Planetary Sciences

Updated Second Edition

An authoritative introduction for graduate students in the physical sciences, this award-winning textbook explains the wide variety of physical, chemical, and geological processes that govern the motions and properties of planets.

This updated second edition has been revised and improved with data that are current as of mid 2014, while maintaining its existing structure and organization. Many data tables and plots have been updated to account for the latest measurements. A new Appendix G focuses on recent discoveries since the second edition was first published (2010), compiled in chapter order. These include results from Cassini, Kepler, MESSENGER, MRO, LRO, Dawn at Vesta, Curiosity, and others, as well as many ground-based observatories.

With over 300 exercises to help students apply the concepts covered, this textbook is ideal for graduate courses in astronomy, planetary science, and earth science, and well suited as a reference for researchers. Color versions of many figures, movie clips supplementing the text, and other resources are available at www.cambridge.org/depater.

Imke de Pater is a Professor in the Astronomy Department and the Department of Earth and Planetary Science at the University of California, Berkeley, and is affiliated with the Faculty of Aerospace Engineering at the Delft University of Technology, the Netherlands. She began her career observing and modeling Jupiter's synchrotron radiation followed by detailed investigations of the planet's atmosphere. In 1994, she led a worldwide campaign to observe the impact of Comet D/Shoemaker–Levy 9 with Jupiter. Currently, she is exploiting adaptive optics techniques in the infrared range to use high angular resolutions data to study the giant planets with their ring and satellite systems.

Jack J. Lissauer is a Space Scientist at NASA's Ames Research Center in Moffett Field, California, and a Consulting Professor at Stanford University. His primary research interests are the formation of planetary systems, detection of extrasolar planets, planetary dynamics and chaos, planetary ring systems, and circumstellar/protoplanetary disks. He is lead discoverer of the six-planet Kepler-11 system and co-discoverer of the first four planets found to orbit about faint M dwarf stars, and co-discovered two broad tenuous dust rings and two small inner moons orbiting the planet Uranus.

Planetary Sciences received the Chambliss Astronomical Writing Award for 2007. This is an award given by the American Astronomical Society (AAS) for astronomy books for an academic audience, specifically textbooks at either the upper division undergraduate level or the graduate level.

Front Cover: Artist's conception of a protoplanetary disk. A growing giant planet appears in the foreground (lower right). This planet has a massive atmosphere, and it has partially cleared a gap around its orbit via gravitational torques (see Chapters 11 and 13). It is accreting both gas and small planetesimals; the latter shed material as they fall into the planet's atmosphere and look like comets. Numerous lunar-sized planetary embryos within the disk are visible through the gravitational wakes that they create in the disk of small planetesimals; such wakes have been observed in Saturn's rings. A pair of these bodies has just collided and glows red. The star at the center of the disk is in its final stages of accretion, and is expelling gas through a bipolar wind. The disk near the star is warmed by both starlight and viscous dissipation within the disk itself; both processes provide more energy closer to the center of the disk. The blue shading of the outer disk is intended to give the impression of cool temperatures, but in reality such regions would appear dark red; similarly, the radially symmetric structure in the disk has been exaggerated in order to convey the impression of rotation. The top of the painting shows other young stars and interstellar gas and dust that inhabit the same stellar nursery as the star/disk system seen close-up. Painted by Lynette Cook (www.lynettecook.com) in 1999, with scientific consultation of Jack Lissauer.

Back Cover: Panoramic view of Saturn, created by combining 165 images taken by the wide-angle camera on the Cassini spacecraft in September 2006. The mosaic images were acquired while the spacecraft was in Saturn's shadow, and hence the rings are seen in forward scattered light. Such a viewing geometry enhances light from microscopic grains. The G ring is easily seen here, outside the bright main rings, while the extended E ring encircles the entire system. Enceladus appears as a white dot within the E ring in the lower left portion of the figure, and Earth is the pale blue dot just inside the G ring in the upper left. (Based on PIA08329, NASA/JPL/SSI/M. Hedman, M. Fondeur and F. van Breugel.)

Planetary Sciences

Updated Second Edition

Imke de Pater
University of California, Berkeley
& Delft University of Technology

and

Jack J. Lissauer
NASA – Ames Research Center
& Stanford University

Winner of the AAS Chambliss Astronomical Writing Award 2007

CAMBRIDGE
UNIVERSITY PRESS

Shaftesbury Road, Cambridge CB2 8EA, United Kingdom

One Liberty Plaza, 20th Floor, New York, NY 10006, USA

477 Williamstown Road, Port Melbourne, VIC 3207, Australia

314–321, 3rd Floor, Plot 3, Splendor Forum, Jasola District Centre, New Delhi – 110025, India

103 Penang Road, #05–06/07, Visioncrest Commercial, Singapore 238467

Cambridge University Press is part of Cambridge University Press & Assessment, a department of the University of Cambridge.

We share the University's mission to contribute to society through the pursuit of education, learning and research at the highest international levels of excellence.

www.cambridge.org
Information on this title: www.cambridge.org/9781107091610

First published 2001
Second edition published 2010
Updated second edition published 2015 (version 5, August 2022)

Printed in the United Kingdom by TJ Books Limited, Padstow Cornwall

A catalog record for this publication is available from the British Library

Library of Congress Cataloging in Publication data
De Pater, Imke, 1952–
Planetary sciences / Imke de Pater, University of California, Berkeley & Delft University of Technology, and Jack J. Lissauer, NASA-Ames Research Center, & Stanford University. – Updated second edition.
    pages cm.
"Winner of the AAS Chambliss Astronomical Writing Award 2007."
Includes bibliographical references and index.
1. Planetology. 2. Solar system. 3. Astronomy.
I. Lissauer, Jack Jonathan. II. Title.
QB601.D38 2015
523.2–dc23 2014028553

ISBN 978-1-107-09161-0 Hardback

Additional resources for this publication at www.cambridge.org/depater

Contents

List of Tables

Preface

Preface to the First Edition

The study of Solar System objects was the dominant branch of Astronomy from antiquity until the nineteenth century. Analysis of planetary motion by Isaac Newton and others helped reveal the workings of the Universe. While the first astronomical uses of the telescope were primarily to study planetary bodies, improvements in telescope and detector technology in the nineteenth and early twentieth centuries brought the greatest advances in stellar and galactic astrophysics. Our understanding of the Earth and its relationship to the other planets advanced greatly during this period. The advent of the Space Age, with lunar missions and interplanetary probes, has revolutionized our understanding of our Solar System over the past forty years. Dozens of planets in orbit about stars other than our Sun have been discovered since 1995; these massive extrasolar planets have orbits quite different from the giant planets in our Solar System, and their discovery is fueling research into the process of planetary formation.

Planetary Science is now a major interdisciplinary field, combining aspects of Astronomy/Astrophysics with Geology/Geophysics, Meteorology/Atmospheric Sciences, and Space Science/Plasma Physics. We are aware of more than ten thousand small bodies in orbit about the Sun and the giant planets. Many objects have been studied as individual worlds rather than merely as points of light. We now realize that the Solar System contains a more dynamic and rapidly evolving group of objects than previously imagined. The cratering record on dozens of imaged bodies shows that impacts have been quite important in the evolution of the Solar System, especially during the epoch of planetary formation. Other evidence, including the compositions of meteorites and asteroids and the high bulk density of the planet Mercury, suggests that even more energetic collisions have disrupted objects. More modest impacts, such as the collision of comet D/Shoemaker–

Levy 9 with Jupiter in 1994, continue to occur in the current era. Dynamical investigations have destroyed the regular 'clockwork' image of the Solar System that had held prominence since the time of Newton. Resonances and chaotic orbital variations are now believed to have been important for the evolution of many small and possibly some large planetary bodies.

The renewed importance of the Planetary Sciences as a subfield of Astronomy implies that some exposure to Solar System studies is an important component to the education of astronomers. Planetary Sciences' close relationship to Geophysics, Atmospheric and Space Sciences means that the study of the planets offers the unique opportunity for comparison available to Earth scientists.

The amount of material contained in this book is difficult to cover in a one year graduate-level course. Moreover, many professors will prefer to cover their favorite topics at greater depth using supplemental materials. Most students using this book are likely to be taking one semester classes, and many will be undergraduates. Although many superficially differing aspects of the Planetary Sciences are interconnected, and we have included extensive cross-referencing between chapters, we have also attempted to organize the text in a manner that allows for courses to focus on more limited topics. Chapter 1 and the first sections of Chapters 2 and 3 should be covered by all students. The remainder of Chapter 2 is particularly useful for Chapters 9–13, and is essential for Chapter 11 and portions of Chapter 12. The remainder of Chapter 3 is essential for Chapter 4 and useful for Chapters 5, 6, 9, and 10. Portions of Chapter 5 are needed for Chapter 6. Chapter 7 is probably the most technical. Chapter 8 contains necessary material for Chapters 9 and 12, and parts of Chapters 9 and 10 are closely related. Although details of observing techniques are beyond the scope of this book, we think it is important that the students are familiar with the variety of observational methods. We have therefore included a general summary of observational techniques in Chapter 9.

Various symbols are commonly used to represent variables and constants in both equations and the text. Some variables have a unique correspondence with standard symbols in the literature, whereas other variables are represented by differing symbols by different authors and many symbols have multiple uses. The interdisciplinary nature of the Planetary Sciences exacerbates the problem because standard notation differs between fields. We have endeavored to minimize confusion within the text and provide the student with the greatest access to the literature by using standard symbols, sometimes augmented by non-standard subscripts or printed using calligraphic fonts in order to avoid duplication of meanings whenever practical. A list of the symbols used in this book is presented as Appendix A.

Inclusion of high-quality color figures within the main text would have added substantially to this book's production costs and consequently to its price. We have thus used monochrome illustrations wherever possible and included color plates in a separate section. However, to facilitate the flow of figures in the book, we have included a monochrome representation and the figure caption within the main text, with the color image and figure number presented in the plates.

We feel that the learning of concepts in the physical sciences, as well as obtaining a feel for Solar System properties, is greatly enhanced when students get their 'hands dirty' by solving problems. Thus, we have included an extensive collection of exercises at the end of each chapter in this text. We rank these problems by degree of conceptual difficulty: The easiest problems, denoted by E, should be accessible to most upper level undergraduate science majors; indeed, some are simply plugging numbers into a given formula. Intermediate (I) problems involve more sophisticated reasoning, and are geared towards graduate students. Some of the difficult (D) problems are quite challenging. Note that these rankings are not related to the number of calculations required, and some E problems take most graduate students longer to solve than some of the I problems.

The breadth of the material covered in the text extends well beyond the area of expertise of the authors. As such, we benefited greatly from comments by many of our colleagues. Especially useful suggestions were provided by Michael A'Hearn, James Bauer, Alice Berman, Donald DePaolo, John Dickel, Luke Dones, Martin Duncan, Stephen Gramsch, Russell Hemley, Bill Hubbard, Donald Hunten, Andy Ingersoll, Raymond Jeanloz, David Kary, Monika Kress, Typhoon Lee, Janet Luhmann, Geoffrey Marcy, Jay Melosh, Bill Nellis, Eugenia Ruskol, Victor Safronov, Mark Showalter, David Stevenson, John Wood, and Dorothy Woolum. Our special thanks go to Catherine Flack, our initial editor, who helped make the book more readable. We enjoyed discussions and comments by the students who were taught with drafts of book chapters, and who worked through half-baked problem sets. This book, like the rapidly evolving field of Planetary Sciences, is a work in progress; as such, we welcome corrections, updates, and other suggestions that we may use to improve future editions. Cambridge University Press has set up a web page for this book on their website: www.cup.cam.ac.uk/scripts/textbook.asp. This page includes errata, various updates, color versions of some of the figures that appear in black and white in this volume, and links to various Solar System information sites.

We dedicate this book to our parents and teachers, and to family, friends, and colleagues who provided us encouragement and support over many years, and to Floris van Breugel, now a teenager, who has never known his mother *not* to be working on this book.

Imke de Pater and Jack J. Lissauer
Berkeley, California
December, 1999

Preface to the Second Edition

Humanity's knowledge (and, hopefully, understanding) of our Solar System has increased by leaps and bounds since the first edition of *Planetary Sciences* was written and published, and data on extrasolar planets have increased manifold. We have thus substantially revised and updated many parts of this text. But the primary purpose of this book remains the same as when we first conceived of it two decades ago: To provide the student/reader with a broad-based introduction to planetary sciences at a level sufficiently high to understand relationships between planetary sciences and related disciplines, to make most of the research literature accessible, and to have the background for teaching planetary sciences at the introductory level. Many researchers (including ourselves!) also find it very useful as a ready reference to planetary processes and data.

When we submitted the manuscript of the first edition for publication, we tabulated the basic physical and orbital properties of all known planetary satellites; by the time that we were reviewing proofs, almost two dozen more small outer moons of Jupiter and Saturn had been discovered, and we referred to the new bodies only as groups. The number of newly discovered small moons is now so large that, even though we have increased the length of the table of satellite orbital properties, only about half of the known moons are included.

Hundreds of Kuiper belt objects have now been studied as individual physical bodies, and therefore we have moved our discussion of this population from Chapter 10 (Comets) to Chapter 9 (formerly called Asteroids, now named Minor Planets), although the dynamics of the transfer of these bodies to the inner Solar System where they exhibit cometary activity remains in Chapter 10. Oort cloud objects have not yet been observed *in situ*, so we continue to include the Oort cloud in Chapter 10.

The first extrasolar planets were discovered while we were writing the first edition of this text. We included this topic as an afterthought in a short final chapter. Substantially more about exoplanets is now known, and this body of knowledge now comprises important clues and constraints about the process of planetary formation. Thus, we have moved the discussion of exoplanets to Chapter 12, prior to covering planetary formation, which is now done in Chapter 13.

The appendices have been substantially enhanced for this edition. Acronyms are common in our field, so we now list the ones used in this book in Appendix B. Some key observing techniques are discussed in the new Appendix E; emphasis is placed on methods used more frequently in studying Solar System bodies apart from the Sun and Earth than in astronomy of more distant objects or geology; techniques specific to certain types of objects, e.g., small bodies and extrasolar planets, are discussed within the appropriate chapters. As the resurgence in planetary studies during the past half century is due primarily to spacecraft sent to make close-up observations of distant bodies, we introduce rocketry and list the most significant lunar and planetary missions in Appendix F. Last but not least, as planetary science is a rapidly advancing field, Appendix G shows a selection of Solar System images released in 2009; we plan to update this appendix with a summary of recent developments in future printings.

The use of the world wide web has increased substantially in recent years. For the first edition, we posted only an erratum on the web. The book's website, www.cambridge.org/depater now includes downloadable versions of many of the figures in this book, many of which are in color, as well as some in movie format. Captions to figures with associated movies on the website are indicated in the margin by 🎬, and those for figures in color on the website by 🖼.

We have benefited greatly from comments on the first edition and draft chapters of the second edition from many students and colleagues. Substantial parts of the text, and many of the problems at the ends of the chapters, have been revised, clarified, and/or updated. Particularly helpful suggestions were provided by Dana Backman, Bill Bottke, Dave Brain, Mike DiSanti, Tony Dobrovloskis, Denton Ebel, Alison Farmer, Bill Feldman, Jonathan Fortney, Richard French, Pat Hamill, Joop Houtkooper, Olenka Hubickyj, Wing Ip, Margaret Kivelson, Rob Lillis, Mark Marley, Paul Mahaffy, Hap McSween, Julie Moses, Francis Nimmo, Larry Nittler, Dave O'Brien, Kaveh Pahlevan, Derek Richardson, Adam Showman, Steve Squyres, Glen Stewart, Chad Trujillo, Len Tyler, Bert Vermeersen, Kees Welten, Josh Winn, Kevin Zahnle and many of the individuals acknowledged above for their assistance in preparing the first edition.

Imke de Pater and Jack J. Lissauer
Berkeley, California
1 May 2009

Preface to the Updated Second Edition (2015)

Planetary sciences is an active research field, and our knowledge of the planets and smaller bodies in the Solar System is increasing very rapidly. The newer discipline of exoplanet research is expanding at an even faster pace. Thus, no compendium on this subject can be completely up to date. In this updated printing, we have corrected errors, revised tables, and in some cases provided updated figures within the main text. Substantial new material, which would require repagination of the main text leading to higher textbook cost, is presented in Appendix G.

Figure 1.3a

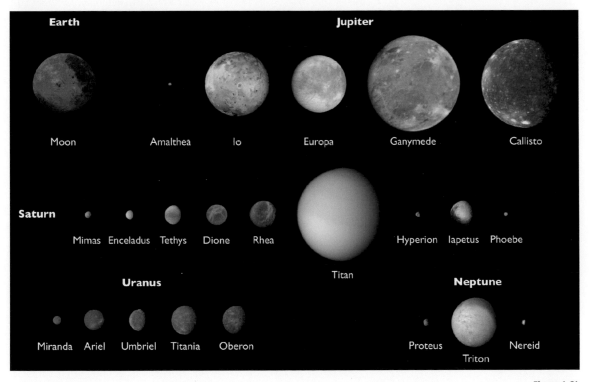

Figure 1.3b

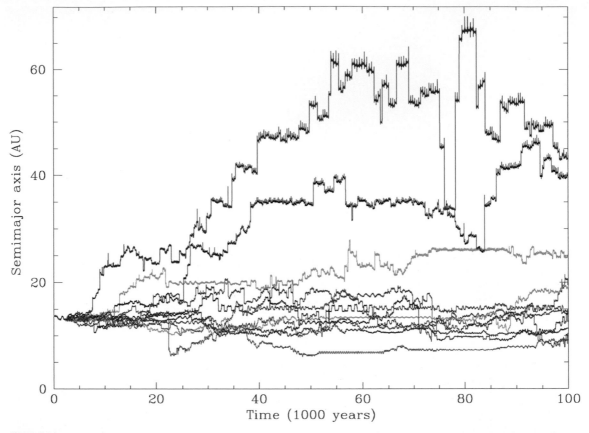

Figure 2.11

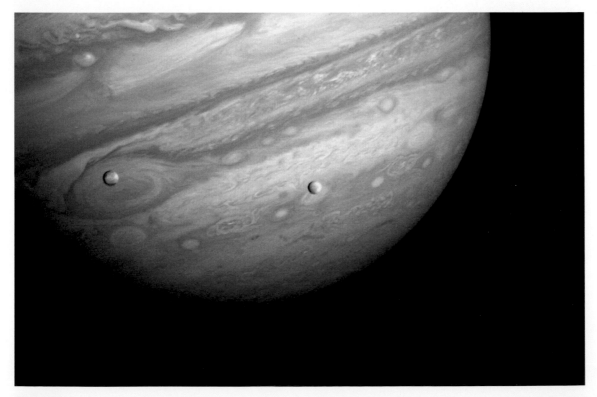

Figure 4.34

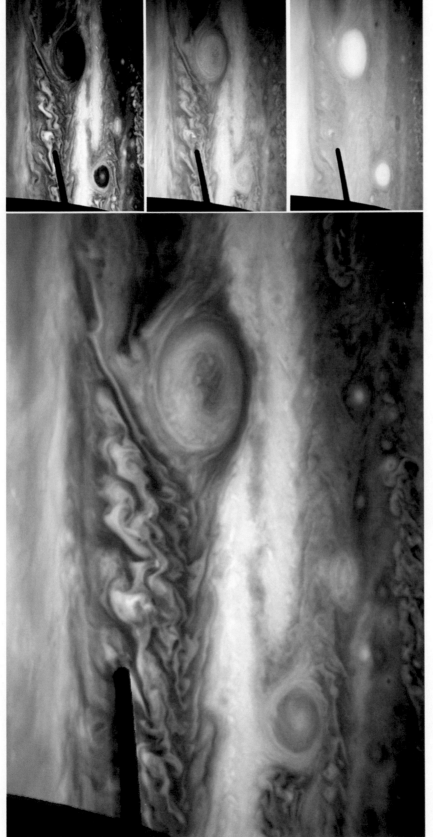

Figure 4.36a

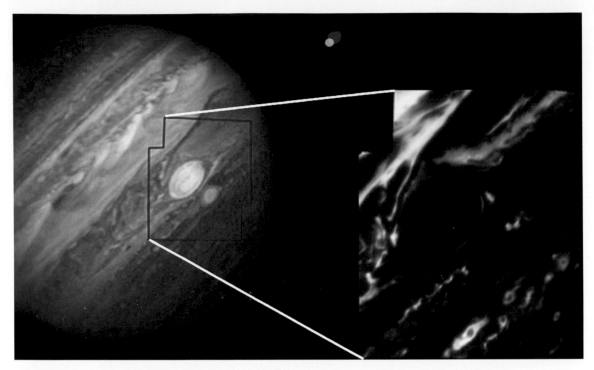

Figure 4.36b

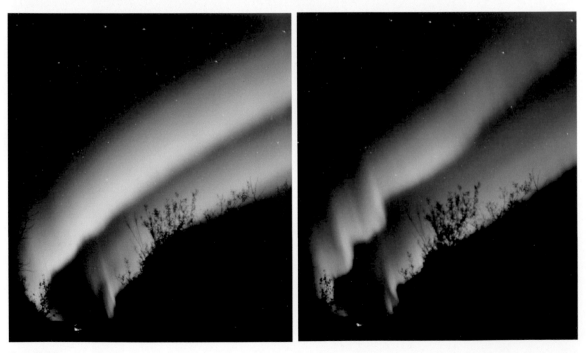

Figure 4.44a

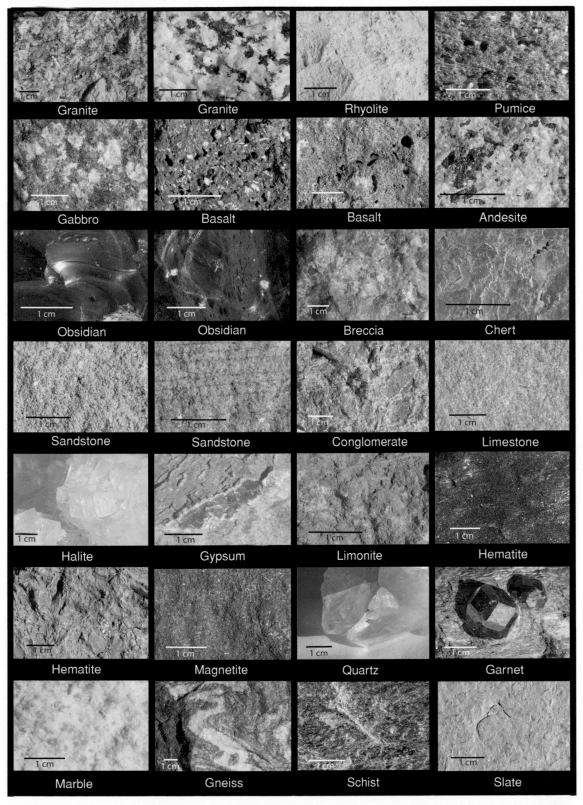

Granite   Granite   Rhyolite   Pumice

Gabbro   Basalt   Basalt   Andesite

Obsidian   Obsidian   Breccia   Chert

Sandstone   Sandstone   Conglomerate   Limestone

Halite   Gypsum   Limonite   Hematite

Hematite   Magnetite   Quartz   Garnet

Marble   Gneiss   Schist   Slate

Figure 5.3

Figure 5.21b

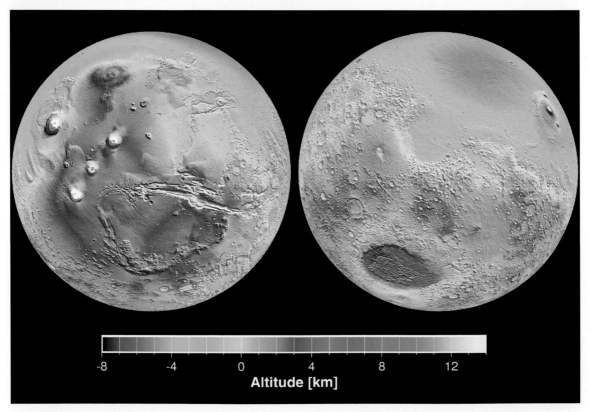

Figure 5.58

Lower-Limit of Water Mass Fraction on Mars

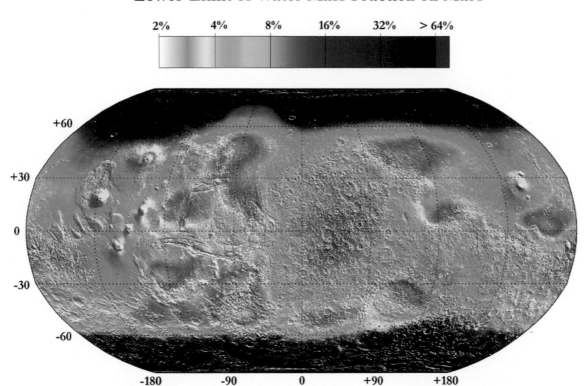

Figure 5.59

Figure 5.70

Figure 5.71a

Figure 5.71c

Figure 5.73

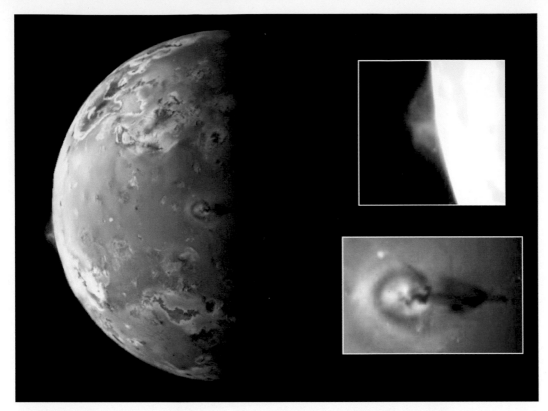

Figure 5.75a

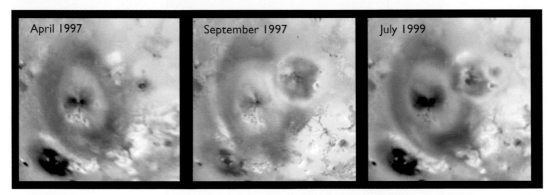

April 1997   September 1997   July 1999

Figure 5.75b

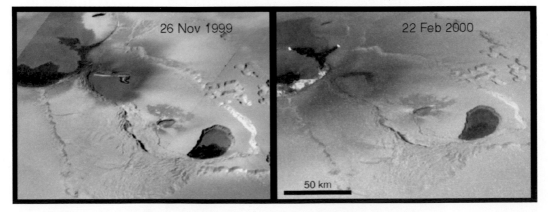

26 Nov 1999   22 Feb 2000

50 km

Figure 5.75c

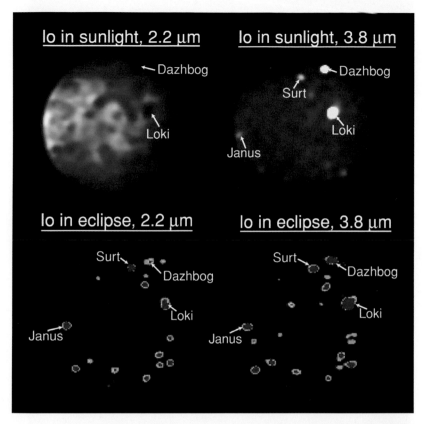

Io in sunlight, 2.2 μm
←Dazhbog
Loki

Io in sunlight, 3.8 μm
→Dazhbog
Surt
Loki
Janus

Io in eclipse, 2.2 μm
Surt
Dazhbog
Loki
Janus

Io in eclipse, 3.8 μm
Surt
Dazhbog
Loki
Janus

Figure 5.78b

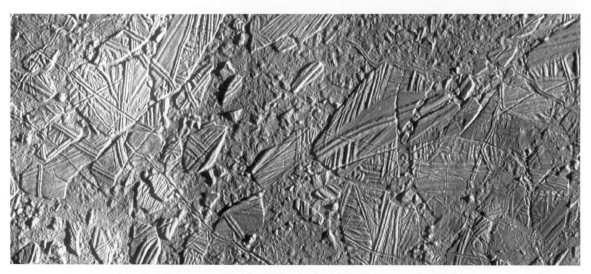

Figure 5.82c

Figure 5.98a

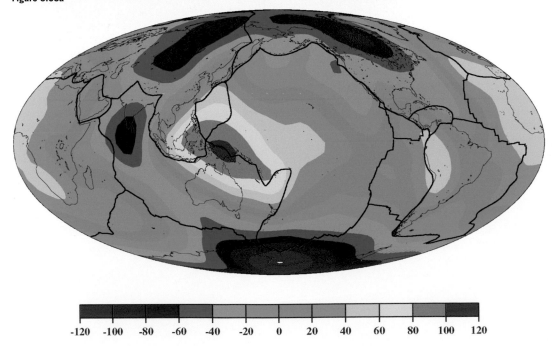

-120 -100 -80 -60 -40 -20 0 20 40 60 80 100 120

Figure 6.10

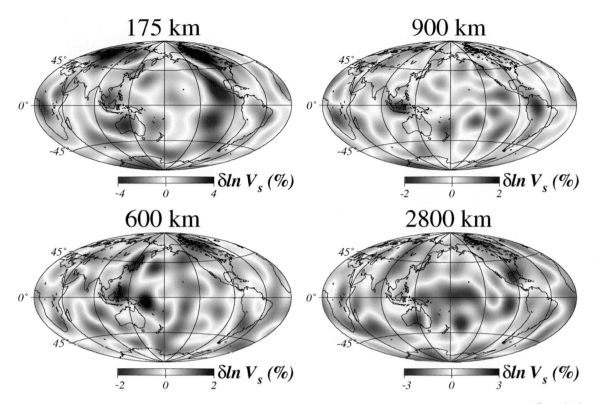

175 km

900 km

$\delta ln\, V_s\ (\%)$

$\delta ln\, V_s\ (\%)$

-4   0   4

-2   0   2

600 km

2800 km

$\delta ln\, V_s\ (\%)$

$\delta ln\, V_s\ (\%)$

-2   0   2

-3   0   3

Figure 6.18a

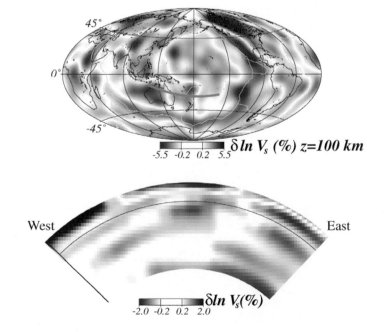

$\delta ln\, V_s\ (\%)\ z{=}100\ km$

-5.5 -0.2 0.2 5.5

West

East

$\delta ln\, V_s(\%)$

-2.0 -0.2 0.2 2.0

Figure 6.18b

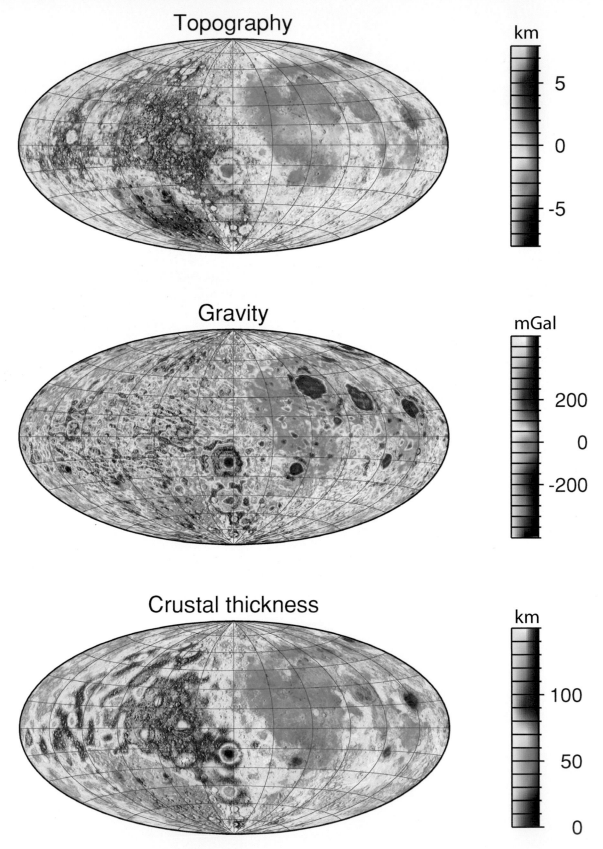

Topography

km

5

0

-5

Gravity

mGal

200

0

-200

Crustal thickness

km

100

50

0

Figure 6.19

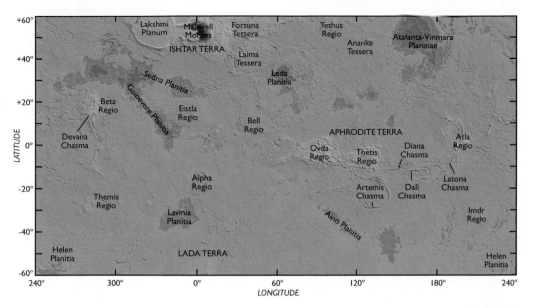

Figure 6.22a

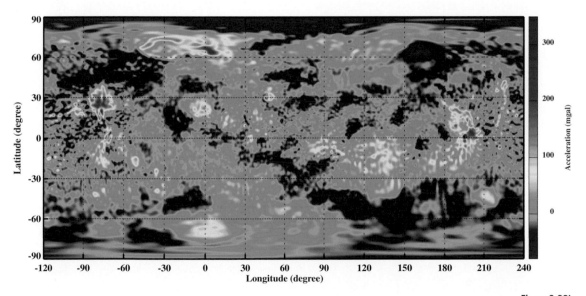

Figure 6.22b

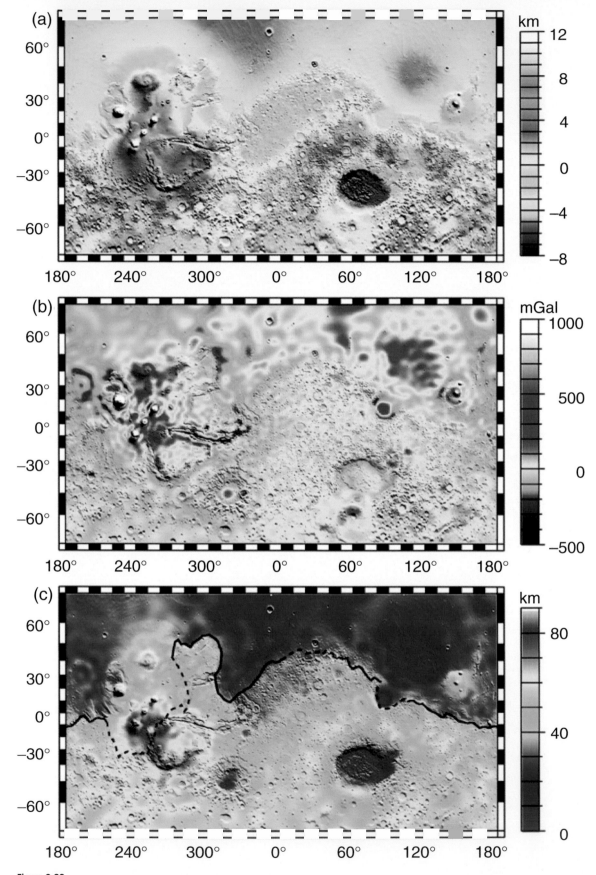

Figure 6.23

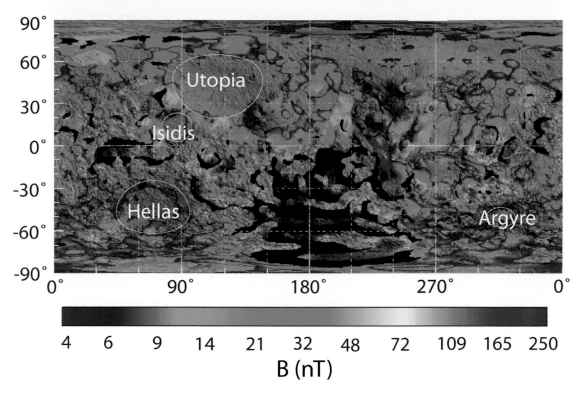

Figure 7.28

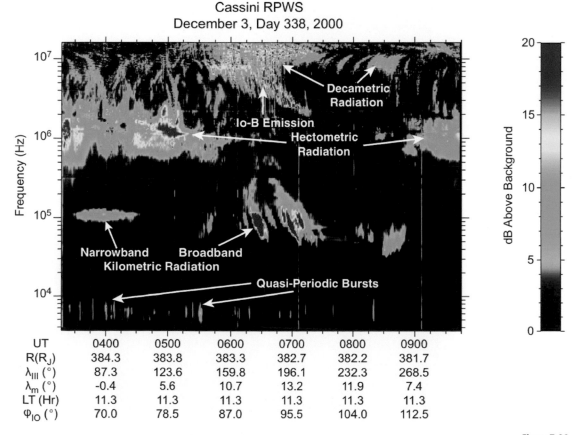

Figure 7.38

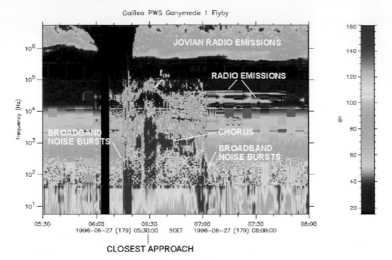

Figure 7.40

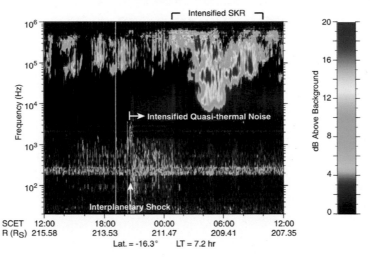

Figure 7.42a

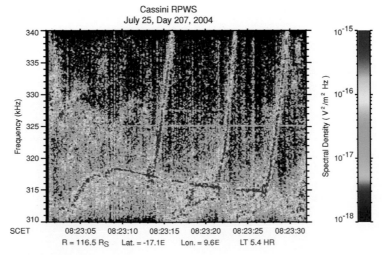

Figure 7.42b

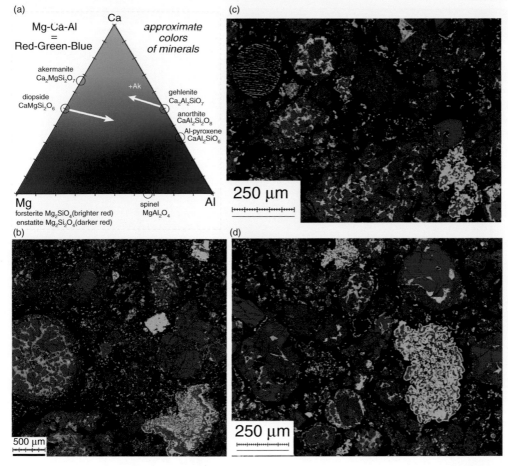

(a) Mg-Ca-Al = Red-Green-Blue

approximate colors of minerals

Ca

akermanite Ca$_2$MgSi$_2$O$_7$

diopside CaMgSi$_2$O$_6$

+Ak

gehlenite Ca$_2$Al$_2$SiO$_7$

anorthite CaAl$_2$Si$_2$O$_8$

Al-pyroxene CaAl$_2$SiO$_6$

Mg

forsterite Mg$_2$SiO$_4$ (brighter red)
enstatite Mg$_2$Si$_2$O$_4$ (darker red)

spinel MgAl$_2$O$_4$

Al

(c)

250 μm

(b)

500 μm

(d)

250 μm

Figure 8.17

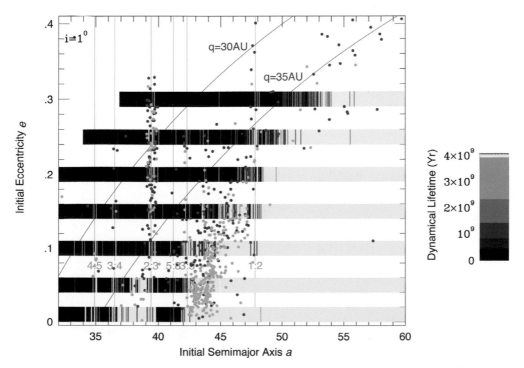

$i = 1^0$

q = 30 AU

q = 35 AU

4:5 3:4 2:3 5:3·5 1:2

Initial Eccentricity e

Initial Semimajor Axis a

Dynamical Lifetime (Yr)

4×10^9
3×10^9
2×10^9
10^9
0

Figure 9.3b

Figure 10.1a

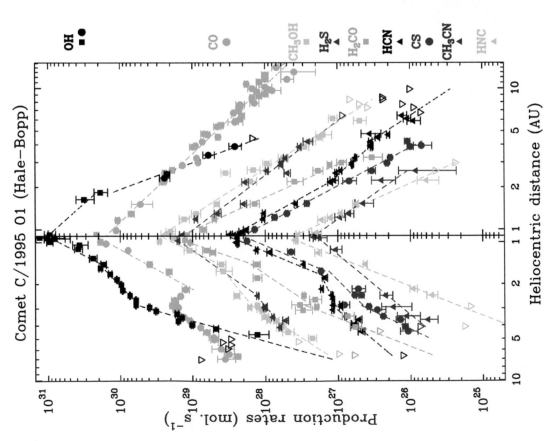

Figure 10.7

Figure 11.4

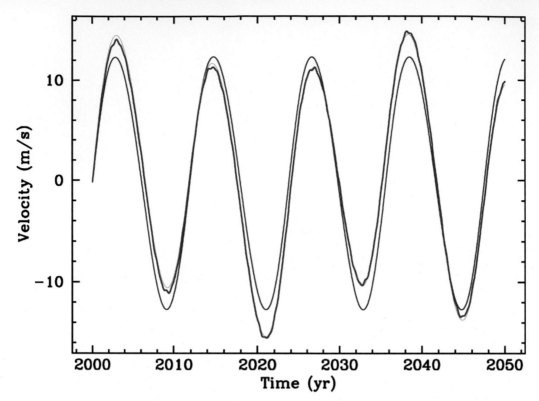

Figure 12.4

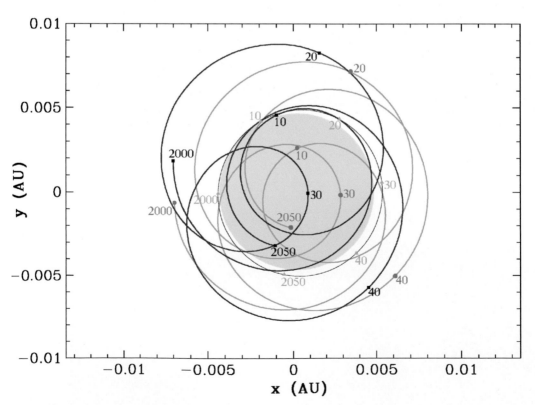

Figure 12.5

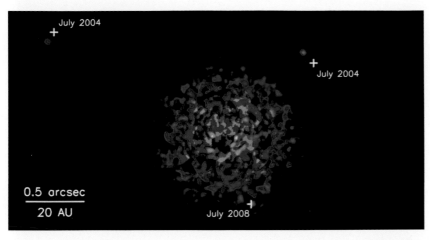

Figure 12.8

Figure 12.25

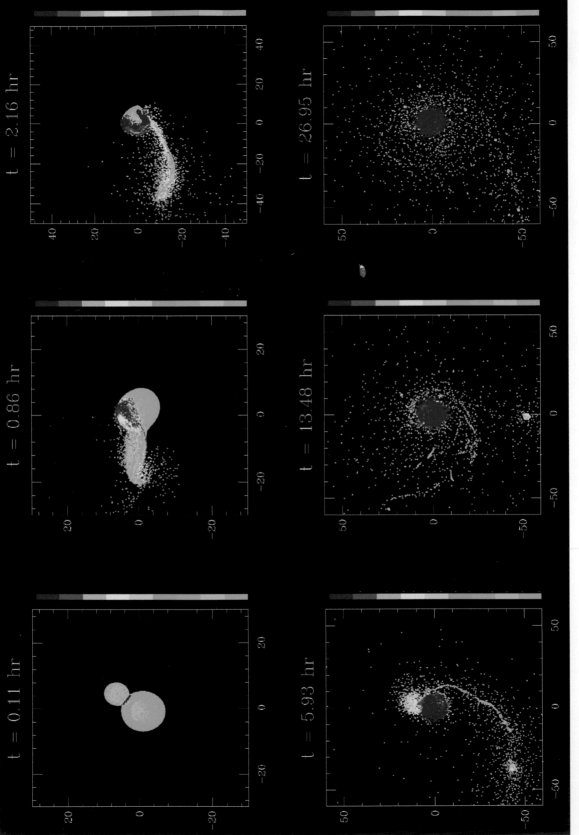

Figure 13.26

1 Introduction

SOCRATES: *Shall we set down astronomy among the subjects of study?*

GLAUCON: *I think so, to know something about the seasons, the months and the years is of use for military purposes, as well as for agriculture and for navigation.*

SOCRATES: *It amuses me to see how afraid you are, lest the common herd of people should accuse you of recommending useless studies.*

Plato, *The Republic VII*

The wonders of the night sky, the Moon and the Sun have fascinated mankind for many millennia. Ancient civilizations were particularly intrigued by several brilliant 'stars' that move among the far more numerous 'fixed' (stationary) stars. The Greeks used the word $\pi\lambda\alpha\nu\eta\tau\eta\zeta$, meaning wandering star, to refer to these objects. Old drawings and manuscripts by people from all over the world, such as the Chinese, Greeks, and Anasazi, attest to their interest in comets, solar eclipses, and other celestial phenomena.

The Copernican–Keplerian–Galilean–Newtonian revolution in the sixteenth and seventeenth centuries completely changed humanity's view of the dimensions and dynamics of the Solar System, including the relative sizes and masses of the bodies and the forces that make them orbit about one another. Gradual progress was made over the next few centuries, but the next revolution had to await the space age.

In October of 1959, the Soviet spacecraft Luna 3 returned the first pictures of the farside of Earth's Moon (Appendix F). The age of planetary exploration had begun. Over the next three decades, spacecraft visited all eight known terrestrial and giant planets in the Solar System, including our own. These spacecraft have returned data concerning the planets, their rings and moons. Spacecraft images of many objects showed details which could never have been guessed from previous Earth-based pictures. Spectra from ultraviolet to infrared wavelengths revealed previously undetected gases and geological features on planets and moons, while radio detectors and magnetometers transected the giant magnetic fields surrounding many of the planets. The planets and their satellites have become familiar to us as individual bodies. The immense diversity of planetary and satellite surfaces, atmospheres, and magnetic fields has surprised even the most imaginative researchers. Unexpected types of structure were observed in Saturn's rings, and whole new classes of rings and ring

systems were seen around all four giant planets. Some of the new discoveries have been explained, whereas others remain mysterious.

Six comets and eleven asteroids have thus far been explored by close-up spacecraft, and there have been several missions to study the Sun and the solar wind. The Sun's gravitational domain extends thousands of times the distance to the farthest known planet, Neptune. Yet the vast outer regions of the Solar System are so poorly explored that many bodies remain to be detected, possibly including some of planetary size.

Thousands of planets are now known to orbit stars other than the Sun. While we know far less about any of these extrasolar planets than we do about the planets in our Solar System, it is clear that many of them have gross properties (orbits, masses, radii) quite different from any object orbiting our Sun, and they are thus causing us to revise some of our models of how planets form and evolve.

In this book, we discuss what has been learned and some of the unanswered questions that remain at the forefront of planetary sciences research today. Topics covered include the orbital, rotational, and bulk properties of planets, moons, and smaller bodies; gravitational interactions, tides, and resonances between bodies; chemistry and dynamics of planetary atmospheres, including cloud physics; planetary geology and geophysics; planetary interiors; magnetospheric physics; meteorites; asteroids; comets; and planetary ring dynamics. The new and rapidly blossoming field of extrasolar planet studies is then introduced. We conclude by combining this knowledge of current Solar System and extrasolar planet properties and processes with astrophysical data and models of ongoing star and planet formation to develop a model for the origin of planetary systems.

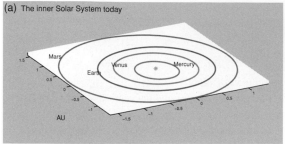

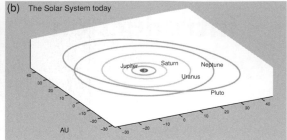

Figure 1.1 The orbits of (a) the four terrestrial planets, and (b) all eight major planets in the Solar System and Pluto, are shown to scale. Two different levels of reduction are displayed because of the relative closeness of the four terrestrial planets and the much larger spacings in the outer Solar System. The axes are in AU. Note the high inclination of Pluto's orbit relative to the orbits of the major planets. The movies show variations in the orbits over the past three million years; these changes are due to mutual perturbations among the planets (Chapter 2). Figure 2.14 presents plots of the variations in planetary eccentricities from the same integrations. (Illustrations courtesy Jonathan Levine)

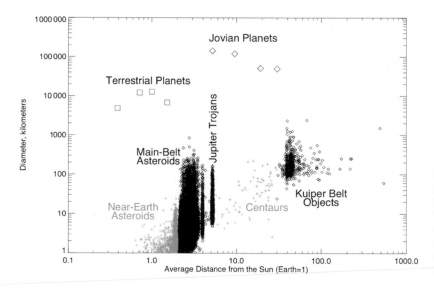

Figure 1.2 Inventory of objects orbiting the Sun. The jovian planets dominate the outer Solar System and the terrestrial planets dominate the inner Solar System. Small objects tend to be concentrated in regions where orbits are stable, or at least long-lived. (Courtesy John Spencer)

1.1 Inventory of the Solar System

What is the *Solar System*? Our naturally geocentric view gives a highly distorted picture, thus it is better to phrase the question as: What is seen by an objective observer from afar? The *Sun*, of course; the Sun has a luminosity 4×10^8 times as large as the total luminosity (reflected plus emitted) of Jupiter, the second brightest object in the Solar System. The Sun also contains >99.8% of the mass of the known Solar System. By these measures, the Solar System can be thought of as the Sun plus some debris. However, by other measures the planets are not insignificant. Over 98% of the angular momentum in the Solar System lies in orbital motions of the planets. Moreover,

the Sun is a fundamentally different type of body from the planets, a ball of plasma powered by nuclear fusion in its core, whereas the smaller bodies in the Solar System are composed of molecular matter, some of which is in the solid state. This book focuses on the debris in orbit about the Sun. This debris is comprised of the giant planets, the terrestrial planets, and numerous and varied smaller objects (Figs. 1.1–1.3).

1.1.1 Giant Planets

Jupiter dominates our planetary system. Its mass, 318 Earth masses ($M_\oplus$), exceeds twice that of all other known Solar System planets combined. Thus as a second

(a)

(b)

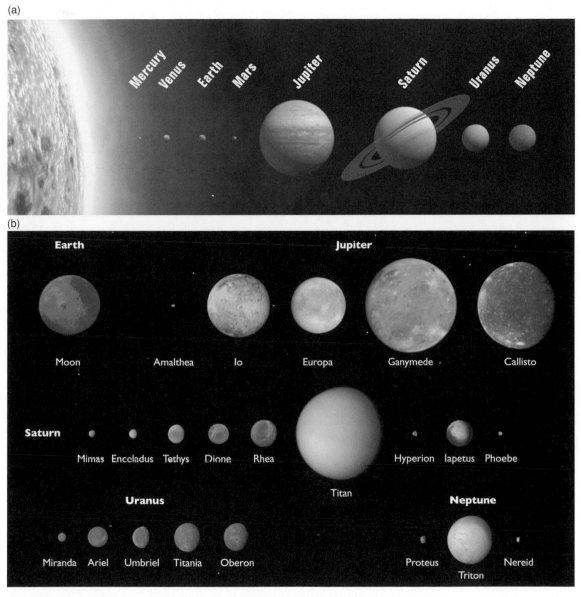

Figure 1.3 COLOR PLATE (a) Images of the planets with radii depicted to scale, ordered by distance from the Sun. (Courtesy International Astronomical Union/Martin Kornmesser) (b) Images of the largest satellites of the four giant planets and Earth's Moon, which are depicted in order of distance from their planet. Note that these moons span a wide range of size, albedo, and surface characteristics; most are spherical, but some of the smallest objects pictured are quite irregular in shape. (Courtesy Paul Schenk)

approximation, the Solar System can be viewed as the Sun, Jupiter, and some debris. The largest of this debris is *Saturn*, with a mass of nearly $100\,M_\oplus$. Saturn, like Jupiter, is made mostly of hydrogen (H) and helium (He). Each of these planets probably possesses a heavy element 'core' of mass $\sim 10\,M_\oplus$. The third and fourth largest planets are *Neptune* and *Uranus*, each having a mass roughly one-sixth that of Saturn. These planets belong to a different class, with most of their masses provided by a combination of three common astrophysical 'ices', water (H_2O), ammonia (NH_3), methane (CH_4), together with 'rock', high temperature condensates consisting primarily of silicates and metals, and yet most of their volumes are occupied by relatively low mass (1–4 $M_{\oplus}$) H–He dominated atmospheres. The four largest planets are known collectively as the *giant planets*; Jupiter and Saturn are called *gas giants*, with radii of $\sim$70 000 km and 60 000 km respectively, whereas Uranus and Neptune are referred to as *ice giants* (although the 'ices' are present in fluid rather than solid form), with radii of $\sim$25 000 km. All four giant planets possess strong magnetic fields. These planets orbit the Sun at distances of approximately 5, 10, 20, and 30 AU, respectively. (One *astronomical unit*, 1 AU, is defined to be the semimajor axis of a massless (test) particle whose orbital period about the Sun is one year. As our planet has a finite mass, the semimajor axis of Earth's orbit is slightly larger than 1 AU.)

1.1.2 Terrestrial Planets

The mass of the remaining known 'debris' totals less than one-fifth that of the smallest giant planet, and their orbital angular momenta are also much smaller. This debris consists of all of the solid bodies in the Solar System, and despite its small mass it contains a wide variety of objects that are interesting chemically, geologically, dynamically, and, in at least one case, biologically. The hierarchy continues within this group with two large *terrestrial*[1] planets, *Earth* and *Venus*, each with a radius of about 6000 km, at approximately 1 and 0.7 AU from the Sun, respectively. Our Solar System also contains two small terrestrial planets: *Mars* with a radius of $\sim$3500 km and orbiting at $\sim$1.5 AU and *Mercury* with a radius of $\sim$2500 km orbiting at $\sim$0.4 AU. All four terrestrial planets have atmospheres. Atmospheric composition and density varies widely among the terrestrial planets, with Mercury's atmosphere being exceedingly thin. However, even the most massive terrestrial planet atmosphere, that of Venus, is minuscule by giant planet standards. Earth and Mercury each have an internally generated magnetic field, and there are signs that Mars possessed one in the distant past.

1.1.3 Minor Planets and Comets

The *Kuiper belt* is a thick disk of ice/rock bodies beyond the orbit of Neptune. The two largest members of the Kuiper belt to have been sighted are *Eris*, whose *heliocentric distance*, the distance from the Sun, oscillates between 38 and 97 AU, and *Pluto*, whose heliocentric distance varies from 29 to 50 AU. The radii of Eris and Pluto exceed 1000 km. Pluto is known to possess an atmosphere.

[1] In this text, the word 'terrestrial' is used to mean Earth-like or related to the planet Earth, as is the convention in planetary sciences and astronomy. Geoscientists and biologists generally use the same word to signify a relationship with land masses.

Numerous smaller members of the Kuiper belt have been cataloged, but the census of these distant objects is incomplete even at large sizes. *Asteroids*, which are minor planets that all have radii <500 km, are found primarily between the orbits of Mars and Jupiter.

Smaller objects are also known to exist elsewhere in the Solar System, for example as moons in orbit around planets, and as comets. Comets are ice-rich objects that shed mass when subjected to sufficient solar heating. Comets are thought to have formed in or near the giant planet region and then been 'stored' in the *Oort cloud*, a nearly spherical region at heliocentric distances of $\sim$1 – 5 $\times$ 10^4 AU, or in the Kuiper belt or the *scattered disk*. Scattered disk objects have moderate to high eccentricity orbits that lie in whole or in part within the Kuiper belt. Estimates of the total number of comets larger than one kilometer in radius in the entire Oort cloud range from $\sim$10^{12} to $\sim$10^{13}. The total number of Kuiper belt objects larger than 1 km in radius is estimated to be $\sim$10^8–10^{10}. The total mass and orbital angular momentum of bodies in the scattered disk and Oort cloud are uncertain by more than an order of magnitude. The upper end of current estimates place as much mass in distant unseen icy bodies as is observed in the entire planetary system.

The smallest bodies known to orbit the Sun, such as the dust grains that together produce the faint band in the plane of the planetary orbits known as the *zodiacal cloud*, have been observed collectively, but not yet individually detected via remote sensing.

1.1.4 Satellite and Ring Systems

Some of the most interesting objects in the Solar System orbit about the planets. Following the terrestrial planets in mass are the seven major moons of the giant planets and Earth. Two planetary satellites, Jupiter's moon Ganymede and Saturn's moon Titan, are slightly larger than the planet Mercury, but because of their lower densities they are less than half as massive. Titan's atmosphere is denser than that of Earth. Triton, by far the largest moon of Neptune, has an atmosphere that is much less dense, yet it has winds powerful enough to strongly perturb the paths of particles ejected from geysers on its surface. Very tenuous atmospheres have been detected about several other planetary satellites, including Earth's Moon, Jupiter's Io and Saturn's Enceladus.

Natural satellites have been observed in orbit about most of the planets in the Solar System, as well as many Kuiper belt objects and asteroids. The giant planets all have large satellite systems, consisting of large and/or medium-sized satellites and many smaller moons and rings (Fig. 1.3b). Most of the smaller moons orbiting

Table 1.1 Planetary mean orbits and symbols.

Planet	Symbol	a (AU)	e	i (deg)	Ω (deg)	ϖ (deg)	λ_m
Mercury	☿	0.387 098 80	0.205 631 75	7.004 99	48.3309	77.4561	252.2509
Venus	♀	0.723 332 01	0.006 771 77	3.394 47	76.6799	131.5637	181.9798
Earth	⊕	1.000 000 83	0.016 708 617	0.0	0.0	102.9374	100.4665
Mars	♂	1.523 689 46	0.093 400 62	1.849 73	49.5581	336.6023	355.4333
Jupiter	♃	5.202 758 4	0.048 495	1.303 3	100.464	14.331	34.351
Saturn	♄	9.542 824 4	0.055 509	2.488 9	113.666	93.057	50.077
Uranus	⛢	19.192 06	0.046 30	0.773	74.01	173.01	314.06
Neptune	♆	30.068 93	0.008 99	1.770	131.78	48.12	304.35

λ_m is mean longitude. All data are for the J2000 epoch and were taken from Yoder (1995).

Pluto, which was classified as a planet from its discovery in 1930 until 2006, also has an official symbol, ♇. The symbol for Earth's Moon is ☾.

close to their planet were discovered from spacecraft flybys. All major satellites, except Triton, orbit the respective planet in a *prograde* manner (i.e., in the direction that the planet rotates), close to the planet's equatorial plane. Small, close-in moons are also exclusively in low-inclination, low-eccentricity orbits, but small moons orbiting beyond the main satellite systems can travel around the planet in either direction, and their orbits are often highly inclined and eccentric. Earth and Pluto each have one large moon: our Moon has a little over 1% of Earth's mass, and Charon's mass is just over 10% that of Pluto. These moons probably were produced by giant impacts on the Earth and Pluto, when the Solar System was a small fraction of its current age. Two tiny moons travel on low-inclination, low-eccentricity orbits about Mars.

The four giant planets all have ring systems, which are primarily within about 2.5 planetary radii of the planet's center. However, in other respects, the characters of the four ring systems differ greatly. Saturn's rings are bright and broad, full of structure such as density waves, gaps, and 'spokes'. Jupiter's ring is very tenuous and composed mostly of small particles. Uranus has nine narrow opaque rings plus broad regions of tenuous dust orbiting close to the plane defined by the planet's equator. Neptune has four rings, two narrow ones and two faint broader rings; the most remarkable part of Neptune's ring system is the ring arcs, which are bright segments within one of the narrow rings. As with interplanetary dust, individual ring particles have not been observed directly via remote sensing.

1.1.5 Tabulations

The orbital and bulk properties of the eight 'major' planets are listed in Tables 1.1–1.3. Table 1.4 gives orbital elements and brightnesses of all inner moons of the eight planets, as well as those outer moons whose radii are estimated to be $\gtrsim$10 km. Many of the orbital parameters listed in the tables are defined in §2.1. Rotation rates and physical characteristics of these satellites, whenever known, are given in Table 1.5. Properties of some the largest 'minor planets', asteroids, and Kuiper belt objects are given in Tables 9.1 and 9.2, and minor planet satellites are discussed in §9.4.4.

1.1.6 Heliosphere

All planetary orbits lie within the *heliosphere*, the region of space containing magnetic fields and plasma of solar origin. The *solar wind* consists of plasma traveling outwards from the Sun, at supersonic speeds. The solar wind merges with the interstellar medium at the *heliopause*, the boundary of the heliosphere.

The composition of the heliosphere is dominated by solar wind protons and electrons, with a typical density of 5 protons cm^{-3} at 1 AU (decreasing as the reciprocal distance squared), and speed of ~400 km s^{-1} near the solar equator but ~700–800 km s^{-1} closer to the solar poles. In contrast, the local interstellar medium, at a density of less than 0.1 atoms cm^{-3}, contains mainly hydrogen and helium atoms. The Sun's motion relative to the mean motion of neighboring stars is roughly 26 km s^{-1}. Hence, the heliosphere moves through the interstellar medium at about this speed. The heliosphere is thought to be shaped like a teardrop, with a tail in the downwind direction (Fig. 1.4). Interstellar ions and electrons generally flow around the heliosphere, since they cannot cross the solar magnetic field lines. Neutrals, however, can enter the heliosphere, and as a result interstellar H and He atoms

Table 1.2 Terrestrial planets: Geophysical data.

	Mercury	Venus	Earth	Mars
Mean radius R (km)	2440 ± 1	$6051.8(4 \pm 1)$	$6371.0(1 \pm 2)$	$3389.9(2 \pm 4)$
Mass ($\times 10^{27}$ g)	0.3302	4.8685	5.9736	0.64185
Density (g cm^{-3})	5.427	5.204	5.515	$3.933(5 \pm 4)$
Flattening ϵ			1/298.257	1/154.409
Equatorial radius (km)			6378.136	3397 ± 4
Sidereal rotation period	58.6462 d	-243.0185 d	23.934 19 h	24.622 962 h
Mean solar day (in days)	175.9421	116.7490	1	1.027 490 7
Equatorial gravity (m s^{-2})	3.701	8.870	9.780 327	3.690
Polar gravity (m s^{-2})			9.832 186	3.758
Core radius (km)	~ 1600	~ 3200	3485	~ 1700
Figure offset ($R_{CF} - R_{CM}$) (km)		0.19 ± 0.01	0.80	2.50 ± 0.07
Offset (lat./long.)		$11°/102°$	$46°/35°$	$62°/88°$
Obliquity to orbit (deg)	~ 0.1	177.3	23.45	25.19
Sidereal orbit period (yr)	0.240 844 5	0.615 182 6	0.999 978 6	1.880 711 05
Escape velocity v_e (km s^{-1})	4.435	10.361	11.186	5.027
Geometric albedo	0.106	0.65	0.367	0.150
$V(1,0)^a$	-0.42	-4.40	-3.86	-1.52

All data are from Yoder (1995).

[a] $V(1,0)$ is the visual equivalent magnitude at 1 AU and 0° phase angle.

The apparent visual magnitude at phase angle ϕ, m_v, can be calculated from: $m_v = V(1,0) + C\phi + (5 \log_{10})(r_{\odot AU} r_{\Delta AU})$, with C the phase coefficient in magnitudes per degree, $r_{\odot AU}$ the planet's heliocentric distance (in AU) and $r_{\Delta AU}$ the distance from the observer to the planet (in AU).

Table 1.3 Giant planets: Physical data.

	Jupiter	Saturn	Uranus	Neptune
Mass (10^{27} g)	1898.6	568.46	86.832	102.43
Density (g cm^{-3})	1.326	0.6873	1.318	1.638
Equatorial radius (1 bar) (km)	$71\,492 \pm 4$	$60\,268 \pm 4$	$25\,559 \pm 4$	$24\,766 \pm 15$
Polar radius (km)	$66\,854 \pm 10$	$54\,364 \pm 10$	$24\,973 \pm 20$	$24\,342 \pm 30$
Volumetric mean radius (km)	$69\,911 \pm 6$	$58\,232 \pm 6$	$25\,362 \pm 12$	$24\,624 \pm 21$
Flattening ϵ	0.064 87	0.097 96	0.022 93	0.0171
	$\pm 0.000\,15$	$\pm 0.000\,18$	± 0.0008	± 0.0014
Sidereal rotation period	$9^h 55^m 29\overset{s}{.}71$	$10^h 32^m 35^s \pm 13^a$	-17.24 ± 0.01 h	16.11 ± 0.01 h
Hydrostatic flattening[b]	0.065 09	0.098 29	0.019 87	0.018 04
Equatorial gravity (m s^{-2})	23.12 ± 0.01	8.96 ± 0.01	8.69 ± 0.01	11.00 ± 0.05
Polar gravity (m s^{-2})	27.01 ± 0.01	12.14 ± 0.01	9.19 ± 0.02	11.41 ± 0.03
Obliquity (deg)	3.12	26.73	97.86	29.56
Sidereal orbit period (yr)	11.856 523	29.423 519	83.747 407	163.723 21
Escape velocity v_e (km s^{-1})	59.5	35.5	21.3	23.5
Geometric albedo	0.52	0.47	0.51	0.41
$V(1,0)$	-9.40	-8.88	-7.19	-6.87

Most data are from Yoder (1995).

[a] Saturn's rotation period is from Anderson and Schubert (2007); the true uncertainty in its value is far greater than the formal error listed because different measurement techniques yield values that differ from one another by up to tens of minutes.

[b] Hydrostatic flattening as derived from the gravitational field and magnetic field rotation rate.

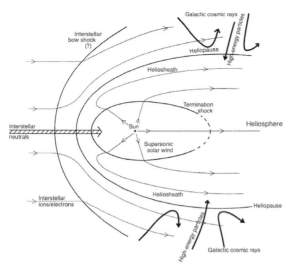

Figure 1.4 Sketch of the teardrop-shaped heliosphere. Within the heliosphere, the solar wind flows radially outward until it encounters the heliopause, the boundary between the solar wind dominated region and the interstellar medium. Weak cosmic rays are deflected away by the heliopause, but energetic particles penetrate the region down to the inner Solar System. (Adapted from Gosling 2007)

move through the Solar System, in the downstream direction, with a typical speed of 22 (for H) – 26 (for He) km s^{-1}.

Just interior to the heliopause is the *termination shock*, where the solar wind is slowed down. Due to variations in solar wind pressure, the location of this shock moves radially with respect to the Sun, in accordance with the 11-year solar activity cycle. The Voyager 1 spacecraft crossed the termination shock in December 2004 at a heliocentric distance of 94.0 AU; Voyager 2 crossed the shock (multiple times) in August 2007 at ~83.7 AU. While Voyager 2 is still in the *heliosheath*, between the termination shock and the heliopause, Voyager 1 crossed the heliopause in 2012 and entered the interstellar medium (§G.7).

1.2 Planetary Properties

All of our knowledge regarding specific characteristics of Solar System objects, including planets, moons, comets, asteroids, rings, and interplanetary dust, is ultimately derived from observations, either astronomical measurements from the ground or Earth-orbiting satellites, or from close-up (often *in situ*) measurements obtained by interplanetary spacecraft. One can determine the following quantities more or less directly from observations:

(1) Orbit
(2) Mass, distribution of mass
(3) Size
(4) Rotation rate and direction
(5) Shape
(6) Temperature
(7) Magnetic field
(8) Surface composition
(9) Surface structure
(10) Atmospheric structure and composition

With the help of various theories, these observations can be used to constrain planetary properties such as bulk composition and interior structure, two attributes which are crucial elements in modeling the formation of the Solar System.

1.2.1 Orbit

In the early part of the seventeenth century, Johannes Kepler deduced three 'laws' of planetary motion directly from observations:

(1) All planets move along elliptical paths with the Sun at one focus.

(2) A line segment connecting any given planet and the Sun sweeps out area at a constant rate.

(3) The square of a planet's orbital period about the Sun, P_{orb}, is proportional to the cube of its semimajor axis, a, i.e., $P_{\mathrm{orb}}^2 \propto a^3$.

A Keplerian orbit is uniquely specified by six orbital elements, a (semimajor axis), e (eccentricity), i (inclination), ω (argument of periapse; or ϖ for the longitude of periapse), Ω (longitude of ascending node), and f (true anomaly). These orbital elements are defined graphically in Figure 2.1 and discussed in more detail in §2.1. The first of these elements are more fundamental than the last: a and e fully define the size and shape of the orbit, i gives the tilt of the orbital plane to some reference plane, the longitudes ϖ and Ω determine the orientation of the orbit, and f (or, indirectly, t_ϖ, the time of periapse passage) tells where the planet is along its orbit at a given time. Alternative sets of orbital elements are also possible, for instance an orbit is fully specified by the planet's location and velocity relative to the Sun at a given time (again six independent scalar quantities), provided the masses of the Sun and planet are known.

Kepler's laws (or more accurate versions thereof) can be derived from Newton's laws of motion and of gravity, which were formulated later in the seventeenth century (§2.1). Relativistic effects also affect planetary orbits, but they are small compared to the gravitational perturbations that the planets exert on one other (Problem 2.15).

All planets and asteroids revolve around the Sun in the direction of solar rotation. Their orbital planes generally lie within a few degrees of each other and close

Table 1.4 Principal planetary satellites: Orbital data and visual magnitude at opposition.

Planet		Satellite	a (10^3 km)	Orbital period (days)	e	i (deg)	m_v
Earth		Moon	384.40	27.321 661	0.054 900	5.15[a]	−12.7
Mars	I	Phobos	9.375	0.318 910	0.015 1	1.082	11.4
	II	Deimos	23.458	1.262 441	0.000 24	1.791	12.5
Jupiter	XVI	Metis	127.98	0.294 78	0.001 2	0.02	17.5
	XV	Adrastea	128.98	0.298 26	0.001 8	0.054	18.7
	V	Almathea	181.37	0.498 18	0.003 1	0.388	14.1
	XIV	Thebe	221.90	0.674 5	0.017 7	1.070	16.0
	I	Io	421.77	1.769 138	0.004 1f	0.040	5.0
	II	Europa	671.08	3.551 810	0.010 1f	0.470	5.3
	III	Ganymede	1 070.4	7.154 553	0.001 5f	0.195	4.6
	IV	Callisto	1 882.8	16.689 018	0.007	0.28	5.6
	XIII	Leda	11 160	241	0.148	27[a]	19.5
	VI	Himalia	11 460	251	0.163	28.5[a]	14.6
	X	Lysithea	11 720	259	0.107	29[a]	18.3
	VII	Elara	11 737	260	0.207	28[a]	16.3
	XII	Ananka	21 280	610	0.169	147[a]	18.8
	XI	Carme	23 400	702	0.207	163[a]	17.6
	VIII	Pasiphae	23 620	708	0.378	148[a]	17.0
	IX	Sinope	23 940	725	0.275	153[a]	18.1
Saturn	XVIII	Pan	133.584	0.575 05	0.000 01	0.000 1	19.4
	XXXV	Daphnis	136.51	0.594 08	0.000 03	0.004	21
	XV	Atlas	137.670	0.601 69	0.001 2	0.01	19.0
	XVI	Prometheus	139.380	0.612 986	0.002 2	0.007	15.8
	XVII	Pandora	141.710	0.628 804	0.004 2	0.051	16.4
	XI	Epimetheus	151.47[b]	0.694 590[b]	0.010	0.35	15.6
	X	Janus	151.47[b]	0.694 590[b]	0.007	0.16	16.4
	I	Mimas	185.52	0.942 421 8	0.020 2	1.53f	12.8
	XXXII	Methone	194.23	1.009 58	0.000	0.02	23
	XLIX	Anthe	197.7	1.037	0.02	0.02	24
	XXXIII	Pallene	212.28	1.153 7	0.004	0.18	22
	II	Enceladus	238.02	1.370 218	0.004 5f	0.02	11.8
	III	Tethys	294.66	1.887 802	0.000 0	1.09f	10.3
	XIV	Calypso (T−)	294.66[b]	1.887 802[b]	0.000 5	1.50	18.7
	XIII	Telesto (T+)	294.66[b]	1.887 802[b]	0.000 2	1.18	18.5
	IV	Dione	377.71	2.736 915	0.002 2f	0.02	10.4
	XII	Helene (T+)	377.71[b]	2.736 915[b]	0.005	0.2	18.4
	XXXIV	Polydeuces (T−)	377.71[b]	2.736 915[b]	0.019	0.18	23
	V	Rhea	527.04	4.517 500	0.001	0.35	9.7
	VI	Titan	1 221.85	15.945 421	0.029 2	0.33	8.4
	VII	Hyperion	1 481.1	21.276 609	0.104 2f	0.43	14.4
	VIII	Iapetus	3 561.3	79.330 183	0.028 3	7.52	11.0[c]
	IX	Phoebe	12 952	550.48	0.164	175.3[a]	16.5
	XX	Paaliaq	15 198	687	0.36	45[a]	21.2
	XXVI	Albiorix	16 394	783	0.48	34[a]	20.4
	XXIX	Siarnaq	18 195	896	0.3	46[a]	20.0

(cont.)

Table 1.4 (cont.)

Planet		Satellite	a (10^3 km)	Orbital period (days)	e	i (deg)	m_v
Uranus	VI	Cordelia	49.752	0.335 033	0.000	0.1	24.2
	VII	Ophelia	53.764	0.376 409	0.010	0.1	23.9
	VIII	Bianca	59.166	0.434 577	0.000 3	0.18	23.1
	IX	Cressida	61.767	0.463 570	0.000 2	0.04	22.3
	X	Desdemona	62.658	0.473 651	0.000 3	0.10	22.5
	XI	Juliet	64.358	0.493 066	0.000 1	0.05	21.7
	XII	Portia	66.097	0.513 196	0.000 5	0.03	21.1
	XIII	Rosalind	69.927	0.558 459	0.000 6	0.09	22.5
	XXVII	Cupid	74.393	0.612 825	~0	~0	25.9
	XIV	Belinda	75.256	0.623 525	0.000	0.0	22.1
	XXV	Perdita	76.417	0.638 019	0.003	~0	23.6
	XV	Puck	86.004	0.761 832	0.000 4	0.3	20.6
	XXVI	Mab	97.736	0.922 958	0.002 5	0.13	25.4
	V	Miranda	129.8	1.413	0.002 7	4.22	15.8
	I	Ariel	191.2	2.520	0.003 4	0.31	13.7
	II	Umbriel	266.0	4.144	0.005 0	0.36	14.5
	III	Titania	435.8	8.706	0.002 2	0.10	13.5
	IV	Oberon	582.6	13.463	0.000 8	0.10	13.7
	XVI	Caliban	7 231	580	0.16	141[a]	22.4
	XX	Stephano	8 004	677	0.23	144[a]	24.1
	XVII	Sycorax	12 179	1288	0.52	159[a]	20.8
	XVIII	Prospero	16 256	1978	0.44	152[a]	23.2
	XIX	Setebos	17 418	2225	0.59	158[a]	23.3
Neptune	III	Naiad	48.227	0.294 396	0.00	4.74	24.6
	IV	Thalassa	50.075	0.311 485	0.00	0.21	23.9
	V	Despina	52.526	0.334 655	0.00	0.07	22.5
	VI	Galatea	61.953	0.428 745	0.00	0.05	22.4
	VII	Larissa	73.548	0.554 654	0.00	0.20	22.0
	VIII	Proteus	117.647	1.122 315	0.00	0.55	20.3
	I	Triton	354.76	5.876 854	0.00	156.834	13.5
	II	Nereid	5 513.4	360.136 19	0.751	7.23[a]	19.7
	IX	Halimede	15 686	1875	0.57	134[a]	24.4
	XI	Sao	22 452	2919	0.30	48[a]	25.7
	XII	Laomedeia	22 580	2982	0.48	35[a]	25.3
	XIII	Neso	46 570	8863	0.53	132[a]	24.7
	X	Psamathe	46 738	9136	0.45	137[a]	25.1

Data are from Yoder (1995), with updates from Showalter and Lissauer (2006), Jacobson *et al.* (2009), Nicholson (2009), Jacobson, (2010), http://ssd.jpl.nasa.gov, and other sources.

i = orbit plane inclination with respect to the parent planet's equator, except where noted.

Abbreviations: T, Trojan-like satellite which leads (+) or trails (−) by ~60° in longitude the primary satellite with same semimajor axis; f, forced eccentricity or inclination.

[a] measured relative to the planet's heliocentric orbit, because the Sun (rather than the planetary oblateness) controls the local Laplacian plane of these distant satellites.

[b] varies due to coorbital libration; value shown is long-term average.

[c] varies substantially with orbital longitude; average value is shown.

Table 1.5 Planetary satellites: Physical properties and rotation rates.

Satellite	Radius (km)	Mass (10^{23} g)	Density (g cm^{-3})	Geom. albedo	Rot. period (days)
Earth	$6378^2 \times 6357$	59 742	5.515	0.367	0.997
Moon	1737.53 ± 0.03	734.9	3.34	0.12	S
Mars	$3396^2 \times 3376$	6419	3.933	0.150	1.026
MI Phobos	$13.1 \times 11.1 \times 9.3 (\pm 0.1)$	1.063×10^{-4}	1.90	0.06	S
MII Deimos	$(7.8 \times 6.0 \times 5.1)(\pm 0.2)$	1.51×10^{-5}	1.50	0.07	S
Jupiter	$71\,492^2 \times 66\,854$	1.8988×10^7	1.326	0.52	0.414
JXVI Metis	$(30 \times 20 \times 17)(\pm 2)$			0.06	S
JXV Adrastea	$(10 \times 8 \times 7)(\pm 2)$			0.1	S
JV Amalthea	$(125 \times 73 \times 64)(\pm 2)$			0.09	S
JXIV Thebe	$(58 \times 49 \times 42)(\pm 2)$			0.05	
JI Io	1821.3 ± 0.2	893.3 ± 1.5	3.53 ± 0.006	0.61	S
JII Europa	1565 ± 8	479.7 ± 1.5	3.02 ± 0.04	0.64	S[a]
JIII Ganymede	2634 ± 10	1482 ± 1	1.94 ± 0.02	0.42	S
JIV Callisto	2403 ± 5	1076 ± 1	1.85 ± 0.004	0.20	S
JVI Himalia	85 ± 10	0.042 ± 0.006			0.324
JVII Elara	40 ± 10				0.5
Saturn	$60\,268^2 \times 54\,364$	5.6850×10^6	0.687	0.47	0.44
SXVIII Pan	$17 \times 16 \times 10$	5×10^{-5}	0.41 ± 0.15	0.5	S
SXXXV Daphnis	$(4.5 \times 4.3 \times 3.1)(\pm 0.8)$	8×10^{-7}	0.34 ± 0.21		
SXV Atlas	$21 \times 18 \times 9$	7×10^{-5}	0.46 ± 0.1	0.9	S
SXVI Prometheus	$68 \times 40 \times 30$	0.0016	0.48 ± 0.09	0.9	S
SXVII Pandora	$52 \times 41 \times 32$	0.001 37	0.49 ± 0.09	0.9	S
SXI Epimetheus	$65 \times 57 \times 53$	0.0053	0.64	0.8	S
SX Janus	$102 \times 93 \times 76$	0.019	0.63 ± 0.06	0.8	S
SI Mimas	$208 \times 196 \times 191$	0.38	1.15	0.5	S
SXXXIII Pallene	$3 \times 3 \times 2$				
SII Enceladus	$257 \times 251 \times 248$	0.65	1.61	1.0	S
SIII Tethys	533 ± 2	6.27	0.99	0.9	S
SXIV Calypso	$15 \times 11.5 \times 7$			0.6	
SXIII Telesto	$16 \times 12 \times 10$			0.5	
SIV Dione	561.7 ± 0.9	11.0	1.48	0.7	S
SXII Helene	$22 \times 19 \times 13$			0.7	
SXXXIV Polydeuces	$(1.5 \times 1.2 \times 1.0)(\pm 0.4)$				
SV Rhea	764 ± 2	23.1	1.24	0.7	S
SVI Titan	2575 ± 2	1345.7	1.88	0.21	$\sim$S
SVII Hyperion	$(180 \times 133 \times 103)(\pm 4)$	0.054	0.6	0.2–0.3	C
SVIII Iapetus	$746 \times 746 \times 712$	18.1 ± 1.5	1.09	0.05–0.5	S
SIX Phoebe	$109 \times 109 \times 102$	0.083	1.64	0.08	0.387
Uranus	$25\,559^2 \times 24\,973$	8.6625×10^5	1.318	0.51	0.718
UVI Cordelia	13 ± 2			0.07	
UVII Ophelia	16 ± 2			0.07	
UVIII Bianca	22 ± 3			0.07	
UIX Cressida	33 ± 4			0.07	
UX Desdemona	29 ± 3			0.07	

(cont.)

Table 1.5 (*cont.*)

Satellite	Radius (km)	Mass (10^{23} g)	Density (g cm^{-3})	Geom. albedo	Rot. period (days)
UXI Juliet	42 ± 5			0.07	
UXII Portia	55 ± 6			0.07	
UXIII Rosalind	29 ± 4			0.07	
UXIV Belinda	34 ± 4			0.07	
UXV Puck	77 ± 3			0.07	
UV Miranda	$240(0.6) \times 234.2(0.9)$ $\times 232.9(1.2)$	0.659 ± 0.075	1.20 ± 0.14	0.27	S
UI Ariel	$581.1(0.9) \times 577.9(0.6)$ $\times 577.7(1.0)$	13.53 ± 1.20	1.67 ± 0.15	0.34	S
UII Umbriel	584.7 ± 2.8	11.72 ± 1.35	1.40 ± 0.16	0.18	S
UIII Titania	788.9 ± 1.8	35.27 ± 0.90	1.71 ± 0.05	0.27	S
UIV Oberon	761.4 ± 2.6	30.14 ± 0.75	1.63 ± 0.05	0.24	S
Neptune	$24\,764^2 \times 24\,342$	1.0278×10^6	1.638	0.41	0.671
NV Despina	74 ± 10			0.06	
NVI Galatea	79 ± 12			0.06	
NVII Larissa	$104 \times 89(\pm 7)$			0.06	
NVIII Proteus	$218 \times 208 \times 201$			0.06	
NI Triton	1352.6 ± 2.4	214.7 ± 0.7	2.054 ± 0.032	0.7	S
NII Nereid	170 ± 2.5			0.2	0.48

Most data are from Yoder (1995), with updates from http://ssd.jpl.nasa.gov, Porco *et al.* (2007), Jacobson *et al.* (2008), Thomas *et al.* (1998, 2007), Thomas (2010), and Pilcher *et al.* (2012).

Abbreviations: S, synchronous rotation; C, chaotic rotation.

[a] Europa's ice crust may rotate slightly faster than synchronous.

to the solar equator. For observational convenience, inclinations are usually measured relative to the Earth's orbital plane, which is known as the *ecliptic plane*. Dynamically speaking, the best choice would be the *invariable plane*, which passes through the center of mass and is perpendicular to the angular momentum vector of the Solar System. The Solar System's invariable plane is nearly coincident with the plane of Jupiter's orbit, which is inclined by $1.3°$ relative to the ecliptic. In this book, we follow standard conventions and measure inclinations of heliocentric orbits with respect to the ecliptic plane and inclinations of planetocentric orbits relative to the planet's equator. The Sun's equatorial plane is inclined by $7°$ with respect to the ecliptic plane. Among the eight major planets, Mercury's orbit is the most tilted, with $i = 7°$. (However, because inclination is effectively a vector, the similarity of these two inclinations does not imply that Mercury's orbit lies within the plane of the solar equator. Indeed, Mercury's orbit is inclined by $3.4°$ relative to the Sun's equatorial plane.) Similarly, most major satellites orbit their planet close to its equatorial plane. Many smaller objects that

orbit the Sun and the planets have much larger orbital inclinations. In addition, some comets, minor satellites, and Neptune's large moon Triton orbit the Sun/planet in a *retrograde* sense (opposite to the Sun's/planet's rotation). The observed 'flatness' of most of the planetary system is explained by planetary formation models that hypothesize that the planets grew within a disk which was in orbit around the Sun (Chapter 13).

1.2.2 Mass

The mass of an object can be deduced from the gravitational force that it exerts on other bodies.

- Orbits of moons: The orbital periods of natural satellites, together with Newton's generalization of Kepler's third law (eq. 2.11), can be used to solve for mass. The result is actually the sum of the mass of the planet and moon (plus, to a good approximation, the masses of moons on orbits interior to the one being considered), but except for the Earth/Moon and various minor planets, including Pluto/Charon, the secondaries' masses are very small compared to that of the primary. The major source of uncertainty in this method results from measurement errors in the semimajor axis; timing errors are negligible.

• What about planets without moons? The gravity of each planet perturbs the orbits of all other planets. Because of the large distances involved, the forces are much smaller, so the accuracy of this method is not high. Note, however, that Neptune was discovered as a result of the perturbations that it forced on the orbit of Uranus. This technique is still used to provide the best (albeit in some cases quite crude) estimates of the masses of some large asteroids. The perturbation method can actually be divided into two categories: short-term and long-term perturbations. The extreme example of short-term perturbations are single close encounters between asteroids. Trajectories can be computed for a variety of assumed masses of the body under consideration and fit to the observed path of the other body. Long-term perturbations are best exemplified by masses derived from periodic variations in the relative positions of moons locked in stable orbital resonances (Chapter 2).

• Spacecraft tracking data provide the best means of determining masses of planets and moons visited, as the Doppler shift and periodicity of the transmitted radio signal can be measured very precisely. The long time baselines afforded by *orbiter* missions allow much higher accuracy than *flyby* missions. The best estimates for the masses of some of the outer planet moons are those obtained by combining accurate short-term perturbation measurements from Voyager images with Voyager tracking data and/or resonance constraints from long timeline ground-based observations.

• The best estimates of the masses of some of Saturn's small inner moons were derived from the amplitude of spiral density waves they resonantly excite in Saturn's rings or of density wakes that they produce in nearby ring material. These processes are discussed in Chapter 11.

• Crude estimates of the masses of some comets have been made by estimating nongravitational forces, which result from the asymmetric escape of released gases and dust (Chapter 10), and comparing them with observed orbital changes.

The gravity field of a mass distribution that is not spherically symmetric differs from that of a point source of identical mass. Such deviations, combined with the knowledge of the rotation period, can be used to estimate the degree of central concentration of mass in rotating bodies (Chapter 6). The deviation of the gravity field of an asymmetric body from that of a point mass is most pronounced, and thus most easily measured, closest to the body (§2.5). To determine the precise gravity field, one can

make use of both spacecraft tracking data and the orbits of moons and/or eccentric rings.

1.2.3 Size

Bodies in the Solar System exhibit a wide range of sizes and shapes. The size of an object can be measured in various ways:

• The diameter of a body is the product of its angular size (measured in radians) and its distance from the observer. Solar System distances are simple to estimate from orbits; however, limited resolution from Earth results in large uncertainties in angular size. Thus, other techniques often give the best results for bodies not imaged at close distances by interplanetary spacecraft.

• The diameter of a Solar System body can be deduced by observing a star as it is occulted by the body. The angular velocity of the star relative to the occulting body can be calculated from orbital data, including the effects of the Earth's orbit and rotation. Multiplying the duration of an occultation as viewed from a particular observing site by its angular velocity and its distance gives the length of a chord of the body's projected silhouette. Three well-separated chords suffice for a spherical planet. Many chords are needed if the body is irregular in shape, and observations of the same event from many widely spaced telescopes are necessary. This technique is particularly useful for small bodies which have not been visited by spacecraft. Occultations of sufficiently bright stars are infrequent and require appropriate predictions as well as significant observing campaigns in order to obtain enough chords (even if some sites are clouded out); thus, occultation diameters exist only for a miniscule fraction of the known small bodies in our Solar System.

• Radar echoes can be used to determine radii and shapes (§E.7). The radar signal strength drops as $1/r^4$ ($1/r^2$ going to the object and $1/r^2$ returning to the antenna), so only relatively nearby objects may be studied with radar. Radar is especially useful for studying solid planets, asteroids, and cometary nuclei.

• An excellent way to measure the radius of an object is to send a lander, and triangulate using an orbiter. This method, as well as the radar technique, also works well for terrestrial planets and satellites with substantial atmospheres.

• The size and the albedo of a body can be estimated by combining photometric observations at visible and infrared wavelengths. At visible wavelengths one measures the sunlight reflected off the object, while at infrared wavelengths one observes the thermal radiation from the body itself (see Chapter 3 and §E.3 for a detailed discussion).

The mean density of an object can be trivially determined once its mass and size are known. The density of an object gives a rough idea of its composition, although compression at the high pressures which occur in planets and large moons must be taken into account, and the possibility of significant void space should be considered for small bodies. The low density ($\sim 1 \mathrm{\,g\,cm^{-3}}$) of the four giant planets, for example, implies material with small mean molecular weight. Terrestrial planet densities of 3.5–5.5 $\mathrm{g\,cm^{-3}}$ imply rocky material, including some metal. Most of the medium and large satellites around the giant planets have densities between 1 and 2 $\mathrm{g\,cm^{-3}}$, suggesting a combination of ices and rock. Comets have densities of the order of 1 $\mathrm{g\,cm^{-3}}$ or less, indicative of rather loosely packed dirty ices.

In addition to the density, one can also calculate the escape velocity using the mass and size of the object (eq. 2.16). The escape velocity, together with temperature, can be used to estimate the ability of the planetary body to retain an atmosphere.

1.2.4 Rotation

Simple rotation is a vector quantity, related to spin angular momentum. The *obliquity* (or *axial tilt*) of a planetary body is the angle between its spin angular momentum and its orbital angular momentum. Bodies with obliquity <90° are said to have *prograde* rotation, whereas planets with obliquity >90° have *retrograde* rotation. The rotation of an object can be determined using various techniques:

• The most straightforward way to determine a planetary body's rotation axis and period is to observe how markings on the surface move around with the disk. Unfortunately, not all planets have such features; moreover, if atmospheric features are used, winds may cause the deduced period to vary with latitude, altitude, and time.

• Planets with sufficient magnetic fields trap charged particles within their magnetospheres. These charged particles are accelerated by electromagnetic forces and emit radio waves. As magnetic fields are not uniform in longitude, and as they rotate with (presumably the bulk of) the planet, these radio signals have a periodicity equal to the planet's rotation period. For planets without detectable solid surfaces, the magnetic field period is viewed as more fundamental than the periods of cloud features (see, however, §7.5.5.1).

• The rotation period of a body can often be determined by periodicities observed in its *lightcurve*, which gives the total disk brightness as a function of time. Lightcurve

variations can be the result of differences in albedo or, for irregularly shaped bodies, in projected area. Irregularly shaped bodies produce lightcurves with two very similar maxima and two very similar minima per revolution, whereas albedo variations have no such preferred symmetry. Thus, ambiguities of a factor of two sometimes exist in spin periods determined by lightcurve analysis. Most asteroids have double-peaked lightcurves, indicating that the major variations are due to shape, but the peaks are distinguishable from each other because of minor variations in hemispheric albedo and local topography.

• The measured Doppler shift across the disk can give a rotation period and a crude estimate of the rotation axis, provided the body's radius is known. This can be done passively in visible light, or actively using radar.

The rotation periods of most objects orbiting the Sun are of the order of three hours to a few days. Mercury and Venus, both of whose rotations have almost certainly been slowed by solar tides, form exceptions with periods of 59 and 243 days respectively. Six of the eight planets rotate in a prograde sense with obliquities of 30° or less. Venus rotates in a retrograde direction with an obliquity of 177°, and the rotation axis of Uranus lies very close to the plane of Uranus's orbit. Most planetary satellites rotate synchronously with their orbital periods as a result of planet-induced tides (§2.6.2).

1.2.5 Shape

Many different forces together determine the shape of a body. Self-gravity tends to produce bodies of spherical shape, a minimum for gravitational potential energy. Material strength maintains shape irregularities, which may be produced by accretion, impacts, or internal geological processes. As self-gravity increases with the size of an object, larger bodies tend to be rounder. Typically, bodies with mean radii larger than $\sim 200 \mathrm{\,km}$ are fairly round. Smaller objects may be quite oddly shaped (Fig. 1.5a).

There is a relationship between a planet's rotation and its oblateness, since the rotation introduces a centrifugal pseudoforce, which causes a planet to bulge out at the equator and to flatten at the poles. A perfectly fluid planet would be shaped as an oblate spheroid. Polar flattening is greatest for planets which have a low density and rapid rotation. In the case of Saturn, the flattening parameter, $\epsilon \equiv (R_e - R_p)/R_e$, where R_e and R_p are the equatorial and polar radii, respectively, is ~ 0.1, and polar flattening is easily discernible on some images of the planet (e.g., Fig. 1.5b).

(a)

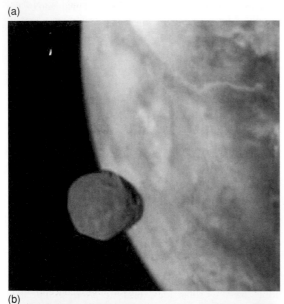

(b)

Figure 1.5 (a) Image of the small irregularly-shaped moon Phobos against the background of the limb of the nearly spherical planet Mars. Phobos appears much larger relative to Mars than it actually is, because the Soviet spacecraft Phobos 2 was much closer to the moon than to the planet when it took this image. (b) Hubble Space Telescope image of Saturn taken on 24 February 2009, less than five months prior to saturnian equinox passage. The rings are seen at a low tilt angle, with the ring shadow appearing across the planet just above the rings. Four moons are seen to be transiting (partially eclipsing the planet); from left to right they are Enceladus, Dione, Titan, and Mimas; the shadows of Enceladus and Dione can also be seen. Note the pronounced oblateness of the low-density, rapidly rotating planet. (NASA/STSci/Hubble Heritage)

The shape of an object can be determined from:

- Direct imaging, either from the ground or spacecraft.
- Length of chords observed by stellar occultation experiments at various sites (see §1.2.3).
- Analysis of radar echoes (see §E.7).
- Analysis of lightcurves. Several lightcurves obtained from different viewing angles are required for accurate measurements (Fig. 9.4).
- The shape of the *central flash*, which is observed when the center of a body with an atmosphere passes in front of an occulted star. The central flash results from the focusing of light rays refracted by the atmosphere and can be seen only under fortuitous observing circumstances (§E.5).

1.2.6 Temperature

The equilibrium temperature of a planet can be calculated from the energy balance between solar insolation and reradiation outward (Chapter 3). However, for many planets internal heat sources provide a significant contribution. Moreover, there may be diurnal, latitudinal, and seasonal variations in the temperature. The *greenhouse effect*, a thermal 'blanket' caused by an atmosphere which is more transparent to visible radiation (the Sun's primary output) than to infrared radiation from the planet, raises the surface temperature on some planets far above the equilibrium blackbody value. For example, because of the high albedo of its clouds, Venus actually absorbs less solar energy per unit area than does Earth; thus (as internal heat sources on these two planets are negligible compared to solar heating), the effective radiating temperature of Venus is lower than that of Earth. However, as a consequence of the greenhouse effect, Venus's surface temperature is raised up to $\sim$730 K, well above the surface temperature on Earth.

Direct *in situ* measurements with a thermometer can provide an accurate estimate of the temperature of the accessible (outer) parts of a body. The thermal infrared spectrum of a body's emitted radiation is also a good indicator of the temperature of its surface or cloud tops. Most solid and liquid planetary material can be characterized as a nearly perfect blackbody radiator with its emission peak at near- to mid-infrared wavelengths. Analysis of emitted radiation sometimes gives different temperatures at differing wavelengths. This could be due to a combination of temperatures from different locations on the surface, e.g., pole to equator differences, albedo variations, or volcanic hot spots such as those seen on Io. Also, the opacity of an atmosphere varies with wavelength, which allows us to remotely probe different altitudes in a planetary atmosphere.

1.2.7 Magnetic Field

Magnetic fields are created by moving charges. Currents moving through a solid medium decay quickly (unless the medium is a superconductor, which is unreasonable to expect at the high temperatures found in planetary interiors). Thus, internally generated planetary magnetic fields must either be produced by a (poorly understood) *dynamo* process, which can only operate in a fluid region of a planet, or be due to *remanent ferromagnetism*, which is a result of charges that are bound to atoms of a solid locked in an aligned configuration. Remanent ferromagnetism is not viewed to be a likely cause of large fields because, in addition to the fact that it is expected to decay away on timescales short compared to the age of the Solar System, it would require the planet to have been subjected to a nearly constant (in direction) magnetic field during the long period in which the bulk of its iron cooled through its Curie point. (At temperatures below the Curie point of a ferromagnetic material, the magnetic moments are partially aligned within magnet domains.) Magnetic fields may also be induced through the interaction between the solar wind (which is composed predominantly of charged particles) and conducting regions within the planet or its ionosphere.

A magnetic field may be detected directly using an *in situ* magnetometer or indirectly via effects of accelerating charges which consequently produce radiation (radio emissions). The presence of localized *aurorae*, luminous disturbances caused by charged particle precipitation in a planet's upper atmosphere, is also indicative of a magnetic field. The magnetic fields of the planets can be approximated by dipoles, with perturbations to account for their irregularities. All four giant planets, as well as Earth, Mercury, and Jupiter's moon Ganymede, have magnetic fields generated in their interiors. Venus and comets have magnetic fields induced by the interaction between the solar wind and charged particles in their atmosphere/ionosphere, while Mars and the Moon have localized crustal magnetic fields. Perturbations in Jupiter's magnetic field near Europa and Callisto are indicative of salty oceans in the interiors of these moons (§7.5.4.9). Geyser activity on Enceladus perturbs Saturn's magnetic field (§7.5.5.1).

1.2.8 Surface Composition

The composition of a body's surface can be derived from:

- Spectral reflectance data. Such spectra may be observed from Earth; however, spectra at ultraviolet wavelengths can only be obtained above the Earth's atmosphere.

- Thermal infrared spectra and thermal radio data. Though difficult to interpret, these measurements contain information about a body's composition.

- Radar reflectivity. Such observations can be carried out from Earth or from nearby spacecraft.

- X-ray and γ-ray fluorescence. These measurements may be conducted from a spacecraft in orbit around the planet (or, in theory, even a flyby spacecraft) if the body lacks a substantial atmosphere. Detailed measurements require landing a probe on the body's surface.

- Chemical analysis of surface samples. This can be performed on samples brought to Earth by natural processes (meteorites) or spacecraft, or (in less detail) by *in situ* analysis using spacecraft. Other forms of *in situ* analysis include mass spectroscopy and electrical and thermal conductivity measurements.

The compositions of the planets, asteroids and satellites show a dependence on heliocentric distance, with the objects closest to the Sun having the largest concentrations of dense materials (which tend to be *refractory*, i.e., have high melting and boiling temperatures) and the smallest concentration of ices (which are much more *volatile*, i.e., have much lower melting and boiling temperatures).

1.2.9 Surface Structure

The surface structure varies greatly from one planet or moon to another. There are various ways to determine the structure of a planet's surface:

- Structure on large scales (e.g., mountains) can be detected by imaging, either passively in the visible/IR/radio or actively using radar imaging techniques. It is best to have imaging available at more than one illumination angle in order to separate tilt-angle (slope) effects from albedo differences.

- Structure on small scales (e.g., grain size) can be deduced from the radar echo brightness and the variation of reflectivity with *phase angle*, the angle between the illuminating Sun and the observer as seen from the body. The brightness of a body with a size much larger than the wavelength of light at which it is observed generally increases slowly with decreasing phase angle. For very small phase angles, this increase can be much more rapid, a phenomenon referred to as the *opposition effect* (§E.1).

1.2.10 Atmosphere

Most of the planets and some satellites are surrounded by significant atmospheres. The giant planets Jupiter, Saturn, Uranus, and Neptune are basically huge fluid balls, and their atmospheres are dominated by H_2 and He. Venus has a very dense CO_2 atmosphere, with clouds so thick that one cannot see its surface at visible wavelengths; Earth has an atmosphere consisting primarily of N_2 (78%) and O_2 (21%), and Mars has a more tenuous CO_2 atmosphere. Saturn's satellite Titan has a dense nitrogen-rich atmosphere, which is intriguing since it contains many kinds of organic molecules. Pluto and Neptune's moon Triton each have a tenuous atmosphere dominated by N_2, while the atmosphere of Jupiter's volcanically active moon Io consists primarily of SO_2. Mercury and the Moon each have an extremely tenuous atmosphere ($\lesssim 10^{-12}$ bar); Mercury's atmosphere is dominated by atomic O, Na, and He, while the main constituents in the Moon's atmosphere are He and Ar. The gaseous components of cometary comae are essentially temporary atmospheres in the process of escaping.

The composition and structure (temperature–pressure profile) of an atmosphere can be determined from: spectral reflectance data at visible wavelengths, thermal spectra and photometry at infrared and radio wavelengths, stellar occultation profiles, *in situ* mass spectrometers and attenuation of radio signals sent back to Earth by atmospheric/surface probes (Appendix E).

1.2.11 Interior

The interior of a planet is not directly accessible to observations. However, with help of the observable parameters discussed above, one can derive information on a planet's bulk composition and its interior structure.

The *bulk composition* is not an observable attribute, except for extremely small bodies such as meteorites which we can actually take apart and analyze (Chapter 8). Thus, we must deduce bulk composition from a variety of direct and indirect clues and constraints. The most fundamental constraints are based on the mass and the size of the planet. Using only these constraints, together with material properties derived from laboratory data and quantum mechanical calculations, it can be shown that Jupiter and Saturn are composed mostly of hydrogen, simply because all other elements are too dense to fit the constraints (unless the internal temperature is much higher than is consistent with the observed effective temperature in a quasi-steady state). However, this method only gives definitive results for planets composed mostly of the lightest element. For all other bodies, bulk composition is best estimated from models which include mass and radius as well as the composition of the surface and atmosphere, the body's heliocentric distance (location is useful because it gives us an idea of the temperature of the region during the planet-formation epoch, and thus which elements were

likely to condense), together with reasonable assumptions of cosmogonic abundances (§1.4 and Chapters 8 and 13).

The *internal structure* of a planet can be derived to some extent from its gravitational field and the rotation rate. From these parameters, one can estimate the degree of concentration of the mass at the planet's center. The gravitational field can be determined from spacecraft tracking and the orbits of satellites or rings (Chapter 2). Detailed information on the internal structure of a planet with a solid surface may be obtained if seismometers can be placed on its surface, as was done for the Moon by Apollo astronauts. The velocities and attenuations of seismic waves propagating through the planet's interior depend on density, rigidity and other physical properties (Chapter 6), which in turn depend on composition, as well as on pressure, temperature, and time. Reflection and refraction off internal boundaries provides information on layering. The free oscillation periods of gaseous planets can, in theory, also provide clues to internal properties, just as helioseismology now provides important information about the Sun's interior. Evidence of volcanism and plate tectonics constrain the thermal environment below the surface. Energy output provides information on the thermal structure of a planet's interior.

The response of moons that are subject to significant time-variable tidal deformations depends upon their internal structure. Repeated observations of such moons can reveal internal properties, including in some cases the presence of a subterranean fluid layer. Combining altimetry with satellite gravitometry could give indications about lateral inhomogeneities under the surface of icy moons and thus, for instance, indicate volcanic sources and tectonic structures.

Magnetic fields are produced by moving charges. While a small magnetic field such as the Moon's may be the result of remanent ferromagnetism, substantial planetary magnetic fields are thought to require a conducting fluid region within the planet's interior. Centered dipole fields are probably produced in or near the core of the planet, whereas highly irregular offset fields are likely to be produced closer to the planet's surface.

1.3 Stellar Properties and Lifetimes

Planets are intimately related to stars, their larger and much more luminous companions. Stellar gravity dominates planetary motions (Chapter 2). The luminosity of our star, the Sun, is the primary source of energy for most planets (Chapter 3), and solar energy inputs dominate planetary weather (Chapter 4). The solar wind affects and controls planetary magnetospheres (Chapter 7), and solar heating is responsible for cometary activity (Chapter 10). Moreover, past generations of stars produced most of the elements out of which terrestrial planets are composed (§13.2.2), and stars and planets form together (§1.4 and Chapter 13). Thus, understanding a few basic properties of stars is quite useful for learning about planets. More details on the physical distinctions between stars, planets, and intermediate mass objects known as brown dwarfs are presented in §12.1.

Stars are huge balls of gas and *plasma* (ionized gas) that radiate energy from their surfaces and liberate energy via thermonuclear fusion reactions in their interiors. The internal structure of a star is determined primarily by a balance between gravity and pressure. Fusion reaction rates are extremely sensitive to temperature (§13.2.2), so a very small warming of the interior enables the star to release nuclear energy much more rapidly. A quasi-equilibrium state is maintained because if the interior gets too cool, the star's core contracts and heats up, whereas if it becomes too hot, pressure builds and the core expands and cools.

During the star's long-lived *main sequence* phase, hydrogen in its core is gradually 'burned up' (fused into helium) to maintain the pressure balance. High-mass stars are much more luminous than low-mass stars, as greater pressure and hence higher temperature is required to balance their larger gravity. Along the main sequence, stellar luminosity, $\mathcal{L}_\star$, is roughly proportional to the fourth power of a star's mass, $M_\star$:

$$\mathcal{L}_\star \propto M_\star^4. \tag{1.1}$$

As the amount of hydrogen fuel in the core increases approximately linearly with the star's mass, stellar lifetime varies inversely with the cube of the star's mass. Figure 1.6 shows the relationship between stellar mass and luminosity more accurately. More-massive stars are larger than low-mass stars, but an equally important reason that they are able to radiate much more energy is that they are hotter (bluer) than low-mass stars. The physics of energy radiation is discussed in §3.1.1.

Stars range in mass from 0.08 $M_\odot$ (solar masses) to a little more than 100 $M_\odot$. Smaller objects cannot sustain sufficient fusion in their cores to balance gravitational contraction (§12.1), whereas radiation pressure resulting from high luminosity would blow away the outer layers of more massive bodies. Low-mass stars are much more common than high-mass stars, but because high-mass stars are vastly more luminous, they can be seen from much further away, and the majority of stars visible to the naked eye are more massive than the Sun.

Although stellar luminosities vary by many orders of magnitude (Figure 1.7), in other respects stars are a

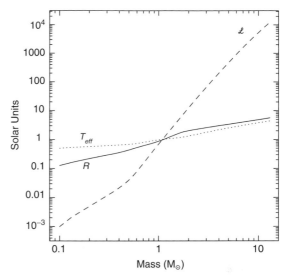

Figure 1.6 Logarithm of the radius (solid line), radiating temperature (dotted line), and luminosity (dashed line) of a zero-age main sequence star as functions of the star's mass. All stars are taken to have solar composition, and all quantities are ratioed to those of the Sun at the present epoch. The models range from 0.1 to 13 $M_\odot$ and were generated with the CESAM code (Morel 1997). (Courtesy Jason Rowe)

relatively homogeneous class of objects. Stars are supported against gravitational collapse by thermal pressure that is maintained by fusion reactions in their interiors;[2] they range in mass by a little over 3 orders of magnitude. In contrast, even the most conservative definition of planets encompasses a more diverse family of objects within our Solar System alone. These include Mercury, a condensed body composed primarily of iron and other heavy elements, and Jupiter, a fluid object that is almost 4 orders of magnitude more massive than Mercury and is made mostly of hydrogen and helium. And some definitions of planets include Pluto and extrasolar objects more than ten times as massive as Jupiter (Chapter 12), extending the range to over 6 orders of magnitude in mass.

A star's luminosity grows slowly during the star's main sequence phase, as fusion increases the mean particle mass in the core and greater temperature is required for pressure to balance gravity. Once hydrogen in the core is completely used up, the core shrinks. Hydrogen burning occurs in a shell of material surrounding the helium-rich core. The star's luminosity increases significantly, and its outer layers swell and cool, turning the star into a *red giant*. If the star is at least ~60% as massive as our Sun, the core

---

[2] Stellar mass objects such as *white dwarfs* and *neutron stars* that have exhausted their nuclear fuel are properly referred to as *stellar remnants* rather than stars.

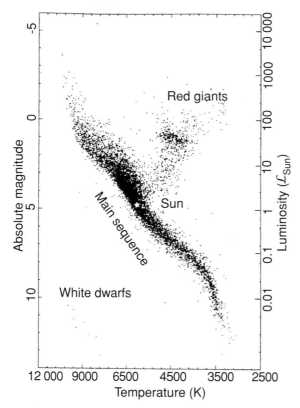

Figure 1.7 *Hertzsprung–Russell* (H–R) *diagram* for single stars from the Hipparcos catalog with the most accurately known distances and magnitudes, i.e., nearby bright stars. (Adapted from Perryman *et al.* 1995)

gets hot and dense enough for helium to fuse into carbon and oxygen. (*Electron degeneracy pressure*, which results from the inability of two electrons to occupy the same quantum state, stops the contraction at too low a temperature in smaller stars.) A helium-burning equilibrium analogous to the main sequence exists, but it lasts a much shorter time, as the star is more luminous (more thermal pressure is needed to balance gravity in the star's denser core) and helium burning liberates far less energy than does hydrogen burning (Fig. 13.1). Thus, the helium fuel is exhausted more rapidly than was hydrogen.

In solar mass stars, electron degeneracy pressure prevents the star from attaining temperatures required for fusion to produce elements more massive than carbon and oxygen. *White dwarfs* are remnants of small- and medium-sized stars in which electron degeneracy pressure provides the primary support against gravitational collapse. But in very massive stars, fusion continues until iron, the most stable nucleus, is produced in the core. No energy can be liberated by nuclear fusion beyond iron, so

the core collapses, rapidly releasing an immense amount of gravitational energy. This energy can fuel a *supernova* explosion, freeing some of the heavy elements that the star has produced to be incorporated into subsequent generations of stars and planets, and leaving a neutron star or black-hole remnant.

1.4 Formation of the Solar System

Questions concerning the formation of the Solar System are among the most intellectually challenging in planetary science. Observations provide direct information on the current state of the Solar System, but only indirect clues to its origin. Thus, even though placing a chapter on Solar System formation at the beginning of this book would make sense from a chronological perspective, we have chosen to defer such a discussion to the end so that the reader can have more clues in hand when we attempt to piece together the puzzle. Nonetheless, it is useful for the student to begin with a brief overview of the currently accepted model of planetary formation because it provides a framework for interpreting unobservable planetary properties, such as the compositions of planetary interiors (Chapter 6) and of extrasolar planets (Chapter 12), and it motivates the study of objects like meteorites (Chapter 8). While there is some component of circular reasoning to this arrangement, it emphasizes that scientific development is not linear, and that placing a new piece in the puzzle is aided by the perspectives we have from the pieces already in place, but also occasionally requires revising some previously accepted ideas.

The nearly planar and almost circular orbits of the planets in our Solar System argue strongly for planetary formation within a flattened circumsolar disk. Astrophysical models suggest that such disks are a natural byproduct of star formation from the collapse of rotating cores of molecular clouds. Observational evidence for the presence of disks of Solar System dimensions around young stars has increased substantially in recent years, and infrared excesses in the spectra of young stars suggest that the lifetimes of protoplanetary disks range from 10^6 to 10^7 years.

Our galaxy contains many molecular clouds, most of which are several orders of magnitude larger than our Solar System. *Molecular clouds* are the coldest and densest regions of the interstellar medium. They are inhomogeneous, and the densest regions of molecular clouds are referred to as *cores*. These are the sites in which star formation occurs at the current epoch. Even a very slowly rotating molecular cloud core has far too much spin angular momentum to collapse down to an object of stellar dimensions, so a significant fraction of the material in a collapsing core falls onto a rotationally-supported disk orbiting the pressure-supported (proto)star. Such a disk has the same initial elemental composition as the growing star. At sufficient distances from the central star, it is cool enough for ∼1–2% of this material to be in solid form, either remnant interstellar grains or condensates formed within the disk. This dust is primarily composed of rock-forming compounds within a few AU of a $1\,M_\odot$ star, whereas in the cooler, more distant regions, the amount of ices (H_2O, CH_4, CO, etc.) present in solid form is comparable to that of rocky solids.

During the infall stage, the disk is very active and probably highly turbulent, as a result of the mismatch of the specific angular momentum of the gas hitting the disk with that required to maintain Keplerian rotation. Gravitational instabilities and viscous and magnetic forces may add to this activity. When the infall slows substantially or stops, the disk becomes more quiescent. Interactions with the gaseous component of the disk affect the dynamics of small solid bodies, and the growth from micrometer-sized dust to kilometer-sized planetesimals remains poorly understood. Meteorites (Chapter 8), minor planets (Chapter 9), and comets (Chapter 10), most of which were never incorporated into bodies of planetary dimensions, best preserve a record of this important period in Solar System development.

The dynamics of larger solid bodies within protoplanetary disks are better characterized. The primary perturbations on the Keplerian orbits of kilometer-sized and larger planetesimals in protoplanetary disks are mutual gravitational interactions and physical collisions. These interactions lead to accretion (and in some cases erosion and fragmentation) of planetesimals. Eventually, solid bodies agglomerated into the terrestrial planets in the inner Solar System, and planetary cores several times the mass of the Earth in the outer Solar System. These massive cores were able to gravitationally attract and retain substantial amounts of gaseous material from the solar nebula. In contrast, terrestrial planets were not massive enough to attract and retain such gases, and the gases in their current thin atmospheres are derived from material that was incorporated in solid planetesimals.

The planets in our Solar System orbit close enough to one another that the final phases of planetary growth could have involved the merger or ejection of planets or planetary embryos on unstable orbits. However, the low eccentricities of the orbits of the outer planets imply that some damping process, such as accretion/ejection of numerous small

planetesimals or interactions with residual gas within the protoplanetary disk, must also have been involved.

As researchers learn more about the individual bodies and classes of objects in our Solar System, and as simulations of planetary growth become more sophisticated, theories about the formation of our Solar System are being revised and (we hope) improved. The detection of planets around other stars has presented us with new challenges to develop a unified theory of planet formation which is applicable to all stellar systems. We discuss these theories in more detail in Chapters 12 and 13.

Further Reading

A good nontechnical overview of our planetary system, complete with many beautiful color pictures, is given by:

Beatty, J.K., C.C. Peterson, and A. Chaikin, Eds., 1999. *The New Solar System*, 4th Edition. Sky Publishing Co., Cambridge, MA and Cambridge University Press, Cambridge. 421pp.

A terse, but detailed overview including reproductions of paintings of various Solar System objects by the authors, is provided by:

Miller, R., and W.K. Hartmann, 2005. *The Grand Tour: A Traveler's Guide to the Solar System*, 3rd Edition. Workman Publishing, New York. 208pp.

An overview of the Solar System emphasizing atmospheric and space physics is given by:

Encrenaz, T., J.-P. Bibring, M. Blanc, M.-A. Barucci, F. Roques, and Ph. Zarka, 2004. *The Solar System*, 3rd Edition. Springer-Verlag, Berlin. 512pp.

Two good overview texts aimed at undergraduate nonscience majors are:

Morrison, D., and T. Owen, 2003. *The Planetary System*, 3rd Edition. Addison Wesley Publishing Company, New York. 531pp.

Hartmann, W.K., 2005. *Moons and Planets*, 5th Edition. Brooks/Cole, Thomson Learning, Belmont, CA. 428pp.

Short summaries of a multitude of topics, ranging from mineralogy to black holes, at a level of sophistication comparable to that of this book, are provided by:

Cole, G.H.A., and M.M. Woolfson, 2002. *Planetary Science: The Science of Planets Around Stars*, Institute of Physics Publishing, Bristol and Philadelphia. 508pp.

Chemical processes on planets and during planetary formation are covered in some detail by:

Lewis, J.S., 2004. *Physics and Chemistry of the Solar System*, Second Edition. Elsevier, Academic Press, San Diego. 684pp.

The following encyclopedia forms a nice complement to this book:

Spohn, T., D. Breuer, and T.V. Johnson, Eds., 2014. *Encyclopedia of the Solar System*, 3rd Edition. Academic Press, San Diego. 1311pp.

Extensive planetary data tables can be found in:

Yoder, C.F., 1995. Astrometric and geodetic properties of Earth and the Solar System. In *Global Earth Physics: A Handbook of Physical Constants*. AGU Reference Shelf 1, American Geophysical Union, 1–31.

For updated information see: http://ssd.jpl.nasa.gov.

Problems

1.1.**E** Because the distances between the planets are much larger than planetary sizes, very few diagrams or models of the Solar System are completely to scale. However, imagine that you are asked to give an astronomy lecture/demonstration to your niece's second-grade class, and you decide to illustrate the vastness and near emptiness of space by constructing a scale model of the Solar System using ordinary objects. You begin by selecting a (1 cm diameter) marble to represent the Earth. What other objects can you use, and how far apart must you space them? Proxima Centauri, the nearest star to the Solar System, is 4.2 light years distant; where, in your model, would you place it?

1.2.**E** The satellite systems of the giant planets are often referred to as 'miniature solar systems'. In this problem you will make some calculations comparing the satellite systems of Jupiter, Saturn, and Uranus to the planetary system.

(a) Calculate the ratio of the sum of the masses of the planets with that of the Sun, and similar ratios for the jovian, saturnian, and uranian systems, using the respective planet as the primary mass.

(b) Calculate the ratio of the sum of the orbital angular momenta of the planets to the rotational angular momentum of the Sun. You can assume circular orbits at zero inclination for all planets, and ignore the effects of planetary rotation and the presence of satellites.

(c) Repeat the calculation in (b) for the jovian, saturnian, and uranian systems, using the respective planet as the primary mass.

(d) Calculate the orbital semimajor axes of the planets in terms of solar radii, and the orbital semimajor axes of Jupiter's moons in jovian radii. How would a scale model of the jovian system compare to the model of the planetary system in Problem 1.1?

1.3.**I** A planet which keeps the same hemisphere pointed towards the Sun must rotate once per orbit in the prograde direction.

(a) Draw a diagram to demonstrate this fact. The rotation period (in an inertial frame) or *sidereal day* for such a planet is equal to its orbital period, whereas the length of a *solar day* on such a planet is infinite.

(b) Earth rotates in the prograde direction. How many times must Earth rotate per orbit in order for there to be 365.24 solar days per year? Verify your result by comparing the length of Earth's sidereal rotation period (Table 1.2) to the length of a mean solar day.

(c) If a planet rotated once per orbit in the retrograde direction, how many solar days would it have per orbit?

(d) Determine a general formula relating the lengths of solar and sidereal days on a planet. Use your formula to determine the lengths of solar days on Mercury, Venus, Mars, and Jupiter.

(e) For a planet on an eccentric orbit, the length of either the solar day or the sidereal day varies on an annual cycle. Which one varies and why? Calculate the length of the longest such day on Earth. This longest day is how much longer than the mean day of its type?

1.4.**I** For the same reasons that the length of a mean solar day is not exactly equal to the sidereal rotation period of Earth (Problem 1.3), the length of the month is not equal to the sidereal orbital period of the Moon about the Earth. What, physically, does a month refer to? Calculate the length of an average (astronomical) month.

1.5.**I** As you may have already guessed by now if you have read the two previous problems, the length of the year is not exactly equal to the time it takes Earth to complete one orbit about the Sun. The common usage of year is the mean length of time over which seasons repeat, this is called the *tropical year*. The change in seasons is primarily a result of Earth's motion about the Sun, but seasons are also affected by a gradual change in the direction of Earth's spin axis. The principal cause of the precession of Earth's spin axis is torques on Earth's equatorial bulge exerted by the Moon and the Sun. The resulting *lunisolar precession* has a period of ~26 000 years. (Torques exerted by the other planets also affect the direction of Earth's spin axis. These torques are important because they induce the quasi-periodic variations in Earth's obliquity seen in Figure 2.19, but their influence on the precession rate of Earth's axis is small.)

(a) Draw a diagram of the system and use it to derive a formula relating the lengths of the tropical year, the *sidereal year* and the precession period. The sidereal year is longer than the tropical year. Use this fact to deduce the direction of Earth's lunisolar precession.

(b) Compute the length of the sidereal year in terms of tropical years and in terms of days. Note that although the fractional difference between the tropical and sidereal years is much smaller than the difference between solar and sidereal days, it is still almost twice as large as the difference between the Julian and Gregorian calendars.

1.6.E A *total solar eclipse* occurs when the Moon blocks the entire disk (*photosphere*) of the Sun, allowing the observer to view only the Sun's extended atmosphere, the *corona*. An *annular eclipse* occurs when the Moon obscures the central portion of the Sun, but a narrow annulus of the Sun's photosphere can be seen around the Moon.

(a) Using the data in Tables 1.1, 1.4, 1.5, and C.5, show that the eccentricities of the orbits of Earth about the Sun and the Moon about the Earth make it possible for both types of eclipse to be viewed from the surface of Earth.

(b) Which occur more frequently, total solar eclipses or annular eclipses? Why?

2 Dynamics

No human investigation can be called real science if it cannot be demonstrated mathematically.

Leonardo da Vinci

In 1687, Isaac Newton showed that the relative motion of two spherically symmetric bodies resulting from their mutual gravitational attraction is described by simple conic sections: ellipses for bound orbits and parabolas and hyperbolas for unbound trajectories. However, the introduction of additional gravitating bodies produces a rich variety of dynamical phenomena, even though the basic interactions between pairs of objects can be straightforwardly described. In this chapter, we describe the basic orbital properties of Solar System objects (planets, moons, minor bodies, and dust) and their mutual interactions. We also provide several examples of important dynamical processes which occur in the Solar System and lay the groundwork for describing some of the phenomena which are discussed in more detail in other chapters of this book.

2.1 The Two-Body Problem

2.1.1 Kepler's Laws of Planetary Motion

By careful analysis of the observed orbits of the planets, Johannes Kepler deduced his three 'laws' of planetary motion:

(1) All planets move along elliptical paths with the Sun at one focus. We can express the *heliocentric distance*, $r_\odot$ (i.e., the planet's distance from the Sun), as

$$r_\odot = \frac{a(1 - e^2)}{1 + e \, \cos f},$$ (2.1)

with a the *semimajor axis* (average of the minimum and maximum heliocentric distances). The *eccentricity* of the orbit, $e \equiv (1 - b_\mathrm{m}^2/a^2)^{1/2}$, where $2b_\mathrm{m}$ is the minor axis of the ellipse. The *true anomaly*, f, is the angle between the planet's *perihelion* (closest heliocentric distance) and its instantaneous position. These quantities are displayed graphically in Figure 2.1a.

(2) A line connecting any given planet and the Sun sweeps out area, $\mathcal{A}$, at a constant rate (Fig. 2.2):

$$\frac{d\mathcal{A}}{dt} = \text{constant}.$$ (2.2)

The value of this constant rate differs from one planet to the next.

(3) The square of a planet's orbital period about the Sun (in years), P_yr, is equal to the cube of its semimajor axis (in AU), a_AU:

$$P_\mathrm{yr}^2 = a_\mathrm{AU}^3.$$ (2.3)

2.1.2 Newton's Laws of Motion, Gravity

Although Kepler's laws were originally deduced from careful observation of planetary motion, they were subsequently shown to be derivable from Newton's laws of motion together with his universal law of gravity. Consider a body of mass m_1 at instantaneous location $\mathbf{r}_1$ with instantaneous velocity $\mathbf{v}_1 \equiv d\mathbf{r}_1/dt$ and hence momentum $m_1\mathbf{v}_1$. The acceleration produced by a net force $\mathbf{F}_1$ is given by Newton's second law of motion:

$$\frac{d(m_1\mathbf{v}_1)}{dt} = \mathbf{F}_1.$$ (2.4)

Newton's third law states that for every action there is an equal and opposite reaction; thus, the force on each object of a pair due to the other object is equal in magnitude but opposite in direction:

$$\mathbf{F}_{12} = -\mathbf{F}_{21},$$ (2.5)

where $\mathbf{F}_{ij}$ represents the force exerted by body j on body i. Newton's universal law of gravity states that a second body of mass m_2 at position $\mathbf{r}_2$ exerts an attractive force on the first body given by

$$\mathbf{F}_{\mathrm{g}12} = -\frac{Gm_1m_2}{r^2}\hat{\mathbf{r}},$$ (2.6)

(a)

(b)

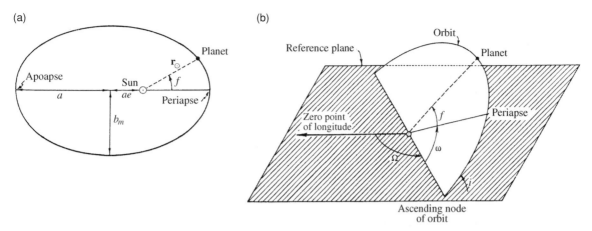

Figure 2.1 (a) Geometry of an elliptical orbit. The Sun is at one focus, and the vector $\mathbf{r}_\odot$ denotes the instantaneous heliocentric location of the planet (i.e., $r_\odot$ is the planet's distance from the Sun). The semimajor axis of the ellipse is a, e denotes its eccentricity, and b_m is the ellipse's semiminor axis. The true anomaly, f, is the angle between the planet's perihelion and its instantaneous position. (b) Geometry of an orbit in three dimensions; i is the inclination of the orbit, Ω is the longitude of the ascending node, and ω is the argument of periapse. (Adapted from Hamilton 1993)

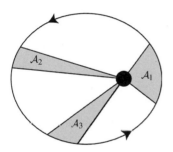

Figure 2.2 Schematic illustration of Kepler's second law. (Murray and Dermott 1999)

where $\mathbf{r} \equiv \mathbf{r}_1 - \mathbf{r}_2$ is the vector distance from particle 2 to particle 1, G is the gravitational constant, and $\hat{\mathbf{r}} \equiv \mathbf{r}/r$.

The equation for the relative motion of two mutually gravitating bodies can be derived from Newton's laws to be (Problem 2.1)

$$\mu_r \frac{d^2\mathbf{r}}{dt^2} = -\frac{G\mu_r M}{r^2}\hat{\mathbf{r}},$$ (2.7)

where μ_r is the reduced mass and M is the total mass:

$$\mu_r \equiv \frac{m_1 m_2}{m_1 + m_2},$$ (2.8a)

$$M \equiv m_1 + m_2.$$ (2.8b)

Thus, the relative motion is completely equivalent to that of a particle of reduced mass μ_r orbiting a *fixed* central mass M. Newton's generalization of Kepler's laws for the two-body problem is as follows:

(1) The two bodies move along elliptical paths, with one focus of each ellipse located at the center of mass (CM) of the system,

$$\mathbf{r}_{CM} = \frac{m_1 \mathbf{r}_1 + m_2 \mathbf{r}_2}{M}.$$ (2.9)

(2) A line connecting two bodies (as well as lines from each body to the center of mass) sweeps out area at a constant rate. This is a consequence of the conservation of angular momentum, $\mathbf{L}$:

$$\frac{d\mathbf{L}}{dt} = 0,$$ (2.10a)

where

$$\mathbf{L} = \mathbf{r} \times m\mathbf{v}.$$ (2.10b)

(3) The orbital period of a pair of bodies about their mutual center of mass is given by

$$P_{orb}^2 = \frac{4\pi^2 a^3}{G(m_1 + m_2)}.$$ (2.11)

Equation (2.11) reduces to Kepler's third law in the limit as $m_2/m_1 \to 0$. The derivation of (the Newtonian generalization of) Kepler's laws is the topic of Problem 2.2.

2.1.3 Orbital Elements

The Sun contains more than 99.8% of the mass of the known Solar System. The gravitational force exerted by a body is proportional to its mass (eq. 2.6), so to an excellent first approximation we can regard the motion of the planets and many other bodies as being solely influenced by a fixed central point-like mass. For objects such as the planets, which are bound to the Sun and hence cannot go arbitrarily far from the central mass, the general solution for the orbit is the ellipse described by equation (2.1). The orbital plane, although fixed in space, can be arbitrarily oriented with respect to whatever reference plane we have chosen. This reference plane is usually taken to be either the Earth's orbital plane about the Sun, which is called the *ecliptic*, or the equatorial plane of the largest body in the system, or the *invariable plane* (the plane perpendicular to the total angular momentum of the system). The *inclination*, i, of the orbit is the angle between the reference plane and the orbital plane; i can range from $0°$ to $180°$. Conventionally, secondaries orbiting in the same direction as the primary rotates are defined to have inclinations from $0°$ to $90°$ and are said to be on *prograde* (or *direct*) orbits. Secondaries orbiting in the opposite direction are defined to have $90° < i \le 180°$ and said to be on *retrograde* orbits. For heliocentric orbits, the Earth's orbital plane rather than the Sun's equator is usually taken as the reference. The intersection of the orbital and reference planes is called the *line of nodes*, and the orbit pierces the reference plane at two locations – one as the body passes upward through the plane (the *ascending node*) and one as it descends (the *descending node*). A fixed direction in the reference plane is chosen, and the angle to the direction of the orbit's ascending node is called the *longitude of the ascending node*, Ω. The angle between the line to the ascending node and the line to the direction of *periapse* (the point on the orbit when the two bodies are closest, which is referred to as *perihelion* for orbits about the Sun and *perigee* for orbits about the Earth) is called the *argument of periapse*, ω. For heliocentric orbits, Ω and ω are measured eastward from the *vernal equinox*.[1] Finally, the true anomaly, f, specifies the angle between the planet's periapse and its instantaneous position. Thus, the six *orbital elements*, a, e, i, Ω, ω, and f, uniquely specify the location of the object in space (Fig. 2.1). The first three quantities, a, e, and i, are often referred to as the *principal orbital elements*, as they describe the size, shape, and tilt of the orbit.

For two bodies with known masses, specifying the elements of the relative orbit and the positions and velocities of the center of mass is equivalent to specifying the positions and velocities of both bodies. Alternative (sets of) orbital elements are often used for convenience. For example, the *longitude of periapse*,

$$\varpi \equiv \Omega + \omega, \tag{2.12a}$$

is often used in place of ω. The time of perihelion passage, t_ϖ, is commonly used instead of f as an alternative

way in which to specify the location of the particle along its orbital path. The *mean motion* (average angular speed),

$$n \equiv \frac{2\pi}{P_{\text{orb}}}, \tag{2.12b}$$

and the *mean longitude*,

$$\lambda = n(t - t_\varpi) + \varpi, \tag{2.12c}$$

are also used frequently to specify orbital properties. See Danby (1988) for a more detailed discussion of the various sets of commonly used orbital elements.

2.1.4 Bound and Unbound Orbits

For a pair of bodies to travel on a circular orbit about their mutual center of mass, they must be pulled towards one another enough to balance inertia. Quantitatively, gravity must balance the centrifugal pseudoforce that is present if the problem is viewed as a steady state in the frame rotating with the angular velocity of the two bodies, n. The *centripetal force* necessary to keep an object of mass μ_r in a circular orbit of radius r with speed v_c is

$$\mathbf{F}_c = \mu_r n^2 \mathbf{r} = \frac{\mu_r v_c^2}{r}\hat{\mathbf{r}}. \tag{2.13}$$

Equating this to the gravitational force exerted by the central body of mass M, we find that the speed of a circular orbit is

$$v_c = \sqrt{\frac{GM}{r}}. \tag{2.14}$$

The total energy of the system, E, is a conserved quantity:

$$E = \frac{1}{2}\mu_r v^2 - \frac{GM\mu_r}{r} = -\frac{GM\mu_r}{2a}, \tag{2.15a}$$

where the first term in the middle expression is the kinetic energy of the system and the second term is potential energy. For circular orbits, the second equality in equation (2.15a) follows immediately from equation (2.14). If $E < 0$, the absolute value of the potential energy of the system is larger than its kinetic energy, and the system is *bound*: the body orbits the central mass on an elliptical path. Simple manipulation of equation (2.15a) yields an expression for the velocity along an elliptical orbit at each radius r:

$$v^2 = GM\left(\frac{2}{r} - \frac{1}{a}\right). \tag{2.15b}$$

Equation (2.15b) is known as the *vis viva equation*. If $E > 0$, the kinetic energy is larger than the absolute value of the potential energy, and the system is *unbound*. The

[1] The hour circle (i.e., great circle through the celestial poles) of the vernal equinox is the great circle that crosses the equator at the location of the Sun on the first day of spring. This is also the zero-point of the *right ascension*, which is a coordinate used by observers to describe the apparent location of a body in the sky.

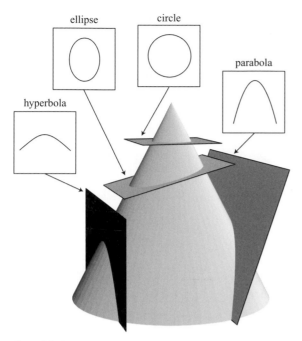

ellipse   circle

parabola

hyperbola

Figure 2.3 Conic sections. (Murray and Dermott 1999)

orbits, the generalized eccentricity is no different from the eccentricity defined in §2.1.1. For a parabola, $e = 1$ and $\zeta = 2q$, where q is the *pericentric separation*, i.e., the distance of closest approach. For a hyperbola, $e > 1$ and $\zeta = q(1 + e)$; $e \gg 1$ signifies a hyperbola with only a slight bend, nearly a straight line. For all orbits, the three orientation angles i, Ω, and ω are defined as in the elliptical case.

Whereas the energy of an orbit is uniquely specified by its semimajor axis (eq. 2.15a), the angular momentum is also affected by the orbit's eccentricity:

$$|\mathbf{L}| = \mu_r \sqrt{GMa(1 - e^2)}. \tag{2.18}$$

As with energy, the angular momentum of a circular orbit follows immediately from equation (2.14). For a given semimajor axis, a circular orbit contains the maximum possible amount of angular momentum (eq. 2.18). This occurs because when $r = a$ for an eccentric orbit, the magnitude of the velocity is the same as that for a circular orbit (by conservation of energy), but not all of this velocity is directed perpendicular to the line connecting the two bodies.

orbit is then described mathematically as a hyperbola. If $E = 0$, the kinetic and potential energies are equal in magnitude, and the orbit is a parabola. By setting the total energy (eq. 2.15a) equal to zero, we can calculate the *escape velocity* (alternatively referred to as the *escape speed*) at any separation:

$$v_e = \sqrt{\frac{2GM}{r}} = \sqrt{2}\,v_c. \tag{2.16}$$

For circular orbits, it is easy to show that both the kinetic energy and the total energy of the system are equal in magnitude to half the potential energy (Problem 2.5b); the same is true in a time-averaged sense for elliptical orbits (eq. 2.61).

As noted above, the orbit in the two-body problem is either an ellipse, parabola, or hyperbola depending on whether the energy is negative, zero, or positive, respectively. These curves are known collectively as *conic sections*, and are illustrated in Figure 2.3. The generalization of equation (2.1) to include unbound as well as bound orbits is

$$r = \frac{\zeta}{1 + e\cos f}, \tag{2.17}$$

where r and f have the same meaning as in equation (2.1), e is the *generalized eccentricity* and ζ is a constant. Bound orbits have $e < 1$ and $\zeta = a(1 - e^2)$, but the generalized eccentricity can take any non-negative value. For elliptical

2.1.5 Interplanetary Spacecraft

A spacecraft launched towards another body in our Solar System must be given enough energy to escape Earth's gravity (eq. 2.16) and still be moving rapidly enough relative to our planet that its new heliocentric orbit intersects the orbit of the other body. It is also advantageous to approach the target body at low speed, regardless of whether the encounter is a *flyby*, or the spacecraft is to be an *orbiter*, an *atmospheric probe*, or a *lander*. Travel between coplanar circular orbits may be accomplished via a *Hohmann transfer orbit*, an elliptical orbit that is tangent to both circular orbits. The periapse distance of a Hohmann transfer orbit is equal to the radius of the inner orbit, and the apoapse distance to that of the outer orbit. Using the vis viva equation (2.15b), one can calculate the speed at which the spacecraft needs to leave the vicinity of the Earth (Problem 2.4a) and the speed at which it approaches the target object. Under many circumstances, Hohmann transfer orbits require the least amount of rocket fuel for the interplanetary journey.

The trajectories of spacecraft sent to Venus or Mars are typically well approximated by Hohmann transfer orbits, but other destinations have been reached via more complicated routings (Fig. 2.4). As the velocity required to depart Earth on an orbit which crosses that of Mercury is quite large (Problem 2.4e), both spacecraft thus far sent

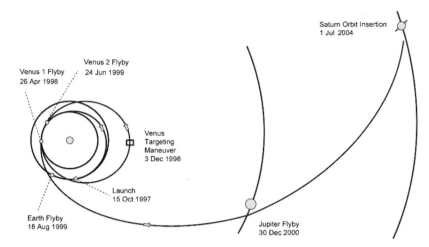

Figure 2.4 Schematic view of the trajectory of the Cassini spacecraft, as viewed in an inertial frame from well above the plane of the Solar System. The initial portion of the trajectory, from launch from Earth to the first Venus flyby, is similar to a Hohmann ellipse. The legs from the Earth flyby to Jupiter and then onwards to Saturn differ substantially from Hohmann ellipses because of the long times that would be required for lowest energy transfers to planets in the outer Solar System. (Courtesy E.J.O. Schrama)

to Mercury have relied on *gravity assists* obtained by initially targeting the spacecraft to Venus, and using Venus's gravity to alter the spacecraft's orbit about the Sun (essentially converting it from an Earth–Venus transfer orbit to a Venus–Mercury one). Gravity assists by Jupiter have been used to send spacecraft to Saturn and beyond. For spacecraft destined to targets in the outer Solar System, the time required to travel to the target on a Hohmann transfer orbit can be excessive (Problem 2.4d), so more eccentric orbits have been used, albeit at a penalty of greater fuel requirements to depart Earth and faster approach speeds at the target.

The above discussion pertains to chemical rockets, which burn (exert their thrust) quickly and thus can change orbits rapidly (§F.1). Electric propulsion rockets can be very fuel efficient, but exert their thrust very slowly, gradually altering the spacecraft's orbital path. The heliocentric trajectory of a spacecraft using electric propulsion typically takes the shape of a spiral.

2.1.6 Gravitational Potential

For many applications, it is convenient to express the gravitational field in terms of a potential, $\Phi_g(\mathbf{r})$, defined as:

$$\Phi_g(\mathbf{r}) \equiv -\int_\infty^{\mathbf{r}} \frac{\mathbf{F}_g(\mathbf{r}')}{m} \cdot d\mathbf{r}'. \qquad (2.19a)$$

By inverting equation (2.19a), one can see that the gravitational acceleration is the gradient of the potential and

$$\frac{d^2\mathbf{r}}{dt^2} = -\nabla\Phi_g. \qquad (2.19b)$$

In general, $\Phi_g(\mathbf{r})$ satisfies Poisson's equation:

$$\nabla^2\Phi_g = 4\pi\rho G. \qquad (2.20a)$$

In empty space, $\rho = 0$, so $\Phi_g(\mathbf{r})$ satisfies Laplace's equation:

$$\nabla^2\Phi_g = 0. \qquad (2.20b)$$

2.2 The Three-Body Problem

Gravity is not restricted to interactions between the Sun and the planets or individual planets and their satellites, but rather all bodies feel the gravitational force of one another. The motion of two mutually gravitating bodies is *completely integrable* (i.e., there exists one independent integral or constraint per degree of freedom), and the relative trajectories of the two bodies are given by simple conic sections, as discussed above. However, when more bodies are added to the system, additional constraints are needed to specify the motion; not enough integrals of motion are available (Problem 2.6), so the trajectories of even three gravitationally interacting bodies cannot be deduced analytically except in certain limiting cases. The general three-body problem is quite complex, and little progress can be made without resorting to numerical integrations. Fortunately, various approximations based upon large differences between the masses of the bodies and nearly circular and coplanar orbits (which are quite accurate for most Solar System applications) simplify the problem sufficiently that some important analytic results may be obtained.

If one of the bodies is of negligible mass (e.g., a small asteroid, a ring particle, or an artificial satellite), its effects on the other bodies may be ignored; the simpler system that results is called the *restricted three-body problem*, and the small body is referred to as a *test particle*. If the relative motion of the two massive particles is a circle, we refer to the situation as the *circular restricted three-body problem*. An alternative to the restricted three-body problem is *Hill's problem*, in which the mass of one of the bodies is much greater than the other two, but there is no restriction on the masses of the two small bodies relative to one another. An independent simplification is to assume that all three bodies travel within the same plane, the *planar three-body problem*. Various, but not all (see Problem 2.7), combinations of these assumptions are possible. Most of the results presented in this section are rigorously true only for the circular restricted three-body problem, but they are valid to a good approximation for many configurations which exist in the Solar System.

2.2.1 Jacobi Constant, Lagrangian Points

Our study of the three-body problem begins by considering an idealized system in which two massive bodies move on circular orbits about their common center of mass. A third body is introduced which is much less massive than the smaller of the first two, so that to good approximation it has no effect on the orbits of the other bodies. Our analysis is performed in a noninertial frame which rotates about the z-axis at a rate equal to the orbital frequency of the two massive bodies. We choose units such that the distance between the two bodies, the sum of the masses and the gravitational constant are all equal to one; this implies that the angular frequency of the rotating frame also equals unity (Problem 2.8). The origin is given by the center of mass of the pair, and the two bodies remain fixed at points on the x-axis, $\mathbf{r}_1 = (-m_1/(m_1 + m_2), 0)$ and $\mathbf{r}_2 = (m_2/(m_1 + m_2), 0)$. By convention, $m_1 \geq m_2$; in most Solar System applications, $m_1 \gg m_2$. The (massless) test particle is located at $\mathbf{r}$, so $|\mathbf{r} - \mathbf{r}_i|$ is the distance from mass m_i to the test particle. The velocity of the test particle in the rotating frame is denoted by v.

By analyzing a modified energy integral in the rotating frame, Jacobi deduced the following constant of motion for the circular restricted three-body problem:

$$C_J = x^2 + y^2 + \frac{2m_1}{|\mathbf{r} - \mathbf{r}_1|} + \frac{2m_2}{|\mathbf{r} - \mathbf{r}_2|} - v^2. \qquad (2.21)$$

The first two terms on the right-hand side of equation (2.21) represent twice the centrifugal potential energy, the next two twice the gravitational potential energy, and the final one twice the kinetic energy; C_J is known as *Jacobi's constant*. Note that a body located far from the two masses and moving slowly in the inertial frame has small C_J because the gravitational potential energy terms are small and the centrifugal potential almost exactly cancels the kinetic energy of the test particle's motion viewed in the rotating frame.

For a given value of Jacobi's constant, equation (2.21) specifies the magnitude of the test particle's velocity (in the rotating frame) as a function of position. As v^2 cannot be negative, surfaces at which $v = 0$ bound the trajectory of a particle with fixed C_J (note that the allowed region need not be finite). Such *zero-velocity surfaces*, or in the case of the planar problem *zero-velocity curves*, are quite useful in discussing the topology of the circular restricted three-body problem (Fig. 2.5e).

Lagrange found that in the circular restricted three-body problem there are five points where test particles placed at rest would feel no net force in the rotating frame. Three of these so-called *Lagrangian points* (L_1, L_2, and L_3) lie along a line joining the two masses m_1 and m_2. Zero-velocity curves intersect at each of the three collinear Lagrangian points, which are saddle points of the total (centrifugal + gravitational) potential in the rotating frame. The other two Lagrangian points (L_4 and L_5) form equilateral triangles with the two massive bodies (Fig. 2.5a–d). The two triangular Lagrangian points together form the zero-velocity 'curve' with the smallest value of C_J. All five Lagrangian points are in the orbital plane of the two massive bodies.

Particles displaced slightly from the three collinear Lagrangian points will continue to move away; hence these locations are unstable. The triangular Lagrangian points are potential energy maxima, but the Coriolis force stabilizes them for $(m_1 + m_2)^2/(m_1 m_2) \gtrsim 27$, which is the case for all known examples in the Solar System that are more massive than the Pluto–Charon system. The precise ratio required for linear stability of the Lagrangian points L_4 and L_5 is $\frac{(m_1+m_2)^2}{m_1 m_2} > (25 + \sqrt{621})/2 \sim 25$. (See Danby (1988) for a derivation and further details.) If a particle at L_4 or L_5 is perturbed slightly, it will start to *librate* about these points (i.e., oscillate back and forth, without circulating past the secondary). Note with irony that particles located at L_4 or L_5 have such a low value of C_J that Jacobi's integral does not exclude them from any location within the plane, nonetheless they remain stable indefinitely at these potential-energy maxima!

The L_4 and L_5 points are important in the Solar System. For example, the *Trojan asteroids* are located near Jupiter's triangular Lagrangian points, several asteroids are known to librate about Neptune's L_4 point, and several small asteroids, including 5261 Eureka, are martian Trojans. There are also small moons in the saturnian system near the triangular Lagrangian points of Tethys and Dione (Table 1.4). The L_4 or L_5 points in the Earth–Moon system have been suggested as possible locations for a future space station.

(a)

(b)

(c)

(d)

(e)

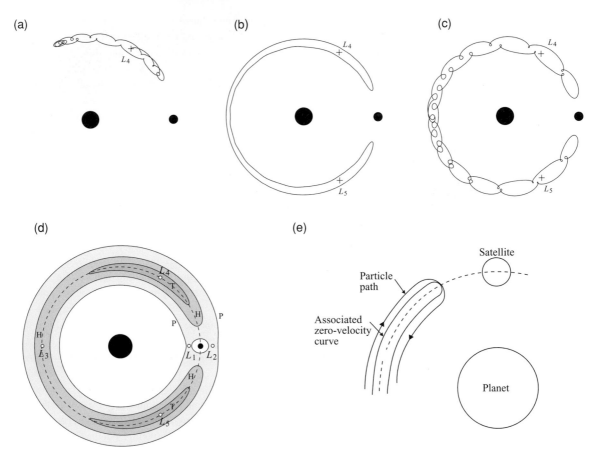

Figure 2.5 Schematic diagrams illustrating various properties of orbits in the circular restricted three-body problem. All cases are shown in the frame that is centered on the primary and rotating at the orbital frequency of the two massive bodies (corotating with the secondary).
(a) Example of a tadpole orbit of a test particle viewed in the rotating frame.
(b) Similar to (a), but for a horseshoe orbit with small eccentricity.
(c) As in (b), but the particle has a larger eccentricity. (Panels a–c adapted from Murray and Dermott 1999)
(d) The Lagrangian equilibrium points and various zero-velocity curves for three values of the Jacobi constant, C_J. The mass ratio $m_1/m_2 = 100$. The locations of the Lagrangian equilibrium points L_1–L_5 are indicated by small open circles. The white region centered on the secondary is the secondary's Hill sphere. The dashed line denotes a circle of radius equal to the secondary's semimajor axis. The letters T (tadpole), H (horseshoe), and P (passing) denote the type of orbit associated with the curves. The regions enclosed by each curve (shaded) are excluded from the motion of a test particle that has the corresponding C_J. The critical horseshoe curve actually passes through L_2, and the critical tadpole curve passes through L_3. Horseshoe orbits can exist between these two extremes. (Courtesy Carl Murray)
(e) Schematic diagram showing the relationship between a horseshoe orbit and its associated zero-velocity curve. The particle's velocity in the rotating frame drops as it approaches the zero-velocity curve, and it cannot cross the curve. (Adapted from Dermott and Murray 1981)

2.2.2 Horseshoe and Tadpole Orbits

Consider a moon on a circular orbit around a planet. A particle just interior to the moon's orbit has a higher angular velocity, and moves with respect to the moon in the direction of corotation. A particle just outside the moon's orbit has a smaller angular velocity, and moves relative to the moon in the opposite direction. When the outer particle approaches the moon, the particle is pulled towards the moon and consequently loses angular momentum. Provided the initial difference in semimajor axis is not too

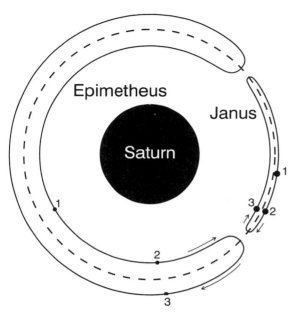

Figure 2.6 Diagram of the librational behavior of the Janus and Epimetheus coorbital system in a frame rotating with the average mean motion of both satellites. The system is shown to scale, apart from the radial extent of the librational arcs being exaggerated by a factor of 500 and the radii of the moons inflated by a factor of 50. The ratio of the radial widths (as well as the azimuthal extents) of the arcs is equal to the Janus/Epimetheus mass ratio (∼0.25). (Tiscareno *et al.* 2009)

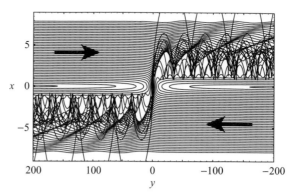

Figure 2.7 The trajectories of 80 test particles in the vicinity of a secondary of mass $m_2 \ll m_1$ are shown in the frame rotating with the secondary's (circular) orbit about the primary. The scale of the plot is expanded in the radial (x) direction relative to that in the azimuthal (y) direction, with numerical values in both directions given in units of the radius of the secondary's Hill sphere. The secondary mass is located at the origin and the L_1 and L_2 points are at $y = 0$, $x = \pm 1$. The particles were all started with $dx/dt = 0$ (i.e., circular orbits) at $y = \pm 200$. The arrows indicate their direction of motion before encountering the secondary. The primary is located at $y = 0$, $x = -\infty$. In an inertial frame, the secondary and the test particles all move from right to left. (Adapted from Murray and Dermott 1999)

large, the particle drops to an orbit lower than that of the moon. The particle then recedes in the forward direction. Similarly, the particle on the lower orbit is accelerated as it catches up with the moon, resulting in an outward motion towards a higher, and therefore slower, orbit. Orbits like these encircle the L_3, L_4, and L_5 points and appear shaped like horseshoes in the rotating frame (Fig. 2.5b), thus they are called *horseshoe orbits*. Saturn's small moons Janus and Epimetheus execute just such a dance, changing orbits every 4 years (Fig. 2.6). As Janus and Epimetheus are comparable in mass, Hill's approximation is more accurate than is the restricted three-body formalism used above, but the dynamical interactions are essentially the same.

Since the Lagrangian points L_4 and L_5 are stable, material can librate about these points individually; such orbits are called *tadpole orbits* after their asymmetric elongated shape in the rotating frame (Fig. 2.5a). The tadpole libration width at L_4 and L_5 is proportional to $(m_2/m_1)^{1/2}r$, and the horseshoe width varies as $(m_2/m_1)^{1/3}r$, where m_1 is the mass of the primary, m_2 the mass of the secondary, and r the distance between the two objects. For a planet of Saturn's mass, $M_\mathrm{h} = 5.7 \times 10^{29}$ g, and a typical moon of mass $m_2 = 10^{20}$ g (a 30 km radius

object with density of ∼1 g cm^{-3}) at a distance of 2.5 R$_\mathrm{h}$, the tadpole libration half-width is ∼3 km and the horseshoe half-width ∼60 km.

2.2.3 Hill Sphere

The approximate limit to a secondary's (e.g., planet's or moon's) gravitational dominance is given by the extent of its *Hill sphere*,

$$R_\mathrm{H} = \left(\frac{m_2}{3(m_1 + m_2)} \right)^{1/3} a, \qquad (2.22)$$

where m_2 is the mass of the secondary and m_1 the primary's (e.g., Sun's or planet's) mass. A test particle located at the boundary of a planet's Hill sphere is subject to a gravitational force from the planet comparable to the tidal difference between the force of the Sun on the planet and that on the test body. The Hill sphere stretches out to the L_1 point, and essentially circumscribes the Roche lobe (§11.1) in the limit $m_2 \ll m_1$. Planetocentric orbits that are stable over long periods of time are those well within the boundary of a planet's Hill sphere; all known natural satellites lie in this region. Stable heliocentric orbits are always well outside the Hill sphere of any planet (Fig. 2.7; eq. 2.28). Comets and other bodies which enter the Hill sphere of a planet at very low velocity can remain gravitationally bound to the planet for some time as *temporary satellites* (Fig. 2.8).

(a)

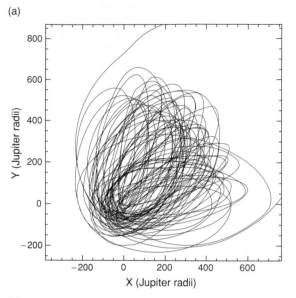

(b)

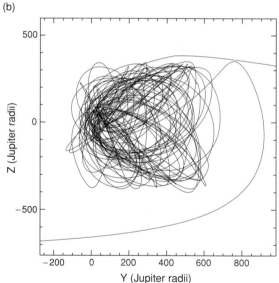

Figure 2.8 Trajectory relative to Jupiter of a test particle initially orbiting the Sun that was temporarily captured into an unusually long duration (140 years) unstable orbit about Jupiter. (a) Projected into the plane of Jupiter's orbit about the Sun. (b) Projected into a plane perpendicular to Jupiter's orbit. (Kary and Dones 1996)

2.2.4 Distant Planetary Satellites and Quasi-Satellites

The orbits of moons that lie in the inner part of a planet's Hill sphere are classified as prograde if the moons move in the sense that the planet rotates and retrograde if they travel in the opposite sense. However, for very distant satellites, the more important dynamical criterion is whether they

travel in the same direction as the planet orbits the Sun (prograde) or the opposite sense (retrograde).

Retrograde orbits are stable to larger distances from a planet than are prograde ones, and moons on retrograde orbits are found at greater distances (Table 1.4). Indeed, at large distances, retrograde orbits transition to a third type of coorbital behavior known as *quasi-satellites*. Quasi-satellites travel around the Sun with the same orbital period as does the planet, but because they have larger orbital eccentricities, as seen from the planet they travel on retrograde orbits beyond its Hill sphere. Note that quasi-satellites librate around 0° (the orbital longitude of the planet), whereas horseshoe orbits librate about 180° and tadpole orbits move about 60° or 300°.

No known quasi-satellites occupy stable orbits. However, a few small asteroids travel on temporary quasi-satellite orbits about either Earth or Venus. Note that, in principle, a planet on an orbit with significant eccentricity can have a quasi-satellite that travels around the Sun/star on a more circular path.

2.3 Perturbations and Resonances

Within the Solar System, one body typically produces the dominant gravitational force on any given object, and the resultant motion can be thought of as a Keplerian orbit about a primary, subject to small perturbations by other bodies. In this section, we consider some important examples of the effects of these perturbations on the orbital motion.

Classically, much of the discussion of the evolution of orbits in the Solar System used perturbation theory as its foundation. Essentially, the method involves writing the potential as the sum of a part that describes the independent Keplerian motion of the bodies about the Sun, plus a part (called the *disturbing function*) that contains the *direct terms* which account for the pairwise interactions among the planets and minor bodies and the *indirect terms* associated with the motion of the Sun in response to the gravitational tugs exerted by the planets. For example, if m_2 and m_3 are two point masses in orbit about a common primary at instantaneous locations $\mathbf{r}_2$ and $\mathbf{r}_3$ relative to the primary, then the disturbing function, $\mathcal{R}$, for the action of m_2 on m_3 may be written as

$$\mathcal{R} = -Gm_2 \left(\frac{1}{|\mathbf{r}_2 - \mathbf{r}_3|} - \frac{\mathbf{r}_2 \cdot \mathbf{r}_3}{r_2^3} \right), \qquad (2.23)$$

where the first quantity in the parentheses is the direct term and the second is the indirect term. The force on m_3 due to m_2 is given by $\mathbf{F} = -m_3 \nabla \mathcal{R}$. In a many-planet system, the disturbing function can be expressed as the sum of

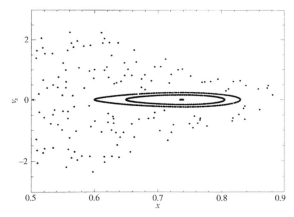

Figure 2.9 *Surface of section* for the trajectories of four different test particles in the planar circular restricted three-body problem. The dots display the x coordinate and x velocity of the particles each time the particles pass through the $y = 0$ plane with positive y velocity. The four particles have the same value of C_J but different initial conditions. Three of the trajectories are regular and produce well-defined quasi-periodic patterns on the plot. The unconnected dots all represent the trajectory of the fourth particle, which is on a chaotic orbit and therefore is less confined in phase space. (Adapted from Duncan and Quinn 1993)

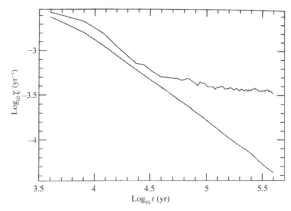

Figure 2.10 Distinction between regular (lower curve, nearly straight) and chaotic trajectories (upper curve) as characterized by the Lyapunov characteristic exponent, γ_c. Both trajectories are near the 3:1 resonance with Jupiter, and they have been integrated using the elliptic restricted three-body problem. For chaotic trajectories, a plot of log γ_c versus log t eventually levels off at a value of γ_c that is the inverse of the Lyapunov timescale for the divergence of initially adjacent trajectories, whereas for regular trajectories, $\gamma_c \to 0$ as $t \to \infty$. (Adapted from Duncan and Quinn 1993)

terms each having the form of the expression on the right-hand side of equation (2.23). A detailed discussion of the disturbing function, Fourier expansions thereof, and applications to planetary dynamics is presented by Brouwer and Clemence (1961).

2.3.1 Regular and Chaotic Motion

In general, one can expand the disturbing function in terms of the small parameters of the problem (such as the ratio of the planetary masses to the Sun's mass, the eccentricities and inclinations, etc.) as well as the other orbital elements of the bodies, including the mean longitudes (i.e., the locations of the bodies in their orbits) and attempt to solve the resulting equations for the time dependence of the orbital elements. However, in the late nineteenth century, Poincaré showed that these perturbation series are often divergent and have validity only over finite time-spans. Direct integrations on computers demonstrate that for some initial conditions the trajectories are *regular* with variations in their orbital elements that seem to be well described by the perturbation series, while for other initial conditions the trajectories are found to be *chaotic* and are not as confined in their motions (Fig. 2.9). The evolution of a system which is chaotic depends so sensitively on the system's precise initial state that the behavior is in effect unpredictable, even though it is strictly determinate in a mathematical sense.

There is a key feature of the chaotic orbits that we will use here as a definition of *chaos*: Two trajectories that begin arbitrarily close in phase space (which can be defined using coordinates such as positions and velocities, or a more complicated set of orbital elements) within a chaotic region will typically diverge exponentially in time. Within a given chaotic region, the timescale for this divergence does not typically depend on the precise values of the initial conditions! The distance, $d(t)$, between two particles having an initially small separation, $d(0)$, increases slowly for regular orbits, with $d(t) - d(0)$ growing as a power of time t (typically linearly). In contrast, for chaotic orbits,

$$d(t) \sim d(0)e^{\gamma_c t}, \tag{2.24}$$

where γ_c is the *Lyapunov characteristic exponent* and γ_c^{-1} is the *Lyapunov timescale* (Fig. 2.10). From this definition of chaos, we see that chaotic orbits show such a sensitive dependence on initial conditions that the detailed long-term behavior of the orbits is lost within several Lyapunov timescales. Even a fractional perturbation as small as 10^{-10} in the initial conditions will result in a 100% discrepancy in about 20 Lyapunov times. However, one of the interesting features of much of the chaotic behavior seen in simulations of the orbital evolution of bodies in the Solar System is that the timescale for large changes in the principal orbital elements is often many orders of magnitude longer than the Lyapunov timescale.

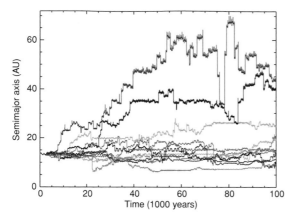

Figure 2.11 COLOR PLATE The future evolution of the semimajor axis of P/Chiron's orbit according to 11 numerical integrations. The initial orbital elements of the simulated bodies differed by about 1 part in 10^6. The orbit of Chiron currently crosses the orbits of both Saturn and Uranus, and is not protected from close approaches with either planet by any resonance. Chiron's orbit is highly chaotic, with gross divergence of trajectories in $<10^4$ years. (Courtesy L. Dones)

In dynamical systems like the Solar System, chaotic regions do not appear randomly but, rather, many of them are associated with trajectories in which the ratios of characteristic frequencies of the original problem are sufficiently well approximated by rational numbers, i.e., near resonances. The simplest of these resonances to visualize are so-called *mean motion resonances*, in which the orbital periods of two bodies are commensurate (e.g., have a ratio of the form $N/(N + 1)$ or $N/(N + 2)$, where N is an integer). Some examples of the consequences of this type of resonance are given below. In §2.4, we define secular resonances and indicate their relationship to the stability of the Solar System.

The above discussion applies to orbits that do not closely approach any massive secondaries. Close approaches can lead to highly chaotic and unpredictable orbits, such as the possible future behaviors of the giant, distant cometary centaur Chiron shown in Figure 2.11. These planet-crossing trajectories do not require resonances to be unstable and generally are not well characterized by a constant Lyapunov exponent.

2.3.2 Resonances

Although perturbations on a body's orbit are often small, they cannot always be ignored. They must be included in short-term calculations if high accuracy is required, e.g., for predicting stellar occultations or targeting spacecraft. Most long-term perturbations are periodic in nature, their directions oscillating with the relative longitudes of the bodies or with some more complicated function of the

bodies' orbital elements. Small perturbations can produce large effects if the forcing frequency is commensurate or nearly commensurate with the natural frequency of oscillation of the responding elements. Under such circumstances, perturbations add coherently, and the effects of many small tugs can build up over time to create a large-amplitude, long-period response. This is an example of *resonance forcing*, which occurs in a wide range of physical systems.

An elementary example of resonance forcing is given by the one-dimensional forced harmonic oscillator, for which the equation of motion is

$$m\frac{d^2x}{dt^2} + m\omega_o^2 x = F_f \cos \omega_f t, \tag{2.25}$$

where m is the mass of the oscillating particle, F_f is the amplitude of the driving force, ω_o is the natural frequency of the oscillator and ω_f is the forcing frequency. The solution to equation (2.25) is

$$x = \frac{F_f}{m(\omega_o^2 - \omega_f^2)} \cos \omega_f t + C_1 \cos \omega_o t + C_2 \sin \omega_o t,$$

$$\tag{2.26a}$$

where C_1 and C_2 are constants determined by the initial conditions. Note that if $\omega_f \approx \omega_o$, a large-amplitude, long-period response can occur even if F_f is small. Moreover, if $\omega_o = \omega_f$, equation (2.26a) is invalid. In this (resonant) case, the solution is given by

$$x = \frac{F_f}{2m\omega_o} t \sin \omega_o t + C_1 \cos \omega_o t + C_2 \sin \omega_o t. \tag{2.26b}$$

The t in the middle of the first term at the right-hand side of equation (2.26b) leads to secular (i.e., steady rather than periodic) growth. Often this linear growth is moderated by the effects of nonlinear terms which are not included in the simple example provided above. However, some perturbations have a secular component.

2.3.2.1 *Examples of Orbital Resonances*
Almost exact orbital commensurabilities exist at many places in the Solar System. Io orbits Jupiter twice as frequently as Europa does, and Europa in turn orbits Jupiter in half of the time that Ganymede takes. *Conjunction* (the moons being at the same longitude in their orbits about the planet) between Io and Europa always occurs when Io is at its perijove. How can such commensurabilities exist? After all, the rational numbers form a set of measure zero on the real line, which means that the probability of randomly picking a rational from the real number line is nil! The answer lies in the fact that *orbital resonances* may be held in place by stable 'locks', which result from

nonlinear effects not represented in the simple mathematical example of the harmonic oscillator. Differential tidal recession (§2.6) brings moons into resonance, and nonlinear interactions between the moons can keep them there. The stabilizing mechanisms are beyond the scope of this book; see Peale (1976) for an explanation.

Other examples of resonance locks include the Hilda and Trojan asteroids with Jupiter, Neptune–Pluto, and several pairs of moons orbiting Saturn, such as Janus–Epimetheus, Mimas–Tethys, and Enceladus–Dione. Scattered disk objects and centaurs can also shuttle back and forth between various temporary mean motion resonance locks with an outer planet. Resonant perturbations can force bodies into eccentric and/or inclined orbits, which may lead to collisions with other bodies; this is believed to be the dominant mechanism for clearing the Kirkwood gaps in the asteroid belt (see below). Several moons of Jupiter and Saturn have significant resonantly produced *forced eccentricities*, which are denoted by the symbol f in Table 1.4.

Spiral density waves can result from resonant perturbations by a moon on a self-gravitating disk of particles. Density waves are observed at many resonances in Saturn's rings; they explain most of the structure seen in Saturn's A ring. The vertical analog of density waves, bending waves, are caused by resonant perturbations perpendicular to the ring plane from a satellite in an orbit which is inclined to the ring. Spiral bending waves excited by the moons Mimas and Titan have been observed in Saturn's rings. We discuss these manifestations of resonance effects in more detail in Chapter 11.

2.3.2.2 *Resonances in the Asteroid Belt*

There are obvious patterns in the distribution of asteroidal semimajor axes that appear to be associated with mean motion resonances with Jupiter (Fig. 9.1a). At these resonances, a particle's period of revolution about the Sun is a small integer ratio multiplied by Jupiter's orbital period. The Trojan asteroids travel in a 1:1 mean motion resonance with Jupiter, as described above. These asteroids execute small amplitude (tadpole) librations about the L_4 and L_5 points 60° behind or ahead of Jupiter and therefore never suffer a close approach to Jupiter. Another example of a protection mechanism provided by a resonance is the Hilda group of asteroids at Jupiter's 3:2 mean motion resonance and the asteroid 279 Thule at the 4:3 resonance. The Hilda asteroids have a libration about 0° of their *critical argument* (the combination of orbital elements that signifies the resonant configuration), $3\lambda' - 2\lambda - \varpi$, where λ' is Jupiter's longitude, λ is the asteroid's longitude, and ϖ

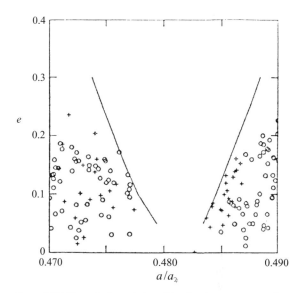

Figure 2.12 The outer boundaries of the chaotic zone surrounding Jupiter's 3:1 mean motion resonance in the a–e plane are shown as lines. Locations of numbered asteroids are shown as circles and Palomar–Leiden survey (PLS) asteroids (whose orbits are less well determined) are represented as plus signs. Note the excellent correspondence of the observed 3:1 Kirkwood gap with theoretical predictions. (Adapted from Wisdom 1983)

is the asteroid's longitude of perihelion. In this way, whenever the asteroid is in conjunction with Jupiter ($\lambda = \lambda'$), the asteroid is close to perihelion ($\lambda' \approx \varpi$) and well away from Jupiter.

Using resonances to explain the Kirkwood gaps in the main asteroid belt and the general depletion of the outer belt proves to be more difficult than understanding the protection mechanisms at other resonances. A feature subject to much investigation has been the gap at the 3:1 mean motion resonance. Early investigations found that most orbits starting at small eccentricity were regular and showed very little variation in eccentricity or semimajor axis over timescales of 5×10^4 yr. In the 1980s, Jack Wisdom showed that an orbit near the resonance could maintain a low eccentricity ($e < 0.1$) for nearly a million years and then have a sudden increase in eccentricity to $e > 0.3$. This illustrates an important feature which often occurs in simulations to be discussed later: A particle can remain in a low-eccentricity state for hundreds of Lyapunov times before 'jumping' relatively quickly to high eccentricity.

The outer boundaries of the chaotic zone coincide well with the boundaries of the 3:1 Kirkwood gap (Fig. 2.12). Since asteroids which begin on near-circular orbits in the gap acquire sufficient eccentricities to cross the orbits of Mars and the Earth, and in some cases become so eccentric

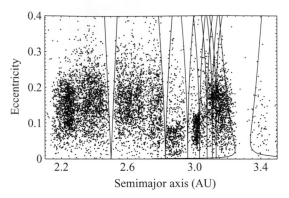

Figure 2.13 The maximum libration widths in *a–e* space of the strong jovian resonances superimposed on the distribution of asteroids in the main belt. Note the correspondence between the widths of the gaps and the widths of the resonances. (Murray and Dermott 1999)

that they hit the Sun, the perturbative effects of the terrestrial planets are believed capable of clearing out the 3:1 gap in a time equivalent to the age of the Solar System. There is also a strong correlation between the libration widths (intervals over which the critical argument can librate) of other resonances and regions of *a–e* space depleted of asteroids (Fig. 2.13).

2.3.3 The Resonance Overlap Criterion and Jacobi–Hill Stability

For nearly circular and coplanar orbits, the strongest mean motion resonances occur at locations where the ratio of test particle orbital periods to the massive body's period are of the form $N:(N \pm 1)$, where N is an integer. At these locations, conjunctions (closest approaches) always occur at the same phase in the orbit, and tugs add coherently. (The locations of these strong resonances are shifted slightly when the primary is oblate; see §2.5 and Chapter 11 for details.) The strength of these first-order resonances increases as N grows, because the magnitudes of the perturbations are larger closer to the secondary. First-order resonances also become closer to one another near the orbit of the secondary (Problem 2.12). Sufficiently close to the secondary, the combined effects of greater strength and smaller spacing cause resonance regions to overlap; this overlapping can lead to the onset of chaos as particles shift between the nonlinear perturbations of various resonances. The region of overlapping resonances is approximately symmetric about the planet's orbit, and has a half-width, Δa_{ro}, given by

$$\Delta a_{ro} \approx 1.5 \left(\frac{m_2}{m_1} \right)^{2/7} a, \tag{2.27}$$

where a is the semimajor axis of the planet's orbit. The functional form of equation (2.27) has been derived analytically, whereas the coefficient 1.5 is a numerical result.

Zero-velocity surfaces can also be used to prove the stability of certain orbits in the circular restricted three-body problem. For example, in the planetary case, where $m_2 \ll m_1$, a test particle initially on a circular planar orbit separated in semimajor axis from the secondary by an amount larger than

$$\Delta a_J = 2\sqrt{3} \left(\frac{m_2}{3m_1} \right)^{1/3} a \approx 3.5 R_H \tag{2.28}$$

can never approach to within the Hill sphere of the planet, and remains in an orbit inferior or superior to that of the planet forever. The relationship between the stability criteria given by equations (2.27) and (2.28) is analyzed in Problem 2.13.

2.4 Stability of the Solar System

We turn now to one of the oldest problems in dynamical astronomy: whether or not the planets will continue indefinitely in almost circular, almost coplanar orbits.

2.4.1 Secular Perturbation Theory

To study the very long term behavior of planetary orbits, a fruitful approach involves averaging the disturbing function over the mean motions of the planets, resulting in what is known as the *secular part of the disturbing function*. Physically, this approximation is equivalent to replacing each of the planets by an eccentric wire of nonuniform thickness, which can be stretched and rotated by perturbations of the other wires. If the disturbing function is further limited to terms of lowest order, the equations of motion of the orbital elements of the planets can be expressed as a coupled set of first-order linear differential equations. This system can then be diagonalized to find the proper modes, which are sinusoids, and the corresponding eigenfrequencies. The evolution of a given planet's orbital elements is, therefore, a sum of the proper modes. With the addition of higher order terms, the equations are no longer linear. It is, however, sometimes possible to find an approximate solution of a form similar to the linear solution, except with shifted proper mode frequencies and terms involving combinations of the proper mode frequencies. Long-term numerical integrations have confirmed and extended many of the results initially derived using secular perturbation theory.

2.4.2 Chaos and Planetary Motions

The above discussion implies that if the mutual planetary perturbations were calculated to first order in the masses, inclinations, and eccentricities, the orbits could be described by a sum of periodic terms, indicating stability. This is still the case if the perturbations are expanded to somewhat higher orders. However, although perturbation expansions are done in powers of small parameters, the existence of resonances between the planets introduces small divisors into the expansion terms (e.g., eq. 2.26). Such small divisors make some high-order terms in the power series unexpectedly large and destroy the convergence of the series. There are two separate points in the construction of the secular system at which resonances can cause nonconvergence of the expansion. The first is in averaging over mean motions. Mean motion resonances between the planets can introduce small divisors, leading to divergences when forming the secular disturbing function. Second, there can be *secular resonances* between the proper mode frequencies, e.g., apse precession rates, leading to problems trying to solve the secular system using an expansion approach.

Mathematical stability (in the sense that planetary orbits would remain well separated and the system would remain bound for infinite time) has been proven for a system of extremely small but nonetheless finite-mass planets with orbits similar to those in our Solar System. However, the set of initial conditions for which this proof does not apply is *everywhere dense*, i.e., there is always a point in phase space arbitrarily close to a given choice of initial conditions for which the proof does not guarantee stability. So a system which satisfies the criteria for mathematical stability might not remain stable if it were subjected to perturbations, even if those perturbations were arbitrarily small.

From an astronomical viewpoint, stability implies that the system will remain bound (no ejections) and that no mergers of planets will occur for the possibly long but finite period of interest, and that this result is robust against (most if not all) sufficiently small perturbations. In the remainder of this discussion, we shall only be concerned with stability in an astronomical sense.

The analytical complexity of the perturbation techniques and the development of ever faster computers has led to the investigation of Solar System stability by purely numerical models. Figure 2.14 shows the behavior of the eccentricities of all eight planets for 3 million years into the past as well as into the future. Mercury's eccentricity reaches higher values on 10^8 year timescales (Fig. 2.15), but the eccentricities of the other planets do not extend much beyond their range shown in Figure 2.14 over this time interval. Variations in the semimajor axis of Earth's orbit over ± 3 Myr are shown in Figure 2.16. The small fractional changes in semimajor axis relative to the variations in eccentricity evident for Earth are characteristic of all eight planets.

Early numerical integrations of the orbits of the giant planets on million-year timescales compared well with perturbation calculations, showing quasi-periodic behavior for the four major outer planets. Pluto's behavior, however, was sufficiently different to inspire further study. It was found that the angle $3\lambda_P - 2\lambda_\Psi - \varpi_P$ is in libration with a period of 20 000 yr, where λ_P and λ_Ψ are the mean longitudes of Pluto and Neptune, respectively, and ϖ_P is the longitude of perihelion of Pluto. This 2:3 mean motion resonance acts to prevent close encounters of Pluto with Neptune and hence protects the orbit of Pluto. However, numerical integrations show that Pluto's orbit is not quasi-periodic. There is evidence for the existence of very long period changes in Pluto's orbital elements, with a Lyapunov exponent of $\sim(20 \text{ Myr})^{-1}$. Nonetheless, no study has shown evidence for Pluto leaving its resonance with Neptune.

Long duration numerical integrations that include the terrestrial planets show a surprisingly high Lyapunov exponent, $\sim(5 \text{ Myr})^{-1}$. Such large Lyapunov exponents certainly suggest chaotic behavior. However, the apparent regularity of the motion of the Earth and Pluto, and indeed the fact that the Solar System has survived for 4.5 billion years, implies that any pathways through phase space that might lead to (highly chaotic) close approaches must be narrow. Nonetheless, the exponential divergence of orbits with a 5 Myr timescale shown by the calculations implies that an error as small as 10^{-8} in the initial conditions will lead to a 100% discrepancy in longitude in 100 Myr. It is also worth bearing in mind the lessons learned from integration of test particle trajectories, namely that the timescale for macroscopic changes in the system can be many orders of magnitude longer than the Lyapunov timescales. Thus, the apparent stability of the current planetary system on billion-year timescales may simply be a manifestation of the fact that the Solar System is in the chaotic sense a dynamically young system. As planetary perturbations appear to be capable of bringing the Solar System to the verge of instability on geological timescales, the planets within our Solar System may be about as closely spaced as can be expected for a mature planetary system containing planets as massive as those orbiting the Sun. Although somewhat more crowded configurations can be long-lived, it may well be that the planet formation process (Chapter 13) is unlikely to produce more densely packed systems of similar planets that survive on gigayear timescales.

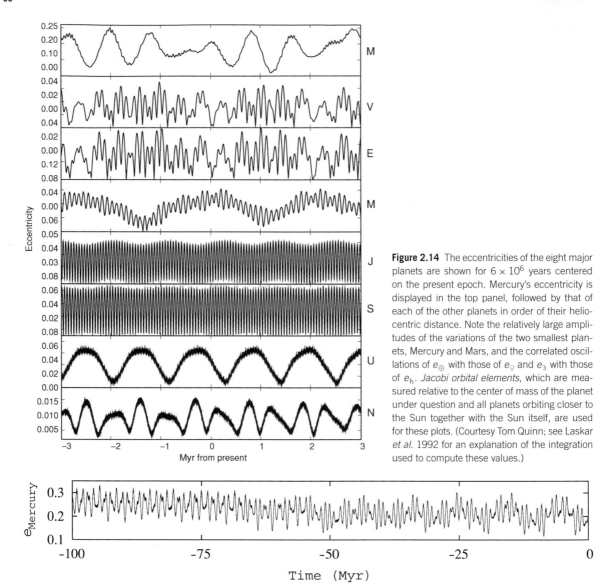

Figure 2.14 The eccentricities of the eight major planets are shown for 6×10^6 years centered on the present epoch. Mercury's eccentricity is displayed in the top panel, followed by that of each of the other planets in order of their heliocentric distance. Note the relatively large amplitudes of the variations of the two smallest planets, Mercury and Mars, and the correlated oscillations of $e_\oplus$ with those of $e_♀$ and $e_♃$ with those of $e_♄$. *Jacobi orbital elements*, which are measured relative to the center of mass of the planet under question and all planets orbiting closer to the Sun together with the Sun itself, are used for these plots. (Courtesy Tom Quinn; see Laskar *et al.* 1992 for an explanation of the integration used to compute these values.)

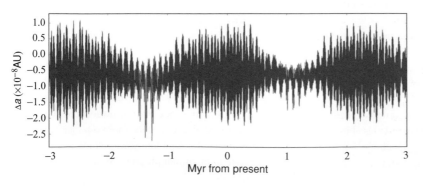

Figure 2.15 Variations in the eccentricity of Mercury's orbit over the past 100 million years. Integrations included the Sun, all eight planets and first-order post-Newtonian effects of general relativity; the eccentricities of all of the planets over the past 3 million years look the same in the integration used to produce this figure as those shown in Figure 2.14. (Courtesy Julie Gayon)

Figure 2.16 Variations in the semimajor axis of Earth's orbit (more precisely, the semimajor axis of the center of mass of the Earth–Moon system) over a time interval of six million years centered on the present epoch. Note the scale of the vertical axis, which indicates that the Earth's semimajor axis varies by only a few kilometers over timescales of millions of years. These data were taken from the integrations used to produce the plot of eccentricities of the planets shown in Figure 2.14. (Courtesy Tom Quinn)

2.4.3 Survival Lifetimes of Small Bodies

Interplanetary space is vast, yet few bodies orbit within this great expanse. And those few bodies are far from randomly distributed. Rather, minor planets are concentrated within a few regions (§9.1): the Kuiper belt beyond Neptune's orbit, the main asteroid belt between the orbits of Mars and Jupiter, the regions surrounding the triangular Lagrangian points of the Sun–Jupiter system (§2.2.1), and probably around the regions surrounding the triangular Lagrangian points of the Sun–Neptune system. Dynamical analyses show that orbits within these regions remain stable for far longer than trajectories passing through most other locations in the Solar System.[2] What causes the removal of bodies from other regions of the Solar System? How rapidly are they removed?

Trajectories crossing the paths of one or more of the major planets are rapidly destabilized by scatterings resulting from close planetary approaches, unless they are protected by some type of resonance (as is Pluto). Small bodies orbiting between a pair of terrestrial planets or a pair of giant planets can be stable for much longer, but most are perturbed into planet-crossing paths in less than the age of the Solar System by the same resonance overlap-induced chaos that makes planetary orbits unpredictable on long timescales. Lifetimes of orbits vary greatly, and collections of test particles spread randomly over even fairly small regions of phase space last for quite diverse amounts of time (Fig. 2.17). Loss rates are rapid early on, but as particles near the stronger resonances are removed, it takes longer and longer for a given fraction of the remaining bodies to be destabilized. This decay rate is more gradual than that of other natural processes, such as radioactivity (§8.6.1), where the population drops exponentially with time.

2.5 Orbits About an Oblate Planet

Thus far we have approximated Solar System bodies as point masses for the purpose of calculating their mutual gravitational interactions. Self-gravity causes most sizable celestial bodies to be approximately spherically symmetric. Newton showed that the gravitational force

exerted by a spherically symmetric body exterior to its surface is identical to the gravitational force of the same mass located at the body's center (Problem 2.16); thus, the point-mass approximation is adequate for most purposes. There are, however, several forces which act to produce distributions of mass which deviate from spherical symmetry. In the Solar System, rotation, physical strength, and tidal forces produce important departures from spherical symmetry in some bodies. The gravitational field of an aspherical body differs from that of a point-mass, with the largest deviation generally being found near the body's surface.

Most planets are very nearly axisymmetric, with the major departure from sphericity being due to a rotationally induced equatorial bulge. Thus, in this section, we analyze the effects of an axisymmetric body's deviation from spherical symmetry on the gravitational force that it exerts. The gravitational field of a nonaxisymmetric body is discussed in §6.1.4.

2.5.1 Gravitational Potential

The analysis of the gravitational field of an axisymmetric planet is most conveniently done by using the Newtonian gravitational potential, $\Phi_g(\mathbf{r})$, which is defined in equation (2.19a). As $\Phi_g(\mathbf{r})$ satisfies Laplace's equation (2.20b) in free space, the gravitational potential exterior to a planet can be expanded in terms of Legendre polynomials (instead of the complete spherical harmonic expansion, which would be required for the potential of a body of arbitrary shape; see eqs. 6.6):

$$\Phi_g(r, \phi, \theta) = -\frac{Gm}{r}\left[1 - \sum_{n-2}^{\infty} J_n P_n(\cos\theta)\left(\frac{R}{r}\right)^n\right].$$

$$(2.29)$$

Equation (2.29) is written in standard spherical coordinates, with ϕ the longitude and θ representing the angle between the planet's symmetry axis and the vector to the particle (i.e., the colatitude). The terms $P_n(\cos\theta)$ are the Legendre polynomials, given by the formula:

$$P_n(x) = \frac{1}{2^n n!}\frac{d^n}{dx^n}\left(x^2 - 1\right)^n. \qquad (2.30)$$

The *gravitational moments*, J_n, are determined by the planet's mass distribution (§6.1.4). The origin is chosen to be the center of mass, thus the gravitational moment $J_1 = 0$. For a nonrotating fluid body in hydrostatic equilibrium, the potential is spherically symmetric and the moments $J_n = 0$. If the planet's mass is distributed symmetrically about the planet's equator, then the J_n are zero for all odd n.

[2] The region well interior to Mercury's perihelion is also quite stable to planetary perturbations. The lack of observed bodies in this zone is attributed to cosmogonic considerations (the difficulty of forming bodies in such a hot region of the protoplanetary disk, §13.4 and §13.5). Additionally, high orbital velocities near the Sun would lead to very destructive impacts when bodies orbiting in this region are struck by the occasional comets and asteroids that reach this near the Sun. Small bodies orbiting interior to Mercury would also have their orbits altered by strong solar radiation forces (§2.7).

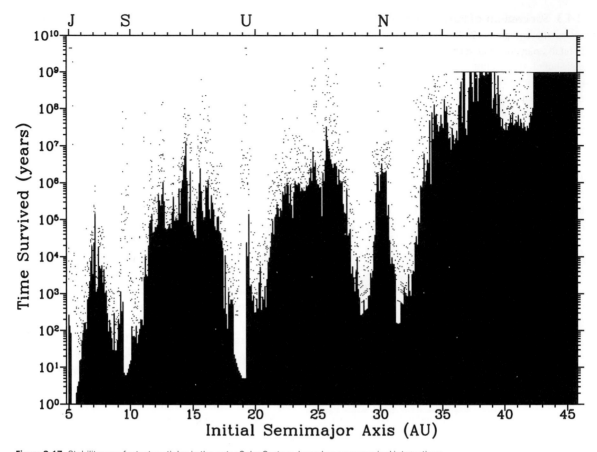

Figure 2.17 Stability map for test particles in the outer Solar System, based upon numerical integrations that include the Sun and the four giant planets. The time that each particle survived is plotted as a function of particle initial semimajor axis. For each semimajor axis bin, six particles were started at differing longitudes. The vertical bars mark the minimum of the six termination times. The points mark the termination times of the other five particles. The scatter of points gives an idea of the spread in particle lifetimes at each semimajor axis. The locations of the planets are denoted on the top of the figure; the spikes in particle lifetimes near these semimajor axes represent particles initially in tadpole or horseshoe orbits. The integrations extend to 4.5×10^9 years for particles initially interior to Neptune and to 10^9 years for those further out. Only a few particles initially interior to Neptune survived the entire integrations, but many particles exterior to 33 AU and all particles beyond about 43 AU remained on non-planet-crossing orbits for the entire time interval simulated. (Courtesy Matt Holman; see Holman 1997 for details on the calculations)

Let us consider a small body, e.g., a moon or ring particle, which travels around a planet on a circular orbit in the equatorial plane ($\theta = 90°$) at a distance r from the center of the planet. The centripetal force must be provided by the radial component of the planet's gravitational force (eq. 2.13), so the particle's angular velocity n satisfies:

$$rn^2(r) = \left.\frac{\partial \Phi_g}{\partial r}\right|_{\theta=90°}. \tag{2.31}$$

If the particle suffers an infinitesimal displacement from its circular equatorial orbit, it will oscillate freely in the horizontal and vertical directions about the reference circular orbit with radial (epicyclic) frequency $\kappa(r)$ and vertical frequency $\mu(r)$ respectively, given by

$$\kappa^2(r) = r^{-3}\frac{\partial}{\partial r}[(r^2 n)^2], \tag{2.32}$$

$$\mu^2(r) = \left.\frac{\partial^2 \Phi_g}{\partial z^2}\right|_{z=0}. \tag{2.33}$$

2.5.2 Precession of Particle Orbits

Using the equations (2.29–2.33), one can show that the orbital, epicyclic, and vertical frequencies can be written as

$$n^2 = \frac{Gm}{r^3}\left[1 + \frac{3}{2}J_2\left(\frac{R}{r}\right)^2 - \frac{15}{8}J_4\left(\frac{R}{r}\right)^4 + \frac{35}{16}J_6\left(\frac{R}{r}\right)^6 - \frac{315}{128}J_8\left(\frac{R}{r}\right)^8 + \cdots\right],$$ (2.34)

$$\kappa^2 = \frac{Gm}{r^3}\left[1 - \frac{3}{2}J_2\left(\frac{R}{r}\right)^2 + \frac{45}{8}J_4\left(\frac{R}{r}\right)^4 - \frac{175}{16}J_6\left(\frac{R}{r}\right)^6 + \frac{2205}{128}J_8\left(\frac{R}{r}\right)^8 + \cdots\right],$$ (2.35)

$$\mu^2 = 2n^2 - \kappa^2.$$ (2.36)

For a perfectly spherically symmetric planet, $\mu = \kappa = n$. Since planets are oblate, μ is slightly larger than the orbital frequency, n, and κ is slightly smaller. The oblateness of a planet therefore causes periapse longitudes of particle orbits in and near the equatorial plane to precess in the direction of the orbit, and lines of nodes of nearly equatorial orbits to regress. Orbits about oblate planets are thus not Keplerian ellipses. However, as the trajectories are nearly elliptical, they are often specified by instantaneous Keplerian orbital elements. Note that

$$\frac{d\varpi}{dt} = n - \kappa,$$ (2.37)

$$\frac{d\Omega}{dt} = n - \mu.$$ (2.38)

2.5.3 Torques Upon an Oblate Planet

The nonspherical distribution of mass within an oblate planet allows the planet to exert torques upon its satellites, changing their orbital angular momenta and thereby precessing the planes of their orbits, as discussed above. Other bodies can exert torques on an oblate planet, via a corresponding back force, thereby changing the direction of its rotational pole in inertial space.

The strongest of torques exerted on an oblate planet are exerted by the Sun. In some cases, such as Earth, large satellites act as intermediaries and their presence can affect the magnitude of the torque. Solar torques cause a planet's rotation axis to precess, resulting in a difference between the length of a tropical year (periodicity of the seasons) and the planet's orbital period (see Problem 1.5), as well as changing the position of the pole on the celestial sphere. Thus, we use a different northern pole star than did the ancient Greeks and Romans. This precession is discussed further in §6.1.4.3.

The torques on a planet's equatorial bulge resulting from gravitational forces of other planets are, of course, much smaller than are the corresponding solar torques.

Nonetheless, interplanetary torques can be more fundamentally important than are solar torques, because they can cause planetary obliquities (axial tilts) to vary. The obliquity of Mars varies chaotically as a result of these perturbations, reaching values as high as 60° (Fig. 2.18). For obliquity in the range 54°–126°, the seasonally averaged flux of solar energy reaching a planet's pole is greater than that reaching its equator.

Earth's obliquity is stabilized by the Moon; without the Moon, Earth's obliquity would also vary considerably, resulting in substantial variations in climate. As it is, the small variations in Earth's obliquity that have occurred over the past few million years (Fig. 2.19) are correlated with ice ages. The quasi-periodic climate variations associated with changes in the Earth's obliquity and with our planet's eccentricity are known as *Milankovitch cycles*.

2.6 Tides

The gravitational force arising from the pull of external objects varies from one part of a body to another. These differential tugs produce what is known as the *tidal force*. The net force on a body determines the acceleration of its center of mass, but tidal forces can deform a body, and can produce torques that alter its rotation state. Time-variable tidal forces such as those experienced by moons on eccentric orbits can result in flexing, which leads to internal heating.

Tidal forces are important to many aspects of the structure and evolution of planetary bodies. For example, on short timescales, temporal variations in tides (as seen in the frame rotating with the body under consideration) cause stresses that can move fluids with respect to more rigid parts of the planet, like the ocean tides with which we are familiar. These stresses can even cause seismic disturbances. (While the evidence that the Moon causes some earthquakes is weak and disputable, it is clear that the tides raised by the Earth are a major cause of moonquakes.) On long timescales, tides change the orbital and spin properties of planets and moons. Tides, along with rotation,

(a)

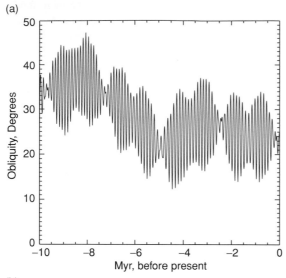

(b)

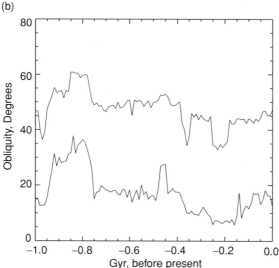

Figure 2.18 Torques by other planets on the martian equatorial bulge produce variations in the obliquity of Mars on a variety of timescales. (a) The martian obliquity over the past 10 Myr. (b) Maxima and minima of the martian obliquity during 10 Myr time intervals are shown for the past 1 Gyr. Because the Lyapunov time for this system is far less than 1 Gyr, the obliquity of Mars cannot be calculated accurately for this timescale, but this integration yields one possible, randomly selected, realization. (Courtesy John Armstrong; see Armstrong *et al.* 2004 for details on the calculation)

determine the equilibrium shape of a body located near any massive object; note that many materials which behave as solids on short timescales are effectively fluids on very long geological timescales, e.g., the Earth's mantle. In some cases, tidal forces are so strong that they exceed a body's cohesive force, and the body fragments.

2.6.1 The Tidal Force and Tidal Bulges

Consider a nearly spherical body of radius R, centered at the origin, which is subject to the gravitational force of a point mass, m, at $\mathbf{r}_o$, where $r_o \gg R$. At a point $\mathbf{r} = (x, y, z)$, the specific (per unit mass) tidal force is the difference between the pull of m at $\mathbf{r}$ and the pull of m at the origin:

$$\mathbf{F}_T(\mathbf{r}) = \frac{Gm}{|\mathbf{r}_o - \mathbf{r}|^3}(\mathbf{r}_o - \mathbf{r}) - \frac{Gm}{r_o^3}\mathbf{r}_o. \qquad (2.39)$$

For points along the line joining the center of the body to the point mass (which we take to be the x-axis), equation (2.39) reduces to

$$F_T(x) = \frac{Gm}{(x_o - x)^2} - \frac{Gm}{x_o^2} \approx \frac{2xGm}{x_o^3}. \qquad (2.40)$$

The tidal approximation used for the last part of equation (2.40) can be derived by Taylor expansion of the first term in the middle expression and retaining only the first two terms. Equation (2.40) states that, to lowest order, the tidal force varies proportionally to the distance from the center of the stressed body and inversely to the cube of the distance from the perturber. The portion of the body with positive x coordinate feels a force in the positive x-direction and the portion at negative x is tidally pulled in the opposite direction (Fig. 2.20).

Note from Figure 2.20 and equation (2.39) that material off the x-axis is tidally drawn towards the x-axis. If the body is deformable, it responds by becoming elongated in the x-direction. For a perfectly fluid body, the degree of elongation is that necessary for the body's surface to be an equipotential, when self-gravity, centrifugal force due to rotation, and tidal forces are all included in the calculation (§6.1.4.2).

The gravitational attraction of, for example, the Moon and Earth on one another thus causes tidal bulges which rise along the line joining the centers of the two bodies. The near-side bulge is a direct consequence of the greater gravitational attraction closer to the other body, whereas the bulge on the opposite side results from the weaker attraction at the far side of the object than at its center. The differential centrifugal acceleration across the body also contributes to the size of the tidal bulges.

The Moon spins once per orbit, so that the same face of the Moon always points towards the Earth and the Moon is always elongated in that direction. The Earth, however, rotates much faster than the Earth–Moon orbital period. Thus, different parts of the Earth point towards the Moon and are tidally stretched. Water responds much more readily to these varying forces than does the 'solid Earth', resulting in the tidal variations in the water level

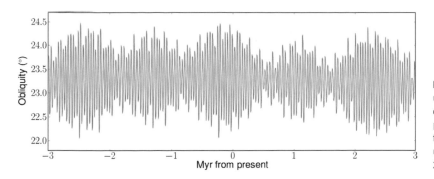

Figure 2.19 Variations in the obliquity of Earth over a time interval of 6×10^6 years centered on the present epoch. These data were taken from the same integrations used to produce Figures 2.14 and 2.16. (Courtesy Tom Quinn)

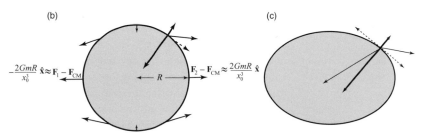

Figure 2.20 Schematic illustration of the tidal forces of a moon on a deformable planet. (a) Gravitational force of the moon on different parts of the planet. (b) Plain solid arrows indicate the differential force of the moon's gravity relative to the force on the planet's center of mass. (c) Response of the planet's figure to the moon's tidal pull.

seen at ocean shorelines (Problem 2.20). As the combined effects of terrestrial rotation and the Moon's orbital motion imply that the Moon passes above a given place on Earth approximately once every 25 hours (Problem 2.21), there are almost two tidal cycles per day, and the principal tide that we see is known as the *semidiurnal tide*. The Sun also raises semidiurnal tides on Earth, with a period of 12 hours and an amplitude just under half those of lunar tides (Problem 2.22a). Tidal amplitudes reach a maximum twice each (astronomical) month, when the Moon, Earth, and Sun are approximately aligned, i.e., when the Moon is 'new' or 'full'. Tides are also larger when the Moon is near perigee and when the Earth is near perihelion (the latter occurs in early January).

Strong tides can significantly affect the physical structure of bodies. Generally, the strongest tidal forces felt by Solar System bodies (other than Sun-grazing or planet-grazing asteroids and comets) are those caused by planets on their closest satellites. Near a planet, tides are so strong that they can rip apart a fluid (or weakly aggregated solid)

body. In such a region, large moons are unstable, and even small moons, which could be held together by material strength and friction, are unable to accrete due to tides. The boundary of this region is known as *Roche's limit*. Interior to Roche's limit, solid material remains in the form of small bodies, and we see rings instead of large moons. The derivation of Roche's limit is outlined in §11.1.

2.6.2 Tidal Torque

Tidal dissipation causes secular variations in the rotation rates and orbits of moons and planets. Although the total angular momentum of an orbiting pair of bodies is conserved in the absence of an external torque, angular momentum can be transferred between rotation and orbital motions via tidal torques. Orbital angular momentum is given by equations (2.10b) and (2.18). The rotational angular momentum of a rigidly rotating body is given by

$$\mathbf{L} = \overset{\leftrightarrow}{\mathbf{I}} \cdot \boldsymbol{\omega}_{\mathrm{rot}}, \qquad (2.41a)$$

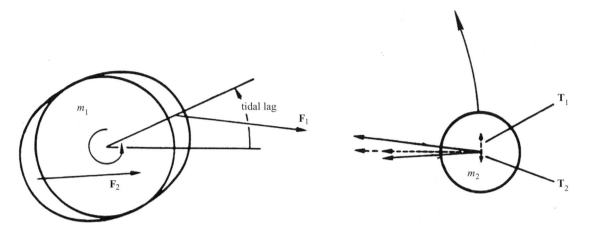

Figure 2.21 Schematic illustration of the tidal torque that a planet exerts on a moon orbiting in the prograde direction with a period longer than the planet's rotation period. Dissipation within the planet causes the tidal bulges that the moon raises on the planet to be located at places on the planet that were nearest to and farthest from the moon at a slightly earlier time. Because there is a temporal lag in the tidal bulges, for a moon on a slow prograde orbit the bulges lead the position of the moon. The asymmetries in the planet's figure imply that its gravity is not a central force, and thus it can exert a torque on the moon. The far-side bulge exerts a retarding torque, T_2, upon the moon, but the near-side bulge exerts a larger positive torque, T_1, on the moon, so the moon receives a net positive torque, and its orbit evolves outwards.

where $\overleftrightarrow{\mathbf{I}}$ is the *inertia tensor* of the body and $\boldsymbol{\omega}_{\rm rot}$ is its spin angular velocity. The kinetic energy of rotation is given by

$$E_{\rm rot} = \frac{1}{2}\boldsymbol{\omega}_{\rm rot} \cdot \overleftrightarrow{\mathbf{I}} \cdot \boldsymbol{\omega}_{\rm rot} = \frac{1}{2}\boldsymbol{\omega}_{\rm rot} \cdot \mathbf{L}. \qquad (2.41b)$$

The components of the inertia tensor are

$$I_{\rm jk} = \iiint \rho(\mathbf{r})(r^2\delta_{\rm jk} - x_{\rm j}x_{\rm k})d\mathbf{r}, \qquad (2.42a)$$

where $\rho(\mathbf{r})$ represents density and $\delta_{\rm jk}$ is the Kronecker δ, i.e., $\delta_{\rm jk} = 1$ if $j = k$; $\delta_{\rm jk} = 0$ if $j \neq k$. The moment of inertia of a body about a particular axis is a scalar given by

$$I = \iiint \rho(\mathbf{r})r_{\rm c}^2 d\mathbf{r}, \qquad (2.42b)$$

where $r_{\rm c}$ is the distance from the axis, and the integral is taken over the entire body. The moment of inertia of a uniform density sphere of radius R and mass m about its center of mass is given by

$$I = \frac{2}{5}mR^2 \qquad (2.43)$$

(Problem 6.4b). Centrally condensed bodies have moment of inertia ratios $I/(mR^2) < 2/5$. The moment of inertia ratios for the planets are listed in Table 6.2.

If planets were perfectly fluid, they would respond immediately to varying forces, and tidal bulges raised by

a satellite would point directly towards the moon responsible. However, the finite response time of a planet's figure causes the tidal bulges to lag 'behind', at locations on the planet which pointed towards the moon at a slightly earlier time (Fig. 2.21). Provided the planet's rotation period is shorter than the moon's orbital period and the moon's orbit is prograde, this tidal lag causes the nearer bulge to lie in front of the moon, and the moon's greater gravitational force on the near-side bulge than on the far-side bulge acts to slow the rotation of the planet. The reaction force upon the moon causes its orbit to expand. Satellites in retrograde orbits (e.g., Triton) and satellites whose orbital periods are less than the planet's rotation period (e.g., Phobos) spiral inwards towards the planet as a result of tidal forces (Problem 2.23).

The tidal torque depends upon the size of the tidal bulge and the lag angle. We denote the *tidal Love number*, which measures the elastic deformation of a body in response to a tidal perturbation, by $k_{\rm T}$. The specific dissipation factor, Q, is the ratio of the peak energy stored in the tidal bulge to that dissipated in one cycle to the energy. The rotational angular velocity is denoted by $\omega_{\rm rot}$, the orbital angular velocity by n, and the distance between primary and secondary by r, and we use the subscripts 1 and 2 to refer to the primary and secondary, respectively. The size of the tidal bulge varies as $k_{\rm T_1}m_2r^{-3}$, the phase lag as Q_1^{-1}; the torque on a bulge of given size and lag angle is also a tidal effect, and is proportional to $Gm_2 r^{-3}$. Accounting for all of these factors as well as the planet's size, the direction of the torque, and an overall constant, yields the following formula for the torque on a satellite from the tidal bulge that it raises on its primary:

$$\dot{L}_{2(1)} = \frac{3}{2} \frac{k_{T_1}}{Q_1} \frac{Gm_2^2 R_1^5}{r^6} \text{sign}(\omega_{rot_1} - n), \qquad (2.44a)$$

where sign $(x) = 1$ if $x > 0$ and sign$(x) = -1$ if $x < 0$. Equation (2.44a) gives the rate at which tides transport angular momentum between planetary rotation and satellite orbits. If no other torques are present, a low-eccentricity orbit ($r \approx a$) expands (or contracts) at the rate

$$\dot{a} = 3\frac{k_{T_1}}{Q_1} \frac{G^{1/2} m_2 R_1^5}{m_1^{1/2} a^{11/2}} \text{sign}(\omega_{rot_1} - n) \qquad (2.44b)$$

(Problem 2.23a).

The above arguments remain valid if 'moon' is replaced by 'Sun', or if moon and planet are interchanged. Indeed, the stronger gravity of planets means that they have a much greater effect on the rotation of moons than vice versa. Most, if not all, major moons have been slowed to a synchronous rotation state, in which the same hemisphere of the moon always faces the planet; thus, no tidal lag occurs for these moons.

Evidence exists for the tidal slowing of Earth's rotation on a variety of timescales. Growth bands observed in fossil bivalve shells and corals imply that there were 400 days per year approximately 350 million years ago. Eclipse timing records imply that the day has lengthened slightly over the past two millennia. Precise measurements using atomic clocks show variations in the Earth's rotation rate; however, care must be taken to separate secular tidal effects from the short-term periodic influences. Most of the secular decrease in Earth's rotation rate is caused by tides raised by the Moon, but at the present epoch $\sim$20% is due to solar tides (Problem 2.22b).

The Pluto–Charon system has evolved even further. Charon is between one-ninth and one-eighth as massive as Pluto, a much larger secondary to primary mass ratio than observed for any of the more massive bodies in the Solar System (Table 1.5). The semimajor axis of their mutual orbit is just 19 636 km, which is only 5% of the Earth–Moon distance and smaller than any planet–moon separation apart from Mars–Phobos (Table 1.4). The Pluto–Charon system has reached a stable equilibrium configuration, in which each of the bodies spins upon its axis in the same length of time they orbit about their mutual center of mass (Problem 2.28). Thus,

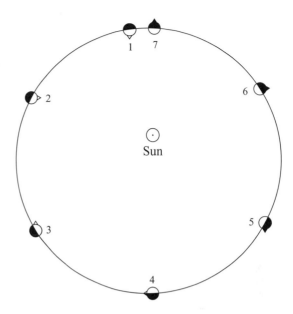

Figure 2.22 Mercury's rotation and solar day. Mercury's 3:2 spin-orbit resonance implies that the same axis is always aligned with the direction to the Sun at perihelion, and that the mercurian (solar) day lasts twice as long as the mercurian year.

the same hemisphere of Pluto always faces Charon, and the same hemisphere of Charon always faces Pluto.

Solar tides have resulted in a stable spin-orbit lock for nearby Mercury, but one that is more complicated than the synchronous state which exists for most planetary satellites. Mercury makes three rotations around its axis every two orbits about the Sun (Fig. 2.22). The reason that equilibrium exists at exactly one and one-half rotations per orbit is that Mercury has a small permanent (nontidal) deformation and a highly eccentric orbit. It is energetically most favorable for Mercury's long axis (the axis with smallest moment of inertia) to point towards the Sun every time the planet passes perihelion, a configuration consistent with the observed 3:2 spin-orbit resonance. It may thus not be by random chance that the *Caloris basin* (Fig. 5.49a,b), which is the largest known geologic feature on Mercury, faces directly towards (i.e., lies at solar noon) or away from the Sun at alternate perihelion passages, although confirmation of a causal connection awaits precise measurements of Mercury's gravity field.

Were Mercury transformed into a fluid planet, its permanent deformation would vanish, and solar tides would slow Mercury's rotation further. However, because of the substantial eccentricity of Mercury's orbit, synchronous rotation would not be achieved. According to Kepler's second law, a planet orbits the Sun much faster at perihelion than at aphelion. The variation is so large for Mercury that for a short time each orbit the planet's angular velocity about the Sun is even faster than its present rate of spin. During this brief interval, the tidal bulge raised on Mercury by the Sun trails the Mercury–Sun line, so the Sun's gravity acts to speed up Mercury's rotation rate. As tidal effects on spin vary inversely with the sixth power of the distance between the two objects (eq. 2.44), the short interval during which Mercury's rotation is rapidly accelerated is almost able to balance the much greater fraction of the time during which the Sun's tides act to slow the planet's spin. For small orbital eccentricity, e, the equilibrium spin rate of a fluid planet is given by

$$\frac{\omega_{rot}}{n} = 1 + 6e^2 + \mathcal{O}(e^4), \tag{2.45}$$

where $\mathcal{O}$ signifies the order of the term in the expansion. Equation (2.45) is more directly applicable to large, close-in (extrasolar) planets (§12.3), which are likely to be fluid, than it is to the planet Mercury. Nonetheless, if our hypothetical fluid Mercury's spin period were increased to roughly 70 days (significantly above its current value of 59 days, yet still well below the planet's 88 day orbital period), a balance between addition of spin angular momentum by solar tides near perihelion and removal during the remainder of the orbit would be achieved. Note that the value of a fluid Mercury's equilibrium rotation rate in the absence of a permanent bulge would vary as Mercury's orbital eccentricity increased and decreased in response to perturbations from the other planets (Fig. 2.14). (Changes in Mercury's semimajor axis in response to planetary perturbations are very small.)

The situation would be more complicated if Mercury were a solid planet lacking a permanent deformation. Solid bodies respond less rapidly to tidal variations than do fluid ones, and as a result the equilibrium rotation rate would be quantized at resonant values, taking the value of 88 days for the present eccentricity, but shifting to 59 days when Mercury's eccentricity was near the high end of its range. Mercury's actual rotation rate, however, would have a more complex behavior, with the rate of dissipation of tidal energy within the planet determining how rapidly the equilibrium rotation rate of the epoch was approached.

Solar gravitational tides are probably the principal reason that Venus rotates very slowly, but they do not explain why our sister planet spins in the retrograde direction. Solar heating produces asymmetries in Venus's massive atmosphere which are known as *atmospheric tides*, and the Sun's gravitational pull on these atmospheric tides probably prevents Venus's solar day from becoming longer than it is at present. Tidal forces slow the rotation rates of the other planets, but at rates too small to be significant, even over geologic time.

2.6.3 Tidal Heating

In addition to transporting angular momentum (eq. 2.44), tidal torques transfer energy. The energy transport rate is n times that of angular momentum. As can be seen by taking the ratio of the derivatives of equations (2.15a) and (2.18) with respect to a, the ratio of change in mechanical energy to that of orbital angular momentum for expanding circular orbits is given by $dE/dL = n$. So torques exerted by the tidal bulges raised upon the planet do not (directly) alter the eccentricity of satellite orbits.

Temporal variations in tidal forces can lead to internal heating of planetary bodies. The locations of the tidal bulges of a moon having nonsynchronous rotation vary as the planet moves in the sky as seen from the moon. A synchronously rotating moon on an eccentric orbit is subjected to two types of variations in tidal forces. The amplitude of the tidal bulge varies with the moon's distance from the planet, and the direction of the bulge varies because the moon spins at a constant rate (equal to its mean orbital angular velocity), whereas the instantaneous orbital angular velocity varies according to Kepler's second law. Since planetary bodies are not perfectly rigid, these variations in tidal forces change the shape of the moon. Since bodies are not perfectly fluid either, moons dissipate energy as heat while they change in shape. Internal stresses caused by variations in tides on a body that is on an eccentric orbit or that is not rotating synchronously with its orbital period can therefore result in significant tidal heating of some bodies, most notably in Jupiter's moon Io. If no other forces were present, the dissipation that results from variations in tides raised on Io by Jupiter would lead to a decay of Io's orbital eccentricity. Io's orbit would approach circularity, the lowest energy state (smallest semimajor axis) for a given angular momentum (eq. 2.18), with the dissipated orbital energy being converted into thermal energy. As Io orbits exterior to the semimajor axis at which its orbital period would be synchronous with Jupiter's spin, the tides raised on Jupiter by Io transfer some of Jupiter's rotational energy to Io's orbit, causing Io to spiral outwards (eq. 2.44). As noted above, these torques do not directly affect the eccentricity of Io's orbit. However, there exists a 2:1 mean motion resonance lock between Io and Europa (Table 1.4 and §2.3.2.1). Io passes on some of the orbital energy and angular momentum it receives from Jupiter to Europa, and because $n_{Io} > n_{Europa}$, Io's eccentricity is increased as a consequence of this transfer (Problem 2.29). This forced eccentricity maintains a high tidal dissipation rate, and consequently there is a large internal heating of Io, which displays itself in the form of active volcanism (§5.5.5.1).

2.7 Dissipative Forces and the Orbits of Small Bodies

The gravitational interactions between the Sun, planets, and moons were described in §§2.1–2.6. In this section, we consider the effects of solar radiation, the solar wind, and gas drag. Gravity exerts itself on the entire volume of a body, whereas these other forces act only at the surface. Therefore, these nongravitational forces most significantly affect the orbits of small bodies, which have the largest surface to volume ratios.

Four dynamical effects of solar radiation can be distinguished:

(1) *Radiation pressure*, which pushes particles (primarily micrometer-sized dust) outwards from the Sun.

(2) *Poynting–Robertson drag*, which causes centimeter-sized particles to spiral inward towards the Sun.

(3) The *Yarkovsky effect*, which changes the orbits of meter to ten-kilometer-sized objects due to uneven temperature distributions across their surfaces.

(4) The *YORP effect*, which can significantly alter the rotation rates of asteroids up to $\sim$20 kilometers in radius.

The solar wind produces a *corpuscular drag* similar in form to the Poynting–Robertson drag; corpuscular drag is most important for submicrometer particles. We discuss each of these processes in the next five subsections and then examine the effect of gas drag on orbital motion. Perturbations on the motion of dust in planetocentric orbits caused by solar radiation are analyzed in §11.5.1. The motion of charged dust particles in planetary rings is discussed in §11.5.2. Nongravitational forces resulting from asymmetric mass loss by comets are considered in §10.2.1.

2.7.1 Radiation Pressure (micrometer grains)

The Sun's radiation exerts a repulsive force, $\mathbf{F}_{rad}$, on all bodies in our Solar System. This force is given by

$$\mathbf{F}_{rad} \approx \frac{\mathcal{L}_{\odot} A}{4\pi c r_{\odot}^2} Q_{pr} \hat{\mathbf{r}}, \qquad (2.46)$$

where A is the particle's geometric cross-section, $\mathcal{L}_{\odot}$ the solar luminosity, $r_{\odot}$ the heliocentric distance, c is the speed of light, and Q_{pr} the dimensionless *radiation pressure coefficient*. The radiation pressure coefficient accounts for both absorption and scattering and is equal to unity for a perfectly absorbing particle. For large particles, Q_{pr} is typically of order unity, but $Q_{pr} \ll 1$ for grains much

smaller than the wavelength of the impinging radiation. Relativistic effects produced by the Doppler shift between the frame of the Sun and that of the particle are generally small, and have been omitted from equation (2.46), but they will be considered in §2.7.2.

The radiation pressure exerted on a body that is large compared to the wavelength of the light impinging upon it varies in proportion to its projected area, πR^2, whereas the gravitational force is proportional to the body's mass, $4\pi \rho R^3/3$, so the ratio of the two goes as $(\rho R)^{-1}$. This ratio is expressed numerically for grains in heliocentric orbit using the dimensionless parameter β, which is defined as the ratio between the forces due to the radiation pressure and the Sun's gravity:

$$\beta \equiv \left| \frac{F_{rad}}{F_g} \right| = 5.7 \times 10^{-5} \frac{Q_{pr}}{\rho R}, \qquad (2.47a)$$

with the particle's radius, R, in cm and its density, ρ, in g cm^{-3}. Because both radiation pressure and gravity fall off as $r_{\odot}^{-2}$, β is independent of heliocentric distance. Solar radiation pressure is only important for micrometer- and submicrometer-sized particles. Extremely small particles are not strongly affected by radiation pressure, because Q_{pr} decreases as the particle radius drops below the (visible wavelength) peak in the solar spectrum (Fig. 2.23). The Sun's effective gravitational attraction is given by

$$F_{g,\text{eff}} = \frac{(1-\beta)GmM_{\odot}}{r_{\odot}^2}, \qquad (2.47b)$$

that is, particles 'see' a Sun of mass $(1-\beta)M_{\odot}$. It is clear that small particles with $\beta > 1$ are repelled more strongly by the Sun's radiation than they are attracted by solar gravity, and thus quickly escape the Solar System, unless they are gravitationally bound to one of the planets. Some 'large' bodies that orbit at the Keplerian velocity shed dust (Fig. 10.1); dust released from large bodies that are on circular orbits about the Sun is ejected from the Solar System if $\beta > 0.5$. Critical values of β for dust released from bodies on eccentric orbits are calculated in Problem 2.32.

The importance of solar radiation pressure can, for example, be seen in comets (§10.3): Cometary tails always point away from the Sun. The ion tails point near the antisolar direction because the ions are dragged along with the solar wind (§10.5.2), which moves rapidly compared to orbital velocities. The dust tails also lie further from the Sun than the nucleus, but are curved. The dust grains initially have the same Keplerian orbital velocities as the comet nucleus. But as a consequence of radiation

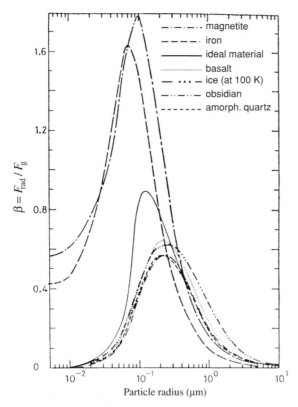

Figure 2.23 The relative radiation pressure force, $\beta = |F_{rad}/F_g|$, as a function of particle size for six cosmically significant substances and a hypothetical ideal material that absorbs all radiation of wavelength $\lambda < 2\pi R$ but is completely transparent to longer wavelengths and has a density $\rho = 3\,\text{g}\,\text{cm}^{-3}$. Most solar energy is radiated in the form of photons of wavelength $0.2-4\,\mu\text{m}$ (Fig. 3.2). For grains much larger than the wavelengths of these photons, the curves are inversely proportional to particle radius, with the constant of proportionality depending upon particle reflectivity. Grains interact weakly with photons whose wavelength is much larger than the grain size, so β values decline precipitously for grains smaller than $\sim 0.1\,\mu\text{m}$. Note that these values are for particles in orbit about the Sun. Grains orbiting stars of different mass, luminosity, and/or spectral type would have different values of β. (Adapted from Burns *et al.* 1979)

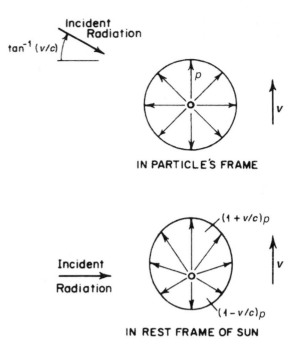

Figure 2.24 A particle in heliocentric orbit that reradiates the solar energy flux isotropically in its own frame of reference, preferentially emits more momentum, p, in the forward direction as seen in the solar frame, because the frequencies and momenta of the photons emitted in the forward direction are increased by the particle's motion. (Adapted from Burns *et al.* 1979)

pressure, they feel a smaller net attraction to the Sun (eq. 2.47), so they drift slowly outwards relative to the nucleus.

2.7.2 Poynting–Robertson Drag (small macroscopic particles)

A particle in orbit around the Sun absorbs solar radiation and reradiates the energy isotropically in its own frame. The particle thereby preferentially radiates (and loses momentum) in the forward direction in the inertial frame of the Sun (Fig. 2.24). This leads to a decrease in the particle's energy and angular momentum and causes

dust in bound orbits to spiral sunward. This effect is called *Poynting–Robertson drag*.

Let us consider a perfectly absorbing, rapidly rotating, dust grain. The flux of solar radiation absorbed by a grain with cross-sectional area A is equal to

$$\frac{\mathcal{L}_\odot A}{4\pi r_\odot^2}\left(1 - \frac{v_r}{c}\right), \qquad (2.48a)$$

where $v_r = \mathbf{v} \cdot \hat{\mathbf{r}}$ is the radial component of the particle's velocity (i.e., the component which is parallel to the incident beam of light). The second term in expression (2.48a) accounts for the Doppler shift between the Sun's rest frame and that of the particle; the transverse Doppler shift is of order $(v_\theta/c)^2 \ll 1$ and will be ignored here. The absorbed flux is reradiated isotropically and can be written as a mass loss rate in the particle's frame of motion (using $E = mc^2$):

$$\frac{\mathcal{L}_\odot A}{4\pi c^2 r_\odot^2}\left(1 - \frac{v_r}{c}\right). \qquad (2.48b)$$

As the particle moves relative to the Sun with velocity $\mathbf{v}$, there is a momentum flux from the particle as seen in the rest frame of the Sun, since the particle emits more momentum in the forward direction than in the backward direction (Fig. 2.24). This flux produces a retarding force on the particle equal to

$$\frac{-\mathcal{L}_\odot A}{4\pi c^2 r_\odot^2}\left(1 - \frac{v_r}{c}\right)\mathbf{v}. \qquad (2.48c)$$

Expression (2.48c) can be generalized to the case in which the particle reflects and/or scatters some of the radiation impinging upon it via multiplication by Q_{pr}. The net force on the particle in this more general case is given by

$$\mathbf{F}_{rad} = \frac{\mathcal{L}_\odot Q_{pr} A}{4\pi c r_\odot^2}\left(1 - \frac{v_r}{c}\right)\hat{\mathbf{r}} - \frac{\mathcal{L}_\odot Q_{pr} A v}{4\pi c^2 r_\odot^2}\left(1 - \frac{v_r}{c}\right)\hat{\mathbf{v}}$$

$$(2.49a)$$

$$\approx \frac{\mathcal{L}_\odot Q_{pr} A}{4\pi c r_\odot^2}\left[\left(1 - \frac{2v_r}{c}\right)\hat{\mathbf{r}} - \frac{v_\theta}{c}\hat{\boldsymbol{\theta}}\right]. \qquad (2.49b)$$

The first term in equation (2.49b) is that due to radiation pressure and the second and third terms (those involving the velocity of the particle) represent the Poynting–Robertson drag.

From the above discussion, it is clear that small dust grains in the interplanetary medium are removed, with (sub)micrometer-sized grains being rapidly blown out of the Solar System (see eqs. 2.47 and surrounding discussion), while centimeter-sized particles slowly spiral inward towards the Sun. The (orbit averaged) decay rate caused by Poynting–Robertson drag is

$$\frac{da}{dt} = -\frac{\mathcal{L}_\odot Q_{pr} A}{4\pi c^2 a}\frac{2 + 3e^2}{(1 - e^2)^{3/2}}, \qquad (2.50a)$$

and the eccentricity damps as:

$$\frac{de}{dt} = -\frac{\mathcal{L}_\odot Q_{pr} A}{4\pi c^2 a^2}\frac{5e}{(1 - e^2)^{1/2}}. \qquad (2.50b)$$

Typical decay times (in years) for particles on circular orbits are given by

$$t_{pr} \approx 400\frac{r_{AU}^2}{\beta}. \qquad (2.50c)$$

The *zodiacal light* is a band centered near the ecliptic plane that appears almost as bright as the Milky Way on a dark night. It is visible in the direction of the Sun, just after sunset or before sunrise. Particles that produce the bulk of the zodiacal light (at infrared and visible wavelengths) are between 20 and 200 μm, so their lifetimes at Earth's orbit are of the order of 10^5 years, which is much less than the age of the Solar System. Dust grains responsible for the zodiacal light are resupplied primarily from the asteroid belt, where numerous collisions occur between countless small asteroids, and from comets. Some grains released by Kuiper belt objects spiral inwards as a result of Poynting–Robertson drag provide a small additional contribution to the zodiacal light.

2.7.3 Yarkovsky Effect (meter–ten-kilometer objects)

Consider a rotating body heated by the Sun. The afternoon/evening hemisphere is typically warmer than the morning hemisphere, by an amount $\Delta T \ll T$. Let us assume that the temperature of the morning hemisphere is $T - \frac{\Delta T}{2}$, and that of the afternoon/evening hemisphere is $T + \frac{\Delta T}{2}$. The radiation reaction upon a surface element dA, normal to its surface, is

$$dF = \frac{2\sigma T^4 dA}{3c}, \qquad (2.51)$$

where σ is the Stefan–Boltzmann constant. For a spherical particle of radius R, the transverse reaction force in the orbit plane due to the excess emission on the evening side is

$$F_Y = \frac{8}{3}\pi R^2\frac{\sigma T^4}{c}\frac{\Delta T}{T}\cos\psi, \qquad (2.52)$$

where ψ the particle's obliquity, i.e., the angle between its rotation axis and orbit pole. This process is referred to as the diurnal *Yarkovsky effect*. The diurnal Yarkovsky force is positive (expands orbits) for an object which rotates in the prograde direction, $0° \leq \psi < 90°$, and negative (causes orbital decay, as does Poynting–Robertson drag) for an object with retrograde rotation, $90° \leq \psi \leq 180°$. There is also an analogous seasonal Yarkovsky effect which is produced by temperature differences between the spring/summer and the autumn/winter hemispheres.

The Yarkovsky effect significantly alters the orbits of bodies in the meter to ten-kilometer-size range. The first direct observational evidence of Yarkovsky forcing was obtained using measurements of the deviations in the orbit of the ~300 meter radius near-Earth asteroid 6489 Golevka from purely gravitational models. Another observed consequence of the Yarkovsky effect is the size-dependent distribution of orbital elements of objects within the Karin cluster of main belt asteroids, which were produced by a disruptive collision six million years ago. The Yarkovsky effect plays an important role in transporting most meteorite parent bodies from the asteroid belt to Earth by helping to sweep them into resonances. Asymmetric outgassing of comets produces a nongravitational force similar to the Yarkovsky force. Nongravitational forces on comets are discussed in §10.2.1.

2.7.4 YORP Torques (Rotation of Asymmetric Bodies)

Sunlight impinging on an asymmetric body can secularly alter the said body's rotation state. Both reflected photons and light that is absorbed and reradiated contribute to these torques, which are referred to as the Yarkovsky–O'Keefe–Radzievskii–Paddack effect, typically abbreviated as the *YORP effect*, after the scientists who first discovered and analyzed it. YORP torques on uniform albedo triaxial ellipsoids average to zero over a rotation period, but such cancelation does not occur for objects with wedge-like asymmetry or nonuniform albedo. These torques can spin up objects in a manner analogous to a windmill spinning in a uniform breeze.

A quantitative analysis is quite involved, but examining the dependence of the magnitude of the YORP effect on the size of the body is straightforward. The torque is proportional to the product of the area of the body ($\propto R^2$) and the moment arm (R), and thus varies as R^3, while the moment of inertia of the body varies with R^5 (eqs. 2.42*b* and 2.43). Taking the ratio, the rate at which the rotation rate changes varies as R^{-2}. Moreover, small bodies tend to be more asymmetric, so the dependence of the YORP effect on size is typically steeper than suggested by this scaling analysis.

Although the magnitude of YORP torques are generally quite small, both because of the weakness of the force associated with reradiation of photons and the near-symmetry of most objects within the Solar System, these torques build up over time. Current estimates suggest that YORP torques significantly affect the rotation states of asteroids with $R \lesssim 20$ km. Spin-up or spin-down of a typical 5 km radius asteroid occurs on a timescale of $\sim 10^8$ years, and some close satellites of asteroids (§9.4.4) were probably produced when YORP torques increased the rotational angular momentum of a single asteroid so much that it shed material from its equator.

2.7.5 Corpuscular Drag (submicrometer dust)

Particles with sizes much smaller than one micrometer are also subjected to a significant corpuscular 'drag' by solar wind particles. This effect can be calculated in a manner similar to the Poynting–Robertson drag above, except that the energy–momentum relation must be replaced by that of nonrelativistic particles:

$$p_{sw} = \frac{2E_{sw}}{v_{sw}}, \tag{2.53}$$

where v_{sw} is the solar wind velocity. The momentum flux densities carried by the solar wind are roughly four orders of magnitude less than that carried by the electromagnetic radiation; hence, since pressure is proportional to the momentum flux density, the pressure of the solar wind is much smaller than the radiation pressure. However, the aberration angle (the change in apparent position of

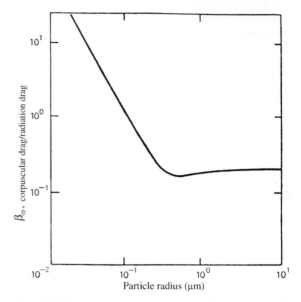

Figure 2.25 Ratio of corpuscular drag (caused by the solar wind) to that resulting from solar radiation, β_{cp}, is plotted against radius for grains composed of obsidian. (Adapted from Burns *et al.* 1979)

the source due to the motion of the receiving body) of the solar wind, $\tan^{-1}(v/v_{sw})$, is much larger than that for solar radiation, $\tan^{-1}(v/c)$, so the solar wind produces a significant drag force. The ratio of this *corpuscular drag* to the radiation drag, β_{cp}, can be expressed as

$$\beta_{cp} = \frac{p_{sw}}{p_r} \frac{c}{v_{sw}} \frac{C_{Dcp}}{Q_{pr}}, \tag{2.54}$$

where C_{Dcp} is the corpuscular drag coefficient.

Note that the first two fractions in equation (2.54) together are equivalent to the ratio between the mass flux of the solar wind and the mass-equivalent flux of solar photons (E/c^2). As the Sun loses about five times as much mass via photon emission as via the solar wind, $\beta_{cp} \approx 0.2$ for grains large enough that the last term in equation (2.54) is approximately unity.

Figure 2.25 shows a graph of β_{cp} as a function of particle radius. Corpuscular drag is more important than radiation drag for particles $\lesssim 0.1$ μm in size, which couple poorly to solar radiation, and it is the primary force behind these very small dust grains' inward spiral towards the Sun.

2.7.6 Gas Drag

Although for most purposes interplanetary space can be considered to be a vacuum, there are certain situations in which interactions with gas can significantly alter the motion of solid particles. Two prominent examples of this process are planetesimal interactions with the gaseous component of the protoplanetary disk during the formation of the Solar System and orbital decay of ring particles as a result of drag caused by extended planetary atmospheres.

In the laboratory, gas drag slows solid objects down until their positions remain fixed relative to the gas. In the planetary dynamics case, the situation is more complicated. For example, a body on a circular orbit about a planet loses mechanical energy as a result of drag with a static atmosphere, but this energy loss leads to a decrease in semimajor axis of the orbit, which implies the body actually speeds up! Other, more intuitive effects of gas drag are the damping of eccentricities and, in the case where there is a preferred plane in which the gas density is the greatest, the damping of inclinations relative to this plane.

Objects whose dimensions are larger than the mean free path of the gas molecules experience *aerodynamic drag*:

$$F_D = -\frac{C_D A \rho_g v^2}{2},\tag{2.55a}$$

where v is the velocity of the body with respect to the gas, ρ_g is the gas density, A is the projected surface area of the body, and C_D is a dimensionless drag coefficient, which is of order unity unless the Reynolds number (eq. 4.48) is very small. Smaller bodies are subject to *Epstein drag*:

$$F_D = -A \rho_g v v_o,\tag{2.55b}$$

where v_o is the mean thermal velocity of the gas. Note that as the drag force is proportional to surface area and the gravitational force is proportional to volume (for constant particle density), gas drag is usually most important for the dynamics of small bodies.

The gaseous component of the protoplanetary disk was partially supported against the gravity of the Sun by a negative pressure gradient in the radial direction. Thus, less centrifugal force was required to maintain equilibrium, and consequently the gas orbited less rapidly than the Keplerian velocity. The 'effective gravity' felt by the gas was

$$g_{eff} = -\frac{GM_\odot}{r_\odot^2} - \frac{1}{\rho_g}\frac{dP}{dr_\odot}.\tag{2.56}$$

For circular orbits, the effective gravity must be balanced by centrifugal acceleration, $r_\odot n^2$. For estimated protoplanetary disk parameters, the gas rotated $\sim 0.5\%$ slower than the Keplerian speed. The implications of gas drag for the accretion of planetesimals are discussed in §13.5.2.

Drag induced by a planetary atmosphere is substantially more effective for a given gas density than is drag in a primarily centrifugally supported disk. Because atmospheres are almost entirely pressure supported, the relative velocity between the gas and orbiting particles is large. As atmospheric densities drop rapidly with height, particle orbits decay slowly at first, but as they reach lower altitudes their decay can become very rapid (see Problem 2.34). Gas drag is the principal cause of orbital decay of artificial satellites in low Earth orbit.

2.8 Orbits About a Mass-Losing Star

At the present epoch, the Sun's mass is decreasing as it expels a bit more than 10^{-14} $M_\odot$ per year in matter through the solar wind, emits about five times as much via photon luminosity (Problem 2.35), and loses a relatively small amount through emission of neutrinos. Post-main-sequence stars can shed substantially more mass, through enormous stellar winds during their distended red giant phases as well as via supernova explosions. Very young stars both accrete and eject substantial quantities of matter. The dynamical consequences of interactions between the photons and massive particles escaping from the Sun were discussed in §2.7. In this section, we consider the direct effects of stellar mass loss on planetary orbits.

The response of planetary orbits to stellar mass loss depends qualitatively on the timescale over which the mass loss occurs. If the mass is lost (passes beyond the orbit of the planet) in a time short compared to the planet's orbital period, then the planet's instantaneous position and velocity are unchanged, but the diminished mass of the star affects the subsequent motion of the planet. The situation is dynamically analogous to the release of a small dust grain from a larger body on a Keplerian orbit, with the fractional decrease in the star's mass being the parameter analogous to β in equation (2.47). Thus, for example, an initially circular orbit becomes unbound if the star suddenly loses more than half of its mass. A smaller amount of 'instantaneous' stellar mass loss results in eccentric bound orbits, provided this loss occurs symmetrically, so that the velocity of the star itself does not change.

If the star loses mass on a timescale long compared to planetary orbital periods, as does the Sun, then planetary orbits expand gradually. No torque is exerted on the planet, so its orbital angular momentum (eq. 2.18) is conserved. The shape (eccentricity) of the planet's orbit also remains unchanged, so its semimajor axis increases according to the formula:

$$\frac{\dot{a}}{a} = -\frac{\dot{M}_\star}{M_\star}.\tag{2.57}$$

Further Reading

A good introductory text:

Danby, J.M.A., 1988. *Fundamentals of Celestial Mechanics*, 2nd Edition. Willmann-Bell, Richmond, VA. 467pp.

An excellent overview of many important aspects of planetary dynamics is presented by:

Murray, C., and S. Dermott, 1999. *Solar System Dynamics*. Cambridge University Press, Cambridge. 592pp.

A somewhat more mathematical treatment is given in:

Morbidelli, A., 2002. *Modern Celestial Mechanics: Aspects of Solar System Dynamics*. Taylor and Francis Cambridge Scientific Publishers, London. 368pp. Out of print. Downloadable from: http://www.oca.eu/morby/

The following book is not for the faint of heart, but it contains some important information that is difficult to locate elsewhere:

Brouwer, D., and G.M. Clemence, 1961. *Methods of Celestial Mechanics*. Academic Press, New York. 598pp.

Legendre expansions and spherical harmonics are covered in:

Jackson, J.D., 1999. *Classical Electrodynamics*, 3rd Edition. John Wiley and Sons, New York. 641pp.

A detailed discussion of the effects of solar radiation and the solar wind on the motion of small particles is given by:

Burns, J.A., P.L. Lamy, and S. Soter, 1979. Radiation forces on small particles in the Solar System. *Icarus*, **40**, 1–48.

Other useful individual articles include:

Duncan, M.J., and T. Quinn, 1993. The long-term dynamical evolution of the Solar System. *Annu. Rev. Astron. Astrophys.*, **31**, 265–295.

Peale, S.J., 1976. Orbital resonances in the Solar System. *Annu. Rev. Astron. Astrophys.*, **14**, 215–246.

Problems

2.1.E Consider two mutually gravitating bodies of masses m_1 and m_2 and positions $\mathbf{r}_1$ and $\mathbf{r}_2$.

(a) Write down the equations which govern the motion of these bodies.

(b) Using Newton's third law, show that the center of mass of the system moves at a constant velocity and that the relative position of the bodies, $\mathbf{r} \equiv \mathbf{r}_1 - \mathbf{r}_2$, changes according to

$$\frac{d^2\mathbf{r}}{dt^2} = -\frac{GM}{r^2}\hat{\mathbf{r}}, \qquad (2.58)$$

where $M \equiv m_1 + m_2$. This reduces the two-body problem to an equivalent one-body problem.

2.2.I In this problem, you will finish the derivation of the Newtonian generalization of Kepler's laws that you began in the previous problem.

(a) Derive the law of conservation of angular momentum for the system, $d(\mathbf{r} \times \mathbf{v})/dt = 0$, by

taking the cross product of $\mathbf{r}$ with equation (2.58) and using various vector identities. By writing the expression for angular momentum in polar coordinates, deduce Kepler's second law and determine the constant rate of sweeping, $d\mathcal{A}/dt$.

(b) Take the dot product of $\mathbf{v}$ with equation (2.58) to deduce the conservation of energy per unit mass. Integrate your result to determine an expression for the specific energy of the system, E. Express your answer in polar coordinates and solve for dr/dt. Take the reciprocal, multiply both sides by $d\theta/dt$ and then use the magnitude of the specific angular momentum, L, to eliminate the angular velocity from your expression, yielding the following purely spatial relationship for the orbit:

$$\frac{d\theta}{dr} = \frac{1}{r}\left(\frac{2Er^2}{L^2} + \frac{2GMr}{L^2} - 1\right)^{-1/2}. \qquad (2.59)$$

Integrate equation (2.59) and solve for r. Set the constant of integration equal to $-\pi/2$, define $r_0 \equiv L^2/(GM)$, and use the relationship $e = \left(1 + (2EL^2)/(G^2M^2)\right)^{1/2}$ to obtain:

$$r = \frac{r_0}{1 + e\cos\theta}. \qquad (2.60)$$

For $0 \le e < 1$, equation (2.60) represents an ellipse in polar coordinates. Thus, Kepler's first law is also precise in the two-body Newtonian approximation, although the Sun itself is not fixed in space. Note that if $E = 0$ then $e = 1$ and equation (2.60) describes a parabola, and if $E > 0$ then $e > 1$ and the orbit is hyperbolic.

(c) Show that the semimajor and semiminor axes of the ellipse given by equation (2.60) are $a = r_0/(1 - e^2)$ and $b = r_0/(1 - e^2)^{1/2}$, respectively. Determine the orbital period, P, by setting the integral of $d\mathcal{A}/dt$ equal to the area of the ellipse, πab. Note that your result, $P = (4\pi^2 a^3/GM)^{1/2}$, differs from Kepler's third law by replacing the Sun's mass, m_1, by the sum of the masses of the Sun and the planet, M.

2.3.E A baseball pitcher can throw a fastball at a speed of $\sim$150 km/hr. What is the largest size spherical asteroid of density $\rho = 3$ g cm^{-3} from which he can throw the ball fast enough that it:

(a) escapes from the asteroid into heliocentric orbit?

(b) rises to a height of 50 km?

(c) goes into a stable orbit about the asteroid?

2.4.I In this problem, you will make some calculations which would be useful for planning spacecraft missions to Jupiter and to other planets. To simplify matters, you may assume that the planets move on circular orbits.

(a) Calculate the velocity (relative to Earth) at Earth's orbit of the Hohmann transfer orbit that is tangent to both Earth's orbit and Jupiter's orbit.

(b) Calculate the minimum velocity necessary to launch a spacecraft from the surface of Earth to Jupiter, ignoring Earth's rotation.

(c) Calculate the minimum velocity necessary to launch a spacecraft from the Earth's equator to Jupiter, including Earth's rotation but ignoring its obliquity.

(d) Calculate the time required for a spacecraft moving along a Hohmann transfer orbit to travel from Earth to Jupiter.

(e) Repeat part (b) for spacecraft sent on Hohmann transfer orbits to Venus and to Mercury.

2.5.I The *virial theorem* states that the time-averaged potential energy of a bound self-gravitating system is equal to two times the negative of the time-average of the kinetic energy of the system,

$$\langle E_G \rangle = -2 \langle E_K \rangle. \tag{2.61}$$

(a) State the virial theorem for the N-body problem in words and also in mathematical form.

(b) Verify the virial theorem for the case of two bodies on a circular orbit.

2.6.I The complete solution to the N-body problem requires knowledge of $6N$ quantities at all times, representing the positions and velocities of every particle or some equivalent set of 'orbital elements'. In general, the system has 10 integrals of motion, six representing the location of the center of mass as a function of time, $\mathbf{x} + \mathbf{v}t$, three representing the angular momentum of the system, $\mathbf{L}$, and one representing the total energy of the system, E.

(a) The longitude at epoch provides an independent constraint for the two-body problem. One more (12−10−1) integral or independent constraint is required to completely specify the solution of the two-body problem. What is it?

(b) The circular restricted three-body problem includes an already-solved two-body problem (as the mass of the third body is ignored). The Jacobi constant is an integral of motion for the test particle in the circular restricted three-body problem. State two additional integrals of motion for the test particle in the planar circular restricted three-body problem.

2.7.I We listed several simplifications for the three-body problem in §2.2: planar, restricted, circular, and Hill. If all of these simplifications were independent, there would be $2^4 = 16$ possibilities. However, there are actually only 12 distinct viable cases. Name these combinations, and state why they are possible, but the other 4 are not. (Note: Hill's original papers assumed coplanar orbits, but his calculations can be generalized to the noncoplanar case. We use this more general definition of Hill's problem here.)

2.8.E Show that if the two-body problem is analyzed using units in which the gravitational constant, the semimajor axis of the mutual orbits of the bodies, and the sum of their masses are all set equal to one, then the orbital period of the two bodies about their mutual center of mass equals 2π.

2.9.I This problem concerns the stability of orbits in the planar circular restricted three-body problem. The ratio of the star's mass to that of the planet is 333. Choose a rotating coordinate system with the center of mass at the origin and the planet located at $x = 1, y = 0$.

(a) Calculate (approximately) the location of the equilibrium points.

(b) Which, if any, of these points are stable?

(c) In what region about the planet may moons have stable orbits? What regions must asteroids avoid in order to have stable orbits about the star? Be quantitative.

2.10.I One of the greatest triumphs of dynamical astronomy was the prediction of the existence and location of the planet Neptune on the basis of irregularities observed in the orbit of Uranus. The motion of Uranus could not be accurately accounted for using only (the Newtonian modification of) Kepler's laws and perturbations of the then-known planets. Estimate the maximum displacement in the position of Uranus caused by the gravitational effects of Neptune as Uranus catches up and passes this slowly moving planet. Quote your results both in kilometers along Uranus's orbital path, and in seconds of arc against the sky as observed from Earth. For your calculations, you may neglect the effects of the other planets (which can be and were accurately estimated and factored into the solution) and assume that the unperturbed orbits of Uranus and Neptune are circular and coplanar, and neglect the influence of Uranus on Neptune.

Note that although the displacements of Uranus in radius and longitude are comparable, the longitudinal displacement produces a much larger observable signature because of geometric factors.

(a) Obtain a very crude result by assuming that the potential energy released as Uranus gets closer to Neptune increases Uranus's semimajor axis and thus slows Uranus down. You may assume that Uranus's average semimajor axis during the interval under consideration is halfway between its semimajor axis at the beginning and the end of the interval. Remember to use the synodic (relative) period of the pair of planets rather than just Uranus's orbital period.

(b) Obtain an accurate result using numerical integration of Newton's equations on a computer.

2.11.E Explain how the Lyapunov characteristic exponent, γ_c, is used to distinguish between regular and chaotic trajectories. What is the value of γ_c for regular trajectories?

2.12.E A small planet travels about a star with an orbital semimajor axis equal to 1 in the units chosen. Calculate the locations of the 2:1, 3:2, 99:98 and 100:99 resonances of test particles with the planet.

2.13.I Two criteria for the stability of the orbits of test particles near the semimajor axis of a planet were introduced in §2.3.3: The resonance overlap criterion for the onset of chaos, $\Delta a_{ro} \approx 1.5(m_2/m_1)^{2/7}$; and the criterion for exclusion from close approaches to the planet resulting from the Jacobi integral, $\Delta a_J \geq 2\sqrt{3}(m_2/3m_1)^{1/3}$.

(a) Comment on the conceptual differences between these two stability criteria.

(b) For what value of m_2/m_1 are the two equal?

(c) Comment on the qualitative difference between orbits near the instability boundary for m_2/m_1 greater or less than this value.

2.14.E Einstein's *general theory of relativity* is conceptually quite different from Newton's theory of gravity, but the predictions of general relativity reduce to those of Newton's model in the low-velocity (relative to the speed of light, c), weak gravitational field (relative to that required for a body to collapse into a black hole) limit.

(a) Calculate v^2/c^2 for the following bodies:
(i) Mercury, using its circular orbit velocity
(ii) Mercury, using its velocity at perihelion
(iii) Earth, using its circular orbit velocity
(iv) Neptune
(v) Io, using its velocity relative to Jupiter

(vi) Metis, using its velocity relative to Jupiter.

(b) The *Schwarzchild radius* of a body,

$$R_{Sch} = \frac{2Gm}{c^2}, \tag{2.62}$$

is the radius inside of which light cannot escape from the body. Calculate the Schwarzchild radii of the following Solar System bodies, and the ratios of these radii to the sizes of the bodies in question and to the semimajor axes of their nearest (natural) satellites:
(i) Sun
(ii) Earth
(iii) Jupiter.

2.15.E As demonstrated in the previous problem, Newton's theory of gravity is quite accurate for most Solar System situations. The most easily observable effect of general relativity is the precession of orbits, because it is nil in the Newtonian two-body approximation. The first-order (weak field) general relativistic corrections to Newtonian gravity imply a precession of the periapse of orbit of a small body ($m_2 \ll m_1$) about a primary of mass m_1 at the rate

$$\dot{\varpi} = \frac{3(Gm_1)^{3/2}}{a^{5/2}(1-e^2)c^2}. \tag{2.63}$$

Calculate the general relativistic precession of the periapse of the following objects:
(a) Mercury
(b) Earth
(c) Io (in its orbit about Jupiter).

Quote your answers in arcseconds per year.

Note: The average observed precession of Mercury's periapse is 56.00″ yr^{-1}, all but 5.74″ yr^{-1} of which is caused by the observations not being done in an inertial frame far from the Sun. Newtonian gravity of the other planets accounts for 5.315″ yr^{-1}. Your answer should be close to the difference between these two values, 0.425″ yr^{-1}. The small difference between this rate and the value calculated by the procedure outlined above is within the uncertainties of the observations and calculations, and may also be affected by a very small contribution resulting from solar oblateness.

2.16.I Use multiple integration to show that the gravitational potential exterior to a spherically symmetric body is identical to that of a point-like particle of the same mass located at the center of the sphere.

(Hint: Divide the sphere into concentric shells, and then subdivide the shells into rings that are oriented perpendicular to the direction from the center of the sphere to the point at which the potential is being evaluated. Determine the potential of each ring, then integrate over angle to deduce the potential of a shell and finally integrate over radius to determine the potential of the sphere.)

2.17.E Saturn is the most oblate major planet in the Solar System, with gravitational moments $J_2 = 1.63 \times 10^{-2}$, $J_4 = -9 \times 10^{-4}$, and $J_6 = 10^{-4}$. Calculate the orbital periods, apse precession rates, and node regression rates for particles on nearly circular and equatorial orbits at 1.5 R_h and 3 R_h:
(a) neglecting planetary oblateness entirely,
(b) including J_2, but neglecting higher order moments,
(c) including J_2, J_4, and J_6.

2.18.E If the Earth–Moon distance was reduced to half its current value then:
(a) Neglecting solar tides, how many times as large as at present would the maximum tide heights on Earth be?
(b) Including solar tides, how many times as large as at present would the maximum tide heights on Earth be?

2.19.E Calculate the semimajor axis of a moon on a synchronous orbit around Mars. Express your answer in martian radii (R_σ) from the center of the planet. Compare this distance with the orbits of Phobos (at 2.76 R_σ) and Deimos (at 6.9 R_σ). Describe the motion of each of these moons on the sky as seen from Mars.

2.20.I Estimate the amplitude of tides that the Moon raises on Earth. (Hint: Integrate the tidal force to compute a tidal potential. Next, calculate the tidal potential of a test particle at the sublunar point on the Earth's surface. Finally, determine the height by which the blob must be raised for the change in its gravitational potential relative to Earth to be equal in magnitude to its tidal potential.)

2.21.E Calculate the mean *synodic period* (time it takes for a relative configuration to repeat) between Earth's rotation and the Moon's orbit. Note that this is equal to the average interval between successive moonrises.

2.22.E (a) Compute the ratio of the height of tides raised on Earth by the Moon to those raised by the Sun.
(b) Compute the ratio of the tidal torque on the Earth due to the Moon and to the Sun.

2.23.I (a) Derive equation (2.44b) from equation (2.44a). (Hint: Use a simplified version of eq. 2.18.)
(b) Using equation (2.44b), show that a moon starting inside synchronous orbit will impact the planet's surface after a time t_{impact}:

$$t_{\text{impact}} = \frac{2}{39} \frac{m_1}{m_2} \frac{Q_1}{k_{T_1}} \frac{a_2^{13/2}(0) - 1}{n_1^*}, \qquad (2.64)$$

where $a_2 \equiv r/R_1$ and $n_1^* \equiv (Gm_1/R_1^3)^{1/2}$. State all assumptions that you make.

2.24.E If the Moon's mass was reduced to half its current value, then:
(a) The rate at which the Moon recedes from Earth would be how many times as large as at present?
(b) Neglecting solar tides, maximum tide heights on Earth would be how many times as large as at present?
(c) Including solar tides, maximum tide heights on Earth would be how many times as large as at present?

2.25.I If a moon and planet are both initially rotating faster than their mutual orbit period, then the tides raised by the planet on the moon slow the moon's rotation until it becomes synchronous. Tides raised by the moon on the planet cause a slowing of the planet's rotation rate. Both tides cause the moon and planet to move further apart.
(a) Go over the details of this argument (using illustrations) and compare timescales for the various processes.
(b) Repeat for the case when a moon and planet are initially spinning more slowly than their mutual orbit period. What is the outcome in this case?
(c) Repeat for the case when the moon is on a retrograde orbit, i.e., its orbital angular momentum is antiparallel to the planet's spin angular momentum.

2.26.I Extrapolate the evolution of the Moon's orbit backwards in time. Use conservation of angular momentum for the Earth–Moon system to derive a variant of equation (2.64). You may neglect solar tides in your calculations, but comment qualitatively on their effects. Assume that k_{T_1}/Q_1 has remained constant and use the data on fossil bivalve shells to determine the value of this constant. State your result in the form $a_2(t)$, where a_2 is in units of $R_\oplus$ and t is in Gyr before present. As tidal evolution was much more rapid when the bodies were closer, your result should imply that the Moon was quite close to the Earth substantially less than 4×10^9 years ago. This was considered a major problem until it was realized that a substantial fraction of the tidal dissipation in Earth today results from sloshing of waters in shallow seas, and that k_{T_1}/Q_1 could have been much less in the past when Earth's continents were configured differently.

2.27.I The Moon is receding from Earth as a result of the lag of the tidal bulge that it raises on Earth. This process will continue until the day is the same length as the lunar month.

(a) Neglecting solar perturbations, what will this day/month period be? You may approximate the Earth by a homogeneous sphere in order to calculate its moment of inertia. You may use other approximations if they induce errors of <5%. Quote your answer in seconds.

(b) Compare the orbital radius of the Moon when the day and month are equal to the radius of the Earth's Hill sphere.

(c) Qualitatively, what will the effects of the Sun be on the tidal evolution of the Earth–Moon system; specifically, will the Moon stop its recession closer to or further from Earth? Will this effect be large or small? Explain.

2.28.D The only possible true equilibrium state for a planet/moon system subject to tidal dissipation is one in which both bodies spin in a prograde sense with a period equal to their mutual orbital period.

(a) Draw a diagram of this system.

(b) Show that for a given total angular momentum of the system there are in general two possible equilibrium states, one in which the bodies are close to each other and most of the angular momentum resides in the planet's rotation, and the other where the bodies are far apart and the moon's orbital motion holds most of the angular momentum.

(c) Such a system may or may not be stable. Under what conditions is such a system stable?

(d) Can both equilibrium states exist for all planet/moon systems? Why, or why not?

2.29.D Two moons are spiraling outwards, away from their planet, due to tidal forces. They become locked in a stable orbital resonance which requires them to maintain a constant ratio of orbital periods (see, e.g., Peale 1976). Calculate the energy available for tidal heating in equilibrium (i.e., assume that the moons' orbital eccentricities do

not change). Your answer should depend on the masses of the two moons, m_I and m_{II}, and of the primary, m_p, on the angular velocities of the moons, n_I and n_{II}, and on (the z-components of) the tidal torques exerted by the planet on each of the moons, $\dot{L}_{I(p)}$ and $\dot{L}_{II(p)}$. (Hint: Use conservation of angular momentum and energy:

$$\frac{d}{dt}(L_I + L_{II}) = \dot{L}_{I(p)} + \dot{L}_{II(p)}, \qquad (2.65a)$$

$$\frac{d}{dt}(E_I + E_{II}) = \dot{L}_{I(p)}n_I + \dot{L}_{II(p)}n_{II} - \mathcal{H}, \qquad (2.65b)$$

where L_I and L_{II} are the orbital[3] angular momenta of the moons, E_I and E_{II} are the orbital energies, and $\mathcal{H}$ is the heating rate.)

2.30.E Calculate the orbital period around the Sun for a dust grain with $\beta = 0.3$ and semimajor axis $a = 1$ AU.

2.31.I A grain with $\beta = \beta_0$ travels on a circular orbit about the Sun. The grain splits apart into smaller grains with $\beta = \beta_n$. What are the eccentricities and semimajor axes of these new grains?

2.32.I (a) Show that dust released at perihelion from a body on an eccentric Keplerian orbit will escape from the Solar System if the ratio of radiation pressure to the solar gravity it feels is

$$\beta \geq \frac{1-e}{2}. \qquad (2.66)$$

(b) Derive an analogous expression for the stability of a dust grain released at aphelion.

2.33.I Consider an icy ($\rho = 1$ g cm^{-3}) particle in Saturn's rings, located at 1.5 R$_h$. The radius of the particle is 0.1 μm.

(a) Calculate the radiation pressure on the particle due to the Sun and due to Saturn's reflected light, at noon (Saturn's albedo is 0.46). (Hint: Use Figure 2.23 to estimate Q_{pr}.)

(b) Compare the radiation forces to the gravitational forces from the Sun and Saturn. Calculate the *appropriate* β. What will happen to the particle (e.g., blown in/out, stay in orbit)?

2.34.I This problem concerns the orbital decay of an artificial satellite in Earth orbit, but it is also applicable to, e.g., the decay of particles in the rings of Uranus

---

[3] Technically, the rotational angular momenta of the moons should also be included; however, their values are so small that you may ignore them.

as a result of that planet's extended atmosphere. The satellite has mass m and drag cross-section πR^2 and is initially on a circular orbit of semi-major axis a_o. The density of the atmosphere is given by the formula $\rho = \rho_o \exp[-(a - a_o)/H]$, where H is the scale height of the atmosphere (§4.1). Assume the drag coefficient is 0.4 and that the aerodynamic drag formula is applicable, i.e., $F_D = 0.4\rho\pi R^2 v^2$.

(a) Assuming F_D is small, calculate the change in semimajor axis during one orbit as a function of F_D.

(b) Calculate F_D as a function of a.

(c) Using the results from (a) and (b), calculate, approximately, the semimajor axis of the satellite's orbit as a function of time.

2.35.E (a) Calculate the rate at which the Sun is losing mass as a result of radiating photons by using Einstein's famous formula,

$$E = mc^2. \tag{2.67}$$

(b) Estimate the Sun's mass loss integrated over the past 4 billion years. You may assume that the solar luminosity and solar wind have remained constant, and that the Sun's neutrino luminosity is negligible.

3 Solar Heating and Energy Transport

Temperature is one of the most fundamental properties of planetary matter, as is evident from everyday experience such as the weather and cooking a meal, as well as from the most basic concepts of chemistry and thermodynamics. For example, H_2O is a liquid between 273 K and 373 K (at standard pressure), a gas at higher temperatures, and a solid when it is colder; silicates undergo similar transitions at substantially higher temperatures and methane condenses and freezes at lower temperatures. Most substances expand when heated, with gases increasing in volume the most; the thermal expansion of liquid mercury allowed it to be the 'active ingredient' in most thermometers from the seventeenth century through the twentieth century. The equilibrium molecular composition of a given mixture of atoms often depends on temperature (as well as on pressure), and the time required for a mixture to reach chemical equilibrium generally decreases rapidly as temperature increases. Gradients in temperature and pressure are responsible for atmospheric winds (and, on Earth, ocean currents) as well as convective motions that can mix fluid material within planetary atmospheres and interiors. Earth's solid crust is dragged along by convective currents in the mantle, leading to continental drift. Temperature can even affect the orbital trajectory and rotation state of a body, as we have seen in our discussions of the Yarkovsky and YORP effects (§2.7.3 and §2.7.4, respectively).

Temperature, T, is a measure of the random kinetic energy of molecules, atoms, ions, etc. The energy, E, of a perfect gas is given by

$$E = \frac{3}{2}NkT, \tag{3.1}$$

where N is the number of particles, and k is *Boltzmann's constant*. The temperature of a (given region of a) body is determined by a combination of processes. Solar radiation is the primary energy source for most planetary bodies, and reradiation to space is the primary loss mechanism. In this chapter, we summarize the mechanisms for solar heating and energy transport. We then use this background in our discussions of planetary atmospheres, surfaces, and interiors in subsequent chapters.

3.1 Energy Balance and Temperature

Planetary bodies are heated primarily by absorbing radiation from the Sun, and they lose energy via radiation to space. While a point on the surface of a body is illuminated by the Sun only during the day, it radiates both day and night. The amount of energy incident per unit area depends both on the distance from the Sun and the local elevation angle of the Sun. As a consequence, most locales are coldest just before sunrise and hottest a little after local noon, and the polar regions are colder than the equator for bodies with obliquity $\psi < 54°$ (or $\psi > 126°$).

Over the long term, most planetary bodies radiate almost the same amount of energy to space as they absorb from sunlight; were this not the case, planets would heat up or cool off. (The giant planets Jupiter, Saturn, and Neptune are exceptions to this rule. These bodies radiate significantly more energy than they absorb, because their interiors are cooling or becoming more centrally condensed.) Although long-term global equilibrium is the norm, spatial and temporal fluctuations can be large. Energy is stored from day to night, perihelion to aphelion, and summer to winter, and can be transported from one location on a planet to another. We begin our discussion with the fundamental laws of radiation in §3.1.1 and factors affecting global energy balance in §3.1.2.

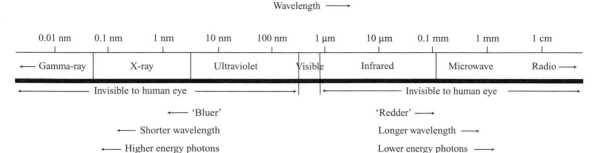

Figure 3.1 The electromagnetic spectrum. (Adapted from Hartmann 1989)

3.1.1 Thermal (Blackbody) Radiation

Electromagnetic radiation consists of photons at many wavelengths (Fig. 3.1). The frequency, ν, of an electromagnetic wave propagating in a vacuum is related to its wavelength, λ, by

$$\lambda\nu = c, \tag{3.2}$$

where c is the *speed of light in a vacuum*, 2.998×10^{10} cm s^{-1}.

Most objects emit a continuous spectrum of electromagnetic radiation. This *thermal emission* is well approximated by the theory of 'blackbody' radiation. A *blackbody* is defined as an object which absorbs all radiation that falls on it, at all frequencies and all angles of incidence; i.e., no radiation is reflected or scattered. A body's capacity to emit radiation is the same as its capability of absorbing radiation at the same frequency. The radiation emitted by a blackbody is described by *Planck's radiation law*:

$$B_\nu(T) = \frac{2h\nu^3}{c^2}\frac{1}{e^{h\nu/(kT)}-1}, \tag{3.3}$$

where $B_\nu(T)$ is the *specific intensity* or *brightness* (erg s^{-1} cm^{-2} Hz^{-1} sr^{-1}), and h is Planck's constant. Figure 3.2a shows a graph of brightness as a function of frequency for various blackbodies with temperatures ranging from 40 to 30 000 K. Note that the brightness curve for a body like our Sun, with a surface temperature 5777 K, peaks at optical wavelengths (Fig. 3.2b), while those of the planets ($\sim$40–700 K) peak at infrared wavelengths. The brightness of most Solar System objects near their spectral peaks can be approximated quite well by blackbody curves.

Two limits of Planck's radiation law can be derived:
(1) *The Rayleigh–Jeans law*: When $h\nu \ll kT$ (i.e., at radio wavelengths for temperatures typical of planetary bodies), the term $(e^{h\nu/(kT)} - 1) \approx h\nu/(kT)$, and equation (3.3) can be approximated by

$$B_\nu(T) \approx \frac{2\nu^2}{c^2}kT. \tag{3.4a}$$

(2) *The Wien law*: When $h\nu \gg kT$:

$$B_\nu(T) \approx \frac{2h\nu^3}{c^2}e^{-h\nu/(kT)}. \tag{3.4b}$$

Equations (3.4) are simpler than equation (3.3), and thus they can be quite useful in the regimes in which they are applicable.

The frequency, $\nu_{\max}$, at which the peak in the brightness $B_\nu(T)$ occurs, can be determined by setting the derivative of equation (3.3) equal to zero, $\partial B_\nu/\partial \nu = 0$. The result is known as the *Wien displacement law*:

$$\nu_{\max} = 5.88 \times 10^{10}T, \tag{3.5a}$$

with $\nu_{\max}$ in Hz. With

$$B_\lambda = B_\nu|\frac{d\nu}{d\lambda}|, \tag{3.5b}$$

the blackbody spectral peak in wavelength can be found by setting $\partial B_\lambda/\partial \lambda = 0$:

$$\lambda_{\max} = \frac{0.29}{T}, \tag{3.5c}$$

with $\lambda_{\max}$ in cm (Problem 3.1). Note that $\lambda_{\max} = 0.57\,c/\nu_{\max}$, i.e., the brightness peak measured in terms of wavelength is blueward of the brightness peak measured in terms of frequency, due to the fact that $B_\lambda \neq B_\nu$.

The *flux density*, $\mathcal{F}_\nu$ (erg s^{-1} cm^{-2} Hz^{-1} or Jy),[1] of radiation from an object is given by (§3.2.3.1)

$$\mathcal{F}_\nu = \Omega_s B_\nu(T), \tag{3.6}$$

where Ω_s is the solid angle subtended by the object. At the surface of a sphere of uniform brightness B_ν (caused, e.g., by a blackbody point source radiator at its center), the flux density is equal to (Problem 3.2)

$$\mathcal{F}_\nu = \pi B_\nu(T). \tag{3.7}$$

[1] 1 Jy (Jansky) $\equiv 10^{-23}$ erg s^{-1} cm^{-2} Hz^{-1}.

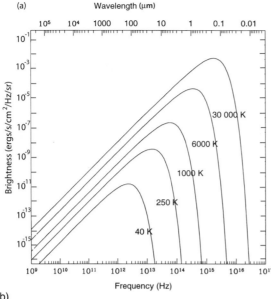

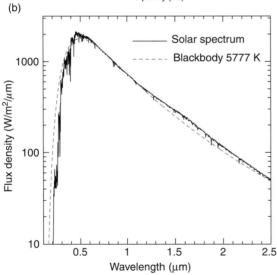

Figure 3.2 (a) Blackbody radiation curves, $B_\nu(T)$, at various temperatures ranging from 40 K up to 30 000 K (eq. 3.3). The 6000 K curve is representative of the solar spectrum. (b) Solar spectrum between 0.1 and 2.5 μm wavelengths. A blackbody spectrum at 5777 K is superposed. The solar data are from Colina *et al.* (1996).

The *flux*, $\mathcal{F}$ (erg s^{-1} cm^{-2}), is defined as the flux density integrated over all frequencies:

$$\mathcal{F} \equiv \int_0^\infty \mathcal{F}_\nu d\nu = \pi \int_0^\infty B_\nu(T) d\nu = \sigma T^4, \qquad (3.8)$$

where σ is the *Stefan–Boltzmann constant*. This relationship is known as the *Stefan–Boltzmann law*. We note that in some texts the flux is defined as a function of frequency,

i.e., what we call flux density (e.g., in Chamberlain and Hunten 1987; Chandrasekhar 1960).

3.1.2 Temperature

One can determine the temperature of a blackbody using Planck's radiation law by measuring a small part of the object's radiation (Planck) curve. This is usually not practical, since most bodies are not perfect blackbodies, but exhibit spectral features that complicate temperature measurements. It is common to relate the observed flux density, $\mathcal{F}_\nu$, to the *brightness temperature*, T_b, which is the temperature of a blackbody that has the same brightness at this particular frequency (i.e., replace T in eq. 3.3 by T_b). Conversely, if the total flux integrated over all frequencies of a body can be determined, the temperature that corresponds to a blackbody emitting the same amount of energy or flux $\mathcal{F}$ is referred to as the *effective temperature*, T_e:

$$T_e \equiv \left(\frac{\mathcal{F}}{\sigma}\right)^{1/4}. \qquad (3.9)$$

The frequency range at which the object emits most of its radiation can be estimated via Wien's displacement law (eq. 3.5). This is typically at mid-infrared wavelengths (10–20 μm) for objects with temperatures of 150–300 K (inner Solar System), and far-infrared wavelengths (~60–70 μm) for 40–50 K bodies in the outer Solar System.

3.1.2.1 *Albedo and Emissivity*

When an object is illuminated by the Sun, it reflects part of the energy back into space (which makes the object visible to us), while the remaining energy is absorbed. In principle, one can determine how much of the incident radiation is reflected into space at each frequency; the ratio between incident and reflected + scattered energy is called the *monochromatic albedo*, A_ν. Integrated over frequency, the ratio of the total radiation reflected or scattered by the object to the total incident light from the Sun is called the *Bond albedo*, A_b. The energy or flux absorbed by the object determines its temperature, as discussed further in §3.1.2.2. With regard to the albedo, it is important to consider how a unit surface element scatters light. The Sun's light is scattered off a planet and received by a telescope. The four angles of relevance are: i, the angle that incident light makes with the normal to the planet's surface; θ, the angle that the reflected ray received at the telescope (i.e., the ray along the line of sight) makes with the normal to the surface (Fig. 3.3); the *phase angle* or angle of reflectance, ϕ, as seen from the object (Fig. 3.4); *scattering angle*, ϕ_{sc}, defined as the change in direction a photon undergoes during a scattering event. The

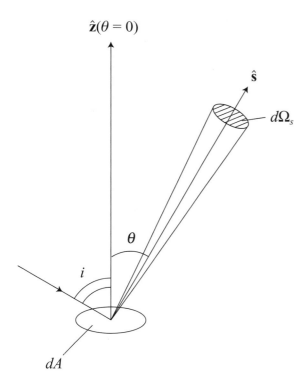

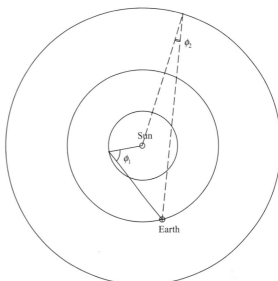

Figure 3.4 Scattering of light by a body that is illuminated by the Sun, with radiation received on Earth. For purely backscattered radiation, the phase angle $\phi = 0$; in the case of forward scattered light, $\phi = 180°$. The scattering angle $\phi_{sc} = 180° - \phi$. Two planets are indicated: one inside Earth's orbit, with phase angle ϕ_1, and one outside Earth's orbit, with phase angle ϕ_2.

Figure 3.3 Sketch of the geometry of a surface element dA: $\hat{z}$ is the normal to the surface, $\hat{s}$ is a ray along the line of sight, and θ is the angle the ray makes with the normal to the surface.

scattering and phase angles are related to one another: $\phi = 180° - \phi_{sc}$.

The *phase integral*, q_{ph}, contains the phase dependence of the scattering:

$$q_{ph} \equiv 2 \int_0^\pi \frac{\mathcal{F}(\phi)}{\mathcal{F}(\phi = 0)} \sin \phi \, d\phi. \qquad (3.10)$$

The phase integral can be measured from Earth for planets with heliocentric distances less than 1 AU (Mercury, Venus) as well as for the Moon, since the angle of reflectance, ϕ, varies between 0° and 180°. The outer planets are observed from Earth at phase angles close to 0°. Additional information on the phase integral can be recovered from Earth by using center-to-limb data, but only with help of spacecraft data can the full phase integral be determined.

We define the Bond albedo:

$$A_b \equiv A_0 q_{ph}, \qquad (3.11)$$

with A_0 the *geometric albedo* or head-on reflectance:

$$A_0 = \frac{r_{\odot AU}^2 \mathcal{F}(\phi = 0)}{\mathcal{F}_\odot}, \qquad (3.12)$$

where $\mathcal{F}(\phi = 0)$ is the flux reflected from the body at phase angle $\phi = 0$. The heliocentric distance, $r_{\odot AU}$, is expressed in AU, and $\mathcal{F}_\odot$, the solar constant, is defined as the solar flux at $r_{\odot AU} = 1$:

$$\mathcal{F}_\odot \equiv \frac{\mathcal{L}_\odot}{4\pi r_\odot^2} = 1.37 \times 10^6 \text{ erg cm}^{-2} \text{ s}^{-1}, \qquad (3.13)$$

with $r_\odot$ the heliocentric distance (in cm) and $\mathcal{L}_\odot$ the solar luminosity. The combination $(\mathcal{F}_\odot/r_{\odot AU}^2)$ is equal to the incident solar flux at heliocentric distance $r_{\odot AU}$ AU.

The geometric albedo can be thought of as the amount of radiation reflected from a body relative to that from a flat *Lambertian surface*, which is a diffuse perfect reflector at all wavelengths. Usually one determines a quantity referred to as $I/\mathcal{F}$ from planetary observations, where I is the reflected intensity at frequency ν, and $\pi\mathcal{F}$ is the incident solar flux density upon the planet at frequency ν. By this definition, $I/\mathcal{F} = 1$ for a flat Lambertian surface when viewed at normal incidence, and $I/\mathcal{F}$ thus equals the geometric albedo at frequency ν when observed at a phase angle $\phi = 0$.

As stated by *Kirchhoff's law*, the reflectivity, A_ν, and emissivity, ϵ_ν, at frequency ν of a smooth, nonscattering sphere are complementary under the same viewing conditions:

$$(1 - A_\nu) = \epsilon_\nu. \qquad (3.14)$$

If scattering is present, the sum of the reflectivity and emissivity remains unity when averaged over 4π steradians (conservation of energy), but not necessarily when viewed from a specific angle.

3.1.2.2 Equilibrium Temperature

Provided the incoming solar radiation (insolation), $\mathcal{F}_{in}$, is balanced, on average, by reradiation outwards, $\mathcal{F}_{out}$, one can calculate the temperature of the object. This temperature is referred to as the *equilibrium temperature*. If indeed the temperature of the body is completely determined by the incident solar flux, the equilibrium temperature equals the effective temperature. Any discrepancies between the two numbers contains valuable information on the object. For example, the effective temperatures of Jupiter, Saturn, and Neptune exceed the equilibrium temperature, which implies that these bodies possess internal heat sources (§4.2 and §6.1.5). Venus's surface temperature is far hotter than the equilibrium temperature of the planet indicates, a consequence of a strong greenhouse effect in this planet's atmosphere (§4.2). The effective temperature of Venus, which is dominated by radiation emitted from that planet's cool upper atmosphere, is equal to the equilibrium temperature, implying that Venus has a negligible internal heat source. We next discuss the average effect of insolation and reradiation for a rapidly rotating spherical object of radius R, using approximate equations. At the end of this section, we provide the more precise equations that can be used for more detailed modeling.

The sunlit hemisphere of a (spherical) body receives radiation from the Sun:

$$\mathcal{P}_{in} = (1 - A_b)\frac{\mathcal{L}_\odot}{4\pi r_\odot^2}\pi R^2, \tag{3.15}$$

with πR^2 the projected surface area for intercepting solar photons. A rapidly rotating planet reradiates energy from its entire surface (i.e., an area of $4\pi R^2$):

$$\mathcal{P}_{out} = 4\pi R^2 \epsilon \sigma T^4. \tag{3.16}$$

Note that the incoming solar radiation is primarily at optical wavelengths (Fig. 3.2), while thermal emission from planets is radiated primarily at infrared wavelengths. The emissivity, ϵ_ν, is usually close to 0.9 at infrared wavelengths, but can differ substantially from unity at radio wavelengths. From a balance between insolation and reradiation, $\mathcal{P}_{in} = \mathcal{P}_{out}$, one can calculate the equilibrium temperature, T_{eq}:

$$T_{eq} = \left(\frac{\mathcal{F}_\odot}{r_{\odot AU}^2}\frac{(1 - A_b)}{4\epsilon\sigma}\right)^{1/4}. \tag{3.17}$$

Even though this simple derivation has many shortcomings, the disk-averaged equilibrium temperature from equation (3.17) gives useful information on the temperature well below a planetary surface. If ϵ is close to unity, it corresponds well with the actual (physical) temperature of subsurface layers that are below the depth where diurnal (day/night) and seasonal temperature variations are important, typically a meter or more below the surface. These layers can be probed at radio wavelengths, and the brightness temperature observed at these long wavelengths can be compared directly with the equilibrium temperature. In the derivation of equation (3.17), we omitted latitudinal and longitudinal effects of the insolation pattern. The magnitude of these effects depends on the planet's rotation rate, obliquity, and orbit. Latitudinal and longitudinal effects are large, for example, on airless planets that rotate slowly, have small axial obliquities, and/or travel on very eccentric orbits about the Sun.

In another limit for equilibrium temperatures, one can consider the subsolar point of a slowly rotating body. In this case, the surface areas πR^2 in equation (3.15) and $4\pi R^2$ in equation (3.16) should both be replaced by a unit area dA. It follows that the equilibrium temperature at the subsolar point of a slowly rotating body is $\sqrt{2}$ times the disk-average equilibrium temperature for a rapidly rotating body. The subsolar temperature calculated in this way corresponds well with the measured subsolar surface temperature of airless bodies.

For more detailed modeling, consider the solar flux incident per unit surface area dA at location (α, δ) on a planet's surface:

$$\frac{\mathcal{F}_{in}}{dA} = \int_0^\infty (1 - A_\nu)\frac{(\mathcal{F}_\odot)_\nu}{4\pi r_{\odot AU}^2}$$

$$\times \cos[\alpha_\odot(t) - \alpha]\cos[\delta_\odot(t) - \delta]\,d\nu, \tag{3.18a}$$

where $(\alpha_\odot, \delta_\odot)$ are the coordinates of the Sun, and A_ν is the reflectivity at frequency ν. A surface element dA emits radiation according to the Stefan–Boltzmann law:

$$\mathcal{F}_{out} = \epsilon\sigma T^4\,dA. \tag{3.18b}$$

The emissivity, ϵ_ν, depends upon wavelength; it usually varies from a few tenths up to unity for bodies that are large compared to the wavelength considered. Objects much smaller than the wavelength ($R \lesssim 0.1\lambda$) do not radiate efficiently.

3.2 Energy Transport

The temperature structure in a body is governed by the efficiency of energy transport. There are three principal mechanisms to transport energy: *conduction*, *radiation*, and *mass motion*. Usually, one of these three mechanisms dominates and determines the thermal profile in any given region. Energy transport in a solid is usually dominated by conduction, while radiation typically dominates in space and tenuous gases. Mass motion is important in fluids and dense gases, and can transport energy and other properties via a process known as *advection*. The most important advective process for planets is *convection*, the vertical motion produced by buoyancy. In atmospheric sciences, the term advection is typically reserved for (predominantly) horizontal transport.

All three energy transport mechanisms are experienced in everyday life, e.g., when boiling water on a stove: the entire pan, including the handle (in particular if metal), is heated by conduction, while the water in the pan is primarily heated through 'convection', up and down motions in the water. These motions are visible in the form of bubbles of vaporized water, which rise upwards because they are lighter than the surrounding water. Heat is transported from the Sun to planets, moons, etc. via radiation. Although it is obvious in these examples which transport mechanism dominates, this is not always easy to determine; in some parts of a planet's interior energy transport is dominated by convective motions, while in other parts conduction is by far the most efficient. In a planet's atmosphere we typically encounter all three mechanisms, though a particular mechanism is usually dominant in a certain altitude range. Almost all of the energy transport to and from planetary bodies occurs via radiation. (Jupiter's moon Io is an exception to this rule; Io receives a substantial amount of energy from its orbit via tidal dissipation, see §2.6.3 and §5.5.5.1.) In this section, we discuss all three principal mechanisms for energy transport and derive equations for the thermal profile in a surface or an atmosphere under the assumption that a particular heat transport mechanism is dominant.

3.2.1 Conduction

Conduction, i.e., when the transfer of energy is primarily via collisions between molecules, is important in a solid body, as well as in the tenuous upper part of an atmosphere (the upper thermosphere). In the latter situation, the mean free path is so long that atoms exchange locations very rapidly and the conductivity is therefore large. This high conductivity tends to equalize temperatures in this part of an atmosphere.

Sunlight heats a planet's surface during the day, and the heat is transported downwards from the surface mainly by conduction. The rate of flow of heat, the *heat flux*,

$\mathbf{Q}$ (erg s^{-1} cm^{-2}), is determined by the *temperature gradient*, ∇T, and the *thermal conductivity*, K_T:

$$\mathbf{Q} = -K_T \nabla T. \tag{3.19}$$

The thermal conductivity is a measure of the material's physical ability to conduct heat. The *thermal heat capacity* is the amount of heat that is needed to raise the temperature of a substance by 1 degree kelvin. We define the *molecular* or *molar heat capacity*, C_P, as the amount of heat, Q, necessary to raise the temperature of one mole of matter by one degree kelvin without changing the pressure (C_P: $dP = 0$) or, alternatively, the volume (C_V: $dV = 0$). The *specific heat*, c_P (or c_V), is the amount of energy necessary to raise the temperature of one gram of material by one degree kelvin without changing the pressure (or volume).

$$m_{gm} c_P \equiv C_P \equiv \left(\frac{dQ}{dT} \right)_P, \tag{3.20a}$$

$$m_{gm} c_V \equiv C_V \equiv \left(\frac{dQ}{dT} \right)_V, \tag{3.20b}$$

where m_{gm} is a gram-mole.[2] The rate at which the subsurface layers gain heat is given by

$$\rho c_P \frac{\partial T}{\partial t} = -\frac{\partial Q}{\partial z}, \tag{3.21}$$

where ρ is the density of the material. Equations (3.19) and (3.21) together lead to the *thermal diffusion equation*:

$$k_d \frac{\partial^2 T}{\partial z^2} = \frac{\partial T}{\partial t}, \tag{3.22}$$

with the *thermal diffusivity*:

$$k_d \equiv \frac{K_T}{\rho c_P}. \tag{3.23}$$

Equations (3.21–3.23) are used in §6.1.5.2 (eq. 6.31) to estimate the depth in a body over which changes in the temperature gradient become significant over a time t.

The amplitude and phase of the diurnal temperature variations, and the temperature gradient with depth in the crust, are largely determined by the *thermal inertia*,

$$\gamma_T \equiv \sqrt{K_T \rho c_P}, \tag{3.24}$$

which measures the ability of the surface to store energy, and the *thermal skin depth* of the material,

$$L_T \equiv \sqrt{\frac{2K_T}{\omega_{rot} \rho c_P}}. \tag{3.25}$$

[2] A gram-mole is the mass of a mole of molecules in units of grams. A mole contains $N_A \equiv 6.022 \times 10^{23}$ molecules (N_A is Avogadro's number), and its mass is numerically equal to its weight in atomic mass units (amu). Thus, a mole of the lightest and most common isotope of carbon atoms has a mass of 12 g.

(a)

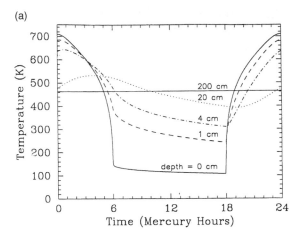

(b)

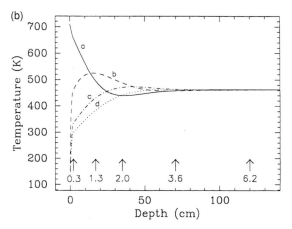

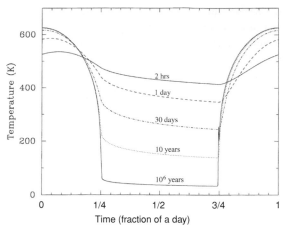

Figure 3.6 The surface temperature of a solid coherent rocky body which orbits the Sun in a circular orbit at a heliocentric distance of 0.4 AU and has zero obliquity. Curves are shown for bodies with rotation periods, P_{rot} (defined with respect to the Sun), of 2 hours, 1 day, 30 days, 10 years, and 10^6 years. The calculations were performed assuming the following parameters: $A_b = 0.1$, $\epsilon_{ir} = 0.9$, $\rho = 2.8$ g cm^{-3}, $\gamma_T = 36.9$ mcal cm^{-2} K^{-1} s$^{-1/2}$, $\epsilon_r = 6.5$, and $\tan \Delta = 0.02$. ϵ_r and $\tan \Delta$ are defined in §3.4. (Courtesy David L. Mitchell)

Figure 3.5 (a) The equatorial temperature structure of Mercury as a function of time after local noon, at various depths, for the subsolar longitude at perihelion (i.e., a 'hot' longitude). (Mitchell 1993) (b) Vertical temperature profiles of the same region of Mercury at four different local times of day: a, noon; b, dusk; c, midnight, and d, dawn. Different depths are probed at different radio wavelengths, as shown by the arrows at the bottom of the graph (wavelengths are in cm). (Mitchell 1993)

The amplitude of diurnal temperature variations is largest at the surface, and decreases exponentially into the subsurface, with an e-folding scale length equal to the thermal skin depth. Moreover, since it takes time for the heat to be carried downwards, there is a phase lag in the diurnal heating pattern of the subsurface layers. The peak temperature at the surface is reached at noon, or soon thereafter, while the subsurface layers reach their peak temperature later in the afternoon. At night the surface cools off, becoming cooler than the subsurface layers. Heat is then transported upwards from below. However, since the conductivity is a function of temperature (eq. 6.30), the surface acts like an insulator at night, preventing the subsurface from cooling off very rapidly. Examples of this effect are illustrated in

Figure 3.5, which shows the temperature as a function of depth and local time for Mercury. Figure 3.6 shows the surface temperature as a function of local time for a hypothetical large rocky body in a circular orbit about the Sun at a heliocentric distance of 0.4 AU. Curves for various rotation periods from 2 hours up to 10^6 years are shown. This figure clearly demonstrates that the peak temperature is primarily determined by the heliocentric distance, while the night-side temperature also depends on the planet's rotation rate (and thermal inertia). Note the time delay in peak temperature from local noon when $P_{rot} = 2$ hours. This delay is caused by the rapid rotation rate combined with the relatively high thermal inertia.

Note that the thermal inertia (eq. 3.24) depends upon the product $K_T c_P$, and the thermal skin depth (eq. 3.25) depends upon the ratio K_T/c_P as well as the angular rotation of the object, ω_{rot}. A typical value for the thermal skin depth on Mars and the Moon is ~4 cm, and on Mercury ~15 cm. When the thermal conductivity is low, the amplitude of the temperature wave is large, but it does not penetrate deeply into the crust. If the thermal conductivity is high, temperature variations are smaller near the surface, but penetrate to greater depths in the subsurface layers. Although the amplitude of diurnal variations in the subsurface layers may be small, seasonal effects can still be large on planets with significant axial obliquities, such as Earth and Mars. Mercury has an interesting variation in temperature with longitude well below the depth where diurnal heating is important. Due to the 3/2 resonance between Mercury's rotation and orbital periods, in combination with Mercury's large orbital eccentricity, the average diurnal insolation varies significantly with longitude (as well as with latitude). Regions along Mercury's equator near longitudes $\lambda = 0°$ and $180°$ (the subsolar longitudes when the planet is at perihelion) receive on average approximately 2.5 times as much sunlight as at longitudes $90°$ and $270°$. The nighttime surface temperature is approximately 100 K, independent of longitude, but the peak (noon) surface temperature near Mercury's equator varies between 700 K at $\lambda = 0°$ and $180°$ to 570 K at $\lambda = 90°$ and $270°$. This nonuniform heating pattern produces longitudinal variations in the subsurface temperature, such that the temperature at depth, where the diurnal variations in temperature can be ignored, is higher at longitudes $0°$ and $180°$ ($T \sim 470$ K) than at $90°$ and $270°$ ($T \sim 350$ K). This effect can best be observed at radio wavelengths longwards of ~ 3 cm, where fluctuations of the diurnal heating cycle are minimal (Fig. 5.50a).

3.2.2 Convection

In dense atmospheres, molten interiors of planets, and protoplanetary disks, transport of heat is often most efficient by large-scale fluid motions, in particular by convection. Convection is the motion in a fluid caused by density gradients which result from temperature differences. Consider a parcel of air in a planet's atmosphere that is slightly warmer than its surroundings. In order to re-establish pressure equilibrium, the parcel expands, and thus its density decreases below that of its surroundings. This causes the parcel to rise. Since the surrounding pressure decreases with height, the rising parcel expands and cools. If the temperature of the environment drops sufficiently rapidly with height, the parcel remains warmer than its surroundings, and thus continues to rise, transporting heat upwards. This process is an example of convection. For convection to occur, the temperature has to decrease with decreasing pressure (thus outwards in a planetary environment) at a sufficiently rapid rate that the parcel remains buoyant.

The temperature structure of an atmosphere in which energy transport is dominated by convection follows an adiabatic lapse rate (eq. 3.34). The temperature gradient in a planet's troposphere, that region in the atmosphere where we live and most clouds form, is usually close to an adiabat (§3.2.2.3). Since the derivation of this lapse rate follows from the equation of hydrostatic equilibrium and the first law of thermodynamics, we discuss these two concepts in the next two subsections. The atmospheric structure, i.e., adiabatic lapse rate, is subsequently derived in §3.2.2.3.

3.2.2.1 Hydrostatic Equilibrium

The relationship between temperature, pressure, and density in a planetary atmosphere and in a planet's interior is governed by a balance between gravity and pressure; this balance is referred to as *hydrostatic equilibrium*. Consider a 'slab' of material of thickness Δz and density ρ. The z-coordinate is taken to be positive going outward (decreasing pressure). This slab exerts a force due to its weight on the slabs below it. Per unit area, this force becomes a pressure. So the change in pressure across the slab, ΔP, is simply how much a column of height Δz and density ρ weighs:

$$\Delta P = -g_\mathrm{p} \rho \Delta z. \tag{3.26a}$$

In general, both the density and gravitational acceleration, g_p, change with altitude, and the equation of hydrostatic equilibrium in differential form is

$$\frac{dP}{dz} = -g_\mathrm{p}(z)\rho(z). \tag{3.26b}$$

The relationship between temperature, pressure, and density (the *equation of state*) in a planetary atmosphere is usually well approximated by the *ideal gas law* (alternatively referred to as the *perfect gas law*):

$$P = NkT = \frac{\rho R_{\mathrm{gas}} T}{\mu_\mathrm{a}} = \frac{\rho kT}{\mu_\mathrm{a} m_{\mathrm{amu}}}, \tag{3.27}$$

where N is the particle number density (cm^{-3}), R_{gas} the universal gas constant ($R_{\mathrm{gas}} = N_\mathrm{A} k$, with N_A Avogadro's number), μ_a the mean molecular mass (in atomic mass units), and $m_{\mathrm{amu}} \approx 1.67 \times 10^{-24}$ g the mass of an atomic mass unit, which is slightly less than the mass of a hydrogen atom.

3.2.2.2 Thermodynamics: First Law

The first law of thermodynamics is an expression for the conservation of energy:

$$dQ = dU + P\,dV, \tag{3.28}$$

where dQ is the amount of heat absorbed by the system from its surroundings, dU the change in internal energy (sum of potential plus kinetic energy), and $P\,dV$ is the work done by the system on its environment, such as an expansion of the system; P is the pressure and dV the change in volume V. The thermal molar heat capacities, C_P and C_V, were defined above (eq. 3.20), and become

$$C_\mathrm{V} = \left(\frac{\partial U}{\partial T}\right)_\mathrm{V}, \tag{3.29a}$$

$$C_\mathrm{P} = \left(\frac{\partial U}{\partial T}\right)_\mathrm{P} + P\left(\frac{\partial V}{\partial T}\right)_\mathrm{P}. \tag{3.29b}$$

In the following we assume that V is the *specific volume*, containing one gram of molecules. Differentiating the ideal gas law gives

$$dV = \frac{k}{\mu_a m_{amu} P} dT - \frac{kT}{\mu_a m_{amu} P^2} dP. \quad (3.30)$$

In an ideal gas, the difference between the two thermal heat capacities (erg mole^{-1} K^{-1}) or specific heats (erg g^{-1} K^{-1}) is given by

$$C_P - C_V = R_{gas}, \quad (3.31a)$$

$$m_{gm}(c_P - c_V) = R_{gas}, \quad (3.31b)$$

with R_{gas} the *universal gas constant* and m_{gm} the mass of a gram-mole (see eq. 3.20). If, in an ideal gas, a parcel of air moves *adiabatically*, i.e., no heat is exchanged between the parcel of air and its surroundings ($dQ = 0$), the first law of thermodynamics requires that (Problem 3.15)

$$c_V \, dT = -P \, dV, \quad (3.32a)$$

$$c_P \, dT = \frac{1}{\rho} dP. \quad (3.32b)$$

Integration of equation (3.32b) yields the following relations for a dry adiabatic gas:

$$TP^{-(\gamma-1)/\gamma} = \text{constant}, \quad (3.33a)$$

$$P\rho^{-\gamma} = \text{constant}, \quad (3.33b)$$

where γ is defined as the *ratio of the specific heats*, $\gamma \equiv c_P/c_V = C_P/C_V$. Typical values for γ are 5/3, 7/5, 4/3, for monatomic, diatomic, and polyatomic gases, respectively.

3.2.2.3 Adiabatic Lapse Rate
For an atmosphere that is marginally unstable to convection, we can use the thermodynamic relations above with the equation of hydrostatic equilibrium (eq. 3.26) to obtain its temperature structure, or the *dry adiabatic lapse rate* (Problem 3.16):

$$\frac{dT}{dz} = -g_p/c_P = -\frac{\gamma-1}{\gamma} \frac{g_p \mu_a m_{amu}}{k}. \quad (3.34)$$

The dry adiabatic lapse rate on Earth is roughly 10 K km^{-1}. We show in §4.4.1 that the latent heat of condensation acts to decrease the adiabatic gradient in cloud-forming regions of moist atmospheres.

Convection is extremely efficient at transporting energy whenever the temperature gradient or lapse rate is *superadiabatic* (larger than the adiabatic lapse rate). Energy transport via convection thus effectively places an upper bound on the rate at which temperature can increase with depth in a planetary atmosphere or fluid interior. Substantial superadiabatic gradients are only possible when convection is suppressed by gradients in mean molecular mass or by the presence of a flow-inhibiting boundary, such as a solid surface.

3.2.3 Radiation
The transport of heat in a planetary atmosphere is typically dominated by radiation in regions where the optical depth of the gas is neither too large nor too small. This is usually the case in a planet's upper troposphere and stratosphere (§4.2). The radiation efficiency depends critically upon the emission and absorption properties of the material involved. To develop equations of radiative transfer, one needs to be familiar with atomic structure and energy transitions in atoms and molecules, and with the radiation 'vocabulary' like specific intensity, flux density, and mean intensity. In §3.2.3.1, we summarize a few basic definitions before developing the equations of radiative transfer in §3.2.3.4. Essential background of atomic and molecular structure, energy transitions, and the Einstein A and B coefficients, is given in §§3.2.3.2–3.2.3.3. The thermal structure and the greenhouse effect for an atmosphere that is in radiative equilibrium are derived in §3.3. Equations of radiative transfer for solid bodies are summarized in §3.4.

3.2.3.1 Definitions
The energy and momentum of a photon are given by

$$E = h\nu, \quad (3.35a)$$

$$\mathbf{p} = \frac{E}{c}\hat{\mathbf{s}}, \quad (3.35b)$$

where h is Planck's constant, c the speed of light, ν the frequency, and $\hat{\mathbf{s}}$ is a unit vector pointing in the direction of propagation.

The amount of energy crossing a differential element of area $\hat{\mathbf{s}} \cdot d\mathbf{A} = dA \cos\theta$, where θ is the angle between the surface normal, $\hat{\mathbf{z}}$, and the direction of propagation, into a solid angle $d\Omega_s$ in time dt and frequency range $d\nu$, is given by (Fig. 3.3)

$$dE = I_\nu \cos\theta \, dt \, dA \, d\Omega_s \, d\nu, \quad (3.36)$$

where I_ν is the *specific intensity*, which has the dimensions of erg s^{-1} cm^{-2} sr^{-1} Hz^{-1}. The specific intensity of radiation at frequency ν emitted by a blackbody is

$$I_\nu = B_\nu(T). \quad (3.37)$$

The solid angle, $d\Omega_s$, expressed in steradians (sr, i.e., radians^2) is defined such that, integrated over a sphere, it is

$$\oint d\Omega_s = \int_0^{2\pi} \int_0^\pi \sin\theta \, d\theta \, d\phi = 4\pi \text{ sr.} \quad (3.38)$$

The *mean intensity*, J_ν, or the zeroth *moment* of the radiation field, is equal to (using the conversion $\mu_\theta \equiv \cos\theta$)

$$J_\nu \equiv \frac{\oint I_\nu \, d\Omega_s}{\oint d\Omega_s} = \frac{1}{2}\int_{-1}^{1} I_\nu \, d\mu_\theta. \tag{3.39}$$

The *energy density*, u_ν, of the radiation is the amount of radiant energy per unit volume at frequency ν. Since, in vacuum, photons travel at the speed of light,

$$u_\nu = \frac{1}{c}\oint I_\nu \, d\Omega_s = \frac{4\pi}{c}J_\nu. \tag{3.40}$$

The net flux density, $\mathcal{F}_\nu$, in the direction $\hat{\mathbf{z}}$ at frequency ν can be obtained by integrating over all solid angles (§3.1.1):

$$\mathcal{F}_\nu = \oint I_\nu \cos\theta \, d\Omega_s. \tag{3.41a}$$

Note that for an isotropic radiation field the net flux density $\mathcal{F}_\nu = 0$, since $\oint \cos\theta \, d\Omega_s = 0$. Changing variables to $\mu_\theta \equiv \cos\theta$ and performing the integration over ϕ, equation (3.41a) becomes

$$\mathcal{F}_\nu = 2\pi \int_{-1}^{+1} I_\nu \mu_\theta \, d\mu_\theta. \tag{3.41b}$$

The momentum flux, integrated over all frequencies along a ray's path in direction $\hat{\mathbf{s}}$, is equal to $d\mathcal{F}/c$. The radiation pressure is the component of momentum flux in the $\hat{\mathbf{z}}$ direction:

$$p_r = \frac{\mathcal{F}\cos\theta}{c} = \frac{2\pi}{c}\int_{-1}^{+1} I\mu_\theta^2 \, d\mu_\theta. \tag{3.42a}$$

In an isotropic radiation field, $I_\nu(\mu_\theta) \equiv I_\nu$, and the radiation pressure is equal to

$$p_r = \frac{4\pi}{3c}I_\nu. \tag{3.42b}$$

3.2.3.2 Energy Transitions

Emission and absorption of photons by atoms or molecules involve a change in energy state. Each atom consists of a nucleus (protons plus neutrons) surrounded by a 'cloud' of electrons. In the semiclassical Bohr theory, the electrons orbit the nucleus such that the centrifugal force is balanced by the Coulomb force:

$$\frac{m_e v^2}{r} = \frac{Zq^2}{r^2}, \tag{3.43}$$

where m_e and v are the mass and velocity of the electron, respectively, r the radius of the electron orbit (assumed circular), Z the atomic number, and q the electric charge. Electrons are in orbits such that the angular momentum,

$$m_e v r = n\hbar, \tag{3.44a}$$

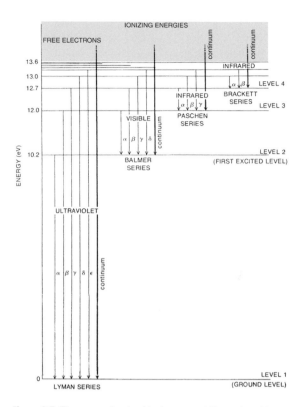

Figure 3.7 The energy levels of hydrogen and the series of transitions among the lowest of these energy levels. (Adapted from Pasachoff and Kutner 1978)

and the radius,

$$r = \frac{n^2\hbar^2}{m_e Zq^2}, \tag{3.44b}$$

where n, an integer, is the *principal quantum number*, and $\hbar \equiv h/2\pi$. The radius of the lowest energy state ($n = 1$) for the hydrogen atom ($Z = 1$) is called the *Bohr radius*: $r_{Bohr} = \hbar^2/(m_e q^2)$. The principal quantum number is the sum of the radial and azimuthal quantum numbers: $n = n_r + k$. These quantum numbers define the semimajor and semiminor axes, a and b, of an electron's orbit: $a/b = n/k$. The energy of orbit n is given by

$$E_n = -\frac{Zq^2}{r} + \frac{Zq^2}{2r} = -\frac{Zq^2}{2r} \approx -\frac{\mathcal{R}Z^2}{n^2}, \tag{3.45}$$

where $\mathcal{R} \equiv \mu_r e^4/h^2$, which is the *Rydberg constant* in the case of the hydrogen atom. The reduced mass, μ_r, was defined in equation (2.8a), where m_1 and m_2 in this case represent the mass of the electron and nucleus, respectively. The frequency of various transitions can be calculated using equations (3.35) and (3.45). An example of energy levels in the hydrogen atom is given in Figure 3.7. The transitions between the ground state and higher levels are called the Lyman series, where Ly α is the transition between levels 1 and 2, Ly β between levels 1 and 3, etc. The Balmer, Paschen, and Brackett series indicate transitions between levels 2, 3, and 4 with higher levels, respectively. If the electron is unbound, the atom is *ionized*. For hydrogen in the ground state, photons with energies ≥ 13.6 eV, or wavelengths shorter than 91.2 nm (the *Lyman limit*) may *photoionize* the atom (Problem 3.17).

Pauli's exclusion principle states that each electron orbit is specified by a unique set of quantum numbers, so that the total number of sublevels for energy level n, which is referred to as the *statistical weight* or *degeneracy* of level n, is given by

$$g_n = 2n^2. \tag{3.46}$$

The energy levels of molecules are more numerous than those of isolated atoms, since rotation and vibration of the nuclei with respect to each other require energy. This multiplicity leads to numerous molecular lines. For a complete treatment of atomic and molecular structure the reader is referred to, e.g., Herzberg (1944).

Transitions between energy levels may result in the absorption or emission of a photon with an energy ΔE_{ul} equal to the difference in energy between the two levels u and l. However, transitions are only possible between certain levels: they follow specific selection rules. The energy difference between electron orbits and therefore the frequency of the photon associated with the transition decrease with increasing n (Fig. 3.7). Whereas electronic transitions involving the ground state ($n = 1$) may be observed at ultraviolet or optical wavelengths, transitions at high n, (hyper)fine structure in atomic spectra and molecular rotation and rotation–vibration transitions all occur at infrared or radio wavelengths, since the spacing between energy levels is much smaller. Because each atom/molecule has its own unique set of energy transitions, one can use measurements of absorption/emission spectra to identify particular species in an atmosphere or surface. According to the *Heisenberg uncertainty principle*, a photon can be absorbed/emitted with an energy slightly different from ΔE_{ul}, which results in a finite width line profile, Φ_ν, discussed more fully in §4.3.2.

3.2.3.3 Einstein A and B Coefficients

How does a gas of atoms/molecules maintain the population of its electron excitation levels in equilibrium with a radiation field? An atom may absorb a photon, bringing it to a higher energy state, and if it is in an excited state, it may spontaneously emit a photon. Additionally, as an oscillator absorbs or emits energy depending on its phase relative to a driving force, a radiation field may cause

atoms in the upper energy state to transition to a lower energy state, a process referred to as *stimulated emission*.

The probability per unit time for emission or absorption of a photon can be expressed by the Einstein A and B coefficients. The coefficient A_{ul} is the probability per unit time for spontaneous emission from the upper energy level, u, to the lower one, l; $B_{lu}J_\nu$ is the probability per unit time for absorption, and $B_{ul}J_\nu$ the probability per unit time for stimulated emission. All probabilities are expressed as frequencies. Because of the finite width of the line profile, Φ_ν, the mean intensity, J_ν, for the probabilities given above should be integrated over the line profile: $\int J_\nu \Phi_\nu \, d\nu$.

In *thermodynamic equilibrium*, the following equations are valid:

(1) The radiation field obeys:

$$I_\nu = J_\nu = B_\nu(T). \tag{3.47a}$$

(2) There is equilibrium between the rate of absorption and emission:

$$N_l B_{lu} J_\nu = N_u A_{ul} + N_u B_{ul} J_\nu. \tag{3.47b}$$

(3) The number density of atoms in energy state i, N_i, is determined by the temperature of the gas:

$$N_i \propto g_i e^{-E_i/kT}, \tag{3.47c}$$

with g_i the statistical weight of level i.
The ratio N_l/N_u is given by *Boltzmann's equation*:

$$\frac{N_l}{N_u} = \frac{g_l}{g_u} e^{\Delta E_{ul}/kT}, \tag{3.48a}$$

where the energy difference between the upper and lower energy levels is given by $\Delta E_{ul} = E_u - E_l$. At low temperatures, most atoms are in the ground state, while higher energy levels are populated at higher temperatures. A more general form of the Boltzmann law is

$$N_i = \frac{N g_i}{Z_p} e^{-E_i/kT}, \tag{3.48b}$$

with N the total number of atoms and Z_p the *partition function*:

$$Z_p = \sum_j g_j e^{-E_j/kT}. \tag{3.48c}$$

Planck's radiation law (eq. 3.3) can be derived from the above formulae together with the *Einstein relations*:

$$g_l B_{lu} = g_u B_{ul} \tag{3.49a}$$

$$A_{ul} = \frac{2h\nu^3}{c^2} B_{ul}. \tag{3.49b}$$

In contrast to equations (3.47) and (3.48), the Einstein relations do not depend on temperature, so are valid whether or not the medium is in thermodynamic equilibrium.

The mass absorption and emission coefficients, κ_ν and j_ν, (erg g^{-1} s^{-1} sr^{-1} Hz^{-1}) become

$$\kappa_\nu \rho = \frac{\Delta E_{ul}}{4\pi}(N_l B_{lu} - N_u B_{ul})\Phi_\nu \qquad (3.50a)$$

$$j_\nu \rho = \frac{\Delta E_{ul}}{4\pi}N_u A_{ul}\Phi_\nu, \qquad (3.50b)$$

where Φ_ν is the line profile. Note that stimulated emission is added by the term with a negative absorption coefficient in equation (3.50a).

3.2.3.4 Equations of Radiative Transfer

When the primary mechanism of energy transport in an atmosphere is the absorption and re-emission of photons, the temperature–pressure profile is governed by the equations of radiative energy transport. The change in intensity, dI_ν, due to absorption and emission within a cloud of gas is equal to the difference in intensity between emitted and absorbed radiation:

$$dI_\nu = j_\nu \rho \, ds - I_\nu \alpha_\nu \rho \, ds. \qquad (3.51)$$

In equation (3.51), j_ν is the emission coefficient due to scattering and/or thermal excitation: $j_\nu = j_\nu(\text{scattering}) + j_\nu(\text{thermal excitation})$. The quantity α_ν is the mass extinction coefficient. Absorption (including stimulated emission) and scattering both contribute to the extinction: $\alpha_\nu = \kappa_\nu + \sigma_\nu$, where κ_ν and σ_ν are the mass absorption and mass scattering coefficients, respectively.

Denoting the coordinate in the direction of the normal to the planet's surface by $\hat{z}$, and the angle between $\hat{s}$ and $\hat{z}$ by θ (Fig. 3.3), we get $ds = \sec\theta \, dz$. With $\mu_\theta \equiv \cos\theta$, equation (3.51) becomes

$$\mu_\theta \frac{dI_\nu}{d\tau_\nu} = -I_\nu + S_\nu, \qquad (3.52)$$

where the *optical depth*, τ_ν, is defined as the integral of the extinction coefficient (along z):

$$\tau_\nu \equiv \int_{z_1}^{z_2} \alpha_\nu(z)\rho(z)\,dz. \qquad (3.53)$$

The *source function*, S_ν, is defined as

$$S_\nu \equiv \frac{j_\nu}{\alpha_\nu}. \qquad (3.54)$$

The formal solution to equation (3.52) is (for $\mu_\theta = 1$)

$$I_\nu(\tau_\nu) = I_\nu(0)e^{-\tau_\nu} + \int_0^{\tau_\nu} S_\nu(\tau_\nu')e^{-(\tau_\nu - \tau_\nu')}\,d\tau_\nu', \qquad (3.55a)$$

where $I_\nu(0)$ is the 'background' radiation that gets attenuated when propagating through the absorbing medium. In

radiative transfer calculations of planetary atmospheres, where the observer probes into the atmosphere, one usually defines the optical depth to increase into the atmosphere, i.e., $\tau_\nu = 0$ at the top of the atmosphere. Equation (3.55a) then becomes

$$I_\nu(0) = I_\nu(\tau_\nu)e^{-\tau_\nu} + \int_0^{\tau_\nu} S_\nu(\tau_\nu')e^{-\tau_\nu'}\,d\tau_\nu'. \qquad (3.55b)$$

For the more general case, the optical depth should be replaced by the slant optical depth, i.e., τ/μ_θ. If S_ν is known, equation (3.55) can be solved for the radiation field. In practice the situation is often more complicated because S_ν usually depends upon the intensity I_ν (e.g., through scattering) and/or the temperature of the medium, which may, in part, be determined by I_ν. If S_ν does not vary with optical depth, equation (3.55a) reduces to

$$I_\nu(\tau_\nu) = S_\nu + e^{-\tau_\nu}(I_\nu(0) - S_\nu). \qquad (3.56)$$

If $\tau_\nu \gg 1$, then $I_\nu = S_\nu$; i.e., the intensity of the emission is completely determined by the source function. If $\tau_\nu \ll 1$, then $I_\nu \to I_\nu(0)$; i.e., the intensity of the radiation is defined by the incident radiation.

In the remainder of this subsection we examine the equation of radiative transfer, or, more explicitly, the source function, S_ν, for four 'classic' cases. These examples help to elucidate the theory of radiative transfer.

(1) Consider a nonemitting cloud of gas along the line of sight, thus $j_\nu = 0$. Suppose that there is a source of radiation behind the cloud. The incident light $I_\nu(0)$ is reduced in intensity according to equation (3.56), and the resulting observed intensity becomes

$$I_\nu(\tau_\nu) = I_\nu(0)e^{-\tau_\nu}. \qquad (3.57)$$

This relation is called *Lambert's exponential absorption law*, also known as *Beer's law*, or various combinations thereof. If the gas cloud is optically thin ($\tau_\nu \ll 1$), equation (3.57) can be approximated by: $I_\nu(\tau_\nu) = I_\nu(0)(1 - \tau_\nu)$. If the cloud is optically thick ($\tau_\nu \gg 1$), the radiation is reduced to near zero.

(2) Assume the material to be in *local thermodynamic equilibrium* (LTE), and the scattering coefficient $\sigma_\nu = 0$ and $\kappa_\nu = \alpha_\nu$. If the material is in equilibrium with the radiation field, the amount of energy emitted must be equal to the amount of energy absorbed, as described by Kirchhoff's law:

$$j_\nu = \kappa_\nu B_\nu(T), \qquad (3.58)$$

where Planck's function, $B_\nu(T)$, describes the radiation field in thermodynamic equilibrium. In this case, the source function is given by

$$S_\nu = B_\nu(T). \tag{3.59}$$

The energy levels of the atoms/molecules are populated according to Boltzmann's equation (3.48).

(3) In this example, j_ν is due to scattering only: $j_\nu = \sigma_\nu I_\nu$. We receive sunlight that is reflected towards us (see Fig. 3.4). In general, scattering removes radiation from a particular direction and redirects or introduces it into another direction. If photons undergo only one encounter with a particle, the process is referred to as *single scattering*; multiple scattering refers to multiple encounters. The angular distribution of the scattered radiation is given by the *scattering phase function*, $\mathcal{P}(\cos\phi_{sc})$, which depends on the scattering angle ϕ_{sc}. The phase function is normalized such that integrated over a sphere,

$$\frac{1}{4\pi} \int \mathcal{P}(\cos\phi_{sc})\, d\Omega_s = \frac{\sigma_\nu}{\alpha_\nu} \equiv \varpi_\nu. \tag{3.60}$$

The term ϖ_ν is referred to as the *albedo for single scattering*, and represents the fraction of radiation lost due to scattering. The single scattering albedo is equal to unity if the mass absorption coefficient, κ_ν, is equal to zero. The source function can be written:

$$S_\nu = \frac{1}{4\pi} \int I_\nu \mathcal{P}(\cos\phi_{sc})\, d\Omega_s. \tag{3.61}$$

Common scattering phase functions are:

- Isotropic scattering:

$$\mathcal{P}(\cos\phi_{sc}) = \varpi_\nu, \tag{3.62a}$$

in which case the source function becomes: $S_\nu = \varpi_\nu J_\nu$.

- Rayleigh scattering phase function:

$$\mathcal{P}(\cos\phi_{sc}) = \frac{3}{4}(1 + \cos^2\phi_{sc}), \tag{3.62b}$$

which is representative for scattering by particles much smaller than the wavelength of light, such as scattering of sunlight by air molecules.

- First-order anisotropic scattering:

$$\mathcal{P}(\cos\phi_{sc}) = \varpi_\nu(1 + q_{ph}\cos\phi_{sc}), \tag{3.62c}$$

with $-1 \leq q_{ph} \leq 1$. The scattering is isotropic if $q_{ph} = 0$; the radiation is backscattered if $q_{ph} < 0$, and scattered in the forward direction if $q_{ph} > 0$. Radiation is scattered predominantly in the forward direction if the particles are similar in size or slightly larger than the wavelength of the scattered light.

- Henyey–Greenstein phase function:

$$\mathcal{P}(\cos\phi_{sc}) = \frac{\varpi_\nu(1 - g_{hg}^2)}{(1 + g_{hg}^2 - 2g_{hg}\cos\phi_{sc})^{2/3}}, \tag{3.62d}$$

where the asymmetry parameter, g_{hg}, represents the expectation value for $\cos\phi_{sc}$: $g_{hg} \equiv <\cos\phi_{sc}>$. For isotropic scattering, $g_{hg} = 0$; for forward (backward) scattering, $g_{hg} > 0$ (< 0). This one-parameter phase function yields good empirical fits to scattering by small nonspherical particles, and is widely used in planetary science. *Mie scattering*, derived analytically for spherical particles from Maxwell's equations (see Van de Hulst 1957), is also often used.

(4) In the situation of LTE and isotropic scattering, the source function, S_ν, becomes (Problem 3.20)

$$S_\nu = \varpi_\nu J_\nu + (1 - \varpi_\nu)B_\nu(T), \tag{3.63}$$

where $1 - \varpi_\nu = \kappa_\nu/\alpha_\nu$, the fraction of radiation absorbed by the medium.

Problems 3.20–3.23 contain exercises related to radiative transfer in an atmosphere. For example, in Problem 3.22 the student is asked to calculate the hypothetical brightness temperature of Mars at different frequencies. The temperature depends upon the optical depth in the planet's atmosphere; if the optical depth is near zero or infinity, the problem essentially simplifies to case (1) discussed above. If the optical depth is closer to unity, the radiation from the planetary disk as well as the atmosphere contributes to the observed intensity.

3.3 Atmosphere in Radiative Equilibrium

Energy transport in a planet's stratosphere, the region above the tropopause (§4.2), is usually dominated by radiation. If the total radiative flux is independent of height, the atmosphere is in *radiative equilibrium*. In this section, we derive the thermal profile for an atmosphere that is in radiative equilibrium.

3.3.1 Thermal Profile
An atmosphere is said to be in radiative equilibrium if the total flux, $\mathcal{F} = \int \mathcal{F}_\nu\, d\nu$, is constant with depth:

$$\frac{d\mathcal{F}}{dz} = 0. \tag{3.64}$$

The temperature structure in such an atmosphere can be obtained from the *diffusion equation*, an expression for the radiative flux at altitude z. In the following, we derive the diffusion equation in an optically thick atmosphere that is approximately in LTE: $I_\nu \approx S_\nu \approx B_\nu(T)$. We assume the atmosphere to be in monochromatic radiative equilibrium: $d\mathcal{F}_\nu/dz = 0$.

Integration of equation (3.52) over a sphere yields (Problem 3.24a):

$$\frac{d\mathcal{F}_\nu}{d\tau_\nu} = 4\pi(B_\nu - J_\nu),\qquad(3.65)$$

where $\mathcal{F}_\nu$ is the flux density across a layer in a stratified atmosphere. Both $\mathcal{F}_\nu$ and the mean intensity J_ν were defined in §3.2.3.1. Multiplying equation (3.52) by μ_θ and integrating over a sphere yields the following relationship between J_ν and $\mathcal{F}_\nu$ (Problem 3.24b):

$$\frac{4\pi}{3}\frac{dJ_\nu}{d\tau_\nu} = -\mathcal{F}_\nu.\qquad(3.66)$$

Setting $d\mathcal{F}_\nu/d\tau_\nu = 0$ in equation (3.65), and using equation (3.66), we find (Problem 3.24c)

$$\frac{dB_\nu}{d\tau_\nu} = -\frac{3}{4\pi}\mathcal{F}_\nu.\qquad(3.67)$$

Integrating over frequency yields the total radiative flux, or the *radiative diffusion equation*:

$$\mathcal{F}(z) = -\frac{4\pi}{3\rho}\frac{\partial T}{\partial z}\int_0^\infty \frac{1}{\alpha_\nu}\frac{\partial B_\nu(T)}{\partial T}d\nu.\qquad(3.68)$$

Equation (3.68) can be simplified by the use of a mean absorption coefficient, such as the *Rosseland mean absorption coefficient*, α_R:

$$\frac{1}{\alpha_R} \equiv \frac{\int_0^\infty \frac{1}{\alpha_\nu}\frac{\partial B_\nu}{\partial T}d\nu}{\int_0^\infty \frac{\partial B_\nu}{\partial T}d\nu}.\qquad(3.69)$$

With this simplification, we write the radiative diffusion equation as

$$\mathcal{F}(z) = -\frac{16}{3}\frac{\sigma T^3}{\alpha_R\rho}\frac{\partial T}{\partial z}.\qquad(3.70)$$

Note that flux travels upwards in an atmosphere if the temperature gradient dT/dz is negative (i.e., where the temperature decreases with altitude).

Applying equations (3.9) and (3.70), the atmospheric temperature profile is

$$\frac{dT}{dz} = -\frac{3}{16}\frac{\alpha_R\rho}{T^3}T_e^4.\qquad(3.71)$$

If an atmosphere is both in hydrostatic and radiative equilibrium and its equation of state is given by the perfect gas law, then its temperature–pressure relation is

$$\frac{dT}{dP} \approx -\frac{3}{16}\frac{T}{g_p}\left(\frac{T_e}{T}\right)^4\alpha_R.\qquad(3.72)$$

Equations (3.71) and (3.72) are approximate, since they use a mean absorption coefficient. Both T and $B_\nu(T)$, as well as the abundances of many of the absorbing gases, vary with depth in a planetary atmosphere. The best approach to solving for the temperature structure is

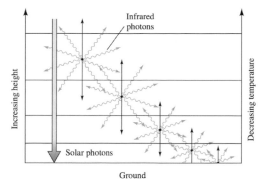

Figure 3.8 Schematic view of the transmission and scattering of radiation by an atmosphere that contains greenhouse gases. The short-wavelength sunlight passes through the atmosphere, delivering energy to the surface. In contrast, energy is reradiated by the planet at longer wavelengths, and much of this radiation is scattered back downwards by the greenhouse gases in the atmosphere. (Lunine 2005)

to solve the transport equation (3.52) at all frequencies, together with the requirement that the flux, $\mathcal{F}$, is constant with depth (eq. 3.64).

3.3.2 Greenhouse Effect

The surface temperature of a planet can be raised substantially above its equilibrium temperature if the planet is overlain by an atmosphere that is optically thick at infrared wavelengths, a situation referred to as the *greenhouse effect*. Sunlight, which has its peak intensity at optical wavelengths (Fig. 3.2 and eq. 3.5 for a blackbody of temperature $\sim$5700 K), enters the atmosphere, which is relatively transparent at visible wavelengths, and heats the surface. The warm surface radiates its heat at infrared wavelengths. This radiation does not immediately escape into interplanetary space, but is absorbed by air molecules, especially CO_2, H_2O, and CH_4. When these molecules de-excite, photons at infrared wavelengths are emitted in a random direction. The net effect of this process is that the atmospheric (and surface) temperature is increased until equilibrium is reached between solar energy input and the emergent planetary flux (Fig. 3.8). In this subsection, we calculate the greenhouse effect in an atmosphere that is in radiative equilibrium. Since we are concerned with atmospheric heating at infrared wavelengths, we consider the planet itself as the energy source. Thus the atmosphere is heated from below, and direct solar illumination of atmospheric gases is ignored.

For this analysis, it is convenient to use the *two-stream approximation*:

$$I_\nu = (I_\nu^+ + I_\nu^-),\qquad(3.73)$$

where I_ν^+ is the upward and I_ν^- the downward radiation at frequency ν. The net flux density across a layer becomes

$$\mathcal{F}_\nu = \pi(I_\nu^+ - I_\nu^-). \tag{3.74}$$

We consider an atmosphere in monochromatic radiative equilibrium $(d\mathcal{F}_\nu/dz = 0)$ and LTE, which is heated from below (i.e., $I_\nu^- \equiv 0$ at the top of the atmosphere). The upward intensity at the ground, $I_{\nu g}^+$, can be expressed with help of equations (3.73), (3.74), and (3.65) (Problem 3.25):

$$I_{\nu g}^+ \equiv B_\nu(T_g) = B_\nu(T_1) + \frac{1}{2\pi}\mathcal{F}_\nu, \tag{3.75}$$

where T_1 is the air temperature just above the ground. The downward and upward intensities at the top of the atmosphere are

$$I_{\nu 0}^- \equiv 0 = B_\nu(T_0) - \frac{1}{2\pi}\mathcal{F}_\nu, \tag{3.76a}$$

$$I_{\nu 0}^+ = B_\nu(T_0) + \frac{1}{2\pi}\mathcal{F}_\nu = 2B_\nu(T_0), \tag{3.76b}$$

where T_0 is the temperature of the upper boundary, usually referred to as the *skin temperature*. Thus, the upward intensity at the top of the atmosphere is twice as large as that emitted by an opaque blackbody at temperature T_0. We can derive the brightness, and therefore temperature, at τ by solving equation (3.67), and using equation (3.76a):

$$B_\nu(\tau) = B_\nu(T_0)\left(1 + \frac{3}{2}\tau_\nu\right). \tag{3.77}$$

Integration over frequency and conversion to temperature via the Stefan–Boltzmann law (eq. 3.8) yields:

$$T^4(\tau) = T_0^4\left(1 + \frac{3}{2}\tau\right). \tag{3.78}$$

The total radiant flux from a body can be obtained by integrating equation (3.76b) over frequency. This flux translates into the effective temperature, T_e:

$$T_e^4 = 2T_0^4. \tag{3.79}$$

If the temperature of a body is determined exclusively by the incident solar flux, the effective and equilibrium temperatures are equal, i.e., $T_e = T_{eq}$. At the top of the atmosphere, where the optical depth is zero, the temperature $T_0 \approx 0.84 T_e$. By combining equations (3.78) and (3.79), we see that the temperature, $T(\tau)$, is equal to the effective temperature, T_e, at an optical depth $\tau(z) = 2/3$. Thus, continuum radiation is received from an effective depth in the atmosphere where $\tau = 2/3$ (Problem 3.26).

The ground or surface temperature, T_g, can be obtained from equation (3.75):

$$T_g^4 = T_1^4 + \frac{1}{2}T_e^4. \tag{3.80a}$$

Note that there is a discontinuity: The surface temperature, T_g, is higher than the air temperature, T_1, just above it. In a real planetary atmosphere, conduction reduces this difference. Equation (3.80a) can be rewritten using equations (3.77) and (3.79):

$$T_g^4 = T_e^4\left(1 + \frac{3}{4}\tau_g\right), \tag{3.80b}$$

where τ_g is the optical depth to the ground. Equation (3.80b) shows that the surface temperature in a radiative atmosphere can be very high if the infrared opacity of the atmosphere is high.

The greenhouse effect is particularly strong on Venus, where the surface temperature reaches a value of 733 K, well above the equilibrium temperature of $\sim$240 K. The greenhouse effect is also noticeable on Titan and Earth and, to a lesser extent, Mars.

However, greenhouse warming on Titan is partially compensated by cooling produced by small haze particles in the stratosphere that block short-wavelength sunlight, but are transparent to long-wavelength thermal radiation from Titan; this process is known as the *anti-greenhouse* effect. Similar effects are observed on Earth after giant volcanic eruptions, such as the 1991 explosion of Mount Pinatubo in the Philippines, which injected huge amounts of ash into the stratosphere.

Calculations performed for an atmosphere assumed to be in radiative equilibrium may produce superadiabatic lapse rates (e.g., in a planet's troposphere). In such cases, convection develops and drives the atmospheric structure to an adiabat, and the temperature structure can better be calculated assuming *radiative–convective equilibrium*, where the convective layer (troposphere) supplies the same amount of upward radiative flux as would have been produced under radiative equilibrium, while the temperature structure in the convective layer follows an adiabat. The temperature thus calculated is somewhat cooler near the surface compared to radiative equilibrium calculations and warmer at higher altitudes.

Icy material allows sunlight to penetrate several centimeters or more below the surface, but is mostly opaque to reradiated thermal infrared emission. Thus, the subsurface region can become significantly warmer than the equilibrium temperature would indicate. In analogy with atmospheric trapping of thermal infrared emission, this process is known as the *solid-state greenhouse effect*. This process may be important on icy bodies, such as the Galilean satellites and comets.

3.4 Radiative Transfer in a Surface

The equilibrium temperature of a planet can be determined from a balance between incoming solar radiation and reradiation outwards (eqs. 3.15, 3.16). The heat is transported downwards primarily by conduction (§3.2.1), and the thermal structure in the crustal layers can be determined if the albedo, emissivity, thermal inertia, and thermal skin depth of the material are known. The brightness temperature of a planet, i.e., the temperature of a blackbody that would emit the same amount of energy at a particular wavelength, can be calculated by integrating the equation of radiative transfer (eq. 3.52) through the crustal layers. The mass absorption coefficient, κ_ν, is usually expressed as

$$\kappa_\nu = \frac{2\pi \sqrt{\epsilon_r} \tan \Delta}{\rho \lambda}, \tag{3.81}$$

where ϵ_r is the real part of the complex dielectric constant. The *loss tangent* of the material, $\tan \Delta$, is the ratio of the imaginary part of the dielectric constant to the real part. Note that the dielectric constant is usually wavelength dependent. It has been determined empirically that the loss tangent increases approximately linearly with the density of the material. For the Moon, $\tan \Delta / \rho \approx 0.007$–$0.01$ at wavelengths between a few mm and 20 cm.

The *electrical skin depth* of a material, L_e, is equivalent to the depth at which unit optical depth is reached:

$$L_e = \frac{\lambda}{2\pi \sqrt{\epsilon_r} \tan \Delta}. \tag{3.82}$$

The electrical skin depth is typically of the order of ten wavelengths. Thus, at infrared wavelengths one probes the surface layers, while at radio wavelengths one can probe up to a few meters down into the crust. Radio observations therefore sample the entire region of diurnal temperature variations, and by modeling the solar heating and outward radiation one can, through comparison with radio data, constrain the thermal and electrical properties in the upper few meters of the crust. For example, via such methods it was noticed that Mercury's surface is largely devoid of basalt (§5.5.2).

We discussed in §3.2.1 how the subsurface layers of a planet heat up during the day, when heat is transported downwards through the crust by conduction. The emission from the subsurface layers is transported upwards and transmitted through the surface into space. If θ_i is the angle with respect to the surface normal at which radiation from

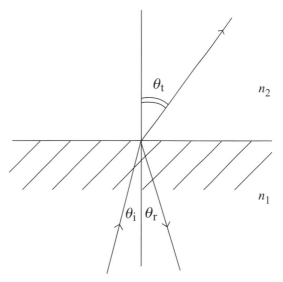

Figure 3.9 Geometry of refraction and reflection of radiation at the interface of two media with refractive indexes $n_2 < n_1$. Measured relative to the surface normal, θ_i and θ_r are the angles of incidence and reflection, respectively; θ_t is the angle for a ray transmitted (refracted) through the medium.

below the surface impinges upon the surface (Fig. 3.9), one can use *Snell's law of refraction* to relate θ_i to θ_t, the transmission or emission angle (toward the observer), and θ_r, the direction of propagation of the reflected component:

$$\theta_i = \theta_r, \tag{3.83a}$$

$$\frac{\sin \theta_t}{\sin \theta_i} \equiv n = \frac{n_1}{n_2} = \sqrt{\frac{\epsilon_{r1}}{\epsilon_{r2}}}, \tag{3.83b}$$

with n_1 and n_2 the indexes of refraction, and ϵ_{r1} and ϵ_{r2} the real parts of the dielectric constants of the two media. For radiation from below a planet's surface into space ($n_2 = 1$), $n = n_1 = \sqrt{\epsilon_r}$, and $\sin \theta_t > \sin \theta_i$. The subsurface radiation transmitted through the surface into space is equal to $\left(1 - R_{0,p}(\theta_t)\right)$, where the *Fresnel reflection coefficient* at the crust–vacuum interface, $R_{0,p}(\theta_t)$, is the ratio of the amount of energy in the reflected wave to that in the incident wave at frequency ν.

The Fresnel reflection coefficients for each sense of polarization of the outgoing thermal emission, assuming a perfectly smooth surface, are given by

$$R_\parallel = \frac{\tan^2(\theta_i - \theta_t)}{\tan^2(\theta_i + \theta_t)}, \tag{3.84a}$$

$$R_\perp = \frac{\sin^2(\theta_i - \theta_t)}{\sin^2(\theta_i + \theta_t)}, \tag{3.84b}$$

where $R_\parallel$ and $R_\perp$ are linearly polarized in and normal to the plane of incidence, respectively. Assuming that the subsurface emission is unpolarized, the emissivity at frequency ν, $\epsilon_\nu(\theta_t)$, is given by

$$\epsilon_\nu(\theta_t) = 1 - \frac{1}{2}R_\perp(\theta_t) - \frac{1}{2}R_\parallel(\theta_t). \tag{3.85}$$

Using the equations of radiative transfer, assuming that the source function, S_ν, is given by the Planck function $B_\nu(T)$, the brightness temperature of polarization p at frequency ν can be expressed as

$$T_{B_p}(\theta_t) = \left(1 - R_{0,p}(\theta_i)\right) \int_0^\infty \frac{\rho_\nu(z)\alpha_\nu(z)T_b(z)\, e^{(-\tau_\nu(z)/\sqrt{1-\epsilon_r(z)^{-1}\sin^2\theta_i})}}{\sqrt{1 - \epsilon_r(z)^{-1}\sin^2\theta_i}}\, dz, \tag{3.86}$$

with $T_b(z)$ the subsurface brightness temperature at depth z:

$$T_b(z) = \frac{h\nu/k}{e^{(h\nu/kT(z))} - 1}, \tag{3.87}$$

and $T(z)$ the physical temperature at depth z. Although the thermal radiation from a planet's subsurface layers is usually unpolarized, the emergent radiation is polarized. Polarization increases strongly towards the limb of the planet, and the total emergent radiation decreases sharply at viewing angles larger than $\sim 70°$, an effect known as *Fresnel limb darkening*. Surface roughness decreases the polarization and limb darkening.

The Fresnel coefficient at normal incidence for a wave which hits the surface from free space becomes (Problem 3.29)

$$R_0 = \left(\frac{1 - \sqrt{\epsilon_r}}{1 + \sqrt{\epsilon_r}}\right)^2. \tag{3.88}$$

The reflected radiation for a wave from free space is linearly polarized in a plane normal to the plane of incidence at the *Brewster angle* of incidence ($R_\parallel = 0$):

$$\tan\theta_i = n = \sqrt{\epsilon_r}. \tag{3.89}$$

If the incoming plane wave is linearly polarized in the plane of incidence, no wave is reflected at the Brewster angle.

Further Reading

Good books on spectroscopy:

Bernath, P.F., 2005. *Spectra of Atoms and Molecules*. Oxford University Press, Oxford. 439pp.

Herzberg, G., 1944. *Atomic Spectra and Atomic Structure*. Dover Publications, New York. 257pp.

Townes, C.H., and A.L. Schawlow, 1955. *Microwave Spectroscopy*. McGraw-Hill, New York. 698pp.

Books that discuss radiative transfer in detail:

Chandrasekhar, S., 1960. *Radiative Transfer*. Dover, New York. 392pp.

Rybicki, G.B., and A.P. Lightman, 1979. *Radiative Processes in Astrophysics*. John Wiley and Sons, New York. 382pp.

Shu, F.H., 1991. *The Physics of Astrophysics. Vol. I: Radiation*. University Science Books, Mill Valley, CA. 429pp.

Thomas, G.E., and K. Stamnes, 1999. *Atmospheric and Space Science Series: Radiative Transfer in the Atmosphere and Ocean*. Cambridge University Press, Cambridge. 517pp.

A classic book on light scattering:

Van de Hulst, H.C., 1957. *Light Scattering by Small Particles*, Wiley, New York. (Dover edition, 1981, 470pp.)

Problems

3.1.E (a) Write Planck's radiation law (eq. 3.3) in terms of λ rather than ν.

(b) Use your expression from part (a) to derive equation (3.5b).

3.2.E A sphere of radius R, at a distance r from the observer, has a uniform brightness B. The specific intensity is equal to B if the ray intersects the sphere, and zero otherwise. Use equation (3.41) to express the flux $\mathcal{F}$ in terms of brightness B and θ_c, the angle at which a ray from the observer is tangent to the sphere (i.e., $2\theta_c$ is the angle subtended by the sphere as seen from the observer). Show that on the surface of the sphere, $\mathcal{F} = \pi B$ (eq. 3.7).

3.3.I (a) Find the unidirectional energy flux in space of the 2.7 K background radiation. (Hint: A blackbody in thermal equilibrium with the radiation field has the same temperature as the radiation.) Express your answer in cgs units (erg cm^{-2} s^{-1}). Note that as the background radiation is (almost) isotropic, the net flux summed over all directions is zero.

(b) Galactic stars (other than the Sun) occupy a total solid angle of $\sim 10^{-14}$ sr as seen from the Solar System. Assuming a typical stellar effective temperature of 10 000 K (this is weighted towards the blue stars, which radiate most of the energy), what is the unidirectional energy flux from galactic stars?

(c) Calculate the solar energy flux at Earth's orbit, at Neptune's orbit, and at a typical Oort cloud distance of 25 000 AU.

3.4.I (a) A dust grain is heated by the interstellar radiation field, which has a temperature $T_{\rm ISM}$. If the grain were a perfect blackbody, its temperature would also be $T_{\rm ISM}$. Do you expect the grain to be warmer or colder than this value, and why?

(b) Assume that a dust grain is heated entirely by the interstellar radiation field at UV wavelengths, and that the UV flux is 2×10^6 photons cm^{-2} s^{-1} nm^{-1}. The bandwidth of the radiation is 100 nm, and the mean energy is 9 eV per photon. If the grain radiates with an efficiency of 0.1%, calculate its temperature. Compare your answer with part (a).

3.5.E Calculate the equilibrium temperature for the Moon:

(a) Averaged over the lunar surface. (Hint: Assume the Moon to be a rapid rotator.)

(b) As a function of solar elevation, assuming the Moon to be a slow rotator.

3.6.I Calculate the equilibrium temperature of the Moon as a function of latitude, assuming the Moon to be a rapid rotator with zero obliquity.

3.7.E Is the spectrum of emitted thermal radiation from a planet broader or narrower than the blackbody spectrum? Why?

3.8.E A body emits and absorbs radiation of any given frequency with the same efficiency. Given this fact, why does the temperature of a planet depend upon its albedo?

3.9.E Thermal emissions by solid bodies can be analyzed to provide information on the temperature at a few wavelengths below the surface. Observations of the night hemisphere of Mercury longward of ~10 cm yield temperatures close to the diurnal equilibrium temperature. As the radiation is obviously able to escape directly from the regions being observed, why does it not get much colder than this during the long mercurian night?

3.10.I (a) Assuming slow rotation and neglecting any internal heat sources, calculate the equilibrium temperature of each planet as a function of solar elevation angle. Assume the emissivity $\epsilon = 1$.

(b) Assuming rapid rotation and neglecting both internal heat sources and planetary obliquity, calculate the equilibrium temperature of each planet as a function of latitude.

3.11.E (a) Neglecting internal heat sources and assuming rapid rotation, calculate the average equilibrium temperature for all eight planets, using the data provided in Tables 1.1, 4.1, and 4.2.

(b) At which wavelength would you expect the blackbody spectral peak of each planet?

3.12.E Jupiter's effective temperature is observed to be 125 K. Compare the observed temperature with the equilibrium temperature calculated in the previous problems. What could be responsible for the difference? (Hint: Consider the assumptions that are involved in the derivation of the equilibrium temperature.)

3.13.E Mercury's observed surface temperature is 100 K at night, and 700 K at the subsolar point (noon and on the equator) at perihelion. Compare these temperatures with the values that you calculated in Problems 3.10 and 3.11, and comment on your result.

3.14.I Calculate the expected increase in the global average temperature of the Earth at a full Moon compared to a new Moon (neglecting eclipses). Which effect is larger, the change in position of the Earth or radiation reflected and emitted from the Moon?

3.15.I (a) Show that, in an ideal gas, equation (3.31) holds.

(b) Derive equations (3.32) and (3.33) from equations (3.28)–(3.31).

3.16.I Derive equation (3.34) for the dry adiabatic lapse rate, using the thermodynamic relations together with the equation for hydrostatic equilibrium.

3.17.E (a) Calculate the wavelength and energy of photons corresponding to the Lyman α, Balmer β, and Brackett α emission from a hydrogen atom.

(b) Calculate the wavelength and energy of photons necessary to ionize a hydrogen atom from the electronic ground state.

3.18.I Show that in an isotropic radiation field $I_\nu = J_\nu$.

3.19.I (a) Using equation (3.51) for a cloud of atomic hydrogen gas in LTE, show that the absorption coefficient can be written as:

$$\kappa_\nu \rho = \frac{\Delta E_{\rm ul}}{4\pi} N_{\rm l} B_{\rm lu} \left(1 - e^{-\Delta E_{\rm ul}/kT}\right) \Phi_\nu. \qquad (3.90)$$

(b) Calculate the relative importance of stimulated emission in the Ly α line ($n = 2 \rightarrow 1$) if the temperature of the gas is 100 K. (Hint: Which energy levels are populated?)

(c) Calculate the relative importance of stimulated emission in the Ly α line if the temperature of the gas is 10^4 K.

(d) Calculate the relative importance of stimulated emission in the Ly α line if the temperature of the gas is 10^6 K.

3.20.I Derive an expression for the source function, S_ν, and intensity, $I_\nu(\tau_\nu)$, for an atmosphere in LTE:

(a) where scattering can be ignored,

(b) in which scattering is isotropic.

3.21.I (a) Consider an optically thick ($\tau \gg 1$), spherical, cloud at a geocentric distance $d \gg R_\oplus$. The cloud is in LTE. Express the observed intensity, $I_\nu(\tau_\nu)$, at the center of this cloud, as a function of its brightness, $B_\nu(T)$, and optical depth, τ.

(b) The brightness temperature, T_b, is the temperature of a blackbody which has the same brightness at this frequency. Derive the relationship between T_b, τ, and T using the Rayleigh–Jeans approximation. Under what conditions does this approximation hold?

(c) If θ is the angle between the line of sight and the normal to the surface, determine the center-to-limb variation in the observed intensity.

(d) Answer parts (a)–(c) for an optically thin cloud, $\tau \ll 1$.

3.22.E The solid surface of Mars is opaque at all wavelengths and is surrounded by an optically thin atmosphere. The atmosphere absorbs in a narrow spectral region: its absorption coefficient is large at a wavelength $\lambda_0 = 2.6$ mm and negligibly small at other radio wavelengths; thus for most radio wavelengths λ_1: $\alpha_{\lambda_0} \gg \alpha_{\lambda_1}$. Assume that the temperature of Mars's surface $T_s = 230$ K, and of its atmosphere $T_a = 140$ K. (Note: For both parts of this problem you may use approximations appropriate for radio wavelengths, even though they are not fully correct at mm wavelengths.)

(a) What are the observed brightness temperatures at λ_1 and at λ_0?

(b) What is the observed brightness temperature at λ if the optical depth of the atmosphere at this wavelength is equal to 0.5?

3.23.E Consider a hypothetical planet of temperature T_p, surrounded by an extensive atmosphere of temperature T_a, where $T_a < T_p$. The atmosphere absorbs in a narrow spectral line; its absorption coefficient is large at frequency ν_0 and is negligibly small at other frequencies, such as ν_1: $\alpha_{\nu_0} \gg \alpha_{\nu_1}$. The planet is observed at frequencies ν_0 and ν_1. Assume that the Planck function does not change much between ν_0 and ν_1.

(a) When observing the center of the planet's disk, at which frequency, ν_0 or ν_1, is the brightness temperature higher? Is the same true when observing near the limb, where the limb is defined

as the atmosphere beyond the outer edge of the solid planet? Make a sketch of the 'observed' spectra.

(b) Repeat for $T_a > T_p$.

3.24.I (a) Derive equation (3.65) for an optically thick atmosphere that is approximately in LTE ($S_\nu \approx B_\nu$). (Hint: Integrate equation (3.52), and use the definitions for $\mathcal{F}_\nu$ and J_ν as given in equations (3.39) and (3.41).)

(b) Derive equation (3.66) by multiplying equation (3.52) by μ_θ and integrating over a sphere. (Hint: Use the definitions for all three moments of the radiation field (eqs. 3.39–3.42). Assume that the radiation is isotropic, so you can use equation (3.42b) and the relation $I_\nu = J_\nu$ (Problem 3.15).)

(c) Assuming monochromatic radiative equilibrium ($d\mathcal{F}_\nu/dz = 0$), and equations (3.65) and (3.66), derive equation (3.67).

3.25.I (a) Derive equation (3.75) for an atmosphere in LTE and in monochromatic radiative equilibrium. (Hint: Use equation (3.65) and the two-stream approximation.)

(b) Prove that the intensity at the top of a radiative atmosphere is twice that emitted by an opaque blackbody at the temperature of the top of the atmosphere (eq. 3.76b).

3.26.I Consider a rapidly rotating planet with an atmosphere in radiative equilibrium. The planet is located at a heliocentric distance $r_\odot = 2$ AU; its Bond albedo $A_b = 0.4$, and emissivity $\epsilon = 1$ at all wavelengths. Assume that the planet is heated exclusively by solar radiation.

(a) Calculate the effective and equilibrium temperatures.

(b) Calculate the temperature at the upper boundary of the atmosphere, where $\tau = 0$.

(c) Show that continuum radiation from the planet's atmosphere is received from a depth where $\tau = 2/3$.

(d) If the optical depth of the atmosphere $\int_0^\infty \tau(z)dz = 10$, determine the surface temperature of this planet.

3.27.E The radiation pressure coefficient, Q_{pr}, is given by:

$$Q_{pr} = Q_{abs} + Q_{sca}(1 - \langle\cos\phi_{sc}\rangle), \qquad (3.91)$$

where $\langle\cos\phi_{sc}\rangle = 1$ in the case of pure forward scattering, and $\langle\cos\phi_{sc}\rangle = -1$ for backscattered (reflected) radiation. Write down expressions for Q_{pr} and the net force on the particle due to radiation pressure and Poynting–Robertson drag for a particle in orbit around the Sun for the following (Hint: See §2.7.):

(a) Perfect absorber.

(b) Particle which does not absorb radiation and scatters in the forward direction only.

(c) Particle which is a perfect reflector (no absorption, backscattered light only).

3.28.E Plot the scattering phase functions given in equations (3.62a–d). You can normalize your plot to ϖ_ν (i.e., $\varpi_\nu = 1$). For the Henyey–Greenstein asymmetry parameter, use values of: $h_{hg} = 0$, $+0.7$ (appropriate for small dust grains), and -0.7. Comment on the similarities and differences between the various scattering functions.

3.29.I Derive the Fresnel coefficient at normal incidence for a wave which hits the surface from free space (eq. 3.88), and determine the emissivity $\epsilon_\nu(\theta_t)$ at normal incidence. (Hint: Use equations (3.83)–(3.85).)

3.30.I Assume that you are observing a point on the surface of the asteroid Ceres, from viewing angles θ of 0° up to 90° (θ is the angle between the line of sight and the normal to the surface). The dielectric constant of the surface material is $\epsilon_r = 6$ (note that $\epsilon_r = 1$ for free space). Plot the Fresnel coefficient for each sense of polarization, as well as the total emissivity as a function of θ. Comment on your results.

4 Planetary Atmospheres

In the small hours of the third watch, when the stars that shone out in the first dusk of evening had gone down to their setting, a giant wind blew from heaven, and clouds driven by Zeus shrouded land and sea in a night of storm.

Homer, *The Odyssey*, ~800 BCE

The *atmosphere* is the gaseous outer portion of a planet. Atmospheres have been detected around all planets and several satellites, and each is unique. Some atmospheres are very dense, and gradually blend into fluid envelopes which contain most of the planet's mass. Others are extremely tenuous, so tenuous that even the best vacuum on Earth seems dense in comparison. The composition of planetary atmospheres varies from the solar-like hydrogen/helium envelopes of the giant planets to atmospheres dominated by nitrogen, carbon dioxide, or esoteric gases such as sulfur dioxide or sodium for terrestrial planets and satellites of giant planets. However, even though all atmospheres are intrinsically different, they are governed by the same physical and chemical processes. For example, clouds form in many atmospheres, but with vastly different compositions since the gases available to condense differ. The upper layers of an atmosphere are modified by photochemistry, with the particulars depending on atmospheric composition. Variations in temperature and pressure lead to winds, which can be steady or turbulent, strong or weak. The various processes operating in planetary atmospheres are discussed in this chapter, and the characteristics of the atmospheres of bodies within our Solar System are summarized.

4.1 Density and Scale Height

The relationships between temperature, pressure, and density in a planetary atmosphere are governed by a balance between gravity and pressure: to first approximation, atmospheres are in hydrostatic equilibrium (§3.2.2.1). Using the equations for hydrostatic equilibrium (eq. 3.26) and the ideal gas law (eq. 3.27), the atmospheric pressure varies with altitude as

$$P(z) = P(0)e^{-\int_0^z dr/H(r)}. \tag{4.1}$$

The *pressure scale height*, H, is given by

$$H(z) = \frac{kT(z)}{g_p(z)\mu_a(z)m_{amu}}, \tag{4.2}$$

where $g_p(z)$ is the acceleration due to gravity at altitude z, $\mu_a m_{amu}$ is the molecular mass and k is Boltzmann's constant. Thus for a constant pressure scale height, H is equal to the distance over which the pressure decreases by a factor e. Small values of H imply a rapid decrease of atmospheric pressure with altitude. The scale height, however, usually varies with altitude. We can similarly express the density as a function of altitude:

$$\rho(z) = \rho(0)e^{-\int_0^z dr/H^*(r)}, \tag{4.3}$$

where $\rho(0)$ is the number density at altitude $z = 0$.[1] The *density scale height*, H^*, is

$$\frac{1}{H^*(z)} = \frac{1}{T(z)}\frac{dT(z)}{dz} + \frac{g_p(z)\mu_a(z)m_{amu}}{kT(z)}, \tag{4.4}$$

where we have neglected the (usually small) terms that result from gradients in μ_a and g_p. Note that for an isothermal (region of an) atmosphere, $H^*(z) = H(z)$. Approximate pressure scale heights for the planets, Titan, the Moon, and Pluto are shown in Tables 4.1–4.3. It is interesting to note that H is of the order of 10–25 km for most planets, since the ratio $T/(g_p\mu_a)$ for the giant and terrestrial planets is similar. Only in the tenuous atmospheres of Mercury, Pluto, and various moons is the scale height larger (Problem 4.1).

4.2 Thermal Structure

The *thermal structure* of a planet's atmosphere, dT/dz, is primarily governed by the efficiency of energy transport, as discussed in §3.2. This process depends largely on the

[1] The location of the $z = 0$ 'plane' is selected for convenience. For Earth, mean sea level is often used. For giant planets, the 1 bar pressure level is usually chosen.

Table 4.1 Basic atmospheric parameters for the giant planets.

Parameter	Jupiter	Saturn	Uranus	Neptune	References
Mean heliocentric distance (AU)	5.203	9.543	19.19	30.07	1
Geometric albedo $A_{0,v}$	0.52	0.47	0.51	0.41	1
Geometric albedo $A_{0,ir}$	0.274 ± 0.013	0.242 ± 0.012	0.208 ± 0.048	0.25 ± 0.02	3
Bond albedo	0.343 ± 0.032	0.342 ± 0.030	0.290 ± 0.051	0.31 ± 0.04	3
Phase integral	1.25 ± 0.10	1.42 ± 0.10	1.40 ± 0.14	1.25 ± 0.10	3
Effective temperature (K)	124.4 ± 0.3	95.0 ± 0.4	59.1 ± 0.3	59.3 ± 0.8	2
Equilibrium temperature (K)	110.0	81.3	58.4	46.3	Calc.[b]
Temperature ($P = 1$ bar) (K)	165.0	134.8	76.4	71.5	5
Tropopause temperature (K)	111	82	53	52	5
Mesosphere temperature (K)	160–170	150	140–150	140–150	4
Exobase temperature (K)	900–1300	800	750	750	6, 7
Tropopause pressure (mbar)	140	65	110	140	5
Scale height (at 1 bar) (km)	24	47	25	23	Calc.[c]
Dry adiabat (K/km, ~1 bar)	2.1	0.9	1.0	1.3	Calc.
Energy balance[a]	1.63 ± 0.08	1.87 ± 0.09	1.05 ± 0.07	2.68 ± 0.21	Calc.

[a] Ratio (energy radiated into space)/(solar energy absorbed).

[b] Calculated with eq. (3.17), Table 1.1, and $\epsilon = 1$.

[c] Calculated with eq. (4.2).

1: Yoder (1995). 2: Hubbard *et al.* (1995). 3: Conrath *et al.* (1989b). 4: Chamberlain and Hunten (1987). 5: Lindal (1992). 6: Atreya (1986). 7: Bishop *et al.* (1995).

Table 4.2 Basic atmospheric parameters for Venus, Earth, Mars, and Titan.

Parameter	Venus	Earth	Mars	Titan	Reference
Mean heliocentric distance (AU)	0.723	1.000	1.524	9.543	1
Geometric albedo $A_{0,v}$	0.84	0.367	0.15	0.21	1, 3, 4
Bond albedo	0.75	0.306	0.25	0.20	1, 2, 3, 4, 5
Surface temperature (K)	737	288	215	93.7	1, 2, 3
Equilibrium temperature (K)	232	255	210	85	Calc.[a]
Exobase[b] temperature (K)	270–320	800–1250	200–300	149	2, 6, 7
Surface pressure (bar)	92	1.013	0.00636	1.47	1, 2, 3
Scale height at surface (km)	16	8.5	11	20	Calc.[c]
Dry adiabat (K/km)	10.4	9.8	4.4	1.4	Calc.

[a] Calculated with eq. (3.17) and $\epsilon = 1$. The globally and wavelength-averaged emissivity for Earth is 0.96–0.98.

[b] A range of values is given for Venus, Earth, and Mars, appropriate for a range of solar (low to high) activities.

[c] Calculated with eq. (4.2).

1: Yoder (1995), and http://nssdc.gsfc.nasa.gov/planetary/. 2: Chamberlain and Hunten (1987). 3: Fulchignoni *et al.* (2005). 4: Moroz (1983). 5: Hunten *et al.* (1984). 6: Waite *et al.* (2005). 7: Forbes *et al.* (2008).

optical depth of the atmosphere, which is determined by a variety of physical and chemical processes. Stellar atmospheres are heated from below and most are so hot that the elements are primarily in atomic form. In contrast, planetary atmospheres consist of molecular gases and are in part heated from the top. To determine the thermal structure in such an atmosphere one has to consider all possible processes that may, directly or indirectly, affect its temperature:

(1)   The top of the atmosphere is irradiated by the Sun. Some of this radiation is absorbed and scattered in the atmosphere. This process, together with other heating

Table 4.3 Basic atmospheric parameters for Mercury, the Moon, Triton, and Pluto.

Parameter	Mercury	Moon	Triton	Pluto	Reference
Mean heliocentric distance (AU)	0.387	1.000	30.069	39.48	1
Geometric albedo $A_{0,v}$	0.138	0.113	0.76	0.44–0.61	1, 2, 3, 5
Bond albedo	0.119	0.123	0.85	∼0.3–0.7	1, 2, 3, 4, 5
Surface temperature (K)	100–725	277	38	∼40–60	1, 2, 3, 5
Equilibrium temperature (K)	434	270	32	39	Calc.[a]
Exobase temperature (K)	600	270–320	100	58	4, 6
Surface pressure (bar)	few $\times 10^{-15}$	3×10^{-15}	1.4×10^{-5}	1.5×10^{-5}	1, 2, 3, 4, 5
Scale height at surface (km)	13–95	65	14	33	Calc.[b]

[a] Calculated using eq. (3.17) and $\epsilon = 1$.

[b] Calculated using eq. (4.2).

1: Yoder (1995), and http://nssdc.gsfc.nasa.gov/planetary/. 2: Veverka *et al.* (1988). 3: Stern (2007). 4: Chamberlain and Hunten (1987). 5: McKinnon and Kirk (2007). 6: Krasnapolsky *et al.* (1993).

processes (§4.2.1.1), radiative losses, and conduction, basically defines the temperature profile in the upper part of the atmosphere.

(2)   Energy from internal heat sources (the giant planets)[2] and reradiation of absorbed sunlight by a planet's surface or dust in its atmosphere modify (in some cases dominate) the temperature profile.

(3)   Chemical reactions in an atmosphere change its composition, which leads to changes in opacity and hence thermal structure.

(4)   Clouds and/or photochemically produced haze layers not only change the atmospheric opacity, but also change the temperature locally through release (cloud formation) or absorption (evaporation) of latent heat.

(5)   Volcanoes and geyser activity on some planets and satellites may modify their atmospheres substantially.

(6)   On the terrestrial planets and satellites, chemical interactions between the atmosphere and the crust or ocean influence their atmosphere.

(7)   The Earth's atmospheric composition, opacity, and thermal structure are influenced by biochemical and anthropogenic processes.

Even though the composition of an atmosphere varies drastically from one planet/satellite to another, aside from the most tenuous atmospheres the temperature structure is qualitatively similar, as shown in Figure 4.1. The profiles in this figure are discussed in detail in §4.2.2; here we merely summarize the terminology. Moving upwards from the surface or, for the giant planets, from the deep

[2]   Throughout this book we use the term 'internal heat source' for a giant planet's loss of primordial energy, i.e., a slow cooling of the planet from its initially hot condition, supplemented in some cases by the energy released from He differentiation (§6.1.5).

atmosphere, the temperature decreases with altitude: this part of the atmosphere is called the *troposphere*. It is in this part of the atmosphere that condensable gases, usually trace elements, form clouds. The atmospheric temperature typically reaches a minimum at the *tropopause*, near a pressure level of ∼0.1 bar. Above the tropopause the temperature structure is inverted. This region in the atmosphere is called the *stratosphere*. At higher altitudes one finds the *mesosphere*, characterized by a temperature gradient that decreases with altitude. The *stratopause* forms the boundary between the stratosphere and mesosphere. On Earth, Titan, and perhaps Saturn the *mesopause* forms a second temperature minimum. Above the mesopause, in the *thermosphere*, the temperature increases with altitude, up to the *exosphere*, which is the outermost part of an atmosphere. Collisions between gas molecules in the exosphere are rare, and the rapidly moving molecules have a relatively large chance to escape into interplanetary space. The *exobase*, at the bottom of the exosphere (∼500 km on Earth), is the altitude above which the mean free path length exceeds the atmospheric scale height H (eq. 4.73).

4.2.1 Sources and Transport of Energy
4.2.1.1 *Heat Sources*
All planetary atmospheres are subject to solar irradiation, which heats an atmosphere through absorption of solar photons. Since the solar 5700 K blackbody curve peaks near 500 nm, most of the Sun's energy output is in the visible wavelength range. These photons heat up a planet's surface (terrestrial planets) or layers in the atmosphere where the optical depth is moderately large (typically near the cloud layers). Reradiation of sunlight by a planet's surface or atmospheric molecules, dust particles,

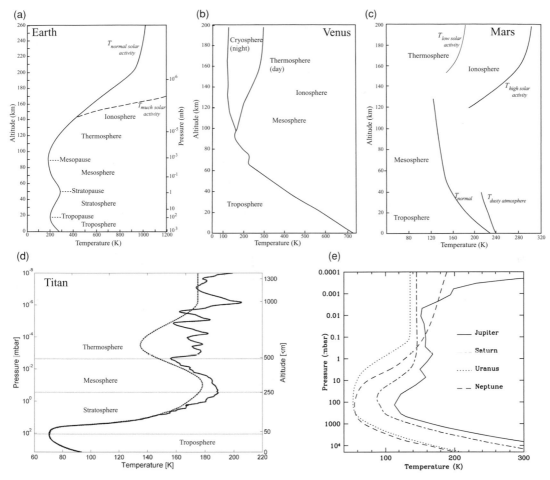

Figure 4.1 The approximate thermal structure of the atmospheres of (a) Earth, (b) Venus, (c) Mars, (d) Titan, and (e) Jupiter, Saturn, Uranus, and Neptune. The temperature–pressure profile in each planet's atmosphere is shown as a function of altitude (Venus, Mars) or pressure (gaseous planets) or both (Earth, Titan). The profile for Venus is derived from Seiff (1983); for Mars from Barth *et al.* (1992); for Titan from Fulchignoni *et al.* (2005). The solid line in the latter graph shows the measurements by the Huygens Atmospheric Structure Instrument while the probe descended through Titan's atmosphere. The dashed line shows the Yelle *et al.* (1997) engineering model. The profiles for the giant planets were constructed with help of Lindal's (1992) radio occultation data and an adiabatic extrapolation to deeper levels, plus for Jupiter: the Galileo probe data (Seiff *et al.* 1998), Saturn: Cassini CIRS retrievals (Fletcher *et al.* 2007), Uranus: Spitzer retrievals (Fletcher *et al.* 2008), Neptune: Spitzer retrievals (Fletcher *et al.* 2008).

or cloud droplets occurs primarily at infrared wavelengths, and forms a source of heat imbedded within or below the atmosphere. Internal heat sources may also heat the atmosphere from below; this is important for the giant planets.

Solar heating of the upper atmosphere is very efficient at EUV (extreme ultraviolet) wavelengths, even though the number of photons in this wavelength range (100 to 10 nm) is very low. Typical EUV photons have energies between 10 and 100 eV, which is enough to ionize several of the atmospheric constituents (§4.6.2). The excess energy from ionization is carried off by electrons freed in

the process; these electrons are referred to as *photoelectrons*. Photoelectrons collide with and excite/ionize other particles, either directly or via *bremsstrahlung*[3] induced by Coulomb collisions.

In addition to heating processes triggered by sunlight, an upper atmosphere can be heated substantially by *charged particle precipitation*: charged particles that enter the atmosphere from above (solar wind or planetary

---

[3] Bremsstrahlung or free–free emission is electromagnetic radiation resulting from the acceleration of a charged particle in the Coulomb field of another charged particle.

magnetosphere). On planets with intrinsic magnetic fields, charged particle precipitation is confined to high magnetic latitudes, the *auroral zones* (§4.6.4). Direct particle precipitation can heat an atmosphere substantially more than the photoelectrons mentioned above. Although most of the heating is localized, thermospheric winds may distribute the heat over the entire globe.

Joule heating, resulting from electric currents in a planet's ionosphere, can be important in the thermosphere as well. The dissipation of electrical energy occurs through charged particle collisions (§4.6.3).

4.2.1.2 *Energy Transport*
The temperature structure in an atmosphere is governed by energy transport. There are three distinct mechanisms to transport energy: conduction, mass motion (e.g., convection), and radiation. Each of these mechanisms is discussed in detail in §3.2. *Conduction* is important in the very upper part of the thermosphere and in the exosphere, and very near the surface if one exists. Collisions tend to equalize the temperature distribution, resulting in a nearly isothermal profile in the exosphere.

In the troposphere, energy transport is usually driven by *convection*, and the temperature profile is therefore close to an adiabat. The *dry adiabatic lapse rate* was derived in §3.2.2.3 (eq. 3.34); the formation of clouds decreases the temperature gradient due to the latent heat of condensation (§4.4.1). Convection thus effectively places an upper bound to the rate at which the temperature can decrease with height. Substantial superadiabatic gradients are possible, however, when convection is suppressed, for example by gradients in the mean molecular weight or by the presence of a flow-inhibiting boundary, such as a surface. Superadiabatic gradients may also exist under extreme heating conditions, such as over a desert on Earth on hot summer days, where the surface may limit the near-surface convective velocities and hence the ability to move heat away from the surface via convection. Such superadiabatic gradients may extend up to $\sim$300 m above the surface.

When the most efficient way for energy transport is via absorption and re-emission of photons, i.e., *radiation*, the thermal profile is governed by the equations of radiative energy transport. The temperature structure for an atmosphere in radiative equilibrium was derived in §3.3 (eq. 3.71).

The thermal structure in any part of an atmosphere is governed by the most efficient mechanism to transport energy. In Chapter 3, the thermal structure was calculated for an atmosphere in which energy transport was either by convection (eq. 3.34) or radiation (eq. 3.71). Which

process is most efficient depends upon the temperature gradient, dT/dz. In the tenuous upper parts of the thermosphere, energy transport is dominated by conduction. At deeper layers, down to a pressure of $\sim$0.5 bar, an atmosphere is usually in radiative equilibrium, and below that convection dominates.

4.2.2 Observations of Thermal Profiles
A body's temperature can be determined from observations of its thermal energy flux (eq. 3.9). It is interesting to compare these effective (observed) temperatures with their equilibrium values (eq. 3.17), as discussed in §3.1.2.2. Tables 4.1–4.3 show a comparison of these numbers for many bodies. The observed effective temperatures of Jupiter, Saturn, and Neptune are substantially larger than the equilibrium values, which implies the presence of internal heat sources (see footnote 2 in §4.2). For Venus, Earth, and Titan (as well as the day side of Mars) the observed surface temperature exceeds the equilibrium value because of a greenhouse effect (§3.3.2).

The thermal structure in an atmosphere can be determined via observations at different wavelengths. Different wavelengths probe different depths in a planet's atmosphere, because opacity is a strong function of wavelength. Although the precise altitudes probed differ from planet to planet, we can make a few generic statements here. At optical and infrared wavelengths, the radiative part of an optically thick atmosphere is probed. Convective regions at $P \gtrsim 0.5$–1 bar can be investigated at infrared and radio wavelengths. The tenuous upper levels, at $P \lesssim 10$ μbar, are typically probed at UV wavelengths, or via stellar occultations at UV, visible, and IR wavelengths. Atmospheric profiles for several bodies are shown in Figure 4.1. For the terrestrial planets and Titan these have been derived from *in situ* measurements by probes and/or landers, as well as inversion of IR and microwave spectra. For the giant planets, the temperature–pressure profiles have been derived via inversion of IR spectra, combined with UV and radio occultation profiles from the Voyager and other spacecraft. At deeper levels in the atmosphere, typically at pressures $P > 1$–5 bar, where no direct information on the temperature structure can be obtained via remote observations, one usually assumes the temperature to follow an adiabatic lapse rate. *In situ* observations by the Galileo probe in Jupiter's atmosphere showed the temperature lapse rate to be close to that of a dry adiabat.

4.2.2.1 *Terrestrial Planets and Titan*
Earth
The average temperature on Earth is 288 K, and the average surface pressure at sea level is 1.013 bar. This

temperature is 33 K above the equilibrium value, a difference which can be attributed to the greenhouse effect (§3.3.2), due most notably to the presence of water vapor, carbon dioxide (CO_2), and a variety of trace gases including ozone (O_3), methane (CH_4), and nitrous oxide (N_2O).

Earth's troposphere extends up to an altitude of $\sim$20 km at the equator, decreasing to $\sim$10 km above the poles. Above the tropopause, the temperature in the stratosphere increases with altitude as a result of the formation and presence of ozone, which absorbs both at UV and IR wavelengths. Above the stratopause at $\sim$50 km the temperature decreases with altitude, caused by a decrease in O_3 production and an increase in the CO_2 cooling rate to space. This region is referred to as the mesosphere. A second temperature minimum is found at the mesopause, at altitudes near $\sim$80–90 km. The temperature structure in Earth's stratosphere–mesosphere is unusual; massive atmospheres other than Earth's and Titan's (and perhaps Saturn's) show a single temperature minimum. Above the mesosphere lies the thermosphere. Thermospheres are often hot because they are dominated by atoms or homonuclear molecules that do not radiate efficiently. In Earth's thermosphere, the temperature increases with altitude, due in part to absorption of UV sunlight (O_2 photolysis and ionization), but primarily because there are too few atoms/molecules to cool the atmosphere efficiently through emission of IR radiation. Most of the IR emission originates from O and NO, molecules which radiate less efficiently than CO_2. At the base of the thermosphere, there is enough CO_2 gas to cool the atmosphere. The upper thermosphere heats up to 1200 K or more during the day, and cools to $\sim$800 K at night.

Venus

Venus's lower atmosphere, or troposphere, extends from the ground up to the level of the visible cloud layers, at $\sim$65 km, which is also the altitude of the tropopause. The surface temperature and pressure are 737 K and 92 bar, respectively. The equilibrium temperature for a rapidly rotating body at Venus's heliocentric distance, with a Bond albedo equal to that of Venus, is only $\sim$240 K, about 500 K less than the observed surface temperature. This difference results from a strong greenhouse effect due primarily to the planet's massive CO_2 atmosphere. At infrared wavelengths, where the top of the cloud layers are probed, the observed temperature is $\sim$240 K, as expected for a body in equilibrium with solar irradiation. The mean lapse rate from the surface up to the base of the cloud layers at $\sim$45 km is $\sim$7.7 K km^{-1}, slightly less than the mean adiabatic lapse rate of 8.9 K km^{-1}. In the lower atmosphere,

no temperature variations (diurnal, latitudinal, or temporal) have been measured in excess of $\sim$5 K.

Venus's middle atmosphere, the mesosphere, extends from the top of the cloud layers up to $\sim$90 km. The temperature lapse rate decreases sharply at $\sim$63 km, and the atmosphere is nearly isothermal between 63 and 75 km altitude.

At higher altitudes, in the thermosphere, there is a distinct difference in temperature between the day and night sides. Above about 100 km on the day side the temperature begins to rise, reaching about 300 K at 170 km. The night side is much colder, 100–130 K, and is often referred to as the *cryosphere*. A relatively warm (180–220 K) layer is present at the base of the cryosphere (90–120 km altitude), probably the result from adiabatic heating caused by day-to-night winds (§4.5). The temperature rise in the thermosphere is much less than that on Earth, despite Venus's proximity to the Sun. The relatively small rise in temperature is attributed to the large concentration of CO_2 gas, a very efficient cooling agent.

Mars

The average surface pressure on Mars is 6 mbar, and the mean temperature $\sim$215 K. However, due to the planet's low atmospheric pressure and hence low heat capacity, its obliquity and orbital eccentricity, the surface temperature displays large latitudinal, diurnal, and seasonal variations. At mid-latitudes, the surface temperature drops to $\sim$200 K at night and peaks at $\sim$300 K during the day. The temperature at the winter pole is $\sim$130 K, while in the summer the temperature at the pole may reach $\sim$190 K. The adiabatic lapse rate for a clear CO_2 atmosphere in radiative–convective equilibrium is 5 K km^{-1}, but Mars's observed lapse rate is seldom steeper than 3 K km^{-1}. Pronounced pressure variations are induced by the condensation of a significant fraction of Mars's CO_2 dominated atmosphere onto the planet's seasonal polar caps (§4.5.1.3).

Figure 4.2 displays the globally averaged day and nighttime temperature of Mars's surface and atmosphere (altitude 25 km) as a function of a martian year. The temperature is usually highest during perihelion season (at Mars solar longitude $L_s = 180°$–$360°$), and lowest when Mars is near aphelion. Dust storms are common near perihelion, and cause large interannual variability in temperature. During a dust storm, the atmosphere is always (day and night) warmer than under dust free conditions, while the surface is colder during the day, but warmer at night. The atmospheric temperature just above the surface, up to an altitude of $\sim$2 km, was retrieved (via inversion of the radiance in the 15 μm CO_2 band) with the Miniature Thermal Emission Spectrometers (Mini-TES) on board

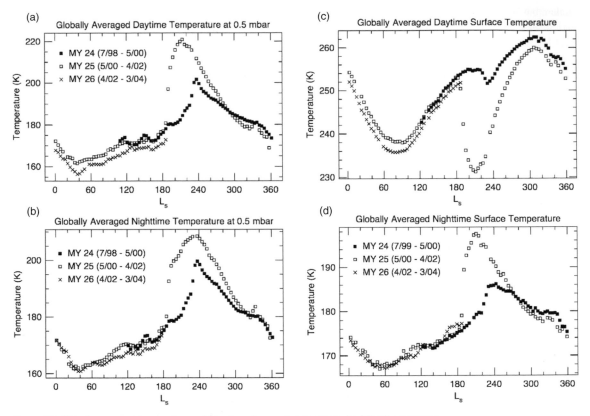

Figure 4.2 Globally averaged temperatures of Mars's surface and atmosphere (at 0.5 mbar pressure, or 25 km altitude) during the day and at night as retrieved from thermal IR sounding observations with the Thermal Emission Spectrometer (TES) on the Mars Global Surveyor (MGS) orbiter over 3 martian years. A $\sim$20 K seasonal variation (with Mars solar longitude, L_s) is forced by Mars's eccentric orbit, through an enhanced solar flux near perihelion ($L_s = 251°$). Dust storms, prevalent near perihelion, cause additional heating in the atmosphere (panels a and b) and nighttime surface temperature (panel d), while the surface temperature during the day drops considerably (panel c). As shown, the globally averaged surface temperature was 23 K cooler at the peak of the planet-encircling 2001a dust storm (MY 25, $L_s = 210°$) during the day than it was one martian year earlier, while at night the temperature was 18 K higher. (Smith 2004)

the Mars rovers Spirit and Opportunity. The atmosphere here interacts most strongly with the surface. In the afternoon, the temperature gradient is superadiabatic just above the surface, while an inversion layer has been observed at night. The surface temperature closely follows the solar input, while the atmospheric temperature, driven by solar radiation and surface heat, continues to warm during the day until it is equal to the surface temperature. Depending on altitude, the maximum atmospheric temperature lags maximum soil temperatures by several hours (2.75 hr at 1 m altitude, 4.5 hr at 1 km).

Like Venus, Mars lacks a stratosphere. At altitudes above $\sim$120 km, in Mars's thermosphere, the temperature is almost isothermal with a value of $\sim$160 K. As on Venus, the low temperature can be explained by the efficiency of CO_2 as a cooling agent.

Titan

The thermal structure in Titan's atmosphere has been determined by the Huygens probe. At the surface, the temperature and pressure were measured at 93.65 K and 1.467 bar, respectively. This temperature results from the competing greenhouse and anti-greenhouse effects (§3.3.2). Since photochemically produced haze in Titan's stratosphere (§4.6.1.4) absorbs much sunlight and is transparent at some infrared wavelengths, it creates an anti-greenhouse effect, reducing the surface temperature by 9 K. The greenhouse effect alone, created through pressure-induced absorption by N_2, CH_4, and H_2, would raise the temperature by 21 K. The two effects together yield a net warming of 12 K.

The temperature profile in Titan's troposphere follows a dry adiabat up to about 7 km, above which dT/dz is smaller, but larger than the moist adiabat. The tropopause, at a temperature of 70.4 K, is located at an altitude of 44 km (0.115 bar), above which the temperature rises (the stratosphere) to 186 K at 250 km altitude. In this region of the atmosphere, radiation in the rotation–vibration bands of hydrocarbons cannot escape, and hence inhibits cooling, resulting in high stratospheric temperatures. In contrast, at higher altitudes, in the mesosphere, radiative cooling into space cools the atmosphere more effectively, although the temperature is not as low as predicted. A second temperature minimum (152 K) is encountered at an altitude of ∼500 km, which might mark the mesopause. Between 500 and ∼1000 km, in the thermosphere, solar EUV heating is important; however, the temperature remains relatively low due to radiative cooling by HCN, which is produced in the thermosphere as a byproduct of ionospheric chemistry. The fluctuations in temperature reveal inversion layers, perhaps indicative of dynamic phenomena, such as gravity waves or gravitational tides. In the exosphere, at altitudes above ∼1200 km, the temperature profile is isothermal at ∼160 K.

4.2.2.2 Giant Planets

The various parameters which characterize the thermal structure of the giant planets' atmospheres are summarized in Table 4.1. The observed effective temperatures of Jupiter, Saturn, and Neptune are significantly higher than expected from solar insolation alone. This excess emission, tabulated as an energy balance, implies that Jupiter, Saturn, and Neptune emit roughly twice as much energy as they receive from the Sun. The excess heat escaping from these planets is attributed to a slow cooling off of the planets since their formation, combined (in particular for Saturn) with He differentiation (§6.1.5 and §6.4). For Uranus, the upper limit to excess heat is 14% of the solar energy absorbed by the planet. It is not known why Uranus's internal heat source is so different than those of the other three giant planets.

The thermal profiles for all four giant planets likely follow adiabats in the troposphere; the tropopause occurs at a pressure level between ∼50 and 200 mbar. Tropopause temperatures vary from ∼50 K for Uranus and Neptune to 110 K for Jupiter. At higher altitudes, in the stratosphere, the temperature increases with height. The stratopause is reached at a pressure of ∼1 mbar and temperature of ∼150 K. Above the stratopause lies a near-isothermal region, the mesosphere. At pressures ≲1 μbar, the temperature increases substantially with altitude, but the heat source for the high temperature in the thermospheres of the giant planets is not completely understood.

All four giant planets have similar mesospheric temperatures despite the solar and planetary heat sources differing by factors of ∼30. Methane gas and dust or smog of photochemical origin in the stratosphere absorb at IR and UV wavelengths, while methane (CH_4), ethane (C_2H_6), and acetylene (C_2H_2) are efficient coolants in the stratosphere/mesosphere at wavelengths between 8 and 14 μm. At 150 K, the peak of the Planck function is near 19 μm, and barely overlaps with the 12.2 μm ethane band. If the atmosphere were colder, there would be no overlap, so cooling would be less efficient. As the 12.2 μm transition is closer to the peak of the Planck function for an atmosphere warmer than 150 K, cooling becomes more efficient as the temperature rises. Thus, there is an effective thermostat which keeps the mesosphere close to 150 K. Similarly, H_3^+ prevents the thermospheres of the giant planets from becoming much hotter than ∼1100 K.

4.3 Atmospheric Composition

The composition of a planetary atmosphere can be measured either via remote sensing techniques, or in situ using mass spectrometers on a probe or lander. In a mass spectrometer, the atomic weight and number density of the gas molecules are accurately measured. However, molecules are not uniquely specified by their mass (unless it is measured far more accurately than is currently possible using spacecraft), and isotopic variations further complicate the situation. Hence, atmospheric composition is deduced from a combination of in situ measurements, observations via remote sensing techniques, and/or theories regarding the most probable atoms/molecules to fit the mass spectrometer data. In situ measurements have been made in the atmospheres of Venus, Mars, Jupiter, the Moon, and Titan (and, of course, Earth). These data contain a wealth of information on atmospheric composition, since trace elements and atoms/molecules which do not exhibit observable spectral features, such as nitrogen and the noble gases, can be measured with great accuracy. A drawback of such measurements, besides the cost, is that they are performed only along the path of the probe at one specific moment in time. Landers can measure the composition for a longer time, though again at only one location. So in situ data, though extremely valuable, may not be representative of the atmosphere as a whole at all times.

Spectral line measurements are performed either in reflected sunlight or from a body's intrinsic thermal emission. The central frequency of a spectral line is indicative of the composition of the gas (atomic and/or molecular) producing the line, while the shape of the line contains information on the abundance of the gas, as well

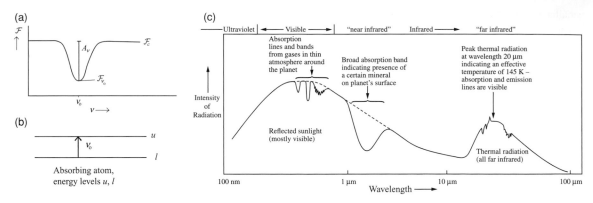

Figure 4.3 (a) Sketch of an absorption line profile. The flux density at the continuum level is $\mathcal{F}_c$; at the center of the absorption line at frequency ν_0, the flux density is $\mathcal{F}_{\nu_0}$. The absorption depth is A_ν. (b) Sketch of the upper, u, and lower, ℓ, energy levels in an atom giving rise to the absorption line in (a). (c) Example of a spectrum from a hypothetical planet with an effective temperature of 145 K. The spectrum is shown from ultraviolet through far-infrared wavelengths. At the shorter wavelengths, the Sun's reflected spectrum is shown. The dashed line shows the spectrum if there were no absorption lines and bands. The spectrum is already corrected for the Sun's Fraunhofer line spectrum. At infrared wavelengths, the planet's thermal emission is detected, where both absorption and emission lines might be present. Note the hyperfine structure of the molecular bands. (Adapted from Hartmann 1989)

as the temperature and pressure of the environment. The strongest spectral lines can be employed to detect small amounts of trace gases (volume mixing ratios of $\lesssim 10^{-9}$ in the giant planet atmospheres) and the composition of extremely tenuous atmospheres (Mercury: $P \lesssim 10^{-12}$ bar). At the same time, high angular resolutions can be obtained ($\lesssim 0.5''$ from the ground via conventional observing techniques, and up to an order of magnitude better via adaptive optics, speckle, and interferometric techniques, and from HST), so the spatial distribution of the gas over the disk can be measured. In addition, the line profile may contain information on the altitude distribution of the gas (through its shape) and the wind velocity field (through Doppler shifts).

In this section we discuss spectra and spectral line profiles, followed by a description of the atmospheric composition of the planets and various satellites. The basic principles of atomic and molecular line transitions were discussed in §3.2.3.

4.3.1 Spectra

Spectra include emission and absorption lines resulting from transitions between energy levels in atoms or molecules (§3.2.3). In astrophysics, one generally sees absorption lines when atoms/molecules absorb photons at a particular frequency from a beam of broadband radiation, and emission lines when they emit photons. The intensity, $\mathcal{F}_{\nu_0}$, at the center of an absorption line is less than the intensity from the background continuum level, $\mathcal{F}_c$: $\mathcal{F}_{\nu_0} < \mathcal{F}_c$ (Fig. 4.3); for emission lines $\mathcal{F}_{\nu_0} > \mathcal{F}_c$. For

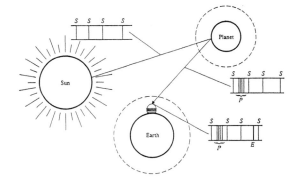

Figure 4.4 A sketch to help visualize the various contributions to an observed planetary spectrum. Sunlight, with its absorption spectrum (indicated by lines S) is reflected off a planet, where the planet's atmosphere may produce additional absorption/emission lines, P. Finally, additional absorption may occur in the Earth's atmosphere before the spectrum is recorded at a telescope, indicated by line E. (Adapted from Morrison and Owen 2003)

planets, we see the effect of atomic and molecular line absorption both in spectra of reflected sunlight (at UV, visible, and near-infrared wavelengths) and in thermal emission (at infrared and radio wavelengths) spectra (Fig. 4.3). Planets, moons, asteroids, and comets are visible because sunlight is reflected off their surface, cloud layers, or atmospheric gases (see schematic in Fig. 4.4). Sunlight itself displays a large number of absorption lines, the *Fraunhofer absorption spectrum*, since atoms in the outer layers of the Sun's atmosphere (photosphere) absorb part of the sunlight coming from the deeper, hotter layers. If all of the sunlight hitting a planetary surface is reflected back into space, the planet's spectrum is shaped like the solar spectrum, aside from an overall Doppler shift induced by the planet's motion (eq. 4.13); the spectrum thus exhibits the solar Fraunhofer line spectrum. Atoms and molecules in a planet's atmosphere or surface may absorb some of the Sun's light at specific frequencies, producing additional absorption lines in the planet's spectrum. For example, Uranus and Neptune are greenish-blue because methane gas, abundant in these planets' atmospheres, absorbs in the red part of the visible spectrum, so primarily bluish sunlight is reflected back into space.

As in the case of the Sun, most of the thermal emission from a planetary atmosphere comes from deeper warmer layers and may be absorbed by gases in the outer layers. In the Sun's photosphere the temperature decreases with altitude, and the Fraunhofer absorption lines are visible as a decrease in the line intensity. Similarly, spectral lines formed in a planet's troposphere are also visible as absorption profiles. In a spectrum of a planet's atmosphere, the optical depth at the center of the line is always largest, (much) larger than in the far wings or with respect to the continuum background. The line profile reflects the temperature and pressure at the altitude probed. In the troposphere, the temperature decreases with altitude, so that lines forming in the troposphere are seen in absorption against the warm continuum background. In contrast, if a line is formed above the tropopause, where the temperature is increasing with altitude, the line may be seen in emission against the cooler background. Thus, whether a spectral line is seen in emission or absorption depends upon the temperature–pressure profile in the region of line formation. Therefore, rather than speaking about emission or absorption lines, in atmospheric sciences one says that spectral lines are seen *in emission* when $\mathcal{F}_{v_0} > \mathcal{F}_c$ and *in absorption* when $\mathcal{F}_{v_0} < \mathcal{F}_c$.

Energy transitions triggered by electronic excitation/de-excitation of atoms and molecules are observed primarily at visible and UV wavelengths. Energy transitions induced by vibrations of atoms can be observed at infrared and submillimeter wavelengths, and those resulting from rotation of molecules at radio wavelengths. The detailed shape of a line profile is determined by the abundance of the element or compound producing the line, as well as the pressure and temperature of the environment. The absorption depth is defined as

$$A_v \equiv \frac{\mathcal{F}_c - \mathcal{F}_{v_0}}{\mathcal{F}_c}, \tag{4.5a}$$

with $\mathcal{F}_c$ the flux density of the continuum background and $\mathcal{F}_{v_0}$ the flux density at the center of the absorption line

(Fig. 4.3). With help of equation (3.56) and $S_v = 0$, this can be written as

$$A_v = 1 - e^{-\tau_v}, \tag{4.5b}$$

where τ_v is the optical depth (eq. 3.53).

Unfortunately, it is not always possible to resolve the line profile. For unresolved lines one measures the *equivalent width*, EW:

$$EW = \int_0^\infty A_v dv = \int_0^\infty (1 - e^{-\tau_v}) dv. \tag{4.6}$$

The equivalent width is equal to the area between the line and the continuum. For an absorption line, EW is equal to the width of a totally black line ($\mathcal{F}_v = 0$), with the same total flux absorbed as by the line. The absorption depth, A_v, of the line is determined by the optical depth τ_v:

$$\tau_v = \tau_{v_0} \Phi_v. \tag{4.7}$$

The optical depth at the center of the line, τ_{v_0}, is determined by the extinction coefficient at the line center, α_{v_0}:

$$\tau_{v_0} = \int_0^L N\alpha_{v_0} dl = N_c \alpha_{v_0}, \tag{4.8}$$

with $N_c \equiv \int N dl$ the *column density* of the absorbing material. The *line shape*, Φ_v, is defined by

$$\Phi_v \equiv \frac{\alpha_v}{\alpha_{v_0}}, \tag{4.9}$$

where α_v corresponds to the extinction coefficient at frequency v (§3.2.3.4). (Note that $\alpha_v = \kappa_v$ when the scattering coefficient $\sigma_v = 0$.)

If the thermal structure in an atmosphere is known, spectral line observations can be used to derive the integrated density of the absorbing material. If the line itself is not resolved but its shape is known, measurements of EW can be used. For an optically thin line ($\tau \ll 1$), EW can be written with help of equations (4.6)–(4.9):

$$EW \approx \int_0^\infty \tau_v dv = N_c \alpha_{v_0} \int_0^\infty \Phi_v dv. \tag{4.10}$$

The equivalent width increases linearly with N_c as long as $\tau_v \ll 1$. When the optical depth increases, the line profile becomes saturated, and the equivalent width cannot continue to increase linearly with the column density. When $\tau \gg 1$, EW is proportional to the square root of N_c (Problems 4.3–4.5). A graph of the equivalent width as a function of column density is called a *curve of growth*, and can be used to determine the abundance of an element from the observed EW. In between the linear and square root regime, the curve of growth is almost flat, i.e., EW is nearly independent of column density.

4.3.2 Line Profiles

The shape of emission and absorption lines is determined by the abundance of the element or compound, and by the pressure and temperature of the environment. Spectral lines are therefore used to determine the abundance of an element or compound at a specific altitude or its distribution over altitude, as well as the thermal structure of an atmosphere over the altitude range probed. Since the observed spectral line center depends upon the observed radial (along the line of sight) velocity of the gas molecules (Doppler shift, eq. 4.13), spectral line observations can also be used to measure the wind velocity field. In this section, we discuss the most common line profiles encountered in planetary atmospheres.

4.3.2.1 Natural Damping: Lorentz Profile

Emission and absorption lines always display some width. The narrowest profile that a line can have is given by the natural damping profile, which results from the finite lifetime of excited states. A natural broadened line profile is given by a *Lorentz* line shape:

$$\alpha_\nu = \alpha \frac{4\Gamma}{(4\pi)^2(\nu - \nu_0)^2 + \Gamma^2}, \tag{4.11}$$

where Γ is the reciprocal of the lifetime of all states giving rise to emission or absorption at frequency ν ($\Gamma \propto 1/\Delta t \propto \Delta \nu$), ν_0 is the central frequency, and α is the spectrally integrated extinction coefficient:

$$\alpha = \int \alpha_\nu d\nu. \tag{4.12}$$

4.3.2.2 Doppler Broadening: Voigt Profile

When an atom has a velocity v_r along the line of sight, the frequency of its emission and absorption lines is Doppler shifted by the amount

$$\Delta\nu = \frac{\nu v_r}{c}. \tag{4.13}$$

The Doppler shift is positive (*blue shifted*) if the atom moves towards the observer, and negative (*red shifted*) if the atom moves in the opposite direction. In an atmosphere, atoms and molecules move in all directions: the radial velocity can generally be expressed by a *Maxwellian velocity distribution*, i.e., the probability $P(v_r)dv_r$ of finding an atom with a radial velocity between v_r and $v_r + dv_r$ is given by

$$P(v_r)dv_r = \frac{1}{\sqrt{\pi}} e^{-(v_r/v_{r_0})^2} \frac{dv_r}{v_{r_0}}, \tag{4.14}$$

where $v_{r_0} = \sqrt{2kT/(\mu_a m_{amu})}$ and the product $\mu_a m_{amu}$ is the molecular mass. Each particle absorbs at frequency

ν in its own reference frame, and hence the absorption profile measured in the inertial frame is Doppler shifted in frequency to $(\nu - \frac{\nu v_r}{c})$. The net effect of such a velocity distribution on the line shape is a broadening of the line profile, which can be computed by convolving the Lorentzian line profile with the velocity (Maxwellian) distribution:

$$\alpha_\nu = \int_{-\infty}^{\infty} \alpha(\nu - \frac{\nu v_r}{c}) P(v_r) dv_r. \tag{4.15}$$

This results in the *Voigt profile* for absorption:

$$\alpha_\nu = \alpha \frac{1}{\sqrt{\pi}\Delta\nu_D} H(a,x), \tag{4.16}$$

with the *Doppler width* $\Delta\nu_D$:

$$\Delta\nu_D \equiv \frac{v_{r_0}\nu_0}{c}. \tag{4.17a}$$

The Doppler width is the full width of the line at half power divided by a factor of $2\sqrt{\ln 2}$ (Problem 4.8). The *Voigt function*, $H(a,x)$, is defined:

$$H(a,x) \equiv \frac{a}{\pi} \int_{-\infty}^{\infty} \frac{e^{-y^2} dy}{(x-y)^2 + a^2}, \tag{4.17b}$$

with $x \equiv (\nu - \nu_0)/\Delta\nu_D \equiv (\lambda - \lambda_0)/\Delta\lambda_D$; $y \equiv v_r/v_{r_0}$; and $a \equiv \Gamma/(4\pi\Delta\nu_D)$. Note that $H(a, x = 0) = 1$ and $\int_{-\infty}^{\infty} H(a,x)dx = \sqrt{\pi}$.

When Doppler broadening dominates, the Voigt profile can be represented schematically by:

$$\alpha_\nu \approx e^{-x^2} + \frac{a}{\sqrt{\pi}x^2}. \tag{4.18}$$

The first term in equation (4.18) represents the core of a Doppler broadened line, which is Gaussian up to a width of $\sim 3\Delta\nu_D$. The second term is due to the wings from the naturally broadened profile.

4.3.2.3 Pressure or Collisional Broadening

In a dense gas, collisions between particles dominate, and perturb the energy levels of the electrons such that photons with a slightly lower or higher frequency can cause excitation/de-excitation. This leads to a broadening of the line profile. The line shape can be expressed by a Lorentzian profile (eq. 4.11), with Γ the reciprocal of the lifetime of all states giving rise to emission/absorption at frequency ν. In a collision-dominated environment $\Gamma = 2/t_c$, where t_c is the mean time between molecular collisions. In contrast to the narrow Lorentz profile, where Γ is the reciprocal of the lifetime in absence of collisions, the collision-broadened profile is sometimes referred to as the *Debye line shape*.

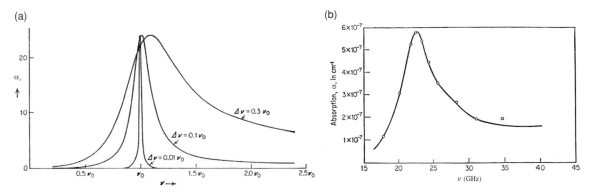

Figure 4.5 (a) Calculations of the Van Vleck–Weisskopf (VVW) line profile for various line widths $\Delta \nu$, where $\Delta \nu$ is the half-width at half-power of the line $(1/(2\pi t_c))$. (From Townes and Schawlow 1955). (b) Comparison of the observed line profile for water vapor in air (open circles: 10 g of H_2O per cubic meter) with a VVW line profile. The VVW profile fits the data quite well, although there is a small discrepancy at the higher frequencies. (Adapted from Becker and Autler 1946)

When pressure increases, the relative time molecules spend in collisions with other molecules increases, and the effects of these collisions become more important. When a molecule undergoes a collision, its geometry is temporarily altered, which changes the central frequency of the range of photons that the molecule can absorb. In an ensemble of molecules, the frequency shifts in each molecular absorption cause a broadening in the line width of each transition. As the line widths increase, individual lines may (partially) overlap one another; when the line widths become comparable to the average frequency spacing between individual lines, the individual line characters are lost. Instead, the molecular absorption can be represented by one broad absorption line. In 1945, Van Vleck and Weisskopf derived an expression for this absorption, based upon a quantum mechanical treatment of absorbing molecules in a collisional environment. The *Van Vleck–Weisskopf* line profile is given by:

In the derivation of their line shape, Van Vleck and Weisskopf assumed that the relative time between collisions is much larger than the time spent in collisions. This assumption breaks down at pressures above $P \gtrsim 0.5$–1 bar, as shown by an apparent mismatch between the theory and observed line profiles. In the mid-sixties, Ben Reuven attempted to improve the line profile through a much more complex quantum mechanical treatment, in which he assumed the molecules to undergo collisions constantly, and by including the effects of coupling between adjacent transitions, which becomes important when the individual lines overlap. Since intermolecular forces are still poorly known, the various coefficients in the *Ben Reuven line profile* are usually determined empirically.

We note that the center of a spectral line is generally formed higher up in an atmosphere than the wings of the line. The observed profile, therefore, may consist of

$$\alpha_\nu = \alpha \left(\frac{\nu}{\nu_0} \right)^2 \times \left(\frac{4\Gamma}{(4\pi)^2(\nu - \nu_o)^2 + \Gamma^2} + \frac{4\Gamma}{(4\pi)^2(\nu + \nu_o)^2 + \Gamma^2} \right). \tag{4.19}$$

Note that the Van Vleck–Weisskopf profile at high frequencies is equal to the Debye (or Lorentzian) line shape. At low frequencies (radio wavelengths) the last term, caused by negative resonance terms, produces an asymmetry in the line shape (see Fig. 4.5), such that the absorptivity in the high frequency tail is larger than in the low frequency tail at equal distances from the center frequency. This asymmetry in the Van Vleck–Weisskopf line shape mimics observed line profiles. The Van Vleck–Weisskopf line profile is widely used to model line profiles in planetary atmospheres if the line width is determined by pressure broadening.

a combination of profiles: a Doppler broadened profile from high altitudes (typically at pressures $P < 0.1$ mbar), a Van Vleck–Weisskopf profile from down to about the 1 bar level, and a Ben Reuven shape from larger depths. The precise pressure level at which the Doppler or pressure broadened profile dominates is determined for each molecular species and planetary atmosphere separately.

4.3.3 Observations

The atmospheric composition of various bodies is given in Tables 4.4–4.6. We specify *abundances* as *volume mixing*

Table 4.4 Atmospheric composition of Earth, Venus, Mars, and Titan.[a]

Constituent	Earth[b]	Venus	Mars	Titan[c]	References
N_2	0.7808	0.035	0.027	~0.95	1, 2, 3, 4
O_2	0.2095	0–20 ppm	0.0013		1, 2, 5, 6
CO_2	400 ppm	0.965	0.953	10 ppb	1, 2, 4
CH_4	2 ppm		10–250 ppb	0.014–0.049[d]	2, 3, 4, 7, 8, 9
H_2O	<0.03[f]	30 ppm[e]	<100 ppm[f]	0.4 ppb	1, 10, 11
H_2O_2	1 ppb		18 ppb		2, 12
Ar	0.009	70 ppm	0.016	28 ppm	2, 3, 4, 8
CO	0.2 ppm	20 ppm[e]	700 ppm	45 ppm	2, 11, 13
O_3	~10 ppm[g]		0.01 ppm		1, 2
C_2H_2	8.7 ppb			189 ppm	14, 15
C_2H_6	13.6 ppb			121 ppm	14, 15
C_3H_8	18.7 ppb			0.6 ppm	14, 15, 16
C_2H_4	11.2 ppb			40 ppm	14, 15
C_3H_4				3.9 ppm	14
NO	<0.5 ppb		3 ppm		2
N_2O	0.35 ppm				1
SO_2	<2 ppb[f]	100 ppm			11, 15
H_2SO_4		1–2.5 ppm			17
H_2	0.5 ppm		10 ppm	0.004	2, 14
HCl		0.1 ppm[h]			11
HF		2 ppb[h]			11
COS		4 ppm			11
He	5 ppm	12 ppm			2
Ne	18 ppm	7 ppm	2.5 ppm	<0.01	2
Kr	1 ppm	0.2 ppm	0.3 ppm	<0.1 ppb	2, 5, 8
Xe	0.09 ppm	<0.1 ppm	0.08 ppm	<0.1 ppb	2, 8

[a] All numbers are volume mixing ratios (ppm: part per million; ppb: part per billion).

[b] There are numerous species in Earth's atmosphere at $\lesssim$ ppb levels in addition to those listed.

[c] Upper limits of 5 ppm for HCN, HC_3N, C_4H_2, C_2N_2, C_6H_6 (ref. 14).

[d] 0.0141: Titan stratosphere; 0.049: Titan troposphere.

[e] H_2O is 0.8 ppm in mesosphere; CO increases to >200 ppm in mesosphere.

[f] Variable.

[g] Varies with altitude (Fig. 4.40).

[h] In the mesosphere.

1: Salby (1996). 2: Chamberlain and Hunten (1987). 3: Yelle (1991). 4: Coustenis and Lorenz (1999). 5: Hunten (2007). 6: Nair *et al.* (1994). 7: Dowling (1999). 8: Niemann *et al.* (2005). 9. Formisano *et al.* (2004); Mumma *et al.* (2004). 10: Coustenis *et al.* (1998). 11: Svedhem *et al.* (2007). 12. Clancy *et al.* (2004). 13: Flasar *et al.* (2005). 14. Waite *et al.* (2005) (at 1200 km altitude). 15: Seinfield and Pandis (2006). 16. Roe *et al.* (2003) (at 90–250 km). 17. Butler *et al.* (2001).

ratios, i.e., the fractional number density of particles (or mole fraction) of a given species in a given volume.

4.3.3.1 *Earth, Venus, and Mars*

Earth's atmosphere consists primarily of N_2 (78%) and O_2 (21%). The most abundant trace gases are H_2O, Ar, and CO_2, but many more have been identified (Table 4.4). The composition of the atmospheres of Mars and Venus is dominated by CO_2, roughly 95–97% on each planet; nitrogen gas contributes approximately 3% by volume; the most abundant trace gases are Ar, CO, H_2O, and O_2. On Venus we also find small amounts of SO_2, H_2SO_4

Table 4.5 Composition of tenuous atmospheres.

Body	Constituent	Number density at surface (cm^{-3})	References	Body	Constituent	Number density at surface (cm^{-3})	References
Mercury	O	4×10^4	1	Io[a]	SO_2	$10^{11}-10^{12}$	7
	Na	3×10^4	1		SO	trace	7
	He	6×10^3	1		Na	trace	8
	K	500	1		O	trace	8
	H	23 (suprathermal)	1		H	trace	9
		230 (thermal)	1		S	trace	10
	Ca	$\sim$30	2		S_2	trace	11
	Mg		17		K	Na/K = 10	12
Moon	He	2×10^3 (day)	1, 3, 4		Cl	trace	13
		4×10^4 (night)			NaCl	trace	14
	Ar	1.6×10^3 (day)	1, 3, 4		H_2S	trace	15
		4×10^4 (night)		Enceladus[b]	H_2O	91%	16
	Na	70	1, 3, 4		CO_2	3%	16
	K	16	1, 3, 4		CO or N_2	4%	16
Pluto	N_2		5		CH_4	1.6%	16
	CO	tracc	5		C_2H_2	<1%	16
	CH_4	trace	5		C_3H_8	<1%	16
Triton	N_2		6				
	CH_4	trace	6				

[a] Only neutral species are listed; see Table 7.3 for ionized species in Io's plasma torus.

[b] Species detected by the INMS on Cassini in the plume emanating from Enceladus's south pole. Abundances are given as volume mixing ratios. Primary mass peaks at 18, 44, 28, and 16 amu were measured, giving some uncertainty to identification.

1: Hunten *et al.* (1988). 2: Bida *et al.* (2000). 3: Sprague *et al.* (1992). 4: Strom (2007). 5: Stern (2007). 6: Stone and Miner (1989). 7: Lellouch *et al.* (1990, 1995). 8: Bouchez *et al.* (2000). 9: Strobel and Wolven (2001). 10: Feaga *et al.* (2002). 11: Spencer *et al.* (2000). 12: Brown (2001). 13: Feaga *et al.* (2004). 14: Lellouch *et al.* (2003). 15: Russell and Kivelson (2001). 16: Waite *et al.* (2006). 17: McClintock *et al.* (2009).

(sulfuric acid), and some hydrogen halides (HCl, HF). Ozone, abundant in Earth's stratosphere, has also been identified on Mars. The differences in composition among the three planets must result from differences in their formation and evolutionary processes, such as differences in temperature, volcanic and tectonic activity, and biogenic evolution.

Figure 4.6 shows coarse thermal infrared spectra of Earth, Venus, and Mars between 5 and 100 μm; all three spectra were taken from space, although with different spacecraft. Each spectrum displays a broad CO_2 absorption band at ~15 μm (wavenumber 667 cm^{-1}). Note that the width of the absorption profile is similar for the three planets, despite vast differences in pressure, since a molecular absorption band consists of numerous transitions (Fig. 4.3). These bands can be used to investigate different depths in the atmospheres. Under clear conditions, when no other absorbers are present, the surface of Earth and Mars is probed in the far wings (continuum) of the band. For Venus, the cloud deck rather than surface is probed. Since the optical depth increases towards the center of the band, higher altitudes are probed closer to the center of the band. Since the profile is seen in absorption, the temperature must decrease with altitude on all three planets. On Earth and Mars, therefore, CO_2 must be present in their tropospheres. On Earth, there is a small emission spike at the center of the CO_2 absorption profile, indicative of some CO_2 in Earth's stratosphere, where, in contrast to the troposphere, the temperature is increasing with altitude. Other prominent features in the terrestrial spectrum are ozone at 9.6 μm (1042 cm^{-1}) and methane at 7.66 μm (1306 cm^{-1}). Note the emission spike

Table 4.6 Atmospheric composition of the Sun and the giant planets.[a]

Gas	Element[b]	Protosolar[c]	Jupiter	Saturn	Uranus	Neptune	References
Major gases							
H_2	H	0.835	0.864	0.88	~0.83	~0.82	1, 2, 3, 4, 5
He	He	0.162	0.136	0.119	~0.15	~0.15	1, 3, 5, 6
Condensable gases							
H_2O	O	8.56×10^{-4}	$>4.2 \times 10^{-4}$	?	?	?	7
	in stratosphere		1.5×10^{-9}	2–20×10^{-9}	5–12×10^{-9}	1.5–3.5×10^{-9}	8
	Galileo, 18–21 bar		4.2×10^{-4}				9
CH_4	C	4.60×10^{-4}	2.0×10^{-3}	4.5×10^{-3}	0.023	0.03	4, 5, 9, 10
NH_3	N	1.13×10^{-4}					
	microwave data[d]		7×10^{-5}	5×10^{-4}	$<1.5 \times 10^{-4}$	$<1.5 \times 10^{-4}$	11, 12
	at pressures		1–2 bar	> few bar	>10 bar	>10 bar	
	Galileo, >8 bar		7×10^{-4}				7
H_2S	S	2.59×10^{-5}					
	microwave data[d]			4.6×10^{-4}	3×10^{-4}	0.001	11
	at pressures			>few bar	>10 bar	>10 bar	
	Galileo, 12–16 bar		7.7×10^{-5}				7
Noble gases							
^{20}Ne	Ne	1.29×10^{-4}	2.0×10^{-5}				7
^{36}Ar	Ar	2.84×10^{-6}	1.6×10^{-5}				7
^{84}Kr	Kr	3.33×10^{-9}	7.6×10^{-9}				7
^{132}Xe	Xe	3.26×10^{-10}	7.6×10^{-10}				7
Disequilibrium species							
PH_3	P	4.29×10^{-7}	5×10^{-6}	6×10^{-6}			13
GeH_4			6×10^{-9}	3.5×10^{-10}			13
AsH_3			2×10^{-10}	2.6×10^{-9}			13
CO[e]			1.3×10^{-9}	1.8×10^{-9}	2.5×10^{-8}	1×10^{-6}	8
CO_2[e]			2.5×10^{-10}	2.5×10^{-10}	4×10^{-11}	4×10^{-10}	8, 14
HCN[e]			detected[f]			2.5×10^{-10}	8
Photochemical species in stratosphere (~1 μbar – 10 mbar)							
CH_3			detection	3×10^{-7}		7×10^{-10}	15
C_2H_2			2–200×10^{-8}	2–30×10^{-7}	1–200×10^{-8}	4–300×10^{-9}	15
C_2H_4			5×10^{-10} – 1×10^{-6}	3×10^{-9}	detection	3–50×10^{-10}	15
C_2H_6			2–9×10^{-6}	3–10×10^{-6}	2×10^{-8}	1–3×10^{-6}	15
C_3H_4			3×10^{-9}	2×10^{-9}	2×10^{-10}	few $\times 10^{-10}$	15
C_3H_8			$<1 \times 10^{-7}$	3×10^{-8}			15
C_4H_2			detection	3×10^{-10}	2×10^{-10}	detection	15
C_6H_6			2×10^{-10}	4×10^{-12}		detection?	15

[a] All numbers are volume mixing ratios (i.e., mole fractions).

[b] The elements O, C, N, S, and P are in the form of H_2O, CH_4, NH_3, H_2S, and PH_3 on the giant planets, respectively.

[c] The protosolar values for the elements are from Grevesse *et al.* (2007).

[d] Values derived via models of radio spectra (e.g., Fig. 4.15).

[e] Species in stratosphere (~1 μbar – 10 mbar).

[f] After the impact of Comet Shoemaker–Levy 9 on Jupiter (§5.4.5).

1: Niemann *et al.* (1998). 2: Atreya (1986). 3: Conrath *et al.* (1989b). 4: Gautier *et al.* (1995). 5: Flasar *et al.* (2005). 6: Burgdorf *et al.* (2003). 7: Taylor *et al.* (2004). 8: Encrenaz (2005). 9: Wong *et al.* (2004). 10: Gautier and Owen (1989). 11: de Pater and Mitchell (1993). 12: de Pater *et al.* (2001). 13: Atreya *et al.* (1999, 2003). 14: Burgdorf *et al.* (2006). 15: Moses *et al.* (2005).

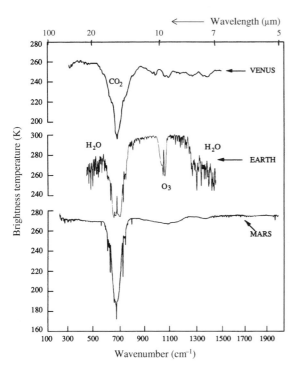

Figure 4.6 Thermal infrared emission spectra of Venus, Earth, and Mars. The Venus spectrum was recorded by Venera 15, the spectrum of the Earth by Nimbus 4, and that of Mars by Mariner 9. (Adapted from Hanel *et al.* 1992)

at the center of the ozone profile, as in the CO_2 absorption band. Numerous water lines are visible in the spectrum, which make the Earth's atmosphere almost opaque in some spectral regions (e.g., at IR wavelengths longer than 20 μm, and at ~5–7.7 μm; note that the CO_2 band prevents transmission near 15 μm). Water lines, though of lesser strength, are also visible in the spectra of Mars and Venus.

Since the emission/absorption lines in planetary atmospheres depend strongly on the temperature structure probed, spectra at different locations may appear very different, even if the concentrations of the absorbing gases are similar. An example is shown in Figure 4.7 for Mars. At mid-latitudes there is a pronounced CO_2 absorption band, as in Figure 4.6. In the polar regions, however, CO_2 shows up in emission. Assuming that the planet's surface is probed in the wings of the lines, the surface temperature can be determined by fitting blackbody curves to the background level. As shown, the background level of the spectrum at mid-latitudes is considerably depressed from a blackbody curve at 280 K, the temperature expected for Mars's surface. This depression is caused by dust in the martian atmosphere, which absorbs sunlight and heats up the atmosphere (by conduction), while it partially shields

the surface from direct sunlight (see Fig. 4.2). The CO_2 feature, therefore, is not as pronounced in this figure as it would be under dust-free conditions. The broad absorption feature is caused by suspended dust particles. The background level of the north pole spectrum can be fitted with a ~140 K blackbody curve, the condensation temperature of CO_2 under martian conditions. Since the atmosphere above the surface is warmer, CO_2 gas appears in emission. The continuum background temperature for the south (almost summer) polar spectrum cannot be fitted with a single blackbody curve; it can be matched quite well using the sum of two curves: one at ~140 K, which covers about 65% of the field of view, and one at 235 K covering the rest. Apparently, CO_2-ice has sublimated away over roughly 1/3 of the surface area probed, while 60–70% was still covered by CO_2-ice. In addition to CO_2, the south polar spectrum also shows several water lines in emission.

Microwave observations of carbon monoxide (CO) on Mars and Venus form an important probe of the thermal structure of their atmospheres. The ^{12}CO line is optically thick, whereas the ^{13}CO line is optically thin. Disk-averaged line profiles for the two planets are shown in Figure 4.8. The ^{12}CO line on Mars is formed high up in the atmosphere, where it is cold; hence the core of the line is seen in absorption against the continuum background. Emission wings appear at either side of the line, where the atmosphere just above the surface is probed. As a consequence of the physical temperature and surface emissivity ($\epsilon < 1$) at the depth of the surface probed, the brightness temperature of the surface is somewhat less than the kinetic temperature of the atmosphere just above it. The wings of the line are therefore seen in emission against the continuum background. On Venus, the continuum emission at millimeter wavelengths arises from within the planet's main cloud deck. The CO lines on Venus are formed in the mesosphere, well above the cloud layers, where the temperature is decreasing with altitude. The CO lines on Venus are therefore also seen in absorption against the warm continuum background.

Observations of different line transitions and/or different isotopes of CO allow retrieval of both the CO abundance and atmospheric temperature structure. The CO abundance in Mars's atmosphere appears to be quite stable over time, in contrast to the thermal structure, which varies much, e.g., in response to the amount of dust in the atmosphere. Microwave observations of the CO line in Venus's atmosphere reveal little variability in the temperature structure, but there are substantial diurnal variations in the CO abundance, as displayed in

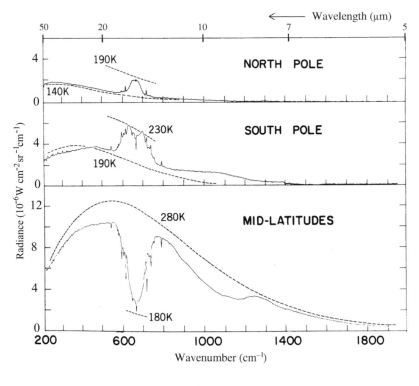

Figure 4.7 Thermal infrared spectra of the poles and mid-latitudes on Mars. The data were taken by the Mariner 9 spacecraft when it was late spring in the southern hemisphere. Blackbody curves at various temperatures (not necessarily best fits) are indicated for comparison. Note that the CO_2 feature is seen in emission in the polar spectra, while in absorption at mid-latitudes. (Adapted from Hanel *et al.* 1992)

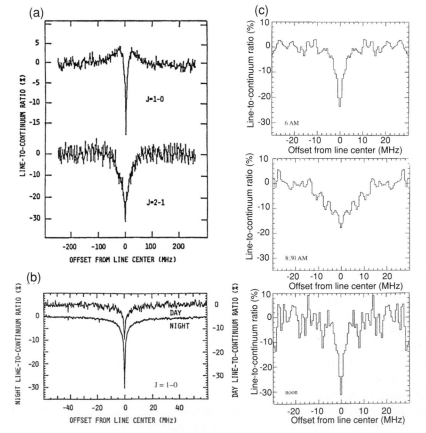

Figure 4.8 Radio spectra of carbon monoxide on Mars and Venus. (a) Full disk radio spectra of Mars in the CO $J = 1-0$ and $J = 2-1$ transitions. (Schloerb 1985) (b) Full disk radio spectra of Venus in the CO $J = 1-0$ line. Spectra of Venus's day and night sides are shown. (Schloerb 1985) (c) Carbon monoxide spectra at different locations on Venus. The locations are indicated by local venusian times. (de Pater *et al.* 1991a)

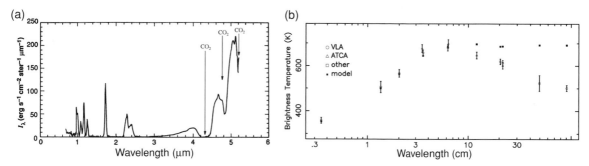

Figure 4.9 (a) A near-infrared spectrum of the dark side of Venus (near the center of the disk), taken by the Galileo spacecraft. At wavelengths $\lambda > 2.8$ μm, Venus's sulfuric acid clouds are opaque and thermal emission from the clouds ($T_b \approx 235$ K) is received. The CO_2 absorption bands reduce the intensity of the blackbody curve at ~4.3, 4.8, and 5.2μm. At several specific wavelengths shortwards of 2.8μm, the clouds are rather transparent, allowing one to probe deeper warmer layers in the atmosphere, shown in the form of emission lines. (Adapted from Carlson *et al.* 1991) (b) A microwave spectrum of Venus. At millimeter wavelengths the planet's cloud layers are probed, while the surface is probed longwards of ~7 cm. (Butler and Sault 2003)

Figure 4.8c. The CO line is deep and narrow on the night side, but broad and shallow on the day side, except at local noon, where the spectrum is similar to that at night. This suggests that the line is formed at lower pressures on the night side and at local noon than at other locations on the day side. Thus, the CO abundance must be largest at high altitudes on the night side and at local noon, despite the fact that CO is formed upon photodissociation of CO_2 on the day side (§4.6.1.2). This apparent discrepancy between expectations and observations may be caused by rapid day-to-night winds, transporting the CO from the day to the night side (§4.5.5.2).

Although Venus's atmosphere and cloud deck are optically thick, one can probe through the clouds at radio wavelengths and, on Venus's night side, at several infrared wavelengths shortwards of ~2.5 μm (Fig. 4.9a). At these wavelengths Venus's atmosphere and clouds are relatively transparent, so deep warm atmospheric layers or the planet's surface are probed. At wavelengths longwards of 2.8 μm, thermal emission from Venus's clouds is observed, and shows strong CO_2 absorption bands (Fig. 4.6). In the absorption bands one probes higher, cooler altitudes (see also §4.4.3.1).

Figure 4.9b shows a microwave spectrum of Venus. Roughly half of the microwave opacity is attributed to CO_2 gas, while gaseous sulfuric acid (H_2SO_4) and sulfur dioxide (SO_2) gas provide the remaining opacity. At wavelengths longwards of ~7 cm, Venus's atmosphere is transparent and its surface is probed. Both the SO_2 and H_2SO_4 abundances are largely confined to the lower atmosphere, within and below the cloud layers. Sulfur dioxide gas may provide a source for Venus's sulfuric acid cloud layers (§4.6.1.2). Both ground-based and spacecraft measurements indicate that the SO_2 abundance near the

cloud tops varies over time, by over an order of magnitude. These variations may be correlated with volcanic eruptions (§5.5.3), or the variations may hint at changes in the eddy diffusion coefficient (§4.7).

The detection of methane gas in Mars's atmosphere by Mars Express and several ground-based telescopes is puzzling. Observed abundances range from 10 ppb up to 250 ppb, and reveal large variations both in time (over time scales of weeks–months) and place. With an average lifetime of 300–600 years, the presence and reported variations in CH_4 require both a strong source and sink. Hot debates continue as to whether or not the detections are real; and if so, what might be its cause: volcanic activity, microbial life, or low-temperature 'serpentinization', a metamorphic process (§5.1.2.3) where (ultra)mafic rocks (e.g., olivine) are converted into serpentinite via hydration and oxidation, thereby releasing methane gas. Potential sinks for CH_4 are oxidation and condensation, both of which are difficult to reconcile with the observations.

4.3.3.2 *Titan*

In 1944, Gerard Kuiper discovered absorption bands of methane gas on Titan, which led to the definitive discovery of an atmosphere surrounding this relatively small body. Since the Voyager flybys we know that CH_4 makes up only a few percent (Table 4.4) of Titan's atmosphere, which, like Earth's, is dominated by N_2 gas, visible in the form of strong UV line emissions. Voyager, and more recently Cassini, further revealed the presence of numerous hydrocarbons and nitriles, which are displayed as prominent emission lines at infrared wavelengths (Fig. 4.10). Since these lines show up in emission, they must form in Titan's stratosphere, where the temperature is rising with increasing altitude.

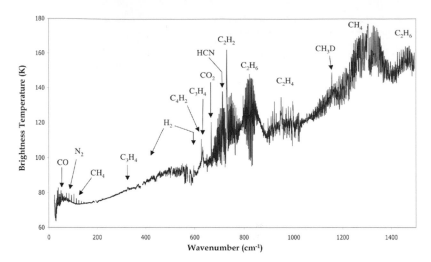

Figure 4.10 Thermal infrared spectrum of Titan, obtained with Cassini/CIRS. Note the numerous emission (i.e., stratospheric) lines from hydrocarbons and nitriles, superposed on a smooth continuum. (Coustenis *et al.* 2007)

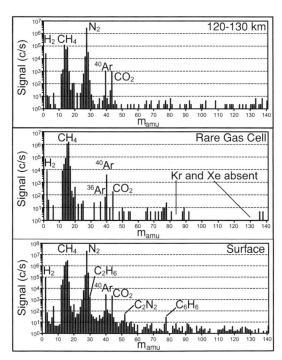

Figure 4.11 Mass spectra (in counts/second) at different altitudes in Titan's atmosphere as measured with the GCMS on the Huygens probe while descending through Titan's atmosphere. Spectra are averaged over 120–130 km altitude (top), rare-gas cell measurements over 75–77 km (middle), and a surface spectrum averaged over 70 minutes after impact (bottom). (Niemann *et al.* 2005)

The Gas Chromatograph Mass Spectrometer (GCMS) on the Huygens probe measured Titan's atmospheric composition while descending through the atmosphere. Results at different altitudes are shown in Figure 4.11. Note the very low abundance of ^{36}Ar, and nondetections

of ^{38}Ar, Kr, and Xe. The implications of these results are discussed in §4.9. Not many heavy hydrocarbons were detected below 130 km altitude, in agreement with expectations (§4.6.1.4). Clear signatures of numerous hydrocarbons and nitriles were, however, detected on the surface upon landing, after heating the material by ~70 K above the background temperature. These gases probably come from haze particles that precipitated out of the atmosphere; the accumulation of such particles over several billion years may have built up a few-hundred-meter thick layer of hydrocarbons on the surface. The methane abundance increased by ~40% upon landing, indicative of liquid CH_4 mixed with surface material.

It is clear from both remote sensing and *in situ* data that Titan's atmosphere presents a richness in atmospheric chemistry that is unique within our Solar System.

4.3.3.3 *Bodies with Tenuous Atmospheres*

All planets and major satellites possess an atmosphere of some kind, though many of these atmospheres are so tenuous that to us these regions are essentially a vacuum. Continuous 'bombardment' by energetic particles (solar wind, magnetospheric plasma) and micrometeorites kick up atoms and molecules from a planet's surface in a process called 'sputtering' (see §4.8.2). The particles kicked up from the surface usually have too low a velocity to escape the body's gravitational field, and form an extended 'corona' or atmosphere around the body. We find such atmospheres around Mercury and the Moon, many of the icy satellites, and Saturn's rings. Other processes which can lead to the formation (or modification) of an atmosphere are volcanoes (e.g., Io), geysers (e.g., Triton, Enceladus), and sublimation of ices (e.g., Mars, Pluto, Triton, Io). Below we briefly summarize observations of tenuous atmospheres around several of the smaller planets and larger satellites. Because of the low gravity of these bodies, the atmospheric scale height is usually large (Table 4.3, Problem 4.1), resulting in a voluminous atmosphere.

Mercury and the Moon

Mercury has an extremely tenuous atmosphere with a surface pressure $\lesssim 10^{-12}$ bar. The atmosphere was discovered when oxygen, helium, and hydrogen atoms were observed using the airglow spectrometer on board the Mariner 10 spacecraft. Later, ground-based telescopes detected sodium, potassium, and calcium atoms, all of which have strong resonance lines at visible wavelengths. The major constituents observed in Mercury's atmosphere are O, Na, and He, with number densities near Mercury's surface of a few thousand atoms cm^{-3} for He, and up to a few tens of thousands of atoms cm^{-3} for O and Na (Table 4.5). Both K and H have been detected at levels of a few hundred atoms cm^{-3} and calcium is another order of magnitude below this. On its first encounter with Mercury in January 2008, the Messenger spacecraft measured *in situ* a mass-per-charge spectrum that suggests ions such as Mg^+, Si^+, Fe^{2+}, and S^+ in Mercury's ionized exosphere and plasma environment. Most intriguing was the identification of water-group ions, as H_2O^+ and OH^+. During the spacecraft's second encounter, 6 October 2008, a firm detection of neutral Mg in Mercury's exosphere was obtained. Sodium, potassium, calcium, magnesium, and oxygen probably come from the planet's surface, after having been kicked up into the atmosphere through sputtering (§4.8.2). In contrast, H and He, major constituents of the solar wind, are probably captured therefrom (§4.8.2). Neutral species that are ejected from the surface with sufficient energy are accelerated by solar radiation pressure and form an extended tail in the antisolar direction (§10.4.2.3), as observed. Sodium emissions (and possibly Mg) show clear maxima at high latitudes, while Ca peaks near the equator. Many species display local enhancements and asymmetries (e.g., north–south and dawn–dusk), indicative of multiple source and loss processes.

Mass and UV spectrometers on the Apollo spacecraft detected He and Ar on our Moon, with a surface density of a few thousand atoms cm^{-3} on the day side and an order of magnitude larger on the night side (Table 4.5). Ground-based spectroscopy revealed Na and K at levels of a few tens of atoms cm^{-3}. As on Mercury, the Moon's atmosphere is in part formed from sputtering by micrometeorites and energetic particles, and by capturing particles from the solar wind.

Pluto and Triton

Pluto and Triton are in many respects quite similar: they are alike in size, and both extremely cold (Pluto: ~40 K for ice-covered regions, up to 55–60 K for darker surface areas; Triton: frost at ~38 K, dark regions at ~57 K). The surface temperature, however, is still high enough to partially sublime N_2, CH_4, and CO_2, constituents detected on the surfaces of both bodies via infrared spectroscopy. The atmospheric abundance of these gases can be calculated from the equations for vapor pressure equilibrium (§4.4.1). When Pluto occulted a 12th magnitude star in 1988, a year prior to its closest approach to the Sun, the gradual rather than abrupt disappearance and later reappearance of the star revealed the presence of an atmosphere, with a surface pressure between 10 and 18 μbar. Because nitrogen ice is about 50 times more abundant than methane and carbon dioxide ice, one might expect nitrogen gas to be the dominant constituent of Pluto's atmosphere. This was indeed confirmed when spectroscopic measurements revealed traces of CH_4 (<1%) and CO (<0.5%) gases in Pluto's atmosphere. Pluto occulted other bright stars in 2002 and 2006. Observations of these occultations revealed significant changes in its atmosphere. Contrary to predictions, the pressure and density in Pluto's atmosphere had increased by a factor of 2, while the atmospheric temperature had remained constant. These changes are likely due to seasonal effects, combined with Pluto's thermal inertia.

Airglow and occultation measurements by the UV spectrometer on board the Voyager spacecraft revealed an atmosphere around Triton, with a surface pressure 14 ± 1 μbar (in 1989). At the time, Triton's south pole was illuminated by the Sun. Triton's atmosphere likely formed in the same way as that of Pluto, from subliming ices and perhaps, as detected on Triton, from geyser activity (§5.5.8). Triton's atmosphere is dominated by N_2 gas, with a trace of CH_4 near the surface (mixing ratio ~10^{-4}). Stellar occultations in the 1990s showed a ~5 μbar increase in atmospheric pressure, despite the less favorable geometry wih respect to the Sun. This increase in pressure suggests either a high thermal inertia for Triton's surface, or changes in Triton's albedo and emissivity as caused by the seasonal insolation cycle.

Io

Io is the only body in our Solar System with an atmosphere dominated by sulfur dioxide. Whereas this gas was first detected above a volcanic hot spot by the Voyager spacecraft, the presence of a global atmosphere was established from the ground at radio wavelengths (Fig. 4.12a). Typical column densities are a few $\times 10^{16}$ cm^{-2}, i.e., a tenuous

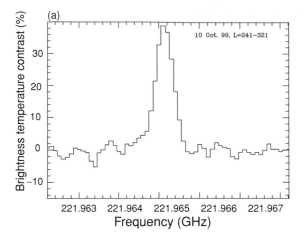 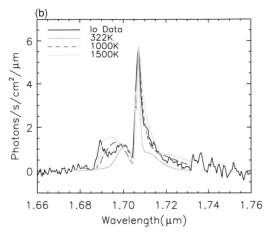

Figure 4.12 Observations of SO_2 and SO on Io taken in September–October 1999. (a) The 221.965 GHz SO_2 line, averaged over its disk, observed with the IRAM telescope on 10 October 1999. (Lellouch *et al.* 2000) (b) Disk-averaged SO emission band at 1.7 μm, taken on 24 September 1999 while the satellite was in Jupiter's shadow. Overplotted are model spectra for an atmosphere at 322, 1000, and 1500 K. The observed photons are probably emitted by SO molecules upon ejection from a volcanic vent (presumably Loki, which was extremely active during that period). (Adapted from de Pater *et al.* 2002)

but collisionally thick atmosphere. Densities tend to be higher (2–5×) in volcanic plumes. The radio data suggest an areal coverage of ∼25% over the leading hemisphere, with an atmospheric temperature of ∼200 K, while the trailing hemisphere appears to be hotter (∼400 K) with a smaller fractional coverage (∼8%). Debates regarding the primary source of Io's atmosphere are still ongoing, although it is clear that both volcanoes and subliming SO_2 frost on Io's surface (global vapor pressure equilibrium) play a role. At night, the temperature drops almost instantaneously and the global SO_2 atmosphere collapses. Since only Io's day-side hemisphere can be viewed from the ground, such events can only be observed during an eclipse, i.e., when Io is in Jupiter's shadow, or by (interplanetary) spacecraft. Infrared spectra taken when Io was in eclipse revealed SO emission in a forbidden (§10.4.3.1) electronic transition at 1.71 μm, at a rotational temperature of ∼1000 K (Fig. 4.12b). This emission band, clearly from a volcanic source, is highly variable over time.

Sulfur monoxide was first detected at radio wavelengths at a global abundance ∼3–10% of that of SO_2, in agreement with photochemical models of SO_2 (§4.6). Such models also predict relatively large quantities of molecular oxygen, as of yet undetected. Both atomic oxygen and sulfur have been detected, as well as Na, K, and Cl. With the detection of NaCl, it appears that the latter atomic species are likely volcanic in origin (see also §5.5.5.1).

Icy satellites

Europa, its surface covered by water-ice (§5.5.5.2), has an oxygen atmosphere. Sputtering processes knock off H_2O molecules from Europa's surface, which, upon dissociation, break up into hydrogen and oxygen. Hydrogen escapes the low gravity field of Europa, leaving an oxygen-rich atmosphere behind. HST measurements suggest excitation of atomic oxygen via electron impact dissociation of molecular oxygen, which has an inferred column abundance of $(1-10) \times 10^{14}$ cm^{-2} ($P \sim$ few picobar). Both Na and K have been detected as well, but the Na/K ratio is higher than that seen on Io and in meteorites. The alkali elements on Europa may originate in a subsurface ocean, and fractionation during transport up to the surface may have led to a relative loss of potassium.

HST observations revealed oxygen on *Ganymede*, with a similar O_2 column density as on Europa, but on Ganymede the oxygen emissions are strongly peaked near the poles. No alkalis have been detected, which despite the low (relative to Europa) sputtering rate, is still indicative of a lower alkali concentration on Ganymede's surface compared to Europa. Galileo observed Lyman α emissions, corresponding to an H column density of ∼10^4 cm^{-2}.

Infrared data from Galileo show a CO_2 atmosphere around *Callisto*, with a column density similar to the O_2 columns on Europa and Ganymede. Although no oxygen has been detected, a Galileo radio occultation experiment revealed an ionosphere with an electron density of

$\sim 2 \times 10^4$ cm^{-3}, which is interpreted as caused by an O_2 column density ~ 100 times higher than that on Europa and Ganymede.

As discussed in §11.3.2, *Enceladus* is the source of Saturn's E ring, a dusty ring which consists predominantly of micron-sized material. Ever since the Voyager flyby of the saturnian system, Enceladus's bright and smooth (young) surface had evoked suggestions of geyser activity supplying the E ring's material. Cassini detected plumes emanating from hot cracks at Enceladus's south pole, and measured their composition *in situ* with the ion and neutral mass spectrometer (INMS) (§5.5.6). The plumes are dominated by H_2O ($\gtrsim 90\%$), with $\sim 3\%$ CO_2, 1.6% CH_4, and 3% of a gas with a mass of 28 amu, which could be either CO and/or N_2. Trace amounts of acetylene and propane are also present.

High signal-to-noise HST spectra at UV wavelengths reveal oxidized gases on several satellites, trapped as microscopic inclusions in the icy surfaces. In particular, SO_2 is seen on Callisto, Ganymede, and Europa, while absorption bands near 280 nm suggest SO_2 and/or OH on several uranian satellites and Triton. Ganymede shows trapped ozone, which like the oxygen emissions is strongest at the poles. Ozone has further been detected on the saturnian satellites Rhea and Dione. Saturn's rings, which consist predominantly of water-ice (§11.3.2.4), are surrounded by a hydroxyl (OH) atmosphere. HST observations of the rings during ring plane crossing (when the rings were seen edge-on) suggest an OH density of ~ 500 molecules cm^{-3}. The Cassini INMS instruments detected atomic and molecular oxygen ions near the A ring.

4.3.3.4 Giant Planets

All four giant planets have deep atmospheres, composed primarily of molecular hydrogen (~ 80–90% by volume) and helium (~ 15–10% by volume). Observations at different wavelengths probe different altitudes, determined by the atmospheric opacity at the particular wavelength used for the observations. For example, at wavelengths $\lambda < 110$ nm, H_2 dominates the absorption, and nanobar pressure levels are probed. At longer UV wavelengths, as opacity by H_2 decreases, absorption by hydrocarbons, Rayleigh scattering, and absorption/scattering by aerosols becomes dominant, and progressively deeper layers are probed. At $\lambda > 160$ nm, one becomes sensitive to the lower stratosphere and upper troposphere. Reflected sunlight from the cloud tops (and above) dominates at visual and near-IR wavelengths (Fig. 4.13), while at longer wavelengths the thermal (blackbody) emission is observed. Thermal IR radiation at ~ 10 μm, sensitive

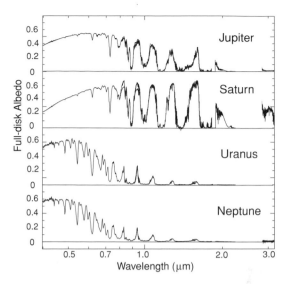

Figure 4.13 Full-disk albedo spectra of Jupiter, Saturn, Uranus, and Neptune. All spectra show strong CH_4 absorption bands. The data used to construct the figure were taken with the 1.5-m ESO observatory at 0.4–1.05 μm by Karkoschka (1994), and with the 3-m IRTF telescope at 0.8–3 μm by Rayner *et al.* (2009).

to hydrocarbons, comes from altitudes at pressure levels between 1 μbar and 1 mbar, while the stratosphere just above the tropopause is probed near 20 μm (H_2 absorption) (Figs. 4.14, 4.37c). Near 5 μm, the gaseous opacity is very low, and one can probe depths of a few bars in cloud-free regions (Figs. 4.35a, 4.37b,e). The continuum opacity at far-IR and (sub)millimeter wavelengths is mainly provided via collision-induced absorption by molecular hydrogen gas, and via absorption/scattering by cloud particles. The latter depends on particle size: If the particles are much smaller than the wavelength, they are relatively transparent, and if they are comparable to or larger than the wavelength, the clouds become more opaque. Deeper layers of the atmospheres can be probed at radio wavelengths. The main source of opacity at radio wavelengths is ammonia gas, which has a broad absorption band at 1.3 cm. At wavelengths longwards of 1.3 cm, the opacity decreases roughly with λ^{-2}, and ever-increasing depths are probed (Fig. 4.15).

The compositions of the giant planets' atmospheres, together with the elemental protosolar mixing ratios, are shown in Table 4.6. If the giant planets had, like the Sun, formed via a gravitational collapse in the primitive solar nebula, one would expect these planets to have a composition similar to the protosolar values quoted in the table (§13.7). But regardless of whether the giant planets formed via a gravitational collapse, or via core accretion followed by gravitational accumulation of gas, one would expect

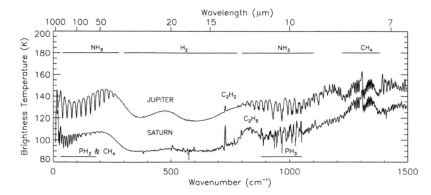

Figure 4.14 Thermal infrared spectra of Jupiter and Saturn taken with CIRS on Cassini. (Courtesy Conor Nixon and NASA/GSFC/UMCP)

the atmospheres of the planets to contain 83–84% H_2 and $\sim$16% He, unless they acquired a late veneer of heavy elements from planetesimals, parts of their cores were mixed upwards, and/or separation of He from H has occurred within their interiors (§§6.1.2.1, 6.4.2). Under equilibrium conditions, the elements O, C, N, and S should be present as water vapor, methane, ammonia, and hydrogen sulfide. Precise estimates of the abundances of these equilibrium species, helium, the noble gases, and isotopic ratios, would help refine formation scenarios of the giant planets.

Although He has only been measured directly using the Galileo probe on Jupiter, its abundance has been estimated by combining thermal infrared spectra and radio occultation profiles (§E.5). As shown in Table 4.6, the detailed mixing ratios for He vary from planet to planet. Helium is similar in abundance to the protosolar values on Uranus and Neptune, but appears to be depleted on Jupiter (by $\sim$20%) and Saturn (by $\sim$35%). This depletion is attributed to the immiscibility of He in metallic hydrogen at pressures of 1–3 Mbar, resulting in a 'raining out' of He towards the core (§6.4).

Although the trace gases H_2O, CH_4, NH_3, and H_2S have energy transitions at visible, UV, and/or IR wavelengths, it has proven difficult to determine the global abundances of these species. All these gases, with the exception of CH_4 on Jupiter and Saturn, condense in the upper troposphere of the giant planets (§4.4). Therefore, to obtain representative values of their mixing ratios, one needs to probe below the cloud layers, which is quite impractical at most wavelengths. As a result, Table 4.6 contains many question marks or empty spaces. Reliable measurements have only been obtained on all four planets for CH_4; all four of these trace gases have been measured on Jupiter *in situ* by the Galileo probe. The probe, however, entered an *infrared hot spot*, a very 'dry' region in Jupiter's atmosphere, which is far from representative for the planet as whole. The

methane abundance on all four planets has been derived from spectra in reflected sunlight (Fig. 4.13), and/or from thermal infrared observations (Fig. 4.14), and, for Uranus and Neptune, in an indirect way via radio occultation measurements (§E.5). On Jupiter, the abundance is about 4 times higher than the protosolar value, and the enhancement is larger for planets at larger heliocentric distances. The Galileo probe made *in situ* measurements of Jupiter's atmosphere down to pressures of 15–20 bar. At these deep levels CH_4, NH_3, H_2S, Ar, Kr, and Xe were detected at levels several (3–6) times the protosolar value. As expected from their immiscibility in metallic hydrogen, He and Ne were below solar. Somewhat puzzling is the low mixing ratio of H_2O ($\sim$1/2 protosolar O), while all other heavy elements are enhanced. It is likely that H_2O is also enhanced at much deeper layers, and that the apparent depletion is caused by dynamics (downdrafts) in the infrared hot spot into which the probe made its descent.

Altitudes within and well below the visible cloud layers can be probed at radio wavelengths, where the opacity is controlled by NH_3 gas, and, to a lesser extent, H_2O vapor. On Uranus and Neptune, absorption by H_2S is noticeable, and perhaps also by PH_3. Microwave spectra for Jupiter and Uranus are shown in Figure 4.15, with various model calculations superposed. The absorption lines are pressure-broadened to such an extent that the radio measurements form a quasi-continuum dataset. Assuming the thermal profile in a giant planet's troposphere is adiabatic, the microwave data can be inverted to yield an altitude profile of NH_3 gas. If the atmosphere is in thermochemical equilibrium, the microwave data can also provide, indirectly, an estimate for the mixing ratio of H_2S gas (§4.4.3.4). For example, the apparent large depletion of NH_3 gas on Uranus and Neptune has been explained through the formation of clouds of NH_4SH. If there is enough H_2S, ammonia gas can be almost entirely removed from the atmosphere above this cloud layer.

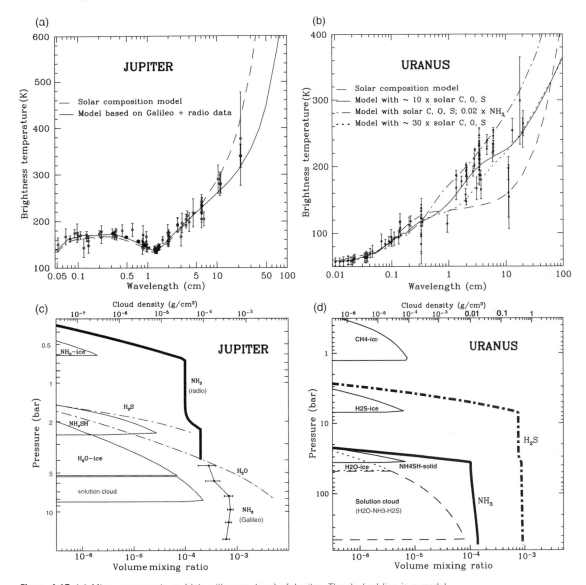

Figure 4.15 (a) Microwave spectrum (dots with error bars) of Jupiter. The dashed line is a model atmosphere in thermochemical equilibrium with a composition that is approximately solar. The solid line is a model atmosphere with abundances ~5 times solar for H_2O and H_2S in Jupiter's deep atmosphere, and an NH_3 profile that agrees with the Galileo probe and radio measurements (panel c). Absorption by clouds was ignored, and an NH_3 humidity of 1% was adopted for the ammonia ice cloud (after de Pater *et al.* 2001, 2005, and Gibson *et al.* 2005). (b) Microwave spectrum of Uranus. The dashed line is a model atmosphere in thermochemical equilibrium with a composition that is approximately solar. The solid and dotted lines are thermochemical equilibrium models with enhanced (factor 10 and 30, respectively) abundances of H_2O, H_2S, and CH_4 in the deep atmosphere. NH_3 was kept at the solar N value in the deep atmosphere (panel d; §4.4.3.4). The top curve is for a solar composition atmosphere, except that NH_3 was decreased by a factor of 50. Absorption by clouds was ignored in all calculations. The spread in data points at 3.5 cm (taken from Klein and Hofstadter 2006) is caused by variations in Uranus's brightness temperature over the course of a uranian year (see Fig. 4.38d) (after de Pater and Mitchell 1993). (c) Sketch of the cloud layers in Jupiter's atmosphere, as suggested by model atmosphere calculations (panel a). The horizontal axis indicates the cloud density (top) and gas abundance (bottom); the vertical axis the pressure. The cloud densities are maximum values, based upon thermochemical equilibrium calculations. The actual densities are lower due to precipitation (after de Pater *et al.* 2001). (d) Sketch of the cloud layers in the atmosphere of Uranus, where the abundances of CH_4, H_2S, and H_2O in the planet's deep atmosphere were enhanced by a factor of 30 above solar.

Whether atmospheres are in thermochemical equilibrium, however, is not clear. A comparison of radio observations with *in situ* measurements by the Galileo probe on Jupiter shows that the observed abundance profiles of NH_3 and H_2S cannot be reconciled with simple chemistry, but are probably strongly influenced by dynamics. Hence, the H_2S numbers listed in Table 4.6 should be taken with caution.

Thermal infrared spectra (Fig. 4.14) show emissions from CH_4 on all four planets, a clear sign that methane gas is also present in the stratospheres of these bodies. Although this may be expected on Jupiter and Saturn, where the temperature in the troposphere never gets low enough for CH_4 to condense out, one would expect the low temperatures in the upper tropospheres of Uranus and Neptune to form an effective *cold trap* to prevent CH_4 from rising up into the stratosphere (§4.4.3.4). Solar photons dissociate CH_4 in the stratosphere, and subsequent chemical reactions lead to the formation of hydrocarbons (§4.6.1.3). As shown in Figure 4.14, emission lines of acetylene (C_2H_2) and ethane (C_2H_6) have been detected on all four planets, and are particularly strong on Neptune. Emissions from more complex hydrocarbons have been identified as well (Table 4.6).

In addition to hydrocarbon emissions, thermal spectra show many other absorption and emission features. For example, the prominent broad absorption features at 28.2 μm (354 cm^{-1}) and 16.6 μm (602 cm^{-1}) in Figure 4.14 are caused by collision-induced absorptions of molecular hydrogen. Ammonia absorptions are prominent in the jovian and saturnian spectra; as expected, these gases are not seen in spectra of Uranus and Neptune. As summarized in Table 4.6 and identified in Figure 4.14, several gases have been detected which would not have been present if the atmospheres were in thermochemical equilibrium. For example, Saturn's spectrum is dominated by absorption features of phosphine (PH_3) at 8.3–11.8 μm (850–1200 cm^{-1}). One of these lines (8.9 μm or 1118 cm^{-1}) is also clearly visible in Jupiter's spectrum, but the rest of the PH_3 lines on Jupiter are masked by NH_3 lines. Absorption bands of germane (GeH_4) and arsine (AsH_3) are visible in areas on Jupiter and Saturn that are relatively clear of clouds and absorbing gases. In addition to these tropospheric constituents, CO, CO_2, H_2O, and HCN have been detected in the stratospheres of several planets (Table 4.6). Prominent emission lines of CO and HCN were first detected in the early 1990s in the 1 and 2 mm wavelength bands on Neptune, with abundances ~ 1000 times higher than predicted from thermochemical equilibrium models.

Disequilibrium species in a planet's stratosphere could have been brought up from below (fast vertical transport, §4.7.2), or have fallen in from outside. Water vapor, detected in the stratospheres of all four planets, cannot have come up from their deep atmospheres because the temperature in the upper troposphere is much too low. It is much more likely H_2O molecules have come in from outside. All four planets are surrounded by rings and moons; these, as well as interplanetary dust and meteoritic material, probably supply the water to these planets' upper atmospheres. The CO abundance is somewhat higher in the stratospheres of Jupiter and Neptune than in their tropospheres, which, one might argue, suggests that CO, at least in part, may fall in from the outside. The relatively high HCN abundance in Neptune's stratosphere may be explained via *in situ* formation, from CH_3 radicals with nitrogen atoms, where the latter may come from Triton, or from N_2 in Neptune's deep interior. Alternatively, both the large CO and HCN abundances in Neptune's stratosphere might be indicative of a large cometary impact sometime in the past.

4.4 Clouds

Earth's atmosphere contains a small amount of water vapor (Table 4.4). The air is said to be *saturated* if the abundance of water vapor (or, in general, any condensable species under consideration) is at its maximum vapor partial pressure. Under equilibrium conditions, air cannot contain more water vapor than indicated by its *saturated vapor pressure curve*, sketched in Figure 4.16a. Water at a partial pressure of ~ 10 mbar in a parcel of air to the right of the solid curves (e.g., point A) is all in the form of vapor, while liquid water is present in parcels between the two solid curves (e.g., at point B), and we find water-ice on the left side of the solid curves (e.g., at point C). The solid lines indicate the saturated vapor curves for liquid (on the right) and ice (on the left). Along these lines evaporation (called *sublimation* if ice transforms directly into gas) is balanced by condensation (sometimes referred to as *deposition* if gas condenses directly in the solid form). The symbol T_{tr} indicates the triple point of water where ice, liquid, and vapor coexist.

Consider a parcel of air at point A, with a vapor pressure of 10 mbar and a temperature of 15 °C. If the parcel is cooled, condensation starts when the solid line is first reached, at point D. Upon further cooling, the partial vapor pressure decreases along the curve D–D'–T_{tr}. At 3 °C (point D') the water vapor pressure is 7.6 mbar. Further chilling to below 0 °C results in the

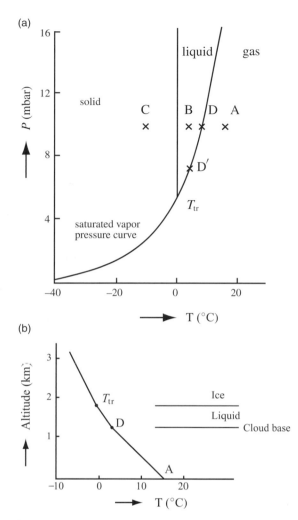

(a)

(b)

Figure 4.16 (a) Saturation vapor pressure curve for water. The vertical axis indicates the partial pressure for H_2O vapor at the temperature (in °C) indicated along the horizontal axis (see text for a detailed discussion). (b) Idealized sketch of the temperature structure in the Earth's atmosphere. In the lower troposphere, the air follows a dry adiabat. A wet air parcel rising up through the atmosphere starts to condense when the water vapor inside the air parcel exceeds the saturated vapor curve (at point D in panel a). The temperature profile in the atmosphere follows the wet adiabat between D and T_{tr}, and changes again at T_{tr}, when the ice line is crossed.

formation of ice, where ice first forms upon crossing the second solid line. As an example, the vapor pressure at $-10°C$ (point C) is 2.6 mbar.

The numerous water droplets and/or ice crystals that form this way make up clouds. Clouds on other planets are composed of various condensable gases, e.g., in addition to H_2O, we find NH_3, H_2S, and CH_4 clouds on the giant planets, and CO_2 clouds on Mars. Clouds on Venus

consist of H_2SO_4 droplets. Clouds may modify the surface temperature and atmospheric structure considerably by changing the radiative energy balance. Clouds are highly reflective; thus they decrease the amount of incoming sunlight, cooling off the surface. Clouds absorb incoming sunlight, and thus heat the immediate environment. Clouds can also block the outgoing infrared radiation, increasing the greenhouse effect (Fig. 4.2). The thermal structure of an atmosphere is influenced by cloud formation through these types of radiative effects, and through the release of latent heat of condensation. Clouds further play a major role in the meteorology of a planet, in particular in the formation of storm systems (§4.5). Cloud formation is discussed in the following subsections.

4.4.1 Wet Adiabatic Lapse Rate

Figure 4.16b shows an idealized sketch of the temperature structure in the Earth's troposphere. The temperature gradient in the lower troposphere is ~ 10 K km^{-1}, while the pressure drops according to equation (4.1). The symbols A, D, and T_{tr} correspond roughly to the same points as in Figure 4.16a. A moist parcel of air rising upward in the Earth's troposphere cools adiabatically as it rises from A to D. At point D, the air parcel is saturated, and liquid water droplets condense out. The condensation process releases heat: the *latent heat of condensation*. This decreases the atmospheric lapse rate, as shown by the change in slope from D to T_{tr}. At T_{tr} the atmospheric temperature is $0°C$ (273.16 K), and water-ice forms, reducing the lapse rate even more, because the latent heat of fusion is added to that of condensation.

The saturated vapor curve is calculated assuming a balance between evaporation and condensation between the vapor and the liquid or solid. The calculations are based upon a flat interface between the vapor and the liquid/ice. If the surface is curved, like for small droplets, molecules evaporate/sublimate more readily, in part because of an increase in surface tension. The saturated vapor curve is therefore somewhat higher above tiny droplets. Air is therefore often *supersaturated*, unless condensation nuclei are present. On Earth, air is often supercooled by up to 20 K before clouds form, unless there are condensation nuclei present at lower altitudes.

Relative humidity is the ratio of the measured partial pressure of the vapor relative to that in saturated air. The relative humidity in terrestrial clouds is usually $100 \pm 2\%$, although considerable departures from this value have been observed. The humidity can be as low as 70% at the edge of a cloud, caused by turbulent mixing or entrainment of drier air. In the interior layers, the humidity can be as high as 107%.

The saturation vapor pressure at temperature T is given by the *Clausius–Clapeyron equation of state*:

$$P = C_L e^{-L_s/(R_{gas}T)}, \tag{4.20}$$

where L_s is the latent heat, R_{gas} the gas constant, and C_L a constant. The thermodynamic equations discussed in §3.2.2.2 (eqs. 3.32) are altered slightly by the inclusion of the release of latent heat:

$$c_V dT = -PdV - L_s dw_s, \tag{4.21a}$$

$$c_P dT = \frac{1}{\rho} dP - L_s dw_s, \tag{4.21b}$$

with w_s the mass of water vapor that condenses out per gram of air. The temperature gradient in a convective atmosphere becomes

$$\frac{dT}{dz} = -\frac{g_p}{c_P + L_s dw_s/dT}. \tag{4.22}$$

The latent heat is effectively added to the specific heat c_P, resulting in a decrease from the dry adiabatic lapse rate. The lapse rate in the presence of clouds is commonly referred to as the *wet adiabatic lapse rate*. On Earth (in the tropics), the wet lapse rate is 5–6 K km^{-1} (Problem 4.14), slightly more than half the dry rate. Note that the wet adiabatic gradient can never exceed the dry lapse rate. Values for L_s and C_L for various gases can be found in, e.g., Atreya (1986) and the *CRC Handbook of Physics and Chemistry* (Lide 2005).

4.4.2 Clouds on Earth

A parcel of air in the Earth's troposphere can be lighter than surrounding air because it is warmer or because it is humid compared to its surroundings (Problem 4.12). Such a parcel convects upwards (§3.2.2 and §4.5.4.2). When the temperature of this rising air drops below the condensation or freezing temperature of water vapor, water droplets or ice crystals form. The tiny (typically up to 10 μm across) water droplets and/or ice crystals form a cloud. Clouds form primarily in the troposphere, that region in the atmosphere where the temperature decreases with altitude. Since condensation/freezing occurs at a particular temperature, the bottom of a cloud is usually flat.

4.4.2.1 *Shape*

Clouds come in a large variety of shapes (Fig. 4.17), determined primarily by the degree of stability of atmospheric air. When the air is stable, extensive flat (stratified) cloud layers are formed, referred to as *stratus* in the lower troposphere, or *altostratus* in the middle troposphere. *Cumulus*

clouds form in unstable air. Since individual packets of air are rising, cumulus clouds look puffy. Small puffy clouds form in shallow layers of unstable air, whereas towering cumulus clouds, which can produce thunderstorms, form in deep layers of unstable air. *Stratocumulus* (stratified cumulus) clouds sometimes occur in the lower troposphere. *Cirrus* and/or *cirrocumulus* are found at high tropospheric altitudes, above ~6 km. These clouds consist of ice crystals, and sometimes show *mares' tails*, which are long, extended patterns formed by falling ice crystals. Clouds cannot rise through the tropopause into the stratosphere, since convection stops when the temperature rises with altitude. If a strong convective storm pokes through the tropopause into the stratosphere, air flows away laterally and forms an *anvil* shape.

4.4.2.2 *Formation and Precipitation*

Clouds are often related to precipitation: droplets and ice crystals fall under the influence of gravity, while atmospheric viscosity resists free fall. When these forces balance, the water droplets fall at the *terminal velocity*, v_∞. For particles that are larger than the mean free path of the gas molecules, the terminal (or *sedimentation*) velocity is given by equating aerodynamic drag to gravity (§2.7.6):

$$v_\infty \approx \frac{2g_p R^2 \rho_d}{9\nu_v \rho_g}, \tag{4.23}$$

where ρ_d and ρ_g are the density of the particle and the atmosphere, respectively, R the particle's radius, and $\nu_v \rho_g$ is the dynamic viscosity of the atmosphere (ν_v is the kinematic viscosity). The terminal velocity is proportional to the size (R^2) of the rain drops. Particles much smaller than the mean free path of the gas molecules feel less drag, while the fall velocity of typical (few mm in size) rain drops is strongly affected by turbulence, so that eq. (4.23) should be modified (see, e.g., Jacobson 1999).

The simplest way for a cloud droplet to form is through direct condensation of vapor, where several H_2O molecules collide by chance (*homogeneous nucleation*). Such droplets usually evaporate immediately (§4.4.1). Instead, droplets form via condensation on atmospheric aerosols, or cloud condensation nuclei (*heterogeneous nucleation*). After a droplet has formed, it grows relatively rapidly (~0.1 μm s^{-1}) through condensation up to a radius of ~20 μm, which is still much smaller than droplets that hit the ground. Growth to larger sizes occurs through collisions with smaller particles, while the droplet falls. The rate at which a droplet grows is proportional to its projected surface area and its velocity with respect to other droplets (e.g., §13.5). A few droplets grow much faster than the rest of them, and reach the ground in the form

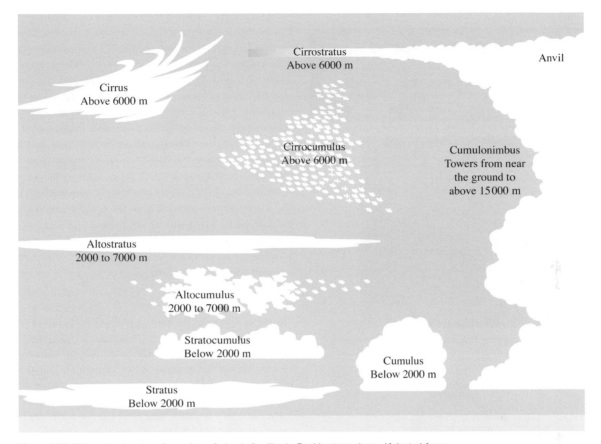

Figure 4.17 Schematic drawing of a variety of clouds familiar in Earth's atmosphere. (Adapted from Williams 1992)

of rain, hail, or snow. Although collisions help droplets to grow, they also limit the size of precipitating droplets to a few millimeters, since larger droplets typically break up during collisions.

4.4.3 Clouds on Other Planets

Most terrestrial clouds are made of water droplets and ice crystals. Even though water vapor is a minor constituent of the Earth's atmosphere, it is the dominant, though not only, constituent of the clouds. Clouds on other planets also form from trace gases, each of which freezes or condenses out when its saturated vapor pressure at the atmospheric temperature is exceeded.

4.4.3.1 Venus

At visible, UV, and most infrared wavelengths, Venus's clouds are so thick ($\tau \gg 1$) that the planet's surface is completely hidden. The cloud particles consist of sulfuric acid, H_2SO_4, with some contaminants. At visible wavelengths, Venus appears as a bright yellow featureless disk. Distinct markings with an overall V-shaped morphology

are discerned at UV wavelengths (Fig. 4.18). Although the precise composition of UV absorbers in the cloud layers has yet to be identified, one expects sulfur- and chlorine-bearing gases together with hazes to dominate the UV absorptions. Venus's main cloud layers span the altitude range between 45 and 70 km, with additional hazes up to 90 km and down to 30 km. Based upon the microphysical properties of the cloud particles, the main cloud deck can be subdivided into a lower, middle, and upper cloud layer. Particle sizes range from a few tenths of a micrometer up to $\sim$35 μm. Most particles, however, are about 0.5 μm (dominant in upper cloud) or 1 μm in radius; in addition the middle and lower clouds contain some particles with radii of $\sim$4 μm. The droplets are formed at high altitudes (80–90 km), where solar UV light photodissociates SO_2 (§4.6.1.2), and chemical reactions (with, e.g., H_2O) lead to production of H_2SO_4. When the droplets fall, they grow. However, since the temperature increases at lower altitudes, the droplets tend to evaporate below 45 km. Below 30 km altitude the temperature is too high for the droplets to exist.

(a)

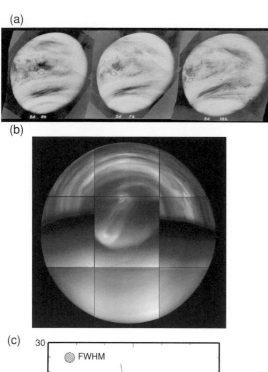

(b)

(c)

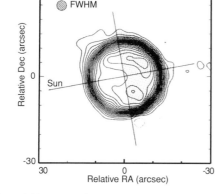

Figure 4.18 Images of Venus at different wavelengths. (a) Three images of Venus taken in ultraviolet light 7 hours apart. A right-to-left motion of the cloud features can be seen (indicated by the tiny arrow). (Mariner 10/NASA: P14422) (b) A mosaic of the south pole of Venus on 16 May 2006. The night side (at the top) is constructed from images at 1.74 μm, sensitive to the cloud deck at ~45 km altitude. The bright regions reveal thermal radiation leaking through from below. The day side (at the bottom) consists of images at 480 nm, and shows the cloud tops at ~65 km altitude. Part of the image at the center was obtained at 3.8 μm, and shows the double vortex at the south pole, at an altitude of ~60 km, surrounded by a collar of 'cold' air. (Venus Express/VIRTIS/ESA, R. Hueso, University of Bilbao; ID number: SEMN8273R8F) (c) A radio image at a wavelength of 3 mm. On the night side one probes deeper, warmer layers than on the day side, presumably because the relative humidity is lower. This image represents an average over several Earth days. The spatial resolution is indicated by the beam width (FWHM). (de Pater *et al.* 1991)

At infrared (1–22 μm) and millimeter (~3 mm) wavelengths the spatial distribution of Venus's brightness temperature is inhomogeneous (see Fig. 4.18). At millimeter wavelengths one probes down into Venus's cloud layers, where the night side appears to be ~10% brighter than the day side. This has been attributed to spatial variations in the H_2SO_4 vapor (i.e., cloud humidity). At near-infrared wavelengths, one is sensitive to Venus's thermal emission on the night-side hemisphere, while reflected sunlight dominates the emission on the day side. Bright markings in the night-side images reveal inhomogeneities in the venusian cloud deck, since at places where the clouds are thinner, warm thermal emission from below the clouds leaks through (§4.3.3.1). At mid-infrared wavelengths one measures the temperature above Venus's cloud layers. At these wavelengths the poles are generally bright, surrounded by a colder 'collar' (Fig. 4.18b).

4.4.3.2 Mars
The surface pressure on Mars is well below the saturated vapor pressure curve for liquid water, so water is present either as a vapor or as ice. The small amount of water vapor in the martian air forms water-ice clouds at altitudes of ~10 km above the equatorial regions. Such clouds are often seen near the martian volcanoes (Fig. 5.57a). At higher altitudes, typically near ~50 km, the temperature is low enough (~150 K) for CO_2-ice clouds to form.

4.4.3.3 Titan
Based upon the temperature structure and CH_4 mixing ratio of several percent, one might expect the formation of methane clouds in Titan's middle troposphere. However, like Titan's surface, such clouds would be hidden by the smog of hydrocarbons in Titan's stratosphere (§4.6.1.4). It is possible to probe through the smog at infrared wavelengths away from the methane absorption bands to 'see' the surface and any low-altitude clouds (§5.5.6.1). In the mid 1990s, disk-averaged infrared spectral measurements indicated, indirectly, the presence of methane clouds in the troposphere, covering ≲1% of Titan's total surface area, although occasionally large storms covered up to ~10%. A decade later, images obtained with the 10 m Keck telescope equipped with adaptive optics (§E.6) revealed clouds near Titan's south pole, where it was summer at the time (Fig. 4.19). Given the Saturn system's obliquity, the south pole is the warmest region on Titan during summer solstice, which may drive convection with cloud formation here.

Analysis of the methane relative humidity profile at the Huygens probe entry site, in combination with ground-based images, suggests that Titan is covered globally by

(a)   (b)   (c)   (d)

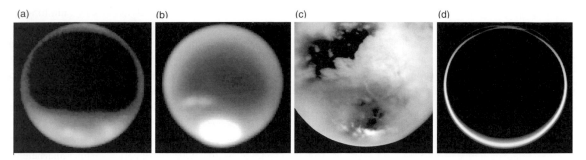

Figure 4.19 (a) Clouds on Titan imaged with the adaptive optics system on the Keck telescope on 21 December 2001. The data were taken with a filter that probes just the troposphere (2.111–2.145 μm). Titan's disk is limb-brightened, due to aerosol scattering in its atmosphere. Several small clouds are visible near Titan's south pole. These clouds are distinct from the haze, which is ubiquitous at latitudes south of −45°. (Adapted from Roe *et al.* 2002) (b) A giant storm at Titan's south pole imaged with Keck on 2 October 2004, using the same set-up as in panel (a). This image also shows clouds at southern mid-latitudes. (Courtesy W.M. Keck Observatory Adaptive Optics Team) (c) Clouds on Titan's south pole imaged by Cassini on 25 October 2004, at 0.9 μm, with a pixel size of 4.2 km/pixel. (NASA/JPL/Space Science Institute, PIA06125) (d) High-altitude hazes are clearly visible when Titan is illuminated from behind. This detached haze layer is visible over the entire globe, and reveals an unusual (not understood) structure over the northern hemisphere. The image was taken at a wavelength of 460 nm and a resolution of ~8 km/pixel. (NASA/JPL/Space Science Institute, PIA06184)

an optically thin methane-ice cloud at 25–35 km altitude. A persistent light methane drizzle below this cloud is present at least near the Xanadu mountains. Distinct clouds have further been seen regularly at southern mid-latitudes, near −40°. These may be confined latitudinally by Titan's geography and/or by its global atmospheric circulation pattern (§4.5.5.4). Large (but short-lived) clouds have been observed over tropical latitudes near the epoch of saturnian equinox. The Cassini spacecraft identified a large cloud of ethane over Titan's north (winter) pole in the upper troposphere, and some smaller (presumably methane) clouds at lower altitudes, which have been hypothesized as lake-effects. At much higher altitudes, in the stratosphere and mesosphere, distinct haze layers are present (Fig. 4.19d; §4.6.1.4).

4.4.3.4 *Giant Planets*

The atmospheres of the giant planets are dominated by H_2 and He. As on Earth, Venus, and Mars, clouds result from the condensation of trace gases, in particular CH_4, NH_3, H_2S, and H_2O. Unfortunately, the compositions of clouds are not easily determined by remote sensing techniques. Nor can we 'see' clouds directly below the upper cloud deck, unless there is a clearing in the upper cloud deck or significant convective penetration of the lower clouds. Our discussion about the cloud layers on these planets is therefore largely theoretical, based upon the (usually) known composition of the atmospheres, together with laboratory measurements of the saturated vapor pressure curves of

the condensable gases. Model calculations of the giant planet atmospheres suggest the presence of the following cloud layers (see Fig. 4.15, and Table 4.6):

(1) In each planet's deep atmosphere H_2O forms an *aqueous solution cloud*: liquid water with NH_3 and H_2S dissolved into it. This cloud is expected to form at temperatures above 273 K. The cloud forms when the water partial pressure exceeds the saturated vapor pressure. Hence, the precise altitude of the bottom or base level of the cloud depends upon the mixing ratio of water. If water is present at an abundance five times the solar oxygen value, the base levels of the aqueous solution clouds on Jupiter and Saturn are at ~305 K and 325 K, respectively; on Uranus and Neptune the water abundance is more likely enhanced by a factor of 10–30 above solar oxygen, resulting in a base level for the solution cloud near 400–450 K. Note that the aqueous cloud cannot exist at temperatures over 650 K, the *critical point* of water; above this temperature there is no first-order phase transition between gaseous and liquid H_2O. It becomes a supercritical fluid, a phase that is neither gas nor liquid.

(2) At $T \leq 273$ K, water-ice forms. Thus, the aqueous solution cloud is topped off with a water-ice layer.

(3) At $T \sim 230$ K, NH_3 and H_2S condense via a heterogeneous reaction:

$$NH_3 + H_2S \rightarrow NH_4SH. \qquad (4.24)$$

The precise altitude or temperature/pressure level of the base of the NH_4SH cloud layer depends upon both the

NH$_3$ and H$_2$S abundances. The gas that is least abundant of the two, NH$_3$ or H$_2$S, is effectively removed from the atmosphere through this reaction (eq. 4.24); hence either NH$_3$ or H$_2$S is present above the NH$_4$SH cloud. This may explain why H$_2$S, in contrast to NH$_3$, has never been detected on the planets Jupiter and Saturn via remote sensing techniques (Fig. 4.14). H$_2$S gas has been detected *in situ* at deeper levels of Jupiter's atmosphere by the Galileo probe.

(4) At temperatures close to 140 K, NH$_3$ or H$_2$S, whichever remains, condenses into its own ice cloud. Given our best estimates for the atmospheric composition of the four giant planets (Table 4.6), we expect NH$_3$-ice to form on Jupiter and Saturn and H$_2$S-ice on Uranus and Neptune. The composition of the NH$_3$-ice cloud on Jupiter has been confirmed by direct spectral measurements obtained with the ISO satellite and the Cassini spacecraft.

(5) On Uranus and Neptune, temperatures in the upper troposphere are low enough for CH$_4$-ice to form (at a temperature ~80 K). Thus, the upper visible cloud decks on Jupiter and Saturn are likely composed of NH$_3$-ice, while the upper cloud deck on Uranus and Neptune should consist of CH$_4$-ice. However, model simulations of infrared spectra show that, globally, the methane cloud must be optically thin, while a much thicker cloud is present at pressures of 3–4 bar. The latter cloud is likely the H$_2$S-ice cloud. The white wispy clouds seen in Voyager photographs and the discrete cloud features visible in near-infrared images (Figs. 4.38, 4.39) are methane-ice clouds.

Although the composition of the clouds cannot easily be measured directly via remote sensing techniques, altitude profiles of condensable gases or the mean molecular weight in an atmosphere give indirect information on the composition of cloud layers. Radio occultation experiments on Voyager suggest a decrease in the atmospheric mean molecular weight (gaseous) on Uranus and Neptune at 1.2–1.3 bar, which probably is caused by the formation of a CH$_4$-ice cloud. The altitude of the base of this cloud has been used to derive a CH$_4$ mixing ratio ~30 times the solar C value (Problem 4.16). Since no global optically thick cloud has been detected at these wavelengths, as mentioned above, the particles must have precipitated out.

Since at radio wavelengths the opacity is mainly provided by NH$_3$ gas, analysis of a microwave spectrum yields an altitude profile of this gas. If the abundance changes at particular altitudes, this may signify cloud formation, either the NH$_3$-ice or NH$_4$SH cloud (e.g., Fig. 4.15c, d). Note that, in principle, the NH$_3$ altitude profile can thus

also be used to determine the H$_2$S mixing ratio (assuming thermochemical equilibrium), since one molecule of NH$_3$ combines with one other molecule of H$_2$S gas to form NH$_4$SH (eq. 4.24). In practice it has been difficult to relate *in situ* measurements by the Galileo probe to microwave observations and cloud formation under thermochemical equilibrium assumptions (perhaps because the probe descended in a dry region).

Multi-wavelength images of clouds are discussed in §4.5.6, together with atmospheric dynamics and meteorology.

4.5 Meteorology

Everyone is familiar with 'weather', usually caused by a combination of Sun, winds, and clouds. On Earth we have different seasons, and each season is associated with particular weather patterns, which vary with geographic location. One sometimes experiences long periods of dry sunny weather, while at other times we are threatened by long cold spells, periods of heavy rain, huge thunderstorms, blizzards, hurricanes, or tornadoes. What is causing this weather, and what can we infer about weather on other planets? In this section, we summarize the basic motions of air as caused by pressure gradients (induced by, e.g., solar heating) and the rotation of the planetary body. We further discuss the vertical motion of air that causes its temperature to change by adiabatic expansion or contraction.

4.5.1 Winds Forced by Solar Heating

Differential solar heating induces pressure gradients in an atmosphere, which trigger winds. Some examples of wind flows triggered directly by solar heating are the Hadley circulation, thermal tidal winds, and condensation flows. Each of these topics is discussed below. The effects of planetary rotation on the winds are discussed in §4.5.3.

4.5.1.1 *Hadley Circulation*

If the planet's rotation axis is approximately perpendicular to the ecliptic plane, the planet's equator receives more solar energy than do other latitudes. Hot air rises and flows towards regions with a lower pressure, thus towards the north and south. The air then cools, subsides, and returns back to the equator at low altitudes. This atmospheric circulation is called the *Hadley cell circulation*. For a slowly rotating or nonrotating planet, such as Venus, there is one Hadley cell per hemisphere. If the planet rotates, the meridional winds are deflected (Fig. 4.20; §4.5.3), and the circulation pattern breaks up. On Earth, there are three mean-meridional overturning cells per hemisphere, and the cell closest to the equator is called the Hadley cell

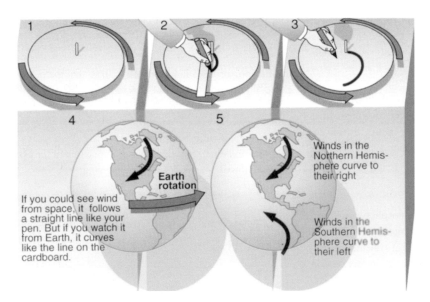

Figure 4.20 A schematic explanation of the Coriolis force: (1) A turntable which rotates counterclockwise. (2) Hold a ruler at a fixed position in inertial space and draw a 'straight' line on the turntable. (3) Even though you drew a straight line, the line on the turntable is curved. This is caused by the 'Coriolis' force. (4, 5) The Coriolis force on the rotating Earth. The rotation of the Earth is indicated by the thick arrow. (Williams 1992)

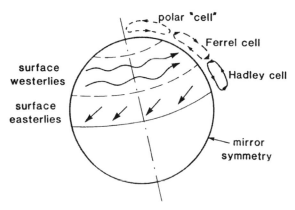

Figure 4.21 Sketch of the Hadley cell circulation on Earth. Three cells are indicated, with the surface winds caused by the Earth's rotation. The winds are indicated as easterly and westerly winds, that is winds blowing from the east and west, respectively. (Ghil and Childress 1987)

(Fig. 4.21). The middle, or *Ferrel, cell* in each hemisphere circulates in a thermodynamically indirect sense: the air rises at the cold end and sinks at the warm end of the pattern. The third cell, closest to the poles, is referred to as the *polar cell*. The giant planets rotate very rapidly, and latitudinal temperature gradients lead to a large number of zonal winds. If the planet's rotation axis is not normal to the ecliptic plane, the Hadley cell circulation is displaced from the equator, and weather patterns can vary with season. Moreover, a planet with a large obliquity on an eccentric orbit (e.g., Mars) may have large orbit-averaged differences between the two polar regions. Such difference can be enhanced by topography and other surface properties.

4.5.1.2 *Thermal Tides*

If there is a large difference in temperature between the day and night hemispheres of a planet, air flows from the hot day side to the cool night side. Such winds are called *thermal tidal winds*. A return flow is present at lower altitudes. The presence of such winds thus depends on the fractional change in temperature over the course of a day, $\Delta T/T$. To estimate this number we compare the solar heat input, $\mathcal{P}_{in}$ (eq. 3.15), with the heat capacity of the atmosphere (Problem 4.18).

The fractional change, $\Delta T/T$, is typically less than 1% for planets with substantial atmospheres, such as Venus and the giant planets. But it can be large for planets which have tenuous atmospheres. For example, it is $\sim$20% for Mars (Problem 4.18). We therefore expect the strong thermal winds near the surface only on Mars and planets/satellites with tenuous atmospheres. On Venus and Earth, thermal tides are strong in the thermosphere, well above the visible cloud layers, where the density is low and day–night temperature difference is large.

4.5.1.3 *Condensation Flows*

On several bodies, such as Mars, Triton, and Pluto, gas condenses out at the winter pole and sublimes in the summer. Such a process drives *condensation flows*. At the martian summer pole, CO_2 sublimes from the surface, thus enhancing the CO_2 content of the atmosphere. At the winter pole it condenses, either directly onto the surface, or onto dust grains which then fall down due to their increased weight. Mars's atmospheric pressure varies by $\sim$20% from one season to the next (Mars's eccentric orbit contributes to the large annual variation). On Triton and

Pluto, the condensable gases nitrogen and methane may induce condensation flows on these bodies. Such flows may explain why a fresh layer of ice overlays most of Triton's (cold) equatorial regions, while no ice cover is seen in the warmer areas. The observed decrease in Pluto's albedo as it approached perihelion may also be evidence for evaporation of a substantial amount of ground frost (e.g., §4.3.3.3). Sulfur dioxide on Io sublimates on the day side and condenses at night, which may drive fast (supersonic) day-to-night winds.

4.5.2 Wind Equations

Winds are induced by gradients in atmospheric pressure, and they are deflected by a planet's rotation. This is the basic concept behind the winds and storm systems encountered on Earth and seen on other planets. In this section we summarize the equations which describe air motions. In §4.5.3 we discuss specific examples of steady flows, while turbulent motions are discussed in §4.5.4.

4.5.2.1 *Inertial Frame*

Euler's equation describes the motion of an incompressible, inviscid fluid that results from pressure gradients and the gravity field:

$$\rho \frac{D\mathbf{v}}{Dt} = -\nabla P + \rho \mathbf{g}_\mathrm{p}. \tag{4.25}$$

In equation (4.25) we have used the *material* or *advective derivative, D/Dt*, which is the total time derivative for an observer in an inertial frame of reference:

$$\frac{D}{Dt} \equiv \frac{\partial}{\partial t} + \mathbf{v} \cdot \nabla. \tag{4.26}$$

The first term on the right-hand side of equation (4.26) is the local derivative, caused by temporal changes in the fluid, and the second term is the advective contribution, which is the transport of material by the bulk motion of the wind. If viscosity is important, Euler's equation must be replaced by the *Navier–Stokes equation*, which for an incompressible fluid with a constant kinematic viscosity, ν_v (cm^2 s^{-1}), becomes

$$\frac{D\mathbf{v}}{Dt} = -\frac{1}{\rho}\nabla P + \mathbf{g}_\mathrm{p} + \nu_\mathrm{v}\nabla^2\mathbf{v}. \tag{4.27}$$

In addition to the above equation(s) of momentum, the temporal evolution of the density, pressure, temperature, and velocity of atmospheric winds are related through the (hydrodynamic) equations of continuity and energy. Mass conservation is described by the equation of continuity:

$$\frac{\partial \rho}{\partial t} = -\nabla \cdot (\rho \mathbf{v}), \tag{4.28a}$$

which can be rewritten using the material derivative:

$$\frac{D\rho}{Dt} = -\nabla \cdot (\rho \mathbf{v}) + \mathbf{v} \cdot \nabla \rho = -\rho \nabla \cdot \mathbf{v}. \tag{4.28b}$$

In an incompressible fluid, $D\rho/Dt = 0$, so that the divergence of the velocity is equal to zero, $\nabla \cdot \mathbf{v} = 0$.

If there is no interchange of heat by conduction or radiation, the flow of gas is isentropic and adiabatic. Under more general circumstances, when heat is exchanged, we need to complement the equations with one that describes conservation of energy in a moving element:

$$\rho c_\mathrm{P} \frac{DT}{Dt} = \nabla \cdot (K_\mathrm{T}\nabla T) + \frac{DP}{Dt} + Q, \tag{4.29}$$

with K_T the thermal conductivity, c_P the specific heat, and Q the heat as arising from, e.g., viscous dissipation and radiation.

4.5.2.2 *Rotating Frame*

Since all planets rotate, it is convenient to express the equation of motion in a rotating frame of reference. The velocity, $\mathbf{v}$, in the inertial frame is equal to

$$\mathbf{v} = \mathbf{v}' + \omega_\mathrm{rot} \times \mathbf{r}, \tag{4.30}$$

with $\mathbf{v}'$ the velocity in the rotating frame of reference, at distance $\mathbf{r}$ ($= R\sin\theta$, with R the radius of the planet and θ the colatitude) from the rotation axis, while the planet (i.e., frame of reference) rotates with an angular velocity ω_rot. Differentiating equation (4.30) yields

$$\frac{D\mathbf{v}}{Dt} = \left(\frac{D\mathbf{v}}{Dt}\right)' + \omega_\mathrm{rot} \times \mathbf{v}, \tag{4.31}$$

with $(D\mathbf{v}/Dt)'$ the material derivative relative to the rotating observer. Euler's equation in a rotating frame of reference becomes (Problem 4.19)

$$\rho\left(\frac{D\mathbf{v}'}{Dt}\right)' = -2\rho\omega_\mathrm{rot} \times \mathbf{v}' - \nabla P + \rho \mathbf{g}_\mathrm{eff}, \tag{4.32}$$

with the effective gravity:

$$\mathbf{g}_\mathrm{eff} = \mathbf{g}_\mathrm{p} + \omega_\mathrm{rot}^2 \mathbf{r}. \tag{4.33}$$

For most planets $\omega_\mathrm{rot}^2 r \ll g_\mathrm{p}$, so $g_\mathrm{eff} \approx g_\mathrm{p}$.

The *vorticity* of a velocity field is the measure of spin around an axis, and is defined by

$$\varpi_\mathrm{v} \equiv \nabla \times \mathbf{v}. \tag{4.34}$$

For a particle or fluid element stationary relative to a planet's surface ($\mathbf{v}' = 0$), the vorticity is equal to twice the angular velocity (Problem 4.20):

$$\varpi_\mathrm{v} = 2\omega_\mathrm{rot}. \tag{4.35}$$

In the following sections, we drop the quotes from the quantities in the rotating frame.

4.5.3 Horizontal Winds

Let us consider a thin (vertical scale length, h, much smaller than the horizontal scale length, ℓ, i.e., $h/\ell \ll 1$) layer of fluid that is incompressible and inviscid. Such an approximation is often valid in atmospheres of rotating bodies, where in addition to vertical density gradients, the rotation of the body itself induces vertical stability. We adopt a locally Cartesian coordinate system, in which y is the coordinate on the surface to the north, x is to the east, and z is upwards, perpendicular to the surface. The wind velocities are generally expressed as u, v, and w along the x, y, and z coordinates, respectively. Atmospheres are approximately in hydrostatic equilibrium, and hence $\partial P/\partial z \approx -\rho g_p$. This gradient is much larger than pressure gradients in the horizontal direction, $\partial P/\partial z \gg \partial P/\partial x, \partial P/\partial y$. For a nonrotating planet, scaling arguments can be used to show that $W/h \sim U/\ell$, where W and U stand for the characteristic velocities in the vertical and horizontal directions, respectively.

4.5.3.1 Coriolis Force

Because planets rotate, winds cannot blow straight from a high pressure region to an area of low pressure, but rather follow a curved path. This phenomenon can be visualized with help of a turntable (see Fig. 4.20). Draw a line along a ruler held fixed in inertial space while the platform rotates: The line comes out curved in the direction opposite to the platform's rotation. According to the same principle, winds on Earth (or any other prograde rotating planet) are deflected to the right on the northern hemisphere and to the left on the southern hemisphere. (On retrograde rotating planets the opposite pairing occurs.) This is called the *Coriolis effect*, and the 'fictitious force' causing the wind to curve is referred to as the *Coriolis force*.

The Coriolis effect follows from the conservation of angular momentum about the rotation axis. A parcel of air at colatitude θ has an angular momentum:

$$L = (\omega_{rot}R\sin\theta + u)R\sin\theta, \tag{4.36}$$

where R is the planet's radius and u the wind velocity along the x coordinate. If an air parcel initially at rest relative to the planet moves poleward while conserving angular momentum, then u must grow in the direction of the planet's rotation to compensate for the decrease in $\sin\theta$. Hence, a planet's rotation deflects the wind perpendicular to its original direction of the motion, with an acceleration equal to $f_C\sqrt{u^2 + v^2}$. The *Coriolis parameter*, f_C, is defined as the planet's vorticity, ϖ_v, normal to the surface at the latitude of interest:

$$f_C \equiv 2\omega_{rot}\cos\theta = \varpi_v\cos\theta. \tag{4.37}$$

The direction of the wind is changed, but, since the acceleration is always perpendicular to the wind direction, no work is done and the speed of the wind is not altered.

The Hadley cells on Earth cause the well known *easterly* (from the east) trade winds in the tropics, as the return Hadley cell flow near the surface is deflected to the west by the Coriolis force. Similarly, one might expect westerlies at mid-latitudes on the low-altitude return flow in the Ferrel cell (as indicated in Figure 4.21); however, in reality the situation is more complex (§4.5.5.1). On the giant planets, the large gradient in the Coriolis force with latitude (referred to as the β-effect, where on a spherical planet $\beta \equiv \varpi_v\sin\theta/R$) leads to a large number of zonal winds. Winds on individual planets are described in more detail in §4.5.5 and §4.5.6.

The importance of the Coriolis force can be judged from the *Rossby number*, $\Re_o$, which is a measure of the ratio of the characteristic horizontal wind speed, U, and Coriolis term:

$$\Re_o \equiv \frac{U}{f_C\ell}, \tag{4.38}$$

where ℓ is a length scale. Thus, when the Coriolis term is important, the Rossby number is small. In analogy to using scaling arguments for a stationary planet (see above), one can show for a rotating planet that $W/h \approx \Re_o U/\ell$. Hence, for small Rossby numbers, $W \ll U$, and the winds are essentially horizontal:

$$\frac{D\mathbf{v}}{Dt} - f_C\mathbf{v} \times \hat{\mathbf{z}} - \frac{1}{\rho}\nabla P. \tag{4.39}$$

4.5.3.2 Geostrophic, Cyclostrophic Balance

It is convenient to write the wind equations in terms of forces tangential, $\hat{\mathbf{t}}$, and normal, $\hat{\mathbf{n}}$, (positive towards the left) to the flow, respectively, in a plane parallel to the surface (Fig. 4.22). Consider a steady flow ($\partial\mathbf{v}/\partial t = 0$), where the pressure gradient and the Coriolis force just balance each other, a situation referred to as *geostrophic balance*. The wind flows along *isobars*, lines of constant pressure, in a direction perpendicular to the pressure gradient (Fig. 4.23). Under these circumstances, the velocity $\mathbf{v}$ follows directly from equation (4.39):

$$\mathbf{v} = \frac{1}{\rho f_C}(\hat{\mathbf{z}} \times \nabla P), \tag{4.40}$$

with ∇P the horizontal pressure gradient. The geostrophic approximation is usually valid if the Rossby number $\Re_o \ll 1$. Prime examples of geostrophic winds are the trade winds and westerly jets in the Earth's troposphere (§4.5.5.1) and the zonal winds on the giant planets (§4.5.6).

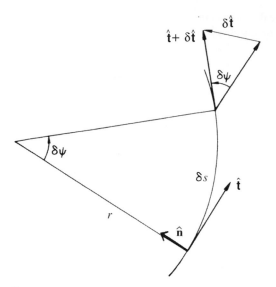

Figure 4.22 Curvilinear coordinates, showing the differential change in the unit tangent vector $\hat{t}$. (Adapted from Holton 1972)

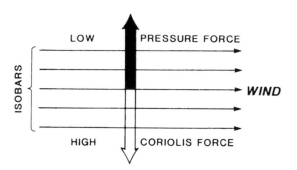

Figure 4.23 Geostrophic balance: the pressure and Coriolis forces balance each other and the wind flows along isobars. (Kivelson and Schubert 1986)

Equations (4.39) and (4.40) can be rewritten in terms of tangential and normal components (Fig. 4.22):

$$f_C \mathbf{v} \times \hat{\mathbf{z}} = -f_C v_h \hat{\mathbf{n}}, \tag{4.41}$$

$$\frac{D\mathbf{v}}{Dt} = \frac{Dv_h}{Dt}\hat{\mathbf{t}} + v_h \frac{Ds}{rDt}\hat{\mathbf{n}} = \frac{Dv_h}{Dt}\hat{\mathbf{t}} + \frac{v_h^2}{r}\hat{\mathbf{n}}, \tag{4.42}$$

where $v_h = \sqrt{u^2 + v^2}$ is the horizontal wind speed. The equations for the forces tangential and normal to the flow become:

$$\frac{Dv_h}{Dt} = -\frac{1}{\rho}\frac{\partial P}{\partial s}, \tag{4.43}$$

$$\frac{v_h^2}{r} = -f_C v_h - \frac{1}{\rho}\frac{\partial P}{\partial n}. \tag{4.44}$$

Equation (4.44) describes the centrifugal acceleration, present in any circular motion (Fig. 4.24).

For a pure zonal flow, the velocity is along the x coordinate, and $u = v_h$. At the equator, the centrifugal force for a zonal wind is perpendicular to the surface. At other latitudes the centrifugal force can be written as a radial plus a tangential term, where the latter is directed towards the equator: $u^2/(R\tan\theta)$. Usually this term is very small compared to the geostrophic term $f_C u$, and the flow is in geostrophic balance. When the centrifugal force $u^2/r \gg f_C u$, it may balance the force induced by a meridional pressure gradient (Fig. 4.25), known as *cyclostrophic balance*.

The only terrestrial planet on which 'planet-wide' cyclostrophic balance is important is Venus, where the predominantly horizontal pressure gradient above the cloud tops is north–south, with pressure decreasing towards the poles. The winds near Venus's cloud tops are predominantly east to west, with a period of 4 days. Since Venus rotates only once in 240 days, these winds are in *superrotation*. In this case, the centrifugal force cannot be neglected, and in fact balances the pressure force, so the winds move along isobars. Cyclostrophic balance is also important on Saturn's satellite Titan, where the atmosphere rotates much faster than the body, and near the equator of some other bodies. Cyclostrophic balance is sometimes satisfied on small scales, such as in storm systems and eddies.

4.5.3.3 Thermal Wind Equation

By combining the equations of hydrostatic equilibrium and geostrophic balance, one can relate vertical *wind shear*, i.e., changes in the wind's velocity with altitude, to variations in temperature along isobars. For this purpose it is convenient to use pressure, P, rather than altitude, z, as the vertical coordinate. For an atmosphere in hydrostatic equilibrium (eq. 3.26), we write the *geopotential*, Φ_g:

$$\Phi_g = \int_0^z g_p dz = -\int_{P_0}^P \frac{dP}{\rho}. \tag{4.45}$$

In isobaric coordinates, the geostrophic wind equation (4.40) becomes:

$$\mathbf{v} = \frac{1}{f_C}\hat{\mathbf{z}} \times (\nabla \Phi_g)_P. \tag{4.46}$$

The vertical gradient of the geostrophic velocity can be obtained by differentiating equation (4.46) with respect to P:

$$\frac{\partial \mathbf{v}}{\partial \ln P} = \frac{R_{gas}}{\mu_a f_C}\hat{\mathbf{z}} \times (\nabla T)_P, \tag{4.47}$$

where R_{gas} is the universal gas constant and μ_a the mean molecular mass (in amu). In this derivation we further made use of the ideal gas and barometric laws.

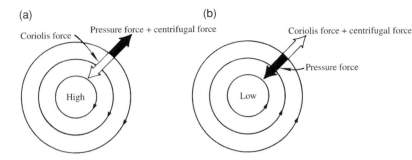

(a) Coriolis force Pressure force + centrifugal force

(b) Coriolis force + centrifugal force Pressure force

High

Low

Figure 4.24 Isobars and wind flows around (a) a high-pressure region (anticyclone) in the Earth's northern hemisphere and (b) a low-pressure region (cyclone) in the Earth's northern hemisphere. (Kivelson and Schubert 1986)

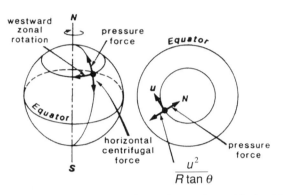

Figure 4.25 Cyclostrophic balance: the equatorward horizontal centrifugal force is balanced by a poleward pressure force. The zonal wind velocity is given by u, planet radius by R, and colatitude by θ. (Adapted from Kivelson and Schubert 1986)

Equation (4.47) is known as the *thermal wind equation*. The thermal wind equation can be integrated vertically to calculate the difference in zonal wind velocity over a range in altitudes. The thermal wind equation is used to calculate changes in the zonal wind velocity with altitude induced by a meridional gradient in temperature, $(\partial T/\partial y)_P$. It can also be used to derive the vertical extent of the wind if the temperature gradient and wind speed along a particular isobar are known (Problem 4.21). Even if the horizontal temperature gradient is small, winds can be fierce if they extend to great depths; if the vertical extent of the winds is small, winds can only be strong if the meridional temperature gradient is large. Note that in a *barotropic* atmosphere, where density is a function of pressure only, density and temperature are constant along an isobaric surface, and there is no thermal wind, i.e., no vertical shear. The geostrophic wind is constant with depth. Only in a *baroclinic* atmosphere, where the density depends on pressure and temperature, do we find vertical wind shear.

4.5.4 Storms
4.5.4.1 *Turbulence*
Thus far we have discussed steady, large-scale fluid motions. Laboratory experiments show that turbulent

motions tend to arise when the dimensionless *Reynolds number*, $\mathfrak{R}_e$, is large:

$$\mathfrak{R}_e \equiv \frac{\ell U}{\nu_v} > 5000, \qquad (4.48)$$

where ℓ and U are a characteristic length scale and velocity, respectively, and ν_v is the kinematic viscosity. A typical value for ν_v in the Earth's lower troposphere is ~ 0.1 cm^2 s^{-1}. Hence, for a length scale of one meter, the critical value for $\mathfrak{R}_e$ is already exceeded for velocities of the order of 5 cm s^{-1}. Thus atmospheres inevitably display some turbulence, in particular in regions of high wind shear, vigorous convection, or near variations in surface 'topography'. The flows can be modeled by allowing variations, perturbations, and wave motions away from geostrophic balance.

4.5.4.2 *Convection*
The *potential temperature*, Θ, is a quantity that is conserved along an adiabatic path in (T, P) coordinates. It represents the temperature that a parcel of unsaturated air would have if it were compressed or expanded adiabatically to a pressure $P_0 = 1$ bar:

$$\Theta \equiv T \left(\frac{P_0}{P} \right)^{(\gamma-1)/\gamma}, \qquad (4.49)$$

with γ the ratio of specific heats, c_P/c_V (assumed to be constant). The vertical gradient in the potential temperature is the difference between the actual lapse rate and the dry adiabatic lapse rate:

$$\frac{\partial \Theta}{\partial z} = \frac{\Theta}{T} \left[\frac{\partial T}{\partial z} - \left(\frac{\partial T}{\partial z} \right)_{ad} \right]. \qquad (4.50)$$

If $\partial \Theta/\partial z < 0$, the lapse rate is superadiabatic, and the atmosphere is unstable against convection. Small-scale turbulence in such an atmosphere is caused by *free convection*. If the gradient in potential temperature is close to zero, large-scale wind flows dominate local effects, and turbulence can only be induced by *forced convection*, i.e., transfer of heat under such conditions is caused by motions

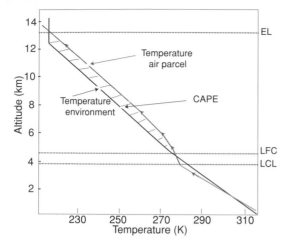

Figure 4.26 A sketch of the temperature–pressure profile of a rising parcel of air in Earth's atmosphere, visualizing the LFC (Level of Free Convection), LCL (Lifting Condensation Level), and CAPE (Convective Available Potential Energy). The temperature of the air parcel follows a dry adiabat below LCL, and a wet adiabat above. CAPE is equal to the area bounded by the temperature curve of the environment and the moist adiabat between LFC and EL, as indicated on the figure. Above EL, the Equilibrium Level (near the tropopause), the parcel is cooler than the environment, and stops its ascent. (After Salby 1996)

in the atmosphere that are not themselves triggered by a 'heating' (buoyancy) process.

As discussed in §4.4.1, condensation of gas decreases the atmospheric lapse rate, and we define the vertical gradient in the *equivalent potential temperature*, Θ_e, analogous to equation (4.50) as the difference between the actual lapse rate and the wet adiabat. If $\partial\Theta_e/\partial z > 0$, the (layer in the) atmosphere is absolutely stable against convection. A layer with a lapse rate in between the dry and wet adiabats is conditionally unstable. An example of a rising air parcel in Earth's atmosphere under conditionally unstable conditions is sketched in Figure 4.26. Initially, the rising parcel of air cools at the dry adiabatic lapse rate, i.e., more rapidly than its surroundings. Once it reaches the *lifting condensation level*, the parcel cools more slowly, with temperature gradient equal to that of the wet adiabat (note that the slope of the wet adiabat is altitude dependent, since it depends on the available latent heat of condensation). Above the *level of free convection*, the parcel is warmer than its surroundings, and thus continues to ascend on its own account. The buoyant energy that can be converted into kinetic energy of moist convection for this rising air parcel is known as the *convective available potential energy*, typically referred to as *CAPE*.

The stability of an atmosphere can also be evaluated from the Rossby deformation radius and the

Brunt–Väisälä frequency. The *Rossby deformation radius*, L_D, is a typical horizontal scale length in a geostrophic flow over which rotational effects become as important as buoyancy:

$$L_D = \frac{Nh}{f_C}, \qquad (4.51)$$

with h the vertical thickness of the flow structures under consideration and N the *buoyancy* or *Brunt–Väisälä frequency*, the frequency of up-and-down oscillation of a fluid element about its equilibrium value:

$$N = \sqrt{\frac{g_P}{\Theta}\frac{\partial\Theta}{\partial z}}. \qquad (4.52)$$

Since the Brunt–Väisälä frequency depends upon the restoring force of buoyancy, it provides a measure of atmospheric stability. Larger values of N are indicative of a more stable atmosphere. In the Earth's stratosphere, $N \approx 0.02$ s^{-1}, while in the troposphere it is a factor of $\sim$2 smaller.

4.5.4.3 Eddies and Vortices

Local topography on a terrestrial planet may induce *stationary eddies*, which are storms that do not propagate in an atmosphere. Stationary eddies are seen over mountains on Earth and Mars, and on Earth at the interface between oceans and continents where large differences in temperature exist.

Baroclinic eddies may form in an atmosphere with geostrophic flows, if such a flow becomes baroclinically unstable. To evaluate this, we write the equation that governs the fluid vorticity in a geostrophically balanced fluid where friction is negligible, in a frame rotating with the planet:

$$(2\omega_{rot}\cdot\nabla)\mathbf{v} - 2\omega_{rot}\nabla\cdot\mathbf{v} = -\left(\frac{\nabla\rho\times\nabla P}{\rho^2}\right), \qquad (4.53)$$

where the term on the right-hand side of equation (4.53) is the baroclinic term. If the atmosphere is barotropic, i.e., the density does not vary along isobars, the baroclinic term is equal to zero. If the density varies along isobaric surfaces, the baroclinic term is nonzero, and baroclinic eddies may form. Such conditions occur, for example, in transition layers between two flows. Only *prograde* (rotating in the direction of flow motion) baroclinic eddies survive (see, e.g., Fig. 4.27). The winds in such eddies flow along isobars, where the pressure force is balanced by the combined Coriolis and centrifugal forces (eq. 4.44, Fig. 4.24). In *cyclones*, the wind blows around a region of low pressure, while in an *anticyclone* the wind blows around a high-pressure region. Cyclones and anticyclones

(a)

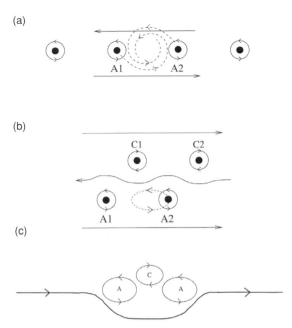

(b)

(c)

Figure 4.27 (a) A row of counterclockwise rotating anticyclones (A) embedded in a shear east–west flow with an eastward moving jet stream (solid arrow) south of the row and a westward moving jet stream to the north. The configuration is an unstable equilibrium. Broken lines show the movement of the vortices. (b) Vortices in a Kármán vortex street consisting of a row of clockwise cyclones (C) north of and parallel to a row of anticyclones (A). This configuration of vortices is stable to vortex mergers. Solid and broken curves are as in panel (a). (c) Schematic of two anticyclones trapped in the trough of a Rossby wave on an eastward moving jet stream. Between the anticyclones is a cyclone, which prevents the anticyclones from merging. (Adapted from Youssef and Marcus 2003)

are observed in the atmospheres of, e.g., Earth, Mars, and the giant planets.

The *potential vorticity* of an eddy, ϖ_{pv}, is proportional to angular momentum around the vertical axis, and is a conserved quantity under frictionless, adiabatic conditions:

$$\varpi_{pv} \equiv \frac{\varpi_v + f_C}{h_i}, \qquad (4.54a)$$

where h_i is a measure of the spacing between *isentropes* (surfaces of constant Θ):

$$h_i = \frac{1}{g_p} \frac{\partial P}{\partial \Theta}. \qquad (4.54b)$$

Since potential vorticity is conserved, storm systems change when they move in latitude to compensate for the change in the Coriolis term, f_C. Equation (4.54) shows that either the storm's vertical extent or its spin must change if the storm changes latitude. This may, in part, explain the disappearance of Neptune's *Great Dark Spot* (GDS) (Fig. 4.39a), which during the Voyager era was observed

to slowly drift in the direction of the equator. If the storm stays at the same latitude but its vertical extent changes, the storm must spin up or down. This is, for example, seen when a storm meets a mountain: The bottom of the storm is forced upwards, so the storm is compressed in altitude and expands horizontally, resulting in a spin-down. When the system has passed over the mountain, it descends and forms a tall, rapidly spinning column of air.

Kármán Vortex Street

Observations of Jupiter's atmosphere (Fig. 4.34) reveal zonal winds and rows of anticyclones that appear to be stable over long periods of time. However, computer simulations show that this cannot be the case. A steady zonal flow, with a north–south wind profile such as measured for Jupiter, is unstable and breaks up into a series of small eddies. Once a series of eddies is present, such as the white ovals in Jupiter's southern hemisphere, they will merge, as shown schematically in Figure 4.27a. The anticyclones A1 and A2 are located midway between two jet streams. Even the smallest perturbation on this system leads to a merger. For example, if the storm A2 moves slightly upward in latitude, it is carried to the left in longitude by the jet stream above it. Within weeks, A2 and A1 merge into a single storm. The system is not stable until all vortices at this latitude have merged into a single storm. This may explain the formation and longevity of Jupiter's *Great Red Spot* (GRS).

The above result is in stark contrast to observations of apparently stable rows of white anticyclonic ovals in Jupiter's southern hemisphere. The persistence of these ovals calls for a mechanism that prevents mergers. If a row of cyclones is added to the system, alternating with the anticyclones as sketched in Figure 4.27b, the cyclone will repel any anticyclone that comes too close. This configuration is known as a *Kármán vortex street*, where anticyclones are longitudinally staggered with cyclones. Under such circumstances, if an anticyclone is perturbed upward, the jet stream above it moves it towards the left, as before, but this time it is carried downward by the clockwise flow until the jet stream below it moves it back towards the right, to its original position. Computer simulations show that this configuration is stable for long periods of time, with storms oscillating back and forth in longitude. In a situation as sketched in Figure 4.27c, and shown in Figure 4.28, a minor perturbation may dislodge cyclone C, in which case the anticyclones, A, quickly merge (§4.5.6).

4.5.4.4 Hurricanes

Hurricanes on Earth form in tropical regions above warm oceans, and dissipate when they move over land, where

(a)

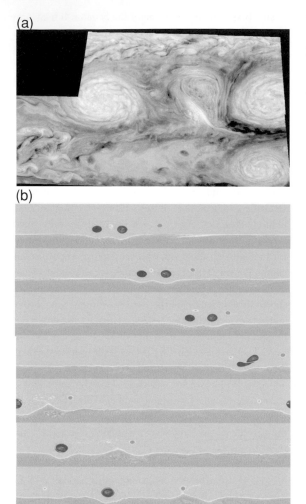

(b)

Figure 4.28 (a) Galileo image of Jupiter's white anticyclonic ovals with a cyclonic feature in between. This mosaic was constructed from images at 756 nm, 727 nm, and 889 nm, taken in 1998. (NASA/JPL-Caltech, PIA00700) (b) Seven frames from a movie showing two anticyclones (large – red in movie) with an intervening cyclone (small – blue in movie) trapped (as in Fig. 4.27c) in the trough of a Rossby wave on a jet stream (indicated by the interface between the light and dark colors). The trio of vortices travels together as a stable unit to the east until they encounter a rogue cyclone on their eastern side that is drifting more slowly than the trio. The trio repels the cyclone but is sufficiently perturbed by it that their cyclone is ejected allowing the two trapped anticyclones to merge. The image is computed with a periodic domain, so that the left-hand edge of the image wraps around to join the right-hand edge. (Adapted from Youssef and Marcus 2003)

the winds and temperatures change. Their fierce winds, rain, and high waves can cause a considerable amount of damage in the coastal areas. Hurricanes are cyclones, i.e., low pressure weather systems. A parcel of humid air

rises, because it is lighter than the surrounding air (Problem 4.12). Such rising air leaves behind an area of low pressure. Winds rush in to equalize pressure, causing both air and water to flow towards the hurricane 'eye'. Water is carried off (return flow) at large depths in the ocean. While the humid air is rising, it cools and water droplets and/or ice crystals form when the temperature drops below the condensation temperature of water. The release of latent heat warms up the air parcel, thereby increasing the gradient in the potential temperature, $|\partial\Theta/\partial z|$, 'fueling' the upward motion of air. The storm system can reach great heights if there is no vertical wind shear in the atmosphere; otherwise the storm is ripped apart. The temperature gradient is inverted at the tropopause and convection stops. The high-pressure build-up at the tropopause helps to pump the air away, aided by stratospheric winds.

4.5.4.5 *Lightning*

Bright flashes of light, attributed to *lightning*, have been observed on Earth, Jupiter, Saturn, and possibly Venus. The electrostatic discharges that have been observed on Saturn (SED) and Uranus (UED) at radio wavelengths may also be caused by lightning (§7.4.3). The basic mechanism for lightning is collisional charging of cloud droplets, followed by a (gravitational) separation of oppositely charged small and large particles, so that a vertical potential gradient develops. The amount of charge that can be separated this way is limited; once the resulting electric field becomes strong enough to ionize the intervening medium, a lightning stroke or discharge occurs, releasing the energy stored in the electric field. This only happens if the electric field is large enough that a typical free electron gains sufficient energy while traversing the medium (i.e., a mean free path) that it ionizes the molecule it collides with, producing an exponential cascade of electrons and collisions. This requires a potential difference of about 30 V over an electron's mean free path.

In Earth's atmosphere, lightning is almost always associated with precipitation, although large-scale electrical discharges occur occasionally during volcanic eruptions (and nuclear explosions). In analogy, lightning on other planets is only expected in atmospheres where both convection and condensation take place. Moreover, the condensed species, such as water, must be able to undergo collisional charge exchange. Lightning on other planets (such as Venus) might be triggered by active volcanism. Electrostatic discharges have been measured in terrestrial dust storms and on Mars (most likely).

4.5.4.6 *Waves*

Disturbances in atmospheric pressure propagate in the form of waves, whose mathematical description can be derived from the hydrodynamic equations by adding small perturbation terms to the equations. Waves are important, since they 'communicate' changes in one region to another, and hence they may cause planet-wide weather patterns. There are several types of waves, most of which are discussed in detail in books on fluid dynamics. The simplest waves are *sound* or *acoustic waves*, which are compressional waves, known as p modes in astronomy and P waves in seismology (§6.2.1.1).

Surface gravity waves are the familiar oscillations that propagate on the surface of water when a stone is dropped in. They are known as f modes in astronomy. While the compressibility of air provides the restoring force in case of a sound wave, the buoyancy provides the restoring force for gravity waves, with the buoyancy or Brunt–Väisälä frequency given in equation (4.54). If the wavelength of gravity waves is short compared to the radius of the planet, they behave like ordinary waves on the surface of a deep ocean. *Internal gravity waves* (g modes) are confined to a stably stratified portion of an atmosphere. All gravity waves propagate vertically as well as horizontally. Various *seismic waves* may travel through atmospheres as well (§6.2.1). *Atmospheric tides* are driven by solar heating (§4.5.1), as well as by the gravitational effects from the Sun and the Moon.

The *Rossby* or *planetary wave* results from changes in the balance between the relative vorticity of an air parcel and its planetary vorticity, through conservation of the potential vorticity (eq. 4.54). Such waves are triggered when an air parcel moves to higher or lower latitudes, so that its Coriolis term changes. For example, when an eastward moving parcel of air in the northern hemisphere (on a prograde planet) is deflected towards the equator, the Coriolis term decreases, so the parcel spins up cyclonically (eq. 4.54). This induces a deflection to the north in the flow ahead of the parcel, so its trajectory is then deflected poleward, back to its original latitude. If the parcel overshoots its latitude, it spins up anticyclonically, and is deflected equatorward. The variation in the Coriolis term with latitude exerts a torque on the displaced air, which provides the restoring force and enables air to move back and forth about its undisturbed latitude. The Rossby wave usually propagates westward (on a prograde rotating planet) with respect to the mean zonal flow; although its speed is usually low, few m s^{-1}, velocities of hundreds of meters per second can be reached. Rossby waves commonly travel along eastward-going jet streams, such as seen on Earth and Jupiter, and cause the flows to meander about lines of constant latitude.

4.5.5 Observations: Bodies with Surfaces

4.5.5.1 *Earth*

Earth's global atmospheric wind system is characterized by a Hadley cell circulation with three cells per hemisphere. Note that the middle cell, the Ferrel cell, circulates in a thermodynamically indirect sense (see Fig. 4.21). The Coriolis force adds an east–west component to the meridional movement of the circulation, which is, in part, the reason why there are three cells rather than one per hemisphere. In general, air rises near the equator and descends in the subtropics. The descending dry air dries the troposphere, inhibits convection and thus maintains the deserts common at subtropical latitudes. In the tropics, the low-altitude return flows from the Hadley cell are deflected westward by the Coriolis force, which leads to the easterly (from the east) trade winds in the tropics. The equatorward flow of the trade winds of the northern and southern hemispheres results in a convergence of the two air streams in a region known as the intertropical convergence zone (ITCZ). This zone extends over a ∼10° latitude range, centered at the equator, and is characterized as a zone of calm and weak winds with no prevailing wind direction, often referred to as *doldrums*. Deep convective clouds, showers, and thunderstorms occur along the ITCZ.

At northern and southern mid-latitudes (roughly between 20° and 60°), the global atmospheric circulation[4] in the troposphere is characterized by westerly jet streams in the upper troposphere. These jets are zonal winds which flow from the west towards the east and increase in strength with altitude up to the tropopause. The equatorward portion of this jet is driven by an eastward acceleration induced by the Coriolis force on the poleward flow in the Hadley cell. Closer to the surface the Hadley cell flow is towards the equator, and hence the Coriolis force induces a westward acceleration; the vertical wind shear is therefore very large at these latitudes. The poleward portion of the jet coincides with the Ferrel cell. The Coriolis force in this cell induces a westward acceleration in the upper troposphere, and eastward near the surface. These latitude bands, however, are baroclinically unstable, and eddies transport a net eastward momentum to the flow, so that the zonal flow is eastward (westerly) at all altitudes. In fact, if the Earth would rotate faster,

---

[4] Globally and temporally averaged wind patterns are rarely seen on individual days. The day-to-day wind patterns deviate significantly from the average global circulation, as a consequence of local pressure highs and lows.

there might be two distinct mid-latitude westerly jets per hemisphere (one driven by the Coriolis force, and one by baroclinic eddies) at all times. At present, one can sporadically distinguish two distinct jets, at specific longitudes and times.

The obliquity of Earth causes the global circulation pattern to vary between summer and winter. The jet stream in the upper troposphere weakens with altitude above the tropopause. In the winter, westerly winds increase again with altitude above $\sim$25 km (or $P \lesssim 30$ mbar), up to $\sim$70 km altitude ($P \approx 0.05$ mbar). This stratospheric/mesospheric jet is known as the *polar-night jet*, and reaches a speed of 60 m s^{-1} in the lower mesosphere. This jet forms a large cyclonic vortex over the winter pole. In the summer, the westerly flow in the upper troposphere also weakens with altitude above the tropopause, and above $\sim$25 km the westerly flow is replaced by easterly winds which intensify up to $\sim$70 m s^{-1} in the mesosphere. The jets thus have opposite directions in the two hemispheres.

Planetary waves cause large-scale disturbances in the zonal flows, both in the troposphere and stratosphere/mesosphere, away from zonal symmetry.

Along the equator the meridional temperature gradient is relatively small and the Coriolis force weak. The trade winds drive the ocean currents, causing surface water to flow from east to west along the equatorial regions of the Pacific Ocean. The water is heated by the Sun, causing warm surface water to build up across the western Pacific. The water level here is typically 0.5 m higher than in the eastern Pacific. Evaporation of the warm moist air in the west leads to the formation of storm systems. The latent heat release inside organized convection cells is the primary source of energy for the circulation in the tropics. In addition to the zonal-mean Hadley cell, there are monsoon and Walker circulations in the tropics. *Monsoon* circulations are driven by horizontal gradients in surface temperature, where during the summer the subtropical land masses, such as India and northern Australia, are warmer than the surrounding oceans, a situation which is reversed during the winter. Thus air is rising above the land masses during the summer (bringing rain), and subsiding during the winter (dry air). The *Walker circulation* in the tropics is driven by nonuniform heating, with air rising at longitudes where heating takes place (near Africa, Indonesia, and South America) and sinking at longitudes where it is cooler (oceans).

The Walker circulation changes on interseasonal timescales. Every 3–5 years the trade winds weaken and the Walker cell circulation breaks down. The pressure difference between the eastern and western Pacific decreases, and less water is transported westward. Eventually, the warm water in the west, at a higher elevation, begins to move eastward, resulting in a warming of the sea surface temperature in the eastern and central Pacific, $\gtrsim$1.5 K above its normal temperature of $\sim$298 K. This phenomenon is known as *El Niño* (Spanish for little boy, named after the infant Jesus, since the warming of the ocean usually starts around Christmas time). Small changes in temperature produce large variations in the evaporation and latent heat release, a consequence of the exponential dependence in the Clausius–Clapeyron equation (eq. 4.20). This brings about a large eastward shift of the convection cell patterns, causing air to rise above the central Pacific. This is accompanied by a change in surface pressure between the western (low $\rightarrow$ high pressure) and eastern Pacific (high $\rightarrow$ low pressure), a phenomenon known as the *El Niño Southern Oscillation*, which is propagated worldwide through planetary waves. El Niño has marked consequences for the global air circulation and weather patterns on Earth, as experienced through, e.g., abnormally wet or dry seasons.

In addition to these large-scale circulations, weather on Earth is characterized by baroclinic eddies, which can cause (large) storms on our planet, sometimes resulting in hurricanes and tornadoes. The development and movement of baroclinic eddies is also modified by El Niño. In contrast to moving baroclinic eddies, stationary eddies develop at places of varying local topography, such as mountain ranges and volcanoes, and from temperature differences between oceans and continents.

Earth's climate, and changes therein over time, in particular as it relates to global warming, can best be investigated with help of general circulation models (GCM). Such models solve the equations of fluid dynamics, incorporating thermodynamics (cloud formation), chemistry (such as photochemistry), and the coupling between the atmosphere and surface (continents versus oceans). Although the models are becoming more and more sophisticated, it remains challenging to accurately predict our climate (global warming) 50–100 years into the future.

4.5.5.2 Venus

On slowly rotating Venus, we find a 'classical' Hadley cell circulation, i.e., one cell per hemisphere. Air is rising above the equator, and subsiding at latitudes near $\sim$60°, the edge of the polar collar (§4.3.3.1). Strong westward (in the same direction as the planet's rotation) zonal winds are observed in Venus's cloud deck at altitudes of $\sim$60 km. The winds circle the planet in 3–5 days ($\sim$100 m s^{-1}), and hence are superrotating. These winds decrease linearly in strength with decreasing altitude, and are only $\sim$1 m s^{-1}

Figure 4.29 Dust devils observed by Mars rover Spirit in Gusev crater on 13 July 2005. Such dust devils are common in the mid-afternoons before summer solstice. (Image credit: NASA/JPL/Texas A&M)

at the surface. The superrotating zonal winds are in cyclostrophic balance.

At higher altitudes, in the thermosphere, strong day-to-night winds prevail, as a consequence of the large temperature gradient between Venus's thermosphere and cryosphere (Fig. 4.1b). Typical wind speeds are $100 \, \mathrm{m \, s^{-1}}$. The transition region between the retrograde winds in the troposphere and the day-to-night winds in Venus's thermosphere is not well studied. This transition takes place in the mesosphere, a region which can be probed at radio wavelengths in various line transitions of the CO molecule and at infrared wavelengths using heterodyne spectrometry in the line core of the CO_2 molecule. Observations have confirmed the theoretically predicted thermal tides (day-to-night wind) in the upper mesosphere. These winds may also explain the observed day-to-night variations in the CO abundance (§4.3.3.1).

4.5.5.3 *Mars*

Air rises over Mars's summer hemisphere and subsides above the winter hemisphere. Since the warmest latitude does not usually coincide with the equator, the Hadley cells are not confined to the northern and southern hemispheres, but are displaced. Local topography with extreme altitude variations, from the deep Hellas basin up to the top of the Olympus and Tharsis ridge, leads to the formation of stationary eddies, while baroclinic eddies form over the winter hemisphere. Mars has substantial condensation flows, where CO_2 freezes out over the winter pole and sublimates above the summer pole.

Since Mars's atmosphere is tenuous, it responds rapidly to the solar heating, leading to strong winds across the terminator, the day–night line. These are the *thermal tide winds*, strong day-to-night winds similar to the winds in Venus's thermosphere. If winds near the surface of a body have a preferred direction and speed, one might expect the formation of dunes. Dune fields have been observed on the terrestrial planets Earth, Venus, and

Mars, as well as Titan (§5.5). Such fields yield clues to the prevailing local wind direction and speed. On arid planets, when such winds exceed $\sim$50–100 $\mathrm{m \, s^{-1}}$, they may start local dust storms, either initiated by *saltation*, where grains start hopping over the surface, or when the dust is raised up in *dust devils*, due to convection in an atmosphere with a superadiabatic lapse rate. Such dust devils are frequently seen in desert areas on Earth. They resemble funnel-like chimneys (narrow at the base, broadening out towards the top), through which hot air and dust rises up while spinning. Dust devils and their tracks have been photographed regularly both by Mars orbiters and landers/rovers (Figs. 4.29, 4.30). Dust devils can be hundreds of meters in diameter, and several km high. They occur most frequently during spring and summer. They usually leave a dark streak behind where dust has been removed (light colored streaks have been seen too), sometimes kinked and curved due to their swirly motion (Fig. 4.30a). Once in the air, dust fuels the tidal winds, since the grains absorb sunlight and heat the atmosphere locally. Within just a few weeks, dust storms may grow so large that they envelop the entire planet (Fig. 4.30b). Such global storms may last for several months, and have pronounced effects on Mars's climate (e.g., Fig. 4.2).

Martian GCM models have been used to simulate seasonal variations in the dust, CO_2, and H_2O. Coupling between different atmospheric regions helps studies of the photochemistry in the upper atmosphere and formation of ice on and below the surface. GCM models are key to reconstructing the climate on early Mars, and/or how the climate may have evolved over time while the atmospheric composition and the orbit and/or obliquity of the planet changed.

4.5.5.4 *Titan*

Winds in Titan's atmosphere at altitudes above $\sim$100 km have been measured using disk-resolved heterodyne and microwave spectroscopy from the ground. A

(a)

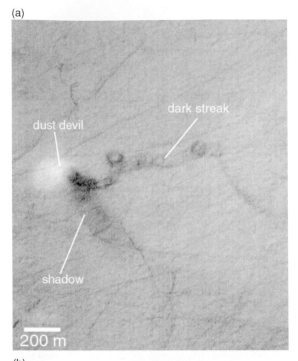

(b)

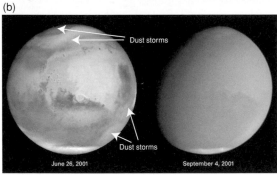

Figure 4.30 (a) Dust devil (∼100 m diameter) and track observed with Mars Global Surveyor. Note the curlicue shape of the track, indicative of the path and spin of the dust devil. (MOC image M1001267, NASA/JPL/Malin Space Science Systems) (b) HST images of Mars at the onset of spring in the southern hemisphere. In June (left image), the seeds of the storm were caught brewing in the giant Hellas Basin, and in another storm near the north pole. Over the months following, surface features got obscured, and by September (right image) the surface could no longer be distinguished. (HST/NASA, J. Bell, M. Wolff, STScI/AURA)

limb-to-limb Doppler shift in the line frequency of the species observed (C_2H_6 at 12 μm, HC_3N at 227.4 GHz, and CH_3CN at 220.7 GHz) indicate high winds (100–200 m/s) in the upper stratosphere (∼200–400 km altitude), dropping down to 60 ± 20 m/s in the lower mesosphere (∼350 − 550 km). At lower altitudes the winds were determined via the Doppler Wind Experiment on the Huygens

probe. While the probe descended through Titan's atmosphere, the radio signal from the probe (communication to the Cassini Orbiter) was recorded by the Very Long Baseline Interferometry (VLBI) network. Winds in Titan's atmosphere affected the horizontal velocity of the probe during its descent, which was measured by the VLBI network through a shift in the probe's transmitted frequency (Doppler shift). These measurements revealed weak prograde winds near the surface, rising to ∼100 m/s, still prograde, at 100–150 km altitude, with a substantial drop (down to a few m/s at most) near 60–80 km altitude.

GCM models for Titan are key to understanding the present, future, and past climate. Titan's surface most likely has a low enough thermal inertia that its surface temperature is regulated by seasonal forcing, which in turn controls Titan's global circulation. A meridional circulation takes place in the stratosphere, where air is rising above the summer hemisphere and descending at the winter pole, so that aerosols accumulate in a polar hood above the winter pole. Several chemical species (such as ethane) condense onto aerosols in the lower stratosphere when brought down in the downwelling branch of the circulation pattern. These condensed species form clouds above the winter pole (such as the ethane cloud at altitudes of ∼40 km, §4.4.3.3). This circulation pattern reverses between the two poles, such that the descending branch is always over the winter pole. Above the summer pole, GCM models show a secondary circulation cell at altitudes between 50 and 200 km, where the circulation is in the opposite direction.

The meridional circulation extends down into the troposphere, with descending air above the winter pole. The circulation above the summer hemisphere is more complicated, in part because methane condensation plays a role. In the lower 10–20 km, air rises along slant-upwards paths from near the equator to a latitude of ∼45°; at this latitude air rises vertically upwards above ∼20 km, and connects to the stratospheric meridional circulation at higher altitudes. Air descends at a latitude ∼60°. Additional secondary circulations are seen over the poles. The 40–60° latitude region is somewhat analogous to the intertropical convergence zone (ITCZ) on Earth. The GCM models correctly predict clouds to appear over the pole near summer solstice, when the surface temperature reaches a maximum and triggers vigorous convection. Clouds are also expected, and have been observed, at mid-latitudes (near 40°S). Based on GCM calculations for a body with a limited supply of liquid methane on its surface, there could be a net drying of the surface at low latitudes, analogous to what is seen in the deserts on Earth just polewards of the tropics. This might explain the general absence of clouds

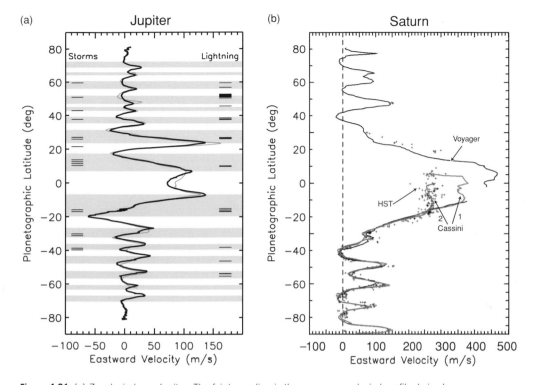

Figure 4.31 (a) Zonal winds on Jupiter. The faint gray line is the average zonal wind profile derived from Voyager 2 images (1979), and the heavy black line is the profile derived from Cassini data (2001). The gray regions represent the belts, white the zones. Also indicated are the occurrence of lightning and convective storms. (Vasavada and Showman 2005) (b) Zonal winds on Saturn. One of the solid lines is the averaged profile as derived from Voyager 1 and 2 images (1980–1981; wavelength: broad-band green). The crosses mark the wind velocities derived from HST data between 1995 and 2002 (wavelength: broadband red). The two other solid lines are data as measured by Cassini (wavelengths: 1: methane band and 2. broadband red). The wind velocities are based on the Voyager period (vertical dashed line). (Adapted from Del Genio *et al.* 2009)

and liquids at the equatorial and mid-latitude regions at the present time. We note, though, that data keep pouring in, and our picture of Titan with regard to these latitudes may change in the near future. Several clouds have already been spotted near the equator on several days in April 2008.

4.5.6 Observations: Giant Planets

Wind velocities on terrestrial planets are measured with respect to the planet's surface. Giant planets lack such a solid surface. The winds on these planets are measured with respect to the rotation rate of the planets' magnetic fields. It is assumed that the fields are 'anchored' in the planet's interior, and that the rotation rates represent the 'true' rotation of the planet's interior. The rotation period is determined from low-frequency radio emissions, as discussed in Chapter 7.

High-velocity zonal winds have been observed on all four giant planets (Figs. 4.31, 4.32). Jupiter and Saturn

have several jets in each hemisphere (5–6 on Jupiter, 3–4 on Saturn), of which the equatorial jet stream is by far the strongest: $\sim$100 m s^{-1} for Jupiter, and 470 m s^{-1} for Saturn; the equatorial jets on both of these planets move in the eastward direction (faster than the planet rotates). On Uranus and Neptune the wind near the equator lags behind the planet's rotation, and blows westward on Neptune and eastward on Uranus (Uranus rotates in the retrograde sense). On Uranus, the wind speed is $\lesssim$100 m s^{-1}, but on Neptune speeds of $\sim$350 m s^{-1} are reached. At higher latitudes ($\gtrsim$20° on Uranus, $\gtrsim$50° on Neptune) the winds become prograde, and reach velocities of $\sim$200 m s^{-1}.

The zonal wind patterns on the giant planets are very stable over time. For example, on Jupiter, although the jets appear correlated with its white zones and brown belts (Fig. 4.31a), the banded structure sometimes changes morphology rather drastically (entire bands may disappear or change color), while the zonal winds remain relatively stable. Saturn looks different. As shown in Figure 4.31b,

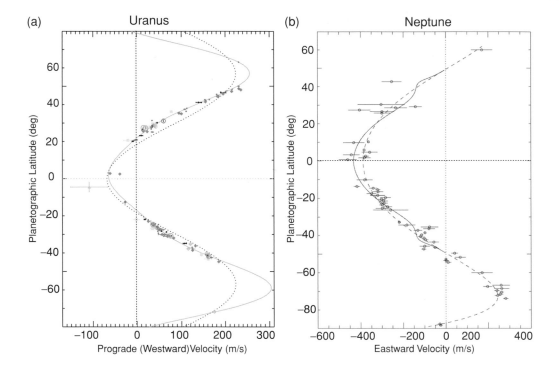

Figure 4.32 (a) Zonal winds on Uranus. A compilation of data derived from HST, Keck, and Voyager data from 1986 to 2008. The dotted curve is a symmetric fit to the Voyager zonal wind profile. The solid line is an average high-order polynomial fit to all the data. (Adapted from Sromovsky *et al*. 2009) (b) Zonal winds on Neptune. The data are bin-averaged wind velocities based on cloud features in Voyager images and the Voyager radio occultation results. The dashed curve is an empirical fit to the Voyager observations, and the solid line is an empirical fit to HST data from 1995–1998. (Adapted from Sromovsky *et al*. 2001)

the wind speed in the equatorial jet appears to have slowed since the Voyager flybys. The measurements were obtained at different wavelengths, though, which probe different hazes/clouds in the atmosphere. Moreover, the altitude of the haze layers has changed over time: it was at a higher altitude during the HST–Cassini era (∼70 mbar) than when Voyager flew by (∼200 mbar). Hence the various measurements in Figure 4.31b indicate wind speeds at different altitudes, and reveal a decay in velocity with altitude that is more or less consistent with the thermal wind equation (and as measured on Jupiter).

The wind profile on Uranus has now been measured almost from pole-to-pole, thanks to changes in viewing geometry (Fig. 4.32a). During the Voyager era, the south pole of Uranus was facing the Sun, and in 2007–2008 the rotational equator and ring plane were viewed edge-on. The wind profile has remained remarkably stable over the past decades. On Neptune, after binning the data within each latitude band, the profile appears to also be well defined. Within individual latitude bands, though, clouds

move substantially with respect to each other, a phenomenon not seen as such on the other three giants.

The vertical extent of the winds and the mechanisms that create them (forcing models) are still topics of debate. The winds may be confined to a relatively thin weather layer, as on Earth, or may extend very deep in the atmosphere. The winds may be forced from below (*deep-forcing models*), or from above by solar heating (*shallow-forcing models*).

It has been suggested that the interiors of Jupiter and Saturn may consist of large cylinders, *Taylor columns*, each of which rotates at its own speed (Fig. 4.33). The zonal winds would be the surface manifestation of these cylinders. Laboratory experiments in the 1920s showed the tendency of fluids in a rotating body to indeed align with the rotation axis. Theoretically, the existence of such columns can be derived for a geostrophic flow from equation (4.53) (small Rossby number, no friction), if the fluid is barotropic. This model was applied to Jupiter and Saturn by F.J. Busse in the 1970s, and has gained much support,

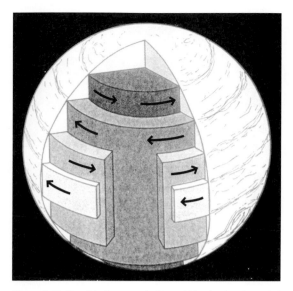

Figure 4.33 Possible large-scale flow within the interiors of the giant planets. Each cylinder has a unique rotation rate, and the zonal winds on the giant planets may be the surface manifestation of these flows. (Adapted from Ingersoll 1999)

Figure 4.34 COLOR PLATE Image of Jupiter at visible wavelengths, taken by the Voyager spacecraft in 1979. The white zones, brown belts, and the Great Red Spot are prominent features, as well as the satellites Io (near the GRS) and Europa. Two of the large White Ovals are visible at latitudes immediately south of the GRS, and a Kármán vortex street of white anticyclonic ovals staggered with filamentary cyclones just south of Europa. (Voyager 1/NASA, PIA00144)

since the winds on both Jupiter and Saturn are somewhat symmetric around the equator. Numerical simulations show, though, that jets similar to those observed on Jupiter can only arise from these deep convection layers if, in fact, these layers are thin shells pertaining to the outer 10% of the planet. As discussed in Chapter 6, interior to $\sim 0.9\ R_{2\downarrow}$, hydrogen takes on a phase referred to as fluid metallic hydrogen, and the convection model may naturally be confined to the outer 10% of Jupiter's radius.

The winds may also be entirely triggered by solar radiation combined with Jupiter's rapid rotation, as in the shallow-forcing model. In this case the zonal winds may be confined to the upper atmospheric layers, as in a thin weather layer, or they may extend as deep as in the deep-forcing model. In either model, numerical simulations can produce the large number of jets as observed on Jupiter and Saturn. The Galileo probe measured an increase in wind speed with depth, indicative of a deep structure of the jets. Unfortunately, the structure of the jets cannot be used to constrain the forcing model. The jets may result from either the shallow- or deep-forcing model, or some combination of the two.

4.5.6.1 *Jupiter*

Figure 4.34 shows an image of Jupiter at visible wavelengths, taken by the Voyager spacecraft in 1979. The white zones, brown belts, and the GRS feature prominently in this view of Jupiter. Immediately south of the

GRS are three large White Ovals, and in the South Equatorial Belt one can discern a row of smaller white ovals. The circulation in these eddies is almost always clockwise in the northern hemisphere and anticlockwise in the south, indicative of high-pressure systems. Many more details can be discerned on this image, such as thunderstorms, filaments, anticyclonic and cyclonic features, all signs of a highly dynamic atmosphere.

A comparison of visible-light images, where Jupiter is seen in reflected sunlight, with images of the planet's thermal emission at a wavelength of 5 μm (Fig. 4.35), shows a strong correlation between color on optical and temperature on 5 μm images. The white zones are generally slightly colder than the brownish belts, suggesting more opaque clouds in the zones than the belts. The opacity at radio wavelengths is mainly provided by ammonia gas, and the clouds are (almost) transparent. Overall, the belts are radio-bright, implying a relatively low ammonia gas abundance, so deep warm layers are probed. All observations together indicate gas rising in the zones, with subsidence in the belts. As a parcel rises, ammonia gas freezes out (NH_4SH, NH_3-ice) when its partial pressure exceeds the saturated vapor curve. Dry air subsides in the belts, and this forces a simple convection pattern. Because the air above the belts is relatively dry, the ammonia-ice cloud layer is either absent or rather thin, and hence not seen at infrared wavelengths. The rising/subsiding air motions produce a latitudinal temperature gradient between belts and zones, which drives the zonal winds on the planet (through geostrophic balance with the Coriolis force).

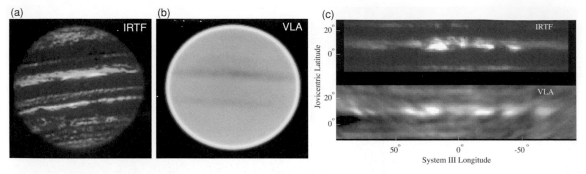

(a) IRTF

(b) VLA

(c)

Jovicentric Latitude

System III Longitude

Figure 4.35 Images of Jupiter's thermal emission at wavelengths of (a) 5 μm taken with the IRTF telescope (Courtesy Glenn Orton), and (b) 2 cm, taken with the VLA (de Pater *et al.* 2001). The IRTF image was taken on 3 October 1995, and the VLA image on 25 January 1996. The Galileo probe entered Jupiter's atmosphere on 7 December 1995. The VLA image, with an angular resolution $\sim 1.4'' \approx 0.044 R_5$ is integrated over 6–7 hours, so that longitudinal structure is smeared out. In this presentation, the 'colors' are inverted: the dark bands across the disk represent a high temperature, while the light 'ring' around the planet represents a low temperature (limb darkening). (c) Detailed comparison of Jupiter's North Equatorial Belt on the two images; the radio data were processed in a novel way to reveal longitudinal structure. (Sault *et al.* 2004)

We note, though, that the simple picture of rising motions in the zones and subsidence in the belts is oversimplified. The Cassini spacecraft detected numerous small-sized (typically a few hundred km diameter) convective storms predominantly in the belts. Lightning also appears to be confined to the belts (Fig. 4.31a). Hence within regions dominated by subsiding motions, there is vigorous convection from deep layers in localized areas. This may in fact be key to explaining the observed altitude profiles of condensable gases on Jupiter (Fig. 4.15c).

Figure 4.35c shows a detailed comparison between a longitude-resolved radio and infrared image of the North Equatorial Belt, featuring several infrared *hot* (or dry) *spots*. The correlation between hot spots at radio (dry air; devoid of NH_3 gas) and 5 μm (devoid of clouds) is perfect, confirming that the hot spots are regions of downdrafts. The Galileo probe entered Jupiter's atmosphere in the brightest of these hot spots. Hence the abundances of the gases measured *in situ* at the deepest levels probed (10–20 bar) are likely to be more representative of Jupiter's envelope as a whole than are abundances measured at higher levels.

The Great Red Spot was first seen by G.D. Cassini in 1665. There is some debate whether Cassini saw the present-day GRS, since during a period in the nineteenth century no red spots were seen. However, we know from observations throughout the 1980s and 1990s that the appearance of the GRS can change drastically in prominence, size, and shape to the point of essential disappearance, so the system could be very long-lived. The 3 large White Ovals south of the GRS trace back to the 1930s,

and formed part of a Kármán vortex street. The cyclonic eddies stabilizing such streets are visible as elongated filamentary features staggered in between, and slightly north of, the anticyclonic white ovals. A trio of vortices, two anticyclones and one cyclone, moved together in a tight packet between 1996 and 1998 (Fig. 4.28a), rather than oscillating back and forth in longitude. Numerical simulations show that such groups of vortices can move together as a single unit if they are trapped in the trough of a Rossby wave, as shown schematically in Figure 4.27c. The sides of the trough squeeze the three trapped vortices so hard that any small perturbation, for example, a collision with a vortex outside the trough, will cause the cyclone to be pushed out of the trough, leaving the two anticyclones in direct contact. Once contact is made between the two anticyclones, they merge within days. Figure 4.28b shows a series of computer simulations visualizing this process. This scenario of vortex trapping followed by cyclone ejection and then merger is probably what happened in 1998 and again in 2000, when the three White Ovals, seen so prominently in the Voyager era (Fig. 4.34), merged.

After the second merger of the White Ovals, the resulting storm remained white, but in late 2005 it turned red. This new Red Oval is similar in color to the GRS (Fig. 4.36). The panels on the right of Figure 4.36a show images at 330, 550, and 892 nm from top to bottom, respectively. The image at 892 nm, in a methane absorption band, shows that both red spots are bright, indicative of material at high altitudes. Clouds at lower altitudes are dark, since sunlight at this wavelength is absorbed by methane gas in Jupiter's atmosphere, and hence much attenuated on its

(a)　　　　　　　　　　　　　　　　　　　　(b)

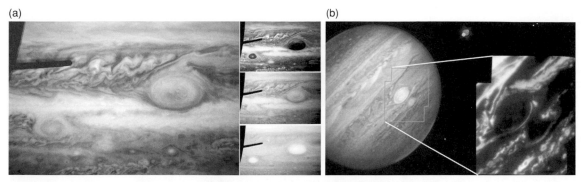

Figure 4.36 COLOR PLATE (a) Deprojected image of the GRS and Red Oval on Jupiter obtained with the high-resolution camera ACS/HRC on HST, on 25 April 2006. Each pixel spans 0.05 deg in latitude and longitude, with the top of the image lying just along the equator. This visual image is constructed from images in a red (F658N), green (F502N), and blue (F435W) filter. The small black-and-white images on the side are deprojected HST images of the GRS and Red Oval from 24 April 2006, at different wavelengths (330, 550, and 892 nm from top to bottom, respectively). The black protrusion on the images is the occulting finger in the ACS camera. (b) [Left] A false-color composite near-infrared image of Jupiter and its moon Io, taken UT 21 July 2006 with the Keck II telescope on Mauna Kea, Hawaii, using adaptive optics to sharpen the image. Images taken in narrow band filters centered at 1.29 and 1.58 microns (shown in gold in this image) detect sunlight reflected off Jupiter's upper cloud deck – the same clouds that are seen in visible light. The narrow band image at 1.65 micron (shown in blue) shows sunlight reflected back from hazes lying just above these clouds. [Right] A close-up of the two red spots through a 5 micron filter, which samples thermal radiation from deep in the cloud layer. (Adapted from de Pater *et al.* 2010)

way into, and, upon reflection, out of the atmosphere. The image at 330 nm show both spots as dark features, due to absorption of sunlight likely by this same high-altitude material. Figure 4.36b is an image at infrared wavelengths that shows the spots when they passed each other. The 5 μm insert reveals both spots to be cold, and thus confirms their high altitudes. As appears to be typical for anticyclonic eddies, both storm systems are surrounded by narrow 'clear' regions, where the thermal heat from deeper layers leaks through. We note, though, that in contrast to the large ovals (GRS, new Red Oval), small eddies are usually tightly surrounded by 5 μm rings.

While the white color of the White Ovals could probably be attributed to ammonia ice, no one knows for certain what makes the Great Red Spot or the new Red Oval, red. Since both spots rise up well above the surrounding cloud deck, the color may be tied to the altitude of the spots, and caused by a coloring agent contaminating the ice particles. The contaminant or *chromophore* could be a minor constituent of the upper atmosphere only encountered by high clouds, or it could be associated with material dredged up from deeper in the planet's atmosphere by the vortices. Potential sources for red chromophores are phosphine gas (PH_3) or ammonium hydrosulfide particles (NH_4SH), which, after been carried upwards by the storm systems, would be broken down by ultraviolet sunlight

and undergo a series of chemical reactions which produce the red chromophores (§4.6.1.3). Another hypothesis is that the red particles (perhaps the chromophores discussed above) are usually coated with white ammonia ice. If the temperature exceeds the sublimation temperature of the ice, the true color of the condensation nuclei is revealed. Finally, it is also possible that the color is merely caused by the particle size distribution.

Every 10–40 years, Jupiter appears to go through a *global upheaval*, a period of dramatic turmoil in its atmosphere that lasts for a few years. It has been proposed that such upheavals may be manifestations of climate cycles, and that the change in color from white to red is just one of the more dramatic changes that have beeen witnessed.

4.5.6.2 *Saturn*

Cassini images of Saturn show that this planet, like Jupiter, displays a large variety of atmospheric phenomena, though the features are not as prominent as on Jupiter (Fig. 4.37). Clear zonal bands can be discerned, usually with a mottled texture and sometimes eddies on their edges. In the southern hemisphere, near $\sim35°$S, bright eruptions have been detected, indicating moist convection. The Radio and Plasma Wave Science (RPWS) instrument has detected simultaneous broadband (2–40 MHz) bursts, referred to

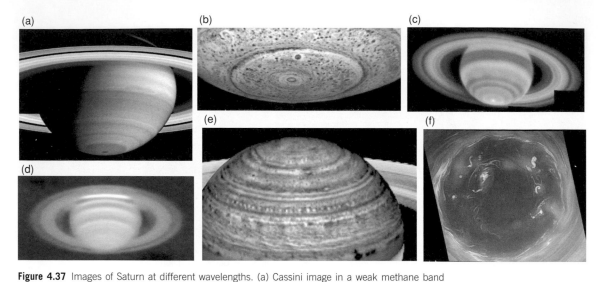

Figure 4.37 Images of Saturn at different wavelengths. (a) Cassini image in a weak methane band at 727 nm, highlighting bands of clouds at high altitudes. Note the dark spot at the south pole. (NASA/JPL/SSI, PIA5391) (b) Thermal emission from Saturn's south polar region, imaged by Cassini VIMS at 5 μm. (NASA/JPL/University of Arizona, PIA11214) (c) Thermal emission from Saturn and its rings at a wavelength of 17.65 μm, imaged with the Keck telescope. Note the bright dot at the south pole, in contrast to the dark spot in panel (a). Particles in Saturn's ring which rotate into daylight are cold; they heat up during the day, being warmest (brightest) on the evening side. (NASA/JPL, PIA07008) (d) Thermal emission from Saturn's deep atmosphere observed with the VLA on UT 27 September 2002 at a wavelength of 3.6 cm. To bring out structures on the disk, and simultaneously see Saturn's rings (in reflected Saturn light, §11.3.2.2), a uniform disk with a brightness temperature of ~130 K was subtracted. (Adapted from Dunn *et al.* 2007) (e) Thermal 5 μm emission from Saturn's northern hemisphere, imaged by Cassini VIMS, revealing a hexagon-shaped vortex (see also Fig. 4.46b) near a latitude of 78°. Near 40°, note the well-defined bright bands and the 'string of pearls' spaced every 3.5° in longitude over a 60 000 km stretch. This may be caused by a large planetary wave. (NASA/JPL, PIA01941) (f) High-resolution Cassini image of Saturn's south polar vortex, constructed from images at 617 and 750 nm taken on 14 July 2008, at a resolution of 2 km/pixel. The image spans ~2500 km. Vigorous convective storms are visible inside the brighter ring, or 'eyewall' of the hurricane. (NASA/JPL/Space Science Institute, PIA11104)

as *Saturn Electrostatic Discharges* (SED), coming from lightning associated with the storms.

Images at 5 μm are sensitive to thermal emission from Saturn's deep atmosphere, which can be detected in areas devoid of clouds. The patterns as observed by Cassini/VIMS in both the southern and northern hemisphere are striking (see Fig. 4.37b and e). Each pole is characterized by a giant vortex. Above the north pole, the vortex is shaped like a hexagon, probably a hexagonal wave pattern. The south polar vortex resembles a hurricane (cyclone), with a clearly defined 'eyewall', which on Earth is made up of a ring of towering thunderstorms. A high-resolution Cassini image (Fig. 4.37f) reveals numerous small storm systems inside the eye, although the overall clear air within the vortex eye indicates that most of the air is descending.

At mid-infrared (8–24 μm) wavelengths, thermal emission from Saturn's stratosphere and upper troposphere is observed. Data shown in Figure 4.37c, taken just after Saturn's southern summer solstice, show a distinctive warming trend from the equator to the south pole, as a result of strong seasonal forcing. The hot spot at the south pole coincides with the dark (i.e., cloud-free) spot at visible and near-infrared wavelengths (Fig. 4.37a). Both radiative forcing and atmospheric dynamics (downdrafts at the pole) appear to be required to explain the high stratospheric temperature and low tropospheric cloud coverage.

Radio observations (2–20 cm) during Saturn's northern summer showed enhanced brightness temperatures at mid-northern latitudes, indicative of a decreased opacity, most likely NH_3 gas, which is the dominant source of opacity at radio wavelengths. Together with Voyager infrared observations, which indicated a thinner ammonia ice cloud at mid-northern latitudes compared to other latitudes, the radio data suggest subsiding air motions at mid-latitudes, down to at least the ~5 bar level. Radio data

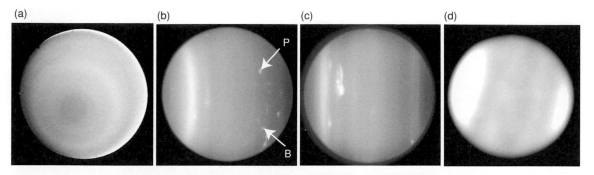

(a) (b) (c) (d)

Figure 4.38 Images of Uranus at different wavelengths. (a) Visible-light Voyager 2 image from 1986, with enhanced variations of albedo (NASA/JPL). (b) Image of Uranus at a wavelength of 1.6 μm, taken in July 2004 with the Keck telescope, equipped with adaptive optics. Numerous small clouds, as well as the moons Belinda and Portia (indicated by arrows), are visible in the northern hemisphere (right side of image). The faint straight line is caused by reflection off Uranus's rings. (H.B. Hammel and I. de Pater) (c) Keck 1.6 μm image from August 2007, a few months before equinox. Note the bright storm in the southern hemisphere. (Sromovsky *et al.* 2009) (d) Thermal emission observed with the VLA at a wavelength of 1.3 cm, on 26 May 2005. This map was constructed from ~7 hours of data, so longitudinal features are smeared by the planet's rotation. The bright regions near the poles imply a relative lack of absorbing gases (H_2S, NH_3) above both poles. (Courtesy Mark Hofstadter and Bryan Butler)

taken during southern summer show a series of warmer and colder bands at southern latitudes (Fig. 4.37d). A detailed analysis of all datasets combined will help determine the global dynamics and seasonal changes in Saturn's atmosphere.

4.5.6.3 *Uranus*

During the Voyager encounter, Uranus seemed a rather 'boring' planet, without prominent cloud features or convective storms. Since Uranus is tipped on its side (Fig. 4.38a), it was expected (based upon a simple insolation model) that Uranus's pole would be warmer than its equator by ~6 K. However, Voyager infrared measurements indicate that the temperature is remarkably similar at all latitudes, which suggests a redistribution of heat in Uranus's atmosphere or interior. A similar redistribution of heat takes place in the other three giant planets, where the equatorial region receives most of the heat, yet the infrared emission from the poles and equator are very similar. However, because Uranus's internal heat source is so small (§6.1.5), it is difficult to explain strong convection on this planet. This view was amplified by the apparent lack of convective motions in the form of eddies and storm systems during the Voyager era. During the 1990s and into the twenty-first century, however, prominent cloud features became visible in the northern hemisphere when it rotated into sunlight after spending ~42 years in darkness (Fig. 4.38b). During the last few years before the 2007–2008 equinox, eye-catching clouds developed in the southern hemisphere.

The cloud-complex featured in Figure 4.38c may have been around since 1994, but since 2005 it changed drastically in morphology. Around the same time, it started moving towards the equator, where it likely will dissipate (eq. 4.54).

Large-scale atmospheric motions have always been present, as infrared observations from Voyager suggested rising air at latitudes between 20° and 40° on the southern hemisphere, and subsiding air at other latitudes. Radio observations are consistent with this global circulation, and imply subsiding motions at both poles, down to depths of ~50 bar (Fig. 4.38d).

4.5.6.4 *Neptune*

At visible and near-infrared wavelengths, Neptune's appearance is strikingly different from that of Uranus. During the Voyager era (1989) Neptune was characterized by the Great Dark Spot (GDS), a smaller dark spot (DS2) in the south, and a small bright cloud feature, referred to as Scooter, moving faster than either dark spot (Fig. 4.39a). When Neptune was imaged several years later by HST, all three features had vanished. Since the late 1990s, the planet has been imaged regularly by HST and the 10-m Keck telescope equipped with adaptive optics (§E.6). These data show that Neptune has changed dramatically since 1989, and that its appearance continues to change over time (Fig. 4.39). These changes may reflect seasonal variations, as suggested by changes in mid-IR spectra, in particular in the ethane emission band. The mid-IR observations suggest a steady ~20 K increase in

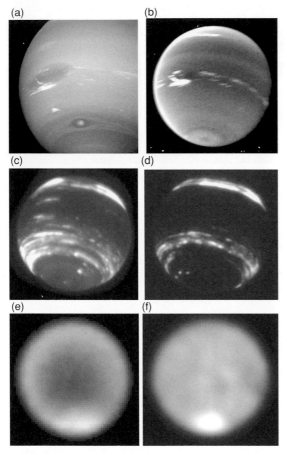

Figure 4.39 (a) Voyager 2 image of Neptune at visible wavelengths, taken in 1989, revealing the Great Dark Spot (GDS), the Little Dark Spot and Scooter (the bright white cloud feature in between the dark spots). The GDS is located relatively deep in the atmosphere, in contrast to the white hazes which are at higher altitudes. (NASA/JPL) (b) A Voyager 2 image in the methane band (890 nm) shows that the white clouds are at high altitudes. (NASA/JPL) (c) Keck adaptive optics image at a wavelength of 1.6 μm, in a methane absorption band, taken on 5 October 2003. (d) Keck adaptive optics image at a wavelength of 2.2 μm, where methane gas is strongly absorbing, taken on 3 October 2003. (Panels c, d from: de Pater *et al.* 2005) (e) Thermal infrared image at 11.7 μm, in the ethane absorption band, obtained with the Long Wavelength Spectrometer (LWS) on the Keck telescope. The disk is limb brightened with a bright south pole. (f) Thermal emission from Neptune's deep atmosphere, as observed with the VLA at a wavelength of 2 cm. (Panels e, f from: Martin *et al.* 2008)

the effective stratospheric temperature between 1985 and 2003; after 2003 the temperature started to decrease.

Disk-resolved images at mid-infrared and radio wavelengths (Fig. 4.39e,f) reveal a bright south pole on Neptune, i.e., the pole appears to be warm both in the stratosphere (probed at mid-infrared wavelengths) and in

the troposphere, down to tens of bars (as observed at radio wavelengths). This may be caused by a global circulation pattern in the atmosphere: as on Uranus, subsidence of dry air above the pole allows one to probe deep warm levels at radio wavelengths, while a similar subsidence in the stratosphere might lead to adiabatic heating there. In addition, in analogy with Saturn, the stratosphere above the south pole may simply be warm because it is summer in the south, and the pole is bathed in sunlight. Regardless of the origin of the relatively high temperature in the stratosphere and near the tropopause, it allows methane gas to convect upwards through the tropopause into the stratosphere, without condensing into clouds (§4.7.2.2).

4.6 Photochemistry

All planetary atmospheres are subjected to solar irradiation, which both heats the atmosphere and can change the atmosphere's composition. Typically, absorption of photons at far-infrared and radio wavelengths ($\lambda \gtrsim 100$ μm) induces excitation of a molecule's lowest quantum states, i.e., of the rotational levels. Photons at infrared wavelengths ($\lambda \sim 2$–20 μm) can excite vibrational levels, while photons at visible and UV wavelengths may excite electrons to higher quantum states within atoms and molecules. Photons with $\lambda \lesssim 1$ μm may break up molecules, a process referred to as *photodissociation* or *photolysis*. Photons at higher energies, $\lambda \lesssim 100$ nm, may *photoionize* atoms and molecules. The solar blackbody curve (Fig. 3.2b) peaks at visible wavelengths, and the number of UV and higher energy photons drops significantly with decreasing wavelength. The penetration depth of any solar photon into an atmosphere depends upon the optical depth at the particular wavelength of radiation, which depends on clouds, hazes, Rayleigh scattering (§3.2.3.4), and absorption by molecules/atoms. Since the optical depth is particularly large for high-energy photons, most of the photochemical reactions occur at high altitudes. If the production rate of a particular species created via photochemistry is balanced by its loss rate, there is *photochemical equilibrium*. In the subsections below we assume the species to be in photochemical equilibrium, and under these conditions we derive information on the altitude distribution of such species.

4.6.1 Photolysis and Recombination

Photodissociation typically takes place at high altitudes, whereas the reverse reaction, *recombination*, proceeds faster at lower altitudes. Therefore, the balance between dissociation and recombination may well be affected by vertical transport. In this section we discuss the most important photochemical reactions on Earth, Venus, Mars, Titan, and the giant planets.

4.6.1.1 Oxygen Chemistry on Earth

The reactions for photodissociation (reaction 1 below) and recombination (reactions 2–4) for oxygen in the Earth's atmosphere can be written as:

Photodissociation:

(1) $O_2 + h\nu \rightarrow O + O$, $\lambda < 175$ nm.

The production rate at altitude z:

$$\frac{d[O]}{dt} = 2[O_2]J_1(z), \tag{4.55}$$

where a compound in square brackets, e.g., [O], refers to the number of O atoms per unit volume, and the subscript 1 in $J_1(z)$ refers to reaction 1. The photolysis or photodissociation rate, $J(z)$, is given by

$$J(z) = \int \sigma_{x_\nu} \mathcal{F}_\nu e^{-\tau_\nu(z)/\mu_\theta} d\nu, \tag{4.56}$$

where σ_{x_ν} is the photon absorption cross-section at frequency ν, μ_θ is the cosine of the angle between the solar direction and the local vertical, and $\mathcal{F}_\nu$ is the solar flux density impinging on the atmosphere (expressed in photons cm^{-2} s^{-1} Hz^{-1}). The mean photolysis rate for oxygen at an altitude of 20 km is $J_1(20 \text{ km}) = 4.7 \times 10^{-14}$ s^{-1}; at an altitude of 60 km: $J_1(60 \text{ km}) = 5.7 \times 10^{-10}$ s^{-1}. Since the number density of solar photons decreases exponentially with optical depth (thus with decreasing altitude), and the number of oxygen molecules decreases with increasing altitude according to the barometric law (eq. 4.1), the concentration of oxygen atoms increases with altitude.

Recombination: the direct, two-body reaction

(2) $O + O \rightarrow O_2 + h\nu$,

is very slow, and oxygen recombination is therefore dominated by the three-body processes:

(3) $O + O + M \rightarrow O_2 + M$
(4) $O + O_2 + M \rightarrow O_3 + M$,

where M is an atmospheric molecule which takes up the excess energy liberated in the reaction. Since the abundance of M and O_2 follows the barometric altitude distribution, reactions (3) and (4) are most effective at low altitudes, provided atomic oxygen is present.

The production rate of [O_2] in (3) is given by

$$\frac{d[O_2]}{dt} = [O]^2[M]k_{r3}, \tag{4.57}$$

where the reaction rate, k_{r3}, (cm^6 s^{-1}) is the rate at which the three-body reaction proceeds (the subscript $r3$ stands

for reaction 3). Reaction rates depend upon the collisional rates between molecules, which in turn depend on the temperature of the medium. A two-body reaction rate (cm^3 s^{-1}), as in reaction (2), is usually expressed as

$$k_{r2} = c_1 \left(\frac{T}{300}\right)^{c_2} e^{-E_o/kT}, \tag{4.58}$$

where E_o is the *activation energy* to overcome the potential barrier that reaction (2) might have. Note that the exponential in equation (4.58) is essentially a Boltzmann factor (§3.2.3.3), and recombination does not take place if $E < E_o$. The temperature, T, is expressed in kelvin; c_1 and c_2 are constants. An upper limit to k_{r2} is given by the gas-kinetic rate of collisions, k_{gk} (where every atom sticks upon collision):

$$k_{gk} = \sigma_x \bar{v}_o \approx 2 \times 10^{-10} \sqrt{\frac{T}{300}}, \tag{4.59}$$

where the collisional cross-section, σ_x, for atmospheric molecules is typically a few $\times 10^{-15}$ cm^2, and $\bar{v}_o$ is the mean (most probable) thermal velocity of the particle ($\sqrt{2kT/m}$).

For a three-body interaction as in reactions (3) and (4), rate k_{r2} has to be multiplied by the chance a third molecule collides at the same time with the oxygen atoms/molecules. The duration of a collision is typically of the order of $2R/\bar{v}_o$, where R is the molecular radius, typically a few tenths of a nm. The three-body reaction rate, k_{r3} for reaction (3), becomes

$$k_{r3} = \frac{2R}{\bar{v}_o} k_{r2}^2 \approx 10^{-12} k_{r2}^2. \tag{4.60}$$

One can show that for gas-kinetic collisions (assuming $k_{r2} = k_{gk}$), reactions (2) and (3) above would proceed equally fast if the atmospheric density at Earth's surface were $\sim$200 times larger than its measured value, i.e., *Loschmidt's number* ($n_o = 2.686 \times 10^{19}$ cm^{-3}) (Problem 4.30). Three-body interactions can therefore usually be ignored, unless the 2-body reaction rate is unusually low, i.e., if $k_{r2} \ll k_{gk}$. Since the reaction rate for oxygen recombination, reaction (2), is very low, $k_{r2} < 10^{-20}$ cm^3 s^{-1}, oxygen recombination usually proceeds via the three-body reaction (3) ($k_{r3} = 2.76 \times 10^{-34} e^{-(710/T)}$ cm^6 s^{-1}).

Ozone, O_3, is produced in chemical reaction (4), where $k_{r4} = 6 \times 10^{-34} \left(\frac{T}{300}\right)^{-2.3}$ cm^6 s^{-1}. Ozone is very important for life on Earth, since it effectively blocks penetration of UV sunlight to the ground. To deduce the vertical distribution of ozone in our atmosphere, we must consider both the processes which lead to its formation and to its destruction. In a pure oxygen atmosphere the relevant processes are the *Chapman reactions* (reactions 1, 4, 5,

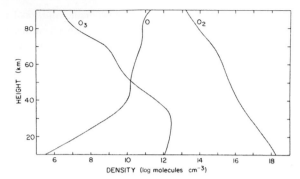

Figure 4.40 Graph of calculated densities of O, O_2, and O_3 in the Earth's atmosphere. (Chamberlain and Hunten 1987)

and 6), where ozone is formed in reaction (4) and destroyed by:

Photodissociation:

(5) $O_3 + h\nu \rightarrow O_2 + O$,
$\lambda \lesssim 310$ nm, $J_5(60$ km$) = 4.0 \times 10^{-3}$ s^{-1},
$J_5(20$ km$) = 3.2 \times 10^{-5}$ s^{-1},

or the reaction:

(6) $O + O_3 \rightarrow O_2 + O_2$,
$k_{r6} = 8.0 \times 10^{-12} e^{-2060/T}$ cm^3 s^{-1}.

The net change in the atomic oxygen and ozone number densities is:

$$\frac{d[O]}{dt} = 2J_1(z)[O_2] + J_5(z)[O_3] - k_{r6}[O][O_3] - k_{r4}[O][O_2][M] \tag{4.61}$$

$$\frac{d[O_3]}{dt} = k_{r4}[O][O_2][M] - k_{r6}[O][O_3] - J_5(z)[O_3], \tag{4.62}$$

where the subscripts to J and k_r refer to reactions (1)–(6). In chemical equilibrium, the net change [O] and [O_3] is equal to zero. This leads to the altitude profiles of atomic oxygen and ozone (Problem 4.28):

$$[O] = \frac{J_1(z)[O_2]}{k_{r6}[O_3]} \tag{4.63}$$

$$[O_3] = \frac{k_{r4}[O][O_2][M]}{k_{r6}[O] + J_5(z)}. \tag{4.64}$$

The altitude distributions of atomic and molecular oxygen in our atmosphere together with ozone are shown in Figure 4.40. As expected from photolysis arguments, the number of oxygen atoms increases with increasing altitude up to a certain altitude, while the number of oxygen molecules is decreasing with altitude. Ozone peaks in number density at altitudes near 30 km.

In reality our atmosphere is not a pure oxygen atmosphere, and *catalytic* destruction of ozone takes place in addition to reactions (5) and (6) given above. Given that

ozone protects life on Earth from harmful solar UV photons, much research is currently devoted to the chemistry of production and destruction of ozone. Free hydrogen atoms are highly reactive with ozone, and readily destroy O_3:

(7) $H + O_3 \rightarrow OH + O_2$,
$k_{r7} = 1.4 \times 10^{-10} e^{-470/T}$ cm^3 s^{-1}.

Free atomic hydrogen on Earth is produced by chemical reactions starting from H_2O and CH_4, but it is not very abundant in the Earth's atmosphere.

Nitric oxides, chlorine, and halomethanes form 'important' catalysts for the destruction of ozone. These molecules react with O_3, and are immediately regenerated through a reaction with atmospheric oxygen. The abundance of these molecules is thus constant:

(8a) $NO + O_3 \rightarrow NO_2 + O_2$
$k_{r8a} = 1.8 \times 10^{-12} e^{-1370/T}$ cm^3 s^{-1}
(8b) $NO_2 + O \rightarrow NO + O_2$
$k_{r8b} = 9.3 \times 10^{-12}$ cm^3 s^{-1}
(9a) $Cl + O_3 \rightarrow ClO + O_2$
$k_{r9a} = 2.8 \times 10^{-11} e^{-257/T}$ cm^3 s^{-1}
(9b) $ClO + O \rightarrow Cl + O_2$
$k_{r9b} = 7.7 \times 10^{-11} e^{-130/T}$ cm^3 s^{-1}.

Reactions with fluorine atoms are not important, since F is transformed rapidly into HF via a reaction with CH_4; it is not regenerated, in contrast to Cl. Bromine atoms may, like Cl, destroy ozone. Nitric oxides (NO_x) in the Earth's stratosphere lead to a rapid decrease in the ozone abundance. Nitric oxides are generated by biological activities, lightning storms, engine exhausts, and industrial fertilizers close to the ground. Before these nitric oxides can destroy ozone, they must be brought up into the stratosphere. It is therefore not clear how harmful the production of NO_x in the troposphere is; it actually leads to a local production of ozone in the troposphere, through the formation of atomic oxygen. However, O_3 is not desirable in the lower troposphere because it is highly reactive/corrosive.

The penetration of galactic and solar cosmic rays over the polar caps leads to free atomic nitrogen in an excited state, which causes a rapid enhancement in the stratospheric NO abundance. Observations show that the O_3 abundance over the polar caps is indeed anticorrelated with NO. During high solar activity, the solar wind and heliomagnetic field are strong, and shield the Earth from galactic cosmic rays. This may explain a suspected 11-year cycle in the ozone abundance. Solar cosmic rays, however, occur in bursts during solar flares, and are out of phase with the galactic cosmic rays.

4.6.1.2 Venus and Mars

The primary atmospheric constituent on Venus and Mars is CO_2, which is dissociated into CO and O under the influence of solar UV light:

Photodissociation:

(10) $CO_2 + h\nu \rightarrow CO + O$, $\qquad \lambda < 169$ nm

Recombination:

(11) $CO + O + M \rightarrow CO_2 + M$.

Since the recombination rate k_{r11} is very slow, one expects large quantities of CO in the atmospheres of Venus and Mars. In addition, individual oxygen atoms may recombine into molecular oxygen. The reaction rate for the latter process is roughly 10^3–10^4 larger than k_{r11}. The creation of molecular oxygen is balanced by photodissociation (reaction 10). Despite the low recombination rate of CO (reaction 11), the observed CO, O, and O_2 densities on both Venus and Mars are very low. An explanation for this observation on Mars may be rapid downward transport of CO, O, and O_2, where recombination of CO and O proceeds faster in the presence of OH chemistry. Essentially, water vapor provides OH and H:

Photodissociation:

(12) $H_2O + h\nu \rightarrow H + OH$, $\qquad \lambda < 210$ nm

where OH oxidizes CO, so that the net product is restoration of CO_2:

Oxidation:

(13) $CO + OH \rightarrow CO_2 + H$.

Regular photolysis of water in the martian atmosphere may be too slow, however. New ideas being pursued include the creation of OH by electric fields in martian dust storms. The hydrogen atoms liberated by reactions (12) and (13) may eventually lead to the production of hydrogen peroxide (H_2O_2), which may condense (e.g., onto dust grains) and precipitate out of the atmosphere. On the martian surface H_2O_2 oxidizes materials and removes both CO and CH_4 from the atmosphere. On Venus, recombination of CO may be aided by catalytic reactions resulting from chlorine and sulfur chemistry on this planet. Since oxygen is formed on both Venus and Mars, one might expect some

ozone. Destruction of ozone is tied to the H_xO_y chemistry, resulting in a near total absence of ozone on the two planets, except near Mars's winter pole. Above Mars's winter pole, temperatures are so low that H_xO_y products freeze out, allowing ozone to accumulate.

Venus's clouds consist of sulfuric acid, formed by the reaction of sulfur dioxide with (photochemically produced) oxygen and water:

(14) $SO_2 + O \rightarrow SO_3$,

(15) $SO_3 + H_2O \rightarrow H_2SO_4$.

Sulfuric acid readily condenses out in the upper troposphere of Venus, producing an optically thick smog of sulfuric acid droplets.

4.6.1.3 Giant Planets

In the 'visible' part of the H_2-rich atmospheres of the giant planets, N, C, S, O, and P are present in the form of NH_3, CH_4, H_2S, H_2O, and PH_3. Since Rayleigh scattering provides a large source of opacity at UV wavelengths, photodissociation of these species is only important in the upper troposphere and stratosphere, at pressures $P \lesssim 0.3$ bar. Below, we summarize the effects of photodissociation and subsequent chemical reactions for NH_3, CH_4, H_2S, and PH_3.

Ammonia, which is photolyzed by photons with $\lambda < 230$ nm, produces amidogen radicals, $NH_2(X)$. Roughly 30% of these radicals recycle back to NH_3, and the remainder form hydrazine gas (N_2H_4) via a self-reaction (i.e., $NH_2(X)$ with $NH_2(X)$). Hydrazine gas has a low saturation vapor pressure, and consequently we expect it to condense in the upper troposphere, thus contributing to the dust and hazes in the upper atmospheres of the giant planets. Hydrazine is expected to photodissociate and/or react with H to form the condensable gas hydrazyl (N_2H_3). Reactions involving N_2H_3 ultimately lead to N_2, which is stable against photolysis. The observed NH_3 abundance in Jupiter and Saturn's atmospheres near and above the tropopause is well below saturation levels, as expected from photochemistry. On Uranus and Neptune, the tropopause temperatures are so low that NH_3 is completely frozen out well before an air parcel has risen up to the tropopause. Therefore, we do not expect nor see evidence of NH_3 photochemistry on the outer giants. The detection of NH_3 gas in Jupiter's stratosphere after the impact of Comet D/Shoemaker–Levy 9 presented a nice test of the NH_3 photodissociation rate. After the impact of one of the largest fragments (fragment K), NH_3 gas was detected above the impact site. Its subsequent decay was consistent with the photochemical destruction rate, a half-life of ~ 3 days.

Methane gas is dissociated by UV photons with $\lambda < 160$ nm, and the photolysis products undergo a complicated series of chemical reactions. In the 10 mb–10 μb region, the hydrocarbons acetylene (C_2H_2), ethylene (C_2H_4), and ethane (C_2H_6) form. Ethylene is quickly lost by photolysis into C_2H_2 or recycled back to CH_4. Ethane is also converted to C_2H_2 or CH_4, but the reaction rate is about 10 times slower than that of C_2H_4. Hence, the mixing ratio of C_2H_4 is expected to be over an order of magnitude smaller than that of C_2H_6 and C_2H_2. Hydrocarbons have been detected on all four giant planets. On Jupiter, C_2H_6 has been measured at a few ppm, while C_2H_2 and C_2H_4 have mixing ratios 1–2 orders of magnitude less. Relatively copious amounts of hydrocarbons are present on Neptune (see §4.7.2.2), which has been attributed to rapid vertical transport (CH_4 is expected to condense near the tropopause; §4.4.3). In agreement with chemical reactions, propane (C_3H_8) is the most abundant amongst higher order hydrocarbons.

Hydrogen sulfide: Photons at $\lambda < 317$ nm break up H_2S molecules, if there are any present in the upper troposphere and stratosphere. Photolysis and subsequent chemical reactions eventually lead to the formation of different sulfur allotropes, that is different molecular configurations, including chains and rings, of pure sulfur, ranging from S_3 to S_{20}. Ammonium polysulfides ($(NH_4)_xS_y$) and hydrogen polysulfides (H_xS_y) are expected to form as well. Colors of these compounds vary from red to yellow, and they may be the chromophores for the brownish bands on Jupiter and Saturn, and perhaps Jupiter's red spots.

Phosphine (PH_3) should not be present in the atmospheres of the giant planets, since it should oxidize to P_4O_6 deep in their atmospheres (at $300 < T < 800$ K), and dissolve in water. Phosphine is, however, detected on Jupiter and Saturn. At higher altitudes the gas is photolyzed by photons with $160 < \lambda < 235$ nm. Subsequent chemical reactions may eventually lead to the formation of red phosphorus, P_4. A more likely outcome of the photochemistry, though, may be that phosphine combines with ammonia and hydrocarbons (derived from methane) to produce complex polymers or compounds.

4.6.1.4 *Titan, Icy Satellites, and Io*

Titan's atmosphere consists predominantly of nitrogen gas, with a small percentage of methane. As on Earth, N_2 is quite inert. The main source for dissociation of N_2 is impact by charged particles, as opposed to photolysis. Methane gas, on the other hand, is easily dissociated by solar photons. The methane photochemistry is in principle similar to that seen on the giant planets, although it is more efficient since there is far less free atomic hydrogen, as it readily escapes Titan's gravity field. In fact, without a semi-continuous supply of methane gas, photolysis would destroy these molecules within $\sim 10^7$ years, suggestive of a continuous source of methane gas (§§4.9.1.2, 5.5.6.1). As on the giant planets we expect the production of numerous hydrocarbons, including C_2H_2, C_2H_4, C_2H_6, and higher order ones as C_3H_8, C_4H_{10}, etc. Ion chemistry, as interactions between ions and molecules, also leads to the production of hydrocarbons, such as, e.g., benzene.

Reactions between N and CH_4 should lead to the formation of HCN, CN, and more complex nitriles, such as cyanogen (C_2N_2), cyanoacetylene (HC_3N), ethylcyanide (C_2H_3CN), and possibly HCN polymers. Because of the low stratospheric temperature, ethane and other photochemically produced complex molecules condense to form a dense layer of smog in Titan's atmosphere. Laboratory measurements by Carl Sagan and coworkers in a simulated Titan atmosphere show the formation of such 'gunk', a reddish-brown powder, referred to as *tholins*. Above Titan's winter pole, due to the near-absence of solar photons and the low temperatures, complex molecules in gaseous and condensed phases (HC_3N, C_4H_2, polyacetylenes, and HCN polymers) build up over time, and form a polar cap of smog in the atmosphere.

The smog particles ultimately sediment out and fall to the ground. Over time they might have built up a few-hundred-meter thick layer of hydrocarbons. The detection of numerous hydrocarbons and nitriles by the probe GCMS after landing and heating the surface may be indicative of such a layer (§4.3.3.2).

Icy satellites must have atmospheres which contain a small amount of water vapor (§4.3.3.3), which readily dissociates to produce oxygen. Io's atmosphere must contain oxygen as well, from photodissociation of SO_2. In analogy with Earth, we therefore expect oxygen photochemistry on all these satellites. Indeed, HST has detected ozone on Ganymede and several saturnian satellites, although this ozone may be a direct product of the interaction of energetic magnetospheric particles with the water-ice on satellites' surfaces.

4.6.2 Photoionization: Ionospheres

UV photons at $\lambda \lesssim 100$ nm can ionize atoms and molecules. Radiative recombination of atomic ions (e.g., $O^+ + e^- \rightarrow O + h\nu$) is very slow compared to molecular recombination. Atomic ions are usually converted to molecular ions by ion–neutral reactions, and the molecular ions recombine. Photolysis in a tenuous atmosphere

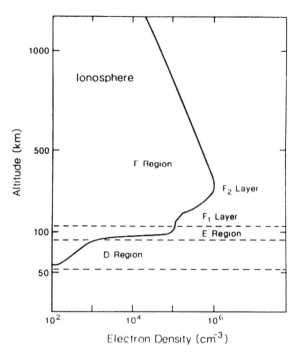

Figure 4.41 Sketch of the electron density in Earth's day-side atmosphere, with the approximate locations of the ionospheric layers. (Russell 1995)

therefore leads to the formation of the *ionosphere*, a region characterized by the presence of free electrons. The electron density in the ionosphere is determined by both the ionization rate and how rapidly the ions recombine, whether directly, or indirectly via charge exchange. Each planet with a substantial atmosphere is expected to have an ionosphere. In this section we first discuss the ionosphere of Earth, followed by a discussion and comparison of the ionospheres of Venus, Mars, and the giant planets.

4.6.2.1 *Earth*

On Earth there are four distinct ionospheric layers: the D, E, F_1, and F_2 layers (Fig. 4.41). The nominal altitude of the D layer is at $\sim$90 km, with a peak electron density of $\sim 10^4$ cm^{-3}. The E layer is concentrated around 110 km, and the peak electron density is an order of magnitude larger than in the D layer. The F_1 layer peaks at 200 km, with an electron density of $\sim 2.5 \times 10^5$ cm^{-3}, and the F_2 layer at 300 km, with a maximum electron density of $\sim 10^6$ cm^{-3}. The height and electron density of the various layers is highly variable in time, since they depend sensitively on the solar UV flux, which varies strongly during the day. The D and F_1 layers are generally absent at night, while the electron densities in the E and F_2 layers are usually smaller at night than during the day. Typical neutral densities at these altitudes can be calculated from

the barometric law with allowance for diffusive separation (§4.7), and are many orders of magnitude larger than the electron densities.

The ionospheric layers are distinct from each other, since the ionization and recombination processes are different for each layer. This is caused by variations in the composition and absorption characteristics of the atmosphere with altitude. In analogy with the physical processes that govern each ionospheric layer on Earth, ionospheric regions on other planets are sometimes denoted with similar letters. Below we summarize the various processes which govern the different layers.

The *E layer* is characterized by direct photoionization of molecular oxygen:

$$(16) \quad O_2 + h\nu \rightarrow O_2^+ + e^-, \qquad \lambda < 103 \text{ nm.}$$

Solar coronal X-rays contribute as well, by ionizing O, O_2, and N_2, which leads to the production of O_2^+ and NO$^+$ ions via rapid charge exchange (16) or atom–ion interchange (17):

$$(17) \quad N_2^+ + O_2 \rightarrow N_2 + O_2^+,$$
$$(18) \quad N_2^+ + O \rightarrow NO^+ + N.$$

Recombination occurs principally through dissociative recombination:

$$(19) \quad O_2^+ + e^- \rightarrow O + O,$$
$$(20) \quad NO^+ + e^- \rightarrow N + O.$$

During the daytime the O_2^+ and NO$^+$ densities in the E layer are roughly equal.

The main ions formed in the F_1 *region* are from atomic oxygen and molecular nitrogen:

$$(21) \quad O + h\nu \rightarrow O^+ + e^-, \qquad \lambda < 91 \text{ nm}$$
$$(22) \quad N_2 + h\nu \rightarrow N_2^+ + e^-, \qquad \lambda < 80 \text{ nm.}$$

Dissociative recombination of N_2^+ ($N_2^+ + e^- \rightarrow N + N$) is very rare, since N_2^+ is rapidly converted into N_2 and NO$^+$ via reactions (17) and (18) above. Radiative recombination of atomic oxygen ions (O$^+ + e^- \rightarrow O + h\nu$) is very slow (reaction rate $k_r \approx 3 \times 10^{-12}$ cm^3 s^{-1}) compared to atom–ion interchange (roughly ten times faster):

$$(23) \quad O^+ + O_2 \rightarrow O_2^+ + O,$$
$$(24) \quad O^+ + N_2 \rightarrow NO^+ + N,$$

followed by rapid dissociative recombination of O_2^+ and NO$^+$ (reactions (19) and (20) above, where $k_r \approx 3 \times 10^{-7}$ cm^3 s^{-1}).

The F_2 *region* is optically thin to most ionizing photons. Reaction (21) is the dominant ionization process. Since radiative recombination of O$^+$ is very slow, and the molecular density is low at these high altitudes (so reactions (23) and (24) are unimportant), the electron density is high in the F_2 region.

The dominant ionization process in the *D layer* is photoionization of O_2 and N_2 by X-rays, and of NO by Ly α photons. There is a pronounced peak in the electron density at an altitude near 90 km, presumably caused, in part, by metallic ions (Fe^+, Mg^+, Na^+, Al^+), which also peak near that altitude. The D layer is further characterized by the presence of O_2^- and more complex negative ions. Negative ions are formed via electron attachment, such as in the three-body reaction between oxygen molecules:

(25) $\quad O_2 + e^- + O_2 \rightarrow O_2^- + O_2,$

at a rate of $\sim 5 \times 10^{-31}$ cm^6 s^{-1}. Although N_2 could serve as a catalyst in place of O_2, the attachment rate would be much lower. These negative ions are destroyed, i.e., the electron is removed from the negative ion, by sunlight (*photodetachment*) or collisions (*collisional detachment*).

Although one would expect the ionosphere to be absent at night due to a lack of photoionizing photons, observations indicate the continued presence of an ionosphere (in particular in the F_2 and E regions), though with reduced electron and ion number densities. A night-side ionosphere is, in part, explained by the fact that the recombination rate, especially for O^+, is slow relative to Earth's rotation. However, ionization is also triggered by other processes, in particular precipitating electrons, micrometeorite bombardment, and UV photons from stars. These processes are an important source of ionization at night and in the polar regions.

4.6.2.2 *Venus and Mars*

The electron density in Mars's ionosphere reaches a maximum of $\sim 10^5$ cm^{-3} at an altitude of ~ 140 km, and it is a factor of 3–5 higher in Venus's ionosphere (Fig. 4.42). Since the dominant constituent in both atmospheres is CO_2, this gas is the main source for ionization, a process which, like in the E layer of Earth's ionosphere, happens through direct photoionization by sunlight:

(26) $\quad CO_2 + h\nu \rightarrow CO_2^+ + e^-, \qquad \lambda < 90$ nm.

The dominant ambient ion, however, is not CO_2^+, but O_2^+, which can be formed through various processes, all of which happen quickly:

(27) atom–ion interchange: $O + CO_2^+ \rightarrow O_2^+ + CO,$
(28) or charge transfer: $O + CO_2^+ \rightarrow O^+ + CO_2,$
(29) quickly followed by: $O^+ + CO_2 \rightarrow O_2^+ + CO.$

Both O_2^+ and CO_2^+ disappear via dissociative recombination:

(30) $\quad CO_2^+ + e^- \rightarrow CO + O,$
(31) $\quad O_2^+ + e^- \rightarrow O + O.$

Peak electron densities on Mars's night-side hemisphere are $\sim 5 \times 10^3$ cm^{-3}. In analogy with Earth, these relatively high densities are caused, in part, by the relatively rapid rotation of the planet, although direct ionization by precipitating electrons and meteor bombardment may play a role as well.

Despite Venus's slow rotation, it has a considerable night-side ionosphere, with peak electron densities of $\sim 10^4$ cm^{-3}. At night, the dominant ion is O_2^+ at ~ 150–170 km, with O^+ at higher altitudes. Models suggest that fast horizontal transport (day-to-night winds, see §4.5.5.2) may carry O^+ from the day to the night side, where it descends to lower altitudes. Chemical reactions with CO_2 may produce the observed densities of O_2^+ (reaction (29)). When the solar wind pressure is high, the ionopause is low and the nightward ion flow is choked off. At these times the ionosphere is confined to altitudes less than 200 km, and precipitating low-energy (~ 30 eV) electrons may be the dominant ionization agent.

4.6.2.3 *Giant Planets*

Direct photoionization of molecular hydrogen, the main constituent of a giant planet atmosphere, yields:

(32) $\quad H_2 + h\nu \rightarrow H_2^+ + e^-, \qquad \lambda < 80$ nm
(33) $\quad H_2 + h\nu \rightarrow H + H^+ + e^-$
(34) $\quad H + h\nu \rightarrow H^+ + e^-.$

The H_2^+ concentration in the giant-planet ionospheres, however, is very low since H_2^+ undergoes rapid charge transfer interactions with molecular hydrogen:

(35) $\quad H_2^+ + H_2 \rightarrow H_3^+ + H.$

H^+ could form H_3^+ when interacting with two hydrogen molecules at once. However, at altitudes above the peak electron density the molecular densities are too low for reaction (35). Since radiative recombination of H^+ is very slow, H^+ remains in the ionosphere as a *terminal ion*. It also accounts for the large number of electrons that are observed. Dissociative recombination of H_3^+ ions is very rapid, so these ions are not expected to stay in the atmosphere for very long. Detections of H_3^+ emissions have been reported, however, in particular in the auroral regions where the production is enhanced due to energetic-particle precipitation. In addition to hydrogen atoms and molecules, the upper atmospheres of the giant planets

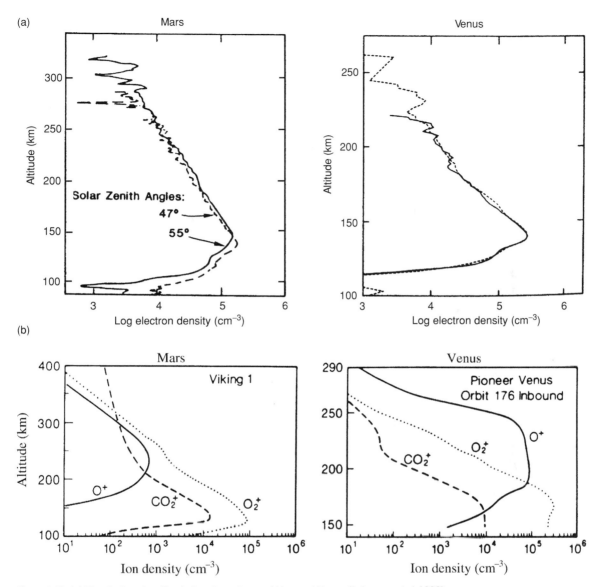

Figure 4.42 (a) The electron densities in the atmospheres of Mars and Venus. (Luhmann *et al.* 1992) (b) Observed ion densities on Mars and Venus. (Adapted from Luhmann 1995)

contain many hydrocarbons and helium. Photoionization thresholds of these molecules and subsequent chemical reactions are discussed in, e.g., Atreya (1986).

The ionospheric structure of the giant planets has been measured using radio occultation experiments on spacecraft (§E.5). Typical electron densities in Jupiter's ionosphere are of the order of $5–20 \times 10^4$ cm^{-3}. When the ionospheric density is plotted as a function of altitude (Fig. 4.43a), many layers can be distinguished, some of which are extremely narrow and dense. The data show considerable latitudinal, diurnal, and temporal variations

in the location and magnitude of the peak electron densities. On Saturn, typical peak electron densities are a few $\times 10^4$ cm^{-3}, up to $\sim 10^5$ cm^{-3} in sharp layers. On Uranus typical electron densities range from a few thousand electrons cm^{-3} up to a few $\times 10^5$ cm^{-3} in a few sharply defined layers. The ionosphere of Neptune is similar to that of Uranus, with local ionization layers enhanced in electron densities by up to an order of magnitude over the nominal values. The sharp ionization layers detected in most ionospheric profiles suggest the presence of long-lived (atomic, perhaps metallic) ions.

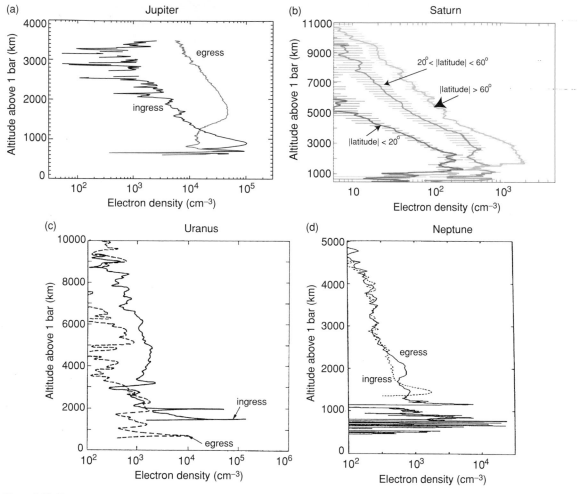

Figure 4.43 Electron density in the ionospheres of the four giant planets. Adapted from (a) Yelle and Miller (2004), (b) Kliore *et al.* (2009), (c) Lindal *et al.* (1987), (d) Lindal (1992).

4.6.3 Electric Currents

As discussed in the previous sections, there are numerous charged particles in an ionosphere, both positively charged ions and negatively charged electrons. These particles gyrate around magnetic field lines of a planet's internal or induced magnetic field (Chapter 7). Charge separation of electrons and ions (due to, e.g., diffusion, §4.7) leads to electric fields. In the absence of collisions, both ions and electrons move (together) under the influence of electric and magnetic fields, at a velocity v as given by equation (7.69). At high altitudes where the lifetimes of ions are large (e.g., in the F region of the Earth's ionosphere), this type of transport is important. At lower altitudes, the drift motion of charged particles is only momentary, until a collision with another particle deviates the electron/ion's path. The bulk motion thus depends upon the ratio of the collision to cyclotron frequencies (§7.3). On Earth the drift or bulk motion for ions typically breaks down at altitudes below 200 km. As the ion–neutral collision cross-section is much larger than that between electrons and neutrals, the bulk motion for electrons breaks down at lower altitudes, ~100 km on Earth.

The relative drift motion of ions and electrons under the influence of magnetic and electric fields induces a current, **J**, described by Ohm's law (eq. 7.15). If collisions between particles are important, there is a finite conductivity, which consists of three components:

(1)   the normal, direct, or longitudinal conductivity (σ_{o}) parallel to the magnetic field:

$$\sigma_{\mathrm{o}} = \left(\frac{n_{\mathrm{i}}}{m_{\mathrm{i}} \nu_{\mathrm{i}}} + \frac{n_{\mathrm{e}}}{m_{\mathrm{e}} \nu_{\mathrm{e}}} \right) q^2, \tag{4.65}$$

(2) the Pederson conductivity (σ_p) perpendicular to the magnetic field:

$$\sigma_p = \left[\frac{n_i \nu_i}{m_i(\nu_i^2 + \omega_i^2)} + \frac{n_e \nu_e}{m_e(\nu_e^2 + \omega_e^2)} \right] q^2, \qquad (4.66)$$

(3) the Hall conductivity (σ_h) perpendicular to both the magnetic field and applied electric field:

$$\sigma_h = \left[\frac{n_e \omega_e}{m_e(\nu_e^2 + \omega_e^2)} - \frac{n_i \omega_i}{m_i(\nu_i^2 + \omega_i^2)} \right] q^2, \qquad (4.67)$$

where q is the electric charge, n_i the number density, ν_i the collisional frequency, and ω_i the cyclotron frequency (the subscript i stands for ion, and e for electron). These conductivities induce the Birkeland, Pederson, and Hall currents, respectively. Generally, σ_o, which is parallel to the magnetic field, is much larger than the Pederson and Hall conductivities, and $\mathbf{E}_\parallel \ll \mathbf{E}_\perp$.

Electric currents lead to *Joule* or *frictional heating* of the ionosphere through the dissipation of electrical energy via charged particle collisions. The Joule heating rate, Q_J, due to currents transverse to the magnetic field is given by

$$Q_J = \mathbf{J}_\perp \cdot \mathbf{E}_\perp = \frac{J_\perp^2}{\sigma_c}, \qquad (4.68)$$

with the Cowling conductivity:

$$\sigma_c = \frac{\sigma_p^2 + \sigma_h^2}{\sigma_p}. \qquad (4.69)$$

The height-integrated Pederson and Hall conductivities on Earth are about 20 mho on the day side and 1 mho at night. Similar numbers are expected on Jupiter and Saturn, although the uncertainties are very large. This implies that Joule heating may be comparable in magnitude to that caused by charged particle precipitation.

4.6.4 Airglow and Aurora
4.6.4.1 Airglow

Airglow results from emissions by atoms/molecules that have been excited (directly or indirectly) by EUV solar photons or cosmic rays (minor contribution). Airglow is confined to high altitudes, and is relatively uniform over the entire globe. A *dayglow*, seen on a body's sunlit hemisphere and dominated by resonant scattering from atomic hydrogen, has been detected from many planets, as well as from Saturn's rings and a torus enveloping Saturn at Titan's orbit. Hydrogen is the most abundant species at very high altitudes because diffusive separation leaves the lightest element at the highest altitudes (§4.7). Resonant Ly α scattering, therefore, gives a planet an extensive *corona*. The Earth's corona can be viewed from outside the atmosphere.

Jupiter, in addition to being bright in Ly α emission, displays a brightening over the equator above that expected from simple resonant scattering. This brightening peaks at a longitude of 100° (in the jovian magnetic longitude System III, §7.5.4.1), and is referred to as the *hydrogen bulge*. This bulge is seen on the day side and, at a reduced intensity, also on the night side. The hydrogen bulge implies a local enhancement in H atoms, which may be caused by magnetospheric effects (§7.5.4.2).

In addition to atomic hydrogen, airglow has been detected from other species, such as nitric oxide (NO), carbon monoxide (CO), and molecular hydrogen, oxygen, and nitrogen. *Nightglow*: Nitric oxide emissions have been detected on Earth, Mars, and Venus. Venus also displays a non-LTE oxygen nightglow at 1.27 μm that originates from the recombination of oxygen atoms in the descending branch of the solar-to-antisolar winds in the thermosphere (§4.5.5.2).

4.6.4.2 Aurora

Aurora, commonly referred to as *northern* or *southern lights*, are seen as oval-shaped regions roughly centered about the magnetic north and south poles of a planet (Figs. 4.44–4.46). In contrast to airglow, auroral emissions are triggered when charged particles from 'outer space' excite atmospheric particles (through collisions). These charged particles enter the atmosphere along magnetic field lines, i.e., as field-aligned currents (§7.5.1.2). Since the footpoints of field lines which thread through the 'storage place' of charged particles form an oval about the magnetic poles, the emissions occur in an oval-shaped region, referred to as the *auroral zone*. Atmospheric atoms, molecules, and ions are excited through interactions with the precipitating particles, or by photoelectrons produced in the initial 'collision' with these particles. Upon de-excitation, the atmospheric species emit photons which can be observed at optical, infrared, UV, and, on Jupiter and Earth, at X-ray wavelengths.

Earth

On Earth, auroral phenomena are truly spectacular (Fig. 4.44). They are regularly seen in the night sky at high latitudes (auroral zones). Rapidly varying, colorful displays fill large fractions of the sky. The lights may appear diffuse, in arcs with or without ray patterns, or may resemble draperies. The auroral emissions are studied both from the ground and from space, at wavelengths varying from X-rays up to radio wavelengths. Aurora are clearly related to disturbances of the geomagnetic field (§7.5.1.4), usually induced by fluctuations in the solar wind, such as those triggered by solar flares or coronal

Figure 4.44 (a) COLOR PLATE Photographs of the northern (auroral) lights. Left photograph taken on 29 September 1996 at 11:25, from Fairbanks, Alaska. One can see a double arc with a developing ray below the Big Dipper. A few minutes later (right image) the arcs have become unstable and develop curtains/draperies. Both exposures were for 10 seconds. (Courtesy J. Curtis, Geophysical Institute, UAF) (b) Aurora Australis (southern lights) imaged by the satellite IMAGE on 11 September 2005, four days after a strong solar flare went off. The image is taken over Antarctica, at ultraviolet wavelengths, and superposed on an image obtained with NASA's Earth Observing System. (NASA) (c) Aurora Australis as photographed from the space shuttle. (NASA)

mass ejections. From the ground, the northern lights are generally only visible at high latitudes, but at times of strong disturbances the emissions are sometimes seen at more moderate latitudes. About once per century they can be seen from near-equatorial regions. Aurora display discrete and diffuse components. The diffuse component is visible over a large region of the sky, while discrete 'arcs' are localized phenomena. The emissions generally vary rapidly in color, form, and intensity. Although the emissions look quite disorderly, a specific pattern, referred to as *auroral substorms*, can be discerned in most auroral displays. An aurora usually starts out as a relatively dim arc oriented in the east–west direction. After a while, maybe a few hours, the arc moves towards the equator, becomes brighter and may develop rays. All of a sudden the emissions spread over the entire sky, and move rapidly (up to tens of kilometers per second) while changing dramatically in form and intensity. This phase typically lasts a few minutes, after which the recovery phase begins, where the emissions weaken and the aurora becomes more diffuse.

The most prominent lines and bands in the terrestrial aurora are due to molecular and atomic nitrogen and oxygen. The visible display is dominated by the green (forbidden; §10.4.3.1) and red atomic oxygen lines at 557.7 nm and 630.0/636.4 nm. Since the 557.7 nm transitions are forbidden, they originate from high atmospheric altitudes, where collisions with other particles are relatively rare (above ~200 km). Blue emissions are produced by nitrogen (427.8 nm and 391.4 nm). Strong lines at UV wavelengths are also caused primarily by nitrogen and oxygen.

Giant Planets

Aurora have been detected on all four giant planets. The aurora on Jupiter have been studied extensively at wavelengths from infrared to X-rays (Fig. 4.45), as well as through auroral decametric emissions (§7.5.4.8). Emissions from ammonia, methane, H_3^+, and numerous hydrocarbons are enhanced within the auroral regions, as a result of differences in temperature and/or composition with the surrounding regions. Several molecular species

Figure 4.45 Composite HST image of Jupiter at visible wavelengths, with superposed northern and southern aurora at UV wavelengths. Note the 'trail of light' produced by Io, just outside the aurora. See also Figure 7.39. (Courtesy John Clarke, NASA/HST)

display spatial variations within the auroral zone. In particular, an auroral infrared bright spot is centered near 180° in the northern hemisphere (fixed in Jupiter's magnetic coordinate III system; §7.5.4.1), and another near 0° in the southern hemisphere. The UV and H_3^+ aurora seem to be well correlated. The H_3^+ aurora occur at high

magnetic latitudes and are linked to distant regions in Jupiter's magnetic field ($\gtrsim 30$ R$_{2\!\!\!/}$). The UV aurora occur at all longitudes and extend equatorward of the H_3^+ aurora. They appear to be triggered by charged particle precipitation from regions in the magnetosphere which are at least ~ 15 R$_{2\!\!\!/}$ away. HST images (Figs. 4.45, 7.39) show faint UV emissions extending roughly 60° in the wake or plasma flow direction beyond Io's magnetic footprint. Auroral emissions are also associated with the footprints of Ganymede, and Europa. These images provide direct evidence of charged particles entering Jupiter's ionosphere along the magnetic field lines that connect Io, Ganymede, and Europa with Jupiter's ionosphere, since the emissions are intensified at and along the footprints of these magnetic field lines (§7.5.4). Such currents at Io's orbit have been detected directly by the Galileo spacecraft.

Aurora on Saturn differ morphologically from those on both Earth and Jupiter. The UV aurora vary slowly (Fig. 4.46), and are most prominent in the morning sector, where some patterns appear fixed with respect to the solar wind direction, but others are in partial corotation with the planet. The aurora brighten (considerably) in response to (large) increases in the solar wind dynamic pressure, and show a strong correlation with Saturn's kilometric radiation (SKR; §7.5.5.4). The latter originates in the auroral region and appears to be fixed near local noon. Cassini images of Saturn's infrared aurora show a bright ring, as well as localized emissions inside and outside of this ring (Fig. 4.46b). The infrared auroral emissions may be

(a)   (b)

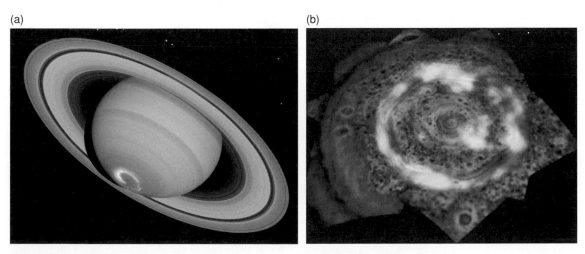

Figure 4.46 (a) HST image of Saturn's UV aurora. The planet and rings were imaged at visible wavelengths, and the bright circle of light at the south pole is an image of the aurora at UV wavelengths. A sequence of images taken over several days shows the day-to-day changes in brightness in response to large changes in the solar wind dynamic pressure. (Courtesy John Clarke and Z. Levay, NASA/ESA) (b) Thermal 5 μm emission from Saturn's north polar hexagon, with superposed auroral emission at 4 μm, as imaged with VIMS on Cassini in November 2006. (NASA/JPL/University of Arizona, PIA11396)

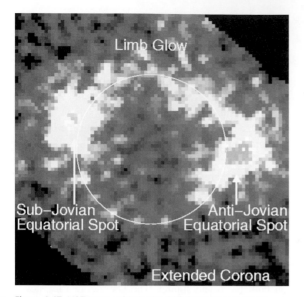

Figure 4.47 HST image of Io in the 135.6 nm line of atomic oxygen. The equatorial emissions result from the interaction between electrons flowing along magnetic field lines and Io's atmosphere. (Adapted from Retherford *et al.* 2003)

indicative of a higher temperature across the pole, which would only affect the infrared auroral brightness. Such an increase in temperature could be caused by increased heating due to particle precipitation.

Smaller Solid Bodies

Glowing spots of UV emissions are observed (Fig. 4.47) in the equatorial plane at the tangent points between field-aligned electrons and Io's atmosphere (Birkeland currents) (§4.6.3). The spots 'wobble' up and down about Io's equator, due to the motion of the tilted jovian magnetic field past Io. The spot on the antijovian hemisphere is brighter than the subjovian spot, which may be caused by an asymmetry in the directions of the electron and ion convection patterns, as a result of the Hall effect, that produce hotter electrons on the antijovian side. A faint glow along the limb and an extended corona are detected as well. Since the HST image in Figure 4.47 was taken in the 135.6 nm transition of atomic oxygen, this image shows only the excited oxygen gas, in contrast to Figure 5.79b which shows all emissions integrated between 350 and 850 nm, which complicates distinction between volcanoes and auroral emissions.

HST and the 10 m Keck telescope detected aurora of atomic oxygen on Ganymede at UV and visible wavelengths, respectively. These emissions, confined to Ganymede's auroral regions, are produced by dissociative excitation of O_2 by electrons traveling along Ganymede's own magnetic field lines.

Most surprising, perhaps, is the Mars Express detection of an aurora on Mars. CO and CO_2^+ emissions were observed over a region where the crustal magnetic field is maximum (§7.5.3.2).

4.7 Molecular and Eddy Diffusion

In the previous section we discussed how the atmospheric composition is changed by photochemical reactions. We showed how to deduce the altitude distribution of photochemically derived constituents if the reaction and photolysis rates of the relevant reactions are known. Theoretically deduced altitude profiles for these constituents, however, seldom agree exactly with observations. This is, at least in part, caused by vertical movements of air parcels, known as *eddy diffusion*, and of individual molecules within the air, referred to as *molecular diffusion*.

4.7.1 Diffusion

The net vertical flux, Φ_i, for minor constituent i in an atmosphere that is in hydrostatic equilibrium can be written as:

$$\Phi_i \equiv N_i v_u = -N_i D_i \left(\frac{1}{N_i} \frac{\partial N_i}{\partial z} + \frac{1}{H_i^*} + \frac{\alpha_i}{T(z)} \frac{\partial T(z)}{\partial z} \right) - N\mathcal{K} \frac{\partial (N_i/N)}{\partial z}, \quad (4.70)$$

where N is the atmospheric number density, N_i is the number density of constituent i, D_i is the molecular diffusion coefficient, $\mathcal{K}$ the eddy diffusion coefficient, and H_i^* is the density scale height for each atmospheric constituent i (eq. 4.4). The coefficient α_i is the thermal diffusion parameter, and $T(z)$ is the atmospheric temperature at altitude z. The physical interpretation of the first three terms on the right-hand side of equation (4.70) is as follows:

$(1/N_i)(\partial N_i/\partial z)$: Molecular diffusion caused by a gradient in the density N_i tends to smooth out density gradients, driving the mixing ratio of a constituent towards a constant with altitude. As an example, consider the O/O_2 mixing ratio in the Earth's atmosphere. The altitude profile of O/O_2 can be calculated according to the chemical reactions described in §4.6. Molecular diffusion is most effective at high altitudes ($z > 100$ km on Earth), where O atoms can be carried downwards to equalize the O/O_2 ratio with height. At lower altitudes, the oxygen atoms combine into molecules. This process causes the O/O_2

ratio at $z > 100$ km to be less than expected from the chemistry described in §4.6.

$(1/H_i^*)$: Molecular buoyancy diffusion, with H_i^* the density scale height (eq. 4.4) for individual constituents i, drives an atmosphere towards a barometric height distribution for each species i. Since the scale height varies with $1/\mu_a$, heavy molecules tend to concentrate at lower altitudes. Collisions between particles slow the diffusion process, so diffusion is only effective at high altitudes. As mentioned above, on Earth molecular diffusion is important at $z \gtrsim 100$ km. In contrast to molecular diffusion induced by $\partial N_i/\partial z$, molecular buoyancy diffusion enhances the O/O_2 ratio at these high altitudes, above that expected from local photochemical equilibrium considerations alone. The net effect of the two processes is that the O/O_2 mixing ratio is less than expected from photochemical considerations alone at altitudes around 100 km, but larger at altitudes above this level.

$(\alpha_i/T(z))(\partial T(z)/\partial z)$: Thermal molecular diffusion, with α_i the thermal diffusion parameter, is triggered by gradients in temperature (but note that there is a temperature dependence in $1/H_i^*$ as well).

The molecular diffusion coefficient D_i is inversely proportional to the atmospheric number density N: $D_i = b_i/N$, with b_i the binary collision parameter, which can best be determined empirically. The maximum rate of diffusion occurs for complete mixing, $\partial(N_i/N)/\partial z = 0$, which leads to the concept of *limiting flux*, Φ_ℓ. For an atmosphere with small or no temperature gradients, where a light gas flows through the background atmosphere, the limiting flux can be written:

$$\Phi_\ell = \frac{N_i D_i}{H}\left(1 - \frac{\mu_{a_i}}{\mu_a}\right) \approx \frac{N_i D_i}{H}, \qquad (4.71a)$$

with H the atmospheric (pressure) scale height. Using $D_i = b_i/N$, equation (4.71a) can be approximated by

$$\Phi_\ell \approx \frac{b_i(N_i/N)}{H}. \qquad (4.71b)$$

The binary collision parameter (cm^{-1} s^{-1}) is given by

$$b_i = C_b T^q, \qquad (4.72)$$

where the parameters C_b and q for some gases are given in Table 4.7. The net outward flux is limited by the diffusion rate, and cannot exceed the limiting flux. The limiting flux depends only on the mixing ratio of constituent i and the pressure scale height. To calculate the limiting flux for hydrogen atoms in Earth's atmosphere (Problem 4.32), one needs to consider the upward flux of all hydrogen-bearing molecules (H_2O, CH_4, H_2) just below the homopause (i.e., ~100 km). The mixing ratio

Table 4.7 Parameters C_b and q (cgs units) in equation (4.72) for various gases[a].

Gas 1	Gas 2	C_b	q
H	H_2	145×10^{16}	1.61
	air	65×10^{16}	1.7
	CO_2	84×10^{16}	1.6
H_2	air	26.7×10^{16}	0.75
	CO_2	22.3×10^{16}	0.75
	N_2	18.8×10^{16}	0.82
H_2O	air	1.37×10^{16}	1.07
CH_4	air	7.34×10^{16}	0.75
Ne	N_2	11.7×10^{16}	0.743
Ar	air	6.73×10^{16}	0.749

[a] All parameters from Chamberlain and Hunten (1987).

of all hydrogen-bearing molecules at the homopause is $\sim10^{-5}$, which results in a limiting flux of $\sim2 \times 10^8$ cm^{-2} s^{-1}.

4.7.2 Eddy Diffusion Coefficient

The last term on the right-hand side of equation (4.70) is the *eddy* or *turbulent diffusion*, with the *eddy diffusion coefficient*, K. Eddy diffusion is a macroscopic process, in contrast to molecular diffusion and mixing, discussed above. Eddy diffusion may occur if an atmosphere is unstable against turbulence, which occurs when Reynolds number $\Re_e > 5000$ (eq. 4.48). The product $v\ell$ gives a crude estimate for the diffusion coefficient K. Assuming $\ell \approx H$ and v is between 1 and 10^4 cm s^{-1}, K is of the order of 10^6 to 10^{10} cm^2 s^{-1}.

We find that eddy diffusion dominates the atmosphere below the *turbopause* or *homopause*, while molecular diffusion dominates above the homopause. On Earth, the homopause is located at an altitude of ~100 km. As mentioned above, at higher altitudes molecular diffusion becomes important. In (super)adiabatic atmospheres, free convection is usually the dominant motion involved in vertical mixing of the atmosphere. In subadiabatic regions, eddy diffusion might be driven by internal gravity waves or tides. The eddy diffusion coefficient is usually estimated from observed altitude distributions of trace gases.

Venus and Mars

The observed CO, O, and O_2 abundances in the upper atmospheres of Venus and Mars are much smaller than estimated using local photochemistry (§4.6.1.2). On Mars, recombination of CO and O might proceed more rapidly via catalytic reactions with OH if CO and O are transported downwards via an efficient eddy diffusion mechanism. The observed abundances of CO and O can be matched if the eddy diffusion coefficient $\mathcal{K} \approx 10^8$ cm^2 s^{-1}, roughly two orders of magnitude larger than in the Earth's stratosphere. On Venus, the observations can be matched if $\mathcal{K} \approx 10^5$ cm^2 s^{-1} between 70 and 95 km, and if $\mathcal{K}$ is increasing at higher altitudes.

Giant Planets

The eddy diffusion coefficient on the giant planets is often estimated from the altitude profile of methane gas and/or of compounds in the upper atmosphere that are not expected to be there under equilibrium situations. Lyman α emission from the planets usually results from resonance scattering of solar photons by atmospheric hydrogen. Since methane gas is a strong absorber of these photons, the Ly α emission arises from above the methane homopause, the altitude of which is determined by the eddy diffusion coefficient: a large value raises the homopause, and hence diminishes the Ly α emission because of the low number of hydrogen atoms.

Compounds like GeH$_4$ and PH$_3$ are stable deep ($T >$ 1000 K) in the giant planet atmospheres. Since these species have been detected in Jupiter and Saturn's tropospheres, there must be rapid vertical mixing to bring these species up against other transformation processes, such as oxidation. On Uranus and Neptune the temperature at the tropopause is well below the condensation point of methane gas, so not much methane is expected above the tropopause. Any methane brought up through this tropopause 'cold trap' undergoes photochemical reactions, leading to the production of hydrocarbons such as C$_2$H$_2$ and C$_2$H$_6$. A comparison between the observed and calculated mixing ratio of the hydrocarbons yields an estimate for the diffusion coefficient across the tropopause.

The much larger concentration of hydrocarbons in Neptune's stratosphere compared to Uranus is suggestive of much stronger convection on Neptune. Alternatively, Neptune's warm south pole as revealed in high spatial resolution images at mid-infrared wavelengths (Fig. 4.39e), may form a pathway for methane gas to be brought up from deeper levels into the stratosphere.

Estimates for the eddy diffusion coefficient near the homopause of planets and Titan are listed in Table 4.8.

4.8 Atmospheric Escape

A particle may escape a body's atmosphere if its kinetic energy exceeds the gravitational binding energy *and* it

Table 4.8 Eddy diffusion coefficient near homopause.[a]

Planet	Eddy diffusion (cm^2 s^{-1})	Pressure (bar)
Earth	$(0.3–1) \times 10^6$	3×10^{-7}
Venus	10^7	2×10^{-8}
Mars	$(1–5) \times 10^8$	2×10^{-10}
Titan	$\sim 10^8$	6×10^{-10}
Jupiter	$\sim 10^6$	10^{-6}
Saturn	$\sim 10^8$	5×10^{-9}
Uranus	$\sim 10^4$	3×10^{-5}
Neptune	$\sim 10^7$	2×10^{-7}

[a] All parameters from Atreya (1986) and Atreya *et al.* (1999).

moves along an upward trajectory without intersecting the path of another atom or molecule. The region from which escape can occur is referred to as the *exosphere*, and its lower boundary is called the *exobase*. The exobase is located at an altitude z_{ex} at which

$$\int_{z_{ex}}^{\infty} \sigma_x N(z) dz \approx \sigma_x N(z_{ex}) H = 1. \qquad (4.73)$$

We assume the scale height, H, to be constant in the exosphere. Since the mean free path

$$\ell_{fp} = 1/(\sigma_x N), \qquad (4.74)$$

the exobase is located at an altitude z_{ex}, where $\ell_{fp}(z_{ex}) = H$. Within the exosphere the mean free path for a molecule/atom is thus comparable to, or larger than, the atmospheric scale height, so an atom with sufficient upward velocity has a reasonable chance of escaping. In addition to the thermal or Jeans escape, there are various nonthermal processes which can also lead to atmospheric escape.

4.8.1 Thermal (Jeans) Escape

For a gas in thermal equilibrium, the velocities follow a Maxwellian distribution function:

$$f(v) dv = N \left(\frac{2}{\pi} \right)^{1/2} \left(\frac{m}{kT} \right)^{3/2} v^2 e^{-mv^2/(2kT)} dv, \qquad (4.75)$$

with v the particle's velocity, m its mass, and N is the local particle (number) density. At and below the exobase, collisions between particles drive the velocity distribution into a Maxwellian distribution. Above the exobase, collisions are essentially absent and particles in the tail of the Maxwellian velocity distribution which have a velocity $v > v_e$, may escape into space. A Maxwellian distribution formally extends up to infinite velocities, but due to the steep dropoff in the Gaussian distribution, there are practically no particles with velocities larger than about four times the mean (most probable) thermal velocity $\bar{v}_o = \sqrt{2kT/m}$.

The ratio of the potential to kinetic energy is referred to as the *escape parameter*, λ_{esc}:

$$\lambda_{esc} = \frac{GMm}{kT(R+z)} = \frac{(R+z)}{H(z)} = \left(\frac{v_e}{v_o}\right)^2. \tag{4.76}$$

Integrating the upward flux in a Maxwellian velocity distribution above the exobase results in the *Jeans formula* for the rate of escape (atoms cm^{-2} s^{-1}) by thermal evaporation:

$$\Phi_J = \frac{N_{ex}v_o}{2\sqrt{\pi}}(1 + \lambda_{esc})e^{-\lambda_{esc}}, \tag{4.77}$$

where the subscript 'ex' refers to the exobase and λ_{esc} is the escape parameter at the exobase. Typical parameters for Earth are $N_{ex} = 10^5$ cm^{-3} and $T_{ex} = 900$ K. For atomic hydrogen $\lambda_{esc} \approx 8$, and $\Phi_J \approx 6 \times 10^7$ cm^{-2} s^{-1}, which is a factor of 3–4 smaller than the limiting flux (eq. 4.71) for hydrogen atoms on Earth (Problem 4.32). Note that lighter elements/isotopes are lost at a much faster rate than heavier ones. Jeans escape can thus produce a substantial isotopic fractionation.

To first approximation, calculations of Jeans escape can be used to predict whether or not an object has an atmosphere (Problem 4.33). It can also be used to evaluate which volatile ices one might find on a body, since the presence of such ices depends on both temperature and gravity (§9.3.4).

4.8.2 Nonthermal Escape

Jeans escape gives a lower limit to the escape flux from a body's atmosphere; *nonthermal processes* often dominate the escape rate. In such processes (1–6 discussed below), neutral particles can gain sufficient energy to escape into space. In the following, we adopt the notation: i_2 = molecule; i, j = atoms; i^+, j^+ = ions; e$^-$ = electron, and * indicates excess energy.

(1) *Dissociation and dissociative recombination*: when a molecule is dissociated by UV radiation or an impacting electron, or when an ion dissociates upon recombination, the end products may gain sufficient energy to escape the body's gravitational attraction:

$$i_2 + h\nu \rightarrow i^* + i^*,$$

$$i_2 + e^{-*} \rightarrow i^* + i^* + e^-,$$

$$i_2^+ + e^- \rightarrow i^* + i^*.$$

(2) *Ion–neutral reaction*: when an atomic ion interacts with a molecule, a molecular ion and a fast atom may result:

$$j^+ + i_2 \rightarrow ij^+ + i^*. \tag{4.78}$$

(3) *Charge exchange*: when a fast ion meets a neutral, charge exchange may take place, where the ion loses its charge, but retains its kinetic energy. The new neutral (former ion) may have sufficient energy to escape the body's gravitational attraction:

$$i + j^{+*} \rightarrow i^+ + j^*. \tag{4.79}$$

This process plays an important role on Io, where fast sodium atoms are created by charge exchange with magnetospheric plasma (§7.5.4.3).

(4) *Sputtering*: when a fast atom or ion hits an atmospheric atom, the atom may gain sufficient energy to escape the body's gravitational attraction. Since it is much easier to accelerate an ion than an atom, sputtering is usually caused by fast ions. In a single collision, the atom is accelerated in the forward direction (conservation of momentum), a process generally referred to as *knock-on*. Sputtering usually refers to a multiple sputtering process, involving a cascade of collisions. Such processes are important in both thick and tenuous atmospheres, as well as on airless bodies. In the latter cases, the fast ions/atoms hit the surface directly, ejecting one or several atoms from the crust into space. Such atoms may not be fast enough to escape into interplanetary space, but rather get trapped in a 'corona', i.e., they stay gravitationally bound to the body. The atmospheres of the Moon and Mercury are partly formed by sputtering processes, and by meteoroid impacts (the latter probably dominate). The exobase for Mercury and the Moon is at their surface. Sputtering is also important on outer planet satellites and rings. The atoms can gain sufficient energy to escape into interplanetary space:

$$i + j^{+*} \rightarrow i^* + j^{+*},$$

$$i + j^* \rightarrow i^* + j^*.$$

(5) *Electric fields*: ionospheres contain electric fields, since any motion of ionized air (due to solar heating, tides, etc.) induces such fields. Electric fields accelerate charged particles, which may transfer momentum to a neutral species upon collision. Of particular note is the molecular diffusion process (§4.7.1), which leads to a separation in altitude between the heavier ions and lighter electrons, inducing a potential difference or electric field in the upper atmosphere that is directed normal to the 'surface'. Such fields lead to an upward acceleration of ions,

which in the polar regions induce a *polar wind*. The depletion of H^+ and He^+ above the Earth's magnetic poles is attributed to such ion escape along open magnetic field lines.

(6) *Solar wind sweeping*: in the absence of an internal magnetic field, charged particles may interact directly with the solar wind, a process referred to as *solar wind sweeping*. The giant planets and Earth have strong intrinsic magnetic fields, and the trajectories of solar wind particles are deflected to flow around the field (§7.1.4). Thus, there is no direct interaction between these planets and the solar wind. If the body has an ionosphere but no intrinsic magnetic field, such as Venus and comets, particle exchange between the solar wind and the body's atmosphere occurs. Particles are captured from the solar wind at the subsolar point, and lost to the wind near the limbs. Via this process, an airless body may temporarily capture solar wind particles. For example, the hydrogen and helium atoms in Mercury's atmosphere, as well as the helium atoms on the Moon, are captured from the solar wind. Satellites embedded in a planet's magnetic field interact in a similar way with magnetospheric plasma.

4.8.3 Blowoff and Impact Erosion

Atmospheric losses in the early Solar System were, in part or occasionally, dominated by impact erosion and hydrodynamic escape, or blowoff. In the present era, hydrodynamic escape can be important for some bodies, such as Pluto.

Hydrodynamic escape occurs when a planetary wind comprised of a light gas (e.g., H) entrains heavier gases, which by themselves would not escape according to Jeans equation. The planetary wind can be compared to the solar wind, in which the initially subsonic flow goes through a critical point where the velocity equals the speed of sound (§7.1.1). At larger heliocentric distances the solar wind is supersonic. Chamberlain and Hunten (1987) derived an expression for hydrodynamic escape in an atmosphere. They assume that the light gas moves at speeds approaching the sonic value, in which case there are large drag forces with other constituents. By ignoring the terms in dT/dz and in $\mathcal{K}$ in equation (4.70), the outgoing flux of the heavier gas, Φ_2, becomes

$$\Phi_2 = \frac{N_2}{N_1}\left(\frac{m_c - m_2}{m_c - m_1}\right)\Phi_1, \tag{4.80}$$

where the subscripts 1 and 2 refer to the light and heavy gas, respectively, and m_c is given by:

$$m_c = m_1 + \frac{NkT\Phi_1}{bg_p}, \tag{4.81}$$

with b the binary collision parameter (§4.7.1). Escape requires Φ_2 to be positive, and thus $m_2 < m_c$. If Φ_1 is equal to the limiting flux (eq. 4.71), $m_c = 2m_1$.

To maintain an atmosphere in a blowoff state requires a large input of energy to the upper atmosphere. Solar energy is usually not large enough to maintain any present-day terrestrial-type atmosphere in a blowoff state. Calculations show that the early atmospheres of Venus, Earth, and Mars may have experienced periods of hydrodynamic escape, triggered by intense solar UV radiation and a strong solar wind. With Pluto's atmospheric temperature of ~ 100 K, as derived from stellar occultation experiments, Pluto may experience hydrodynamic escape. However, because Pluto's orbit is highly eccentric, the atmosphere will be frozen, or collapsed, during most of a plutonian year.

Impact erosion can occur during or immediately following a large impact on a body that has a substantial atmosphere. For an impactor that is smaller than the atmospheric scale height, shock-heated air flows around the impactor and the energy is dispersed over a relatively large volume of the atmosphere. If the impactor is larger than an atmospheric scale height, however, a large fraction of the shock-heated gas can be blown off, since the impact velocity exceeds the escape velocity from the planet. The mass of the atmosphere blown into space, M_e, is given by

$$M_e = \frac{\pi R^2 P_0 \mathcal{E}_e}{g_p}, \tag{4.82}$$

where R is the radius of the impactor, and P_0/g_p is the mass of the atmosphere per unit area. The atmospheric mass that can escape is thus the mass intercepted by the impactor multiplied by an enhancement factor $\mathcal{E}_e$:

$$\mathcal{E}_e = \frac{v_i^2}{v_e^2(1 + \mathcal{E}_v)}, \tag{4.83}$$

where v_i and v_e are the impact and escape velocities, respectively, and $\mathcal{E}_v$ is the evaporative loading parameter, which is inversely proportional to the impactor's latent heat of evaporation. A typical value for $\mathcal{E}_v$ is ~ 20 (for meteors). Significant escape occurs when $\mathcal{E}_e > 1$. If $\mathcal{E}_e < 1$, evaporative loading is much larger than the energy gained by impact heating, and the gas does not have enough energy left to escape into space. In the case of a colossal cratering event, the ejecta from the crater may also be large enough and contain sufficient energy to accelerate atmospheric gas to escape velocities. Such large impactors can remove all of the atmosphere above the horizon at the location of the impact.

4.9 History of Secondary Atmospheres

4.9.1 Formation

The initial stages of planetary growth involve the accumulation of solid materials, but gas can be trapped within some solids; chemical alteration can produce volatiles and radioactive nuclides can decay into volatiles. Moreover, if a planet becomes massive enough, it may gravitationally trap gases. The formation of planets and their atmospheres is discussed in detail in Chapter 13.

The atmospheres of the giant planets are composed primarily of hydrogen and helium, with traces of C, O, N, S, and P in the form of CH_4, H_2O, NH_3, H_2S, and PH_3, respectively. In contrast, the atmospheres of the terrestrial planets and satellites are dominated by CO_2, N_2, O_2, H_2O, and SO_2. The main difference between the giant and terrestrial planets is gravity, which allowed the giant planets to accrete large quantities of common species (e.g., H_2, He; see Table 8.1) which remain gaseous at Solar System temperatures. The light elements H and He (if present originally) would have escaped the shallow gravitational potential wells of the terrestrial planets.

In this section we argue that the atmospheres of the terrestrial planets and Titan cannot be remnants of gravitationally trapped primordial atmospheres, but must have formed from outgassing of bodies accreted as solids. In §4.9.2 we discuss the subsequent evolution of the 'climate' on Earth, Mars, and Venus.

4.9.1.1 *Terrestrial Planets*

The following chemical reactions can occur in an atmosphere between H_2 and other volatiles:

$$CH_4 + H_2O \longleftrightarrow CO + 3H_2$$

$$2NH_3 \longleftrightarrow N_2 + 3H_2$$

$$H_2S + 2H_2O \longleftrightarrow SO_2 + 3H_2$$

$$8H_2S \longleftrightarrow S_8 + 8H_2$$

$$CO + H_2O \longleftrightarrow CO_2 + H_2$$

$$CH_4 \longleftrightarrow C + 2H_2$$

$$4PH_3 + 6H_2O \longleftrightarrow P_4O_6 + 12H_2.$$

A loss of hydrogen shifts the equilibrium towards the right, hence oxidizing material. We refer to an atmosphere as *reducing* if a substantial amount of hydrogen is present, as on the giant planets, and as *oxidizing* if little hydrogen is present, as on the terrestrial planets.

If the atmospheres of the terrestrial planets were primordial in origin (accreted from a solar composition gas, similar to the formation of the giant planets; §13.6), and all of the hydrogen and helium subsequently escaped, the

most abundant gases in these atmospheres would be CO_2 (~63%), Ne (~22%), and N_2 (~10%), with a small fraction of carbonyl sulfide (OCS, ~4%; this molecule does not form via one simple reaction as those listed above). In addition, one would expect solar concentrations for Ar, Kr, and Xe. These abundances are quite different from what is observed. In particular, neon on Earth is present in minuscule amounts, about ten orders of magnitude less than predicted from this model. Similarly, nonradiogenic Ar, Kr, and Xe are present but at abundances over six orders of magnitude less than expected for a solar composition atmosphere. These three noble gases are too heavy to escape via thermal processes if initially present, and could not be chemically confined to the condensed portion of the planet, as CO_2 is on Earth (see below). The observed small abundances of the noble gases form a major argument in the conclusion that the atmospheres of the terrestrial planets are secondary in origin. A secondary atmosphere could have been produced (*i*) during the accretion phase of the planet, when impacts caused intense heating, and/or by (late) accreting volatile-rich asteroids and comets, (*ii*) at the time of core formation, when the entire planet was molten, and/or (*iii*) through 'steady' outgassing via volcanic activity.

The measured ratio of argon isotopes, $^{40}Ar/^{36}Ar$, in Earth's atmosphere and volcanic glasses can be used to deduce when gases were released into the atmosphere. ^{36}Ar is a primordial isotope, incorporated into planetesimals only at extremely low temperatures ($\lesssim 30$ K). ^{40}Ar, in contrast, originates from radioactive decay of potassium, ^{40}K, which has a half-life of 1.25 Gyr (Table 8.3). Both the stable and radioactive isotopes of potassium are incorporated in rock-forming minerals. Upon decay of ^{40}K, the resulting argon gets released only when the mineral melts. The $^{40}Ar/^{36}Ar$ ratio seen in bubbles within volcanic glasses is about one hundred times as large as the atmospheric value of 300. This implies that the vast majority of the ^{36}Ar now in the atmosphere either never resided in the mantle or was outgassed from the mantle within the first few tens of million years after the bulk of Earth's accretion.

4.9.1.2 *Titan*

It is puzzling why Titan has such a dense nitrogen atmosphere, while the similarly sized satellites of Jupiter have practically no atmosphere. Since the abundances of the noble gases are many orders of magnitude below solar values, Titan's atmosphere must be secondary in origin, just like the atmospheres of the terrestrial planets. Would the atmosphere have formed during the accretion phase of Titan, via delivery of cometary impacts during the late bombardment era, or via steady volcanic outgassing over Titan's history? A related question is whether the nitrogen in the planetesimals that formed Titan was in the form of N_2 or as a mixture of nitrogen compounds, in particular NH_3-ice. Direct condensation or trapping of N_2 in amorphous ice, or in clathrate hydrates, requires such low temperatures (<45 K) that noble gases would have been captured as well. Hence, the absence of noble gases from Titan's atmosphere means that N_2 must have been delivered in the form of nitrogen compounds and ammonia-ice. This conclusion, however, does not solve the question of when the atmosphere formed, nor why only on Titan.

Calculations show that the energetics in delivering volatiles via cometary impacts on average add to an atmosphere around Titan, while eroding pre-existing atmospheres around Ganymede and Callisto. The D/H ratio measured on Titan, however, is similar to that on Earth, and smaller than observed on comets (Fig. 10.24). Hence, Titan's atmosphere cannot have formed during the late bombardment era via impacts. This, together with the detection of the radiogenic isotope ^{40}Ar, although at much lower levels than in Earth's atmosphere, suggests that some outgassing from Titan's interior has taken place.

The ^{14}N/^{15}N ratio, being much smaller than the terrestrial value, was explained via fractionation during atmospheric escape. Estimates suggest Titan's early atmosphere to have been 5–10 times more dense. Such losses and fractionation should have depleted ^{12}C relative to ^{13}C as well, but instead the ^{12}C/^{13}C ratio was similar to the terrestrial value. Since photolysis would destroy all methane in Titan's atmosphere within $\sim 10^7$ years (§4.6.1.4), these observations suggest that there must be a continuous supply of methane gas to the satellite's atmosphere. Despite this supply, the C/N ratio is well (3–4 times) below the solar value, indicative of a large loss of carbon over time. If carbon was not lost to space, it must still be hidden somewhere, in the form of ancient deposits of aerosols or below the surface (on Earth it is hidden as carbonate rocks, §4.9.2.1). This could be the reservoir that supplies the methane gas, either continuously or episodically via cryovolcanic (ice) volcanism or geyser eruptions. In fact, if nitrogen was indeed delivered in the form of ammonia-ice, this constituent, mixed in the water, would lower the freezing point of water, facilitating cryovolcanism on Titan.

4.9.2 Climate Evolution

As discussed in Chapter 3, a planet's surface temperature is determined mainly by solar insolation, its Bond albedo, and atmospheric opacity. The Sun's luminosity has slowly increased during the history of the Solar System. Early on, $\sim$4.5 Gyr ago, the solar luminosity was probably 25–30% smaller than it is nowadays, implying

(a)

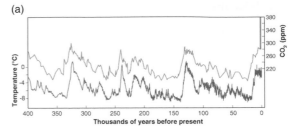

(b)

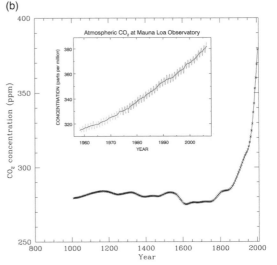

Figure 4.48 (a) Long-term variations in temperature (bottom graph) and in the atmospheric concentration of carbon dioxide (top graph) over the past 400 000 years as inferred from Antarctic ice-core records. (Fedorov *et al.* 2006) (b) CO_2 concentration in the Earth's atmosphere over the past 1000 years. (Adapted from Etheridge *et al.* 1996) The insert is the CO_2 concentration from data obtained at a single location in Hawaii. This graph reveals a steady increase of the CO_2 concentration, in addition to the seasonal oscillations. The data before 1974 were obtained by C.D. Keeling; the 1974–2006 data are from K.W. Thoning and P.P. Tans. (NOAA Earth System Research Laboratory)

lower surface temperatures for the terrestrial planets. In contrast, although the Sun's total energy output was less than it is at the present time, its X-ray and UV emission were much larger, and the solar wind was stronger and probably erratic during the Sun's T-Tauri phase. In addition to changes in the solar flux, the planetary albedos may have fluctuated, as variations in a planet's cloud deck, ground-ice coverage, and volcanic activity may significantly alter its albedo. Changes in atmospheric composition, in particular with regard to greenhouse gases, also have a profound effect on climate. Further, periodic variations (i.e., the Milankovitch cycles discussed in §2.5.3) or sudden modifications in a planet's orbital eccentricity and the obliquity of its rotation axis (e.g., due to a large impact) play an important role in climate evolution. In particular, the $\sim$40 000 and 100 000 year cycles of Earth's ice ages (Fig. 4.48) have been attributed to the Milankovitch cycles.

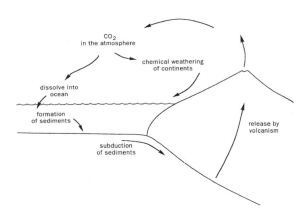

Figure 4.49 Schematic of the CO_2 cycle on Earth. Carbon dioxide is removed from the atmosphere by the Urey weathering reaction, transported down into the mantle via plate tectonics, and recycled back into the atmosphere by volcanic activity. (Jakosky 1998)

4.9.2.1 Earth

Even though the luminosity of the young Sun was smaller, the presence of sedimentary rocks and possible absence of glacial deposits on Earth about 4 Gyr ago suggest that the Earth may have been quite warm, maybe even warmer than today. If true, this could be attributed to an increased greenhouse effect, since the atmospheric H_2O, CO_2, CH_4, and possibly NH_3 content was likely larger during the early stages of outgassing. In fact, the long-term cycling of CO_2 may be regulated in such a way that the Earth's surface temperature does not change too much over long time periods. Carbon dioxide is removed from the atmosphere–ocean system by silicate weathering, the *Urey weathering reaction*, a chemical reaction between CO_2, dissolved in water, with silicate minerals in the soil (Fig. 4.49). The reaction releases Ca and Mg ions and converts CO_2 into bi-carbonate (HCO_3^-). The bi-carbonate reacts with the ions to form other carbonate minerals.[5] An example of such a chemical reaction for calcium is given by

$$CaSiO_3 + 2CO_2 + H_2O \rightarrow Ca^{2+} + SiO_2 + 2HCO_3^-$$

$$(4.84a)$$

$$Ca^{2+} + 2HCO_3^- \rightarrow CaCO_3 + CO_2 + H_2O. \qquad (4.84b)$$

Carbonate sediments on the ocean floor are carried downwards by plate tectonics (§5.3.2.2), and are transformed back into CO_2 in the high temperature/pressure environment of the Earth's mantle:

$$CaCO_3 + SiO_2 \rightarrow CaSiO_3 + CO_2. \qquad (4.84c)$$

Volcanic outgassing returns the CO_2 to the atmosphere. The weathering rate increases when the surface

[5] On Earth, organisms in the oceans make shells of calcium carbonate.

temperature is higher, while the surface temperature is related to the CO_2 content of the atmosphere through the greenhouse effect. This effectively causes a self-regulation in the atmospheric CO_2 abundance on Earth. The role of this carbonate–silicate weather cycle during the recovery from glaciations is well established.

The abundance of O_2 in the Earth's atmosphere is primarily due to photosynthesis of green plants, together perhaps with a small contribution from past photodissociation of H_2O and subsequent escape of the hydrogen atoms. The O_2 in our atmosphere rose to significant levels about 2.2 Gyr ago. The presence of iron carbonate ($FeCO_3$) and uranium dioxide (UO_2) in sediments that date back to over 2.2 Gyr ago (Fig. 4.50), and its absence in younger sediments, provides evidence for the presence of free oxygen in Earth's atmosphere over the past 2.2 Gyr, since oxygen destroys these compounds today. The most striking evidence of a low oxygen abundance on Earth is provided by *banded iron formations* (BIFs, Fig. G.1) on the ocean floor, a sediment which consists of alternating (few cm thick) layers of iron oxides (such as hematite and magnetite) and sediments as, e.g., shale and chert (§5.1.2). Banded iron formations, which were formed by precipitation, imply that iron was able to accumulate within seawater, something that cannot occur in today's oxygen-rich oceans. Since BIFs are common in sediments laid down prior to 1.85 Gyr ago, but very rare in more recently formed rocks, free oxygen must have been a rare commodity over 2 Gyr ago. The rise of oxygen coincided with the first large ice age. The increase in O_2 may have eliminated much of the methane gas, hypothesized to have been a major greenhouse gas on early Earth, by reducing its photochemical lifetime and constraining the environments in which methanogens (methane-producing archaebacteria) could survive.

The influence of mankind on the evolution of the Earth's atmosphere at the present time should not be underestimated. The CO_2 levels in our atmosphere are rising at an alarming rate (Fig. 4.48), which is leading to a global warming through the greenhouse effect. Although the increased temperature will also increase the weathering rate, these geophysical processes occur on much longer timescales than the present (largely human-induced) rapid rate of CO_2 accumulation in the atmosphere. An increased absorption (dissolution) of CO_2 in seawater, however, has been measured already, and causes the ocean to become less basic, i.e., this decreases the ocean's pH. This process is referred to as *ocean acidification*, and may have serious consequences for marine ecosystems. In addition, chemical reactions between atmospheric gases and pollutants may also influence the atmospheric composition, the consequences of which are difficult to predict

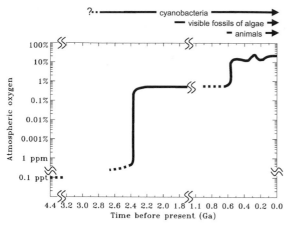

Figure 4.50 The abundance of oxygen in the Earth's atmosphere is shown as a function of time. Oxygenation of our planet's atmosphere appears to have occurred in a highly nonuniform manner. While the general trend of increasing oxygen is well established, quantitative estimates of the abundance are quite uncertain, especially in the more distant past. The best constrained epochs are represented by the solid portions of the curve, gaps in the curve denote epochs with no or very weak constraints, and dashes represent values which are moderately uncertain. (Courtesy David Catling)

with any certainty. We have embarked on a giant inadvertent experiment with our home planet's atmosphere, which may have dire consequences for the future of life on Earth.

4.9.2.2 Mars

Mars's small size may be as important a cause for the difference in climate between Mars and Earth as is the difference in heliocentric distance of these two planets. Although liquid water cannot exist in significant quantities on the surface of Mars at the present time, the numerous channels on the planet, layers of sandstone, and minerals that can only form in the presence of water, are suggestive of running water in the past (§5.5.4). This implies that Mars's atmosphere must have been denser and warmer in the past. Since the runoff channels are confined to the ancient, heavily cratered terrain, the warm martian climate did not extend beyond the end of the heavy bombardment era, about 3.8 Gyr ago. Estimates of Mars's early atmosphere suggest a mean surface pressure of the order of 1 bar, and temperature close to 300 K. Widespread volcanism, impacts by planetesimals, and tectonic activity must have provided a large source of CO_2 and H_2O, whereas impacts by very large planetesimals may also have led to (repeated) losses of atmospheric gases through impact erosion. In addition to atmospheric escape into space, Mars has probably lost most of its CO_2

via carbonaceous (weathering) processes, adsorption onto the regolith, and/or condensation onto the surface. Since Mars does not show current tectonic activity, the CO_2 cannot be recycled back into the atmosphere. Without liquid water on the surface, weathering has ceased, and Mars has retained a small fraction of its CO_2 atmosphere. The present abundance of H_2O on Mars is largely unknown. Most of the H_2O might have escaped, but recent theories invoke large amounts of subsurface water-ice on the planet (§5.5.4). A potential problem with the weathering theory is the apparent lack of carbonates on the martian surface.

Climatic changes may also be caused by changes in Mars's orbital eccentricity and obliquity. On Earth, these parameters vary periodically on timescales of $\sim 10^4$–10^5 yr, and may be responsible for the succession of ice ages and ice-free epochs during the past million years. For Mars, these parameters have periods about ten times larger than for Earth, and departures from the mean values are also much larger. The polar regions receive more sunlight when the obliquity is large, and large eccentricities increase the relative amount of sunlight falling on the summer hemisphere at perihelion. The layered deposits in Mars's polar region and glaciations in the tropics and at mid-latitudes (§5.5.4) suggest that such periodic changes have taken place on Mars.

The Tharsis region of Mars contains many volcanoes that appear to be roughly the same age. The eruptions of these volcanoes must have enhanced the atmospheric pressure and, via the greenhouse effect, the surface temperature. However, the sparsity of impact craters implies that the volcanic eruptions occurred well after the formation of the runoff channels on Mars's highlands.

4.9.2.3 Venus

Venus is very dry at the present time, with an atmospheric H_2O abundance of only 100 parts per million. This is about 10^5 times less H_2O than is present in the Earth's oceans. Various theories have been offered to explain the lack of water on Venus. It could have simply formed with very little water, because the minerals which condensed in this relatively warm region of the solar nebula lacked water. However, mixing of planetesimals between the accretion zones and asteroid/cometary impacts may have provided similar amounts of volatiles to Venus and Earth, in which case it seems probable that there was an appreciable fraction of a terrestrial ocean on early Venus. The D/H ratio on Venus is ~ 100 times larger than on Earth, which is persuasive evidence that Venus was once much wetter than it is now. But where did the water go? Water can be dissociated into hydrogen and oxygen, either by photodissociation

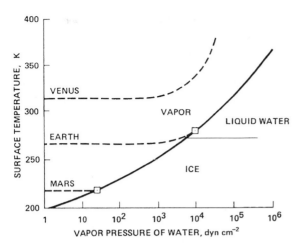

Figure 4.51 Evolution of the surface temperatures of Venus, Earth and Mars for a pure water-vapor atmosphere. (Goody and Walker 1972)

or chemical reactions, and the hydrogen will escape into space. However, the current escape rate is only 10^7 H cm^{-2} s^{-1}, implying that only 9 meters of ocean water could have escaped during the planet's entire lifetime.

The classical explanation for Venus's loss of water is via the *runaway greenhouse effect*. Figure 4.51 shows the evolution of the surface temperatures for Earth, Mars, and Venus, where each of the planets was assumed to have outgassed a pure water-vapor atmosphere, starting from an initially airless planet. While the vapor pressure of water in the atmosphere is increasing, the surface temperature increases due to the enhanced greenhouse effect. There is thus a positive feedback between the increasing temperature and increasing opacity. On Venus, the temperature stayed well above the saturation pressure curve, which led to a runaway greenhouse effect, and all of Venus's water accumulated in the atmosphere as steam. If one postulates an effective mixing process, the water is distributed throughout the atmosphere, and is photodissociated at high altitudes, with subsequent escape of the hydrogen atoms from the top of the atmosphere.

An alternative model to explain Venus's loss of water is the *moist greenhouse effect*. This model, in contrast to the runaway greenhouse model, relies on moist convection. Venus may have had a surface temperature near 100 °C. Convection transported the saturated air upwards. Since the temperature initially decreases with altitude, water condensed out, and the latent heat of condensation led to a decrease in the atmospheric lapse rate, and increased the altitude of the tropopause. In this scenario, water vapor naturally reached high altitudes, where it was dissociated and escaped into space. With the accumulation of water vapor in the atmosphere, the atmospheric pressure increased, preventing the oceans from boiling. Because of the rapid loss rate of water from the top of the atmosphere, the oceans continued to evaporate. As long as there was liquid water on the surface, CO_2 and O_2 were removed from the atmosphere by weathering processes. Once the liquid water was gone, CO_2 could not form carbonate minerals, and accumulated in the atmosphere.

4.9.3 Summary

Although the atmospheres of the terrestrial planets and satellites are all different in detail, common processes probably led to their formation. All atmospheres were predominantly formed via outgassing, while the mass (gravity) of the individual bodies, together with their atmospheric temperature and composition, are key to retaining the atmosphere. Subsequent evolution led to large variations in composition and atmospheric pressure. The carbon–silicate weathering process probably played a crucial role on Venus, Earth, and Mars, while surface temperature together with tectonic activity ultimately led to the observed differences. Late-accreting planetesimals may have eroded and supplied the planets with additional volatiles. Satellites around the giant planets formed within the planets' local subnebulas, where temperatures were well above those where planetesimals could have trapped N_2 and noble gases. The nitrogen in these planetesimals was therefore incorporated in the form of nitrogen compounds, in particular ammonia-ice. Outgassing of this constituent led to Titan's dense nitrogen atmosphere, while methane gas must be supplied continuously or episodically either via cryovolcanism, or perhaps via a meteorological cycle akin to the water cycle on Earth. Because of the large kinetic energies involved, any late-impacting planetesimals would have eroded any pre-existing atmosphere around the large jovian satellites Ganymede and Callisto, while such impacts on Titan, being less energetic, might have led to an increase, rather than a decrease, in the satellite's volatile budget.

Further Reading

A good book on weather at the nonscience major level is written by:
Williams, J., 1992. *The Weather Book*. Vintage Books, New York. 212pp.

Up-to-date descriptions of the atmospheres of individual planets are in:
McFadden, L., P. R. Weissman, and T.V. Johnson, Eds., 2007. *Encyclopedia of the Solar System*, 2nd Edition. Academic Press, San Diego. 982pp.

A few books on atmospheric sciences (graduate student level) we recommend are:

Atreya, S.K., 1986. *Atmospheres and Ionospheres of the Outer Planets and Their Satellites.* Springer-Verlag, Heidelberg. 224pp.

Chamberlain, J.W., and D.M. Hunten, 1987. *Theory of Planetary Atmospheres.* Academic Press, Inc., New York. 481pp.

Jacobson, M.Z., 1999. *Fundamentals of Atmospheric Modeling.* Cambridge University Press, New York. 656pp.

Salby, M.L., 1996. *Fundamentals of Atmospheric Physics.* Academic Press, New York. 624pp.

Seinfeld, J.H., and S.N. Pandis, 2006. *Atmospheric Chemistry and Physics: From Air Pollution to Climate Change,* 2nd Edition. John Wiley and Sons, New York. 1203pp.

Papers reviewing the origin and evolution of planetary atmospheres can be found in:

Atreya, S.K., J.B. Pollack, and M.S. Matthews, Eds., 1989. *Origin and Evolution of Planetary and Satellite Atmospheres.* University of Arizona Press, Tucson, Arizona. 881pp.

A classical text on fluid dynamics is given by:

Pedlovsky, J., 1987. *Geophysical Fluid Dynamics*, 2nd Edition. Springer-Verlag, New York. 710pp.

An excellent review paper on jovian atmospheric dynamics:

Vasavada, A.R., and A.P. Showman, 2005. Jovian atmospheric dynamics: An update after Galileo and Cassini. *Rep. Prog. Physics,* **68**, 1935–1996.

Problems

4.1.E Estimate the pressure scale height near the surfaces of Earth, Venus, Mars, Pluto, and Titan, and at the 1 bar levels of Jupiter and Neptune. Comment on similarities and differences.

4.2.E Although in some respects Earth and Venus are 'twin planets', they have very different atmospheres. For example, the surface pressures on Earth and Venus are 1 bar and 92 bar, respectively. Calculate the mass of each atmosphere both in grams and as a fraction of each planet's total mass. Recalculate these values for Earth including Earth's oceans as part of its 'atmosphere'. (If all of the water above Earth's crust were spread evenly over the planet, this global ocean would be $\sim$3 km deep.) Compare the values for the two planets and comment.

4.3.E (a) Show that in an optically thin medium, the equivalent width is proportional to the column density of absorbing material, regardless of the shape of the line profile.

(b) Assume the line profile can be represented by a Voigt profile. Show that the equivalent width in an optically thin medium is

$$EW \propto \Delta\nu_D N_c. \qquad (4.85a)$$

4.4.I Derive an expression for the equivalent width in a saturated line. Assume a Voigt profile, with the difference in optical depth between the center of the line and the wings being $\sim 10^4$. The wings of the line can be ignored. Define a frequency $x_1 = (\nu_1 - \nu_o)/\Delta\nu_D$, where the optical depth $\tau_\nu = 1$. Inside of x_1 the line is fully saturated, and outside x_1 the line is optically thin. Show that the equivalent width is

$$EW \propto \Delta\nu_D \sqrt{\log N_c}. \qquad (4.85b)$$

Note that the equivalent width is practically insensitive to the number density of absorbing material.

4.5.I In an optically thick medium, the wings in the Voigt profile become important. Define the frequency $x_1 = (\nu_1 - \nu_o)/\Delta\nu_D$, such that the product of τ_{ν_o} with the absorbing wings (second term in equation 4.18) is equal to unity. Express the frequency x_1 in terms of a and τ_{ν_o}. The medium is optically thick at frequencies $|x| < |x_1|$, while at $|x| > |x_1|$ the medium is optically thin. Show that the equivalent width is

$$EW \propto \sqrt{N_c \Delta\nu_D}. \qquad (4.85c)$$

4.6.E If you were to observe Jupiter's thermal emission at radio wavelengths, where one probes down to well below the planet's tropopause, would you expect limb brightening, darkening, or no change in intensity when you scan the planet from the center to the limb? Explain your reasoning.

4.7.E Assume that you observe Saturn at infrared wavelengths in the line transitions of C_2H_2 and PH_3. The C_2H_2 line is seen in emission, and the PH_3 line in absorption.

(a) Where in the atmosphere are these gases located?

(b) Do you expect the limb of the planet to be brighter or darker than the center of the planet in these lines?

4.8.I (a) The intensity of a Doppler broadened line profile is proportional to $e^{-(\Delta\nu/\nu_o)^2}$, where $\nu_o^2 = 2kT/m$, $\Delta\nu = |\nu - \nu_o|$, m is the mass of the molecules, T the temperature, k Boltzmann's constant, and ν the velocity. Derive the equation for the full width at half power for the line profile:

$$\Delta\nu = \frac{\nu_o \nu_o}{c} 2\sqrt{\ln 2}. \qquad (4.86)$$

(b) Compare equation (4.84) to the Doppler width, $\Delta\nu_D$, in equation (4.17).

4.9.E In the following you can assume that the lines are Doppler broadened.

(a) Calculate the full line width at half the maximum intensity (FWHM) in km s^{-1} for the hydrogen atoms in the upper atmosphere of Jupiter. You may assume a temperature $T = 10^3$ K, and that the upper atmosphere consists of H atoms only.

(b) Calculate the full line width at half power in km s^{-1} for Mars's atmosphere. Assume the atmosphere to consist entirely of CO_2 molecules, and the temperature to be 140 K.

(c) CO is a minor species in Mars's atmosphere; the mixing ratio CO/CO_2 is about 10^{-4}. What is the full width at half power in km s^{-1} for the CO lines?

4.10.E Why do we believe that we can calculate fairly precisely the temperature vs. altitude profiles well below the observable clouds for Jupiter and Saturn? Sketch one of these profiles and describe how it is derived. Why would the assumptions made in deriving this profile be questionable if applied to Uranus?

4.11.I The SO_2 line on Io is observed in emission at 222 GHz. The contrast between the peak of the line and the continuum background is 18 K. The FWHM (full width at half maximum) is 600 kHz. Io's surface temperature is 130 K; the emissivity $\epsilon = 0.9$. The line strength is approximately $\alpha_{\nu_0} = 3.2 \times 10^{-22}(\frac{300}{T})^{5/2}$ cm per molecule. In the following you will derive approximate values of the optical depth, number density, temperature, and surface pressure, under the assumption that the atmosphere is optically thin. You may assume that the observations pertain to the center of the disk, and that the plane parallel atmosphere approximation applies.

(a) Assume that line broadening is due to Doppler broadening. Calculate the atmospheric temperature, T_A.

(b) Calculate the optical depth in Io's atmosphere, assuming the Rayleigh–Jeans approximation (eq. 3.4a) to be valid.

(c) Calculate the Doppler width $\Delta\nu_D$ from the FWHM.

(d) Using the answers above, determine the column density and surface pressure. (Hint: Convert the Doppler width $\Delta\nu_D \rightarrow \Delta\lambda_D$.) Your estimates of column density and surface pressure should be roughly an order of magnitude below published values (§4.3.3.3). The reason for this discrepancy lies in the assumptions; better agreement can be reached for a more optically thick atmosphere and lower atmospheric temperatures.

4.12.E Consider a parcel of dry air in the Earth's atmosphere. Show that if you replace some portion of the air molecules (80% N_2, 20% O_2) by an equivalent number of water molecules, the parcel of air becomes lighter and rises.

4.13.E Calculate the dry adiabatic lapse rate (in K km^{-1}) in the atmospheres of the Earth, Jupiter, Venus, and Mars. Assume the atmospheres of Venus and Mars to consist entirely of CO_2 gas; Earth is 20% O_2 and 80% N_2; Jupiter is 90% H_2 and 10% He. Make a reasonable guess for the value of γ in each atmosphere. (Hint: See §3.2.2.3.)

4.14.I Estimate (crudely) the wet adiabatic lapse rate (in K km^{-1}) in the Earth's lower troposphere, following steps (a)–(d) below.

(a) Determine c_P from the dry adiabatic lapse rate (see previous problem).

(b) Set $T = 280$ K and $P = 1$ bar. The saturation vapor pressure of water near 280 K is roughly approximated by the Clausius–Clapeyron relation, with $C_L = 3 \times 10^7$ bar, and $L_S = 5.1 \times 10^{11}$ erg mole^{-1}. Calculate the partial pressure of H_2O in a saturated atmosphere at 280 K.

(c) As the concentration of water in a saturated atmosphere decreases with height much more rapidly than the total pressure, you may estimate the value of w_s (grams of water per gram of air) by multiplying the value of the partial pressure of water in bars by the ratio of the molecular mass of water to the mean molecular mass of air. Determine the value of w_s.

(d) Estimate the wet adiabatic lapse rate. Note: Watch your units. The latent heat in equation (4.20) is given in erg mole^{-1}, whereas that in equation (4.22) is in ergs g^{-1}.

4.15.I The saturated vapor pressure curve for NH_3 gas is given by equation (4.20) with $C_L = 1.34 \times 10^7$ bar and $L_s = 3.12 \times 10^{11}$ erg mole^{-1}.

(a) Calculate the temperature at which ammonia gas condenses out if the NH_3 volume mixing ratio is 2.0×10^{-4}. Assume the atmosphere to consist of 90% H_2 and 10% He, and that the pressure is 1 bar. (Hint: Convert the volume mixing ratio to partial pressure.)

(b) Calculate the temperature at which ammonia gas condenses out if the NH_3 volume mixing ratio is 1.0×10^{-3}.

4.16.I The base of the methane cloud in Uranus's atmosphere is at a pressure level of 1.25 bar and temperature of 80 K. The saturation vapor pressure curve is given by equation (4.20), with

$C_L = 4.658 \times 10^4$ bar and $L_s = 9.71 \times 10^{10}$ erg mole^{-1}. Derive the CH_4 volume mixing ratio in Uranus's atmosphere, assuming the composition of the atmosphere is 83% H_2 and 15% He. Compare your answer with the solar volume mixing ratio for carbon.

4.17.**E** Determine the terminal velocity of rain droplets in the Earth's atmosphere, assuming the viscosity $\nu_v = 0.134$ cm^2 s^{-1} and $\rho_{air} = 1.293 \times 10^{-3}$ g cm^{-3}.

(a) Determine the terminal velocity for rain drops 10 μm in radius.

(b) Determine the terminal velocity for rain droplets 1 mm in radius.

(c) Compare your answer to the escape velocity from Earth. Is the terminal velocity a realistic fall velocity for droplets several mm in size?

4.18.**I** (a) Derive an expression for the fractional increase in temperature for an atmosphere that is heated only by the Sun. (Hint: Equate the solar heat input (eq. 3.15) per rotation to the heat needed to raise the temperature of an atmosphere by ΔT.)

(b) Calculate the fractional change in temperature, $\Delta T / T$, near the surface on Venus and Mars, between local noon and midnight.

(c) Calculate the fractional change in temperature, $\Delta T / T$, in the thermospheres of Venus and Mars.

4.19.**I** Derive the Navier–Stokes equation for an incompressible fluid in the rotating frame of reference. (Hint: Consider equations (4.26) and (4.30); and note that the fluid is incompressible.)

4.20.**E** Show that a planet's vorticity is equal to twice its angular velocity (eq. 4.35).

4.21.**I** The zonal wind velocity on Jupiter can be measured at the cloud tops by following features in Jupiter's cloud deck. The cloud tops are at a pressure level of ~400 mbar. The wind speed is measured to be 100 m s^{-1} at a latitude of $\theta = 30°$. The meridional temperature gradient at this latitude $\frac{\partial T}{\partial \theta} \approx 3$ K deg^{-1}. Assume Jupiter's atmosphere to consist of 90% H_2 and 10% He. The average temperature in the range 0.4–4 bar can be taken as 150 K.

(a) Use the thermal wind equations to derive the depth (in bars) at which the zonal wind vanishes.

(b) How far (in km) below the cloud tops is this location?

4.22.**I** Consider a planet whose atmosphere can be approximated by an ideal gas. The planet's obliquity is small, and the surface temperature varies smoothly from equator to pole. If the atmospheric density is a function of altitude only, then the pressure varies over the surface, and the gas is accelerated in the poleward direction.

(a) Show that the acceleration is $dv/dt = -R_{gas} \nabla T / \mu_a$, where R_{gas} is the gas constant and μ_a is the mean molecular mass.

(b) Using the parameters of Earth, with an equator-to-pole temperature difference of 60 K, calculate the time it would take a parcel to 'free fall' from the equator to the pole.

4.23.**I** Consider the planet described in the previous problem. Planetary rotation produces a Coriolis acceleration, which measured in the rotating frame of reference is given by: $(d\mathbf{v}/dt) = -2\omega_{rot} \times \mathbf{v}$, where ω_{rot} is the planet's rotation rate. The Coriolis acceleration has a horizontal component everywhere except at the equator. Thus, moving air masses tend to go in circles of radius $\sim v/\omega_{rot}$. Using the free-fall velocity derived in the previous problem, estimate the characteristic radius of such motions on Earth, at a latitude $\theta = 30°$. Comment on your results; do you think the free-fall velocity is characteristic for wind velocities on Earth?

4.24.**I** Consider a storm on Earth at a geocentric latitude $\theta = 20°$. The height of the storm $\ell = 1$ km, and the radius is 100 km.

(a) Assume that the storm rotates with a velocity of 50 km per hour; i.e., the winds in the storm blow with this speed. Calculate the potential vorticity of the storm system.

(b) If the storm moves northwards to a latitude $\theta = 45°$ and both the vertical and horizontal scale of the storm stays the same, calculate its new vorticity (wind speeds).

4.25.**E** Neptune's Great Dark Spot was prominent during the Voyager era (1989), but the spot had disappeared by the mid 1990s. If the storm system indeed moved towards the equator, explain why the storm may have disappeared.

4.26.**E** (a) Calculate the minimum speed of air in the Earth's atmosphere to allow turbulence to develop at scales exceeding the scale height of the atmosphere (take $\nu_v = 0.134$ cm s^{-1}).

(b) If the flow velocity is 10 cm s^{-1}, what is the characteristic length scale on which we can expect turbulent motions?

4.27.**E** Explain briefly, qualitatively, why the density of ozone in Earth's atmosphere peaks near an altitude of 30 km. List the relevant reactions.

4.28.I (a) Using the Chapman reactions (reactions 1, 4, 5, and 6 in §4.6.1.1), derive equations (4.61) and (4.62).

(b) Assuming chemical equilibrium, derive the altitude profiles given by equations (4.63) and (4.64).

(c) Calculate the number density of O_3 molecules in Earth's atmosphere at altitudes of 20 and 60 km, using the number densities of [O] and [O_2] from Figure 4.40. The number density for [M] at $z = 0$ km is equal to Loschmidt's number.

4.29.E Explain why the NH_3 mixing ratio in the stratospheres of Saturn and Jupiter is well below the mixing ratio based upon the saturated vapor curve.

4.30.I In the following, assume that recombination of atoms proceeds at the gas-kinetic rate of collisions, i.e., for direct recombination of two oxygen atoms into a molecule $k_{r2} = k_{gk}$. Compare the recombination rates for a two-body and three-body reaction, such as in the formation of O_2, reactions (2) and (3). (Hint: You may want to use equations analogous to eq. 4.57 for the production of O_2, and calculate the number density of molecules needed for equal rates $k_{r2} = k_{r3}$.)

4.31.E Cloud or haze layers of hydrocarbons are observed in the stratospheres of Uranus and Neptune. Explain why we see these hazes above rather than below the tropopause.

4.32.I The limiting flux is the maximum diffusion rate through a planetary atmosphere. The limiting flux can be calculated by assuming the same value for N_i/N for that part of the atmosphere which is well mixed (i.e., below the homopause, which for Earth is at $z < 100$ km). Consider hydrogen atoms in the Earth's atmosphere in all forms (H_2O, CH_4, H_2), at a fractional abundance $N_i/N \approx 10^{-5}$.

(a) Calculate the limiting flux of hydrogen-bearing molecules (and thus hydrogen) from the Earth's atmosphere. You can approximate the binary diffusion parameter using the quantities for H_2 in air (Table 4.7). (Hint: Calculate the limiting flux for an altitude $z = 100$ km; why?)

(b) Calculate the Jeans rate of escape for hydrogen atoms from Earth.

(c) Compare your answers from (a) and (b) and comment on the results.

4.33.E The presence or absence of an atmosphere is to first order determined by Jeans escape. To verify this statement, plot a normalized escape parameter for atomic hydrogen, λ_{esc}, as a function of heliocentric distance for all eight planets, Io, Ganymede, Titan, Enceladus, the asteroid Ceres, and the Kuiper belt objects Pluto, Eris, and Varuna. To calculate relative temperatures, simply use the equilibrium temperature (Chapter 3) at perihelion, with an albedo of zero, and emissivity of unity. Assume the exobase altitude $z = 0$. Discuss your findings. (Hint: You may want to use Tables 1.1–1.5, 9.1, 9.2, and 9.5.)

4.34.E Suppose a body with a radius of 15 km hits a planet at a velocity of 30 km s^{-1}. Assume the evaporative loading parameter is $\sim$20.

(a) Calculate the mass of the atmosphere blown into space if the impactor hits the Earth.

(b) Calculate the mass of the atmosphere blown into space if the impactor hits Venus.

(c) Calculate the mass of the atmosphere blown into space if the impactor hits Mars.

(d) Express the masses of escaping gas calculated in (a)–(c) as a fraction of each planetary atmosphere's mass. Comment on your results.

4.35.E Why is the $^{15}N/^{14}N$ ratio larger in Mars's atmosphere than it is in Earth's atmosphere?

4.36.I (a) Explain how the carbon cycle moderates climatic variations on Earth.

(b) Explain qualitatively why, if there were to be a major nuclear war, the temperature on Earth may drop to levels well below freezing, a scenario referred to as 'nuclear winter'.

5 Planetary Surfaces

I believe this nation should commit itself to the goal, before this decade is out, of landing a man on the Moon and returning him safely to the Earth.

USA President John F. Kennedy, in a speech before Congress, 25 May 1961

That's one small step for man, one giant leap for mankind.

Astronaut Neil Armstrong, 20 July 1969, as he became the first human to set foot on the Moon

The four largest planets in our Solar System are gas giants, with very deep atmospheres and no detectable solid 'surface'. All of the smaller bodies, the terrestrial planets, asteroids, moons, and comets, have solid surfaces. These bodies display geological features that yield clues about their formation, as well as past and current geological activity. The surface reflectivity varies dramatically from one body to another; some surfaces have very low albedos (such as the maria on the Moon, carbonaceous asteroids, comet nuclei), while others are highly reflective (Europa, Enceladus). Large albedo variations may even be seen on a single object (Iapetus). Some bodies are almost completely covered by impact craters (Moon, Mercury, Mimas), while others show little or no sign of impacts (Io, Europa, Earth). The terrestrial planets and many of the larger moons show clear evidence of past volcanic activity, and some (Earth, Io, Enceladus, Triton) are active even today. Past volcanic activity may be seen in the form of volcanoes of different shapes and size (Earth, Mars, Venus) or large solidified lava lakes (Moon). Most bodies, even small asteroids, display linear features like faults, ridges, and scarps that are suggestive of past tectonic activity. None of the other objects, however, displays the motion of tectonic plates as the Earth does. Why are planetary surfaces so different superficially, and what similarities do they share? In this chapter, we review planetary geological processes. We start with a basic review of rocks and minerals, and discuss the crystallization of *magma* (molten rock at depth). Processes that 'shape' the surfaces (gravity, volcanism, tectonism, impacts) are discussed in §5.3 and §5.4, and the surface characteristics of individual bodies are summarized in §5.5.

Although the interior structure of planets is discussed in Chapter 6, some basic terminology is required for the current chapter. Figure 5.1 shows a sketch of the interior structure of the Earth, as it has been deduced from seismological data. The Earth consists of a solid iron–nickel

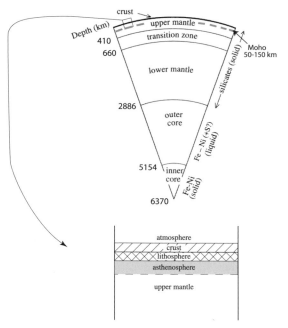

Figure 5.1 Sketch of the interior structure of the Earth. (Adapted from Putnis 1992)

inner core, surrounded by a fluid metallic *outer core*. The core extends over roughly half the Earth's diameter. The outer ~3000 km is the primarily rocky *mantle*, which itself is divided into a lower and upper mantle, separated by a transition zone. A cool elastic *lithosphere* sits on a hot, highly viscous 'fluid', the *asthenosphere*. Although the asthenosphere is highly viscous, the mantle below it is even more so, and the asthenosphere, therefore, is like a 'lubricating' layer between the rigid lithosphere and highly viscous mantle below it. The lithosphere is an elastic layer which responds to a 'load', i.e., it bends down under the load of an ocean island. The terminology for the exact description of the lithosphere is somewhat confusing, however, because the precise definition depends on the subfield in geophysics. For example, in addition to the elastic lithosphere, one can also refer to the mechanical or the thermal lithosphere, where the former is defined as the material (rocks) that remain a coherent part of the plates over geological timescales ($\sim 10^8$ yr). Since deformation of material is determined by its viscosity, which decreases with increasing temperature, only that part of the lithosphere that has a temperature below ~ 1400 K is cold enough to stay rigid. Under oceans, the lithosphere varies in thickness from ~ 0 km (mid-ocean ridge) up to ~ 100 km, and it is ~ 200 km thick under the continents. Because the lithosphere is so mucher thicker under the continents, the asthenosphere is found primarily under oceans.

The outer 'skin' of the planet is the *crust*, a rather brittle layer, which is the topic of the present chapter. Earth's crust has a mean thickness of ~ 6 km under the oceans, and a mean thickness of ~ 35 km under the continents. Oceanic crust, primarily of basaltic composition, is denser than the more silicic (granitic) continental crust.

5.1 Mineralogy and Petrology

Petrology is the study of the composition, structure, and origin of rocks. Since solid planetary material consists of rock and ice, and rocks are made up of different minerals, it is essential for a planetary scientist to have some basic knowledge of rocks and minerals. We begin this section with a short summary of the basics of mineralogy, before reviewing the various types of rocks, how rocks are formed, and where they are found.

5.1.1 Minerals

Minerals are solid chemical compounds that occur naturally and that can be separated mechanically from other minerals that make up a rock. Each mineral is characterized by a specific chemical composition and a specific regular architecture of the atoms from which it is made. The forces that hold molecules together depend upon the electronic structure of the constituent atoms. Some atoms, for example, Si, Mg, Fe, Ti, tend to give up electrons in their outer shells, creating a positively charged ion (a *cation*), while other atoms, in particular O, may adopt electrons, creating a negatively charged ion (an *anion*). Whether an atom gains or gives up electrons depends upon the electronic structure of the element. In particular, most atoms have one or more loosely bound *valence electrons* that can be 'shared' with other atoms to fill up electron shells, thereby lowering the energy state of the formed compound. Such interactions determine the chemical behavior of the elements. The valence of each atom is indicated in the

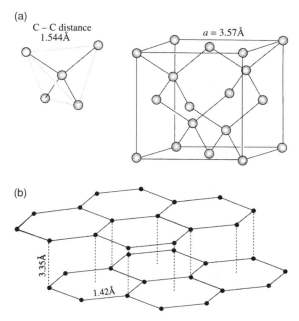

Figure 5.2 (a) Mineral structure of diamond: each carbon atom is surrounded by four others in a regular tetrahedron (shown by the dashed line, left figure). The diamond (right figure) is built up from these tetrahedra. (Putnis 1992) (b) The structure of graphite is made up of layers in which the carbon atoms lie at the corners of a hexagonal mesh. Within the layers the atoms are strongly bonded, while the layers are weakly bonded to one another. (Putnis 1992)

periodic table of elements (Appendix D). The chemical bond between cations and anions is referred to as an *ionic bond*, caused by the electrostatic attraction between oppositely charged particles (Coulomb's law). Examples are the mineral halite (Na^+ with Cl^-) and magnesium oxide (Mg^{2+} with O^{2-}). Another type of bonding is *covalent bonding*, where the atoms share electrons in their outer shells. In a diamond, each carbon atom is surrounded by four others in a regular tetrahedron (Fig. 5.2a), and these atoms are held together by a covalent bond. A third, but much weaker, bonding is due to the *van der Waals force*, a weak electrical attractive force that exists between all ions and atoms in a solid. The strength of the bonds, often a combination of the three types mentioned above, determines the hardness of a mineral. The *Mohs scale of hardness* (Table 5.1) runs from 0 to 10, where talc is 1 and diamond 10.

A mineral is characterized by a combination of its chemical composition and crystalline structure. A different spatial arrangement of the atoms that make up one material can lead to a very different mineral, even if the chemical composition is the same. A classic example is graphite versus diamond, each of which consists exclusively of C atoms. Diamond, where the atoms are

Table 5.1 Mohs scale of hardness.[a]

Mineral	Scale number	Common objects
Talc	1	graphite (0.6)
Gypsum	2	fingernail (2.5)
Calcite	3	copper coin
Fluorite	4	
Apatite	5	teeth
Orthoclase	6	window glass (5.5)
Quartz	7	steel file (6.5)
Topaz	8	
Corundum	9	
Diamond	10	

[a] After Press and Siever (1986), and others.

Table 5.2 Chemical classes of minerals.[a]

Class	Defining anions	Example
Native elements	none	copper Cu, gold Au
Sulfides and similar compounds	S^{2-} similar anions	pyrite FeS_2
Oxides and hydroxides	O^{2-} OH^-	hematite Fe_2O_3 brucite $Mg(OH)_2$
Halides	$Cl^-, F^-,$ Br^-, I^-	halite $NaCl$
Carbonates and similar compounds	CO_3^{2-}	calcite $CaCO_3$
Sulfates and similar compounds	SO_4^{2-} similar anions	barite $BaSO_4$
Phosphates and similar compounds	PO_4^{3-} similar anions	apatite $Ca_5F(PO_4)_3$
Silicates and similar compounds	SiO_4^{4-}	pyroxene $MgSiO_3$

[a] After Press and Siever (1986).

bonded in a covalent bond, is an extremely hard mineral, whereas graphite is very soft. Graphite is made up of layers within which the C atoms form a hexagonal mesh, and van der Waals forces bond the layers to one another (Fig. 5.2b).

Minerals in the field can be identified through a combination of their *hardness*, *cleavage* or breakage along certain planes (e.g., mica), *fracture*, *density, color, luster*, and *streak* (color of the powder that comes off when a mineral is scraped). Although there are several thousand known minerals, each with its own unique set of properties, most of them can be classified within a few major chemical classes. These classes are listed in Table 5.2. In addition to the native elements, such as Cu, Fe, Zn, etc., minerals can be made up of several different atoms organized in a regular crystalline structure, such as quartz, SiO_2, or olivine, $(Fe,Mg)_2SiO_4$. The notation (Fe,Mg) indicates that the elements Fe and Mg can be substituted for one another. Substitution of atoms in minerals takes place between elements of the same size and valence.

The most abundant types of minerals on terrestrial planets are the *silicates*, minerals which contain silicon and oxygen, such as quartz, olivine, feldspar $((K,Na)AlSi_3O_8, CaAl_2Si_2O_8)$, and pyroxene $((Mg,Fe)SiO_3)$. Feldspars make up about 60% of the surface rocks on Earth. They have a typical density of ~ 2.7 g cm^{-3}, and thus are relatively light. They, therefore, tend to float upwards in a magma, and end up relatively close to a planet's surface. Potassium-rich feldspars are referred to as orthoclase feldspars, while plagioclase feldspars are rich in sodium and/or calcium. Quartz is, like feldspars, very abundant on the Earth. Quartz has a density of ~ 2.7 g cm^{-3}, thus is also present on or near a planet's surface. Pyroxenes

make up $\sim 10\%$ of the Earth's crust. They contain a relatively large fraction of heavy elements, such as Mg and Fe, which makes them denser than feldspars ($\rho \approx$ 2.8 to 3.7 g cm^{-3}). Common minerals in this class are, e.g., augite $(Ca(Mg,Fe,Al)(Al,Si)_2O_6)$ and enstatite $(MgSiO_3)$. Olivine $((Fe,Mg)_2SiO_4)$, an olive-colored mineral, is denser than the pyroxenes, so it sinks in magmas. Olivine is therefore an important constituent of rocks formed at depth, and believed to be a major constituent of the Earth's mantle. Amphibole is a group of (Mg,Fe,Ca)-silicates, which are slightly less dense than pyroxenes, but have a more amorphous structure. They make up $\sim 7\%$ of the Earth's crustal minerals. An example is hornblende $((Ca,Na)_{2-3}(Mg,Fe,Al)_5(Si,Al)_8O_{22}(OH)_2)$. Micas are sheet silicates of K, Al, and/or Mg. The most common examples of micas are biotite (black to dark brown; $K(Mg,Fe)_3AlSi_3O_{10}(OH,F)_2$) and muscovite (silvery, colorless or white, translucent; $KAl_2(AlSi_3O_{10})(OH,F)_2$).

After silicates, the most abundant minerals on Earth are *oxides*, which are composed primarily of metals (in particular Fe) and oxygen. Common iron oxides include magnetite (Fe_3O_4), which has a black, metallic luster, hematite (Fe_2O_3), which is often reddish-brown or steely-gray and black, and limonite $(HFeO_2)$. The color of limonite is yellowish-brown to dark brown, similar to rust. These minerals are believed to redden the surface of Mars. Ilmenite ($(Fe,Mg)TiO_3$), a black opaque mineral, and spinel ($MgAl_2O_4$), provide most of the opacity in the Moon's maria.

Other common minerals on Earth are pyrite (fool's gold, FeS_2) and troilite (FeS), both of which are probably abundant in planetary interiors (§6.1.2.3), since their high density ($\sim$5 g cm^{-3}) causes them to sink down in magma. Clay minerals are hydrous aluminum silicates, major erosion products on Earth and Mars, which also have been detected on carbonaceous asteroids (e.g., Ceres). Clay minerals may hold substantial amounts of chemically bound water.

In the outer Solar System, beyond a heliocentric distance of $\sim$4 AU, *ices* made up over half the mass of material which condensed in the solar nebula, and are thus important constituents of minerals in the outer Solar System. Important ices include water (H_2O), carbon dioxide (CO_2), ammonia (NH_3), and methane (CH_4). Much of the water is also found in the form of *hydrate* minerals (such as the hydrous aluminum silicates mentioned above), while water-ice can exist in the form of *clathrates*, where a guest molecule occupies a cage in the water-ice lattice. Other low-temperature condensates important in the outer Solar System are carbonaceous minerals, which color the surfaces of objects blackish and reddish-black (albedos $\sim$2–8%). *In situ* measurements of Comet P/Halley by the Giotto spacecraft revealed the presence of CHON particles, dust grains which are dominated by combinations of the elements H, C, N, and O (§10.3.5).

5.1.2 Rocks

Planetary surfaces are composed of solid material, which is generally referred to as 'rocks', assemblages of different minerals. Rocks are classified on the basis of their formation history. We distinguish four major groups, each of which is discussed in more detail in the following subsections: primitive, igneous, metamorphic, and sedimentary rocks. Within these groups, the rocks can be further subdivided on the basis of the minerals of which they are composed, and/or on the basis of their texture, such as the size of the grains which make up the rock. Some rocks, such as breccias (§5.1.2.5), can include material from various groups.

5.1.2.1 *Primitive Rocks*

Primitive rocks are formed directly from material that condensed out of the primitive solar nebula. These rocks have not undergone transformations in interiors of objects like the planets and larger moons and asteroids, where materials are altered significantly due to the high temperatures and pressures prevailing there. These primitive

rocks have never been heated much, although some of their constituents (e.g., chondrules) may have been quite hot early in the history of our Solar System. Primitive rocks are common on the surfaces of many asteroids, and the majority of meteorites are primitive rocks (§8.1).

5.1.2.2 *Igneous Rocks*

Igneous rocks are the most common rocks on Earth and other bodies that have undergone melting. Igneous rocks are formed when a *magma*, i.e., a large amount of hot molten rock, cools. The physics and chemistry of a cooling magma, and the crystallization of minerals therein, is discussed in §5.2. Here we describe the end products of the melt, i.e., the various types of rocks that result from the cooling process. Rocks form from the magma either underground (*intrusive* or *plutonic* rocks) or above ground (*extrusive* or *volcanic* rocks). Magma deep underground cools slowly, and crystals have plenty of time to grow. The resulting intrusive rocks are therefore coarse grained, and the minerals can easily be distinguished with the naked eye (e.g., common granite, Fig. 5.3). When magma erupts through the planetary crust, it cools rapidly through radiation into space. Volcanic rocks thus show a fine-grained structure, in which individual minerals can only be seen through a magnifying glass. In cases of extremely rapid cooling, the rock may 'freeze' into a glassy material. Obsidian is a volcanic rock that cooled so fast that it shows no crystalline structure (Fig. 5.3). The minerals in these rocks are no longer in the form of crystals, but show an amorphous, glassy structure. The texture of the rock thus depends on how rapidly the magma cools, while the composition of the rock depends upon the minerals which crystallize from the melt.

Although the classification of the major rock groups is based upon their chemical and mineralogical composition, for practical purposes one can simply use the silica content of the rock. The two basic rock types are basalts and granites, where basalts contain 40–50% silica (by weight) and granites much more ($\sim$70% by weight). In addition to silica, basalts consist largely of heavy minerals, such as pyroxenes and olivines. Basalts are sometimes referred to as *basic* or *mafic* (from Mg, Fe) rocks. *Ultrabasic* (*ultramafic*) rocks (e.g., peridotite, the primary constituent of planetary mantles) have a very large percentage of heavy elements. In contrast, feldspars, in particular orthoclase (K-rich feldspars), and quartz are the dominant minerals in granite. Granites are therefore also referred to as *felsic* (from feldspar) or *silicic* rocks. Granites are usually light-colored, whereas basalts, and in particular the ultramafic basalts, are dark. Basaltic rocks are probably the most common rocks on planetary surfaces, as they make up

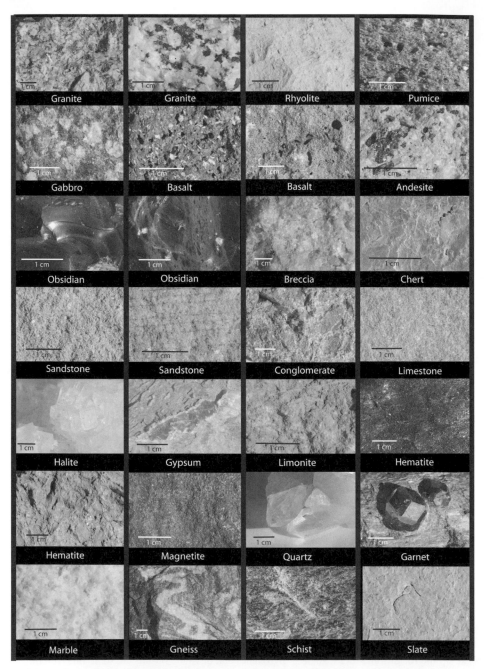

Figure 5.3 COLOR PLATE Examples of different types of rocks. This collection is by no means complete, but gives an idea of the differences between important rock types. An approximate scale (1 cm bar) is indicated on each frame. The rocks are grouped by type (igneous, sedimentary, metamorphic) and grain size (see Fig. 5.4). In some cases, we show two examples, e.g., where colors may differ substantially (such as a white and pink granite, a dark-gray and a reddish hematite). Individual crystals (e.g., quartz, biotite, muscovite, plagioclase) can be identified in the granite samples, while in rhyolite only small specks of grains are visible. Pumice is a very light and froth-like rock. In gabbro, individual crystals are large, as in granite, whereas the crystals are smaller in basalt and andesite. No crystal structure is present in obsidian, a glassy rock. Sandstones, sedimentary rocks, can be clearly recognized from their granular appearance (a gray sandstone and Arizona pinkish one are shown), while conglomerates are much coarser grained. Chert is a relatively hard sediment. Halite and gypsum are representative samples of evaporites. Magnetite is an iron oxide. Limonite and hematite are sediments formed from clay minerals. Minerals are sometimes present in the form of large crystals, such as the quartz and garnet crystals shown. Several metamorphic rocks are displayed (bottom row): Marble, gneiss, schist, and slate. (Photographs taken by Floris van Breugel, using rock samples provided by K. Ross, Museum of Geology, UC Berkeley)

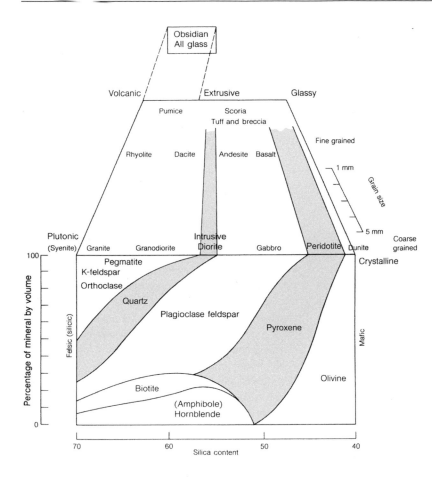

Figure 5.4 A classification cube for igneous rocks. The horizontal axis shows the silica content of the rocks (percent by volume), and the vertical axis the percentage of a given mineral. The texture of the rocks is indicated as a function of grain size along the receding axis, at the top of the cube. A granite with a silica content of 70% contains about 25% quartz (SiO_2), less than 10% of each biotite, hornblende, and plagioclase feldspar, and 50% K-feldspar or orthoclase. Fine-grained granites are called rhyolite. Rocks with a lower silica content are more mafic. Peridotite consists of varying fractions of pyroxene and olivine. Rocks that contain more than 90% olivine are called dunite. (Press and Siever 1986)

the lava (solidified magma) flows on bodies like the Earth and Moon. Although abundant on Earth, granite is less common on other planetary bodies.

Figure 5.4 shows a classification scheme for the various rock types in the form of a data cube. The horizontal axis shows the silica content of the rock, while the vertical axis indicates the mineral content. The receding axis (towards the back of the cube) indicates grain size. The silica content determines whether a rock is felsic (granitic) or mafic (basaltic). The grain size increases with the time it took the rock to cool. Intrusive rocks consist of large grains, while volcanic rocks are fine grained. Some pictures of rocks from the field are shown in Figure 5.3. The minerals that make up granite are typically several millimeters in size, and consist primarily of feldspars and quartz. Volcanic forms of granite are rhyolite, pumice, and obsidian, in order of decreasing grain size (obsidian, a glass, does not contain crystals). Basalt is a fine-grained mafic volcanic rock, which consists primarily of the mineral pyroxene. The plutonic, coarse-grained mafic rock is called gabbro. Dunite is a very ultramafic rock, which consists almost entirely of the mineral olivine.

Extrusive rocks are directly correlated with volcanic activity. The type of eruption is determined by the viscosity (a measure of resistance to flow in a fluid) of the magma, and depends on the temperature, composition (in particular silica content), and the gas content of the melt. Typically, magma/lava with a high silica content, i.e., felsic lava, has a high viscosity, which is increased even more if the gas content is high. Such eruptions are explosive, with typical temperatures of $\sim$1050–1250 K, and they form thick local deposits. In contrast, basaltic melts are very fluid, erupt with temperatures of 1250–1500 K, and flow fast and far; they form large lava beds and fill in lowlands, such as the maria on the Moon and the lava beds on Hawaii. These lava flows are usually dark in color, in contrast to the lighter felsic deposits.

Volcanic rocks can vary widely in appearance and density. These rocks are usually identified by a combination of their texture and composition. Violent explosions, where lava is ejected into the air and shattered in the process, usually due to a sudden release of gases, are more common in the silicic than basaltic magmas. Rocks that are blasted out during such a violent explosion are called *pyroclasts*, and vary in size from micrometer-sized *dust*, to millimeter-sized *ash*, and larger *bombs*. Many pyroclasts are glassy or fine grained, caused by the rapid cooling of the suddenly ejected magma. The silicic melts may have a high gas content, and because the gases usually cannot escape, pyroclastic fragments can be extremely light. Examples of such *vesicular rocks* are *pumice* or *volcanic foam*, a sponge-like glassy rock with numerous bubbles or cavities (*vesicles*), formed by gas in the melt. Pumice is so light that it floats in water. When pyroclasts hit the ground, the rocks may get cemented together under extreme heat into volcanic *tuff* and *breccias*.

5.1.2.3 *Metamorphic Rocks*

As there has been much reworking of the Earth's surface by great forces from within the interior, many rocks have been altered when subjected to high temperature and pressure, or when introduced to other chemically active ingredients. Rocks that have been so altered are called *metamorphic* rocks. Such rocks are often named for a mineral constituent that is predominant in the rock, such as marble (from limestone or other carbonate rock), quartzite (from quartz), and amphibolite (amphiboles) (see Fig. 5.3 for some examples). Metamorphism can act either on a *regional* or a *local* scale. On the regional scale, rocks (igneous, sedimentary, metamorphic) are transformed many kilometers below the surface by extremely high temperatures and pressures. Large areas or regions of rock can be metamorphosed this way. Regional metamorphic rocks show *foliation*, a platy structure caused by a parallel alignment of the minerals, often perpendicular to the stress exerted on them. Examples are coarse-grained gneiss, transformed from granite, medium-grained schist from shale and/or granite, and fine-grained slate from shale. Serpentinite, composed of serpentine minerals (hydrous magnesium iron silicates, $(Mg, Fe)_3Si_2O_5(OH)_4$), forms through oxidation of the magnesium-rich minerals olivine and pyroxene in peridotite in the oceanic mantle, at relatively low temperatures. Tectonic plate motions have brought serpentinite onto the California coast, where the rock is known as the 'California state rock'. On a local scale, rocks are transformed near an igneous intrusion, largely by heat. Magma forces its way into layered rocks or penetrates cracks or cavities. If the stress on the rocks and the temperature are high enough, the rock is altered, metamorphosed. An example is hornfels, a very fine-grained silicate rock. Rocks formed this way are sometimes referred to as *contact* metamorphic rocks. Though not as common on Earth as regional or contact metamorphic rocks, rocks may also get altered as a result of impact-induced shocks at impact sites. Shocked quartz has been found at several impact sites.

5.1.2.4 *Sedimentary Rocks*

On planets that possess an atmosphere, material may be transported by winds, rain, and liquid (e.g., water) flows. Sedimentation is the final stage of this process, where material is deposited at some other location. These sediments may form new *sedimentary* rocks (Fig. 5.3). *Detrital* sediments have physically been transported from one place to another, for example by winds or water after erosion of rocks. The components of detrital sediments are fragments of pre-existing rocks and minerals, and are referred to as *clastic* (which, in Greek, means to break) fragments. While the rock fragments are transported, the minerals are sorted by size and weight with a variable efficiency. Due to the sorting process, rocks form with different textures, varying from coarse to very fine grained rocks. Coarse-grained fragments, such as gravel, form *conglomerates* when cemented together. Medium-grained sands form *sandstone*, and fine-grained clays and silt may be cemented together into *mudstone* or *shale*. The individual grains in shale and sandstones are fairly round, as a result of the erosion. Shale and sandstone are the two most abundant sedimentary rocks on Earth. Shale makes up ~70% of the sediments on Earth, while sandstone makes up ~20%. The remaining 10% consists predominantly of limestone, a chemical sediment discussed briefly below.

The composition of rocks may be altered through the interaction with other chemical constituents, such as those present in an atmosphere. Such sediments are referred to as *chemical* sediments. Prime examples are limestone ($CaCO_3$) and dolomite ($CaMg(CO_3)_2$), derived from the *Urey weathering reaction* (§4.9.2), wherein CO_2, dissolved in water, reacts with silicate minerals in the soil to form calcium carbonate or calcite, $CaCO_3$. We note that on Earth, however, most carbonates are from biological deposits, fossils of animal shells produced by organisms in the oceans. *Evaporite* is a rocky material from which liquid evaporated, leaving behind sediments such as halite or common salt (NaCl) and sulfate minerals, such as gypsum ($CaSO_4 \cdot 2H_2O$). Evaporites may bond other rocks together into a loose crumbly rock. Another type of sediment is formed by clay minerals, hydrous aluminum silicates such as hematite and limonite. They are abundant in erosion products on both Earth and Mars, as well as in carbonaceous material on asteroids and bodies in the outer Solar System.

5.1.2.5 *Breccias*

Breccias are 'broken rocks' that consist of sharp angular fragments which are cemented together. These rocks may originate from meteoroid impacts, where the pieces are 'glued together' under the high temperature and pressure during and immediately following the impact. They therefore cover the bottom of many impact craters. Breccias that consist of pieces of a single type of rock are called *monomict*; *polymict breccias* consist of a mix of fragments from different types of rocks. Breccias may also form tectonically, e.g., along fault zones.

5.2 Cooling of a Magma

The composition, pressure, and temperature of a magma determine which minerals ultimately form when the magma cools. In this section, we discuss phase diagrams of a melt and a general sequence of reactions that take place in a cooling magma.

5.2.1 Phases of the Magma

The states that the magma goes through as it cools and crystallizes can be shown on a *phase diagram*, similar to the phase diagrams discussed for water and other condensable gases in §4.4. For simplicity, let us assume the magma to crystallize under equilibrium conditions. The phase changes that occur in the magma are best predicted by making use of the *Gibbs free energy*, G, of the system:

$$G = H - TS, \tag{5.1a}$$

or the change in Gibbs free energy, ΔG:

$$\Delta G \equiv \Delta H - T\Delta S, \tag{5.1b}$$

where T is the temperature and S the entropy (defined below) of the system.

The *enthalpy*, H, is defined as the sum of the internal energy, U (potential energy stored in the interatomic bonding plus kinetic energy of the atomic vibrations), and the work done on the system, PV, with P the pressure and V the volume:

$$H = U + PV. \tag{5.2}$$

The enthalpy of the pure elements is defined to be zero, and the enthalpy of a mineral is the change in enthalpy or 'heat' that is required to form the mineral from the individual elements. If it takes energy to form the mineral, the reaction is *endothermic*, and the formation enthalpy is positive. If the reaction frees up energy, the reaction is *exothermic*, and the formation enthalpy is negative. The formation enthalpy is related to the thermal heat capacity of the system, C_P, defined as the amount of heat needed to raise the temperature of 1 mole of material by 1 K while keeping the pressure constant (eqs. 3.20, 3.29):

$$\left(\frac{\partial H}{\partial T}\right)_P = \left(\frac{\partial Q}{\partial T}\right)_P \equiv C_P. \tag{5.3}$$

The enthalpy of the system at temperature T_1 is thus equal to

$$H = H_0 + \int_0^{T_1} C_P dT, \tag{5.4}$$

with H_0 the enthalpy at $T = 0$ K.

Entropy, S, is a quantity which measures the change in a mineral's state of order when it changes from one phase or structure to another. For a thermodynamically reversible process, the change in entropy is equal to the ratio of the amount of heat absorbed by the system, dQ, and T:

$$dS = \frac{dQ}{T}. \tag{5.5}$$

The entropy of such a system at temperature T_1 is then given by

$$S = S_0 + \int_0^{T_1} \frac{C_P}{T} dT, \tag{5.6}$$

with S_0 the entropy at $T = 0$ K.

For a perfect crystal with $S_0 = 0$, all atoms are in the ground state. After heating a thermodynamically reversible sample and cooling it back down to its original temperature, the change in entropy would be zero. However, for any natural process, the change in entropy is always positive ($dS > 0$), which implies that if a mineral becomes more ordered in a transformation process (i.e., $dS < 0$), then the heat liberated in the process must increase disorder in the environment.

The Gibbs free energy is often used to evaluate phase transitions in samples. Figure 5.5 shows a graph of the Gibbs free energy as a function of temperature for the liquid, l, and solid, s, phase of a melt. The phase with the lowest free energy is the stable phase. At the critical temperature, T_c, the curves cross, and upon further heating or cooling of the melt a phase transformation takes place. The phase transition brings about a change in the enthalpy, ΔH, which is the *latent heat of transformation* (see also §4.4.1):

$$\Delta H = T\Delta S. \tag{5.7}$$

At a constant pressure, transformations or phase transitions only take place upon cooling or heating the sample, since the free energies of the two phases are equal at T_c. In practice, transformations may be *reversible*, i.e., the phase changes upon cooling the sample and changes

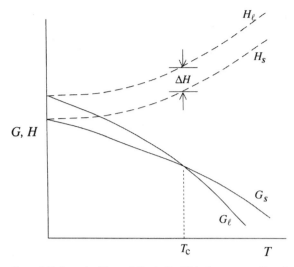

Figure 5.5 A graph of the variation in the Gibbs free energy, G, and the enthalpy, H, of the liquid (ℓ) and solid (s) phase of a solution as a function of temperature. The phase with the lowest free energy is the stable phase. At temperatures $T > T_c$, the mixture is in a liquid phase, while at $T < T_c$ the solution is solid. (Adapted from Putnis 1992)

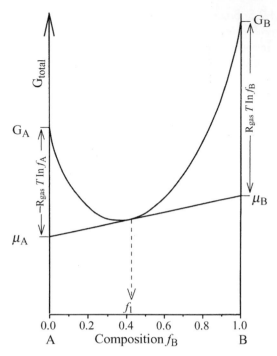

Figure 5.6 The Gibbs free energies curve for an ideal solid solution, consisting of components A and B. The x-axis shows the solution's fraction of component B. The Gibbs free energy for the end members, G_A and G_B, are plotted along the y-ordinate, where the solution is 100% A (i.e., 0% B) and 100% B, respectively. The tangent to the curve at any composition f_B (such as f_1 shown in the figure) intercepts the free energy axes at the chemical potential of the component, μ_A and μ_B, respectively. The total free energy is then given by the sum $\mu_A f_A + \mu_B f_B$ (eq. 5.11). (Adapted from Putnis 1992)

back when heating it, or *irreversible*, when the sample does not convert back to its original state when reversing the temperature gradient. Irreversible transformations occur because transformations depend upon the kinetics or reaction rate of the processes involved in addition to the temperature. A prime example of an irreversible process is the case of diamond and graphite. Diamonds grow deep within the Earth's crust where the high pressure and temperature favors this compact configuration; when brought up to the surface, they are not transformed into graphite, even though at surface conditions the Gibbs free energy of graphite is lower than that of diamond. The reason for this is that the rate of transformation of diamond into graphite at room temperature and pressure is so small that little change occurs, even over billions of years.

Most melts consist of a variety of constituents. The total Gibbs free energy of such a melt is determined by the energies of the individual components:

$$G = \Delta G_{mix} + \sum (f_i G_i)$$

$$= \Delta H_{mix} - T \Delta S_{mix} + \sum (f_i G_i), \tag{5.8}$$

where f_i is the fractional concentration of the ith component. The mix term, ΔG_{mix}, comes from a change in the entropy and enthalpy of the system when the various constituents are mixed. From statistical arguments it can be shown that

$$\Delta S_{mix} = -n_s R_{gas} \sum (f_i \ln f_i), \tag{5.9}$$

where n_s is the number of structural sites over which substitution of atoms/molecules can take place, and R_{gas} is the universal gas constant. Since $f_i < 1$, $\Delta S_{mix} > 0$, and the Gibbs free energy is decreased, which thus favors the formation of a solid solution. In the ideal solid solution, $\Delta H_{mix} = 0$, and ΔG_{mix} is entirely determined by the entropy of mixing. By defining the *chemical potential* or *partial mole free energy* of constituent i in the solid phase as

$$\mu_i = G_i + R_{gas} T \ln f_i, \tag{5.10}$$

the Gibbs free energy in an ideal solid solution ($\Delta H_{mix} = 0$) becomes

$$G = \sum (\mu_i f_i). \tag{5.11}$$

Figure 5.6 shows a free energy curve for a two-component solution, with $G = \mu_A f_A + \mu_B f_B$. If we consider the free energy curves for a liquid and a solid solution, phases can only coexist if the chemical potential of each component is

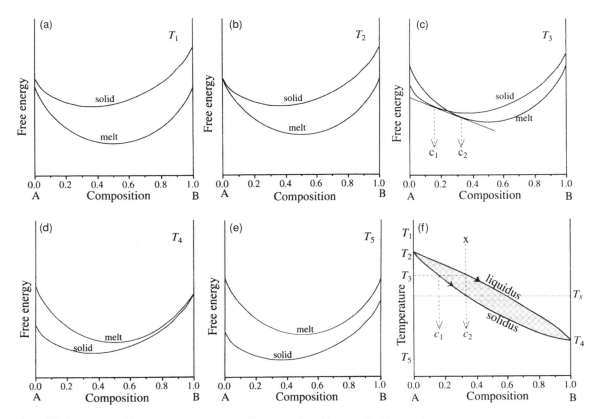

Figure 5.7 A sequence of free energy versus composition curves for a binary melt, at temperatures T_1 (highest) to T_5 (lowest). The resulting phase diagram is shown in panel (f). At high temperatures, T_1 in panel (a), the free energy of the liquid is lower everywhere than for the solid, and the magma is completely molten. At low temperatures, T_5 in panel (e), the free energy of the solid is lower everywhere and the mixture is completely solid. Upon cooling the melt, starting at temperature T_1 in panel (a), solid A first condenses out when temperature T_2 is reached (panel b). At lower temperatures, T_3 (panel c), G_{solid} is below G_{melt} for a range of compositions. The coexisting compositions of the solid and liquid are defined by the common tangent, since here $\mu_{c_1} = \mu_{c_2}$ (panel c). The equilibrium phase diagram in panel (f) consists of the locus of the common tangent points, and defines the composition of the coexisting phases at each temperature. The liquidus and solidus curves define the 'boundaries' above or below which the mixture is entirely liquid or solid, respectively. Consider, for example, a melt of composition c_2 at point x in panel (f). Upon cooling of the melt, the vertical dashed line intersects the liquidus curve at temperature T_3, and a solid of composition c_1 crystallizes out. Upon further cooling of the melt, the composition of both the solid and liquid are adjusted or re-equilibrated continuously, and the compositions are given by the intersection of the horizontal line at temperature T with the curves liquidus and solidus. The composition of the liquid and solid follow essentially the curves liquidus and solidus to the right. At temperature T_x the entire melt has solidified into a solid with a composition equal to that of the original melt (c_2). If the original melt starts out with a higher concentration of B, the temperature of the melt has to be lowered more before the entire melt has solidified. (Adapted from Putnis 1992)

the same in each phase. An example of crystallization of a binary melt is given in Figure 5.7, where the free energy of the liquid and solid solution is shown at different temperatures during the cooling process (panels a–e). The phase diagram (panel f) shows the *liquidus* and *solidus* curves, which define the 'boundaries' above or below which the mixture is entirely liquid or solid, respectively.

Phase diagrams can be very complicated; the solid and/or melt may become immiscible (insoluble in each other) at certain temperatures, solid phases may exist in different forms at different temperatures, the melting point of a substance may be lowered in the presence of a particular melt (*eutectic* behavior), intermediate products may be formed, etc. For example, the MgO–SiO$_2$

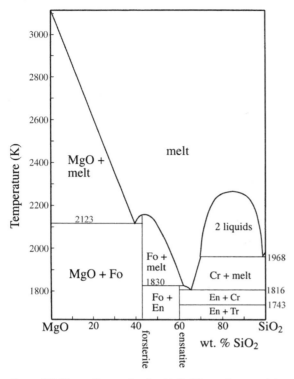

Figure 5.8 Phase diagram for the MgO–SiO$_2$ system, containing the phases periclase (MgO), forsterite (Mg$_2$SiO$_4$), enstatite (MgSiO$_3$) and silica (SiO$_2$). At very high temperatures, $T > 2270$ K, the melt exists as a single phase over the entire compositional range. At $T < 2270$ K, the melt becomes immiscible at the silica-rich end, so that silica-rich and magnesium-rich liquids coexist. At the magnesium-rich end in the diagram, periclase (MgO) crystallizes out at $T \lesssim 3070$ K. At $T \lesssim 2120$ K, in addition to periclase, forsterite crystallizes out in melts with $\lesssim 40\%$ SiO$_2$, and at 1830 K, in 40–60% SiO$_2$ melts, forsterite and enstatite appear without periclase. Note the eutectic behavior of forsterite: Pure forsterite melts at 2170 K, while in a mixture the melting point is lowered by 50 K. In more silica-rich melts, silica crystallizes as different polymorphs of quartz (cristobalite, Cr, and tridymite, Tr) around 1770 K. (Adapted from Putnis 1992)

system sketched in Figure 5.8 contains several distinct solid phases as periclase (MgO), forsterite (Mg$_2$SiO$_4$), enstatite (MgSiO$_3$), and the silica (SiO$_2$) polymorphs, crystobalite and tridymite. Depending on the composition and temperature, the melt may exist as one single liquid phase (high temperatures), or a number of different solid and/or liquid phases. Which solid phase is present depends upon the temperature and original composition of the cooling melt.

5.2.2 Crystallization and Differentiation

In the previous section, we showed that the physics and chemistry of a cooling melt is complex. As in binary systems, magmas or molten 'rocks' crystallize over a

large range of pressures and temperatures. While the melt is cooling, the composition of the crystallizing material as well as that of the magma is changing continuously. The minerals that crystallize from the magma, and the sequence in which they crystallize, depend on pressure, temperature, and composition and how these vary as the system changes from a liquid to a solid phase. While the magma is cooling, existing crystals or nucleation seeds grow. The rate of growth is regulated, since the latent heat of crystallization warms the local environment. This heat has to be removed from the crystal to allow it to grow. On the other hand, if the temperature is reduced too rapidly, the magma becomes very viscous and does not provide the crystal with enough material to continue to grow. Geologists study cooling magmas through analyses of igneous rocks and semi-molten lavas, and by conducting laboratory experiments on melting of rock and cooling of melts.

To complicate matters further, a magma usually does not cool under equilibrium conditions. The crystallized matter may not equilibrate with the magma, or heavy crystals may sink down in the magma, a process referred to as *differentiation*. In the latter case, the crystals are essentially removed from the melt. In the former case, if the crystals do not equilibrate, zoned crystals may form. Zoned crystals consist of a core rich in composition A, surrounded by a mantle (referred to as *rim*) which gradually grades to a composition richer and richer in B. For example, in a plagioclase melt, the mineral anorthite (CaAl$_2$Si$_2$O$_8$) has the higher melting temperature (like the binary melt of composition A in Fig. 5.7) than albite (NaAlSi$_3$O$_8$), resulting in plagioclase feldspars with calcium-rich cores and sodium-rich rims.

Based on laboratory experiments, Bowen derived in 1928 a general scheme for the cooling of a magma with *fractional crystallization* and *magmatic differentiation*. His reaction series are summarized in Figure 5.9. Starting with a high-temperature (ultra)mafic magma, olivine crystals are the first that condense out. Since these crystals are heavy, they sink to the bottom of the magma chamber, and are thus removed from the melt through magmatic differentiation. The remaining melt is of basaltic composition. Upon further cooling, pyroxenes condense and differentiate out, leaving a melt of a more andesite (more silica-rich; Fig. 5.4) composition. Then amphiboles and biotite micas appear, and the magma left behind is more and more silicic in composition. If the crystals had remained in the cooling melt, they may have reacted with the magma, and consequently been changed. For example, if the olivine crystals had not settled out, they would have been converted to pyroxenes through interaction with the cooling

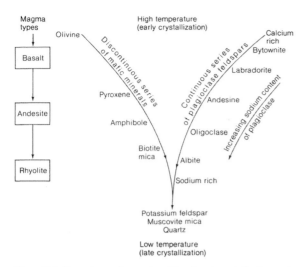

Figure 5.9 Bowen's reaction series of fractional crystallization and magmatic differentiation. The magma follows a double path of fractional crystallization: a discontinuous mafic series and a continuous series of plagioclase feldspars. (Press and Siever 1986)

melt, in which case there would not be much olivine on Earth.

Upon cooling a more silica-rich melt, crystallization starts with calcium-rich plagioclase (anorthite), and at lower temperatures the crystals gradually become more and more sodium-rich (albite). Since feldspars are relatively light, they tend to float on top of the magma. The cooling magma becomes gradually more and more silicic, ending with granitic material (potassium feldspar, muscovite mica, and quartz). At any time during the cooling process, if the melt reaches the surface, the resulting rocks have the composition of the melt 'frozen'. So, if the melt surfaces early on in the cooling process, basaltic rocks are formed. If the magma surfaces later during the cooling process, it contains relatively more silica.

Bowen's reaction series of fractional crystallization and magmatic differentiation is based upon laboratory experiments of melting and subsequent cooling of igneous rocks. The sequence of reactions taking place in a cooling magma in the Earth's mantle, however, is often far more complicated, and research is still going on. Usually there is partial rather than complete melting of rocks, and there is a wide range of temperatures even within one magma chamber. The differences in temperature may create chemical separation. While convective motions may mix magmas of different composition, some melts are immiscible, so that there may be melts with different compositions within one magma chamber. These melts each give rise to their own crystallization products.

In addition to the crystallization of a magma, it is important to consider the melting process of materials. For example, on Earth in regions of subducting tectonic plates, sediments on these subducting plates melt at a certain temperature and pressure. The presence of water lowers the melting temperature of some materials considerably. Water-rich silicic magmas may be produced by a remelting of the crust, which upon cooling may produce more granitic-type rocks.

5.3 Surface Morphology

The surfaces of planets, asteroids, moons, and comets show distinct morphological features, such as mountain chains, volcanoes, craters, basins, (lava) lakes, canyons, faults, scarps, etc. Such features can result from *endogenic* (within the body itself) or *exogenic* (from outside) processes. In this section we summarize endogenic processes that are common on planetary bodies. In §5.4 we discuss exogenic processes, in particular, impact cratering. The last section of this chapter gives a brief summary of the features observed on a variety of Solar System bodies, and what these features tell us about the formation and evolution of these objects.

5.3.1 Gravity and Rotation

Gravity is ubiquitous. It wants to pull everything 'down', shaping a planet into a perfect sphere, the equipotential of a stationary fluid body. On the largest scales, the force which best competes with gravity is rotation, through the centrifugal force. Polar flattening caused by rotation is the largest deviation from sphericity for planet-sized (radii exceeding a few 100 km) bodies (see §6.1.4). Only small bodies, where the gravity field is weak, can have very irregular shapes (Figs. 9.5, 9.15, 9.24, and Figs. 5.86, 5.94). The surface that is generated through the rotation of an ellipse about its minor axis is an equipotential surface called an *ellipsoid*. The *geoid* (§6.1.4) is the equipotential surface that best matches mean sea level on Earth. Changes in sea level due to tidal effects from the Moon and the Sun are typically ± 1–1.5 m. Such daily amplitude variations may be tens of meters on Jupiter's moon Io.

The surface *topography* of a planet is measured with respect to the planet's geoid. Whether or not local structures on planetary surfaces survive the gravitational pull depends on the density and strength of the material. Although downhill movements of material are induced by gravity, whether or not such movements occur is determined by the steepness of the slope as compared to the *angle of repose*, which is the greatest slope that a particular material can support. The angle of repose depends mainly on friction. If one piles up sand in a sand box, the slope of the resulting hill is the same for small and large hills, but the slope is different if the mound is built out of fine sand, gravel, or pebbles. The angle of repose thus depends upon the type of material, the size and shapes of the 'granules', water and air content, and temperature. If the slope of a hill is steeper than the angle of repose, *mass movements* or *mass wasting* such as landslides, mudflows, or rockslides will occur. But even on slopes less steep than the angle of repose, material can migrate downhill in *slumping* motions (e.g., landslides, avalanches), or as a slow continuous *creeping* motion (e.g., glaciers, lava flows). Such downhill migrations can be triggered by seismic activity (earthquakes on Earth, moonquakes on the Moon, etc.) caused by either internal or external processes. Precipitation and the presence of a liquid, such as water on Earth and methane on Titan, also play a major role in downhill motions. Spacecraft observations have revealed that mass wasting is a widespread surface process, and the behavior or characteristics of landslides (e.g., length/height ratios) seem to be remarkably insensitive to gravity and the material properties.

5.3.2 Tectonics

Tectonic activity is controlled by the rheology or rheidity of a planet's materials. Technically, *rheology* is the study of the deformation and flow of materials, while *rheidity* is the capacity of a material to flow. In practice, rheology is often used synonymously with rheidity, and we use it here as an empirically derived quantity which determines the stress/strain relation of a material. *Stress*, like pressure, is defined as force per unit area. The deformation of a material due to an applied stress, such as a load, can be characterized by the dimensionless quantity *strain*:

$$\epsilon_{ij} = \frac{1}{2}\left(\frac{\partial u_i}{\partial x_j} + \frac{\partial u_j}{\partial x_i}\right), \qquad (5.12)$$

with u_i the deformation component or displacement. *Elastic* material responds to stress, and if the load is removed, an elastic material will regain its original properties. If a stress is applied to a viscous material, the material will deform or flow in a slow, smooth way as long as the stress is exerted. When the stress is removed, the flow stops, because of the material's intrinsic resistance (*viscosity*) against deformation. A material's viscosity depends on the applied stress, the strain rate, and temperature. A material typically behaves as a 'fluid' if the ratio between viscous and elastic strain is >1000. Whether a material is elastic or viscous also depends upon the timescales involved. The Earth's mantle, for example, behaves in an elastic manner on short timescales, but is viscous on (geologically)

long timescales. Such materials are *viscoelastic*. The ratio between a material's (dynamic) viscosity and its rigidity is known as the (exponential) *viscoelastic relaxation time*, t_{rx}:

$$t_{rx} \approx \frac{\nu_v \rho}{\mu_{rg}}, \qquad (5.13)$$

with $\nu_v \rho$ the dynamic viscosity and μ_{rg} the material's rigidity. At low temperatures, materials are usually brittle, but behave elastically until they break. At high temperatures, material behaves in a *ductile* fashion, which means that it can undergo considerable deformation before it fractures.

5.3.2.1 Tectonic Features

Any crustal deformation caused by motions of the surface, including those induced by extension or compression of the crust, is referred to as *tectonic* activity. Many planetary bodies (terrestrial planets, most major satellites and asteroids) show evidence of crustal motions due to shrinking and/or expanding of the surface layers, commonly caused by heating or cooling of the crust. Consider a forming planet as a hot ball of fluid magma. The outer layers are in direct contact with cold outer space and thus cool off first by radiating the heat away, so that a thin crust forms over the hot magma. While the crust cools, it shrinks. Convection in the mantle may move 'hot plumes' around and heat the crust locally, leading to a local expansion of the crust. The interior cools off through convection and conduction, with volcanic eruptions at places where the crust is thin enough that the hot magma can burst through. The added weight of the magma on the crust may lead to local depressions, such as the 'coronae' on Venus.

Extensional and compressional forces on the crust result in folding and faulting. Common tectonic deformations are illustrated in Figure 5.10a. *Folding* refers to an originally planar structure that has been bent. *Faulting* involves fractures. If the crust is cracked because it moved in response to a compression or expansion of the crust, the cracks are called *faults*. The movement can be up and down as in *normal faults*, which result from tensional stresses, and in *reverse faults*, which result from compressional forces. In *strike-slip faults* the crustal motion is primarily in horizontal directions, such as seen when two tectonic plates slide alongside each other (e.g., as across the San Andreas fault in California). Fault displacements can be of order ~ 30 m after a very large earthquake. If there is no crustal motion involved, cracks are called *joints*. Along such joints, rocks are relatively weak, and therefore particularly susceptible to erosion and weathering. A spectacular example are the columns of columnar

(a)

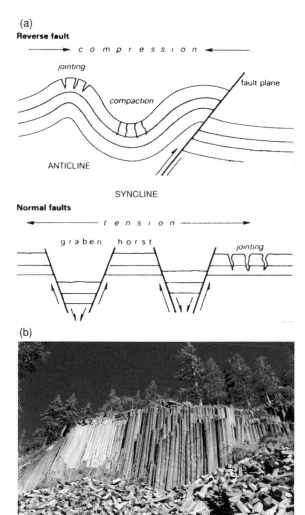

(b)

Figure 5.10 (a) Cartoon showing common tectonic deformations such as faults, and the formation of grabens and horsts. (Greeley 1994) (b) Vertical basalt columns in Devils Postpile National Monument (California, USA). (Courtesy Cooper, Wikimedia Commons)

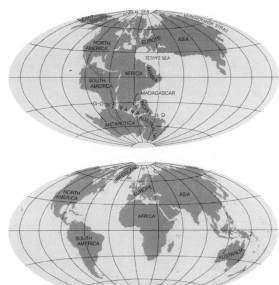

Figure 5.11 Continental drift on Earth: 200 million years ago the continents fit together as a jig-saw puzzle, a supercontinent named Pangaea (top panel). (Press and Siever 1986)

basalts in Devils Postpile National Monument in the USA (Fig. 5.10b).

Folding and faulting help shape a planet's surface, and the effects are visible as distinct geological features. Typical examples are *grabens* and *horsts*. A graben is an elongated fault block that has been lowered in elevation relative to surrounding blocks (Fig. 5.10a), whereas a horst is a fault block that has been uplifted. *Scarps* are steep cliffs that can be produced by faulting or by erosion processes. Many bodies display *rilles*, which are elongated trenches, either sinuous in shape or relatively linear, both of which can be tectonic (from, e.g., faulting) or volcanic (e.g.,

collapsed lava tubes, as the Hadley Rille on the Moon, Fig. 5.21a) in origin. Folding and faulting processes can also lead to the formation of mountain ridges, such as found on many continents on Earth.

5.3.2.2 *Plate Tectonics*

A study of the shape and motions of the continents on Earth has led to the concept of *plate tectonics*. The various continents seem to fit together as a jig-saw puzzle (Fig. 5.11), and current theories suggest that roughly 200 Myr ago there was only one large landmass, *Pangaea*. Since that time the continents have moved away from each other, a process known as *continental drift*. This motion is induced by 'plate tectonics'. The lithosphere consists of 15 large plates, which move with respect to each other by a few, in some cases up to nearly 20, centimeters per year. The current motion of the plates can be measured either by using very long baseline interferometry (VLBI), which uses quasars (highly luminous objects billions of lightyears from Earth) as fixed radio sources, or by using the Global Positioning System (GPS), which is based upon a satellite ranging technique. These techniques agree with the velocities derived by geologic means from the magnetic field of the oceanic lithosphere (§7.6). Hence the motion of the plates must have hardly changed over the past million years! Before the existence of the supercontinent Pangaea, there must have been more continental assemblages and

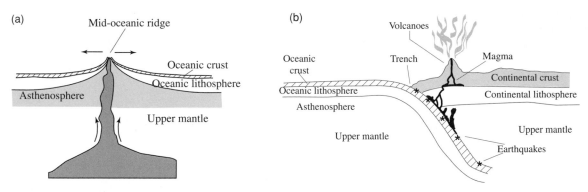

Figure 5.12 Schematic presentations of plate tectonics. (a) Sea floor spreading: Plates recede from each other at the mid-oceanic rift, where magma rises and fills the void. (b) Subduction zones: At convergent boundaries between two lithospheric plates, at least one of which is oceanic, the heavier oceanic plate is subducted. Volcanoes form near such subduction zones.

breakups. These cycles of opening and closing of ocean basins (complementary to the formation and breakup of supercontinents) is known as the *Wilson cycle*. The Wilson cycle has the longest period in the geologic time cycle on Earth. We note that, although the typical period of the Wilson cycle is 300–500 million years, the formation of supercontinents over 600 million years ago did not happen at regular time intervals.

Plate tectonics is caused by convection in the mantle, which induces a large-scale circulation pattern, where the plates 'ride' on top. This circulation is sometimes compared to a conveyor belt. Although the driving force for mantle convection and plate tectonics is still a topic of debate, it is clear that the plates recede from each other at the mid-oceanic rift (Fig. 5.12), where hot magma rises and fills the void. When this magma reaches the surface, it solidifies and becomes part of the oceanic plates. On its journey away from the mid-ocean ridge, the material cools further, which causes a thickening in the oceanic lithospheric plates. When reaching a continental plate at the plate boundary, the cold plates subduct. Such an *active plate margin* is called a *subduction zone*. This subduction (negative buoyancy) 'pulls' on the plates, and this *slab-pull* is probably the driving force of the mantle circulation, since subducting plates have been detected all the way into/through the lower mantle (§6.2.2.1). As a consequence, plates are pulled apart at the mid-ocean ridge. Since the source of the rising magma here is in the upper, rather than lower, mantle, the *ridge-push* mechanism (plates pushed apart by hot rising magma) does not provide the primary driving force for the mantle circulation. There is a passive ridge-push component, however, in the sense that the hot upwelling mantle material creates an elevated height (ridge, Fig. 5.12a),

from which newly formed rigid plates slide away through gravity.

Regardless of the slab-pull or ridge-push mechanism, new ocean floor is created at mid-oceanic ridges, and the recession of plates is referred to as *sea floor spreading*. Where plates meet, they bump into each other (compare a river full of logs), or slip past one another at transform faults. These 'collisions' result in *earthquakes*. The energy, E, released by earthquakes can be quantified using the *Richter magnitude* scale, $\mathcal{M}_R$:

$$\log_{10}E = 12.24 + 1.44\mathcal{M}_R. \quad (5.14)$$

A (large) earthquake of magnitude 7.5 on the Richter scale releases 10^{23} ergs.

Oceanic plates (lithosphere + crust) are thinner (0–100 km) than continental plates ($\sim$200 km), and also denser because they are composed of basalt, while the continental crust is made of lighter (more granitic) material. When an oceanic and continental plate are pushed against each other, the oceanic plate will therefore *subduct*, or dive under, the continental plate, at least at active margins. At passive margins, oceanic and continental plates are juxtaposed without subduction. Mountain ranges and volcanoes form on the continental side of a subduction zone, and ocean trenches on the oceanic side (Fig. 5.12b). While an oceanic plate is brought down, sediments on its 'surface' (e.g., bones and shales from marine animals, sands from rivers) are squeezed and heated, so that new metamorphic rocks may form, and at greater depths this material, which has a low melting temperature, melts. This forms a continuous source of magma for (and initially formed) the volcanoes along the fault line. Since water lowers the melting temperature of rocks, solidification of the rising magma (Fig. 5.12b) results in more granitic-type rocks. The Cascade mountain range in North California–Oregon–Washington is formed this way, and the entire Pacific Ocean boundary is surrounded by volcanoes of this type, known collectively as the *Ring of Fire*. Recycling of oceanic lithosphere typically occurs on timescales of $\sim 10^8$ years (Problem 5.5).

Tectonic plates are not strictly separated into oceanic and continental plates; many consist of a combination of the two types. When two tectonic plates meet, they can also slide alongside each other, as they do at the San Andreas fault zone in California. A similar earthquake prone fault line exists in Israel, where earthquakes have repeatedly destroyed old 'biblical' towns, such as Bet She'an and Jericho. Since oceanic plates are heavy enough to subduct, one will go down when two oceanic plates meet. This may create (volcanic) islands, such as the Aleutan islands in Alaska. When two continental plates are pushed against each other, being buoyant they buckle up, resulting in the formation of mountain ranges, such as the Himalayas.

Plate tectonics is unique to Earth, and is not seen on any other body in our Solar System. The smaller planets and large satellites, such as Mercury, Mars, and the Moon, cooled off rapidly, and developed one thick lithospheric plate. Tectonic features on these planets involve primarily vertical movements, with features such as grabens and horsts. Venus shows evidence of local lateral tectonic movements, but not of plates. The absence of plate tectonics has been attributed to a lack of water on Venus, a consequence of its high surface temperature, making its outer shell too stiff to break into plates. Water in Earth's interior is thought to play a major role in plate tectonics by reducing the strength of rocks, leading to a breakup of the lithosphere. Based upon the geology, petrology, and magnetic field measurements of Mars (§7.5.3.2), it has been suggested that this planet may have had plate tectonics early in its history, although these theories are highly controversal. Perhaps the jovian satellite Europa shows features that most closely resemble the mid-oceanic ridges seen on Earth (Fig. 5.81). The bands on Europa, however, formed in a different way and by other forces (tidal), when ice or water rose to fill fractures created by regional stresses in the surface.

5.3.3 Volcanism

Many planets and several moons show signs of past volcanism, while a few bodies, in particular Earth, Io, and Enceladus, are still active today (Figs. 5.13, 5.75, 5.91). Volcanic eruptions can change a planet's surface drastically, both by covering up old features and by creating new ones. Volcanism can also affect, and even create, atmospheres (§4.3.3.3). In this section we discuss what

(a)

(b)

(c)

Figure 5.13 Examples of volcanic activity on Earth. (a, b) Photographs are shown of Mount St. Helens before (a) and after (b) the big explosion in 1980. This volcano is located in Washington state, along the boundary of the Pacific and North American plates. (Courtesy USGS/Cascades Volcano Observatory) (c) PuùÒò–Kupaianaha eruption of the Kilauea volcano on Hawaii on 6 September 1983, during its 8th episode, about 5 months after the eruption began. (Courtesy J.D. Griggs and USGS/Hawaii Volcano Observatory)

volcanic activity is, where it is found and how it changes the surface. We focus our discussion on Earth, a volcanically active planet that has been studied in detail.

A prerequisite for volcanic activity is the presence of buoyant material, like magma, below the crust. There are several sources of heat to accomplish this (§6.1.5): (*i*) Heat

Figure 5.14 Lava flows on Hawaii: Glowing 'a'a flow front advancing over pahoehoe on the coastal plain of Kilauea Volcano, Hawaii. (USGS Volcano Hazards program)

can be generated from accretion during the planet's formation and continuing differentiation of heavy and light material. (*ii*) Tidal interactions between bodies can lead to substantial heating, such as is the case for Jupiter's moon Io. (*iii*) Radioactive nucleides form an important source of heat for all of the terrestrial planets.

Earth's upper mantle consists of hot, primarily unmolten, rock under pressure, and it behaves as a highly viscous fluid. That is, when considering timescales of at least a few hundred years, (upper) mantle material flows, but on shorter timescales it does not move much (Problem 5.8). The solid lithosphere and crust overlying the mantle can be compared to a lid on a pressure cooker or espresso machine. Magma formed within the hot rock, being less dense than the solid rocks surrounding it, is buoyant and hence rises. It is pushed out through any cracks or weakened structures in the surface.

On Earth, we distinguish three types of volcanism. Two of these three types are associated with plate tectonics: eruptions along the mid-oceanic ridges and in subduction zones (§5.3.2; Fig. 5.12). A third type of volcanism is found above hot thermal mantle 'plumes' at places where the crust is weak and the magma can break through (§6.2.2.1). Such magma is very hot – the plumes are thought to originate close to the core–mantle boundary, and do not seem to be connected to plate tectonics (although there is some debate about this). As tectonic plates move over the hot magma plume, the magma forms islands in the ocean, such as the chain of Hawaiian islands.

The style of volcanic eruptions is governed by the (chemical) composition and the physical properties of the magma. Volatile compounds such as water, carbon dioxide, and sulfur dioxide are dissolved in the magma at high temperatures and pressures. These volatiles come out of solution when the temperature and pressure drops, i.e., when the magma approaches the surface. The type of volcanism, *explosive* versus *effusive* (the nonexplosive extrusion of magma at the surface), depends primarily on the viscosity of the magma, which is determined by its composition, especially its silica content. The higher the silica content, the lower the melting temperature and the higher the viscosity. In low-viscosity basaltic melts, volatiles can rise and escape freely into space as gas bubbles, while in high-viscosity silicic magmas bubbles cannot rise but are carried upwards with the magma, resulting in explosive eruptions at the surface.

Basaltic magma erupts at temperatures of 1250–1500 K. Due to its low viscosity, basaltic magma is very fluid, and can cover vast (many square km) areas within hours, resulting in extensive lava flows, such as seen in Hawaii, on the Moon and the planets Venus and Mars. In addition to viscosity and topography, the characteristics of lava flows depend on the *eruption* rate, defined as the instantaneous lava output by a vent, and the *effusion* rate, which is the total volume of lava emplaced since the beginning of the eruption, divided by the time since the eruption began. The basaltic flows in Hawaii are referred to as *'a'a* or *pahoehoe*. 'A'a is an Hawaiian word, which sounds like ah ah, outcries made when walking barefoot over this jagged form of lava. Pahoehoe is an Hawaiian word which means 'ropy'. A photograph of both lava flows is shown in Figure 5.14; pahoehoe is the glistening smoother surface, in part covered by 'a'a, a very rough, jagged and broken lava flow.

In contrast, the highly viscous silicic magma, such as rhyolite, erupts at temperatures of 1050–1250 K. This lava flows very slowly, oozing out of a vent like toothpaste from a tube. Such volcanic events create *domes*, which may largely consist of obsidian. Such 'glass mountains' can, for example, be seen near Mono Lake and Mount Shasta in California.

Volcanic eruptions can emanate from long narrow fissures, or from a central vent or pipe. Most spectacular eruptions are *fire fountains*, which occur in basaltic magmas that are rich in dissolved gases. On Earth such fountains may be tens to hundreds of meters high. Fissure eruptions, such as occur along oceanic ridges, cause large lava floods that can cover extensive areas. Such areas are known as *lava plains* or *lava plateaus*. A prime example of such plains are the lunar maria (Fig. 5.15). Localized eruptions form volcanic 'mounts' and are often accompanied by large lava flows that gush or ooze out and flow downhill.

Volcanic activity can create many different types of features. *Cinder cones* and *spatter cones* are cone-shaped hills up to a few hundred meters high, built up around vents that eject pyroclastic material as cinders, ash, and boulders. The profile of the cone is determined by the angle of repose. A *shield volcano* is a gently sloping volcanic mountain, built by low-viscosity lava flowing out from a central vent (Figs. 5.16, 5.17). These volcanoes may be very large; the largest known shield volcano is Olympus Mons on Mars, with a height of ~25 km and base diameter ~600 km. The largest shield volcano on Earth, Mauna Loa on Hawaii, measures about 9 km from its peak to the bottom of the ocean floor and has a base diameter of ~100 km (Fig. 5.18). *Composite cones* are built when a volcano emits (alternately) lava as well as pyroclasts. This is the most common type of large continental volcanoes, such as Vesuvius, Mount Etna, and Mount St. Helens.

Water vapor is the main constituent (60–95%) of volcanic gases on Earth, followed by carbon dioxide (10–40%). Sulfur is usually present in the form of SO_2, although at lower temperatures ($\lesssim$700 K) sulfur may be in the form of H_2S. Volcanic gases usually carry traces of, for example, nitrogen, argon, helium, and neon, and sometimes metals, as iron, copper, zinc, and/or mercury. Although water vapor is the main constituent of volcanic gases on Earth, volcanic explosions and plumes on other bodies may be driven by other gases, such as sulfur dioxide on Io.

Emissions of gas and vapor without the eruption of lava or pyroclastic matter often mark the last stages of volcanic activity. Vents that emit only gas and steam are referred to as *fumaroles*. Groundwater that is heated by magma can produce *hot springs* and *geysers* (Fig. 5.19). Such springs and geysers are found in volcanic areas on

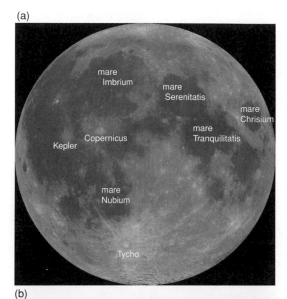

(a)

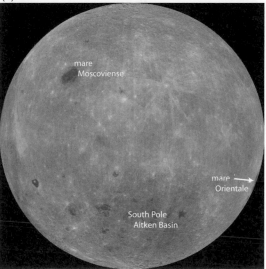

(b)

Figure 5.15 Images of the near (a) and far (b) side of the Moon taken by the Clementine spacecraft. (Courtesy USGS)

Earth, such as Yellowstone National Park. They have been discovered by the Cassini spacecraft on Saturn's moon Enceladus (Fig. 5.91), and may occur on Europa, one of Jupiter's Galilean moons. The Voyager spacecraft imaged geysers of liquid nitrogen on Triton, Neptune's largest moon (Fig. 5.98).

Craters are found on the summits of most volcanoes. A volcanic crater is centered over a vent, and is produced when the central area collapses as the pressure that caused the eruption dissipates. Since the original crater walls are steep, they usually cave in after an eruption, enlarging the crater to several times the vent diameter. Craters can

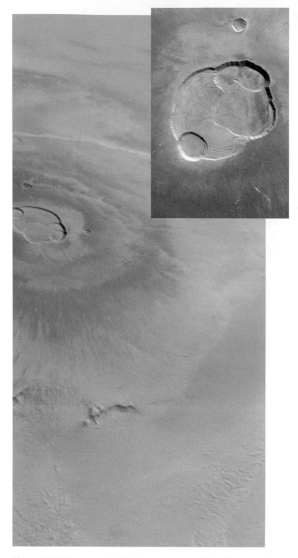

Figure 5.16 The largest shield volcano in our Solar System: Olympus Mons on the planet Mars. (NASA/Mars Global Surveyor) The inset shows a Mars Express image of the caldera at the summit. (ESA/DLR/FU Berlin; G. Neukum)

Figure 5.17 Three-dimensional, computer-generated view of the surface of Venus showing Maat Mons, an 8-km high volcano. This view is based upon radar data obtained with the Magellan spacecraft. Lava flows extend for hundreds of km across the fractured plains in the foreground. The vertical scale has been exaggerated 22.5 times. (NASA/JPL, PIA00254)

be hundreds of meters deep. A volcano in Italy, Mount Etna, has a central vent that is 300 m in diameter and over 850 m deep. Small craters are termed *pit craters* ($\lesssim 1$ km) or, if even smaller, *collapse depressions*. *Calderas* are large basin-shaped volcanic depressions, varying in size from a few kilometers up to 50 km in diameter. Such volcanic depressions are caused by collapse of the underlying magma chamber. After a volcanic eruption, when the magma chamber is empty, its roof, which is the crater floor, may collapse. Over time the crater walls erode, and lakes may form within the depression. Many years later (this may take 10^5–10^6 yr) new magma may enter the

chamber, push up the crater floor, and the entire process may start again. Several calderas show evidence of multiple eruptions. Examples of such *resurgent calderas* are the Yellowstone caldera in Wyoming, the Valles caldera in Mexico, Crater Lake in Oregon, and the Aniakchak Caldera in Alaska (Fig. 5.20). After an explosion, when the lava flow cools and contracts, shrinkage cracks may form. Sometimes, when the source of a lava flow is cut off, and the outer layers have solidified, the lava drains out, and *lava tubes* or *caves* result. Such caves are found, for example, in Hawaii and northern California. A sinuous rille, the Hadley Rille in Mare Imbrium on the Moon (Fig. 5.21a), might be a lava tube where the roof has collapsed. Such volcanic structures, characterized as steep-sided troughs, are also known as *lava channels*.

The characteristics and morphology of volcanic features are determined by the viscosity of the magma, its temperature, density, and composition, the planet's gravity, lithospheric pressure, and strength, and the presence and properties of the atmosphere. Earth possesses a wide variety of volcanic features, and many of these plus additional forms of volcanism have been identified on other planets, moons, and asteroids.

5.3.4 Atmospheric Effects on Landscape

An atmosphere can profoundly alter the landscape of a planetary body. If the atmospheric pressure and temperature at the surface are high enough for liquids, such as water, to exist, there may be oceans, rivers, and precipitation, which modify the landscape through both

Heights of mountains on Mars, Venus, Earth

Olympus Mons

Mt. Everest

Maxwell Mountains

Mauna Loa

Sea level

(Vertical scale exaggerated × 2)

Figure 5.18 A comparison of volcanoes/mountains on Mars (Olympus Mons), Earth (Mauna Loa and Mount Everest), and Venus (Maxwell Mountains). (Morrison and Owen 1996)

Figure 5.20 The Aniakchak Caldera in Alaska, USA, is an example of a resurgent volcanic caldera. The caldera formed about 3500 years ago; it is 10 km in diameter and 500–1000 m deep. Subsequent eruptions formed domes, cinder cones and explosion pits on the caldera floor. (Courtesy M. Williams, National Park Service 1977)

Figure 5.19 One of the numerous geysers in Yellowstone National Park, USA. (Courtesy Wil van Breugel)

mechanical and chemical interactions. In 'dry' areas, e.g., currently on Mars and in deserts on Earth, winds displace dust grains and erode rocks. Over time, these processes 'level' a planet's topography: high areas are gradually worn down, and low areas filled in. This process is called *gradation*, and material is displaced by *mass wasting*. The main driving force for gradation is gravity (§§5.3.1, 6.1.4). Although gradation and mass wasting occur on all solid bodies, the presence of an atmosphere and liquids on the surface enhance these processes and give rise to particular surface features, as discussed in more detail below. In addition to mass wasting, there are numerous chemical interactions between the crust and the atmosphere. These

vary from planet to planet, since they depend on atmospheric and crustal composition, temperature, and pressure, as well as the presence of life. In addition to the changes a massive atmosphere can make to surface morphology, it also protects a planet's surface from impacting debris, especially small and/or fragile projectiles, as well as from cosmic rays and ionizing photons.

In this section we summarize the most common morphological features on Earth as caused by water and wind. These features serve as a basis for comparative studies with other planets.

5.3.4.1 *Water*
The atmospheric temperature and pressure on Earth are close to the triple point of water, so H_2O exists as a vapor, a liquid, and as ice. Although at present Earth is unique in this respect, water may have flowed freely over Mars's surface during the first 0.5–2 Gyr after its formation, and Titan's surface shows clear evidence of fluvial features, though produced by liquid hydrocarbons rather

(a)   (b)

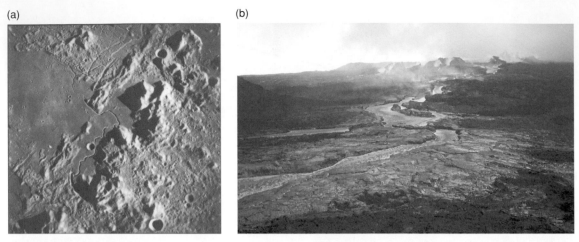

Figure 5.21 (a) Hadley Rille, a typical sinuous rille on the Moon, close to the base of the Apennine mountains. The rille starts at a small volcanic crater and 'flows' downhill. The picture is approximately 130 × 150 km. (NASA/Lunar Orbiter IV-102H3) (b) COLOR PLATE Lava pours down a well-developed lava channel near the erupting vents (background) in Hawaii (on 28 March 1984). (Courtesy R.W. Decker and USGS)

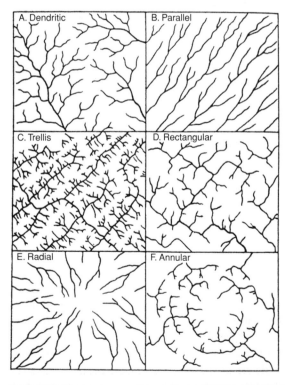

Figure 5.22 Short summary of Howard's classification scheme of drainage patterns on Earth. Dendritic: gentle regional slope at the time of drainage. Parallel: moderate to steep slopes. Trellis: areas of parallel fractures. Rectangular: joints and/or faults at right angles. Radial: volcanoes, domes, and residual erosion features. Annular: structural domes and basins. (Adapted from Howard 1967)

than water. Europa most likely has a large water ocean under its ice crust. Liquids on planetary surfaces, whether water, hydrocarbons or lava, tend to flow downhill at a velocity determined by the flow's viscosity, the terrain, and the planet's gravity. In addition to the liquid itself, the flow transports solid materials, such as sediments eroded from rocks. The faster a river flows, the larger the particles or rocks it can transport. The largest particles usually stay close to the bottom, and may roll and slide over the surface, while the finest particles (clay in the case of water flows) may be suspended throughout the flow. The finest particles are carried the farthest, a sifting process which produces the various sedimentary rocks described in §5.1.2.

Flow patterns contain information on the local topography and the surface characteristics of the underlying rocks. Various patterns are associated with specific terrain (Fig. 5.22). Dendritic patterns indicate gentle slopes, where water trickles down the small channels and accumulates in the larger rivers. Radial patterns are associated with dome-like features, often volcanic in origin. Annular features are associated with domes or basins. Classification of the morphology of flow features on other planets yields clues to the origin of the flow patterns and the local surface topography.

In addition to running water on the surface, groundwater just below the surface may leave profound marks as well. Some rocks are dissolved in water (e.g., limestone, gypsum, salt), leading to *karst topography*, which may show its appearance in the form of various-sized sinkholes, solution valleys, or as haystacks or pinnacles. Groundwater seeping up from below also leads to particular drainage patterns. Finally, dry lake beds or *playas* are associated with former lakes, swamps, or oceans, while sea cliffs and beaches mark the shorelines of oceans.

The temperature on most bodies in our Solar System is well below freezing, which makes ice an important constituent of planetary surfaces. The thawing and freezing of permafrost leaves a characteristic pattern of polygons on the surfaces of some bodies, e.g., Earth and Mars (Fig. 5.64). This pattern is caused by a contraction of the ground when it is freezing cold (in the winter), which creates spaces that fill with melt water in the summer. During the winter this water freezes, thereby widening the cracks.

Although water-ice is the dominant type of ice throughout the Solar System, in the far outer reaches of the planetary system, ices of methane, ammonia, or carbon dioxide may have similar roles as water-ice on Earth. For example, carbon dioxide on Mars freezes out above the winter pole, while it sublimes during the summer. It is cold enough above Mars's poles for permanent water-ice caps, whereas water vapor freezes near this planet's equator at night, subliming again during the day. The temperature on Titan's surface is near the triple point of methane, so methane vapor, liquid, and ice coexist. On Pluto and Triton, nitrogen-ice forms during the winter, and sublimes in the summer. Ice may be considered a 'pseudo-plastic fluid' which moves downhill. On Earth we find *valley glaciers* and *ice sheets*, as well as the morphological features of past glaciation, such as U-shaped valleys, grooves, and striations parallel to the flow, and amphitheater-shaped *cirques* at the head of the flow. Usually the ice contains dust and rocks, which are left behind when the ice melts or sublimes. The dust and rocks may be deposited, carried away by melt water, or blown away by winds. The morphology of the deposits contains information on the glaciers, and hence the surface topography and past climate.

5.3.4.2 *Winds*

Most planets with atmospheres and surfaces show the effects of *aeolian* or wind processes. On Earth, such processes are most pronounced in desert and coastal areas. The winds transport material. The smallest particles, such as clay and silt ($\lesssim 60$ μm) are *suspended* in the atmosphere. Larger dust and sand grains (~ 60–2000 μm) are transported via *saltation*, an intermittent 'jumping' and 'bounding' motion of the dust grains. Still larger grains are transported via *surface creep*, where particles are rolled or pushed over the ground. The amount of dust that the winds

can displace depends upon the atmospheric density, viscosity, temperature, and surface composition and roughness. When the wind blows over a large sandy area, it first ripples the surface and then builds dunes. Because the turbulence created by the wind increases with increasing surface roughness, the winds get stronger during this process (positive feedback). The amount of dust that winds can transport depends upon the atmospheric density and wind strength. On Earth, half a ton of sand can be moved per day over a meter-wide strip of sand if the wind blows at ~ 50 km per hour; the amount of sand transported by stronger winds increases more rapidly than the increase in wind speed. In large terrestrial dust storms, one cubic kilometer of air may carry up to 1000 tons of dust, and thus many millions of tons of dust can be suspended in the air if the dust storm covers thousands of square kilometers. On planets with a low-density atmosphere, the winds need to be much stronger to transport material. The winds on Mars need to be about an order of magnitude stronger than on Earth to transport the same amount of material.

Winds both erode and shape the land. They erode the land through removal of loose particles, thereby lowering or *deflating* the surface. When winds are loaded with sand, they can wear away and shape rocks through *sandblasting*. This causes erosion and rounding of rocks. The best known example of a wind-blown landform is the *dune* (Fig. 5.23a,b). Any obstacle to the wind, such as a large rock, can start the formation of a dune, where sand grains are deposited at the lee-side of the obstacle. The shape of dunes can be used to determine the local wind patterns. Dunes are found in desert and coastal areas, where winds are strong and there is an abundance of particulate material. Dunes have also been found on Mars, Venus, and Titan, and *wind streaks* on Mars and Venus (Fig. 5.23c).

5.3.4.3 *Chemical Reactions*

The interaction between a planet's atmosphere and its surface can lead to *weathering*, a process which depends upon the composition of both the atmosphere and surface rocks. Weathering on Earth is usually a two-part process, consisting of *mechanical weathering* or fragmentation of rocks, together with *chemical weathering* or decay of the rock fragments. Many iron silicates, such as pyroxene, weather or oxidize slowly through the interaction with oxygen and water and get a rusty-iron color. Hydration is a more general process on planetary surfaces, since it only requires the presence of water. It is therefore not too surprising that hydrated minerals have been detected on several asteroids. Some minerals, such as feldspar, partially dissolve when in contact with water and leave behind a layer of clay.

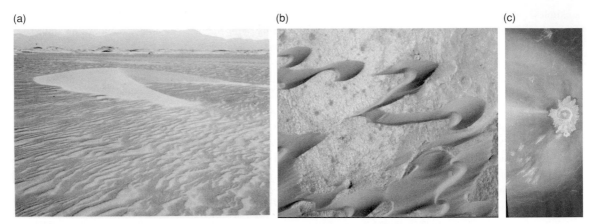

Figure 5.23 (a) Small barchan sand dune on Earth, in Peru. Prevailing wind is from the left to the right. The dune is moving over a surface of dark coarse granule ripples. The light sinuous streaks on the desert surface are the slip faces of small ripples, where light, finer grained sand is temporarily trapped. (USGS Interagency Report, 1974) (b) Aerial photograph of the dark sand dunes of Nili Patera, Syrtis Major, on Mars. The shape of the dunes indicates that the wind has been steadily transporting dark sand from the right/upper right towards the lower left. The width of the picture is 2.1 km. (NASA/Mars Global Surveyor, MOC2-88) (c) Magellan radar image of the 30-km diameter Adivar crater and surrounding terrain on Venus. Crater ejecta appear bright due to the presence of rough fractured rock. A much broader area has been affected by the impact, particularly to the west of the crater. Radar-bright materials, including a jet-like streak just west of the crater, extend for over 500 km across the surrounding plains. A darker streak, in a horseshoe or paraboloidal shape, surrounds the bright area. These unusual streaks, seen only on Venus, probably result from the interaction of crater materials (the meteoroid, ejecta, or both) and high-speed winds in the upper atmosphere. The precise mechanism that produces the streaks is poorly understood. (NASA/Magellan, PIA00083)

Calcite and some mafic minerals may completely dissolve away. In the desert, the products of chemical weathering, such as silica, calcium carbonate, and iron oxides, may make up a hard surface crust, called *duricrust*. This hardened soil has also been seen on the surface of Mars, at the landing sites of the Viking and Mars Exploration Rover missions. The cementation process of these grains usually goes together with the formation of evaporates like salts (§5.1.2.4). Salts are transported with liquid water, which migrates up (from a brine or ice reservoir deeper down) or down (from frost or dew on the surface), often assisted by capillary action. Upon evaporation of the water the salts are left behind, cementing the soil.

The presence of life may have profound effects on the surface morphology of a planet, as we well know from our own planet, Earth. Plants cover large fractions of continental crust, changing the albedo (even varying with season), atmospheric and soil composition as well as the local climate. Plants add matter to the soil and influence erosion. Mankind has large effects on the surface morphology, e.g., through building and mining projects, and by altering the composition of the atmosphere. Even micro-organisms change the atmospheric and soil composition (locally), e.g., through metabolism. Since these

effects are (at present) only applicable to Earth, we will not discuss them further in this book.

5.4 Impact Cratering

Despite the large distances between bodies (compared to their sizes) within our Solar System, collisions occur frequently on a geological timescale. Collision speeds are typically high enough that impacts are very violent events, with *ejecta* thrown outwards from the location of the impact, often leaving a long-lasting depression in the surface. *Impact craters* are produced on all bodies in the Solar System that have a solid surface, and have been observed on the four terrestrial planets, many moons, asteroids, and comets. They are the dominant landform on geologically inactive bodies without substantial atmospheres, which includes most small bodies. We only have to look at the Moon through a modest telescope or a good pair of binoculars to see that the lunar surface is covered by craters, which were formed by meteoroids which hit the Moon over the past 4.4 Gyr. Earth has been subjected to a somewhat higher flux of impacts (due to gravitational focusing), but most craters have disappeared on our planet as a result of plate tectonics and erosion.

(a)

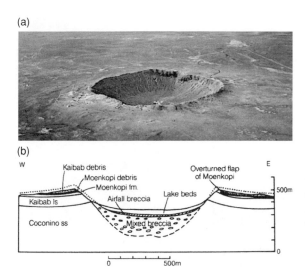

(b)

Figure 5.24 (a) Meteor crater in Arizona. The crater has a diameter of 1 km and is 200 m deep. (Courtesy D. Roddy, USGS/NASA) (b) A cross-section through Meteor crater. (Melosh 1989, as derived from Shoemaker 1960)

Figure 5.25 Microcrater with a diameter of 30 μm. This is a scanning electron microprobe photograph of a glass sphere from the Moon, brought to Earth by Apollo 11. (Courtesy D. McKay, NASA S70-18264)

Impact cratering involves the nearly instantaneous transfer of energy from the impactor to the target. If the target has a substantial atmosphere, as does Earth, the impactor is seen as a fireball, or *bolide*, prior to impact. High-velocity collisions are common, with impact energy being provided from the kinetic energy of the relative orbital motion of the two colliding bodies augmented by the gravitational potential energy released as they approach one another. Typical impact velocities of large meteoroids (which are not significantly slowed by the atmosphere) on Earth are 10–40 km s^{-1}, although long-periodic comets may impact at speeds up to 73 km s^{-1} (Problem 8.1). So characteristic impact energies are of the order of $\sim 10^{12}$ erg g^{-1}. This is more than an order of magnitude larger than the specific energy of chemical explosives (TNT releases $\sim 4 \times 10^{10}$ erg g^{-1}) and the foods that we consume, but about six orders of magnitude smaller than the specific energy of nuclear explosives. For example, a nickel–iron meteoroid, 30 m in diameter, would impart an energy a few times 10^{23} ergs, or the equivalent of several million tons (megatons) of TNT. This energy is comparable to that of a (large) earthquake with Richter magnitude 8 (eq. 5.14), but note that the impact energy is not fully transferred into kinetic energy of material motion within the target body. Nonetheless, collisions can be very energetic, and the impacting body typically creates a hole (crater) much larger than its own size. Meteor Crater in Arizona (Fig. 5.24), with a diameter of about 1 km and 200 m deep, was formed

(within one minute!) by an impacting 30 m nickel–iron meteoroid.

Impact cratering has been studied using astronomical and geological data pertaining to the crater morphology seen on our planet, on the Moon, and on more remote Solar System objects. *Hypersonic* (much faster than the speed of sound) impact experiments using small (mm–cm) projectiles have been conducted in the laboratory. Studies of craters produced by conventional and nuclear explosions sample a broad range of energies. Numerical computer simulations are of great value in studies concerning impacts and impact craters. Impact theories were tested and refined after astronomers witnessed a series of large impacts of Comet D/Shoemaker–Levy 9 with Jupiter in 1994 (§5.4.5). The Deep Impact mission smashed a 370 kg 'spacecraft' into Comet Temple 1 at 10.3 km s^{-1} during 2005; this event was monitored by a companion spacecraft and various telescopes on and near Earth (§10.4).

5.4.1 Crater Morphology

Craters can be 'grouped' according to their morphology into four classes:

(1) *Microcraters* or *pits* (Fig. 5.25) are subcentimeter craters caused by impacts of micrometeoroids or high-velocity cosmic dust grains on rocky surfaces. Pits are only found on airless bodies. The central hole is often covered with glass.

(2) Small or *simple craters* (Fig. 5.26), typically up to several kilometers across, are bowl-shaped. The depth (bottom to rim) of a simple crater is $\sim 1/5$ its diameter,

Figure 5.26 This photograph shows the 2.5 km diameter simple crater Linné in western Mare Serenitatis on the Moon. (NASA/Apollo panoramic photo AS15-9353)

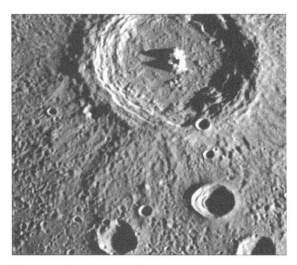

Figure 5.27 A close-up of a 98 km diameter complex mercurian crater, characterized by a relatively flat crater floor, a central peak and terraced walls. Note that the smaller craters in the foreground (25 km diameter) also are terraced. This image (FDS 80) was taken during Mariner 10's first encounter with Mercury. (NASA/JPL/Northwestern University)

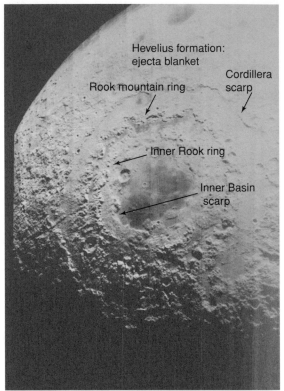

Figure 5.28 A photograph of the lunar multiring basin Mare Orientale. (NASA/Lunar Orbiter IV 194 M)

although variations do occur, depending upon the strength of the surface material and the surface gravity.

(3) Large craters are more complex. They usually have a flat floor and a central peak, while the inside of the rim is characterized by terraces (Fig. 5.27). *Complex craters* have diameters of a few tens up to a few hundred km. The transition size between simple and complex craters is ∼12 km on the Moon and scales inversely with the gravitational acceleration, g_p, because gravity is responsible for the modifications that convert transient simple craters into

complex craters. The minimum size for a crater to acquire a complex morphology also depends on the strength of the target's surface material. Craters with dimensions between 100 and 300 km on the Moon, Mars, and Mercury show a concentric ring of peaks, rather than a single central peak. The inner ring diameter is typically half the rim-to-rim diameter. The crater size at which the central peak is replaced by a peak ring scales in the same way as the transition diameter between small and complex craters. More details on the attributes of complex craters, together with a discussion of their formation, are provided in §5.4.2.3.

(4) *Multiring basins* (Fig. 5.28) are systems of concentric rings, which cover a much larger area than the complex craters mentioned above. The inner rings often consist of hills in a rough circle, and the crater floor may be partly flooded by lava. In some cases, it is not clear which of the outer rings is the true crater rim.

Craters on icy planets/satellites show systematic differences with craters on rocky bodies, such as the minimum size of a crater when central peaks form. Also, no peak-ring craters have been seen on icy satellites. These differences are presumably due to the difference in material properties.

5.4.2 Crater Formation

The formation of a crater consists of a rapid sequence of phenomena, which starts when the impactor first hits the target and ends when the last debris around the crater has fallen down. It helps to understand the process by identifying three stages: An impact event begins with the *contact and compression stage*, is followed by an *ejection* or *excavation stage*, and ends with a *collapse and modification stage*. These three stages are sketched in Figure 5.29, and are discussed below for an airless planet. The influence of an atmosphere is summarized in §5.4.3.

5.4.2.1 *Contact and Compression Stage*

Upon collision of a meteoroid with a planet, the relative kinetic energy is transferred to the bodies in the form of shock waves, one of which propagates into the planet, and another one into the projectile. The impact velocity of a typical meteoroid with a planet like the Earth is of the order of 10 km s^{-1} (equal to or larger than the escape velocity from the planet). Since the velocity of seismic waves in rocks is only a few km s^{-1}, the impact velocity is hypersonic.

The propagation of shock waves can be modeled using the *Rankine–Hugoniot conditions*, which relate the density, velocity, pressure, and energy of the material across the shock front. The Rankine–Hugoniot equations are derived from the conservation of mass, momentum, and energy, and can be written:

$$\rho(v - v_p) = \rho_0 v, \tag{5.15a}$$

$$P - P_0 = \rho_0 v_p v, \tag{5.15b}$$

$$E - E_0 = \frac{P + P_0}{2}\left(\frac{1}{\rho_0} - \frac{1}{\rho}\right), \tag{5.15c}$$

where ρ and ρ_0 are the compressed and uncompressed densities; P_0 and P are the pressures in front of and behind the shock; v is the shock velocity and v_p the particle velocity behind the shock, and E_0 and E the internal energies per unit mass in front of and behind the shock, respectively. This situation is sketched schematically in Figure 5.30.

The pressure involved in a collision of a meteoroid with a solid surface can be derived from the Rankine–Hugoniot equations, and it follows that in low-velocity impacts:

$$P \approx \frac{1}{2}\rho_0 c_s v, \tag{5.16}$$

with c_s the sound speed:

$$c_s = \sqrt{\frac{K_m}{\rho_0}}, \tag{5.17}$$

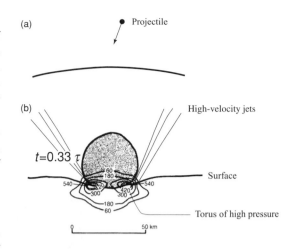

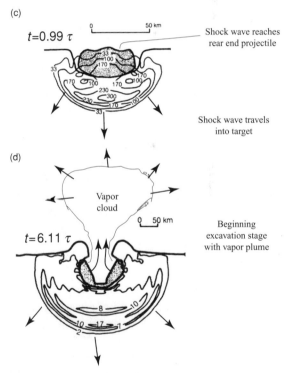

Figure 5.29 Schematic presentation of a hypervelocity (15 km/s) impact of a 23.2 km radius iron projectile onto a rocky surface. Times subsequent to impact are given in units of τ, the ratio of the diameter of the projectile to its initial velocity, which is ~3 s in this case, and pressure contours are labeled in GPa (10^9 Pa = 10^{10} dyne/cm^2). (a) Projectile approaching the target. (b) A torus of extra-high pressure is centered on the circle of contact between the projectile and the target (perpendicular impact). Heavily shocked material squirts or jets outwards at velocities of many km/s. (c) Shock waves propagating into the target and projectile. The latter wave has reached the rear end of the projectile. The projectile melts or vaporizes (depending on the initial pressure) when decompressed by rarefaction waves. (d) Beginning of the excavation stage, preceded by the vapor plume leaving the impact site. (Adapted from Melosh 1989)

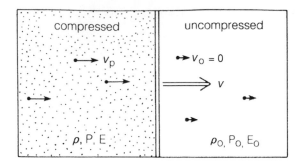

Figure 5.30 A sketch of a medium traversed by a shock front. The various quantities in the compressed and uncompressed medium are indicated (eq. 5.15). (Melosh 1989)

where K_m is the bulk modulus of material (see §6.2.1.1). The sound speed in rock is a few km s^{-1}. In high-velocity impacts:

$$P \approx \frac{1}{2}\rho_0 v^2. \qquad (5.18)$$

Rocks are typically compressed to pressures well over a few Mbar. The shock waves originate at the point of first contact, and compress the target and projectile material to extremely high pressures. A region of extra-high pressure (Fig. 5.29b) is centered on the point of contact between the projectile and target.

The geometry of the shock wave system is modified by the presence of free surfaces on the target and the meteoroid, i.e., the outer surface of the material that is in contact with air or the interplanetary medium. Free surfaces cannot sustain a state of stress, and therefore *rarefaction* (release) *waves* develop behind the shock wave. When the shock wave in the projectile reaches its rear surface, a rarefaction wave is reflected from the surface. This rarefaction wave travels at the speed of sound through the shocked projectile, thereby decompressing the material to near-zero pressure. The contact and compression phase lasts as long as it takes the shock wave and subsequent rarefaction wave to traverse the projectile, which is typically 1–100 ms for meteoroids with sizes between 10 m and 1 km (Problem 5.10).

As soon as the heavily shocked mixture of target and projectile material is decompressed by rarefaction waves, it squirts or *jets* outwards at velocities of many km s^{-1}. This jetting occurs nearly instantaneously with the projectile hitting the target, and is often finished by the time the projectile is fully compressed. The shock, propagating hemispherically into the target, can be detected as a seismic wave (§6.2).

Most rocks vaporize when suddenly decompressed from pressures exceeding $\sim$600 kbar. Since the initial pressure of the shocked material can be very high, the projectile may nearly completely melt or vaporize upon decompression and, shortly after the passage of the rarefaction wave, remnants of the projectile leave the crater as a *vapor plume* or *fireball*.

5.4.2.2 Ejection or Excavation Stage

Upon decompression, the projectile and target region vaporize, provided the initial pressure was high enough. The vapor plume or fireball expands adiabatically upward and outward, where a parcel of gas at distance r is accelerated:

$$\frac{d^2 r}{dt^2} = -\frac{1}{\rho_g}\frac{dP}{dr}, \qquad (5.19)$$

with ρ_g the gas density. Meanwhile, the shock wave expands and weakens as it propagates into the target. The shock wave gradually degrades to a stress wave, propagating at the local speed of sound. The rarefaction waves behind the shock decompress the material, and initiate a subsonic *excavation flow*, which opens up the crater. The excavation of material may last for several minutes, depending upon the size of the crater and the surface gravity. A rough estimate of the time it takes to excavate the crater, i.e., the crater formation time t_{cf}, can be obtained from the period of a gravity wave with a wavelength equal to the crater diameter, D (for craters whose excavation is dominated by gravity, i.e., craters larger than a few km):

$$t_{cf} = \left(\frac{D}{g_p}\right)^{1/2}. \qquad (5.20)$$

Material down to $\sim$1/3 the depth of the transient cavity is excavated. Target material below the excavation depth is pushed downwards, whereas the strata above this depth are bent upwards, and are either excavated or lifted upwards to form the crater walls or *rim*. The rim height on relatively small craters (diameter $\lesssim$15 km on the Moon) is typically $\sim$4% of the crater diameter:

$$h_{rim} \approx 0.04D \quad \text{(small craters)}, \qquad (5.21a)$$

and the depth (bottom to rim):

$$d_{br} \approx 0.2D \quad \text{(small craters)}. \qquad (5.21b)$$

For larger craters these relationships break down, as gravity causes various morphological changes, including rim and crater collapse.

Figure 5.31 The excavation flow forms an outward expanding conical ejecta curtain. The excavation flow in a large planetary impact is thought to be very similar to that from this small impact in a laboratory. (Courtesy P.H. Schultz)

Rocks and debris excavated from the crater are ejected at velocities that are much lower than those of the initial 'jets' of fluid-like material during the compression stage. The ejecta are thrown up and out along ballistic, nearly parabolic, trajectories (Fig. 5.33; see also §2.1) that re-impact at a distance from the crater given by

$$r = \frac{v_{ej}^2}{g_p} \sin 2\theta, \qquad (5.22)$$

where r is the distance, v_{ej} is the ejection velocity, g_p the gravitational acceleration, and θ the ejection angle with respect to the ground. The rock is in the air (or space) for a time t:

$$t = \frac{2v_{ej}}{g_p} \sin \theta. \qquad (5.23)$$

Equations (5.22) and (5.23) are 'flat Earth' approximations, only valid for ejecta traveling much slower than the planet's escape speed, $v_{ej} \ll v_e$. The curvature of the planet needs to be considered for higher velocity ejecta.

The rarefaction waves cause material to move approximately upwards, while the original material velocities are directed radially away from the impact site. The excavation flow, therefore, forms an outwardly expanding *ejecta curtain*, which has the shape of an inverted cone, whose sides make an angle of $\sim 45°$ with the target surface (Fig. 5.31). The ejection velocities are highest (up to several km s^{-1}) early in the excavation process, thus near the impact site. The sides of the crater continue to expand until all of the impact energy is dissipated by viscosity and/or carried off by the ejecta. The resulting crater is many times larger than the projectile that produced it. The crater is nearly hemispherical until the maximum depth is reached, after which it only grows horizontally. The ejecta form an *ejecta blanket* around the crater (Figs. 5.28, 5.32, 5.33, 5.36), up to one or two crater radii from the rim, covering up the old surface. When the ejecta fall down, the outward momentum is retained, so that the material skids along

horizontally before it stops, altering the surface. The morphology of the ejecta blanket is defined by the subsurface material.

Some ejecta blankets on Mars look morphologically like mudflows, suggestive of a fluid substance. Various fluids have been suggested, including subsurface ice which liquefies when heated by an impact, adsorbed CO_2 released by the impact, and trapped air.

Some of the excavated rocks may, when they hit the surface, create *secondary craters*. Because the ejecta move along ballistic trajectories, the secondaries are closer to the primary crater on more massive planets (eq. 5.22; e.g., Problems 5.11, 5.12). The sizes of the secondary craters depend on the mass and impact velocity of the ejecta, as well as on the target material. Because the impact velocities of the ejecta are lower than those of the original impactors, the morphology of secondary craters is somewhat different, but the differences are often subtle. The secondary craters are usually seen outside the ejecta blanket from the primary crater, and may be found many crater radii away from the primary crater. Relatively young craters on the Moon ($\lesssim 10^9$ yr) display bright *rays* emanating outwards from the primary. These rays may be visible over a large surface area (Fig. 5.34), extending to ten or more crater radii from the primary. Many, but not all, secondary craters are associated with bright rays. The rays are composed primarily of local material that has been overturned by the re-impact of ejecta from the primary crater. Over time the rays disappear, probably due to radiation damage from the solar wind (space weathering, §9.3.2).

At the end of the excavation stage, the crater is referred to as the *transient cavity*. The shape of this crater depends upon the meteoroid's size, speed, composition, the angle at which it struck, the planet's gravity, and the material and structure of the surface in which the crater formed. The energy required for excavation varies approximately as D^4, because the mass of excavated material is

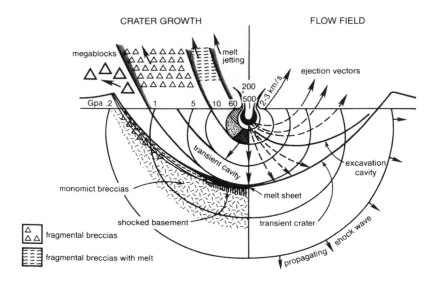

CRATER GROWTH   FLOW FIELD

Figure 5.32 Schematic cross-section through a growing impact crater before crater modification takes place. The left side shows various regimes of physical modification. Ejecta from the location of the impact is largely molten; material ejected from further away is primarily solid, with larger blocks emanating from greater distances. Material remaining below the crater also suffers various degrees of modification, including melting and the formation of breccias (§5.1.2.5). The right side shows the particles' flow field. (Taylor 1992)

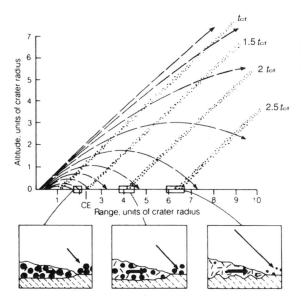

Figure 5.33 The trajectories (ballistic) of debris ejected from the crater at times 1, 1.5, 2, and $2.5t_{cf}$, with t_{cf} the crater formation time given by equation (5.20). (Adapted from Melosh 1989)

Figure 5.34 Rays from the young lunar crater Tycho. The rays are visible over nearly an entire hemisphere. (Courtesy UCO/Lick Observatory)

proportional to the third power of the crater diameter, and the distance that it must be moved to be clear of the crater adds the fourth power. The crater dimension thus scales approximately with the meteoroid's kinetic energy:

$$D \propto E^{1/4}. \qquad (5.24)$$

Equation (5.24) describes *energy scaling*, meaning that the size depends only on the energy of the event. Energy is, indeed, the most important factor in determining crater size, but other factors also matter.

A more general scaling law, derived empirically, is given by (in mks units)

$$D \approx 2\rho_m^{0.11} \rho_p^{-1/3} g_p^{-0.22} R^{0.11} E_K^{0.22} (\sin\theta)^{1/3}, \qquad (5.25)$$

where ρ_p and ρ_m are the densities of the planet (= target) and the meteoroid, respectively, R the physical radius of the projectile, E_K the impact (kinetic) energy, and θ the angle of impact from the local horizontal. Some key terms in equation (5.25) have simple physical interpretations beyond the simple energy scaling given in

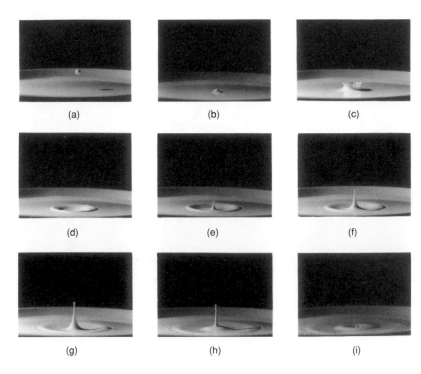

(a)   (b)   (c)

(d)   (e)   (f)

(g)   (h)   (i)

Figure 5.35 A series of photographs from the impact of a milk drop into a 50/50 mixture of milk and cream. An ejecta curtain forms immediately after impact (panel c), which creates the 'crater' wall. After the initial 'crater' forms (panel d), the central peak appears (panel e) and grows (panels f, g). Collapse of the peak (panel h) triggers the formation of a second, outwardly expanding ring (panel i). (Courtesy R.B. Baldwin; photographs by Gene Wentworth of Honeywell Photograph Products)

equation (5.24). For fixed projectile energy, crater size is larger for more massive (higher momentum) projectiles, leading to the ρ_m and R terms in equation (5.25), as well as the slightly smaller explicit dependence on E. The energy is required to work against the planet's gravity, leading to the g_p term. The mass of material excavated varies as $D^3 \rho_p$. The shallower the impact angle, the less effectively the kinetic energy of the impact is coupled to the surface; although only very oblique impacts (within $\sim 10°$ of horizontal) produce asymmetric craters, smaller departures from verticality can lead to significant reduction in crater size, as well as asymmetric ejecta blankets. It follows from equation (5.25) that a typical crater on Earth is roughly ten times as large as the size of the impacting meteoroid.

5.4.2.3 *Crater Collapse and Modification*

After all of the material has been excavated, the crater is modified by geological processes induced by the planet's gravity, which tends to pull excess mass down, and by the relaxation of compressed material in the crater floor. The shape of the final crater depends upon the original morphology of the crater, its size, the planet's gravity, and the material involved. The four basic morphological classes of craters were summarized in §5.4.1. The transition size between simple and complex craters is inversely proportional to the planet's gravity, and depends upon, e.g., the material strength, melting point, and viscosity.

Complex craters are characterized by central peaks and terraced rims. Shortly after the excavation process ends, the debris remaining in the crater moves downwards and back towards the center, while the crater floor undergoes a rebound of the compressed rocks, which probably leads to the formation of the central peak or mountain rings. The formation of the central peak might be analogous to the impact of a droplet into a fluid, shown in Fig. 5.35. After the inital crater has formed (panel d), the central peak comes up (panel e) and grows (panels e–g). Collapse of the peak (panel h) triggers the formation of a second concentric ring (the first being the crater rim), which propagates outwards. The possibly analogous process on a solid surface is sketched in Fig. 5.36. The rebound starts before the crater has been completely excavated, and the central uplift 'freezes' to form the central peak. For larger craters the peak becomes too high and collapses, triggering an outwardly propagating ring, which 'freezes' into a ring of mountains. Although the details of this process are not completely understood, the peak ring must form before the crater material comes to rest.

After material has been excavated, the crater rim collapses or *slumps*, moving outwards and thereby increasing the crater diameter, filling in the floor of the crater, and shaping the wall in the form of terraces. Observations indicate that these terraces form before the rock, which acts like a liquid because of the 'shaking' from the impact, has

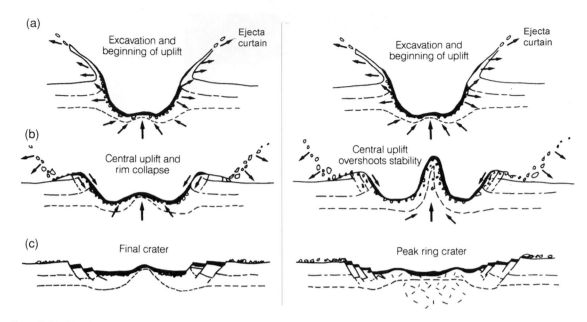

Figure 5.36 (a)–(c) Schematic illustration of the formation of central peaks (left side) and peak rings (right panels). (Adapted from Melosh 1989)

'solidified'. The entire collapse process typically takes several minutes.

Some crater properties vary with crater size in a well-defined manner. For craters with sizes between about 15 and 80 km on the Moon, the height of the central peak, h_{cp} (in km), typically increases with crater diameter, D (in km), as

$$h_{cp} \approx 0.0006D^2. \qquad (5.26)$$

For larger craters, the central peak on the Moon tops out at about 3 km, so the central peak is usually lower than the rim (eq. 5.21*a*). The central peak width is approximately 20% of the crater diameter. In craters with $D > 140$ km, a ring of mountains develops in the crater, replacing the central peak. This ring develops about halfway between the center and crater rim. In some craters both a central peak and peak ring are seen. Like the transition between simple and complex craters, the transition from craters with a central peak to peak rings is gravitationally induced and consequently varies from planet to planet in proportion to $1/g_p$. Thus, this transition also occurs at substantially smaller crater diameters on Venus and Earth than it does on the Moon.

Some complex craters have a central *pit* rather than a peak. These are seen on Ganymede and Callisto for craters over 16 km in diameter. The formation of central pits might be caused by the properties of icy surfaces and subsurface regions.

The Orientale basin on the Moon (Fig. 5.28) is the youngest and best-preserved multiring basin. The center of the basin is surrounded by four rings: the Inner Basin Ring, with a diameter $D \approx 320$ km; the Inner Rook Ring ($D \approx 480$ km); the Rook Mountains ($D \approx 620$ km), and the Cordillera Mountains ($D \approx 920$ km). There may be an additional mountain range with $D \approx 1300$ km. The Rook and Cordillera Mountains are about 6 km high. The Rook Mountains may form the original crater rim, although the rim might have been located interior to this ring. Rings appear to form outside the original crater rim in multiring basins, whereas peak rings in complex craters always form inside the crater rim. Moreover, peak rings are symmetric on the inside and outside slopes, whereas the outer rings on multiring basins are asymmetric, with steep inward-facing scarps and gentle backslopes. The ratio of the radii of successive rings is about $\sqrt{2}$, although no consensus has been reached about the generality or importance of this number. Multiring basins on other planets and satellites are similar, but differ in details. The Valhalla structure on Callisto has more than a dozen rings, and the steep scarps face outwards rather than inwards. No theory to explain the formation of multiring basins is universally accepted. One possibility is that the rings form from a ripple effect, just like the peak rings in complex craters. Another theory attributes the rings to crater collapse in a layered medium wherein the material strength decreases with depth into the planet, such as for a planet where lithospheric plates overlie a fluid-like medium, like the Earth's asthenosphere.

Further modification of craters happens on long time-scales, i.e., months–years–aeons. Erosion and micrometeoroid impacts slowly erode the rim away, and smooth out or flatten the crater. The lifetime of a terrestrial 1 km impact crater against erosion is $\lesssim 10^6$ years. Isostatic adjustments (§6.1.4.4) may be important in large craters, where the crater floor may be uplifted to account for the deficiency in mass caused by the excavation of the crater. On icy satellites, craters are flattened or slowly disappear by plastic-like ice flows. Many large craters on Ganymede and Callisto may have disappeared and left vague discolored circular patches on the surface, called *palimpsests*. Volcanism and tectonic forces within the general area of the crater can modify a crater at a (much) later stage. Many impact basins on the Moon have been flooded with basaltic lava. However, this flooding happened at much later times, and is not connected to the impact events. The lava flows have a very low viscosity, about one-tenth that of flows on Earth. They therefore can cover large areas with a thin (10–40 m) layer of lava. Some small volcanic domes have been spotted, but most mare floors are very smooth. There may have been more volcanic eruptions in the maria than elsewhere because the crust here is thinnest (§6.3.1).

Analysis of craters on other bodies provides information regarding material on and below the surface of that body. The most direct examples are the central peak and ejecta blankets around the craters, which were formed from material originally below the surface. The number, shape, and size of craters also yield information regarding the surface composition/materials and the impacting bodies, as discussed further below and in §5.5.

5.4.2.4 Regolith

Large impacts may fracture the crust down to ~30 kilometers below the surface, while small impacts affect only the upper few millimeters to centimeters of the crust. The cumulative effect of meteoritic bombardment over millions or billions of years pulverizes the bedrock so that a thick layer of rubble and dust is created on airless bodies. This layer is called *regolith* (the Greek word for rocky layer), also termed 'soil' in a terrestrial analog, though the properties of regolith are very different from those of terrestrial soil. Most planetary bodies are covered by a thick layer of regolith. Since the population of meteoroids display a steep size distribution, the rate of 'gardening' or turnover decreases rapidly with depth on a planet. With present-day meteoritic bombardment rates, within the past

10^6 years half of the regolith on the Moon has been over-turned to a depth of 1 cm, while in most locations the uppermost millimeter has been turned over several tens of times. The thickness of the regolith depends upon the age of the underlying bedrock. In the ~3.5 Gyr old lunar maria the regolith is typically several meters thick, while it is well over ten meters deep in the 4.4 Gyr old lunar highlands. The ejecta from larger impacts have formed a 2–3 km thick *mega-regolith*, which consists of (many) meter-sized boulders. The regolith may be bonded at depth as a consequence of the higher (past or present) pressure and temperature.

5.4.2.5 Summary of Cratering

The following features are recognized in connection with cratering events:

- Primary crater: the crater formed upon impact.
- Ejecta blanket: debris ejected from the crater up to roughly one crater diameter beyond the rim. The appearance of the ejecta blanket depends upon the subsurface properties of the target. For example, ejecta blankets on the Moon consist largely of boulders; in contrast, some ejecta blankets on Mars display evidence of fluid motions.
- Secondary impact craters: these are caused by impacts of high-velocity chunks of rock ejected from the primary crater.
- Rays: bright linear features 'radiating' outwards from the impact crater. They extend about 10 crater diameters out. Secondary craters may be clustered around the rays.
- Crater chains: linear arrays of secondary craters, often similar in size and overlapping, sometimes form along rays emanating out from a primary crater. Chains of primary craters form when a meteoroid is tidally disrupted by a planet and impacts into a moon before the fragments disperse.
- Breccia (Fig. 5.37b,c) and melt glasses: high temperature and pressure minerals, melt glasses, and breccias (rocks of broken fragments cemented together, §5.1.2.5) form and line the inside of a crater.
- Regolith: the rocks that were broken or ground down by (micro)meteoroids and secondary impacts.
- Focusing effects: large impacts can produce surface and body waves of sufficient amplitude that they may propagate through the entire planet. In planets with a seismic low-velocity core, the waves are 'focused' at the *antipode* (the point directly opposite the impact site). If the impact is very large, the waves still have enough energy to substantially modify terrain. Pre-existing landforms in regions antipodal to very large impact features

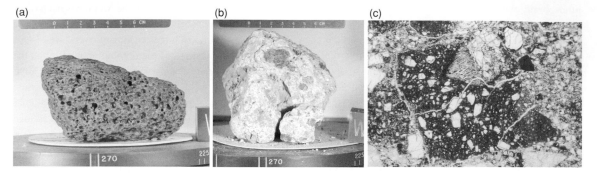

Figure 5.37 (a) Basaltic rock from the lunar mare brought to Earth by Apollo 15. This rock, sample 15016, crystallized 3.3 billion years ago. The numerous vesicles (bubbles) were formed by gas that had been dissolved in the basaltic magma before it erupted. (NASA/Johnson) (b) A breccia from the lunar highlands, sample 67015, collected by Apollo 16 astronauts. This rock is termed *polymict* because it contains numerous fragments of pre-existing rocks, some of which were themselves breccias. It was compressed into a coherent rock about 4.0 billion years ago. (NASA/Johnson) (c) Petrographic thin section (2–3 mm across) of the sample in panel (b) in polarized transmitted light. (Courtesy Paul Spudis)

on these rocky bodies are overprinted and modified with rugged equidimensional hills and narrow linear troughs. This effect is, for example, clearly seen on Mercury and the Moon. No unequivocal evidence of antipodal effects from impacts has yet been identified on any icy body.

• Erosion and disruption: sufficiently energetic impacts can substantially erode or even catastrophically disrupt the target. The impact energy required to disrupt and disperse a planet-sized target body is substantially larger than the energy required to produce a crater whose diameter is about equal to the diameter of the target body.

5.4.3 Impact Modification by Atmospheres

Hitherto we discussed impacts on airless bodies. If the target is enveloped by a dense atmosphere, such as Earth and Venus, impacts may be modified extensively. Projectiles can be completely vaporized while plunging through the atmosphere, and never hit the ground. Or the projectile might break up into many pieces. Such fragmentation is thought to have led to the formation of *crater clusters* on Venus. No analogous close groupings are observed on the lunar surface. Atmospheric drag can slow down a small meteoroid, so that it merely hits the surface at the terminal velocity (§4.4.2.2). Larger bodies can explode in the air, never creating an impact crater. These processes are described quantitatively in §8.3. In this section, we consider the effects of atmospheric transit on the cratering process.

Not only does the atmosphere affect the projectile, the projectile can noticeably perturb the atmosphere during and after the explosion. Exploding meteors are the source of dust and vapor that is not in chemical equilibrium with

surrounding atmospheric gases, and they can also release considerable amounts of energy, most of which is immediately converted into heat. Both dynamical and chemical interactions can be substantial.

5.4.3.1 *Passage through Atmosphere*

A meteoroid is slowed down by atmospheric drag, with the aerodynamic force being proportional to the atmospheric ram pressure ($0.5\rho v^2$). Atmospheric drag may be substantial compared to a meteor's internal strength. Meteors, therefore, often break up and fall down in clusters. A body that is broken apart by ram pressure is slowed very rapidly in the atmosphere, although the cluster of remnant bodies can hit the ground and produce either a crater or a strewn field (Figure 8.15). The effect of such a breakup is usually reflected in the crater morphology.

The mass displaced during a meteor's fall through the atmosphere is $\sigma_\rho A / \sin\theta$, where σ_ρ is the mass of the atmosphere per unit area, A is the body's cross-sectional area and θ is the angle that the trajectory makes with the surface. The meteor is slowed substantially when it encounters an atmospheric mass about equal to its own mass. Significantly smaller bodies are slowed close to the terminal velocity (eq. 4.23) prior to hitting the surface. It follows that the radius of a typical iron meteor must be over one meter for it to hit Earth in a hypersonic impact (Problem 5.17).

However, actual meteors do not remain constant in size as they pass through the atmosphere. As their exteriors are heated to melting and even vaporization temperatures, they shed material. This process is known as *ablation*, and is discussed in greater detail in §8.3. Meteors which hit the Earth at speeds $\lesssim 100$ m s^{-1} produce a small hole or pit equal in size to the diameter of the meteor itself. At somewhat higher speeds, the hole produced is larger than that of the impacting meteor, while impacts at hypersonic speeds produce craters as discussed in §5.4.2.

When a meteor plunges through the atmosphere at supersonic speeds, a bow shock forms in front of it and gases are considerably compressed. The shock waves of such meteors, even for projectiles which do not hit the surface, can be devastating and leave obvious marks. On 30 June 1908, a $\sim$40 m radius meteor that disintegrated in an airburst produced a shock wave that flattened about 2000 km^2 of forest near the Tunguska river in Siberia. On Venus, the Magellan spacecraft detected several radar-dark features, some of which have been connected to impact events. Such features probably result from the blast waves from meteors which never hit the ground.

5.4.3.2 Fireball

Immediately after a high-velocity impact, a hot plume of gases, known as the *fireball*,[1] leaves the impact site, preceded by a shock wave in the lower atmosphere. The fireball expands adiabatically, and its radius can be calculated from its initial pressure, P_i, and volume, V_i, assuming $PV^\gamma = $ constant, where γ, the ratio of specific heats, usually equals 1.5. Since the vapor is hotter and therefore less dense than its surroundings, the fireball rises, driven by buoyancy forces. Fine dust brought up in the plume or by ejecta may stay suspended in the Earth's atmosphere for many months to a year, which may have profound climatic consequences through blocking of sunlight and trapping of outgoing infrared radiation (§5.4.6). If the size of the impactor exceeds the atmospheric scale height (i.e., $\sim$10 km on Earth; §4.8.3), a large fraction of the shock-heated air may be blown off into space.

5.4.4 Spatial Density of Craters

Although most airless bodies show evidence of impact cratering, the crater density (number of craters per unit area) varies substantially from object to object. The *size–frequency distribution* of craters quantifies the number of craters per unit area as a function of crater size. Examples of size–frequency distributions for various bodies are shown in Figure 5.38. Some surfaces, including regions of the Moon, Mercury, and Rhea, appear *saturated* with craters, i.e., the craters are so closely packed that, on average, each additional impact obliterates an existing crater.

The surface has reached a 'steady state'. Other bodies, like Io and Europa, have very few, if any, impact craters. Why is there such a large range in crater densities? The variations must be caused by the combined effect of impact frequency and crater removal, two topics that are discussed below.

5.4.4.1 Cratering Rate

The surface of the Moon shows the cumulative effect of impact cratering since the time that the Moon's crust solidified. The size–frequency distribution of lunar craters is shown in Figure 5.38a. For the Moon, we can determine the ages of some regions of the surface accurately using radioisotope dating (§8.6) on rock samples returned to Earth by the various Apollo and Luna landers. Nine different missions returned a total of 382 kg of rocks and soil to Earth (Table F.1). A comparison of these absolute ages with graphs such as shown in Figure 5.38a have been used to determine the historic cratering rate. The very oldest regions, the lunar highlands, $\sim$4.45 Gyr, are saturated with craters, while younger regions (lunar maria, $\sim$3–3.5 Gyr) show a much lower crater density. This difference can be understood when considering planetary accumulation processes (Chapter 13). Since there were many more stray bodies around during the early history of the Solar System, the impact frequency was highest during the planet-forming era. This period is referred to as the *early bombardment era*. The impact frequency dropped off rapidly during the first billion and a half years, to a roughly constant cratering flux during the past 3 Gyr.

The present cratering rate on the Moon for craters with sizes $D > 4$ km is $\sim$2.7 $\times 10^{-14}$ craters km^{-2} yr^{-1}, a rate that can be accounted for within a factor of $\sim$3 by the observed and extrapolated distributions of Earth-crossing asteroids and comets. This rate, extrapolated to Mars, agrees with the $\sim$20 craters between 2 and 150 meters in diameter that were observed (by Mars Global Surveyor) to have formed between 1999 and 2006 in a region of area $\sim$20 million km^2.

The cratering rate was clearly larger prior to 3.4 Gyr ago, and much larger prior to 3.8 Gyr ago. But the data are not adequate to uniquely constrain the cratering rate in the first 700 Myr of the Solar System. One monotonic model suggests that the cumulative crater density for the Moon for craters with $D > 4$ km can be crudely approximated by

$$N_{\text{cum}} \sim 2.7 \times 10^{-5} \left(t + 4.6 \times 10^{-7} (e^{qt} - 1) \right), \quad (5.27)$$

where N_{cum} is expressed in km^{-2}, t is the age of the surface in Gyr, and $q = 4.5$ Gyr^{-1}. However, most of the impact melts in rocks returned from the Moon by the Apollo and

---

[1] This term, unfortunately, is also used to describe a large meteor blazing its way through a planetary atmosphere on its way towards the surface.

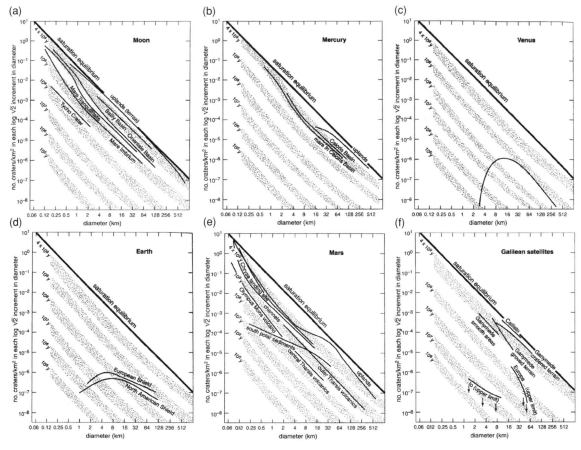

Figure 5.38 Graphs of the crater density (differential crater size–frequency) as a function of crater diameter for (a) the Moon, (b) Mercury, (c) Venus, (d) Earth, (e) Mars, (f) Galilean satellites. (Hartmann 2005)

Luna missions, as well as lunar meteorites, date from 3.8–3.9 Gyr before the present. These dates suggest that the Moon may have been subjected to a terminal cataclysm during this epoch, known also as the *late heavy bombardment*, an era when the cratering rate substantially exceeded that of the previous few hundred million years (Fig. 5.39). Assuming that it occurred, the late heavy bombardment produced a spike in the cratering rate, but cratering was less than given by equation (5.27) at earlier (albeit not the very earliest) times. The size distribution of craters (Figure 5.38) also provides a historical record of the size–frequency distribution of impactors.

Table 5.3 summarizes when an impactor of a given size last hit the Earth – some numbers are based upon data concerning particular events, others (indicated by ∼ preceding time) on statistical arguments based upon the mean impact frequency. The impact rate at the current epoch is shown in Figure 5.47. Based upon the observed

size distribution of asteroids and comets, one expects the present-day Earth to be hit by a body 5 km in radius roughly once every 10^8 years. Smaller bodies hit more frequently: An object with a 2.5 km radius hits Earth on average once every ∼10^7 yr, and a body with a radius of 200 m once every ∼10^5 yr. Note that most impact frequencies shown in Figure 5.47 are less than the inverse time since the last impact of a body of a given size given in Table 5.3. A factor of two difference is to be expected, as typically we would expect to be halfway between two impacts, and some uncertainties are inherent in the estimation techniques used for the two compilations. However, examining the individual differences provides interesting insights. It appears to be chance that an impact as energetic as the Tunguska event occurred just a century ago. The K–T impactor that wiped out the dinosaurs 65 Myr ago (§5.4.6) seems to have been one of the most energetic impact events on Earth within the past couple of billion

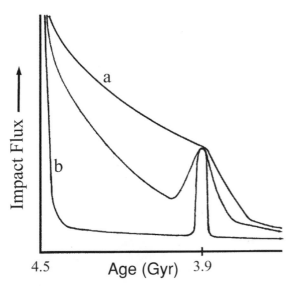

Figure 5.39 Schematic diagram showing possible evolution curves for the early impact flux at the Moon and Earth. Curve 'a' is for a relatively long period of heavy bombardment following planetary accretion. Curve 'b' shows a sharp decline in the impact cratering rate after planetary accretion, followed by an intense cataclysmic period of bombardment 3.9 Gyr ago (known as the late heavy bombardment). Note that no scale is given for the vertical axis, and the variations shown may be orders of magnitude. While it is generally agreed that there was a decrease in the impact rate 3.85 Gyr ago, the precise 'shape' of the variation in impact flux over time before ∼3.85 Gyr ago is not well known, as indicated by the variety of curves on this graph. (Adapted from Kring 2003)

years, but this is difficult to confirm because the majority of Earth's crust is oceanic, and therefore gets subducted into the mantle in ∼200 Myr (§5.3.2.2). It is also possible that long-period comets, not included in the NEO data, make a significant contribution to impacts of this energy and above. Finally, the impact rate during the first billion years of Solar System history was much larger than it is at the current epoch, accounting for the very large impacts that occurred ≳3.8 Gyr ago.

5.4.4.2 *Crater Removal*

The gravity and material strength of the target body influence the size and shape of craters produced by impacts. *Viscous relaxation* sets an upper limit on how long a surface feature can be recognized as an impact crater, although this time can exceed the age of the Solar System. There are numerous other processes that play a larger role in shaping, modifying, and obliterating/removing craters. Volcanic activity is an important endogenic process that effectively removes craters from sight by bending, breaking, or covering up part of the crust. No impact

craters have been seen on Jupiter's satellite Io; their absence has been attributed to Io's extremely active volcanism. Melting of (sub)surface material, which may be triggered by an impact, a volcanic outburst, or by a change in atmospheric temperature (e.g., via the greenhouse effect, §3.3.2), can shorten the lifetime of an impact crater considerably. A major removal process on Earth is plate tectonics, which 'recycles' the Earth's oceanic crust on timescales of ∼10^8 years (§5.3.2.2). Non-plate tectonic processes have affected craters on various planets and moons. Other endogenic changes in the crust, for example through (local) shrinkage or expansion of the crust, and/or the formation of mountains, also can remove or modify existing craters. Atmospheric (and oceanic) weathering slowly erodes craters, both mechanically (e.g., water and wind flows) and through chemical interactions.

In addition to endogenic removal processes, craters can be destroyed and eroded by exogenic processes. Impacts may hit a pre-existing crater and destroy it, or can cover up old craters with ejecta blankets. *Seismic shaking* resulting from an impact can obliterate craters either locally or globally. Seismic shaking has the greatest influence on the smallest bodies.

Small craters are easier to obliterate than larger ones. The more densely cratered a surface is, the more craters are likely to be destroyed by a given impact. Eventually, the surface reaches a statistical equilibrium, where one new crater destroys, on average, one old crater. Further bombardment by the same population of projectiles does not cause any further secular changes in the size–frequency distribution of craters. Such a surface is saturated with craters. Because small craters are easier to destroy than large ones, saturation is usually first reached for the smallest craters. However, on some bodies, such as Jupiter's moon Callisto (§5.5.5.3), endogenic processes are much more efficient at destroying small craters than large ones, leaving the surface saturated with moderate to large craters, but with far less area covered by small craters.

Micrometeoritic impacts have an eroding effect on airless bodies, which is referred to as *sandblasting* and *gardening*. A similar effect, *sputtering*, is caused by low-energy ions (keV range) (solar wind, magnetosphere) (§4.8.2). As discussed in §5.4.2.4, micrometeoroid impacts play a role in the formation of regolith. Impacts by charged particles and UV photons also contribute to this process, though these are more important in changing the chemical composition of the surface locally through radiolysis and photolysis, respectively. If the surface consists of water-ice contaminated with carbon products, then charged particles and/or energetic photons may dissociate the molecules and/or knock off atoms/molecules from the ice into an atmosphere or corona (§4.8.2). The lightest atoms, such as H, may gain sufficient energy in this process to escape the gravitational attraction from the body, leaving the darker material behind. Icy surfaces, therefore, tend to darken over time, an effect clearly seen on the outer planets' satellites, centaur asteroids, and comets.

5.4.4.3 Stratigraphy

Stratigraphy is the study of the time sequence of geological events and of dating the events with respect to each other. The simple counting of craters on a given surface yields a crude estimate of its age. Surfaces of bodies that formed ~4.5 Gyr ago are saturated with craters. As all large bodies in our Solar System formed ~4.5 Gyr ago, we know that something happened to a body if its surface is less than 4.5 Gyr old. For example, when the plains on the Moon (maria) were flooded by lava, any pre-existing craters were covered and disappeared. When the lava solidified, new craters started to accumulate, and crater counts yield the age of the surface after it last solidified. We can thus date less heavily cratered surfaces relative to more densely cratered areas. Several bodies display linear features that are clearly caused by an endogenic process. Sometimes these features cross or dissect a pre-existing crater, while others appear partially obliterated by an impact crater (Fig. 5.40). The stratigraphy of these features yields information on the chronology of the various events. Also, the study of the appearance of the craters themselves yields information on the sequence of events involved in the impact. For example, the crater walls and central peak may have collapsed over time, and a good photograph of the crater reveals the sequence of these events.

The word stratigraphy originates from a study of the sequence and correlation of stratified rock layers. On other bodies there are uplifts of strata (e.g., the walls of an impact crater), and such morphologies yield information not only on the dating of the event, but also on the crustal layers and the forces involved.

5.4.4.4 Dating Cratered Surfaces

The time that has passed since a rock has solidified can often be determined in a laboratory using radiometric methods (§8.6). While this technique has proven extremely useful in determining the ages of terrestrial rocks, meteorites, and lunar samples that have been brought to Earth, it cannot be applied to most planetary surfaces at present.

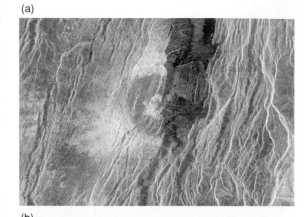

(a)

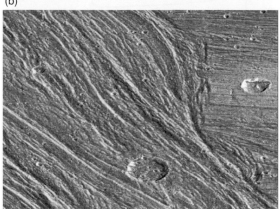

(b)

Figure 5.40 Examples of stratigraphy. (a) A Magellan radar image of a 'half crater' on Venus, located in the rift between Rhea and Theia Montes in Beta Regio. The crater is 37 km in diameter, and has been cut by many fractures or faults since it was formed. The eastern half of the crater was destroyed during the formation of a fault valley that is up to 20 km wide and apparently quite deep. (NASA/Magellan, PIA00100) (b) The Nippur Sulcus region, an example of bright terrain on Ganymede, shows a complex pattern of multiple sets of ridges and grooves. The intersections of these sets reveal complex age relationships. The Sun illuminates the surface from the southeast (lower right). In this image, a younger sinuous northwest–southeast trending groove set cuts through and has apparently destroyed the older east–west trending features on the right of the image. The area contains many impact craters; the large crater at the bottom of the image is about 12 km in diameter. (NASA/Galileo Orbiter, PIA01086)

The relative ages of some rocks and planetary surfaces can be estimated by the stratigraphic techniques discussed above, but the most common technique applied to the determination of relative ages of planetary and satellite surfaces uses the spatial density of craters on the surface. Such cratering ages can be deduced from remote imaging observations.

The longer a surface has been solid, the greater the integrated flux of impactors that it has been exposed to, and thus the larger the expected number of craters on its surface. However, there are many complications inherent in estimating the age of a surface from the size–frequency distribution of craters observed thereupon. As discussed in §5.4.4.1, cratering has occurred at a very nonuniform rate over the history of the Solar System. Craters can be removed by various processes, including subsequent cratering (§5.4.4.2). Different surface materials can affect the production and destruction of craters.

Crater production rates vary substantially from body to body and even across the surface of an individual object. For example, most regular moons rotate synchronously with their orbit (Table 1.5), so the same hemisphere always faces the planet, and one location on the surface, the *apex*, always faces forward along the moon's orbital path. If the primary source of impacts is from bodies on heliocentric orbits (asteroids and comets), then the surface near the moon's apex is likely to be cratered much more frequently than the opposite side of the moon, near the *antapex*. Theoretical apex–antapex asymmetries can be factors of several (Problem 5.18). However, little evidence of such asymmetries are apparent in the cratering record. An explanation for this apparent inconsistency may be that some moons have been reoriented subsequent to their last global resurfacing.

Cratering rates can differ substantially from body to body. The trajectories of heliocentric bodies passing near giant planets are substantially altered by these planets' gravitational perturbations. Inner moons are subjected to more impacts by this gravitational focusing, and characteristic impact velocities are also higher. Additionally, apex–antapex asymmetries are larger on moons closer to the planet. However, the magnitudes of these effects depend upon the dynamical properties of the impacting population. For instance, long-period comets on highly eccentric orbits are less strongly affected by planetary perturbations than are centaurs that approach the planet at a slower speed.

Estimating the relative cratering rates at different planets is even more challenging, especially when comparing the inner planets with outer Solar System objects. For instance, the breakup of a large main belt asteroid located near a strong resonance with Jupiter can produce a spike in the cratering rate in the inner Solar System, but won't substantially increase the rate on outer planet satellites, which are impacted primarily by objects coming from beyond the orbit of Neptune.

Assuming a model for the density and size distribution of impactors over time and throughout the Solar System, one can estimate a body's age from graphs such as that shown in Figure 5.38, and using the Moon as the 'groundtruth'. This *crater dating* technique is widely used, since it typically provides the only means to assign an age to a body's surface. The simplest assumption regarding impact frequency and size distributions contains no variations with location or time. This assumption is clearly too simplistic. The population and velocity distribution of impactors varies within the Solar System. Less than one-fifth of the known Earth-crossing asteroids pass also inside Mercury's orbit. However, gravitational focusing by the Sun, the increased velocity of the impactor and Mercury's weaker gravitational field cause the cratering rate per unit area on Mercury's surface to be roughly half that on Earth. Since more asteroids cross Mars's orbit, the cratering rate on Mars is somewhat higher than that on Earth. While the vast majority of the impact craters on Earth and Mars can be attributed to impactors emanating from the asteroid belt, most impact craters on the satellite systems around the giant planets are caused by objects that have spent most of the history of the Solar System beyond the orbit of Neptune. Gravitational focusing (see eq. 13.23) by the Earth causes the number of meteoroids per unit area impinging on the top of our planet's atmosphere to be somewhat greater than the collision rate with the Moon. The gravitational focusing effect of the planets causes the impact rate to be largest for the satellites closest to the planet. The present cratering rate on Io is estimated to be 2–3 times as large as that on Callisto, while that on Callisto presently is roughly 1–2 times that on the Moon. The cratering rates on the largest satellites of Saturn and Uranus are similar to those on the Galilean satellites. In addition to variations in impact frequency with location, the lunar cratering frequency and size distributions of the impactors have changed drastically over time (Figs. 5.39, 5.38a); it may have changed differently for other bodies.

Even when cratering rates are expected to be the same, or relative cratering rates can be estimated, other processes complicate efforts to determine surface ages from crater counts. Saturation is approached gradually, not achieved all at once. Uniform cratering probabilities produce random cratering rates, and random implies statistical variations (Problem 5.19). Secondary craters can indeed be clustered, i.e., less uniformly distributed than random. Thus, nonuniform crater spatial distributions don't necessarily imply nonuniform surface ages. Indeed, in the portions of the surface of Saturn's mid-size moon

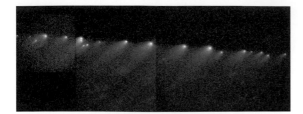

Figure 5.41 An HST image of Comet D/Shoemaker–Levy 9 (SL9) about 2 months before the comet crashed into Jupiter. Note that each fragment, A–W, is a small comet, with its own tail. (Courtesy Hal Weaver and T. Ed Smith, HST/NASA)

Rhea imaged at high resolution by Voyager, moderately large craters are more uniformly distributed than random, implying saturation effects dominate over any possible age differences.

The size distribution of craters can also yield information on the population of impactors. The oldest, most heavily cratered regions of the Moon and Mars have crater size distributions consistent with impacts by a population of bodies that was derived from the asteroid belt in a size-independent manner. In contrast, Venus, and less heavily cratered regions of the Moon and Mars have crater size distributions consistent with impacts by the current population of stray bodies in the inner Solar System (extrapolated to larger sizes); while this population is also primarily derived from the asteroid belt, it is enhanced in small objects relative to large ones because such objects are more transportable as collisional ejecta (§9.4) and by the Yarkovsky force (§2.7.3).

5.4.5 Comet SL9 Impacts Jupiter

In 1993, Carolyn and Gene Shoemaker, in collaboration with David Levy, discovered their ninth comet, Comet Shoemaker–Levy 9 (or SL9). Comet SL9 was very unusual: it consisted of more than 20 individual cometary fragments (Fig. 5.41) that orbited Jupiter, rather than the Sun. Orbital calculations showed that the comet was captured into an unstable jovicentric orbit around 1930. In July 1992, the comet's perijove passage was $\sim$1.3 $R_{\jupiter}$ from Jupiter's center, well within Roche's limit (§11.1). As a consequence, SL9 was torn apart by Jupiter's strong tidal forces, after which the individual fragments developed cometary characteristics (coma and tail) and continued their journey around the planet, each fragment in its own quasi-Keplerian orbit. The comet's next perijove, two years later, would have been below Jupiter's surface, so each fragment crashed into the planet. The impacts occurred over a 6 day period, on the hemisphere facing away from Earth, a few degrees behind the east (dawn) limb. The Galileo spacecraft, at 1.6 AU from Jupiter on its

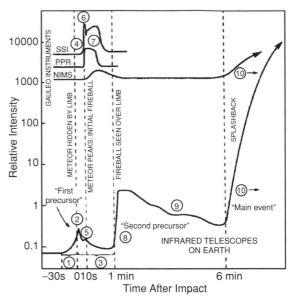
Figure 5.42 Schematic of the SL9 lightcurves during the first 6–8 minutes. 1–3: coma meteor shower (sometimes called leader emission); 4 and 6: bolide entry; 5: possible reflection of bolide emission on trailing coma dust; 7: fireball; 8: plume becoming visible over the limb; 9: cooling plume; 10: plume re-entry. (Harrington *et al.* 2004)

way to the planet, had a direct view of the impact sites. It is the only large impact mankind has witnessed directly. In this section, we give a brief summary of the observations.

By utilizing both ground-based and Galileo data at different wavelengths, together with elaborate numerical simulations of the impacts, we identify a sequence of events as sketched in the form of lightcurves in Figure 5.42, and as a cartoon in Figure 5.43. The first observed event was a short flash, lasting a few tens of seconds, the *first precursor*, interpreted as a meteor shower when the fragment's coma first impacted Jupiter's atmosphere (Figs. 5.42, 5.43a, 5.44). While the fragment fell through the atmosphere, it disrupted and vaporized. Most of its kinetic energy was deposited near its terminal atmospheric depth, probably at a few bars. While Galileo was not sensitive enough to pick up the meteor trail, it did detect the fireball immediately after the explosion, an event hidden from Earth. Galileo measured a temperature of $\sim$8000 K, consistent with an explosion temperature of well over 10 000–20 000 K.

The second observed event was hot thermal emission radiated by the *plume* of hot material, the fireball (preceded by a shock front), rising back up the 'chimney', or entry path, of the comet (Figs. 5.42, 5.43b). This hot gas contained a fair fraction of the original cometary gas, mixed with a similar mass of shocked jovian air, followed

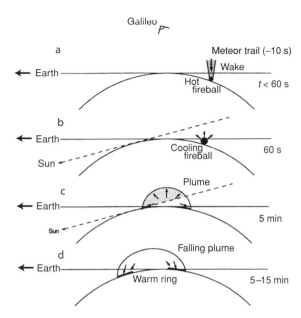

Figure 5.43 A cartoon illustrating the geometry during the SL9 impacts. For each panel, the time, t, after impact is indicated. (Adapted from Zahnle 1996)

by a trail of nearly pure jovian air from a depth of several bars. The fireball's thermal emission was seen at visible and infrared wavelengths as soon as the plume rose above the 'limb' of Jupiter, i.e., the local line of sight to the Earth; at this point the fireball was still in the pre-dawn shadow (Figs. 5.43b, 5.44). The onset of this second flash was abrupt, $\sim$7 s, as expected from a fireball with a diameter of 100–200 km rising at a velocity of 10–15 km s^{-1}. The decay in intensity of the flash represents the rapid cooling (mostly adiabatic, enhanced by radiative cooling) of the rising, expanding fireball.

The plumes reached a height of $\sim$3000 km above Jupiter's cloud tops, before falling back down. Once a plume had risen high enough to become visible in reflected sunlight, both the plume and subsequent collapse onto the atmosphere was imaged with HST (Fig. 5.44b). The splashback of material onto the atmosphere is characterized by a dramatic brightening at infrared wavelengths, six minutes after the first flash (Figs. 5.42, 5.44a), which lasted for about ten minutes. This was seen simultaneously by Galileo and ground-based observers, and signifies a dramatic heating of the atmosphere. The heating was produced by plume material 'raining' down onto the atmosphere, where the vertical component of its kinetic energy shock-heated the atmosphere.

Since the tangential velocity component of the material falling down onto the atmosphere is conserved across the re-entry shock, plume material slides horizontally on

the atmosphere, so that the 'impact regions' expanded rapidly in the radial direction and, within minutes of time, covered areas well over 10 000 kilometers in extent. Immediate re-entry of fast plume material on near-horizontal trajectories drove a lateral shock, which gave rise in a few cases to a *third precursor* and an outward propagating (thermal) 'ring' at scales much larger than the impact sites imaged by HST (Fig. 5.45). This ring was visible at 3.08 μm (often referred to as McGregor's ring, after its discoverer).

Following an impact, HST images (Fig. 5.45) revealed the impact sites to have a special morphology. Each site showed a brown dot at the point of entry, surrounded by one or two dark rings propagating outwards at a velocity of 400–500 m s^{-1} (the inner of the two rings, if present, was fainter and moved much slower). A large crescent is visible to the southwest, sometimes referred to as 'impact ejecta'. Although the impact phenomena appeared dark against Jupiter's clouds at visible wavelengths, they were bright at near-infrared wavelengths (Fig. 5.45), indicative of material at high altitudes (well above the visible clouds; §4.5.6.1). Much of this material has been attributed to submicron-sized dust ($\lesssim$0.3 μm), likely carbonaceous material or 'soot' generated by shock-heating atmospheric methane. The impact morphology can be accounted for by a combination of the ballistic trajectories of the plume material, the Coriolis force (§4.5.3.1), and the horizontal 'sliding' of material for about 20–30 minutes after impact. The impacts were followed by several *bounces*, periodic brightenings in the emission, caused by material 'bouncing' off the top of the atmosphere and re-entering and shock-heating the atmosphere for a second or third time.

The effects of the impacts did not stop in the atmosphere. The fireball rose up through the atmosphere into the ionosphere. Coupling between the atmosphere, ionosphere, and magnetosphere took place via shocks, plumes, and precipitation of particles trapped in Jupiter's radiation belts (§7.5). This coupling led to modifications of the trapped radiation belts, placement of ionized material onto magnetic field lines, and a variety of emissions from the upper atmosphere at wavelengths ranging from the infrared to UV and X-rays (§7.5).

Computer models of the tidal disruption of the parent body and the impacts showed that the individual pieces must have been $\lesssim$1 km across, with densities of $\sim$0.5– 0.6 g cm^{-3} in order to match the number of fragments and the length of the comet chain as a function of time. The kinetic energy involved in an impact by one fragment is a few times 10^{27} ergs (5.9), equivalent to the explosive energy of about a million megatons of TNT.

(a)

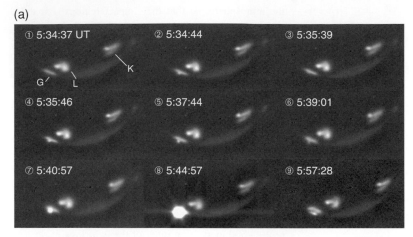

(b)

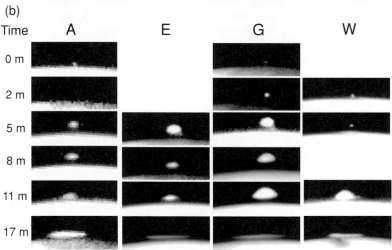

Figure 5.44 (a) Impact of SL9 fragment R with Jupiter as observed at 2.3 μm by the Keck telescope. Each panel, labeled with UT time, is a frame from the movie. Only Jupiter's southern hemisphere is shown, with the former impact sites G (coming into view on the dawn limb), L, and K indicated. The planet is dark at 2.3 μm since methane and hydrogen gases in the atmosphere absorb (incoming and reflected) sunlight. The impact sites are bright, since some of the impact material is located at high altitudes, above most of the absorbing gases. (Graham *et al.* 1995) (b) A sequence of HST images showing the meteor shower (at time $t = 0$ min), thermal emission from the plume ($t = 2$ min), plume (in sunlight) rising ($t = 5$ min), spreading ($t = 8$–11 min) and collapsing ($t = 17$ min) after the impacts of SL9 fragments A, E, G, and W. The plumes continued to slide after they had fully collapsed. Note emission from the hot ejection tube at $t = 5$ min images of the E and G impacts. (Adapted from Hammel *et al.* 1995; HST/NASA)

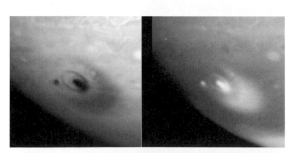

Figure 5.45 Two HST images of Jupiter, on 18 July 1994, approximately 1.75 hours after the impact of SL9 fragment G, one of the largest fragments. The left image was taken through a green filter (555 nm) and the right image through a near-infrared methane filter (890 nm). Note that the impact sites appear bright at near-infrared wavelengths in methane absorption bands and dark at visible wavelengths. The G impact site has concentric rings around it, with a central spot 2500 km in diameter. The thick outermost ring's inner edge has a diameter of 12 000 km. The small spot to the left of the G impact site was created by the impact of the smaller sized fragment D, about 20 hours earlier. (Courtesy Heidi B. Hammel, HST/NASA)

The largest nuclear bomb ever exploded had a yield of $\sim$60 megatons TNT equivalent, and most were well under one megaton. Large volcanic eruptions on Earth perhaps reach 100–1000 megatons of TNT. These were powerful impacts!

Forensic evidence implies Jupiter was hit by an object of radius $\sim$100–200 meters in July 2009. The scar on Jupiter produced by this event had a morphology very similar to those left by medium-sized SL9 fragments, and it was spotted by amateur astronomers within a day after the impact (neither the impactor nor the impact event were detected). Analysis of spectroscopic observations revealed that the impacting body was most likely a small asteroid.

5.4.6 Mass Extinctions

The well-preserved history of impacts on the Moon, together with the widely observed impact of Comet

Shoemaker Levy 9 with Jupiter, makes us uncomfortably aware of the dangers of being hit by meteoroids. Because the Earth's crust is continuously renewed, impact craters are relatively rare and difficult to find or recognize. Yet studies of asteroids, comets, and interplanetary dust suggest that Earth 'sweeps up' 10 000 tons of micrometeoritic material each year. Meteors, primarily centimeter-sized and smaller material falling through Earth's atmosphere, are a familiar sight, especially around August 10 (Perseids), November 17 (Leonids), December 11 (Geminids), and January 1 (Quadrantids). One calculation suggests that ∼7240 meteorites over 100 g in mass ($\gtrsim 2$ cm in radius) make it to the ground each year, which translates into one fall per square kilometer every 100 000 years.

The smallest meteoroids to hit the Earth's atmosphere are slowed to benign speeds by gas drag or vaporized before they hit the ground. Extremely large impactors can melt a planet's crust and eliminate life entirely.

An airburst creates a shock wave whose destructive reach varies roughly as the one-third power of the energy released by the explosion. However, the size of the area devastated by an airburst depends upon the burst's height as well as its energy, because shock waves created by high airbursts can dissipate before reaching the ground and shocks from low airbursts approach the ground very obliquely aside from immediately below the explosion. The Tunguska event in 1908, an airblast of ∼4 Mt caused by a ∼40 m radius stony body, flattened about 2000 km² of forests. This area is about as large as could be destroyed by an explosion of this energy, since it occurred near the optimal height for producing destructive effects. In addition to the direct collisional destruction, an impact can have dramatic effects on climate worldwide.

The Moon likely formed about 4.5 Gyr ago in a disk that was produced by an impact of a Mars-sized or larger body with the proto-Earth (§13.11). At that time, just after the formation of the planets, there still were many planetesimals around that battered the surfaces of planets, and moderately large impacts were probably not uncommon. To assess the impact probabilities at the current epoch, we first investigate the evidence of moderately large (several kilometers in size) impactors that hit our planet over the past billion years.

It is well known from fossil records in rock strata that many species of animal and plant life have been wiped out nearly simultaneously on a few occasions during the past half billion years, in processes referred to as *mass extinctions*. One of the most dramatic mass extinctions occurred ∼65 million years ago, in which the dinosaurs and large numbers of other animal and plant species suddenly disappeared from Earth. This event marks the end

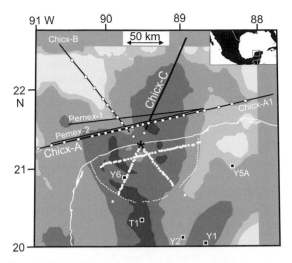

Figure 5.46 The Chicxulub seismic experiment. Solid lines show offshore reflection lines, white dots show wide-angle receivers. Shading shows the Bouguer gravity anomaly (§6.1.4.4); the crater is marked by a ∼30 mGal circular gravity low. The dashed white lines mark the positions of the sink-holes in the carbonate 'platform'. Squares show well locations; Y6 is ∼1.6 km deep, and T1, Y1, Y2, and Y5a are 3–4 km deep. All radii are calculated using the asterisk as the nominal center. (Morgan *et al.* 1997)

of the Cretaceous (K) time period and the beginning of the Tertiary (T) period, and is known as the *K–T boundary*. The discovery of much higher (10–100 times) than normal levels of iridium in sediments deposited at the K–T boundary provided a strong indication that this mass extinction had an extraterrestrial origin. The subsequent discovery of the Chicxulub crater on the Yucatán Peninsula in Mexico (Fig. 5.46) was convincing evidence that the K–T boundary and the extinction of the dinosaurs was indeed triggered by an impact of an ∼10 km sized object. The Chicxulub crater is no longer visible at the surface, but can be studied by using gravity anomaly measurements (§6.1.4). The crater appears to have a multiring basin morphology: a peak ring with a diameter $D \approx 80$ km, an inner ring with $D \approx 130$ km, and an outer ring with $D \approx 195$ km.

The energy involved in an impact of an $R \approx 10$ km sized body with a velocity of ∼15 km s^{-1} is about two orders of magnitude larger than the energy involved in the impact of individual fragments of Comet Shoemaker–Levy 9. At the instant the bolide hit the surface, two shock waves must have propagated away from the impact site, as discussed in §5.4.2. One shock wave should have propagated into the bedrock, while the other must have gone backwards, into the impactor. Immediately following this, a colossal plume of vaporized rock, the fireball, must have risen upward into space, launching dust and rocks on ballistic trajectories which carried them far around the Earth. In this particular case, this fireball is thought to have been followed by a second plume, driven by the sudden release of CO_2 gas from a layer of shocked limestone located about 3 km below the surface. The cavity itself may have reached a depth of about 40 km before the center rose into a central peak. The peak grew so large and high that it collapsed, thereby triggering several outward expanding rings and ridges. Meanwhile, the crater walls continued to expand outwards. The transient cavity is thought to have had a diameter of $\sim$100 km.

The heat from ejecta re-entering the atmosphere probably ignited global forest fires. Also large amounts of nitric acid (HNO_3) and sulfuric acid (H_2SO_4) were likely formed, raining down (acid rain), killing plants and animals and dissolving rocks over a large area around the impact site. Because the impact happened on a peninsula, a large tsunami wave (recorded in rocks found in Mexico and Cuba) spread outward and, upon hitting Florida and the Gulf coast, must have destroyed vast areas in what is now Mexico and the United States. Fine dust, which had been brought up by the fireballs, stayed suspended in the atmosphere for many months before reaching the surface. This could have turned the sky dark over the entire Earth and prevented sunlight from reaching the surface, so the surface temperature dropped to well below freezing for many months. Once the sky cleared, temperatures may have risen to uncomfortably high levels due to enhanced levels of greenhouse gases such as H_2O and CO_2. This global cycle of extreme hot–cold–hot temperatures would have destroyed many animal and plant species in areas quite remote from the impact site. An alternative model of the consequences of the K–T impact suggests that the greenhouse warming effect is small, but that the slow (over many years) production of sulfuric acid kept the temperature low (by tens of degrees kelvin) for many decades, which could have had a similarly destructive effect on plant and animal life.

Impacts of increasingly larger sizes have greater potential for killing individual organisms and wiping out entire species; such impacts also become increasingly rare. If the amount of dust injected into the stratosphere is sufficient to produce an optical depth >2 worldwide for several months, the surface temperature can be suppressed by $\sim$10 K globally. To produce an optical depth of 2 requires about 10^{16} g of dust, about a hundred times more than has been lofted by any of the large volcanic eruptions from the past century (e.g., Mt. Pinatubo in 1991). This much dust is injected into the stratosphere by an impact with an energy of 10^5–10^6 Mt, i.e., an impact by a stony asteroid 500 m–1

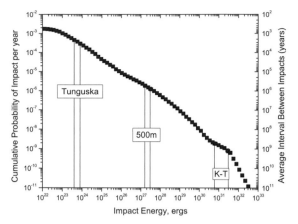

Figure 5.47 Frequency of projectiles of various kinetic energies impacting on the top of Earth's atmosphere at the present era. Bodies 500 m in radius and with the energies of K–T and Tunguska impactors are highlighted. These estimates were made using the observed distribution of NEOs (near-Earth objects). For additional information, see §9.4.1 and Stuart and Binzel (2004). (Courtesy Scott Stuart)

km in radius at a velocity of 20 km s^{-1}. Such impacts occur roughly once per million years (Fig. 5.47). The effect of the impact on the planet and on life are indicated in the last two columns of Table 5.3.

5.5 Surface Geology of Individual Bodies

A combination of remote sensing techniques from the ground and space have led to quite detailed geological views of many bodies in our Solar System. The surface characteristics have been determined via remote sensing techniques such as imaging, photometry, polarimetry, thermal and reflectance spectra, radio and radar observations. These techniques are discussed in more detail in Appendix E. Although the spatial resolution from the ground has been relatively poor ($\gtrsim 0.3''$) compared to the size of the object ($<1''$ up to $1'$ for the largest body, Venus), one can now achieve spatial resolutions of order $\sim$0.05$''$ with the Hubble Space Telescope (HST), or using speckle imaging and/or adaptive optics techniques on large ground-based telescopes. Our highest quality images, however, usually come from spacecraft flybys, orbiters, or landers (Appendix F). Spectacular images have been returned by, e.g., Cassini for the saturnian system, Galileo for the jovian system, the Mars Exploration Rovers roaming the surface of Mars and by Mars Express and the Mars Reconnaissance Orbiter (MRO) for global coverage. Deep Impact revealed the surface of Comet C/Tempel 1, and the Near-Earth Asteroid Rendezvous (NEAR) and the Japanese probe Hayabusa returned exquisite views

Table 5.3 Impacts and life.[a]

Impactor size[b]	Example(s)	Most recent	Planetary effects	Effects on life
Super colossal $R > 2000$ km	Moon-forming event	4.5×10^9 yr ago	Melts planet	Drives off volatiles Wipes out life on planet
Colossal $R > 700$ km	Pluto 1 Ceres (borderline)	$\gtrsim 4.3 \times 10^9$ yr ago	Melts crust	Wipes out life on planet
Mammoth $R > 200$ km	4 Vesta (large asteroid)	$\sim 3.9 \times 10^9$ yr ago	Vaporizes oceans	Life may survive below surface
Jumbo $R > 70$ km	95P/Chiron (largest active comet)	3.8×10^9 yr ago	Vaporizes upper 100 m of oceans	Pressure-cooks photic zone May wipe out photosynthesis
Extra large $R > 30$ km	Comet Hale–Bopp	$\sim 2 \times 10^9$ yr ago	Heats atmosphere and surface to ~ 1000 K	Continents cauterized
Large $R \gtrsim 10$ km	K–T impactor 433 Eros (largest NEO)	65×10^6 yr ago	Fires, dust, darkness Atmosphere/ocean chemical changes Large temperature swings	Half of species extinct
Medium $R > 2$ km	1620 Geographos	$\sim 5 \times 10^6$ yr ago	Optically thick dust Ozone layer threatened Substantial cooling	Photosynthesis interrupted Significant extinction
Small $R > 500$ m	~ 1000 NEOs Lake Bosumtwi	$\sim 500\,000$ yr ago	High altitude dust for months Some cooling	Massive crop failures Many individuals die, but few species extinct Civilization threatened
Midget $R > 200$ m	99942 Apophis (borderline)	$\sim 50\,000$ yr ago	Tsunamis	Coastal damage
Peewee $R > 30$ m	Tunguska event Meteor crater	30 June 1908	Major local effects Minor hemispheric dusty atmosphere	Newspaper headlines Romantic sunsets increase birth rate

[a] Adapted from Lissauer (1999) and Zahnle and Sleep (1997).

[b] Based upon USDA classifications for olives, pecan halves, and eggs.

of the surface of asteroids 433 Eros and 25143 Itokawa, respectively.

Each Solar System body that has been studied in detail has its own peculiarities. Every solid surface has its own geology, which often looks quite different from that of any other known body. Yet there are similarities too, including many features known from Earth, such as volcanic and tectonic structures, atmospheric effects such as winds and condensation flows, impact craters, etc. In this section, we discuss the surfaces of the terrestrial planets and of many moons. Geological properties of the few asteroids that have been imaged at high resolution are reviewed in §9.5, and the surface characteristics of comets visited by spacecraft are summarized in §10.6.5.

When viewed from space, Earth looks much like other planets, though distinctly different in color because of the overwhelming presence of the oceans (Fig. 5.48). The continents are visible as brown land masses, and white ice sheets dominate the polar regions. The northern polar cap is rapidly shrinking due to the global warming that is a consequence of increasing abundance of greenhouse gases in Earth's atmosphere (§4.9.2.1). Our planet has been mapped from space at different wavelengths and also with radar techniques. These data are useful in comparison with images of other planets obtained via very similar techniques. Since we have used planet Earth as a prototype throughout this chapter, and have already shown many photographs of terrestrial features, we only discuss other

Figure 5.48 Image of the Earth taken by the Galileo spacecraft on 11 December 1990, from a distance of ~2.5×10^6 km. India is near the top of the picture, and Australia is to the right of center. The white, sunlit continent of Antarctica is below. Picturesque weather fronts are visible in the South Pacific, at the lower right. (NASA/Galileo, PIA00122)

Solar System bodies here, comparing them to Earth where applicable.

5.5.1 Moon

One can discern two major types of geological units on the Moon with the naked eye: the bright *highlands* or *terrae* that account for over 80% of the Moon's surface area and have an albedo of 11–18%, and the darker plains or *maria* with an albedo of 7–10% that cover 16% of the lunar surface (Fig. 5.15). The maria are concentrated on the hemisphere facing Earth. The dominant landforms on the Moon are impact craters. The highlands appear saturated with craters, varying in diameter from micrometers (Fig. 5.25) up to hundreds of kilometers in size (e.g., Orientale basin, Fig. 5.28). Some of the large younger craters show the bright rays and patterns of secondary craters. The highlands clearly date back to the early bombardment era ~4.4 Gyr ago (Fig. 5.38a). In contrast, the maria are less heavily cratered and must therefore be younger.

Some maria cover parts of seemingly older impact basins, e.g., Mare Imbrium within the Imbrium basin. Radioisotope dating (§8.6) of rocks brought back by the various Apollo and Luna missions indicates that the maria are typically between 3.1 and 3.9 Gyr old. The lunar

samples further indicate that the maria consist of fine-grained, sometimes glass-like basalt, rich in iron, magnesium, and titanium. All of this together suggests that the mare basalts originated hundreds of kilometers below the surface, and must have been brought up by volcanic activity, 3.1–3.9 Gyr ago. The lava lakes cooled and solidified rapidly, as evidenced by the glassiness and small grain size of the minerals in the rocks. Data from the Clementine spacecraft show that the lava-flooded maria are extremely flat, with slopes of less than 1 part in 10^3; they are typical topographic lows. Since the large impact basins were created by energetic impacts, the crust was probably fractured down to many kilometers below the surface. The hot magma may have oozed its way up through the cracks and flooded the low-lying regions of the impact basins. A few small volcanic domes and cones can be discerned on the surface.

The most pronounced topographic stucture on the Moon is the South Pole Aitken Basin, the oldest discernible impact feature. It is 2500 km in diameter, with a maximum depth of 8.2 km below the reference ellipsoid, the Moon's geoid (§6.3.1), or 13 km from the rim crest to the crater floor. This is the largest and deepest impact basin known in the entire Solar System. The highest point on the Moon, 8 km above the reference ellipsoid, is in the highlands on the far side of the Moon, adjacent to the South Pole Aitken Basin.

Some regions near the lunar poles are in permanent shadow, and may remain as cold as 40 K. In 1998, measurements made with a neutron spectrometer on board the Lunar Prospector spacecraft demonstrated the presence of hydrogen, presumably contained in water-ice, in cold craters at the lunar poles (§E.9). Radar observations from Earth, however, show no sign of thick ice deposits (compare Mercury, §5.5.2). In addition, when a sensitive camera onboard JAXA's KAGUYA mission peered into permanently shadowed craters using solar radiation scattered from the rim, no concentrations of high albedo material was detected at a resolution of ~10 m. Hence ice, if indeed present, must be in the form of tiny crystals mixed in with the soil, at an estimated concentration of ~1.5% by weight. The presence of such ice crystals appears to have been confirmed by the Lunar Crater Observation and Sensing Satellite, LCROSS, which impacted a permanently shadowed region in the Cabeus crater near the Moon's south pole on 9 October 2009. Spectra of the vapor plume created by the impact revealed several absorption features, including those of water and hydroxyl molecules. The volatiles in this and other permanently shadowed craters were presumably brought in by comets and asteroids well after the Moon had formed, because the Moon, in general, is very dry. Several months before the LCROSS impact, however, the Moon Mineralogy Mapper (M^3) on ISRO's Chandrayaan-1 detected absorption features near 2.8–4 μm, indicative of widespread OH- and H_2O-bearing material in the upper few mm of the Moon's regolith. This suggests that ongoing bombardment by cosmic rays may play a role in supplying H to form ice in the permanently shadowed craters.

The Apollo missions showed that the lunar crust in the highlands has been pulverized by numerous impacts. Continued micrometeoroid impacts on the broken rocks created a fine-grained layer of regolith, more than 15 m deep. The maria are also covered by regolith, but because the maria are younger this layer is only 2–8 m deep. Many of the lunar rocks are polymict breccias (Fig. 5.37), different types of rock cemented together by impact processes. The breccias, as well as the lunar soil, contain smooth glass spherules or dumbbells, created during impacts. The lunar highlands generally lack rocks abundant in heavy minerals such as iron and titanium. The rocks are dominated by *anorthosites*, composed of calcium-rich plagioclase feldspars, or anorthites (§5.2.2). The Apollo missions revealed an unusual chemical component in the lunar rocks, referred to as *KREEP*, named after its mineral components: potassium (K), rare-earth elements (REE; elements with atomic numbers 57–70, see Appendix D), and phosphorus (P).

At some locations, KREEP is enriched in REE by up to 1000 times the chondritic value. During crystallization of a magma, REE are excluded from the major mineral phases, as olivine, pyroxene, and plagioclase, so KREEP may be the final crystallization product of a global magma system. This concept is reinforced by the anomalies observed in the abundance of europium ($_{63}$Eu) compared to other REE. Both KREEP and the mare basalts are depleted in Eu, whereas Eu is enhanced compared to the other REE in the highlands. In contrast to the other REE and other incompatible elements (§6.2.2.1) such as K, U, Th, P, europium is readily incorporated in plagioclase. Since plagioclase is light, it floats upwards in a magma (§5.2.2) to form the lunar crust. Hence the presence of KREEP, together with the Eu anomalies, provides geochemical evidence of crystallization in an initially global lunar magma ocean. Although there still is some debate whether the entire Moon, or only its outer portions, was molten and differentiated, the relative depletion of siderophile elements (e.g., Fe, Mg) and enhancement in lithophile elements (e.g., Ca, Al, Ti) as compared to the Earth's crust and mantle is suggestive of a global magma ocean and differentiation leading to the formation of an iron-rich core (§6.3.1). Crystallization of the main phases, e.g., the anorthosite

rocks that make up the highlands, was probably complete 4.44 Gyr ago, while the final KREEP residue was solid ∼4.36 Gyr ago.

Although impact cratering is by far the most important geological process on the Moon, there is also clear evidence of volcanism and tectonics. Volcanism is most evident in the maria, but there are a few other features which show evidence of volcanic flows. The sinuous (Hadley) rille shown in Figure 5.21 has been interpreted as a collapsed lava channel. Tectonic features such as linear rilles, similar to graben faults, likely formed by expansion or contraction. There is no evidence for (past or present) plate tectonics. Geological activity on the Moon tailed off ∼3 Gyr ago, although results by the SELENE (Kaguya) orbiter suggest that volcanic activity on the far side continued, perhaps episodically, until 2.5 Gyr ago. Tidally and thermally induced moonquakes continue to occur on the Moon.

5.5.2 Mercury

Superficially Mercury's surface resembles that of the Moon, because craters are the dominant landform on both bodies (Fig. 5.49c). However, the two bodies differ significantly in detail, as discussed below. Craters on Mercury are shallower than like-sized craters on the Moon, and secondary craters and ejecta blankets are closer to the primary craters of a given size. Both of these differences are caused by the greater surface gravity on Mercury. Mercury's heavily cratered terrain is interspersed with smooth *intercrater plains*, resembling in some ways the maria on the Moon; the least cratered areas on Mercury, however, are bright, in contrast to the Moon's maria.

By far the largest feature observed on Mercury is the 1550-km diameter Caloris basin (Fig. 5.49a), a huge ring basin analogous to the large impact basins on the Moon. This basin is directly facing the Sun at the perihelion of every other orbit, a consequence of Mercury's 2:3 spin-orbit resonance (§2.6.2). The basin is surrounded by a 2-km high ring of mountains. Shock waves from the impact that produced the Caloris basin are thought to be responsible for the irregular or 'weird' terrain antipodal to Caloris, which consists of a chaotic formation of rock-blocks and hills.

Mercury shows unique scarps, or *rupes*, on its surface: linear features hundreds of kilometers long, which range in height from a few hundred meters up to a several kilometers (Fig. 5.49d,e). These lobate scarps are the most prominent tectonic features on the planet; they cut across all terrain, at seemingly random orientations. These scarps probably were produced by planet-wide contraction due

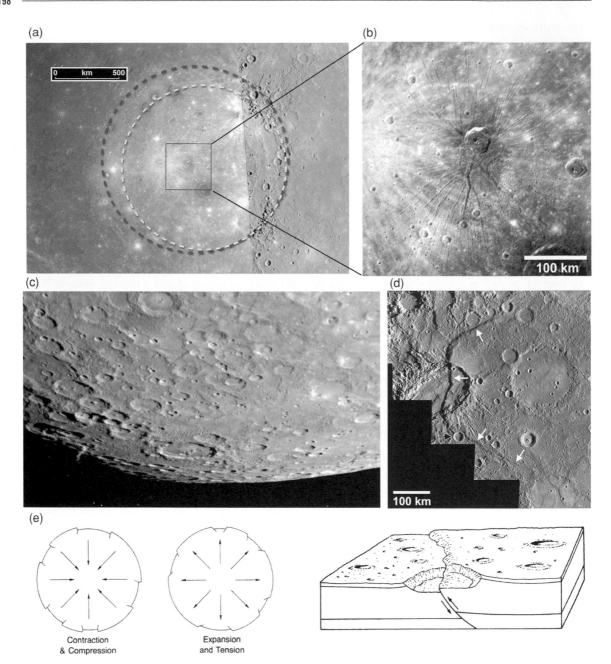

Figure 5.49 (a) A composite image of Caloris basin, the largest impact basin on Mercury. The eastern half of the basin was photographed in 1974 by Mariner 10; it was the only part in sunlight at that time. During its first flyby of Mercury, on 14 January 2008, the MESSENGER spacecraft imaged the western half of the basin. This composite image shows that Caloris basin is larger (outer dashed circle, diameter of 1550 km) than originally derived from the Mariner 10 data (inner dashed circle, diameter of 1300 km). The black box at the center is enlarged in panel (b),which shows a detailed image of the center of Caloris basin. The radial troughs probably result from an extension of the floor materials that filled the Caloris basin after its formation. (NASA/MESSENGER, PIA10383; Murchie *et al*. 2008) (c) MESSENGER image of Mercury's south pole, taken on 14 January 2008, from a distance of 200 km. The southern limb of the planet is visible in the bottom right. Towards the left is the terminator, where a raised crater rim is just catching the last glint of sunlight. (NASA/MESSENGER, PIA10187) (d) Mercury shows prominent lobate scarps, such as Beagle Rupes, an ~600 km long scarp (white arrows) that offsets the floor and walls of an ~220 km diameter impact crater. Lava appears to have flooded this crater, which subsequently was deformed by wrinkle ridges before the scarp developed. In contrast, the black arrow points at a ~30 km diameter crater, that must have formed afterwards. (NASA/MESSENGER, Solomon *et al*. 2008) (e) Schematic of the formation of scarps. (Hamblin and Christiansen 1990)

to cooling, including partial core solidification, like the wrinkles on the skin of a dried-out apple. They suggest a global decrease of Mercury's radius by up to ~4 km.

As on the Moon, the smooth plains formed after the late heavy bombardment era, i.e., they are not more than 3.8 Gyr old, while the cratered highlands stem from before the early bombardment era. Even though the surface reflectance between the plains and highlands are similar, the MESSENGER (MErcury Surface, Space ENvironment, GEochemistry, and Ranging) spacecraft revealed conclusive evidence that the plains were formed volcanically. For example, a variety of volcanic features were seen, such as vents, deposits, as well as a ~100-km diameter shield volcano along the inner margin of the southern Caloris basin. At the center of Caloris basin is a radial graben structure, Pantheon Fossae (Fig. 5.49b), showing hundreds of grabens, up to several km wide and over 100 km long. This structure may result from rising magma just below the surface, where melting may have been triggered by the excavation of material. Such excavations release the pressure on the underlying rocks, which reduces the melting temperature. The structure is reminiscent of volcanic constructs on Venus (§5.5.3).

Volcanism appears to be widespread on Mercury. Smooth plains seem to fill craters and embay crater rims. Volcanic plains can be several km thick. The composition of the volcanic flows, however, is quite different from basaltic flows on the Moon and Earth, which explains its brighter color. Mercury lacks basaltic material that is rich in heavy elements such as iron and titanium (Fig. 5.50). The mineral ilmenite ($(Fe,Mg)TiO_3$) that makes up most of the opaque material in the Moon's maria appears to be largely absent on Mercury.

The most likely explanation for Mercury's high uncompressed density (§6.3.2) is that part of the planet's mantle was stripped away by a giant impact (or many somewhat smaller high-velocity impacts) early in its formation history (§13.6.1). Such an impact would also 'boil away' the more volatile elements, leaving the crust dominated by very refractory material. Since Mercury's surface appears to lack heavy elements, the lava which covered or created the smooth plains must have originated close to the crust, in contrast to the Moon's maria, which were flooded by basalt from deep below the Moon's surface.

Since Mercury's obliquity is 0°, the poles do not receive much sunlight. In fact, the crater floors in some regions are in permanent shadow, and the temperature stays well below 100 K. Radar echoes indicate an unusual large reflection from the poles, which has been attributed to the presence of water-ice. On its first flyby of Mercury

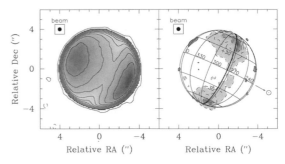

Figure 5.50 Radio image of Mercury (left panel) at a wavelength of 3.6 cm, as observed with the VLA. Contours are at 42 K intervals (10% of maximum), except for the lowest contour, which is at 8 K. (Dashed contours are negative.) The beam size is 0.4″, equal to 1/10 of Mercury's radius. The geometry of Mercury during the observation, the direction to the Sun, and the morning terminator (dashed line) are superimposed on the image on the right. This image shows the residuals after subtracting a model image from the observed map. Note that the darker grays show peaks in emission in the image on the left, but minima on the right. One effectively probes a depth of ~70 cm at this wavelength (Fig. 3.5). The two 'hot' longitudes (those facing the Sun at perihelion) are clearly seen in the left image (§3.2.1). The image on the right shows the thermal depressions at both poles and along the sunlit side of the morning terminator. Contour intervals (right image) are in steps of 10 K, which is roughly three times the rms noise in the image. The loss tangent of the material determined from these images appears to be much less than that on the Moon (§3.2.1). This may be evidence of a lack of basaltic material on Mercury's surface. (Mitchell and de Pater 1994)

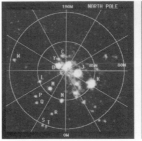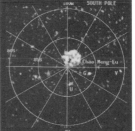

Figure 5.51 Radar images of the north (left) and south (right) polar regions of Mercury, obtained with the Arecibo radio observatory. The radar-bright features are interpreted as layers of ice. Their locations correspond to craters on the poles. The floors of these craters are in permanent shadow, and hence cold enough that ice is stable over billions of years. (Harmon *et al.* 1994)

(January 2008), MESSENGER revealed water-derived ions in Mercury's exosphere, perhaps related to the radar detections of water-ice. Detailed radar images (Fig. 5.51) show a high correlation of the icy regions with visual craters near the pole. It may seem counterintuitive to have water-ice on a planet so close to the Sun, which otherwise

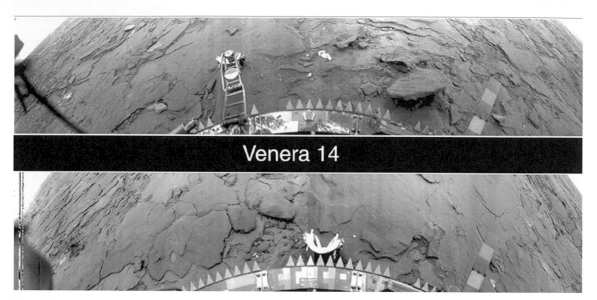

Figure 5.52 Venera 14 Lander images of the surface of Venus. The lander touched down at 13° S, 310° E on 5 March 1982. It transmitted from the surface for 60 minutes before succumbing to the planet's heat. Chemical analyses performed by the Venera landers indicate that most venusian rocks, including those shown here, are basaltic and therefore black or gray. They appear slabby or platy, and are separated by minor amounts of soil. Parts of the lander can be seen at the bottom of each picture (a mechanical arm in the upper picture, a lens cover on the lower one). The landscape appears distorted because Venera 14's wide-angle camera scanned in a tilted sweeping arc. The horizon is seen in the upper left and right corners of both images. (Courtesy Carle Pieters and the Russian Academy of Sciences)

is very dry like the Moon. However, comets and volatile-rich asteroids have continued to impact Mercury. Although the impactors' volatile material rapidly evaporates, some of it may not have escaped Mercury's gravitational attraction. Water molecules might 'hop' over the surface, until they hit the polar regions where they freeze and remain stable for long periods of time. At temperatures $T < 112$ K, water-ice is stable to evaporation over billions of years. Calculations do show, however, that a very large source of icy material is needed to build up a many meters ($\gtrsim 50$ m) thick layer of ice, as required to explain the radar data. Assuming an impacting distribution of meteoritic material as expected at Mercury during the early and late heavy bombardment era, one might only build up a layer ~20 cm thick! However, a few giant comets or volatile-rich asteroids might significantly alter these results.

5.5.3 Venus

Although Venus is covered by a thick cloud deck, impenetrable to visible light, the surface can be probed at a few specific infrared wavelengths and at radio wavelengths longwards of a few centimeters (§4.4). Global maps of Venus's surface have been constructed using radar

experiments from the ground and from space. Several Venera spacecraft have landed on the surface and sent back photographs thereof (Fig. 5.52). The color of the landscape in these images is orange-like, because the dense, cloudy atmosphere scatters and absorbs the blue component of sunlight. The photographs reveal a dark surface and slightly eroded rocks; the rocks are not as smooth, however, as typical terrestrial rocks. The compositions of rocks at the landing sites are similar to various types of terrestrial basalts.

The most detailed (spatial resolutions of ~0.2–1 km) almost global dataset of Venus's surface has been obtained by the Magellan spacecraft using radar. A small fraction of the surface is covered by highlands, large continent-sized areas of volcanic origin that are well (3–5 km) above the average surface level (Fig. 6.22a). The four major highlands (Ishtar and Aphrodite Terra, Alpha and Beta Regio) together cover about 8% of the surface. Roughly 20% of the surface consists of lowland plains, and ~70% of rolling uplands. Overall, most of the surface lies within a kilometer of the mean planetary radius. The difference in elevation between the highest and lowest features is ~13 km, which is similar to the elevation contrast on Earth (Himalayas ~8 km above sea level, oceans ~5 km

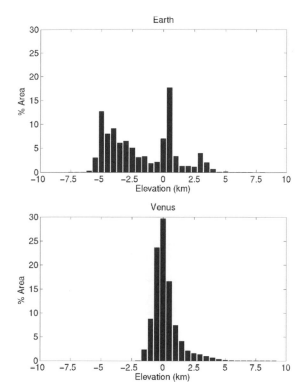

Figure 5.53 Histograms of the elevation (in 0.5 km bins) for Earth and Venus, normalized by area. Note the multiple peaks for Earth and the single peak for Venus. (Smrekar and Stofan 2007)

below sea level; volcano Mauna Loa ∼9 km above the sea floor). However, histograms of surface area as a function of elevation show very different distributions for the two planets (Fig. 5.53): Earth shows a bimodal distribution, reflecting the division between oceans and continents, while Venus shows one peak centered near zero km altitude. Such a unimodal histogram argues strongly against the presence of plate tectonics.

The radar reflection coefficient varies from roughly 0.14 in the lowlands to ∼0.4 in the highlands. Regions with a high radar reflectivity usually correlate with low radio brightness temperatures, or low radio emissivities. The radio emissivities and radar reflectivities can be related to the dielectric constant of the material (§3.4). The disk-averaged value for the dielectric constant is ∼5, a typical value for solid rocks (granite, basalt). There is no evidence of a dielectric constant as low as ∼2, a value expected for porous surface material. This suggests that Venus's surface consists mostly of dry solid rock, and (parts of) the surface might be overlain at most by a few centimeters of soil or dust. The highlands show dielectric constants well over 20–30, indicative of inclusions of metallic and/or sulfide material, such as iron pyrites. Alternatively, the

high dielectric constants may also be caused by scattering effects of rocks embedded in dry soils.

Volcanism

In addition to the four highlands mentioned above, Magellan identified more than a thousand volcanic constructs on Venus. These include numerous small dome-like hills, which are probably shield volcanoes, many circular flattened domes with a small pit at their summit, as well as peculiar structures, including pancake-like domes, coronae, and arachnoids. The low-lying plains themselves are covered by volcanic deposits, likely caused by massive floods of basalt-like outpourings, volumetrically comparable to, e.g., the Deccan Traps in India. The Deccan Traps are composed of a ≳2 km thick plateau that covers an area of about half a million km^2.

Pancake-like domes (Fig. 5.54a) have diameters of ∼20–50 km, and their heights range from ∼100 to ∼1000 m. Morphologically, they resemble terrestrial rhyolite–dacite domes ('glass' mountains, made of obsidian), which are composed of very thick (silica-rich) lava flows originating from an opening on relatively level ground, so that the lava flows outwards in all directions. Although the venusian domes may not be composed of silica-rich lavas, the lava flows must be highly viscous and hence differ from the basalt flows seen at other places on Venus. The complex fractures on top of these domes suggest that the outer layer cooled before magma activity below had completely stopped, resulting in a stretching (and hence fracturing) of the surface. It is also possible that the domes were formed by magma pushing the surface upwards. The near-surface magma then withdrew to deeper levels, causing the collapse and fracturing of the dome surface. The radar-bright margins suggest the presence of rock debris on the slopes of the domes. Some of the fractures on the plains cut through domes, while others appear to be covered by domes. This indicates that active processes both predate and postdate the formation of the dome-like hills.

Coronae (Fig. 5.54c) are large circular or oval structures, with concentric multiple ridges and diameters ranging from ∼100 km to over 1000 km. They are located primarily within the volcanic plains, and are thought to form over hot upwellings of magma within the venusian mantle. *Arachnoids* (Fig. 5.54d) look like spiders sitting on webs of interconnecting fractures. The central region, 50–150 km in diameter, is usually depressed in altitude, as if collapsed, merging with radial lineaments on the outer flanks. Arachnoids are similar in form but generally smaller than coronae. They may be a precursor to coronae formation. The radar-bright lines extending for many

(a)

(b)

(c)

(d)

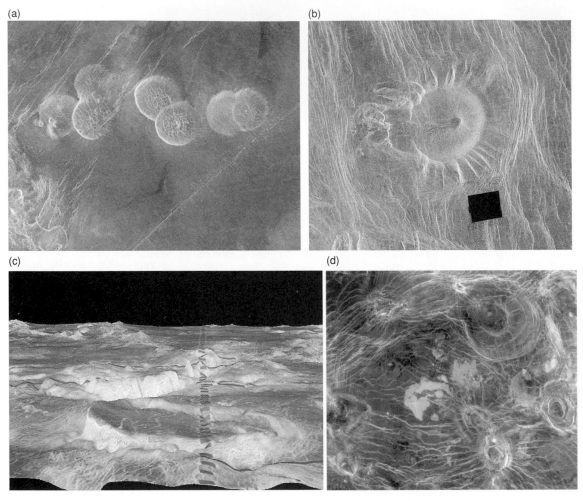

Figure 5.54 A variety of volcanic features on Venus. (a) Radar image showing seven pancake-shaped domes on Venus's surface, each averaging ~25 km in diameter with maximum heights of 750 m. (NASA/Magellan, PIA00215) (b) This peculiar volcanic construct on Venus, nicknamed 'the tick', is approximately 66 km across at the base and has a relatively flat, slightly concave summit 35 km in diameter. The sides of the edifice are characterized by radiating ridges and valleys with a fluted appearance. To the west, the rim of the structure is infiltrated by dark lava flows emanating from a shallow summit pit. A series of coalescing, collapsed pits are located 10 km west of the summit. The edifice and western pits are circumscribed by faint, concentric lineaments up to 70 km in diameter. A series of north–northwest trending graben are deflected eastward around the edifice; the interplay of these graben and the fluted rim of the edifice produce a distinctive scalloped pattern in the image. (NASA/Magellan, PIA00089) (c) A perspective view of Venus with Atete Corona in the foreground, an ~600 × 450 km oval volcano-tectonic feature. (NASA/Magellan/JPL/USGS, PIA00096) (d) This image features 'arachnoids' on radar-dark plains on Venus. These arachnoids range in size from approximately 50 km to 230 km in diameter. (NASA/Magellan, P37501)

kilometers may have resulted from an upwelling of magma from Venus's mantle which pushed the surface up to form 'cracks'. *Novae* have radial fracture patterns, without the central corona.

Magellan images further reveal hundreds of lava channels on volcanic plains. Some of these are hundreds of kilometers long. Others are long narrow (≲2 km wide) features, similar to the sinuous rilles on the Moon. These sometimes end in delta-like distributaries in the plains. One particular channel, Baltis Vallis, is 6800 km in length, the longest sinuous rille yet identified in the Solar System (Fig. 5.55).

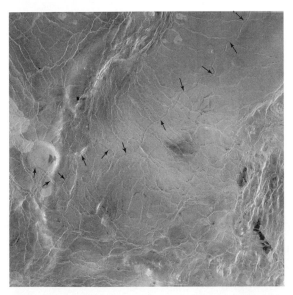

Figure 5.55 A 600 km segment of sinuous channel Baltis Vallis (indicated by arrows) on Venus, the longest channel known in our Solar System. (NASA/Magellan, PIA00245)

It is not known if Venus remains active today. Large temporal variations in the abundance of atmospheric sulfur dioxide had been attributed to possible volcanic outbursts, but Venus Express data cast doubt on this hypothesis (§4.3.3.1).

Tectonics

Magellan radar images show that Venus's surface has undergone numerous episodes of volcanic and tectonic deformations. The distribution of impact craters suggests that volcanic resurfacing is locally very efficient, but also quite episodic. Prominent tectonic features include long linear mountain ridges and strain patterns, which can be parallel to one another, or they may crosscut each other. They can extend over hundreds of kilometers. These tectonic deformations reflect the crustal response to dynamical processes in the mantle. The tectonic deformations clearly both pre- and postdate episodes of volcanic activity. Although there is no evidence that a planet-wide system of tectonic plates exist on Venus, there may have been some local 'tectonic plate' activity.

Erosion

The formation of sediments and erosion is less important on Venus than on Earth and Mars, due to Venus's extremely dense and hot atmosphere. The atmosphere prevents small meteoroids from hitting the ground, a process that on airless bodies is the main source of erosion and regolith formation. The lack of water and thermal cycling on Venus limits weathering processes, and because there is little wind near the surface (wind velocities are $\lesssim 1$ m/s), there is little erosion by wind. Weathering, winds, and water erode rocks on Earth, while on Mars the main erosion process at present is high-speed winds. Even though the wind velocities near the surface of Venus are extremely low, Magellan revealed several morphological features which must be caused by winds, such as wind streaks at the lee side of obstacles (Fig. 5.23c) and a dune field in the neighborhood of a large crater. The wind streaks and dune field yield information on the prevailing venusian winds. We note that, because Venus's atmosphere is 90 times as dense as the Earth's atmosphere, even slow winds may be able to displace a considerable amount of sand.

Impact Craters and Surface Age

The number and size distribution of impact craters (Fig. 5.38) implies that Venus's surface is younger than that of Mars, but older than Earth's. Typical age estimates range from a few hundred million up to one billion years. Craters seem to be randomly distributed over the planet's surface, suggesting that most areas have quite similar ages. Thus, Venus may have undergone major global resurfacing. Venus's lithosphere may be very thick (§6.3.3), perhaps ~200 km, so that heat produced in the planet's mantle by radiogenic processes cannot escape as fast as it is generated. At some point the thick lithosphere may rupture and sink through the overheated buoyant mantle. Once subduction starts, slabs of crust may sink at rates of 20–50 cm per year, so that the crust is completely renewed within 10^8 yr. Alternatively, the crust can be thin, where slices of crust occasionally peel off and sink down, while the crust is slowly renewed by global volcanic resurfacing events.

Impact craters larger than ~15 km are generally circular complex craters with (multiple) central peaks and peak rings (Fig. 5.56). Smaller craters show multiple floors, and often appear in clusters, suggestive of a breakup of meteors in Venus's dense atmosphere. No craters with diameters less than 3 km have been seen. The projectiles which would have created such small craters must have been broken up or substantially slowed down in the atmosphere. The ejecta blankets around impact craters typically extend out to ~2.5 crater radii. In radar images, they look like a bright pattern of flower petals. Such patterns are seen only on Venus, whose atmosphere is so thick that ejecta cannot travel very far. Still, the ejecta extend much further than predicted from simple ballistic emplacements. The ejecta patterns are often asymmetric as a result of oblique

(a)   (b)

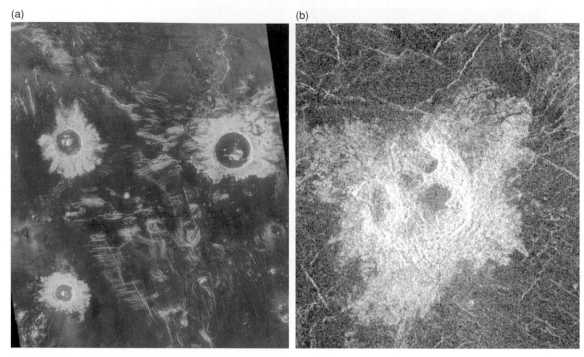

Figure 5.56 Examples of impact craters on Venus. (a) Three impact craters with diameters that range from 37 to 50 km, located in a region of fractured plains, show many features typical of meteoroid impact craters: rough (bright) material around the rim resembling asymmetric flower petals, and terraced inner walls and central peaks. Numerous domes, 1 to 12 km in extent, which are probably caused by volcanic activity, are seen in the southeastern corner of the mosaic. (NASA/Magellan, PIA00214) (b) An irregular crater on Venus, approximately 14 km in diameter. The crater is actually a cluster of four separate craters that are in rim contact. The noncircular rims and multiple, hummocky floors are probably the result of the breakup and dispersion of an incoming meteoroid during passage through the dense venusian atmosphere. After breaking up, the meteoroid fragments impacted nearly simultaneously, creating the crater cluster. (NASA/Magellan, PIA00476)

impacts, where the missing sector is in the uprange direction. In several cases, lava flows are connected with the ejecta. In a number of places radar-dark streaks are seen, some of which surround an impact crater. These streaks are rather smooth areas, and may be caused by the deposition of fine material or pulverization of the surface by atmospheric shocks or pressure waves produced by an incoming meteor, which itself may have broken up completely in the atmosphere, as was the case in the Tunguska event on Earth (§§5.4.3.1, 5.4.6).

5.5.4 Mars

Some detailed drawings from the late nineteenth and early twentieth century telescopic observations of the surface of Mars contained long straight linear features that some astronomers referred to as 'channels' or 'canals' (Fig. 5.57b). Around the same time, the martian polar caps were discovered and were observed to vary seasonally, reminiscent of the polar ice caps on our own planet.

The ice caps, canals, and other seasonal changes convinced some scientists that life existed on Mars. Although we now know that the surface of the red planet is not currently inhabited, the questions of whether there is life underground or has been life in the past are central to Mars exploration programs today.

5.5.4.1 *Global Appearance of Mars*

Mars has been mapped in detail by numerous orbiting spacecraft, and these reveal a striking asymmetry between the northern and southern hemispheres. The Mars Orbiter Laser Altimeter (MOLA) experiment on the Mars Global Surveyor (MGS) spacecraft reveals this asymmetry most vividly (Fig. 5.58). One half of the planet, mostly in the southern hemisphere, is heavily cratered and elevated 1–4 km above the 'nominal' surface level, the martian geoid, where zero altitude is taken as the MOLA-derived mean equatorial radius of 3396.0 ± 0.3 km. The other hemisphere is relatively smooth, and lies at or below this level.

(a)

(b)

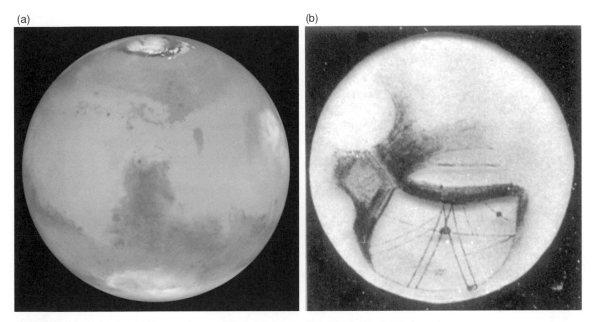

Figure 5.57 (a) Hubble Space Telescope image of Mars, taken in April/May 1999. The dark feature at the center is Syrtis Major. To the south of Syrtis, near the limb of the planet, a large circular feature, Hellas basin, is visible. On this picture it is partly filled with surface frost and water-ice clouds. Towards the planet's right limb, late afternoon clouds have formed around the volcano Elysium. Note also the ring of dunes around Mars's north pole. (Steve Lee, Jim Bell, Mike Wolff, and HST/NASA) (b) One of Percival Lowell's sketches of Mars, showing details of his 'canals'. He thought that most canals were in pairs of two, as shown here.

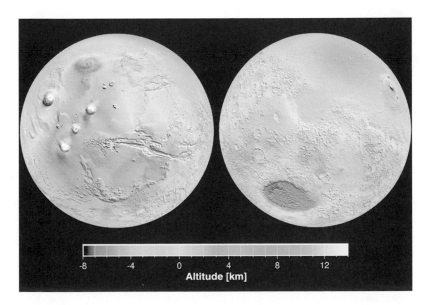

Figure 5.58 COLOR PLATE Global topographic views of Mars at different orientations, constructed from Mars Orbiter Laser Altimeter (MOLA) data. The image on the right features most strikingly the crustal dichotomy (division between the northern plains and heavily cratered southern hemisphere) and the Hellas impact basin (dark blue). The left-hand image shows the Tharsis topographic rise and Valles Marineres. (D. Smith, NASA/MGS-MOLA, PIA02820)

The geologic division between these two hemispheres is referred to as the *crustal dichotomy*, characterized by complex geology and prominent scarps. Superficially the highlands, being saturated with craters and interspersed with younger intercrater plains, resemble the Moon and smaller bodies in appearance. In detail, however, the martian terrain and its craters are quite different.

In addition to the global asymmetry, Mars's appearance is characterized by four massive shield volcanoes in the Tharsis region, including Olympus Mons, and a giant canyon system, Valles Marineris (Fig. 5.66). Volcanism and tectonics have clearly been important in the planet's history. Although Mars is small, the scale of these martian features dwarfs similar structures on Earth. Figure 5.18 shows a photograph of Olympus Mons, and a comparison in size with Mauna Loa, the largest volcano on Earth. Mars's relatively low surface gravity and cold thick lithosphere enables the existence of such high mountains, which on Earth and Venus would have collapsed due to the larger surface gravity (and tectonic plate movements on Earth). The Tharsis region is about 4000 km wide and rises 10 km above Mars's mean surface level. Three giant shield volcanoes rise another 15 km higher, while Olympus Mons, the largest volcano in our Solar System with a base of ~600 km, rises 18 km above the surrounding high plains to a total height of 27 km. The Tharsis rise is surrounded by linear rilles and fractures. Naturally, this huge volcanic area is located near the equator. If such a huge bulge had formed elsewhere, the large excess of mass on a rotating planet would cause a change in the rotation axis of Mars (true polar wander, §6.1.4.3), to bring it back into rotational equilibrium. Calculations on the rotational stability of Mars prior to the rise of Tharsis show that the position of the spin axis could have been changed substantially (tens of degrees).

Valles Marineris is a tectonically formed canyon system extending eastwards of Tharsis for 4000 km. The canyons of the Valles Marineris system are 2–7 km deep, and over 600 km wide at its broadest section. The system was probably created by the tectonic activity that accompanied the formation, or uplifting, of the Tharsis region via magma pushing the ground up from below. While the Tharsis area was rising upwards, surrounding crust was stretched, resulting in faults and fractures, and accompanying landslides. This process may explain the formation of Valles Marineris, with its long straight walls or fault scarps. At the same time it may have opened up pathways for subsurface water to enter the canyons, which could explain phenomena in the canyons that appear to result from erosion and sedimentation, as shown on many photographs taken by various spacecraft orbiting Mars (see Figs. 5.65, 5.66).

Crater density studies show the martian highlands to be ancient (~4.45 Gyr), and the smooth northern hemisphere to be much younger (3–3.5 Gyr). Based upon a variety of observations, however, it appears now that the plains originally formed around the same time as the highlands, and were covered up later, perhaps in part with material that eroded from the southern highlands, and/or via volcanic resurfacing, like the lunar maria. Obviously, the age of the crust underneath the resurfaced plains, and

the timing of the resurfacing, yield clues to Mars's geological evolution, including the formation of the crustal dichotomy, the Tharsis rise and the northern plains. Two leading theories on the formation of the dichotomy are a giant impact, or large-scale mantle convection, with upwelling in one hemisphere and subsidence in another. Since the northern plains have an elliptical shape, and not circular, the giant impact hypothesis had originally been ruled out. However, more recent calculations show that oblique (30°−60° angles) planet-scale impacts with velocities up to twice Mars's escape velocity can produce a basin that matches the observed basin's eccentricity. If true, this *Borealis* basin would be the largest impact basin in the Solar System.

Although the presence of catastrophic outflow channels (see below) that drained into the smooth plains are suggestive of a vast ocean once covering the northern lowlands, this hypothesis is controversial. Several >1000 km long paleoshorelines have been identified in the topographic data, but these do not appear to follow equal gravitational potential surfaces, as vast oceans should. However, calculations show that the shorelines would have followed equipotential surfaces if, at the time, the planet was oriented in a different way. These calculations suggest a reorientation of the planet after the formation of the Tharsis rise such that the pole moved along a great circle connecting the present poles, ~90° east of the Tharsis rise. Moving along this line would keep the Tharsis rise on the equator (§6.4.3). Some questions remain about how to explain the subtle topographic relief of the plains, if the area was covered by an ocean.

5.5.4.2 *Frost, Ice, and Glaciers*

The atmospheric pressure at the martian surface is about 6 mbar, and the temperature varies between ~130 K and 300 K (§4.2.2.1). Due to the low surface pressure, H_2O exists only in vapor or ice form (§4.4.3.2). Water vapor, with an annual mean column density between 10 and 20 precipitable microns, freezes out at night and forms a thin layer of frost on Mars's surface (see Fig. 5.69). This frost sublimes immediately after sunrise. At the poles water is permanently frozen (Fig. 5.59). In the winter, the temperature above the poles drops below the freezing point of carbon dioxide, so CO_2 condenses out to form a (seasonal) polar cap of dry ice (§4.5.1.3). The CO_2 gas either condenses directly onto the surface or in the air on condensation nuclei, such as dust grains, which then fall down to the surface. In the winter the ice sheet extends out to a latitude of ~60°. During the summer the ice sublimes, leaving the dust behind. On 25 May 2008, the Phoenix spacecraft made a successful landing at a latitude of 68°, north of

(a) (b) (c)

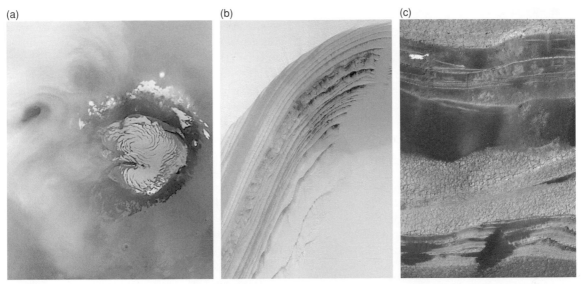

Figure 5.60 The martian north and south polar regions are covered by large areas of layered deposits that consist of a mixture of ice and dust. (a) This picture shows Mars's north polar cap in its entirety, surrounded by dunes, photographed by MGS–MOC in October 2006. The image has a resolution of about 7.5 km/pixel. Annular clouds, visible in the upper left corner, are common here in mid-northern summer. They typically dissipate later in the day. The summer caps exhibit many exposed layers and steep cliffs, some of which are shown at higher resolution in adjacent panels. (NASA/JPL/Malin Space Science Systems) (b) A springtime view of frost-covered layers, as revealed by MGS–MOC on an eroded scarp in the martian north polar cap. Some layers are known to be a source for the dark sand seen in nearby dunes. The picture covers an area about 3 km wide, and is illuminated by the Sun from the lower left. (NASA/JPL/Malin Space Science Systems) (c) This image reveals the basal layers of Mars's north polar layered deposits at Chasma Boreale. The image is taken by the High-Resolution Imaging Science Experiment (HiRISE) on NASA's Mars Reconnaissance Orbiter (MRO) in October 2006. The resolution is 64 cm/pixel, and the imaged region is 568 meters wide. (NASA/JPL/University of Arizona, PIA01925)

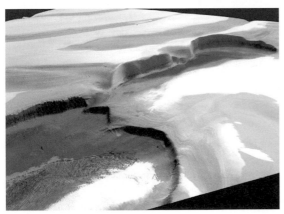

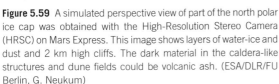

Figure 5.59 A simulated perspective view of part of the north polar ice cap was obtained with the High-Resolution Stereo Camera (HRSC) on Mars Express. This image shows layers of water-ice and dust and 2 km high cliffs. The dark material in the caldera-like structures and dune fields could be volcanic ash. (ESA/DLR/FU Berlin, G. Neukum)

the Tharsis region. Over a period of ~5 months, Phoenix observed the arctic region while the season changed from summer to autumn.

Over time the sublimation and condensation of CO_2 has produced a layered structure of dust and ice, as shown in spacecraft images (Fig. 5.60). The dry ice in the northern ice cap sublimes completely away during the summer, leaving behind a permanent cap of water-ice, ~1000 km in diameter. In the south, the CO_2 never completely sublimates away, leaving a permanent southern cap ~350 km in diameter. Mixed in with the permanent CO_2-ice is ~15% water-ice, while the steep scarps consist almost entirely of

H_2O-ice. This residual south polar cap displays a history indicative of depositional and ablational events unique to Mars's south pole. The northern cap is surrounded by large dune fields, indicative of differences in dust storms between the north and the south poles. The difference between the ice caps on the two poles has been attributed to periodic variations in the orbital eccentricity, obliquity, and season of perihelion of Mars. At the current epoch, the northern summer is hotter but shorter than the southern summer.

While the presence of ice deposits in the polar regions was never surprising, when the idea of glacial deposits at mid-latitudes was first brought up, several decades ago after analysis of Viking images, it was considered highly

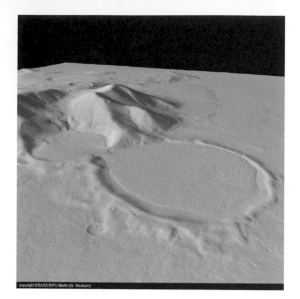

Figure 5.61 Although ice deposits are found in the polar regions, as might be expected, evidence for glaciers at low latitudes is also accumulating. This Mars Express HRSC image shows a simulated perspective view of a 3.5–4 km high massif in the Hellas region, with a viscous flow of material from one crater down to the next one, through a narrow notch. The image was taken from an altitude of 590 km, at a resolution of 29 m/pixel. (ESA/DLR/FU Berlin, G. Neukum)

controversial. Evidence for such past glaciations, however, is accumulating with images from Mars Express (obtained with the High Resolution Stereo Camera). These images provide less-ambiguous evidence for geologically recent and recurring glaciation in the tropics and at mid-latitudes. Figure 5.61 shows a simulated perspective view of a 3.5–4 km high massif with a viscous flow of material from one crater down to the next one, through a narrow notch. Mars Express images further reveal rock–glacier features at the base of the Olympus Mons scarp that are on top of older, debris-covered glaciers. The images suggest a high ice-to-debris ratio in the flows, which under the cold dry martian conditions suggests that the ice most likely is of atmospheric origin. Crater counts provide evidence of multiple eras of glaciation, the most recent of which, at Olympus Mons, occurred only a few million years ago. These periods of glaciation at such low latitudes may be caused by changes in the planet's obliquity (Fig. 2.18, §4.9.2), where excursions from the nominal value can be much larger than for Earth. In addition to evidence for recurring glaciations, the Mars Express data also revealed, via size–frequency distributions of craters, that volcanoes have been active up to only 2 million years ago.

As detailed in §5.5.4.3, there is strong evidence that, despite the low atmospheric pressure at present, water may have flowed on Mars's surface. If true, where did the water go? Did it escape into space, or is it still present as permafrost below the surface? Data from the Neutron Spectrometer on 2001 Mars Odyssey revealed that the residual ice caps at the poles extend out to latitudes of $\pm 50°$, with H_2O mass fractions of 20–100%. The Phoenix lander did indeed encounter some water-ice while digging. Additional ice reservoirs were found at two mid-latitude locations, at H_2O mass fractions of 2–10% (Fig. 5.62).

5.5.4.3 Water on Mars

Much current research is focused on the search for water, liquid and frozen, on the surface or immediate subsurface of Mars. As mentioned above, under current climate conditions liquid water cannot exist on Mars (§§4.4.3.2, 4.9.2) but, since the Mariner 9 mission in 1971, evidence is growing based upon morphological features that water once flowed on this cold planet, and some scientists suggest that this may happen even today. We summarize below the morphological evidence that Mars was once much wetter than today.

Impact craters

Ejecta blankets of many martian craters appear to have 'flowed' to their current positions (Fig. 5.63a), rather than traveled through space along ballistic trajectories. This suggests that the surface was fluidized by the impacts. Craters with fluidized ejecta blankets are referred to as *rampart craters*. Hence, in contrast to the Moon and Mercury, Mars must have a significant fraction of water-ice in its crust, or at least have had subsurface ice during the early bombardment era. In addition, martian craters are usually shallower than those seen on the Moon and Mercury, and the craters (rocks and rim) show signs of atmospheric erosion, though not as much as on Earth. Mars Global Surveyor (MGS) images revealed evidence of seepage at the edge of some crater walls (see gullies below), and of (past) 'ponding', the accumulation of water in ponds on some crater floors. The observed polygon structures on Mars (Fig. 5.64a) are typical of ice-wedge polygons on Earth that form via seasonal (or episodic) melting and freezing of water in and on the surface (Fig. 5.64b).

Channels

The oldest martian terrain contains numerous channels, similar in appearance to dendritic river systems on Earth (Fig. 5.63b), where water acts at slow rates over long

Lower-Limit of Water Mass Fraction on Mars

2%   4%   8%   16%   32%   > 64%

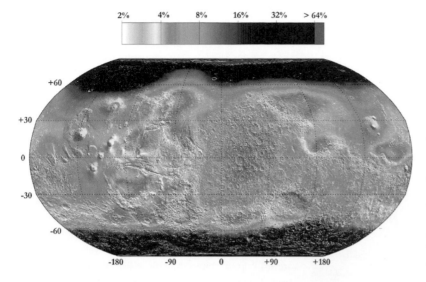

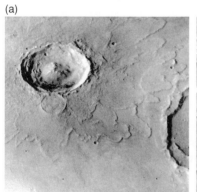 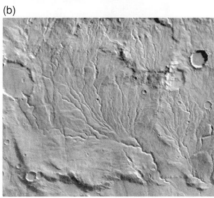 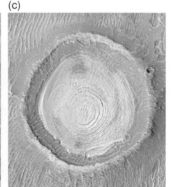

Figure 5.62 COLOR PLATE This map shows the estimated lower limit of the water content in Mars's surface. The estimates are obtained from the epithermal neutron flux, as measured with the neutron spectrometer component of the gamma-ray spectrometer on Mars Odyssey. The epithermal neutron flux is sensitive to the amount of hydrogen in the upper meter of the soil (see §5.5.1). (NASA/JPL/Los Alamos National Laboratory)

(a)   (b)   (c)

Figure 5.63 (a) The ejecta deposits around this martian impact crater Yuty (18 km in diameter) consist of many overlapping lobes. This type of ejecta morphology is characteristic of many craters at equatorial and mid-latitudes on Mars, but is unlike that seen around small craters on the Moon. (NASA/Viking Orbiter image 3A07) (b) These channels on Mars resemble dendritic drainage patterns on Earth, where water acts at slow rates over long periods of time. The channels merge together to form larger channels. Because the valley networks are confined to relatively old regions on Mars, their presence may indicate that Mars once possessed a warmer and wetter climate in its early history. The area shown is about 200 km across. (Courtesy Brian Fessler, image from the Mars Digital Image Map, NASA/Viking Orbiter) (c) Layers of sedimentary rocks in an old impact crater in the Schiaparelli Basin. The crater is 2.3 km wide. With the Sun shining from the left, one can see that the mesa top in the middle of the crater stands higher than the other stair-stepped layers. (NASA/JPL/Malin Space Science Systems)

periods of time. Individual segments are usually less than 50 km long and up to 1 km wide, while the entire dendritic system may be up to 1000 km long. Crater rims and volcanoes are sometimes eroded by such channels. In addition to these dendritic systems, there are other fluvial features on Mars, such as the immense channel systems, or *outflow channels*, starting in the highlands and draining into the low northern plains (Fig. 5.65). Some of these channels are many tens of kilometers wide, several kilometers deep and hundreds-to-thousands of kilometers long. The presence of teardrop-shaped 'islands' in the outflow channels suggests that vast flows of water have flooded the plains. While some of the martian channels may have been formed by lava flows, the morphology of most (dendritic and outflow) channels suggests that they must have been carved out by water, i.e., by a sustained fluid flow. Since the

(a)

(b)

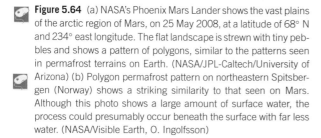

Figure 5.64 (a) NASA's Phoenix Mars Lander shows the vast plains of the arctic region of Mars, on 25 May 2008, at a latitude of 68° N and 234° east longitude. The flat landscape is strewn with tiny pebbles and shows a pattern of polygons, similar to the patterns seen in permafrost terrains on Earth. (NASA/JPL-Caltech/University of Arizona) (b) Polygon permafrost pattern on northeastern Spitsbergen (Norway) shows a striking similarity to that seen on Mars. Although this photo shows a large amount of surface water, the process could presumably occur beneath the surface with far less water. (NASA/Visible Earth, O. Ingolfsson)

martian drainage systems lack small-scale streams feeding into the larger valleys, they may have been carved by groundwater rather than by runoff of rain.

Gullies

Figure 5.67a shows perhaps the most tantalizing images of groundwater flow, morphological features which likely result from fluid seepage and surface runoff. The 'head alcove', located just below the brink of a slope (e.g., on the wall of a crater, valley, or hill) seems to be the 'source' of a depositional apron just below it. Most of these aprons show a main and some secondary channels emanating from the downslope apex of the alcove. These channels start broad and deep at their highest point, taper downslope, and are diverted around obstacles. The features are clearly distinct from landforms involving 'dry' mass movements, such as granular flows or avalanches. In analogy to terrestrial landforms, the martian gullies must involve a low-viscosity fluid like water, either pure, salty, acidic, or alkaline, the source of which may be groundwater or ice, perhaps through a capillary action as in the Antarctic Dry Valleys. Though these gullies are not the only landforms requiring liquid water to form, intriguing is the observation that all gullies must be very young: there are no impact craters superposed on these features, and some features partly obscure aeolian landforms – indicative of the (geologically) very recent past. Comparison of images taken over the decade that MGS was active shows that at least two new light-toned gully deposits formed at mid-southern latitudes (Fig. 5.67b). The morphology of these gullies suggests transport of sediments. Hence, perhaps even today subsurface liquid water reservoirs may break through the crust in short-lived outbursts, enabling downhill transport of debris. We note that these light-toned deposits are quite different from the dark slope streaks often seen on dust-covered slopes (Fig. 5.68), which are probably caused by a dry granular flow. The darkest streaks on these slopes are usually the youngest; they get covered by dust over time.

5.5.4.4 Geology at Lander Sites

In situ photographs of the martian surface taken by the Viking landers in the late 1970s (Fig. 5.69), and decades later by various Mars rovers and landers (Table F.2), show a reddish landscape where big boulders are scattered on a surface of finer grained soil, an iron-rich 'clay' (like rust; §5.3.4.3), which gives Mars its reddish color. Near the surface this clay is sometimes cemented together by evaporite materials, such as salt, to form a hard crust, called *duricrust*. The micrometer-sized regolith particles are carried around the planet by fierce winds during giant dust storms (§4.5.5.3). Hence the presence of dunes and ripples on the sandy areas is not surprising (Fig. 5.23b). The rovers also imaged *flutes*, scallop-shaped depressions, and narrow longitudinal grooves in rocks, likely caused by dust-loaded wind.

In 1997, Mars Pathfinder landed in Ares Vallis, a 'flood plain', where rover Sojourner moved around to

(a)

(b)

(c)

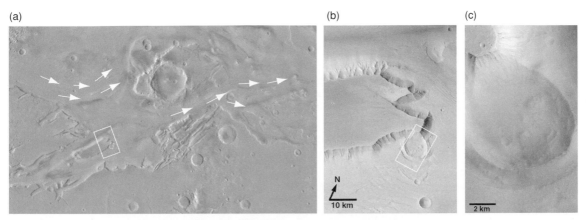

Figure 5.65 Images of a portion of the Kasei Vallis outflow channel system. (a) Large-scale view from Viking showing flow patterns (see arrows) that created 'islands'. The large white box shows the outline of the Viking 1 image shown in panel (b), while the small white box outlines the area imaged by MGS–MOC, shown in panel (c). The large crater in the upper center of this overview scene is 95 km in diameter. Panel (c) shows a 6 km diameter crater that was once buried by about 3 km of martian 'bedrock'. This crater was partly excavated by the Kasei Valles floods over a billion years ago. The crater is poking out from beneath an 'island' in the Kasei Valles. The mesa was created by a combination of the flood and subsequent retreat via small landslides of the scarp that encircles it. (USGS Viking 1 mosaic; Viking 226a08; MOC34504)

(a)

(b)

(c)

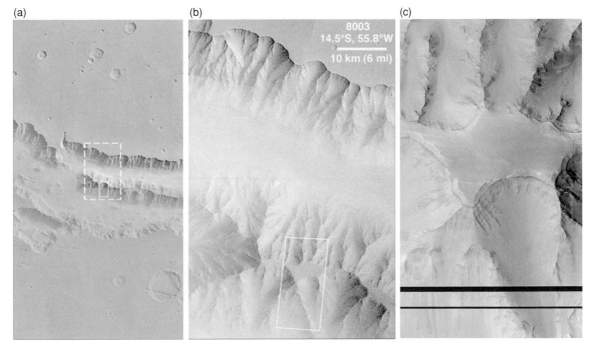

Figure 5.66 Close-up photographs of Coprates Chasma in the eastern Valles Marineris system which show many layered outcrops in various locations across the surface of the red planet. The images in (a) and (b) give context images of the high-resolution MGS–MOC image in (c). The white boxes give the approximate size and location. The highest terrain in the image is the relatively smooth plateau near the center of the right frame. Slopes descend to the north and south from this plateau in broad, debris-filled gullies with intervening, rocky spurs. Multiple rock layers, varying from a few meters to a few tens of meters thick, are visible in the steep slopes on the spurs and gullies. (NASA/MGS–MOC 8003)

(a)

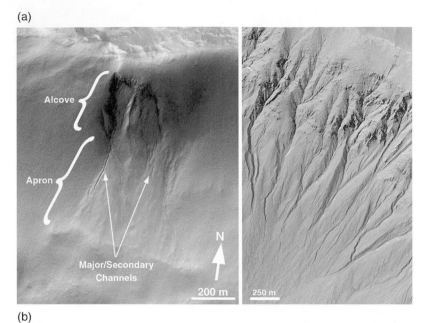

(b)

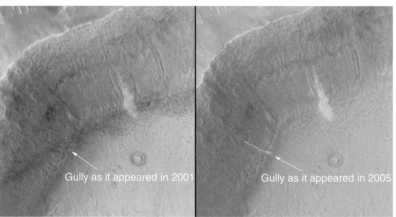

Figure 5.67 (a) Examples of land-forms that contain martian gullies. These features are characterized by a theater-shaped 'alcove' that tapers downslope, below which is an apron. The apron appears to be made of material that has been transported downslope through the channels or gullies on the apron. On the right is a larger scale view of some such channels. (M03_00537, M07_01873; Malin and Edgett 2000) (b) Comparison of two MGS–MOC images of gullies in a crater in Terra Sirenum, taken in 2001 (left) and 2005 (right). A new light-toned deposit had appeared in what was otherwise a nondescript gully. (NASA/JPL/Malin Space Science Systems, PIA09027)

analyze rocks. The area indeed resembled depositional plains such as seen on Earth after catastrophic floods, with (semi-)rounded pebbles and here and there a possible conglomerate. Many of the rocks show albedo variations, where the bottom 5–7 cm is lighter than the top, such as could result from a higher soil level in the past, swept away by catastrophic floods.

The Mars Exploration Rovers (MER) Spirit and Opportunity arrived on Mars in January 2004. Although the nominal mission was designed for 90 sols (martian days), Spirit lasted for more than 2000 sols and Opportunity is still roaming around after 4000 sols. Both rovers were equipped with a full suite of instruments, amongst them a rock abrasion tool to grind into rocks, and various spectrometers to analyze the rocks. We note the advantage for both atmospheric and geological investigations of

having rovers on the surface and simultaneously several spacecraft observing the planet from orbit using remote sensing techniques. Spirit landed in Gusev crater, a flat-floored, 160-km diameter crater, which was most likely a lake ~4 Gyr ago, connected to the northern lowlands via a channel. Surprisingly, no sedimentary rocks were found. Since these would have been deposited over 3 Gyr ago, they may all have been buried since then by more recent volcanic and/or aeolian processes. Indeed, the rocks in Gusev crater are mostly basaltic in composition, with a texture that also points at a volcanic origin. The rocks are weathered primarily by impacts and wind. The soil contains weakly bound agglomerates of dust, and is rich in olivine. This suggests that physical weathering dominates over chemical processes. Spirit set course, via some small impact craters, to the 'Columbia Hills'

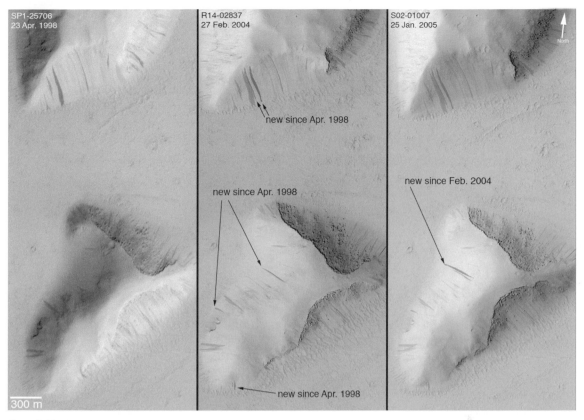

Figure 5.68 New slope streaks on Mars are frequent events. New streaks are always dark. This image shows several examples on slopes of hills in the Elysium/Cerberus region, taken from April 1998 to January 2005. (NASA/JPL/Malin Space Science Systems, PIA09030)

Figure 5.69 Photograph of the surface of Mars taken by Viking Lander 2 at its Utopia Planitia landing site on the early morning (local martian time) of 18 May 1979. It shows a fresh coating of water-ice on the rocks and soil. The ice seen in this picture is extremely thin, perhaps only tens of micrometers thick. (NASA/Viking lander 2, PIA00533)

(Fig. 5.70). On approach, the rocks and soil changed. The rocks became largely granular in appearance, and both the rocks and soil in the hills are relatively rich in salts, suggestive of significant aqueous alteration compared to the rocks near the landing site.

Opportunity landed in Eagle crater on Meridiani Planum, a landing site that was selected because the thermal emission spectrometer (TES) on MGS had discovered that 15–20% of the area contained the mineral hematite. Although hematite can form in various ways, it often involves the action of liquid water. With the rovers' prime goal of searching for evidence of liquid water, in the past or present, this appeared to be an opportune area for closer investigation. Opportunity landed near a 30–50 cm high bedrock outcrop (Fig. 5.71a), characterized by a fine (mm-sized) lamination. The bedrock is mostly sandstone, composed of materials derived from weathering of basaltic rocks, with several tens of percent (by weight) sulfate minerals, as magnesium and calcium sulfates and the iron sulfate jarosite ($(K,Na,X^{+1})Fe_3(SO_4)_2(OH)_6$), as well as hematite. Also Cl and Br were detected, at ratios that vary by over two orders of magnitude from rock to rock, indicative of variations in the evaporative processes. Scattered throughout the outcroppings, and partly embedded within, Opportunity discovered small (4–6 mm across) gray/blue-colored spherules ('blueberries'), sometimes

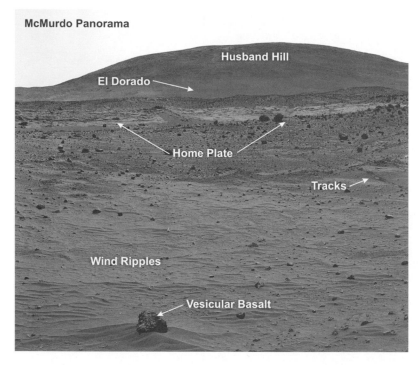

McMurdo Panorama

Figure 5.70 COLOR PLATE A view from Mars Exploration Rover Spirit, taken during its winter campaign in 2006 (the COLOR PLATE shows a larger view than the black/white version). In the distance (850 m away) is 'Husband Hill', behind a dark-toned dune field and the lighter-toned 'home plate'. In the foreground are wind-blown ripples, along with a vesicular basalt rock. (NASA/JPL-Caltech/Cornell)

multiply fused, composed of >50% hematite by mass. These are likely concretions that formed when minerals precipitated out of water-saturated rocks. In the same outcrops, small voids or *vugs* in the rocks also hint at the past presence of water; soluble materials, like sulfates, dissolved within the rocks, leaving vugs behind. While rocks partially dissolved or weathered away, the hematite concretions fell out of the bedrock, covering the plains. The sulfate-rich sedimentary rocks at Meridiani Planum, underneath a meter-thick layer of sand, probably preserve a historic record of a climate that was very different than the martian conditions we know today. Liquid water most likely covered Mars's surface, at least intermittently, where wet episodes were followed by evaporation and desiccation.

5.5.4.5 *Signs of (Past) Life?*

As discussed above and in §4.9.2, Mars must have had a very different climate early in its history. When Mars had running water on its surface, the climate may have been suitable for life to develop. The Viking landers, therefore, searched for life via a number of different experiments. In addition to simple cameras, some instruments looked for organic chemicals and metabolic activity in the atmosphere and soil, for example, through the addition of nutrients to the soil and looking for chemical byproducts resulting from living organisms (life as we know it). No signs of life were detected. The soil was completely devoid

of organic molecules. In retrospect, this should have been expected, since the martian soil is directly exposed to solar UV radiation, which breaks up any organic molecules. The experiments designed to search for metabolic activity gave some positive results, which are now attributed to unfamiliar reactive chemical states in martian minerals that were produced by solar UV radiation.

The meteorite ALH 84001, an igneous rock found in Antarctica that formed on Mars 4.5 Gyr ago (§8.2), has some chemical and morphological features which initially were attributed to possible traces of microbial life on ancient Mars. The rock was probably ejected into space by an impact about 16 Myr ago, and fell in Antarctica 13 000 yr ago, as judged from the time that cosmic ray exposure stopped. This rock contains tiny globules of carbonate minerals, which may have been deposited in the cracks by martian groundwater laden with CO_2. In and near these globules are PAHs (polycyclic aromatic hydrocarbons; mothballs are examples of these), constituents which can be formed through the transformation of dead organisms when exposed to mild heat. However, PAHs are abundant in the interstellar medium, and form easily through a reaction of CO/CO_2 with hydrogen in the presence of minerals, such as magnetite. Tiny (<0.1 μm long) perfect crystals of pure magnetite, such as are made by bacteria on Earth, have been found in the carbonate globules. Microbial action was also suggested from the coexistence of seemingly incompatible minerals as iron

Figure 5.71 (a) COLOR PLATE A panoramic view from Mars Exploration Rover Opportunity of the 'Payson' outcrop on the western edge of Erebus Crater on 26 February 2006. One can see layered rocks in the ~1 m thick crater wall. To the left of the outcrop, a flat, thin layer of spherule-rich soil lies on top the bedrock. (NASA/JPL-Caltech/USGS/Cornell, PIA02696) (b) Small (millimeter-sized) spherules, dubbed 'blueberries', are scattered throughout the rock outcrop near Rover Opportunity's landing site. The rocks show finely layered sediments, which have been accentuated by erosion. The blueberries are lining up with individual layers, showing that the spherules are concretions, which formed in formerly wet sediments. (NASA/JPL/Cornell, PIA05584) (c) COLOR PLATE This series of pointy features, a few cm high and less than 1 cm wide, stick up at the edge of flat rocks in Endurance Crater. These features probably formed when fluids migrated through fractures, depositing minerals. The minerals that filled the fractures would have formed veins that were composed of a harder material, which eroded more slowly than the rock slabs. Scattered throughout the area are blueberries. (NASA/JPL/Cornell, PIA06692)

oxide and iron sulfide, unless the globules are formed under extreme high temperatures. Recent studies show that the planes of atoms in the martian magnetites are aligned with the atomic planes of the surrounding carbonate minerals, which suggests that the magnetites likely formed in the rock, via shock heating during impact, and not inside micro-organisms.

5.5.4.6 *Phobos and Deimos*

Mars's two moons, Phobos and Deimos, were discovered in 1877 by Asaph Hall, Sr., at the US Naval Observatory. Their visual albedos, $A_v \sim 0.07$, and spectral properties are similar to those of carbonaceous asteroids. Their densities,

1.9 g cm^{-3} for Phobos and 2.1 g cm^{-3} for Deimos, are similar to that of Ceres, indicative of a mixture of rock and ice. Phobos orbits Mars at a distance of 2.76 R$_\sigma$, which is well inside the synchronous orbit, while Deimos orbits Mars outside synchronous orbit at 6.92 R$_\sigma$. Both satellites are in synchronous rotation. Images of these two moons are shown in Figure 5.72. It is not surprising that both objects, being so small (Table 1.5), are very irregular in shape.

Phobos is heavily cratered, close to saturation. The shape of the craters is very similar to that of lunar craters. The largest crater is ~10 km in diameter, almost equal to the radius of the satellite; if the impactor had been

(a)

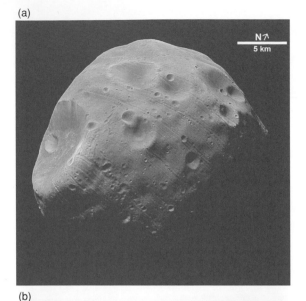

N↗
5 km

(b)

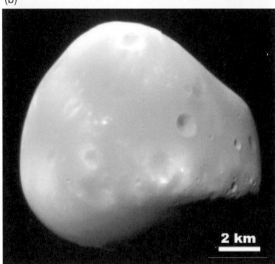

2 km

Figure 5.72 (a) Image of Phobos, the inner and larger of the two moons of Mars, taken by Mars Express in 2004. The spatial resolution is 7 m/pixel. (ESA/DLR/FU Berlin, G. Neukum) (b) Image of Deimos taken 21 February 2009 at a spatial resolution of 20 m/pixel. (HiRISE/MRO NASA/JPL/University of Arizona, PIA11826)

slightly larger, Phobos would have been shattered to pieces. Intriguing are the linear depressions or grooves, typically 10–20 m deep, 100–200 m across, and up to 20 km long. These grooves are centered on the leading apex of Phobos in its orbit, and are probably formed as (secondary) crater chains from material ejected into space from impacts on the surface of Mars. The thermal inertia derived from thermal infrared measurements by Viking suggests that Phobos is covered by very loose fine-grained

regolith, similar to lunar soil. The depth of some grooves suggests that the regolith may be over 100 m deep at places. Although initially some or all the ejecta may have been lost from this tiny moon ($v_e \approx 8 \text{ m s}^{-1}$), only micronsized dust may have escaped recapture by the moon. The orbits of such dust grains are altered by radiation forces (§2.7), and as such these grains might escape re-impacting the moon. Grains eroded from Phobos and Deimos thus should produce rings around Mars (compare §11.6). The lack of detection of any such rings imply that they would need to be quite tenuous.

Deimos's surface is rather smooth, and shows prominent albedo markings, varying from 6 to 8%. The images also show a concavity 11 km across, twice as large as the mean radius of the object. It is not clear whether this feature is a crater, or evidence that Deimos may consist of two or more objects, i.e., a rubble pile. The craters on Deimos are partly or totally filled by sediments, material which moved downhill into the lower lying craters. Many craters may have been buried this way, which may explain why Deimos appears less heavily cratered than Phobos.

5.5.5 Satellites of Jupiter

Jupiter's four largest moons range in size from Europa, which is slightly smaller than Earth's Moon, to Ganymede, the largest moon in our Solar System (Fig. 5.73). They are collectively referred to as the Galilean satellites, named after Galileo Galilei who discovered them in 1610. Ganymede is slightly larger than the planet Mercury, but less than half as massive. The satellites Io, Europa, and Ganymede are locked in a 4:2:1 orbital (Laplace) resonance. Compositionally the Galilean satellites represent a diverse grouping, ranging from rocky Io, which has a bulk composition similar to that of the terrestrial planets, to Ganymede and Callisto, which are ~50% rock and 50% water-ice by mass. Geologically, the Galilean satellites are even more diverse: Io is the most volcanically active body in the Solar System, Europa is covered by a vast ocean topped off by a layer of water-ice, Ganymede has a diverse and complex geological history plus generates its own internal magnetic field, while Callisto's surface is saturated with craters over 10 km in size, but has few small craters.

5.5.5.1 Io

The mass and density of Io are very similar to those of Earth's Moon, but the surfaces of the two bodies look vastly different. While the Moon's surface is essentially saturated with impact craters, not a single impact crater has been identified on Io, suggestive of an extremely young ($\lesssim$ a few Myr) surface. This contrast in appearance is due

Figure 5.73 COLOR PLATE Galilean satellites: Io, Europa, Ganymede and Callisto, shown (left to right) in order of increasing distance from Jupiter. All satellites have been scaled to a resolution of 10 km/pixel. Images were acquired in June 1996 (Io and Ganymede), September 1996 (Europa), November 1997 (Callisto). (NASA/Galileo orbiter PIA01299)

primarily to the dynamical environments that the Moon and Io occupy. Another important difference is that Io is far richer in moderately volatile elements, such as sodium and sulfur, than is the Moon. (Despite containing fewer volatiles, the Moon is less dense than Io because it is also depleted in iron, see §§6.3.1, 13.11.) Reflectance spectra (Fig. 5.74) show that Io's crust is dominated by sulfur-bearing species, in particular SO_2-frost. Many of the colors that give Io its spectacular visual appearance (Fig. 5.75) may be attributed to a variety of sulfur allotropes (different molecular configurations, including chains and rings, of pure sulfur, such as S_2–S_{20}) and metastable polymorphs of elemental sulfur mixed in with other species. In contrast to these volatile materials, spectra of many of the dark calderas indicate the presence of (ultra)mafic minerals, as pyroxene and olivine.

The Moon occupies a relatively benign environment, with negligible external stresses or tidal heating. The only major resurfacing process currently operating on the Moon is impact cratering. The Moon's surface is dominated by impact craters together with endogenic features (volcanic basins and mountains) created billions of years ago when the Moon was still warm. In contrast, Io is constantly flexed and heated by varying tidal forces from Jupiter (§2.6.2). The satellite is deformed into a triaxial ellipsoid (radii $\sim$1830 × 1819 × 1816 km) by Jupiter's tidal force, with the long axis pointed towards Jupiter and the shortest axis aligned with Io's spin axis. Io's permanent tidal bulge is thus $\sim$11 km. Because of Io's eccentric orbit, the bulge moves with respect to Io itself, and its amplitude varies with radial distance, being largest at perijove. These variations cause daily distortions in Io's shape that are many tens of meters in amplitude. Because Io is not perfectly elastic, this leads to dissipation of massive amounts of

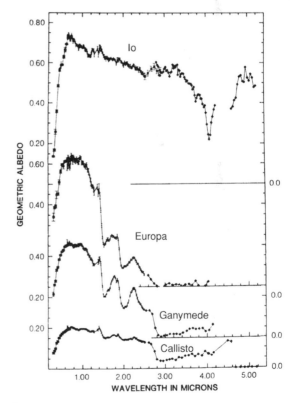

Figure 5.74 Spectra of the Galilean satellites. (Clark *et al.* 1986)

energy in its interior, so much that Io's global heat flux is $\sim$25 times larger than the terrestrial value. The tidal heat that is deposited in Io's interior is too large to be removed by conduction or solid-state convection. Melting therefore occurs, and lavas erupt through the surface via giant volcanoes. More than 400 volcanic calderas, varying in size from a few up to $\gtrsim$200 km in size, are distributed over

(a)

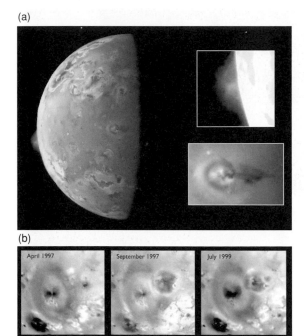

(b)

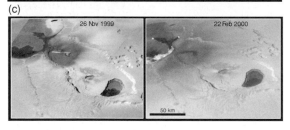

(c)

Figure 5.75 COLOR PLATE (a) A 140-km high plume is seen on the bright limb of Io (see inset at upper right) on 28 June 1997, erupting over Pillan Patera. A second plume, Prometheus, is seen near the terminator (see inset at lower right). The shadow of the 75 km high plume extends to the right of the eruption vent, near the center of the bright and dark rings. The blue color of the plumes is caused by light scattering off micron-sized dust grains, which makes the plume shadow reddish. (NASA/Galileo Orbiter, PIA00703) (b) Images of Pele taken on 4 April 1997, 19 September 1997, and 2 July 1999 show dramatic changes on Io's surface. Between April and September 1997, a new dark spot, 400 km in diameter, developed surrounding Pillan Patera, just northeast of Pele. The plume deposits to the south of the two volcanic centers also changed, perhaps due to interaction between the two large plumes. The image from 1999 shows further changes, such as the partial covering of Pillan by new red material from Pele. A new eruption took place in Reiden Patera, north-west of Pillan, that deposited a yellow ring. (NASA/Galileo, PIA02501) (c) A pair of images taken of Tvashtar Patera. The eruption site has changed locations over a period of a few months in 1999 and early 2000. The red and yellow lava flow in the left image is an illustration based upon (saturated) imaging data. The image on the right is a composite based on 5 colors. (NASA/Galileo, PIA02584)

Io's surface. The lava flows from the calderas can be hundreds of km long, which implies that the lavas have low viscosities, like basaltic lavas on Earth. Io further displays a variety of geological features (Fig. 5.76), all of which are probably connected to the satellite's strong volcanic activity. However, mountains on Io do not resemble volcanoes seen elsewhere in the Solar System. Instead, they are rugged kilometers-high ridges (the highest is 17 km high), which appear to have formed by the uplift of large blocks of crust (Fig. 5.76), a process that is probably enabled by Io's extremely active volcanism. Other geological features clearly volcanic in origin are the plateaus formed by layers of materials, and many irregular depressions or paterae (volcanic calderas), which can be several km deep. There are also numerous dark lava flows and bright deposits of SO_2 frost and/or other sulfurous materials that have no discernible topographic relief (Fig. 5.77).

Ground-based as well as spacecraft observations of Io at (near-)infrared wavelengths reveal a body covered by numerous *hot spots*. The Near Infrared Mapping Spectrometer (NIMS) on Galileo imaged more than 150 hot spots in the vicinity of volcanoes (Fig. 5.77b), while ground-based speckle and adaptive optics images at infrared wavelengths show over two dozen hot spots on Io when the satellite is in eclipse (in Jupiter's shadow) (Fig. 5.78b). Spacecraft images at visible and UV wavelengths reveal in addition also volcanic plumes and auroral glows when Io is in eclipse (Fig. 5.79b). These glows are atomic (likely neutral oxygen and sodium) and molecular (probably SO_2) emissions in Io's atmosphere and plumes, caused by electron excitation in Jupiter's magnetosphere (§7.5.4). Hot spots are usually associated with low-albedo regions at visible wavelengths, and seem to appear and disappear at random times. Only some of the hot spots are associated with volcanic plumes: the Voyager and Galileo spacecraft combined observed a total of 17 volcanic plumes, four of which may be permanent features, having been seen by several spacecraft many years apart. Plumes are usually dominated by SO_2 gas and often also dust. Small amounts of SO, S_2, S, and NaCl have been detected as well. The tiny, ~10 nm sized, grains from Io's largest plumes create the dust streams that have been detected in interplanetary space far from Jupiter (§9.4.7).

Blackbody fits to near-infrared (1–5 μm) spectra show that individual hot spots are composed of areas at different temperatures, with a small (up to a few km² at most) hot 'core', and cooler regions which can be hundreds of kilometers in extent. The highest temperature detected is indicative of the melting temperature of the magma. Terrestrial basaltic flows typically have temperatures of

Figure 5.76 Global view (upper left) of Tohil Mons on Io, as photographed by the Galileo spacecraft in October 2001. This mountain rises 5.4 kilometers above Io's surface. Two volcanic calderas lie to the northeast of Tohil's peak. High-resolution views of the mountain and calderas, illuminated from the right, are shown in the mosaic, printed in the same orientation (diagonal) as on the low-resolution picture. The total area covered is 280 km from upper right to lower left, at a resolution of 50 m/pixel. The bottom left image shows evidence of numerous landslides from the mountain. However, despite the closeness of the small, dark-floored patera to the mountain walls, no landslide debris is shown on the floor of the patera. Perhaps the caldera floor has recently been resurfaced with lava, or it is, in fact, a lava lake in which debris would sink through. The image at the bottom right shows Mongibello Mons, the jagged ridge at the left of the image. This ridge rises 7 km above the plains of Io. These angular mountains are thought to be relatively young, while older mountains have a more subdued topography, such as the rise near the top center of this image. A 250 m high scarp is visible in between these mountain ranges. The image covers an area of 265 km east–west, and has a resolution of 335 m/pixel. (NASA/Galileo, PIA03527, PIA03886)

1200–1500 K. Contrary to early (Voyager) reports that low temperature ($\lesssim$650 K) sulfur volcanism may be widespread on Io, Galileo and high resolution ground-based images revealed that typical hot spot temperatures were too high ($\gtrsim$900 K) for sulfur volcanism, and instead implied that volcanism on Io is similar to silicate volcanism on Earth. Several low-temperature volcanic areas may be dominated by secondary sulfur volcanism, where sulfur deposits are melted and mobilized by injection of hot silicates from below. Interestingly, some observations of hot spots on Io suggest temperatures exceeding 1700 K, indicative of volcanism driven by ultramafic

(a)

(b)

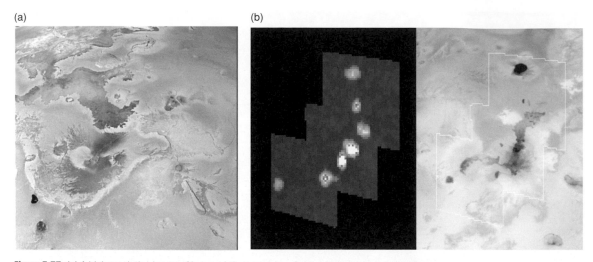

Figure 5.77 (a) A high-resolution image of Io reveals immense lava flows and other volcanic landforms. Several high-temperature volcanic hot spots have been detected in this region, suggesting active silicate volcanism in lava flows or lava lakes. The large dark lava flow in the upper left region of the image is more than 400 km long, similar to ancient flood basalts on Earth and mare lavas on the Moon. The image was taken 6 November 1996, covers an area 1230 km wide, and the smallest features that can be discerned are 2.5 km in size. (NASA/Galileo Orbiter, PIA00537) (b) The 300 km long Amirani lava flow on Io (image on the right) consists of many individual flows; the youngest flows are hottest, and are revealed by the bright spots in the 5 μm infrared image on the left. In this 5 μm image, Amirani includes the two brightest spots and two others closest to that pair. The image also shows three other active volcanoes on Io, where the one on the lower left and that at the top both correlate with dark, roughly circular areas in the image on the right. (NASA/Galileo, PIA03533)

magmas (e.g., komatiites), a style of volcanism that has not occurred for billions of years on Earth. In addition to the temperature, one can estimate the age and the eruption style of new hot spots by comparing observed spectra with a time sequence of synthetic spectra for a simulated volcanic eruption. Hot spots thus yield important clues to the satellite's heating and cooling mechanisms, the driving force behind the plumes, and the composition of Io's interior, its surface, and subsurface layers.

One can distinguish three types of ionian eruptions: flow-dominated or Promethean, explosion-dominated or Pillanean, and intra-Patera or Lokian volcanism.

(*i*) The Prometheus and Amirani volcanic centers are typical examples of flow-dominated eruptions. These eruptions are characterized by a long-lived flow field, fed by insulated lava tubes or sheets, where several km^2 day^{-1} of old lava may be covered by fresh flows (Fig. 5.77). Global resurfacing rates based upon persistent hot spots are estimated to be ~0.1–1 cm yr^{-1}. They often are associated with small explosive plumes, <200 km high and a few hundred km across at their base. They may last for years at a time, and have been seen to migrate laterally over the surface. The Prometheus plume moved ~100 km between the Voyager encounters in 1979 and

the first Galileo images in 1996. These plumes probably result from a volatilization of SO$_2$-ice by the slowly advancing hot lavas. The plume morphology suggests that material is ejected on ballistic trajectories, at velocities of ~0.5 km s^{-1}, a factor of ~5 higher than typical vent velocities on Earth.

(*ii*) Explosion-dominated eruptions are characterized by episodes of highly energetic, high-temperature ($\gtrsim$1400 K) eruptions, that usually last for days–weeks, although activity can continue at a lower level of intensity for months. Examples include, e.g., Pillan, Tvashtar, Surt, and Pele (Fig. 5.75b,c, and Fig. 5.78). These eruptions usually show plumes >200 km high and >1000 km wide at their base. They leave extensive pyroclastic deposits. Most notable for such plumes is a ring of red deposits, several hundred kilometers in diameter, which may be composed of red sulfur allotropes, produced via polymerization of S$_2$, a gas that has been detected in Pele's plume. These high-temperature events may be driven by liquid sulfur, which could be heated to over 1000 K by hot, possibly molten silicates at depths of a few kilometers. The high temperature brings about a phase change in the sulfur (liquid → gas), driving the volcano. Detailed images of the Tvashtar plume were obtained in 2007 by the New Horizons spacecraft (Fig. 5.79). The plume morphology is not consistent

(a)

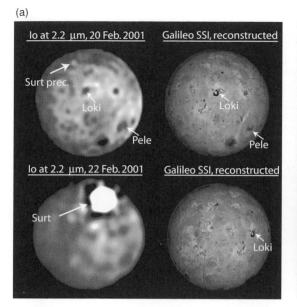

(b)

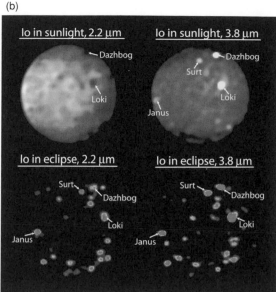

Figure 5.78 Various images of Io at infrared wavelengths taken with the adaptive optics system on the Keck telescope. All images have been deconvolved ('sharpened') using various deconvolution algorithms. Typical spatial resolutions are ∼140 km. (a) Images of Io at 2.2 μm taken on UT 20 February 2001 (top), and two days later (bottom), compared with visible-light images based on Galileo–Voyager data at the same viewing aspect (created via the IMMCE website). Several volcanoes are indicated. A giant eruption, the most energetic ever witnessed on Io, took place at the volcano Surt between 20 and 22 February. The energy released from this volcano is 6500 times that released by Mt. Etna in 1992. (Adapted from Marchis *et al.* 2002) (b) COLOR PLATE Images of Io at 2.2 μm (left) and 3.8 μm (right), taken on UT 18 December 2001. (top): Io in sunlight. At both wavelengths, Io's emission is dominated by sunlight reflected off the satellite. Because the Sun's intensity is lower at 3.8 μm than at 2.2 μm, and 3.8 μm is closer to the peak of a typical hot spot's blackbody curve, hot spots are easier to recognize at a wavelength of 3.8 μm than at 2.2 μm. Note that some volcanoes (Loki, Dazhbog) show up as hot spots at 3.8 μm, but as low-albedo features at 2.2 μm. (bottom): Io in eclipse. Images of Io taken 2 hours later, after the satellite had entered Jupiter's shadow. Without sunlight reflecting off the satellite, even faint hot spots can be discerned. The difference in brightness between the two wavelengths gives an indication of the temperature of the spot. Both Loki and Dazhbog, very bright at 3.8 μm, are low-temperature (∼500 K) hot spots. Surt and Janus, on the other hand, are also very bright at 2.2 μm, indicative of higher temperatures (∼800 K). (Adapted from de Pater *et al.* 2004a)

with ballistic trajectories; it can be simulated well using hydrodynamic models, including a gas shock front at the top of the plume.

(*iii*) A few volcanic centers are best characterized as intra-Patera eruptions, such as lava lakes, where eruptions are triggered when (part of) the lake is overturning. Pele and Loki are the best examples of such lakes. Loki, a prominent feature on visible-light images of Io (Figs. 5.78, 5.80), is thermally the most persistent energetic hot spot, with modeled eruption rates of ∼10^4 m^3 s^{-1}. Variations in Loki's infrared brightness have been modeled by a basaltic lava lake whose crust overturns when it becomes buoyantly unstable, a process that starts at the southwest end of the lake and propagates around the horseshoe-shaped

caldera. Measured temperatures at Loki rarely exceed 1000 K. In contrast, temperatures of over 1700 K have been reported for Pele, a much smaller lava lake with a lower eruption rate (∼300 m^3 s^{-1}), but a huge, faint, plume that deposits the bright red material mentioned under (*ii*). Over time this veneer of red material fades, and hence such rings are indicative of recent (weeks–months) plume activity.

5.5.5.2 *Europa*

Europa is slightly smaller and less dense than the Moon. Its surface is very bright and has the spectral properties of nearly pure water-ice. The combination of spectral and density information suggests that Europa is a mostly

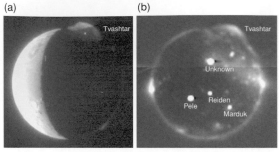

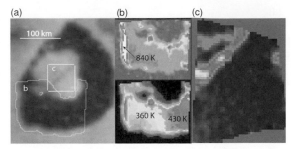

Figure 5.79 (a) An image of the 2006–2007 Tvashtar eruption, captured by the Long Range Reconnaissance Imager (LORRI) on the New Horizons spacecraft, 28 February 2007. Io's day side is overexposed to bring out faint details in the plumes and on the moon's night side. On the night side, at the 'center' of the eruption, the glow of the hot lava is visible as a bright point of light. Another plume, likely from the volcano Masubi, is illuminated by Jupiter just above the lower right edge, and a third very faint plume can be seen at the 2 o'clock position. A high (4.5 km) plateau is visible just beyond the terminator (the bright vertical line), as well as a mountain south of it. (b) A LORRI New Horizons image of Io in eclipse, showing only glowing hot lava (the brightest points of light), as well as auroral displays in Io's tenuous atmosphere and the moon's volcanic plumes. The edge of Io's disk is outlined by the auroral glow produced as intense radiation from Jupiter's magnetosphere bombards the (patchy) atmosphere. Tvashtar's plume is visible on the limb, and several smaller ones show up as diffuse glows scattered across the disk. Bright glows at the edge of Io on the left and right sides of the disk mark regions where electrical currents connect Io to Jupiter's magnetosphere. Both images are composites of images taken at wavelengths between 350 and 850 nm. (NASA/APL/SWRI, PIA09250, PIA09354)

rocky body with an H_2O 'crust' ∼100–150 km thick. Part of the lower portion of the H_2O layer must be liquid, as suggested by details in surface topology and, most convincingly, from measurements of Jupiter's magnetic field near Europa (§7.5.4.9). This liquid ocean is maintained by tidal heating, and decouples the ice shell from Europa's interior. As such, the shell may rotate slightly faster than synchronously. Observations show that the timespan over which the shell may perform one extra rotation may be of order 50 000 years.

How thick is Europa's ice shell? The characteristics of impact craters suggest ∼20 km, a thickness that is consistent with estimates based upon convective upwelling of warm buoyant ice in *diapirs*, forming 'lenticulae' (Fig. 5.81, and below). In contrast, in some places, as in 'chaos' terrain (Fig. 5.82, and below), it looks like liquid ocean water rose up to the surface, i.e., melted through the ice shell, in which case a thicknesss ≲6 km is implied. Depending on the thickness of the ocean and ice shell, temporal variations of the tidal deformation may reach amplitudes of 1–30 m, and hence if this deformation can

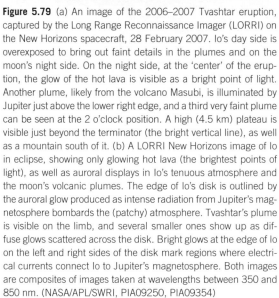

Figure 5.80 Detailed views of the 200 km diameter volcanic caldera and lava lake Loki at different wavelengths. (a) Image taken in visible light (with Galileo/SSI), outlining the areas of the infrared NIMS images on the right. (b) Two NIMS temperature maps of the southern portion of Loki taken of Io's night side on 16 October 2001. The lower right image is at a wavelength of 4.4 μm, and the upper right at 2.5 μm. The spatial resolution is 2 km/pixel. (c) This thermal 4.7 μm map (taken of Io's night side) shows that heat is also emitted from the dark 'channel' on the island in the SSI image. (NASA/JPL Galileo, PIA02595 and PIA02514)

be measured, it would give a direct measurement of the thickness of the ice shell.

Europa's surface is relatively flat compared to the surfaces of, e.g., Io and the Moon. The Galileo spacecraft found only a few impact craters, suggestive of a very young surface (tens to at most a few hundred million years), which may still be undergoing active resurfacing. Central peaks in these craters rise up to no more than 500 m, with the youngest features typically showing the highest relief. Most of the geologic features that have been seen on the satellite's surface result from the diurnal tidal stresses. The oldest terrain is characterized by *ridged plains*, which are often crisscrossed by younger *bands*. At higher resolutions, the ridges (Figs. 5.81, 5.82) often appear as two parallel ridges separated by a V-shaped trough. Sometimes they appear as triple bands. They can be thousands of km long and typically are 0.5–2 km wide. The formation of these features is not understood. They may have formed by an expansion of the crust, or when two ice plates pulled slightly apart. Warmer, slushy or liquid material may have been pushed up through the crack, forming a ridge. The brownish color suggests the slush consists in part of rocky material, hydrated minerals or clays, or salts. Alternatively, a compression of two plates pushing against each other may have formed a ridge in a similar way, by pushing up the edges of the plates. There is some evidence of (past) geyser-like or volcanic activity along these ridges, causing one or more lines of fresh ice on the ridges.

Although most ridges are linear in shape, some of them appear curved, like cycloids (Fig. 5.81a). The cycloidal

(a)

(b)

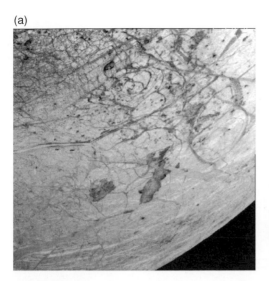

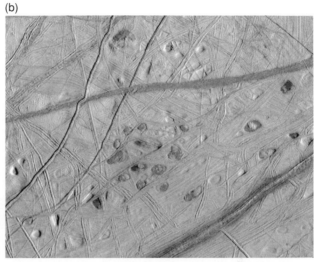

Figure 5.81 (a) Europa's southern hemisphere: The upper left portion of the image shows the southern extent of the 'wedges' region, an area that has undergone extensive disruption. Thera and Thrace Macula are the dark irregular features southeast of the ~1000 km long Agenor Linea. The image covers an area approximately 675 by 675 km, and the finest details that can be discerned are about 3.3 km across. (NASA/Galileo Orbiter, PIA00875) (b) Reddish spots and shallow pits pepper the surface of Europa. The spots and pits on this image are about 10 km across. (PIA03878; NASA/JPL/University of Arizona/University of Colorado)

shape results from the propagation of a crack in the surface due to diurnal stresses, which produces a curved rather than straight feature. Dark and gray bands that crisscross the surface (Fig. 5.83) have been compared with the mid-ocean ridge on Earth, as the bands appear to have been pulled apart while darker material from below the surface has filled the voids. In contrast to the tectonic plate movement on Earth, no subduction zones have been seen on Europa. One of the youngest features on the satellite's surface is the chaotic terrain, displayed in Figure 5.82. The morphology of these features resembles km-scale blocks or sheets of ice 'floating' on softer/slushy ice below. Some of these broken-up plates have been rotated, tilted, and/or moved. They can be reassembled like a jig-saw puzzle. In these areas, ocean water may have reached the surface and destroyed part of the original icy crust.

Other intriguing extremely young features are *lenticulae*, the Latin term for freckles (Fig. 5.81b). Their morphology suggests that they originate from convective upwelling of warm buoyant ice in diapirs, somewhat analogous to lava lamps. This could lead to the observed 'domes' if the diapirs reach the surface, or depressions if the diapir does not break through to the surface, but instead weakens/melts the ice above it, which subsequently sags down. Obviously, the presence of a liquid ocean on Europa has led to a surface with an intriguing morphology. When

adding the possibility of (cryo)volcanic activity, one may not be surprised that Europa is a prime target for speculations on the possible existence of a variety of life-forms, in analogy with early microbial life-forms on Earth that may have thrived on hot vents in the deep ocean.

5.5.5.3 *Ganymede and Callisto*

The outer two Galilean satellites represent a fundamentally different type of body from the inner two. Their low densities imply they contain substantial amounts of water-ice. We note, however, that the ice near the surfaces of these bodies is at such low temperatures that it behaves more like rock than the ice we are familiar with on Earth. The large sizes of Callisto and Ganymede imply high enough pressures that the ice must be significantly compressed, thus a larger ice fraction is suspected than would be for smaller bodies of the same density. Ice is also observed in the spectra of both bodies (Fig. 5.74), but it is far more contaminated than the ice on Europa's surface, which is regularly 'refreshed' from below. The difference in density between Ganymede (1.94 g cm^{-3}) and Callisto (1.85 g cm^{-3}) is greater than can be accounted for by compression alone, so Callisto probably contains a slightly higher fraction of water-ice. The surfaces of the two bodies are vastly different (Fig. 5.73), which might be related to differences between their interiors. Ganymede

(a)   (b)   (c)

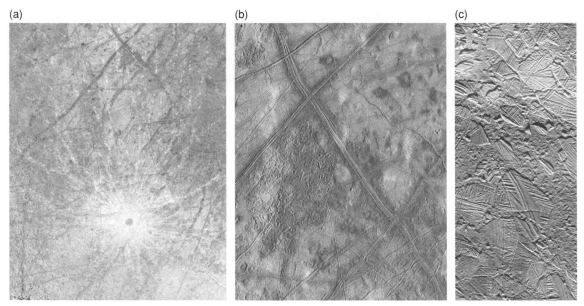

Figure 5.82 (a) The 26 km diameter impact crater Pwyll, just below the center of the image, is likely one of the youngest major features on the surface of Europa. The central dark spot is ~40 km in diameter, and bright white rays extend >1000 km in all directions from the impact site. One can also discern several dark lineaments, called 'triple bands' because they have a bright central stripe surrounded by darker material. The order in which these bands cross each other can be used to determine their relative ages. The image is 1240 km across. (NASA/Galileo Orbiter, PIA01211) (b) This image of Europa shows surface features such as domes and ridges, as well as a region of disrupted terrain including crustal plates which are thought to have broken apart and 'rafted' into new positions. The image covers an area of Europa's surface ~250 × 200 km (the X-shaped ridges are seen north of Pwyll crater, panel a). (NASA/Galileo Orbiter, PIA01296) (c) COLOR PLATE View of a small region of the thin, disrupted, ice crust in the Conamara region of Europa, the disrupted terrain displayed in panel b. This image shows the interplay of surface (enhanced) color with ice structures. The white and blue colors outline areas that have been blanketed by a fine dust of ice particles ejected at the time of formation of the large crater Pwyll. A few small craters, ≲500 m in diameter, are visible. These craters were probably formed by large, intact, blocks of ice thrown up in the impact that formed Pwyll. The unblanketed surface has a reddish brown color that has been painted by mineral contaminants carried and spread by water vapor released from below the crust when it was disrupted. The original color of the icy surface was probably the deep blue seen in large areas elsewhere on Europa. The image covers an area of 70 × 30 km; north is to the right. (NASA/Galileo Orbiter, PIA01127)

is clearly differentiated, whereas Callisto may be more homogeneous (§6.3.5.1).

Impact craters are ubiquitous on both satellites. They are usually flatter than on the Moon, and some show unique features, which have been attributed to the relatively low (compared to rock) viscosity of the icy crust. Craters ≲2–3 km in diameter reveal the classic bowl-shaped morphology (§5.4.1), while at larger sizes central peaks appear. Craters over ~35 km in diameter, however, possess central pits rather than peaks, and at even larger sizes (≳60 km) the younger craters exhibit round domes in their center. Surface fractures suggest an origin like lava domes – a rapid extrusion of warm viscous ice in the center of the craters just after their formation. In some cases, all that can be distinguished is a large bright circular patch, some with and others without concentric rings around them. These features, called *palimpsests*, are similar to large, several hundred km across, impact basins, but completely lack any topographic relief, likely caused by relaxation due to flowing subsurface ice. Both satellites also show many bright impact craters. These may be relatively young, with their high albedo resulting from fresh ice that was ejected from the impact site. In contrast, some craters on Ganymede show an unusually dark floor, and others show dark ejecta. The dark component on the floor of a crater may be residual material from the impactor that formed the crater, or the impactor may have punched through the bright surface to reveal a dark layer beneath.

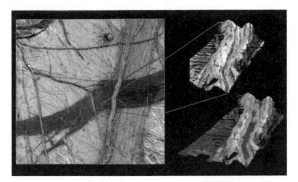

Figure 5.83 An east–west (north is to the right) oriented double ridge on Europa, with a deep intervening trough that cuts across older background plains and the dark wedge-shaped band. At the numerous cracks and bands the crust may have been pulled apart. Sometimes dark material from below the surface may have welled up and filled the cracks. A computer generated three dimensional perspective (upper right) shows that bright material, probably pure water-ice, prevails at the ridge crests and slopes, while most dark material (perhaps ice mixed with silicates or hydrated salts) is confined to lower areas such as valley floors. The ridges reach elevations of over 300 m above the surrounding plains. The two ridges are separated by a valley about 1.5 km wide. The image on the left has a spatial resolution of ∼26 m/pixel. (NASA/Galileo Orbiter, PIA01664)

Callisto's surface shows signs of weakness/crumbling at small scales, which may be produced by sublimation of a volatile component of the crust. This degradation appears to bury and/or destroy craters, and may explain the apparent dearth of (sub)kilometer-sized craters on Callisto's surface (Fig. 5.84).

The geology of Ganymede is very complex. At low resolution Ganymede resembles the Moon, in that both dark and light areas are visible (Fig. 5.73). However, in contrast to the Moon, the dark areas on Ganymede's surface are the oldest regions, being heavily cratered, nearly to saturation. The lighter terrain is less cratered, though more than the lunar maria; so it must be younger than the dark terrain, though probably still quite old. The light-colored terrain is characterized by a complex system of parallel ridges and grooves (Fig. 5.85a), up to tens of kilometers wide and maybe a few hundred meters high. These features are clearly of endogenic origin and may be of a similar origin (tensional) as the grabens on Earth. It is thought that this form of tectonics is typical for icy satellites, since similar patterns have been seen on a few other icy bodies (e.g., Europa, Enceladus, and Miranda). One can generally see younger grooves or faults overlaying older sets of grooves, while smooth bands on the surface may be evidence of cryovolcanism, i.e., volcanic activity involving ices rather than silicates.

5.5.5.4 *Jupiter's Small Moons*

Jupiter's other moons are all very small compared to the Galilean satellites. Their combined mass is about 1/1000 that of Europa, the smallest of the Galilean satellites.

Four moons have been detected inside the orbit of Io. Amalthea, the largest of these moons (Fig. 5.86), with a mean radius of 83.5 km, was discovered in 1892 by Edward Barnard. It is distinctly nonspherical in shape (Table 1.5), dark and red, and heavily cratered. It possesses two large craters, 90 and 75 km in diameter, which are probably 8–15 km deep, in addition to two mountains. Its low density, 0.86 ± 0.1 g cm^{-3}, of a presumably rocky object, is suggestive of a 'rubble pile' composition. Since impacts on such a porous body are quickly damped (§9.4.2), it is a little easier to understand how Amalthea can have several craters almost half its size. The low density itself hints at a violent collisional history. The other inner satellites of Jupiter, Thebe, Metis, and Adrastea, are also dark and red. Thebe, located exterior to Amalthea, is smaller (mean radius of 49.3 km) and much rounder than Amalthea. Metis and Adrastea are substantially smaller still, and are located near the outer edge of Jupiter's main ring. All four of these moons are obviously associated with, and likely form Jupiter's dusty rings (§§11.3.1, 11.6.3).

The orbital properties of these small moons suggest that they condensed from a disk of dust and gas surrounding proto-Jupiter, rather than being captured at a later stage. Judging from the close proximity of Metis and Adrastea, and the low density of Amalthea, and their intimate connectivity with the jovian ring, we expect that the present moons may be fragments from an initially larger body, an hypothesis strengthened by the observation that ∼15% of the jovian ring optical depth is in parent bodies $\gtrsim 5$ cm in radius (§11.3.1). The nonzero inclinations of the orbits of Amalthea and Thebe have been explained via past resonance interactions with Io. In this scenario, Io originally formed between ∼4 and 5 R, and subsequently migrated outwards to its present position due to tidal interactions with Jupiter. During this process, its resonance locations moved outwards as well, and calculations show that when Io's 3:1 resonance location passed over Amalthea, it would have led to an increase in Amalthea's inclination, to a value consistent with observations.

Jupiter's outer moons are much further away from Jupiter than the Galilean satellites. They are in highly eccentric, inclined, and often retrograde orbits. Collectively they are referred to as the 'irregular' satellites. As of early 2009, 54 of such satellites had been discovered. The orbital elements of these objects are not

(a)

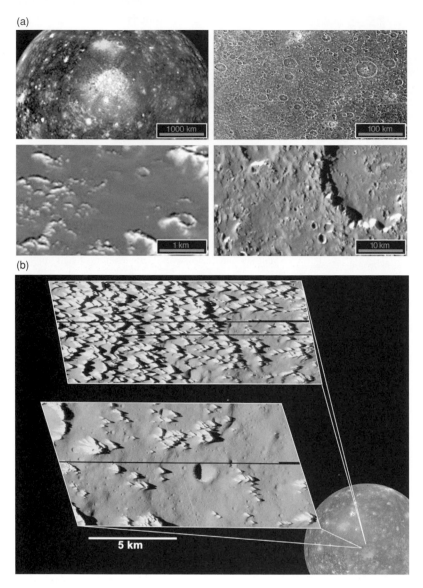

(b)

Figure 5.84 (a) Four views of Callisto, at increasing resolution. In the hemispheric view (top left; 4400 × 2500 km) the surface shows many small bright spots, with the Valhalla basin seen at the center. The regional view (top right; 10× higher resolution) reveals the spots to be impact craters. The local view (bottom right; again 10× higher resolution) not only brings out smaller craters and detailed structure of larger craters, but also shows a smooth dark layer of material that appears to cover much of the surface. The close-up frame (bottom left) presents a surprising smoothness in this highest resolution (30 m per pixel; area covered 4.4 × 2.5 km) view of Callisto's surface. (NASA/Galileo Orbiter, PIA01297) (b) A region just south of the multiring impact crater Asgard on Callisto reveals numerous bright, sharp knobs, approximately 80–100 m high. They may consist of material thrown outward from a major impact billions of years ago. These knobs, or spires, are very icy but also contain some darker dust. As the ice erodes, the dark material appears to slide down and accumulate in low-lying areas. The lower image shows somewhat older terrain, judging from the number of impact craters. This image suggests that the spires erode away over time. (NASA/JPL/Arizona State University, PIA03455)

randomly distributed, but reveal the presence of dynamical groupings. Five such 'families' have been identified. Each of these families, most likely, resulted from the breakup of a body (likely an asteroid, judging from spectra) after capture by Jupiter. The giant planet is also known to have captured Jupiter-family comets in the past. Some such bodies are known to have orbited Jupiter for decades before being ejected from the system, or colliding with the planet (e.g., Comet D/Shoemaker–Levy 9; §5.4.5) or with a satellite (Fig. 5.85c).

5.5.6 Satellites of Saturn

Saturn has a total of 62 satellites (counted as of November 2009), many of which are discussed in the following

subsections. Titan is by far the largest satellite, with a radius of 2575 km. In addition to Titan, we also devote an entire subsection to Enceladus, which surely is a most enigmatic moon in the saturnian system.

5.5.6.1 *Titan*

Titan, Saturn's largest satellite, was discovered in 1655 by Christiaan Huygens. Titan is similar in size to Ganymede, Callisto, and Mercury. With a mean density of $1.88 \, \mathrm{g \, cm^{-3}}$, Titan belongs to the 'icy' satellites. The moon is surrounded by a dense atmosphere (1.44 bar surface pressure) composed primarily of nitrogen and a small but significant amount of methane. The atmosphere contains a dense smog layer of photochemical origin (§4.6.1.4),

(a)

(b)

(c)

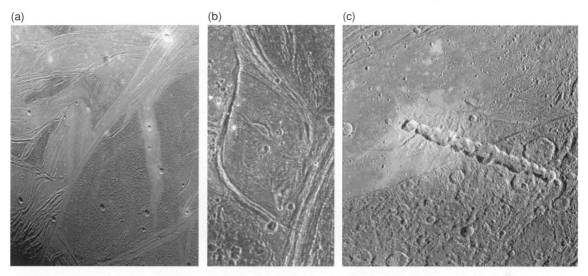

Figure 5.85 (a) View of the Marius Regio and Nippur Sulcus area on Ganymede showing the dark and bright grooved terrain which is typical on this satellite. The older, more heavily cratered dark terrain is rutted with furrows, shallow troughs perhaps formed as a result of ancient giant impacts. Bright grooved terrain is younger and was formed through tectonics, probably combined with icy volcanism. The image covers an area ~664 × 518 km at a resolution of 940 m/pixel. (NASA/Galileo Orbiter, PIA01618) (b) The 80-km wide lens-shaped feature in the center of the image is located along the border of Marius Regio, a region of ancient dark terrain near Nippur Sulcus, displayed in panel (a). The tectonics that created the structures in the bright terrain nearby has strongly affected the local dark terrain to form unusual structures, such as the one shown here. The lens-like appearance of this feature is probably due to shearing of the surface, where areas have slid past each other and also rotated slightly. The image covers ~63 × 120 km, at 188 m/pixel. (NASA/Galileo Orbiter, PIA01091) (c) This chain of 13 craters on Ganymede was probably formed by a comet that was pulled into pieces by Jupiter's tidal force as it passed too close to the planet. Soon after this breakup, the 13 fragments crashed onto Ganymede in rapid succession. The craters formed across the sharp boundary between areas of bright terrain and dark terrain. It is difficult to discern any ejecta deposit on the dark terrain. This may be because the impacts excavated and mixed dark material into the ejecta and the resulting mix is not apparent against the dark background. The image covers 214 × 217 km, at 545 m/pixel. (NASA/Galileo orbiter, PIA01610)

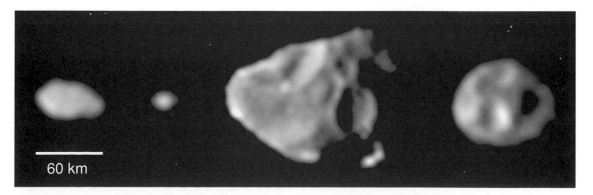

60 km

Figure 5.86 The four small, irregularly shaped 'ring'-moons that have orbits within Jupiter's ring system. The moons are shown in their correct relative sizes. From left to right, arranged in order of increasing distance from Jupiter (with north up), are Metis, Adrastea, Amalthea, and Thebe. (NASA/Galileo Orbiter, PIA01076)

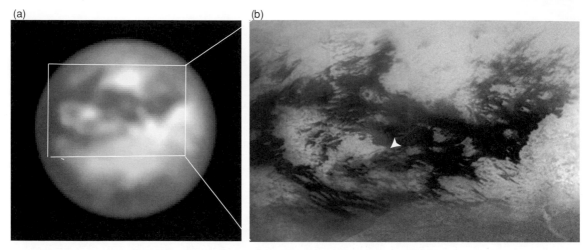

Figure 5.87 (a) An image of Titan's surface at a wavelength of 2.06 μm, obtained with the adaptive optics system on the W.M. Keck telescope one day after the descent of the Huygens probe on 14 January 2005. (Adapted from de Pater *et al.* 2006c) (b) Cassini map at a wavelength of 938 nm of a region on Titan's surface as outlined (approximately) in panel (a). Approximate longitude and latitude coverage is between 120° and 240° west longitude and from −45° to +30° in latitude. The Huygens probe landing site is indicated by an arrow. (NASA/JPL Cassini orbiter, PIA08399)

which makes it impossible to remotely probe the satellite's surface at visible wavelengths. The smog is transparent, however, at longer wavelengths, so that the surface can be imaged at infrared wavelengths outside of the methane absorption bands. Such images (Fig. 5.87) display significant surface albedo variations.

When Cassini arrived at Saturn and the Huygens probe made its descent through Titan's atmosphere, the combined orbiter/probe observations revealed a surface that has been etched by fluids, and possibly been resurfaced by cryovolcanism (Fig. 5.88). Numerous channels cut across different types of terrain. Radar-bright rivers may be filled with boulders, while radar-dark channels suggest the presence of either liquids or smooth deposits. Some rivers show tributaries, and the system photographed by the Huygens probe resembles a delta (Fig. 5.88a). These features suggest formation via rainfall, likely in the form of liquid hydrocarbons. No liquids were seen by the probe, however. Upon landing, a penetrometer under the probe measured the force as a function of penetration depth (Fig. 5.88c). Comparison of the graphs with laboratory experiments suggest that the probe landed in a substance similar to wet clay, sand, or snow. Upon landing, the Gas Chromatograph Mass Spectrometer (GCMS) measured an increase in methane gas, also indicative of a 'moist' surface (see also discusssion in §4.3.3.2).

Although no liquids have been seen on Titan's surface at low latitudes, there is clear evidence of lakes filled

with hydrocabon liquids at high latitudes (Fig. 5.89). The depth and extent of these lakes have been observed to vary over time. Radar-dark features over both poles combined cover over 600 000 km², which is about 1% of Titan's total surface area. However, even if all of these radar-dark features were filled with liquids, it would not be enough to explain Titan's methane cycle in a manner analogous to the hydrological cycle on Earth (§§4.6.1.4, 4.5.5.4).

The Cassini radar instrument detected only five features that are clearly impact craters on images obtained through December 2007 (22% of Titan's surface). An additional few dozens of features might also be impact craters, but interpretation of these is not unambiguous. A comparison with models of the impact crater production rate suggests Titan's surface to be young, perhaps a few ×10⁸ yr, and certainly less than 10⁹ yr. Hence Titan clearly is a geologically active moon, with a high resurfacing rate. This agrees well with the many smooth-looking radar-bright 'flows' on its surface, which are reminiscent of cryovolcanic lava flows. One ~180 km wide potential shield volcano has been identified, with a ~20 km diameter caldera at its center, and sinuous channels and/or ridges radiating away from the caldera. Numerous longitudinal dunes, dark both in radar echoes and at infrared wavelengths, have been identified in the equatorial region (Fig. 5.90). These dunes are all oriented in the east–west direction and are up to thousands of km long. They appear to bend around radar-bright, likely elevated, features. The orientation of the dunes has been used to derive the wind

(a)

(b)

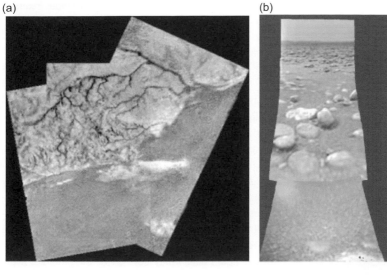

(c)

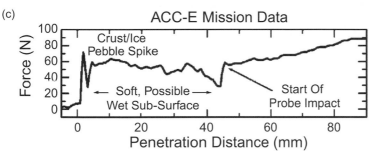

Figure 5.88 (a) Mosaic of three frames from the Descent Imager/ Spectral Radiometer (DISR) instrument on the Huygens probe shows a remarkable view of a 'shoreline' and channels, from an altitude of 6.5 km. The bright 'island' is about 2.5 km long. (NASA/JPL/ESA/University of Arizona, PIA07236) (b) After landing, the DISR instrument on the Huygens probe obtained this view of Titan's surface, including 10–15 cm sized rocks, presumably made of ice. (ESA/NASA/JPL/University of Arizona, PIA06440) (c) Graph of the force on the probe's penetrometer as a function of penetration depth. The initial spike near 0 mm depth may be caused by a pebble or an icy crust, while the shape of the curve suggests a surface that is neither hard nor fluffy. The surface most likely is similar to wet (methane) clay, sand, or snow. (Adapted from Zarnecki *et al.* 2005)

direction, which contrary to expectation is towards the east rather than the west (§4.5.5.4).

To ultimately solve the overarching question of how methane gas is resupplied to Titan's atmosphere, one needs to determine Titan's surface composition, and whether or not there is ongoing cryovolcanism. Titan's bulk density suggests an icy surface. However, to verify this spectrally is challenging because of the intervening atmosphere. In addition, if aerosols have been settling out over hundreds of millions to billions of years, the surface may be covered with a few hundred meters of hydrocarbons or tholins, which may cover up the water-ice (§§4.6.1.4, 4.3.3.2).

5.5.6.2 Enceladus

Enceladus, in orbit around Saturn between Mimas and Tethys, is a most remarkable and enigmatic satellite. Parts of this moon are heavily cratered, but large regions on the surface show virtually no impact craters at all. The youngest parts are probably no more than one Myr old, while the oldest terrain is likely a few billion years old. Enceladus's surface reflectivity is very high, about 100%, implying fresh, uncontaminated ice. With a bulk density

of 1.6 g cm^{-3}, the satellite probably has a rocky core ($R \approx 170$ km, $\rho \approx 3$ g cm^{-3}) and an ~80 km thick icy crust. Like Ganymede and Europa, the crust displays regions of grooved terrain, indicative of tectonic processes, and smoother parts, possibly resurfaced by water flows.

Although geyser activity and venting from Enceladus had been suggested as a possible source for E ring material based upon Voyager data (§11.3.2), Cassini's discovery of giant plumes of vapor, dust, and ice emanating from Enceladus's south pole was unexpected. The plumes emanate from 'cracks', dubbed *tiger stripes* in the satellite's south polar region (Fig. 5.91), an area where Cassini's Composite Infrared Spectrometer (CIRS) detected temperatures of at least 180 K along some of the brightest tiger stripes, well above the 72 K background temperatures at other places in the south polar region. When flying through the plume, the Ion and Neutral Mass Spectrometer (INMS) on Cassini measured a gas composition of $91 \pm 3\%$ H_2O, ~3% CO_2, 4% N_2 or CO, and 1.6% CH_4. Particle velocities in the plume are ~60 m s^{-1}, well below the escape speed (235 m s^{-1}).

Figure 5.89 Cassini radar images of lakes of liquid hydrocarbons near Titan's north pole. The lakes are darker than the surrounding terrain, indicative of regions of low (sometimes zero) backscatter. The strip of radar imagery is foreshortened to simulate an oblique view of the highest latitude region, seen from a point to its west. (NASA/JPL/USGS, PIA09102)

Only about 1% of the particles escape, but these escapees supply material to Saturn's tenuous E ring (§11.3.2). These plumes are likely the source of most of the material in the E ring.

The observed geyser activity on Enceladus requires a substantial heat source, the cause of which is still a puzzle. Primordial heat or radioactive decay are not sufficient, while tidal heating resulting from orbital eccentricities excited by its 2:1 orbital resonance with Dione may only be marginally adequate. Because the plume is composed primarily of water, it may erupt from chambers of liquid water just below the surface, at temperatures over 273 K. Dissolution of ammonia in water could lower the temperature of a putative liquid near-surface ocean. The plume composition further suggests that degassing of clathrate hydrates (§5.1.1) on a (putative) seafloor may play a role. Since decomposition of clathrate hydrates liberates energy, this process may help solve the energy puzzle. In addition, when (some of) the gases condense on the surface, this energy reappears as latent heat of condensation. A liquid ocean may not be essential, though, if the jets are 'driven' by *diapirs*, as discussed in §6.3.5.2.

5.5.6.3 *Other Mid-sized Saturnian Moons*

Apart from Enceladus, Saturn has five other mid-sized, nearly spherical satellites (Figs. 5.92, 5.93). These moons range in radius from a little under 200 km (Mimas) up to 750 km (Rhea), with densities from just under 1 g cm^{-3} (Tethys) up to 1.6 g cm^{-3} (Enceladus), i.e., clearly icy bodies. Most of Saturn's satellites are quite bright, with albedos, $A_{\rm v}$, ranging from about 0.3 up to 1.0. All of the regular satellites show water-ice in their surface spectra. All of these moons are relatively spherical, suggestive of relatively low viscosities in their interiors at some point in their histories.

Detailed images taken by various spacecraft show that each satellite has its own unique characteristics. The

Figure 5.90 Part of Titan's surface as mapped by the Cassini radar instrument on 7 September 2006. The impact crater on the left side is ∼30 km in diameter. This crater shows a central peak, while the dark floor indicates smooth and/or highly absorbing materials. Towards the right on the figure are longitudinal dunes, which make up most of Titan's equatorial dark regions. These ∼100 km long features run east–west on the satellite, are 1–2 km wide and spaced similarly, and roughly 100 m high. They curve around the bright features in the image – which may be high-standing topographic obstacles – following the prevailing wind pattern. Unlike Earth's sand (silicate) dunes, these may be solid organic particles or ice coated with organic material. (NASA/JPL Cassini orbiter, PIA09172)

(a)

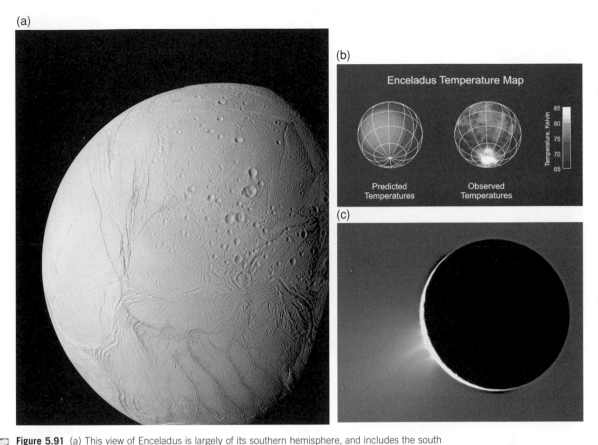

(b)

(c)

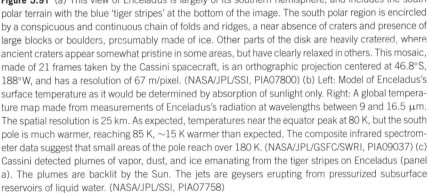

Figure 5.91 (a) This view of Enceladus is largely of its southern hemisphere, and includes the south polar terrain with the blue 'tiger stripes' at the bottom of the image. The south polar region is encircled by a conspicuous and continuous chain of folds and ridges, a near absence of craters and presence of large blocks or boulders, presumably made of ice. Other parts of the disk are heavily cratered, where ancient craters appear somewhat pristine in some areas, but have clearly relaxed in others. This mosaic, made of 21 frames taken by the Cassini spacecraft, is an orthographic projection centered at 46.8°S, 188°W, and has a resolution of 67 m/pixel. (NASA/JPL/SSI, PIA07800) (b) Left: Model of Enceladus's surface temperature as it would be determined by absorption of sunlight only. Right: A global temperature map made from measurements of Enceladus's radiation at wavelengths between 9 and 16.5 μm. The spatial resolution is 25 km. As expected, temperatures near the equator peak at 80 K, but the south pole is much warmer, reaching 85 K, ~15 K warmer than expected. The composite infrared spectrometer data suggest that small areas of the pole reach over 180 K. (NASA/JPL/GSFC/SWRI, PIA09037) (c) Cassini detected plumes of vapor, dust, and ice emanating from the tiger stripes on Enceladus (panel a). The plumes are backlit by the Sun. The jets are geysers erupting from pressurized subsurface reservoirs of liquid water. (NASA/JPL/SSI, PIA07758)

surfaces of Mimas, Tethys, and Rhea are heavily cratered. *Mimas* is characterized by one gigantic crater near the center of its leading hemisphere, about 135 km in diameter, one-third the moon's own size. The crater is about 10 km deep, and the central peak ~6 km high. The impacting body must have been ~10 km across. *Tethys* displays an ~2000 km long complex of valleys or troughs, Ithaca Chasma, which stretches three-quarters of the way around the satellite. This system was produced by tectonic activity, which might have been triggered by the large impact that produced the 400 km diameter crater, Odysseus, on Tethys's leading hemisphere. The craters on Tethys tend to be flatter than on the Moon or Mimas, probably because of viscous relaxation of its icy surface. *Dione* exhibits variations in surface albedo of almost a factor of 2, which is much larger than those seen on *Rhea*, but much less extreme than Iapetus's hemispheric asymmetry. The trailing hemispheres of both satellites are relatively dark and covered by wispy, white streaks, perhaps snow or ice. Their leading hemispheres

Figure 5.92 Images of Saturn's inner mid-sized satellites other than Enceladus. The cratered surface of Mimas shows its 140 km diameter crater Herschel (PIA06258). Tethys shows its anti-Saturn facing hemisphere. The rim of the 450 km diameter impact basin Odysseus lies on the eastern limb, making the limb appear flatter than elsewhere. Other large craters seen here are Penelope (left of center) and Melanthius (below center) (PIA08870). The trailing hemisphere of Dione shows many bright cliffs. At lower right is the feature called Cassandra, exhibiting linear rays extending in multiple directions (PIA08256). Rhea's crater-saturated surface shows a large bright blotch and radial streaks, which were likely created when a geologically recent impact sprayed bright, fresh ice ejecta over the moon's surface (PIA08189). (All images taken by the Cassini spacecraft; NASA/JPL/SSI)

are bright, bland, and heavily cratered, although on Dione the crater density varies quite substantially from one region to another. This implies that extensive resurfacing must have taken place. Mimas, Dione, and Rhea all have fractures on their surface which manifest themselves as narrow shallow troughs. The wispy streaks may have formed by frost and ice extruded along such fractures.

Iapetus is a bizarre body, with its trailing hemisphere $\sim$10 times as bright as the leading hemisphere ($A_v \approx 0.5$ versus $A_v \approx 0.05$). Its icy trailing hemisphere and polar regions are very similar to the cratered surface of Rhea. The black material on its leading hemisphere still puzzles scientists. This material is reddish, and might consist of organic, carbon-bearing compounds. It looks like a coat of dark material without any brighter markings on top of it. Radar measurements show the coating to be quite thin (up to several tens of cm). It is still unknown whether the material is of internal or external origin. Being on the leading side, Iapetus may just sweep up 'dirt' from Saturn's magnetosphere, such as dust from the dark satellite Phoebe. However, several craters appear to have black floors on the bright side of this moon, which does not appear to be broken up by yet other craters; i.e., no light material is uncovered by additional impacts. This argues for an internal origin. Iapetus's most remarkable topographic feature is a mysterious $\sim$1300 km long ridge, up to 20 km high at places, that coincides almost exactly with its geographic equator (Fig. 5.93). Crater counts suggest the ridge to be ancient. Isolated peaks are observed at many of the places where segments of the ridge are absent.

5.5.6.4 Phoebe

Phoebe is by far the largest of Saturn's irregular moons, and the only one for which we have resolved images. The moon is very dark ($A_v \approx 0.06$), similar to that of C-type asteroids and comets. Phoebe's density (1.6 g cm^{-3}) suggests a body composed of ice and rock. Water-ice was detected via ground-based spectroscopy, and Cassini identified CO_2-ice and organic materials on its surface. These findings, together with Phoebe's retrograde orbit, which points at a capture origin, suggest that the moon probably originated in the Kuiper belt. Cassini images of the moon (Fig. 5.96) show an unusual variation in brightness, where some crater slopes and floors display bright material – probably ice – on what is otherwise an extremely dark body. Some craters exhibit several layers of alternating bright and dark material, which might be caused during crater formation when ejecta thrown out from the crater bury pre-existing surface that was itself covered by a relatively thin, dark deposit over an icy mantle.

Saturn, like the other giant planets, has a large number of irregular satellites at relatively large distances (up

Figure 5.93 The leading side of Iapetus, displayed in this image, is about ten times darker than its trailing side. An ancient, 400 km wide impact basin shows just above the center of the disk. Along the equator is a conspicuous, 20 km wide topographic ridge that extends from the western (left) side of Iapetus almost to the day/night boundary on the right. On the left horizon, the peak of the ridge rises at least 13 km above the surrounding terrain. (Cassini, NASA/JPL/SSI, PIA06166)

to $\sim$20 $\times$ 10^6 km) from the planet. Most of these have been discovered by ground-based observations using large fields of view. These moons typically move in highly eccentric and inclined orbits, often retrograde, suggestive of captured objects rather than satellites that formed within Saturn's subnebula.

5.5.6.5 Small Regular Satellites of Saturn

In addition to the seven satellites described above, Saturn has a large number of small moons. Most of Saturn's small regular moons were discovered by the Voyager and Cassini spacecraft, and some of them during ring plane crossings from ground-based observations. When the Earth travels through Saturn's ring plane, the rings are seen 'edge-on' and are therefore practically invisible. During these times it is possible to detect tiny moons near Saturn's main ring system.

All of Saturn's small inner moons are oddly shaped, heavily cratered, and as reflective as Saturn's larger satellites (Fig. 5.94). Two of the small moons, *Janus* and *Epimetheus* share the same orbits, and change places every 4 years (§2.2.2). *Calypso* and *Telesto* are located at the L_4 and L_5 Lagrangian points of Tethys's orbit, while *Helene* and *Polydeuces* reside in Dione's Lagrangian points. *Atlas* is a small moon orbiting just outside the A ring. *Prometheus* and *Pandora* are the inner and outer

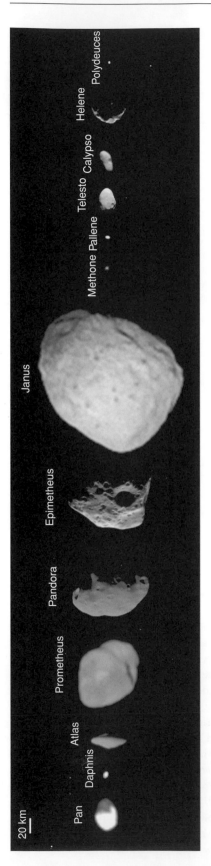

shepherds of the F ring, and play a key role in 'shaping' the kinky appearance of the F ring (§11.4). The Cassini spacecraft discovered the satellites *Pallene* and *Methone* between the orbits of Mimas and Enceladus. Pallene is embedded within a faint ring of material. *Pan* and *Daphnis* orbit within the Encke and Keeler gaps, respectively. The densities of these inner moons are very low, less than that of water (Table 1.5). Such low densities imply that the moons are very porous.

Hyperion is oddly shaped, $\sim 400 \times 250 \times 200$ km, and saturated with craters that appear to be deeply eroded (Fig. 5.95). Its irregular shape implies that it is a collisional remnant of a larger body. Hyperion is the only satellite that displays a chaotic rotation.

5.5.7 Satellites of Uranus

Uranus's five mid-sized 'classical' moons (Table 1.5, Fig. 5.97) were discovered prior to the space age. They orbit in or near the plane of the planet's equator, which is tilted by 98° with respect to Uranus's orbit around the Sun. Their radii vary from 235 km for Miranda, the innermost large moon, to almost 800 km for Titania.

Figure 5.95 Chaotically tumbling and seriously eroded by impacts, Hyperion is one of Saturn's more unusual satellites. The moon may be quite porous; in fact in this view it resembles a sponge rather than a solid rocky object. Its color is unusual as well, being rosy tan, perhaps from debris from moons further out. (Cassini, NASA/JPL/SSI, PIA07740)

Figure 5.96 This mosaic of Phoebe reveals the satellite's irregular topography. Unusual variations in brightness are visible on some crater slopes and floors, showing evidence of layered deposits of alternating bright and dark material. (Cassini, NASA/JPL/SSI, PIA06073)

The smallest and innermost of the five classical moons, *Miranda*, is almost as large as Enceladus, and its surface is bizarre. Some areas are extremely heavily cratered, as expected for a small cold object. Other regions, however, have only a few craters and a surprising endogenic terrain, referred to as coronae, characterized by subparallel sets of bright and dark bands, scarps and ridges, with very sharp boundaries between differing types of terrain. There is no good explanation for this great diversity in terrain types. One theory invokes tidal heating, possibly due to chaotic excitation of orbital eccentricities upon passage through resonances eons ago. The surface features may result from an incomplete differentiation and convection pattern. It has also been suggested that Miranda was disrupted by a catastrophic impact early in its history and then re-accreted with some of the initial rocky core pieces falling on top of the icy mantle. The differences in terrain could then have resulted from a sinking of material in regions where the heavy core material re-accreted on the outside and subsequently sank to the center.

Next in distance from Uranus is *Ariel*. Ariel shows clear signs of local resurfacing, but nothing as striking

as Miranda. The age of terrain varies significantly, but the entire surface appears to be younger than the oldest terrains on each of the three outer classical moons. There exists a global system of faults and evidence of flows attributed to ice volcanism. Although *Umbriel* is similar in size to Ariel, this moon is heavily cratered and appears to have the oldest surface in the uranian system. There is little or no evidence of tectonic activity. Most of *Titania* is heavily cratered; however, some patches of smoother material with fewer craters imply local resurfacing. An extensive network of faults cuts the surface of Titania. *Oberon*'s surface is dominated by craters, but there are several high-contrast albedo features and signs of faulting.

The densities of the four largest satellites of Uranus are significantly higher than those of Saturn's satellites of comparable size. Uncompressed densities of these moons are estimated to be $\sim$1.45–1.5 g cm^{-3}. These high densities have led to the suggestion that the circum-uranian disk out of which these moons presumably formed had most of its oxygen in the form of CO rather than H_2O, and thus was depleted in water relative to the saturnian nebula. While plausible, this suggestion must be regarded as quite tentative owing to all the bizarre ways in which other bodies are thought to have gotten their anomalous densities. Note especially the wide range of densities of Saturn's mid-sized moons (Table 1.5).

Thirteen small moons are known to orbit Uranus inside Miranda's orbit. Eleven of these tiny satellites were discovered on Voyager 2 images, and the two smallest satellites, Mab and Cupid, were found using HST. *Cordelia* and *Ophelia* are two ring shepherds which control the inner and outer edges of the ϵ ring, the brightest and outermost of the planet's main rings (§11.3.3). The largest of the small satellites is *Puck*, with $R \approx 80$ km. Puck is darker than any of the five large satellites, slightly irregular in shape, and heavily cratered. Nine moons, collectively referred to as the 'Portia group' after their largest member, have orbits between 59 200 km and 76 400 km from Uranus. Orbital calculations show that this family of satellites is chaotic and dynamically unstable; they may be remnants of a larger satellite. All of these objects are quite dark, and neutral to slightly reddish in color, indicative of carbonaceous materials. Their surfaces may have darkened over time as a result of micrometeoroid impacts and sputtering, subliming ices away but leaving the darker material behind (§4.8.2). *Mab* is a particularly intriguing satellite, since its orbit is centered in the outermost ring of Uranus, the μ ring, a ring which shows some similarities to Saturn's E ring (§11.3.3).

Irregular satellites have beeen discovered up to $\sim$20 $\times$ 10^6 km from Uranus. Like the irregular satellites of the other giant planets, many are on retrograde and/or highly

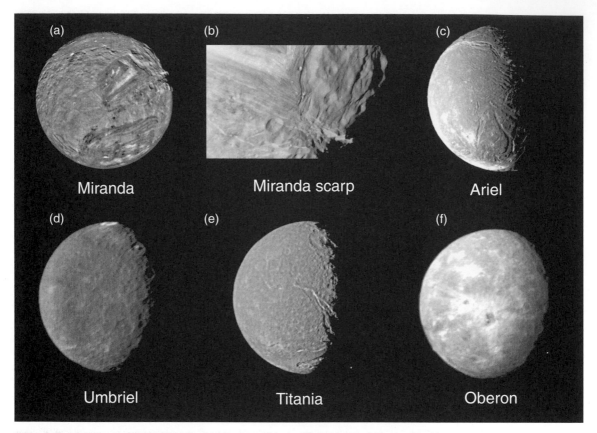

Figure 5.97 (a) Uranus's satellite Miranda displays two strikingly different types of terrain. An old, heavily cratered rolling terrain with relatively uniform albedo, contrasted by a young, complex terrain that is characterized by sets of bright and dark bands, scarps, ridges, and cliffs up to 20 km high, as seen most distinctly in the 'chevron' feature displayed in panel (b). (PIA01490) (c) Most of Ariel's visible surface consists of relatively heavily cratered terrain transected by fault scarps and fault-bounded valleys (graben), as shown on this mosaic of four high resolution (2.4 km/pixel) images. Some of the largest valleys are partly filled with younger deposits. (PIA01534) (d) Umbriel is the darkest of Uranus's larger moons and the one that appears to have experienced the least geological activity. Note the bright ring at the top, ~140 km across, which lies near the satellite's equator. This may be a frost deposit, perhaps associated with an impact crater. Just below this feature, on the terminator, is an ~110 km diameter crater which has a bright central peak. (PIA00040) (e) Titania, Uranus's largest satellite, is heavily cratered and displays prominent fault valleys up to 1500 km long and 75 km wide. In valleys seen at right-center, the sunward-facing walls are very bright, suggestive of younger frost deposits. A prominent impact crater is visible at the top. (PIA00039) (f) Oberon's icy surface is covered by impact craters, many of which are surrounded by bright rays. Near the center of the disk is a large crater with a bright central peak and a floor partially covered with very dark material. This may be icy, carbon-rich material erupted onto the crater floor sometime after the crater formed. Another striking topographic feature is a large mountain, about 6 km high, peeking out on the lower left limb. (PIA00034) (All images were taken with the Voyager 2 spacecraft, NASA/JPL)

eccentric orbits. These moons are most likely captured objects.

5.5.8 Satellites of Neptune

Before the Voyager flyby only two moons of Neptune, *Triton* and *Nereid*, were known. Both of these bodies occupy 'unusual' orbits. *Triton* orbits Neptune at $14.0\,R_{\Psi}$, has a very small eccentricity ($e < 0.0005$), but its orbit

is inclined 159° with respect to Neptune's equator, so that the satellite orbits in a retrograde manner. This odd orbit is the reason that it is generally accepted that Triton must have been captured from the Kuiper belt. With Neptune's rotation axis inclined by 28.8° relative to the planet's orbit about the Sun, and Triton's inclined orbit precessing about Neptune's equatorial plane, Triton undergoes a complicated cycle of seasons within seasons, which

(a)   (b)   (c)

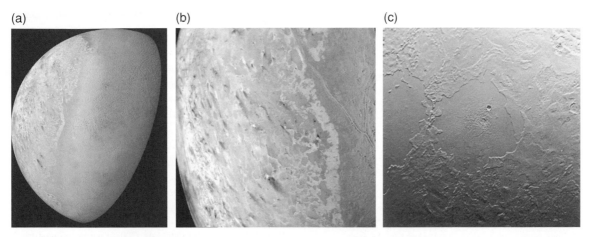

Figure 5.98 (a) COLOR PLATE A global color mosaic of Neptune's largest moon, Triton. Color was synthe-
sized by combining images taken through orange, violet, and ultraviolet filters, displayed in red, green,
and blue, and combined to create this color version. Triton's surface is covered by nitrogen-ice, while
the pinkish deposits on the south polar cap (on the left) may contain methane-ice, which would have
reacted under sunlight to form pink or red compounds. The dark streaks may be carbonaceous dust
deposited from huge geyser-like plumes. The bluish-green band extends all the way around Triton near
the equator; it may consist of relatively fresh nitrogen-frost deposits. The greenish areas include what
is called the cantaloupe terrain, and a set of 'cryovolcanic' landscapes. (NASA/Voyager 2, PIA00317)
(b) This image of the south polar terrain of Triton reveals about 50 dark 'wind streaks'. A few plumes
are observed to originate at dark spots; these are several km in diameter and some are more than 150
km long. The spots may be vents or geysers where gas has erupted from beneath the surface, carrying
dark particles into Triton's atmosphere. Southwesterly winds then transported this dust, which formed
gradually thinning deposits to the northeast of most vents. (NASA/Voyager 2, PIA00059) (c) This view of
Triton is about 500 km across. It encompasses two depressions, possibly old impact basins, that have
been extensively modified by flooding, melting, faulting, and collapse. Several episodes of filling and partial removal of material appear to have occurred. The rough area in the middle of the bottom depression
probably marks the most recent eruption of material. Only a few impact craters dot the area, which shows
the dominance of internally driven geologic processes on Triton. (NASA/Voyager 2, PIA01538)

lasts about 600 years. *Nereid* orbits Neptune on a prograde
orbit, with a semimajor axis of 219 R_Ψ, inclined by $\sim$27°
to Neptune's equator and, more importantly, by 7.2° to its
Laplace plane (essentially Neptune's orbital plane). The
orbit has the largest eccentricity of any known moon, with
$e = 0.76$. Nereid is a small moon, with a radius of $\sim$170
km, and is reasonably round.

Triton is by far the largest moon in the neptunian satellite system (Fig. 5.98), with a size somewhat smaller than
Europa. It has a tenuous atmosphere of nitrogen, with a
trace of methane gas (mixing ratio $\sim$10^{-4}; §4.3.3.3). Triton has the lowest observed surface temperature of any
planetary satellite, 38 ± 4 K. The polar cap on the southern hemisphere is bright, with an albedo of $\sim$0.9, while
the equatorial region is somewhat darker and redder. Most
of the surface is covered with a thin layer of nitrogen-
and methane-ice, although it does not completely hide the
underlying terrain. The western (trailing) hemisphere of
Triton looks like a 'cantaloupe'; a dense concentration of
pits or dimples, crisscrossed by ridges or fracture systems.

This terrain may have a long history of repeated fracturing
and some form of viscous icy volcanism. It is the oldest
terrain on Triton, but, since it is much less cratered than
the satellites of Uranus and Saturn, it must be geologically
young. The leading hemisphere consists of a smoother surface, with large calderas and/or lava lakes. This terrain is
probably a few billion years old. The icy substance which
produced the smooth plains was less viscous than that seen
on the trailing hemisphere, which suggests a chemically
different composition, perhaps ammonia. Multiple levels
of cooling and stagnation are apparent near the ice lava
lakes. At several places it appears as if a volcanic fluid
cooled to form a solid lid, after which the fluid underneath drained away causing parts of the lid to founder and
melt. When the newer, lower lying, lava lakes undergo a
similar process, layers of 'lids' are formed which look like
collapsed calderas. It is not clear if the features seen on Triton are formed by drainage of lava lakes or from collapsed
calderas. The polar regions are covered with N_2-ice, which
evaporates in the spring (§4.5.1.3). The ice has a slightly

reddish tint, indicative of organic compounds. In these regions a large number of relatively dark (10–20% lower albedos than the surroundings) streaks are seen. At least two of these streaks appear to be active geyser plumes, likely driven by liquid nitrogen. The plumes rise $\sim$8 km and then are swept westwards by the winds. Such plumes have been observed to reach lengths exceeding 100 km. Although the heating mechanism for the geysers has not yet been understood, sunlight might play a role, since all four geysers detected are in the south polar region where the surface was continuously illuminated by the Sun. Perhaps sunlight combined with a solid-state greenhouse effect in the subsurface (§3.3.2) heats the ice by a few degrees, leading to enhanced subsurface nitrogen vapor pressure that produces geyser-like eruptions of gas and dust.

Voyager discovered six satellites within and near Neptune's ring system. The largest of these, *Proteus*, with a radius of 200 km, is slightly bigger than Nereid. The radii of the other satellites are between 27 and 100 km. All of these satellites are dark and irregular in shape.

Irregular satellites have been found out to 50×10^6 km from the planet.

Further Reading

General books on our Solar System were listed in Chapter 1.

Background material on mineral phases and cooling of a magma is summarized by, e.g.:

Putnis, A., 1992. *Mineral Science*. Cambridge University Press, Cambridge. 457pp.

Examples of books on general geology/geophysics of the Earth and other planetary bodies are:

Fowler, C.M.R., 2005. *The Solid Earth: An Introduction to Global Geophysics*. 2nd Edition. Cambridge University Press, New York. 685pp.

Greeley, R., 1994. *Planetary Landscapes*, 2nd Edition. Chapman and Hall, New York, London. 286pp.

Grotzinger, J., T. Jordan, F. Press, and R. Siever, 2006. *Understanding Earth*, 5th Edition. W.H. Freeman and Company, New York. 579pp.

Lopes, R.M.C., and T.K.P. Gregg, Eds., 2004. *Volcanic Worlds*, Springer-Praxis, New York. 236 pp.

Turcotte, D.L., and G. Schubert, 2002. *Geodynamics*, 2nd Edition. Cambridge University Press, New York. 456pp.

An excellent monograph on impact cratering is written by:

Melosh, H.J., 1989. *Impact Cratering: A Geologic Process*. Oxford Monographs on Geology and Geophysics, No. 11. Oxford University Press, New York. 245pp.

Post-Galileo views of the jovian system:

Lopes, R.M.C., and J.R. Spencer, Eds., 2007. *Io after Galileo: A New View of Jupiter's Volcanic Moon*. Springer, Praxis Publishing, Chichester, UK. 342pp.

Bagenal, F., T. Dowling and W. McKinnon, Eds., 2004. *Jupiter: The Planet, Satellites, and Magnetosphere*. Cambridge University Press, Cambridge. 719pp. We recommend in particular the chapters by:

Harrington, J., *et al.*: Lessons from Shoemaker–Levy 9 about Jupiter and planetary impacts.

Burns, J.A. *et al.*: Rings and inner small satellites.

McEwen, A.S. *et al.*: Lithosphere and surface of Io.

Greeley, R. *et al.*: Geology of Europa.

Pappalardo, R.T. *et al.*: Geology of Ganymede.

Moore, J.M. *et al.*: Callisto.

Schenk, P.M. *et al.*: Ages, interiors, and the cratering record of the Galilean Satellites.

The detailed geology of planets and satellites is discussed in several chapters in the encyclopedia by McFadden, L., P. R. Weissman, and T.V. Johnson, Eds., 2007. *Encyclopedia of the Solar System*, 2nd Edition. Academic Press, San Diego. 982pp.

A pleasant 'novel' about the history of the development of the impact theory on the extinction of the dinosaurs is written by:

Alvarez, W., 1997. *T. Rex and the Crater of Doom*. Princeton University Press, Princeton, NJ. 185pp.

Problems

5.1.**E**   (a) Determine the approximate mineral composition and texture of dacite with a silica content of 60%. (Hint: Use Figure 5.4.)

(b) Determine the approximate silica content, mineral composition, and texture of obsidian, pumice, rhyolite, and granite. Explain the similarities and differences.

(c) Determine the approximate silica content, mineral composition and texture of gabbro and basalt. Compare with your answer in (b) and comment on your result.

5.2.**I**   (a) With help of the first law of thermodynamics (eq. 3.28), show that the enthalpy can be written as

$$dH = dQ + VdP. \tag{5.28}$$

(b) Show that for a (idealized) thermodynamically reversible process, changes in the Gibbs free energy can be expressed as a function of pressure and/or temperature as

$$dG = VdP - SdT. \tag{5.29}$$

5.3.**E**   Calcium carbonate, $CaCO_3$, exists in two polymorphic forms: calcite and aragonite. In the following you will calculate which form (or phase) of calcium carbonate exists at a temperature of 298 K and pressure of 1 atm. The enthalpy at a temperature of 298 K and pressure of 1 atm for calcite is -12.0737×10^{12} erg mole^{-1} and for aragonite it is -12.0774×10^{12} erg mole^{-1}. The entropy for

Table 5.4 Constants at $P = 1$ atm for the system jadeite + quartz → albite.[a]

Mineral	$S_{298} \times 10^9$	$c_1 \times 10^{10}$	$c_2 \times 10^5$	$c_3 \times 10^{10}$	$c_4 \times 10^{10}$
Albite	2.074	0.4521	−1.336	−1276	−3.954
Jadeite	1.335	0.3011	1.014	−2239	−2.055
Quartz	0.415	0.1044	0.607	34	−1.070

[a] From Putnis (1992).

calcite is 91.7×10^7 erg mole^{-1} K^{-1}; the entropy for aragonite is 88×10^7 erg mole^{-1} K^{-1}.

(a) Calculate the change in enthalpy for the transformation aragonite → calcite at a temperature of 298 K and 1 atm pressure. Is the reaction endothermic or exothermic?

(b) Calculate the change in entropy for this same transformation (aragonite → calcite).

(c) Determine which form of $CaCO_3$ is the stable form at a temperature of 298 K and pressure of 1 atm.

5.4.I Consider the system jadeite + quartz → albite ($NaAlSi_2O_6 + SiO_2 \rightarrow NaAlSi_3O_8$). The temperature dependence of the thermal heat capacity is usually expressed by

$$C_p = c_1 + c_2 T + c_3 T^{-2} + c_4 T^{-1/2}, \qquad (5.30)$$

with c_1, c_2, c_3, and c_4 constants. For the above system these constants, as well as the entropy at room temperature and 1 atm pressure, are given in Table 5.4 (in units of erg mole^{-1} K^{-1}). The change in enthalpy for the transformation jadeite + quartz → albite at 298 K is $+12.525 \times 10^{10}$ erg mole^{-1}.

(a) Calculate the stable phase(s) for the system at a temperature of 298 K.

(b) Calculate the stable phase(s) for the system at a temperature of 1000 K.

5.5.E If the lithospheric plates move, on average, at a speed of 6 cm yr^{-1}, what would be a typical recycling time of terrestrial crust? (Hint: Calculate the recycling time based upon the motion of one plate over the Earth's surface. How would your answer change if you have several plates moving over the surface?)

5.6.E The Voyager 1 spacecraft detected nine active volcanoes on Io. If we assume that, on average, there are nine volcanoes active on Io, and that the average eruptive rate per volcano is 50 km^3 yr^{-1}, calculate:

(a) the average resurfacing rate on Io in cm yr^{-1},

(b) the time it takes to completely renew the upper kilometer of Io's surface.

5.7.E Suppose a child leaves a toy truck on a deserted sandy beach facing the ocean, and returns many years later to retrieve it. The beach is characterized by strong winds, which usually blow inland. Sketch the dune formed around the toy truck.

5.8.E Calculate a typical timescale over which material in the Earth's upper mantle flows. The dynamic viscosity is $\sim 10^{21}$ Pa s, and the shear modulus is ~ 250 GPa.

5.9.E (a) Calculate the kinetic energy and pressure involved when the Earth gets hit by a stony meteoroid ($\rho = 3.4$ g cm^{-3}) that has a diameter of 10 km, and $v_\infty = 0$.

(b) Calculate the kinetic energy were the same meteoroid to hit Jupiter instead of Earth, assuming the body has zero velocity at a large distance from Jupiter.

(c) Calculate the kinetic energy involved when a fragment of Comet D/Shoemaker–Levy 9 ($\rho = 0.5$ g cm^{-3}, $R = 0.5$ km) hits Jupiter at the planet's escape velocity.

(d) Express the energies from (a)–(c) in magnitudes on the Richter scale, and compare these with common earthquakes.

5.10.E (a) The compression stage typically lasts a few times longer than the time required for the impacting body to fall down a distance equal to its own diameter. Calculate the duration of the compression stage for $R = 10$ m and $R = 1$ km meteoroids that impact Earth at $v = 15$ km s^{-1}.

(b) Estimate the pressure involved in these collisions, assuming the meteoroids are stony bodies with density $\rho = 3$ g cm^{-3}.

5.11.E Consider the impact between an iron meteoroid ($\rho = 7$ g cm^{-3}) with a diameter of 300 m and the Moon.

(a) Calculate the kinetic energy involved if the meteoroid hits the Moon at $v = 12$ km s^{-1}.

(b) Estimate the size of the crater formed by a head-on collision, and one where the angle of impact with respect to the local horizontal is 30°.

(c) If rocks are excavated from the crater with typical ejection velocities of 500 m s^{-1}, calculate how far from the main crater one may find secondary craters.

5.12.E Repeat the same questions as in Problem 5.11 for Mercury. Comment on the similarities and differences.

5.13.E After the Moon has been hit by the meteoroid from Problem 5.11, many rocks are excavated from the crater during the excavation stage.

(a) If the ejection velocity is 500 m s^{-1}, calculate how long the rock remains in flight, if its ejection angle with respect to the ground is 25°, 45°, and 60°.

(b) Calculate the maximum height above the ground reached by the three rocks from (a).

5.14.E The secondary craters related to a primary crater of a given size on Mercury typically lie closer to the primary crater than do the secondary craters of a similarly sized primary on the Moon. Presumably, this is the result of Mercury's greater gravity reducing the distance that ejecta travel.

(a) Verify this difference quantitatively by calculating the 'throw distance' of ejecta launched at a 45° angle with a velocity of 1 km s^{-1} from the surfaces of Mercury and the Moon.

(b) Typical projectile impact velocities are greater on Mercury than they are on the Moon. Why doesn't this difference counteract the surface gravity effect discussed above?

5.15.E (a) Determine the diameter of the crater produced when a stony ($\rho = 3$ g cm^{-3}) meteoroid, 1 km across, hits the Earth at a velocity of 15 km s^{-1}, at an angle of 45° (ignore the Earth's atmosphere). Use a density of 3.5 g cm^{-3} for the Earth's surface layer.

(b) Determine the crater diameter from (a) if the meteoroid is 10 km across.

(c) Determine the diameter of craters produced by both meteoroids if they had hit the Moon ($v = 15$ km s^{-1}) instead of Earth.

5.16.I (a) Calculate the average crater density (km^{-2}) for craters over 4 km in size for a portion of the lunar highlands that is 4.44 Gyr old.

(b) Calculate the average crater density (km^{-2}) for craters over 4 km in size for a region on the lunar surface that is 1 Gyr old.

(c) Calculate the approximate age of the lunar maria if the average crater density for craters over 4 km in diameter is 9×10^{-5} craters km^{-2}.

5.17.I (a) Determine the minimum radius of an iron meteoroid ($\rho = 8$ g cm^{-3}) to impact the Earth at hypersonic speed.

(b) Determine the minimum radius for a similar meteoroid of similar composition to make it through Venus's atmosphere.

(c) Calculate the approximate crater size the meteoroids produce on both planets, if the impacting velocity is equal to the escape velocity from the planet. Have craters smaller than this size been observed on Venus and Earth?

5.18.E (a) Calculate the ratio of collisions per unit area near the apex of Callisto's motion to that at the antapex for a population of impactors that approaches Jupiter with $v_\infty = 5$ km s^{-1}. You may assume that Callisto is on a circular orbit and neglect the moon's gravity, but don't neglect Jupiter's gravitational pull on the impactors.

(b) Calculate the ratio of kinetic energies per unit mass at impact for the situation considered in (a).

(c) Repeat your calculations for Io in place of Callisto.

(d) Repeat your calculations in (a) and (b) for long-period comets that approach Jupiter with $v_\infty = 15$ km s^{-1}.

5.19.E (a) Draw a 5×5 square grid on which 25 craters are uniformly distributed at the centers of each grid square. You may represent the craters as either circles (which are substantially smaller than the grid squares) or points.

(b) Draw the same grid, but place the center of one crater at random within each grid square. Use a random number generator to compute the coordinates of each grid point.

(c) Place the 25 craters at random within the grid, as the production of craters would occur under most circumstances.

(d) Place the 25 craters in a clustered distribution, as would be expected if one part of the surface is much older than another part.

(e) Comment on using the spatial distribution of craters to determine the relative ages of planetary surfaces.

6   Planetary Interiors

at somewhere between 0.6 and 0.5 of the radius, measured from the surface, a very marked and remarkable change in the nature of the material, of which the Earth is composed, takes place.

R.D. Oldham, 1913

In the previous two chapters, we discussed the atmospheres and surface geology of planets. Both of these regions of a planet can be observed directly from Earth and/or space. But what can we say about the deep interior of a planet? We are unable to observe the inside of a planet directly. For the Earth and the Moon we have seismic data, revealing the propagation of waves deep below the surface and thereby providing information on the interior structure (§6.2). The interior structure of all other bodies is deduced through a comparison of remote observations with observable characteristics predicted by interior models. The relevant observations are the body's mass, size (and thus density), its rotational period and geometric oblateness, gravity field, characteristics of its magnetic field (or absence thereof), the total energy output, and the composition of its atmosphere and/or surface. Cosmochemical arguments provide additional constraints on a body's composition, while laboratory data on the behavior of materials under high temperature and pressure are invaluable for interior models. Quantum mechanical calculations are used to deduce the behavior of elements (especially hydrogen) at pressures inaccessible in the laboratory.

In this chapter we discuss the basics of how one can infer the interior structure of a body from the observed quantities. As expected, there are large differences between the interior structure of the giant planets, the terrestrial planets, and the icy moons. Moreover, even within each of these groupings there are noticeable differences in interior structure.

6.1 Modeling Planetary Interiors

Key observations used to extract information on the interior structure of a body are its mass, size, and shape. The mass and size together yield an estimate for the average density, which can be used directly to derive some first-order estimates on the body's composition. For small bodies, a density $\rho \lesssim 1$ g cm^{-3} implies an icy and/or porous object, while large planets of this density consist primarily of hydrogen and helium. A density $\rho \approx 3$ g cm^{-3} suggests a rocky object, while higher (uncompressed)[1] densities indicate the presence of heavier elements, in particular iron, one of the most abundant heavy elements in the cosmos (Table 8.1). The shape of a body depends upon its size, density, material strength, rotation rate, and history (including tidal interactions for moons). An object is approximately spherical if the weight of the mantle and crust exerted on its inner parts is large enough to deform the body. Any nonrotating 'fluid-like' body will take on the shape of a sphere, which corresponds to the lowest energy state. Note that the term 'fluid-like' in this context means deformable over geologic time (i.e., $\gtrsim$ millions of years), also referred to as *plasticity*. The shape or *figure* of a planet depends upon the *rheology* of the material (§5.3.2) and the rotation rate of the body. Rotation flattens a deformable object somewhat, changing its figure to an *oblate spheroid*, the equilibrium shape under the combined influence of gravity and centrifugal forces. A rocky body with a typical density $\rho = 3.5$ g cm^{-3} and material strength $\mathcal{S}_{\mathrm{m}} = 2 \times 10^9$ dyne cm^{-2} is approximately round if $R \gtrsim 350$ km; the maximum radius is $\sim$220 km for iron bodies to be oddly shaped (Problem 6.2).

In this chapter, we discuss the interior structure of bodies large enough to be in *hydrostatic equilibrium* (§3.2.3.1). To calculate the balance between gravity and pressure, one must know the gravity field as well as an equation of state which relates the temperature, pressure, and density in a planet's interior. The equation of state

---

[1] Uncompressed density: the density that a solid or liquid planet would have if material was not compressed by the weight of overlying layers.

Table 6.1 Densities and central properties of the planets and the Moon.

Planet	Radius (equatorial) (km)	Density (g cm^{-3})	Uncompressed density (g cm^{-3})	Central pressure (Mbar)	Central temperature (K)
Mercury	2 440	5.427	5.3	~0.4	~2 000
Venus	6 052	5.204	4.3	~3	~5 000
Earth	6 378	5.515	4.4	3.6	6 000
Moon	1 738	3.34	3.3	0.045	~1 800
Mars	3 396	3.933	3.74	~0.4	~2 000
Jupiter	71 492	1.326		~80	~20 000
Saturn	60 268	0.687		~50	~10 000
Uranus	25 559	1.318		~20	~7 000
Neptune	24 766	1.638		~20	~7 000

Data from Hubbard (1984), Lewis (1995), Hood and Jones (2000), Guillot (1999), and Yoder (1995).

depends upon the constituent relations of the various materials the planet is made of. In addition, the sources, losses, and transport mechanism(s) of the heat inside a planet are crucial to determining the object's thermal structure, which in turn is an important parameter in the derivation of a body's interior structure.

6.1.1 Hydrostatic Equilibrium

To first order, the internal structure of a spherical body is determined by a balance between gravity and pressure, and we assume hydrostatic equilibrium (see eq. 3.26):

$$P(r) = \int_r^R g_p(r')\rho(r')dr'. \tag{6.1}$$

Equation (6.1) can be used to calculate the pressure throughout the planet provided $\rho(r)$ is known (Table 6.1 provides a summary of central pressures and temperatures for the planets and our Moon). If the density is constant throughout the planet's interior, the pressure at the center of a planetary body, P_c, is given by (Problem 6.3b)

$$P_c = \frac{3GM^2}{8\pi R^4}. \tag{6.2}$$

Equation (6.2) provides a lower limit to the central pressure, since the density usually decreases with distance r. This method yields good estimates for relatively small bodies, with a nearly uniform density, such as the Moon, which has a central pressure of (only) 45 kbar. An alternative quick estimate can be obtained by assuming the planet consists of one slab of material, in which case the central pressure is a factor of two larger than the value obtained in the previous estimate (Problem 6.3b). Since the single slab technique overestimates the gravity over most of the region of integration, the actual pressure at the center of the planet usually lies between these two values.

On the other hand, if the planet is extremely centrally condensed, the density increases sharply towards the center of the planet, and the pressure calculated using the single slab model may still be too low compared to the actual value. We find that the central pressure of Earth calculated according to the single slab model agrees quite well with the actual value of 3.6 Mbar (Problem 6.3). The Earth is differentiated, and the increase in density towards the center just about compensates for our overestimate in gravity. Using the single slab model, Jupiter's central pressure is still underestimated by a factor of ~4, since this planet is very dense near its center (realistic models show that Jupiter's central pressure is roughly 80 Mbar).

An accurate estimate of a planet's internal structure requires assumptions regarding the planet's composition, as well as knowledge of the equation of state and constituent relations of the material. It is also crucial to know the temperature structure throughout the interior, which is determined by internal and external (such as tidal friction) heat sources, heat transport, and heat-loss mechanisms. The sources of heat are strongly tied to the planet or moon's formation history. All of this information must be used to compute interior models, which can then be checked against observations and refined in an iterative manner.

6.1.2 Constituent Relations

In order to develop realistic models of a planet's interior, one needs to know the phases of the materials inside a planet as functions of temperature and pressure. In §5.2 we showed that the state of the material, i.e., whether the material is in a solid, liquid, or vapor phase, depends upon the temperature and pressure of the environment. In principle, one can determine the melting temperature, T_m,

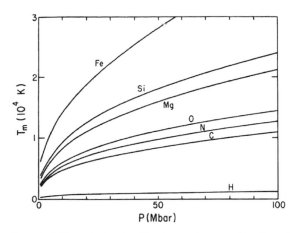

Figure 6.1 The calculated melting temperature as a function of pressure for various common elements. (Hubbard 1984)

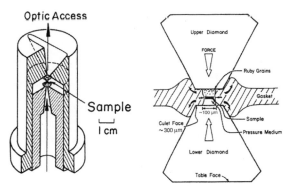

Figure 6.2 Left: Sketch of a diamond anvil press. Two single-crystal gem-quality diamonds are compressed in a piston-cylinder assembly. The sample is placed between the opposed points of the diamonds, and because of the small surface area, extremely high pressures (~1 Mbar) can be reached and maintained over long periods of time. Because the diamonds are transparent, the sample can be 'seen' at wavelengths extending from far-infrared to hard X-ray and γ-ray energies. Right: Details of the sample and pressure medium. Spectra of embedded fine-grained ruby-powder ($\lesssim 5$ μm grain size) allow precise determination of the temperature and pressure. (Jeanloz 1989)

as a function of pressure for any component by solving the equation (§5.2.1):

$$G_\ell(T_\mathrm{m}, P) = G_\mathrm{s}(T_\mathrm{m}, P), \tag{6.3}$$

where G_ℓ and G_s are the Gibbs free energies for the liquid and solid phase of the material, respectively. The *Lindemann criterion* states that melting occurs when the thermal oscillations of the ions in the material lattice become a significant fraction (~10%) of the equilibrium spacing of the ions in the lattice. With the thermal oscillations being proportional to T/Z^2, the melting temperature, T_m, can be approximated by

$$T_\mathrm{m} \approx \frac{Z^2}{150 r_\mathrm{s}}. \tag{6.4}$$

where r_s is a measure of the equilibrium spacing of ions in the lattice (in atomic units) and Z is the atomic number. Lindemann's criterion holds quite well for simple crystals which have structures with closely packed atoms. Figure 6.1 shows the approximate relationship between T_m and pressure, P, for various elements.

The derivation of empirical constituent relations is relatively easy at low pressure, where the chemical reactions and phase transitions are well known for many materials. However, the pressures and temperatures in planetary interiors can be very high, and in such an environment it is difficult to predict whether a material will be in a solid or liquid phase. Moreover, as discussed in §5.2, mixtures of elements undergo chemical reactions that depend upon the temperature and pressure of the environment, eutectic behaviors play a role and phase diagrams may become quite complicated. A stable system is one where the Gibbs free energy is at a minimum. Typically, above some critical temperature, a solution is in a single liquid phase, while

below this temperature several liquid and solid phases of different composition may coexist. Under the temperature and pressure conditions encountered in the interiors of terrestrial planets, one usually expects a chemical separation of the compounds (such as metals/siderophiles from the silicates/lithophiles).

Experiments under high pressure can be conducted with various techniques. By using a hydraulic press, rocks in the laboratory can be squeezed to pressures of ~100 kbar, and heated to roughly 1200–1400 K. Static pressures of a little over one Mbar can be reached using a diamond anvil press (Fig. 6.2). In these devices, material is squeezed between two diamonds, each ~350 μm across. The high pressure can be kept constant for weeks, months, or longer. Since the diamonds are transparent, the sample can be seen while being compressed and heated, and the temperature can be regulated. Measurements at pressures $\gtrsim 1$ Mbar are conducted by means of shock wave experiments (Fig. 6.3) or powerful lasers. Unfortunately, the duration of the high-pressure state in shock wave experiments typically lasts only a fraction of a microsecond, and the rock sample is destroyed in the process. In these experiments, however, temperatures of many thousand K can be reached, close to the temperatures prevailing in planetary interiors. The magnetic 'Z' accelerator at Sandia National Laboratories is a giant X-ray generator where such extremely high temperatures and pressures have been reached that fusion of deuterium occurred. However, at

Figure 6.3 Photograph of the 60 foot long, two-stage light-gas (usually H₂) gun at Lawrence Livermore National Laboratory. This gun is used to obtain the equation of state of various materials through shock wave experiments and to investigate impact events. (Courtesy Lawrence Livermore National Laboratory)

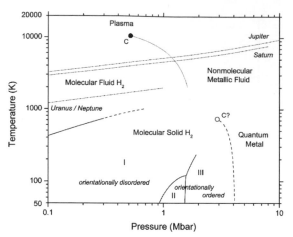

Figure 6.4 Phase diagram of hydrogen at high pressures, showing the transition from molecular hydrogen into metallic hydrogen. At low temperatures (unrealistic for planets within our Solar System) hydrogen is present as an electrically insulating solid (I, II, or III) below ~4 Mbar. At pressures $\gtrsim 4$ Mbar, molecular hydrogen is believed to change to (solid) molecular metallic hydrogen, and at still higher pressures to a quantum metal. Whether this is a fluid or a solid is not known, and the transformation temperatures at higher pressures, if any, are not determined at this time. At much higher temperatures, hydrogen breaks up into a plasma (or becomes highly degenerate), a transition which is probably continuous rather than discontinuous. A critical point, C, exists near 0.5 Mbar and 10 000 K. A second critical point may exist near 3 Mbar, 1000 K. The adiabats of the giant planets are superposed as dot-dashed curves. (Diagram courtesy of Stephen A. Gramsch, Carnegie Institution of Washington; adiabats for the planets were provided by William B. Hubbard)

present one largely relies on theoretical arguments to estimate conditions at pressures much above 5–10 Mbar.

In the following subsections, we discuss the various phases of materials which make up the planets and relate our findings to what is known about planetary interiors. We give detailed descriptions of individual planetary interiors in §6.2, §6.3, and §6.4.

6.1.2.1 *Hydrogen and Helium*

Typical temperatures and pressures in the giant planets range from about 50–150 K at a planet's tropopause up to 7000–20 000 K at their centers, while the pressure varies from near zero in the outer atmosphere up to 20–80 Mbar at the planet's center. A phase diagram for hydrogen over a wide range of pressures and temperatures is shown in Figure 6.4. This diagram is based upon experiments and theoretical calculations by many different groups, and it is far from complete. At pressures below ~4 Mbar, hydrogen exists in its molecular form. Above ~4 Mbar, and below approximately 1000 K, there are probably continuous transitions from insulating molecular H₂ to metallic molecular H₂, and finally to a quantum metal. The nature of the quantum metal, and whether it is a fluid or a solid is not known. In Jupiter, dissociation (H₂ → 2H) begins near 0.95 R_♃ and goes to completion at an estimated radius of ~0.8 R_♃. Because of the energy required to dissociate H₂, the temperature is nearly constant over a relatively wide range of pressures and radii. The fluid is so densely packed at these high (Mbar) pressures that the separation between the molecules becomes comparable to the size of the molecules, so that their electron clouds start to overlap and an electron can hop or percolate from one molecule

to the next. Shock wave experiments at pressures between 0.1 and 1.8 Mbar and temperatures up to ~5000 K show that the conductivity of fluid molecular hydrogen increases monotonically and reaches the minimum conductivity of a metal at a pressure of 1.4 Mbar. Thus at $T \gtrsim 2000$ K and $P \gtrsim 1.4$ Mbar fluid hydrogen behaves like a metal, a phase referred to as *fluid metallic hydrogen* (Fig. 6.4). In Jupiter, this transition occurs at a radius of ~0.90 R_♃. Convection in this fluid is thought to create the magnetic fields observed to exist around Jupiter and Saturn. At $P \gtrsim 4$ Mbar, it is expected that the molecular hydrogen dissociates into an atomic metallic state. No measurements exist on the dissociation of metallic molecular hydrogen into metallic atomic hydrogen, however. At much higher temperatures, theory predicts that hydrogen may either become highly degenerate, or a plasma. Whether this transition is continuous, as expected for ionization resulting from high temperatures, or a first-order phase transition with discontinuities in densities and entropy, is unknown. If there would be a phase transition, it would create a

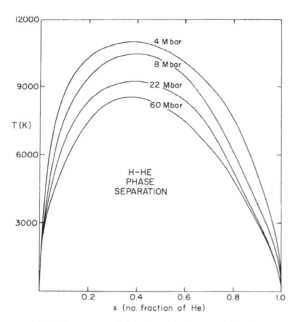

Figure 6.5 Phase separation between hydrogen and helium as a function of temperature at different pressures. The miscibility gap is below the curve in each case. (Stevenson and Salpeter 1976)

barrier for convection between the metallic and molecular hydrogen regions, affecting mixing of chemical species, so that observed atmospheric abundances may not be indicative of the planet's bulk composition.

The giant planets consist primarily of a mixture of hydrogen and helium. Hydrogen is expected to be in liquid metallic form in the interiors of both Jupiter and Saturn. Helium transforms into a liquid metallic state only at pressures much higher than encountered in the giant planets. Hydrogen and helium are fully mixed only if the temperature and pressure are high enough. At lower temperatures and pressures, the liquid phases of hydrogen and helium do not mix. Calculations on the miscibility of hydrogen and helium, as a function of temperature and helium abundance, are shown in Figure 6.5. Curves for four different pressures are shown. Above the line the two phases are completely mixed, but below the line the phases separate out. Given the temperatures and pressures expected to prevail in the interiors of Jupiter and Saturn, we expect that helium and hydrogen are not fully mixed. The observed He/H ratio in Jupiter's atmosphere is slightly less than the solar ratio, and helium is even more depleted in Saturn's atmosphere. These depletions have been attributed to He separation within the metallic hydrogen region. Because the temperature at a given pressure in Saturn's interior is always less than in Jupiter, the effect of immiscibility is stronger within Saturn than within Jupiter (see §6.4.2).

6.1.2.2 Ices

Water-ice is a major constituent of bodies in the outer parts of our Solar System. Depending on the temperature and pressure of the environment, water-ice can take on at least 15 different crystalline forms, many of which are shown in the water phase diagram in Figure 6.6. Although the water molecule does not change its identity, the molecules are more densely packed at higher pressures, so that the density of the various crystalline forms varies from 0.92 g cm^{-3} for common ice (form I) up to 1.66 g cm^{-3} for ice VII near the triple point with ices VI and VIII. Temperatures and pressures expected in the interiors of the icy satellites range from $\sim$50–100 K at the surface up to several hundred K at pressures of up to a few tens kbar in their deep interiors. So one might expect a wide range of ice-forms in these satellites. The adiabats of the giant planets are superposed on Figure 6.6. At the higher temperatures, above 273 K at moderate pressures, water is a liquid. The *critical point* of water ($T = 647$ K, $P = 221$ bars) is indicated by a C; above this temperature there is no first-order phase transition between gaseous and liquid H_2O. Water becomes a supercritical fluid, characterized as a substance that is neither gas nor liquid, with properties that are very different from ambient water.

Pure water is slightly ionized, with H_3O^+ and OH^- ions. At higher temperatures and pressures, the ionization is enhanced. Shock wave data on 'synthetic Uranus', which is a mixture of water, isopropanol, and ammonia, show that the mixture ionizes at pressures over 200 kbar to form an electrically conductive fluid. The conductivity of this mixture is essentially the same as that for pure water under the same high P conditions. The conductivity is high enough to explain the existence of the observed magnetic fields of Uranus and Neptune. At pressures over 1 Mbar the ice constituents dissociate, and the fluid becomes rather 'stiff', i.e., the density is not very sensitive to pressure.

In addition to water-ice, one would expect the outer planets and moons to contain substantial amounts of other 'ices', such as ammonia, methane, and hydrogen sulfide. Phase diagrams of these ices might be as complex as those for water. Although experiments up to 0.5–0.9 Mbar have been carried out for some of these constituents, they are not as well studied under high pressures as is water-ice, and mixtures of these various ices are characterized even less.

6.1.2.3 Rocks and Metals

The phase diagrams of magmas, i.e., molten rocks, were discussed in §5.2, and were shown to be very complex. The various elements and compounds interact in different

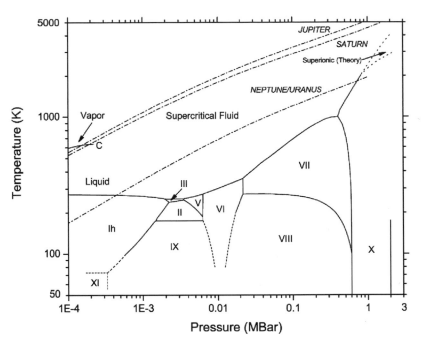

Figure 6.6 Phase diagram for water at temperatures and pressures relevant for the icy satellites as well as the giant planets (adiabats for the giant planets are superposed). The various crystal forms of ice are indicated by roman numerals I–XII, where 'h' in Ih refers to the hexagonal crystal form of ordinary ice (all natural snow and ice on Earth is in the Ih form). The metastable ices IV and XII are not shown (they fall in the top corner of II, and middle of V, respectively). The critical point of water is indicated by a C. (Diagram courtesy of Stephen A. Gramsch, Carnegie Institution of Washington; adiabats for the planets were provided by William B. Hubbard)

ways, depending on the temperature, pressure, and composition of the magma. From terrestrial rocks, models of phase diagrams, laboratory experiments, and discontinuities in the density profile of the Earth's interior (§6.2), we have obtained a reasonable idea of the composition of the Earth's mantle and core. The primary minerals of the Earth's upper mantle are olivine, $(Mg,Fe)_2SiO_4$, and pyroxene, $(Mg,Fe)\,SiO_3$, which together make up the rock *peridotite*. It is dominated by Mg, with a molar concentration, [Mg/(Mg+Fe)], of 89. Experiments have shown that at higher pressure the molecules get more closely packed, which leads to a reorganization of the atoms in the lattice structure. This rearrangement of atoms/molecules into a more compact crystalline structure involves an increase in density, and characterizes a phase change. A common example is carbon, which at low pressure is present in the form of graphite, and at high pressure as diamond. At a depth of $\sim$50 km ($\sim$15 kbar) basalt changes into *eclogite*, i.e., the structure of some minerals in the basalt changes, such as the conversion of pyroxene to garnet. Since this transformation takes place only at several 'spots' near the Earth's surface, where cold oceanic lithospheric plates descend (§6.2.2), there is no global seismic discontinuity measured at this depth. Global seismic discontinuities are measured at a depth of $\sim$400 km, where the mineral olivine changes into a spinel structure (i.e., ringwoodite, $MgSiO_4$) via an exothermic reaction, and at a depth of $\sim$660 km (0.23 Mbar), where ringwoodite is decomposed into magnesiowüstite, $(Mg,Fe)O$, and the perovskite

phase $(Mg,Fe)SiO_3$ via an endothermic reaction. Technically, although we refer to this mineral as perovskite, it has a composition similar to pyroxene but a structure like the mineral perovskite, $CaTiO_3$. Perovskite is stable up to pressures of $\sim$1.25 Mbar (near the core–mantle boundary, §6.2.2), and is probably the dominant 'rock' in the Earth's interior. At higher pressures, the perovskite, with an orthorhombic or cubic-like crystal structure, changes into postperovskite, characterized by a more octahedral sheet structure, with a $\sim$1% increase in density.

The primary constituents in the Earth's core are almost certainly iron and nickel (§6.2.2.2), though the density of the Earth's core is $\sim$5–10% lower than the density of an Fe–Ni alloy. This suggests that the core is composed of iron and nickel mixed with a lower density material, possibly sulfur, oxygen, or hydrogen (§6.2.2.2). The phase diagram of iron is well known for pressures up to $\sim$200 kbar; four different solid phases can be distinguished. The melting curve for iron has been determined up to 2 Mbar using laser heating in a diamond anvil cell. Shock compression experiments have been used at higher pressures. Although much progress has been made in understanding the behavior of iron under high pressure, details of the various experiments are not (yet) in perfect agreement.

In the protoplanetary nebula at temperatures below $\sim$700 K, Fe reacts with H_2O and H_2S to form FeO and FeS. Within a larger body, iron oxide is largely incorporated with magnesium silicates in rocks as olivine and pyroxene. Iron sulfide, however, is expected to settle in the planet's core with iron. This could explain the lower density of the Earth's core compared to that of pure iron. The alloy of iron and sulfur has a *eutectic* behavior, which means that the mixture melts completely into Fe + FeS at a temperature well below the melting temperature of each mineral separately. At the 1 bar pressure level, Fe melts at a temperature of 1808 K, FeS at 1469 K, while an iron–sulfur alloy (mixture) of 27% sulfur and 73% iron (solar mixture, which is close to a eutectic mixture) melts at 1262 K. At pressures as high as 100 kbar, the melting temperature of this eutectic mixture is depressed by nearly 1000 K compared to the melting temperature of pure iron.

6.1.3 Equation of State

The equation of state is an expression which relates the pressure, density, temperature, and composition, $P = P(\rho, T, f_i)$. In planetary atmospheres at pressures below ~50 bar, one can use the perfect (ideal) gas law, equation (3.27). At higher temperatures and pressures this simple equation is not adequate, because the molecules can no longer be treated as infinitesimally small spheres. When intermolecular spacings decrease to ~0.1–0.2 nm, van der Waals forces become important, and the atoms/molecules start to interact. At higher pressures, liquids and solids may be formed, mineral phases may change, and the electronic structure of individual atoms and molecules may be modified. The equation of state is usually derived from measurements at room temperature, augmented by data at higher temperature and pressure based upon, e.g., shock wave, laser, and/or diamond anvil experiments.

In a hydrogen-rich environment, like the giant planets, the electron clouds of the hydrogen molecules start

Over limited ranges in pressure, the equation of state of material in planetary interiors is usually well-approximated by that of a *polytrope*:

$$P = K_{po}\rho^{1+n_{po}}, \tag{6.5}$$

where K_{po} and n_{po} are the *polytropic constant* and *polytropic index*, respectively. At very low pressures, $P \to 0$, $1/n_{po} \approx \infty$, while in the limit of high pressures, $1/n_{po} = 3/2$ and $P \propto \rho^{5/3}$.

For planets composed of incompressible material, the relation between planetary mass and radius is given by $R \propto M^{1/3}$. When a sufficient amount of matter is added to the planet, the material is compressed and the radius increases more slowly. There is a maximum size that a sphere of matter can reach, and adding more mass makes it shrink (§12.1, Problem 12.1, Fig. 6.25). Jupiter's size is near-maximal for a solar composition planet. Cool massive brown dwarfs, discussed in §12.1, are slightly smaller than Jupiter.

6.1.4 Gravity Field

The gravity field of a planet or moon contains information on the internal density structure. The gravity field can be determined to quite high accuracy by tracking the orbits of spacecraft close to the body or from the rate of precession of the periapses of moons and rings orbiting the planet. In the following we show how to extract information on a body's internal structure from its gravity field.

The gravitational potential of a body can be obtained by solving Laplace's equation (2.20b):

$$\Phi_g(r,\phi,\theta) = -\left(\frac{GM}{r} + \Delta\Phi_g(r,\phi,\theta)\right), \tag{6.6a}$$

where $\Delta\Phi_g$ represents any deviations in the gravitational potential from that corresponding to a nonrotating fluid body in hydrostatic equilibrium:

$$\Delta\Phi_g(r,\phi,\theta) = \frac{GM}{r} \sum_{n=1}^{\infty} \sum_{m=0}^{n} \left(\frac{R}{r}\right)^n (C_{nm}\cos m\phi + S_{nm}\sin m\phi)P_{nm}(\cos\theta). \tag{6.6b}$$

to overlap at high pressures, which increases the conductivity. At pressures of ~1.4 Mbar, hydrogen enters a molecular metallic state, which is referred to as *metallic hydrogen* (§6.1.2.1). In the extreme limit of a fully pressure ionized gas ($P \gtrsim 300$ Mbar), hydrogen is presumably present in atomic form, the electrons become degenerate, and the pressure is independent of temperature. Although the pressures where electrons become degenerate are much higher than encountered in planetary interiors, a short discussion on the mass–density relation for this situation is enlightening for our discussion of the giant planets in §6.4 (see also §12.1).

Equation (6.6) is written in standard spherical coordinates, with ϕ the longitude (or azimuthal position) and θ the colatitude. R is the mean radius, M the mass of the body, and the terms $P_{nm}(\cos\theta)$ are the associated Legendre polynomials, of degree n and order m, given by

$$P_{nm}(x) = \frac{\left(1-x^2\right)^m}{2^n n!} \frac{d^{n+m}}{dx^{n+m}}\left(x^2-1\right)^n. \tag{6.7}$$

Note that $P_{n0}(x) = P_n(x)$ as given in equation (2.30). The Stokes coefficients, C_{nm} and S_{nm}, are determined by the internal mass distribution, a distribution which is, in part, determined by the body's rotation and tidal deformation effects:

$$C_{nm} = \frac{2 - \delta_{0m}}{MR^n} \frac{(n-m)!}{(n+m)!} \int_E \rho r^n P_{nm} \cos(\theta) \cos m\phi \, dV$$

$$(6.8a)$$

$$S_{nm} = \frac{2 - \delta_{0m}}{MR^n} \frac{(n-m)!}{(n+m)!} \int_E \rho r^n P_{nm} \cos(\theta) \sin m\phi \, dV,$$

$$(6.8b)$$

with ρ the density, δ_{0m} the Kronecker δ, i.e., $\delta_{0m} = 1$ if $m = 0$ and $\delta_{0m} = 0$ if $m \neq 0$, and E the entire volume of the planet. With equations (2.41) and (2.42), the Stokes coefficients can be rewritten in the moments and products of inertia. For example, it can be shown that the Stokes coefficients C_{20} and C_{22} are related to the moments of inertia I_A, I_B, and I_C along the three (orthogonal) axes with $A > B > C$, with the short axis parallel to the rotation axis:

$$C_{20} = \frac{I_A + I_B - 2I_C}{2MR^2},$$

$$(6.9a)$$

$$C_{22} = \frac{I_B - I_A}{4MR^2}.$$

$$(6.9b)$$

Most planets are very nearly axisymmetric, with the major departure from sphericity being due to a rotationally induced equatorial bulge. Under these circumstances, and with the coordinate system centered at the center of mass, all $S_{nm} = 0$, and for $m \neq 0$ the coefficients $C_{nm} = 0$. This simplifies equation (6.6) to equation (2.29), where we define the zonal harmonics or *gravitational moments*, J_n:

$$J_n \equiv -C_{n0}.$$

$$(6.10)$$

For a nonrotating fluid body in hydrostatic equilibrium, the moments $J_n = 0$, and the gravitational potential reduces to $\Phi_g(r, \phi, \theta) = -GM/r$. Rotating fluid bodies in hydrostatic equilibrium have $J_n = 0$ for all odd n. This is a very good approximation for the giant planets. However, to accurately model the gravity fields of the terrestrial planets and moons we may need to use more terms in the spherical harmonic expansion (eq. 6.6). For example, the Earth's gravity field has been modeled with harmonic degrees and orders >360.

6.1.4.1 *Equipotential Surface*
Within rotating bodies, the effective gravity is less than the gravitational attraction calculated for a nonrotating planet, since the centrifugal force induced by rotation is directed outwards from the planet (§2.1.4). The equipotential surface on a planet, i.e., the *geoid*[2] on Earth and *areoid* on

[2] On Earth, the geoid is measured at the mean sea level.

Mars, must thus be derived from the sum of the gravitational potential, Φ_g, and the rotational, or centrifugal, potential Φ_c:

$$\Phi_g(r, \phi, \theta) + \Phi_c(r, \phi, \theta) = \text{constant},$$

$$(6.11)$$

where $\Phi_g = -GM/r$ and Φ_c is defined as

$$\Phi_c = -\frac{1}{2} r^2 \omega_{rot}^2 \sin^2 \theta.$$

$$(6.12a)$$

We can rewrite equation (6.12a) into a radial term and one that describes the equatorial flattening, to make it easier to relate the Stokes coefficients in equations (6.6) or (2.29) to the planet's rotation (Problem 6.6):

$$\Phi_c = \frac{1}{3} r^2 \omega_{rot}^2 (1 - P_2(\cos\theta)).$$

$$(6.12b)$$

The planetocentric distance of the equipotential surface, $r(\theta)$, and the surface gravity, $g_p(\phi, \theta)$, as a function of colatitude θ can then be obtained by solving equation (6.11):

$$g_p(\theta) = (1 + C_1 \cos^2 \theta + C_2 \cos^4 \theta) g_p(\theta = 90°).$$

$$(6.13)$$

Equation (6.13) is known as the *reference gravity formula*, with $g_p(\theta = 90°)$ the equatorial gravity. The coefficients C_1 and C_2 can be determined accurately for Earth ($C_1 = 5.278\,895 \times 10^{-3}$ and $C_2 = 2.3462 \times 10^{-5}$), since both Φ_g and Φ_c are known to high accuracy. Determining the coefficients for other planets is not straightforward, since both the gravity field and rotation rate need to be well known. The latter may be difficult to determine, because measurements from the rotation of features on planets with optically thick atmospheres (giant planets, Venus) yield the rotation period of atmospheric winds rather than the interior. For the giant planets, observations of the nonthermal radio emissions are useful diagnostics of the rotation period of the interior of a planet (but note the uncertainties in Saturn's rotation rate, §7.5), while radar techniques can be used to determine the rotation rate for solid bodies covered by a dense atmosphere.

6.1.4.2 *Gravitational Moments*
Axisymmetric Planet in Hydrostatic Equilibrium
An axisymmetric rotating planet which consists of an incompressible fluid of uniform density takes on the form of a *Maclaurin spheroid*. The moment of inertia of a Maclaurin spheroid about its short axis is larger than that of a sphere of equal volume, enabling a rotating body with this shape to have a lower total energy (gravitational potential energy plus kinetic energy of rotation) than does

a sphere with the same rotational angular momentum.[3] The polar flattening is determined by the rotation rate and the rheology (§5.3.2) of the material, quantified by the *fluid* or *tidal Love number*, a quantity originally derived by Love in 1944 for a homogeneous elastic body:

$$k_T = \frac{3}{2} \left(1 + \frac{19\mu_{rg}}{2\rho g_p R} \right)^{-1}, \qquad (6.14)$$

where R is the radius, ρ the density, g_p the gravitational acceleration, and μ_{rg} the *rigidity* or *shear modulus*. Note that for fluids, $\mu_{rg} = 0$, and k_T becomes equal to 3/2. It can be shown that the second harmonic, J_2, is proportional to k_T and q_r:

$$J_2 = \frac{1}{3} k_T q_r. \qquad (6.15)$$

The nondimensional quantity q_r is the ratio of the centrifugal to the gravitational force at the body's surface:

$$q_r \equiv \frac{\omega_{rot}^2 R^3}{GM}. \qquad (6.16)$$

For a uniform density distribution, $k_T = 3/2$, and $J_2 = 0.5 q_r$. Usually the density in a planet increases towards the center, so that $J_2 < 0.5 q_r$ (Problem 6.7). The ratio between J_2 and q_r is known as the *response coefficient* Λ_2, which measures the response of the planet to its own rotation:

$$\Lambda_2 \equiv \frac{J_2}{q_r} = \frac{1}{3} k_T. \qquad (6.17)$$

The coefficient Λ_2 contains information on the spatial distribution of the mass in a planet's interior: Rotating planets with high-density cores have small values of Λ_2, whereas bodies with a more homogeneous density distribution have larger values of Λ_2. For an incompressible fluid of uniform density, $\Lambda_2 = 0.5$ (Problem 6.7).

From a solution of the equipotential surface for a planet in hydrostatic equilibrium, it follows that the geometric oblateness, ϵ, is related to the rotation period and the second harmonic (Problem 6.8):

$$\epsilon \equiv \frac{R_e - R_p}{R_e} \approx \frac{3}{2} J_2 + \frac{q_r}{2}, \qquad (6.18)$$

where R_e and R_p represent the planet's equatorial and polar radius, respectively. The oblateness, ϵ, and J_2 are of order q_r (eqs. 6.15, 6.18). For rapidly rotating planets in hydro-

[3] A very rapidly rotating self-gravitating incompressible fluid takes the form of a *Jacobi ellipsoid*. A Jacobi ellipsoid is triaxial, and thus has an even greater moment of inertia than does a Maclaurin spheroid of comparable volume and gravitational potential energy.

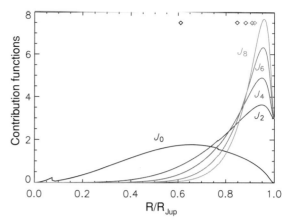

Figure 6.7 Contribution functions of the gravitational moments for Jupiter, as a function of jovian radius. The diamonds at the top show the median radius corresponding to each moment. J_2 has the largest contribution from the planet's interior, and each higher order moment is more and more sensitive to the outer layers only. J_0 is equivalent to the planet's mass. The discontinuities in each gravitational moment are caused by the core/envelope ($\sim$0.07 R$_{\jupiter}$) and helium rich/helium poor (metallic/molecular) transition ($\sim$0.77 R$_{\jupiter}$). (Adapted from Guillot 2005)

static equilibrium, higher order zonal harmonics are proportional to q:

$$J_{2n} \propto q_r^n. \qquad (6.19)$$

The higher order zonal harmonics for such planets are therefore small compared to J_2, and hence more difficult to determine. However, because progressively higher order moments reflect the mass distribution closer and closer to a planet's surface (Fig. 6.7), the J_2 and J_4 moments are important parameters for the determination of a body's interior structure.

Since the Stokes coefficients, and hence J_2, can be written in terms of the products and moments of inertia, one can derive an approximate algebraic formula to illustrate the relations between the moment of inertia, the rotation, and J_2, which is known as the *Radau–Darwin approximation*:

$$\frac{I}{MR^2} = \frac{2}{3} \left(1 - \frac{2}{5} \sqrt{\frac{5q_r}{2\epsilon} - 1} \right). \qquad (6.20)$$

In the above we used the moment of inertia along the polar axis, which is the largest inertia value. If the density ρ is uniform throughout the planet, $I = 0.4MR^2$ (Problem 6.4). $I/(MR^2) = 0.667$ for a hollow sphere. The ratio $I/(MR^2) < 0.4$ if ρ increases with depth in the planet. Usually the density increases towards a planet's center,

Table 6.2 Gravitational moments and the moment of inertia ratio.

Body	J_2 $(\times 10^{-6})$	J_3 $(\times 10^{-6})$	J_4 $(\times 10^{-6})$	J_6 $(\times 10^{-6})$	q_r	Λ_2	I/MR^2	C_{22} $(\times 10^{-6})$	Refs.
Mercury	60 ± 20				1.0×10^{-6}	60	0.33		1
Venus	4.46 ± 0.03	-1.93 ± 0.02	-2.38 ± 0.02		6.1×10^{-8}	73	0.33		1
Earth	$1\,082.627$	-2.532 ± 0.002	-1.620 ± 0.003	-0.21	3.45×10^{-3}	0.314	0.331		1
Moon	203.43 ± 0.09				7.6×10^{-6}	26.8	0.393	22.395	1, 2
Mars	$1\,960.5 \pm 0.2$	31.5 ± 0.5	-15.5 ± 0.7		4.57×10^{-3}	0.429	0.365		1
Jupiter	$14\,696.4 \pm 0.2$		-587 ± 2	34 ± 5	0.089		0.165	0.254	1
Saturn	$16\,290.7 \pm 0.3$		-936 ± 3	86 ± 9	0.151		0.108	0.210	4
Uranus	$3\,343.5 \pm 0.1$		-28.9 ± 0.2		0.029		0.114	0.23	1
Neptune	$3\,410 \pm 9$		-35 ± 10		0.026		0.136	0.23	1
Io	$1\,860 \pm 3$				1.7×10^{-3}	1.08	0.378	558.8	3
Europa	436 ± 8				5.02×10^{-4}	0.87	0.346	131.5	3
Ganymede	128 ± 3				1.91×10^{-4}	0.67	0.312	38.3	3
Callisto	33 ± 1				3.67×10^{-5}	0.90	0.355	10.2	3

1: Yoder (1995) and http://ssd.jpl.nasa.gov/. 2: Konopliv *et al.* (1998). 3: Schubert *et al.* (2004). 4: Anderson and Schubert (2007).

both because dense compounds tend to sink and because material gets compressed at higher pressure.

Table 6.2 shows the J_n, q_r, Λ_2, and $I/(MR^2)$ values for the planets and several large moons. As mentioned above, the relation between the gravitational moments and the internal density distribution was derived for a rotating planet in hydrostatic equilibrium. Using the measured J_2 and q_r for Earth, we find that the 'actual' $I/(MR^2)$ as listed in Table 6.2 (derived using a 'best' approximation to the density profile in Earth – §6.2.2) is close to the equilibrium value calculated from equation (6.20). For Mars the equilibrium value is slightly higher (0.375). The discrepancy has been attributed to the Tharsis uplift, which is not in isostatic equilibrium (§6.3). The giant planets are in hydrostatic equilibrium and rapid rotators; the small values for Λ_2 and $I/(MR^2)$ suggest a pronounced increase in density towards their centers.

The situation is different for Mercury and Venus. These planets have long rotation periods (and therefore small values of q_r), so that nonhydrostatic effects (e.g., mantle convection) have a much larger contribution to the J_2 than does the rotational effect. Indeed, the $I/(MR^2)$ values as calculated from equation (6.20) are very different from the 'actual' values (Problem 6.5).

Triaxial Bodies

Our Moon and the Galilean satellites are in synchronous rotation with their orbital period, so that in addition to

rotation, tidal forces affect their shape and hence gravity fields (§2.6.2). The primary harmonic coefficients of relevance are the quadrupole C_{20} (or J_2) (dynamic polar flattening) and C_{22} (dynamic equatorial flattening) coefficients in equations (6.6) and (6.9). A synchronously rotating satellite takes on a (nearly) triaxial shape with axes $A > B > C$, where the long axis is directed along the planet–satellite line and the short axis is parallel to the rotation axis. If the satellite is in hydrostatic equilibrium, it can be shown that the quadrupole coefficients depend on k_T, q_r, and q_T:

$$C_{22} = -\frac{1}{12} k_T q_T, \tag{6.21a}$$

$$J_2 = \frac{1}{3} k_T \left(q_r - \frac{1}{2} q_T \right), \tag{6.21b}$$

where the tidal coefficient, q_T, is defined:

$$q_T \equiv -3 \left(\frac{R_s}{a^3} \right)^3 \frac{M_p}{M_s}, \tag{6.22}$$

with R_s and M_s the satellite's radius and mass, respectively, M_p the planet's mass, and a the satellite–planet distance, or semimajor axis of the satellite's orbit if the eccentricity is small. For a body in hydrostatic equilibrium and synchronous rotation, $q_T = -3q_r$ and $J_2 = \frac{10}{3} C_{22}$ (Problem 6.9). Measurements of both J_2 and C_{22} can thus be used to determine whether a satellite is in hydrostatic equilibrium or not. For example, our Moon is not in equilibrium; its gravity field is dominated by internal mass distributions that are not isostatically compensated.

6.1.4.3 *Effects of Mass Anomalies*
Precession

As discussed in the previous sections, rotating bodies develop an equatorial bulge. External forces, such as gravitational interactions with the Sun, produce a net torque on the body's equator, which leads to a rotation of the rotational axis, i.e., the axis will change its position relative to distant stars; this spinning of the rotation axis is referred to as *precession* (see also §2.5.3). For example, the Sun, Moon, and other planets produce a net torque on the Earth's equator, which leads to a precession of the Earth's rotation axis with a period of ∼26 000 years. Because of variations in the torque due to the motions of the Sun and Moon relative to the Earth's equatorial bulge, there are wobbles superposed on the spinning axis, known as *nutation*. The precession of the Moon's orbit has the largest effect on the nutation, with a period of 18.6 years.

The angular rate at which the body is precessing, Ω_{rot}, is equal to the torque applied to it divided by its spin angular momentum, ω_{rot}, which is related to the body's moments of inertia:

$$\Omega_{rot} = \frac{3Gm\sin(2\psi)}{2r^3\omega_{rot}}\left(\frac{I_C - I_A}{I_C}\right), \qquad (6.23)$$

with m the mass of the perturbing body (Moon, Sun), r the distance to the perturbing body, and ψ the angle the rotational equator makes with the orbital plane of the perturber (e.g., obliquity in the Sun–Earth case). A measurement of the precession and J_2 for an axisymmetric planet ($I_B = I_A$) thus yields the moments of inertia I_C and I_A, parameters which are essential to extract information on a body's internal structure.

Polar Wander

The previous subsections dealt with bodies which were in rotational equilibrium. We considered primarily the polar and equatorial flattening of objects as induced by rotation and tides, which all depend on the moments of inertia along the principal axes. For solid objects, even when confining ourselves only to Stokes coefficients of degree 2, we find that, e.g., C_{21}, S_{21}, and S_{22} are related to moments of inertia which are not along the principal axes. Bodies with nonzero values for these coefficients are not in rotational equilibrium, and undergo a *torque-free precession*, i.e., they *wobble* (see also §9.4.6).

In this torque-free precession, angular momentum is conserved, and it is the planet, e.g., Earth, that is reorienting itself in space, while the rotational axis stays fixed with respect to distant stars. In the reference frame of Earth, the rotation axis appears to 'wander' across the globe, a phenomenon known as *polar wander*. On Earth we distinguish between *apparent* and *true* polar wander, where true polar wander is most significant (and important on other planets and moons) as it is measured with respect to the deep mantle (e.g., using hot mantle plumes as a reference system, §6.2.2). Large mass anomalies would lead to a complete reorientation of the polar axis with respect to the globe, while periodicities in polar wander must be caused by periodic displacements of mass. For example, on Earth the growth and decay of ice sheets, changes in the convection of the mantle and/or core, tectonic plate motions, and (seasonal) changes in atmospheric winds and ocean currents all contribute to polar wander, with different magnitudes (displacement in meters per year) and periodicities. For example, seasonal changes in atmospheric pressure lead to an annual wobble of ∼5 m yr^{-1}, while post-glacier rebound (§6.1.4.4) is the primary cause of a secular drift in the polar wander with a magnitude of ∼1 m yr^{-1}.

6.1.4.4 *Isostatic Equilibrium*

Deviations in the measured gravity with respect to the geoid provide information on the structure of the crust and mantle. In the eighteenth century, it was already recognized that the measured surface gravity field of Earth does not deviate substantially from an oblate spheroid, even in the proximity of high mountains, despite the large land masses which make up the mountains. This observation led to the concept of *isostatic equilibrium*, which is based upon *Archimedes principle* and the theory of hydrostatic equilibrium (§3.2.2.1).

Figures 5.1 and 5.12 showed schematic presentations of the outermost layers or shell of the Earth. The rigid surface layers, the *lithosphere*, sit on a hot, highly viscous 'fluid' layer, the *asthenosphere* and *upper mantle* (§5.3.2). The lithosphere itself is topped off with a lighter *crust*, which is relatively light and thick for continents (20–80 km, $\rho = 2.7$ g cm^{-3}), and denser and thinner under the oceans (mean ∼6 km, $\rho = 3.0$ g cm^{-3}). We can compare this picture with an iceberg floating in water: the iceberg floats, because the volume submerged is lighter than the volume of water displaced. Archimedes principle states that any object (partially) submerged in a fluid feels a net upward (buoyancy) force that is equal to the weight of the fluid that is displaced by the object, i.e., $g_p\rho_m b$ for continental crust in Figure 6.8, with ρ_m the density of the

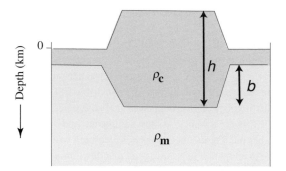

Figure 6.8 Schematic representation of a mountain illustrating the concept of isostatic equilibrium.

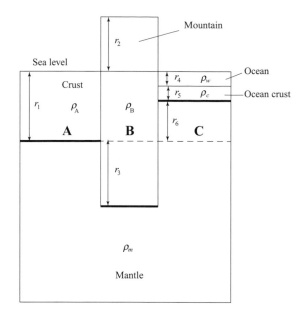

Figure 6.9 This figure illustrates Airy's and Pratt's hypotheses. Three columns are shown: A, B, and C. In Airy's hypothesis, the mantle rises up to the heavy lines in each column, with a density ρ_m that is higher than that found in the crust. The densities in the crust, $\rho_A = \rho_B = \rho_c$. The mountain is compensated by a mass deficiency of height r_3, and the ocean crust, r_5, by a layer of extra mass, r_6. In Pratt's hypothesis the mantle is at a constant level represented by the dashed line, and the densities $\rho_A \neq \rho_B \neq \rho_c$ (see Problems 6.10, 6.11).

mantle. This is also equal to the weight of the object itself, i.e., $g_p \rho_c h$ for the continental crust, assuming a density ρ_c for the crust (hydrostatic equilibrium).

Isostatic equilibrium simply states that a floating object displaces its own weight of the substance on which it floats. Like an iceberg floating in water, a mountain in isostatic equilibrium is compensated by a deficiency of mass underneath, since the part of the mountain that is submerged in the upper mantle is lighter than the mantle material displaced. Similarly, ocean and impact basins in isostatic equilibrium have extra mass deeper down. The deficiency or addition of mass at deeper layers can be calculated by assuming isostatic equilibrium (see Fig. 6.9). *Airy's hypothesis* assumes one and the same density for all the crustal layers, $\rho_A = \rho_B = \rho_c$, and a larger value for the fluid mantle material, ρ_m ($\rho_c < \rho_m$). Isostatic equilibrium is reached by varying the height of the crust in the various regions A, B, and C (heavy solid line) (Problem 6.10). In *Pratt's hypothesis* it is assumed that the depth of the base level of the crust is the same for all regions (dashed line in Fig. 6.9), and that isostasy is reached because the densities in the columns A, B, and C are different ($\rho_A \neq \rho_B \neq \rho_c$, Problem 6.11).

Before one can use gravity measurements to determine if a region is in isostatic equilibrium, a number of corrections need to be applied to the data. In addition to the variations in surface gravity as a function of latitude (eq. 6.13), one must take the height, h, above the geoid into account, i.e., the altitude at which the measurements were made, $g_p(obs)$ ($h \ll R$, with R the radius of the planet). If one assumes there is only air present above sea level up to altitude h, this correction is known as the *free-air correction*, Δg_{fa}, and is equal to $2GMh/R^2$ (Problem 6.12). The *free-air gravity anomaly* is defined as

$$g_{fa} = g_p(obs) - g_p(\theta)\left(1 - \frac{2h}{R}\right), \tag{6.24}$$

with $g_p(\theta)$ the gravity at the reference geoid (eq. 6.13). If there is a large slab of rock at the latitude of the measurements around the entire planet, the free-air correction factor should be modified by the gravitational attraction of the rocks, $2\pi G\rho h$, where ρ is the density and h the height of the slab above the reference geoid. This correction is known as the *Bouguer correction*, Δg_B. Deviations from this uniform slab of rock can be taken into account by using a terrain correction factor, δg_T, based upon a topographic map. With these corrections we define the *Bouguer anomaly*, g_B, as the observed gravity minus the theoretical value at the observing location:

$$g_B = g_p(obs) - g_p(\theta) + \frac{2MGh}{R^2} - 2\pi \rho Gh + \delta g_T. \tag{6.25}$$

Lateral variations in the density distribution result in gravity anomalies, i.e., as the free-air and Bouguer anomalies. In addition, such variations cause deviations in the measured geoid with respect to the reference geoid (eq. 6.13). This difference, the radius of measured geoid minus radius of reference geoid, is known as the *geoid height anomaly*, Δh_g, which is related to the measured anomaly in the gravitational potential, $\Delta \Phi_g$ (eq. 6.6):

$$g_p(\theta)\Delta h_g = -\Delta \Phi_g. \qquad (6.26a)$$

For an isostatic density distribution, the geoid height anomaly becomes

$$\Delta h_g = -\frac{2\pi G}{g_p(\theta)} \int_0^D \Delta\rho(z) z \, dz, \qquad (6.26b)$$

with $\Delta\rho(z)$ the anomalous density at depth z and D the compensation depth, below which no horizontal gradients in density are presumed. Depth is measured positive going down, and $z = 0$ corresponds to the geoid surface. The effective gravitational attraction is always normal to the geoid, so that there is a trough in the geoid where there is a negative gravity anomaly or positive potential anomaly (mass deficit), and there is a bulge in the geoid if there is a positive gravity anomaly or negative potential anomaly (mass excess).

The surface gravity can be estimated from small changes in the orbital parameters of artificial or natural satellites and radar altimetry measurements. A gravity map of a planet is usually represented as a contour plot of the geoid height anomaly, and shows the elevations and depressions of the surface equipotential with respect to the mean planetary surface (sea level on Earth).

As mentioned in the beginning of this section, the surface gravity field as measured for the Earth does not deviate much from the reference geoid, despite the presence of large topographic features. The free-air anomaly is therefore close to zero ($g_{fa} \approx 0$), as it should be for a planet that is in hydrostatic equilibrium. The Bouguer anomaly over large landmasses is negative, however, since the Bouguer anomaly 'corrects' for excess mass above sea level, but does not take the mass deficit below sea level into account. It is this mass deficit that compensates for the excess mass above if the features are isostatically compensated. The degree to which surface topography and gravity are correlated can be interpreted in terms of how much or little isostatic compensation is present, information which can then be used to derive information on a planet's lithosphere and mantle.

Before a geoid map can be used to extract such information, a number of assumptions need to be made. For example, the local topography needs to be known to high accuracy in order to calculate the Bouguer anomaly, even if the small terrain factor δ_T is ignored. Also, the density of the underlying structures must be known (Airy and Pratt hypotheses; see Problems 6.10 and 6.11). Finally, hitherto we have tacitly assumed static structures; however, we know that there is convection in the mantle, with regions

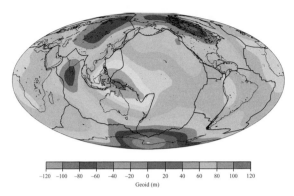

Figure 6.10 COLOR PLATE The observed geoid (degrees 2–15) superposed on a map of the planet Earth. Contour levels are from −120 meters (near the South Pole) to +120 meters (just north of Australia). (Lithgow-Bertelloni and Richards 1998)

of upwelling and subsiding motions. The columns of rising material are hotter, and therefore less dense than neighboring regions, while the opposite is true in columns of sinking material. Additionally, the surface is influenced: there are ridges or mountains above columns of rising material, whereas we find depressions above columns of sinking material. The combined effect usually shows up as small positive gravity and geoid height anomalies above regions of rising material, because the surface deflections introduce a larger positive anomaly than the negative one introduced by the low density. At locations where material is sinking, however, the effect on gravity and geoid height is strongly influenced by the viscosity in the mantle. So whether the net geoid anomaly is positive or negative depends upon the subtle cancelation of the 'dynamic' surface topography and the density effect in the mantle, whereby the surface topography is influenced by the viscosity structure in the mantle.

Figure 6.10 shows the geoid shape for Earth superposed on a map of the Earth. This is a relatively low-order ($J_2 - J_{15}$) gravity map, comparable to that obtained for other planets. Although there clearly is structure in the geoid, there is no correlation with the topography. However, the structure does seem to correlate well with tectonic features, such as mid-ocean ridges and subduction zones, and must be caused by mantle convection and deep subduction ('dynamic isostasy').

A low-order harmonic gravity map, as shown in Figure 6.10, essentially reveals how dynamic a planet is. When including higher order harmonics, gravity and topography show a somewhat higher correlation, due to the fact that topography over small areal scales is usually not well compensated. Such maps yield information on the mantle viscosity and thickness of the lithosphere.

Table 6.3 Heat-flow parameters.

Body	T_e (K)	T_{eq} (K)	H_i (erg cm^{-2} s^{-1})	L/M (erg g^{-1} s^{-1})	References
Sun	5770		6.2×10^{10}	1.9	1
Carbonaceous					
chondrites				4×10^{-8}	1
Mercury		446			3
Venus		238			3
Earth		263	75	6.4×10^{-8}	1, 3, 4
Moon		277	~18	~10^{-7}	3, 5
Mars		222	40	9×10^{-8}	1, 3, 4
Io		92	1500–3000	~10^{-5}	
Jupiter	124.4	113	5440	1.8×10^{-6}	1, 2, 3
Saturn	95.0	83	2010	1.5×10^{-6}	1, 2, 3
Uranus	59.1	60	<42	<4×10^{-8}	1, 2, 3
Neptune	59.3	48	433	3.2×10^{-7}	1, 2, 3

1: Hubbard (1984). 2: Hubbard *et al.* (1995). 3: Tables 4.1, 4.2. 4: Carr (1999). 5: Turcotte and Schubert (2002).

Although changes on the Earth's surface usually lead to isostatic adjustments in the fluid mantle below, such adjustments may take many thousands of years because of the high viscosity of the mantle. A good example is the *post-glacial rebound*, a general rise of the land masses that were depressed by the large weight of ice sheets during the last ice age, ~21 000 years ago. With the melting and hence removal of the ice, the land started to rise. This effect is indeed visible on the low-order harmonic gravity map (e.g., under Canada).

As we will see in our discussion of the individual planets (§6.3), the poor correlation between topography and low-order harmonic gravity maps of Earth is quite unique. Most other bodies reveal a correlation between gravity field and topography.

6.1.5 Internal Heat: Sources and Losses

In §4.2.2 we compared the equilibrium temperatures of the giant planets, i.e., the temperature the planet would have if heated by solar radiation only, with their observed effective temperatures (Table 4.1). This comparison showed that Jupiter, Saturn, and Neptune are warmer than can be explained from solar heating alone, which led to the suggestion that these planets possess internal heat sources. The Earth must also have an internal source of heat, as deduced from its measured heat flux of 75 erg cm^{-2} s^{-1}. In this section we discuss possible sources of internal heat, as well as mechanisms to transport the energy and ultimately lose it to space. Heat-flow parameters for all planets are summarized in Table 6.3.

6.1.5.1 Heat Sources
Gravitational
Accretion of material during the formation of planets is likely one of the largest sources of heat (§13.6; Problems 13.27, 13.28). Bodies hit the forming planet with roughly the escape velocity, yielding an energy for heating of GM/R per unit mass. The increase in energy per unit volume is equal to $\rho c_P \Delta T$, with c_P the specific heat (per gram material), ρ the density, and ΔT the increase in temperature. The gain in energy at the surface of the planet, whose radius is $R(t)$, must be equal to the difference between the gravitational energy acquired at R, $GM(R)/R$, and the energy which is radiated away, σT^4 (§3.1), over a time dt during which the body accreted a layer of thickness dR:

$$\frac{GM(R)\rho}{R}\frac{dR}{dt} = \sigma(T(R)^4 - T_0^4) + \rho c_p(T(R) - T_0)\frac{dR}{dt},$$

(6.27)

with T_0 the initial temperature of the accreting material. If accretion is rapid, much of the heat is 'stored' inside the planet before it has time to radiate away into space, since subsequent impacts 'bury' it. The ultimate temperature structure inside a planet further depends on the size of accreting bodies and the internal heat transfer (§13.6.2, Problems 13.19–13.22).

Giant Planets. The internal heat sources of the giant planets Jupiter, Saturn, and Neptune are attributed to gravitational energy, either from gradual escape of primordial heat generated during the planet's formation, and/or from previous or ongoing differentiation. The luminosity of the planetary body, L, consists of three components: L_v is reflected sunlight (mainly at optical wavelengths), L_{ir} is incident sunlight absorbed by the planet and re-emitted at infrared wavelengths, and L_i is the planet's intrinsic luminosity. For the terrestrial planets L_i is very small, but for the gas giants, Jupiter, Saturn, and Neptune, the internal luminosity is comparable to L_{ir}. The effective temperature T_e is obtained by integrating the emitted energy over all infrared wavelengths, and it thus consists of both L_{ir} and L_i. The equilibrium temperature T_{eq} is the temperature the planet would have in the absence of internal heat sources (§3.1.2.2). The intrinsic luminosity of the planet is thus equal to

$$L_i = 4\pi R^2 \sigma (T_e^4 - T_{eq}^4). \tag{6.28}$$

If we assume that the internal heat flux merely represents the leakage of the primordial heat stored during the planet's formation period, we can express the rate of change in the mean internal temperature dT_i/dt:

$$\frac{dT_i}{dt} = \frac{L_i}{c_V M}, \tag{6.29}$$

where M is the planet's mass and c_V the specific heat at constant volume. For metallic hydrogen, $c_V \approx 2.5k/m_{amu}$ erg g^{-1} K^{-1}, with k Boltzmann's constant and m_{amu} the atomic mass unit.

Jupiter's excess luminosity is consistent with the energy released from gravitational contraction/accretion in the past. In contrast, detailed models of Saturn's interior structure, including evolutionary tracks (such as those discussed below for Uranus and Neptune), show that primordial heat alone is not sufficient to explain Saturn's excess heat. The additional heat loss can be accounted for using differentiation of helium from hydrogen, a process which at the same time explains the observed depletion in the helium abundance in Saturn's atmosphere compared to solar values. Since Saturn is less massive and therefore colder in its deep interior than is Jupiter, the temperature in Saturn's metallic hydrogen dropped a few billion years ago to levels where the hydrogen and helium phases separated out. In contrast, helium became only 'recently' immiscible in Jupiter's interior. Helium, therefore, has steadily 'rained out' of the metallic hydrogen region towards Saturn's core. The energy release from this process can explain Saturn's observed L_i.

For Uranus and Neptune, the specific heat can be approximated by $c_V \approx 3k/(\mu_a m_{amu})$ erg g^{-1} K^{-1}, with μ_a the mean atomic weight, which is ~5 (amu) for icy material. The drop in temperature for Neptune over the age of the Solar System is thus ~200 K (Problem 6.18), which is small compared to the internal temperature expected for

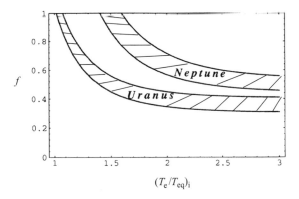

Figure 6.11 The relationship between the fraction, f, of the initial heat reservoir that may drive the observed luminosity of Uranus and Neptune, and the planets' initial internal thermal energy. The latter quantity is characterized by the initial ratio between the effective and equilibrium temperatures, $(T_e/T_{eq})_i$. The upper and lower curves for each planet correspond to upper and lower bounds on their observed present-day luminosity, and hence the parameter space indicated by the hashed areas gives acceptable fits to each planet's observed luminosity. (After Hubbard et al. 1995)

an adiabatic planet (a few thousand kelvin). Thermal evolution is characterized by equation (6.29), which can be used to determine which fraction of the internal heat reservoir gives rise to the observed luminosity. This fraction, f, is less than unity if convection is (partly) inhibited, such as may happen if there is a stable stratification in the interior. The relationship between f and the initial ratio $(T_e/T_{eq})_i$, just after the planet formed, is shown in Figure 6.11 for Uranus and Neptune, where the upper and lower curves correspond to the upper and lower limits to the observed present-day luminosities. Thus, the lower curve for Uranus corresponds to a planet which reached a zero heat flow 4.5 Gyr after it formed. Any parameter choice to the left of this curve is unacceptable, because this corresponds to states that cannot sustain the dynamo required to explain the existence of Uranus's magnetic field (§6.4.3). It seems impossible to find a parameter choice which satisfies the thermal evolution curves for both Uranus and Neptune.

Owing to planetary accretion, the internal temperature for both planets was probably very high initially, so that f must be substantially less than unity, of the order of 0.4 for Uranus and 0.6 for Neptune. Such small fractions imply that convection is inhibited inside a radius of 0.6 R_{δ} for Uranus and inside 0.5 R_{ψ} for Neptune. This difference between the two planets seems small, but nevertheless could lead to the observed differences in the intrinsic luminosities for the two planets.

Terrestrial Planets. As shown by equation (6.27), the interior of a planet can become very hot if accretion is rapid. For the 10^8 year accretion times estimated for the terrestrial planets, there is, however, little accretional heating (Problem 13.19), unless large impacts buried heat below the surface (§13.6.2). Accretional heating alone, however, cannot account for the heat budget of the terrestrial planets and smaller bodies, even after specifically including the extra energy released from differentiation and latent heat release from condensation of, e.g., a liquid outer core. Gravitational energy alone has never been sufficient for the smaller asteroids and satellites to cause a separation of heavy and light materials, yet the interiors of some of these bodies are differentiated (§5.2). As discussed below, the decay of radioactive elements and, for some solid bodies, tidal and ohmic heat may provide significant additional energy sources.

Radioactive Decay

Radioactive decay has been proposed as an important source of heat in the interiors of the terrestrial planets, satellites, asteroids, and the icy bodies in the outer Solar System. If radioactive decay of elements is a major contributor to the heat flow measured today, the elements must have long half-lives, roughly of the order of a billion years. ^{235}U, ^{238}U, ^{232}Th, and ^{40}K have lifetimes approximately 0.71, 4.5, 13.9, and 1.4 Gyr respectively. These isotopes are present in the Earth's crust at levels of a few parts per million, and produce on average an energy of ~ 10 erg cm^{-2} s^{-1}. These elements are about two orders of magnitude less abundant in the mantle, but because the volume of the mantle is so much larger than that of the crust, the heat produced in the mantle significantly influences the total heat output. Only $\sim 20\%$ of the Earth's radioactive heating occurs in the crust. The heat generation from radioactive elements was larger in the past, and it appears that the total energy liberated by these elements over the first ~ 1–2 Gyr would have been enough to melt Earth and Venus (this does not imply that Earth differentiated as a result of radioactive heating, just that radioactive heating must have been important early on). The heat generated by short-lived radionuclides, in particular ^{26}Al with a half-life of 0.74 Myr, was high at an early stage, but disappeared quickly. At present the total heat released by radioactive decay in carbonaceous chondrites is about 4×10^{-8} erg g^{-1} s^{-1}, which is only slightly less than the total heat flow from Earth (6.4×10^{-8} erg g^{-1} s^{-1}). The heating from long-lived radionucleides was an order of magnitude larger at the time of planet formation (Problems 13.24, 13.25). At present, roughly half of the Earth's heat output is thought to be due to radioactive heat production, while the remainder is from secular cooling.

Tidal and Ohmic Heat

Temporal variations in tidal forces can lead to internal heating of planetary bodies, as discussed in detail in §2.6.2. For most objects, this potential source of energy is much smaller than that due to gravitational contraction or radioactive heating. For the moons Io, Europa, and likely Enceladus, however, tidal heating is a major source of energy (§§2.6.2, 5.5). Tidal heating was probably important in the past for Triton and possibly also for Ganymede.

Ohmic heating results from the dissipation of an induced electric current, such as the processes discussed in Chapter 7. Ohmic heating might have been sufficient to melt asteroid-sized planetesimals in some regions of the protoplanetary disk during the young Sun's active T-Tauri phase (§13.3.4).

6.1.5.2 Energy Transport and Loss of Heat

Energy transport determines the temperature gradient in a planet's interior, just like it does in a planetary atmosphere. The temperature gradient is determined by the process which is most effective in transporting heat. The three mechanisms by which heat is transported are conduction, radiation, and mass motion, primarily convection (see §3.2). Conduction and convection are important in planetary interiors, while radiation is important in transporting energy from a planet's surface into space, and in a planet's atmosphere where the opacity is small but finite.

Conduction and Radiation

The heat flux, $\mathbf{Q}$ (erg s^{-1} cm^{-2}), is given by the Fourier heat law (eq. 3.19; §3.2.1). Conduction is the most effective way to transport energy in solid materials, such as the crustal layers of the terrestrial planets and throughout the interiors of smaller bodies like asteroids and satellites. The energy may be transported by free electrons or heavier particles, or by photons or *phonons*. The latter correspond to waves excited by vibrations in the crystal lattice. Good heat conductors are materials with a large number of free electrons, such as metals. In metal-poor materials, such as silicates, heat transport is dominated by phonons, where the energy is carried by propagating elastic lattice waves. The propagation of these waves decreases with increasing temperature, because the anharmonicity in the crystal increases at higher temperatures, which causes the waves to scatter. At high temperatures, radiative transport of heat by photons becomes more significant. At low temperatures, the photon mean free path is small and the total energy in the photon field is small compared to the vibrational thermal energy; but the energy in the photon field is proportional to T^3, so that it becomes appreciable at high temperatures. The thermal conductivity, K_T, in uncompressed silicate is therefore written as the sum of K_L, the lattice thermal conductivity, and K_R, the radiative thermal conductivity (in erg cm^{-2} s^{-1} K^{-1})

$$K_T = K_L + K_R, \tag{6.30a}$$

where

$$K_L = \frac{4.184 \times 10^7}{30.6 + 0.21T}, \tag{6.30b}$$

$$K_R = 0 \quad \text{(for } T < 500\,\text{K)}, \tag{6.30c}$$

$$K_R = 230(T - 500) \quad \text{(for } T > 500\,\text{K)}. \tag{6.30d}$$

A quick calculation of the importance of temperature changes due to conduction can be obtained via the thermal diffusivity, k_d (§3.2.1):

$$\ell \approx \sqrt{k_d t}, \tag{6.31}$$

where ℓ is a length scale over which changes in the temperature gradient become significant over a timescale t and the thermal diffusivity is given by equation (3.23). In Problem (6.20) the thermal diffusivity and scale length ℓ are calculated for a typical rocky body. Over the age of the Solar System, the temperature gradient in such a body has been affected by conduction over only a few hundred kilometers. Hence, the temperature structure in small bodies, such as asteroids and satellites, can be modified by thermal conduction over the age of the Solar System, but planet-sized bodies cannot have lost much of their primordial heat via conduction.

Convection

As discussed in §3.2, convection is caused by rising and sinking motions of the material: hot material rises towards cooler regions at higher altitudes, and cold material sinks down. Convection can only proceed if the rate at which energy is liberated by buoyant forces exceeds that at which energy is dissipated by viscous forces. This criterion is expressed in terms of the *Rayleigh number*, $\mathfrak{R}_a$:

$$\mathfrak{R}_a = \frac{\alpha_T \Delta T g_p \ell^3}{k_d \nu_v} > \mathfrak{R}_a^{crit}, \tag{6.32}$$

with α_T the thermal expansion coefficient, and ΔT the temperature difference in excess of the adiabatic gradient across the layer of thickness ℓ. $\mathfrak{R}_a^{crit}$ is a critical value, usually of the order 500–1000. The Rayleigh number can only exceed this critical value if the kinematic viscosity, ν_v, is finite. Since the viscosity is small in fluids and gases,

convection is likely the primary mode of energy transport in regions where the planet is gaseous or fluid, as in giant planets and the liquid outer core of Earth. Many materials, including rocks, can deform under an applied rate of strain (§5.3.2). Energy may therefore be transported by solid-state convection in planetary mantles, if the characteristic timescale for convection is small compared to geological timescales. It has been shown empirically that, if $\mathfrak{R}_a \gg \mathfrak{R}_a^{crit}$, the ratio of the heat flux carried by convection plus conduction to that carried by conduction alone, the *Nusselt number*, is approximately equal to

$$\mathcal{N}_u \equiv \frac{Q(\text{convection} + \text{conduction})}{Q(\text{conduction})}$$

$$\approx \left(\frac{\mathfrak{R}_a}{\mathfrak{R}_a^{crit}}\right)^{0.3}. \tag{6.33}$$

The viscosity of rocks depends strongly on temperature. At low temperatures, the viscosity is essentially infinite, and the material behaves as a solid. At temperatures above roughly 1100–1300 K, the viscosity of rock is low enough that the material 'flows' over geological timescales. This explains why the cold outer layers of a planet, the *lithosphere*, are made of solid rock where energy is transported by conduction, while deeper in the planet (mantle), energy is mainly transported by convection.

Heat Loss

Solid bodies generally lose heat by conduction upwards through the crust and radiation from the surface into space. At deeper levels, for planets that are large enough that solids 'flow' over geological timescales, the energy may be transported by convection, a process which, when it occurs, typically transports more heat than conduction. In the upper 'boundary' layers heat transport is again by conduction upwards through the rigid lithosphere and crust. However, conduction alone may not be sufficient to 'drain' the incoming energy from below. On Earth, additional heat is lost through tectonic activity along plate boundaries, and via hydrothermal circulation along the mid-ocean ridge. Heat may also be lost during episodes of high volcanic activity, either in volcanic eruptions, or through vents or hot spots. The latter source of heat loss is dominant on Io (and Enceladus); in fact, the total heat outflow through Io's hot spots may, at the present time, even exceed generation of heat due to tidal dissipation.

In the giant planets, heat is transported by convection throughout the mantle and most of the troposphere, while radiation to space plays an important role at higher altitudes, in the stratosphere and lower thermosphere. From measurements of the effective temperature, we know that the giant planets Jupiter, Saturn, and Neptune have an internal source of energy that is comparable in magnitude to the energy these planets receive from the Sun (§6.1.5.1).

Measurements of the heat flow from solid bodies are much more difficult, since the outer parts of a planet from which we can measure the temperature or luminosity, are also heated up by the Sun during the day, so that the crustal layers display the effects of the diurnal solar insolation, as discussed in §3.2.1. If the thermal conductivity in these layers is known, the heat flow can in principle be determined via measurements of the temperature gradient in the upper layers of a planet's crust. This temperature gradient can, in principle, be obtained by drilling holes (Earth, Moon), or by using remote sensing techniques. A typical value for the heat flux from Earth is 75 erg cm^{-2} s^{-1}, which corresponds to an intrinsic luminosity $L_i = 3.84 \times 10^{20}$ erg s^{-1}, roughly 5000 times smaller than L_{ir} and L_v combined (Problem 6.17). The heat flux from the Moon is roughly half that of Earth, as determined from boreholes in the lunar surface. In contrast to boreholes or mines on Earth, which can go down 100–1000 m, remote sensing techniques only sample the upper few meters of the crust at best, which makes it impossible to directly determine the temperature gradient below the region sensitive to diurnal heating. It is therefore not surprising that we usually do not know the heat flux from solid bodies (Table 6.3), except for the most volcanically active ones (Io, Enceladus), and those where *in situ* measurements have been made.

6.2 Interior Structure of the Earth

Whereas the structure of the Earth's crust can be determined by *in situ* experiments such as drilling, details on the Earth's deep interior have to be obtained via indirect means. The most powerful method to obtain information on the Earth's interior comes from *seismology*, the study of the passage of elastic waves through the planet. The combination of seismic studies and gravity measurements has led to detailed models of the Earth's interior.

6.2.1 Seismology

Seismic waves can be induced by, e.g., *earthquakes*, meteoritic impacts, and volcanic or man-made explosions. The waves are detected by *seismometers*, sensitive instruments that measure the motion of the ground on which they are located. A seismograph recording of vertical and horizontal ground motions is shown in Figure 6.12. With a number

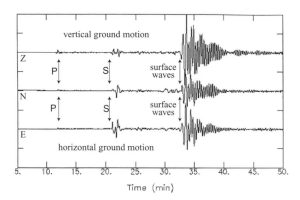

Figure 6.12 A recording from a seismograph showing the onset of P, S, and surface waves from a distant earthquake. Vertical (Z) and horizontal (N = to the north, E = to the east) motions are shown. (UC Berkeley Seismological Laboratory)

of seismometers spread around Earth, the waves are studied from many different 'viewing' points, and all data together can be used to derive the structure of the Earth's interior via *seismic tomography*, like CAT-scanning the interior.

6.2.1.1 *Seismic Waves*

Body waves are seismic disturbances that travel through a planet's interior, whereas *surface waves* propagate along the surface. Body waves obey Snell's law, and are reflected and transmitted at interfaces where the material density changes. We distinguish between P and S waves (see Fig. 6.13). *P waves* are Primary, Push, or Pressure waves, where the individual particles of the material are oscillating back and forth in the direction of wave propagation. They are longitudinal waves, similar to ordinary sound waves, and involve compression and rarefaction of the material as the wave passes through it. The first P waves travel rapidly and arrive at a seismic station well before the first S waves. *S waves* are the Secondary, Shake, or Shear waves, which have their oscillations transverse to the direction of propagation. They are analogous to the waves you can make on a rope, and to electromagnetic waves. They involve shearing and rotation of the material as the wave passes through it.

Surface waves are confined to the near-surface layers on Earth. These waves have a larger amplitude and longer duration than body waves, and because their velocity is lower than that of body waves, they arrive later at a seismograph (Fig. 6.12). The motion in *Love waves* is entirely horizontal, but transverse to the propagation of the waves (Fig. 6.14a). In *Rayleigh waves* the particle motion is a vertical ellipse, so that the motion can be described as a

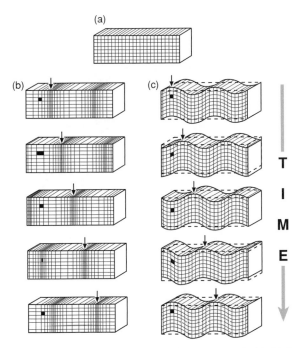

Figure 6.13 Seismic P and S waves: (a) Unperturbed grid. (b) A propagating P wave, in which the particles oscillate back and forth along the direction of motion. (c) The particle motion in S waves is transverse to the direction of motion. (Phillips 1968)

'ground roll' (Fig. 6.14b). These waves show many similarities to waves on water, and their amplitude decreases exponentially with depth.

The equations for P and S waves can be deduced from the theory of elasticity (see, e.g., Fowler 2005). The compressional wave equation for P waves is

$$\frac{\partial^2 \Phi}{\partial t^2} = v_P^2 \nabla^2 \Phi, \tag{6.34a}$$

and the rotational wave equation for S waves is

$$\frac{\partial^2 \bar{\psi}}{\partial t^2} = v_S^2 \nabla^2 \bar{\psi}, \tag{6.34b}$$

where v_P and v_S are the P wave and S wave velocities. The displacement of the medium, $\mathbf{x}$, can be expressed as the sum of the gradient of the scalar potential, Φ, and the curl of the vector potential, $\bar{\psi}$:

$$\mathbf{x} = \nabla \Phi + \nabla \times \bar{\psi}. \tag{6.35}$$

The P wave and S wave velocities are related to the thermodynamic properties of the medium (for a detailed treatment, see Fowler 2005):

$$v_P = \sqrt{\frac{K_m + \frac{4}{3}\mu_{rg}}{\rho}}, \tag{6.36a}$$

and

$$v_S = \sqrt{\frac{\mu_{rg}}{\rho}}, \tag{6.36b}$$

where ρ is the density, μ_{rg} the shear modulus (§6.1.4.2), and K_m the bulk or (adiabatic) *incompressibility modulus* of the material at constant entropy S:

$$K_m \equiv \rho \left(\frac{\partial P}{\partial \rho} \right)_S. \tag{6.37a}$$

If a planet's interior is adiabatic and chemically homogeneous, the bulk modulus of the material becomes

$$K_m \approx \rho \frac{dP}{d\rho}. \tag{6.37b}$$

From the equations above it follows that K_m, v_P, and v_S are related as follows:

$$\frac{K_m}{\rho} = v_P^2 - \frac{4}{3} v_S^2. \tag{6.38}$$

The bulk modulus is a measure of the stress or pressure needed to compress a material, thus it involves a change in volume of the material. The shear modulus is a measure of the stress needed to change the shape of the material, without changing its volume. Note that v_P depends both on K_m and μ_{rg}, since P waves involve a change of both volume and shape of the material, while v_S depends only on μ_{rg}, since S waves do not involve a change in volume (see Fig. 6.13). Since $K_m > 0$, P waves travel faster than S waves: $v_P > v_S$. The wave velocities are measured by timing how long it takes the wave to travel from the source (*hypocenter*) of the earthquake to

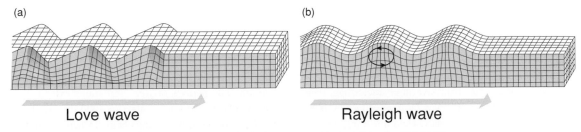

(a)   (b)

Love wave   Rayleigh wave

Figure 6.14 Seismic surface waves: (a) Love waves, (b) Rayleigh waves. (Fowler 2005, after Bolt 1976)

the seismograph. (The *epicenter* is the point on the surface vertically above the hypocenter.) In a liquid, $\mu_{rg} = 0$, and therefore, in contrast to P waves, the S waves cannot propagate through liquids. Because there are no detectable seismic phases corresponding to S wave propagation in the Earth's outer core, we know that the Earth's outer core is liquid (Fig. 6.15). In fact, when an S wave is incident on the outer core, part of it is reflected and part is transmitted as a P wave (S → P wave conversion). Both K_m and μ_{rg} depend on the density, but increase faster than ρ. Since the density increases towards the center of a planet, the velocities of both P and S waves increase with depth. In addition, because the density changes along a wave's path, the path is curved (upwards, since ρ increases with depth) according to Snell's law. Since v_P and v_S can be determined as a function of depth from seismic data, equation (6.38) allows for a direct determination of the density as a function of depth.

6.2.1.2 *Free Oscillations*

When a rope is fastened at one end and moved up and down with an appropriate frequency, one can create stationary *standing waves*. Such waves can be induced in any elastic body. In contrast to other waves, standing waves may persist for a long time. In daily life one makes use of this physical phenomenon when playing a musical instrument, like a violin, an organ, or just ringing a bell. In a planet, standing waves can be induced by a large quake, causing the planet to 'ring as a bell' for many days to months. Such vibrations display a large number of *modes*, some of which are shown graphically in Figure 6.16. The motion is either radial (up and down) or tangential to the body's surface, causing *spheroidal* ($_nS_m$) and *toroidal* ($_nT_m$) modes, respectively, where n indicates the overtone (n = 0 is the fundamental mode), and the harmonic degree m is the number of *nodes* in latitude (i.e., locations which do not move at all). The simplest spheroidal oscillation is a purely radial expansion and contraction of the Earth as a whole, as indicated by the $_0S_0$ mode. The fundamental surface spheroidal oscillations are equivalent to the standing waves resulting from interference of Rayleigh waves. Fundamental toroidal oscillations are equivalent to the interference of Love waves, and involve twisting of the surface in opposite directions. On Earth, free oscillations have periods between 100 s and 1 hr; they have been recorded over several months after large ($\mathcal{M}_R > 8$) earthquakes.

Surface waves are dispersive in character, i.e., their velocities depend on frequency. *Dispersion curves*, plots of the velocity versus frequency, yield much information on the velocity structure in the crust and upper mantle,

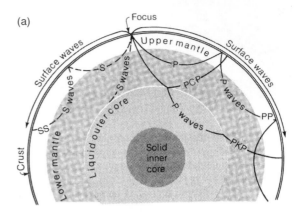

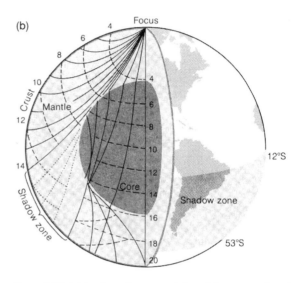

Figure 6.15 (a) A schematic representation of the propagation of seismic waves on Earth superimposed on a sketch of the interior structure as determined from seismological experiments. Surface waves propagate along the near-surface layers, while the S and P waves propagate through the Earth's interior. Waves reflected from the Earth's crust are called SS or PP waves. When an S wave is incident on the outer core, part of it is reflected and part is transmitted as a P wave; S waves themselves cannot propagate through liquid material. P waves are refracted (PKP waves) as well as reflected (PCP waves) at the core–mantle interface. Since changes in the wave velocity take place continuously throughout the Earth's interior (in addition to abrupt changes at interfaces), the wave paths are curved rather than straight. (b) A different representation of seismic waves, which shows the presence of a *shadow zone*, a region which cannot be reached by P waves because they are deflected by the Earth's core. (Press and Siever 1986)

and hence density and rigidity as a function of depth in the Earth. The periods of free oscillations extend these dispersion curves to much longer (~3000 s rather than a few 100 s) periods, which thus helps to better constrain models of the interior structure of the entire Earth.

(a)

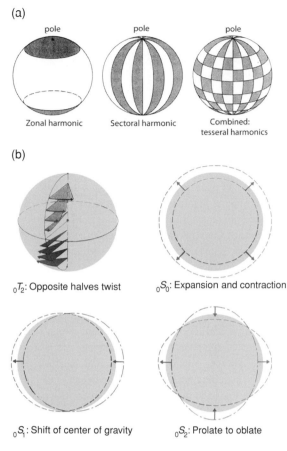

pole	pole	pole
Zonal harmonic	Sectoral harmonic	Combined: tesseral harmonics

(b)

$_0T_2$: Opposite halves twist

$_0S_0$: Expansion and contraction

$_0S_1$: Shift of center of gravity

$_0S_2$: Prolate to oblate

Figure 6.16 (a) Surface movements of some free oscillations on Earth. The light and dark areas have displacements which are in opposite senses at any instant. They are separated by *nodes*, lines where there is never any movement. (b) Examples of modes of oscillation. In toroidal modes, T, the movements are tangential to the Earth's surface, and in spheroidal modes, S, the movement is predominantly radial. (Adapted from Brown and Mussett 1981)

Usually, seismic waves on Earth are related to earthquakes. However, the Earth appears to be 'humming' continuously, even if there are no earthquakes, as was revealed after stacking seismic data taken over many years when *no* earthquakes had occurred. This is apparently caused by the continuous flow of air over the surface.

In addition to Earth, free oscillations have been observed on the Sun and the Moon. Despite many searches for such oscillations on Jupiter, in particular during and following the impacts of Comet D/Shoemaker–Levy 9 (§5.4.5), none have been detected.

6.2.2 Density Profile

The density profile in a planet, $\rho(r)$, can be derived from the equation of hydrostatic equilibrium (eq. 6.1) and

the definition of the bulk modulus, K_m, in an adiabatic and chemically homogeneous planet (eq. 6.37b) (Problem 6.21):

$$\frac{d\rho}{dr} = -\frac{GM\rho^2(r)}{K_m r^2}. \tag{6.39}$$

Since K_m/ρ can be determined from the seismic wave velocities v_P and v_S (eq. 6.36), one can integrate the *Adams–Williams equation* (eq. 6.39) from the surface down to determine the Earth's density structure, assuming a *self-compression model*, where the density at each point is assumed to be due only to compression by the layers above it. The relevant mass is the mass within radius r, $M(r)$, which is given by

$$M(r) = M_\oplus - 4\pi \int_r^{R_\oplus} \rho(r)r^2 dr, \tag{6.40}$$

with $M_\oplus$ and $R_\oplus$ the mass and radius of the Earth, respectively. The result of this calculation depends on the density in the top layer. It may not be too surprising that it appears impossible to find a density structure with this model which satisfies the seismic wave velocities in detail. These velocities show clear 'jumps' at particular depths, such as at the core–mantle boundary, indicative of abrupt changes in density. Seismic data (e.g., travel-time-distance data, surface-wave-dispersion curves, and free oscillation periods), together with the gravitational potential of the Earth (e.g., mass, moments of inertia) have been used to derive the 'Preliminary Reference Earth Model' (PREM) shown in Figure 6.17.

In the outer 3000 km, the density as well as v_P and v_S increase with depth. One may notice a small decrease with depth in the outer 200 km, which is probably not real. It might be caused by insufficient resolution to properly invert the data. An abrupt boundary appears at ~3000 km, where the S waves disappear altogether, and P waves slow down considerably. This boundary is interpreted as the *interface between the solid mantle and liquid outer core*, since S waves cannot propagate through liquids. In addition to the disappearance of S waves, analysis of Earth's free oscillations excited by large earthquakes also suggests the presence of a liquid outer core. Model fits to geodetic observations of the nutation of the Earth's rotation axis form a third piece of evidence that the outer core is a liquid, since the viscosity of the outer core must be relatively low to explain these data. At ~5200 km depth, there is another discontinuity, interpreted as the *boundary between the liquid outer core and a solid inner core*. At this boundary the P wave velocity increases. Analysis of seismic free oscillations shows conclusively that the inner core must have some rigidity, i.e., be solid, with a density about $0.5\,\mathrm{g\,cm^{-3}}$

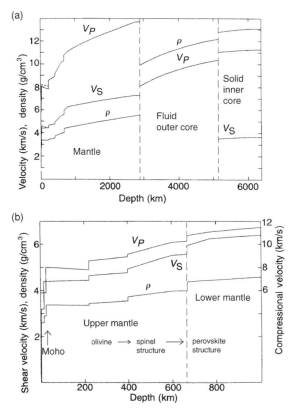

(a)

(b)

Figure 6.17 The PREM model, as derived by Dziewonski and Anderson (1981). (a) The seismic P and S wave velocities, together with the density are shown throughout the Earth's interior. (b) An expansion of the uppermost 1000 km from (a). (Adapted from Pieri and Dziewonski 1999)

higher than that of the outer core. Although S waves disappear at the outer core boundary, they can travel through the inner core (this has been observed as PKIKP waves in seismograms).

The boundary between the crust and the mantle, within the lithosphere, is called the *Mohorovičić* or *Moho discontinuity*. The continental crust consists mostly of granitic rocks, with gabbro at the bottom, while the oceanic crust is composed entirely of basaltic rocks. The thickness of the continental crust varies from roughly <20 km under active margins up to 70–80 km under the Himalayas. The oceanic crust is zero km at the mid-ocean ridges, and has a mean value of ∼6 km. The crust is the upper part of the *lithosphere*, which sits on a plastic-like layer called the *asthenosphere*. The lithosphere is cool and rigid, while the asthenosphere has a relatively low viscosity compared to the mantle underneath. The lithosphere is often modeled as an elastic layer; it bends under loads, such as glacier caps, whereas the asthenosphere flows. The shear wave velocity decreases somewhat at the transition between the lithosphere and asthenosphere. As further shown in

Figure 6.17b, the wave velocities in the outer 670 km of the Earth increase stepwise with depth. The velocities change where the density changes, due to, for example, phase changes of the material: olivine → ringwoodite → perovskite (§6.1.2.3).

6.2.2.1 *Mantle Dynamics*

Earth's lithosphere is divided into ∼15 quasi-rigid plates, and the movement of these tectonic plates with sea floor spreading and continental drift suggests that these plates 'ride on top' of a convection pattern in the Earth's mantle. The subject of plate tectonics, together with the forces that drive the mantle circulation, was discussed in §5.3.2.2. Not discussed in Chapter 5 is the still ongoing debate as to what depth mantle convection extends: does it go down to the core–mantle boundary (i.e., whole mantle convection), as suggested by some seismic velocity models, or are there separate convection patterns in the upper and lower mantle, as suggested by geochemical data?

It is generally assumed that when Earth formed, its composition was largely homogeneous and similar to that found in volatile-poor chondritic meteorites. The upper mantle, however, seems depleted in incompatible elements,[4] including some rare earth elements and noble gases, compared to the crust and the lower mantle. Estimates of the concentration of these elements in the upper mantle come from the mid-oceanic ridge, while the lower mantle is assumed to be sampled from certain ocean island basalts. These oceanic islands, including, e.g., the Hawaiian islands, are known as volcanic 'hot spots'. They formed from hot mantle material that is rising up in a narrow (≲100 km) plume. These 'mantle' plumes appear not to move much relative to one another, while the lithospheric plates move over them. The plate movement leads to the formation of island chains, rather than a single volcanic island. An example is the chain of Hawaiian islands, where the southeast end of the youngest island, the Big Island, is still volcanically active. The next Hawaiian island, Loihi, is already to the southeast (its top is still 1000 m under water).

In contrast, the magma which fills the oceanic ridge is less hot and slowly oozes up to fill the void left by the receding oceanic plates. Seismic data show that this material originates in the upper mantle, while some volcanic islands formed from mantle plumes that might originate in the lower mantle, possibly near the core–mantle boundary. If global mixing between the lower and upper mantle

---

[4] An incompatible element is an element that cannot be incorporated in a mineral, either because its valence or its ionic radius is not compatible with the major element that makes up the mineral. Such elements will partition into a melt during partial melting.

were to take place, one would expect the same concentration of rare earth elements and noble gases throughout the mantle, and thus also between the mid-oceanic ridge and ocean island basalts. Since this has not been observed, the geochemical data appear to support theories on convection cells which are separate between the lower and upper mantle. Moreover, variations in the isotope abundances as measured from island to island suggest the lower mantle to be quite heterogeneous. Convection in the lower mantle may thus be much less efficient than in the upper mantle; this difference in convection efficiency can be explained if the viscosity in the mantle increases with depth.

Seismic tomography has led to detailed three-dimensional seismic velocity models of the Earth's mantle. Figure 6.18a shows anomalies in the S wave velocities at different depths. The blue color indicates areas with above-average seismic velocities, and the red areas indicate below-average velocities. Since the seismic speed decreases with increasing temperature, the blue areas are interpreted as cold regions and the red areas as warmer (an interpretation that completely ignores variations in rock types, as observed at the surface). At a depth of 175 km, an extended cold region is seen under North America into South America, and between Europe and Indonesia, across southern Asia. These cold regions correspond to the stable part of the continents ('cratons') which have been tectonically inert for long periods of time ($\gtrsim 10^9$ years) and are therefore cold, and as such 'fast' in the tomography. Clear hot/cold structures are seen at many depths. At some places the structures can be correlated over large depths, as shown in Figure 6.18b. On geological timescales (millions of years) the hot material rises, while cold material (except for the continents) sinks down. The cold areas correspond indeed to subducting plates, in active margins. Figure 6.18b shows a seismic model at 100 km depth (upper panel), and a vertical slice through the Earth's crust and mantle down to the core–mantle boundary, in the South Pacific at the latitude of the 'Tonga Kermadec' subduction zone (along the green line). In the cross-section, the cold regions are associated with the westward dipping slab. At the eastern end of the cross-section, the warm area corresponds to the Pacific 'super-swell'. The line in the cross-section corresponds to the separation between the upper and lower mantle at a depth of 670 km. From this figure it appears as if some of the hot–cold patterns extend all the way down to the core–mantle boundary, where the cold (blue) streaks suggest that lithospheric plates have descended continuously through the mantle over perhaps 40–50 Myr in geologically young areas (e.g., the Mariana subduction zone), and up to 180–190 Myr in older areas (e.g., the Tonga subduction zone). There are also examples of seismic models, however,

(a)

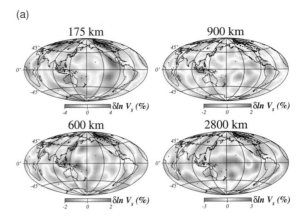

(b)

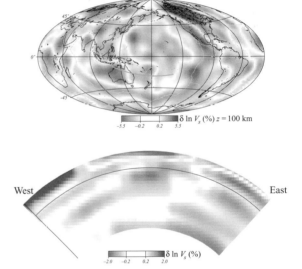

Figure 6.18 COLOR PLATE Seismic tomography maps of the Earth. (a) Four depth sections through the seismic model SAW12D derived by Li and Romanowicz (1996). (b) Plots of the seismic model SAW18B16. The top map shows the model at 100 km depth, and the bottom plot shows a cross-section through the green line in the South Pacific at the latitude of the Tonga Kermadec subduction zone. The line at 670 km in the cross-section corresponds to the separation between upper and lower mantle, while the vertical slice extends down to the core–mantle boundary. (Megnin and Romanowicz 2000)

where descending slabs are stopped ~1300 km above the core–mantle boundary, or at the 670 km discontinuity.

The lowermost ~150–200 km of the mantle, just above the core–mantle boundary, is referred to as the D'' region. This region exhibits a strong lateral heterogeneity in seismic wave velocities, including thin patches of 'ultralow' wave speeds. The overall heterogeneity is suggestive of a chemical heterogeneity (e.g., in the silicate/oxide ratio), temperature anomalies, and/or of a phase transition from perovskite to postperovskite, or melting. Chemical variations may result from the interaction between the rocky mantle and the liquid metal outer core, or from being a 'graveyard' for subducting slabs. Laboratory experiments suggest that the oxides from the rocky mantle undergo vigorous reactions with the liquid metal, so the mantle may be slowly dissolving into the outer core, creating patches of partial melting at the base of the mantle. On the other hand, spatial variations in the metal alloys at the core–mantle boundary, having a high conductivity, might trigger temperature anomalies. Both processes could explain the observed patches of ultralow seismic wave velocities, which appear to indicate narrow ($\lesssim 40$ km wide) hot (velocities of $\gtrsim 10\%$ below average) 'plumes' of magma, rising up from the core–mantle boundary. These plumes would cool off the Earth's core by transferring energy outwards. They probably surface in the form of ocean islands, as mentioned above. This view thus suggests that some processes on the Earth's surface are connected to those happening at the core–mantle boundary.

The models based upon seismic tomography and those based upon geochemical data are thus seemingly incompatible. However, there is a theory which may satisfy both sets of data: Computer models suggest that convection is modified at boundaries where a phase change takes place, very similar to inhibition of convection through a layered atmosphere (§4.2.1.2). The subducting slabs of material sink down to such a boundary, and mass accumulates here. When a critical mass is reached, the material suddenly plunges through the boundary, like an 'avalanche'. Computer models, supported by seismic tomography maps, suggest that such avalanches might have occurred and may fall all the way through the mantle to the core–mantle boundary. There may thus be a partially layered convection, but with some intermittent mixing of materials throughout the entire mantle so that the distinction between layered and whole mantle convection becomes vague. This remains an active area of research.

6.2.2.2 Earth's Core

The core of the Earth consists of a solid inner and liquid outer part, as discussed above, with an atomic weight close to that of iron. Since iron is one of the most abundant heavy elements in space (as a result of stellar nucleosynthesis; §13.2), and since the metal in metallic meteorites is usually iron-dominated, it seems most reasonable that the inner core is an iron-rich alloy. Since the density of the outer core is ~ 5–10% less than that of pure iron (or Ni–Fe), there must be some lighter element present. Possible 'contaminants' are sulfur, oxygen, silicate, carbon, and/or hydrogen (§6.1.2.3), and some of these might infiltrate

the outer core via chemical reactions at the core–mantle boundary. The presence of these lighter elements further lowers the melting temperature of the core, leading to the large liquid outer core as observed. Since overall the Earth is cooling down since its formation 4.5 Gyr ago, one might expect outer core material to condense onto the inner core, a process which should liberate energy (from latent heat) (Problem 6.25). This heat may drive convection in the outer core, essential to the formation of the Earth's magnetic field (§6.2.2.3).

Seismic waves pass through the inner core slightly faster when they follow a north–south track than in the east–west direction. The source of this anisotropy is not known, but it is probably related to the crystalline structure of the inner core. At the high pressures prevailing in the Earth's core, iron forms hexagonal close packed crystals, which may line up in a certain orientation as a result of solid-state convection. The seismic 'fast-track' axis is tilted by $\sim 10°$ compared to the rotation axis, and recent studies have shown that this axis traces out a circle around the geographic north pole, with the inner core appearing to rotate slightly (up to perhaps a few tenth's of a degree per year) faster than the rest of the planet. Why the inner core should spin at a different rate than the rest of the solid Earth is a complex problem, possibly related to connections between the fluid outer core and the Earth's magnetic field. When fluid from the outer core sinks down, it is spun up (conservation of angular momentum), and magnetic field lines in the fluid are dragged forward (compare with the magnetic braking effect, discussed in §13.4.2). Since the field lines thread through the inner core, the inner core is slightly accelerated in its rotation. An alternative model is that the rotation rate of Earth's mantle is gradually slowed by tidal torques from the Moon and the Sun, and that these torques are only weakly coupled to the inner core because the viscosity of the outer core is low.

6.2.2.3 Earth's Magnetic Field

The presence of a magnetic field around a planet poses constraints on interior models of that planet. As discussed in Chapter 7, it is generally assumed that magnetic fields are produced by a magnetohydrodynamic dynamo, where electric currents in a planet's interior generate magnetic fields just like currents in a coiled copper wire. The details of this process, however, are not understood, although computer models have made significant progress in modeling the Earth's magnetic field, including its field reversals.

The geological history of the Earth's magnetic field is most apparent from the orientation of iron-bearing crystals in rocks. To explain this, we consider the mid-ocean ridge, where for millions of years magma has risen to form new crust. Thus the youngest material is found directly on top of the ridge, and the crustal age increases with increasing distance from the ridge. The ocean floor therefore is lined with 'strata', symmetrically around the ridge, and the geological record can be established by investigating the material at different distances from the mid-ocean ridge. Basaltic magma is rich in iron, and any iron-bearing minerals in a cooling magma or lava become magnetically aligned with the Earth's field. Upon cooling below a magnetic 'freezing-in' (Curie) temperature, this orientation gets locked in. Thus, by analyzing the iron-bearing minerals as a function of distance from the mid-ocean ridge, one can (in principle) determine the geological history of the Earth's magnetic field. The orientation of the magnetic crystals produces a striped pattern parallel to the ridge. These variations have been interpreted as caused by changes in the magnetization direction, where the magnetic orientation changes direction from one stripe to the next. These changes in magnetic orientation are indicative of reversals in the orientation of the Earth's magnetic field. Additional evidence for the Earth's 'field reversals' comes from paleomagnetic studies, where the magnetism and age of old rocks in many parts of the world have been determined. From these records scientists have been able to reconstruct the history of the Earth's magnetic field, and it has become apparent that magnetic field reversals were quite common on geological timescales; they appear to have occurred at irregular intervals of between 10^5 and a few million years.

If a planet has an internally generated magnetic field, it must have a region in its interior that is fluid, convective, and electrically conductive. Since the Earth's outer core is fluid, likely convective, and conductive because iron is its major constituent, our planet's magnetic field must originate in the outer core. The presence or absence of an internally generated magnetic field around a planet thus adds constraints to models of its interior.

6.3 Interiors of Other Solid Bodies

Seismic data are only available for the Earth and the Moon. One can use these bodies as prototypes, and develop models for the internal structure of other solid bodies, which mimic the observable parameters as summarized in the introduction of this chapter. In this section we summarize our current understanding of the Moon, the terrestrial planets, and some of the satellites in the outer Solar System. Minor planets and comets are discussed in Chapters 9 and 10, respectively. The interior structure of the giant planets is described in §6.4.

6.3.1 Moon

Measurements of the Moon's moment of inertia show a value $I/(MR^2) = 0.3932 \pm 0.0002$, only slightly less than the value of 0.4 for a homogeneous sphere. A model which fits the measured $I/(MR^2)$ together with the lunar seismic data suggests the average density of the Moon to be 3.344 ± 0.003 g cm^{-3}, with a lower density crust ($\rho = 2.85$ g cm^{-3}) that is on average between 54 and 62 km thick, and an iron core ($\rho = 8$ g cm^{-3}) with a radius $R \lesssim 300$–400 km. The latter estimates on the Moon's core are consistent with measurements of the large localized magnetic field in the Moon's crust (§7.5.3.3).

A laser ranging device and a microwave transmitter on board the Clementine spacecraft, launched in 1994, provided detailed measurements on the Moon's gravity field and surface topography. Gravity and geoid height anomalies are determined from microwave Doppler tracking, and topography is determined from laser ranging. Topography and gravity anomaly maps are shown in Figure 6.19. The lunar gravity models typically show an elevated equipotential surface at low latitudes and gravitational lows closer to the poles. Assuming a crustal density of 2.8 g cm^{-3}, the Bouguer gravity correction factor was calculated and subtracted from the free-air anomalies to allow determination of the subsurface density distribution. If all deviations in the Bouguer anomaly are attributed to variations in the thickness of the crust, these data yield a map of the relative crustal thickness. Such a map for the lunar crustal thickness is shown in Figure 6.19c, where a density of 3.3 g cm^{-3} was assumed for the lunar mantle, and an average reference value of 64 km for the lunar crust.

The maps displayed in Figure 6.19 show that the highlands are gravitationally smooth, indicative of isostatic compensation such as seen over most regions on Earth. The lunar basins, however, show a broad range of isostatic compensations. Some basins on the near side show clear gravity highs (e.g., the Imbrium basin), circular features called *mascons* (from mass concentrations). A photograph of the Moon, annotated with many of the geologically interesting features, is shown in Figure 5.15. The gravity highs in the lunar basins suggest that the Moon's lithosphere here was very strong at and since the time the area was flooded by lava. Both the local uplift of mantle material following the impact and the later additon of dense mare basalt must have caused the gravitational highs. The high-gravity areas over young, large, multiringed basins, such as Mare Orientale with a Bouguer anomaly of +200 mGal,[5] are surrounded by a negative anomaly ring (−100 mGal for Orientale), with an outer positive collar

[5] 1 mGal = 10^{-3} cm s^{-2}.

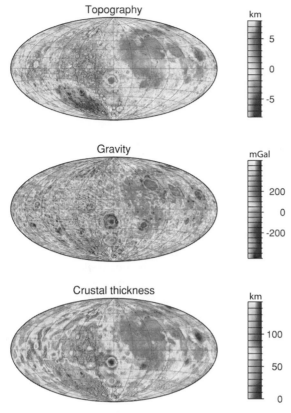

Topography

km

5

0

-5

Gravity

mGal

200

0

-200

Crustal thickness

km

100

50

0

Figure 6.19 COLOR PLATE A map of the lunar topography (in km), free-air gravity anomaly (in mGal) and the crustal thickness (in km). (a) Topography model GLGM-2 based upon Clementine data (Smith *et al.* 1997). (b) Gravity model LP75G (Konopliv *et al.* 1998). This model also includes data from Lunar Prospector. (c) Single layer Airy compensation crustal thickness model using topography model GLTM-2 and gravity model LP75D. This latter model corrects for the presence of mare basalt in the major impact basins in the determination of crustal thickness. (Hood and Zuber 2000)

(+30–50 mGal for Orientale), suggestive of flexure of the lithosphere in response to the flooding and mantle uplift following the impact. The Aitken basin, adjacent to Mare Orientale, on the other hand, is nearly fully compensated, or relaxed. There is no correlation between isostatic compensation of the basins with their size or age, so the lithosphere must have displayed large spatial variations in strength at the time the basins were formed and flooded. This conclusion was reinforced based upon more recent gravity data by the SELENE (Kaguya) orbiter, which revealed that the far-side lithosphere must have been stronger than the near side during the late heavy bombardment era when the mascons were formed. The

gravitational differences can, to first approximation, be interpreted as caused by variations in crustal thickness. Assuming a constant crustal density, the far-side crust is, on average, thicker (68 km) than the near-side crust (60 km). The crust is usually thinner under the maria, with a minimum near 0 km at Mare Crisium, 4 km under Mare Orientale, and 20 km under the Aitken basin. The thickest crust, 107 km, is on the far side, beneath a topographic high.

The interior structure of the Moon has largely been determined from seismic measurements of *moonquakes* at several Apollo landing sites. In contrast to the Earth where quakes originate close to the surface, moonquakes originate both deep within the Moon (down to ~1000 km) and close to its surface. Deep (700–1000 km) moonquakes are usually caused by tides raised by Earth, although such quakes can also occur closer to the surface. Meteoritic impacts are another common source of moonquakes. Note that such impacts are more numerous on the Moon than on Earth since the Moon does not have much of an atmosphere. Moonquakes can also be triggered thermally, by the expansion of the frigid crust when first illuminated by the morning Sun after having been in the cold (night) for two weeks. The lunar free oscillations require a long time to damp, implying that the Moon has little water and also lacks other volatiles. Figure 6.20 shows a lunar seismic velocity profile for P waves (panel a) and a sketch of the interior structure of the Moon (panel b), as deduced from the various seismic experiments. The shallow (near surface) velocities are attributed to ejecta blankets. The velocity of both S and P waves increases steadily down to ~20 km. From 20 down to 60 km the average P wave velocity is ~6.8 km s^{-1}, indicative of an anorthositic composition. This region is the lunar crust, which varies in thickness from less than a few kilometers over maria up to over 100 km over the highlands, as mentioned above. The crust is on average thicker on the far side, which results in an offset of the Moon's center of mass from its geometric center by 1.68 km ± 50 m in the Earth–Moon direction, with the center of mass closer to Earth (Fig. 6.20b). This offset is probably caused by an asymmetry that developed during crystallization of the magma.

There appears to be a slight seismic velocity inversion at a depth of 300–500 km, which might suggest a chemical gradient with depth, such as a relative increase in the ratio Fe/(Fe + rock). The region below the crust down to ~500 km is called the upper mantle. Its composition is dominated by olivine. The middle mantle, down to

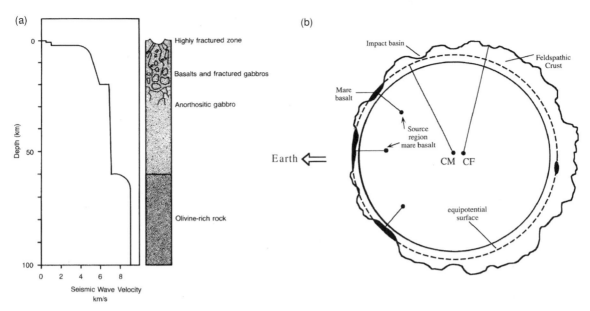

Figure 6.20 (a) Lunar seismic velocity profile of P waves, measured with seismometers deployed by astronauts during Apollo moonwalks. The interpretation of the measurements is shown on the right. (After Hartmann 2005, with data from Taylor 1975) (b) Sketch of the interior structure of the Moon, in the equatorial plane, showing the displacement toward the Earth of the center of mass (CM) (greatly exaggerated) relative to the center of the Moon's figure (CF). (Taylor 2007)

~1000 km, is dominated in composition by olivine and pyroxene, like the Earth's mantle. The average S and P wave velocities down to this depth are $v_S \approx 4.5$ km s^{-1} and $v_P \approx 7.7$ km s^{-1}, respectively. Attenuation of S waves at deeper levels may imply the Moon's lower mantle to be partially molten. No moonquakes have been detected below ~1000 km. The seismic velocity of P waves decreases by a factor of ~2 below ~1400 km, indicative of an iron-rich core. Although this last measurement was derived from only one, relatively weak meteoritic impact, the observation that many of the lunar samples seemed to have solidified in a strong magnetic environment ~3–4 Gyr ago, while the Moon itself does not now have a magnetic field, is also suggestive that the Moon has a small metallic core. A core with a radius $R \lesssim 300$–400 km is consistent with all available data.

6.3.2 Mercury

Unfortunately, there is not much data on Mercury yet that can be used to infer its interior structure. Until recently, photographs and other measurements from three flybys by the Mariner 10 spacecraft, together with radar experiments and other Earth-based observations provided all the available data. Although the Messenger spacecraft passed by Mercury in January and October of 2008 and October

2009, there have not yet been any updates on the internal structure of the planet.

Radar data revealed the unique 3:2 commensurability of Mercury's rotation and orbital period (§2.6.2), with the upshot that Mercury spins around its axis in 59 days. The high bulk density, $\rho = 5.43$ g cm^{-3} (uncompressed density of 5.3 g cm^{-3}), of a planet about 20 times less massive than the Earth, implies that ~60% of the planet's mass consists of iron (Problem 6.27), which is twice the chondritic percentage. Models suggest the iron core extends out to 75% of the planet's radius. The outer 600 km is the mantle, composed primarily of rocky material, where the outermost 200 km, at temperatures below 1100–1300 K, is the lithosphere (Fig. 6.21a). The moment of inertia of such a planet is $I/(MR^2) = 0.325$. A measurement of J_2 and the precession of Mercury's rotation axis would yield values for the moments of inertia (eqs. 6.9a, 6.23), and hence constrain the interior structure of this planet.

The absence of a large rocky mantle has led to the theory that the planet was hit by a large object towards the end of its formation. The impact ejected and possibly vaporized much of Mercury's mantle, leaving an iron-rich planet. When the large iron core started to cool, it contracted and the rigid outer crust collapsed and formed the unique scarps seen all over Mercury (§5.5.2). Far-infrared and radio observations indicate a general lack of

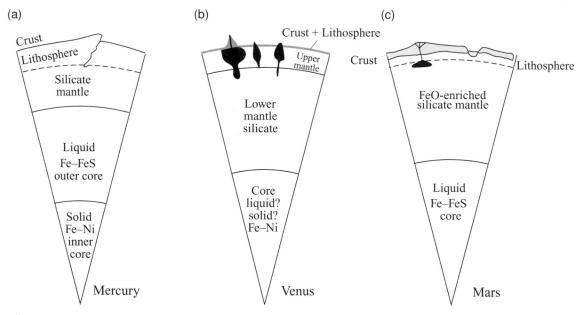

Figure 6.21 Sketch of the interior structure of (a) Mercury, (b) Venus, and (c) Mars. These sketches are based upon our 'best guess' models for the interior structure of the three planets, as discussed in the text.

basaltic iron- and titanium-rich material; the abundance of these elements is even lower on Mercury's surface than in the lunar highlands. This suggests that deep-seated widespread volcanism shut off early in Mercury's formation history. If true, it suggests a very slow cooling of the iron core, so that the core may still be partially molten.

Using high precision radar measurements, the position of Mercury's spin axis has been determined with a high accuracy (2.11 ± 0.1 arcmin). Variations in the planet's spin rate during Mercury's 88 day orbital period indicate a small oscillation or libration, which is forced by solar torques. The amplitude of this forced libration implies that the planet's mantle must be decoupled from the core, so Mercury must have a liquid outer core. The outer core likely consists of a mixture of Fe and FeS. A sulfur content of a few percent would lower the melting temperature of the outer core sufficiently to maintain it as a liquid, yet permit the solidification of an inner core. In analogy with Earth, the solidification releases sufficient energy to keep the outer core convective, a necessary condition to sustain a magnetohydrodynamic dynamo system that could produce the magnetic field detected around Mercury (§7.5.2).

6.3.3 Venus

Venus is very similar to Earth in size and mean density, suggesting similar interior structures for the two planets. This argument is strengthened by observations from the Soviet landers that show Venus's surface to be basaltic in composition. Venus's mean uncompressed density is

slightly ($\sim 3\%$) less than that of the Earth (Table 6.1). Venus, in contrast to the Earth, does not possess an intrinsic magnetic field, which implies the absence of a convective metallic region in its mantle and/or core. Could the core be completely frozen? The slightly lower mean density of Venus compared to Earth hints at the absence of some abundant heavy element. Based upon theories on the origin of the Solar System, Venus may contain less sulfur than Earth, since it formed in a warmer region of the solar nebula. If true, Venus's core contains less iron sulfide than the terrestrial core. Since iron sulfide lowers the melting temperature of iron, the absence of this alloy may have led to a completely frozen core. Alternatively, the difference between planetary densities may be the result of stochastic disruptions of differentiated planetesimals. In contrast to a frozen core, it might be possible that Venus's core is liquid, but that there is no convection. On Earth a major source of energy driving convection in the liquid outer core comes from the phase transition between the solid inner and liquid outer core (Problem 6.25). Solidification of core material liberates energy, driving convection in the Earth's core. If this phase boundary is absent on Venus, there may be no convection, even if the core is completely liquid. Measurements of the venusian gravitational field from Doppler tracking of Magellan and Pioneer Venus Orbiter spacecraft data seem to support the hypothesis that Venus's core is liquid. An alternative explanation for the absence of convection in a liquid core would be provided if the mantle were hotter than the core, a situation which would inhibit

(a)

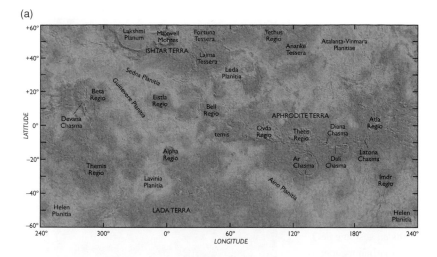

(b)

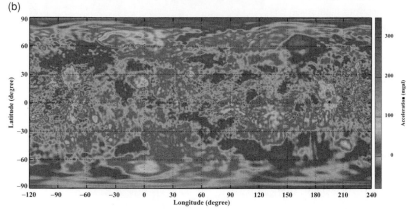

Figure 6.22 COLOR PLATE (a) Mercator-projected view of Venus's surface as derived from Magellan's radar altimeter data. Maxwell Montes, the planet's highest mountain region, rises 12 km above the mean elevation. (Courtesy Peter Ford, NASA/Magellan) (b) Free-air gravity map of Venus: Note the high correlation with Venus's topography in panel (a), such as the gravity and topography highs at the Beta Regio and Alta Regio, while the low lying planitiae show gravity lows. (Konopliv *et al.* 1999)

rather than trigger convection in the core. Although such a situation may sound counterintuitive, below we reason that such may be the case on Venus.

The lowlands and highlands on Venus are sometimes compared to the ocean floors and continents on Earth, even though the composition and detailed topography of the venusian regions are quite different from those on Earth. The largest difference between the two planets is the absence of planet-wide tectonic plate activity on Venus (Fig. 5.53). Water on Earth plays a key role in driving plate tectonics. Water weakens a rock's rigidity, or strength, and lowers its melting temperature. The water in Earth's lithosphere is therefore thought to be essential to enable plate tectonics. Venus's atmosphere and rocks are extremely dry, which most likely is a consequence of the planet's extreme high surface temperature having driven off all the water (§4.9.2.3). The dry rocks retain a high rigidity even at relatively high temperatures, so that Venus's lithosphere does not break up, and may be as thick as that of the Earth, as has been suggested based upon gravity mesurements.

On Earth, plate tectonics is a major avenue of heat loss. In the absence of plate motions, hot spot and volcanic

activity may be more important. Indeed, there are numerous volcanic features on Venus, including giant volcanoes, small domes, and coronae (§5.5.3). However, all these features together can at most account for a heat loss ∼20–30% of that generated in the planet's interior. The interior, therefore, is slowly heating up, and in contrast to Earth, which displays *active lid* mantle convection, Venus may be in a *stagnant lid* convection mode, where a strong cold lithosphere blocks the heat from below. Assuming that most of the radioactive heating occurs in the mantle rather than the core, this may lead to a hot mantle overlying a relatively cooler core. Although pieces of the lithosphere may break off when it becomes too dense, such a process is not a very efficient cooling mechanism. Numerical models such as those leading to tectonic plate convection on Earth show that under venusian conditions it is possible that the lithosphere may completely subduct, i.e., show a 'catastrophic resurfacing' event every few $\times 10^8$ years (§5.5.3).

Venus's gravity field is highly correlated with its topography (Fig. 6.22), suggestive of a lithosphere strong enough to support the topography. Although some highlands may, in part, be isostatically compensated, most are not, and appear to be compensated by large mantle plumes.

6.3.4 Mars

Mars is intermediate in size between the Moon and Earth. Its mean density is 3.93 g cm^{-3}, which is larger than expected for a planet composed entirely of chondritic material. A more interesting comparison is the value of the uncompressed density of the martian mantle, 3.55 g cm^{-3}, with that of Earth, 3.34 g cm^{-3}, which implies a larger abundance of iron in Mars's mantle. Estimates yield mass abundances for FeO in the range 16–21%, or more than twice the value in the Earth's upper mantle (7.8%). Analysis of Mars meteorites (Chapter 8) yields an absolute FeO concentration of 18% by weight for the martian mantle; a similar value is obtained on the surface at various landings sites. In fact, Mars's red color is a consequence of relatively large quantities of rust, Fe_2O_3, something that has not been seen on any other planet. Although the iron content on Mars's surface and in its mantle are much larger than those on Earth, the upper regions on Mars are still depleted compared to chondritic material (the chondrite-normalized iron value is 0.39), presumably caused by a separation of Fe–FeS into the core.

Mars's low surface temperature suggests a relatively thick lithosphere (Problem 6.29), which is consistent with the observed absence of tectonic plate activity. A significant fraction of the early heat loss was probably via volcanic activity, as suggested by the presence of several large shield volcanoes. These large volcanoes further imply that the erupting magma had a low viscosity, such as measured for lavas rich in FeO (25% by mass). The magma probably originated at a depth of ∼200 km (Problem 6.30), a region within the planet's mantle, perhaps akin to the Earth's asthenosphere.

A map of Mars's global topography reveals a striking ∼5 km difference in elevation between the northern and southern hemispheres (Fig. 6.23a; §5.5.4). After removing a best-fit reference ellipsoid, which is displaced ∼3 km to the south from the center of mass, most hemispheric asymmetries disappear. A global gravity map (Fig. 6.23b) shows that the martian geoid, or areoid, is highly correlated with topography, indicating that topographic features are not isostatically compensated. The Tharsis region, for example, shows a clear gravity high of ≳1000 mGal. Localized mass concentrations appear centered at impact basins, and Valles Marineris shows a pronounced mass deficit. Such features are suggestive of a thick rigid lithosphere. Assuming that the Bouguer gravity anomalies are directly correlated with variations in crustal thickness, Figure 6.23c

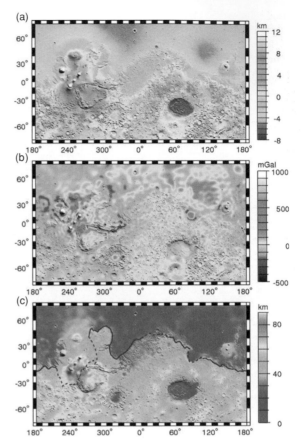

Figure 6.23 COLOR PLATE Relationship between local topography on Mars (a), free-air gravity (b), and crustal thickness (c). Note the good correlation between topography and gravity: topographic highs, as the Tharsis region, show large positive gravity anomalies and a thick crust. In contrast, the Hellas basin, a topographic low, shows a large negative gravity anomaly and the crust is thin. (Zuber *et al.* 2000)

shows that to first order such variations correlate well with Mars's topography. For example, the crust is thinnest in the northern hemisphere and at Hellas, and thickest under the Tharsis region.

It is not a coincidence that the large Tharsis region is on the planet's equator. Even if it would have formed at a different latitude, true polar wander would have reoriented the planet to bring it into rotational equilibrium. As shown from the gravity maps, the Tharsis rise is not in hydrostatic equilibrium. This is also evident from a comparison of the 'true' moment of inertia, $I/(MR^2) = 0.365$, with one calculated directly from the measured gravitational moment, J_2, by assuming hydrostatic equilibrium, $I/(MR^2) = 0.377 \pm 0.001$. The smaller number is essentially based on measurements of the J_2 (by tracking orbiting spacecraft) and of the precession rate of the planet as determined from data taken by the landers and rovers on Mars. A model taking into account Mars's gravity field, moments of inertia, and tidal deformation by the Sun, measured a tidal Love number $k_T = 0.153 \pm 0.017$, which is indicative of a liquid core, at least in part, with a radius between 1520 and 1840 km. Since there is no internal magnetic field, the core is probably completely fluid, as expected from laboratory measurements if it is composed of an Fe–FeS mixture.

As discussed in §5.5.4, there is a debate whether or not the northern plains on Mars may have beeen covered by an ancient ocean. The main evidence against this theory is the fact that the paleoshorelines, which can be traced over many thousands of km, do not follow surfaces of equal gravitational potential. However, if a true polar wander event changed the orientation of Mars, after the Tharsis rise formation and reorientation to the equator, an ancient ocean could have covered the plains such that the shorelines did follow the equipotential surfaces as expected for standing bodies of water.

6.3.5 Satellites of Giant Planets

The density of large (radii over a few hundred km) satellites varies from ~ 1 g cm^{-3} up to ~ 3.5 g cm^{-3}, indicative of almost pure iceballs up to bodies composed mostly of rock. The composition of these bodies clearly depends upon their place of formation. In the outer Solar System, beyond the *ice line*, i.e., the surface in the solar nebula outside of which temperatures remained low enough for water to condense, bodies accreted ice during their formation stage, which is reflected in a relatively low bulk density. In the giant planet subnebulae, however, it may have been too warm close to the planet for ice to form, and hence satellites like Io and Europa near Jupiter are dominated by rock, despite being formed beyond the ice line in the primitive solar nebula itself. In the following we discuss the interior structure of some of the larger satellites in detail.

6.3.5.1 *Galilean Satellites*

Io's mean density of 3.53 g cm^{-3} is indicative of a rock + iron composition. Direct spectral information on surface composition reveals a body devoid of water-ice, and rich in SO_2 frost and other sulfur-bearing species. In addition, mafic minerals such as pyroxene and olivine have been identified. As discussed in §2.6 and §5.5.5, Io is the most volcanically active body in our Solar System, with a heat flow 20–40 times higher than on Earth (Table 6.3), a characteristic that has been attributed to strong tidal interactions with Jupiter, while being in a Laplace resonance with Europa and Ganymede. Io's interior must,

therefore, be very hot and partially molten. Eruption temperatures of 1200–1400 K are common, with occasional temperatures exceeding 1800 K. This, coupled to measurements of Io's moment of inertia by the Galileo spacecraft, $I/(MR^2) = 0.378$, suggests that Io is differentiated, with the heavier elements concentrated in the core. The combination of Io's detailed shape and gravity field, both influenced by tidal and rotational forces, provides further constraints on the satellite's internal properties, assuming the body has relaxed into an equilibrium shape. Indeed, measurements show a triaxial ellipsoid, as expected, with the long axis pointed towards Jupiter and short axis aligned with the rotational axis. Although no unique model for Io's interior structure can be derived from the data, there are sufficient constraints to support the following general picture.

Io's core radius could be as small as $\sim 1/3$ of Io's radius if the core is made of pure Fe–Ni, or as large as $\sim(1/2)R_{\mathrm{Io}}$ if it is composed of an eutectic mixture of Fe–FeS (+Ni). Since Io has no intrinsic magnetic field, the core is probably either completely solid or liquid. Overlying Io's core is a hot silicate mantle, topped off with a, perhaps lower density, crust + lithosphere (Fig. 6.24a). The lithosphere has to be strong enough to support the >10 km high mountains observed on the satellite (Fig. 5.76), yet at the same time it must allow the passage of a high heat flux, i.e., it must be cold enough to sustain elastic stresses for a long time, yet hot enough to explain the heat flow. Io's crust/lithosphere is probably ~ 30–40 km thick, while a 'heat-pipe' mechanism may transport heat in the form of magma rising up through fissures. This process may lead to vast amounts of molten rock breaking through the surface and covering older colder lava flows, which ultimately are mixed back into the, partially molten, mantle. The advection of material corresponding to Io's global resurfacing rate of ~ 1 cm/yr allows indeed a relatively thick cold surface layer to develop, which is strong enough that it can support the >10 km high mountains as observed on the satellite. The observed eruption temperatures are consistent with global melt fractions of 10–20%, and melt fractions could be much higher locally. All the melt might be concentrated in an asthenosphere with a thickness of several tens, perhaps ~ 100 km. However, whether there is an asthenosphere depends in part on where most of the tidal heat is dissipated, throughout the mantle or near the surface. This is not known.

Europa has a mean density of 3.01 g cm^{-3}, indicative of a rock/ice composition where the rocky mantle/core takes up over 90% of the mass (Fig. 6.24b). Europa's moment of inertia ratio, $I/(MR^2) = 0.346$, implies a differentiated, centrally condensed body. The rocky, most

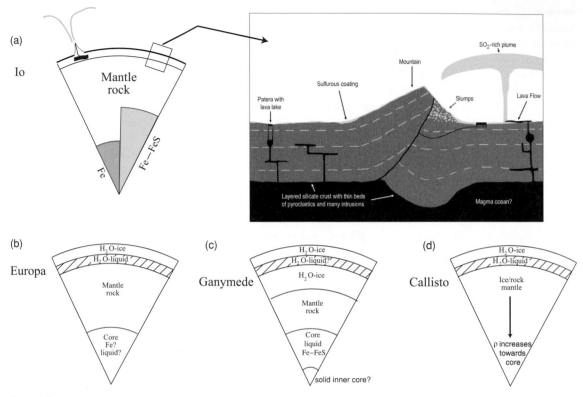

Figure 6.24 Best estimates of the interior structures of the four Galilean satellites. (The Io inset is from McEwen *et al.* 2004)

likely dehydrated, mantle is overlain by a thick H_2O crust, while the satellite probably has an inner metallic core. The H_2O layer is at least 80 km thick, but not more than 170 km, while the core may subtend up to 45% of Europa's radius, or only ~12%. The exact size depends on the composition of the core (Fe–Ni, Fe–FeS–Ni or other mixtures), and the thickness of the H_2O crust. Galileo images (Fig. 5.81c) suggest that the crust consists of thin ice sheets or plates floating on a liquid ocean or a layer of soft ice; the images remind one of the ice in the Arctic regions on Earth. Galileo images are suggestive of nonsynchronous rotation (§5.5.5.2), which implies that the outer shell must be decoupled from the mantle, indicative of a liquid layer in between. The strongest evidence for the existence of a liquid ocean beneath Europa's icy crust comes from Galileo magnetometer data, which revealed disturbances in the magnetic field around Europa consistent with that expected from a body covered by a salty (briny) ocean (§7.5.4.9). Dissipation of tidal energy at the base of the ice shell, just above the liquid ocean, is probably sufficient to keep the ocean liquid.

Ganymede has an average density $\rho = 1.94$ g cm^{-3}, suggestive of a mixture of comparable amounts (by mass)

of rock and ice. Its moment of inertia, $I/(MR^2) = 0.312$, implies that its mass is heavily concentrated towards the center. Ganymede has an intrinsic magnetic field, suggestive of a liquid metallic core, most likely with a small solid inner core (§7.4.9). Best fits to the gravity, magnetic field, and density data are obtained with three-layer internal models, where each layer is ~900 km thick. These models consist of a liquid metallic core, surrounded by a silicate mantle, which is topped off by a thick ice shell (Fig. 6.24c). The ice may be liquid at a depth of ~150 km (2 kbar), where the temperature (253 K) corresponds to the minimal melting point of water. Since water-ice under high pressure takes on many different phases (Fig. 6.6), with different ice densities, a layer of liquid water could be sandwiched in between two layers of different forms of ice. Galileo radio Doppler data revealed several gravity anomalies, perhaps large concentrations of rock at or underneath the ice surface, or perhaps at the bottom of the ocean, if there is a liquid ocean.

Callisto's average density, $\rho = 1.83$ g cm^{-3}, suggests this moon must contain a larger percentage of ice than Ganymede, consistent with the trend of increasing ice abundance from the inner- to the outermost Galilean satellite. Its moment of inertia, $I/(MR^2) = 0.355$, is slightly less than expected for a pressure compressed, yet compositionally homogeneous, mixture of ice and rock, but substantially larger than that of Ganymede. Callisto, if in hydrostatic equilibrium, may be partially differentiated, with an icy crustal layer (a few hundred km) and an ice/rock mantle which is slightly denser towards the center of the satellite. The magnetometer on board the Galileo spacecraft discovered magnetic field disturbances which, as for Europa, suggest the presence of a salty ocean within Callisto (§7.5.4.9). As on Ganymede, such an ocean may exist at a depth of $\sim$150 km.

6.3.5.2 Satellites of Saturn

Saturn's regular satellites display a wide variety of densities (Table 1.5). Titan is by far the largest, and it is also the densest, in part because of compression of ices deep within its mantle. In contrast to Jupiter's Galilean moons, the densities of Saturn's mid-sized regular moons do not show a monotonic trend of density with distance from the planet, nor do their densities correlate strongly with mass. Enceladus, the second densest of Saturn's regular moons, is too small for significant pressure-induced compression; thus, its density of 1.61 g cm^{-3} implies that it has a bit more rock (by mass) than ice. The neighboring (and slightly larger) moon Tethys, by contrast, has a density of only 0.96 g cm^{-3}, implying that it is almost pure H_2O-ice. The alternative of an admixture of water-ice with lower density, more volatile ices and some rock is unlikely since Tethys's surface is primarily H_2O-ice, and Tethys is too massive to have a substantial porosity. Saturn's tiny, irregularly shaped inner moons have even lower densities than does Tethys, but the pressure within these moons is so small that their low densities presumably result from high porosity, i.e., large amounts of void space on small (*microporosity*) or large (*macroporosity*) length scales.

The mean density of Saturn's satellite Titan is 1.88 g cm^{-3}, which implies a rock-to-ice ratio of $\sim$0.55 by mass. The interior should be differentiated into a rocky (+ iron) core, $\sim$60–65% in radius, and an ice-rich mantle. As discussed in §4.9.1.2 and §5.5.6, the presence of methane gas in Titan's atmosphere requires an active source, which could either be in the form of liquids on the surface (through a methanological cycle akin to the hydrology cycle on Earth) or via cryovolcanism. If indeed there is cryovolcanism on Titan, there must be liquids below the surface, e.g., in the form of a subsurface ocean. Evidence for such an ocean is accumulating, despite the initial problem of how to reconcile Titan's 3% orbital

eccentricity if tidal dissipation would be as high as expected for a satellite covered by large global oceans (on the surface or interior). Such a high eccentricity cannot be maintained via resonances with other moons, as in the case of Io.

Cassini radar measurements indicate that Titan's spin period decreased by 0.36°/yr between 2004 and 2007. Such a change could be caused by a seasonal exchange of angular momentum between the surface and Titan's dense atmosphere, provided the satellite's icy crust is decoupled from the core by a subsurface ocean, e.g., a layer of water enriched with liquid ammonia. The ammonia is essential since it decreases the ocean's melting temperature. Its presence is further predicted from formation models of a satellite in Saturn's subnebula (§13.11.1). Recent models for Titan compute the satellite's thermal evolution over time and the coupling between its interior and orbit, including the effects a freezing subsurface ocean has on the tidal dissipation rate and orbital evolution. Such models can reconcile the observed eccentricity with the presence of a subsurface ocean. In one such model, methane on Titan is stored in the form of methane clathrate hydrates (§5.1.1) within an icy shell above the ocean, and released episodically, injecting the atmosphere with fresh methane gas. A typical thickness for the ice shell is of the order 60–100 km, with an ocean several 100 km thick.

The Cassini spacecraft detected extreme geyser activity on the south pole of Enceladus (§5.5.6), where giant plumes of vapor, dust, and ice emanate from 'tiger stripes', large tectonic features near Enceladus's south pole. It is difficult to explain the presence of subsurface oceans on this relatively small ($R \sim 250$ km) moon, even though some tidal heating is expected from its 2:1 mean motion resonance with Dione, augmented perhaps by interactions with Janus and some radiogenic heat from a rocky core. The heating must be sufficient, though, to induce some geological activity. Perhaps, in addition, decompositon of clathrate hydrates on the ocean floor, an exothermic reaction, may drive geysers. Or perhaps the small amount of tidal heating is sufficient to induce ice *diapirs*, where warmer buoyant material (mushy ice or liquid) moves upwards through the ice shell, to 'explode' into jets on the surface. Diapirism may explain why the jets are on the south pole, and not on the equator, since it will lead to a mass deficit at depth. Depending on the rigidity of the ice shell, the mass deficit will be compensated by topographic highs (a lithosphere without much rigidity), or will produce a net negative gravity anomaly (rigid lithosphere, and no compensation). In the latter case, true polar wander would reorient the satellite such that the jets emanate from the (south) pole.

6.3.5.3 *Satellites of Uranus and Neptune*

The five largest satellites of Uranus have mean densities of ~ 1.5 g cm^{-3}, indicative of bodies composed of ice/rock mixtures. Given the sizes of these bodies, the radioactive heat even without the accretional heat from formation would have been sufficient to differentiate the bodies into rocky cores with an ice/silicate mantle and icy crust, unless it was carried outward via solid state convection. Depending on the size of the object and its environment (tidal heating), the mantle could be solid or partially liquid. The likelihood of a liquid layer increases if a small amount of ammonia is present, as the melting temperature of an eutectic H_2O–NH_3 mixture is much lower than that of pure water-ice.

Triton's density of 2.05 g cm^{-3} indicates a significantly larger rock fraction than other large and mid-sized moons of the giant planets, apart from Io and Europa. (Ganymede, Titan, and Callisto are almost as dense as Triton, but ice within these three more massive moons suffers greater pressure-induced compression.) The mass and density of Triton are similar to those of Pluto, consistent with models that suggest Triton formed in the solar nebula and was subsequently captured by Neptune (§13.11.1).

6.4 Interior Structure of the Giant Planets

6.4.1 Modeling the Giant Planets

The bulk density of the giant planets implies that they consist primarily of light elements. The two largest planets in our Solar System are generally referred to as *gas giants*, even though the elements that make up these giants are no longer gases at the high pressures prevailing in the interiors of Jupiter and Saturn. Analogously, Uranus and Neptune are frequently referred to as *ice giants*, even though the astrophysical ices, such as H_2O, CH_4, H_2S, and NH_3 that make up the majority of these planets' mass, are in fluid rather than solid form. Note that whereas H and He *must* make up the bulk of Jupiter and Saturn because no other elements can have such low densities at plausible temperatures, it is possible, though unlikely, that Uranus and Neptune are primarily composed of a mixture of 'rock' and H/He.

In §6.1.3 we noted that there is a maximum size to any sphere of material; if more matter is added, the radius of the body decreases. Below we derive this maximum radius for a sphere of pure hydrogen gas. Consider the equation of state as given by equation (6.5) with a polytropic index $n = 1$, so that $P = K\rho^2$. In this particular case, the radius is independent of mass. Thus, when more mass is added to the body, the material gets compressed such that its radius does not change. With $n = 1$, the equations can be solved analytically, while the results are in good agreement with calculations based upon more detailed pressure–density relations. Integration of the equation of hydrostatic equilibrium (eq. 6.1) leads to the following density profile:

$$\rho = \rho_c \left(\frac{\sin(C_K r)}{C_K r} \right), \tag{6.41}$$

with ρ_c the density at the center of the body, and

$$C_K = \sqrt{\frac{2\pi G}{K}}. \tag{6.42}$$

The radius of the body, R, is defined by $\rho = 0$, thus $\sin(C_K R) = 0$, and $R = \pi/C_K$. With a value for the polytropic constant $K = 2.7 \times 10^{12}$ cm^5 g^{-1} s^{-2}, as obtained from a fit to a more precise equation of state, we find that the planet's radius $R = 7.97 \times 10^4$ km. This number is thus independent of the planet's mass. The radius of this hydrogen sphere is slightly larger than Jupiter's mean radius, 6.99×10^4 km, and significantly larger than Saturn's mean radius, 5.82×10^4 km. This suggests that the Solar System's two largest planets are composed primarily, but not entirely, of pure hydrogen.

Using experimental data at low pressure and theoretical models at very high pressure, it can be shown that the maximum radius for cold self-gravitating spheres of heavier elements can be approximated by

$$R_{\max} = \frac{Z \times 10^5}{\mu_a m_{\text{amu}} \sqrt{Z^{2/3} + 0.51}} \text{ km,} \tag{6.43}$$

where Z is the atomic number and $\mu_a m_{\text{amu}}$ the atomic mass. This equation results in $R_{\max} = 82\,600$ km for a pure hydrogen and $R_{\max} = 35\,000$ km for pure helium. For heavier material $R_{\max}$ is smaller. Figure 6.25 shows graphs of the mass–radius relation for spheres of zero-temperature matter for different materials, as calculated numerically using precise (empirical) equations of state. The approximate locations of the giant planets are indicated on the graph, where the radius is taken as the distance from a planet's center out to the average 1 bar level along the planet's equator. The observed atmospheric species form a boundary condition on the choice of elements to include in models of the interior structure of the giant planets. The atmospheric composition of Jupiter and Saturn is close to a solar composition, and the location of these planets on the radius–mass graphs in Figure 6.25 shows that the composition of Jupiter and Saturn's interiors must indeed be close to solar. Calculations for more realistic (nonzero

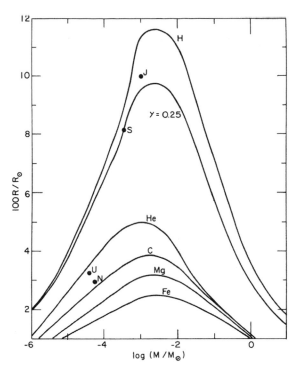

Figure 6.25 Graphs of the mass–radius relation for spheres of different materials at zero temperature, as calculated numerically using precise (empirical) equations of state. The second curve from the top is for a mixture of 75% H, 25% He by mass; all other curves are for planets composed entirely of one single element. The approximate locations of the giant planets are indicated on the graph. (Stevenson and Salpeter 1976)

temperature) models that take contraction and cooling during a planet's formation into account are shown in Chapter 12 (Fig. 12.1).

In precise models the observed rotation rate and inferred density (from measurements of the mass and radius) are used to fit the gravity field. The models are further constrained by the characteristics of the planet's magnetic field, interior heat source, and the nearby presence of massive satellites. The mass of the planets can be accurately determined from spacecraft tracking data, and the radius can be measured from star occultations and/or spacecraft images. The rotation rate of a planet is more complicated to determine, since timing of albedo features gives information on the rotation rate of atmospheric phenomena. The rotation period of the bulk of a gas giant planet is assumed to be equal to the rotation period of its magnetic field, which can usually be determined via ground-based (Jupiter) or spacecraft radio observations (Chapter 7). To first order, these values are consistent with measurements of the planets' oblateness and gravitational moments. The distribution of material inside

the planet, $\rho(r)$, can then be derived from the response coefficient, Λ_2, or the moment of inertia ratio, $I/(MR^2)$, which both indicate that all four giant planets are centrally condensed.

As expected, spacecraft measurements show that all gravitational moments, J_n, for odd n are very small for the giant planets. The second moment, J_2, is related to the planet's rotation, while higher order moments indicate further deviations in a planet's shape from a spinning oblate spheroid. The moments J_2 and J_4 have been measured for all four giant planets, while for Jupiter and Saturn J_6 has also been determined observationally (Table 6.2). These moments provide valuable constraints for detailed models. The observed strong internal magnetic moments require the presence of an electrically conductive and convective fluid medium in the planets' interiors, while any internal heat source for these planets is most likely gravitational in origin (§6.1.5.1).

Since the solution to the modeling technique outlined above is nonunique, there are substantial variations from one model to the next. Overall, the models show that the total amount of heavy ($Z > 2$) material in the giant planets is roughly between 10 and 30 $M_\oplus$, some of which has to reside near the center to satisfy the gravitational moments. Below we discuss the most likely models for the four planets, but keep in mind that research is still ongoing and that details may change in the future.

6.4.2 Jupiter and Saturn

The interior structures of Jupiter and Saturn are constrained by their mass, radius, rotation period, oblateness, internal heat source, and gravitational moments J_2, J_4, and J_6. Most models for Jupiter predict a relatively small dense core containing 5–10 $M_\oplus$. However, there are also models that predict no core, and others that predict a much more massive core. Saturn's core is probably somewhat larger than Jupiter's. Although the masses of the cores alone are difficult to estimate, it is clear that both Jupiter and Saturn have a total of ~15–30 $M_\oplus$ high-Z ($Z > 2$) material distributed throughout the core and surrounding envelope, so that the planets, overall, are enriched in high-Z material by factors of 3–5 for Jupiter and 2–2.5 times more for Saturn, as compared to a solar composition planet. Such enhancements are consistent with observations of the planets' atmospheres (§4.3.3.4). The core is probably composed of relatively large quantities of iron and rock, some of which was incorporated initially when the planet formed from solid body accretion, while more may have been added later via gravitational settling. The mantle material likely contains a relatively large amount of 'ices' of H_2O, NH_3, CH_4, and S-bearing materials. Schematics

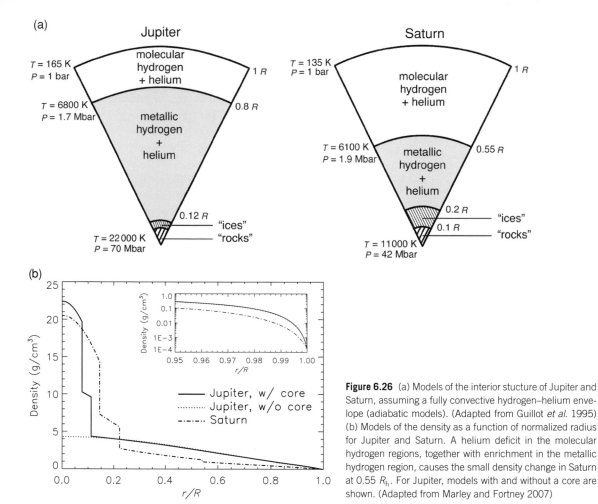

(a)

Figure 6.26 (a) Models of the interior stucture of Jupiter and Saturn, assuming a fully convective hydrogen–helium envelope (adiabatic models). (Adapted from Guillot *et al.* 1995) (b) Models of the density as a function of normalized radius for Jupiter and Saturn. A helium deficit in the molecular hydrogen regions, together with enrichment in the metallic hydrogen region, causes the small density change in Saturn at 0.55 R_h. For Jupiter, models with and without a core are shown. (Adapted from Marley and Fortney 2007)

of the interior structure of Jupiter and Saturn are shown in Figure 6.26.

Jupiter possesses a strong magnetic field, and Saturn is surrounded by a somewhat weaker one (§7.5). As discussed in §6.1.2.1, theories and laboratory experiments suggest hydrogen in the interiors of Jupiter and Saturn to be present in the form of liquid metallic hydrogen, a highly conductive fluid. In analogy to models of the magnetohydrodynamic dynamo theory in the Earth's outer liquid core (§7.6.1), one expects the generation of electromagnetic currents, and hence magnetic field, in the metallic hydrogen region. Since Saturn is less massive than Jupiter, the extent of the metallic hydrogen region is smaller, which readily explains its weaker magnetic field.

Jupiter and Saturn emit much more energy than they receive from the Sun. The excess energy has been attributed to an internal source of heat. As discussed in §6.1.5.1, the source of this internal heat is most likely gravitational. For Jupiter, models suggest that most of the

heat originates from gravitational contraction/accretion in the past. At present, the planet is still contracting by ∼3 cm/yr, and its interior cools by 1 K/Myr. For Saturn a substantial part (roughly half) of the excess heat is attributed to the release of gravitational energy by helium, which, being immiscible in metallic hydrogen, rains out onto the core. This slow 'drainage' of helium also explains the depletion of this element in Saturn's atmosphere compared to the solar value (Table 4.6). This separation process may also have 'recently' started on Jupiter, as shown by the somewhat lower than solar elemental abundance of helium in Jupiter's atmosphere. Since the envelopes of both planets are almost fully convective, the thermal structure is probably very close to adiabatic below the tropopause.

6.4.3 Uranus and Neptune

The mass–density relation for Uranus and Neptune (Fig. 6.27) suggests these planets are quite different from

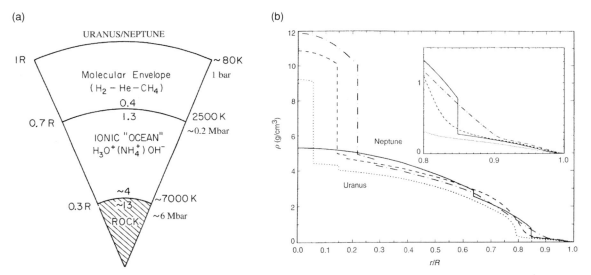

Figure 6.27 (a) Schematic representation of the interiors of Uranus and Neptune. (After Stevenson 1982) (b) Models of the density as a function of normalized radius for three Neptune and one Uranus interior models. The solid, dashed, and dot-dashed curves represent the range in possible Neptune models, where the core could be absent or extend out to a fractional radius of 20%. The dotted curve represents a single Uranus model. Because of Neptune's larger mass, it is denser than Uranus at each fractional radius. The inset shows the region of transition from a hydrogen-rich atmosphere to the icy mantle in more detail. (Marley and Fortney 2007)

Jupiter and Saturn. While the total mass of $Z > 2$ elements is very similar for all four planets, for Uranus and Neptune their abundance relative to hydrogen is enhanced by factors of over 30 compared to solar elemental abundances. The total mass in hydrogen and helium is only a few $M_\oplus$ on these planets, but the light gases dominate Uranus and Neptune by volume.

Neptune is 3% smaller than Uranus, while its mass is 15% larger, resulting in a density that is 24% higher than Uranus's bulk density. Neptune's response coefficient, Λ_2, is somewhat larger that Uranus's, indicative of a planet that is less centrally condensed. Interior models for the two planets are not well constrained at present. As shown in Figure 6.27b, a range of models fit the available data. Some of these models have a core subtending over 20% of the planet's radius, but others have no core at all. The mantle consists of 'ices', and comprises $\sim$80% of the mass of these planets. The outer 5–15% of the planets' radii make up a hydrogen- and helium-rich atmosphere. Since each planet has an internal magnetic field, their interiors must be electrically conductive and convective. Although the pressure and temperature are high in the planets' interiors, the pressure is too low throughout most of the planet interiors for metallic hydrogen to form. The conductivity in Uranus's and Neptune's interiors must thus be attributed to other materials. Because of the high temperature and

pressure, the icy mantles are probably hot, dense, liquid, ionic 'oceans' of water with some methane, ammonia, nitrogen, and hydrogen sulfide. Laboratory shock wave experiments on such a *synthetic Uranus* mixture show a conductivity that is high enough to set up an electromagnetic current system that can generate the observed magnetic fields (see §6.1.2.2).

A comparison of the equilibrium and effective temperatures shows that Neptune must have a large internal heat source, whereas Uranus has only a very small one, if any (§4.2.2.2). The observed upper limit for Uranus is consistent with the heat flow expected from radioactive decay alone. This difference in heat sources is the 'accepted' theory to explain the observed difference in atmospheric dynamics and hydrocarbon species in the planets' atmospheres (§§4.5.6, 4.7.2). An intriguing question is why the two planets are so different in this respect, as their evolution must have been rather similar. Both planets must initially have been hot when they accreted. Perhaps convection is inhibited in the interior of Uranus due to (subtle) differences in density gradients between the two planets. If convection is inhibited within $\sim$60% of Uranus's radius, and within $\sim$50% of Neptune's radius, all observed characteristics can be reconciled. Such compositional gradients might have arisen from late accretion of large (e.g., 0.1–1 $M_\oplus$) planetesimals, which break up upon impact and only partially mix with material already present.

Despite the fact that the ionic oceans on Uranus and Neptune extend deep within the planets' interiors, the absence of convection not only inhibits mass and energy transport from the core to outer layers on the planet, but it also prevents the generation of magnetic fields within 0.5–0.6 planet radii. Hence, the magnetic fields of Uranus and Neptune should originate within a thin shell of the ionic ocean, between 0.5–0.6 and 0.8 planetary radii, via a process similar to that which generates the fields around Earth, Jupiter, and Saturn. This would explain the large offsets and tilts of these planets' magnetic fields (see Chapter 7).

The helium abundance in Uranus's and Neptune's atmospheres is similar to that expected for a solar composition atmosphere, in contrast to the helium abundance on Jupiter and Saturn, where it is smaller than on the Sun. If indeed hydrogen is confined to the outer layers of Uranus and Neptune, metallic hydrogen will not form, and hence helium will not be separated from hydrogen. However, some models indicate substantial amounts of hydrogen and helium mixed in with the icy material deep in the planets' interiors. At pressure levels $\gtrsim$ few Mbar, hydrogen becomes metallic, and helium should separate out. These high-pressure levels are only reached in the deep interior of the planets, a region where we just argued convection may be inhibited. Hence, there is no mass (or energy) transfer between these deep layers and the atmosphere, and the effect of helium differentiation, if present, may not be noticed in the atmosphere.

In contrast to inhibition of convection and mass transfer, suggestions of equilibration between core, mantle, and atmospheric material have been made based upon measurements of the D/H ratio. The observed D/H ratio is $\sim 1.2 \times 10^{-4}$, about an order of magnitude higher than primordial values and D/H ratios on Jupiter and Saturn, and similar to terrestrial numbers and comets. These high values presumably originate from the icy compounds in the planets, either from the ice in the deep interiors of the planets, which requires mass transfer and thus convection (which is inhibited at the present epoch, but could have been different in the past), or in the original icy planetesimals which accreted to form Uranus and Neptune. In the latter case one does not need to invoke efficient mixing processes between the deep interior and the outer layers to explain the high observed D/H numbers, since the planetesimals deposited ice with a high D/H throughout the forming planet. The upshot of both scenarios is that the D/H ratio in Uranus's and Neptune's atmospheres may not be a good indicator of the degree of mixing in these planets' deep interiors.

Further Reading

A short nontechnical review of the Earth as a planet is given by

Pieri, D.C., and A.M. Dziewonski, 2007. Earth as a planet: surface and interior. In *Encyclopedia of the solar system*, 2nd Edition. Eds. L. McFadden, P.R. Weissman, and T.V. Johnson. Academic Press, San Diego. pp. 189–212.

Although this book contains mostly material pertaining to the surface structure of the Earth, the chapters on seismology and the Earth's interior form a good introduction to the more advanced books listed below:

Grotzinger, J., T. Jordan, F. Press, and R. Siever, 2006. *Understanding Earth*, 5th Edition. W.H. Freeman and Company, New York. 579pp.

Standard textbooks on the physics of Earth's interior:

Fowler, C.M.R., 2005. *The Solid Earth: An Introduction to Global Geophysics*, 2nd Edition. Cambridge University Press, New York. 685pp.

Lambeck, K., 1988. *Geophysical Geodesy: The Slow Deformations of the Earth*. Oxford Science Publications. 718 pp.

Schubert, G., D.L.Turcotte, and P. Olsen, 2001. *Mantle Convection in the Earth and Planets*. Cambridge University Press, New York. 456pp.

Turcotte, D.L., and G. Schubert, 2002. *Geodynamics*, 2nd Edition. Cambridge University Press, New York. 456pp.

A book about the Earth's core, magnetic field, and magnetic field reversals:

Jacobs, J.A., 1987. *The Earth's Core*. 2nd Edition. Academic Press, New York. 416pp.

A good book on seismology:

Stein, S., and M. Wysession, 2003. *An Introduction to Seismology, Earthquakes and Earth's Structure*. Wiley-Blackwell, Oxford. 498pp.

A good book (although it is outdated at places) on the interior structure of all planets is:

Hubbard, W.B., 1984. *Planetary Interiors*. Van Nostrand Reinhold Company Inc., New York. 334pp.

Interiors of the giant planets and Galilean satellites are discussed by:

Guillot, T., D.J. Stevenson, W.B. Hubbard and D. Saumon, 2004. The interior of Jupiter. In *Jupiter: Planet, Satellites and Magnetosphere*. Eds. F. Bagenal, T.E. Dowling, and W. McKinnon. Cambridge University Press, Cambridge. pp.19–34.

Hubbard, W.B., M. Podolak, and D.J. Stevenson, 1995. The interior of Neptune. In *Neptune and Triton*. Ed. D.P. Cruikshank. University of Arizona Press, Tucson, pp.109–138.

Marley, M.S., and J.J. Fortney, 2007. Interiors of the giant planets. *Encyclopedia of the Solar System*, 2nd Edition. Eds. L. McFadden, P. Weissman, and T.V. Johnson. Academic Press, San Diego. pp. 403–418.

Moore, W.B., G. Schubert, J.D. Anderson, and J.R. Spencer, 2007. The interior of Io. *Io After Galileo: A New View of Jupiter's Volcanic Moon*. Springer–Praxis, Chichester, UK, pp. 89–108.

Schubert, G., J.D. Anderson, T. Spohn, and W.B. McKinnon, 2004. Interior composition, structure and dynamics of the Galilean satellites. In *Jupiter: Planet, Satellites and Magnetosphere*. Eds. F. Bagenal, T. E. Dowling, and W. McKinnon. Cambridge University Press, Cambridge. pp. 281–306.

Stevenson, D.J., 1982. Interiors of the giant planets. *Annu. Rev. Earth Planet. Sci.,* **10**, 257–295.

Problems

6.1.E Compute the gravitational potential energy of:

(a) A uniform sphere of radius R and density ρ.

(b) A sphere of identical mass and radius to that above, but whose mass is distributed so that its core, whose radius is $R/2$, has twice the density of its mantle.

6.2.E Assume that material can be compressed significantly if the pressure exceeds the material strength. If material can be compressed considerably over a large fraction of a body's radius, the body will take on a spherical shape (the lowest energy state of a nonrotating fluid body).

(a) Calculate the minimum radius of a rocky body to be significantly compressed at its center. Assume an (uncompressed) density $\rho = 3.5$ g cm^{-3} and material strength $S_m = 2 \times 10^9$ dyne cm^{-2}.

(b) Calculate the minimum radius of the rocky body in (a) to be significantly compressed over a region containing about half the body's mass.

(c) Repeat for an iron body, which has an (uncompressed) density of $\rho = 8$ g cm^{-3} and material strength $S_m = 4 \times 10^9$ dyne cm^{-2}.

(d) Calculate the minimum radius for an icy body.

6.3.E Use the equation of hydrostatic equilibrium to estimate the pressure at the center of the Moon, Earth, and Jupiter.

(a) Take the simplest approach, and approximate the planet to consist of one slab of material with thickness R, the planetary radius. Assume the gravity $g_p(r) = g_p(R)$, and use the mean density $\rho(r) = \rho$.

(b) Assume the density of each planet to be constant throughout its interior, and derive an expression for the pressure in a planet's interior, as a function of distance r from the center (Hint: You should get eq. 6.2).

(c) Although the pressure obtained in (a) and (b) is not quite right, it will give you a fair estimate of its magnitude. Compare your answer with the more sophisticated estimates given in §6.1.1, and comment on your results.

6.4.I (a) The moment of inertia is related to the density via equation (2.42). If ρ is a function of distance r, show that for a spherical planet:

$$I = \frac{8}{3}\pi \int \rho(r) r^4 \, dr. \qquad (6.44)$$

(b) Calculate the moment of inertia expressed in units of MR^2 for planets for which the density distributions are given in Problems 6.1 (a) and (b).

6.5.E (a) Show that the net gravitational plus centrifugal acceleration, $g_{\mathrm{eff}}(\theta)$, on a rotating sphere is:

$$g_{\mathrm{eff}}(\theta) = g_p(\theta) - \omega_{\mathrm{rot}}^2 r \sin^2 \theta, \qquad (6.45)$$

where ω_{rot} represents the spin angular velocity, g_p is the gravitational acceleration, and θ is the planetocentric colatitude.

(b) Calculate the ratio of the centrifugal to the gravitational acceleration for Earth, the Moon, and Jupiter.

6.6.E Rewrite equation (6.11a) in terms of Legendre polynomials.

6.7.E (a) Using equations (6.13) and (6.14) show that $J_2 = 0.5q_r$ for a uniform density distribution.

(b) Show that $J_2 < 0.5q_r$ for a planet in which the density increases towards its center.

(c) Calculate the response coefficient for the Earth, the Moon, and Jupiter, and comment on your results.

6.8.E Show that the geometric oblateness, ϵ, is equal to $1.5J_2 + 0.5q_r$ for a rotating fluid body in hydrostatic equilibrium.

6.9.E (a) Calculate the moment of inertia ratios for Earth, Venus, Jupiter, and the Moon from the Radau–Darwin approximation. Compare your results with the values in Table 6.2, and comment on similarities and discrepancies.

(b) Compare the J_2 and C_{22} values in Table 6.2 for the Galilean satellites and the Moon. Are these bodies in hydrostatic equilibrium?

6.10.I Consider Figure 6.9: the Earth's continental crust with a thickness r_1, a mountain on top with a height r_2, an ocean with a depth r_4, and the oceanic crust with a thickness r_5. Assume that the density

of the crust is equal for continental crust, mountains, and oceanic crust. The density of the crust is ρ_c, that of the mantle ρ_m, and water has a density ρ_w. You can further assume that $\rho_w < \rho_c < \rho_m$.
(a) Derive the relationship between the quantities r_3 and r_2, and between r_6 and r_4, assuming isostatic equilibrium. The mountain is compensated by a mass deficiency of height r_3, and the ocean crust by a layer of extra mass of height r_6. These assumptions are referred to as *Airy's hypothesis*.
(b) If the mountain is 6 km high, calculate the height of the mass deficiency r_3. Assume $\rho_w = 1$ g cm^{-3}, $\rho_c = 2.8$ g cm^{-3}, and $\rho_m = 3.3$ g cm^{-3}.
(c) If both the ocean and oceanic crust are 5 km deep, calculate r_6.

6.11.**I** An alternative to Airy's hypothesis for isostatic equilibrium is *Pratt's hypothesis*. Pratt assumed that the depth of the base of the continental and oceanic crust, including mountains and oceans, is the same, i.e., equal to r_1 (see the dashed line on Figure 6.9), and that isostatic equilibrium is reached by variations in the density between the three columns A, B, and C, i.e., ρ_A, ρ_B, and ρ_C. (The density of water is $\rho_w = 1$ g cm^{-3}.)
(a) Derive the relationship between the densities ρ_A, ρ_B, and ρ_C with the heights r_1, r_2, and r_4.
(b) If the crust is 30 km thick and the mountain 6 km high, calculate the density of the mountain + crust, ρ_B, if $\rho_A = 2.8$ g cm^{-3}.
(c) If the ocean is 5 km deep, calculate ρ_C.

6.12.**I** (a) Derive the expression for the free-air correction, i.e., the difference between the gravity g_p at the reference geoid and the gravity measured at an altitude h above sea level. You may assume $h \ll R$, where R is the radius of the planet.
(b) Show how the free-air correction factor at colatitude θ changes if the gravitational attraction of a horizontally infinite slab of rock at colatitude θ, of density ρ and height h, is included.

6.13.**E** (a) Calculate the free-air correction on Earth, Δg_{fa}, per km elevation above sea level.
(b) Determine the Bouguer correction on Earth, Δg_B, per km elevation above sea level due to mountains with a density of 2.7 g cm^{-3}.

6.14.**I** To calculate the geoid anomaly above mountains and oceans, the density anomalies are measured with respect to a reference structure. Using the Airy hypothesis, calculate the geoid height anomaly at the equator above the mountain and the ocean from Problem 6.10. Use the crust with density ρ_c as the reference structure.

6.15.**I** Approximate the interior structure of the Earth by two layers: a uniform density core, out to a radius of 0.57 R$_\oplus$, surrounded by a uniform density mantle. The Earth's average density is 5.52 g cm^{-3}, and its moment of inertia ratio $I/(MR^2) = 0.331$. The planet is in hydrostatic equilibrium.
(a) Using the equation for the moment of inertia, determine the density of the Earth's core and of the Earth's mantle.
(b) Using the equation of hydrostatic equilibrium for a planet which consists of two layers (core plus mantle), determine the pressure at the center of the Earth, and compare with the values obtained in Problem 6.3.

6.16.**E** (a) Calculate the total amount of internal energy lost per year from Earth, assuming a heat flux of 75 erg cm^{-2} s^{-1}.
(b) Calculate the temperature of Earth if the 'heat flux' had been constant over the age of the Solar System, but none of the heat had escaped the planet. A typical value for the specific heat of rock is $c_P = 1.2 \times 10^7$ erg g^{-1} K^{-1}.

6.17.**E** Compare the Earth's intrinsic luminosity (L_i) or heat flux (75 erg cm^{-2} s^{-1}) with the luminosity from reflection of sunlight (L_v), and from emitted infrared radiation (L_{ir}). Assume the Earth's albedo is 0.36. Note that you can ignore the greenhouse effect, and that the value of the infrared emissivity is irrelevant (Hint: Use §3.1.2). Comment on your results, and the likelihood of detecting Earth's intrinsic luminosity via remote sensing techniques from space.

6.18.**I** (a) Using the effective and equilibrium temperatures, calculate the decrease in the internal temperature of each of the giant planets over the age of the Solar System, assuming the planets' luminosities have not changed. (b) Can you use this technique to estimate the temperature these planets had initially? If so, provide such an estimate, if not, explain why not.

6.19.**I** Calculate the ratio of internal radioactive heat to absorbed solar radiation for Pluto. You may assume that Pluto consists of a 60:40 mixture of chondritic rock to ice, and that its albedo is 0.4.

6.20.**E** If energy transport is dominated by conduction, calculate the depth over which the temperature gradient is influenced significantly in a rocky object for the timescales specified below. Assume the surface layers consist of uncompressed silicates, and estimate the thermal conductivity using equation (6.30). Take the heat capacity to be $c_P = 1.2 \times 10^7$ erg g^{-1} K^{-1}, and the density $\rho = 3.3$ g cm^{-3}.

(a) Assume a temperature of 300 K and a timescale of 1 day (i.e., this gives you an approximate depth of diurnal temperature variations in Earth's crust).

(b) Assume a temperature of 300 K and a timescale equal to the age of the Solar System.

(c) Assume a temperature of 10 000 K and a timescale equal to the age of the Solar System.

(d) Comment on your results above. Do you think energy transport is by conduction only throughout the entire Earth?

6.21.E Derive the Adams–Williams equation (6.39) from the equation of hydrostatic equilibrium and the definition of the bulk modulus K.

6.22.E (a) Use Figure 6.17 to determine the time that it takes a P wave to travel from the epicenter of an earthquake to the furthest point on Earth.

(b) Use Figure 6.17 to estimate the time that P waves and S waves take to travel from the epicenter of an earthquake to a point on the globe that is 60° away. Ignore refraction, i.e., assume that the waves propagate along straight paths.

(c) How, qualitatively, would refraction change the value that you computed in part (b)? Explain using a diagram.

6.23.I Suppose a powerful jolt from below woke you up at 2 a.m. sharp. Exactly 4 seconds later a strong horizontal shaking begins and you are knocked off the bed.

(a) Explain both types of motions.

(b) If the average velocity of P and S waves is 5.8 km s^{-1} and 3.4 km s^{-1}, respectively, calculate your distance from the epicenter. You can assume that the focus was at the surface.

(c) How many measurements like your own do you need to have to pinpoint the epicenter exactly? Explain in words and pictures.

6.24.E The continents on Earth drift relative to each other at rates of up to a few centimeters per year; this drift is due to convective motions in the mantle.

(a) Assuming typical bulk motions in the mantle of Earth are 1 cm/yr, calculate the total energy associated with mantle convection.

(b) Calculate the kinetic energy associated with the Earth's rotation.

(c) Calculate the kinetic energy associated with the Earth's orbital motion about the Sun.

6.25.I Fluid motions in the Earth's outer liquid core are believed to be responsible for the currents which produce Earth's magnetic field. By far the most likely source of energy for these motions is convection within the core. The core contains very little radioactive material, so some other process must be providing the buoyancy necessary for such convection. Three mechanisms have been suggested: (1) Energy loss to the mantle at the top of the core. (2) Latent heat released by the freezing out of nickel and iron at the boundary between the inner solid core and the outer fluid core. (3) Decrease in the density of material at the bottom of the outer core due to freezing out of denser portions of the liquid.

Process (2) is the easiest to study quantitatively, and even here several simplifying assumptions are needed. From modeling which includes the size of Earth's solid inner core and estimates of its age (1–3.6×10^9 years), a growth rate of 0.04 cm/yr is estimated at the current epoch. Making the assumption that the latent heat of fusion is not affected by either megabar pressures or the fact that nickel/iron is freezing out from a solution which also contains more volatile elements, estimate the energy released by this process over the course of a year.

6.26.I (a) What is the current rate at which energy is being generated within Earth via radioactive decay? (You may assume that the Earth is made of chondritic material. State your answer in erg/yr.)

(b) How much solar energy is absorbed by Earth? Again, state your answer in erg/yr.

(c) Compare the energies calculated above with the average energy released in earthquakes per year, which is $\sim 10^{28}$ erg, and comment.

6.27.E Mercury's mean density $\rho = 5.43$ g cm^{-3}. This value is very close to the planet's uncompressed density. If Mercury consists entirely of rock ($\rho = 3.3$ g cm^{-3}) and iron ($\rho = 7.95$ g cm^{-3}), calculate the planet's fractional abundance of iron by mass.

6.28.E Assume the density of 'rock' in the jovian satellites is equal to the mean density of Io and that the density of H_2O-ice is 1 g cm^{-3}.

(a) What proportions of each of the other Galilean satellites consists of rock, and what proportion ice?

(b) In reality, this calculation is good for Europa, but underestimates the ice fraction for Ganymede and Callisto. Why?

6.29.I The lithosphere is defined as the solid outer layer of a planet's upper mantle. The upper mantle is solid if the temperature $T \lesssim 1200$ K. For Earth, the thermal conductivity is $K_T = 3 \times 10^5$ erg cm^{-1} s^{-1} K^{-1}, and the average heat flow is 75 erg cm^{-2} s^{-1}. Assume that these same parameters hold for Venus and Mars. Determine the thickness of the lithosphere for Earth, Venus, and Mars (Hint: What is the surface temperature for the three planets?). Comment on the differences, and explain the lack of tectonic plate activity on Venus and Mars.

6.30.E Suppose that the shield volcanoes on Mars are produced from a partial melting of martian rocks at a temperature of $\sim$1100 K. The density is then $\sim$10% less dense than that of the surrounding rocks, and the magma will rise up through the surface and build up the 20-km high Tharsis region. Assuming the pressure on the magma chamber below the magma column is equal to the ambient pressure at that depth, calculate the depth of the magma chamber below the martian surface.

6.31.E Calculate the rotation period for Neptune from the observed oblateness 0.017, equatorial radius of 24 766 km, and $J_2 = 3.4 \times 10^{-3}$. Compare your answer with a typical rotation period of 18 hours for atmospheric phenomena, and of 16.11 hours for Neptune's magnetic field. Comment on your results.

6.32.I Approximate Jupiter and Saturn by pure hydrogen spheres, with an equation of state $P = K\rho^2$, and $K = 2.7 \times 10^{12}$ cm^5 g^{-1} s^{-2}.

(a) Determine the moment of inertia for the planets.

(b) Assume the planets each have a core of density 10 g cm^{-3}, and the moment of inertia ratio is $I/(MR^2) = 0.254$ for Jupiter and 0.210 for Saturn. Determine the mass of these cores.

7 Magnetic Fields and Plasmas

The secret of magnetism, now explain that to me! There is no greater secret, except love and hate.
Johann Wolfgang von Goethe, in *Gott, Gemüt und Welt*

Most planets are surrounded by huge magnetic structures, known as *magnetospheres*. These are often more than 10–100 times larger than the planet itself, and therefore form the largest structures in our Solar System, other than the heliosphere. The solar wind flows around and interacts with these magnetic 'bubbles'. A planet's magnetic field can either be generated in the interior of the planet via a dynamo process (Earth, giant planets, Mercury), or induced by the interaction of the solar wind with the body's ionosphere (Venus, comets). Large-scale remanent magnetism is important on Mars, the Moon, and some asteroids.

The shape of a planet's magnetosphere is determined by the strength of its magnetic field, the solar wind flow past the field, and the motion of charged particles within the magnetosphere. Charged particles are present in all magnetospheres, though the density and composition varies from planet to planet. The particles may originate in the solar wind, the planet's ionosphere, or on satellites or ring particles whose orbits are partly or entirely within the planet's magnetic field. The motion of these charged particles gives rise to currents and large-scale electric fields, which in turn influence the magnetic field and the particles' motion through the field.

Although most of our information is derived from *in situ* spacecraft measurements, atoms and ions in some magnetospheres have been observed from Earth through the emission of photons at ultraviolet and visible wavelengths. Accelerated electrons emit photons at radio wavelengths, observable at frequencies ranging from a few kilohertz to several gigahertz. Radio emissions at $\sim$10 MHz were detected from Jupiter in the early 1950s, and formed the first evidence that planets other than Earth might have strong magnetic fields.

7.1 The Interplanetary Medium

7.1.1 Solar Wind

The presence of corpuscular radiation from the Sun, the *solar wind*, was first suggested by L. Biermann in 1951 based on the observation that cometary ion tails always point away from the Sun. The angle between the ion tail and the Sun–comet line, ϕ, is determined by the ratio of the comet's (transverse) orbital velocity, v_θ, and the speed of the solar wind particles, v_{sw}:

$$\tan\phi = \frac{v_\theta}{v_{sw} - v_r}, \tag{7.1}$$

where v_r is the radial component of the comet's velocity. One can show that $\phi \lesssim 5°$ (Problem 7.1).

Observations of the Sun during a solar eclipse reveal the solar *corona* (Fig. 7.1), which consists of highly variable magnetically controlled *loops* and *streamers*, containing hot plasma that becomes visible in EUV and X-ray images. Closer to the Sun's surface, in the low corona and *chromosphere*, are bright *prominences*, elongated clouds of cooler material that are held aloft in a 'hammock' of sheared or twisted magnetic fields. *Sunspots* are pairs of dark spots on the Sun's surface. They appear dark because their temperature is lower ($T \approx 4000$ K) than the average photospheric temperature ($T \approx 5750$ K). The magnetic field strength in the spots is higher than in the surrounding regions, so that there is approximate pressure equilibrium between sunspots and the surrounding regions. The two spots in each pair show opposite polarity, as expected for loop-like magnetic structures anchored to the sunspots. Sometimes only one spot is visible; in such situations, the opposite polarity is present nearby, but is not strong enough to make a sunspot. Some sunspots occur in complex groups, called *active regions*.

Figure 7.1 The total solar eclipse of 29 March 2006. The dark center is the Moon as it passes between us and the Sun. The bright white streamers extending outwards from the Sun are visible because sunlight is scattered to us by electrons in the streamers. This picture is a composition of 63 images at visible wavelengths taken by Hana Druckmüllerová near Göreme, Cappadocia, Turkey. Image processing has brought out the faint plumes of coronal plasma outlining the magnetic field lines, both the loop-like structures in the streamers and the poloidal magnetic field near the poles of the Sun. For more images and close-up views see: http://www.zam.fme.vutbr.cz/~druck/Eclipse/Index.htm. (Courtesy Hana Druckmüllerová and Miloaslav Druckmüller)

The number of sunspots varies on an 11 year cycle (Fig. 7.2), and the leading sunspot of a pair tends to show the same polarity throughout the cycle. This polarity as well as the overall magnetic field of the Sun reverses every

11 years. The largest scale field of the Sun is dominated by the diffuse field in the polar regions which is made up of the remnants of the decayed sunspots that emerged during the cycle. An average tilt of the sunspot pairs ensures the existence of opposite polarities at the poles throughout most of the cycle. In addition to the 11 year periodic variation in the number of sunspots, the number of sunspots during *solar maximum* (i.e., maximum number of sunspots during a solar cycle) and *solar minimum* also varies over time (Fig. 7.2b). Very few sunspots were seen between about 1645 to 1715, indicative of a period that the Sun was relatively inactive. This period, referred to as the *Maunder minimum*, coincides with a climatic period known as the *Little Ice Age* (§4.9).

X-ray images of the Sun reveal a tremendous amount of structure within the solar corona (Fig. 7.3a). The luminosity from the X-ray bright areas is produced by a dense hot thermal plasma, which is heated by energetic electrons trapped on closed magnetic field loops in the low corona. The dark areas indicate regions largely devoid of hot X-ray emitting plasma, and are referred to as *coronal holes*. Magnetic field lines in these regions have opened up into interplanetary space, so that particles can freely escape into space, forming and/or 'feeding' the solar wind. At Earth's orbit and beyond, the solar wind has an average speed of $\sim$400 km s^{-1} during solar maxima. During solar minima the average solar wind speed is $\sim$400 km s^{-1} near the ecliptic region, but at higher solar latitudes the speed is typically 750–800 km s^{-1}. During this time there are large coronal holes near the Sun's poles from

(a)

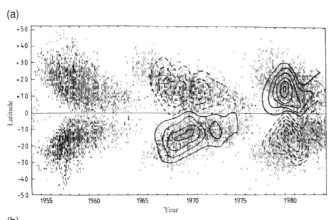

Figure 7.2 (a) The spatial distribution of sunspots as a function of time, which shows the familiar butterfly diagram. The 11 year solar sunspot cycle is clearly visible. The contours outline the radial magnetic field, at multiples of 0.27 G. Solid lines indicate positive magnetic field values, dashed lines negative values. (Stix 1987) (b) The number of sunspots as a function of time, from 1750 to 2007. Note the 11 year periodicity and the variations therein from period to period. (Compiled from NOAA/NESDIS website)

(b)

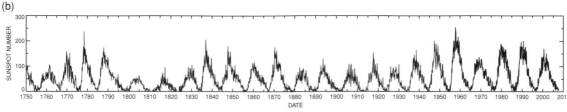

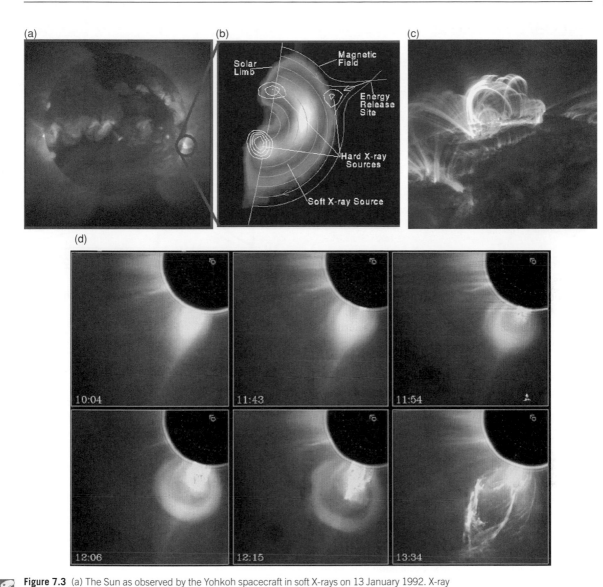

Figure 7.3 (a) The Sun as observed by the Yohkoh spacecraft in soft X-rays on 13 January 1992. X-ray bright regions indicate heating to temperatures in excess of 2×10^6 K. These regions usually overlie sunspots or active regions. The very dark regions are the coronal holes, which are usually located above the solar poles. These coronal holes sometimes extend down to lower latitudes. (Yohkoh Science Team) (b) Expanded view of the extraordinary heating associated with this solar flare. The Sun's magnetic field traps the plasma in loops, and the X-ray bright regions represent plasma that has been heated and accelerated in the solar flare. (Solar Data Analysis Center, Goddard Space Flight Center) (c) This image of prominences on the Sun was taken by the Transition Region and Coronal Explorer (TRACE) in FeIX. It clearly shows the magnetic loop structures such as visualized in (b); the details on this picture are startling. (NASA) (d) Sequence of images from 18 August 1980, showing the formation of a coronal mass ejection (CME) through the distortion of a coronal streamer, triggered by the disruption of a solar prominence beneath it. The prominence material is blown outward along with the original streamer material; the bright, filamentary structures on the 13:34 frame are the remnants of the prominence. The dark circle in the upper right corner of each frame is the occulting disk of the Solar Maximum Mission (SMM) coronagraph, which is 60% larger than the solar disk. (Courtesy J. Burkepile, High Altitude Observatory, NCAR)

Table 7.1 Solar wind properties at 1 AU.[a]

	Most probable value	5–95% range
Density (protons cm^{-3})	5	3–20
Velocity (km s^{-1})	375	320–710
Magnetic field (γ)	5.1	2.2–9.9
Electron temp. (10^5 K)	1.2	0.9–2
Proton temp. (10^5 K)	0.5	0.1–3
Sound speed (km s^{-1})	59	41–91
Alfvén speed (km s^{-1})	50	30–100

[a] After Gosling (2007).

which the wind emanates. The solar wind consists of a roughly equal mixture of protons and electrons, with a minor proportion of heavier ions. The density decreases roughly as the inverse square of the heliocentric distance. A typical ion density at Earth's orbit is 6–7 protons cm^{-3}; the temperature is $\sim 10^5$ K and the magnetic field strength a typically $5-7 \times 10^{-5}$ G (Table 7.1).

Sometimes a large filament or prominence erupts in connection with a *coronal mass ejection* (CME), such as shown in Figure 7.3d. During a typical CME, $\sim 10^{15}-10^{16}$ g of solar material is injected into the solar wind at an ejection speed of $\sim$50 km s^{-1} for a very slow event, up to over 2500 km s^{-1} for an extremely fast one. Particles ejected during a slow event are accelerated up to the local solar wind speed, and those from a fast CME are decelerated. A CME that has a speed much larger than the prevailing solar wind is often referred to as ICME, or interplanetary coronal mass ejection. An ICME is preceded by a shock front that compresses the solar wind ahead of it. CMEs often occur in association with *solar flares*, events where the coronal X-ray and UV emissions suddenly intensify by several orders of magnitude in a relatively localized region in the low corona and chromosphere. A flare is likely triggered by a reconnection of magnetic field lines (§7.2; Fig. 7.10), which releases energy that accelerates particles. Major solar flares may occur approximately once per week during years of maximum sunspot activity, with weak flares and CMEs happening a few times daily. During years of minimum solar acitivity, such weaker events may happen once a week.

7.1.1.1 The Parker Model

In 1958, Eugene Parker predicted the existence of the continuous solar wind flow, assuming that particles flowed radially outward from the Sun, carrying the solar magnetic field as if it were 'frozen in'. The outward acceleration of

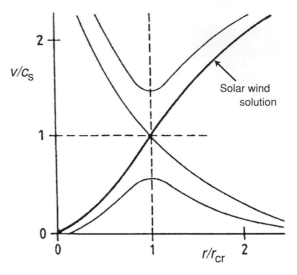

Figure 7.4 Parker's four solutions to the equations which describe the solar wind outflow. The actual solar wind solution is indicated. (Adapted from Parker 1963)

the solar wind is primarily caused by the pressure difference between the corona and the interplanetary medium. Parker used the equations of continuity and momentum, with the ideal gas law as the equation of state (assuming constant temperature), to calculate the solar wind velocity. His family of solutions is shown in Figure 7.4. Two of the solutions start at a supersonic velocity in the corona, which disagrees with observations. A third solution stays subsonic regardless of distance from the Sun. In this solution, the velocity reaches a maximum at a critical radius, r_{cr}, and decreases at larger distances (see, e.g., Hundhausen's chapter in Kivelson and Russell (1995) for a detailed discussion):[1]

$$r_{cr} = \frac{GM_\odot}{2c_s^2}, \qquad (7.2)$$

where c_s is the *speed of sound*:

$$c_s = \sqrt{\frac{\gamma P}{\rho}} = \sqrt{\frac{\gamma kT}{\mu_a m_{amu}}}, \qquad (7.3)$$

γ is the ratio of specific heats ($\gamma \approx 5/3$ for the solar wind), and T the temperature ($T \approx 2 \times 10^6$ K in the solar corona, and $T \approx 10^5$ K at $r = 1$ AU). The sound speed at Earth's orbit is thus approximately 60 km s^{-1}, and $r_{cr} \approx R_\odot$. Since the solar wind velocity does not decrease with distance beyond r_{cr}, the only valid solution is the fourth curve, which starts at a low (subsonic) velocity and turns supersonic at the critical radius.

[1] Note that, in contrast to most texts in space physics, the equations in this chapter, as in the entire book, are in cgs units.

(a)

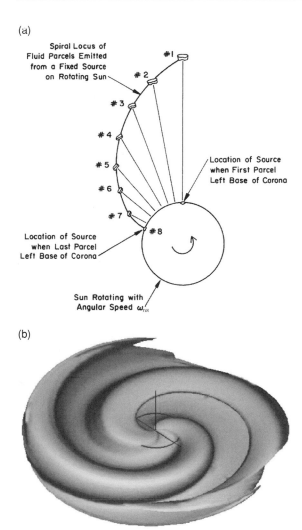

Spiral Locus of
Fluid Parcels Emitted
from a Fixed Source
on Rotating Sun

#1
#2
#3
#4
#5
#6
#7
#8

Location of Source
when First Parcel
Left Base of Corona

Location of Source
when Last Parcel
Left Base of Corona

Sun Rotating with
Angular Speed ω_{rot}

(b)

Figure 7.5 (a) Archimedean (or Parker) spiral of solar wind particles streaming away from the Sun. (Hundhausen 1995) (b) When the Sun's large-scale dipole is inclined with respect to the ecliptic plane, the rotation of the Sun causes the heliospheric current sheet to become wavy, and this waviness is carried out into interplanetary space by the solar wind. The Parker spiral is clearly visible. (Constructed from http://imhd.net/stereo)

The expansion of the solar wind is radially outwards, while the Sun rotates 'underneath'. Each fluid-like element of the wind effectively carries a specific magnetic field line, which is rooted at the Sun. Consequently, the solar wind magnetic field takes on the approximate form of an Archimedean spiral, as displayed in Figure 7.5. The radial and azimuthal components of the field are roughly equal at Earth's orbit, with a strength of a few $\times$ 10^{-5} G each. Since the total magnetic flux through any closed surface around the Sun must be zero, inward and outward magnetic fluxes must balance each other. Spacecraft

measurements have shown that the inward and outward fluxes are distributed in a systematic way such that there are interplanetary sectors with predominantly outward fluxes and others with predominantly inward fluxes. The different sectors are magnetically connected to different regions on the solar surface – generally different coronal holes. However, because of the predominantly dipolar nature of the Sun's largest scale magnetic field, the field in the solar wind or heliosphere is generally in the form of a positive (outwards) hemisphere and a negative (inwards) hemisphere. The *heliospheric current sheet* separates the hemispheres of opposite magnetic polarity. This current sheet is effectively the extension of the solar magnetic equator into the heliosphere. During solar minimum, the solar magnetic dipole is approximately aligned with the Sun's rotation axis. During other phases of the sunspot cycle, in particular the declining phase of the activity cycle, the dipole is tilted substantially. In these situations, the Sun's rotation causes the heliospheric current sheet to be warped, like a ballerina skirt (Fig. 7.5b).

The flows from different coronal holes often have different speeds, and hence 'collide' and produce spiral-shaped compressions in the solar wind. The entire magnetic field and stream structure rotates with the Sun. The magnetic sector structure seen in 1963/1964 by the spacecraft IMP-1 is illustrated in Figure 7.6. This sector structure changes over time as conditions on the Sun change. While the Sun rotates, the different magnetic sectors sweep by the various bodies in our Solar System. The sudden reversals of magnetic field direction and stream structure are possibly responsible for the 'disconnection' events seen in cometary ion tails (§10.5.2), as well as for certain magnetospheric disturbances.

As discussed in §1.1.6 and shown in Figure 1.4, the heliosphere forms a 'bubble' in the interstellar medium. Its boundary, the *heliopause*, is at the location where the solar wind pressure is balanced by that in the interstellar medium (analogous to the magnetopause, discussed in §7.1.4). Interior to the heliopause is the *termination shock*, where the solar wind is slowed down to subsonic velocities. The precise location of the various boundaries varies in response to the highly dynamic solar wind pressure. One of Voyager 2's several crossings of the termination shock is shown in Figure 7.7.

7.1.1.2 *Space Weather*

Interplanetary space is a region full of complicated and sometimes violent processes. Interplanetary shocks, preceding high-speed winds and ICMEs, accelerate particles locally to very high energies. While it takes an ordinary solar wind particle typically a few days to reach

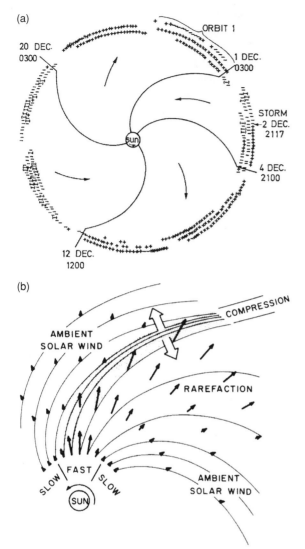

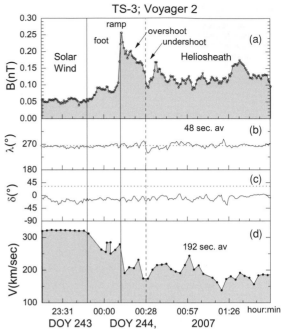

Figure 7.7 The termination shock in the heliosphere was crossed several times by the Voyager 2 spacecraft on 31 August – 1 September 2007 at 83.7 AU. Changes in the magnetic field strength (a) and velocity (d) during the third crossing (TS-3) are shown. No changes were seen in the magnetic field direction (azimuthal angle λ and elevation angle δ in panels b and c). The time resolution in the magnetic measurements is 48 seconds; in the solar wind velocity it was 192 seconds. The magnetic field strength profile shows the classical features of a supercritical quasi-perpendicular shock: a 'foot', 'ramp', 'overshoot', 'undershoot', and smaller oscillations, in that order. Plasma waves were observed on the ramp of TS-3. (Burlaga *et al.* 2008)

Figure 7.6 (a) Sector structure in the interplanetary field as observed by IMP-1 in 1963/1964. Plus signs indicate magnetic field directed outward from the Sun. (Wilcox and Ness 1965) (b) The interaction between a region of high-speed solar wind with the ambient solar wind. The high-speed solar wind is less tightly bound. (Hundhausen 1995)

the Earth, the most energetic cosmic-ray-like particles injected through a solar flare may reach Earth less than an hour after the event. The response of Earth's environment to solar activity and the continuously varying interplanetary medium is known as *space weather*, and the effects of coronal mass ejections and solar flares on the interplanetary medium and Earth's environment are referred to as *space weather storms*. Space weather, a plasma effect, should not be confused with space weathering, a material processing effect, discussed in §9.3.2.

Space weather is of particular importance because of its effects on spacecraft operations and communication systems and on some ground-based electronics. Particles accelerated as a result of fast coronal mass ejections may damage exposed electronic equipment directly, and may disrupt communications indirectly. Such solar wind disturbances trigger magnetic storms (§7.5.1.4) which energize the entire magnetosphere and drive large-scale currents that can bring down power systems. The Earth's upper atmosphere gets heated by enhanced auroral precipitation and electric field-driven 'Joule heating' and expands, which increases the drag on near-Earth satellites and 'space junk', leading to changes in their orbits. Such changes may result in a (temporary) 'loss' of spacecraft. Radio communication systems on the ground rely on the reflection of radio waves by the Earth's ionosphere. Such communications are temporarily disrupted when the ionosphere is 'changed' by a solar flare or coronal mass ejection. Transformers in power grids and long conductivity cables can be damaged by currents induced by the related changes in the magnetic field on the ground.

7.1.2 Modeling the Solar Wind
7.1.2.1 Fluid Behavior

To first approximation, the solar wind consists of a fully ionized hydrogen plasma with roughly equal numbers of protons and electrons, and hence is electrically neutral over large spatial scales. Ideally one would like to model the collective effect of the particles rather than tracing each particle individually. Since we are dealing with charged particles, long-range electromagnetic forces are important, and in principle particles separated by long distances may influence each other such that we cannot model the collective behavior of the solar wind. We thus need to determine over what length scale electrical neutrality is valid, or at which distance particles are effectively 'shielded' from each other.

In a vacuum, the electrostatic potential around a charged test particle with charge q is described by: $\Phi_V = q/(4\pi r)$. In a plasma the interaction of this test charge with other protons and electrons distorts the potential:

$$\Phi_V = \frac{q e^{-r/\lambda_D}}{4\pi r}, \tag{7.4a}$$

with the *Debye length*, λ_D:

$$\lambda_D = \sqrt{\frac{kT}{N_e q^2}} = 6.9 \sqrt{\frac{T}{N_e}} \text{ cm}. \tag{7.4b}$$

In equations (7.4), N_e and q are the electron number density and charge, respectively; k is the Boltzmann parameter, T the temperature, and r the distance from the charged test particle. When r/λ_D is large, the potential reduces to near zero, and the particle's effects are not felt. The effect of charged particles near the test charge thus reduces the sphere of influence of the test charge in an exponential way. The relevant length scale over which a charged particle influences its surroundings is the Debye length. For the solar wind, $\lambda_D \approx 10$ m (near 1 AU), which is very small compared to typical solar wind length scales. For shielding to be effective, the number of particles within a Debye sphere, $N_D = \frac{4}{3}\pi N_e \lambda_D^3$, must be large. In the solar wind, $N_D \approx 10^{10}$. Hence the solar wind can indeed be considered an ideal plasma, or ionized gas in which near-neighbor interactions can be ignored. Since it is a gas, the plasma temperature, number density, and pressure are simply related via the ideal gas law: $P = NkT$, with N the total number of particles, electrons + protons (i.e., equal to $2N_e$). Yet, because the Debye spheres of neighboring charges partially overlap, the plasma also exhibits a fluid

behavior, and its collective behavior can be modeled using the equations of hydrodynamics. The solar wind can thus be considered a collisionless flow. However, since plasmas consist of charged particles, electric and magnetic fields are important. In fact, the interplanetary magnetic field also 'binds' the solar wind particles together, strengthening the fluid behavior. The hydrodynamic equations, therefore, need slight alterations, and we use the equations of *magnetohydrodynamics (MHD)* instead. MHD equations, in general, describe the macroscopic behavior of a plasma, i.e., the response to electric, magnetic, and gravitational fields, including the plasma density and bulk-flow velocity.

7.1.2.2 Magnetohydrodynamics (MHD)

The equation of continuity (conservation of mass) reads

$$\frac{\partial \rho}{\partial t} + \nabla \cdot (\rho \mathbf{v}) = 0, \tag{7.5}$$

where ρ is the total mass density. The equation of motion (conservation of momentum) is

$$\rho \left(\frac{\partial \mathbf{v}}{\partial t} + \mathbf{v} \cdot \nabla \mathbf{v} \right) = -\nabla P + \mathbf{J} \times \mathbf{B} + \rho \mathbf{g}_p, \tag{7.6}$$

with P the thermal pressure of the plasma, $\mathbf{J}$ the current, $\mathbf{B}$ the magnetic field, and $\mathbf{g}_p$ the gravitational acceleration. If the plasma is not neutral, an electrical force term should be added to the right-hand side of equation (7.6).

Conservation of energy in the plasma implies

$$\frac{\partial U}{\partial t} + \nabla \cdot \mathbf{J}_u = 0, \tag{7.7}$$

with U the energy density, which consists of the kinetic, magnetic, and thermal energies, respectively:

$$U = \frac{1}{2}\rho v^2 + \frac{B^2}{8\pi} + \frac{P}{\gamma - 1}, \tag{7.8}$$

with γ the ratio of specific heats. The energy flux vector $\mathbf{J}_u$ is given by

$$\mathbf{J}_u = \mathbf{v} \left(\frac{1}{2}\rho v^2 + \frac{\gamma P}{\gamma - 1} \right) + \frac{c}{4\pi} \mathbf{E} \times \mathbf{B}, \tag{7.9}$$

with $\mathbf{E}$ the electric field. The various contributions to this flux vector are the transport of kinetic and thermal energy, and the work done by hydrostatic and radiation pressure (Poynting vector), respectively.

7.1.2.3 Shocks

It is apparent from §7.1.1 that the solar wind is not a steady continuous flow of particles. Numerous discontinuities and shocks, such as those induced by ICMEs and solar flares, propagate through the medium. A shock wave changes the state of the medium through which it travels. In the frame of reference of the shock, the upstream velocity is supersonic, while downstream from the shock the velocity is subsonic and the density of the medium is higher. In an unmagnetized plasma, the density downstream of the shock can be up to four times higher than in the upstream medium (Problem 7.5). Since the solar wind is a collisionless plasma, the shocks are also collisionless. Mass, momentum, and energy flux are conserved across a shock:

$$[\rho v_\perp] = 0, \tag{7.10a}$$

$$\left[\rho v_\perp \mathbf{v} + P + \frac{B_\parallel^2}{8\pi} - \frac{B_\perp B_\parallel}{4\pi}\right] = 0, \tag{7.10b}$$

$$\left[\rho v_\perp \left(\frac{\gamma}{\gamma - 1}\frac{P}{\rho} + \frac{1}{2}v^2\right) - \frac{B_\parallel}{4\pi}(B_\perp v_\parallel - B_\parallel v_\perp)\right] = 0, \tag{7.10c}$$

$$[B_\perp v_\parallel - B_\parallel v_\perp] = 0, \tag{7.10d}$$

$$[B_\perp] = 0, \tag{7.10e}$$

The brackets in equations (7.10a–e) indicate the difference between the enclosed quantity from one side of the shock to the other. The flow across the shock is indicated by a ⊥ sign, and parallel to the shock by a ∥ sign. These shock or jump equations are called the *Rankine–Hugoniot relations* (compare with §5.4.2.1), which thus relate the density, pressure, temperature, and magnetic field strength before and after the shock. Both *discontinuities* and *shocks* are described by the Rankine–Hugoniot relations.

7.1.3 Maxwell's Equations

In order to discuss solar wind–planet interactions, we need to review the relevant electromagnetic equations. In a magnetized plasma, the motion of charged particles and the presence of electric and magnetic fields are all related through *Maxwell's equations*. Protons and electrons spiral around magnetic field lines in opposite directions. If the magnetic field points towards the observer, protons appear to spiral around the field in a clockwise motion; electrons gyrate anticlockwise. The relative flow of such charged particles forms a current. (The current in a wire is caused by the flow of electrons from a negative to a positive voltage; however, the current through the wire is defined as a flow in the opposite direction, i.e., as given by a flow consisting of positively charged particles moving to the negative voltage.)

Maxwell's equations read as follows:

(*i*) *Poisson's equation* relates the electric field, **E**, to the charge density, ρ_c:

$$\nabla \cdot \mathbf{E} = 4\pi \rho_c. \tag{7.11}$$

(*ii*) There are no magnetic charges, so the divergence of the magnetic field, **B**, is zero:

$$\nabla \cdot \mathbf{B} = 0, \tag{7.12}$$

(*iii*) *Faraday's law* describes the relationship between time-varying magnetic and spatially varying electric fields:

$$\nabla \times \mathbf{E} = -\frac{1}{c}\frac{\partial \mathbf{B}}{\partial t}. \tag{7.13}$$

Faraday's law is related to *Lenz's law*, which states that the current induced in a circuit flows in a direction that produces a magnetic field opposing the change in flux that produces the current. Lenz's law, however, only applies to rigid conductors, and not to deformable systems like space plasmas, and hence is not used in this chapter.

(*iv*) *Ampere's law*, as modified by Maxwell, relates spatial variations in a magnetic field to currents, **J**, and time-variable electric fields:

$$\nabla \times \mathbf{B} = \frac{4\pi}{c}\mathbf{J} + \frac{1}{c}\frac{\partial \mathbf{E}}{\partial t}. \tag{7.14a}$$

Usually, in planetary magnetic fields the term $(\partial E/\partial t)$ is very small compared with either of the other terms and Ampere's law can be approximated by

$$\nabla \times \mathbf{B} \approx \frac{4\pi}{c}\mathbf{J}. \tag{7.14b}$$

The motion of a conductive body or plasma through the interplanetary magnetic field induces a current, **J**, as given by *Ohm's law*:

$$\mathbf{J} = \sigma_0\left(\mathbf{E} + \frac{\mathbf{v} \times \mathbf{B}}{c}\right), \tag{7.15a}$$

where σ_0 is the conductivity of the plasma, **v** the plasma velocity, and **B** the interplanetary magnetic field strength. In a highly conducting plasma like the solar wind, $\mathbf{J}/\sigma_0 \approx 0$, and thus:

$$\mathbf{E} \approx -\frac{\mathbf{v} \times \mathbf{B}}{c}. \tag{7.15b}$$

Hence, if there is an electric field, the plasma must flow. Conversely, if the plasma flows, an electric field must be present.

Maxwell's equations (7.13) and (7.14) can be combined with Ohm's law (eq. 7.15) to yield the *magnetic induction equation*:

$$\frac{\partial \mathbf{B}}{\partial t} = \nabla \times (\mathbf{v} \times \mathbf{B}) + \frac{c^2}{4\pi\sigma_0}\nabla^2\mathbf{B}. \tag{7.16}$$

The induction equation relates time variations in the magnetic field strength to convection and diffusion of the field through the fluid. The first term on the right is convection of field lines with the moving fluid, and the second term

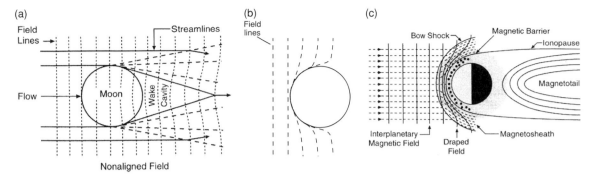

Figure 7.8 Interaction of the solar wind with various types of planetary bodies that do not possess internal magnetic fields. (a) A nonconducting body, (b) a conducting body, and (c) a body with an ionosphere. (Adapted from Luhmann 1995)

is diffusion of the field through the fluid. If the convection term is much smaller than the diffusion term, the induction equation reduces to the (magnetic) *diffusion equation*:

$$\frac{\partial \mathbf{B}}{\partial t} = \frac{c^2}{4\pi\sigma_0}\nabla^2\mathbf{B}, \tag{7.17}$$

which yields the characteristic *Ohmic dissipation time* over which the magnetic field will gradually decay away:

$$t_d = \frac{4\pi\sigma_0\ell^2}{c^2}, \tag{7.18}$$

with ℓ a characteristic length scale. According to this equation, the Earth's magnetic field would have dissipated away in 10^4–10^5 years (Problem 7.29). In the other extreme situation, if the conductivity of the medium is very large, the convection term dominates:

$$\frac{\partial \mathbf{B}}{\partial t} = \nabla \times (\mathbf{v} \times \mathbf{B}). \tag{7.19}$$

It can be shown (see, e.g., Boyd and Sanderson 1969) that this statement is equivalent to saying that the total amount of magnetic flux that passes through a closed loop circuit, A, that moves with the local fluid is constant over time:

$$\Phi_B = \int_A \mathbf{B} \cdot d\mathbf{A} = \text{constant}. \tag{7.20}$$

This means that regardless of changes in the surface area A, such as shrinking or stretching, the flux through this area remains unchanged. So field lines threading this surface may move closer together or further apart, but they will move with the fluid through the medium. This is the same as saying that the magnetic field lines are *frozen into* the plasma. The magnetic field in the interplanetary medium is thus carried away with the solar wind.

In analogy with the Reynolds number in fluid dynamics (eq. 4.48), the *magnetic Reynolds number*, $\mathfrak{R}_m$, can be used to assess the importance of the diffusion versus convection terms:

$$\mathfrak{R}_m \equiv \frac{t_d v}{\ell}, \tag{7.21}$$

with t_d the characteristic diffusion, or Ohmic dissipation, time (eq. 7.18), ℓ a length scale, and v the fluid velocity. When $\mathfrak{R}_m$ is large, the Ohmic diffusion times are long, and the convection term dominates.

7.1.4 Solar Wind–Planet Interactions

All planetary bodies interact to some extent with the solar wind, as schematically indicated in Figure 7.8. For bodies without intrinsic magnetic fields, the interaction depends on the conductivity of the body and/or its atmosphere. Rocky objects, such as the Moon and most asteroids, are poor conductors. In such circumstances, the solar wind particles hit the body directly and are absorbed. The interplanetary magnetic field lines simply diffuse through the body. The wake immediately behind the object is practically devoid of particles (Fig. 7.8a).

The interaction of the solar wind with a body that is conducting is more complicated. If a planetary body is highly conductive, the interplanetary magnetic field lines drape around the body because the plasma flows around the conductor (Fig. 7.8b). The electric field (eq. 7.15) induces a current ($\mathbf{J} = \sigma\mathbf{E}$) in the body, which in turn disturbs the interplanetary magnetic field. The perturbations in the magnetic field and plasma flow propagate away as Alfvén waves (§7.4.2).

If a poorly conducting body has an atmosphere but no internal magnetic field, the solar wind interacts with the atmosphere. The interaction is mainly between charged particles, i.e., with the planet's ionosphere, but charge exchange also operates (§4.8.2). In charge exchange, a solar wind proton or other ion transfers its positive charge to a planetary atom, usually of the same species. The planetary atom then becomes the charged particle that the solar wind interacts with, while the solar wind ion that lost its charge flies away as an energetic neutral atom (ENA).

Since the magnetic field is moving with the solar wind plasma, atmospheric ions exposed to it are accelerated and picked up by the solar wind. If the body has an extensive ionosphere (e.g., is a good conductor, $\sigma_o \neq 0$), currents are set up, which try to prevent the magnetic field from diffusing through the body (Fig. 7.8c). This situation gives rise to a magnetic configuration very similar to the magnetic 'cavity' created by the interaction of the solar wind with a magnetized planet, discussed below. Venus, Mars, and comets are objects which reveal such *induced* magnetospheres. However, all these bodies show solar wind field penetration into the ionosphere when solar wind pressure is larger than the ionospheric thermal plasma pressure, so the shielding is only partial and varies with the interplanetary conditions. Titan, in Saturn's magnetosphere, also exhibits similar behavior.

At the Sun-facing side of the planet, the interaction of the supersonic solar wind with the object's charged particle population induces a *bow shock*. The creation of this shock is analogous to the formation of a bow wave in front of a speedboat on a lake. In both cases the quiescent flow (solar wind or water) does not 'know' about the obstacle, since the relative velocity between the obstacle and medium is larger than the speed of the relevant waves (surface waves in the case of water, and magnetosonic waves in the case of the solar wind). At the bow shock, the solar wind plasma transitions from supersonic to subsonic, slowing and becoming more dense while the entrained interplanetary magnetic field strength increases. The region downstream of the bow shock is called the *magnetosheath*. The bow shock and the magnetosheath within it may be asymmetric, depending on the orientation of the solar wind magnetic field.

Interior to the bow shock, in the inner portion of the magnetosheath, we find a boundary referred to as a *magnetic pileup boundary*, *mantle boundary*, or *depletion layer*. This boundary is characterized by strong, highly organized magnetic fields as a result of the pileup and draping of the interplanetary magnetic field. The innermost boundary for planets without an internal magnetic field is the *ionopause*, above which the density of planetary thermal ions and electrons decreases rapidly. This boundary can in a sense be compared to the magnetopause discussed below, although the magnetic pileup region has also been compared to the magnetopause, because solar wind protons appear not to penetrate this boundary. The ionopause is located where the ionospheric pressure is balanced by the outside plasma pressure (magnetic plus thermal plus ram pressure), which becomes purely magnetic pressure in the pileup region. The solar wind wake that forms behind the planetary obstacle is filled with inner magnetosheath fields that sink into the wake after picking up some of the ionospheric plasma. This region is called an *induced magnetotail* (Fig. 7.8c).

If the body has an internal magnetic field, as Mercury, Earth, and the giant planets do, the solar wind interacts with the field around the object. It confines the magnetic field to a 'cavity' in the solar wind, referred to as a *magnetosphere*. A planetary magnetic field resembles to first approximation a dipole field, similar to that generated by a bar magnet. A sketch of the Earth's magnetic field is given in Figure 7.9. At the Sun-facing side, the interaction of the supersonic solar wind with the magnetic field resembles the interaction with the atmospheric/ionospheric obstacle. At the bow shock, the solar wind plasma is decelerated to subsonic speeds, and flows smoothly around the magnetospheric obstacle. The shape of the magnetosphere depends upon the strength of the magnetic field and the solar wind flow past the field. The magnetospheric boundary is called the *magnetopause*. The solar wind pressure shapes the 'nose' of the field, while the solar wind flow stretches the field out into a tail, also called the *magnetotail*. The magnetotail in a magnetosphere consists of two lobes of opposite polarity, separated by a neutral sheet in the plane of the dipole equator if the dipole axis is oriented perpendicular to the solar wind flow. A plasma or current sheet is located at the interface of the two lobes.

The approximate position of the magnetopause near the subsolar equator can be calculated from a balance between the ram pressure of the solar wind, $P_{\rm sw}$, and the pressure inside the magnetosphere, $P_{\rm m}$:

$$P_{\rm sw} = P_{\rm m}, \tag{7.22}$$

where

$$P_{\rm sw} \approx \rho v^2, \tag{7.23}$$

with ρ the ion density in the solar wind, and

$$P_{\rm m} = \frac{B^2}{8\pi} + P, \tag{7.24}$$

with P the thermal gas pressure. Since the pressure in the magnetosphere is dominated by the magnetic field strength, a balance in pressure leads to the approximate equality:

$$\left(\rho v^2\right)_{\rm sw} \approx \left(\frac{B^2}{8\pi}\right)_{\rm m}, \tag{7.25}$$

(a)

(b)

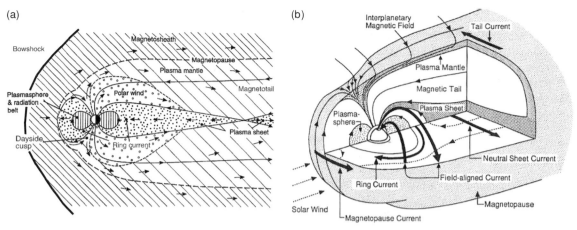

Figure 7.9 Sketches of the Earth's magnetic field in 2D and 3D, showing currents, fields, and plasma regions. (a) 2D (side) view: The solid arrowed lines indicate magnetic field lines, the heavy long-dashed line the magnetopause, and the arrows represent the direction of the plasma flow. Diagonal hatching indicates plasma in, or directly derived from, the solar wind/magnetosheath. Outflowing ionospheric plasma is indicated by open circles; the solid dots indicate hot plasma accelerated in the tail, and the vertical hatching shows the corotating plasmasphere. (Adapted from Cowley 1995) (b) 3D cutaway view: The heavy arrows indicate currents, the plasma sheet and plasmasphere are indicated by different hatchings, and the magnetic field lines are shown by lines with arrows to indicate the field direction. (Russell 1995)

where the parameters on the left-hand side pertain to the solar wind, those on the right-hand side to the magnetosphere. A typical *standoff distance* of the magnetopause for Earth is 6–15 $R_\oplus$ (Problem 7.4).

The standoff distance of the Earth's bow shock, $\mathcal{R}_{bs}$, is determined empirically:

$$\mathcal{R}_{bs} = \mathcal{R}_{mp}\left(1 + 1.1\frac{(\gamma - 1)\mathcal{M}_o^2 + 2}{(\gamma + 1)\mathcal{M}_o^2}\right), \qquad (7.26)$$

with $\mathcal{R}_{mp}$ the standoff distance of the magnetopause, γ the ratio of specific heats, and $\mathcal{M}_o$ the *magnetosonic Mach number*, defined as

$$\mathcal{M}_o \equiv \frac{v_{sw}}{\sqrt{c_A^2 + c_s^2}}, \qquad (7.27)$$

with c_s the speed of sound (eq. 7.3) and c_A the *Alfvén speed*:

$$c_A = \frac{B}{\sqrt{4\pi\rho}}. \qquad (7.28)$$

Typical values for the standoff distance of the Earth's bow shock are 10–20 $R_\oplus$ (Problem 7.4). The minimum empirically determined distance, Δ, between the bow shock and the magnetopause is given by

$$\Delta = 1.1\frac{\rho_{sw}}{\rho_{bs}}\mathcal{R}_{mp}, \qquad (7.29)$$

with ρ_{bs} the density just behind the shock.

As discussed earlier, countless shocks and discontinuities propagate through the interplanetary medium. The density and velocity in such a shocked (or disturbed) solar wind are higher than in the quiescent wind. When a solar wind shock hits a magnetized planet's bow shock and magnetopause, multiple shocks and waves are induced. Energy is transferred from the solar wind to the planet's magnetic field via waves that propagate into/through the magnetosphere. Due to the numerous fluctuations in the solar wind (Table 7.1), the interaction between the solar wind and a magnetosphere is very dynamic.

7.2 Magnetic Field Configuration

The solar wind pressure causes a compression of Earth's magnetosphere at the Sun-facing side, while the solar wind flow past the Earth's magnetic field stretches the field lines radially behind the planet, and forms the magnetotail (Fig. 7.9). Close to the planet where the magnetic field is hardly deformed by the solar wind, we find the *radiation belts* or *Van Allen belts*, regions in the magnetosphere where energetic charged particles are trapped. These particles gyrate around and bounce up and down magnetic field lines, while they drift around the Earth (§7.3.1). Under normal equilibrium circumstances these particles cannot escape from the magnetosphere.

The thin region in the magnetotail which separates field lines of opposite polarity is referred to as the *neutral*

Table 7.2 Characteristics of planetary magnetic fields.

	Mercury	Earth	Jupiter	Saturn	Uranus	Neptune
Magnetic moment ($\mathcal{M}_\oplus$)	4×10^{-4}	1^a	20 000	600	50	25
Surface B at dipole equator (gauss)	0.0033	0.31	4.28	0.22	0.23	0.14
Maximum/minimum[b]	2	2.8	4.5	4.6	12	9
Dipole tilt and sense[c]	$+14°$	$+10.8°$	$-9.6°$	$0.0°$	$-59°$	$-47°$
Dipole offset (R)		0.08	0.12	$\sim$0.04	0.3	0.55
Obliquity	$0°$	$23.5°$	$3.1°$	$26.7°$	$97.9°$	$29.6°$
Solar wind angle[d]	$90°$	$67–114°$	$87–93°$	$64–117°$	$8–172°$	$60–120°$
Magnetopause distance[e] (R)	1.5	10	42	19	25	24
Observed size of magnetosphere (R)	1.4	$8–12$	$50–100$	$16–22$	18	$23–26$

After Kivelson and Bagenal (2007).

[a] $\mathcal{M}_\oplus = 7.906 \times 10^{25}$ G cm^3.

[b] Ratio of maximum to minimum surface magnetic field strength (equal to 2 for a centered dipole field).

[c] Angle between the magnetic and rotation axis.

[d] Range of angles between the radial direction from the Sun and the planet's rotation axis over an orbital period.

[e] Typical standoff distance of the magnetopause at the nose of the magnetosphere, in planetary radii.

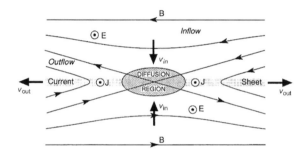

Figure 7.10 Magnetic reconnection according to Sweet's mechanism, occurring on an X-type magnetic neutral line. Plasma and magnetic field flow in from the top and bottom of the figure (v_{in}), and out towards the sides (v_{out}). Plasma is not tied to the magnetic field in the diffusion region. (Adapted from Hughes 1995)

sheet or *current sheet*. Due to the field line reversal, the magnetic field strength is minimal near the neutral sheet, and to maintain pressure equilibrium across the magnetosphere, the plasma density is maximal. Moreover, *field line reconnection* or *annihilation* may take place in the neutral sheet (Fig. 7.10). Energy released from magnetic reconnection accelerates/energizes the plasma. Magnetic field lines in the *polar caps* are connected to the interplanetary magnetic field (IMF), where the amount of interconnection depends on the IMF orientation with respect to the planetary field. On average, reconnection in the magnetotail is balanced by day-side reconnection at the plasmapause (§7.3.3.2, Fig. 7.17).

The magnetic field strength of a planet is usually given in terms of its *magnetic dipole moment*, $\mathcal{M}_B$, expressed in gauss cm^3. Typical parameters for the Earth and other

magnetospheres are summarized in Table 7.2. Note that the 'surface' magnetic field strengths (the 1 bar level for the giant planets) for Earth, Saturn, Uranus, and Neptune are all $\sim$0.3 G; however, because the radii of the giant planets are much larger than that of Earth, their magnetic dipole moments are 25–500 times larger than the terrestrial moment. Jupiter has by far the strongest magnetic dipole moment, nearly 20 000 times stronger than that of the Earth. The north magnetic poles on Earth and Mercury are near the planets' geographic south poles; thus field lines exit the planets in the south, and enter in the northern hemisphere. The north magnetic poles of Jupiter and Saturn are in the northern hemispheres.

The standoff distance of the magnetopause for Earth and the giant planets is typically larger than 6–10 planetary radii. For Mercury, however, the magnetopause is thought to, at times, be pushed down to the surface, in which case the solar wind interacts directly with the planet's surface.

7.2.1 Dipole Magnetic Field

The magnetic field of a dipole in polar coordinates can be described by (Fig. 7.11):

$$B_r = -\frac{2\mathcal{M}_B}{r^3} \cos\theta, \tag{7.30a}$$

$$B_\theta = \frac{\mathcal{M}_B}{r^3} \sin\theta, \tag{7.30b}$$

$$B_\phi = 0, \tag{7.30c}$$

with r, θ, and ϕ the coordinates in the radial, latitudinal (colatitude), and azimuthal (to the east) directions,

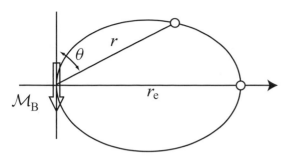

Figure 7.11 Sketch of a field line in a dipole magnetic field. (Adapted from Roederer 1970)

respectively. A field line lies in the meridional plane and is completely specified by the distance to its equatorial point, r_e, and its longitude or azimuth, ϕ (Fig. 7.11):

$$r = r_e \sin^2 \theta, \tag{7.31a}$$

$$\phi = \phi_o = \text{constant}. \tag{7.31b}$$

A small arc length, ds, along the field line is given by

$$ds = \sqrt{dr^2 + (rd\theta)^2} = r_e \sin \theta \sqrt{4 - 3\sin^2 \theta} \, d\theta. \tag{7.32}$$

The magnetic field strength along a field line is

$$B(\theta) = B_e \frac{\sqrt{4 - 3\sin^2 \theta}}{\sin^6 \theta}, \tag{7.33}$$

where the magnetic field strength in the magnetic equator, B_e, is determined by the magnetic dipole moment, $\mathcal{M}_B$:

$$B_e = \mathcal{M}_B / r_e^3. \tag{7.34}$$

7.2.2 Multipole Expansion

Although a planet's magnetic field can to first order be approximated by a dipole, deviations from a simple dipole field are apparent for all magnetospheres. In analogy with the mathematical description of a planet's gravitational field, the planet's internal magnetic field can be described as the gradient of a scalar potential $\Phi_V(r, \theta, \phi)$:

$$\mathbf{B} = -\nabla \Phi_V, \tag{7.35}$$

where r, θ, and ϕ are the planetocentric coordinates of the point in question: radial distance, planetocentric colatitude, and east longitude, respectively. The potential is expressed by

$$\Phi_V = R \sum_{n=1}^{\infty} \left(\frac{R}{r}\right)^{n+1} T_n, \tag{7.36}$$

where R is the planet's radius, and the function T_n is

$$T_n = \sum_{m=0}^{n} [g_n^m \cos(m\phi_i) + h_n^m \sin(m\phi_i)] P_n^m(\cos \theta), \tag{7.37}$$

with g_n^m and h_n^m the *Gauss coefficients*, which define the field configuration, and $P_n^m(\cos \theta)$ are the Schmidt-normalized associated Legendre polynomials:

$$P_n^m(x) \equiv N_{nm}(1 - x^2)^{m/2} \frac{d^m P_n(x)}{dx^m}, \tag{7.38}$$

where $N_{nm} \equiv 1$ if $m = 0$ and $N_{nm} \equiv \left(\frac{2(n-m)!}{(n+m)!}\right)^{1/2}$ if $m \neq 0$.

The terms with $n = 1$ are called dipole terms, those with $n = 2$ quadrupole terms, $n = 3$ octupole terms, etc. Note that the terms decrease with distance from the planet as r^{-n}, so that the higher order terms are most important near a planet's 'surface'. For a magnetic field which consists of only the three dipole terms g_1^0, g_1^1, and h_1^1, its magnetic moment, $\mathcal{M}_B$, tilt angle with respect to the rotational axis, θ_B, and the longitude of the magnetic north pole, λ_{np}, are

$$\mathcal{M}_B = R^3 \sqrt{(g_1^0)^2 + (g_1^1)^2 + (h_1^1)^2}, \tag{7.39}$$

$$\tan \theta_B = \sqrt{\left(\frac{g_1^1}{g_1^0}\right)^2 + \left(\frac{h_1^1}{g_1^0}\right)^2}, \tag{7.40}$$

$$\lambda_{np} = 180° - \tan^{-1}\left(\frac{h_1^1}{g_1^1}\right). \tag{7.41}$$

If the frame of reference is chosen such that the $\theta = 0$ axis is aligned with the magnetic dipole axis and the $\phi = 0$ meridian is at the longitude of the magnetic north pole, the $(1, 1)$ terms in equation (7.37) are zero, and deviations from a pure dipole field are more easily identifiable. The quadrupole terms $(2, 0)$ and $(2, 1)$ in this new frame of reference can be interpreted as a displacement of the main dipole. The quadrupole term $(2, 2)$ causes an apparent longitude-dependent tilt of the field, i.e., a warping of the magnetic equatorial plane, with a periodicity of 180°. The higher order terms have similar effects on the magnetic field configuration, although with different strengths and periodicities.

The above equations show how to mathematically approximate a magnetic field caused by current systems within the planet. In reality the field deviates from this ideal configuration, since currents in and near the magnetosphere itself induce electric and magnetic fields (§7.1.3) which distort the field lines. These effects increase with increasing r. So the total magnetic field is better represented by

$$\mathbf{B}_{total} = -\nabla \Phi_V + \mathbf{B}_{external}. \tag{7.42}$$

Characteristic parameters for the magnetospheres of the planets are shown in Table 7.2. The magnetic field geometries of Jupiter and Saturn are similar to that of the Earth,

Table 7.3 Plasma characteristics of planetary magnetospheres.

	Mercury	Earth	Jupiter	Saturn	Uranus	Neptune
Maximum density (cm^{-3})	1	1000–4000	>3000	~100	3	2
Composition	H^+	$O^+, H^+, N^+,$ He^+, He^{2+}	$O^{n+}, S^{n+},$ SO_2^+, Cl^+	O^+, H_2O^+, H^+	H^+	N^+, H^+
Dominant source	solar wind	ionosphere solar wind	Io	rings, Enceladus, Tethys, Dione	atmosphere	Triton
Production rate (ions s^{-1})	?	2×10^{26}	$>10^{28}$	10^{26}	10^{25}	10^{25}
Ion lifetime	minutes	days,[a] hours[b]	10–100 days	1 month–years	1–30 days	1 day
Plasma motion controlled by:	solar wind	rotation[a] solar wind[b]	rotation	rotation	solar wind + rotation	rotation (+ solar wind?)

After Kivelson and Bagenal (2007).

[a] Inside plasmasphere.

[b] Outside plasmasphere.

where the rotation and magnetic axes are within 10° of each other, and the magnetic field is to first approximation dipolar. Uranus and Neptune have quite irregular magnetospheres, and their magnetic axes make large angles with the rotation axes of the planets. The large-scale structure of individual planetary magnetospheres is discussed in detail in §7.5.

7.3 Particle Motions in Magnetospheres

Magnetospheres are populated with charged particles: protons, electrons, and ions. The dominant species detected in each magnetosphere are summarized in Table 7.3. We find primarily oxygen and hydrogen ions in the Earth's magnetosphere; in Jupiter's magnetosphere these are augmented with sulfur ions, in Saturn's magnetosphere with water-group ions, and in Neptune's magnetosphere both H^+ and N^+ have been detected. Note the large variations in ion densities: in Uranus's and Neptune's magnetospheres the maximum ion densities are only 2–3 protons cm^{-3}, while in the magnetospheres of Earth and Jupiter the densities measure over a few thousand cm^{-3} in some regions. All magnetospheres also contain large numbers of electrons; on average magnetospheric plasma is approximately neutral in charge. The particle distribution functions are almost Maxwellian (eq. 4.14), or thermal, though there is a pronounced high-energy tail to this distribution, with particle energies ranging up to several hundreds of MeV in some magnetospheres.

The spatial distribution of plasma is determined by the sources and losses of the plasma, as well as the motions of the particles in the planet's magnetic field. In equilibrium situations, a charged particle's motion is completely defined by the magnetic field configuration, the planet's gravity field, centrifugal forces, large-scale electric fields, and the particle's charge-to-mass ratio, q/m. In this section, we give a detailed treatment of the motions of charged particles in a stable magnetosphere, and then summarize the sources and sinks of plasma.

7.3.1 Adiabatic Invariants

The general motion of nonrelativistic charged particles of mass m in a magnetic field is governed by

$$m\frac{d^2\mathbf{r}}{dt^2} = \mathbf{F} + \frac{q\mathbf{v} \times \mathbf{B}}{c}, \quad (7.43)$$

with q the elemental charge and c the speed of light. In the absence of external forces ($\mathbf{F} = 0$), this equation simplifies to the *Lorentz force*:

$$\mathbf{F}_L = \frac{q\mathbf{v} \times \mathbf{B}}{c}. \quad (7.44)$$

The Lorentz force leads to rapid circulation of the particle centered on a location known as the *gyrocenter*. External forces change the particle's simple circular motion into three components (Fig. 7.12): (a) a gyration around field lines, (b) a bounce motion along field lines, and (c) a drift motion perpendicular to the field lines. The trajectory of the particle can be approximated by a circular motion

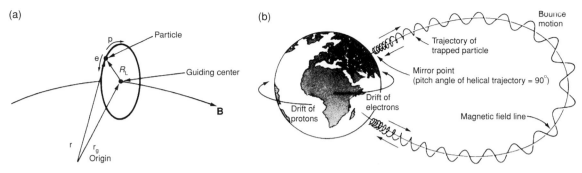

Figure 7.12 Basic motion of charged particles: (a) cyclotron (gyro) motion, (b) bounce motion along a field line and drift motion around the Earth. (Lyons and Williams 1984)

of radius R_L around the instantaneous gyrocenter, plus a displacement of the gyrocenter, referred to as the *guiding center* $\mathbf{r}_g$:

$$\mathbf{r} = \mathbf{r}_g + \mathbf{R}_L. \tag{7.45}$$

7.3.1.1 *First Adiabatic Invariant*

The cyclotron or gyro motion around a field line is a circular motion around the guiding center. The Lorentz force balances the centripetal force of the particle's motion around the field line. The *gyro* or *Larmor* radius is (Problem 7.8)

$$|\mathbf{R}_L| = \left| \frac{c\mathbf{p}_\perp}{qB} \right| = \left| \frac{mc\mathbf{v} \times \mathbf{B}}{qB^2} \right|, \tag{7.46}$$

where $\mathbf{p}_\perp$ is the momentum perpendicular to the field line, $p_\perp = mv \sin\alpha$, with m the mass and α the instantaneous *pitch angle* of the particle, i.e., the angle between the direction of motion and the local magnetic field line. The particle 'orbits' the field line at a frequency n:

$$n = \frac{v_\perp}{2\pi R_L}. \tag{7.47a}$$

The *cyclotron, gyro*, or *Larmor frequency* is defined by (Problem 7.8)

$$\omega_B \equiv 2\pi n = \frac{v_\perp}{R_L} = \left| \frac{qB}{mc} \right|. \tag{7.47b}$$

Typical values for the Larmor radius and period for a $\sim$100 keV electron in the Earth's inner radiation belt are $\sim$100 m and a few µs, respectively; for a proton these numbers are larger by a factor of 1836, the ratio of their masses.

If changes in the magnetic field are small over one gyro radius and during one gyro period, the particle gyrates in a nearly static magnetic field, and the magnetic flux, Φ_B, through a particle's orbit is constant: $d\Phi_B/dt = 0$. From

this one can derive a constant of motion, known as the *first adiabatic invariant*, μ_B (Problem 7.9)

$$\mu_B = \frac{p_\perp^2}{2m_o B}, \tag{7.48}$$

with m_o the particle's rest mass. The above equation is valid for nonrelativistic as well as relativistic particles; for the latter, the mass m (in $p_\perp$) is the relativistic mass:

$$m = \gamma_r m_o, \tag{7.49a}$$

with γ_r the relativistic correction factor:

$$\gamma_r \equiv \frac{1}{\sqrt{(1 - v^2/c^2)}}. \tag{7.49b}$$

For nonrelativistic particles, the first adiabatic invariant is equal to the particle's magnetic moment, μ_b, induced by its circular motion around the magnetic field lines. For relativistic particles, $\mu_B = \gamma_r \mu_b$ (Problem 7.9c).

The first adiabatic invariant can be written in terms of the particle's kinetic energy, E:

$$\mu_B = \frac{E \sin^2 \alpha}{B}. \tag{7.50}$$

For relativistic particles the first adiabatic invariant in terms of a particle's energy becomes (Problem 7.9)

$$\mu_B = \frac{(E^2 + 2m_o c^2 E)}{2m_o c^2 B} \sin^2 \alpha. \tag{7.51}$$

Note that in the nonrelativistic limit equation (7.51) reduces to equation (7.50).

7.3.1.2 *Second Adiabatic Invariant*

Consider a particle's motion along the magnetic field line in the absence of electric fields, so that the particle's kinetic energy, E, is conserved. Conservation of the first adiabatic invariant shows that the pitch angle, α, increases when the particle moves to larger field strengths, until $\alpha = 90°$ at the *mirror point*. At this point the particle turns around, i.e., is 'reflected' back along the field line,

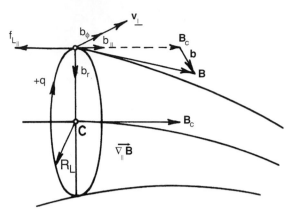

Figure 7.13 Effect of a field-aligned Lorentz force on a particle's helical motion along a field line. (Roederer 1970)

as explained in physical terms below. The pitch angle is minimal at the magnetic equator. For particles confined to the planet's magnetic equator, the pitch angle $\alpha_e = 90°$. Particles which move along field lines with $v_\perp \neq 0$ are subject to a gradient in the field strength directed along the field lines, $\nabla_\parallel \mathbf{B}$ (Fig. 7.13). Since the (dipole-like) field lines bunch up at higher latitudes, the particle experiences a field-aligned Lorentz force:

$$\mathbf{f}_{L_\parallel} = \frac{q\mathbf{v}_\perp \times \mathbf{b}_r}{c}, \tag{7.52}$$

where $\mathbf{b}_r$ is directed along the Larmor radius of the particle (Figure 7.13). The force $\mathbf{f}_{L_\parallel}$ is directed along $\mathbf{B}_c$, always in a direction opposite to $\nabla_\parallel \mathbf{B}$. The Lorentz force reduces the particle's velocity along the field line, until the velocity reaches zero at the particle's mirror point, at which point the particle turns around and moves in the opposite direction. This instantaneous field-aligned Lorentz force causes the particles to bounce up and down the field lines, between the two mirror points.

If changes in the magnetic field are small over the time of one bounce period, one can derive a second constant of motion, known as the *second adiabatic invariant*, J_B, which is the integral of the particle's momentum along the field line, $p_\parallel = mv\cos\alpha$, between the mirror points s_m and s'_m:

$$J_B = 2\int_{s_m}^{s'_m} p_\parallel \, ds. \tag{7.53}$$

A typical bounce period for a $\sim$100 keV electron in the Earth's inner radiation belt is $\sim$0.1 s. We define the integral $I_B \equiv J_B/(2mv)$, which is an invariant if the velocity v is constant:

$$I_B \equiv \frac{J_B}{2mv} = \int_{s_m}^{s'_m} \left(1 - \frac{B_s}{B_m}\right)^{1/2} ds, \tag{7.54}$$

with B_s and B_m the local field strength and the field strength at the particle's mirror point, respectively.

When a particle diffuses radially inwards, its energy increases since the first adiabatic invariant is constant. In equilibrium situations, the second invariant, $J_B = 2mvI_B$, is also constant. Thus, since the particle's kinetic energy, and hence v, changes, the integral I_B changes as well. Since both the first and second adiabatic invariants are constant, the following combination should also be invariant:

$$K_B = \frac{J_B}{2\sqrt{m_0\mu_B}} = I_B\sqrt{B_m}$$

$$= \int_{s_m}^{s'_m} [B_m - B_s]^{1/2} \, ds = \text{constant}. \tag{7.55}$$

To calculate the equatorial pitch angle α_{e2} at position r_2, knowing α_{e1} at $r_1 > r_2$, one needs to evaluate the integral in equation (7.55). This can best be done by field line tracing or a numerical calculation. However, it is fairly easy to show that the pitch angle must increase while the particle is diffusing radially inwards (Problem 7.12). Hence, closer to the planet the equatorial pitch angles of the particles are generally closer to 90°.

7.3.1.3 Third Adiabatic Invariant

Particles which are confined to the magnetic equator ($\alpha_e = 90°$) drift around a planet along contours of constant magnetic field strength (§7.3.2). If changes in the magnetic field are small over timescales of one drift period, then the magnetic flux, Φ_B, enclosed by the particle's orbit is a constant of motion. This constant is known as the *third adiabatic invariant* (eq. 7.20). In planetary magnetospheres, the first and second adiabatic invariants are usually conserved, but the third invariant is often violated.

7.3.2 Drift Motions in a Magnetosphere

Using equations (7.44) and (7.45), the position of a particle's guiding center, $\mathbf{r}_g$, can be written as

$$\mathbf{r}_g = \mathbf{r} - \frac{mc}{qB^2}\mathbf{v} \times \mathbf{B}. \tag{7.56}$$

Any instantaneous force $\mathbf{F}_i$ changes a particle's velocity:

$$d\mathbf{v} = \frac{1}{m}\int \mathbf{F}_i \, dt, \tag{7.57}$$

which results in a change in the position of the gyrocenter:

$$d\mathbf{r}_g = \frac{mc}{qB^2}d\mathbf{v} \times \mathbf{B} = -\frac{c}{qB^2}\mathbf{B} \times \int \mathbf{F}_i \, dt. \tag{7.58}$$

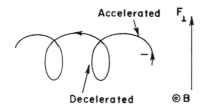
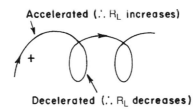

Figure 7.14 General motion of charged particles in a magnetic field caused by a force perpendicular to the field line. The magnetic field is perpendicular to the paper, coming out of the paper. (Roederer 1970)

If **F** is the time average of **F**$_i$, a change in the position of the guiding center leads to a drift in the particle's motion, with a drift velocity **v**$_F$:

$$\mathbf{v}_F = \frac{c\mathbf{F} \times \mathbf{B}}{qB^2}. \tag{7.59}$$

Note that the drift velocity is perpendicular to both the force exerted on the particle and the magnetic field direction. In addition, if F is charge independent, protons and electrons drift in opposite directions for a given direction of **F**. The general motion of a proton and an electron in a magnetic field under the influence of a force perpendicular to the field line is sketched in Figure 7.14. The drift motion of the particle is caused by the subsequent increases and decreases of the particle's Larmor radius.

7.3.2.1 Gradient **B** Drift

The field strength in a planetary magnetosphere decreases with increasing distance from the planet. The overall gradient in field strength thus consists of components parallel and perpendicular to the field lines: $\nabla\mathbf{B} = \nabla_\parallel\mathbf{B} + \nabla_\perp\mathbf{B}$. This gradient in field strength induces a force:

$$\mathbf{F} = -\mu_b\nabla\mathbf{B}, \tag{7.60}$$

with μ_b the magnetic moment, $\mu_b \equiv \mu_B$. The force caused by the gradient $\nabla_\perp B$ causes charged particles to drift around the planet. If changes in the magnetic field are small during a drift period, the magnetic flux enclosed by the drift orbit is conserved (the third adiabatic invariant, eq. 7.20). The drift velocity, **v**$_B$, is equal to

$$\mathbf{v}_B = \frac{\mu_b c\mathbf{B} \times \nabla\mathbf{B}}{qB^2}. \tag{7.61}$$

Particles move in a direction perpendicular to the lines of force and perpendicular to the magnetic gradient.

Consider a dipole field with particles confined to the magnetic equatorial plane, i.e., the equatorial pitch angle $\alpha_e = 90°$. Since the field strength is proportional to r^{-3}, the particles are subject to a force induced by the gradient in the field strength, $\nabla_\perp\mathbf{B}$. This situation is sketched in Figure 7.15; the drift motion is caused by the subsequent increases and decreases in the Larmor radius of the particle due to the changes in field strength (note the difference with Figure 7.14). Protons and electrons drift in opposite directions,

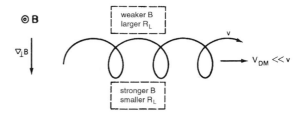

Figure 7.15 Drift motion of an equatorial ($\alpha_e = 90°$) proton due to a gradient in the magnetic field. (Lyons and Williams 1984)

each along contours of constant magnetic field strength. In the Earth's magnetic field, electrons drift around the Earth in the eastward direction, protons towards the west. The drift motion of these particles produces an electric current system known as the *ring current*, which peaks at $\sim 5R_\oplus$. A typical drift period for ring current particles is about 2 hours. The magnetic field induced by the ring current (inside of the current) is oriented to the south, and thus reduces the magnetic field strength on the surface of the Earth (eq. 7.14; Problem 7.17).

The maximum distance of a magnetic field line to the center of a planet is indicated by *McIlwain's parameter*, which in a dipole field is defined as

$$\mathcal{L} \equiv \left(\frac{\mathcal{M}_B}{R^3 B_e}\right)^{1/3}, \tag{7.62}$$

with $\mathcal{M}_B$ the magnetic dipole moment and B_e the magnetic field intensity in the magnetic equator, i.e., the minimum magnetic field strength along a given field line. For a centered dipole field, $\mathcal{L}$ represents the actual distance in planetary radii from the planet's center to the equatorial point of a field line. If radial diffusion is absent or very small and the field is approximately axisymmetric, particles bounce and drift around the planet thereby tracing out *drift shells*. In the case of a pure dipole field, these drift shells are alike for particles at the same planetocentric distance, even if they have different equatorial pitch angles α_e. For a multipole magnetic field this is not the case, and drift shells depend both on α_e and on the particle's starting point. This effect is called *shell degeneracy* or *shell splitting* (§7.5.1.2).

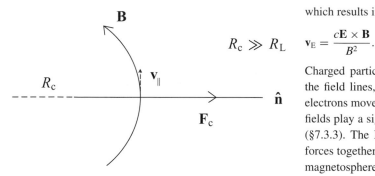

$$R_c \gg R_L$$

Figure 7.16 Geometry of field line curvature (§7.3.2.2).

7.3.2.2 Field Line Curvature Drift

Particles which move along field lines are subject to a field line curvature drift, in addition to the field-aligned Lorentz force which causes the particle to bounce up and down field lines. If the guiding center follows the curved field line (Fig. 7.16), the centripetal force is equal to

$$\mathbf{F}_c = \frac{m\mathbf{v}_\parallel^2}{R_c}\hat{\mathbf{n}}, \tag{7.63}$$

where $\hat{\mathbf{n}}$ is a unit vector outwards, along the direction of the field line's radius of curvature, R_c. The field line's radius of curvature is usually much larger than the particle's Larmor radius, $R_c \gg R_L$. This results in the curvature drift:

$$\mathbf{v}_C = \frac{mcv_\parallel^2}{qR_cB^2}\hat{\mathbf{n}} \times \mathbf{B}. \tag{7.64}$$

This drift motion, being perpendicular to both the field line's radius of curvature and the field line itself, is in the same direction as the gradient **B** drift.

7.3.2.3 Drift Induced by Gravity Fields

Particles in a planetary magnetic field are also subject to forces from the planet's gravitational field, $\mathbf{F} = m\mathbf{g}_p$:

$$\mathbf{v}_g = \frac{mc}{qB^2}\mathbf{g}_p \times \mathbf{B}. \tag{7.65}$$

This gravitational 'perturbation' causes particles to move perpendicular to both the gravitational force and the magnetic lines of force. This drift motion is usually small compared to that due to electric fields and gradients in the magnetic field strength.

7.3.2.4 Drift Induced by Electric Fields

When there is a drop in voltage or an electric field in a magnetosphere, particles are influenced by a force:

$$\mathbf{F} = q\mathbf{E}, \tag{7.66}$$

which results in a drift velocity:

$$\mathbf{v}_E = \frac{c\mathbf{E} \times \mathbf{B}}{B^2}. \tag{7.67}$$

Charged particles move in a direction perpendicular to the field lines, **B**, and the electric field, **E**; protons and electrons move in the same direction. Large-scale electric fields play a significant role in planetary magnetospheres (§7.3.3). The $\mathbf{E} \times \mathbf{B}$ force, the $\nabla\mathbf{B}$, and curvature drift forces together dominate the drift motion of particles in a magnetosphere.

7.3.3 Electric Fields

The overall drift motion of charged particles in a magnetosphere is governed by the gradient in the magnetic field strength, field line curvature, and by the presence of electric fields. The first two forces cause equatorially confined particles to drift around the planet on $B = $ constant contours. Electric fields aligned along magnetic field lines accelerate electrons and protons/ions in opposite directions, resulting in a field-aligned *current*, which usually quickly reduces any $E_\parallel$. Electric fields perpendicular to magnetic field lines cause both ions and electrons to drift in the same direction; thus the potential difference is not decreased by a current, and such large-scale electric fields can be stable. Two large-scale electric fields that are present in each planetary magnetosphere are the corotational and convection electric fields.

7.3.3.1 Corotational Electric Field

A planet's rotation induces an electric field in its magnetosphere, which is directed radially inwards or outwards: the *corotational electric field* (Ohm's law, eq. 7.15b)

$$\mathbf{E}_{cor} = -\frac{\mathbf{v} \times \mathbf{B}}{c} = -\frac{(\omega_{rot} \times \mathbf{r}) \times \mathbf{B}}{c}. \tag{7.68}$$

In the above equation, the plasma velocity, **v**, is assumed to be equal to the cross product of the spin angular velocity of the planet, ω_{rot} and the planetocentric distance **r**. Since a planet's ionosphere rotates with the planet, field-aligned currents will link the rotating plasmas in the ionosphere and magnetic equator (see also §7.5.4.6). The direction of the electric field depends upon the direction of the magnetic field and the sense of rotation of the planet. For the Earth, the corotational electric field is directed inwards, for the giant planets outwards.

7.3.3.2 Convection Electric Field

The solar wind flowing past the Earth's magnetosphere pulls the field lines back, forming the magnetotail. The interaction between the solar wind and Earth's

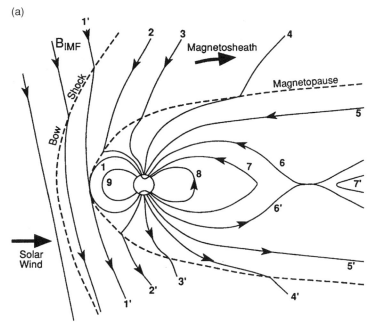

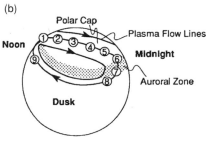

Figure 7.17 Interaction of the solar wind (southward directed magnetic field) with the Earth's magnetosphere. Reconnection takes place at the 'nose' of the magnetosphere, between field lines 1 and 1', and in the geomagnetic tail between 6 and 6'. Panel (b) shows the path of the feet of the numbered field lines in the northern hemisphere. (Hughes 1995)

magnetosphere is sketched in Figure 7.17 for situations where the interplanetary magnetic field is directed southwards, thus approximately antiparallel to the Earth's field. Reconnection (Fig. 7.10) must take place between the interplanetary and geomagnetic field lines. This produces open field lines, which have one end attached to one of the polar regions on Earth, while the other end stretches out into interplanetary space. The interplanetary part of the field line is swept back around the Earth's magnetic field by the solar wind. The plasma on this flux tube senses an electric field $\mathbf{E} \propto \mathbf{v}_{sw} \times \mathbf{B}_{sw}$. The field lines move in the antisolar direction through the locations numbered in Figure 7.17, and form the magnetotail. The return flow of the magnetic flux is achieved through reconnection in the tail. In this noon–midnight meridian view, two originally open field lines (#6 on Fig. 7.17) reconnect in the tail and form a new closed field line. The Earth's rotation carries the field lines back to the Sun-facing side, while the open field lines (#7' on Fig. 7.17) continue to flow down the tail. The path of the numbered flux tube feet is shown in Figure 7.17b. Because this circulation resembles thermal convection cells, the dawn-to-dusk electric field generated by this process is referred to as the *convection electric field*. On Earth, this field is directed from the dawn to the dusk side; on planets such as Jupiter, where the magnetic field is directed northwards, the convection electric field is directed from dusk to dawn.

This large-scale convection electric field causes charged particles in Earth's magnetosphere to drift in the direction of the Sun at low magnetic latitudes and towards the center plane (current sheet) at high latitudes (Fig. 7.9). The dawn-to-dusk electric field thus induces a large-scale global circulation of magnetospheric plasma, where plasma moves from (near-)Earth to the current sheet in the tail, and once it reaches the current sheet it moves back towards Earth, in the direction of the Sun. While particles drift towards a planet, their first and second adiabatic invariants are conserved, while the third invariant is usually violated. Hence a particle's energy increases when moving towards regions of higher magnetic field strength (eqs. 7.50, 7.51). The increase in energy is tapped from the electric field, and thus from the solar wind. Particles can gain a considerable amount of energy this way. A typical solar wind proton has an energy of $\sim$10 eV ($T \sim 10^5$ K), while the interplanetary magnetic field strength at Earth is $\sim$5 × 10^{-5} G. This translates into a first adiabatic invariant $\mu_B \approx 0.2$ MeV G^{-1}, which implies that solar wind electrons entering the Earth's magnetosphere gain energy by adiabatic diffusion alone up to nearly 0.2 MeV at the Earth's surface (Problem 7.14).

7.3.3.3 Particle Drift

A comparison of the strength of the corotational and convection electric fields shows whether magnetospheric circulation is driven primarily by the solar wind or the planet's rotation. In addition, the gradient in the magnetic field and field line curvature give rise to another global

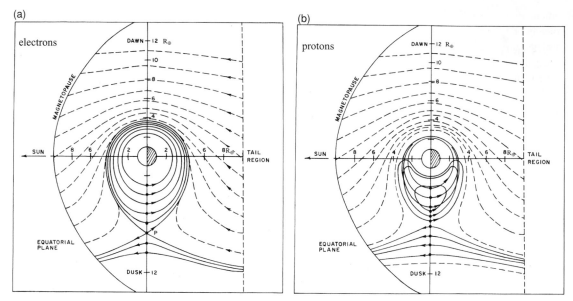

Figure 7.18 Motion of charged particles with $\alpha_e = 90°$ in the Earth's magnetic field. Broken lines: equipotentials for the dawn-to-dusk electric field. These curves represent the drift paths (in the direction of the arrows) for 'zero energy' particles in the convection region. Solid lines: drift paths of electrons (panel a) and protons (panel b) injected with an energy of 1 keV along the dusk meridian (at the dots). (Roederer 1970)

particle drift. We can express the total drift for magnetospheric particles in the form:

$$\mathbf{v}_D = \frac{\mathbf{B} \times \nabla \Phi_{eff}}{B^2}, \tag{7.69}$$

where Φ_{eff} is the effective potential due to the convection and corotational electric fields and ∇B, respectively:

$$\Phi_{eff} = \Phi_{conv} + \Phi_{cor} + \Phi_{\nabla B}, \tag{7.70}$$

where

$$\Phi_{conv} = -E_0 r \sin\phi, \tag{7.71}$$

$$\Phi_{cor} = \frac{-\omega_{rot} B_0 R^3}{r}, \tag{7.72}$$

$$\Phi_{\nabla B} = \frac{\mu_B B_0 R^3}{q r^3}. \tag{7.73}$$

In the above equations, B_0 is the surface magnetic field strength at the magnetic equator, R is the planet's radius, r is the planetocentric distance of the particle, and E_0 is the dawn-to-dusk electric field. The coordinate ϕ is measured in the magnetic equator from the planet–Sun direction to the dusk side. Since the convection potential, Φ_{conv}, is increasing with increasing distance ($\propto r$) while the potential due to the planet's rotation is decreasing ($\propto r^{-1}$), corotation dominates near the planet and solar wind induced

convection is more important at larger distances. In magnetospheres of rapidly rotating planets with strong magnetic fields, such as Jupiter and Saturn, plasma circulation is likely dominated by the planet's rotation, while the solar wind controls the plasma flow in the smaller fields around more slowly rotating planets, such as Mercury (Problem 7.15). At Earth, the inner magnetosphere is controlled by rotation, while the outer magnetosphere is driven by the solar wind.

Given the various drift motions of charged particles in the Earth's magnetosphere, one can predict a particle's trajectory once its energy and initial location are known. Drift paths for protons and electrons that are injected into the Earth's magnetosphere at the dusk meridian with an energy of 1 keV are displayed in Figure 7.18. Electric fields cause protons and electrons to drift in the same direction. The dawn-to-dusk electric field causes a general motion towards the Sun, while the corotational electric field causes the particles to drift around the Earth on closed equipotential contours. The gradient **B** drift causes protons to move westward and electrons eastward around the Earth. The gradient **B** drift motion of the electrons is in the same direction as the drift due to the corotational electric field. For low-energy particles (energy <1 keV), the electric field drift dominates, while for high-energy particles (energy >100 keV), the gradient **B** drift is more important. Protons with intermediate energies may not completely orbit the Earth. The orbits in Figure 7.18b are for protons, injected with an energy of 1 keV at several locations along the dusk meridian. Starting at <5 $R_\oplus$, the corotational field takes them eastwards around the Earth in orbits similar to those of electrons. The energy-dependent gradient **B** drift is negligible at all times. Protons which start between 5 and 7 $R_\oplus$ get sufficiently accelerated in their eastward drift that eventually the gradient **B** drift takes over and turns them around westwards, on the same evening side of the Earth. The proton is then decelerated, and at some point the electric field drift takes over again. These particles thus follow closed drift paths which do not encircle the Earth. At larger distances the convection field always dominates (Problem 7.16).

7.3.4 Particle Sources and Sinks
7.3.4.1 *Sources of Plasma*
There are several sources for magnetospheric plasma, and the relative contributions of each source vary from planet to planet. Charged particles can originate in cosmic rays, the solar wind, the planet's ionosphere, or on satellites/rings which are partially or entirely embedded in the magnetosphere. Although ionospheric particles are usually gravitationally bound to the planet, some charged particles escape along magnetic field lines into the magnetosphere (§4.8.2). Sputtering by micrometeorites, charged particles, and high-energy solar photons may cause ejection of atoms and molecules from moons/rings (§4.8.2); if such particles become ionized, they enrich the magnetospheric plasma.

A planetary magnetosphere is embedded in the solar wind; simple entry of solar wind particles into the magnetosphere would populate a planet's magnetic field with solar wind plasma. The detection of protons, electrons, and helium nuclei in Earth's magnetosphere suggests the solar wind to be a rich source of plasma. Both solar and galactic cosmic rays can also enter the magnetosphere. Their access is mainly at high latitudes, and is energy dependent.

Interplanetary particles can enter a magnetosphere through drift or reconnection processes:
(*i*)   As discussed in §7.1.4, the solar wind pulls the planet's magnetic field lines back, in the antisolar direction, to form the magnetotail, with the fast flowing solar wind on the interplanetary side of the magnetotail, and magnetospheric plasma on the inside. When two fluids in contact try to slip past each other via a tangential discontinuity, the *Kelvin–Helmholtz instability* can be induced, which manifests itself as ripples on a surface, similar to a flag waving in the wind, and the ripples on a lake induced

by the wind blowing over its surface. When the fast solar wind flows past the magnetosphere, it induces ripples in the magnetospheric boundary or magnetopause. These ripples induce a component in **B** perpendicular to the solar wind flow; hence, locally an electric field is set up, which enhances the convection electric field. This is an example of a mechanism that allows solar wind particles to diffuse into the planetary magnetosphere. Charged particles can also diffuse or gradient/curvature drift from the solar wind into the magnetosphere.

(*ii*)   Whenever the interplanetary field has a component antiparallel to the planetary field, magnetic reconnection may occur (Fig. 7.10): field lines merge together, and solar wind particles can enter the magnetosphere through the magnetic neutral points. Not only do they enter this way, they are accelerated in the process. Field line reconnection may occur at the day-side magnetopause and in the magnetotail neutral sheet. Usually, particles enter at the 'nose' of the day-side magnetopause when the interplanetary magnetic field is antiparallel to the planet's field. Contrary to this generally accepted view, however, the THEMIS fleet of 5 satellites (Time History of Events and Macroscale Interactions during Substorms) in Earth's magnetosphere discovered a tremendous influx of particles following a CME when the solar IMF was *parallel* to Earth's field. The particles probably entered at high latitudes, where the fields, locally, are antiparallel, and hence reconnection can occur.

7.3.4.2 *Particle Losses*
Moons, rings, and atmospheres are both sources and sinks of magnetospheric plasma. Particles which hit the surface of a solid body are generally absorbed and lost from the magnetosphere. Similarly, if a particle enters the collisionally thick part of an atmosphere, it gets 'captured' and won't return to the magnetosphere. Charged particles carry out a helical bounce motion along magnetic field lines (Fig. 7.12). The particle is reflected at the mirror point, and if this point lies in or below the ionosphere/atmosphere, the numerous collisions with atmospheric particles 'trap' the magnetospheric particle in the atmosphere. The location of this mirror point depends upon the particle's initial pitch angle, α_e, which is the angle between the direction of motion of a particle in the magnetic equator and the local magnetic field line there. We define a particle's *loss cone*, α_l, as the smallest pitch angle an equatorial particle can have without being absorbed. Particles with equatorial pitch angles $|\alpha_e| \leq |\alpha_l|$, or $\sin^2 \alpha_e < \sin^2 \alpha_l$, have their projected mirror points within or below the atmosphere. These particles are thus lost from the magnetosphere.

Another loss process is induced by charge exchange of magnetospheric ions (§4.8.2). The ions roughly corotate with the planet. Their velocity at large planetocentric distances is therefore much higher than the Keplerian velocity of a neutral particle (Problem 7.11). If such an ion undergoes a charge exchange with a neutral, the newly formed ion picks up the rotation speed of the ambient plasma, and stays trapped in the magnetosphere. The former ion, however, becomes a fast neutral and escapes the system if the corotation speed exceeds the escape velocity.

7.3.5 Particle Diffusion

Diffusion of particles in and through a magnetosphere is an important process: without particle diffusion the radiation belts would be empty, unless there is an *in situ* source. *Radial diffusion* displaces particles across field lines, while *pitch angle diffusion* moves a particle's mirror point along field lines. Whereas the first mechanism transports particles from their place of origin to other regions in a magnetosphere, pitch angle scattering can be regarded as a principal means to lose particles. It causes the particles' pitch angle distribution to spread, forcing some particles into the loss cones.

Radial diffusion is driven by large-scale electric fields in a magnetosphere, such as the solar wind induced convection field, which dominates diffusion in the quiescent terrestrial magnetic field. The corotational electric field dominates diffusion in the magnetospheres of Jupiter and Saturn, although we should note here that much transport in Jupiter and Saturn's magnetospheres takes place through *flux tube* or *centrifugally driven interchange*, where denser flux tubes slip outwards, changing places with less-dense flux tubes. This can be compared with the *Rayleigh–Taylor instability*, where a fluid, when overlain with a heavier one, becomes unstable and 'fingers' of heavier fluid penetrate or sink through the lighter one. In regions of a magnetosphere where rotation dominates over gravity, flux tube interchange leads to an outward transport of mass, 'visible' in the form of plasma fingers. Such signatures have been identified in Jupiter's magnetosphere, near the Io plasma torus.

Plasma instabilities in the plasma sheet, temporal variations in the magnetic field, as well as winds in a planet's upper atmosphere/ionosphere, induce stochastically varying electric fields (§7.1.3), which affect particle diffusion. As a result, particles random walk through the magnetosphere, so that they may diffuse both inwards and outwards. With a particle source somewhere in the outer magnetosphere, and a sink in the planet's atmosphere, the overall diffusion of protons and electrons is inwards, as has been observed *in situ* by spacecraft. For heavy ions in Jupiter's magnetosphere, outward diffusion dominates.

In a dipole-like field (i.e., inner magnetospheres), when particles diffuse radially inwards, conservation of the first and second adiabatic invariants shows that the particle's equatorial pitch angle increases; thus we expect particles to be more confined to the magnetic equator closer to the planet. In contrast to this slow change in a particle's pitch angle, stochastic variations have also been observed. Such a stochastic pitch angle diffusion can be caused by collisions with other particles (Coulomb scattering with atmospheric particles, charge exchange) or via wave–particle interactions.

7.4 Magnetospheric Wave Phenomena

Perturbations in a medium can induce waves which propagate at specific velocities, and which transfer energy from one region to another. Magnetospheres contain copious amounts of plasma, threaded by large-scale magnetic fields, so numerous waves are induced, and have indeed been observed in the magnetospheres of Earth and the four giant planets. Despite the large differences in size, magnetic orientation, energy sources, and plasma content, many of the same types of wave modes are present in all five magnetospheres, although the relative and absolute intensities of the emissions differ from planet to planet.

Waves can grow or decay by taking energy out of a plasma or increasing its energy. Since waves perturb the electric and/or magnetic fields permeating the plasma, they can be detected *in situ* using, e.g., dipole antennas, search coils, or other types of magnetometers. Plasma waves generally have frequencies that are at or below characteristic frequencies in the plasma (such as the plasma and hybrid frequencies discussed below). Such waves usually do not propagate very far. Plasmas can also generate electromagnetic waves at much higher frequencies, exceeding the highest characteristic frequency of the medium (electron plasma frequency). These waves are referred to as *radio waves*. Radio waves of sufficiently high frequency can escape the plasma in which they are produced, and hence these can be detected remotely. Waves which have frequencies that are well below the natural plasma frequencies are referred to as *magnetohydrodynamic (MHD) waves*; these are the lowest frequency waves in a plasma. Both *in situ* and remote observations of plasma and radio waves contain a wealth of information about the medium where they originate and through which they propagate. In this section we briefly discuss some properties of these waves. We do not derive equations or dive into the plasma physics, but merely give the reader a flavor of the rich variety of waves encountered in magnetospheres and the solar wind.

7.4.1 General Wave Theory

The dynamics of a magnetized plasma is controlled both by the electromagnetic field and the thermal pressure. For convenience, wave equations are often derived separately for cold and warm plasmas. In a *cold* plasma the thermal pressure can be ignored (which simplifies the equations), in contrast to that in a *warm* plasma. Usually perturbations in the electromagnetic field and/or the gas are small, and the governing equations can be linearized and solved for the wave properties. High-frequency perturbations propagate as plane waves of the form:

$$e^{i(\mathbf{k}\cdot\mathbf{x}-\omega_o t)} = \cos(\mathbf{k}\cdot\mathbf{x}-\omega_o t) + i\sin(\mathbf{k}\cdot\mathbf{x}-\omega_o t), \quad (7.74)$$

where $\mathbf{k}$ is the wave vector ($k = 2\pi/\lambda$, with λ the wavelength) and ω_o the angular wave frequency ($\omega_o = 2\pi\nu$, with ν the frequency of the wave). The argument of the exponential is constant if

$$x = x_o + \frac{\omega_o t}{k_x}. \quad (7.75)$$

Thus the solutions are constant at a position which moves with the *phase velocity* of the wave:

$$v_{\mathrm{ph}} = \frac{dx}{dt} = \frac{\omega_o}{k_x}. \quad (7.76)$$

The wave energy is carried at the *group velocity*:

$$v_g = \nabla_k\,\omega_o = v_{\mathrm{ph}} + \nabla_k v_{\mathrm{ph}}. \quad (7.77)$$

In a nondispersive medium, where the phase velocity does not depend on wavelength, the phase and group velocity of a wave are identical.

7.4.2 MHD, Plasma, and Radio Waves

Dispersion relations relate the angular wave frequency, ω_o, to the wave vector, $\mathbf{k}$. A familiar example of a dispersion relation is that of electromagnetic waves in free space, which propagate at the speed of light:

$$\omega_o = kc. \quad (7.78)$$

Similarly, sound waves in a gas propagate at the speed of sound, c_s.

In a cold plasma the dispersion relations can be derived from Maxwell's equations (§7.1.3) together with the MHD relations (§7.1.2.2). In a warm plasma one needs to use kinetic theory, where the dynamics of individual particles is taken into account. The dispersion relations show that a plasma can support a large variety of electromagnetic (waves involving perturbations in both $\mathbf{E}$ and $\mathbf{B}$), electrostatic (waves involving perturbations in $\mathbf{E}$ only), and

magnetoacoustic waves. Many waves are excited at or in resonance with the natural plasma frequencies, i.e., the electron and ion cyclotron frequencies ω_{Be} and ω_{Bi} given in equation (7.47), the electron and ion plasma frequencies ω_{pe} and ω_{pi}, which is the frequency at which electrons and ions oscillate about their equilibrium positions in the absence of a magnetic field:

$$\omega_p = \left(\frac{4\pi N q^2}{m}\right)^{1/2}, \quad (7.79)$$

with N the ion or electron density, m the ion or electron mass (m_i, m_e), and q the electric charge. A third set of natural frequencies is given by the upper and lower hybrid resonance frequencies, ω_{UHR} and ω_{LHR}:

$$\omega_{UHR} = \sqrt{\omega_{pe}^2 + \omega_{Be}^2}, \quad (7.80a)$$

$$\omega_{LHR} = \sqrt{|\omega_{Be}\omega_{Bi}|}. \quad (7.80b)$$

The maximum frequency at which a plasma can respond to waves is the electron plasma frequency, ω_{pe} (Problem 7.27). Waves at frequencies $\omega_o > \omega_{pe}$ escape, and can be observed remotely at radio wavelengths (§7.4.3). The dispersion relation for such high-frequency (radio) waves becomes

$$\omega_o^2 = \omega_{pe}^2 + k^2 c^2. \quad (7.81)$$

In free space, equation (7.81) reduces to equation (7.78). The phase velocity of radio waves depends upon the dielectric properties of the medium, since v_{ph} is inversely proportional to the index of refraction of the medium, n:

$$n = \frac{ck}{\omega_o} = \frac{c}{\lambda\nu}. \quad (7.82)$$

Using equations (7.81) and (7.82) one can relate the electron plasma frequency, and hence electron density, to the propagating radio wave (Problem 7.28):

$$n = \sqrt{1 - \frac{\omega_{pe}^2}{\omega_o^2}}. \quad (7.83)$$

The latter equation shows that radio waves do not propagate at or below the electron plasma frequency, when $n \to 0$, since waves cannot propagate when $n^2 < 0$. Hence by sweeping a receiver through frequency, one can determine the plasma frequency of the medium. If a wave moves through a medium in which ω_{pe} varies with position, such as altitude in an ionosphere, the wave will be reflected when $n = 0$. An *ionosonde*, transmitting and receiving radio signals at different frequencies (from the ground), uses this technique to determine the altitude profile of the electron density in Earth's ionosphere.

Plasma waves have frequencies at or below the characteristic plasma frequencies. These waves are generated and trapped locally. Plasma waves can be observed *in situ* by spacecraft. The waves are often named after the region where they occur (e.g., auroral hiss), the natural frequency of the plasma (e.g., upper and lower hybrid waves, electron and ion cyclotron waves), or the 'sound' they make when processed as audio signals and played through a speaker (whistlers, chorus, hiss, lion roar). In contrast, radio waves are usually named after their frequency range (e.g., kilometric radiation).

The lowest frequency waves that occur in a plasma are the magnetohydrodynamic (MHD) waves, which have frequencies that are well below the natural plasma frequencies. In a cold plasma, we find two modes of MHD waves. The first of these is the *shear Alfvén wave*, which is the regular Alfvén wave that causes field lines to bend. The perturbations are all perpendicular to **B**, like the propagation of a wave along a guitar string. These Alfvén waves propagate at a (phase) velocity:

$$v_{ph} = \frac{\omega_o}{k} = c_A \cos\theta, \qquad (7.84a)$$

with c_A the Alfvén velocity (eq. 7.28), and θ the angle between the wave vector **k** and the magnetic field **B**. In contrast to the phase velocity, the group velocity is directed along the magnetic field. A second wave, a *compressional wave*, propagates at the Alfvén velocity:

$$v_{ph} = \frac{\omega_o}{k} = c_A. \qquad (7.84b)$$

In contrast to the Alfvén wave, the compressional wave changes the density of the fluid and the magnetic pressure, like a sound wave. Since in a warm plasma, the plasma pressure may be comparable in strength to the magnetic pressure, the dispersion relation depends on both the sound speed and Alfvén velocity. This changes equation (7.84b) into two modes:

$$\left(\frac{\omega_o}{k}\right)^2 = \frac{1}{2}\left[c_s^2 + c_A^2 \pm \left((c_s^2 + c_A^2)^2 - 4c_s^2 c_A^2 \cos^2\theta\right)^{1/2}\right], \qquad (7.84c)$$

referred to as the *fast* (positive sign) and *slow* (negative sign) MHD or magnetoacoustic modes, respectively.

7.4.3 Radio Emissions

All four giant planets and Earth are strong radio sources at low frequencies (kilometric wavelengths). The strongest planetary radio emissions usually originate near the auroral regions and are intimately related to auroral processes. The average normalized spectra of the radio emissions from the four giant planets and Earth are shown in Figure 7.19. Jupiter is the strongest low-frequency radio source, followed by Saturn, Earth, Uranus, and Neptune.

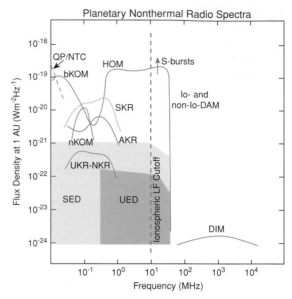

Figure 7.19 A comparison of the peak flux density spectrum of the radio emissions of the four giant planets and Earth. All emissions were scaled to that observed at a distance $r_\Delta = 1$ AU from the planet. Jovian emissions shown include quasi-periodic bursts (QP), nonthermal continuum (NTC), broadband and narrowband kilometric radiation (bKOM, nKOM), hectometric radiation (HOM), decametric radiation (DAM), and decimetric radiation (DIM). Saturn's kilometric radiation is designated SKR, and its electrostatic discharge emissions are labeled SED. Terrestrial auroral kilometric radiation is designated AKR. UKR and NKR refer to kilometric radiation from Uranus and Neptune, respectively. Uranus's electrostatic discharges are labeled UED. (Adapted from Zarka and Kurth 2005)

In this section we discuss general aspects (or background) of radio emissions. Emissions specific to each planet are summarized in §7.5.

7.4.3.1 *Low-Frequency Radio Emissions*
Cyclotron Maser Emissions

Radio emission at frequencies of a few kHz to 40 MHz (for Jupiter) is usually attributed to electron cyclotron maser radiation, emitted by keV (nonrelativistic) electrons in the auroral regions of a planet's magnetic field. The radiation is emitted at the cyclotron or Larmor frequency, ω_B (eq. 7.47). Propagation of the radiation depends on the interaction of the radiation with the local plasma. The oscillation of these particles, as caused by the electromagnetic properties of the plasma, leads to a complex interaction between the propagating radiation (the electromagnetic waves) and the local plasma. For example, the radiation can escape its region of origin only if the local cyclotron frequency is larger than the electron plasma frequency, ω_{pe} (eq. 7.83). This similarly sets the limit for propagation through Earth's ionosphere at ~10 MHz. If the local cyclotron frequency is less than the electron plasma frequency, the waves are locally trapped and amplified, until they reach a region from where they can escape. The cyclotron maser instability also requires a large ratio of ω_B/ω_{pe}. The auroral regions in planetary magnetospheres are characterized by such conditions. The mode of propagation (or polarization) of auroral radio emissions is in the so-called *extra-ordinary (X)* sense,[2] and the polarization (direction of the electric vector of the radiation) depends upon the direction of the magnetic field. The emission is right-handed circularly polarized (RH) if the field at the source is directed towards the observer, and left-handed circularly polarized (LH) if the field points away from the observer.[3]

Cyclotron radiation is emitted in a dipole pattern, where the lobes are bent in the forward direction. The resulting emission is like a hollow cone pattern, as displayed in Figure 7.20. The radiation intensity is zero along the axis of the cone, in the direction of the particle's parallel motion, and reaches a maximum at an angle Ψ. Theoretical calculations show that Ψ is very close to 90°. Observed opening angles, however, are often much smaller, down to ~50°, caused by refraction of the electromagnetic waves as they depart from the source region.

The cyclotron maser instability derives energy from few-keV electrons which have distribution functions with a positive slope towards the direction perpendicular to the magnetic field. Recent observations in the source of Earth's auroral kilometric radiation reveal 'horseshoe' shaped electron distributions that provide a highly efficient (of order 1%) source of free energy for the generation of the radio waves. This distribution is thought to be the result of parallel electric fields in the auroral acceleration region, the loss of small pitch-angle electrons to the planetary atmosphere, and trapping of reflected electrons. Radio emissions generated in planetary magnetospheres by this mechanism often display a bewildering array of frequency–time spectral peaks, including narrowband

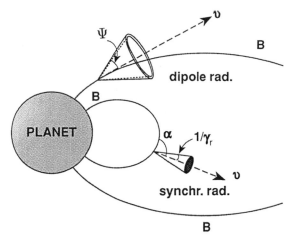

Figure 7.20 Radiation patterns in a magnetic field. The hollow cone pattern caused by cyclotron (dipole) radiation from nonrelativistic electrons near the auroral zone is indicated. The electrons spiral down along the planet's magnetic field lines. The hollow cone opening half-angle is given by Ψ. At low magnetic latitudes, in the radiation belts, the instantaneous radiation cone of a single relativistic electron is indicated. The angle between the particle's instantaneous direction of motion and the magnetic field, commonly referred to as the particle's pitch angle, α, is indicated on the sketch. The emission is radiated into a narrow cone with a half-width of $1/\gamma_r$.

emissions which rise or fall in frequency, sharp cutoffs, and more continuum-like emissions. While it is generally accepted that emissions that rise or fall in frequency are related to tiny sources moving down or up the magnetic field line (hence, to regions with higher or lower cyclotron frequencies), there is no generally accepted theoretical explanation for the fine structure.

Other Types of Low-Frequency Radio Emissions

While the radio emissions generated by the cyclotron maser instability are, by far, the most intense in any planetary magnetosphere, there are other types of low-frequency radio emissions that are also of interest. Perhaps the most ubiquitous of these is the so-called nonthermal continuum radiation which arises from the conversion of wave energy in electrostatic waves near the source plasma frequency to radio waves, usually propagating in the ordinary mode. The term 'continuum' was originally assigned to this class of emissions because they can be generated at very low frequencies and be trapped in low-density cavities in the outer portions of the magnetosphere when the surrounding solar wind density is higher. The mixture of multiple sources at different frequencies and multiple reflections off the moving walls of the magnetosphere tend to homogenize the spectrum.

At higher frequencies, emissions are created at the upper hybrid resonance (UHR) frequency on density

[2] We distinguish between propagation in the ordinary (O) and extra-ordinary (X) mode. The dispersion relation for O mode propagation does not depend on the magnetic field, whereas that in the X mode does.

[3] Circular polarization is in the RH sense when the electric vector of the radiation in a plane perpendicular to the magnetic field direction rotates in the same sense as an RH screw advancing in the direction of propagation. Thus, rotation is counterclockwise when propagation is toward and viewed by the observer. RH polarization is defined as positive; LH as negative. In some cases the radio emissions propagate in the ordinary (O) magneto-ionic mode. In this mode the polarization is reversed. The theory of the cyclotron maser instability does admit the possibility of emission in the ordinary mode. However, it is less common.

gradients in the inner magnetosphere. These emissions can propagate directly away from the source, and reveal a complex narrowband spectrum. They were first discovered at Earth and have been found at all of the magnetized planets, as well as Ganymede's magnetosphere.

Another type of planetary radio emission is closely related to a common solar emission mechanism, and involves *Langmuir waves*, which are electron plasma oscillations at the plasma frequency. Langmuir waves can be converted to radio emissions at either the plasma frequency or its harmonic, resulting in weak narrowband emissions. The Langmuir waves are common features of the solar wind upstream of a planetary bow shock.

Atmospheric Lightning

Radio emissions from planets are sometimes associated with atmospheric lightning. The lightning discharge, in addition to producing a visible flash, also produces broad, impulsive radio emissions. If the spectrum of this impulse extends above the ionospheric plasma frequency and absorption in the atmosphere is not too great, a remote observer can detect the high-frequency end of the spectrum. The 'interference' detected with an AM radio on Earth during a thunderstorm is the same phenomenon.

The most common waves that can be triggered by lightning discharges in an atmosphere are *whistler mode waves*, which have been observed in all five magnetospheres, as well as in Ganymede's magnetic field. Whistler mode waves have phase velocities that almost match the gyro motion of electrons. Since the highest frequency waves propagate fastest, the frequency of the waves reaching an observer decreases over time (Fig. 7.21), reminiscent of a whistling sound with a decreasing pitch. The rate at which the frequency decreases contains information on the plasma density.

7.4.3.2 *Synchrotron Radiation*

In contrast to the low-frequency emissions that are produced by keV electrons, synchrotron radiation is produced by relativistic (MeV energies, $v \approx c$) electrons gyrating around magnetic field lines. In essence, the emission consists of photons emitted by the acceleration of electrons as they execute their helical trajectories about the field lines. The radiation is strongly beamed in the forward direction (Fig. 7.20) within a cone of opening angle $1/\gamma_r$:

$$\frac{1}{\gamma_r} = \sqrt{1 - \frac{v^2}{c^2}}, \tag{7.85}$$

with v the particle's velocity and c the speed of light. The relativistic beaming factor $\gamma_r = 2E$, with E the energy in MeV. The radiation is emitted over a wide range of

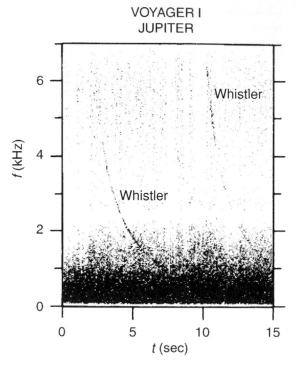

Figure 7.21 Spectrogram of whistler mode waves in Jupiter's magnetosphere, as observed by Voyager 1. More intense waves are represented by darker shading in the spectrogram. (Kurth 1997)

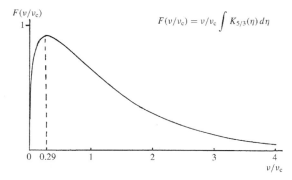

$$F(v/v_c) = v/v_c \int K_{5/3}(\eta)\, d\eta$$

Figure 7.22 Power spectrum of the synchrotron radiation emitted by a single electron trapped in a magnetic field. The $K_{5/3}$ is a modified Bessel function. (Ginzburg and Syrovatskii 1965)

frequencies (Fig. 7.22), but shows a maximum at 0.29 v_c, with the critical frequency, v_c, in MHz:

$$v_c = \frac{3}{4\pi} \frac{q\gamma_r^2 B_\perp}{m_e c} = 16.08 E^2 B_\perp, \tag{7.86}$$

where the energy E is in MeV and the field strength $B_\perp$ (component perpendicular to the line of sight) in G. The emission is polarized, with the direction of the electric vector depending on the direction of the local magnetic field. Jupiter is the only planet for which this type of emission has been observed. It has been mapped by ground based radio telescopes and by Cassini to provide some of the most comprehensive, though indirect, information about Jupiter's intense radiation belts.

7.5 Magnetospheres of Individual Bodies

7.5.1 Earth

Earth's magnetosphere is sketched in Figure 7.9. The bow shock, which results from the interaction of the super-magnetosonic solar wind with the Earth's magnetosphere, is located at ~15 $R_\oplus$ on the side facing the Sun. The turbulent subsonic region behind the bow shock is the magnetosheath. The magnetosheath is shielded from the Earth's magnetic field by the magnetopause, a boundary at ~10 $R_\oplus$ (along the Earth–Sun direction) which separates the solar wind plasma from the terrestrial magnetic field. The solar wind drags the field lines back and forms the magnetotail. The midplane, where magnetic fields of opposite polarity meet, is a region of field reversal, where reconnection takes place. The near-zero magnetic pressure here is balanced by a higher plasma pressure. This region is referred to as the current sheet, embedded in the plasma sheet. Closer to Earth we find regions of stable particle trapping, the plasmasphere and the radiation (Van Allen) belts, the latter containing energetic particles.

As described in detail in §7.3, charged particles in a magnetosphere follow helical paths around field lines, being reflected at their mirror points. The particles' drift around the Earth is caused by gradients in the magnetic field strength and curvature of the field lines. Their drift orbits are modified by the presence of electric fields, in particular the corotational and convection fields. For a pure dipole field, the particle trajectories can be expressed analytically; as planetary magnetic fields are observed to be complex, the particle trajectories need to be described numerically. In addition, the magnetosphere responds continuously to changes in the solar wind, complicating detailed modeling of the field and plasma therein.

7.5.1.1 *South Atlantic Anomaly*

In the inner radiation belts, the most important departures from a dipole field result from the higher order moments in the Earth's internal magnetic field. On the Earth's surface, the magnetic field is weakest near the east coast of South America, a feature of the Earth's field which is known as the *South Atlantic Anomaly*. Because particles drift around the Earth on paths along which the magnetic field is constant, they venture much closer to Earth's surface in the South Atlantic Anomaly than at other longitudes.

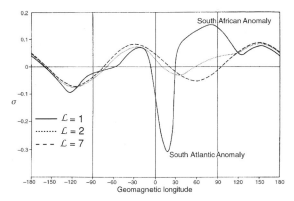

Figure 7.23 Deviation of a particle's drift path in the Earth's magnetic field as compared to the drift orbit in a pure dipole field, at different distances as indicated. The deviation is expressed by $\sigma = \mathcal{L}^2 \left(\frac{\partial^2 B/\partial s^2}{\partial^2 B/\partial s^2|_{\text{dipole}}} - 1 \right)$. (Adapted from Roederer 1972)

Figure 7.23 shows the deviation of drift contours of equatorially confined particles in the Earth's magnetosphere (including only Earth's internal magnetic field) compared to that in a pure centered dipole field, at distances $\mathcal{L} = 1$, 2, and 7. It is clear that the largest deviations occur close to the planet, and the region of the South Atlantic Anomaly stands out as one where the particle drift orbits are at much lower altitudes. The fact that energetic Van Allen belt particles pass Earth at such low altitudes in the South Atlantic Anomaly can substantially affect the performance of spacecraft in low Earth orbit as they pass through this region (Problem 7.18). At $\mathcal{L} = 7$, the largest variations are caused by the offset of the dipole field from the center of the Earth (note: this ignores the field modifications by magnetospheric currents, which are substantial at this distance).

As discussed in §7.3, particles may get lost in the atmosphere when they venture too close to Earth. Along a particle's drift orbit (Fig. 7.23), its loss cone varies with geocentric longitude. A particle's loss cone is largest near the South Atlantic Anomaly, and therefore all particles which have an equatorial pitch angle which is less than the loss cone at the South Atlantic Anomaly are removed from the magnetosphere during their drift around Earth. When a particle drifts around a planet in an offset dipole field, the particle's mirror point may lie well above the planet's atmosphere at certain longitudes, but within the atmosphere at others. If the field is displaced towards the north or south, the atmospheric loss cone is asymmetric. The largest loss cone a particle encounters during its drift orbit regulates the trapped particle distribution; this loss cone is referred to as the *drift loss cone*.

7.5.1.2 Magnetospheric Currents

Major current systems in the Earth's magnetosphere are indicated in Figure 7.9b. The magnetic gradient and curvature drift cause protons in the Earth's magnetic field to drift westwards and electrons eastwards around Earth. This drift induces a large-scale current, known as the *ring current*. Partial ring currents flow partway around the Earth in the middle magnetosphere. The ends of the partial rings are connected to the ionosphere via *field-aligned* or *Birkeland currents*, where currents in the ionosphere complete the circuit. In the magnetopause, at latitudes below the polar cusp, we find the *Chapman–Ferraro currents*, with the *tail current* at higher latitudes in the tail (Fig. 7.9b). The convection electric field induces a dawn-to-dusk current across the magnetotail, near the equatorial plane, referred to as the *neutral sheet current* in Figure 7.9b. Since electric currents induce magnetic fields, the various current systems can cause significant perturbations in the geomagnetic field compared to the internal field structure. For example, the ring current induces a southward field inside the particle orbits, which weakens the magnetic field here, as verified from measurements on Earth's surface. The Chapman–Ferraro current induces a northward field, which strengthens the geomagnetic field in particular on the day side, whereas the westward tail current tends to weaken the magnetospheric field in the tail. These latter two currents thus cause a systematic difference between the day- and night-side magnetosphere, which affects particle drift orbits. Moreover, the effect is different for particles with different pitch angles. Particles with small equatorial pitch angles drift further from the Earth on the night side than on the day side, while the opposite is true for particles which are confined to the magnetic equatorial plane. This phenomenon of *drift-shell splitting* was discussed in §7.3.2.1.

7.5.1.3 Magnetospheric Plasma

The Earth's magnetic field is populated mainly by protons, electrons, and ions of oxygen, helium, and nitrogen. This composition indicates that the particles originate from both the solar wind and the ionosphere. The various plasma regions are indicated in Figure 7.9. Magnetosheath plasma penetrates to low altitudes in the polar cusps, which are effectively dents or gaps in the magnetopause. The open field lines in the polar caps provide a route for magnetosheath plasma to enter the magnetosphere and thereby populate the high-latitude regions. Upon reflection at their mirror points, cusp particles can enter the geomagnetic tail and form the plasma mantle (together with magnetosheath plasma which 'leaks' through the tail magnetopause). As a result of the solar wind induced convection electric field,

there is a large-scale magnetospheric 'circulation', such that particles at low latitudes are driven in the direction of the Sun, with an antisolar flow at higher latitudes. The particles in the plasma mantle thus drift towards the midplane, while the particles in this current sheet either drift towards the Earth (when Earthwards from the neutral X point, where tail reconnection is occurring – Fig. 7.17), or away from it. In the latter case, the particles ultimately join the solar wind. Plasma in the current sheet is accelerated and fills the plasma sheet with hot plasma, as discussed below. A *polar wind*, consisting of low-energy protons and singly ionized oxygen atoms, flows out from the high-latitude ionosphere. Like the polar wind, the *plasmasphere* inside $\mathcal{L} \lesssim 4$ is filled with cold ionospheric plasma.

Plasma Sheet and Current Sheet

Typical plasma parameters in the current sheet, the central part of the plasma sheet, are $\sim$0.3 particles cm^{-3}, with ion energies of a few up to a few tens of keV, and electron energies a factor of 2–3 lower. This is much higher than the typical energy of magnetosheath (a few hundred eV) or suprathermally escaped ionospheric (a few eV) plasma. Hence, the plasma in the plasma and current sheets must be energized somehow. Since the convection electric field is parallel to the neutral sheet current, it energizes the particles carrying this current. Thus, the particles in the current sheet derive their energy from a slowing down of the solar wind. The energy input is $\sim$2–20% of the total energy flux carried by the solar wind through an area equal to the cross-section of the day-side magnetopause (§7.3.3.2; Problem 7.13). To understand how the energy is transferred to the magnetospheric particles in the plasma sheet, consider the particle trajectories in the magnetotail. Since the magnetic field changes dramatically across the current sheet (**B** changes direction and is $\approx$ 0 at the center), conservation of the adiabatic invariants is no longer valid. Above and below the current sheet, the particles undergo their normal helical motion around the field lines. However, when they approach the region of field reversal, their circular motion changes depending on the direction of the field. This causes an oscillatory motion of the particle within the current sheet, whereby the particle gets essentially trapped in this region. Electrons move towards the dawn and protons towards the dusk side. The dawn-to-dusk convection electric field accelerates the particles in this direction, thus energizing them. If **B** is indeed zero in the midplane, the particles are trapped and continually energized, until they reach the edge of the current sheet and are lost from the magnetotail. Energetic particles are then only found in the current sheet, not in the broader plasma sheet. However, the magnetic field has a small northward component in the midplane, which deflects the particles in the current sheet towards the Earth and out of the midplane. The energy input of particles in the current sheet thus forms an important energization mechanism for all particles in the plasma sheet. Only particles entering the current sheet at the Earth side of reconnection points are energized and trapped in the magnetosphere; particles entering the current sheet in the tail side of the reconnection point are carried down the tail and lost from the magnetotail.

Plasmasphere

The *plasmasphere* is located inside $\mathcal{L} \lesssim 4$, the same region in space as occupied by the radiation or Van Allen belts. The plasmasphere is separated from the plasma sheet by the *plasmapause*. The plasmasphere is filled with cold dense plasma from the ionosphere. The low-energy ($\lesssim 1$ keV) particle motion is dominated by the corotational and convection electric fields, while the motion of high-enery ($\gtrsim 100$ keV) particles is dominated by the gradient **B** drift (§7.3.3). On the dusk side, when the velocity of particles orbiting the Earth becomes equal to the velocity of particles in the direction of the Sun, as induced by the convection electric field, a 'bulge' arises in the plasmasphere (Fig. 7.18), as observed. The strength of the convection electric field fluctuates in response to changes in the interplanetary magnetic field and solar wind. Therefore, the size and shape of the plasmasphere are not constant over time. If the convection electric field is suddenly increased, the plasmapause moves inwards, so that particles which were originally inside the plasmasphere are now on trajectories which cause them to drift towards the day-side magnetopause. It takes about 20 hours for an outer layer to 'peel off'. During this time, the bulge of the plasmasphere rotates somewhat towards local noon.

The more energetic particle drifts are dominated by the gradient in the magnetic field, which drives positively charged particles towards the west and electrons towards the east. This gives rise to the ring current, as discussed above (§§7.3.2.1, 7.5.1.2). The more energetic particles therefore cannot penetrate as close to the Earth as the lower energy (colder) particles, so that in general the hotter plasma from the plasma sheet does not penetrate the plasmasphere. However, radial diffusion of plasma, caused by fluctuations in electric fields (e.g., the convection electric field), tends to decrease any radial gradients in the particle populations.

In addition to a reduction in size of the plasmasphere in response to an increase in the convection electric field,

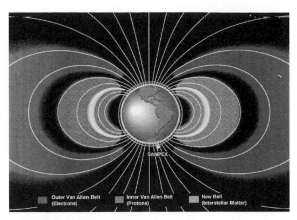

Figure 7.24 Illustration of the Earth's Van Allen belts. The inner (centered at $\mathcal{L} \approx 1.5$) and outer (centered at $\mathcal{L} \approx 4$) belts are characterized by a bimodal distribution in electron energies. In this particular representation, a narrow belt within the inner belt (at $\mathcal{L} = 2$), comprised of anomalous cosmic rays (ACR), is indicated, as seen following the intense solar storm in 1991. (Mewaldt *et al.* 1997)

the plasmapause is deformed, moving in closer to Earth at the night side, but further away from Earth at the dawn and dusk sides. This deformation is coupled to the ionosphere through field-aligned (Birkeland) currents, which flow up from the ionosphere at the dawn side, and down into the ionosphere at dusk (Fig. 7.9). These form the partial ring current in the Earth's middle magnetosphere, and induce a dusk-to-dawn directed electric field across the inner magnetosphere, which effectively shields the inner magnetosphere from the convection (dawn-to-dusk) electric field.

Van Allen Belts

The Earth's radiation belts were discovered in 1958 by James A. Van Allen from data obtained by Explorer 1, the first satellite launched by the USA. The Van Allen belts are a region in space filled with energetic particles which can penetrate deep into dense materials and thus cause damage to spacecraft instruments and humans. All the particles in these belts contribute to the ring current, discussed above. The Van Allen belts consist of two main belts (Fig. 7.24): The inner belt, centered around $\mathcal{L} \approx 1.5$, is characterized by highly energetic protons and electrons (Fig. 7.25), whereas the outer belt, centered near $\mathcal{L} \approx 4$ contains no energetic protons. There is a region at $\mathcal{L} \approx 2.2$ where the number density of energetic electrons is a minimum, caused by an interaction of the electrons with whistler mode waves, where the electrons (otherwise trapped in the Van Allen belts) are effectively 'scattered' into the atmosphere (Figs. 7.24, 7.25).

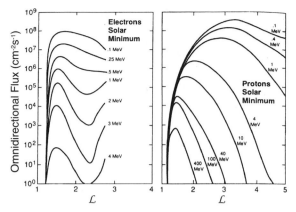

Figure 7.25 Spatial distribution of energetic electrons and protons in the Van Allen belts during a quiescent phase in solar wind activity. (Adapted from Wolf 1995)

There are a variety of sources which contribute to the particle population in the Earth's radiation belts, the ionosphere and the solar wind chief among them. The Earth's ionosphere is the primary source of oxygen and nitrogen, while both the ionosphere and solar wind contribute to the proton density. During times of low solar activity, the O^+/H^+ ratio in the plasma sheet is low, while during periods of enhanced magnetic activity, the concentration of oxygen ions is nearly as high as that of H^+. This suggests that the particle population in the magnetosphere is dominated by solar wind particles during quiescent times, while during active times a large fraction of the particles comes from the Earth's ionosphere.

As pointed out earlier, typical solar wind particles have too low an energy to contribute to the most energetic particles (hundreds of MeV) in the inner magnetosphere. Protons with such high energies may originate from energetic neutrons, produced in the atmosphere through collisions with cosmic rays. When these neutrons leave the atmosphere, they may decay while en route through the magnetosphere. This decay process creates highly energetic protons (few 100 MeV) and electrons of lower energies (few 100 keV).

High-energy protons and electrons may also be injected and/or accelerated by shocks, such as those induced by ICMEs or solar flares (§7.1). Following the extreme intense solar storm in March 1991, the Solar Anomalous Magnetospheric Particle Explorer (SAMPEX) discovered a 'new' radiation belt at $\mathcal{L} \approx 2$, filled with highly energetic multiply charged ions (primarily O, with some N, Ne, and a little C) and electrons. The relative abundances of the O, N, Ne, and C ions were similar to that of interplanetary anomalous cosmic rays (ACR), which are singly charged ions with energies up to several

tens MeV/nucleon (galactic cosmic rays are fully stripped, and have much higher energies). This 'new' radiation belt persisted for many months before it slowly dissipated.

ACRs may be produced when interstellar atoms near the Sun get ionized (by solar UV radiation or by collisions with energetic interplanetary particles). Once ionized, the particles are swept away from the Sun by the solar wind. At the heliospheric boundary they may be accelerated again, after which they can re-enter the heliosphere as low-energy (up to a few 100 MeV) cosmic rays. The Voyager spacecraft, however, did not observe a change in the energy spectrum or in the intensity of the ACR flux near the termination shock at ~90 AU (§1.1.6), so the precise acceleration mechanism of ACRs remains mysterious. The correlation between these ACRs and heavy ions in the magnetosphere, however, seems to be well established. In addition to similar C:N:O abundance ratios, the heavy-ion density in the Earth's magnetosphere correlates well with the interplanetary ACR flux. Both vary over time by factors of up to 3–4, and are anticorrelated with the solar sunspot number.

7.5.1.4 Magnetospheric (Sub)Storms

The highly variable solar wind and polarity changes in the interplanetary magnetic field (IMF) make the Earth's magnetosphere a dynamic environment. In particular when an ICME approaches Earth while the IMF is turned southwards (i.e., opposite to the Earth's field, Fig. 7.17), the magnetosphere may undergo major, albeit temporary, changes. The ordered sequence of changes in the magnetosphere during such times is referred to as a *magnetospheric substorm*, which typically lasts for about an hour, and may repeat on a ~few-hour timescale. One of the first visible phenomena during a substorm is enhanced auroral activity, with the auroral zone moving to lower latitudes (§4.6.4).

When a strong ICME hits the Earth's field while the IMF is turned southward (and keeps this polarity for hours to days), the Earth's surface field strength may decrease by up to several percent. Such geomagnetic or magnetospheric disturbances are known as *magnetic storms*. A magnetic storm is usually characterized by the D_{st} index, the instantaneous worldwide average of the change in magnetic field strength. A typical example of a magnetic storm signature is shown in Figure 7.26. Initially, the Earth's surface field strength increases slightly when the interplanetary shock, followed by the ICME (enhanced dynamic pressure), first hits the magnetosphere (and induces an increase in the magnetopause current). Reconnection of magnetic field on the day side (southward IMF) causes the solar wind to transport magnetic flux to

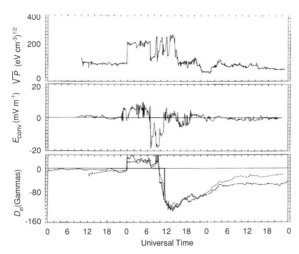

Figure 7.26 Effects of a magnetic storm on Earth (solid line, bottom panel), as recorded in the D_{st} index on 15–17 February 1967. The top panel shows the solar wind dynamic pressure, and the middle panel the solar wind dawn-to-dusk electric field. The dashed line in the bottom panel is a prediction of the D_{st} index based upon the velocity and density of the solar wind and the north–south component of the interplanetary magnetic field. (Adapted from McPherron 1995)

the night side, which puts much stress on the magnetotail. Reconnection in the tail (Fig. 7.17) triggers a flow towards the Earth (restoring magnetic flux on the day side) and in the antisolar direction. Sometimes such flows contain 'bubbles' of plasma embedded in the magnetic field, known as *plasmoids*.

Several hours after the shock hits, the D_{st} decreases. This is known as the storm's *main phase*, which typically lasts for about a day. The decrease in D_{st} is caused by an increase in the ring current, resulting from the enhanced plasma flow towards the Earth. The subsequent recovery phase can last for many days, and is caused by a gradual loss of particles from the radiation belts, due, e.g., to pitch angle scattering into the loss cone by enhanced wave activity and radial diffusion. Although these large-scale phenomena can be explained in a qualitative way, no theories have yet been developed which explain the detailed sequence of all events triggered during a magnetospheric substorm.

An exceptionally strong solar storm occurred on 11 November 2003. During this time the outer radiation belt was displaced inwards from about $\mathcal{L} \approx 4$ to 2.5, for a period that lasted 2 weeks. The inner radiation belt got filled with high-energy electrons to a degree that had last been seen during the 1991 solar storm (§7.5.1.3). The plasmapause, as expected during times of high magnetic activity, was much closer to the Earth than at

quiescent times. During the recovery phase, the outer plasmasphere refilled gradually, over a period of a couple of days. Although the conventional view has been that such strong magnetic (sub)storms happened during a southward-oriented IMF, the THEMIS mission discovered (in June 2007) that a tremendous flux of particles entered the Earth's magnetosphere while the Sun's field was aligned with the Earth's field (§7.3.4.1).

7.5.1.5 Radio Observations

Earth is also a source of nonthermal radio emissions, which have been studied both at close range and larger distances by many Earth-orbiting satellites. *Auroral kilometric radiation* (AKR), the terrestrial version of cyclotron maser emission (§7.4.3.1), is very intense; the total power is 10^7 W, sometimes up to 10^9 W. The intensity is highly correlated with geomagnetic substorms, thus it is indirectly modulated by the solar wind. It originates in the night-side auroral regions and in the day-side polar cusps at low altitudes and high frequencies, and spreads to higher altitudes and lower frequencies. Typical frequencies are between 100 and 600 kHz. As AKR is generated by auroral electrons, it can serve as a proxy for auroral activity. And, since numerous *in situ* studies of the terrestrial auroral electron populations and the resulting radio emissions have been carried out, we can apply our understanding of this emission process to similar emissions at other planets where *in situ* studies have not yet been carried out.

Earth is also a source of nonthermal continuum radiation. Below the solar wind plasma frequency this radiation is trapped within the magnetosphere (§7.4.3.1). The spectrum is relatively smooth down to the local plasma frequency, typically in the range of a few kHz. The low-frequency limit of the continuum radiation at the plasma frequency provides an accurate measure of the plasma density, a quantity that is often difficult to measure directly because of spacecraft charging effects. Above the solar wind plasma frequency, the continuum radiation spectrum exhibits a plethora of narrowband emissions, from a few tens of kHz sometimes up to several hundred kHz.

7.5.2 Mercury

Based on two close encounters by the Mariner 10 spacecraft, we know that Mercury possesses a small Earth-like magnetosphere. The magnetic axis is within 10° of Mercury's rotational axis. Under usual solar wind conditions, Mercury's intrinsic field is strong enough to stand off the solar wind well above its surface (1.3–2.1 $R_{\mathcal{Q}}$). However, at times of increased solar wind pressure, the interplanetary particles may impinge directly onto Mercury's surface.

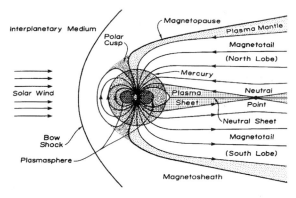

Interplanetary Medium
Polar Cusp
Solar Wind
Bow Shock
Plasmasphere
Magnetopause
Plasma Mantle
Magnetotail
(North Lobe)
Mercury
Plasma Sheet
Neutral Point
Neutral Sheet
Magnetotail
(South Lobe)
Magnetosheath

Figure 7.27 Mercury and Earth, scaled such that their magnetospheres occupy the same volume. A large fraction of an Earth-like inner magnetosphere and radiation belts would reside below Mercury's surface. (Russell *et al.* 1988)

In contrast to the similarity in magnetosphere morphology between Earth and Mercury, both being 'shaped' by the interaction with the solar wind and in which plasma flows are dominated by the solar wind induced convection, the differences make Mercury's field unique. Tiny Mercury occupies a much larger fractional volume of its magnetosphere than do Earth and the giant planets This implies that the stable trapping regions we see in other planetary magnetospheres, the radiation belts, cannot form. A sketch of Mercury's magnetosphere is shown in Figure 7.27, where, in dotted lines, the radiation belts and plasmasphere in the Earth's magnetic field are overlain. As shown, these would lie below Mercury's surface.

Another important difference between Mercury and Earth is the near absence of an atmosphere and ionosphere. An ionosphere usually affects a planet's electric and magnetic fields, thus indirectly the transport of charged particles. It further serves as a source of plasma in a magnetosphere (§7.3.4). The solar wind is expected to be the primary source of Mercury's magnetospheric plasma, although planetary ions released or sputtered from the surface make some contribution according to ground-based observations of emissions from sodium and other elements (§§4.3.3.3, 4.8.2). The plasma densities in the plasma sheet are higher than in the Earth's magnetosphere by roughly the difference in solar wind density at Mercury and Earth's orbits, i.e., by a factor of ~10. The plasma sheet almost touches Mercury's surface near midnight. The convection electric field may cause particles to diffuse from the magnetotail towards the planet, where it is swept out into the solar wind through the magnetopause at the day side. However, these theories are just extensions from our knowledge of the terrestrial magnetosphere. We will not be able to accurately describe or fully understand

the physical processes within Mercury's magnetic field until the spacecraft Messenger and BepiColombo map out Mercury's environment.

7.5.3 Venus, Mars, the Moon
7.5.3.1 *Venus*
Venus does not possess an internal magnetic field, but the interaction of the solar wind with Venus's ionosphere induces a magnetic field that produces an obstacle to the solar wind flow, as discussed in §7.1.4. Charge exchange deposits solar wind energy in the atmosphere, and mass (heavy ions) in the solar wind. Photoionization by UV sunlight ionizes atmospheric atoms, which in turn 'mass load' the solar wind, slowing it down. The interplanetary field lines in the magnetosheath are draped around Venus and form an induced magnetotail behind the planet. The tail consists of two lobes of opposite polarity, and is very similar in appearance to the Earth's magnetotail, except that its 'polarity' is controlled by the draped IMF orientation, like a comet's tail. Without a dipole field, there are no durably trapped particles around Venus.

7.5.3.2 *Mars*
The interaction of the solar wind with Mars is quite similar to that with Venus in some respects, but differs in others. The most surprising result from the magnetic field experiment on Mars Global Surveyor (MGS) was the detection of localized very intense magnetic fields. The strongest field measured ~16 mG at an altitude of 100 km, which, in combination with the ambient ionospheric pressure, is strong enough to stand off and deflect the solar wind at Mars. Several plasma boundaries are formed in this interaction, including a bow shock where solar wind plasma is slowed, a magnetopause-like boundary that shields the upper atmosphere from solar wind protons, and an ionopause below which the planetary ion density increases rapidly in nonmagnetized regions. As at Venus, solar wind magnetic field lines are compressed and drape around the planetary obstacle below the bow shock.

The localized magnetic fields on Mars are caused by remanent crustal magnetism. Most of the strong sources are located in the heavily cratered highlands (Fig. 7.28), south of the crustal dichotomy. There is evidence for little or no crustal magnetization inside some of the younger giant ($\gtrsim1000$ km) impact basins (e.g., Hellas, Utopia and Argyre), while older similarly large basins (e.g., Ares, Daedalia) are quite strongly magnetized. The strong crustal fields perturb the location of the obstacle boundary as they rotate through the sunlit side of the planet.

By estimating the age of these basins using the density of smaller subsequently formed craters, it can be argued

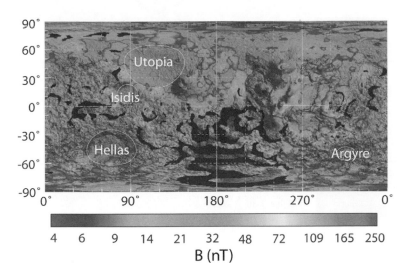

Figure 7.28 COLOR PLATE Smoothed magnetic map of Mars constructed from Electron Reflectometer data from Mars Global Surveyor (MGS). The logarithmic color scale represents the crustal magnetic field magnitude at an altitude of 185 km, overlaid on a topography map as derived from laser altimeter data on MGS. The lower limit of the color scale is the threshold for unambiguously identified crustal features, while the scale saturates at its upper end. Black represents sectors with fewer than 10 measurements within a 100 km radius. These regions are areas where there is closed crustal magnetic field and so the solar wind electrons cannot penetrate to the altitude of the spacecraft where they can be detected. The four largest visible impact basins are indicated (dotted circles). (Adapted from Lillis *et al.* 2008)

that during the first few hundred million years of its history, Mars had a geodynamo with a magnetic moment comparable to, or larger than, Earth's dynamo at present. While this dynamo still operated, ancient martian crust (formed from cooling magma or post-impact melts) acquired a substantial magnetization when it cooled below the Curie point of its magnetic minerals. When Mars's core-to-mantle heat flux decreased sufficiently to stop turbulent convection in the core, dynamo action ceased and the global magnetic field dissipated within ~50 000 years, between 3.9 and 4.1 Gyr ago. Subsequent reheating and shocking of the crust by volcanism and large impacts erased any prior crustal magnetization in those regions (e.g., the Tharsis volcanic province in Fig. 7.28).

Figure 7.28 further shows an interesting 'striped' magnetic pattern at west longitudes between about 120° and 210°, where bands of alternating magnetic polarity up to 2000 km long are seen. On Earth similar magnetic lineations are found, though at sea level, along the mid-ocean rift, and are associated with sea floor spreading and repeated reversals of the Earth's dipole field. Hence, this observation has been interpreted by some researchers as suggesting that Mars may have had plate tectonics during the first few hundred million years where, like on Earth, magma from below filled the void created by two plates moving apart. Upon cooling through the Curie point, the new crust would have Mars's magnetic polarity imprinted. The horizontal scale length of the magnetic lineations on Mars is about 100 km. If the plates moved apart at a rate of about 8 cm yr^{-1}, the frequency of magnetic field reversals on Mars would be comparable to that seen on Earth. But of course, both the rate of plate movement and frequency of field reversals may have been very different 4.5

Gyr ago, both for Mars and the Earth. A drawback of this interpretation is that no spreading center has been found, such as that along the mid-ocean ridge where the magnetic polarity is symmetric at either side.

An alternative explanation for the stripy magnetic field pattern is that the once intact magnetized crust 'broke up' into a series of long narrow 'plates', similar perhaps in appearance to the linear fractures or rilles near the Tharsis region (§5.5.4.1). Magma from below would fill the cracks and dipole magnetic field patterns would bridge the gaps between broken plates (like when breaking a bar magnet in pieces). One thus would end up with hundreds to thousands of kilometer long 'tracks' of dipolar-like fields. The regions in between these tracks are similar to magnetospheric 'cusp' regions (as in the Earth's polar regions) at the altitude of the solar wind obstacle boundary. Solar wind plasma can enter these cusp regions freely, while it is excluded from the crustal dipole-like magnetic field regions. Results from MGS and the more recent Mars Express spacecraft indeed suggest that solar wind plasma penetrates down to low altitudes (below 200 km) in magnetic cusp regions, while no such plasma is present at these altitudes above areas of strong crustal fields. On the night side, penetration of energized electrons to low altitudes creates aurora, as observed by Mars Express.

7.5.3.3 *The Moon*

The Moon shows strong localized patches of surface magnetic fields of a few tens up to 2500 µG. These patches appear to be correlated with the antipodal regions of large young impact basins, such as the Crisium, Serenitatis, and Imbrium basins (§5.5.1). There is also an interesting association between these patches and 'swirl' albedo markings.

Figure 7.29 An image of Io's plasma torus imaged in S⁺ (top) and neutral sodium cloud (bottom). Jupiter is shown in the center of each image; the planet was imaged through a neutral density filter. (Courtesy Nick M. Schneider and John T. Trauger)

The latter have been hypothesized to be caused by partial shielding of the surface from space weathering effects by solar wind ions (though the relative importance in space weathering of micrometeorites versus ions is still being debated).

The presence of magnetic field patches in regions antipodal to impact basins suggests that the crustal magnetization is associated with the formation of large impact basins. The hypervelocity impacts that formed these basins likely produced a plasma cloud that surrounded the Moon within minutes of impact, which would have compressed and amplified any pre-existing magnetic field at the antipode. Seismic waves and impact ejecta arrived at the antipode within tens of minutes, upon which the crust got shocked and magnetized in the amplified field. This model requires the Moon to have had a global magnetic field, and hence a magnetic dynamo, 3.9–3.6 Gyr ago. Paleomagnetic data obtained from returned Apollo samples are also indicative of an early magnetic field with a surface field strength $\sim$0.1–1 G. Given recent estimates on the presence and size of the Moon's iron core (§6.3.1), it appears as if a lunar magnetic dynamo almost 4 Gyr ago is quite plausible.

7.5.4 Jupiter

The presence of a magnetic field surrounding Jupiter was first postulated in the late 1950s after the detection of non-thermal radio signals from the planet (§7.4.3). Later, but well before spacecraft traversed Jupiter's magnetosphere, neutral sodium atoms were detected in the vicinity of Io through optical emissions, soon followed by ground-based detections of potassium and ionized sulfur (Fig. 7.29). As of 2010, seven spacecraft have flown past Jupiter (Pioneer

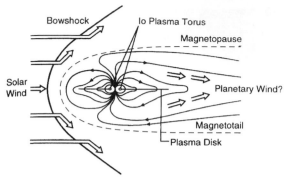

Figure 7.30 A sketch of Jupiter's magnetosphere. (Kivelson and Bagenal 1999)

10 and 11, Voyager 1 and 2, Ulysses, Cassini, New Horizons) and one, the Galileo spacecraft, orbited the planet for almost 8 years. These have studied the planet's magnetic field and the plasma environments of the Galilean satellites in detail, either through remote sensing and/or *in situ*. Intensive monitoring programs from near-Earth and ground-based telescopes enriched the available database. In contrast to Earth (and Mercury), where large-scale plasma flows are primarily driven by the solar wind induced convection electric field, plasma flows in the magnetospheres of Jupiter and Saturn are dominated by the corotational electric field.

7.5.4.1 *Magnetic Field Configuration*

The general form of Jupiter's magnetosphere resembles that of Earth, with a primarily dipole-like field tilted by $\sim$10° with respect to the rotation axis, but its dimensions are over three orders of magnitude larger. If visible to the eye, Jupiter's magnetosphere would appear several times larger in the sky than the Moon. The magnetotail extends to beyond Saturn's orbit; at times, Saturn is engulfed in Jupiter's magnetosphere. A graphical representation of the magnetosphere is shown in Figure 7.30 and quantities characterizing the field are listed in Table 7.2. The magnetosphere is usually divided into an inner ($\lesssim$10 R$_{2\!\!\!/}$), middle (10–40 R$_{2\!\!\!/}$), and outer ($\gtrsim$40 R$_{2\!\!\!/}$) magnetosphere. Analogous to the Earth's Van Allen belts, there are *radiation belts* close to the planet, at $\mathcal{L} \lesssim 2.5$. These are filled with electrons, protons, and helium ions. Coinciding with the orbit of Io is a plasma torus (§7.5.4.4). At around 20 R$_{2\!\!\!/}$, in the middle magnetosphere, corotation of plasma breaks down; the plasma lags more and more behind corotation with increasing distance from the planet. The outer magnetosphere is a large disk-shaped region, characterized by plasma that is flowing outwards due to centrifugal forces, forming the plasma- or magnetodisk.

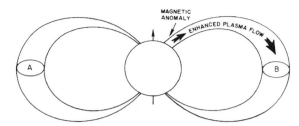

Figure 7.31 Flux tubes emanating from Jupiter's 'surface': the flux tubes A and B have the same cross-sectional areas at the equator. Flux tube B is anchored in a region of weak magnetic field, A to a strong magnetic field. (Hill *et al.* 1983)

When subtracting a best-fit displaced dipole from Jupiter's magnetic field, deviations from a dipole field can readily be identified. There are regions of weak and strong magnetic field strength on its 'surface'. Several field lines are sketched in Figure 7.31. The *flux tubes* A and B have the same cross-sectional areas in the magnetic equatorial plane. However, flux tube B is anchored in a region of weak magnetic field, while flux tube A is connected to a strong field. The footprint of flux tube B is therefore much larger than that of flux tube A, so that the flow of ionospheric plasma through flux tube B is enhanced compared to that through flux tube A. This is the essence of the *magnetic anomaly model* developed to explain many observed phenomena in Jupiter's magnetosphere. In this model, the magnetic anomaly is the depression in magnetic field strength in Jupiter's northern hemisphere, centered near a jovian longitude,[4] $\lambda_{III} \approx 260°$. This region is referred to as the *active sector*.

Jupiter's magnetic anomaly affects the distribution of plasma in the jovian magnetosphere. As mentioned above, the plasma flow from the ionosphere is enhanced in the active sector compared to that at other longitudes, as well as the height-integrated conductivity of the jovian ionosphere. In addition, the ionization rate due to particle bombardment is largest in the active sector because of the reduced mirror altitude, which leads to a larger particle loss cone. This explains why jovian aurora are usually brightest at longitudes in the active sector. In addition to the longitudinal asymmetry of aurora, the magnetic anomaly model explains many phenomena with observed longitudinal asymmetries. For example, there is a well-known 'clock' modulation of relativistic electrons in the interplanetary medium. These electrons seem to originate at Jupiter. The 'clock' modulation can be explained since release of electrons into interplanetary space occurs

primarily through the tail. There is a maximum in the energetic electron population escaping into interplanetary space when the active sector faces the tail.

7.5.4.2 *Hydrogen Bulge*

The resonantly scattered hydrogen Ly α line is enhanced at longitudes about 180° away from the active sector, near the magnetic equator (or more specifically the particle's drift equator). This suggests the presence of a 'mountain of atomic hydrogen', generally referred to as the *hydrogen bulge*. This feature is unique to Jupiter. Ionization of neutrals and thus mass-loading into the Io plasma torus is largest in the active sector. Centripetal forces push the plasma outwards, inducing an electric field pattern that corotates with the planet. This field induces a corotating convection pattern, such that plasma moves outwards in the active sector and inwards at longitudes ~180° away from it. The inward convection causes hot magnetospheric plasma to impinge on Jupiter's atmosphere and dissociate CH_4 and H_2 into atomic hydrogen, leading to the formation of the hydrogen bulge.

7.5.4.3 *Io's Neutral Clouds*

Although sodium and potassium are only trace elements of Io's neutral cloud, these atoms are easily excited by resonant solar scattering and were therefore the first to be detected. The main constituents of the neutral clouds are oxygen and sulfur atoms, but observations of these are more difficult. The sodium cloud is shaped like a banana and pointed in the forward direction (Fig. 7.32). Since the neutral atoms follow Keplerian orbits, those closest to Jupiter travel fastest, which results in the forward pointed banana-shaped cloud.

Sublimation of SO_2 frost from Io's surface, volcanism and sputtering create a tenuous, yet collisionally thick atmosphere around the satellite (Fig. 7.32; §4.3.3.3). Ions corotating with Jupiter's magnetosphere have typical velocities of 75 km s^{-1}, and hence readily overtake Io, which orbits Jupiter at the Keplerian speed of 17 km s^{-1}. The ions interact with Io's atmosphere through a collisional cascade process. Some of the atmospheric molecules escape directly into the neutral cloud, others first form a 'sputter corona' around the satellite (see also §4.8.2).

The extent of the neutral cloud is determined by the lifetime of the individual atoms (Problem 7.20). A neutral can get ionized through photoionization, electron impact ionization, elastic collisions, or charge exchange. Photoionization of neutrals near Io is very slow: typical lifetimes for O and S are a few years, for SO_2 it is about a year, and for Na it is about one month. These times

---

[4] Jupiter's central meridian longitude system, CML III, is based on the rotation period of its magnetic field (§7.5.4.8), and the longitude of the magnetic north pole, λ_{np}, is at $\lambda_{III} = 201°$.

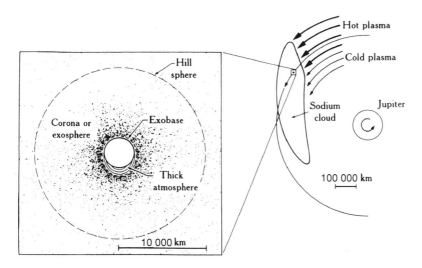

Figure 7.32 A schematic representation of Io's sodium cloud and atmosphere/corona. The right side shows the sodium cloud and the corotating plasma, indicated by arrows. The left side (enlarged by a factor of 40 compared to the right side) illustrates Io's immediate vicinity: the collisionally thick atmosphere is shown by contours (on the sunlit hemisphere). The thin corona or exosphere is indicated by dots. The dashed line shows the location of the Hill sphere. (Schneider et al. 1987)

are much longer than the survival times against the other processes, which are collisional in nature. Although it is relatively straightforward to calculate the collisional rates in the low-density regions far from Io, it is more complicated in the immediate vicinity of the satellite. Lifetimes of neutrals against electron impact ionization depend upon the density and temperature of the electrons. The particle lifetimes are generally quite short; sodium and potassium typically last for about an hour, oxygen for 55 hours, and sulfur for about 10 hours. The newly created ion is accelerated to corotational velocities and, although lost from the neutral cloud, adds to the mass and density of the Io plasma torus, discussed in detail in the next section. Since the plasma is cooler inside of Io's orbit, electron impact ionization is less effective here, so that Io's neutral cloud lies preferentially inside Io's orbit.

Depending on the plasma density, elastic collisions and charge exchange can also terminate the life of a neutral particle. An oxygen atom is more likely to be ionized by charge exchange (typical lifetime is 18 hr) or an elastic collision (13 hr) than by electron impact ionization (55 hr). In an elastic collision, a fast ion impinges on a slow neutral, sending it off at a high velocity nearly perpendicular to the plasma flow (the exact outcome depends upon the impacting angle). In a charge-exchange reaction the corotating ion strips off an electron from a neutral. The neutral is ionized and accelerated to corotational speed, while the former ion becomes a very fast neutral, flying out through the magnetosphere.

Detailed images in the neutral sodium line show jets, fans, and loops; these are created by fast sodium atoms, particles traveling at tens of kilometers per second. On a much larger scale, Jupiter is enveloped by a giant disk-shaped sodium cloud, extending out to a few hundred

$R_?$. This cloud is likely formed by fast neutrals flung out from the magnetosphere via charge exchange or elastic collisions, between particles in the Io plasma torus and neutral clouds (§7.5.4.4). Thus, although electron impact ionization, elastic collisions, and charge exchange form sinks for slow neutrals, the latter two processes also provide a source of fast neutrals. Another major source of fast neutrals may be dissociative recombination and dissociation of sodium-bearing molecules (such as, perhaps, NaCl).

7.5.4.4 Io Plasma Torus
Both *in situ* and ground-based observations at visible and UV wavelengths have provided a wealth of information on the heavy ions in Jupiter's magnetosphere. The ions are concentrated in a torus around the planet, centered near Io's orbit at 5.9 $R_?$ (Fig. 7.29). The plane of symmetry is the centrifugal equator, which goes through the field lines where they are furthest away from Jupiter. This plane makes roughly an angle of 3° with the magnetic equator and 7° with the rotational equator. The exact equilibrium point depends upon the mass and energy of the ion. The vertical extent of the torus depends upon the temperature and mass of the ions. To a first approximation the plasma density $N(z)$ decreases exponentially with distance, z, away from the equator:

$$N(z) \approx N_o e^{-(z^2/H^2)}. \tag{7.87}$$

In equation (7.87), the scale height $H \approx \sqrt{2kT_i/(3m_i n^2)}$, n is the orbital rotation rate, and m_i and T_i the mass and temperature of ion species i, respectively.

The main species detected in Io's plasma torus are ions of oxygen (up to O^{3+}), sulfur (up to S^{4+}), chlorine (Cl^+, Cl^{2+}), and sulfur dioxide (SO_2^+). The excitation process, through which the ions can be detected, is through collisions with electrons. Both forbidden[5] (at optical wavelengths) and allowed transitions have been observed, and used to determine the electron density and temperature in the torus. Typical maximum electron densities are a few thousand cm^{-3}. Estimates to maintain the plasma torus as observed suggest a production rate of $\sim 10^{28}$–10^{29} ions s^{-1}.

Measurements of the particle density and temperature reveal a cold inner torus (a few eV) inside of 5.7 R$_{\jupiter}$ (dropping off sharply inside 5.3 R$_{\jupiter}$) and the hot outer torus (~ 80 eV) up to 7–8 R$_{\jupiter}$. The hot outer torus is much broader in latitude than the cold inner torus (Problem 7.21). While emissions from the cold inner torus are confined to the optical wavelength range, both optical and UV emissions have been observed from the hot outer torus. With Io being the source of this plasma, material must be transported both inward and outward from the satellite.

Radial transport outward from Io is rapid (10–100 days), aided by the centrifugal forces on the plasma. Details of this process are not understood. Some researchers favor centrifugally driven interchange, where denser flux tubes slip outwards, changing places with less-dense flux tubes (§7.3.5). A relatively thin sheet of warm (10–100 eV) plasma, dominated by sulfur and oxygen ions, fills the plasma sheet. Densities decrease from several thousand particles cm^{-3} in the torus to a few particles cm^{-3} near 20 R$_{\jupiter}$. The outward-flowing plasma in the plasma sheet generates a dawn-to-dusk directed electric field over Jupiter's magnetosphere (note: opposite to solar wind induced dusk-to-dawn field), which has a number of observable effects, as discussed below. Radial transport inward from Io is slow, allowing ample time for the ions to cool. This diffusion is probably driven by currents in Jupiter's ionosphere.

Ground-based observations of the torus via emissions from S^{+} ions (Fig. 7.29) show temporal variations in intensity and phenomenology. Part of the variability can be attributed to the changing viewing geometry of the planet's magnetic field, but the time-averaged brightness distribution of the torus shows a clear east–west asymmetry in intensity and location. The torus exhibits a maximum in intensity at the dusk (west, receding) side, and a minimum at the dawn (east, approaching) side. The peak intensity at the dawn side is always located on Io's average $\mathcal{L}$-shell, while the maximum intensity at the dusk side is shifted inwards towards Jupiter by ~ 0.4 R$_{\jupiter}$. These features are interpreted as being due to the dawn-to-dusk electric field, generated by the anti-sunward flow of plasma that

originates in the Io torus and flows down the tail. The electric field accelerates particles during one half of their drift orbit, causing an inward motion of the dusk end of the torus. The torus brightens here due to adiabatic compression of the electrons that excite the torus emissions. The particles lose their energy during the second half of their drift orbits, so the dawn side stays unaffected.

In addition to the east–west asymmetries in the torus, both the intensity and phenomenology of the torus change over time. The peak in torus density near 5.7 R$_{\jupiter}$, forming the 'ribbon' (left side in Fig. 7.29), shows up very clearly in some years, while it is nearly absent in other years. The torus and ribbon are always brightest in the 'active sector'. The intensity of the neutral torus and its jet-like features of fast sodium are also highly time variable.

7.5.4.5 Positive Feedback Mechanism?

Increased volcanic activity on Io would enhance the neutral cloud density, and hence the Io plasma torus. The charged particles in the torus form one component of particles which collide with the satellite and replenish the neutral clouds through sputtering. An increased plasma density thus increases the sputtering rate, leading to a positive feedback mechanism, or runaway model to supply the plasma torus with material. However, even though the ion density in the plasma torus has been observed to increase following an outburst in the sodium emission, the magnetosphere somehow imposes a stabilization mechanism, preventing the plasma torus from runaway growth.

7.5.4.6 Coupling to Ionosphere

Coupling between the magnetosphere, plasma torus, and Jupiter's ionosphere occurs via field-aligned currents, which connect the plasma torus to Jupiter's ionosphere, as shown graphically in Figure 7.33 (see also Fig. 7.41). The existence of such currents has been 'shown' directly via HST images of auroral emissions along the wake of the footpoint of Io's flux tube (Fig. 7.39), and they have been detected *in situ* by the Galileo spacecraft (§4.6.4.2). Through these currents, the ionosphere tries to enforce corotation throughout the magnetosphere. Indeed, inside of ~ 5 R$_{\jupiter}$ the magnetosphere is observed to be in rigid corotation with the planet, but at larger distances there is a significant departure from corotation. At jovian distances of 6–10 R$_{\jupiter}$ the plasma lags behind corotation by ~ 1–10%, while beyond 20 R$_{\jupiter}$ the azimuthal flow is at a constant speed of about 200 km s^{-1}.

There is a complicated three-way coupling between the ionosphere, magnetosphere, and plasma torus. Heating by sunlight causes the weakly ionized gases in the

[5] See footnote [5] §10.4.3.1.

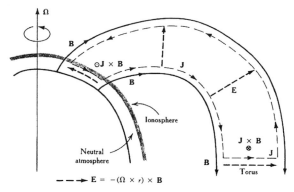

$$E = -(\Omega \times r) \times B$$

Figure 7.33 Sketch of the coupling between Jupiter's magnetosphere and ionosphere along Io's flux tube: the Birkeland current system between Io and Jupiter's ionosphere. (Bagenal 1989; adapted from Belcher 1987)

ionosphere to flow, resulting in currents and electric fields which affect particle motions in the magnetosphere. Conversely, mass-loading and radial motions of particles in the magnetosphere also drive currents in the ionosphere, and particle precipitation modifies the ionospheric conductivity. Any currents in the magnetosphere and ionosphere can be expressed by Ohm's law (eq. 7.15a). In the ionosphere the electrical conductivity, σ_0, perpendicular to the magnetic field lines is large, but just above the ionosphere in the magnetosphere $\sigma_0 \approx 0$. The conductivity along the magnetic field lines, however, is large, and the field lines can be considered equipotentials ($E \cdot B = 0$). Any plasma injected into the magnetosphere is accelerated up to corotation with the planet, since the newly charged particles feel the corotational electric field (directed outwards for Jupiter), which induces an $E \times B$ drift around the planet. Thus the particles are accelerated up to corotational speeds.

The corotational electric field also induces a radial current outwards, which brings about a small radial separation between electrons and positively charged ions, thus inducing an electric field opposing the corotational field. The (outwards) radial current forms the Birkeland current, and must close in Jupiter's ionosphere (Fig. 7.33); the $J \times B$ force brings the plasma into corotation. However, the ionosphere is not a perfect conductor, so an electric field is required to force the current through. This field maps into the magnetosphere as a field opposing the corotational electric field, which is thus equivalent to reducing the corotation speed.

In general, whenever J/σ_0 becomes significant in an ionosphere, corotation may break down. Such situations may be induced, for example, if the plasma density in a magnetosphere increases significantly locally, such as

caused by a mass-loading process, or if the conductivity in the ionosphere changes significantly, e.g., by particle precipitation or a meteorite impact. Interestingly, the impacts of Comet D/Shoemaker–Levy 9 with Jupiter in 1994 (§5.4.5) did not cause any observable changes in the torus emissions.

7.5.4.7 Jupiter's Synchrotron Radiation

Synchrotron radiation from Jupiter is received at wavelengths between a few cm and a few m (frequencies $\gtrsim 40$ MHz). The variation in total intensity and polarization characteristics during one jovian rotation (so called beaming curves, Fig. 7.34) indicated already in the 1960s that Jupiter's magnetic field is approximately dipolar in shape and inclined by $\sim 10°$ with respect to the rotation axis, with most electrons confined to the magnetic equatorial plane. The total flux density of the planet was found to vary significantly over time (Fig. 7.35). To some extent these variations appear to be correlated with solar wind parameters, in particular the solar wind ram pressure, suggesting that the solar wind may influence the supply and/or loss of electrons into Jupiter's inner magnetosphere. In addition to variations in the total flux density, the radio spectrum changes as well (Fig. 7.36).

An image of Jupiter's synchrotron radiation obtained with the Very Large Array (VLA) in 1994 is shown in Figure 7.37a. This image was obtained at a wavelength of 20 cm and has a spatial resolution of $\sim 6''$ or 0.3 R_J. The two main radiation peaks, L and R, result from the line-of-sight integration through the ring of radiating electrons (sketched in Figure 7.34b). Since Jupiter's synchrotron radiation is optically thin, one can use tomography to extract the three-dimensional distribution of the radio emissivity from data obtained over a full jovian rotation. The example in Figure 7.37b shows that most of the synchrotron radiation is concentrated near the magnetic equator, which, due to the higher order moments in Jupiter's field, is warped like a potato chip. The secondary emission regions, apparent at high latitudes in Figure 7.37a, show up as rings of emission north and south of the main ring. These emissions are produced by electrons at their mirror points, and reveal the presence of a rather large number of electrons bouncing up and down field lines that thread the magnetic equator at $\mathcal{L} \sim 2.5$. This emission may be 'directed' by the moon Amalthea. A fraction of the electrons near Amalthea's orbit undergo a change in their direction of motion, caused perhaps by interactions with low-frequency plasma waves near Amalthea (such plasma noise was detected by the Galileo spacecraft when it crossed Amalthea's orbit), and through interactions with dust in Jupiter's rings, while regular

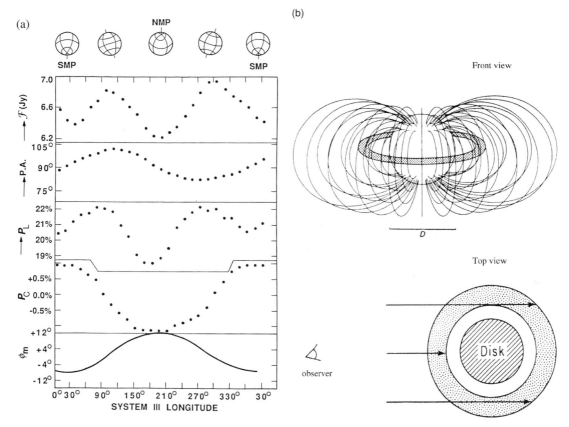

Figure 7.34 (a) An example of the modulation of Jupiter's synchrotron radiation due to the planet's rotation. The orientation of the planet is indicated at the top; the different panels show the total flux density $\mathcal{F}$, the position angle P.A. of the electric vector, the degree of linear and circular polarization P_L and P_C, and the magnetic latitude of the Earth, ϕ_m. This latitude can be calculated with: $\phi_m = D_E + \theta_B \cos(\lambda - \lambda_{np})$, with D_E the declination of the Earth, θ_B the angle between Jupiter's magnetic and rotational axes, λ the central meridian longitude, and λ_{np} the central meridian longitude of the magnetic north pole. (de Pater and Klein 1989) (b) A schematic representation of the energetic electrons in Jupiter's magnetic field, as seen from the front and top. (de Pater 1981)

synchrotron radiation losses also lead to small changes in an electron's direction of motion.

Figure 7.36 shows radio spectra of Jupiter's synchrotron radiation from 74 MHz up to $\gtrsim 20$ GHz. The energy distribution of electrons is usually expressed by a power law:

$$N(E)dE \propto E^{-\zeta} dE, \tag{7.88}$$

so that the flux density of the observed radiation depends on frequency:

$$\mathcal{F}_\nu \propto \nu^{-(\zeta-1)/2}. \tag{7.89}$$

Jupiter's radio spectra show that the electrons in Jupiter's radiation belts do not follow a simple $N(E) \propto E^{-\zeta}$ power law. Well outside the synchrotron radiation region, beyond Io's orbit at 6 R_2, the electron energy

spectrum appears to follow a double power law, $N(E) \propto E^{-0.5}(1 + E/100)^{-3}$, consistent with *in situ* measurements by the Pioneer spacecraft. Radial diffusion, pitch angle scattering, synchrotron radiation losses, and absorption by moons and rings all change the electron spectrum. The radio spectra superposed on the data were derived from such models.

As shown in Figure 7.35, Jupiter's total synchrotron flux density varies significantly over time. A significant ($\sim$20%) increase was seen when Comet D/Shoemaker–Levy 9 collided with Jupiter (§5.4.5). During the same time, the radio spectrum hardened and the spatial brightness distribution of the radio emissions changed considerably. These changes were essentially brought about by the atmospheric explosions of the cometary fragments. These explosions triggered the propagation of shocks and

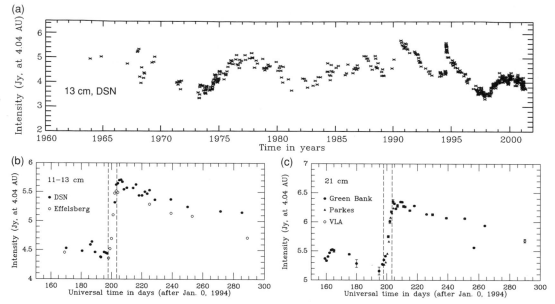

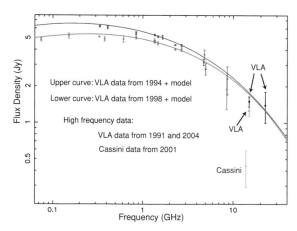

Figure 7.35 Time variability in Jupiter's radio emission. (a) Graph of the radio intensity at a wavelength of 13 cm from 1963 through early 2001. The data after 1969.2 are NASA/JPL Jupiter Patrol observations made by M.J. Klein using the antennas of NASA's Deep Space Network (see Klein *et al.* 2001). The data up to 1968.7 were taken with the Parkes and Nancey radio telescopes. The spike in 1994 was caused by the impact of Comet D/Shoemaker–Levy 9 with Jupiter. (b, c) An expanded view of the SL 9 impact period is shown in the two lower panels, where data taken at 11–13 cm (panel b) and 21 cm (panel c) are shown. Note that the intensity of the 11 cm data is slightly less than that at 13 cm (see Figure 7.36). The impact of Comet D/Shoemaker–Levy 9 with Jupiter occurred 16–22 July 1994, and is indicated by vertical dashed lines on panels (b) and (c). (The data in the latter panels were taken by Klein *et al.* (1995), Bird *et al.* (1996), and Wong *et al.* (1996))

Figure 7.36 Jupiter's radio spectrum as measured in June 1994, September 1998, and July 2004. Superposed are various model calculations. (Adapted from Kloosterman *et al.* 2007)

electromagnetic waves into the magnetosphere, which in turn affected the radiating electrons, e.g., by causing a sudden million-fold enhancement in particle diffusion (see, e.g., Harrington *et al.* 2004, and references therein for details).

7.5.4.8 *Jupiter at Low Frequencies*

Jupiter has the most complex low-frequency radio spectrum of all the planets. Examples are shown in Figure 7.38 and are discussed below.

Decametric (DAM) and Hectometric (HOM) Emissions

Jupiter's DAM (decametric) emission, confined to frequencies below 40 MHz, has routinely been observed from the ground since its discovery in the early 1950s. Frequencies as low as 4 MHz have been observed occasionally. The upper frequency cutoff is determined by the local magnetic field strength in the auroral regions: 40 MHz for RH (right-handed circularly polarized) emissions translates into ~14 gauss in the north polar region, and 20 MHz for LH into ~7 gauss in the south.

The dynamic spectra in the frequency–time domain are extremely complex, but well ordered. On timescales of minutes, the emission displays a series of arcs, like 'open' or 'closed parentheses' (Fig. 7.38). Within one storm, the arcs are all oriented the same way, and have been interpreted as coherent cyclotron emissions. The

(a)

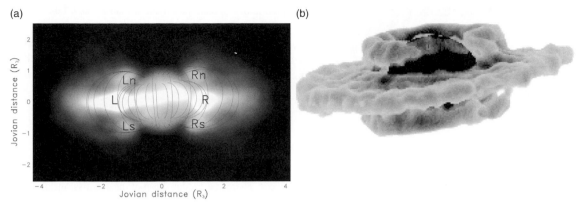

Figure 7.37 (a) Image of Jupiter's flux density at a wavelength of 20 cm, taken with the VLA at a longitude $\lambda_{III} \sim 312°$. The spatial resolution is 0.3 R_J. Several magnetic field lines (at $\mathcal{L} = 1.5$ and $\mathcal{L} = 2.5$) are superposed. (Adapted from de Pater et al. 1997) (b) A 3-D representation of the apparent radio emissivity of the planet. The data were taken with the VLA at a wavelength of 20 cm. The central meridian longitude is 140°. (de Pater and Sault 1998)

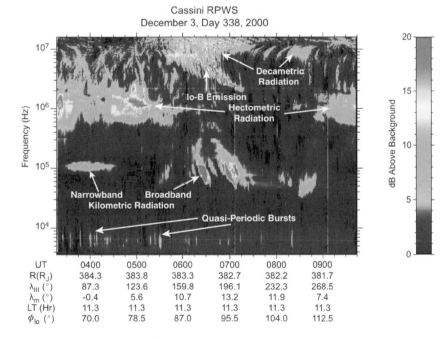

Figure 7.38 COLOR PLATE Representative dynamic spectra of Jupiter's low-frequency radio emissions, taken by the Cassini spacecraft. Color is used to show the intensity of the emission in the time–frequency plane. Indicated along the x-axis are: UT: UT time; R(R$_J$): distance in R$_J$; λ_{III}: jovian central meridian longitude (in degrees); λ_m: magnetic latitude (in degrees); LT (Hr): Local Time; ϕ_{Io}: Io phase (in degrees). (Adapted from Lecacheux 2001)

satellite Io appears to modulate some of the emissions: both the intensity and the probability of the occurrence of bursts increase when Io is at certain locations in its orbit with respect to Jupiter and the observer. The non-Io emission originates near Jupiter's aurora, and is produced by electrons that travel along magnetic field lines from the middle-to-outer magnetosphere towards Jupiter's ionosphere. Particles that enter the atmosphere are 'lost'. These may locally excite atoms and molecules through collisions, which upon de-excitation are visible as aurora at UV and infrared wavelengths (§4.6.4.2). Other electrons

are reflected back along the field lines, and produce DAM, where their motion along the field line is reflected in the form of arcs in the radio emission, i.e., a drift with frequency. The Io-dependent emissions are produced at or near the footprints of the magnetic flux tube passing through Io (similar, but much weaker, emissions originate along the flux tubes passing through Ganymede, and perhaps Callisto) (Fig. 7.39).

HOM emissions are, in many ways, indistinguishable from DAM except that they are found at lower frequencies, from a few hundred kHz to a few MHz, with a local

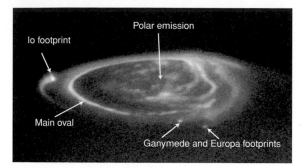

Figure 7.39 HST image of Jupiter showing the auroral footprints from the Galilean satellites at UV wavelengths, as indicated. Also the main oval and polar emissions are labeled. The image was taken on 26 November 1998. (NASA/ESA, John Clarke)

maximum near 1 MHz. The source region of HOM must be further from Jupiter than the DAM source. Otherwise, like DAM, HOM is predominantly emitted in the extraordinary mode and is likely generated by the cyclotron maser instability.

Since the dipole moment of Jupiter is tilted by ∼10° from the rotational axis, most jovian radio emissions exhibit a strong rotational modulation. Given that Jupiter is a gas giant, this modulation is thought to be the best indicator of the rotation of the deep interior of the planet. The rotation period of the interior is important, for example, because this provides a rotating coordinate system against which the atmospheric winds can be measured. Based upon early radio observations, in 1965 the IAU defined Jupiter's internal rotation period to be 9 h 55 m 29.71 s; this is referred to as the System III (1965) period (System I refers to the rotation period of clouds near Jupiter's equator ($-10° <$ latitude $< +10°$), and System II to the period at latitudes $< -10°$ and $> +10°$). In 1997 this period was refined, based upon 35 years of ground-based DAM observations (9 h 55 m 29.685 s $\pm$ 0.004 s); in 2009 a rotation period of 9 h 55 m 29.704 s $\pm$ 0.003 s was derived from the rotation of the (magnetic) dipole field, using data from the Pioneer, Voyager, Ulysses, and Galileo spacecraft over a timespan of 25 years.

Kilometric Radio Emissions (KOM)

Between a few kHz up to 1 MHz, various spacecraft detected both broadband (bKOM) and narrowband (nKOM) kilometric radiation from Jupiter (Fig. 7.38). The lower frequency cutoff for bKOM, ∼20 kHz (sometimes down to ∼5 kHz), is likely set by propagation of the radiation through the Io plasma torus. The source of these emissions is at high magnetic latitudes, and appears fixed in local time. The forward lobe near the north magnetic

pole is of opposite polarization than a 'back lobe' of the same source. The nKOM emissions last longer (up to a few hours) than bKOM, are confined to a smaller frequency range, 50–180 kHz, and show a smooth rise and fall in intensity. The recurrence period for nKOM events suggests the source lags behind Jupiter's rotation by 3–5%, which was the first indication that this emission, in contrast to other low-frequency emissions, is produced by distinct sources near the outer edge of the plasma torus. Galileo and Ulysses studies have shown that these emissions occur as a part of an apparently global magnetospheric dynamic event. There is a sudden onset of these emissions, they are visible for a few to several planetary rotations, and finally, they fade away.

Very Low Frequency Emissions (VLF)

The Voyager spacecraft detected continuum radiation in Jupiter's magnetosphere at frequencies below 20 kHz, both in its escaping and trapped form. As discussed in §7.4.3, radiation can be trapped inside the magnetic cavity if it cannot propagate through the high-plasma-density magnetosheath. This trapped emission has been observed from a few hundred Hz up to ∼5 kHz. Occasionally it has been detected up to 25 kHz, suggesting a compression of the magnetosphere caused by an increased solar wind ram pressure. The lower frequency cutoff of the freely propagating radiation outside the magnetosphere corresponds to the plasma frequency in the magnetosheath and appears to be well correlated with the solar wind ram pressure. This escaping component is characterized by a complex narrowband spectrum, attributed to a linear or nonlinear conversion of electrostatic waves near the plasma frequency into freely propagating electromagnetic emissions. The linear mechanism favors ordinary mode radiation, but the trapped emission appears to be a mix of both ordinary and extraordinary radiation, perhaps from the multiple reflections off high-density regions in the magnetosphere and at the magnetopause.

The quasiperiodic (QP), or jovian type III emissions (in analogy to solar type III bursts, because of their similar dispersive spectral shape) often occur at intervals of 15 and 40 minutes as observed by Ulysses; however, neither Galileo nor Cassini found particularly dominant periodicities in this radiation (Fig. 7.38). The emission likely originates near the poles. Simultaneous measurements by the Galileo and Cassini spacecraft, both in the solar wind but at different locations, observed similar QP characteristics, suggestive of a strobe light pattern rather than a search light rotating with the planet. Within the magnetosphere, the QP bursts can then appear as enhancements of the continuum emission. At the magnetosheath, the lower frequency components of the bursts are dispersed by the higher density plasma, which produces the characteristic type III spectral shape. The 40-minute QP bursts were correlated with energetic (1 MeV) electrons observed by Ulysses. Chandra detected similar periods in X-rays from the auroral region, although not directly correlated with QP bursts themselves. Such observations suggest that the QP bursts are related to an important particle acceleration process, but the details of the relationship and the details of the process remain elusive.

7.5.4.9 Magnetic Field Perturbations by the Galilean Satellites

The Galilean satellites orbit Jupiter while embedded deep in the planet's magnetosphere (distances vary from 6 R_J for Io up to $\sim$27 R_J for Callisto). The immediate environments of the Galilean satellites have been observed *in situ* by the Galileo spacecraft. One major surprise was the discovery of pronounced changes in the jovian magnetic field in the neighborhood of all four Galilean satellites. The spacecraft observed a large depression in the ambient magnetic field strength when crossing Io's plasma wake, which could be reconciled if Io possesses a magnetic field of its own, with a surface magnetic field strength of $\sim$17 mG, anti-aligned with Jupiter's magnetic field. This interpretation is not unambiguous, however, since the magnetometer data would show similar magnetic disturbances induced by current systems in Io's vicinity.

Ganymede, in contrast, is a moon that has been unambiguously shown to possess an intrinsic magnetic field, as attested to by both the Galileo magnetometer and plasma wave results. The magnetometer data reveal a magnetic field with an equatorial surface strength of 7.6 mG, tilted 10° with respect to its spin axis. The dipole moment is anti-aligned with Jupiter's magnetic moment. The Galileo spacecraft revealed a rich plasma wave spectrum (Fig. 7.40), similar to that expected from a planetary magnetosphere. It also is the source of nonthermal narrowband radio emissions at 15–50 kHz, very similar to the escaping continuum emissions from Jupiter. The more intense cyclotron maser emission, seen from the auroral regions of all giant planets and Earth, is absent, however. This is almost certainly because the electron plasma frequency is greater than the cyclotron frequency, hence, the cyclotron maser instability does not operate.

The ambient magnetic field near Europa changed in strength and direction when Galileo passed the satellite. The signature can be modeled as the electromagnetic response to Jupiter's time-varying magnetic field provided

there is a layer of electrically conducting material below Europa's surface. Although Jupiter's magnetic field itself is roughly constant over time, the projection of the field into Europa's equatorial plane as seen from the satellite varies with Jupiter's rotation. Such a time-varying **B** field can induce currents in an electrically conductive medium (eq. 7.13). Such currents, in turn, generate a secondary magnetic field, which is observed as a distortion in Jupiter's **B** field. The magnetic field distortions as measured during multiple close passes of Galileo to Europa, together with images of Europa's cracked icy surface, suggest that the electrically conducting shell is likely a $\sim$100 km deep salty ocean immediately below Europa's crust (§§5.5.5.2, 6.3.5.1).

In contrast to the pronounced signatures detected by the magnetometer when Galileo passed Io, Europa, and Ganymede, only a small change in the field strength was detected near Callisto. Like in the case of Europa, the latter signature can be simulated well if Callisto has a deep salty ocean, a plausible model given constraints on its interior structure (§6.3.5.1).

7.5.5 Saturn

Saturn's magnetosphere is intermediate in extent between those of Earth and Jupiter. The strength of the magnetic field at the equator is a little less than that found on the Earth's surface. However, Saturn is roughly 10 times larger than Earth and it is roughly 10 times further away from the Sun; both of these factors lead to a substantially larger magnetosphere around Saturn than around Earth. Most remarkable is the nearly perfect alignment between Saturn's magnetic and rotational axes. The center of its dipole field is slightly shifted towards the north, by 0.04 R_h.

7.5.5.1 Mass Loading by Enceladus

Five mid-sized icy satellites (Mimas, Enceladus, Tethys, Dione, and Rhea) are located in Saturn's inner magnetosphere, between 3 and 9 R_h. Saturn's main ring system, composed predominantly of water-ice, lies inside 2.27 R_h (§11.3.2). Sputtering by charged particles and meteorite bombardment leads to the formation of an oxygen-rich neutral cloud (H_2O, OH, O, H) and plasma torus surrounding Saturn that was observed by all passing spacecraft. However, the Cassini spacecraft revealed large ($\gtrsim$50%) variations in the oxygen density in the vicinity of Saturn's E ring, which have been attributed to changes in geyser activity on Enceladus (§5.5.6.2).

The first direct evidence for geysers on Enceladus came from Cassini's magnetometer data, which indicated

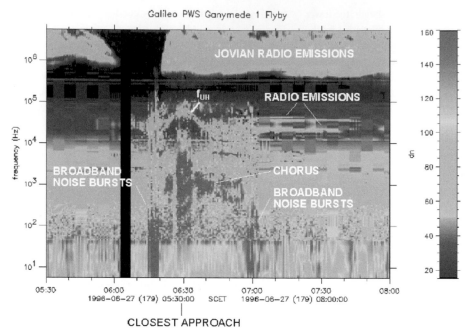

Figure 7.40 COLOR PLATE Dynamic radio spectra (frequency vs. SCET, spacecraft event time) taken during one of Galileo's close approaches to Ganymede. The strong interaction between Ganymede and the jovian magnetosphere provides strong evidence of a small magnetosphere surrounding Ganymede. The band of noise labeled f_{UH} is at the upper hybrid resonance frequency and has been used to determine a plasma density of ∼100 particles cm^{-3}. The broadband bursts at the beginning and end of the interaction period are typical of the plasma wave signature for a magnetopause. The banded emissions after closest approach are electron cyclotron harmonic emissions, which are known at Earth to contribute to the generation of the aurora. The bright, broadband emission centered on closest approach and the emissions identified as 'chorus' in the spectrogram are whistler-mode emissions. The maximum frequency of these emissions translates into a maximum magnetic field strength traversed by Galileo of about 4 mG. The narrowband radio emissions extending primarily to the right of the Ganymede interaction in the spectrogram are the first known radio emissions from a planetary satellite; these are similar to radio emissions studied at Earth and the outer planets, including Jupiter. (Gurnett *et al.* 1996)

a bending or draping of field lines around the moon (Fig. 7.41), indicative of mass-loading processes such as observed on Io (§7.5.4.3). Charge exchange interactions dominate mass loading from the Enceladus plume into magnetospheric plasma; electron impact and photoionization play a much smaller role. The newly formed ions are accelerated up to corotation speed by field-aligned currents (§§7.5.4.4, 7.5.4.6; Fig. 7.33), draining angular momentum from Saturn's ionosphere.

One might expect the integrated mass of ions on flux tubes near the orbit of Enceladus to be larger than that on flux tubes further away, leading to a process known as flux tube or centrifugal interchange. Such an instability near Enceladus may lead to a two-cell convection pattern, as sketched in Fig. 7.41c. Such patterns have been invoked in the past to explain a variety of rotational modulations in magnetospheres (e.g., Jupiter's magnetic anomaly model,

leading to the hydrogen bulge, discussed in §§7.5.4.1–7.5.4.2). A longitudinal asymmetry in the plasma density near Enceladus's orbit, such as might be induced by the above convection pattern, has in fact been observed by the Cassini spacecraft.

The dominant source of the neutral gas cloud in Saturn's magnetosphere is the geyser plume on Enceladus. As magnetospheric plasma moves through the neutral gas (from a to b in Fig. 7.41c), its density increases through ionization of the neutrals (mass loading, dm/dt, Fig. 7.41b). In the frame rotating with the plasma disk, the centrifugal force $F_c(2)$ is therefore larger than $F_c(1)$. This triggers a radial outflow of plasma in the 'heavy' sector of the convection pattern, a flow also referred to as a 'plasma tongue'. The rotation of the plasma disk is coupled to the ionosphere via field-aligned currents (§7.5.4.6; Figs. 7.33, 7.41b). Mass loading and/or changes in

(a)

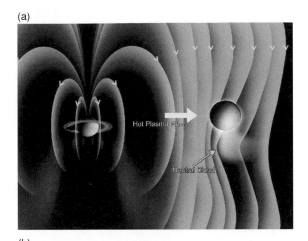

(b)

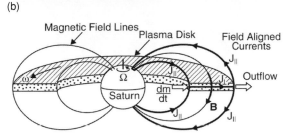

(c)

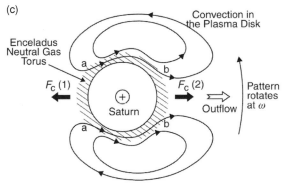

Figure 7.41 (a) Artist's conception of the bending of magnetic field lines around Enceladus, resulting from electric currents generated by the interaction of the Enceladus plume and the magnetospheric plasma. The magnetometer data have been used to infer the presence of a neutral cloud being vented from the south pole of Enceladus. (NASA/JPL, PIA07370) (b) Coupling of Saturn's plasma disk to its ionosphere via field-aligned currents (compare with Fig. 7.33). The plasma disk rotates at an angular rotation rate ω. (c) Corotating (with plasma disk) two-cell plasma convection pattern, that may induce a longitudinal modulation in plasma density. While plasma flows through the neutral gas torus, its density increases due to pickup of ions upon ionization. The difference in centrifugal forces, $F_c(2) - F_c(1)$ (due to the enhanced mass density, $F_c \propto m\omega^2 r$) drives convection, i.e., a radial outflow in the 'heavy' sector of the convection pattern. (b, c from Gurnett *et al.* 2007)

ionospheric conductivity may cause the plasma to lag behind corotation. Several researchers have suggested that the observed variations in the rotation period of Saturn's radio emissions (§7.5.5.4), in sync with fluctuations in the rotation of the magnetic field, can be attributed to a slippage in corotation, caused by a variable mass-loading rate.

7.5.5.2 Radiation Belts

Satellites and rings are both sources and sinks of plasma (§7.3.4). Charged particle data from spacecraft flybys usually reveal clear absorption signatures in the vicinity of a satellite, or when crossing $\mathcal{L}$-shells traversed by a satellite. In fact, the existence of Jupiter's ring was first suggested based upon absorption features in charged particle data from the Pioneer 11 spacecraft. Similarly, one of several interpretations of an ambiguous proton absorption signature in Pioneer 11 data when the spacecraft traversed Saturn's magnetosphere was the presence of a faint ring at $\sim 2.8\ R_h$, where later the Voyager spacecraft located the G ring. Since Saturn's rings lie in the planet's rotational equator, which coincides with its magnetic equator, they are efficient absorbers of magnetospheric plasma. The intensity of trapped particles drops dramatically at the outer edge of the A ring. The main radiation belt is thus outside the rings, and shows clear signatures of the various satellites. There are practically no charged particles interior to the A ring. An intriguing discovery, though, is the presence of an inner radiation belt of energetic neutral atoms (ENA) interior to the D ring, discovered by the Magnetospheric Imaging Instrument (MIMI) on the Cassini spacecraft. This belt has been explained via a double charge-exchange process: ENAs from the main radiation belt which travel towards Saturn, are ionized and trapped when they enter the planet's exosphere. Through charge exchange they can be transformed back into ENAs, and the process of ionization may be repeated. This double (or multiple) charge exchange leads to a low-altitude ENA emission region, i.e., an inner radiation belt.

7.5.5.3 Titan

Titan orbits Saturn at a distance of $\sim 20\ R_h$. Because Saturn's magnetospheric boundary moves in and out with respect to the planet, in response to variations in the solar wind ram pressure, Titan is sometimes embedded in Saturn's magnetosphere, and at other times it is in the solar wind. Under quiet solar wind conditions, Titan is in Saturn's magnetosphere where the flow is submagnetosonic, and we do not expect the formation of a bow shock. The expected interaction of the plasma flow with Titan's dense atmosphere is controlled by atmospheric-ion mass loading of the plasma flow, with a subsequent slowing down of the flow and draping of field lines around Titan. Hence a 'magnetosphere' is induced, just like in the case of comets (§10.5), Venus, and Mars, although its relative dimensions are much smaller. Because of the preferential pickup of ions on one hemisphere (due to electric fields and curvature drift), the magnetic field must be highly asymmetric. The MIMI instrument on Cassini revealed strong ENA emissions near Titan, caused by charge exchange interactions with Titan's upper atmosphere. The emissions are strongest on the downstream side of Titan, probably as a result of the finite gyro radii causing protons on the downstream side to be converted to ENAs without encountering the high-density neutral gas region in Titan's atmosphere. In contrast to expectations, practically no nitrogen has been found in the magnetosphere. This suggests that the nitrogen derived from Titan's atmosphere must completely escape from the magnetosphere.

7.5.5.4 *Radio Observations*
Saturn's nonthermal radio spectrum consists of several components, as displayed in Figure 7.42, and discussed below.

Saturn Kilometric Radio Emissions (SKR)
Saturn's kilometric radiation is characterized by a broad band of emission, 100% circularly polarized, covering the frequency range from 20 kHz up to several hundred kHz. When displayed in the frequency–time domain, it is sometimes organized in arc-like structures, reminiscent of Jupiter's DAM arcs (see Fig. 7.42a). Cassini has revealed complex fine structure, which is also typical of cyclotron maser emissions at Jupiter and Earth (Fig. 7.42b). Like on Earth, the SKR source appears to be fixed at high latitudes primarily in the local morning-to-noon sector, but also appearing at other local times. The SKR intensity is strongly correlated with the solar wind ram pressure, perhaps suggesting a continuous transfer of the solar wind into Saturn's low-altitude polar cusps. A detailed comparison between high-resolution HST images of Saturn's aurora with SKR suggests a strong correlation between the intensity of UV auroral spots and SKR.

Even though the emission is highly variable over time, a clear periodicity at $10^h 39^m 24^s \pm 7^s$ was derived from the Voyager data, which was adopted as the planet's rotation period. Since the emission is tied to Saturn's magnetic field, which is axisymmetric, the cause of the modulation remains a mystery, though may be indirect evidence of higher order moments in Saturn's magnetic field. Even

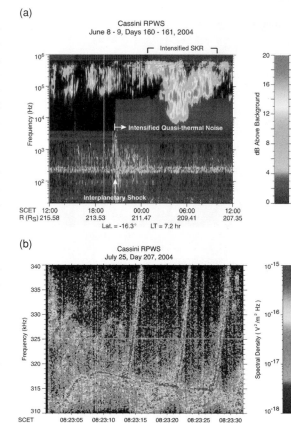

Figure 7.42 COLOR PLATE (a) Dynamic spectra of Saturn's SKR emission reveal a dramatic intensification in response to an interplanetary shock which passed Cassini at about 20:30 on 8 June 2004. (b) This high temporal and spectral resolution record of SKR obtained by Cassini illustrates the complex structure in the SKR spectrum, which is also typical of cyclotron maser emissions at Jupiter and Earth. This spectrogram shows a series of rapidly rising tones ($\sim$10 kHz s^{-1}) evidently triggered by a narrowband tone with a slight negative drift. The time between the apparently triggered emissions is 78 seconds. To hear the sounds, visit http://cassini.physics.uiowa.edu/space-audio/cassini/SKR2/. (Adapted from Kurth *et al.* 2005)

more mysterious, however, is that the SKR modulation period measured by Ulysses and Cassini vary by 1% or more (several minutes) on timescales of a few years or less. This implies that Saturn's radio rotation rate does not accurately reflect the rotation of its interior, in contrast to the radio periods of the other giant planets. The variations may be produced by plasma on field lines that lag behind corotation due to mass loading by Enceladus. Perhaps even more intriguing is the discovery of a north–south asymmetry in the radio rotation period, where SKR originating in the southern auroral region appears to have a period $\sim$0.2 hr shorter than the SKR period from the north.

Both periods (~10.6 and 10.8 hr) are longer than Saturn's internal rotation period as determined from gravity data (Table 1.3).

Very Low Frequency Emissions (VLF)

While Cassini was within Saturn's magnetosphere, it detected low-level continuum radiation ('trapped' radiation) at frequencies below 2–3 kHz. At higher frequencies the emission can escape, and appears to be concentrated in narrow frequency bands. It is believed that both the trapped and narrowband radio emissions are generated by the same mechanism, that is, mode conversion from electrostatic waves near the upper hybrid resonance frequency. However, the source location has not been determined. In particular, one source that has been suggested is related to Saturn's icy moons.

During the passage of the Cassini spacecraft through the inner region of the saturnian system on 1 July 2004, the Radio and Plasma Wave Science (RPWS) instrument detected many narrowband emissions in a plasma density minimum over the A and B rings. It is not clear how these narrowband emissions are related, if at all, to those measured well beyond the planet.

Saturn Electrostatic Discharges (SED)

Saturn electrostatic discharges (SEDs) are strong, unpolarized, impulsive events, which last for a few tens of milliseconds from a few hundred kHz to the upper frequency limit of the Voyager PRA experiment (40.2 MHz), and are also detected by the Cassini spacecraft. Structure in individual bursts can be seen down to the Voyager time resolution limit of 140 microseconds, which suggests a source size less than 40 km. During the Voyager era, episodes of SED emissions occurred approximately every $10^h\ 10^m$, distinctly different from the periodicity in SKR. In contrast to SKR, the SED source is fixed relative to the planet–observer line. The emissions are likely electrostatic discharge events as a counterpart of lightning flashes in Saturn's atmosphere. Some SED episodes have been linked directly to cloud systems observed in Saturn's atmosphere by the Cassini spacecraft. Cassini, however, has found SEDs to be much less common, generally speaking, than Voyager. Cassini can go months without seeing the discharges. Perhaps it may be a seasonal effect or related to the extent of ring shadowing on the atmosphere (or ionosphere, if propagation is an issue).

7.5.6 Uranus and Neptune

Our knowledge of the magnetospheres of Uranus and Neptune is limited to that obtained during brief encounters of these planets by the Voyager 2 spacecraft in 1986 (Uranus)

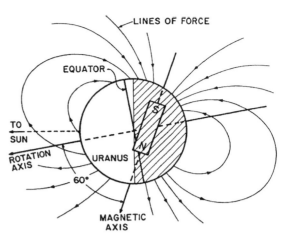

Figure 7.43 A sketch of Uranus's offset dipole magnetic field. (Ness *et al.* 1991)

and 1989 (Neptune). We thus do not have observations of the fields at different seasons, nor could we have detected any other variations on timescales longer than a few weeks.

7.5.6.1 Magnetic Field of Uranus

The axis of Uranus's magnetic field makes an angle of ~60° with the planet's rotation axis. This is much larger than that of Mercury, Earth, Jupiter, and Saturn, which are all aligned to within ~10°. The magnetic center is displaced by ~0.3 $R_{\eth}$ from the planet's center, resulting in a configuration as sketched in Figure 7.43. Note that particles mirroring at the Sun-facing side have a much larger chance of getting lost in the atmosphere than particles mirroring in the opposite hemisphere. Since Uranus's rotation axis is nearly perpendicular to the ecliptic plane, the magnetic field geometry as viewed from the solar wind at the epoch of the Voyager 2 encounter was quite Earth-like. The magnetotail wobbles about the planet–Sun line as the planet rotates, and with it the planet's magnetic polarity, as depicted in Figure 7.44. The stagnation point of the magnetopause is at ~18 $R_{\eth}$, so all five major satellites of Uranus lie in Uranus's magnetosphere, although far from its magnetic equator. Because of the latter, the satellites likely do not contribute much material to Uranus's magnetosphere.

Indeed, Uranus's magnetosphere does not contain much plasma at all. Protons and electrons have been observed at densities ~0.1–1 cm^{-3}. The primary source of plasma is ionization of Uranus's extended neutral hydrogen corona, whereas the solar wind is likely a second source of plasma. Due to the geometry of the planet's rotation and magnetic field direction, plasma motions in the magnetosphere are quite complicated. The plasma

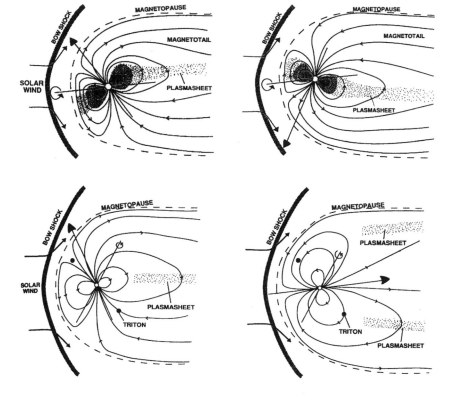

Figure 7.44 A sketch of Uranus's magnetosphere at the epoch of Voyager 2's encounter with the planet in 1986. The left and right panels are separated by half a planetary rotation. (Bagenal 1992)

Figure 7.45 A sketch of two extreme situations of Neptune's magnetic field configuration at the epoch of Voyager 2's encounter with the planet in 1989. The left and right panels are separated by half a planetary rotation. (Bagenal 1992)

corotates with Uranus ($\sim$17 hr) and is circulated throughout the magnetosphere within a couple of days by solar wind driven convection (despite the planet's fast rotation).

7.5.6.2 *Magnetic Field of Neptune*

Neptune's magnetic axis makes an angle of 47° with its rotation axis, which has an obliquity of $\sim$30°. The center of the magnetic dipole is displaced by 0.55 R_Ψ with respect to the planet's center, even more than in Uranus's case. The misalignment between Neptune's rotational and magnetic axes brings about a magnetic field configuration which is unique. While the field is rotating with the planet, two extreme situations are encountered, as sketched in Figure 7.45. At times the field is similar to that of the Earth, Jupiter, and Saturn, where the magnetic field in the tail is separated into two lobes of opposite polarity, separated by the plasma sheet. Half a rotation later the field topology is that of a 'pole-on' configuration, with the magnetic pole directed towards the Sun. The magnetic field topology in this case is very different, with a cylindrical plasma sheet, separating planetward field lines on the outside and field lines pointing away from the planet on the inside. The magnetic pole is facing the Sun, and the solar wind flows directly into the planet's polar cusp. The large offset in the planet's dipole field (which can

be translated into quadrupole moments), together with the presence of high-order moments, result in large variations in the 'surface' magnetic field strength for both Uranus and Neptune (Table 7.2).

Although one would expect Neptune's satellite Triton to be a source of hydrogen and nitrogen ions in Neptune's magnetosphere, observed densities were very low, $\lesssim$0.1 cm^{-3} up to a few tens cm^{-3} close ($\lesssim$2 R_Ψ) to the planet. The H$^+$ and N$^+$ escape Triton via sputtering processes as neutrals and ions, respectively. The N$^+$ moves inward from Triton and was detected even at $\mathcal{L} \approx 1.2$. A hydrogen cloud surrounds Triton with a density of $\sim$500 cm^{-3}, extending inwards to $\mathcal{L} \approx 8$. This cloud serves as a source for H$^+$, which moves inwards from its place of formation.

Theories predict the cumulative effect of solar wind driven convection to result in a net sunward transport of plasma in the magnetic equator. However, since convection is strongest when the field configuration is like that of Earth, the convection is strongly longitude dependent, such that plasma may move either toward or away from the planet, depending on its longitude. Convection also produces longitude-dependent variations in plasma density. The spacecraft trajectory was not well suited for detecting any longitudinal asymmetries in the plasma density.

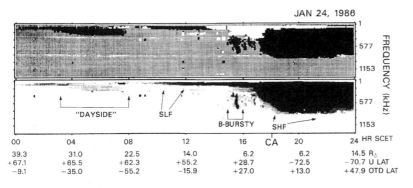

JAN 24, 1986

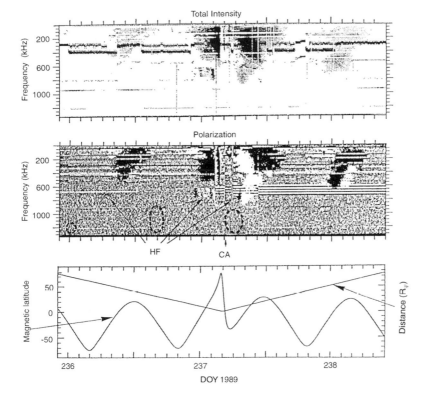

Figure 7.46 Dynamic spectra of Uranus's low-frequency radio emissions. The top panel shows the emission polarization, with white (black) corresponding to RH (LH) polarized signals. The bottom panel shows the intensities of the emissions. The distance to Uranus ($R_{\odot}$), planetocentric latitude (U LAT), and magnetic latitude in an offset-tilted dipole model (OTD LAT) are indicated at the bottom. (Desch *et al.* 1991)

Figure 7.47 Dynamic spectrum from Neptune during 60 hours around Voyager 2's closest approach. The upper panel shows the intensity, where increasing darkness represents increasing intensity. The middle panel shows the polarization: white indicates LH and black RH polarization. The bottom panel shows the magnetic latitude and distance of the spacecraft at the time of the observations. CA stands for closest approach, and HF for higher frequency emissions. (Zarka *et al.* 1995)

However, the observations indicate only inward transport of plasma, inconsistent with any of the convection models proposed to date.

7.5.6.3 *Radio Observations*

Like Saturn's radio emissions, both smooth and bursty components are apparent in the radio emissions from Uranus and Neptune (Figs. 7.46, 7.47), and these emissions probably originate in the southern auroral regions of the planets. Note, though, that the magnetic fields of these planets are inclined by large angles (47° for Uranus, 59° for Neptune) with respect to their rotational axes, and hence the auroral regions are not near the rotation poles. The periodicity of the emissions led to the determination

of the rotation periods of both planets, 17.24 ± 0.01 hr for Uranus and 16.11 ± 0.02 hr for Neptune. The upper bound to the frequency of the emissions is determined by (and indicative of) the planets' surface magnetic field strength.

From Uranus we have also received impulsive bursts, similar to the SED events of Saturn, which are referred to as UED or Uranus electrostatic discharge events. They were fewer in number and less intensive than the SEDs. If these emissions are caused by lightning, the lower frequency cutoff suggests peak ionospheric electron densities on the day side of $\sim 6 \times 10^5$ cm^{-3}. In addition to the broadband emissions, both planets also emit trapped continuum and narrowband radiation.

7.6 Generation of Magnetic Fields

7.6.1 Magnetic Dynamo Theory

Magnetic fields around planets cannot be caused by permanent magnetism in a planet's interior. The Curie point for iron is near 800 K (§6.2.2.3); at higher temperatures iron loses its magnetism. Since planetary interiors are much warmer, all ferromagnetic materials deep inside a planet have lost their permanent magnetism. In addition, a permanent magnetism would gradually decay away (eq. 7.18). The pattern of remanent magnetism observed near mid-ocean rifts implies that Earth's magnetic field has reversed its direction frequently during geologically recent times; a permanent magnet would not behave in this manner. Paleomagnetic evidence from the magnetization of old rocks implies that the Earth's magnetic field has existed for at least $\sim 3.5 \times 10^9$ yr, and that its strength has usually been within a factor of two from its present value (except during field reversals), so there must be a mechanism which continuously produces magnetic field. Electric currents in the interior of a planet form the only plausible source of planetary magnetism. The outer core of the Earth is liquid nickel–iron, thus highly conductive; similarly, the interiors of Jupiter and Saturn are fluid and conductive (metallic hydrogen), and Uranus and Neptune have large ionic mantles. Electric currents are likely to exist in all of these planets, and it is generally assumed that planetary magnetic fields are generated by a *magnetohydrodynamic dynamo*, a process which reinforces an already present magnetic field.

Electric currents are set up in a fluid which moves with respect to a magnetic field, provided the electrical conductivity is nonzero (eq. 7.15). Magnetic field lines tend to convect and distort their shape with the local motion of the fluid while they diffuse through it (Fig. 7.48). Slowly evolving magnetic fields are described by the magnetic induction equation (7.16). Obviously, fluid motions cannot lead to the spontaneous appearance of magnetic field if the field strength $B = 0$ to begin with. The magnetic induction equation, supplemented with the full equations of motion, governs the behavior of the dynamo magnetic field.

Consider a planet as an idealized rotating sphere of conducting fluid with an internal heat source. The heat source sets up a thermal gradient, which causes convection of the fluid. The inner parts of the planet rotate faster than the outer parts, due to the conservation of angular momentum. Hence the fluid sphere rotates differentially. In a convection cell (Fig. 7.48a) there is a radially outward motion of the fluid along the axis of the convective eddy, with a general return flow in the surrounding region. Thus

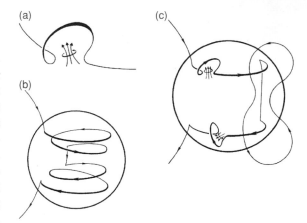

Figure 7.48 (a) Magnetic field lines tend to convect and distort their shapes with the local motion of the fluid. Within the convection cell, a radially outward motion of the fluid occurs along the axis of the eddy, with a general return flow in the surrounding region. Fluid, therefore, is converging at the bottom and diverging at the top. (b) Differential rotation of a planet causes magnetic field lines to coil up. (c) Convective eddies can maintain the overall dipole field by regenerating the poloidal field: The convective fluid motion causes a loop in the toroidal field, in the same direction as in the poloidal field. (Levy 1986)

the fluid is converging to the axis of the cell at the bottom, and diverging at the top. The Coriolis force associated with the converging part of the flow causes the fluid to turn locally around the axis of the convecting eddy, in the same direction as the overall rotation, similar to the motion in a cyclone. The divergence at the top of the eddy reduces the cyclonic motion, and may even reverse its direction.

The magnetic field lines of a planet are rooted in its interior. Differential rotation of the conducting part of the planet causes the magnetic field lines to coil up (Fig. 7.48b): this toroidal field can be a hundred times stronger than the original poloidal field (the polar field lines). Both the poloidal and toroidal fields, however, decay due to resistive dissipation of associated electric currents. The convective eddies can maintain the overall dipole field by regenerating the poloidal field, as sketched in Figure 7.48c: the convective fluid motion causes a loop in the toroidal field, in the same direction as the poloidal field, and therefore reinforces the planet's magnetic field. This is the essence of the magnetic dynamo theory.

Although this theory was proposed near the middle of the twentieth century, it was impossible to solve the complicated nonlinear magnetohydrodynamic equations that govern the geodynamo until the mid 1990s, when two researchers developed a self-consistent three-dimensional computer code that describes the dynamic evolution of Earth's magnetic field, using realistic properties of Earth's core. This simulation shows that indeed the geodynamo model works, including the spontaneous reversals in field polarity (§7.6.2). The geodynamo theory, involving a conductive, convective fluid, tells us that a rotating planet must have an internal magnetic field if its interior contains a heat source to drive convection. We have seen that this statement is true for the giant planets, as well as for Earth and Mercury.

In analogy to the planet's gravitational field, one can further surmise that the dipole terms of a planet's magnetic field are induced deep in the body's interior, whereas the strong high-order terms imply a source closer to the surface (Uranus and Neptune).

7.6.2 Variability in Magnetic Fields

Details in the Earth's magnetic field are best revealed after subtraction of the dipole field. A visualization of this remaining, or higher order moments of, the field shows about a dozen, continent-sized patches of field scattered around the globe. Both the dipole field and these higher order patches change over time. At the current epoch, the magnetic dipole moves westwards and completes a full circle in about 2000 years. The magnetic patches seem to develop and disappear on timescales of about 1000 years. At irregular intervals, between 10^5 and a few million years, the field reverses its polarity, as has been deduced from the magnetic polarity in rock samples of the oceanic ridge and volcanic lava flows (§6.2.2.3). It is clear that the Earth's magnetic field is not a static phenomenon; it is continuously changing, although the timescales involved are long compared to human lifetimes. As the field is generated by electric currents in the Earth's interior, variability in the strength and pattern can be understood, in principle. However, neither the field reversals, nor other excursions of the magnetic pole have been explained adequately, although three-dimensional computer simulations do predict such field reversals to take place. The reversals take place so quickly, on a geological timescale, that it is hard to find rocks preserving a record of such changes. From the (sparse) records, it seems that the intensity of the field decreases by a factor of 3–4 during the first few thousand years of a field reversal, while the field maintains its direction. The field then swings around a few times by $\sim 30°$, and finally moves to the opposite polarity. Afterwards, the intensity increases again to its normal value. The details of the changes in magnetic field configuration during these episodes, however, are not clear.

Whether similar changes occur in magnetic fields around other planets is not known. We do know that the Sun's magnetic field reverses every 11 years, as discussed in §7.1. We also know that Mars must have had a magnetic dynamo during the first few hundred million years after the planet formed; one theory suggests that Mars may have had field reversals as well (§7.5.3.2). Among the planets, only Jupiter's magnetic field can be monitored from Earth, via its radio emissions. However, only half a century has elapsed since the first detection of Jupiter's radio emission; we need a much longer time baseline to see changes in its magnetic field. Moreover, as the radio emissions depend on both the magnetic field configuration and particle distribution, it is not a simple task to disentangle the two.

Further Reading

Good review papers on the solar wind, magnetospheres, and space weather can be found in:

McFadden, L., P. R. Weissman, and T.V. Johnson, Eds., 2007. *Encyclopedia of the Solar System*, 2nd Edition. Academic Press, San Diego. 982pp. We recommend the chapters by:

  Aschwanden, M.J.: The Sun

  Gosling, J.T.: The solar wind

  Luhmann, J.G. and S.C. Solomon: The Sun–Earth connection

  Kivelson, M.G. and F. Bagenal: Planetary magnetospheres

Reviews on Jupiter's magnetosphere and aurora, as well as a complete review on the impacts of Shoemaker–Levy 9 with Jupiter, are in: *Jupiter: Planet, Satellites and Magnetosphere* Eds. F. Bagenal, T. E. Dowling, and W. McKinnon, 2004. Cambridge University Press, Cambridge.

Review papers on the magnetospheres of the giant planets and the martian magnetic field:

Bagenal, F., 1992. Giant planet magnetospheres. *Annu. Rev. Earth Planet. Sci.*, **22**, 289–328.

Brain, D.A., 2006. Mars Global Surveyor measurements of the martian solar wind interaction. *Space Science Reviews*, **126**, 77–112.

Kurth, W.S., and D.A. Gurnett, 1991. Plasma waves in planetary magnetospheres. *J. Geophys. Res.*, **96**, 18 977–18 991.

Basic books on magnetospheric physics and plasma physics:

Boyd, T.J.M., and J.J. Sanderson, 2003. *The Physics of Plasmas* Cambridge University Press, Cambridge. 532pp.

Kivelson, M.G., and C.T. Russell, Eds., 1995. *Introduction to Space Physics*. Cambridge University Press, Cambridge. 568pp.

Lyons, L.R., and D.J. Williams, 1984. *Quantitative Aspects of Magnetospheric Physics*. Reidel Publishing Company, Dordrecht. 231pp.

Roederer, J.G., 1970. *Physics and Chemistry in Space. 2: Dynamics of Geomagnetically Trapped Radiation*. Springer-Verlag, Berlin. 166pp.

Schulz, M., and L.J. Lanzerotti, 1974. *Physics and Chemistry in Space. 7: Particle Diffusion in the Radiation Belts.* Springer-Verlag, Berlin. 215pp.

Problems

7.1.E Consider a comet on a Keplerian orbit around the Sun. The perihelion distance is 0.3 AU, and aphelion distance is 15 AU. Assume a solar wind velocity of 400 km s^{-1}. Calculate the angle between the ion tail and the Sun–comet line both at perihelion and aphelion.

7.2.E Calculate the speed of sound and the Alfvén velocity in the solar wind at Earth's orbit. Compare your answers with the quiet solar wind velocity.

7.3.I (a) Calculate the mean free path between collisions in the solar wind, assuming quiescent solar wind properties (Table 7.1). Compare this number with the size of the Earth's magnetosphere and the thickness of the bow shock.

(b) If the mean free path between collisions is much larger than the typical size of the system, the plasma is collisionless. Is the solar wind a collisionless plasma by this definition? Explain why the solar wind is often treated as a fluid nevertheless.

7.4.I Calculate the approximate standoff distance of the Earth's magnetopause and bow shock, assuming quiescent solar wind properties (Table 7.1). Approximate the Earth's magnetic field as a dipole field, with a surface magnetic field strength of 0.3 G. You may ignore the plasma density in the outer magnetosphere.

7.5.I Show that in an ideal unmagnetized gas with a velocity perpendicular to the shock the following equation holds:

$$\frac{\rho_1}{\rho_2} = \frac{(\gamma + 1)P_1 + (\gamma - 1)P_2}{(\gamma - 1)P_1 + (\gamma + 1)P_2}, \qquad (7.90)$$

where the subscript 1 refers to conditions before the shock (i.e., solar wind) and subscript 2 after the shock (i.e., magnetosheath). Calculate the ratio ρ_2/ρ_1 for a strong shock ($P_2 \gg P_1$) in a monatomic gas.

7.6.I Assume that the magnetic field and velocity in the solar wind parallel to the bow shock are both zero. The solar wind density is 5 protons cm^{-3}, the magnetic field is 5×10^{-5} G, and the temperature is 2×10^5 K. The solar wind velocity is 400 km s^{-1}. Calculate the density, velocity, temperature, and magnetic field strength in the magnetosheath.

7.7.E The Gauss coefficients for Jupiter's magnetic field are: $g_1^0 = 4.218$, $g_1^1 = -0.664$, and $h_1^1 = 0.264$ gauss. Calculate the magnetic dipole moment, the angle between the magnetic and rotational angle, and the longitude of the magnetic north pole.

7.8.I Derive the equations for the Larmor radius and Larmor frequency by balancing the centripetal force with the Lorentz force.

7.9.I (a) If changes in the magnetic field are small over one gyro radius and period, the magnetic field flux through the particle's orbit is constant. Derive the equation for the first adiabatic invariant (eq. 7.48).

(b) Show that for nonrelativistic particles the adiabatic invariant, μ_B, is equal to a particle's magnetic moment, μ_b, induced by its circular motion around a magnetic field line.

(c) Show that for relativistic particles $\mu_B = \gamma_r \mu_b$.

(d) Express the first adiabatic invariant in terms of energy rather than momentum, both for nonrelativistic and relativistic particles (eqs. 7.50 and 7.51).

(e) Show that these two expressions for relativistic and nonrelativistic particles are equal at low energies.

7.10.I Consider a proton in the Earth's magnetic field at a distance $\mathcal{L} = 3$. Assume the Earth's field can be approximated by a dipole with a magnetic moment $\mathcal{M}_B = 7.9 \times 10^{25}$ G cm^3, which is centered at the center of the planet and aligned with the rotation axis.

(a) Calculate the magnetic field strength at the particle's mirror point if the equatorial pitch angle of the particle $\alpha_e = 60°$, $\alpha_e = 30°$, and $\alpha_e = 10°$.

(b) Calculate the latitude of the mirror points for the three cases in (a).

(c) Compare your answers in (a) with the surface magnetic field strength, and comment on the results.

7.11.I Consider a proton and an electron in the Earth's magnetic field at a distance $\mathcal{L} = 2$. Both particles are confined to the magnetic equator, and have an energy of 1 keV. Assume the properties of Earth's magnetic field described in Problem 7.10.

(a) Calculate the gyro period for each particle.

(b) Calculate the drift orbital period for each particle.

(c) Calculate the orbital velocity around the Earth, and compare this with the Keplerian velocity of a neutral particle at the same geocentric distance ($\mathcal{L} = 2$; assume a circular orbit).

Describe what would happen if the ion undergoes charge exchange.

7.12.**I** Consider a particle in a dipole field at a distance $\mathcal{L} = 6$, which diffuses radially inwards until $\mathcal{L} = 2$. The latitude of the mirror points is indicated by θ_m; the location of and field strength at the mirror points by s_{m_1} and B_{m_1}, respectively, at $\mathcal{L} = 6$ and by s_{m_2} and B_{m_2} at $\mathcal{L} = 2$.

(a) Assume the magnetic latitude θ_m is the same for the mirror points s_{m_1} and s_{m_2}. Calculate the relative strength of the magnetic field at the mirror points: B_{m_1}/B_{m_2} and at the equator: B_{e_1}/B_{e_2}.

(b) Calculate the increase in the particle's kinetic energy when diffusing inwards from $\mathcal{L} = 6$ to $\mathcal{L} = 2$.

(c) Write an expression for the invariant $K_B = I_B \sqrt{B_m}$ in terms of B_{m_1} and B_{e_1} at both positions $r_1 = \mathcal{L} = 6$ and $r_2 = \mathcal{L} = 2$. Based upon this expression, do you think B_{m_1} and B_{m_2} are at the same latitude θ_m? Explain your reasoning.

7.13.**E** The total voltage drop, Φ_{conv}, between the dusk and dawn sides on Earth can be estimated:

$$E_{conv} = \frac{\Phi_{conv}}{2R_{pc}} = v_{pc} B_{pc}, \qquad (7.91)$$

where the subscript pc stands for the region threaded by the open field lines across the polar cap, the area around the magnetic pole within the auroral oval. This results in a potential drop of 20–200 kV, with an average value of $\sim$50 kV (Problem 7.16).

(a) Calculate the total potential drop across the Earth's magnetosphere induced by the convection electric field.

(b) Calculate the potential drop over a distance equal to the diameter of the tail (50 $R_\oplus$) in the undisturbed solar wind ($v_{sw} = 400$ km s^{-1}, $B_{sw} = 5\ \gamma$).

(c) Compare your answers in (a) and (b) and determine the fraction of the IMF magnetic flux that reconnects with the geomagnetic field. (Hint: The potential drop would be equal to your answer in (b) if all IMF flux reconnects with the geomagnetic tail.)

7.14.**E** Assume Jupiter's magnetic field can be approximated by a dipole field with a magnetic moment $\mathcal{M}_B = 1.5 \times 10^{30}$ G cm^{-3}. The solar wind magnetic field strength near Jupiter is $\sim 2 \times 10^{-6}$ G. Calculate the gain in energy for a solar wind proton which enters Jupiter's magnetosphere and

diffuses inwards until $\mathcal{L} = 1.5$. (Hint: Calculate the proton's energy assuming a solar wind velocity of 400 km s^{-1}.)

7.15.**E** (a) Estimate the critical radius, r_{cr}, inside of which corotation dominates over convection, for particles near the dawn and dusk sides in the Earth's magnetosphere (see Problem 7.10 for typical geomagnetic parameters). Express your answer in $R_\oplus$.

(b) Calculate the critical radius, r_{cr}, inside of which corotation dominates over convection, for particles near the dawn and dusk sides in Jupiter's magnetosphere, using the magnetic field parameters as specified in Problem 7.14. Express your answer in $R_?$. Comment on the differences between the critical radii for Earth and Jupiter.

7.16.**I** (a) Calculate the effective potential for protons with an energy of 50 keV at the dawn and dusk sides in the Earth's magnetosphere, at $\mathcal{L} = 2$ and $\mathcal{L} = 4$.

(b) Calculate the effective potential for protons with an energy of 200 keV at the dawn and dusk sides in the Earth's magnetosphere, at $\mathcal{L} = 2$ and $\mathcal{L} = 4$.

(c) Calculate the effective potential for electrons with an energy of 200 keV at the dawn and dusk sides in the Earth's magnetosphere, at $\mathcal{L} = 2$ and $\mathcal{L} = 4$.

(d) Calculate the drift velocity for the above mentioned particles, and comment on your results (describe in words what happens to the particles drifting in the geomagnetic field).

7.17.**I** Explain why the ring current tends to weaken the Earth's surface magnetic field.

7.18.**E** Describe how the orbit of a spacecraft which orbits over the Earth's equator at an altitude of 2000 km is influenced by the South Atlantic Anomaly. (Hint: Compare the Keplerian orbital velocity with the orbital velocity of the particles.)

7.19.**I** (a) Calculate the location of Venus's ionopause. Assume the atmospheric pressure to drop with altitude according to the barometric law (Chapter 4), with a surface pressure of 90 bar. Make a reasonable estimate for the temperature based upon the various atmospheric profiles in Chapter 4. Assume a solar wind velocity of 400 km s^{-1}, solar wind density of 10 protons cm^{-3}, and an interplanetary magnetic field strength of 1×10^{-4} G.

(b) Calculate the location of the martian ionopause, assuming a surface pressure of 6 mbar and estimating the atmospheric temperature from profiles given in Chapter 4. Assume that the solar wind density and interplanetary magnetic field strength scale with heliocentric distance as $r_\odot^{-2}$. (What is the value for the solar wind velocity?)

(c) Compare your results from (a) and (b), and comment on the comparison.

7.20.E (a) Calculate the energy involved when an ion in the Io plasma torus impacts Io. (Hint: Calculate the orbital velocity of Io and the ion, assuming both orbit Jupiter at a planetocentric distance of 6 $R_{2\!\!\!/}$.)

(b) Calculate the size of Io's neutral sodium cloud, if the lifetime for sodium atoms is a few hours.

(c) Explain in words why the banana-shaped neutral sodium cloud is pointed in the forward direction.

7.21.I (a) Calculate the scale height for sulfur and oxygen ions in Io's cold inner and hot outer torus.

(b) Estimate the ion production rate from Io if the ion density is 3000 particles cm^{-3} and the torus stretches radially from 5.3 to 7.5 $R_{2\!\!\!/}$; assume the latitudinal extent to be equal to twice the scale height.

7.22.E (a) Calculate the typical energy of electrons which radiate at a wavelength of 20 cm, if the magnetic field strength is 0.8 G.

(b) Calculate the typical energy of electrons which radiate at a wavelength of 6 cm if the magnetic field strength is 0.8 G.

(c) If the energy distribution of the radiating electrons is flat, show the wavelength dependence of the observed radiation.

(d) The above numbers are valid for Jupiter's magnetosphere. Assuming the planet's magnetic field resembles a dipole field with the parameters described in Problem 7.14, calculate the location in the magnetosphere from which the synchrotron radiation in (a) and (b) is emitted.

7.23.E Jupiter emits strong polarized bursts of radiation at frequencies $\nu < 40\,\mathrm{MHz}$. This radiation is thought to originate close to Jupiter's cloud-tops, and is attributed to cyclotron radiation.

(a) Why is the radiation cyclotron and not synchrotron radiation?

(b) Calculate the magnetic field strength at Jupiter's cloud-tops.

7.24.E Structure in SED bursts can be distinguished on timescales of 140 μs. Derive a size scale for the source of this emission.

7.25.E Neptune's magnetic dipole moment is 2.14×10^{27} G cm^3. The cutoff frequency of Neptune's decametric radiation is at 1.3 MHz.

(a) Calculate the magnetic field strength which corresponds to the cutoff frequency.

(b) Compare the magnetic field strength calculated in (a) with Neptune's surface dipole magnetic field strength. Explain the difference.

7.26.I Derive an expression for the frequency of a standing Alfvén wave, as a function of magnetic field strength and plasma density. Assume ℓ is the length of the field line, and n is the number of harmonics.

7.27.E Calculate the six natural plasma frequencies for a magnetosphere in which the magnetic field strength is 0.3 G and plasma density is 20 protons cm^{-3}. Assume the energy of the particles to be 20 kHz, and that there are only protons and electrons ($N_p = N_e$).

7.28.I The dispersion relation for electromagnetic waves is given by equation (7.81).

(a) Express the index of refraction, n, in terms of ω_o and ω_{pe}.

(b) Assume there are two antennas in space. One of the antennas acts as a transmitter, and the other receives the signals. Show that by sweeping the signals in frequency, one can derive the plasma density between the two antennas.

7.29.E If Earth's magnetic field would not be regenerated, calculate how long it would take for the field to dissipate.

8 Meteorites

I could more easily believe two Yankee professors would lie than that stones would fall from heaven.
Attributed (probably incorrectly) to USA President Thomas Jefferson, 1807

A *meteorite* is a rock that has fallen from the sky. It was a *meteoroid* (or, if it was large enough, an asteroid) before it hit the atmosphere and a *meteor* while heated to incandescence by atmospheric friction. A meteor that explodes while passing through the atmosphere is termed a *bolide*. Meteorites that are associated with observations prior to or of the impact are called *falls*, whereas those simply recognized in the field are referred to as *finds*.

The study of meteorites has a long and colorful history. Meteorite falls have been observed and recorded for many centuries (Fig. 8.1). The oldest recorded meteorite fall is the Nogata meteorite, which fell in Japan on 19 May 861. Iron meteorites were an important raw material for some primitive societies. However, even during the Enlightenment it was difficult for many people (including scientists and other natural philosophers) to accept that stones could possibly fall from the sky, and reports of meteorite falls were sometimes treated with as much skepticism as UFO 'sightings' are given today. The extraterrestrial origin of meteorites became commonly acknowledged following the study of some well-observed and documented falls in Europe around the year 1800. The discovery of the first four asteroids, celestial bodies of sub-planetary size, during the same period added to the conceptual framework that enabled scientists to accept extraterrestrial origins for some rocks.

Meteorites provide us with samples of other worlds that can be analyzed in terrestrial laboratories. The overwhelming majority of meteorites are pieces of small asteroids, which never grew to anywhere near planetary dimensions. Primitive meteorites, which contain moderate abundances of iron, come from planetesimals that never melted. Most of the iron-rich meteorites presumably come from the deep interiors of planetesimals that differentiated (§5.2.2) prior to being disrupted, while iron-poor meteorites are samples from the outer layers of differentiated

bodies. As small objects cool more rapidly than do large ones, the parent bodies of most meteorites either never got very hot, or cooled and solidified early in the history of the Solar System. Many meteorites thus preserve a record of early Solar System history that has been wiped out on geologically active planets like the Earth. The clues that meteorites provide to the formation of our planetary system are the focus of most meteorite studies and also of this chapter.

8.1 Basic Classification and Fall Statistics

The traditional classification of meteorites is based upon their gross appearance. Many people think of meteorites as chunks of metal, because metallic meteorites appear quite different from ordinary terrestrial rocks. Museums also tend to specialize in metal meteorites, because most people find these odd-shaped pieces of nickel–iron interesting to look at. Metallic meteorites are made primarily of iron, with a significant component of nickel and smaller amounts of several other *siderophile* elements (elements which readily combine with molten iron, such as gold, cobalt, and platinum); thus, metal meteorites are referred to as *irons*. Meteorites that do not contain large concentrations of metal are known as *stones*. Many stony meteorites are difficult for the untrained eye to distinguish from terrestrial rocks. Meteorites that contain comparable amounts of macroscopic metallic and rocky components are called *stony-irons*.

A more fundamental classification scheme is based on the history of meteorite parent bodies. Most irons and stony-irons, as well as some stones known as *achondrites* (in contrast to chondrites, which are described below), come from *differentiated* parent bodies (i.e., bodies that have undergone density-dependent phase separation, §5.2). Differentiated bodies experienced an epoch in which they were mostly molten, and much of their

Figure 8.1 Woodcut depicting the fall of a meteorite near the town of Ensisheim, Alsace on 7 November 1492. A literal translation of the German caption reads 'of the thunderstone (that) fell in 92 year outside of Ensisheim'. This meteorite is the oldest recorded fall from which material is still available.

iron sank to the center, taking with it siderophile elements. (A small fraction of the analyzed irons were probably produced by impact-induced localized melting.) The bulk compositions of achondrites are enriched in *lithophile* and/or *chalcophile* elements. Lithophile elements tend to concentrate in the silicate phases of a melt, and chalcophile elements tend to concentrate in the sulfide phases of a melt. The abundances of both lithophiles and chalcophiles are also enhanced in the Earth's crust. Relative to a solar mixture of refractory elements, achondrites are significantly depleted in iron and siderophile elements. *Primitive meteorites* have not been differentiated; they are composed of material that formed directly from the solar nebula condensates and surviving interstellar grains, modified in some cases by aqueous processing (implying that liquid water was once present in the parent body) and/or thermal processing (implying that the body was quite warm at some time). Silicates, metals, and other minerals are found in close proximity within primitive meteorites. Primitive meteorites are called *chondrites* because most of them contain small, nearly spherical, igneous inclusions known as *chondrules*, which solidified from melt droplets. Some chondrules are glassy, implying that they cooled extremely rapidly. Apart from the most volatile elements, the composition of all chondrites is remarkably similar to that of the solar photosphere (Fig. 8.2). As meteorite compositions are easier to measure than are abundances in the Sun, analysis of chondrites provides the best estimates of the average Solar System composition of most elements (Table 8.1). Densities of meteorites vary from 1.7 g cm^{-3} for the primitive Tagish Lake carbonaceous chondrite meteorite (which fell in Canada on 18 January 2000) to between 7 and 8 g cm^{-3} for irons.

Most chondrites fit into one of three different *classes* that have been cataloged on the basis of composition and mineralogy; these classes are subdivided into various *groups*. Members of the most volatile-rich class of

chondrites contain up to several percent carbon by mass, and are known as *carbonaceous chondrites*. The carbonaceous chondrite class is divided into eight major groups which differ slightly in composition and are denoted CI, CM, CO, CV, CR, CH, CB, and CK. The most common primitive meteorites are the *ordinary chondrites*, which are subclassified primarily on the basis of their Fe/Si ratio: H (high Fe), L (low Fe), and LL (low Fe, low metal; i.e., most of the iron that is present is oxidized). The third class of primitive meteorites, *enstatite chondrites*, is named after their dominant mineral (MgSiO$_3$). These highly reduced chondrites are also divided on the basis of their iron abundance, and groups are denoted EH and EL. The *Rumuruti chondrite* group, denoted R, does not fit into any of these broad classes. Rumuruti chondrites are the most highly oxidized anhydrous chondrite group, contain fewer chondrules than ordinary chondrites, no metal, and have large amounts of olivine fayalite. The iron abundance and its mineralogical location in various meteorites is diagrammed in Figure 8.3.

Although chondritic meteorites have never been melted, they have been processed to some extent in 'planetary' environments (i.e., in asteroid-like parent bodies) via thermal metamorphism, shock, brecciation (breaking up and reassembly), and chemical reactions often involving liquid water. Chondrites are assigned a *petrographic type* ranging from 1 in the most volatile-rich primitive meteorites to 6 in the most thermally equilibrated chondrites (Fig. 8.5). Type 3 chondrites (Figs. 8.4 and 8.5) appear to be the least altered in planetary environments, and provide the best data on the conditions within the protoplanetary disk. Types 2 and 1 show progressively more aqueous alteration; all known aqueously altered chondrites are carbonaceous. Type 1 are devoid of chondrules, which either never were present or have been completely destroyed by aqueous processing. In contrast, the degree of metamorphic alteration increases in higher numbered

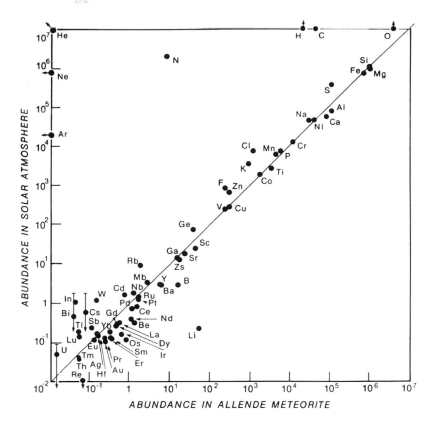

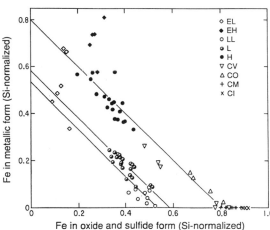

Figure 8.2 The abundance of elements in the Sun's photosphere plotted against their abundance in the Allende CV3 chondrite. Most elements lie very close to the curve of equal abundance (normalized to silicon). Several volatile elements lie above this curve, presumably because they are depleted in meteorites (rather than being enriched in the Sun), while only lithium lies substantially below the curve; lithium is depleted in the solar photosphere because it is destroyed by nuclear reactions near the base of the Sun's convective zone.

Figure 8.3 Plot showing the ratio of metallic iron to silicon as a function of Fe/Si in oxidized phases (mainly silicates) for various chondritic meteorites. The bulk Fe/Si for the meteorites determines the position of the diagonal, while the oxidation state of the Fe determines the location of a given point on the diagonal. (Sears and Dodd 1988)

types above type 3, with type 7 sometimes being included to denote meteorites that have undergone partial melting. Impacts may be responsible for most thermal processing of chondrites.

Iron meteorites are classified primarily on the basis of their abundance of nickel and of the moderately volatile trace elements germanium and gallium. Compositional differences are correlated to observed differences in structure. Crystallization textures of irons also depend upon the rate at which the meteorite cooled. These intriguing structures, the famous *Widmanstätten pattern* (see Fig. 8.6), thus provide information about meteorite parent bodies.

There are two classes of stony-irons: *pallasites*, which are closely related to irons, and *mesosiderites*, which are related to achondrites. Pallasites (Fig. 8.7) consist of networks of iron–nickel alloy surrounding nodules of olivine typically 5–10 mm in size. Pallasites are of igneous origin, and they probably formed at the interface between a region of molten metal and a magma chamber in which olivine could form and sink to the bottom, e.g., at a core–mantle interface. Mesosiderites (Fig. 8.8) contain a mixture of metal and magmatic rocks similar to *eucrite* achondrites.

Several different types of achondrites exist, presumably from different regions of differentiated (or at least locally melted) asteroids of a variety of compositions and sizes. A small percentage of known achondrites are from two larger bodies: the Moon and Mars (§8.2).

Table 8.1 Elemental abundances.[a]

Element		Solar System[b] (atoms/10^6 Si)	CI chondrites (mass fraction)	Element		Solar System[b] (atoms/10^6 Si)	CI chondrites (mass fraction)
1	H	2.431×10^{10}	21.0 mg/g	44	Ru	1.90	692 ng/g
2	He	2.343×10^9	56 nL/g	45	Rh	0.37	141 ng/g
3	Li	55.5	1.46 µg/g	46	Pd	1.44	588 ng/g
4	Be	0.74	25.2 ng/g	47	Ag	0.49	201 ng/g
5	B	17.3	713 ng/g	48	Cd	1.58	675 ng/g
6	C	7.08×10^6	35.2 mg/g	49	In	0.18	78.8 ng/g
7	N	1.95×10^6	2.94 mg/g	50	Sn	3.73	1.68 µg/g
8	O	1.41×10^7	458.2 mg/g	51	Sb	0.33	152 ng/g
9	F	841	60.6 µg/g	52	Te	4.82	2.33 µg/g
10	Ne	2.15×10^6	218 pL/g	53	I	1.00	480 ng/g
11	Na	5.75×10^4	5.01 mg/g	54	Xe	5.39	31.3 pL/g
12	Mg	1.02×10^6	95.9 mg/g	55	Cs	0.37	185 ng/g
13	Al	8.41×10^4	8.50 mg/g	56	Ba	4.35	2.31 µg/g
14	Si	1.00×10^6	106.5 mg/g	57	La	0.44	232 ng/g
15	P	8370	920 µg/g	58	Ce	1.17	621 ng/g
16	S	4.45×10^5	54.1 mg/g	59	Pr	0.17	92.8 ng/g
17	Cl	5240	704 µg/g	60	Nd	0.84	457 ng/g
18	Ar	1.03×10^5	888 pL/g	62	Sm	0.25	145 ng/g
19	K	3690	530 µg/g	63	Eu	0.095	54.6 ng/g
20	Ca	6.29×10^4	9.07 mg/g	64	Gd	0.33	198 ng/g
21	Sc	34.2	5.83 µg/g	65	Tb	0.059	35.6 ng/g
22	Ti	2420	440 µg/g	66	Dy	0.39	238 ng/g
23	V	288	55.7 µg/g	67	Ho	0.090	56.2 ng/g
24	Cr	1.29×10^4	2.59 mg/g	68	Er	0.26	162 ng/g
25	Mn	9170	1.91 mg/g	69	Tm	0.036	23.7 ng/g
26	Fe	8.38×10^5	182.8 mg/g	70	Yb	0.25	163 ng/g
27	Co	2320	502 µg/g	71	Lu	0.037	23.7 ng/g
28	Ni	4.78×10^4	10.6 mg/g	72	Hf	0.17	115 ng/g
29	Cu	527	127 µg/g	73	Ta	0.021	14.4 ng/g
30	Zn	1230	310 µg/g	74	W	0.13	89 ng/g
31	Ga	36.0	9.51 µg/g	75	Re	0.053	37 ng/g
32	Ge	121	33.2 µg/g	76	Os	0.67	486 ng/g
33	As	6.09	1.73 µg/g	77	Ir	0.64	470 ng/g
34	Se	65.8	19.7 µg/g	78	Pt	1.36	1.00 µg/g
35	Br	11.3	3.43 µg/g	79	Au	0.20	146 ng/g
36	Kr	55.2	15.3 pL/g	80	Hg	0.41	314 ng/g
37	Rb	6.57	2.13 µg/g	81	Tl	0.18	143 ng/g
38	Sr	23.6	7.74 µg/g	82	Pb	3.26	2.56 µg/g
39	Y	4.61	1.53 µg/g	83	Bi	0.14	110 ng/g
40	Zr	11.3	3.96 µg/g	90	Th	0.044	30.9 ng/g
41	Nb	0.76	265 ng/g	92	U	0.0093	8.4 ng/g
42	Mo	2.60	1.02 µg/g				

[a] Uncertainties of these estimates are generally larger than suggested by the number of digits quoted. Concentration of noble gases within meteorites is calculated by scaling the values for the principal isotopes from Anders and Grevesse (1989) to all isotopes using the isotopic abundances given in Lodders (2003). All other data from Lodders (2003), who also lists uncertainties.

[b] Protosolar material after radioactive decay to the present epoch.

Figure 8.4 Photographs of various chondritic meteorites. The scale bars are labeled in both centimeters and inches. (a) Brownfield H3.7 ordinary chondrite, which fell in Texas in 1937. Very small chondrules, plus highly reflective metal and sulfide grains, can be picked out. (b) Parnallee LL3 ordinary chondrite, which fell in India in 1857. The cut surface clearly shows well-delineated chondrules and slightly larger clasts. Parnallee is of very similar metamorphic grade to Brownfield, but it has a much coarser texture. (c) Vigarano CV3 carbaceous chondrite, which fell in Italy in 1910. Vigarano has beautifully delineated chondrules. It also contain large CAIs, which appear white in this photograph.

The relative abundances of meteorites that have been collected and cataloged are not representative of the meteoroids whose orbits cross that of Earth. The overwhelming majority of meteorites collected after being observed to fall are stones, most of which are chondrites (Table 8.2). An even larger majority of stones has been found on Antarctic ice sheets, possibly because stones are more fragile than irons and fragments of a disrupted meteor are counted as one body when observed to fall, but presumably they are often multiply counted if found in Antarctica, where ice flows destroy evidence of strewn fields (such as the one illustrated in Figure 8.15). Irons are much more prevalent among non-Antarctic finds, as they are more resistant to weathering and they are also more easily identifiable as meteorites, or at least as unusual objects, to nonexperts.

The total mass of cosmic debris impacting the Earth's atmosphere in a typical year is 10^{10}–10^{11} g. Dust and micrometeorites $\sim 1 - 100$ μm in radius account for most of the mass during most years, although the infrequent impacts of kilometer-sized and larger bodies dominate the flux averaged over very long timescales.

8.2 Source Regions

Meteorites are identified by their extraterrestrial origin. In theory, a rock could be knocked off the Earth, escape

the Earth's gravity and orbit the Sun for a period of time before re-impacting Earth. Such a body would be a meteorite, despite its terrestrial origin. However, such rocks would need to have been accelerated to escape speed from Earth without being vaporized. Also, it would be very difficult to identify such rocks as meteorites, the only evidence being cosmic ray tracks, rare isotopes produced by cosmic rays, and possibly a fusion crust (such as those seen in Figures 8.13 and 8.14); no such meteorite has ever been identified. However, *tektites*, which are rocks formed from molten material knocked off the Earth's surface by impacts, have been found in many locations. The geographic distribution of tektite finds and the lack of cosmic ray tracks in tektites imply that these rocks did not escape, but rather quickly fell back to Earth's surface.

Based upon a comparison with Apollo samples, several dozen achondrite meteorites, many of which are *anorthositic breccias*, are clearly of lunar origin. A similar number of achondrite meteorites, including four falls (representing a total of $\sim 0.4\%$ of all known meteorite falls) are in the *SNC* class (shergotites, nakhlites, and chassignites, named after the places where the first examples were found), which are of martian origin. Figure 8.9 shows an SNC found in Antarctica. These rocks are young, with most having crystallization ages of $<1.3 \times 10^9$ years. (Meteorite ages are measured using radiometric dating techniques, which are described in detail in §8.6.)

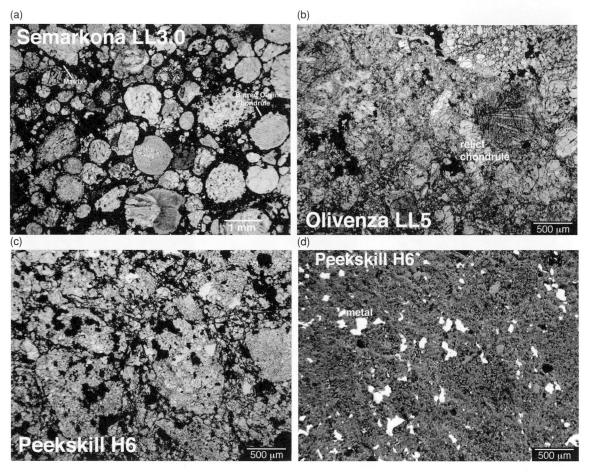

Figure 8.5 Thin-section photomicrographs of representative examples of chondrites showing petro-graphic variations among chondrite groups. (a) The Semarkona LL3.0 ordinary chondrite in plane-polarized light, showing a high density of chondrules with different textures surrounded by opaque matrix. Semarkona fell in India in 1940. (b) The Olivenza LL5 ordinary chondrite showing the texture of a recrystallized LL chondrite in which some relict chondrules are discernible, but chondrule bound-aries are not as sharp as in Semarkona. Olivenza fell in Spain in 1924. (c) The Peekskill H6 chondrite in plane-polarized light. Peekskill is a recrystallized ordinary chondrite in which the primary texture has been essentially destroyed and chondrule boundaries are not easily discernible. Peekskill fell in upstate New York on 9 October 1992. (d) Peekskill shown in reflected light, which emphasizes the 7% (by volume, 20% by mass) metal content of this meteorite. (Weisberg *et al.* 2006)

Convincing evidence for the martian origin of the SNC meteorites came from the similarity between their noble gas abundances and isotopic ratios of both noble gases and nitrogen to those measured in the martian atmosphere by the Viking landers. A 4.5×10^9 year old non-SNC martian meteorite[1] has also been recovered from Antarctica. This

rock, ALH84001, has been extensively studied because it possesses several intriguing characteristics (including magnetite similar to that produced biologically on Earth) which were initially interpreted as evidence for ancient life on Mars (§5.5.4.5). Theoretical studies indicate that it is much easier for impacts to eject unvaporized rocks from Mars than from Earth because the velocity required to escape from Mars is less than half of Earth's escape velocity.

[1] The term 'martian meteorite' is also applied to rocks observed on the surface of Mars which appear to have originated on other bod-ies. Such rocks provide information about the capture of meteorites by planets with tenuous atmospheres as well as surface processes on Mars. But unlike other meteorites, they are not easily accessible for laboratory studies, and they will not be considered further in this book. Likewise, two tiny, otherwise unremarkable, chondrites

were among the lunar samples brought to Earth by the Apollo astronauts.

Figure 8.6 A cut acid-etched surface of the Maltahöhe iron meteorite. This surface shows a Widmanstätten pattern, an intergrowth of several alloys of iron and nickel which formed by diffusion of nickel atoms into solid iron during slow cooling within an asteroid's core. (Courtesy Jeff Smith)

Figure 8.7 Photograph of the Seymchan pallasite, which was found in Siberia in 1967. A continuous network of iron–nickel metal acts as a frame holding grains of Mg-rich olivine. The imaged section is 8.3 centimeters across. (Courtesy Laurence Garvie)

Figure 8.8 Photograph of the Estherville mesosiderite, which fell in Iowa, USA, in 1879. Note Estherville's brecciated structure, with large pods of iron–nickel metal mixed with a seemingly random jumble of stony clasts. The scale bar is labeled both in centimeters and inches, and the size of the meteorite is ~30 cm.

The overwhelming majority (>90%) of meteorites are from bodies of sub-planetary size in the asteroid belt. Two classes of bodies are known in Earth-crossing orbits: comets and asteroids. Although the dividing line between these two sets of objects is not as sharp as once thought, either using orbital or physical characteristics, most objects are clearly in one class or the other. Are meteorites extinct comets depleted of volatiles? A few of the primitive chondrites may be, but 'evolved' objects (achondrites, irons) must come from asteroids (or other rocky objects). Comparison of the spectra of reflected light from several types of meteorites with asteroid spectra yields many close correspondences (Fig. 8.10). This enables us to determine many compositional facts about asteroids that would otherwise be very difficult or impossible to obtain via remote sensing from Earth.

Pre-impact orbits have been determined for only a tiny fraction of meteorite falls, most of which are ordinary chondrites. All of these orbits had perihelia near Earth's orbit; most penetrated the asteroid belt but remained well interior to Jupiter's orbit (Fig. 8.11). The irregularly shaped $R \approx 2$ m rapidly spinning object named 2008 TC$_3$ was detected 20 hours prior to hitting Earth's atmosphere. It exploded at an altitude of about 37 km, and 4 kg of carbonaceous chondrite meteorites from this object were recovered in Almahata Sitta, Sudan. The pre-impact orbit of 2008 TC$_3$ had semimajor axis $a = 1.308$ AU and perihelion distance $q = 0.9$ AU.

Firm identification of individual meteorites with specific asteroids is difficult, but strong cases can be made for some meteorite classes. The spectrum of 4 Vesta is unique among large asteroids and very similar to those of HED achondrite meteorites (Fig. 8.10); Vesta's orbit in the inner asteroid belt (Table 9.1) makes delivery to Earth of debris excavated by impacts easier than is the

Table 8.2 Meteorite Classes and Numbers (as of September 2008).

	Falls	Fall frequency (%)	Non-Antarctic finds	Antarctic finds[a]
Total cataloged	1070	–	9582	15 660
Stones	1009	94.3	8648	15 495
Chondrites	916	85.6	7964	15 082
Carbonaceous chondrites	42	3.8	319	494
Achondrites	87	8.1	684	413
Martian meteorites	4	0.4	53	9
Lunar meteorites	0		93	19
Stony-irons	12	1.1	139	56
Irons	49	4.5	795	109

Data from Meteoritical Bulletin Database (http://tin.er.usgs.gov/meteor/metbull.php).

[a] Lists the well-cataloged ANSMET collection only.

Figure 8.9 A cut surface of the martian meteorite ALHA 77005. This rock contains dark olivine crystals and light-colored pyroxene crystals, plus some patches of impact melt, which were probably produced when it was ejected from the surface of Mars. The cube (W) is 1 cm on a side. (NASA/JSC)

case for most asteroids. Thus, Vesta is the likely source of these interesting and well-studied differentiated meteorites (§8.7.1). Numerous L chondrites with short cosmic ray ages, indicating brief lifetimes, as well as small space debris (§8.6.5), have been found in ∼470 Myr old sedimentary rocks; typical cosmic-ray ages of more recently fallen L chondrites are much longer. Many L chondrites have ^{39}Ar–^{40}Ar gas retention ages (§8.6) of ∼470 Myr, indicating a major shock event at that time. These data suggest that L chondrites share a common origin with either the Flora family asteroids or the Gefion family asteroids. The Flora asteroid family (§9.5, Fig. 9.14) was produced by the breakup of an $R \gtrsim 100$ km asteroid; the parent to the Gefion family had an $R \gtrsim 50$ km. Evidence for linkage

comes from: (i) the dynamical ages from orbit spreading of both asteroid families are ∼500 Myr; (ii) the spectral similarity between L chondrites and many Flora and Gefion family asteroids, and (iii) the location of the Flora family near the ν_6 secular resonance (where perihelion precession rates are equal to that of Saturn, §2.4), and the Gefion family near the 5:2 mean motion resonance with Jupiter; both of these resonances provides an efficient delivery mechanism to Earth-crossing orbits.

The typical time for a large object to reach Earth from the main asteroid belt is much longer than the age of the Solar System, except near certain strong resonances. Poynting–Robertson drag (§2.7.2) can move very small meteoroids to the vicinity of resonances that can transport them into Earth-crossing orbits with characteristic lifetimes of 10^7 years; the Yarkovsky effect (§2.7.3) can transport ∼1 m–10 km bodies into the same resonances. Earth-crossing asteroids provide another source of meteoroids with typical lifetimes as small space rocks of ∼10^7 years. These short intervals are consistent with cosmic-ray exposure ages (§8.6.5), but note that a source is required for Earth-crossing asteroids, whose dynamical lifetime is far less than the age of the Solar System (§9.1).

8.3 Fall Phenomena

Meteoroids encounter the Earth's atmosphere at speeds ranging from 11 km s^{-1} to 73 km s^{-1}, with typical velocities of ∼15 km s^{-1} for bodies of asteroidal origin and ∼30 km s^{-1} for cometary objects (Problem 8.1). At such velocities, meteoroids have substantial kinetic energy per unit mass, enough energy to completely vaporize the bodies if it was converted to heat.

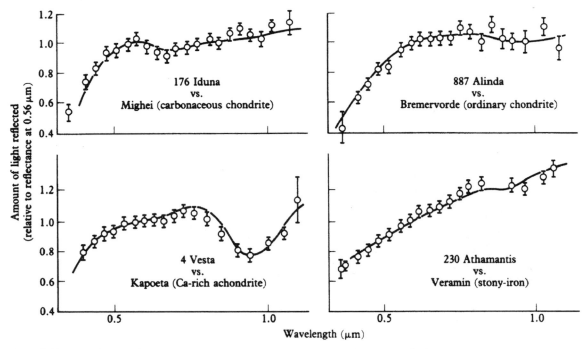

Figure 8.10 A comparison of the reflection spectra of four asteroids (points with error bars) with meteorite spectra as determined in the laboratory (solid curves). (Morrison and Owen 1996)

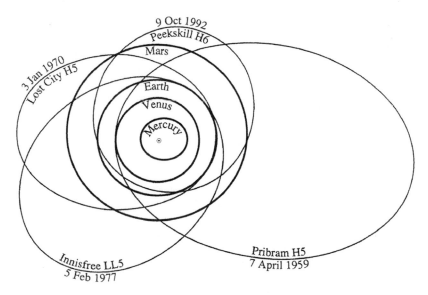

Figure 8.11 Pre-impact orbits of the first four recovered meteorites that were photographed well enough to allow for the calculation of accurate trajectories. All four are metamorphically processed chondrites, with petrographic types shown on the figure. The deduced orbit of Neuschwanstein, which fell on 6 April 2002, is almost identical to that of Pribram, so it was very surprising when laboratory analysis revealed that Neuschwanstein is an EL6 enstatite chondrite. (Adapted from Lipschutz and Schultz 1999)

In the rarefied upper portion of a planet's atmosphere, gas molecules independently collide with the rapidly moving meteoroid. Interactions with this tenuous gas are dynamically insignificant for large meteoroids, but are able to retard the motion of tiny bodies substantially (see eq. 8.3a). *Micrometeoroids* smaller than ∼10–100 μm are able to radiate the heat they acquire from this drag rapidly enough that they can reach the ground. These very small particles are known as *micrometeorites*.

Micrometeorites have been recovered using specially equipped jets flying in the stratosphere and also from sediments found in certain locations in Greenland, Antarctica, and the ocean floor. Elemental and isotopic abundances confirm the extraterrestrial origin of this dust. Most of the highly porous, friable aggregates collected in the stratosphere, such as the one pictured in Figure 8.12a, come from interplanetary dust particles (IDPs) that are likely to have been incorporated in cold, volatile-rich

(a)

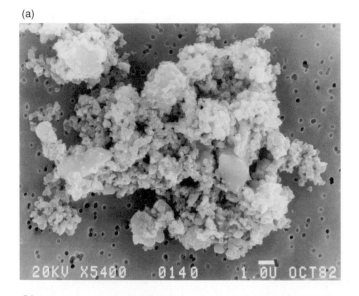

(b)

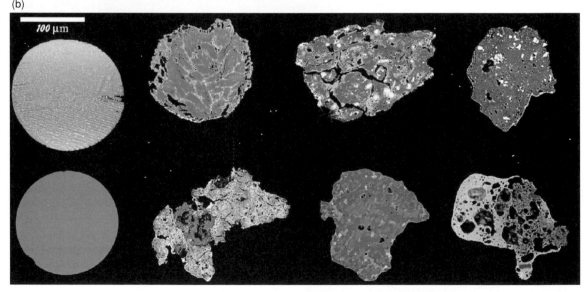

Figure 8.12 (a) Scanning electron microscope image of a micrometeorite that was once an interplanetary dust particle (IDP). Note the fluffy fractal-like structure (compare with Figure 13.15). (Courtesy Donald Brownlee) (b) A selection of micrometeorites and cosmic spherules recovered from Antarctic aeolian sediments. Particles were mounted in epoxy and polished to expose a cross-section. All images are backscattered electron micrographs, where brighter gray means higher atomic number.

First row, left to right: (1) Typical barred-olivine texture cosmic spherule, with dark bars of olivine separated by brighter Fe-rich glass and magnetite crystallites. This diagnostic texture is produced when a melt of chondritic composition quickly crystallizes. These spherules resemble chondrules. (2) A heavily weathered spherule consisting of olivine surrounded by glass and bright magnetite grains. (3,4) Two typical relatively unweathered micrometeorites preserving a fine-grained, glassy, and C-rich matrix, mafic silicates, FeNi metal, and sulfides, with some oxidation along rim and fractures.

Second row, left to right: (5) Glassy cosmic spherule, showing no recrystallization and only a few chips along the exterior surface. (6) Fluffy, thermally-altered micrometeorite showing melting of volatile-rich matrix to produce bright (Fe-rich) glasses surrounding darker, more refractory (and therefore less melted), regions. This grain, like 4, shows bright magnetite-rich glass rims on some surfaces. (7) Igneous micrometeorite consisting of olivine, pyroxene, and marginal melt glass decorated with bright magnetite crystallites. Some vesicles remain, suggesting incomplete melting. (8) Micrometeorite with a highly vesicular texture due to intense thermal alteration, volatilization, and melting of a volatile-rich host (probably a micrometeorite like 3 or 4). Many micrometeoriticists consider particles like 3, 4, and 6 to have been formed when large versions of particles such as that pictured in (a) were thermally altered during atmospheric entry and compacted by inclusion into sediments on Earth. (Courtesy Ralph Harvey) bodies such as comets. The micrometeorites that are collected on the Earth's surface tend to be larger and more coherent (Fig. 8.12b), and most come from IDPs of asteroidal origins.

Micrometeorites are classified according to the extent of atmospheric modification to which they were subjected. *Cosmic spherules* are fully melted objects, whereas micrometeorites that reveal grains have not been melted. Based upon bulk composition, carbon content, and the composition of isolated olivine and pyroxene grains, fine-grained micrometeorites are likely related to carbonaceous chondrites. Micrometeorites of the types related to carbonaceous chondrites appear to be several times as common as those related to ordinary chondrites. This stands in stark contrast to the abundance ratio of larger meteorites in terrestrial collections (Table 8.2).

The surface of a meteor is heated by radiation from the atmospheric shock front that it produces. Meteors rarely get significantly hotter than 2000 K, because this temperature is sufficient to cause iron and silicates to melt. The liquid evaporates or just falls off the meteor, so ablation provides an effective thermostat. Ablation is only important for velocities $v > v_c$, where the critical velocity, v_c, above which radiative cooling cannot release energy rapidly enough to prevent material at the surface of the meteor from melting, is approximately 3 km s^{-1} for iron meteorites. The rate of decrease in the meteor's mass is given by

$$Q\frac{dm}{dt} = -\frac{1}{2}C_H\rho_g Av^3\left(\frac{v^2 - v_c^2}{v^2}\right),\tag{8.1}$$

where Q is the heat of ablation, A is its projected surface area in the direction of motion, and C_H is the heat transfer coefficient. Note the similarity between equation (8.1) and aerodynamic drag, equation (2.55a). The value of C_H is approximately 0.1 above an altitude of 30 km in Earth's atmosphere and varies inversely with atmospheric density at lower altitudes. The heat of ablation is $\sim 5 \times 10^{10}$ erg g^{-1} for stony and iron meteorites. Most visible meteors completely vaporize in the atmosphere. The light emitted by the hot atmospheric gas (plasma) and ablated meteoritic material forms the familiar meteor trails in our atmosphere. Most visible meteors are from millimeter- and centimeter-sized bodies. The initial mass necessary for part of a meteoroid to make it to the ground depends on its initial speed, impact angle, and composition. The surface of a meteor can become hot enough to melt rock to a depth of ~ 1 mm (~ 1 cm for irons, which have a higher thermal conductivity). However, interior to this hot outer skin the temperature stays close to 260 K (Problem 8.6), so a meteorite is cold a few minutes after

Figure 8.13 The 800 g Lafayette (Indiana) nakhlite (martian meteorite) shows an exquisitely preserved fusion crust. During the stone's rapid transit through the atmosphere, air friction melted its exterior. The lines trace beads of melted rock streaming away from the apex of motion. (Courtesy Smithsonian Institution)

hitting the surface of Earth. The crust is melted and greatly altered (Figs. 8.13 and 8.14), becoming re-magnetized, but the interior of the meteorite is basically undisturbed.

Larger cosmic intruders continue to travel at *hypersonic* (substantially supersonic) speeds as they penetrate into denser regions of the planet's atmosphere, and therefore they induce a shock in the gas in front of them. The leading face of a meteor is subject to an average pressure, P, given approximately by the formula:

$$P \approx \frac{C_D\rho_g v^2}{2},\tag{8.2}$$

where v is the velocity of the meteor, C_D is the drag coefficient (which is approximately unity for a sphere) and ρ_g is the local density of the atmosphere. When this pressure exceeds the compressive strength of the meteor, the meteor is likely to fracture and disperse (Problem 8.3), with the resulting debris scattered over a *strewn field* (Fig. 8.15). The meteor is slowed by this pressure, but it continues to

Figure 8.14 Carbonaceous chondrite meteorite ALHA 77307, which was found in Antarctica. The meteorite's rounded surface was sculpted during its passage through Earth's atmosphere. The cracked surface was produced by subsequent cooling. The cube is 1 cm on a side. (NASA/JSC)

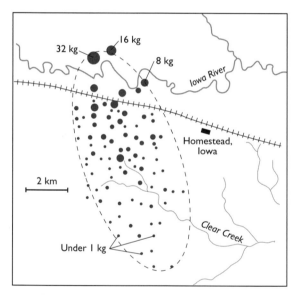

Figure 8.15 Illustration of the strewn field in the Homestead, Iowa, USA, meteorite shower, which occurred on 12 February 1875. The meteor traveled in the upwards direction, slightly towards the left, in the geometry of the figure. Note that larger objects, which are less susceptible to atmospheric drag, landed at the far end of the ellipse, in the upper left of the figure. (Adapted from O. Farrington, 1915)

be accelerated by the planet's gravity, so its velocity varies as

$$\frac{d\mathbf{v}}{dt} = -\frac{C_D \rho_g A v}{2m} \mathbf{v} - g_p \hat{\mathbf{z}}, \tag{8.3a}$$

where m is the mass of the meteor, g_p is the gravitational acceleration, and $\hat{\mathbf{z}}$ is the unit vector pointing in the upwards direction. Equation (8.3a) accounts for aerodynamic drag (eq. 2.55a) in a uniform gravitational field.

A meteor loses a substantial fraction of its initial kinetic energy if it passes through a column of atmosphere with a mass equal to its own.

For shallow angles of incidence, the meteoroid may skip off the atmosphere back into space, just like skipping stones on water. A fireball observed in western North America on 10 August 1972 was caused by such a grazing encounter of a meteor, estimated to be $\sim 4 \times 10^6$ kg in mass. There are unconfirmed reports of a 1996 meteor being observed to skip off our atmosphere, and slightly less than one orbit later re-entering the atmosphere and landing in southern California.

The fate of centimeter- to meter-sized meteoroids depends primarily on their vertical velocity at atmospheric entry, v_{z_0}. Rapidly moving ($v_{z_0} > 15$ km s^{-1}) meteoroids of this size tend to be ablated away, whereas slowly moving ($v_{z_0} < 10$ km s^{-1}) similar objects tend to be aerobraked to the *terminal velocity*, v_∞, at which gravitational acceleration balances atmospheric drag:

$$v_\infty = \sqrt{\frac{2g_p m}{C_D \rho_g A}}. \tag{8.3b}$$

Stony meteors in the ~ 1–10 m size range tend to break up in the atmosphere; the fragments are aerobraked to terminal velocity (eq. 8.3b) and are often recovered in many pieces spread out over strewn fields. Stony bodies between ~ 10 and 100 m in size continue at high speed to deeper and denser levels in the atmosphere, where they can be disrupted by ram pressure of the dense atmosphere. The giant ($\sim 5 \times 10^{23}$ erg ≈ 10 megatons of TNT) explosion that occurred over Tunguska, Siberia in 1908 was probably produced by the disruption of a stony bolide roughly 50–100 m in diameter about 5–10 km above the Earth's surface (see also §5.4.6). Meteoroids larger than ~ 100 m generally reach the surface with high velocity, because even if they are flattened out by ram pressure from their interactions with the atmosphere, they still collide with an amount of gas totaling much less than their own mass. In contrast, iron meteorites are very cohesive, and iron meteors over a broad size range can reach Earth's surface moving with sufficient velocity to produce impact craters, such as the famous Meteor Crater in Arizona (Fig. 5.24). Cratering of planetary surfaces by hypervelocity impacts is described in detail in §5.4. Impact erosion of planetary atmospheres is discussed in §4.8.3.

8.4 Chemical and Isotopic Fractionation

Meteorites provide the oldest and most primitive rocks available for study in terrestrial laboratories. Analysis of meteorites yields important clues as to how, when, and of what type of materials the planets themselves formed. The primary information on early Solar System conditions obtained from meteorites is given by their chemical and isotopic composition, and variations thereof between different parts of individual meteorites and among meteorites as a group. In this section, we provide some of the basic geochemistry background necessary to understand these results. In §8.6, we discuss how some observed variations can be used to determine the ages of meteorites. The mineralogical structure of meteorites also yields information on conditions in the early Solar System, as does remanent ferromagnetism detected in some meteorites; these subjects are reviewed in §8.5 and §8.7.

8.4.1 Chemical Separation

In a sufficiently hot gas or plasma, atoms become well mixed. Provided diffusion has had sufficient time to erase initial gradients and turbulence is large enough to prevent gravitational settling of massive species, a gas is generally well mixed at the molecular level. Solid bodies, however, tend to mix very little, retaining the molecular composition that they acquired when they solidified. When solids form from a gas or from a melt, molecules tend to group with mineralogically compatible counterparts, and distinct minerals are formed. Such minerals have elemental compositions that can be considerably different from the bulk composition of the mixture (§5.2). Condensation from a gas can produce small grains, which mix heterogeneously with one another. Crystallization from a melt allows greater separation of materials, producing samples with large-scale heterogeneities. On an even larger scale, the combination of chemical separation in a melt and density-dependent settling results in planetary differentiation (§5.2). However, under most circumstances, the isotopic composition of each element usually remains uniform across mineral phases.

Analysis of the most primitive meteorites known implies that to a good approximation the material from which the planets formed was well mixed over large distance scales on both the isotopic and elemental level. Gross chemical differences result primarily from temperature variations within the protoplanetary disk. Exceptions to isotopic homogeneity have been used to determine the age of the Solar System (§8.6), and to show that at least a small amount of presolar grains survived intact and never melted or vaporized before being incorporated into planetesimals (§§8.5, 8.7, and 10.4.4).

8.4.2 Isotopic Fractionation

Although different isotopes of the same element are almost identical chemically, there are several physical and nuclear

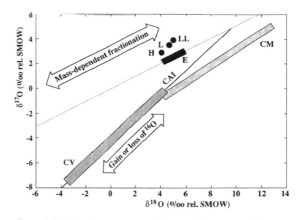

Figure 8.16 Plot showing the distribution of the three stable oxygen isotopes in various Solar System bodies. Isotope abundance ratios are shown relative to the Standard (terrestrial) Mean Ocean Water (SMOW), with units being parts per thousand variations. The dotted line represents the mass-dependent fractionation pattern observed in terrestrial samples. (Kerridge 1993)

processes that can produce isotopic inhomogeneities. Sorting out these different processes is essential to use the isotope data obtained from meteorites. One explanation for isotopic differences between grains is that they came from different reservoirs that were never mixed, e.g., interstellar grains which formed in distinct parts of the galaxy from material with different nucleosynthesis histories. This explanation can have profound consequences for our understanding of planetary formation, so other processes must also be considered.

Isotopes can be separated from one another by mass-dependent processes. These processes can rely on gravitational forces, such as the preferential escape of lighter isotopes from a planet's atmosphere, or be the result of molecular forces, such as the preference for deuterium (as opposed to ordinary hydrogen) to bond with heavy elements, which is a consequence of a slightly lower energy resulting from deuterium's greater mass. Mass-dependent fractionation is easy to identify for elements such as oxygen, which have three or more stable isotopes, because the degree of fractionation is proportional to the difference in mass. One can plot differences in the $^{17}O/^{16}O$ ratio against the corresponding difference in $^{18}O/^{16}O$. Mass-dependent fractionation of oxygen leads to points along a line of slope 0.52 (Fig. 8.16; the fractionation slope is slightly larger than 1/2 because 17/16 > 18/17). The deviations of meteorite data from this line may result from *self-shielding* of the abundant (and thus optically thick) isotope $^{12}C^{16}O$ from photodissociation (§4.6). An alternative explanation involves *chemical mass-independent fractionation*, which can occur because the stability of some gaseous compounds containing multiple oxygen atoms differs between the symmetric case (all ^{16}O atoms) and the asymmetric one (i.e., containing a ^{17}O or ^{18}O atom).

Nuclear processes can also lead to isotopic variations. Paramount among these is radioactive decay, which transforms radioactive parents into stable daughter isotopes, thereby producing isotopic differences. Cosmic rays produce a variety of nuclear reactions. Energetic particles from local radioactivity may also induce nuclear transformations.

8.5 Main Components of Chondrites

Chondrites contain very nearly (within a factor of two) solar abundance ratios of refractory nuclides (Fig. 8.2). Carbonaceous chondrites have even closer to solar abundances of refractories and moderately volatile elements. CI chondrites are the most similar to the Sun in elemental composition. Note, however, that even the most volatile-rich carbonaceous chondrites are depleted relative to the Sun in the highly volatile elements oxygen, carbon, and nitrogen (as well as, of course, the extremely volatile noble gases and hydrogen). Isotopic ratios are even more strikingly regular; almost all differences can be accounted for by radioactive decay (excesses of daughter nuclides), cosmic ray-induced *in situ* nucleosynthesis, or mass fractionation (§8.4.2). However, slight deviations from this rule show that the material within the solar nebula was not completely mixed at the atomic level.

Small condensates that clearly predate the protosolar nebula, some of which formed in outflows from stars and others that may have accreted within the interstellar medium, represent a volumetrically insignificant but scientifically critical component of chondrites. Many *presolar grains* are carbon-rich stardust, which occurs as *nanodiamonds*, graphite, and silicon carbide (SiC). Other common types of presolar grains include silicates and the oxides corundum (Al_2O_3), hibonite ($CaAl_{12}O_{19}$), and spinel ($MgAl_2O_4$).

Chondrites have not been melted since their original accretion $\sim 4.56 \times 10^9$ years ago. Although these primitive meteorites represent well-mixed isotopic and elemental (except for volatiles) samples of the material in the protoplanetary disk, they are far from uniform on small scales. In addition to chondrules, many chondrites contain *CAIs*, which are refractory inclusions that are rich in calcium and aluminum. Chondrules and CAIs are embedded within a dark fine-grained *matrix* that is present in all chondrites. Chondrites formed with different percentages of inclusions (CAIs, chondrules) and differing amounts of moderately volatile elements. Up to $\sim 20\%$ of the mass of some chondrites is comprised of Fe–Ni metal.

Chondrules are small (typically ~ 0.1–2 mm) rounded igneous rocks (i.e., they solidified from a melt) composed primarily of refractory elements (Fig. 8.17). They range from 0 to 80% of the mass of a chondrite, with abundances depending on compositional class (CI chondrites contain neither chondrules nor CAIs) and petrographic type. Chondrules are totally absent in petrographic type 1 (they may have been destroyed by aqueous processes) and are substantially degraded by recrystallization resulting from thermal metamorphism in types 5 and 6; the most pristine chondrules are found in type 3 chondrites (Figs. 8.4 and 8.5). Mineralogical properties imply that chondrules cooled very quickly, dropping from a peak temperature of ~ 1900 K to ~ 1500 K over a period ranging from 10 minutes to a few hours. Chondrules are diverse, with a wide variety of compositions, mineralogies, and sizes. However, strong correlations of chondrule properties (size and compositions) are observed within individual meteorites. These correlations, combined with the compositional complementarity of chondrules and matrix within individual primitive meteorites (together they are nearly solar in composition apart from volatiles, whereas separately they differ substantially), implies that chondrules were not well mixed within the protoplanetary disk before incorporation into larger bodies.

Compound chondrules appear to be two or more chondrules joined together. Most compound chondrules probably were produced by collisions of partially molten objects. Other compound chondrules consist of a primary chondrule that appears to be entirely entrained in a larger secondary chondrule. The secondary chondrules are thought to have formed from the heating and melting of fine-grained dust that had accreted onto the surface of the primary. Many chondrules have melted rims, providing additional evidence for multiple heating events.

CAIs are light-colored inclusions, typically 1–10 millimeters in size (Fig. 8.4c). They are composed of very refractory minerals, including substantial amounts of Ca and Al and abundances of high-Z elements that are greatly enhanced relative to bulk chondrites. CAIs are most abundant in CV chondrites, and are also seen in most other classes of primitive meteorites. They are among the oldest objects formed in the Solar System. *Type A CAIs* are extremely refractory; most type As have not been melted. *Type B CAIs* are slightly less refractory than type As, and all type Bs were once molten. Nonetheless, CAIs are a much more homogeneous (compositionally, mineralogically, and size) class of objects than are chondrules, and unlike chondrules, CAIs are not compositionally complementary to the other components within their host rocks. Many CAIs have melted rims that formed up to

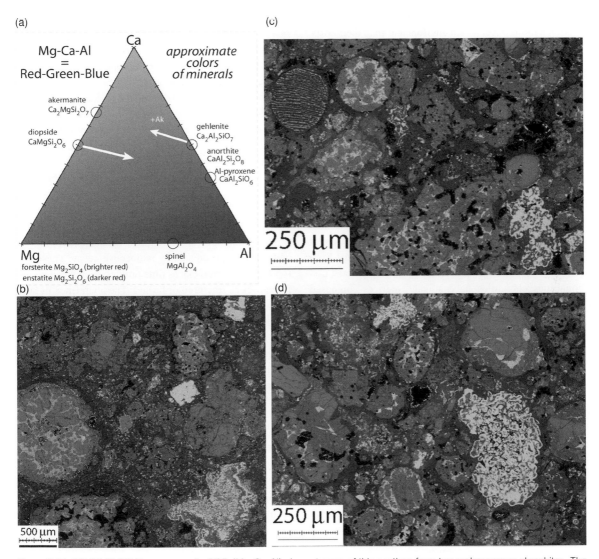

Figure 8.17 COLOR PLATE X-ray composite RGB (Mg–Ca–Al) element maps of thin sections from two carbonaceous chondrites. The images were taken with the electron microprobe at the American Museum of Natural History. (a) Key (triangular diagram) showing the correspondence between the various colors and minerals. Black areas are metals or sulfide grains. (b) Image of Vigarano (CV3) carbonaceous chondrite at 7 μm/pixel resolution. Note the big porphyritic olivine chondrule on the left, fragments of macrocrystalline CAIs in upper left far upper right. The one near the top has spinel (purple) in melilite. The big layered CAI to lower right has a melilite/anorthite/pyroxene interior, spinel MgAl$_2$O$_4$ inner rim, and diopside outer rim beyond that. (c, d) Images of Kainsaz (CO3) carbonaceous chondrite at a resolution of 2 μm per pixel. A barred olivine chondrule in the middle left of panel (c) has FeO-rich olivine (dark red) in glass (green-blue). Right next to it is a (circular) porphyritic olivine–pyroxene chondrule, with brighter (higher Mg/Si ratio) red olivine with darker red (lower Mg/Si) pyroxene, and green crystals of Ca-rich pyroxene (diopside) in blue-green Ca-, Al-rich silicate glass. To the lower right is a nodular CAI with Al-rich phases surrounded by Ca-rich pyroxene. Just above the right corner of the scale bar is a metal grain (metal is black, here) surrounded by a red olivine rim. This texture is not uncommon in AOAs (Amoeboid Olivine Aggregates) in CO chondrites. Image (d) is dominated by a large, nodular CAI containing spinel-rich cores surrounded by diopside and melilite (gehlenite + ackermainte admixture), and also probably anorthite. A similar object in the upper center has a much smaller grain size. A large porphyrytic olivine chondrule, just above the big CAI, is dominated by large crystals of olivine, with minor glasses, and some Mg–pyroxene (enstatite). Three other chondrules lie to the right of the scale bar, the largest one dominated by a large pyroxene grain with included olivine, and, to the left, euhedral small olivines floating in blue-green glass (that was once liquid). Minor Ca-rich pyroxene is also present. Nearly all of the chondrules in these rocks contain metal, such as the large porphyrytic olivine chondrule above the scale bar. (Courtesy Denton Ebel)

300 000 years after the core CAI formed. These rims solidified in a more oxygen-rich environment than their hosts. The interiors of the CAIs were not heated to anywhere near melting during this secondary processing, implying very short duration heating. Clearly, the protoplanetary disk was an active and sometimes violent place! Dust (often similar to the matrix) is seen in CAI rims. Some chondrules are surrounded by analogous rims. Parent-body processing could well have affected some of these rims.

The characteristics of the fine-grained (10 nm–5 μm) matrix material that makes up the bulk of most chondrites vary with petrographic type. Mean grain sizes in the matrix of type 3 chondrites vary from 0.1 to 10 μm. Chondrite matrices appear to contain material from a wide variety of sources, including presolar grains, direct condensates from the protoplanetary disk, and dust from fragmented chondrules and CAIs. A few less-altered carbonaceous chondrites have silicate matrices composed largely of crystalline forsterite (Mg_2SiO_4) and enstatite ($MgSiO_3$) as well as amorphous silicates. These components appear to be solar nebula condensates, with the crystalline minerals having cooled through 1300 K at ~1000 K/hr. In most chondrites, many of these grains have been altered by post-accretional aqueous or thermal processing.

Olivine ($(Mg,Fe)_2SiO_4$) and the pyroxenes orthopyroxene ($(Mg,Fe)SiO_3$) and clinopyroxene (or diopside; a solid solution of $CaSiO_3$, $MgSiO_3$, and $FeSiO_3$) are the most common minerals in the majority of ordinary chondrites and in carbonaceous chondrites of petrographic type 3 and higher; serpentine ($(Mg,Fe)_3Si_2O_5(OH)_4$) is most common in carbonaceous chondrite types 1 and 2; magnetite (Fe_3O_4) is found in a variety of classes and types. Other common minerals in ordinary chondrites are plagioclase feldspar (5–10 wt%), and ~5 wt% troilite (FeS).

Chondritic porous (CP) interplanetary dust particles (IDPs) are extremely fine-grained submicron particles that are compositionally similar to volatile-rich chondrites. Most CP IDPs probably emanate from comets. A common component of CP IDPs is glass with embedded metal and sulfides (*GEMS*); GEMS are submicron particles that contain numerous 10–50 nm Fe–Ni metal and Fe–Ni sulfide grains. Some GEMS show isotopic ratios that differ from chondritic averages and thereby indicate a presolar origin.

8.6 Radiometric Dating

There are several different ages that may be assigned to a given meteorite. All of these meteorite ages are determined by *radionuclide dating*.

The most fundamental age of a meteorite is its *formation age*, which is often referred to simply as the age. The formation age is the length of time since the meteorite (or its components) solidified from a molten or gaseous phase. Most meteorites have ages of 4.55–4.57 × 10^9 years, but a few are much younger. Techniques for determining meteorite formation ages are based upon radioactive decay of long-lived radioactive isotopes. These techniques are presented in §8.6.2. The relative ages of individual chondrites can be measured more precisely than their absolute ages using short-lived extinct radioactive isotopes. This technique, which can be used to determine an upper bound of ~5 × 10^6 to the *formation interval* of chondritic meteorites, is described in detail in §8.6.3.

Some isotopes of noble gases such as helium, argon, and xenon are produced by radioactive decay. These isotopes build up in rocks as time progresses, but can be lost if the rock is fractured or heated. The abundance of such radiogenic gases can be used to determine the *gas retention age* of the rock. Usually this gas retention age is less than the formation age, but for some rocks these two ages are equal. Lighter noble gases diffuse more readily than the heavy ones, implying a lower *closure temperature* to remain bound. So some events lead to the loss of most of a rock's helium, but little of its argon, resulting in a meteorite with a helium retention age younger than its argon retention age.

The time between the final epoch of nucleosynthesis that the material underwent prior to incorporation in the solar nebula and condensation can also be estimated from isotopic measurements. The techniques used for obtaining such estimates are described in §8.6.4. Finally, nuclear reactions produced by cosmic rays can be used to determine how long the meteorite existed as a small body in space, and the time at which it reached Earth (§8.6.5).

8.6.1 Radioactive Decay

Many naturally occurring nuclides are *radioactive*, that is they spontaneously decay into nuclides of other elements that are usually of lesser mass. Radioactive decay rates can be accurately measured; thus the abundances of decay products provide precise clocks that can be used to reconstruct the history of many rocks. Every radioactive decay process releases energy, and the resultant heating can lead to differentiation of planetary bodies. The most common types of radioactivity are *β decay*, whereby a nucleus emits an electron, and *α decay*, in which a helium nucleus (composed of two protons and two neutrons) is emitted.

When an atomic nucleus undergoes *β* decay, a neutron within the nucleus is transformed into a proton, so the atomic number increases by one; the total number of nucleons (protons plus neutrons) remains fixed, so the atomic mass number of the nucleus does not change and the actual mass of the nucleus decreases very slightly. An example of β decay is the following transformation of an isotope of rubidium into one of strontium: $^{87}_{37}\text{Rb} \rightarrow ^{87}_{38}\text{Sr}$; the number shown to the upper left of the element symbol is the atomic weight of the nuclide (number of nucleons), and the number to the lower left (which is usually omitted because it is redundant with the name of the element) is the atomic number (number of protons). Proton-rich nuclei can undergo *inverse β decay* (positron emission), decreasing their atomic number by one; a closely related process is *electron capture*, whereby an atom's inner electron is captured by the nucleus; both of these processes convert a proton into a neutron. An example of a decay that can occur via positron emission or electron capture is $^{40}_{19}\text{K} \rightarrow ^{40}_{18}\text{Ar}$; however, potassium 40 (β) decays into calcium ($^{40}_{20}\text{Ca}$) eight times more frequently than it decays into argon.

When a nucleus undergoes α decay, its atomic number decreases by two and its atomic mass decreases by four; an example of α decay from uranium to thorium is $^{238}_{92}\text{U} \rightarrow ^{234}_{90}\text{Th}$. Some heavy nuclei decay via *spontaneous fission*, which produces at least two nuclei more massive than helium, as well as smaller debris. Spontaneous fission is an alternative decay mode for $^{238}_{92}\text{U}$ (and also of the now almost extinct $^{244}_{94}\text{Pu}$), leading typically to xenon and lighter byproducts.

The time required for an individual radioactive nucleus to decay is not fixed; however, there is a characteristic lifetime for each radioactive nuclide. The probability that a nucleus will decay in a specified interval of time does not depend on the age of the nucleus, so the number of atoms of a given radioactive nuclide remaining in a sample drops exponentially if no new atoms of this nuclide are produced. The timescale over which this process occurs can be characterized by the *mean lifetime* of a species, t_m, or the *decay constant*, t_m^{-1}. The abundance of a 'parent' species at time t is related to its abundance at t_0 as

$$N_p(t) = N_p(t_0)e^{-(t-t_0)/t_m}. \tag{8.4}$$

Alternatively, the *half-life* of the nuclide, $t_{1/2}$, which represents the time required for half of a given sample to decay, can be used to quantify the decay rate. The relationship between these quantities is

$$t_{1/2} = \ln 2\, t_m. \tag{8.5}$$

The half-lives of nuclides commonly used to date events in the early Solar System are given in Table 8.3. Many nuclides produced by radioactive decay are themselves unstable. In many cases, these 'daughter' nuclides have

Table 8.3 Half-lives of selected nuclides.

Parent	Measurable stable daughter(s)	Half-life $t_{1/2}$
Long-lived radionuclides		
^{40}K	^{40}Ar, ^{40}Ca	1.25 Gyr
^{87}Rb	^{87}Sr	48 Gyr
^{147}Sm	^{143}Nd, ^4He	106 Gyr
^{187}Re	^{187}Os	44 Gyr
^{232}Th	^{208}Pb, ^4He	14 Gyr
^{235}U	^{207}Pb, ^4He	0.704 Gyr
^{238}U	^{206}Pb, ^4He	4.47 Gyr
Extinct radionuclides		
^{10}Be	^{10}B	1.4 Myr
^{22}Na	^{22}Ne	2.6 yr
^{26}Al	^{26}Mg	0.72 Myr
^{36}Cl	^{36}Ar, ^{36}S	0.3 Myr
^{41}Ca	^{41}K	0.1 Myr
^{53}Mn	^{53}Cr	3.6 Myr
^{60}Fe	^{60}Ni	1.5 Myr
^{92}Nb	^{92}Zr	35 Myr
^{107}Pd	^{107}Ag	6.5 Myr
^{129}I	^{129}Xe	16 Myr
^{146}Sm	^{142}Nd	68 Myr
^{182}Hf	^{182}W	9 Myr
^{244}Pu	$^{131-136}\text{Xe}$	82 Myr

shorter half-lives than their 'parents'. A sequence of successive radioactive decays leading to a stable or nearly stable nuclide is referred to as a *decay chain*. Two decay chains that are important in meteorite evolution are

$$^{238}_{92}\text{U} \xrightarrow[t_{1/2}=4.47\times10^9\,\text{yr}]{\alpha} {}^{234}_{90}\text{Th} \xrightarrow[21.4\,\text{d}]{\beta} {}^{234}_{91}\text{Pa} \xrightarrow[6.75\,\text{hr}]{\beta} {}^{234}_{92}\text{U}$$

$$\xrightarrow[2.47\times10^5\,\text{yr}]{\alpha} {}^{230}_{90}\text{Th} \xrightarrow[8\times10^4\,\text{yr}]{\alpha} {}^{226}_{88}\text{Ra} \xrightarrow[1600\text{yr}]{\alpha} {}^{222}_{86}\text{Rn}$$

$$\xrightarrow[3.8\text{d}]{\alpha} {}^{218}_{84}\text{Po} \xrightarrow[3\text{m}]{\alpha} {}^{214}_{82}\text{Pb} \xrightarrow[27\text{m}]{\beta} {}^{214}_{83}\text{Bi} \xrightarrow[19.9\text{m}]{\beta} {}^{214}_{84}\text{Po}$$

$$\xrightarrow[1.64\times10^{-4}\text{s}]{\alpha} {}^{210}_{82}\text{Pb} \xrightarrow[21\text{yr}]{\beta} {}^{210}_{83}\text{Bi} \xrightarrow[5\text{d}]{\beta} {}^{210}_{84}\text{Po}$$

$$\xrightarrow[183\text{d}]{\alpha} {}^{206}_{82}\text{Pb}\ (\text{stable}), \tag{8.6a}$$

$$^{235}_{92}\text{U} \xrightarrow[t_{1/2}=7.1\times10^8\,\text{yr}]{\alpha} {}^{231}_{90}\text{Th} \xrightarrow[25.5\text{hr}]{\beta} {}^{231}_{91}\text{Pa} \xrightarrow[3.25\times10^4\,\text{yr}]{\alpha} {}^{227}_{89}\text{Ac}$$

$$\xrightarrow[21.6\text{yr}]{\beta} {}^{227}_{90}\text{Th} \xrightarrow[18.5\text{d}]{\alpha} {}^{223}_{88}\text{Ra} \xrightarrow[11.43\text{d}]{\alpha} {}^{219}_{86}\text{Rn} \xrightarrow[4\text{s}]{\alpha} {}^{215}_{84}\text{Po}$$

$$\xrightarrow[1.8\times10^{-3}\text{s}]{\alpha} {}^{211}_{82}\text{Pb} \xrightarrow[36\text{m}]{\beta} {}^{211}_{83}\text{Bi} \xrightarrow[2.15\text{m}]{\alpha} {}^{207}_{81}\text{Tl}$$

$$\xrightarrow[4.8\text{m}]{\beta} {}^{207}_{82}\text{Pb}\ (\text{stable}). \tag{8.6b}$$

Note that the first decay in each of these chains takes of order 10^9 years, but subsequent decays are much more rapid.

8.6.2 Dating Rocks

With the passage of time, the abundance of radioactive 'parent' nuclides in a rock, $N_p(t)$, decreases, as these atoms decay into 'daughter' species (or 'granddaughter' etc. nuclides if the initial decay products are unstable with short half-lives, e.g., expressions (8.6)). The abundance of the daughter species can be expressed as

$$N_d(t) = N_d(t_0) + \xi(1 - e^{-(t-t_0)/t_m})N_p(t_0), \qquad (8.7)$$

where the *branching ratio*, $0 < \xi \leq 1$, represents the fraction of the parent nuclide that decays into the daughter species under consideration. The branching ratio is a fundamental property of a given nuclide. In most cases, $\xi = 1$.

The current abundances, $N_d(t)$ and $N_p(t)$, are measurable quantities. The initial abundance of the parent, $N_p(t_0)$, can be expressed in terms of the measured abundance and the age of the rock, $t - t_0$, using equation (8.4). However, this combination of equations (8.4) and (8.7) yields a single equation for two unknowns, $(t - t_0)$ and $N_d(t_0)$. If we could determine independently the 'initial' (nonradiogenic) abundance of the daughter nuclide (its abundance when the rock solidified), $N_d(t_0)$, then we could determine both the initial abundance of the parent and the age of the rock, $t - t_0$.

Chemical separation during a rock's solidification epoch can create an inhomogeneous sample that can be analyzed to determine both initial abundances and the age of the rock. Consider two samples within the rock containing different ratios of the parent element to the daughter element. Assume that the system can be considered closed (i.e., no migration since the rock solidified). Analysis of the samples provides relationships between $N_d(t_0)$ and $t - t_0$ for each sample. Presumably $t - t_0$ is the same for both, and at the time the rock solidified each element was isotopically well mixed. Thus, by measuring the abundances of a nonradiogenic isotope of the same element as the daughter isotope (e.g., ^{86}Sr for Rb–Sr dating), we can compute the ratio of $N_d(t_0)$ in the same samples, providing the third equation needed to solve for the three unknowns. In practice, several samples are usually analyzed and solutions are obtained graphically using an *isochron diagram* (Fig. 8.18). Since the abundance ratio of strontium isotopes was the same throughout the meteorite, whereas the abundance ratio ^{87}Rb/^{86}Sr varied between minerals as a result of chemical fractionation, a plot showing the relationship between these abundance ratios would have yielded a constant at $t = t_0$ (dashed line in Figure 8.18a). As the rock aged, however, the abundance ratio ^{87}Sr/^{86}Sr

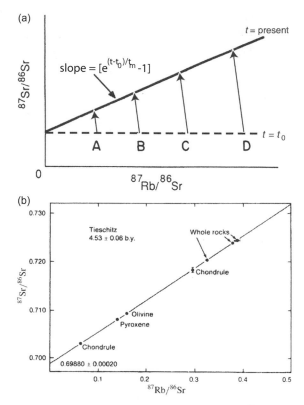

Figure 8.18 (a) Schematic isochron diagram of the ^{87}Rb–^{87}Sr system. Phases A, B, C, and D have identical initial ^{87}Sr/^{86}Sr ratios at $t = 0$, but differing ^{87}Rb/^{86}Sr ratios. Assuming that the system remains closed, these ratios evolve as shown by the arrows to define an isochron for which t is the age of the rock. (b) ^{87}Rb–^{87}Sr isochron for the Tieschitz unequilibrated H3 chondrite meteorite. (Taylor 1992)

has increased in proportion to the ^{87}Rb/^{86}Sr ratio. This has produced a current (measurable) slope on the isochron diagram of

$$\frac{d\left(^{87}\text{Sr}/^{86}\text{Sr}\right)}{d\left(^{87}\text{Rb}/^{86}\text{Sr}\right)} = e^{(t-t_0)/t_m} - 1 \qquad (8.8)$$

(Problem 8.14a). The slope on an isochron diagram thus gives the age of the rock, and the intercept represents the initial abundance of the daughter species. Isochron dating is preferred because if the system was disturbed, then the data do not fall along a straight line, and the age is known to be unreliable.

Operationally, the most difficult part of applying equation (8.8), and the part that leads to the largest uncertainty, is measuring the abundances of daughter and parent nuclides at the present epoch. Additional uncertainties come from imperfect knowledge of $t_{1/2}$ (especially for nuclides that β decay and have long half-lives) and of the branching ratio, ξ.

The long-lived isotopes of uranium, ^{235}U and ^{238}U, initiate decay chains that culminate in the lead isotopes ^{207}Pb and ^{206}Pb, respectively (expressions 8.6). These decay chains can be analyzed individually to estimate ages via the techniques described above for rubidium/strontium dating, using the nonradiogenic lead isotope ^{204}Pb as a stable comparison. However, more accurate dates can be obtained using the two chains in tandem. Two techniques are commonly used for this analysis, one of which only requires measurement of the lead isotopes and is described in Problem 8.14b. Radiometric dates for chondritic meteorites cluster tightly around 4.56×10^9 years. The majority of differentiated meteorites are of similar age, but many are younger, in some cases much younger. These results and their implications are discussed in more detail in §8.7.

8.6.3 Extinct-Nuclide Dating

Absolute radiometric dating requires that a measurable fraction of the parent nuclide remains in the rock. For rocks that date back to the formation of the Solar System, this implies long-lived parents, which due to their slow decay rates cannot give highly accurate ages. The *relative ages* of rocks which formed from a single well-mixed reservoir of material can be determined more precisely using the daughter products of short-lived radionuclides that are no longer present in the rocks. Additionally, extinct nuclei can provide estimates of the time between nucleosynthesis and rock formation (§8.6.4).

Correlations between the Al/Mg ratio and ^{26}Mg/^{24}Mg excess have been detected within chondritic meteorites. This excess cannot be the result of mass-dependent fractionation because the relative abundance of the nonradiogenic isotopes of magnesium, ^{25}Mg/^{24}Mg, is normal (or the excess exists after correcting for any mass-dependent fractionation found in ^{25}Mg/^{24}Mg, see §8.4.2). The stable isotope of aluminum, ^{27}Al, is a common nuclide, but the lighter isotope, ^{26}Al, decays on a timescale much shorter than the age of the Solar System ($t_{1/2} = 720\,000$ years for inverse β decay into ^{26}Mg). If a meteorite or piece thereof contained ^{26}Al when it solidified aeons ago, then decay of this isotope will have produced excess ^{26}Mg, and the amount of this excess is proportional to the local aluminum abundance. The constant of proportionality is the fractional abundance of ^{26}Al at the time of solidification. Provided the ^{26}Al was of presolar origin and the isotopes of aluminum were well mixed within the protoplanetary disk, the relative ages of different samples can be determined with a precision significantly greater than the half-life of ^{26}Al.

The extinct chronometers ^{26}Al and ^{53}Mn provide good estimates of relative times of rock crystallization.

Variations in the derived initial abundances of these nuclides imply that the *formation interval* of CAIs, chondrules, and primitive meteorites was $\lesssim 5$ Myr. Extinct nuclei such as ^{60}Fe and ^{182}Hf imply rapid formation and differentiation of some planetesimals (§8.7.1).

Another extinct chronometer is ^{129}I, which β decays into ^{129}Xe with a half-life of 16 million years. Because xenon is a noble gas, it is much rarer than iodine in meteorites, so even though only a small fraction of the iodine originally in the meteorite was radioactive, the amount of ^{129}Xe in the rock may have undergone a large percentage increase. By taking the ratio of the excess (that beyond the typical Solar System isotopic mix of xenon) ^{129}Xe to the abundance of ^{127}I, one can estimate the fraction of iodine that was radioactive at the time that migration of I and Xe within the meteorite ceased. Assuming that iodine was isotopically well mixed in the protoplanetary disk, this fraction declined with time in a predictable manner. As the ratio ^{129}Xe/^{127}I is the same to within a factor of two for primitive meteorites surveyed, they stabilized within an interval of $\sim 2 \times 10^7$ years. Since iodine and xenon are volatile, they may move subsequent to rock crystallization, and so provide only an upper bound to the actual formation interval.

8.6.4 Time from Final Nucleosynthesis to Condensation

Most atomic nuclei heavier than boron were formed in stars (including supernova explosions) or via radioactive decay of products of stellar nucleosynthesis; bombardment by energetic particles and photons also plays a significant role in transmuting nuclei (§13.2). Some isotopes, such as ^{26}Al, can be formed under many different circumstances, whereas others require very specific conditions. For instance, very heavy nuclides are only produced in supernova explosions. The relative abundances of various isotopes in Solar System materials preserve a record of contributions from the different environments in which nucleosynthesis occurred. Signatures of extinct nuclides place limits on the time between the final episode of nucleosynthesis in a particular environment and the condensation and solidification of Solar System materials. Some short-lived nuclides are particularly useful in constraining the properties of the region of the galaxy in which the Sun and planets formed and timescales of this process.

8.6.4.1 *r-Process Heavy Elements*

The *r-process nucleosynthesis* of heavy elements occurs when there is a sufficiently high flux of neutrons present that slightly unstable nuclei do not have time to β decay before further neutron capture (§13.2.2). Neutron-rich nuclei are thus produced by this rapid process, which is thought to occur in some types of supernovae. The ratio of certain similar nuclides produced in r-process nucleosynthesis can be estimated from models of supernovae.

In order to determine the time that elapsed between the last r-process nucleosynthesis of Solar System material and the condensation of meteorites, consider the initial Solar System abundances of two chemically similar r-process pairs of short-lived and long-lived nuclides which are believed to be produced at the same rate (^{244}Pu, ^{238}U) and (^{129}I, ^{127}I). The abundance ratios at the time of Solar System formation inferred from meteorites are: ^{244}Pu/^{238}U = 0.007 ($t_{1/2}(^{244}$Pu) = 8.2 × 10^7 yr); ^{129}I/^{127}I = 0.0001 ($t_{1/2}(^{129}$I) = 1.6 × 10^7 yr). The ratio of half-lives is 5, thus the ratio of relative abundances should be 5 if nucleosynthesis occurred at a uniform rate prior to formation of the Solar System. The observed ratio of 70 implies that r-process nucleosynthesis stopped ~8 × 10^7 years before condensation (Problem 8.17). This is approximately equal to the time it takes for a molecular cloud at the Sun's distance from the galactic center to pass from one spiral arm to the next. Most stars are formed when molecular clouds are compressed as they pass through spiral arms. The lifetimes of massive stars capable of r-process nucleosynthesis are only a few million years, so presumably the last injection of fresh r-process nuclides occurred during or soon after the Solar System-forming cloud's penultimate passage through a spiral arm.

8.6.4.2 Short-Lived Light Elements

The observed correlations between excess ^{26}Mg and aluminum abundance (§8.6.3) provide strong evidence that ^{26}Al was present in (at least some regions of) the early Solar System with an abundance of ~5 × 10^{-5} that of the stable isotope ^{27}Al. Note that, because aluminum is fairly abundant (Table 8.1), this fraction of radioactive aluminum could have provided a substantial heat source for melting planetesimals in the early Solar System (Problem 8.18).

The ^{26}Mg from ^{26}Al decay has been detected in many large inclusions that are otherwise isotopically normal. The correlation between ^{26}Mg excess and aluminum abundance implies that ^{26}Al was present (not extinct) in the early Solar System, which requires the interval of time between nucleosynthesis of the ^{26}Al and the formation of solid bodies within the Solar System to have been at most a few million years. Note that this is far shorter than the interval since last r-process nucleosynthesis discussed in §8.6.4.1. This is consistent with theoretical models that ^{26}Al is produced in a much wider array of astrophysical environments than are r-process nuclides.

Gamma-ray observations show that ^{26}Al is 'plentiful' in the galaxy today, implying that it can be formed in abundance. The ^{26}Al that we see evidence for in chondrites also may have been produced by high-energy particles released from the active young Sun which bombarded grains that approached to within a few solar radii. Excess ^{41}K, from decay of ^{41}Ca ($t_{1/2} = 10^5$ years), has also been detected in meteorites. The correlation between ^{41}K and ^{26}Mg observed in some meteorite samples indicates a common source for their parent nuclides ^{41}Ca and ^{26}Al.

8.6.4.3 Presolar Grains

Some isotopic anomalies found in chondritic meteorites indicate survival of grains that solidified prior to the formation of the protoplanetary disk rather than radioactive decay within the Solar System. Almost pure concentrations of the rare heavy isotope of the noble gas neon, ^{22}Ne, probably from sodium decay (^{22}Na, $t_{1/2} = 2.6$ years!) have been detected in small, carbon-rich phases within some primitive meteorites. Other pockets with enhanced (relative to Solar System averages) but not pure ^{22}Ne may have been created by implantation of neon from stellar winds. The grains containing heavy neon enhancements probably condensed within outflows from carbon-rich stars, supernova explosions, and possibly nova outbursts. Some SiC-dominated grains contain excess ^{49}Ti correlated with vanadium/titanium ratios; this anomaly may result from the condensation of grains containing ^{49}V ($t_{1/2} = 330$ d) within ejecta from supernova explosions. Some presolar grains show relics of much higher initial abundances of ^{26}Al than do any Solar System condensates. Decay products from the short-lived nuclides ^{41}Ca, ^{44}Ti, and ^{99}Tc have also been identified within presolar grains.

Isotopic abundances in presolar grains that were formed in stellar atmospheres and outflows sometimes indicate the age and mass of the star in which they originated. Some of this stardust clearly formed in supernovae, whereas other presolar grains emanate from post-main-sequence asymptotic giant branch (AGB) stars. Primitive meteorites thus contain information on the nucleosynthesis history of our galaxy. As some of the grains must have formed in stars that were already several billion years of age when our Solar System formed, studies of such grains also yield a lower bound to the age of the galaxy of nine billion years.

8.6.5 Cosmic-Ray Exposure Ages

Galactic cosmic rays are extremely energetic particles (~87% protons, ~12% alpha particles, and ~1% heavier nuclei) that possess enough energy to produce nuclear reactions in particles they collide with. Cosmic rays and the energetic secondary particles that they produce have a mean interaction depth of ~1 meter in rock, thus they do not affect the bulk of material in any sizable asteroid. The amount of cosmic rays that a meteorite has been exposed to indicates how long it has 'been on its own', or at least near the surface of an asteroid.

Cosmic-ray exposure ages are determined using measurements of the abundances of certain rare nuclides that in meteorites are almost exclusively produced by cosmic rays. Some of these nuclides are noble gases, e.g., ^{21}Ne and ^{38}Ar, while others are short-lived radionuclides such as ^{10}Be and ^{26}Al. Production rates of the various nuclides depend upon meteorite composition and depth below the surface. Comparison between various meteorites of the same class are useful in determining production rates. Radionuclides reach equilibrium abundances between production and decay if the meteorite is exposed for much longer than the half-life of the particular nuclide. Together, this information can be used to determine both depth and exposure age of many meteorites.

Typical cosmic-ray exposure ages are 10^5–10^7 years for carbonaceous chondrites, 10^6–10^8 years for other stones, 10^8 years for stony-irons, and 10^8–10^9 years for irons. The differences in age are due to material strength, which governs how quickly a body breaks up or a surface erodes. An additional factor which may contribute to the larger ages of irons and stony-irons is that the Yarkovsky force §2.7.3, which can help move small objects into resonant orbits within which perturbations by the giant planets can increase their eccentricities enough for them to become Earth-crossing (§9.1), is weaker on these meteoroids than it is on stony meteoroids. The relative numbers of meter-sized and smaller meteoroids of different classes that are being formed within the inner Solar System can be estimated by dividing the number of falls by the typical cosmic-ray ages of meteorites of that class. Note that the strength of iron meteorites helps them survive better than chondrites at many stages: in space, during atmospheric entry, and on the ground. Irons are also easier to identify as meteorites (except in Antarctica, where any rock on an ice sheet is distinctive and of suspect origin), making them more over-represented in meteorite collections. However, cosmic-ray ages imply that even fall statistics vastly overestimate the percentage of small fragments produced in the inner Solar System that are composed primarily of iron. The nonrandom cosmic-ray ages deduced imply that certain meteorite groups experienced major breakups that generated a large fraction of the members of each of these groups in a single event.

The *terrestrial age* of a meteorite is the time since it fell, i.e., how long the meteorite has been on Earth.

Weathered appearance is correlated with terrestrial age of hot desert meteorites, but for Antarctic meteorites no such correlation exists. The best method of determining the terrestrial age of a meteorite is to measure the relative abundances of two short-lived cosmic-ray produced radionuclides (two are required in order to eliminate the effects of variations of the cosmic-ray flux with depth). Terrestrial ages of hot desert meteorites are typically <50 kyr, although a few achondrites are up to ~0.5 Myr and some irons have been on Earth even longer. Terrestrial ages of Antarctic meteorites are generally <0.5 Myr, although a few Antarctic chondrites and irons are up to a few Myr old. The distribution in terrestrial ages can be used to constrain possible variations in the influx of meteorites.

8.7 Meteorite Clues to Planet Formation

Small bodies in the Solar System have not been subjected to as much heat or pressure as planet-sized bodies, and they remain in a more pristine state. Meteorites thus provide detailed information about environmental conditions and physical and chemical processes during the epoch of planet formation. This information pertains to timescales, thermal and chemical evolution, mixing, magnetic fields, and grain growth within the protoplanetary disk. Processes identified include evaporation, condensation, localized melting and fractionation, both of solids from gas and among different solids.

Meteorites definitively date the origin of the Solar System to about 1 part in 10^4. Chondritic solids formed within a period of ≲5 million years at the beginning of Solar System history. The age of the Solar System, based on ^{207}Pb/^{206}Pb dating of CAIs in the Efremovka and NWA 2634 CV3 meteorites, is $4.568 \pm 0.001 \times 10^9$ years; other nuclide systems and other CAIs yield similar ages. Some types of chondrules have ages indistinguishable from those of CAIs, whereas others appear to have solidified a few million years later. Meteorites from differentiated parent bodies are often a bit younger, but usually not very much. Almost all meteorites are thus older than known Moon rocks (~3–4.45 × 10^9 yr), and terrestrial rocks (≲4 × 10^9 yr, although some contain grains of the durable mineral zircon up to 4.4 × 10^9 yr old).

The vast majority of elements in most meteorite groups are identical in isotopic composition, aside from variations which may plausibly be attributed to mass fractionation, radioactive decay, or cosmic-ray irradiation; thus, matter within the solar nebula must have been relatively well mixed. Differences in isotopic composition between individual meteorites and between meteorites and the Earth yield information on the place of formation of the individual molecules and grains out of which the meteorites have formed. Some grains have very high D/H ratios (compared with cosmic values); such fractionation implies formation in (very) cold interstellar molecular clouds. Other grains have non-cosmic isotopic ratios in many elements that imply condensation in outflows from stars, such as occurs in the ejecta from supernova explosions (§13.2). Hence, meteorites seem to contain stellar outflow and interstellar condensates in addition to material that formed or was significantly processed within our own Solar System.

8.7.1 Meteorites from Differentiated Bodies

Isotopic anomalies found in some achondrites imply rapid differentiation and recrystallization of planetesimals. Excess ^{60}Ni, which is the stable decay product of ^{60}Fe ($t_{1/2} \sim 1.5 \times 10^6$ yr), is correlated with iron abundances in *HED* (howardite–eucrite–diogenite) achondrite meteorites.[2] HEDs originate from the asteroid 4 Vesta (or possibly another differentiated planetesimal), and were once molten. Live ^{60}Fe must thus have been present when the planetesimal resolidified. This implies differentiation within a few to several ^{60}Fe half-lives after nucleosynthesis. Signatures left by ^{26}Al have also been detected in HED meteorites, and dating using extinct manganese (^{53}Mn) and ^{182}Hf also implies that the HED parent body formed and differentiated within 3–5 Myr of the solidification of the oldest known Solar System materials (CAIs). Some iron meteorites have hafnium–tungsten (^{182}Hf–^{182}W) signatures that date differentiation of their parent bodies to ≲1.5 Myr after CAIs formed, and comparable Pb isotopic ages have been derived for the same differentiated meteorites. The oldest differentiated meteorites thus appear to be at least as old as most chondrules. This suggests that large planetesimals and small grains existed at the same time within the protoplanetary disk, albeit not necessarily in the same location.

The cooling rate of a rock, as well as the pressure and the gravity field that it was subjected to while cooling, can be deduced from the structure and composition of its minerals. Thus, we can estimate the size of a meteorite's original parent body. By knowing what size bodies melted in the early Solar System (and possibly where they accreted relative to other bodies by the presence or lack of volatiles in the meteorite), we can get a better idea of the heat sources responsible for differentiation.

Spontaneous fission of heavy radioactive nuclei, such as ^{244}Pu, causes radiation damage within crystalline materials in the form of *fission tracks*. This damage tends to be annealed out at high temperatures. The annealing temperature varies among minerals. The difference in fission track density among a set of meteoritic minerals with different annealing temperatures can be interpreted in terms of a cooling history. The retention of radiogenic noble gases in meteoritic minerals (§8.6.3) is analogous to fission track retention and can likewise be used to estimate a cooling history.

The composition and texture of differentiated meteorites such as achondrites, irons, and pallasites reflect igneous differentiation processes (i.e., large-scale melting) within asteroid-size parent bodies. It appears as if some small bodies ≲100 km in radius differentiated. Neither accretion energy nor long-lived radioactive nuclides provide adequate sources of heat. Possible heat sources include the extinct radionuclides ^{26}Al and ^{60}Fe, as well as *electromagnetic induction heating*. Electromagnetic induction heating, which occurs when eddy currents are generated within an object and dissipated via Joule heating. Meteorite parent bodies passing through currents that may have been produced by the massive T-Tauri phase solar wind (§13.3) would have been subjected to electromagnetic induction heating. Although there are many uncertainties in this mechanism, maximum heating would occur in bodies nearest the Sun, and probably for bodies between 50 and 100 km in radius. The largest ^{26}Al/^{27}Al concentrations observed in primitive meteorites would be sufficient to melt chondritic composition planetesimals as small as 5 km in radius. (The largest abundance of ^{60}Fe deduced in achondrites is a substantially weaker heat source.) Short-lived radionuclides would not, by themselves, have provided enough energy to melt chondritic planetesimals of any radius formed ≳2 Myr (three half-lives of ^{26}Al) after CAI formation, but they could have led to less drastic thermal processing. The formation ages of many iron meteorites suggests that their parent bodies accreted early enough for melting via ^{26}Al decay. The positive correlation between age of chondrules and degree of metamorphism of the primitive meteorites that contain them is also consistent with ^{26}Al being a major heat source for early-formed planetesimals.

The great age of iron meteorites from the cores of differentiated bodies combined with the paucity of dunite (§5.1.2.2) meteorites from the mantles of such bodies can be explained if the parent bodies of irons formed in the inner Solar System. In this scenario, the mantles of these bodies were collisionally stripped off and pulverized

---

[2] Nickel and several other elements are more siderophile than iron itself, and thus are more enriched in cores and more depleted in mantles and crusts of differentiated bodies.

during the growth of the terrestrial planets, but residual cores of a small fraction of inner Solar System planetesimals were scattered into the asteroid belt, where they have survived to the present epoch in a region dominated by later-formed locally accreted bodies.

8.7.2 Primitive Meteorites

Most meteorites that are observed to fall are chondrites, which have never been molten and thus preserve a better record of conditions within the protoplanetary disk than do the differentiated meteorites. Indeed, some of the grains in chondrites predate the Solar System, and thus also preserve a record of processing in stellar atmospheres, winds, explosions, and the interstellar medium. These grains may have been affected by passage through a hot shocked layer of gas during their entry into the protoplanetary disk. The precursors to chondrules and CAIs formed by agglomeration of presolar grains and solar nebula condensates. These agglomerates were subsequently heated to the point of melting. Many had their rims melted at a later time, and/or were fragmented as the result of high-speed collisions. Ultimately, they were incorporated into planetesimals in which they were subjected to nonhydrous processing at 700–1700 K (especially petrographic types 4–6) and/or hydrothermal processing at lower temperatures (primarily types 1 and 2). Some chondrites also show evidence for shock processing in the geologically recent past.

The differences in bulk composition among chondrites are closely related to the volatilities of the constituent elements. In almost all cases, the more volatile elements are depleted; however, very refractory materials may be depleted relative to silicon in some enstatite and ordinary chondrites. Small inclusions of very refractory material (CAIs) are seen in chondrites, but bulk meteorites of such refractory composition have not been found. Magnesium silicates and iron–nickel metal, which are not significantly depleted in primitive meteorites, condense at ~1300 K. A gradual depletion pattern with increasing volatility is seen in most chondrites (Fig. 8.19). If each meteorite was formed in equilibrium at a unique temperature, then relative abundances of elements with condensation temperatures above this value would be solar, and more volatile elements would be almost absent. The only elements that would be depleted by tens of percent would be those that either condense at temperatures very near the local equilibration temperature (the saturation vapor pressures of most compounds vary rapidly with temperature near their 50% condensation temperature; see §4.4), or those that form refractory compounds exclusively with rarer elements. The gradual depletion patterns which

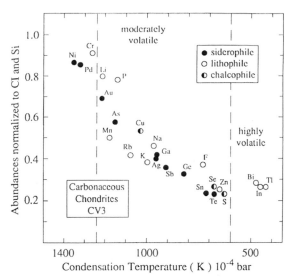

Figure 8.19 The abundances of moderately volatile elements in bulk CV chondrites compared to their abundances relative to silicon in CI chondrites are plotted against the condensation temperatures of the elements in a solar composition gas. The gradual decrease in abundance with decreasing condensation temperature implies that the components of individual meteorites condensed or were altered in a variety of environments. The lack of dependence of abundances on the geochemical character of the elements shows that the meteorites have not been fully melted subsequent to accretion. (Palme and Boynton 1993)

are observed (Fig. 8.19) imply that the constituents of individual meteorites condensed in a broad range of environments. Grains could be brought together from a variety of locations to produce such admixtures, or most of the material that formed the terrestrial planets and asteroids cooled to around 1300 K while the gaseous and solid components remained well mixed, and then gas was subsequently lost as material cooled further. Significant condensation in the asteroid region continued down to ≲500 K before the gas was completely removed. Elemental depletions in bulk terrestrial planets and differentiated asteroids are consistent with this conclusion, although they are less well constrained.

The mineral composition of chondrites implies substantial variations in the *fugacity* (fractional abundance) of oxygen, fO_2, within the solar nebula. Enstatite chondrites formed in a highly reducing environment, whereas CAI rims solidified in environments with fO_2 up to 10^4 times that corresponding to solar composition. Most CAIs appear to have formed within an oxygen-rich gas, whereas most chondrules do not. More reducing conditions could have been created locally by evaporation of C-rich dust (which would have sequestered oxygen in CO) and/or removal of H_2O into icy planetesimals. Oxidizing conditions probably resulted from enhancements in the local ratio of O-rich dust and ice to H-rich gas within the protoplanetary disk or impact-generated vapor clouds and/or photochemical destruction of H_2O.

Remanent magnetism in carbonaceous chondrites suggests that a magnetic field of strength 1–10 G existed at some locations within the protoplanetary disk. The magnetic field recorded in the high-temperature component is anisotropic from chondrule to chondrule, so magnetization presumably occurred prior to the incorporation of the chondrules into their meteorite parent body. Possible sources for such magnetism include a dynamo within the protoplanetary disk, the solar field carried outward by solar wind, or the solar field itself if the chondrules cooled through the Curie point while in the vicinity of the protosun.

The general high degree of uniformity of isotope ratios implies the solar nebula was for the most part well mixed, but the small violations of this rule tell us that some things didn't mix or never vaporized. Oxygen isotopic ratios show relatively large variations that cannot be explained by mass-dependent fractionation within primitive chondrites and between groups of meteorites (Fig. 8.16); these data are usually taken to imply distinct reservoirs that were incompletely mixed during nebular processes, although non-mass-dependent fractionation, via e.g., photochemical processes, can occur in certain circumstances (§8.4.2). Why is oxygen special? Oxygen is the only element common in both high- and low-temperature phases (apart from sulfur, which displays much smaller mass-independent fractionation in meteorites). The oxygen isotopic variations could have been produced if both the gas and the grains were well mixed, but no isotopic equilibrium existed between the two in portions of the solar nebula where the grains never completely evaporated.

There is no strong evidence that any macroscopic ($\gtrsim 10$ μm) grains are of presolar origin. However, tiny carbon-rich grains (most of which are ≪10 μm) that clearly represent surviving interstellar material have been found in some chondrites (Fig. 8.20). Unambiguous proof of the presolar origin of these grains comes from their isotopic compositions, which are anomalous both in trace elements such as Ne and Xe and in the more common elements C, N, and Si. While carbon-rich silicon carbide and graphite grains stand out most clearly, the majority of interstellar grains found in meteorites are silicates. The survival of interstellar grains, and their ability to retain noble gases, constrains the thermal and chemical environments that they experienced on their journey from

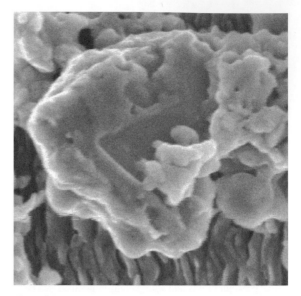

Figure 8.20 Image of a tiny presolar silicon carbide (SiC) grain (1 μm across) extracted from the Murchison meteorite. This very high resolution secondary electron image was obtained using a scanning electron microscope. The worm-like background shows a foil substrate that is not part of the grain. (Courtesy Scott Messenger)

interstellar cloud to meteorite parent bodies. Some of the grains were clearly never heated above 1000 K, and must have been much cooler during any episode in which they were exposed to an oxygen-rich environment.

Some refractory inclusions contain isotopic anomalies in several elements that appear to have been produced by a single (or specific combination of) nucleosynthetic process(es) (§13.2.2); the particular nucleosynthesis environment(s) differs from grain to grain. These anomalies imply that interstellar grains formed in various environments and survived entry into the protoplanetary disk.

The D/H ratios observed in organic material within carbonaceous chondrites are far higher than solar values, and $^{15}N/^{14}N$ is also enhanced. (D/H is more than 1000 times the solar value in some grains.) Fractionation of this magnitude is difficult if not impossible to achieve within the warm solar nebula. Rather, these complex hydrocarbons or their precursor molecules are thought to be formed by ion–molecule reactions in very cold interstellar clouds (which are capable of producing high D/H ratios). Either substantial amounts of interstellar grains survived or the gas containing the hydrocarbon molecules did not get sufficiently hot for isotopic equilibration to occur. This implies that some of the material in meteorites was never heated to the $\gtrsim 1500$ K temperatures experienced by chondrules and CAIs.

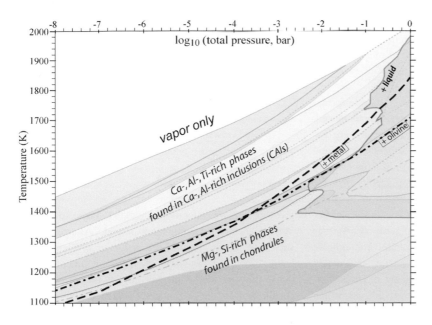

Figure 8.21 Phase diagram showing the principal constituents of chondrules and CAIs. The various lines show the condensation boundaries for different minerals, with the most important constituents labeled. (Ebel 2006)

8.7.3 Chondrule and CAI Formation

A major conceptual gap in meteoritics is how the epoch of global nebular cooling and settling of solids relates to the very rapid heating episodes that produced chondrules and CAIs. The compositions of CAIs imply equilibrium temperatures of ~1400–1500 K (Fig. 8.21). Most chondrules, in contrast, contain various volatile materials, albeit in subsolar abundances relative to silicon. Thus, as with chondrites in bulk, individual chondrules cannot be viewed as solar nebula condensates at any single temperature. The structure of chondrules and the melted rims of CAIs evidence especially high temperatures during transient heating events. The peak temperatures reached by chondrules and CAIs were 1700–2000 K. At nebular densities, molecular hydrogen dissociates at about this temperature range. The energy required to dissociate H_2 could have provided an effective thermostat, preventing the molten chondrules and CAIs from being heated to even higher temperatures.

The presence of significant amounts of moderately volatile elements shows that chondrules were not molten under equilibrium conditions, implying local, brief heating events. The volatile elements would have been lost had the chondrules remained molten for more than a few minutes. If chondrule precursors were fluffy dustballs (as most models assume), then the heating interval must have been similarly short. The textures and mineral compositions imply cooling rates of 50–1000 K per hour for chondrules and 2–50 K per hour for CAIs. The rapidity of chondrule heating and cooling implies local processes were responsible, as large regions of the nebula cannot cool quickly.

The origin of chondrules and CAIs remains enigmatic. Possible mechanisms for chondrule heating include drag during passage through an accretion shock either upon entry into the protoplanetary disk or during subsequent disk processing, flares or lightning within the disk, and heating by intense sunlight coincident with their removal from the vicinity of the Sun by the powerful T-Tauri phase solar wind. Melting during entry into the protoplanetary disk requires pre-existing clumps much larger than observed interstellar grain sizes, and would probably be evidenced by substantially greater isotopic variations than have been detected. The disk mechanisms all suffer from the energetics problem of concentrating a substantial amount of energy more than 1 AU from the Sun. The homogeneity of chondrule properties within individual chondrites compared to the overall diversity of chondrules is most easily explained if chondrules were incorporated into meteorites near where they formed; this homogeneity is difficult to explain if chondrules formed very near the Sun.

Lightning within the protoplanetary disk might have provided the energy to rapidly melt chondrules and CAIs. Models suggest that lightning analogous to that produced in terrestrial clouds (§4.5.4.5) could have occurred within the asteroidal region of the disk. Size-segregated motions of chondritic composition particles could have been produced by gas convection or turbulence, or by the component of solar gravity perpendicular to the midplane of the protoplanetary disk. These moving particles presumably would have transferred charge as do ice particles in thunderstorms within Earth's atmosphere. Large-scale charge separation would have built up until the nebular gas broke down. Observations of lightning in terrestrial volcanic plumes and in outer planet atmospheres (§4.5.6) show that lightning can be produced in a wide variety of environments.

The nebular shock wave model of chondrule formation envisions gas overrun by a shock and abruptly heated, compressed, and accelerated. A *shock front* is a sharp discontinuity between hot, compressed, supersonic gas and cooler, less dense, slower moving gas (§7.1). Shocks may have been produced within the protoplanetary disk by, e.g., energetic outbursts from the protosun or by dense clumps of gas falling onto the disk. The timescales for shock heating are seconds, in accord with observations of chondrules and CAIs. However, radiative cooling in an optically thin region of the protoplanetary disk would also have occurred in seconds, producing different mineral characteristics than those observed. Shocks which produced large numbers of chondrules within dense, dust-rich regions of the disk may have allowed cooling on a slower timescale consistent with observations.

Sufficient energy for the thermal processing required for the formation of CAIs and chondrules was readily available close to the protosun. CAIs, chondrules, and their rims could form when solid bodies were lifted out of the relatively cool shaded region of the protoplanetary disk into direct sunlight by the aerodynamic drag of a magnetocentrifugally driven T-Tauri phase solar wind. Such winds have been theorized to explain the substantial bipolar outflows observed to accompany star formation and the loss of angular momentum by the star and inner disk required for accretion to proceed (§13.3.3). In this *X-wind* model, two-thirds of the matter reaching the inner edge of a protoplanetary disk is accreted by the protostar, whereas one-third is ejected into interstellar space in a powerful stellar wind (Fig. 8.22). For reasonable parameters of the bipolar flow, the peak temperatures reached by solid bodies resemble those needed to melt CAIs and chondrules. The high-velocity expulsion of these bodies from the vicinity of the star yields rapid cooling, although it is not clear whether or not it would be sufficiently quick to be in agreement with the observed chondrule properties. Rims could have formed when very rapidly moving dust grains (which are better coupled to the wind than are larger particles) impacted chondrules and CAIs. Small solid grains would be carried by the wind off into interstellar space. Large particles would fall back to the inner portion of the protoplanetary disk and return to the vicinity of the protostar. Some of these large particles would be accreted by the protostar, with others being cycled again. CAI-sized objects would be

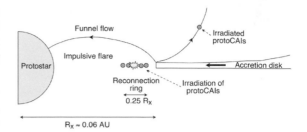

Figure 8.22 Schematic drawing of the magnetic field and gas flow in the stellar X-wind model for the production of CAIs. (Chaussidon and Gounelle 2006)

given just enough velocity for them to fall back to the disk in the planetary region. Chondrule and CAI sizes would vary with position within the disk (the smaller objects being thrown the furthest) and with their time of formation as the protosolar wind evolved, thus the correlation of chondrule properties within individual meteorites could be understood.

A simpler explanation of the correlation of chondrule properties within individual meteorites, together with the compositional complementarity between chondrules and matrix, is that chondrules formed near the locations where they were incorporated into larger bodies. Heating by shocks within the protoplanetary disk is currently the preferred (although by no means consensus) model of chondrule origins. Shocks are favored over stellar heating because they provide for local formation, and preferred over lightning because shock-induced compression produces an environment that has enhanced density (and thus likely less rapid cooling), whereas lightning likely reduces density via thermal expansion. Yet it isn't clear how sufficiently strong shocks can be produced within a protoplanetary disk. The greater homogeneity of CAI ages and physical characteristics, combined with the large amount of heat required to melt these very refractory objects, means that models of heating near the protosun are more likely to apply to CAIs than to chondrules.

8.8 Perspectives

In addition to being fascinating in their own right, meteorites provide us with a wealth of extremely detailed data on conditions during the planet-forming epoch. Some of these data are easy to interpret. For instance, radiometric dates provide an estimate of the age of the Solar System accurate to better than one part in five thousand (200 ppm). Other data give potentially very valuable clues but are subject to more diverse interpretations. The near but not total homogeneity of isotopic composition among meteorites tells us that substantial mixing of presolar material occurred, but some interstellar grains survived, and also that certain short-lived radionuclides were present in the material out of which the planets formed. The local mineralogical and compositional heterogeneity of primitive meteorites implies an active dynamic environment within the protoplanetary disk, but specific models to explain these data remain quite controversial research topics.

Further Reading

Nice nontechnical summaries, including many good photographs can be found in:

Wasson, J.T., 1985. *Meteorites: Their Record of Early Solar-System History.* W.H. Freeman, New York. 274pp.

McSween, H.Y., Jr., 1999. *Meteorites and their Parent Planets*, 2nd Edition. Cambridge University Press, Cambridge. 322pp.

Zinner, E., 1998. Stellar nucleosynthesis and the isotopic composition of presolar grains from primitive meteorites. *Annu. Rev. Earth Planet. Sci.*, **26**, 147–188.

Taylor, S.R., 2001. *Solar System Evolution*, 2nd Edition. Cambridge University Press, Cambridge. 484pp.

Lipschutz, M.E., and L. Schultz, 2007. Meteorites. In *Encyclopedia of the Solar System*, 2nd Edition. Eds. L. McFadden, P.R. Weissman, and T.V. Johnson, Academic Press, San Diego, pp. 251–282.

A very useful collection of review chapters can be found in:

Lauretta, D.S., and H.Y. McSween, Eds., 2006. *Meteorites and the Early Solar System II.* University of Arizona Press, Tucson. 942pp.

A compendium of articles written from various different viewpoints can be found in:

Krot, A.N., E.R.D. Scott and B. Reipurth, Eds., 2005. *Chondrites and the Protoplanetary Disk. ASP Conference Series* **341**, Astronomical Society of the Pacific, San Francisco. 1029pp.

Additional information on radiometric dating can be found in:

Tilton, G.R. 1988. Principles of radiometric dating. In *Meteorites and the Early Solar System*. Eds. J.F. Kerridge and M.S. Matthews. University of Arizona Press, Tucson. pp. 249–258.

More details on dating using specific isotopes are provided in:

Fowler, C.M.R., 2005. *The Solid Earth*, 2nd Edition. Cambridge University Press, Cambridge. 685pp.

Three good articles discussing meteorite clues to the formation of the Solar System are:

Kerridge, J.F., 1993. What can meteorites tell us about nebular conditions and processes during planetesimal accretion? *Icarus*, **106**, 135–150.

Palme, H., and W.V. Boynton, 1993. Meteoritic constraints on conditions in the solar nebula. In *Protostars and Planets III.* Eds. E.H. Levy and J.I. Lunine. University of Arizona Press, Tucson, pp. 979–1004.

Podosek, F.A., and P. Cassen, 1994. Theoretical, observational, and isotopic estimates of the lifetime of the solar nebula. *Meteoritics*, **29**, 6–25.

Two very different models of the formation of chondritic meteorites are given in:

Shu, F.H., H. Shang and T. Lee, 1996. Toward an astrophysical theory of chondrites. *Science*, **271**, 1545–1552.

Scott, E.R.D., and A.N. Krot, 2005. Thermal processing of silicate dust in the solar nebula: Clues from primitive chondrite matricies. *Astrophys. J.*, **623**, 571–578.

Problems

8.1.I Calculate the speed at which meteoroids with the following heliocentric orbits encounter the Earth's atmosphere:

(a) An orbit very similar to that of Earth, so $v_{inf} \ll v_e$, and thus $v_{impact} \approx v_e$.

(b) A parabolic orbit, with perihelion of 1 AU and $i = 180°$.

(c) A parabolic orbit, with perihelion of 1 AU and $i = 0°$.

(d) A parabolic orbit, with perihelion of 1 AU and $i = 90°$.

(e) An orbit with $a = 2.5$ AU, $e = 0.6$, and $i = 0°$.

(f) An orbit with $a = 2.5$ AU, $e = 0.6$, and $i = 30°$.

8.2.I (a) Calculate the kinetic energy of a meteoroid of radius 1 cm and density 1 g cm^{-3} moving at 20 km s^{-1}.

(b) Assume that this meteoroid enters the Earth's atmosphere and radiates away 0.01% of its kinetic energy as visible light over a period of 5 seconds. What is the rate at which it radiates during this period? State your answer both in erg s^{-1} and in watts.

(c) At what visual magnitude would this meteor appear to an observer at a distance of 100 km?

8.3.I (a) Calculate the pressure on a meteor moving at a speed of 10 km s^{-1} at an altitude of 100 km above the Earth's surface.

(b) Repeat for a meteor at the same speed 10 km above Earth's surface.

(c) Repeat your calculations in parts (a) and (b) for a meteor traveling at 30 km s^{-1}.

(d) The tensile strengths of comets are of order 10^4 dyne cm^{-2}, the strengths of chondrites are roughly 10^8 dyne cm^{-2}, stronger stony objects have strengths approximately 10^9 dyne cm^{-2}, whereas iron impactors have effective strengths of about 10^{10} dyne cm^{-2}. Compare these tensile strengths to the pressures calculated in parts (a)–(c) and comment.

8.4.E (a) Calculate the size of an iron meteor (density $\rho = 8$ g cm^{-3}) that passes through an amount of atmospheric gas equal to its own mass en route to the surface of the Earth. You may assume a spherical meteorite, vertical entry into the atmosphere, and neglect ablation.
(b) Repeat your calculation for a chondritic meteorite of density $\rho = 4$ g cm^{-3}.
(c) Repeat your calculation in part (a) for an entry angle of 45°.

8.5.E Calculate the terminal velocity near the Earth's surface for falling rocks of the following sizes and densities:
(a) $R = 10$ cm, $\rho = 8$ g cm^{-3}.
(b) $R = 10$ cm, $\rho = 2$ g cm^{-3}.
(c) $R = 100$ cm, $\rho = 2$ g cm^{-3}.
(d) $R = 100$ μm, $\rho = 2$ g cm^{-3}.

8.6.E (a) Calculate the equilibrium temperature of a meteoroid of mass M, density ρ, and albedo A in the vicinity of the Earth.
(b) Evaluate your result for a chondrite with $M = 10^9$ g, $\rho = 2.5$ g cm^{-3}, and albedo $A = 0.05$ and for an achondrite with $M = 10^6$ g, $\rho = 3$ g cm^{-3}, and albedo $A = 0.3$.

8.7.E Use equation (8.4) to verify that t_m is indeed the average lifetime of the isotope.

8.8.E Calculate the fractional abundance of ^{234}U in naturally occurring uranium ore. (Hint: Use expression (8.6a).)

8.9.I (a) Use tabulated data on nuclear decay (from e.g., the *CRC Handbook* or the web) in order to write the decay chain from ^{244}Pu to ^{232}Th.
(b) Continue this decay chain until a stable isotope is reached.

8.10.E (a) Use the decay chains given in expressions (8.6a,b) to estimate lower bounds on the abundance of elements 84–91 in terrestrial uranium ore.
(b) Why are your values only lower bounds?
(c) Why are your estimates less trustworthy for elements with atomic number ≤86 than they are for elements with higher atomic numbers?

8.11.I Estimate the amount of (naturally occurring) ^{244}Pu present on Earth at the present epoch. (Hint: You may assume that the abundance ratio of the longest-lived isotopes of plutonium and uranium 4.56×10^9 years ago was ^{244}Pu/^{238}U $= 0.005$ and that the Earth has a chondritic abundance of uranium.) Comment on the fact that Pu is often considered not to occur in nature.

8.12.I All elements with atomic numbers 1–118 have either been discovered in nature or produced in the laboratory. In this exercise, you will determine a complicated set of answers to the question of how many of these elements are naturally occurring.
(a) How many elements have at least one stable isotope? What are the atomic numbers of these elements? Note: Ignore the possibility of proton decay, which may make all isotopes unstable on *very* long timescales ($t_{1/2} > 10^{35}$ yr).
(b) Which other elements have isotopes that are so long-lived that they have survived in measurable quantities on Earth?
(c) Which elements not represented above are produced on Earth as the result of radioactive decay of naturally occurring elements? Justify your answer by writing down the appropriate decay chains. Note: Some isotopes have more than one possible decay path.
(d) A very small amount of the plutonium isotope ^{244}Pu has survived since the Earth formed (Problem 8.11). Does ^{244}Pu decay add any more elements to your list of naturally occurring elements?
(e) Spontaneous fission of uranium produces a variety of daughter nuclei. Although the dominant massive element produced is xenon, the distribution is broad and contains small amounts of two elements without stable isotopes. Which elements?
(f) Spontaneous fission releases neutrons which can initiate nuclear reactions in uranium ore. A few billion years ago, the concentration of ^{235}U was still sufficient to initiate chain reactions, producing *natural nuclear reactors*, which have left isotopic signatures that have been detected in some rich uranium ore. Although the concentration of ^{235}U is no longer sufficient to produce chain reactions, some neutrons released by spontaneous fission are absorbed by ^{238}U and others initiate fission in ^{235}U. Can these reactions add any elements to your list? What about neutron capture by ^{244}Pu?

8.13.I Determine the least abundant element represented in each of parts (a)–(d) of the previous problem and the total quantity of these elements on Earth. You may assume chondritic abundances for elements in parts (a) and (b).

8.14.I (a) Derive equation (8.8). (Hint: Differentiate equations (8.4) and (8.7) with respect to $N_p(t_0)$.)
(b) The primary decay modes of uranium isotopes ^{235}U and ^{238}U initiate decay chains which ultimately produce the lead isotopes ^{207}Pb and ^{206}Pb respectively; intermediate decay products

are relatively short-lived and thus contain insignificant amounts of material (see expressions 8.6a,b). At the time that a rock solidifies, an isochron plot of $^{207}Pb/^{204}Pb$ vs. $^{206}Pb/^{204}Pb$ is represented by a point. How does an isochron plot of these ratios appear at subsequent times? Be quantitative, and derive an equation analogous to equation (8.8). You may neglect the small fraction of uranium ($<10^{-4}$) that decays via spontaneous fission. (Hint: The ratio between the abundances of the uranium isotopes is the same in all minerals, but the ratio of uranium to lead varies.) This technique for determining the age of a rock is known as *lead–lead dating*.

8.15.**I** The following isotopic abundances (atoms/10^6 Si atoms) are measured in CI carbonaceous chondrites and meteorite X respectively:

Isotope	CI chondrites	Meteorite X
^{204}Pb	0.0612	0.1224
^{206}Pb	0.603	?
^{207}Pb	0.650	2.63
^{235}U	6.49×10^{-5}	1.59×10^{-2}
^{238}U	8.49×10^{-3}	?

Assume that uranium and its daughters decay exclusively by α and β decay, and that no mass-dependent fractionation has occurred. Determine the values of each of the quantities represented by question marks, and calculate the age of meteorite X.

8.16.**E** In this problem, you will calculate the age of a rock using actual data on the abundances of rhenium and osmium, which are related via the decay

$$^{187}Re \xrightarrow[t_{1/2}=4.16\times10^{10}yr]{\beta} {}^{187}Os.$$

(a) The following list summarizes some measurements of Re and Os isotope ratios for different minerals within a particular rock:

$^{187}Re/^{188}Os$	$^{187}Os/^{188}Os$
0.664	0.148
0.669	0.148
0.604	0.143
0.484	0.133
0.512	0.136
0.537	0.138
0.414	0.128
0.369	0.124

Plot the results on a piece of graph paper with $^{187}Re/^{188}Os$ along the horizontal axis and $^{187}Os/^{188}Os$ along the vertical axis.

(b) Draw a straight line that goes as closely as possible through all the points and extend your line to the vertical axis to determine the initial ratio of $^{187}Os/^{188}Os$.

(c) Draw and label several lines representing theoretical isochrones for a rock with the same initial ratio of $^{187}Os/^{188}Os$ as the rock being studied. Use these lines to estimate the age of the rock.

8.17.**I** Estimate the time interval between the last r-process nucleosynthesis of Solar System material and the condensation of chondrites. (Hint: Write down a formula for the abundance pair ratio of $(^{244}Pu/^{238}U)/(^{129}I/^{127}I)$ as a function of time assuming no ongoing nucleosynthesis. Use your formula to determine the time required for this ratio to grow from its 'steady-state' value of 5 to the observed value of 70.)

8.18.**I** Calculate the abundance of ^{26}Al (in grams per gram of chondritic material and as a ratio to the abundance of ^{27}Al in chondrites) required to generate sufficient heat to melt a chondritic mixture of magnesium silicates and iron initially at 500 K. You may assume that the asteroid is sufficiently large that negligible heat is lost during the period in which most of the ^{26}Al decays.

9 Minor Planets

I have announced this star as a comet, but since it is not accompanied by any nebulosity and, further, since its movement is so slow and rather uniform, it has occurred to me several times that it might be something better than a comet.

Giuseppe Piazzi, 24 January 1801, commenting on the object that he had discovered 23 days earlier, which was later determined to be the first known minor planet, 1 Ceres

In addition to the eight known planets, countless smaller bodies orbit the Sun. These objects range from dust grains and small coherent rocks with insignificant gravity to *dwarf planets* that have sufficient gravity to make them quite spherical in shape. Most are very faint, but some, the comets, release gas and dust when they approach the Sun and can be quite spectacular in appearance (Fig. 10.1); comets are discussed in Chapter 10. In this chapter, we describe the orbital and physical properties of the great variety of non-cometary small bodies ranging in radius from a few meters to over 1000 km that orbit the Sun. We refer to these bodies collectively as *minor planets*.

Minor planets occupy a wide variety of orbital niches (see Fig. 1.2). Most travel in the relatively stable regions between the orbits of Mars and Jupiter (known as the *asteroid belt*), exterior to Neptune's orbit (the *Kuiper belt*), or near the triangular Lagrangian points of Jupiter (the *Trojan asteroids*). The Kuiper belt is by far the most massive of these reservoirs, and contains the largest objects. However, the largest members of the asteroid belt appear much brighter than any Kuiper belt objects (KBOs) by virtue of their proximity to both the Earth and the Sun (see eq. 10.3 with $\zeta = 2$).

Smaller numbers of minor planets are found in unstable regions. Most of these cross or closely approach the orbits of one or more of the eight planets, which control their dynamics. Those that come near our home planet are known as *near-Earth asteroids* (NEAs); those orbiting among the giant planets are called *centaurs*.

Asteroids is a term generally used for rocky minor planets that orbit the Sun at distances ranging from interior to Earth's orbit to a bit exterior to the orbit of Jupiter. Well over 200 000 asteroids have been permanently cataloged, and many more are added each year. Asteroids exhibit a large range of sizes, with the largest asteroid, 1 Ceres, being ∼475 km in radius. The next largest asteroids

are 2 Pallas, 4 Vesta, and 10 Hygiea, ranging in radius from about 260 to 200 km. The 20 largest asteroids are listed in Table 9.1. Smaller asteroids are far more numerous than the larger ones; the number of asteroids with radii between R and $R + dR$ scales roughly as $R^{-3.5}$, implying that most of the mass in the asteroid belt is contained in a few large bodies. The total mass in the asteroid belt is ∼5 × 10^{-4} $M_{\oplus}$.

The Kuiper belt, beyond the orbit of Neptune, is analogous to the asteroid belt, but on a grander scale. *Kuiper belt objects* are icy bodies, and the largest KBOs are an order of magnitude more massive than 1 Ceres. The total mass of the Kuiper belt exceeds that of the asteroid belt by about two orders of magnitude. Yet because the Kuiper belt is located much further from both the Earth and the Sun than is the asteroid belt, far more is known about asteroids than about KBOs.

Pluto, by far the brightest KBO and the first that was discovered, was officially classified as a planet from 1930 until 2006; 1 Ceres, the first detected (in 1801) and by far the largest member of the asteroid belt, was also once considered to be a planet, as were the next few asteroids that were discovered. With the detection of other KBOs, debates began with regard to the classification of Pluto as a planet, and culminated in August 2006 with the resolution by the International Astronomical Union (IAU):

- A *planet* is a celestial body that (1) is in orbit around the Sun, (2) has sufficient mass for its self-gravity to overcome rigid body forces so that it assumes a hydrostatic equilibrium (nearly round) shape, and (3) has cleared the neighborhood around its orbit.

- A *dwarf planet* is a celestial body that (1) is in orbit around the Sun, (2) has sufficient mass for its self-gravity to overcome rigid body forces so that it assumes a hydrostatic equilibrium (nearly round) shape, (3) has not cleared the neighborhood around its orbit, and (4) is not a satellite.

Table 9.1 Twenty largest asteroids ($a < 6$ AU).

#	Name	Tax. class	M_v	Radius^{a,b} (km)	A_0	a (AU)	e	i (deg)	P_{orb} (yr)	P_{rot} (hr)	Axial tilt^{b,c} (deg)
1	Ceres	C/G	3.34	467.6	0.09	2.766	0.080	10.59	4.607	9.075	9
4	Vesta	V	3.20	264.5	0.42	2.362	0.090	7.13	3.629	5.342	32
2	Pallas	B	4.13	256	0.16	2.772	0.231	34.88	4.611	7.811	110
10	Hygiea	C	5.43	203.6	0.07	3.137	0.118	3.84	5.56	27.623	126
511	Davida	C	6.22	163	0.05	3.166	0.186	15.94	5.63	5.130	65
704	Interamnia	F	5.94	158.3	0.07	3.062	0.150	17.29	5.36	8.727	60
52	Europa	C	6.31	158	0.06	3.099	0.104	7.48	5.460	5.631	52
87	Sylvia	P/X	6.94	143.0	0.04	3.489	0.080	10.86	6.52	5.184	35
31	Euphrosyne	C	6.74	127.9	0.05	3.149	0.22	26.32	5.59	5.531	150
15	Eunomia	S	5.28	127.7	0.209	2.644	0.187	11.74	4.30	6.083	157
16	Psyche	M	5.90	126.6	0.12	2.919	0.140	3.09	4.99	4.196	100
65	Cybele	P/X	6.62	118.7	0.07	3.433	0.105	3.55	6.36	4.041	131
3	Juno	S	5.33	117	0.24	2.668	0.258	13.00	4.36	7.210	60
88	Thisbe	C/B	7.04	116	0.067	2.767	0.164	5.22	4.60	6.041	32
324	Bamberga	C	6.82	114.7	0.06	2.683	0.338	11.11	4.39	29.43	148
624	Hektor	D	7.49	112.5	0.025	5.229	0.023	18.19	11.96	6.921	115
451	Patientia	C	6.65	112.5	0.08	3.061	0.077	15.22	5.36	9.727	67
107	Camilla	C/X	7.08	111.3	0.05	3.476	0.078	10.05	6.48	4.844	29
532	Herculina	S	5.81	111.2	0.17	2.770	0.179	16.31	4.61	9.405	75
48	Doris	C	6.90	110.9	0.06	3.110	0.075	6.55	5.49	11.89	63

All orbital data are from http://ssd.jpl.nasa.gov/.

^{a} Mean radius; most asteroids are substantially nonspherical.

^{b} Uncertainties in asteroid radii and axial tilt relative to the ecliptic plane are typically far larger than suggested by the number of digits quoted here.

^{c} Data from http://astro.troja.mff.cuni.cz/projects/asteroids3D/web.php and http://vesta.astro.amu.edu.pl/Science/Asteroids.

More recently, the IAU decided on the name *plutoids* for dwarf planets in orbit around the Sun with a semimajor axis greater than that of Neptune.

Meteorites, asteroids, KBOs, and comets provide unique information regarding the formation of our Solar System. Small bodies can be viewed as remnant planetesimals that have undergone relatively little endogenic geological evolution. KBOs, being at large heliocentric distances, have also experienced minimal heating by sunlight, which enabled preservation of bodies composed largely of ices. A closer look at these 'remnant planetesimals', however, shows that it is not straightforward to develop formation/evolution scenarios of our Solar System based upon our knowledge of these bodies. For example, in Chapter 8 we have seen that some asteroids melted early in the history of the Solar System, and there has been substantial collisional evolution among bodies in both the asteroid belt and the Kuiper belt. This collisional evolution complicates interpretation of the data, but it can also work

to our advantage. For example, during a collision an entire object can be broken up, providing us with samples of the cores of bodies (iron meteorites).

Over the past few decades our knowledge of minor planets has increased dramatically, as a result of the increased sensitivity of optical and infrared detectors, the exploitation of radar techniques, the availability of data from Earth-orbiting observatories, as well as flybys of several asteroids and satellites which resemble asteroids (e.g., Mars's moons Phobos and Deimos) or KBOs (in particular Neptune's moon Triton), plus detailed extended *in situ* investigations of 433 Eros and 25143 Itokawa.

Nomenclature

All minor planets with well-determined orbits are designated by a number, in chronological order, followed by a name, e.g., 1 Ceres, 324 Bamberga, 136199 Eris. After an object is discovered, but before it has a well-determined orbit, it gets a provisional name. The standard designation for this is related to the date of discovery of the object: the name starts with a 4-digit number indicating the year, followed by a space, then a letter to show the half-month (A for January 1–15, K for May 16–31, I is omitted), followed by another letter to show the order of discovery within the half-month (A for 1st, W for 22nd, with I being omitted). This second letter cycles through with a subscript. The first time it cycles through subscript 1 is used, then subscript 2, etc. For example, 2003 UB$_{313}$ means that this object was number $(25 \times 313) + 2 = 7827$ during 2003 October 16.00000–31.99999 UT. In some periods (in particular September and October, with the midnight ecliptic coming higher in the sky in the northern hemisphere), over 12 000 objects have been discovered in a half month (as shown by, e.g., minor planet 2005 UJ$_{516}$). A different nomenclature scheme is used for comets; see §10.1.

9.1 Orbits

Minor planets are often grouped according to orbital properties. Moving outwards from the Sun, near-Earth objects are generally quite small, and most have only been detected because their proximity makes them easier to observe. The asteroid belt lies between the orbits of Mars and Jupiter, and contains the brightest asteroids; the number of asteroids in this region is so much larger than in the neighborhood of our planet that it more than compensates for the greater distance from both the Earth and the Sun in terms of detectability. Next, sharing an orbit with Jupiter, lie the Trojans, which are distinctly more difficult to detect, because of their greater distance from illumination and observer plus their lower albedo. The centaurs occupy unstable trajectories that cross the orbits of one or more giant planets – a fairly rare lot due to their short dynamical lifetimes, but their domain is so immense that some large bodies are included. Although difficult to observe because of their distance, the vast majority of small bodies in our Solar System lie beyond the orbit of Neptune, in the Kuiper belt, the scattered disk, and the even more distant Oort cloud.

9.1.1 Asteroids
Main Belt Asteroids (MBAs)
Figure 9.1a shows the distribution of the semimajor axes for the orbits of the first 100 000 numbered asteroids with absolute magnitude[1] $M_v < 15$. Most asteroids are located in the *main asteroid belt*, at heliocentric distances between

---

[1] The absolute magnitude of a body in our Solar System is equal to the apparent magnitude if the body were at 1 AU from both the observer and the Sun, as seen at $\phi = 0$.

2.1 and 3.3 AU. The spread in the eccentricities of the main belt asteroids (MBAs) appears to be described well by a Rayleigh distribution (like a Maxwellian distribution in one dimension, eq. 4.14), suggesting some kind of quasi-equilibrium situation:

$$N(e) \propto \frac{e}{e_*} \exp\left(\frac{-e^2}{e_*^2}\right), \tag{9.1}$$

where the mean eccentricity $e_* \approx 0.14$. The large eccentricities of many asteroids imply that the perihelia and aphelia of the asteroids occupy a significantly wider zone than do their semimajor axes (Fig. 9.1b). The mean inclination of asteroid orbits to the ecliptic plane is 15°; the standard deviation of asteroidal inclinations is larger than that of a Rayleigh distribution with the same mean value.

Several gaps and concentrations of asteroid semimajor axes can be distinguished in Figure 9.1a. The gaps were first noted in 1867 by Daniel Kirkwood and are known as the *Kirkwood gaps*. The Kirkwood gaps coincide with resonance locations with the planet Jupiter, such as the 4:1, 3:1, 5:2, 7:3, and 2:1 resonances. As discussed in §2.3.2.2, if an asteroid orbits the Sun with a period commensurate to that of Jupiter, the asteroid's orbit is strongly affected by the cumulative gravitational influence of Jupiter. Perturbations by the giant planet produce chaotic zones around the resonance locations, where asteroid eccentricities can be forced to values high enough to cross the orbits of Mars and Earth. These asteroids may then be removed by gravitational interactions and/or collisions with the terrestrial planets. Secular resonances, in particular the ν_6 resonance with Saturn located near the inner edge of the asteroid belt (where the apse precession rate of asteroids is equal to Saturn's apse precession rate), can also excite asteroids onto high-eccentricity orbits. In some cases, eccentricities can be excited to such high values that the asteroids ultimately collide with the Sun, unless they are tidally shredded or thermally vaporized during close approaches to the Sun as their periapses approach the solar photosphere.

The opposite situation occurs in the outer asteroid belt, where asteroid orbits are very strongly perturbed by nearby Jupiter. Asteroids in the outer asteroid belt may acquire such high eccentricities that the asteroid suffers a close approach to Jupiter and is scattered to interstellar space. Asteroids that orbit the Sun at the 3:2 and 4:3 resonances with Jupiter are protected from such encounters, and we find an enhancement in the asteroid population at these resonances: The *Hilda asteroids* complete three orbits during two jovian years, and 279 Thule orbits the Sun four times every three jovian years.

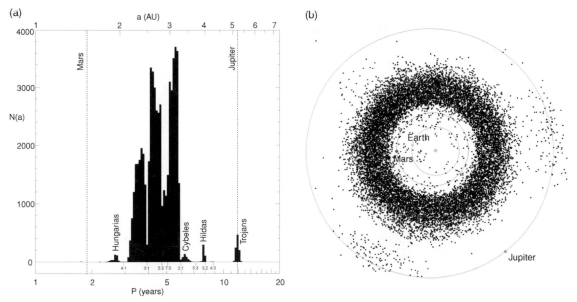

Figure 9.1 (a) Histogram of the first 100 000 numbered asteroids with $M_v < 15$ versus orbital period (with corresponding semimajor axes shown on the upper scale); the scale of the abscissa is logarithmic. The planets Mars and Jupiter are shown by dashed vertical lines. Note the prominent gaps in the distribution for orbital periods 1/4, 1/2, 2/5, 3/7, and 1/3 that of Jupiter. One asteroid, Thule, is located at the 4:3 resonance. (Courtesy A. Dobrovolskis) (b) Locations projected onto the ecliptic plane of approximately 7000 asteroids on 7 March 1997 (many more asteroids are known now, but plotting them all would clutter up the diagram). The orbits and locations of the Earth, Mars, and Jupiter are indicated, and the Sun is represented by the dot in the center. The vast majority of the asteroids depicted are in the main asteroid belt, but Trojans are shown leading and trailing the position of Jupiter; Aten, Apollo, and Amor asteroids are seen in the inner Solar System, crossing the orbits of Mars and Earth. (Courtesy Minor Planet Center)

Near-Earth Objects (NEOs)

Objects that venture close to Earth attract much attention because of the danger posed by the possibility of them colliding with our planet (§5.4.6). Such potential impactors belong to the population of *Near-Earth Objects (NEOs)*, a name collectively given to all asteroids and (inert/dormant) comets that have perihelia inside of 1.3 AU. These objects are sometimes also referred to as planet-crossing objects (as are their kin in the outer Solar System, the centaurs). The 10 000th NEO was detected in 2013. The NEO population is subdivided into four categories based upon the perihelia (q) and aphelia of their orbits (Fig. 9.2). About 40% of the NEO population, with 1.017 AU $< q <$ 1.3 AU, are called *Amor* asteroids, named after one of the prominent members of this group, 1221 Amor; their radii range up to ~15 km. About 50% of the NEOs have $q <$ 1.017 AU and semimajor axes $a >$ 1 AU; these are referred to as *Apollo* asteroids, after the archetype of this group, 1862 Apollo. The largest detected Apollo asteroids have radii of 4–5 km. The *Atens*, $\lesssim$10% of the NEOs, have $a <$ 1 AU and aphelia greater than Earth's perihelion, 0.983 AU.

Objects with orbits completely interior to that of our planet are referred to as *Apohele* asteroids. Apoheles are difficult to detect, and only a few are known.

The dynamical lifetime of NEOs is relatively short, $\lesssim 10^7$ years, and hence the NEO population needs an ongoing replenishment source from more stable orbits. Numerical models show that the primary NEO source regions are the chaotic zones near the resonance locations in the main asteroid belt, discussed above (Kirkwood gaps). These models show that ~35–40% of the NEOs originate at the ν_6 resonance, near the inner edge of the asteroid belt. Asteroids with perihelia $q >$ 1.3 AU and semimajor axes 2.06 AU $< a <$ 2.8 AU may cross the orbit of Mars during a secular oscillation cycle (§2.4.1) of their eccentricity. These asteroids are called Intermediate Mars-Crossing (IMC) objects, and provide about 20–25% of the NEOs. Another ~20–25% are sent inwards at Jupiter's 3:1 resonance. Smaller numbers of NEOs originate in the outer asteroid belt, with perhaps a few percent in the region beyond Neptune. The ultimate fate of most NEOs is either ejection into interstellar space (the

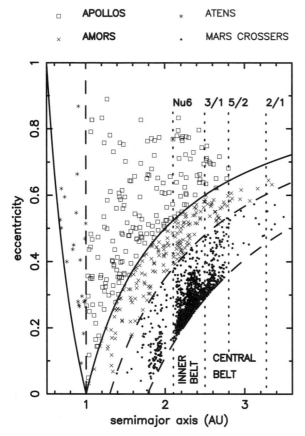

APOLLOS □   **ATENS** *

AMORS ×   **MARS CROSSERS** .

Figure 9.2 The orbital distribution of NEOs. The solid and long-dashed lines mark the boundaries of the different populations amongst the NEOs that are listed above the plot. The short-dash lines mark the location of three major mean motion resonances with Jupiter and the ν_6 secular resonance, as indicated. (Adapted from Morbidelli 2002)

remains balanced. One possibility is resupply via collisions within the asteroid belt. Such collisions are often disruptive (§9.4.2), and the orbits of the smaller fragments are typically altered the most compared to that of the proto-asteroid. Small bodies are also susceptible to the Yarkovsky effect (§2.7.3), which can significantly change their orbital parameters. Typically, kilometer-sized bodies drift in semimajor axis by $\sim 10^{-4}$ AU in one million years. Although this is significant, at the same time it is slow enough that populations of fresh collisional debris may evolve through further collisions into a size distribution as expected for a collisionally evolved population of objects. In §9.4.1 we will see that the slopes of the size distributions of NEOs and main belt asteroids are fairly similar.

Another potential source of NEOs is extinct comets that have developed nonvolatile crusts and ceased activity. A few bodies in Earth-crossing orbits that have been classified as asteroids are associated with meteor streams, suggesting a cometary origin. Also, Comet 2P/Encke is in an orbit typical of Earth-crossing asteroids. The NEO 1979 VA appears to be the same object as the active Comet 107P/1949 Wilson–Harrington. In 1949 this object was identified as having a tail, but in all apparitions since 1979 there has been no evidence of any cometary activity. This object is the first to have been 'seen' to transform from an active cometary state to a dormant asteroidal state. The NEO 3200 Phaeton appears to be the parent of the Geminid meteor shower, suggesting cometary activity in the past. Several other NEOs have orbital elements that, over the past 5000 years, may have been associated with meteor showers. A total of perhaps 10–15% of the NEOs may be inert comets. As much of the NEO population is resupplied by sporadic events, the total number of NEOs may fluctuate significantly over time.

Trojan Asteroids

As of January 2014, more than 6000 asteroids have been discovered near Jupiter's L_4 and L_5 triangular Lagrangian points. These bodies are known as the *Trojan asteroids*. Trojan asteroids are more distant from both the Sun and the Earth than are main belt asteroids, and they have low albedos. Thus, Trojans are much more difficult to detect than are MBAs of comparable size. Extrapolation of the observed distribution suggests that the total population of Trojan asteroids over 15 km in size is roughly half that estimated for MBAs. However, as the Trojans lack bodies comparable in size to the largest main belt objects (624 Hektor, by far the largest Trojan, has a mean radius of ~ 100 km), their total mass is much lower than that of MBAs. Neptune may also have a considerable population

dominant loss mechanism for NEOs originating in the outer asteroid belt), or collisions with or tidal or thermal destructions by the Sun (the dominant loss mechanism for NEOs that originate at the ν_6 and 3:1 resonances). A small but significant fraction of NEOs collide with planets and moons (§5.4.6).

Several objects (27 as of September 2007) are in a 3:2 exterior mean motion resonance with Earth (6 objects), Venus (4 objects), and Mars (17 objects), and orbit the Sun twice while the planet completes 3 orbits. These bodies are thus dynamically protected from close encounters with these planets, just like the Hildas and Thule in the outer asteroid belt, and the plutinos (see below) in the Kuiper belt.

Having identified the resonance zones as source regions for NEOs requires a solution of how to repopulate the resonance zones, so loss and supply of bodies of Lagrange point librators, but because of their faintness, only nine are known.

The dynamics of the Trojans are discussed in §2.2.2. The bodies sharing Neptune's orbit, the Trojans, the 4/3 and 3/2 librators, along with the MBAs that are not near resonances, occupy the only known stable (over the age of the Solar System) orbits between the major planets. Orbits well inside the orbit of Mercury are also stable, and small zones of stability may exist between the orbits of Uranus and Neptune (Fig. 2.17), but no objects have been observed within these orbital niches. Three kilometer-sized asteroids have been found librating on tadpole orbits about the triangular Lagrangian points of Mars. Secular resonances (§2.4.2) rapidly remove objects on high- or low-inclination orbits from the vicinity of Mars's L_4 and L_5 points. All of the observed martian Trojans indeed have inclinations within the range $15° < i < 30°$, the most stable part of the martian L_4 and L_5 regions, and they may date from the early Solar System. Several near-Earth asteroids, e.g., 3753 Cruithne and 2002 AA$_{29}$, are coorbital horseshoe librators (§2.2.2) with Earth, but those orbital locks are most likely of geologically recent origin.

9.1.2 Trans-Neptunian Objects, Centaurs

The overwhelming majority of small bodies within the Solar System are trans-neptunian objects (TNOs), whose orbits lie (entirely or in part) beyond the distance of Neptune. The first known TNO, 134340 Pluto, was discovered in 1930; Pluto's large (and rarely used) minor planet number was not assigned until Pluto was reclassified in 2006. The existence of a disk of numerous small bodies exterior to the major planets was postulated by K.E. Edgeworth in 1949 and (more prominently) by G.P. Kuiper in 1951, based upon a natural extension of the original solar nebula beyond the orbit of Neptune. This ensemble is therefore referred to as either the *Kuiper belt* or the *Edgeworth–Kuiper belt*. No Kuiper belt object other than Pluto and its large moon Charon was known until 1992. The discovery of 1992 QB$_1$ marked the onset of a flurry of search activities, and more than 1000 TNOs have now been detected in multiple years. The orbits of these bodies fall into several dynamical groupings, as shown in Figure 9.3a. Properties of the largest known TNOs are given in Table 9.2.

Classical KBOs (CKBOs)

About half of the known TNOs are *Classical KBOs* (CKBOs), which travel on low eccentricity ($e \lesssim 0.2$) orbits exterior to Neptune. Most CKBOs have semimajor axes between 37 and 48 AU. Two populations, one dynamically cold ($i < 4°$; ~35% of all CKBOs) and the other hot ($i > 4°$; ~65% of all CKBOs), are present, each

with distinct colors (§9.3.4). The total mass of the classical Kuiper belt is a few times that of its largest member, Pluto.

Many of the CKBOs are locked in one or more mean motion resonances with Neptune. Pluto occupies a 2:3 mean motion resonance with Neptune. This resonance appears to be chaotic (§2.4.2), but the chaos is so mild that Pluto's orbit is stable for billions of years, and maybe much longer. In addition to Pluto, many (~10–30% of the CKBO population) small objects are trapped in this same 2:3 resonance with Neptune; these objects are often referred to as *plutinos*. Apart from the 2:3, the most populated resonances are 3:5, 4:7, 1:2, and 2:5 (Fig. 9.3a).

Scattered disk objects (SDOs)

An increasing number of TNOs are being detected on high-eccentricity, nonresonant orbits with perihelia beyond the orbit of Neptune. These objects are referred to as *scattered disk objects* (SDOs). The largest known SDO is 136199 Eris, whose size and mass are slightly larger than those of Pluto (Table 9.2). The number of known SDOs is several times smaller than that of known CKBOs, since many of the SDOs travel on highly eccentric orbits and spend most of the time near aphelion, where they are quite faint. From the observed population, the total mass of SDOs is estimated to be (very roughly) an order of magnitude larger than that of the classical Kuiper belt. An upper limit of ~1 $M_\oplus$ worth of material contained in small bodies distributed in a flattened ring just beyond the orbit of Neptune has been deduced from the lack of detection of the gravitational perturbations of such a disk on Comet P/Halley.

The vast majority of SDOs are on orbits with perihelia 33 AU $\lesssim q \lesssim$ 40 AU. These bodies come close enough to the giant planets that they could have been placed in their current orbits by planetary perturbations, but they are far enough from Neptune that their orbits are stable on billion-year timescales. Nonetheless, some occasionally get close enough to Neptune to be scattered inwards on planet-crossing trajectories (discussed below) and in some cases become comets (see Chapter 10).

The TNO 90377 Sedna, with a perihelion at 76 AU and aphelion of ~900 AU, and possibly 2000 CR$_{105}$ ($q = 44.4$ AU, $a = 221$ AU), could not have been scattered into its present orbit via perturbations of the known planets on their current orbits. The combination of high perihelion, large eccentricity, and lack of resonance with Neptune, implies perturbations from an object exterior to the currently known planets. Sedna is often considered to be a member of the inner Oort cloud (§10.2). The orbits of both objects could have resulted from perturbations by a passing star (probably when the Solar System was very

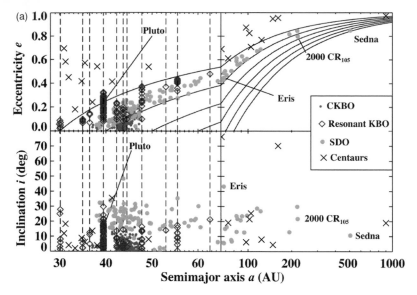

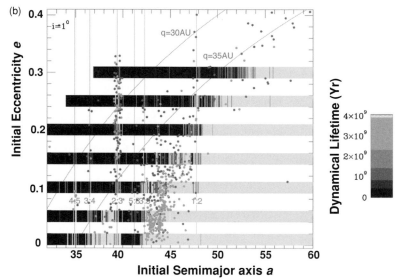

Figure 9.3 (a) Orbital elements for TNOs, time averaged over 10 Myr. The dashed vertical lines indicate mean motion resonances with Neptune which are occupied: the 1:1, 5:4, 4:3, 3:2, 5:3, 7:4, 9:5, 2:1, 7:3, 5:2, 3:1, in order of increasing heliocentric distance. Solid curves trace loci of constant perihelia $q = a(1 - e)$. Several objects are indicated: the two largest KBOs Pluto and Eris (Table 9.2), the high-q Sedna and 2000 CR_{105}. The centaur with semimajor axis near 1000 AU is 2006 SQ_{372}. (Adapted from Chiang *et al.* 2007) (b) COLOR PLATE The dynamical lifetime for small particles in the Kuiper belt derived from 4 billion year integrations. Each particle is represented by a narrow vertical strip, the center of which is located at the particle's initial eccentricity and semimajor axis (initial orbital inclination for all objects was 1 degree). The color of each strip represents the dynamical lifetime of the particle. The yellow strips represent objects that survive for the length of the integration, 4×10^9 years. Dark regions are particularly unstable on these timescales. For reference, the locations of important Neptune mean motion resonances are shown as blue vertical lines and two curves of constant perihelion distance, q, are indicated in red. Observed Kuiper belt objects with well-determined orbits are shown by green dots if they have $i < 4°$ and by magenta dots if their orbital inclinations are larger than four degrees. (Morbidelli and Levison 2007)

young and had not left its crowded stellar nursery) or by an unknown planet that may still orbit in the outer Solar System or may have escaped to interstellar space eons ago. The orbit of the small ($R \sim 25$–50 km) TNO 2006 SQ_{372} is very intriguing, with a perihelion $q = 24.17$ AU and a semimajor axis $a = 1015$ AU, i.e., its aphelion is close to 2000 AU from the Sun (Fig. 9.3a).

Centaurs

Centaurs orbit between (and in some cases also cross) the orbits of Jupiter and Neptune. Dozens of such objects are known, and many are on highly eccentric and/or inclined orbits. Centaurs are in chaotic planet-crossing

orbits, which have dynamical lifetimes of 10^6–10^8 yr (Fig. 9.3b). Dynamical calculations suggest that they are transitioning from the trans-neptunian region (primarily the scattered disk, but also the classical and resonant portions of the Kuiper belt) to becoming short-period comets (§10.2). Centaur 944 Hidalgo ($a = 5.8$ AU) has long been suspected of being a dormant comet. Both its orbit and spectral class (reddish-black, D type)[2] are typical of comets. Centaur 2060 Chiron ($a = 13.7$ AU) has a dark neutral color, like a C type asteroid, and an orbit which crosses those of both Saturn and Uranus. Calculations

[2] Spectral classes are discussed in §9.3.

Table 9.2 Twenty largest distant minor planets (known as of 2009; $a > 6$ AU).

#	Name	Provisional name	Dynamical[a] class	M_v	Radius[b] (km)	A_0^b	a (AU)	e	i (deg)	P_{orb} (yr)	P_{rot} (hr)
136199	Eris	2003 UB_{313}	SDO	−1.17	1163 ± 6	0.96	67.728	0.44	43.97	557.5	
134340	Pluto		RKBO	−0.7	1153 ± 10	0.5^c	39.482	0.249	17.14	247.7	153.3
136472	Makemake	2005 FY_9	RKBO	−0.48	715 ± 30	0.81	45.678	0.16	29.00	308.0	7.77
136108	Haumea	2003 EL_{61}	SDO	0.18	675 ± 125	0.84	43.329	0.19	28.21	284.8	3.92
	Charon		Moon	1.3	606 ± 1.5	0.375	39.482	0.249	17.14	247.7	153.3
90377	Sedna	2003 VB_{12}	IOC	1.56	<800	>0.16	489.6	0.84	11.93	10 718	10.27
84522		2002 TC_{302}	SDO	3.84	575 ± 170	0.03	45.678	0.16	29.00	408.7	
90482	Orcus	2004 DW	RKBO	2.3	450 ± 40	0.28	39.363	0.22	20.59	246.4	
50000	Quaoar	2002 LM_{60}	CKBO	2.67	422 ± 100	0.20	43.572	0.04	7.98	288.0	17.68
55565		2002 AW_{197}	SDO	3.26	367 ± 160	0.12	47.349	0.13	24.39	324.8	
84922		2003 VS_2	RKBO	3.98	363 ± 100	0.06	39.273	0.07	14.79	246.0	
		2002 MS_4	SDO	3.80	363 ± 60	0.08	41.560	0.15	17.72	271.6	
208996		2003 AZ_{84}	RKBO	4.0	343 ± 50	0.12	39.714	0.17	13.52	247.5	13.44
55637		2002 UX_{25}	SDO	3.6	341 ± 60	0.12	42.524	0.14	19.48	277.6	14.38
90568		2004 GV_9	SDO	3.9	338 ± 35	0.08	42.241	0.08	21.95	274.1	
28978	Ixion	2001 KX_{76}	RKBO	3.2	325 ± 130	0.12	39.648	0.24	19.59	250.1	
15874		1996 TL_{66}	SDO	3.2	288 ± 60	0.035	82.756	0.58	24.02	753.1	
38628	Huya	2000 EB_{173}	RKBO	4.7	262 ± 13	0.05	39.773	0.28	15.46	250.6	
20000	Varuna	2000 WR_{106}	CKBO	3.7	250 ± 50	0.16	42.921	0.05	17.20	280.5	6.34
26375		1999 DE_9	RKBO	4.7	231 ± 23	0.07	55.783	0.42	7.62	414.5	

Data from Stansberry *et al.* (2008), Brown *et al.* (2010), and http://ssd.jpl.nasa.gov/.

[a] CKBO: Classical KBO; RKBO: Resonant KBO; SDO: Scattered disk object; IOC: Inner Oort Cloud.

[b] For all bodies, except Pluto/Charon, the radius and albedo were derived from Spitzer (thermal IR) data.

[c] Changes with season between 0.44 and 0.61 as a consequence of variations in ice cover.

show that Chiron must pass close to Saturn every 10^4–10^5 years. When this happens, the orbit is perturbed significantly, and hence Chiron orbits the Sun on a highly chaotic trajectory (Fig. 2.11). In 1987–1988, Chiron developed a coma which was spotted through a brightening of the object (Fig. 10.5). Since then Chiron has also been classified as a comet, 95P/Chiron.

The TNOs, centaurs and other planet-crossers blur the distinction between minor planets and comets. Volatile-rich minor planets can become comets if they are brought close enough to the Sun, and comets look like minor planets if they outgas all of their near-surface volatiles and become dormant or inert. When an object is not outgassing over at least part of its orbit, it is usually considered to be a minor planet. However, many dormant comets may be hidden among the minor planets as, e.g., exemplified by several objects in the main asteroid belt which occasionally eject dust (§10.6.4). In this text, we adopt the traditional observational definition that an object is a comet if, and only if, a coma and/or tail has been observed.

9.2 Determination of Physical Properties

Sizes, masses, and compositions of minor planets are difficult to determine without sending a spacecraft (§1.2). A handful of minor planets have been visited by spacecraft (§9.5); unfortunately, this small sample is far from sufficient to evaluate the overall characteristics of these large reservoirs of bodies. Such analyses require data from large numbers of minor planets, an endeavor that can only be undertaken with ground-based or near-Earth (IRAS, HST, Spitzer) telescopes.

Some of the larger asteroids can be resolved by HST, or from the ground using speckle interferometry or adaptive optics (§E.6), while radar techniques have been used to resolve nearby asteroids (§E.7). Observations of *mutual events*, which occur when the primary and secondary pass in front of and behind each other, have been used to characterize the Pluto–Charon system (§9.5). The majority of the minor planets, however, cannot be resolved by ground-based telescopes or HST. For these bodies, sizes are best estimated using stellar occultations, where the minor planet passes in front of a star and starlight is temporarily blocked out during an interval of time, the duration of which is proportional to the size of the object. A drawback of the star occultation technique is that many chords are required, as most bodies with radii of less than a few hundred km are usually irregular in shape (§6.1; Figs. 5.94, 9.5). Despite these difficulties, the effective size, shape, albedo, and surface structures have now been determined for numerous asteroids and the largest KBOs by using a variety of observing techniques. In the following subsections we discuss what we learned about minor planets (in particular asteroids) by applying specific 'stand-alone' techniques, and via a combination of different observations.

9.2.1 Radius and Albedo

The amount of sunlight reflected off an asteroid is proportional to the body's visual albedo times its projected (onto the sky) surface area (eqs. 3.15, 3.16). In contrast, the total thermal emission is proportional to the product of absorbed insolation, $1 - A_b$, (A_b is the Bond albedo) and the object's projected surface area. To first approximation, the surface of airless bodies is in equilibrium with insolation, so the sum of reflected and emitted radiation is equal to the solar radiation intercepted. Therefore, measurements of the thermal and reflected emissions are sufficient to determine the size and albedo of an object if the relation between visual geometric and Bond albedo is known. The technique of combining measurements at visible and infrared wavelengths to extract the size and albedo of an object is known as *radiometry* (§E.3).

The geometric albedo is related to the Bond albedo via its phase dependence, i.e., phase integral q_{ph} (eq. 3.11), which depends on the photometric and thermal properties of the body's surface. Most minor planets are observed over a limited range of solar phase angles, ϕ. From $r_\odot \approx$ 1 AU, TNOs are only viewed at $\phi \approx 0°$, and most MBAs at $\phi \lesssim 30°$. Only NEOs come close enough to Earth that they may be observed over a large range of phase angles. Hence, for most objects the relationship between geometric and Bond albedo cannot be determined observationally. Since the thermal properties of minor planets are essentially unknown, most researchers usually assumed that asteroids resemble the Moon, a spherical object covered with a thick loose particulate regolith, which has a low thermal inertia (§5.5.1).[3] A body with such properties

Table 9.3 Albedos and phase functions.

Body	$A_{0,v}$	$q_{ph,v}$	A_v	A_b	Ref.
Moon	0.113	0.611	0.069	0.123	1
Mercury	0.138	0.486	0.067	0.119	1
243 Ida	0.21	0.34	0.071	0.081	2
Dactyl	0.20	0.32	0.064	0.073	2
253 Mathilde	0.047	0.280	0.013		3
433 Eros	0.29	0.39	0.11	0.12	4
951 Gaspra	0.23	0.47	0.11		5
25143 Itokawa	0.53	0.13	0.069	0.07	6

$A_{0,v}$: geometric albedo at visible wavelengths, v.

$q_{ph,v}$: phase integral at v.

A_v: albedo at v ($A_{0,v} \times q_{ph,v}$).

A_b: Bond albedo.

1: Veverka *et al.* (1988). 2: Veverka *et al.* (1996). 3: Clark *et al.* (1999). 4: Domingue *et al.* (2002). 5: Helfenstein *et al.* (1994). 6: Lederer *et al.* (2005).

radiates only a small percentage of the absorbed insolation from its dark hemisphere, so that the thermal radiation from large phase angles is small. Assuming lunar-like characteristics for a few large asteroids, however, yields radii and albedos which disagree with radii obtained from stellar occultations, suggestive of differences in the angular distribution of the thermal emission between the Moon and asteroids. The Galileo flyby of 243 Ida provided the first direct measurements of the phase integral and Bond albedo for an asteroid (Table 9.3), and the values are quite different from those of the Moon. To compensate for such differences, a 'beaming' factor, η_v, is usually added to equation (3.16) (at phase angle $\phi \approx 0°$):

$$\mathcal{F}_{out} = \pi R^2 \eta_v \epsilon_v \sigma T^4. \tag{9.2}$$

Setting $\eta_v \approx 0.7$–0.8, the radii and albedos for asteroids with radii over 25 km agree quite well with values derived independently from radiometry, polarimetry, and stellar occultations. This suggests that the surface of asteroids is quite rough compared to that of the Moon.

For smaller objects it is often difficult to derive diameters and albedos from radiometry that agree with values obtained with other techniques, even after inclusion of the beaming factor η_v in equation (9.2). There are a variety of reasons to explain such difficulties: (*i*) the object may be very aspherical, (*ii*) the rate and sense of rotation and obliquity modify the 'observed' temperature, (*iii*) the asteroid may not have a regolith (§9.2.3), in which case

[3] Many modelers use full thermophysical models. However, when the thermal inertia (eq. 3.24) and thermal skin depth of the material (eq. 3.25) are unknown, the reliability of such models may not be

much higher than that of the standard lunar model with beaming factor η_v in equation (9.2).

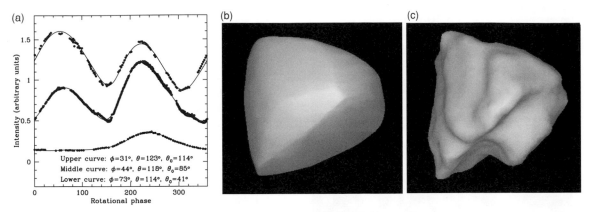

Figure 9.4 (a) Photometric lightcurves of the NEA 6489 Golevka under different viewing conditions, as indicated by the solar phase angle ϕ and the aspect angles of the Earth, θ, and the Sun, θ_0. Superposed are model lightcurves of the shape model that best fits all lightcurve data combined. (b) Shape model of Golevka as derived via an inversion of 30 photometric lightcurves (a subset of which is shown in panel a). (c) Shape model (plane-of-sky views; see Fig. 9.5) of Golevka as reconstructed from Arecibo delay-Doppler observations obtained in May 1999. The NEA is $0.35 \times 0.25 \times 0.25$ km across. (Panel a: Courtesy Mikko Kaasalainen. Panels b, c: Kaasalainen *et al.* 2002)

the thermal inertia is much higher and as much as 50% of the absorbed insolation may be radiated from the night-side hemisphere, and/or (*iv*) the asteroid's surface may be highly metallic. In the latter case, the thermal emissivity may be as low as 0.1, while the thermal conductivity is very high. The low emissivity would raise the surface temperature, and shift the bulk of thermal radiation to shorter wavelengths. Unless the emission is observed at the appropriate wavelengths, this shift may go unnoticed, and the derived asteroid properties may be grossly in error.

9.2.2 Shape
Lightcurve Analysis
Photometric lightcurves (§E.1) have shown that the rotation periods of the majority of minor planets are between 4 and 16 hours. By using lightcurves from different viewing angles, obtained over many years of observing, the shape, pole position, and sense of rotation has been determined for many asteroids (Fig. 9.4). For objects large enough to be resolved from the ground (using adaptive optics techniques or radar), with HST, or asteroids that have been imaged by spacecraft, excellent agreement is reached between lightcurve-derived shapes and high-resolution images (Fig. 9.4).

A comparison of lightcurves from small main belt and near-Earth asteroids shows that the distribution of lightcurve amplitudes is similar, suggesting that the gross distributions in shape of these bodies are alike. In contrast, Trojan asteroids typically show larger mean lightcurve amplitudes than do main belt and near-Earth asteroids,

indicative of more elongated shapes. On the other hand, although we have only a limited sample of KBO data, Kuiper belt objects typically show a higher fraction of low-amplitude lightcurves and longer rotation periods than do similarly sized asteroids (although we note that 136108 Haumea has a rotation period less than 4 hours).

Shape from Radar Data
The shape of an object can also be determined from radar observations, as detailed in §E.7. Radar echoes from NEOs which come very close to the Earth can be very strong. For such objects, high-resolution observations, both in time and frequency (Doppler delay), can be obtained so that inversion of the data recovers the three-dimensional shape of the asteroid. The full radar-signal inversion requires 15 parameter fits to the data; however, despite the large number of free parameters, the results are quite robust. Three-dimensional images of several NEOs and MBAs have been obtained over the past few decades (Fig. 9.5). Many NEOs are quite elongated, and may in fact be contact binaries. In addition to its bifurcated shape, the NEA 4179 Toutatis (Fig. 9.5a) revealed an interesting spin, being composed of two types of motion with periods of 5.4 and 7.3 days. These motions combine in such a way that Toutatis's orientation with respect to the Sun never repeats. It 'tumbles' in space (§9.4.6). The NEA 1620 Geographos (Fig. 9.8) shows peculiar protuberances at the ends of its elongated body. These may result from excavation through impacts, and subsequent deposition of impact ejecta on a body where the ratio of gravitational and centrifugal forces

(a)

4179 Toutatis

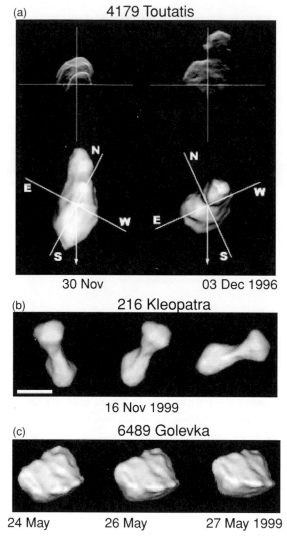

30 Nov   03 Dec 1996

(b)

216 Kleopatra

16 Nov 1999

(c)

6489 Golevka

24 May   26 May   27 May 1999

Figure 9.5 (a) High-resolution delay-Doppler images (top) and the corresponding plane-of-sky appearance of NEA 4179 Toutatis on two dates in 1996. Gray scales have been chosen to emphasize brightness contrast. The crosshairs are 5 km long and centered on Toutatis's center of mass. The radar images are plotted with time delay (range) increasing from top to bottom and Doppler frequency (radial velocity) increasing from left to right. The crosshairs are aligned north–south and east–west on the plane of the sky. In the bottom plane-of-sky frames, the arrow is parallel to the instantaneous spin vector as projected onto the sky. (Ostro *et al.* 2002) (b) Shape model (plane-of-sky views) of the main belt asteroid 216 Kleopatra as reconstructed from Arecibo radar images from 16 November 1999. The scale bar represents 100 km. (Ostro *et al.* 2000) (c) Shape model (plane-of-sky views) of the NEA 6489 Golevka, reconstructed from Arecibo delay-Doppler observations from 24, 26, and 27 May 1999. The asteroid is 0.35 × 0.25 × 0.25 km across. (Chesley *et al.* 2003)

varies drastically over its surface. In contrast, the NEA 6489 Golevka is much rounder (Fig. 9.5c).

M class asteroids have very high radar albedos. For example, the main belt asteroid 216 Kleopatra shows a radar reflectivity of ~0.6. Kleopatra is a dogbone-shaped object (Fig. 9.5b), covered by a metallic regolith with a porosity similar to that of lunar soil. Its shape, like that of other bifurcated bodies, suggests that the object once consisted of two separate bodies which may have 'fused' together via a gentle collision, perhaps from a low-velocity infall of fragments after a disruption event, or from tidal decay of a binary system. With the recent discovery of two moonlets in orbit about this asteroid (§9.4.4), its collisional origin appears to be well established.

9.2.3 Regolith

Most asteroids are covered by a layer of regolith (§5.4.2.4), albeit with different properties than the regolith on our Moon (§9.2.1), as borne out by a variety of observations.

Opposition Effect

Graphs of an asteroid's brightness in reflected sunlight display a strong variation with phase angle, ϕ (Fig. 9.6). In particular, an anomalous increase in intensity is seen at small phase angles, $\phi \lesssim 2°$, which is known as the *opposition effect*. This opposition surge is caused by a combination of 'shadow-hiding' and the coherent-backscatter effect. The latter is indicative of particulate material, i.e., the surface must be covered by a layer of regolith (see §E.1 for details). The characteristics of the opposition effect for asteroids appears to depend on their geometric albedo, and asteroids of different taxonomic (§9.3) classes show different relationships (Fig. 9.6).

Polarization

Figure 9.7a shows a plot of the degree of polarization as a function of phase angle for 1 Ceres. The polarization of most asteroids is negative at small ϕ, and becomes positive at $\phi \gtrsim 15$–$20°$. This is indicative of rough, porous, or particulate surfaces. The minimum degree of polarization, P_{Lmin}, reached at phase angle ϕ_{min}, the phase angle ϕ_0 where $P_L = 0$, and the slope h_p of P_L versus ϕ (near ϕ_0), are diagnostic of the surface texture and optical properties of the object. The curves are steepest for C type asteroids, and flattest (smallest $|P_{Lmin}|$ and h_p) for V and E types. The peak polarization at large ϕ would be diagnostic of the object's surface texture, but is practically impossible to observe from Earth, except for Earth-crossing objects. The slope h_p appears to be directly related to the geometric albedo A_0, irrespective of the nature of the

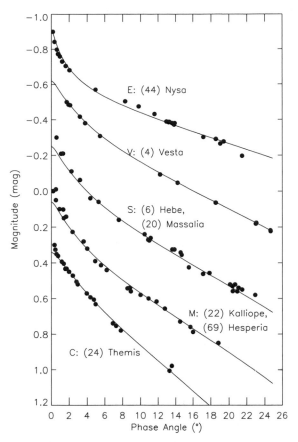

Figure 9.6 Opposition effect on a relative magnitude scale for C, M, S, V, and E class asteroids. The solid lines present models based on both the photometric (shown here) and polarimetric (see Fig. 9.7) data. (Muinonen *et al.* 2002)

surface (Fig. 9.7b). It is thus possible to determine the geometric albedo, and hence diameter of an object, from its polarization–phase curve.

Figure 9.7c shows a plot of P_{Lmin} versus ϕ_0 for lunar and terrestrial rocks (region I) and rock powders and lunar fines (region II); meteorites and rocks with grain sizes between 30 and 300 μm lie between these two regions. Thus, a measurement of the polarization characteristics of an object can be used to determine, in addition to the geometric albedo and radius, also the surface texture of the body. Large bodies, such as Mercury, the Moon, and Mars, lie within region II, suggestive of a fine-grained regolith. Asteroids are located in between regions I and II, suggesting a mixture of pulverized rocks with coarser-grained material.

Radar Echoes

The radar echo from an object is Doppler broadened as a result of the rotation of the body (§E.7). However,

for objects like the Moon and Mercury, the observed bandwidth is usually less than the maximum value (B_{max} in eq. E.7), since the signal is dominated by specular reflections from surface slopes that are tilted in the direction of the radar system. These slopes are concentrated near the center of the projected disk, where Doppler shifts caused by the object's rotation are relatively small. In contrast to radar observations of the Moon, Mercury, or Mars, the observed spectral bandwidth of asteroids is not much smaller than the expected edge-to-edge bandwidth (B_{max}; Fig. 9.8), indicative of objects with a very rough surface. In contrast, the ratio between radar power in the same sense of circular polarization to that in the opposite sense, SC/OC, is generally low. Main belt asteroids usually show $SC/OC < 0.2$. Hence, asteroids appear to be objects that are very irregular on scales exceeding meters, yet smooth on centimeter scales. Near-Earth objects show higher SC/OC ratios, $0.2 \lesssim SC/OC \lesssim 0.5$, and hence must have a rougher surface than MBAs.

Since the radar albedo is related to the Fresnel coefficient (eq. E.5), one can derive the dielectric constant of the material from radar echoes. This parameter yields information both on the composition and the compactness of the body's surface layers (eqs. 3.88, E.6). If meteorites are indeed good analogs to asteroids, the radar albedo thus provides a measure of surface porosity, and hence the presence of a regolith. If one assumes, for example, that the meteoritic analog of a particular asteroid is a stony-iron meteorite, a radar reflectivity of 0.2 suggests a 50% porosity of the regolith material (Fig. E.7). A radar albedo of 0.2 for a C type asteroid, however, would suggest that the asteroid is not overlain by a layer of regolith.

Radio Spectra

Electromagnetic radiation typically probes ∼10 wavelengths deep in a body's surface (§3.4). Thus, observations at radio wavelengths probe much deeper into the asteroid's crust than measurements at visible and infrared wavelengths. A comparison of multi-wavelength thermal-emission data with thermo-physical models provides information on the depth dependence of the density and temperature in the surface layers. Radio spectra of several main belt asteroids suggest that these bodies are typically overlain by a layer of fluffy (highly porous) dust that is a few centimeters thick; similar dust layers have been observed on the Moon and Mercury. This general structure of a fluffy layer of dust overlying a highly compacted regolith is likely caused by bombardment of small meteoroids, which maintains the top layer at a low density while compacting deeper layers.

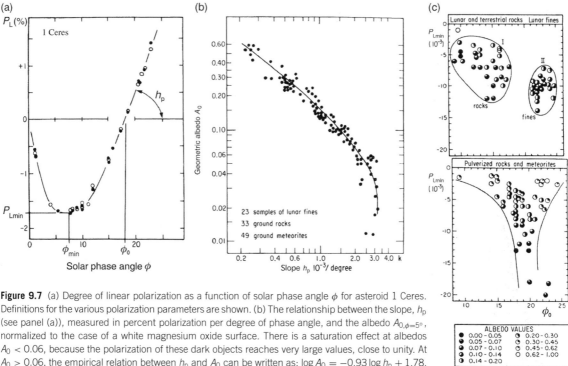

Figure 9.7 (a) Degree of linear polarization as a function of solar phase angle ϕ for asteroid 1 Ceres. Definitions for the various polarization parameters are shown. (b) The relationship between the slope, h_p (see panel (a)), measured in percent polarization per degree of phase angle, and the albedo $A_{0,\phi=5°}$, normalized to the case of a white magnesium oxide surface. There is a saturation effect at albedos $A_0 < 0.06$, because the polarization of these dark objects reaches very large values, close to unity. At $A_0 > 0.06$, the empirical relation between h_p and A_0 can be written as: $\log A_0 = -0.93 \log h_p + 1.78$. (c) A graph of P_{Lmin} versus ϕ_0 for: top, lunar and terrestrial rocks (region I) and rock powders and lunar fines (region II); bottom, meteorites and rocks with grain sizes between 30 and 300 μm lie in between these two regions. (All panels from Dollfus *et al.* 1989)

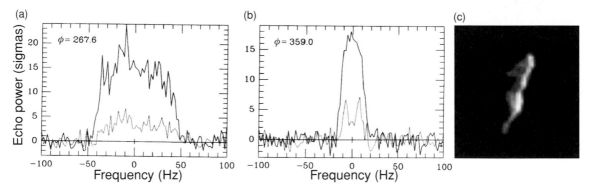

Figure 9.8 Radar echo spectra (a, b) and a radar image (c) of near-Earth asteroid 1620 Geographos obtained in 1994 with the Goldstone antenna at a transmitter frequency of 8510 MHz. The echo power is plotted in standard deviations versus Doppler frequency (in Hz) relative to the estimated frequency of echoes from the asteroid's center of mass. Solid and dotted lines are echoes in the *OC* and *SC* polarizations, respectively. The spectra are at rotational phases, φ, which correspond to bandwidth extrema. The radar image (right panel) is different from a regular visual view: It consist of echoes in time delay (range) increasing towards the bottom, and Doppler frequency (line-of-sight velocity) towards the right. (Ostro *et al.* 1996)

9.3 Bulk Composition and Taxonomy

Asteroids, like meteorites, appear to be a compositionally diverse group of large rocks. Some asteroids contain volatile material as carbon compounds and hydrated minerals, while others appear to be almost exclusively composed of refractory silicates and/or metals. Some asteroids resemble primitive rocks, while others have undergone various amounts of thermal processing. TNOs have probably suffered even less thermal processing, being

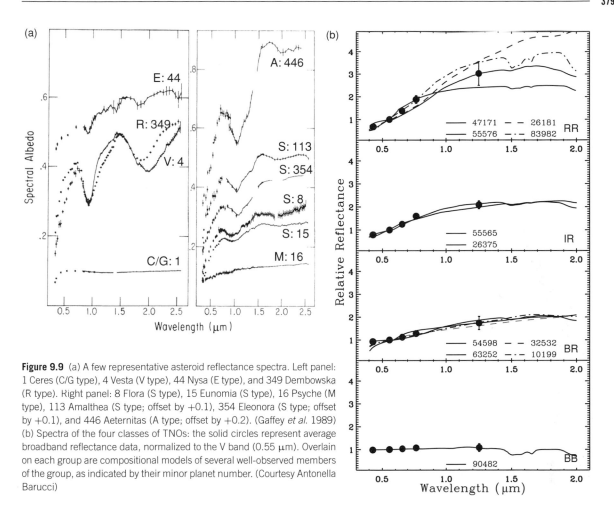

Figure 9.9 (a) A few representative asteroid reflectance spectra. Left panel: 1 Ceres (C/G type), 4 Vesta (V type), 44 Nysa (E type), and 349 Dembowska (R type). Right panel: 8 Flora (S type), 15 Eunomia (S type), 16 Psyche (M type), 113 Amalthea (S type; offset by +0.1), 354 Eleonora (S type; offset by +0.1), and 446 Aeternitas (A type; offset by +0.2). (Gaffey *et al.* 1989) (b) Spectra of the four classes of TNOs: the solid circles represent average broadband reflectance data, normalized to the V band (0.55 μm). Overlain on each group are compositional models of several well-observed members of the group, as indicated by their minor planet number. (Courtesy Antonella Barucci)

at larger distances from the Sun. Based upon cosmic abundance ratios, the composition of TNOs is likely to be an ~50:50 mixture of ice and rock by mass.

The composition of asteroids and TNOs contains clues about the environment in which the planetesimals that ultimately accreted into the eight known planets formed and evolved. Our main means to extract information on compositon is via remote sensing spectroscopy. For many objects, we have reflectance spectra at visible and near-infrared wavelengths, thermal spectra in the infrared, as well as passive radio and/or radar data. In §§9.3.1–3 we summarize the asteroid classification scheme and recent advances in our understanding of the relation between these classes, asteroid composition, and the effect of space weathering. In §9.3.4 we discuss the spectroscopic diversity of TNOs.

9.3.1 Asteroid Taxonomy

Histograms of albedos (at 0.55 μm) for asteroids with radii $R > 20$ km show a bimodal distribution, with pronounced

peaks at $A_0 \approx 0.05$ and 0.18. These albedos together with reflectance spectra at visible and near-infrared wavelengths have historically been used to sort asteroids into *taxonomic classes* (Fig. 9.9a; Table 9.4). The largest class of asteroids contains the *carbonaceous* or *C type* asteroids. Roughly 40% of the known asteroids are C types. They are dark bodies with typical geometric albedos $A_0 \sim$ 0.04–0.06 and flat spectra (neutral in color) longwards of 0.4 μm. Some C type asteroids exhibit an absorption band near 3 μm, indicative of water, which is probably present in the form of hydrated silicates. Carbonaceous chondritic meteorites (CI, CM) have reflectance spectra very similar to those of C type asteroids, so C types are probably carbon-rich (§8.2 and Fig. 8.10). They are low-temperature condensates, primitive objects which have undergone little or no heating. Asteroids with similar, yet quantifiably different, spectra are grouped in classes B, F, and G. In contrast to primitive C type asteroids, B, F, and G asteroids have probably been heated sufficiently to cause some mineralogical changes.

Table 9.4 Asteroid taxonomic types.

Low-albedo ($A_0 < 0.1$) *classes*:

C   Carbonaceous asteroids; similar in surface composition to CI and CM meteorites. Dominant in outer belt (>2.7 AU). Spectrum flat, UV absorption at <0.4 μm; slightly reddish at >0.4 μm. Subclasses: B, F, G.

D   Extreme outer belt and Trojans. Red featureless spectrum, possibly due to organic material.

P   Outer and extreme outer belt. Spectrum is flat to slightly reddish, like M types, but lower albedo.

K   Resembles CV and CO meteorites. Steep absorption feature at <0.75 μm; shallow 1 μm and no 2 μm absorptions.

T   Rare, unknown composition; altered C types? Moderate absorption feature <0.85 μm; flat spectrum >0.85 μm.

Moderate-albedo ($0.1 < A_0 < 0.3$) *classes*:

S   Stony asteroids. Major class in inner-central belt. Absorption feature <0.7 μm. Weak absorptions near 1 and 2 μm.

M   Stony-iron or iron asteroids; featureless flat to reddish spectrum.

W   Like M types, but have an absorption band near 3 μm (indicative of hydration).

Q   Resembles ordinary (H, L, LL) chondrite meteorites. Absorption features shortwards and longwards of 0.7 μm.

A   Very reddish spectrum shortwards of 0.7 μm. Strong absorption feature near 1 μm.

V   Strong absorption feature at <0.7 μm, and near 1 μm. Like basaltic achondrites. Type example: 4 Vesta.

R   Spectrum intermediate between A and V classes. Similar to olivine-rich achondrites. Type example: 349 Dembowska.

High-albedo ($A_0 > 0.3$) *classes*:

E   Enstatite asteroids. Concentrated near inner edge of belt. Featureless, flat to slightly reddish spectrum.

X   Visible spectrum as P, M, or E types, but has absorption bands at 0.49 and 0.60 μm.

The next largest class of cataloged asteroids (30–35%) are the *S type* or *stony* asteroids. These objects are fairly bright, with geometric albedos ranging from 0.14 to 0.17, and reddish. S type asteroids have a strong absorption feature shortwards of 0.7 μm, indicative of iron oxides. S class spectra with their weak to moderate absorption bands near 1 and 2 μm suggest assemblages of iron and magnesium bearing silicates, like pyroxene ((Fe,Mg)SiO_3) and olivine ((Fe, Mg)$_2$ SiO_4), mixed with pure metallic nickel–iron. These bodies likely crystallized from a melt, and consequently S type asteroids are usually classified as igneous bodies. However, as discussed below, it is not yet clear whether S type asteroids are indeed igneous, or if they could be primitive bodies whose surface has been altered by *space weathering*; evidence in favor of the latter is growing (§9.3.2).

About 5–10% of the asteroids have been classified as *D and P types*. The D and P type asteroids are quite dark, with $A_0 \approx 0.02$–0.07, and on average somewhat redder than S types. Neither D nor P types exhibit spectral features. They may represent even more primitive bodies than the carbonaceous C type asteroids. There is no meteoritic analog to their spectra. The red color of P and D types has been attributed to organic compounds, perhaps created via space weathering.

M type asteroids have a spectrum similar to P type asteroids, but with a higher albedo, $A_0 \approx 0.1$–0.2. The M type visible wavelength spectra lack silicate absorption features and are reminiscent of metallic nickel–iron. Their spectra are analogous to iron meteorites and enstatite chondrites, meteorites composed of grains of nickel–iron, embedded in enstatite ($MgSiO_3$), a magnesium-rich silicate.

A few very bright asteroids have been discovered with albedos in the range 0.25–0.6. These are *E type* asteroids, which display linear, flat or slightly reddish spectra. They may consist of enstatite or some iron-poor silicate.

The meteoritic analogs of M and E types suggest these asteroids have undergone substantial thermal processing via a melt phase. They have often been interpreted as fragments of the cores of disrupted large asteroids. This picture became blurred with the discovery of absorption bands near 3 μm on many of the larger ($\sim$75% with $R \gtrsim$ 30 km) M and E type asteroids. These objects have been reclassified into a W class, for 'water' in the form of hydration, features indicative of primitive rather than igneous objects. This apparent mismatch or contrast between M/E and W type asteroids is puzzling.

A group of asteroids, referred to as *X types*, contains asteroids with albedos ranging from low to high, and spectra which resemble P, M, and E types. The spectra of these asteroids differ from the regular P, M, and E types in that they exhibit a strong absorption band near 0.49 μm and a weaker one at 0.60 μm, perhaps produced by the mineral troilite (FeS).

In addition to the large main groups, there are a number of small asteroid classes, some of which contain only a few asteroids. Asteroids whose spectra do not fit into any established class are referred to as *U type* or unclassified. The spectrum of *R type* 349 Dembowska suggests large amounts of olivine with little or no metals. It resembles a silicate residue after metal has been extracted from the upper mantle of an incompletely differentiated body.

Asteroid 4 Vesta (*V type*) is unique among large objects in that it appears to be covered by basaltic material. Its spectrum resembles that of eucrite meteorites (§8.1), displaying pyroxene absorption bands near 1 and 2 μm, as well as a weaker absorption at 1.25 μm attributed to plagioclase feldspar ($(CaAl,NaSi)AlSi_2O_8$). In addition to Vesta, several small V type asteroids have been detected in orbits similar to that of Vesta. These asteroids and eucrite meteorites, may have been blasted off Vesta in the impact that created the 460 km wide crater discernible on high-resolution images of this asteroid.

Practically no Q type asteroids have been found in the main belt, while roughly 1/3 of the NEOs are either Q types, or fall in a 'transition region' between Q and S types. Q type spectra match those of LL ordinary chondrites, and typically have deeper absorption bands and are less red than S type spectra.

The 'alphabet soup' of asteroid classes (still growing and sometimes inconsistent) can be quite confusing to the nonexpert. The most important classes to remember are the very common C type, which are dark, neutral in color, and resemble carbonaceous chondrite meteorites, the brighter reddish S type which are common in the inner asteroid belt, the dark red D type asteroids common at large heliocentric distances, and the M types (at least some of) which appear to be very rich in metals.

9.3.2 Space Weathering

It is curious that no large asteroid class seems to match the spectra of ordinary chondrites, the most common meteoritic samples, while Q type spectra, which match those of LL ordinary chondrites, are relatively common among the NEOs (e.g., 1862 Apollo, 9969 Braille). Where are the parent bodies of the (primitive) ordinary chondrites? Are they hidden in the asteroid belt (i.e., too small to be detected)? An alternative is that *space weathering* may alter the spectra of asteroids: the interaction of solar wind particles, solar radiation, and cosmic rays with planetary bodies may induce chemical alterations in the surface material and therefore 'hide' the real composition of an asteroid. Space weathering may have altered the surface composition of S type asteroids, in which case S type asteroids may be the parent bodies of ordinary chondritic meteorites. This idea seems supported by the observation that the smallest S type asteroids (NEOs) display a greater likeness to the ordinary chondrites than larger ones do; i.e., they appear to fall in a transition region between S and Q types. If the smaller ones are relatively young fragments, they have been exposed to space weathering for less time than larger asteroids, and hence may better resemble spectra of ordinary chondrites.

Observations of a moderately large S type asteroid, the NEA 433 Eros, by the NEAR spacecraft show that its bulk elemental composition and spectrum at visible and near-infrared wavelengths are very similar to those of ordinary chondrites. This observation lends strong support to the theory that space weathering has altered/hidden the original composition of S type asteroids, but it is not easy to unambiguously prove the above hypothesis or to simulate space weathering in the laboratory. Space weathering clearly darkens and reddens a body's surface, but the surface can also be 'rejuvenated' at times by impacts, including micrometeorite impacts which lead to gardening and erosion.

9.3.3 Taxometric Spatial Distribution

There appears to be a strong trend among the taxonomic classes with heliocentric distance (Fig. 9.10). The high-albedo E type asteroids are only found near the inner edge of the asteroid belt. S type asteroids prevail in the inner parts of the main belt, M types are seen in the central regions of the main belt, while the dark C type objects are primarily found near the outer regions of the belt. D and P asteroids are found only in the extreme outer parts of the asteroid belt and among the Trojan asteroids. Figure 9.10b shows the distribution of igneous (assuming S type asteroids to be igneous bodies, see previous subsection) and primitive asteroids as a function of heliocentric distance. This figure shows a correlation with heliocentric distance: Igneous asteroids dominate at heliocentric distances $r_\odot < 2.7$ AU, and primitive asteroids at $r_\odot > 3.4$ AU. Metamorphic asteroids, which must have undergone some changes due to heating as characterized by their spectra, have been detected throughout the main belt. Spectral evidence for hydrated phases is particularly strong for asteroids in the main belt, and is essentially absent beyond 3.4 AU.

The strong correlation of asteroid classes with heliocentric distance cannot be a chance occurrence. It must

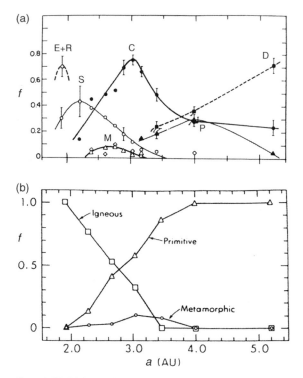

Figure 9.10 (a) Graph showing the relative distribution of the asteroid taxonomic classes as a function of heliocentric distance. The classes E, S, C, M, P, and D are shown. Smooth curves are drawn through the data points for clarity. (Adapted from Gradie *et al.* 1989) (b) Distribution of igneous, primitive, and metamorphic classes as a function of heliocentric distance. (This figure assumes that S type asteroids are igneous bodies.) (Adapted from Bell *et al.* 1989)

be a primordial effect, although it is likely modified by subsequent evolutionary or dynamical processes. If asteroids are indeed remnant planetesimals, their distribution in space might provide insight into the temperatures, pressures, and chemistry of the solar nebula. The difference in spatial distribution between the C type carbon-rich (primitive) and S type carbon-poor (possibly igneous) asteroids is qualitatively consistent with the temperature structure expected in the primitive solar nebula. P and D type asteroids, probably even more primitive than the C types, may have formed at even lower temperatures; unfortunately, not much is known about the composition of P and D type asteroids, as there is only one spectral analog in terrestrial meteorite collections (the Taglish Lake meteorite, which resembles D type asteroids). The highly reflective objects in the inner belt are consistent with a higher temperature than must have prevailed in the inner-central belt.

It is not known to what degree post-formation evolution, including space weathering and dynamical processes, masks the original distribution of asteroid compositions.

However, the orderly arrangement of taxonomic classes with heliocentric distance clearly contains important clues to the formation history of our Solar System. While there is a trend in taxonomic classes with heliocentric distance in the main and outer asteroid belt, NEOs have representatives from almost all the taxonomic classes, indicating that many locations in the asteroid belt feed the NEO population, as expected based upon dynamical arguments (§§9.1, 2.3).

We caution the reader here about statistics of asteroid samples. The samples are (strongly) biased to nearby bright, high-albedo objects. Thus, the S class objects are over-represented relative to the C class asteroids, because they are, on average, brighter and closer to the Sun; an even stronger bias against detection of D class asteroids exists for the same reasons. Indeed, once albedo biases are removed, C types may represent ∼50% of NEAs and inner main belt asteroids.

9.3.4 TNO Spectroscopy

Based upon their visible-light spectra, the TNOs and centaurs can be subdivided into 4 groups: BB, BR, IR, and RR. The BB (blue) group contains objects that are neutral in color with respect to the Sun, while BR, IR, and RR contain progressively redder bodies (Fig. 9.9b). The RR group contains, e.g., the centaur 5145 Pholus, 1992 QB$_1$ (the first TNO detected, other than Pluto/Charon), and the distant object Sedna, while Pluto, Eris, 2060 Chiron, and 90482 Orcus belong to the BB group. We refer to these as groups rather than taxonomic classes to distinguish the TNOs from asteroids where the various classes are clearly related to composition and physical characteristics. It is not clear whether the diverse colors of TNOs reflect different primordial compositions or different degrees of surface processing. Red objects may be red as a consequence of space weathering, while gray bodies may have had their 'clock' reset by collisions or cometary-like activity (as for 2060 Chiron) dredging up primordial materials neutral in color. Surveys of TNOs and centaurs remain inconclusive. All dynamically cold KBOs are red, while the hot population is gray. Centaurs have a bimodal distribution, with essentially BB and RR objects, while SDOs are dominated by BB bodies.

While the reflection spectra of most TNOs and centaurs are relatively featureless at visible wavelengths, they are very rich at near-infrared wavelengths. The largest KBOs, Pluto and Eris, show absorption bands of the most volatile ices CH$_4$ and N$_2$ (Fig. 9.11). Note, though, that nitrogen-ice is very difficult to see, and evidence for its presence on Eris is not as clear as on Pluto. Pluto also exhibits bands of CO-ice. Both 136472 Makemake and

(a)

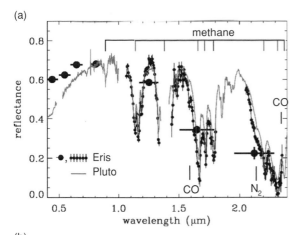

(b)

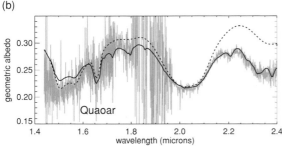

Figure 9.11 (a) Comparison of reflectance spectra of Pluto and Eris, where the Eris spectrum has been scaled to agree with that of Pluto in the I band (0.8 μm). The large filled circles show the photometry points derived from broadband observations. The location of the methane, nitrogen, and carbon monoxide ice bands are indicated. (Adapted from Brown *et al.* 2005a). (b) Near-infrared spectrum of Quaoar obtained with the W.M. Keck telescope. The 1.65 μm feature due to crystalline water-ice is clearly visible. A spectrum that consists of a best-fit water-ice model combined with a red continuum fits the 1–2.1 μm region well, but falls short at 2.2–2.4 μm (dashed line). A good fit to this region can be obtained by including the broad methane (2.32 and 2.38 μm) and (perhaps) narrower ethane (2.37 and 2.32 μm) absorption bands (solid line). (Adapted from Schaller and Brown 2007)

Sedna show CH_4 absorption bands without N_2-ice, while on Haumea, 50000 Quaoar, and Charon (Pluto's large moon) only absorption bands of water-ice are detected (plus NH_3-ice on Quaoar). More exotic ices have been discovered on a few bodies, as methanol (CH_3OH on, e.g., 5145 Pholus and (55638) 2002 VE$_{95}$) and ethane (C_2H_6 on, e.g., Makemake and Quaoar).

Both Pluto and Neptune's largest moon Triton, presumably a captured KBO (§13.11), have tenuous atmospheres dominated by N_2 gas, with traces of CH_4 and CO (§4.3.3.3). It is feasible that other large TNOs have atmospheres as well, but so far have escaped detection. As discussed in §4.8.1, the continued presence of an atmosphere depends upon a body's gravity and on the temperature

of its atmosphere. Similarly, the retention of volatile ices also depends upon gravity and temperature. Calculations of Jeans escape from large KBOs show that Pluto, Triton, Eris, and Sedna are large and cold enough for the continued presence of CH_4, N_2, and CO ices, while Quaoar, Haumea, and Makemake are borderline candidates for volatile retention. Smaller KBOs would have long lost all CO, CH_4, and N_2 via atmospheric escape.

The largest KBOs not covered by methane-ice show deep absorptions by water-ice, while smaller KBOs show spectra that are more neutral or have weak water-ice absorption features. A few of these large KBOs, e.g., Quaoar, Charon, Orcus, and Haumea, reveal signatures of crystalline water-ice (Fig. 9.11b). Since the continuous bombardment by energetic photons from the Sun and cosmic rays converts crystalline ice into its amorphous form within 1–10 Myr, the presence of crystalline ice on these bodies requires an explanation. Radiogenic heat in the large KBOs could, in principle, be sufficient to transform amorphous into crystalline ice, but some mechanism, perhaps micrometeorite gardening (§5.4.2.4) or cryovolcanism, needs to bring the ice up to the surface. While the endogenic processes may be viable on the large KBOs, nonthermal exogenic mechanisms must be invoked to explain crystalline ice features on smaller KBOs, unless such bodies result from a disruption of a larger body (e.g., the collisional family members of Haumea, discussed in §9.4.2).

9.4 Size Distribution and Collisions

Collisions have played a major role in shaping the asteroid and Kuiper belts. Such collisions were frequent events in the early history of our Solar System, and evidence for geologically recent collisions is growing. In the next subsections we discuss phenomena that have shaped our view of the disruptive environment in which minor planets formed and evolved. These topics range from the overall size distribution of minor planets to their bulk densities and the presence of interplanetary dust.

9.4.1 Size Distribution

The size distribution of minor planets can be approximated by a power law valid over a finite range in radius. Size distributions can be given in differential form:

$$N(R)dR = \frac{N_o}{R_o}\left(\frac{R}{R_o}\right)^{-\zeta} dR \quad (R_{min} < R < R_{max}), \qquad (9.3a)$$

where R is the object's radius, and $N(R)dR$ the number of bodies with radii between R and $R + dR$. The

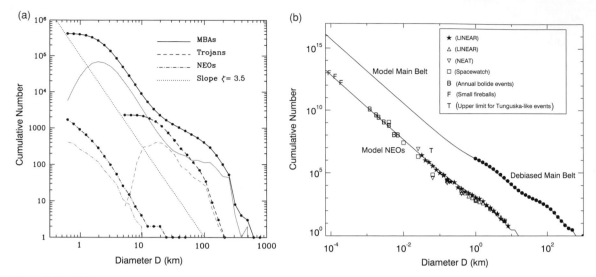

Figure 9.12 (a) Cumulative size distribution, $N_>(D)$, (dark lines with dots) for MBAs, Trojans, and NEOs, with a superposed dotted line that corresponds to a power-law size distribution with $\zeta = 3.5$ (eq. 9.3). The various curves for diameters $D \lesssim 300$ km were produced from the ASTORB catalog (~440 000 bodies in Feb. 2009, ftp://ftp.lowell.edu/pub/elgb/astorb.html); for $D \gtrsim 300$ km we used Table 9.1. The absolute magnitude, M_v, recorded in the database was converted to a diameter using the relation (from Bottke *et al.* 2005a): $D(km) = (1329/\sqrt{A_v})(10^{-M_v/5})$, with $A_v = 0.092$ for MBAs and NEOs and $A_v = 0.04$ for Trojans. The dots show the center of each of our bins (each of which is 0.5 magnitude wide). For comparison, the red curves show the differential size distribution for the three populations. (b) Comparison of a model of the dynamical evolution (solid lines) with the debiased distribution of MBAs and NEOs. Most bodies with $D \lesssim 100$ km are fragments (or fragments of fragments) derived from a limited number of breakups of bodies with $D \gtrsim 100$ km. The NEO model population is compared to estimates derived from telescopic surveys, spacecraft detections of bolide detonations in Earth's atmosphere, and photographs of fireballs. The symbol T is an estimate on the upper limit of 50 m NEOs, derived from the Tunguska airblast in 1908 (§5.4.6). (Adapted from Bottke *et al.* 2005b)

size distribution can also be presented in cumulative form:

$$N_>(R) \equiv \int_R^{R_{max}} N(R')dR' = \frac{N_o}{\zeta - 1}\left(\frac{R}{R_o}\right)^{1-\zeta}, \quad (9.3b)$$

where $N_>(R)$ is the number of bodies with radii larger than R, and R_{max} is the radius of the largest body. In equations (9.3), ζ is the power-law index of the distribution, R_o is the fiducial radius, and N_o is a constant that depends upon the choice of R_o.

Theoretical calculations imply that a population of collisionally interacting bodies evolves towards a power-law size distribution with $\zeta = 3.5$, provided the disruption process is self-similar (i.e., depends only on the speed and size ratio of the colliding bodies). In such a steady state, the number of objects that is destroyed within a certain mass bin (due to grinding and catastrophic disruptions) is equal to the number of objects that is created in this mass bin (due to accretion and agglomeration). A slope of $\zeta = 3.5$ implies that most of the mass is in the largest

bodies and most of the surface area is in the smallest bodies (Problem 9.3).

Figure 9.12a shows the size distributions observed for main belt, near-Earth, and Trojan asteroids. The overall asteroidal size distribution is consistent with the power law expected from a collisional evolution ($\zeta \approx 3.5$). The leveling off in $N_>(R)$ at small sizes (i.e., dropoff in $N(R)$) is caused by an observational bias against the detection of small asteroids. Another bias is introduced based on the size, albedo, and orbital parameters of asteroids. Detectors, the software used to process the data, and the region in the sky surveyed introduce additional biases in the resulting database. Techniques have been developed to 'debias' surveys, so the data can be used in concert with dynamical models to extract information on the past and present conditions in our Solar System. Debiased MBA and NEO populations are shown in Figure 9.12b. The two populations appear to have very similar slopes.

The size distribution of the main belt asteroids appears to be 'wavy'; it shows prominent bumps at $R \sim 2$ and 50 km. This structure, together with the power-law slopes

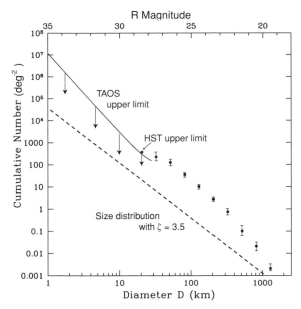

Figure 9.13 Cumulative size distribution for TNOs. The points with error bars are from Bernstein *et al.* (2004), and are binned estimates (Bayesian expectation with 68% credible range) of the TNO number density as a function of brightness, observed through a red filter near the invariable plane (R magnitude refers to the magnitude observed at a wavelength of 680 nm). The solid line is an upper limit as determined from TAOS (Zhang *et al.* 2008). The diameters were determined from the R magnitude assuming $r_\odot = 43$ AU and $A_v = 0.04$. The dashed line corresponds to a power-law size distribution with $\zeta = 3.5$ (eq. 9.3).

of the MBAs and NEOs, constrains models of the dynamical evolution of the asteroid belt. The size–frequency distribution resulting from one such model is reproduced in Figure 9.12b. This model includes the effect of the Yarkovsky force on small bodies in the main asteroid belt, which ultimately resupplies the NEO population and accounts for the similarity in the size distributions of MBAs and NEOs (§9.1.1). This particular model suggests that the bumps in the MBA distribution are 'fossils' produced by the collisional evolution in a primordial main belt that was much more (~200 ×) massive than the present-day asteroid belt. In addition, many of the largest ($R \gtrsim 60$ km) asteroids probably escaped fragmentation, and may date back to the epoch of planet formation.

The cumulative size distribution for TNOs is plotted in Figure 9.13, together with a power law with $\zeta \approx 3.5$ for comparison. This power law is a good match to the size distribution for TNOs at $m_r \gtrsim 23$ ($R \lesssim 130$ km), but the larger objects display a steeper slope. Such a 'double' power-law distribution might be expected if the population is collisionally evolving and has reached a steady state for small objects but not for large ones. There appears

to be a slight difference in the size distribution for the cold CKBOs (steeper slope) and the hot CKBOs + SDOs. The hotter population follows a $\zeta \approx 3.5$ slope towards larger radii than the cold population, i.e., a steady state between destruction and creation has been reached for larger objects in the hot compared to the cold CKBO population. This can be explained by the fact that the hotter TNOs reside in a more violent collisional environment than the cold population. Given enough time and a large enough density of bodies, the entire TNO population should evolve to $\zeta = 3.5$.

Since small (kilometer-sized) objects are too faint for direct detection, several groups carry out stellar occultation surveys to constrain the small size end of the TNO distribution (§E.5). The Taiwanese–American Occultation Survey (TAOS) consists of an array of four small (50 cm) automated telescopes that monitor ~1000 stars at a frequency of 5 Hz. A star's intensity will temporarily be reduced when an object passes in front of it (~0.2 seconds when partially occulted by a KBO). No KBOs were detected during the first two years of operation (with three telescopes), which constrains the power law $\zeta < 4.6$ (Fig. 9.13). Future observations with TAOS and other telescopes will likely put more stringent constraints on ζ.

9.4.2 Collisions and Families

Typical random (relative) velocities in the Kuiper belt are $\gtrsim 1$ km s^{-1}, and for asteroids they are several km s^{-1}. These numbers are much larger than the escape velocities of most minor planets. (The escape velocity from Ceres $v_e \approx 0.5$ km s^{-1}.) Thus, most collisions between minor planets should be erosive or disruptive. The final outcome of a collision depends upon the relative velocity and strength/size of the object (§13.5.3.2). Large minor planets and iron–nickel bodies have the greatest resistance to disruption. In *super-catastrophic collisions*, the colliding bodies are completely shattered, and the fragments are dispersed into independent, yet similar, orbits. Immediately following an impact, the resulting grouping of bodies is tightly clustered in space. However, asteroid fragments rapidly spread out in orbital longitude as a consequence of small differences in orbital period (Problem 9.10), as do the particles within a planetary ring (Problem 11.7). If planetary perturbations were weak, the principal orbital elements, a, e, and i, would not vary substantially, and could be used to identify such clusters. However, Jupiter strongly perturbs the asteroid belt, and induces quasi-periodic variations in the principal orbital elements. These variations can be calculated and subtracted out, yielding *proper orbital elements* that do not vary significantly with time and thus represent more robust dynamical parameters.

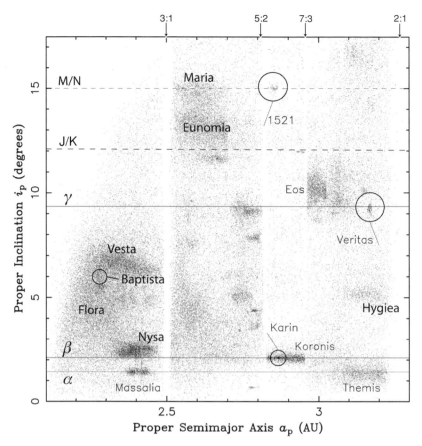

Figure 9.14 This plot of the proper inclination versus proper semimajor axis of ~10^5 main belt asteroids reveals many groupings of asteroids with similar proper orbital elements. These groups, or asteroid families, represent remnants of collisionally disrupted large asteroids. The horizontal lines indicate the model-derived proper inclinations of the IRAS dust bands, with solid lines for the prominent α, β, and γ bands, and dashed lines denoting the fainter M/N and J/K bands. Several asteroid families referred to in the text are indicated, such as the Veritas family, the young Karin cluster, and the Baptista group (§5.4.6). The 3:1, 5:2, 7:3, and 2:1 Kirkwood gaps (indicated at the top) produce the prominent vertical gaps on this graph. (Adapted from Nesvorný *et al.* 2003)

A plot of the proper inclination versus semimajor axis for main belt asteroids (Fig. 9.14) reveals numerous groupings. Such collections of asteroids with nearly the same proper orbital elements are sometimes called *Hirayama families*, in honor of the Japanese astronomer who discovered the first of those. Individual families are named after their largest member. Over 50 families have been discovered in the asteroid belt, as well as some Trojans and one KBO family (Haumea). Members of individual families often share similar spectral properties as well, further supporting a common origin in a single body that has undergone a catastrophic collision. The similarity in spectral properties is particularly striking for the Haumea family in the Kuiper belt. The KBO Haumea and its two satellites (§9.4.4) show unusually deep water-ice absorption features. A subsequent large infrared survey revealed about eight other KBOs with similarly deep water-ice absorptions. All these objects appeared to be dynamically clustered, and hence they probably formed via a giant impact on proto-Haumea. Many of these bodies also show evidence of crystalline water-ice (§9.3.4).

Backward integrations of the orbits of individual family members show that at least seven asteroid families were formed by collisions in the last 10 Myr, including two that formed within the past 250 000 years. Such catastrophic collisions ultimately provide a new influx to the NEO population. The catastrophic disruption of a ~170 km diameter main belt asteroid ~160 Myr ago led to the creation of the Baptista family. Over time, dynamical processes (i.e., Yarkovsky effect, YORP) changed the orbits of fragments produced in this collision such that they could strike the terrestrial planets. It has been suggested that one such fragment may have hit Earth, and led to the extinction of the dinosaurs 65 Myr ago (K–T boundary, §5.4.6). Another striking example of the consequences of catastrophic collisions in the asteroid belt is that of the formation of the Veritas family. This family resulted from the disruption of an $R > 75$ km asteroid 8.3 ± 0.5 Myr ago. This collision was likely the largest main belt asteroid disruption in the past 10^8 years. The median proper inclination of Veritas family members, 9.3°, corresponds to one of the major IRAS dust bands (Fig. 9.14; §9.4.7); this dust is presumably being produced at the present epoch by continued collisional grinding of debris from this disruption. Moreover, there is a four-fold increase in the ^{3}He content of sediments in Earth's oceans ~8.3 Myr ago; the entire event, its rise and fall, lasted ~ 1.5 Myr. This ^{3}He presumably was implanted by the solar wind in dust particles produced by the initial catastrophic disruption as they spiraled inwards towards Earth's orbit due to Poynting–Robertson drag (§2.7.2) and corpuscular drag (§2.7.5).

9.4.3 Collisions and Rubble Piles

The presence of numerous families in the asteroid belt is indicative of a violent collisional environment. Although the (super-)catastrophic collisions that lead to the formation of asteroid families are rare, less energetic impacts, where bodies simply get fractured, or shattered but not dispersed, happen much more frequently (§5.4). The latter case occurs when the velocity of the individual fragments is less than their mutual escape velocity, and some or most of them coalesce back into a single body, forming a *rubble pile*. Rubble piles are composed of a gravitationally bound collection of smaller bodies and internal void spaces. Collisional fragments may also form binary or multiple systems, where bodies are of similar size and mass, or one larger object is orbited by one or more small satellites. Such systems, however, are usually not long lived. Tidal interactions between a bound pair of asteroids lead to orbital evolution on a timescale of $\sim 10^5$ years (eqs. 2.44 and 2.64). Satellites interior to the synchronous orbit evolve inwards, so that the system ultimately becomes a rubble-pile compound object. Satellites outside the synchronous orbit evolve outwards.

As discussed in the next subsections, there is strong evidence that most asteroids in the main belt are rubble piles rather than single coherent objects: (1) Most of those asteroids whose densities have been accurately measured (Table 9.5) are significantly less dense than are meteorites with comparable spectra. (2) There are essentially no asteroids with radii $\gtrsim 100$ m that have a rotation period of less than 2 hours (see Fig. 9.20; a single exception is seen with $R \approx 350$ m). The sharp cutoff of periods near the maximum rotation rate for a strengthless body held together by gravity (Problem 9.11) indicates that all but the smallest asteroids are too weak to hold themselves together against significant centrifugal forces. (3) Many asteroid families have several large bodies of comparable size. Collisional models in which the parent of these families was a rubble pile of similar-sized blocks can reproduce the observed size and velocity distributions of the asteroids in the families, whereas models in which the parent was a single coherent body produce families in which the largest member is much larger than any other member.

Based upon a combination of data and numerical calculations, it is useful to 'sort' minor planets in terms of relative tensile strength and porosity, two parameters which determine a body's reaction to impacts. Bodies with high tensile strength and low porosity, as monolithic rocks (i.e., $\lesssim 100$ m sized bodies) or moderately fractured objects, react to impacts as described in §5.4, and may develop fractures throughout. Highly fractured or shattered bodies with moderate tensile strength but low porosity are harder to breakup, since the faults and joints may suppress propagation of (impact) tensile waves. The energy transferred by impacts on bodies with rubble-pile structures, having a moderate-to-high porosity and low relative tensile strength, is quickly damped, and craters are formed via compaction. As images of 253 Mathilde (see Fig. 9.22a) show, craters look very different on such objects than on a body like our Moon.

Impacts by bodies much smaller than the target object, including (micro)meteoroids, merely breakup or pulverize the near-surface rock, and thereby create a layer of regolith as on our Moon (§5.4.2.4). We hence expect the larger minor planets to be covered by a thick layer of regolith. There probably are systematic differences in the quantity of regolith produced on bodies as a function of size; larger bodies retain a greater fraction of impact ejecta and they have longer lifetimes against super-catastrophic collisions. Calculations by various groups predict asteroids over 100 km in size to be covered by a layer of regolith many meters thick, while bodies less than 10 km in size may be covered by at most a few centimeters of regolith. As discussed in §9.2, photometric observations of the opposition effect, and data obtained with polarimetry, radar, and radio techniques, show that asteroids ($R \gtrsim 25$ km) are indeed generally overlain by a layer of regolith, but with a size distribution that is different from that on the Moon.

9.4.4 Binary and Multiple Systems

Many minor planets orbit the Sun as gravitationally bound pairs. When the bodies are of comparable size, these are known as binaries, whereas when one body is much smaller than the other, the smaller object is generally referred to as a satellite (or moon) of the larger body. There is no precise boundary between satellites and binaries, but the mass ratio required for stability of Lagrange point libration in the circular restricted three-body problem is $m_1/m_2 \gtrsim 25$ (§2.2.1). Hence, we refer to systems with a size ratio $\gtrsim 3$ (for equal density objects) as a primary with a satellite, and if the size ratio <3 we refer to the system as a binary. Most bound pairs in the asteroid belt lie within the satellite regime (widely differing masses), whereas many known Kuiper belt pairs are binaries (similar masses). Relative to physical sizes, asteroid pairs generally occupy closer orbits than do Kuiper belt pairs, whereas orbital separations ratioed to the size of the

Hill spheres (eq. 2.22) are comparable in the two principal minor planet zones within our Solar System.

Bound pairs can be used to determine masses and thus densities of individual minor planets via Kepler's third law (eq. 2.11). Eventually, more detailed observations of orbits may yield information on the internal mass distribution as well (§2.5). The masses of a few of the largest asteroids have been estimated from the perturbations that they cause on the heliocentric orbits of other asteroids, and spacecraft tracking has also yielded some mass estimates (§9.4.5). Table 9.5 lists the properties of selected minor planets with known densities.

The notion that a minor planet may have a satellite goes back to the early 1900s, when a lightcurve of 433 Eros prompted suspicion of a companion. The irregular shapes of some large objects, such as the Trojan asteroid 624 Hektor with radii $\sim 75 \times 150$ km, were suggestive of compound objects, formed when two round bodies collided at low speed. Detailed radar observations show 216 Kleopatra to be shaped like a dumbbell (Fig. 9.5) and several NEOs to resemble contact binaries (§9.2.2). As typical collisions between asteroids occur at $\sim$a few km s^{-1}, such gentle encounters are only likely between fragments produced in a catastrophic collision. Alternatively, a large collision could produce a single coherent large elongated object if most of the ejecta escape.

Roughly 10% of the largest ($\gtrsim 20$ km in diameter) known impact craters on Venus and Earth (e.g., the Clearwater Lakes crater pair, which are 32 and 22 km across) and $\sim$2% on Mars are doublets, and must have been formed by the nearly simultaneous impact of objects of comparable size. The sizes and separations of the craters are too large to be caused by a single asteroid that was tidally disrupted or fragmented just before impact. These doublet craters were likely produced by the impacts of binary asteroids. In contrast, the crater chains on the surfaces of Ganymede (Fig. 5.85c), Callisto, and the Moon are the result of an impact by the remains of a body which was tidally disrupted (by Jupiter or Earth) shortly before impact, indicative of a rubble pile nature of the original body.

The first asteroid satellite to be detected was Dactyl, a small body that orbits 243 Ida, imaged in 1993 by the Galileo spacecraft (Fig. 9.15). This detection renewed searches for asteroid companions, and in 1998 a satellite in orbit around asteroid 45 Eugenia was detected using adaptive optics on the CFHT (Canada–France–Hawaii telescope). Although analyses of asteroid lightcurves have long been suggestive of binary bodies, the first widely accepted binary asteroid resulting from such observations was the NEO 1994 AW$_1$. Its lightcurve showed

Figure 9.15 Galileo image of S type asteroid 243 Ida and its moon Dactyl. Ida is about 56 km long. Dactyl, the small object to Ida's right, is about 1.5 km across in this view, and probably $\sim$100 km away from Ida. (NASA/Galileo PIA000136)

two components with different periods and amplitudes that could be matched well by a model of an eclipsing/occulting binary.

As of January 2013, more than 200 minor planets have been identified as multiple systems. About 2% of all main belt and Trojan asteroids with $R > 10$ km are known binaries or multiplets, where most have been discovered via direct imaging using adaptive optics on 8–10 m telescopes (Fig. 9.16). Over 10% of TNOs are in multiple systems, most of which have been discovered on HST images. The fraction of binary KBOs is much larger for objects with low inclination orbits than for those with high inclinations. The fraction of binaries among the NEOs is largest, $\sim$15% for objects with $R > 100$ m, and most have been discovered via radar techniques (Fig. 9.17) and/or lightcurve analyses. It is possible that binary systems are more common among the NEOs than main belt asteroids because close encounters with Earth and other terrestrial planets could tidally disrupt a compound body, similar to the tidal disruption by Jupiter witnessed for Comet D/Shoemaker–Levy 9 (§5.4.5). Debris from such a disruptive event could evolve into a binary system, and impact the planet on a later return.

The asteroids 87 Sylvia, 45 Eugenia, 93 Minerva, and dogbone-shaped 216 Kleopatra, as well as the third largest known KBO, Haumea, are triple systems (Fig. 9.18). With the discovery of four small moons, Pluto now has five known satellites (Fig. 9.19). These multiple systems reinforce models of a collisional origin.

The distribution of mass ratios and orbital characteristics provide constraints on the origin, collisional history, and tidal evolution of minor planets. Purely two-body gravitational interactions cannot convert unbound orbits to bound ones, nor vice versa. Temporary captures can be caused by three-body interactions with the Sun (Fig. 2.6) or gravitational forces exerted by a nonspherical body,

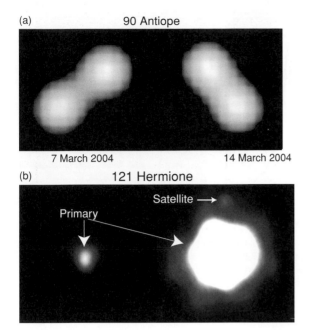

Figure 9.16 Binary systems: (a) The binary C type asteroid 90 Antiope as imaged with the adaptive optics system on the VLT in 2004. The two components are almost equal in size (size ratio of 0.95), with an average radius of 42.9 km, separated by 171 km. The rotation period is 16.1 hr. The derived density of 1.25 g cm^{-3} suggests a macroporosity of 30%. (Adapted from Descamps *et al.* 2007) (b) The C type asteroid 121 Hermione with its small moon imaged with the Keck adaptive optics system on 6 December 2003. Hermione is a member of the Cybele family. The image on the left shows the primary. Although not visible with the viewing geometry in this image, Hermione's shape resembles that of a 'snowman'. The image on the right has been stretched in intensity to reveal the faint companion. In this presentation, scattered light makes the primary appear much larger than its actual size, and with an unrealistic shape. North is up, east is left. (Adpated from Marchis *et al.* 2005a)

but for long-term stability, energy must be dissipated as heat or removed permanently from the system by another body or bodies. Stable bound pairs can be formed when disruptive or large cratering collisions (or strong tidal encounters with a planet) produce debris that subsequently interacts gravitationally or collisionally. The circularizing second stage in such a process is known as the *second burn* in rocketry. Alternatively, bodies that approach one another slowly can become bound when a physical collision or gravitational interaction with a third body removes energy.

Physical tides raised by one body on another (§2.6) and Yarkovsky forces (§2.7.3) can also dissipate energy. Although these processes are too slow to yield significant numbers of captures, they can substantially alter orbits of captured bodies over geologic timescales. Interactions

with interloping bodies can also change mutual orbits of bound pairs. Small (compared to the smaller member of the pair), slowly moving (relative to the mutual orbital speed of the pair) interlopers tend to remove energy, tightening the bound pair, whereas large, rapidly moving interlopers can tear systems apart. Large, slowly moving objects passing within the orbit of the pair can capture one member of the pair, allowing its initial companion to escape; this process is known as an *exchange reaction*. The number of bodies required for these various binary formation mechanisms can be used as constraints on conditions in the early Solar System.

Considering the various scenarios to create binary and multiple systems, and the differences between main belt and Trojan asteroids, NEOs and TNOs, different processes may dominate in the various minor body reservoirs. The impact scenario is the most likely process in the main belt and Trojan asteroid region, while the capture scenario is favored for TNOs. Since the dynamical lifetime of NEOs (∼10 Myr) is much shorter than the collisional timescale against disruption (∼100 Myr), binary formation in the NEO populations may be dominated by rotational disruption, such as caused by, e.g., tidal interactions with planets, cometary jetting, or YORP (§2.7.4).

9.4.5 Mass and Density

To determine a celestial body's mass, one needs to observe a gravitational interaction, such as a natural or artificial satellite orbiting the primary, a binary system, or a gravitational encounter between two bodies. The most efficient gravitational interactions are long-lasting encounters at small distances, or repeated encounters with similar geometries, so that the gravitational effects are accumulated. The gravitational perturbations can best be measured along the line of sight by radar ranging and Doppler shifts in the frequency of the radar echo (or radio transmitter if measured from a spacecraft that encounters or orbits the asteroid). The gravitational perturbation can also be measured from accurate astrometric measurements of the perturbed object in the plane of the sky. Once a body's mass is known, its density can be determined if its size and shape are known. A body's density yields invaluable information on its internal structure if its composition is known (e.g., via spectroscopy). In some cases, e.g., 20 000 Varuna, when the shape and rotational period are known, the object's density can be estimated by assuming the shape to agree with that of a body in hydrostatic equilibrium (§6.1.4).

Although with the ever-increasing number of confirmed binaries our list of asteroid and KBO masses and densities has expanded significantly over the past years,

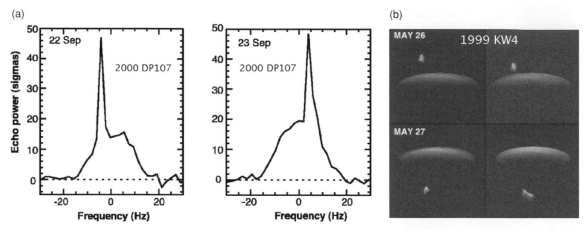

Figure 9.17 (a) Goldstone radar echoes (left panels) of NEA 2000 DP$_{107}$, obtained in September–October 2000. The spectra reveal a narrow spike superposed on a broadband component. The wide-bandwidth echo is indicative of a rapidly rotating primary object. The narrow spike moves at a different rate, and indicates the presence of a smaller and/or slowly spinning secondary. The narrowband echo oscillates between negative and positive frequencies, representing the variations in Doppler shift of an object revolving about the system's center of mass. The spectral resolution of the data is 2 Hz. This was the first binary NEA detected via radar measurements. (Margot *et al.* 2002) (b) Four delay-Doppler radar images of the binary NEA (66391) 1999 KW$_4$ obtained at Arecibo Observatory in 2001. The frame is 5.6 km vertically by 18.6 Hz horizontally. Rotation and revolution are counterclockwise. The oblate primary has a diameter of 1.5 km and a spin period of 2.8 hours. The 0.5 km sized oblong secondary travels around the primary in a roughly circular orbit with a radius of 2.5 km and period of 17.4 hours, with its long axis pointing towards the primary. (Ostro *et al.* 2006)

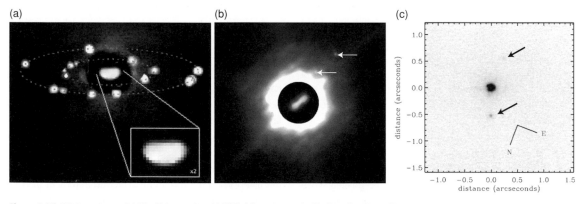

Figure 9.18 Triple systems: (a) The P type asteroid 87 Sylvia as imaged with the adaptive optics system on the VLT in 2004. The dashed lines correspond to the two moons' orbits. This image is a composite of nine individual observations taken on nine nights and filtered with an unsharp mask (a digital sharpening process). North is up and east is left. The inset shows the shape of Sylvia's primary. The dark ring around the primary is an artifact due to the filtering process. The primary is 0.2″ across (long axis). (Adapted from Marchis *et al.* 2005b) (b) M type asteroid 216 Kleopatra with two moonlets (indicated by arrows) observed on UT 19 September 2008 with the adaptive optics system on the W.M. Keck II telescope. The moonlets are just visible against the background of the primary's scattered light. The central dark circle shows the dog-bone shape of the primary (see Fig. 9.5a). (Courtesy Franck Marchis) (c) Image of the triple TNO system Haumea obtained with the adaptive optics system at the Keck telescope. Wavefront sensing was accomplished with a laser guidestar (and the 17.5 magnitude object itself was used as the tiptilt reference star; §E.6). The two moons are indicated with arrows. (Adapted from Brown *et al.* 2005b)

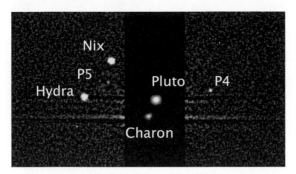

Figure 9.19 Sextuple system: HST image of Pluto, its large moon Charon, the small moons Hydra and Nix, as well as the most recently discovered tiny moons, P4 and P5, as indicated. This image has been processed to substantially enhance the brightness of the four little moons relative to that of the much larger bodies Pluto and Charon. Pluto–Charon can be considered a close binary encircled by four small satellites. The four linear features emanating from Pluto and Charon are diffraction spikes produced by the telescope's optics. (HST/NASA, ESA, Mark Showalter, SETI Institute)

only a tiny fraction of the minor planet population has good density measurements. The parameters for these bodies are summarized in Table 9.5, together with the method used to determine their mass/density. For example, the mass of 4 Vesta was determined from its gravitational perturbation on the orbits of 197 Arete and 17 Thetis; the orbit of the latter asteroid was also observably perturbed by 11 Parthenope, enabling the derivation of Parthenope's density. The masses of 243 Ida, 253 Mathilde, 433 Eros, and 25143 Itokawa have been derived from data obtained during spacecraft encounters.

The densities of minor planets vary substantially from object to object, with measured values ranging from ~0.5 up to almost 4 g cm^{-3}. A comparison of asteroid densities with that of meteorites with similar spectra gives information on the asteroid's porosity. The three largest asteroids, 1 Ceres, 2 Pallas, and 4 Vesta, appear to be coherent objects without substantial porosity, as expected for such large bodies. The measured shape of 1 Ceres is consistent with the hydrostatic shape of a differentiated body. It does not match the shape of an homogeneous object, reinforcing models on early differentiation of asteroidal bodies.

Many of the S type asteroids, e.g., 433 Eros and 243 Ida, have porosities of ~20%, while many C types (e.g., 243 Mathilde, 87 Sylvia) and a few M types (e.g., 16 Psyche) have porosities over 30%, in some cases exceeding 50%. The most porous objects probably have a loosely consolidated rubble-pile structure. Perhaps, relative to the more primitive asteroids, the more 'igneous' asteroids

have an internal strength large enough that the bodies are not completely disrupted in a collision, and instead remain coherent, but highly fractured (low-porosity), bodies. The internal structure of the most porous objects is unknown. When a body is completely shattered, the largest pieces have the lowest initial velocities and largest gravitational attraction. Hence, these accrete first to form the core of a newly assembled body. The core of such an asteroid may thus consist of large irregularly shaped fragments with huge voids (though small pieces could slip through cracks and fill the voids), while the outer layers, made up of smaller pieces, may be less porous. Such structures could be tested observationally for substantially nonspherical objects, once the gravitational moment J_2 can be determined reliably (§6.1.4).

The three largest known KBOs, Eris, Pluto, and Haumea (Table 9.2), which are too massive to retain substantial porosities, have densities of ~2 g cm^{-3}, indicative of coherent bodies composed of comparable amounts of rock and ice. The densities of smaller objects are closer to 1 g cm^{-3}. Since the composition of these bodies is presumably similar to that of the larger objects, with a rock:ice mass ratio of ~0.5, they must be highly porous, like the primitive asteroids. The relatively high observed multiplicity together with the low internal densities of KBOs hint at an early intense collisional epoch. Numerical calculations also show that the Kuiper belt must once have been much (several hundred times?) more massive than at present to explain the accretion of the number of observed large ($R \gtrsim 100$ km) KBOs.

9.4.6 Rotation

Over 80% of all planetary bodies rotate with a period between 4 and 16 hours. Asteroids spinning faster than about 2 hours would throw loosely attached material off their equator, and hence are unlikely to survive (Problem 9.11). There is a clear correlation between rotation period and asteroid size: asteroids with radii less than about 5 km typically spin faster than larger bodies (Fig. 9.20). The shortest rotation period observed for asteroids over 1 km in size is 2.2 hours, essentially equal to the theoretical limit. The asteroidal spin vectors appear to be distributed isotropically.

The spin rates of large ($R \gtrsim 60$ km) asteroids are most likely governed by their collisional history, since their distribution can be fit well by a Maxwellian (eq. 4.75). In contrast, the distribution of spin rates of smaller bodies is highly non-Maxwellian. Asteroids, in particular NEOs with 0.1 km $\lesssim R \lesssim$ 10 km, show a relative excess of slowly rotating bodies, while ~30% of the bodies in this

Table 9.5 Sizes and densities of minor planets.

Body	Class[a]	Mass (10^{19} kg)	R (km)	ρ (g cm^{-3})	Method	Ref.
Near-Earth Asteroids						
2000 UG$_{11}$	R	$9.4^{+2}_{-4} \times 10^{-10}$	0.115 ± 0.15	$1.47^{+0.6}_{-1.3}$	Binary system	1
2000 DP$_{107}$	C	$4.3^{+2}_{-1} \times 10^{-8}$	0.40 ± 0.08	$1.62^{+1.2}_{-0.9}$	Binary system	1
1999 KW$_4$	?	$2.35 \pm 0.1 \times 10^{-7}$	0.6 ± 0.06	1.97 ± 0.24	Binary system	2
1996 FG$_3$	C		0.7	1.4 ± 0.3	Binary system	3
433 Eros	S	$6.7 \pm 0.3 \times 10^{-4}$	18.7	2.67 ± 0.03	Spacecraft encounter	1
25143 Itokawa	S	$3.5 \pm 0.1 \times 10^{-9}$	0.18 ± 0.01	1.9 ± 0.13	Spacecraft encounter	4
Main Belt Asteroids						
1 Ceres	C/G	94.3 ± 0.7	467.6 ± 2.2	2.21 ± 0.04	Orbit perturbation	5
2 Pallas	B	23.9 ± 0.6	256 ± 3	3.4 ± 0.9	Orbit perturbation	21
4 Vesta	V	26.7 ± 0.3	264.5 ± 5	3.44 ± 0.12	Orbit perturbation	1
10 Hygiea	C	10 ± 4	203.6 ± 3.4	2.76 ± 1.2	Orbit perturbation	1
11 Parthenope	S	0.51 ± 0.2	76.7	2.72 ± 0.12	Orbit perturbation	1
15 Eunomia	S	0.84 ± 0.22	127.7	0.96 ± 0.3	Orbit perturbation	1
22 Kalliope	M	0.81 ± 0.02	84.1 ± 1.4	3.35 ± 0.33	Binary system	6
45 Eugenia	C	0.57 ± 0.01	96.5 ± 8	1.1 ± 0.1	Multiple system	7
87 Sylvia	P/X	1.48 ± 0.01	143	1.2 ± 0.1	Multiple system	8
90 Antiope	C	0.083 ± 0.002	42.9 ± 0.5	1.25 ± 0.05	Binary system	9
107 Camilla	C	1.11 ± 0.03	123 ± 7	1.4 ± 0.3	Binary system	7
121 Hermione	C	0.54 ± 0.03	104.5 ± 2.4	1.1 ± 0.3	Binary system	10
130 Elektra	G	0.66 ± 0.04	108 ± 8	1.3 ± 0.3	Binary system	11
216 Kleopatra	M	0.464 ± 0.002	67.5 ± 1	3.6 ± 0.2	Multiple system	12
243 Ida	S	0.0042 ± 0.0006	15.7	2.6 ± 0.5	Spacecraft encounter	1
253 Mathilde	C	0.0103 ± 0.0004	26.5	1.3 ± 0.2	Spacecraft encounter	1
283 Emma	X	0.138 ± 0.003	80 ± 5	0.7 ± 0.2	Binary system	11
379 Huenna	C	0.0383 ± 0.002	49 ± 2	0.8 ± 0.1	Binary system	11
762 Pulcova	C	0.14 ± 0.01	121.5 ± 1	0.9 ± 0.1	Binary system	7
Trojan Asteroids						
617 Patroclus	P	0.136 ± 0.011	61×56	0.8 ± 0.2	Binary system	13
624 Hektor	D	1.0 ± 0.1	$190 \times 100 \times 100$	1.6 ± 0.3	Binary system	14
Trans-Neptunian Objects						
20000 Varuna	CKBO		355	$0.99^{+0.09}_{-0.02}$	Jacobi ellipsoid[b] fit	15
(26308) 1998 SM$_{165}$	RKBO	0.68 ± 0.02	144 ± 18	$0.5^{+0.29}_{-0.14}$	Binary system	16
42355 Typhon	SDO	0.096 ± 0.005	67 ± 7	$0.47^{+0.18}_{-0.10}$	Binary system	17
(47171) 1999 TC$_{36}$	RKBO	1.44 ± 0.25	155 ± 18	$0.5^{+0.3}_{-0.2}$	Binary system	17
65489 Ceto	Centaur	0.541 ± 0.042	87 ± 9	$1.37^{+0.65}_{-0.32}$	Binary system	18
90482 Orcus	RKBO	63.2 ± 0.1	450 ± 40	1.5 ± 0.3	Binary system	20
134340 Pluto	RKBO	1305 ± 62	1153 ± 10	2.03 ± 0.06	Multiple system	17
Charon	Moon	152 ± 6.5	606.0 ± 1.5	1.65 ± 0.06	Multiple system	17
136108 Haumea	SDO	421 ± 10	725	2.9 ± 0.4	Multiple system	17
136199 Eris	SDO	1670 ± 20	1163 ± 6	2.52 ± 0.05	Binary system	17
1998 WW$_{31}$	CKBO	0.27 ± 0.04	67 ± 8	1.5 ± 0.5	Binary system	19
(139775) 2001 QG$_{298}$	RKBO		120	$0.59^{+0.14}_{-0.05}$	Roche model	15

[a] Spectroscopic class for asteroids; dynamical class for TNO (CKBO: Classical KBO; RKBO: Resonant KBO; SDO: Scattered disk object).

[b] Jacobi ellipsoid: see §6.1.4.2.

: Britt *et al.* (2002). 2: Ostro *et al.* (2006). 3: Mottola and Lahulla (2000). 4: Fujiwara *et al.* (2006). 5: Carry *et al.* (2008). 6: Descamps *et al.* (2008). 7: Marchis *et al.* (2008b). 8: Marchis *et al.* (2005b). 9: Descamps *et al.* (2007). 10: Marchis *et al.* (2005a). 11: Marchis *et al.* (2008a). 12: Descamps *et al.* (2010). 13: Mueller *et al.* (2010). 14: Marchis *et al.* (2006). 15: Lacerda and Jewitt (2007). 16: Spencer *et al.* (2006). 17: Noll *et al.* (2008). 18: Grundy *et al.* (2007). 19: Veillet *et al.* (2000). 20: Brown *et al.* (2010). 21: Carry *et al.* (2010).

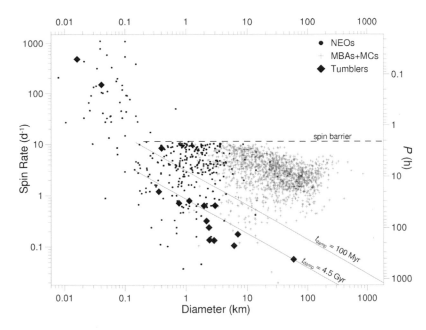

Figure 9.20 Plot of the rotational period (right verical axis) or spin rate (left vertical axis) versus diameter for main belt (MBA) and Mars crossing (MC) asteroids, near-Earth objects (NEO), and tumbling asteroids. The dashed line denotes the maximum rotation rate for objects of density 3 g cm^{-3} that are bound gravitationally. (Adapted from Pravec *et al.* 2007)

size range, including almost all NEA binaries, rotate at periods $P_{rot} \lesssim 4$ hours (but slower than 2.2 hours). Many of the smallest NEOs, $R \lesssim 100$ m, rotate much faster. The tiny (~15 m radius) asteroid 1998 KY$_{26}$ has a rotation period of only 10.7 minutes. Such rapidly rotating bodies must be single coherent bodies without a regolith. A few asteroids with exceptionally long rotation periods may have been tidally despun (§2.6) by (as yet undetected) moons, or the YORP effect (§2.7.4).

Much less is known about the rotation rates of TNOs. As of July 2007, there are ~40 TNOs with a well-determined spin period. For similarly sized TNOs and MBAs, TNOs appear to have longer rotation periods, ~8.2 hours versus 6.0 hours for MBAs. Smaller TNOs may spin somewhat faster than the larger ones, with the notable exception of Haumea, with a 3.9 hr spin period. This high spin rate may have been imparted during the collision that created the Haumea family.

The moment of inertia tensor (eq. 2.42) of any body can be diagonalized (made into the form $I_{jk} = 0$ for $j \neq k$) if the proper coordinate axes are chosen; these axes are referred to as the *principal axes* of the body. The principal axes of a homogeneous triaxial ellipsoid are simply the axes of the ellipsoid itself, and the principal axes of a spherically symmetric body can be chosen to lie along any three mutually orthogonal lines which pass through the center of the body. The lowest energy state for a given rotational angular momentum is simple rotation about a body's axis of greatest moment of inertia (the short axis); see equation (2.41). Simple rotation about a body's axis

of least moment of inertia (long axis) is also possible, but requires more energy and thus is secularly unstable (often on very long timescales) to energy dissipation resulting from rotationally induced stresses within the body. Rotation about the axis of intermediate moment of inertia is unstable on a dynamical (very short) timescale. If a body does not spin about one of its principal axes, then the rotational angular momentum is not parallel to the instantaneous axis of rotation (eq. 2.41*a*), and as a consequence the axis of rotation varies, i.e., the body undergoes *torque-free precession* (§6.1.4.3); in layman's terms, it wobbles. This wobble is described quantitatively by Euler's equations.

Internal stresses caused by precessional wobble damp the rotation of planetary bodies down towards the lowest energy state. A good way to visualize this process is to imagine that the asteroid is a collection of rigid balls connected together by springs. The springs oscillate as variations in the axis of rotation alter internal stresses, and mechanical energy is lost to heat because the springs are damped via friction. The damping timescale, t_{damp}, depends upon the density, radius, and rigidity of the body, ρ, R, and μ_{rg}, a shape-dependent factor K_3^2 that varies from ~0.01 for a nearly spherical body to ~0.1 for a highly elongated one, the ratio of energy contained in the internal oscillations of the body to that lost per cycle, f_Q, and the rotational frequency ω_{rot} as

$$t_{damp} \approx \frac{\mu_{rg} f_Q}{\rho K_3^2 R^2 \omega_{rot}^3}. \tag{9.4a}$$

For nominal asteroidal parameters, the damping timescale in billions of years is given by

$$t_{\text{damp}} \sim \frac{0.7}{R^2}\left(\frac{2\pi}{\omega_{\text{rot}}}\right)^3, \qquad (9.4b)$$

where the radius of the asteroid is in kilometers and its rotation period, $2\pi\omega_{\text{rot}}^{-1}$, is in days. The spread in damping times given by equation (9.4b) is estimated to be a factor of $\sim$10. Precessional motions are thus damped very quickly for large bodies which rotate rapidly, but small slow rotators can remain in complex rotational states for long periods of time (Problem 9.13). Indeed, most identified tumblers have damping times that are at least equal to the age of our Solar System (Fig. 9.20). A few rotate much faster, such as 2000 WL$_{107}$, a small NEA ($R \sim 20$ m) with $P_{\text{rot}} \approx 20$ min. Tumblers with these long damping times, as well as the rapidly rotating bodies, may result from energetic collisions. Nutation of slowly rotating small bodies can also be excited via YORP torques (§2.7.4), or in the case of comets by outgassing. Several small asteroids and comets wobble substantially like, e.g., 4179 Toutatis (§9.2.2).

9.4.7 Interplanetary Dust

The interplanetary medium contains countless microscopic dust grains, visible through faint reflections of sunlight, producing the *zodiacal light* and the *gegenschein*. The zodiacal light is visible in clear moonless skies, in particular in the spring and fall, just after sunset and before sunrise, in the direction of the Sun. Tiny dust particles, concentrated in the local *Laplacian plane* (the mean or reference plane about which satellite orbits precess), scatter sunlight in the forward direction, and the resulting zodiacal light is about as bright as the Milky Way. The gegenschein is visible in the antisolar direction as a faint glow, caused by backscattered light from interplanetary dust. The total volume of the zodiacal dust corresponds to one $\sim$5 km radius sphere.

The presence of interplanetary dust is further 'seen' in the form of meteors, streaks of light in the night sky. Meteors are caused by centimeter-sized dust grains which, when falling down, are heated to incandescence by atmospheric friction (§§5.4.3, 8.3). Under excellent conditions, one may see 5–7 meteors per hour. On rare nights there are many more which appear to come from a single point in space: such events are called *meteor showers*, and are generally named after the stellar constellation that contains the *radiant point*, e.g., the Perseids on 11 August and the Leonids on 17 November. The latter sometimes displays spectacular *meteor storms*, with up to 150 000 meteors per hour! Many of the meteor showers are associated with cometary orbits: The debris left behind by the outgassing comet is intercepted by the Earth when it intersects the comet's path.

Studies show that the Earth 'accretes' about 4×10^7 kg of interplanetary dust per year. Interplanetary dust particles (Fig. 8.12a) have been collected by high-altitude aircraft in the Earth's stratosphere. Analysis in the laboratory shows an overall similarity in composition of these particles with chondritic meteorites. Some particles have a much larger D/H ratio than does the Earth, suggesting that the molecules, and possibly even the dust grains, formed within the interstellar medium and retained their identities when processed within the solar nebula (§§10.7.1, 13.4).

The IRAS satellite, and more recently the Spitzer telescope, sensitive to the thermal emission from 10–100 μm-sized interplanetary dust particles, revealed considerable structure in the zodiacal dust cloud. Bright emission streaks visible on images (Fig. 9.21) have been interpreted as bands of dust encircling the inner Solar System. Many of the *dust bands* are attributed to outgassing by comets,

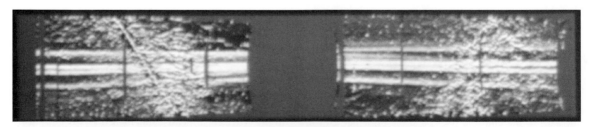

Figure 9.21 IRAS image of the sky near 25 μm showing several zodiacal dust bands in the asteroid belt. Parallel bands are seen above and below the ecliptic, encircling the inner Solar System. The two central bands are connected to the Koronis–Themis families, and the 10° bands may be from the Eos family (Fig. 9.14). The thin band between the center bands and the outer 10° bands is a type 2 dust trail. Type 1 dust trails originate from short-period comets. Type 2 dust trails are not yet understood, and could signify a relatively recent breakup of an asteroid. The diagonally shaped features are the plane of the Milky Way galaxy. (Courtesy Mark V. Sykes)

producing the meteor streams associated with the meteor showers mentioned above. A total of eight of these trails have been positively identified with short-period comets. Several dust bands appear as 'double' structures. Roughly seven pairs of such dust bands straddle the plane of the ecliptic (Fig. 9.21). The IRAS color temperature (~200 K) and parallactic studies place these bands within the asteroid belt. These dust bands are seen as pairs, since the particles which make up the band have the same orbital inclination (to the local Laplacian plane), but their nodes are uniformly distributed over all longitudes. Because each particle spends most of its time near the extrema of its oscillation (as does a swing, pendulum, or sine curve), a collection of dust particles with uniform inclination orbits is seen as a pair of bands straddling the plane of the ecliptic, with a separation depending on the inclination of the particles' orbits (see also §11.3.1). These asteroidal dust bands likely result from erosive collisions in the asteroid belt, or from a catastrophic disruption of an asteroid. Since such disruptions lead to new families as discussed above, it may not be too surprising that several dust bands have indeed been identified with asteroid families (Fig. 9.14).

(Sub)micrometer–to–centimeter sized material is removed from the Solar System by radiation pressure, or by Poynting–Robertson and/or solar wind (corpuscular) drag (§2.7). Thus the orbits of particles in these dust bands change over time, and small grains are lost from the interplanetary medium. Hence, because individual dust bands must eventually fade away, new bands presumably form. Observations of the dust bands and asteroid families, together with collisional theories and numerical simulations, suggest that collisions are frequent events. It is expected that one $R \gtrsim 500$ m asteroid is disrupted every ~100 years, and one $R \gtrsim 5$ km body every ~10^5 years. These collisions, together with refractory particles released as a result of outgassing by comets, form a source of interplanetary dust. The distribution of velocities of meteoroids entering the Earth's atmosphere also implies a mixture of particles of cometary and asteroidal origin. It is not clear, however, which source dominates. In addition, calculations have shown that dust created in the Kuiper belt also contributes to the interplanetary dust in the inner Solar System. Although ~80% of these grains should be ejected from the Solar System by the giant planets, some of them can collide with Earth.

Many spacecraft have made *in situ* observations of the interplanetary dust particles. The Helios 1 and 2 space probes observed them at heliocentric distances between 1 and 0.3 AU, Galileo and Ulysses between 0.7 and 5 AU, and the Pioneers 10, 11, and Cassini out to ~10 AU. Inside ~3 AU the spatial density distribution is roughly inversely proportional with distance, while the dust density is constant outside of Jupiter's orbit. Inside Earth's orbit the dust is distributed into three dynamically distinct populations. Dust in low-eccentricity orbits originates in the asteroid belt, while particles in high-eccentricity orbits, which usually have large semimajor axes, come from short-period comets. A significant flux of tiny particles, the β-meteoroids, appears to come from the solar direction, expelled by radiation pressure.

Closer to Jupiter, Ulysses, Galileo, and Cassini encountered collimated streams of dust grains, ~10 nm in size, moving at speeds of several 100 km s^{-1} away from Jupiter. These particles are ejected from Io via volcanic eruptions, a hypothesis reinforced when mass spectra (§E.10) identified sodium chloride and sulfur compounds. Jupiter's rings form a secondary source of this material. Cassini detected similar dust streams from Saturn's A and E rings.

Several spacecraft have also detected *interstellar* dust grains, which have radii typically between 0.1 and 1 μm. These particles move essentially parallel to the interstellar neutral hydrogen and helium (§1.1.6). There appears to be a deficiency in smaller grains, probably caused by a 'filtering' effect due to radiation pressure and electromagnetic forces.

Numerical integrations of dust particles that originate in the asteroid belt show that 20–25% of these grains are temporarily trapped in corotational resonances just outside the orbits of the terrestrial planets. The Earth is observed to be embedded in an extremely tenuous ring of asteroidal dust particles with a width of a few tenths of an AU. The ring is longitudinally nearly uniform, except for a cavity which contains the Earth. The Earth is closer to the edge of the cavity in the trailing orbital direction than in the leading direction, which explains the observed 3% enhancement in the zodiacal brightness in the trailing direction compared to that in the leading direction.

9.5 Individual Minor Planets

Detailed images, spectra, and other *in situ* data have been obtained for about a half dozen asteroids: 951 Gaspra and 243 Ida have been imaged by the Galileo spacecraft on its way to Jupiter, the NEA 9969 Braille by Deep Space 1, 253 Mathilde and the NEA 433 Eros by NEAR (Near-Earth Asteroid Rendezvous), the NEA 25143 Itokawa by the Hayabusa spacecraft, and 2867 Steins by Rosetta. Rosetta will image a second asteroid, 21 Lutetia, in June 2010, before reaching its final destination in 2014, Comet 67P/Churyumov–Gerasimenko. Dawn is on its way to asteroids 1 Ceres and 4 Vesta, and New Horizons is en route to the Kuiper belt, with a Pluto encounter in 2015.

(a)   Mathilde   (b)   Gaspra

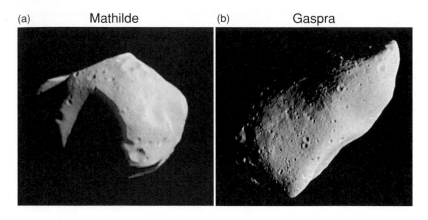

Figure 9.22 (a) Image mosaic (4 images) of C type asteroid 253 Mathilde. The size of the asteroid as shown is 59 × 47 km. (NASA/NEAR, PIA02477). (b) Mosaic (2 images) of S type asteroid 951 Gaspra, which has dimensions of 19 × 12 × 11 km. (NASA/Galileo, PIA00119)

Voyager 2 has investigated Neptune's moon Triton, perhaps a close analog to KBOs (§5.5.8).

The spacecraft observations provide detailed information, and aid substantially in our overall understanding of minor planets, both as individual objects and as a group of bodies/planetesimals that contain information on the formation of our Solar System. Although the most detailed images are obtained by spacecraft during close flybys or when orbiting the body, radar measurements and high angular resolution data on binary/multiple systems have contributed a wealth of information, as discussed in §9.2.2 and §9.4.4. In this section we summarize the characteristics of those minor planets that have been investigated in detail.

951 Gaspra

951 Gaspra is a small ($R = 6.1 \pm 0.4$ km) irregularly shaped S type asteroid (Fig. 9.22b), which is unusually red and olivine-rich. Its geometric albedo is typical for S type asteroids, $A_0 = 0.22$. The object was most likely produced during the catastrophic disruption of the parent body ($R \gtrsim 100$ km) that produced the Flora family ~500 Myr ago (§8.2). Over 600 craters have been identified on Galileo images of Gaspra's surface. Craters with sizes $D \sim 0.2-0.6$ km look relatively young and fresh. They show a size–frequency distribution with $\zeta \sim 4-4.7$, i.e., a steep production function. Larger craters are shallower, with more subdued rims, and a shallower slope, $\zeta \sim 3.6$, consistent with that expected for a collisionally evolved population of impactors. One explanation for the fresh craters and steep slope is that Gaspra was hit by a swarm of fragments produced during the catastrophic disruption of another body. Numerical calculations suggest that the asteroid may have been hit by fragments from the parent body ($R \sim 80-90$ km) that created the Baptista family ~160 Myr ago (§5.4.6).

In addition to craters, Gaspra displays grooves somewhat reminiscent of the fractures seen on Mars's moon Phobos (§5.5.4.6). The grooves and subtle color/albedo variations that cover the surface suggest that Gaspra is covered by a layer of regolith that could be many tens of meters thick.

The interplanetary magnetic field changed abruptly towards a radial orientation in the direction of 951 Gaspra when the Galileo spacecraft made its closest approach to the asteroid. Since an object as small as Gaspra is unlikely to have an active magnetic dynamo, the spacecraft probably detected remanent magnetism, as on Mars (§7.5.3.2). Several meteorites, in particular iron and stony-iron objects, are also strongly magnetized. Gaspra may be a scaled-up version of these meteorites, and show remanent magnetism from the epoch of its formation.

243 Ida and Dactyl

243 Ida, an irregularly shaped S type asteroid, is a member of the Koronis family. The biggest surprise from the Galileo data was the small (mean radius of 0.7 km) almost round satellite Dactyl orbiting Ida at a distance of 85 km (Fig. 9.15). Ida and Dactyl have very similar photometric properties, but distinct differences in spectral properties. Hence their textures must be similar, but Dactyl has a slightly different composition; in particular, it may contain more pyroxene than Ida. The differences are within the range of differences reported for other members of the Koronis family, and argue for compositional inhomogeneities of the Koronis parent body, possibly resulting from differentiation.

The crater density on Ida is roughly five times higher than that on Gaspra for craters over 1 km in size. The craters are severely degraded, which hints at an old age. The density and crater size distribution suggest the surface of Ida to be ~1–2 × 10^9 years old. Dactyl's life expectancy against impact disruption is only a few $\times 10^8$ years old, which is thus much less than Ida's age. Since it is very unlikely that Dactyl would have been captured in the past 10^8 years, the satellite is probably a 'primordial' companion, created at the same time as Ida from the Koronis parent body. However, in principle the moonlet could have formed later from a large impact on Ida. Although it might be possible that the pair is less than 10^8 years old, and was heavily bombarded immediately after the Koronis parent body broke up, this would not explain the degraded condition of craters on Ida's surface, unless the sputtering by micrometeorites is much more severe in Ida's environment. Hence the coexistence of Ida and Dactyl remains puzzling.

Based upon estimates of the thermal inertia, Ida is likely covered by a $\sim$50–100 m thick layer of regolith. Galileo also detected about 17 large (45–150 m across) blocks of rock on its surface. In analogy to Earth, the Moon, Phobos, and Deimos, these blocks could be impact ejecta. On Ida, however, it is also possible that these blocks are fragments of the original Koronis parent body, which survived after the parent body broke up, and re-accreted (gently) onto Ida.

253 Mathilde

253 Mathilde is a C type asteroid, with a geometric albedo of 0.043 ± 0.005 (Fig. 9.22). It is very homogeneous in albedo across its surface, suggesting that the entire body is probably homogeneous and undifferentiated, as expected for primitive bodies. (If the asteroid had a thin dark coating, impact ejecta and craters would result in albedo variations.) Mathilde's color is similar to that of CM carbonaceous chondrites, but Mathilde's density $(1.3 \pm 0.2$ g cm$^{-3})$ is much lower than that of the meteorites, implying a 40–60% internal porosity, i.e., a rubble-pile compound body. Its slow rotation $(P_{\text{rot}} = 17.4$ days$)$ raised suspicion that Mathilde might be a binary system, tidally despun. No satellite was detected by the NEAR spacecraft, however, which would have detected any moon over 300 m in diameter.

About 50% of Mathilde's surface was imaged by NEAR. Four craters with diameters that exceed the asteroid's mean radius of 26.5 ± 1.3 km were found. These large craters look relatively fresh (steep walls, crisp rims), but do not show any evidence of ejecta blankets. Moreover, one might expect impactors that created such large craters to have disrupted the entire body, which apparently did not happen. As discussed in §9.4.3, numerical hydrocode as well as laboratory experiments show that the energy of impacts into bodies as porous as Mathilde is quickly

damped, and that craters form by compaction rather than excavation.

Craters up to 5 km across approach the saturation limit, with a size distribution similar to that seen on Ida, and suggests that Mathilde, like Ida, is $1–2 \times 10^9$ years old. Crater morphologies range from relatively deep fresh craters to shallow degraded ones. Such a range of morphologies is also consistent with a surface in equilibrium with the cratering process.

No water or hydrated features have been seen on Mathilde's surface. The spectral properties of the asteroid match those of the Murchison meteorite (CM) after the meteorite had been heated substantially, so that it had lost all its water and hydrated minerals. This suggests that Mathilde either formed in a region where it was too cold to allow aqueous alterations on its surface, or that the asteroid was heated up so any hydrated material would have been driven off.

433 Eros

The S type near-Earth asteroid 433 Eros (mean $R = 8.4$ km, $A_0 = 0.25 \pm 0.05$) has been studied extensively by the NEAR Shoemaker spacecraft (Fig. 9.23). After orbiting Eros for about a year, NEAR was the first spacecraft to land on such a tiny body. On a global scale, irregularly shaped Eros is dominated by convex and concave forms, including a depression extending over $\sim$10 km (Himeros), and a bowl-shaped crater 5.3 km in diameter (Psyche). Eros's surface is heavily cratered, like Ida's, but craters with diameters <200 m are progressively depleted compared to a lunar-like surface. The asteroid is further covered with about a million ejecta blocks with diameters between 8 and 100 m.

The multiple orientations of sinuous and linear depressions, ridges, and scarps indicate that these features were formed in multiple events. These structures, as well as cliffs with slopes above the angle of repose, imply that Eros must have considerable cohesive strength. Hence Eros, while heavily fractured, must be a coherent body. Groove morphologies, filled-in craters, and benches suggest regolith depths of tens of meters. The saddle-shaped depression, Himeros, exhibits landforms indicative of mass wasting. There are also areas of flat deposits, reminiscent of fluids 'ponding' to an equipotential surface. Perhaps seismic shaking has redistributed fine materials, although this does not seem to be the complete answer.

Eros's elemental composition and spectral properties are similar to ordinary chondrites. Thus, space weathering may indeed 'mask' the surfaces of many asteroids, and the common S type asteroids may be the parent bodies of ordinary chondrites. This would solve the long-standing

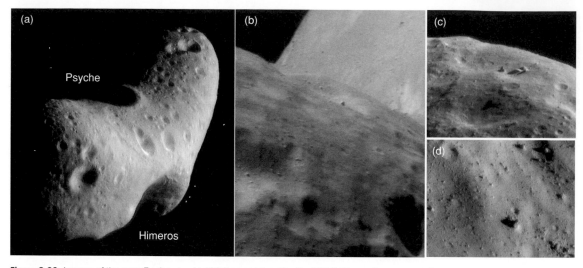

Figure 9.23 Images of the near-Earth asteroid 433 Eros obtained by the NEAR Shoemaker spacecraft. (a) Mosaic taken on 29 February 2000 from an altitude of ~200 km. The ~10 km saddle Himeros and 5.3 km diameter crater Psyche are indicated. (NASA/NEAR, PI02923) (b) A close-up of Himeros, taken from a distance of 51 km, showing a region about 1.4 km across. (NASA/NEAR, PI02928) (c) A view of Eros's horizon sculpted by worn, degraded craters and punctuated by jagged boulders. The angular boulder at the center of the frame is about 60 m tall. The image was taken from an altitude of 50 km; the picture is 1.4 km across, at a resolution of 4 m. (NASA/NEAR, PI02912) (d) This image was taken on 6 October 2000, at a distance of 7 km. The image is ~350 m across, and reveals rocks of all sizes and shapes, yet the floors of some craters are smooth, suggesting accumulation of fine regolith. The large boulder just below and to the right of the center of the picture is about 15 m across. (NASA/NEAR, PI03118)

question of the apparent absence of parent bodies for these common meteorites (§9.3.2). Eros appears to be a primitive undifferentiated body with a density of 2.67 ± 0.03 g cm^{-3}, suggesting a ~20% internal porosity.

25143 Itokawa

In September 2005, the Hayabusa, or 'Falcon', spacecraft went into orbit around the small ($268 \times 147 \times 105$ meters in radius) NEA 25143 Itokawa. For a period of ~3 months, Hayabusa hovered between 7 and 20 km above the asteroid's surface, and performed two touchdowns to attempt to collect materials, which were returned to Earth in 2010.

Remote sensing observations revealed a body with a shape and surface morphology unlike any other seen so far. Itokawa is shaped like a sea otter (Fig. 9.24). The surface shows no evidence of craters, but is instead littered with boulders, a few up to 50 m in size. The largest boulder, about one-tenth the size of the asteroid itself, is located near the end of the body, and a surprising black boulder rests on the otter's head. These boulders may stem from the large impacts which produced Itokawa's odd shape. While 80% of the asteroid's surface is rough and littered with boulders, the remaining 20% is extremely smooth

and featureless. These areas are covered by gravel (mm–cm sized pebbles), without boulders. Barely visible are remnants of craters, almost erased by impact debris and fine materials, which probably filled in the craters due to impact-induced seismic shaking.

Spectra show that Itokawa's bulk composition is similar to that of LL ordinary chondrites. This, combined with its low density (1.9 ± 0.13 g cm^{-3}) suggests a porosity of ~40%, i.e., the body is a rubble pile of material, held together by gravity. As even a small impact would give ejecta velocities well over the escape speed of 10 cm s^{-1}, perhaps most of the smallest grains have left the asteroid over time, while large boulders continued to accumulate.

2867 Steins

While en route to Comet 67P/Churyumov–Gerasimenko, ESA's Rosetta spacecraft flew by asteroid 2867 Steins on 5 September 2008 (Fig. 9.25). Steins, a member of the 434 Hungaria family, is a small ($R \approx 3 \times 2$ km), irregularly shaped E type asteroid, with an average visual albedo of 35%. It spins around its axis in 6.05 hours. Near its northern pole is a 'huge' crater, ~2 km in diameter. A total of 23 craters have been identified, some of which are degraded, perhaps covered by regolith. Most remarkable

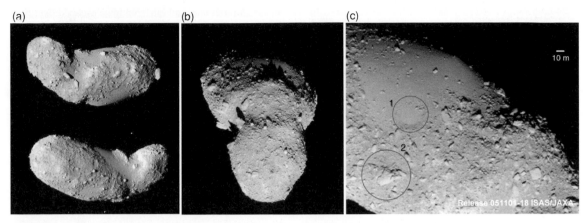

Figure 9.24 Images of 25143 Itokawa obtained with the Hayabusa spacecraft in September 2005. Full views from different perspectives are shown in panels (a) and (b), while panel (c) shows a detailed view from a distance of 4 km (see size bar of 10 m on the image). As shown, the surface is covered with huge boulders and seemingly 'naked' surfaces. At higher resolution these surfaces are covered with finer grained regolith. In panel (c) a subdued crater is encircled (1) as well as a huge boulder (2). (ISAS/JAXA)

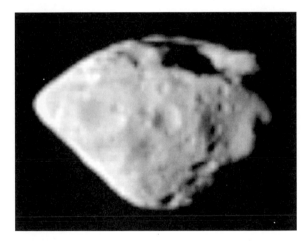

Figure 9.25 Asteroid Steins ($R \approx 3 \times 2$ km) imaged on 5 September 2008 from a distance of 800 km by the Rosetta spacecraft. (ESA, and OSIRIS Team MPS/UPD/LAM/IAA/RSSD/INTA/UPM/DASP/IDA)

is a chain of about seven craters, which may have been produced by a meteoroid stream or by fragments from a shattered object. Prior to the Steins encounter, such crater chains had only been seen on large moons (§5.5.5.3).

134340 Pluto

Because it is by far the brightest KBO, and it has been known for far longer than any other object orbiting the Sun beyond Neptune, we have more information about Pluto than any other TNO. Pluto is about 2/3 the size of our Moon, while its largest moon, *Charon*, is about half

Pluto's size. Pluto's mean density, 2.03 ± 0.06 g cm^{-3}, is larger than expected for a rock/water-ice mixture in cosmic (50:50) proportion, and suggests that Pluto is composed of roughly 70% rock by mass. This dwarf planet is probably differentiated, with the heaviest elements (rock, iron) composing its core, overlain by a mantle of water-ice, and topped off with the most volatile ices. As in the case of the icy satellites (§6.3.5), the water could be liquid ~100–200 km below the surface, in particular if the ice is mixed with ammonia.

Since Pluto and Charon appear so close on an angular scale, in the pre-HST era it was difficult to extract information about each of the bodies separately. In the mid 1980s Charon's orbit was such that the satellite passed in front of and behind Pluto. These mutual events enabled scientists to extract information on Pluto and Charon separately, and to model the spatial distribution of Pluto's surface albedo. Spatially resolved images were later also obtained with HST (Fig. 9.26), which confirmed that Pluto's poles were brighter than its equator, suggestive of poles covered by ice(s) while most of the ice has sublimated off the equatorial regions.

Infrared spectra of Pluto, obtained in the early 1990s, suggest its surface is covered by nitrogen-ice with traces of methane, ethane, and carbon monoxide. Although there are a couple of locations with pure methane-ice, most of the CH$_4$-ice is dissolved in a matrix of nitrogen-ice. High-resolution spectroscopy of the 2.15 μm N$_2$-ice feature revealed a surface temperature of ~40 K over the bright (ice-covered) areas. Thermal infrared data

Figure 9.26 Hubble Space Telescope images of the surface of Pluto, taken over a 6.4 day period in mid 1994. The two smaller inset pictures at the top are the actual images from HST. North is up. At the HST resolution, more than 150 km per pixel, there are 12 major 'regions' where the surface is either bright or dark. The larger images (bottom) are from a global map constructed through computer image processing performed on the Hubble data. (The tile pattern is an artifact of the image enhancement technique.) The two views show opposite hemispheres of Pluto. Most of the surface features, including the prominent northern polar cap, are likely produced by the complex distribution of frosts that migrate across Pluto's surface with its orbital and seasonal cycles and chemical byproducts deposited out of Pluto's nitrogen–methane atmosphere. (Courtesy Alan Stern, Marc Buie, NASA/HST and ESA, PIA00825)

suggest the darker regions have a surface temperature of ∼55–60 K. When Pluto occulted a 12th magnitude star in 1988, a year prior to its closest approach to the Sun, the gradual rather than abrupt disappearance and later reappearance of the star revealed the presence of an atmosphere, with a surface pressure between 10 and 18 μbar (§4.3.3.3). Such a pressure is consistent with an atmosphere in sublimation equilibrium with an icy surface. Contrary to expectations, when Pluto occulted another bright star 13 years after perihelion, the atmospheric pressure had increased by a factor of ∼2, while the atmospheric temperature had remained constant. These changes suggest an increase by 1.3 K in Pluto's surface temperature indicating a time lag in its temperature, caused perhaps by a large thermal inertia.

Both stellar occultations indicated a temperature of 106 K in the upper atmosphere, which combined with Pluto's relatively low gravity might lead to a hydrodynamic escape of gases (§4.8.3.1). The time-averaged rate of escape from Pluto is of the order of $1-5 \times 10^{27}$ mol s^{-1}, which over the age of our Solar System corresponds to a total loss of up to several kilometers of material from Pluto's surface.

In contrast to the nitrogen-ice that covers Pluto's surface, its moon Charon is covered by water-ice, probably

because all methane and nitrogen have escaped over time (§9.3.4). Not much is known about the surface composition of Nix and Hydra, Pluto's two small moons, except that both of them exhibit the same, essentially gray, colors as the larger moon, Charon. The orbits and colors of these small moons reinforce the theory that Charon was produced in a giant impact, similar to the formation of our Moon (§13.11.2). Such an impact could have left Pluto in a 'drier' state than its original composition, and would have left all three moons in coplanar orbits. A subsequent outward migration of Charon might explain the observed orbital resonances of the three moons.

9.6 Origin and Evolution of Minor Planets

Studies of meteorites and minor planets yield valuable information regarding the origin and evolution of our Solar System. However, constructing a generic model of minor planet formation based on the observations is not straightforward, because many 'alterations' have taken place since the first bodies accreted. The size distribution of minor planets, the high porosity of many asteroids, the large number of binary and multiple systems amongst TNOs and asteroids, and the abundance of dust all indicate that collisions were, and still are, common.

Interpretation of surface composition and characteristics of minor planets is muddled by space weathering, which can gradually alter a body's surface composition. For example, it seems likely that S type asteroids are the parents of the primitive chondrites, yet the spectra of these asteroids suggest that they are thermally processed. Differences in spectral shape among the TNOs may also be caused by space weathering. Red spectral slopes, common among some TNOs, are indicative of organic materials, which could be produced by space weathering. In contrast, the many gray bodies amongst the TNOs may have undergone some collisional gardening of their regoliths, resetting their 'clock'. Alternatively, differences in spectra may be genuine differences in the original composition of these bodies. Despite these complications, we summarize below what we have learned about the origin and evolution of the minor planets; the origin and evolution of the Kuiper belt and Oort cloud is continued in §10.7, and all results are folded into the discussion of the origin and evolution of our Solar System as a whole in Chapter 13.

Minor planets presumably grew from planetesimals, as did the terrestrial planets. For those bodies that grew fast and large enough, accretional and radioactive heating led to a (partial) melting of the material, resulting in *differentiation* (§§5.2.2, 13.6.2), i.e., the heavy materials sank down and formed the core of the object. Analysis of meteorites implies that some asteroids are differentiated (§8.7). The asteroids can thus be viewed as remnant planetesimals which failed to accrete into a single body. At present, the mass in the asteroid belt is too small by 3–4 orders of magnitude to allow the formation of a full-sized planet. Numerical simulations of the origin and evolution of the asteroid belt suggest that this region originally had 150–250 times more mass than at present. Many of the original material/bodies have probably been removed from the asteroid belt by the gravitational influence of the planets, in particular of Jupiter (§13.9).

Bodies in the Kuiper belt presumably formed in an analogous manner, and larger bodies are likely differentiated. The TNOs have a higher ice:rock mass ratio than asteroids, because they formed further from the Sun. Yet, as discussed in §10.4.4, the samples returned from the Stardust mission imply that highly refractory processed materials, including CAIs and chondrules, were distributed throughout the solar nebula. Numerical simulations of accretion suggest that the Kuiper belt contained ~100 times more mass during the formation era than it does at present.

The present minor planet population is thus a small fraction of what it once was. Most of the lost material was ejected from the Solar System by gravitational scattering.

Collisions gradually grind the bodies to smaller and smaller fragments, the smallest (dust grains) being removed by Poynting–Robertson drag and radiation forces. The orbits of asteroid fragments $\gtrsim 10$ km across are modified by the Yarkovsky effect, and can replenish the chaotic resonance zones in the main asteroid belt. Once placed into chaotic orbits, they are subsequently driven into planet-crossing trajectories, feeding the NEO population. Modifications to orbits of some TNOs similarly lead to planet-crossing objects, the centaurs, which ultimately feed the population of short-period comets. The few large asteroids that exist today may, by chance, have escaped catastrophic collisions. If such large bodies fragmented at all during their lifespan, they may have coalesced back into a single body. Re-accumulation is more difficult for asteroids $\lesssim 50$ km in size.

It is almost impossible to develop a consistent theory that can explain all aspects of the diversity of minor planets observed today. We note here the existence of the large metallic asteroid 16 Psyche, which could be the remnant of a very large differentiated object, stripped down to its core by numerous impacts. At the opposite extreme, 4 Vesta seems to have preserved its thin basaltic crust.

Collisions between planetesimals may lead to fragmentation of the parent bodies, and may strip the mantles away from differentiated objects, leaving just the metal-rich core of these bodies. Large M type asteroids may be such exposed asteroidal cores. Since the material strength of iron–nickel cores is much larger than that of the silicate mantles, the metal-rich asteroids survived many collisions, while the olivine mantle fragments were likely efficiently destroyed. However, since meteorites from the crusts of differentiated minor planets are also far more common than olivine-rich meteorites, it is difficult to form a coherent scenario unless the metal asteroids were exposed to a much more collisional environment than were bodies like 4 Vesta. One model suggests that metal asteroids, most common in the inner belt, accreted, differentiated, and were stripped of their mantles while in the terrestrial planet region, and were ejected to the asteroid belt towards the end of the planet formation era.

The heliocentric distribution of the various asteroid classes follows the general condensation sequence: High-temperature condensates are found in the inner regions of the asteroid belt, while lower temperature condensates typically orbit the Sun at larger distances. Bound water, e.g., in the form of hydrated silicates, has only been found in a few C, W, and E type asteroids; it has not been seen in D or P types. Hydration of silicates probably took place in the parent asteroidal bodies, rather than in the solar nebula. Aqueous alteration can occur at temperatures as low as 300 K. The lack of hydrated silicates in D and P asteroids may be caused by insufficient heating to melt and mobilize the ice. The dependence on heliocentric distance of igneous, metamorphic, and primitive asteroids (Fig. 9.14) suggests a heating mechanism that declined rapidly in efficiency with heliocentric distance. Although both accretional and radioactive heating are more effective for larger objects, neither accretion nor decay of long-lived radioactive isotopes could have provided enough energy to have melted typical asteroidal-sized bodies (Problems 13.23 and 13.24). Short-lived radioactive nuclei such as ^{26}Al may have provided sufficient energy in the earliest days of Solar System formation. An alternative model involves heating by electromagnetic induction during the Sun's active T-Tauri phase (§13.3.3). Although many aspects of this process remain highly uncertain, calculations suggest that electromagnetic induction heating may melt bodies with $R \gtrsim 50$ km in the inner asteroid belt.

Although much less is known about TNOs, much progress has been made in recent years towards understanding the origin and evolution of this population, in part because of the discoveries of numerous such bodies and detailed follow-up observations of at least the larger bodies with regard to their environment (binary/multiple systems) and surface characteristics. Most intriguing are absorption features of crystalline water-ice, a constituent which is unstable in the Kuiper belt on ~ 10 Myr timescales. Could this be a sign of geyser activity in some large KBOs? On the smaller KBOs it is most likely caused by some nonthermal exogenic mechanisms, such as meteoritic bombardment.

Further Reading

Good review papers on asteroids, TNOs, and interplanetary dust can be found in:

McFadden, L., P. R. Weissman, and T.V. Johnson, Eds., 2007. *Encyclopedia of the Solar System*, 2nd Edition. Academic Press, San Diego. 982pp.

More in-depth topical papers are contained in:

Bottke, W.F., Jr., A. Cellino, P. Paolicchi, and R.P. Binzel, Eds., 2002. *Asteroids III*. University of Arizona Press, Tucson. 785pp.

Barucci, M.A., H. Boenhardt, D.P. Cruikshank, and A. Morbidelli, Eds., 2008. *The Solar System beyond Neptune*. University of Arizona Press, Tucson. 592pp.

The review papers by Cruikshank *et al.* and Chiang *et al.* and Jewitt *et al.* in Reipurth, B. D. Jewitt, and K. Keil, Eds., 2007. *Protostars and Planets V*. University of Arizona Press, Tucson. 951pp.

A good treatment of Euler's equations and other dynamics can be found in:

Goldstein, H., 2002. *Classical Mechanics*, 3rd Edition. Addison Wesley, MA. 638pp.

Problems

9.1.**I** Figure 9.1b shows the instantaneous positions of over 7000 asteroids.

(a) Explain why the Kirkwood gaps cannot easily be detected in this figure, in contrast to their clear visibility in Figure 9.1a.

(b) Discuss the dynamical causes of the radial and longitudinal structure which is apparent in Figure 9.1b.

9.2.**E** Calculate the locations of the 3:1, 5:2, 7:3, 2:1, and 3:2 resonances with Jupiter. (Hint: See Chapter 2.) Use Figure 9.1a to determine which of these resonances produce gaps and which have led to a concentration in the population of asteroids.

9.3.**E** (a) Show that if the exponent $\zeta = 4$ in equation (9.3*a*), then the mass is divided equally among equal logarithmic intervals in radius.

(b) Show that if $\zeta < 3$, then most of the mass in the asteroid belt is contained in a few large bodies, in the sense that the largest factor of 2 in radius contains more mass than all smaller bodies combined.

(c) For which value of ζ does one find equal integrated cross-sectional areas in equal logarithmic size intervals?

9.4.**E** If the differential size–frequency distribution of a group of objects that have equal densities can be adequately described by a power law in radius of the form $N(R)dR \propto R^{-\zeta}$, then it can also be described as a power law in mass of the form $N(m)dm \propto m^{-x}$. Derive the relationship between ζ and x.

9.5.**I** (a) Estimate the number of asteroids in the main belt with radii $R > 1$ km by using the observed number of large asteroids (Table 9.1) and the observed slope of the size–frequency distribution, $\zeta = 3.5$.

(b) How often does a particular 100 km radius asteroid get hit by any asteroid 1 km in radius or larger? You may assume that the asteroids are uniformly distributed in semimajor axis within the main belt.

(c) How often does a particular X km radius asteroid get hit by any asteroid Y km in radius or larger?

(d) How often does any 100 km radius asteroid somewhere in the main belt get hit by any asteroid 1 km in radius or larger?

9.6.**I** (a) Calculate the mean transverse optical depth of the main asteroid belt. (Hint: Divide the projected surface area of the asteroids by the area of the annulus between 2.1 and 3.3 AU. Use the size–frequency distribution given by equation (9.3) and Problem 9.5a, and assume (and justify) reasonable values for $R_{\min}$ and $R_{\max}$ if necessary.)

(b) Calculate the fraction of space (i.e., volume) in the main asteroid belt near the ecliptic that is occupied by asteroids.

9.7.**E** A 'typical' asteroid orbits the Sun at 2.8 AU, has an inclination of $15°$, and an eccentricity of 0.14. Calculate the typical collision velocity of two asteroids. (Hint: Compute the speed of a single asteroid relative to a circular orbit in the Laplacian plane of the Solar System and multiply by $\sqrt{2}$ to obtain the mean encounter velocity.)

9.8.**E** (a) Calculate the gravitational binding energy (in erg) of a spherical asteroid or KBO of radius R (km) and density ρ (g cm^{-3}).

(b) For what size asteroid is the gravitational binding energy equal to the physical cohesion (the fracture stress of rock is $\approx 10^9$ dyne cm^{-2})?

(c) Are nonspherical asteroids more or less tightly bound gravitationally than spherical ones of the same mass?

(d) What is the radius of the largest coherent spherical asteroid that can be disrupted by 1000 Mt of TNT equivalent explosives (1 Mt TNT = 4.18×10^{22} erg)?

(e) What if the asteroid is a rubble pile (no strength)?

(f) Repeat parts (d) and (e) for 1 000 000 Mt TNT.

9.9.**I** Asteroid A has radius R_A, density $\rho = 3$ g cm^{-3}. What is the size, $R_B(R_A)$, of the smallest asteroid B, also of density $\rho = 3$ g cm^{-3}, needed to catastrophically disrupt asteroid A? For your calculation, you may assume a collision velocity of 7.5 km s^{-1} and that disruption requires a kinetic energy equal to gravitational plus physical binding energy. Comment on the sign and magnitude of errors induced by these assumptions.

9.10.**I** (a) Calculate the synodic period (time between conjunctions) of a pair of asteroids with orbital semimajor axes of 2.8 and 2.8028 AU.

(b) Assume that the asteroid with $a = 2.8028$ AU has periapse at 2.8 AU. Calculate its speed relative to local circular orbit when it is at periapse. If the other asteroid is on a circular orbit, this is the relative speed of the two bodies where their paths cross.

(c) Compare the speed that you have calculated in part (b) with typical relative velocities of asteroids and also with the escape velocity from a 100 km radius rocky asteroid ($\rho = 2.5$ g cm^{-3}).

(d) Use your results from the above calculations to explain why members of asteroid families (which were formed via catastrophic collisions) are not necessarily in close proximity to one another. What parameters are most useful in determining membership in asteroidal families?

9.11.**E** Consider a spherical asteroid with a density $\rho = 3$ g cm^{-3} and radius $R = 100$ km. The asteroid is covered by a layer of loosely bound regolith. What is the shortest rotation period this asteroid can have without losing the regolith from its equator?

9.12.**I** (a) Estimate the size of the largest asteroid from which you could propel yourself into orbit under your own power.

(b) What variables other than asteroid size must be considered?

(c) How would you be able to launch yourself into a stable orbit?

9.13.**E** Estimate the damping times for wobbles of asteroids with the following radii and rotation periods:

(a) $R = 200$ km, $2\pi/\omega_{rot} = 16$ hours.

(b) $R = 20$ km, $2\pi/\omega_{rot} = 4$ hours.

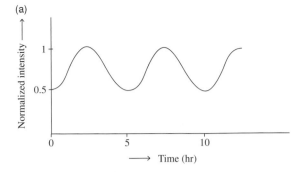

(a)

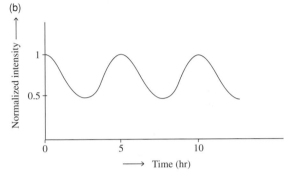

(b)

Figure 9.27 Fictitious lightcurves for Problem 9.15.

(c) $R = 2$ km, $2\pi/\omega_{rot} = 4$ hours.

(d) $R = 2$ km, $2\pi/\omega_{rot} = 4$ days.

(e) $R = 15$ m, $2\pi/\omega_{rot} = 10$ minutes.

9.14.**E** What techniques are used to estimate the diameters of minor planets? Briefly describe each technique, and comment on its advantages and disadvantages.

9.15.**E** (a) Consider the visual lightcurve of an asteroid as sketched in Figure 9.27a. Assume that the shape of the lightcurve at thermal infrared wavelengths is similar. What can you say about the asteroid's shape, size, albedo, and rotation period?

(b) Consider the visual lightcurve in Figure 9.27a and the infrared curve in Figure 9.27b. What can you say about the asteroid's shape, size, albedo, and rotation period?

9.16.**I** Consider an asteroid shaped like a triaxial ellipsoid with axes ratios $a = 2b = 3c$. This asteroid rotates about its short axis and has one small bright (high-albedo) spot. Sketch the visual and infrared lightcurves for this asteroid:

(a) Assuming that the spot is at a pole and the asteroid's rotation axis is perpendicular to the line of sight.

(b) Assuming that the spot is at a pole and the asteroid's rotation axis is parallel to the line of sight.

(c) Assuming that the spot is along the equator at one tip of the long axis and the asteroid's rotation axis is perpendicular to the line of sight.

9.17.I (a) In order to calculate both the radius and the albedo of an unresolved asteroid, one needs observations at infrared as well as visible wavelengths. Explain in one sentence why this is the case.

(b) Assume the Bond and geometric albedos to be equal, the infrared emissivity to be 0.9, the asteroid to be spherical and rapidly rotating. The asteroid is at a heliocentric distance of 2.5 AU, and a geocentric distance of 1.5 AU. The infrared flux (integrated over frequency) is 9.2×10^{-9} erg cm^{-2} s^{-1}, and the flux at visible wavelengths is 1.6×10^{-9} erg cm^{-2} s^{-1}. Determine the temperature of the asteroid, its Bond albedo, and radius. (Hint: See §3.1.)

9.18.E (a) What features distinguish the spectra of S, C, and D type asteroids from one another?

(b) Where do most members of each of these taxonomic classes orbit?

9.19.I Observations of asteroids A1 and A2 show that they are in circular orbits with periods of 4.4 and 6.0 yr, respectively. Assume that each asteroid is spherical, with a radius of 50 km. The geometric albedos are measured to be: $A_{0_1} = 0.245$, and $A_{0_2} = 0.049$, respectively.

(a) Calculate the semimajor axes of the asteroids' orbits.

(b) Assume each asteroid is overlain by a layer of regolith so that the phase integral, $q_{ph,v}$, of the solar reflectivity is equal to that of the Moon. Estimate the visual albedo for the two asteroids from the measured geometric albedos.

(c) Calculate both the average subsurface temperature and the subsolar surface temperature for the two asteroids, assuming they are rapid rotators. (Hint: See §3.1.)

(d) What major taxonomic class would you assign to A1 and A2? Explain your answer in detail.

9.20.I In the previous problem it was assumed that the phase integral for the asteroids A1 and A2 was similar to that of the Moon. If the phase integral has half the value of the Moon's phase integral, calculate the 'beaming factor' $\eta_{v=ir}$, assuming that all other quantities are equal.

9.21.E The observed radar cross-section for asteroid 324 Bamberga is 4500 km^2. Bamberga's radius is 120 km. Assume the radar backscatter gain $g_r = 1$.

(a) Determine Bamberga's radar albedo.

(b) What is the dielectric constant of Bamberga's regolith?

(c) Determine the regolith density and porosity of Bamberga's regolith. Assume the density and dielectric constant of the solid rock in Bamberga's crust are $\rho_0 = 2.6$ g cm^{-3} and $\epsilon_0 = 6.5$.

9.22.I Bamberga's geocentric distance is 0.83 AU and its heliocentric distance 1.78 AU at the time of the observations. Bamberga's Bond albedo is 0.10. The asteroid's total observed emission at 3.6 cm is 1.31 mJy.

(a) Determine Bamberga's brightness temperature T_b at 3.6 cm.

(b) Compare T_b with the expected equilibrium temperature of Bamberga several meters below its surface, and derive the radio emissivity.

(c) Compare the radio emissivity with the radar albedo from Problem 9.21, and comment on your results.

9.23.E The Earth intercepts about 4×10^7 kg of interplanetary dust per year. Estimate the equivalent radius of a spherical body with the above mass (assume a density of 3 g cm^{-3}).

9.24.E (a) Calculate the surface temperature of Pluto and Charon at perihelion, assuming both bodies are rapid rotators and in equilibrium with the solar radiation field. (Hint: See §3.1.2.2.)

(b) Calculate the escape velocity from Pluto and Charon, and compare these numbers with the velocity of N_2, CH_4, and H_2O molecules.

(c) Given your answers in (a) and (b), explain qualitatively the differences in surface ice coverage for Pluto and Charon.

10 Comets

You see therefore an agreement of all the Elements in these three, which would be next to a miracle if they were three different Comets... Wherefore, if according to what we have already said it should return again about the year 1758, candid posterity will not refuse to acknowledge that this was first discovered by an Englishman.

Edmond Halley, 1752, *Astronomical Tables*, London

The generally unexpected and sometimes spectacular appearances of comets have triggered the interest of many people throughout history. A bright comet can easily be seen with the naked eye, and its tail can extend more than 45° on the sky (Fig. 10.1). The name comet is derived from the Greek word $\kappa\omega\mu\eta\tau\eta\zeta$ which means 'the hairy one', describing a comet's most prominent feature: its long *tail*. The earliest records of comets date back to ~6000 BCE in China. In the time of Pythagoras (550 BCE) comets were considered to be wandering planets, but Aristotle (330 BCE) and subsequent natural philosophers thought comets were some kind of atmospheric phenomenon. Comets were therefore scary, and often considered bad omens. An apparition of Comet 1P/Halley is depicted on the Bayeux Tapestry (Fig. 10.2), which commemorates the Norman conquest of England in 1066.

The first detailed scientific observations of comets were made by Tycho Brahe in 1577. Brahe determined that the parallax[1] of the bright Comet C/1577 VI was smaller than 15 arcminutes, and concluded that therefore the comet must be further away than the Moon. Edmond Halley used Newton's gravitational theory to compute parabolic orbits of 24 comets observed up to 1698. He noted that the comet apparitions in 1531, 1607, and 1682 were separated by 75–76 years, and that the orbits were described by roughly the same parameters. He hence predicted the next apparition to be in 1758. It was noticed much later that this Comet Halley, as it was named subsequently, has returned 30 times from 240 BCE to 1986; records of all of these apparitions have been found with the exception of 164 BCE.

[1] The apparent motion of an object compared to the background as seen by an observer moving from one location to another is referred to as its *parallax*. In astronomy, the annual parallax is half the angle over which an object appears to move because of Earth's orbit around the Sun. The diurnal parallax is half the angle over which an abject appears to move because of Earth's rotation about its axis.

Up through the early twentieth century, 3–4 comets were discovered each year; the discovery rate increased to 20–25 comets per year with the development of more powerful cameras (CCDs) in the 1980s, and rose to 40–50 per year from the ground in the first decade of the twenty-first century. In addition, the Solar and Heliospheric Observatory (SOHO) satellite, launched in December 1995, discovered its 1600th comet on 2 January 2009, i.e., an average of over a 100 comets (mostly Sun-grazing, §10.6.4) per year.

A schematic picture of a comet is shown in Figure 10.1c. The small *nucleus*, often only a few kilometers in diameter, is usually hidden from view by the large *coma*, a cloud of gas and dust roughly 10^4–10^5 km in diameter. Not seen with the naked eye is the large *hydrogen coma*, between one and ten million kilometers in extent, which surrounds the nucleus and visible gas/dust coma. Radiation pressure drives the tiny dust particles in a comet's coma outward from the Sun (§2.7.1); these dust particles form the yellowish *dust tail*, which, like the coma, is seen in reflected sunlight. The visible comas of Comet 2P/Encke and a few other comets are dominated by C_2 emissions. As the heliocentric distance of the dust particles grows relative to that of the nucleus, they move less rapidly (conservation of angular momentum), which results in a curvature of the dust tail in a direction opposite to the comet's motion. In contrast to the curved yellowish tail, some comets display a straight, usually bluish, tail in the antisolar direction (Fig. 10.1a). This tail consists of ions that are bound to the interplanetary magnetic field lines, and are dragged along with the solar wind. *Ion tails* can reach lengths up to 10^8 km. The bluish color is produced primarily by emission from CO^+ ions.

Together with meteorites and minor planets, comets provide key information on the origin of our Solar System. Comets originate in the outer Solar System where solar heating is minimal, and they are relatively small ($\lesssim$ tens of km across). Comets, therefore, are among the most primitive objects in our Solar System, and as such

(a)

(b)

(c)

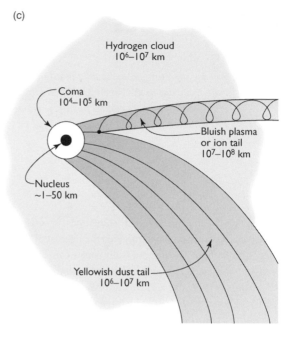

Figure 10.1 (a) COLOR PLATE C/West (1976 VI) as photographed 9 March 1976, by amateur astronomer John Laborde. Both the yellow or 'whitish' dust tail and the blue plasma tail (lower right) are apparent. (Courtesy John Laborde) (b) C/Hale–Bopp (C/1995 O1) as viewed in April 1997 above Natural Bridges National Monument in Utah. (Courtesy Terry Acomb/John Chumack/ PhotoResearchers) (c) Schematic diagram of a comet, showing its nucleus, coma, tails, and hydrogen cloud.

they yield key information about the thermochemical and physical conditions of the regions in which they formed. To determine the composition of a comet nucleus, however, is complicated since it is shrouded in a coma of dust and gas. Moreover, the gas molecules that can readily be observed at optical wavelengths are generally not those housed in the nucleus, but are several steps removed from such *parent molecules* via dissociations and ionizations. Parent molecules are studied through spectroscopy at infrared and radio wavelengths. In this chapter we discuss the physical and chemical properties of comets as derived from observations, and the inferences regarding how and where comets formed and evolved.

10.1 Nomenclature

Comets are named after their discoverer(s). Numbers follow names when one person or group discovers multiple comets.[2] Comets are also given a designation that includes the year of their (re)discovery or perihelion passage. The form of these designations changed in 1995, so the literature contains both formats.

According to the old format, when a comet was discovered or recovered (seen for the first time during an

[2] In recent years these numbers were dropped from the name for long-period and dynamically new comets, but not for short-period comets.

Figure 10.2 Section of the Bayeux Tapestry, commemorating the Norman conquest of England in 1066. 1P/Halley is shown; this comet was very bright at the time William the Conqueror invaded England from Normandy. The comet was considered a bad omen for King Harold of England. The Latin sentence 'Isti Mirant Stella' means 'they marvel at the star'. (Courtesy Beatty *et al.* 1999)

apparition) it was given a provisional designation based on the year of discovery followed by a letter sequentially assigned as comets are discovered, e.g., 1994c represents the third comet discovered in 1994. Several years later, a final designation was given, in which the name of the comet was followed by the year of its perihelion passage, together with a Roman numeral based sequentially on perihelion date to distinguish it from other comets passing perihelion that same year, e.g., Comet C/Kohoutek (C/1973 XII). Short-period comets are preceded by a P/, e.g., P/Halley and P/Encke. Deceased comets, e.g., comets which have collided with the Sun or one of the planets, or simply disintegrated, are preceded by D/; the most famous such object is (was) D/Shoemaker–Levy 9 (§§5.4.5, 10.6.4).

In contrast to the old system, the current system avoids duplication. The year of discovery (or recovery) is followed by a letter indicating the half-month in which the comet was first observed (I is omitted), followed by a number to distinguish it from other comets seen during the same period. The names of long-period comets are now preceded by C/, and short-period comets are given a number according to their initial discovery. Thus, 1P/1682 Q1 represents Halley's comet during its 1682 apparition, and indicates that it was initially spotted in the second half of August of 1682. Note that short-period comets receive a different designation for each apparition in both the old and new systems, but the name and number of a short-period comet remains unchanged.

When a comet splits (§10.6.4), each fragment is given the designation and name of the parent comet, followed by an upper case letter, beginning with the letter A for the fragment that passes perihelion first, e.g., 73P/Schwassmann–Wachmann 3A. If a fragment splits further, the pieces receive numerical indices, as e.g., components Q_1 and Q_2 from Comet D/Shoemaker–Levy 9.

10.2 Orbits and Reservoirs

Many comets travel on extremely eccentric orbits, e.g., $e \approx 0.9999$, so that only a small fraction of their orbital period is spent in the inner planetary region; most of the time comets reside in the cold outer parts of our Solar System (§2.1). Comets with orbital periods >200 years are classified as *long-period (LP) comets*, while those with periods <200 years are called *short-period (SP) comets*. Comets whose orbits suggest that they have entered the inner Solar System, the planetary region, for the first time are referred to as *dynamically new comets*.

In addition to ∼2700 comets recorded only by SOHO and similar spacecraft as of May 2014, the 2008 *Catalogue of Cometary Orbits* (Marsden and Williams 2008) contains 2325 orbits for 2218 cometary apparitions (this difference being because some comets have multiple components) of 1359 specific comets (this difference being because orbits of periodic comets are given for each observed passage through perihelion), of which 415 are SP comets, 200 of these having been observed at more than one passage.

10.2.1 Nongravitational Forces

Although to lowest order comet trajectories are determined by the gravitational pull of the Sun and the planets,[3] the observed orbits of most active comets deviate from these paths in small but significant ways. These variations can advance or retard the time of perihelion passage of a comet by many days from one orbit to the next. The observation of these nongravitational effects was a major motivation for Whipple's *icy conglomerate* model of comet nuclei introduced in 1950 (§10.6). *Nongravitational forces* result from the momentum imparted to a comet's nucleus by the gas and dust which escape as the comet's ices sublime. The process is analogous to rocket propulsion (§F.1), but the magnitude of the effects is much smaller, because only a tiny fraction of the comet's mass is lost per orbit, mass escapes at a slower speed, and forces exerted at differing phases of the comet's orbit produce opposing effects. Nongravitational forces have some similarities to the Yarkovsky effect on small bodies, which is described in §2.7.3.

The standard model (also known as the *symmetric model*) for the nongravitational acceleration of comets is based upon the hypothesis that ice evaporates from a rotating comet nucleus at a specified rate that depends only on heliocentric distance, and thus which is symmetric about perihelion. The equations of motion for a comet are written as

$$\frac{d^2\mathbf{r}_\odot}{dt^2} = -\frac{GM_\odot}{r_\odot^2}\hat{\mathbf{r}}_\odot + \nabla\mathcal{R} + A_1\eta(r_\odot)\hat{\mathbf{r}}_\odot + A_2\eta(r_\odot)\hat{\mathbf{T}} + A_3\eta(r_\odot)\hat{\mathbf{n}},$$ (10.1a)

with

$$\eta(r_\odot) = \eta_1\left(\frac{r_\odot}{r_{\odot 0}}\right)^{-\eta_2}\left(1+\left(\frac{r_\odot}{r_{\odot 0}}\right)^{\eta_3}\right)^{-\eta_4}.$$ (10.1b)

The disturbing function, $\mathcal{R}$, accounts for planetary perturbations (eq. 2.23), A_1, A_2, and A_3 are the nongravitational acceleration coefficients, $\hat{\mathbf{r}}$, $\hat{\mathbf{T}}$, and $\hat{\mathbf{n}}$ are unit vectors pointing in the directions radially outwards from the Sun, perpendicular to $\hat{\mathbf{r}}$ in the plane of the comet's orbit (tangentially along the comet's orbit at perihelion), and normal to the comet's orbit, respectively. The variations of cometary activity with heliocentric distance are approximated by $\eta(r_\odot)$, defined in equation (10.1b). The acceleration is given in AU day^{-2}. For water-ice, the values for the constants in equation (10.1) that are currently being used are: $r_{\odot 0} = 2.808$ AU, $\eta_1 = 0.111\,262$, $\eta_2 = 2.15$, $\eta_3 = 5.093$, $\eta_4 = 4.6142$. The large values of η_3 and η_4 imply that the magnitude of nongravitational forces drops off sharply

around $r_{\odot 0}$; the fact that η_2 is slightly larger than 2 means that nongravitational forces increase slightly more rapidly than solar insolation close to the Sun.

The values of A_1, A_2, and A_3 are determined observationally for individual comets. The value of A_1 is generally much larger than A_2 and A_3, because most of the gas is released by the portion of a comet near the subsolar point and escapes roughly normal to the surface. However, in the standard model, the radial portion of the nongravitational force is symmetric about perihelion, and this component of the force therefore returns to the comet after perihelion the same amount of orbital energy as it removes prior to perihelion. Forcing in the tangential direction is assumed to be a consequence of the thermal lag between cometary noon and the time of maximum outgassing, although it may be more likely caused by asymmetries in the outgassing. In either case it does not suffer from inbound/outbound cancelation. Instead, the tangential component of the nongravitational force produces a component of acceleration along the direction of the comet's motion if the comet rotates in the prograde direction (Problem 10.6). Such an acceleration increases the comet's orbital energy, and thereby increases its orbital period. The situation is reversed for comets with retrograde rotation. Thus, even though A_2 is substantially smaller than A_1, the tangential term is the most important term in the standard theory. For most comets, forces normal to the orbit cancel out over time as a result of symmetries, and they do not produce any significant secular effects, so the last term in equation (10.1a) is generally ignored or found to be negligible in magnitude.

The values of A_1 and A_2 are fitted over both long and short intervals of time, and the solutions are not necessarily constant. Roughly half of the comets with well-studied orbits have values of A_2 that are nearly constant or vary slowly over many apparitions. However, the standard theory clearly breaks down for those comets which show more rapid variations in A_2. Nonuniform distribution of active areas on the surface of a comet can lead to pronounced seasonal effects, as individual vents become more or less active. In such cases, not only is the standard theory inapplicable, but the radial component of the nongravitational force can become far more important than the tangential component (Problem 10.7).

The motion of a comet is clearly intertwined with its evolution as a physical object. Cometary activity depends

[3] Relativistic effects are non-negligible for comets with small perihelion distances, and are included in most high-accuracy orbital calculations.

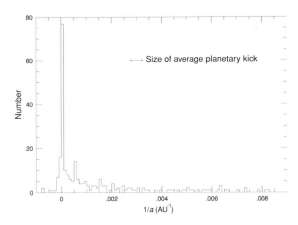

Figure 10.3 Distribution of the original (inbound) inverse semi-major axis, $1/a_0$, of all long-period comets in the 2003 version of Marsden and Williams's *Catalogue of Cometary Orbits*. The typical perturbation on $1/a$ due to a comet's passage through the inner Solar System is indicated on the graph. (Adapted from Levison and Dones 2007)

strongly on heliocentric distance. Outgassing, in turn, changes the orbit of a comet, albeit in a less profound manner.

10.2.2 Oort Cloud

To deduce the source region of dynamically 'new' comets, Jan Oort plotted (in 1950) the distribution of the inverse semimajor axes, $1/a_0$, for 19 long-period comets. Based on this small sample, Oort postulated the existence of about 10^{11} 'observable' comets in what is now known as the *Oort cloud*.

Figure 10.3 shows the distribution of the original $1/a_0$ for a much larger distribution of LP comets. The inverse semimajor axis is a measure of the orbital energy per unit mass, $GM_\odot/(2a_0)$. The original orbit is that of the comet before it entered the planetary region and became subject to planetary perturbations and nongravitational forces. Positive values in $1/a_0$ indicate bound orbits, while negative values denote hyperbolic orbits. The few hyperbolic orbits shown in Figure 10.3, however, are almost certainly not from comets originating in interstellar space (Problems 10.1 and 10.2). The negative $1/a_0$ values are probably caused by errors in the calculation of the orbital elements, possibly induced by unaccounted-for nongravitational forces (§10.2.1).

The evidence for the existence of the Oort cloud at heliocentric distances $\gtrsim 10^4$ AU is based upon the large spike between 0 and 10^{-4} AU^{-1}. The spike, which is much narrower than the typical perturbation on $1/a$ due to a passage through the inner Solar System ($\sim 5 \times 10^4$ AU^{-1}), represents comets from the Oort cloud that have entered

the planetary region for the first time. These comets have semimajor axes of $(1–5) \times 10^4$ AU and are randomly oriented on the celestial sphere, aside from a small signature of the galactic tidal field. The orbits of these comets are highly eccentric ellipses, which appear nearly parabolic when passing through the planetary region of the Solar System. In contrast, the orbits of short-period comets are less eccentric, usually prograde, and appear to be concentrated near the ecliptic plane (inclination angles $i \lesssim 35°$).

The number of comets in the classical Oort cloud can be estimated from the observed flux of dynamically new comets. Counting objects brighter than absolute magnitude $H_{10} = 11$ (for the definition of H_{10}, see §10.3.1), one would expect a total number of $\sim 10^{11} - 10^{12}$ comets in the classical Oort cloud. Comets in the Oort cloud whose orbits are perturbed by the tidal field of the galactic disk, by nearby stars, or by close encounters with giant molecular clouds, may enter the inner Solar System and be observed as dynamically new long-period comets. The galactic tide is the most important perturbation from sending comets from the Oort cloud into the planetary region, but since it alters orbital angular momentum and not orbital energy, it does not remove comets from solar orbit.

The dynamical lifetime of comets in the classical Oort cloud to ejection by passing stars is about half the age of the Solar System. The importance of encounters with giant molecular clouds is much more difficult to estimate due to uncertainties in cloud parameters, but giant molecular clouds could be as much as several times as effective as stars at ejecting comets from the Solar System. This suggests that the Oort cloud is relatively young compared to the age of our Solar System, or that it needs to be replenished, at least occasionally. Replenishment could, in principle, happen from the inside by, e.g., an unseen inner Oort cloud between 10^3 and 10^4 AU, or by capture from the interstellar medium. Due to the rather high interstellar encounter velocities (20–30 km s^{-1}), capture of interstellar comets is exceedingly unlikely. In contrast, the existence of a vast inner Oort cloud is a natural consequence of the formation of our Solar System; dynamical models show that ejection of planetesimals from the planetary region results in an inner Oort cloud which is initially about 5–10 times more populated than the outer (classical) cloud (Fig. 10.4; §10.7.2). Under 'normal' circumstances, the inner Oort cloud contributes essentially nothing to the flux of new comets observable in the inner Solar System because it is virtually unperturbed. However, large perturbations caused by penetrating stellar encounters and giant molecular clouds could produce *comet showers* lasting a few million years about once every 10^8 years, and also repopulate the outer Oort cloud.

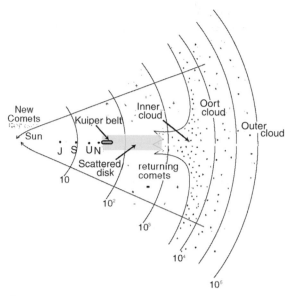

Figure 10.4 Schematic diagram of the structure of the inner and outer Oort cloud. The location of the giant plants and the Kuiper belt are indicated. Note that the distance scale is logarithmic. (Adapted from Levison and Dones 2007)

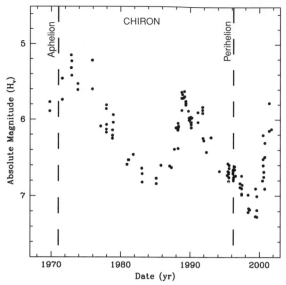

Figure 10.5 This graph of 2060 Chiron's magnitude as a function of time clearly shows dramatic variations in intensity. Interestingly, the object was brighter at aphelion than at perihelion. The data were taken by many different observers, and compiled most recently by Bus *et al.* (2001) and Duffard *et al.* (2002).

When comets pass through the planetary region of the Solar System, gravitational perturbations by the planets scatter their orbits in $1/a$ space. It may take a comet about 400 returns to the planetary region before its orbit is changed to that of a short-period comet. Dynamical calculations show that fewer than 0.1% of long-period comets evolve into short-period comets.

10.2.3 Kuiper Belt

While evolving inwards, comets usually preserve their orbital inclination. Most of the short-period comets have low orbital inclinations and a prograde sense of revolution. It was therefore suggested that most SP comets come from a flattened annulus of objects beyond the orbit of Neptune rather than from the Oort cloud. The subsequent discovery of over 1000 objects in the Kuiper belt, as well as many centaurs which travel on unstable transitional paths crossing the orbits of the giant planets, is convincing evidence for this hypothesis (§9.1).

The giant planets provide the principal perturbations for deflecting Kuiper belt objects sunwards en route to becoming short-period comets. The strong interactions typically begin with a close approach to the planet Neptune. Scattered disk objects are likely the primary source for SP comets, as their paths are much more likely to pass close to Neptune than are those of classical or resonant Kuiper belt objects. However, collisions among KBOs on

relatively stable orbits can produce debris more susceptible to strong perturbations by the giant planets.

Centaurs are in highly unstable orbits that cross the orbits of Saturn, Uranus, and/or Neptune (Figs. 2.11, 9.3). Most centaurs have perihelia between 8 and 11 AU, and aphelia between 19 and 36 AU. Orbital eccentricities are $\sim$0.4–0.6. Typical dynamical lifetimes of the centaurs are 10^6–10^8 years. Because of the difficulty of observing objects at large heliocentric distances, known centaurs are significantly larger than SP comet nuclei. Nonetheless, it is important to study these objects in detail, since they represent the link between Kuiper belt objects (which are technically minor planets, as they do not exhibit a coma) and short-period comets. The first centaur to be discovered was 2060 Chiron, which was subsequently observed to show cometary activity (Fig. 10.5) and thus has also received a comet designation: 95P/Chiron.

10.2.4 Orbits of Comets and Asteroids

Comets have retained their ices because they have spent most of geologic time in a deep-freeze at large distances from the Sun. These ices leave the comet once it is warmed sufficiently. If a comet loses all of its near-surface ices, it can become dormant or extinct. At this point, it is an asteroid. However, its orbit usually betrays its origins in the outer Solar System. Jupiter is the dominant perturber of most objects that cross or pass near its orbit. As Jupiter's

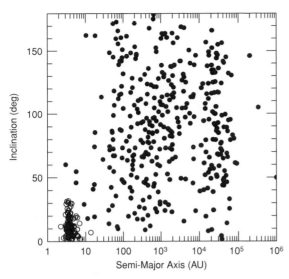

Figure 10.6 The a–i distribution of all comets in the 2003 version of Marsden and Williams's *Catalogue of Cometary Orbits*. Comets with $T > 2$ are denoted by open circles, and those with $T < 2$ are represented by filled circles. (Levison and Dones 2007)

orbital eccentricity is small, the value of the perturbed small body's Jacobi constant (eq. 2.21) is nearly conserved by these interactions. Moreover, as Jupiter's mass is also much less than that of the Sun, there is a simple function of the small body's orbital elements that is nearly constant except when the body is close to Jupiter. This 'constant' is known as the *Tisserand parameter*, and it is given by the formula:

$$C_T \equiv \frac{a_{\jupiter}}{a} + 2\cos i \sqrt{\frac{a}{a_{\jupiter}}(1 - e^2)}, \qquad (10.2)$$

where a, e, and i are the principal heliocentric orbital elements of the small body. As C_T is nearly constant, equation (10.2) is helpful in determining whether two comets observed many years apart may be the same object.

Bodies with $C_T > 3$ cannot cross Jupiter's orbit in the circularly restricted three-body problem. Most asteroid orbits are characterized by $C_T > 3$, and bodies of cometary origin usually have $C_T < 3$; objects with $C_T \approx 3$ can be either asteroidal or cometary. A plot of the inclination versus the semimajor axis of many comets (Fig. 10.6) shows an isotropic distribution in inclination angle for bodies with $C_T < 2$. This population includes the *Halley family* comets (HFC), comets with aphelia between 7.4 and 40 AU, and which have orbital periods between 20 and 200 years. Objects with $2 < C_T < 3$ are usually confined to the ecliptic plane, and are hence referred to as *ecliptic comets* (EC). These comets generally cross Jupiter's orbit; when they have their aphelion near Jupiter's orbit, they are referred to as the *Jupiter family* comets (JFC).

Table 10.1 Orbital elements of main belt comets.

Comet	Minor planet #	a (AU)	e (°)	i (°)
133P/Elst Pizarro	7968	3.157	0.163	1.386
P/2005 U1 (Read)		3.273	0.311	1.188
176P/Linear	118401	3.194	0.193	0.238
P/2008 R1 (Garradd)		2.726	0.342	15.9

This population contains most of the ecliptic comets. Comets with $C_T > 3$ and which are interior to Jupiter's orbit, are classified as *Encke-type* comets, after the prototype 2P/Encke, which has a semimajor axis $a = 2.22$ AU, eccentricity $e = 0.85$, and inclination $i = 11.8°$. Comets with $C_T > 3$ and a semimajor axis $a > 5.2$ AU are called *Chiron-types*, after 95P/Chiron.

Main Belt Comets

The main belt asteroid 7968 Elst Pizarro was renamed as 133P/Elst Pizarro when observations revealed a dust tail. This object ejects dust every 5.6 years when approaching perihelion, at an estimated ejection velocity of $\sim$100 m s^{-1}. It may stay active for weeks–months, and must be driven by sublimation. As of April 2014, about ten such objects are known; some of the first of these to be identified are listed in Table 10.1. Together these bodies are referred to as *main belt comets*. They cannot come from the Kuiper belt or Oort cloud, but must be icy asteroids. Any ice in main belt asteroids that is covered by a layer of regolith 1–100 m thick would be protected against sublimation over the age of the Solar System. Collisions could expose the ice, leading to sublimation inside of 3 AU from the Sun. Elst Pizarro and 176P/Linear are both members of the Themis family of asteroids. Mass loss has been reported from an additional 3 objects, all NEOs (e.g., §9.1.1). Together, these bodies make up a new 'class' of asteroids, referred to as *active asteroids*.

10.3 Coma and Tail Formation

Cometary activity is triggered primarily by solar heating. Comets are usually inert at large heliocentric distances, and only start to develop a coma and tails when they get closer to the Sun. A cometary nucleus is covered with ice, which sublimates (evaporates directly from the solid state) when the comet approaches the Sun. When the sublimating gas *evolves off* the surface, dust is dragged along. The gas and dust form a comet's coma, and hide the nucleus from view. A comet is usually discovered after the coma has formed, when it is bright enough to be seen with relatively small telescopes. Many comets are still inert when they cross Jupiter's orbit, although some show activity even at distances beyond Uranus's orbit. At one extreme, 9P/Tempel 1, with an orbital period of 5.3 years and perihelion and aphelion distances of 1.4 AU and 4.7 AU respectively, is inert over a significant fraction of its orbit, while Comets Shoemaker 1987o (C/1987 H1), Shoemaker 1984f (C/1984 K1), and Cernis (C/1983 O1) still had a dust coma/tail at distances exceeding 20 AU. Comet 29P/Schwassmann–Wachmann 1, which orbits the Sun between 5.4 and 6.7 AU, is known to flare up substantially, sometimes by as much as a factor of 1000. Such *outbursts* may occur if small quantities of ices with low vaporization temperatures are present. On dynamically new comets, outbursts may also be triggered from the conversion of amorphous water-ice to crystalline ice (§10.6.3).

10.3.1 Brightness

The apparent brightness of a comet, B_ν, varies with heliocentric distance, $r_\odot$, and the distance to the observer, r_Δ, a behavior usually approximated by

$$B_\nu \propto \frac{1}{r_\odot^\zeta r_\Delta^2}. \tag{10.3}$$

An inert object, such as an asteroid, has an index $\zeta = 2$, but comets typically show $\zeta > 2$, attributed to the fact that a comet's *gas production rate* (i.e., the amount of gas being released per second) increases with decreasing heliocentric distance. Although some comets indeed appear to follow a power law in $r_\odot$ over a large range of heliocentric distances, there are others which deviate significantly from a power law over many intervals in $r_\odot$. Most comets brighten considerably when their heliocentric distance drops below 3 AU. From the observed relationship between a comet's brightness and its heliocentric distance, it has been deduced that water-ice is the dominant volatile in most comets. The sharp dropoff in the magnitude of nongravitational forces beyond 2.8 AU provides further evidence for the predominance of H_2O-ice. More substantial evidence that comets are composed primarily of water-ice comes from OH emissions, the primary photodissociation product from H_2O: $H_2O + h\nu \rightarrow OH + H$. Hydroxyl emissions at a wavelength of 309 nm are very strong relative to other visible emissions.

C/Hale–Bopp was an exceptionally bright comet, which enabled observations of the evolution in the production rates as a function of heliocentric distance (from 0.9 to 14 AU) for nine different molecules, as displayed in Figure 10.7. The pre-perihelion data were split in three

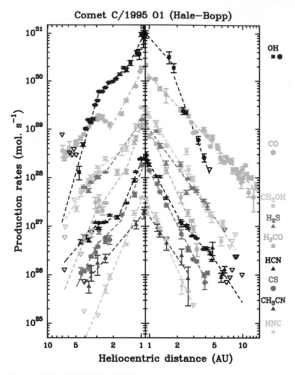

Figure 10.7 COLOR PLATE Time evolution of the production rates of 9 different molecules in C/Hale–Bopp as observed at (sub)mm wavelengths. Power-law fits (dashed lines) are superposed on the data. (Biver *et al.* 2002)

periods, which show different trends with heliocentric distance; the production rates for most molecules followed a single power law after perihelion. As expected, within a heliocentric distance of 3 AU, OH is the most abundant species. At larger distances the comet's brightness was dominated by sublimation of gases more volatile than water. Carbon monoxide is the most volatile species, and dominated the production rate at large distances. CO, as well as the less volatile species CH_3OH, HCN, CH_3CN, and H_2S, showed a moderate increase in production rate at large heliocentric distances inbound, with $\zeta = 2.2$ at $r_\odot > 3$ AU. Between 3 and 1.6 AU inbound the CO production rated stalled or decreased somewhat ($\zeta \approx 0$). Perhaps at that time all the CO had been evaporated from the upper layers of the nucleus, and most of the solar energy went into sublimation of water-ice, rather than heating deeper layers in the cometary crust. A dramatic increase in the production rate of many species was seen at $r_\odot < 1.5$ AU, with $\zeta \approx 4.5$ pre-perihelion, and $\zeta \approx 3.4$ post-perihelion. The evolution in the production rates for some species is suggestive of evaporation of grains in the coma in addition to a nuclear source (e.g., CO, H_2CO, CS), and/or chemical reactions between species in the coma (e.g., HNC).

The brightness of a comet is generally expressed as the *apparent magnitude* at visual wavelengths, m_v. A 6th magnitude star is just visible to the naked eye in a dark sky. The magnitude scale is logarithmic, and a difference of 5 magnitudes equals a factor of 100 in brightness, i.e., a star with $m_v = 0$ is 100 times brighter than one with $m_v = 5$. For a comet, m_v is related to its *absolute magnitude*, M_v:

$$m_v = -2.5 \log B_v$$
$$= M_v + 2.5\zeta \log r_{\odot AU} + 5 \log r_{\Delta AU}, \qquad (10.4)$$

where $r_{\Delta AU}$ is the geocentric distance in AU. A comet's absolute magnitude is equal to its apparent magnitude if the comet were at 1 AU from both the observer and the Sun. The maximum brightness of a comet is usually reached a few days after perihelion, and the brightness variation shows asymmetries between the branches before and after perihelion (the latter is clearly evident in Fig. 10.7). Pronounced differences have been observed in the brightening of old and new comets. Dynamically new comets often brighten gradually on their inbound journey, beginning at large heliocentric distances (at $r_\odot \gtrsim 5$ AU, with $\zeta \approx 2.5$). In contrast, most short-period comets do not brighten much while inbound at large distances, but may 'flare up' when they get closer to perihelion (on average $\zeta \approx 5$). Generally, $\zeta = 4$ is adopted, and the correspondingly derived estimate of the comet's absolute magnitude is denoted as H_{10}.

10.3.2 Gas Production Rate

Assuming only solar heating, the energy balance at the surface of the cometary nucleus is given by

$$(1 - A_b)\frac{\mathcal{F}_\odot e^{-\tau}}{r_{\odot AU}^2}\pi R^2 = 4\pi R^2 \epsilon_{ir}\sigma T^4 + \frac{QL_s}{N_A} + 4\pi R^2 K_T \frac{\partial T}{\partial z}. \qquad (10.5)$$

The term on the left is the energy received from the Sun, with A_b the Bond albedo, $\mathcal{F}_\odot$ the solar constant, τ is the optical depth of the coma, $r_{\odot AU}$ the heliocentric distance in AU, and R the radius of the comet, which is assumed to be spherical. The first term on the right side represents losses caused by thermal infrared reradiation (see §3.1.2), and the last two terms represent losses due to sublimation of ices and to heat conduction into the nucleus, respectively. The thermal infrared emissivity, ϵ_{ir}, is close to unity for most ices. The symbol L_s represents the latent heat of sublimation per mole, and N_A is Avogadro's number. Heat conduction into the surface is represented by the last term on the right, with K_T the thermal conductivity; this term is usually very small for comets and it therefore is often ignored. The gas production rate, Q (molecules

s^{-1}), of water molecules inside $r_\odot < 2.5$ AU can usually be approximated by:

$$Q \approx \frac{1.2 \times 10^{18}\pi R^2}{r_{\odot AU}^2}, \qquad (10.6)$$

with the comet radius, R, in cm. Equation (10.6) was derived assuming $A_b = 0.1$. A better estimate for the mean variation with heliocentric distance would be $Q \propto r_\odot^{-\zeta}$, as exemplified by the power-law fits superposed on Figure 10.7. For most comets, however, the scatter is so large that the simple form given by equation (10.6) is usually considered to be adequate.

Assuming a certain outflow velocity for the gas (e.g., thermal expansion, §10.3.3), both the temperature and coma density can be determined independently. Let us consider two extreme situations: (a) At large heliocentric distances the energy going into sublimation is negligible, and the temperature of the coma can be determined from a balance between insolation and reradiation (equilibrium temperature). (b) At small heliocentric distances, all energy is used for evaporation, and the evaporation rate varies as $r_\odot^{-2}$. The gas coming off the nucleus initially has the temperature of the surface. This is not the equilibrium blackbody temperature, since escaping gas carries away much of the heat as latent heat of sublimation. The temperature is determined by the gas that controls the evaporation. If outgassing is dominated by a single gas, and if the flux received on Earth is proportional to the number of (parent) molecules in the coma, the production rate and latent heat of the parent material can be deduced. If the

sublimation is controlled by CO, the surface temperature of the nucleus is between 30 and 45 K at 0.2 AU $< r_\odot <$ 10 AU; for CO_2 it is between 85 and 115 K; for H_2O it varies from 210 K at 0.2 AU, to 190 K at 1 AU, and to 90 K at 10 AU.

Since the side of the comet facing the Sun is hotter than the anti-sunward side, gas evolves predominantly from the sunward side, as indeed suggested from Earth-based obserations of many comets. Spacecraft encounters with several comets confirm activity primarily from the sunlit side, where release of volatiles and entrained dust is usually confined to a few small areas on the surface (Fig. 10.8). Although water-ice has been detected in a few patchy spots on Comet 9P/Tempel 1 by the Deep Impact spacecraft, these patches did not coincide with the jets, nor

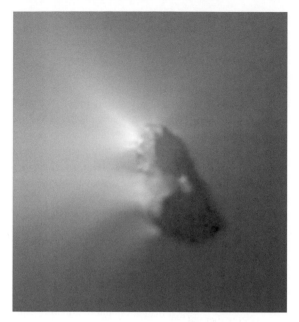

Figure 10.8 The nucleus of 1P/Halley photographed by the Giotto spacecraft from a distance of ∼600 km. The resolution in this composite image varies from 800 m at the lower right to 80 m at the base of the jet at the upper left. The nucleus is ∼16 × 8 km in extent. The (dust) jets visible in the image point in the sunward direction. (Courtesy Halley Multicolor Camera Team; ESA)

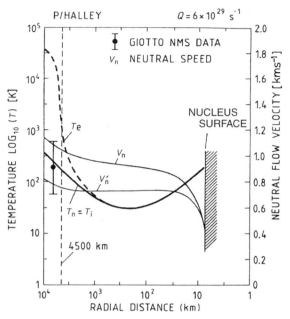

Figure 10.9 A model for various parameters in the inner coma of 1P/Halley: the electron and neutral gas temperatures, T_e and T_n, respectively, and the expansion velocity with (v_n') and without (v_n) IR radiative cooling by water molecules. (Ip and Axford 1990)

did they account for the total H_2O production rate. Perhaps the jets result from sublimation of subsurface ices. Regardless of the precise source, the asymmetric outgassing and jets give rise to the nongravitational forces, which distort a comet's orbit, as discussed in §10.2.1.

10.3.3 Outflow of Gas

Gas comes off the nucleus at the thermal expansion velocity, v_o, where

$$\frac{1}{2}\mu_a m_{amu} v_o^2 = \frac{3}{2}kT, \qquad (10.7)$$

with $\mu_a m_{amu}$ the molecular mass. Typical expansion velocities are ∼0.5 km s^{-1} near $r_\odot \sim 1$ AU, which is much larger than the escape velocity from a comet (Problem 10.12). Since the equilibrium temperature for a body at $r_\odot \approx 3$ AU is roughly equal to the sublimation temperature of water-ice, the coma is usually well developed at heliocentric distances $r_\odot \lesssim 3$ AU. In fact, the appearance of a well-developed coma at heliocentric distances <3 AU was evidence that the volatile component of most comets is dominated by water-ice (§10.3.1).

The gas released from the nucleus expands into a near vacuum and rapidly reaches supersonic velocities. The flow of gas can be calculated using the equations for conservation of mass, momentum, and energy (§7.1.2.2),

including sources and sinks of these quantities. The *terminal velocity*, which is the velocity of the gas at large distances from the comet, is usually reached within several tens of kilometers from the nucleus. Numerical calculations show that terminal velocities of parent molecules are usually between 0.5 and 2 km s^{-1} at heliocentric distances near 1 AU. Figure 10.9 summarizes model calculations for 1P/Halley, as well as a measurement that was obtained with the Giotto spacecraft. Within 100 km of the cometary nucleus, the electron temperature is likely to be a few tens of degrees kelvin, because water molecules are efficient coolants. Since electron–neutral collisions are frequent, electrons and neutrals should have the same temperature.

The gas density in the coma decreases approximately as the square of the distance from the nucleus, and the flow becomes a collisionless free molecular flow rather than a hydrodynamic flow at the *collisional radius*, $\mathcal{R}_c$. The location of $\mathcal{R}_c$ can be defined as the place at which an outward-moving particle has a 50% chance of escaping to infinity without colliding with another particle:

$$\int_{\mathcal{R}_c}^{\infty} \frac{Q\sigma_x}{4\pi r^2 v_o} dr = 0.5, \qquad (10.8)$$

where r is the cometocentric distance, and σ_x is the collisional cross-section. Note that $\mathcal{R}_c$ is analogous in some respects to the height of the exobase, z_{ex} (eq. 4.73), but the background gas flow velocity and the strength of the gravitational field differ greatly between the two cases. Beyond the collisional radius, the electron temperature rises because newly produced electrons have a large excess kinetic energy. The electron temperature may increase to a few hundred degrees kelvin at a cometocentric distance of 4000–5000 km.

More realistic (and much more complex) models are required to explain structures such as jets and shells in a coma and the production rates of gases. Gases may originate both from the nucleus and from grains in the coma, and may also undergo chemical reactions in the coma. Realistic models need to also consider physical conditions at the nucleus–coma interface. For example, the gas outflow near the nucleus is very sensitive to the surface topology and porosity, as well as to its rotation (rate, orientation, changes therein). Dust adds extra complexity to the models. The gas outflow slows down in the first few tens of meters above the nucleus through friction with the dust and energy exchange between the gas and dust. There is also increasing evidence (§10.3.1) that dust is a source of gas (e.g., CO, H_2CO, HCN, CN), and that the chemical composition of the gas may be changed by the presence of dust (e.g., recondensation of H_2O molecules onto dust grains).

The energetics in a comet's coma are further influenced by radiative cooling processes and heating of abundant molecules through photolysis. The gas flow is usually not adiabatic (see §3.2.2.3), as both the gas temperature and outflow velocities are modified by collisions with molecules and radicals (i.e., OH radicals and fast H atoms). Observations and numerical calculations show that the outflow velocity of heavy molecules increases significantly at nuclear distances $\gtrsim 10^3$–10^4 km. For a Halley-like comet at $r_\odot \approx 1$ AU, the outflow of gas can be described by hydrodynamical theory up to a distance of ~1000 km from the nucleus. Collisions are still important at large distances from the cometary nucleus; a free molecular flow is not reached until $\gtrsim 5 \times 10^4$ km from the nucleus (Problem 10.13).

10.3.4 Ultimate Fate of Coma Gas

Except for the inner ~100 km, a comet's coma stays optically thin at most wavelengths. Hence, essentially all molecules in the coma are irradiated by sunlight at visible and UV wavelengths. The mean lifetimes of the molecules and radicals against dissociation and ionization, therefore, varies with $r_\odot^2$. The lifetime, t_ℓ, of a molecular species is thus given by

$$\frac{1}{t_\ell} = \frac{1}{r_{\odot AU}^2} \int_0^{\lambda_T} \sigma_x(\lambda) \mathcal{F}_\odot(\lambda) d\lambda, \tag{10.9}$$

where $\mathcal{F}_\odot(\lambda)$ is the solar flux at 1 AU in the wavelength range $(\lambda, \lambda + d\lambda)$, $\sigma_x(\lambda)$ is the photodestruction cross-section, and λ_T is the threshold wavelength for photodestruction (which is usually in the UV). Although the solar flux at visible wavelengths is nearly constant, it varies much at UV wavelengths both temporally and spatially. Additionally, the cross-section against photodestruction is not always known, and hence it is not easy to constrain lifetimes of individual molecules.

The primary volatile constituent of a cometary nucleus is water-ice. The typical lifetime of water molecules at $r_\odot \approx 1$ AU is ~5–8 $\times 10^4$ s. The main (~90%) initial photodissociation products of water are H and OH. The OH molecules receive an average excess speed of 1 km s^{-1} in the dissociation process, and the H atoms ~18 km s^{-1}. Other dissociation products are excited oxygen (roughly 5%) with H_2, H_2O^+, OH^+, O^+, and H^+. A small fraction of the H_2O molecules dissociate inside the collisional radius. The OH radicals produced in this region are quickly thermalized through numerous collisions with other molecules; the much lighter H atoms are not thermalized. The typical lifetime for OH radicals at $r_\odot = 1$ AU is ~1.6–1.8 $\times 10^5$ s. The OH is dissociated into O and H primarily by solar Lyman α photons (121.6 nm), and the H atoms receive on average an excess velocity of 7 km s^{-1} in this process.

The H atoms produced by dissociation of H_2O and OH form a large *hydrogen coma*, several $\times 10^7$ km in extent. Hydrogen is ultimately lost either by photoionization or by charge exchange reactions (§4.8.2) with solar wind protons. Observations show that the latter is the more efficient of the two processes. Since charge exchange reactions depend on the always varying solar wind properties (§7.1.1), lifetimes for H vary significantly, from ~3 $\times 10^5$ to ~3 $\times 10^6$ s.

Regardless of a molecule's photodissociation cross-section, ultimately all gas species get ionized, and are swept away by the solar wind.

10.3.5 Dust Entrainment

Dust grains of various sizes and compositions are entrained in the outflowing cometary gases; however, there is a maximum size or mass to a grain that can be dragged off the surface of a cometary nucleus. The maximum radius of such a grain, $R_{d,max}$, depends upon the gas production rate, Q, the temperature of the surface, T, the radius of the nucleus (assumed to be spherical), R, and the densities of the nucleus, ρ, and dust grain, ρ_d, as

$$R_{d,max} = \frac{9\mu_a v_o Q}{64 N_A G \rho \rho_d R^3}, \tag{10.10a}$$

where N_A is Avogadro's number, μ_a the molecular weight of the gas, G the gravitational constant, and v_o the outflow velocity (Problem 10.25). Equation (10.10a) is valid if the mean free path of the gas molecules is larger than the dust particle sizes. At heliocentric distances less than a few tenths of an AU, this may not be a valid assumption, and the drag force must be replaced by the aerodynamic drag formula (eq. 2.55). The largest particle that can be dragged off the surface is then

$$R_{d,max} = \left(\frac{27 v_v \rho_g v_o}{8\pi RG\rho\rho_d} \right)^{1/2}, \qquad (10.10b)$$

where the dynamic viscosity, $v_v \rho_g$, between 200 and 300 K is approximately

$$v_v \rho_g \approx 10^{-6} \text{g cm}^{-1} \text{ s}^{-1}. \qquad (10.10c)$$

To get an order of magnitude estimate for the largest size grain that can be dragged off the surface from a comet, imagine a spherical nucleus with a radius $R = 1$ km which consists predominantly of well-compacted water-ice with $\rho = \rho_d = 1$ g cm^{-3}. The maximum size for icy particles dragged off this hypothetical comet at $r_\odot = 1$ AU is ~ 10 cm (Problem 10.26). Particle sizes may be larger during cometary outbursts, where part of the 'dust crust' is thrown into space. Icy particles are subjected to sublimation effects similar to ice at the surface of the cometary nucleus. The lifetime of an icy dust grain depends on the dielectric constant (absorptivity, eq. 3.81) of the grain, as this determines the temperature of the grain, and hence the sublimation rate. Depending on the absorptivity, the lifetime can vary by many orders of magnitude.

Comets thus 'pollute' the interplanetary environment significantly, with both gas and dust. While the gases have a short lifespan, dust particles may stay around longer. In particular, the larger particles, which are affected least by solar radiation pressure, may share the orbital properties of the comet for a long time. These particles are seen as *meteor streams*, or cometary *dust trails* (§9.4.7).

10.3.6 Particle Size Distribution

The particle size distribution in a comet's coma and dust tail can be constrained via observations at different wavelengths. Below we summarize our findings based upon observations at optical and infrared wavelengths in reflected sunlight, including data on the polarization characteristics, and infrared observations of the thermal emission component, as well as radar echoes at cm wavelengths.

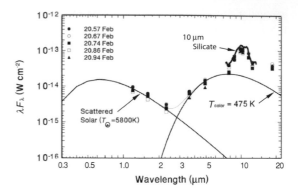

Figure 10.10 Spectrum of Comet C/Hale–Bopp. At wavelengths shortwards of 3 μm we see sunlight reflected off the grains, which can be fit well with a blackbody spectrum of 5800 K, i.e., the temperature of the Sun. On the right, thermal emission from the cometary coma (475 K blackbody curve) dominates. The heavy solid line shows the silicate emission feature. The data were taken in February 1995, at $r_\odot = 1.15$ AU and $r_\Delta = 1.64$ AU. (Adapted from Williams *et al.* 1997)

10.3.6.1 *Optical–Infrared Observations*

Visible and near-infrared spectra of cometary dust tails typically resemble a solar spectrum, including the Fraunhofer absorption lines, because sunlight is scattered off dust particles in the comet's coma and/or tail. The derived blackbody temperature is therefore $\sim$6000 K, the temperature of the Sun. A spectrum of C/Hale–Bopp is shown in Figure 10.10. After dividing by the solar spectrum, cometary spectra are usually neutral to slightly reddish in color. The absence of Rayleigh scattering (§3.2.3.4), i.e., no bluish color, suggests that there are not many particles much smaller than the wavelength of visible light, so a lower limit to the sizes of grains is $\sim$0.1 μm. At a heliocentric distance of 1 AU, thermal emission from dust particles dominates the spectrum at wavelengths longwards of 3 μm. At larger distances, the dust particles are colder, and the 'crossover' wavelength between reflected light and thermal emission shifts to longer wavelengths.

Assuming equilibrium between solar insolation and reradiation outwards (§3.1.2), the equilibrium temperature of a dark, rapidly rotating grain becomes (Problem 10.22)

$$T_{eq} = \frac{280}{\sqrt{r_{\odot AU}}} \left(\frac{1 - A_b}{\epsilon_\nu} \right)^{1/4} \approx \frac{280}{\sqrt{r_{\odot AU}}}, \qquad (10.11)$$

where A_b is the Bond albedo of the grain, ϵ_ν is the emissivity, and $r_{\odot AU}$ is the heliocentric distance in AU. The term $((1 - A_b)/\epsilon_\nu)^{1/4}$ is usually close to unity, except for particles with sizes $2\pi R/\lambda < 1$, with R the radius of the particle and λ the peak radiating wavelength corresponding to T_{eq}. Micrometer-sized grains absorb efficiently at UV and visible wavelengths, near the peak of the Sun's blackbody (Planck) curve, while the emission efficiency is relatively low at infrared wavelengths (§3.1.2.2). For such particles $(1 - A_b)/\epsilon_{ir} > 1$, and the observed, or *color temperature* is higher than the equilibrium temperature. The observed thermal spectrum can thus be used to determine the size of the grains. Many comets show a higher color temperature at 3.5–8 μm than at 8–20 μm, and some show a peak in thermal emission around 7–8 μm. These observations suggest a distribution of cometary grains that are smaller than a few micrometers in size. An independent constraint on grain sizes follows from emission features of silicates, which are near 10 and 18 μm. Such emission features can only be observed if the grains producing them are smaller than ~5 μm in radius. Thus, the bulk of dust grains in a comet's tail are likely between ~0.1 μm and a few μm in radius. Note, though, that although most of the surface area is in these small grains, most of the mass is contained in the larger (cm-sized) grains (see below), even though these are much fewer in number (§9.4.1, Problem 9.3).

Simultaneous observations at visible and infrared wavelengths can be used to estimate the Bond albedo of the dust particles (§E.3); the results depend, however, quite critically on the scattering phase function of the grains. There is some indication that the particle albedos are low, ~0.02–0.05, and that they may increase during periods of high activity. Some observations suggest an increase in albedo with distance from the nucleus, as well as with heliocentric distance.

Light scattered by particles is usually polarized, so observations of the linearly polarized flux density can be used to extract additional information on the dust grains (§E.4). Polarization measurements have the advantage of being normalized to the total intensity, and hence variations in polarization with heliocentric distance and within the coma yield information on the physical properties of the dust particles. Figure 10.11 shows plots of the degree of polarization, P_L (eq. E.1), as a function of the solar phase angle (Sun–comet–observer), ϕ. P_L can be positive or negative. It is generally negative at $\phi \sim 0$–20°, with a minimum of −2% near $\phi \approx 10°$, increasing nearly linearly with a slope $h = 0.2$–0.4% per degree to a maximum positive value near $\phi \approx 90$–100°. The maximum value of P_L increases with increasing wavelength, but decreases again beyond 2 μm. It further varies from comet to comet, with low maxima of ~10–15%, and high values near 20–25%, while C/Hale–Bopp may have had a higher value still. The polarization often increases after outbursts, as exemplified by C/1999 S4 Linear during its perihelion breakup (Fig. 10.11). Variations in P_L within the coma were apparent for, e.g., 1P/Halley and C/Hale–Bopp. The

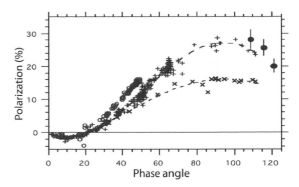

Figure 10.11 Phase dependence of the degree of linear polarization for various comets as measured in a narrowband red filter. (x) comets with a low maximum in polarization. (+) comets with a higher maximum. (o) Comet C/Hale–Bopp. (●) Comet C/1999 S4 Linear when it disrupted. (Kolokolova *et al.* 2004)

latter comet displayed low P_L values near its nucleus (<2000 km), including the extreme low value of −5% at $\phi = 8°$. P_L decreased gradually with distance (at 5000–8000 km from nucleus) for C/Hyakutake (C/1996 B2) and C/Tabur (C/1996 Q1), but increased for 1P/Halley and C/Hale–Bopp. These variations are likely due to a gradient in the physical properties of the dust particles. The polarization measurements rule out significant contributions by grains $\lesssim 0.1$ μm in size, in agreement with the observed lack of Rayleigh scattering.

10.3.6.2 *Radar Measurements*

Larger grains are best observed at longer wavelengths, either via observations of the thermal emission at far-infrared–millimeter wavelengths, or by using radar techniques (§E.7). In cometary radar experiments, one usually transmits a monochromatic signal which is Doppler broadened upon reflection. The echo bandwidth is determined by the size and shape of the object, as well as its rotation rate and the direction of the rotation axis. In addition, large (cm-sized) grains in a coma can broaden the signal substantially, because it is proportional to the radius of the 'object'. About a dozen comets have been detected using these techniques, and for roughly half of these, including e.g., C/IRAS–Araki–Alcock (C/1983 VII), C/Hyakutake, and 8P/Tuttle, a strong narrowband radar echo from the nucleus plus a much weaker broadband echo are observed (Fig. 10.12). The broadband signal is attributed to $\gtrsim$ centimeter-sized grains in a halo around the comet. The signals are polarized, with the OC (§E.7) signal being stronger for both the nucleus and halo, as expected. As shown in Figure 10.12, the spectrum of the nucleus of C/IRAS–Araki–Alcock is broad, similar to that of many asteroids (Fig. 9.8), indicative of a rough surface.

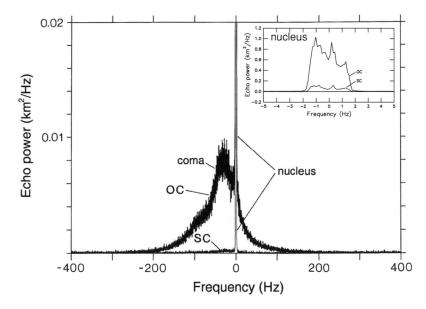

Figure 10.12 Radar echo power from C/IRAS–Araki–Alcock. The observed cross-section (km²/Hz) is plotted as a function of frequency. The strong spike is backscatter from the nucleus, while the 'skirt' is attributed to reflection off large icy grains in the coma. The insert zooms in on the radar echo from the nucleus (note scale on the y-axis). (Adapted from Harmon *et al.* 1989)

The polarization ratio, μ_c, yields information on the surface roughness. The highest ratio has been measured for C/Hyakutake, $\mu_c = 0.5$ at 3.5 cm, suggestive of a surface covered with pebble-sized rubble. The low-level *skirt* in the radar echo from C/IRAS–Araki–Alcock has been attributed to radar reflection off large grains. The small polarization ratio therein, $\mu_c \sim 0.014$, suggests a maximum grain radius of $\sim \lambda/(2\pi)$, i.e., a few cm for $\lambda \sim 13$ cm. Such large grains can still be entrained in the outflowing gas, as discussed in §10.3.5.

10.3.7 Morphology of Dust Tails

The dust-to-gas mass ratio in comets is usually between 0.1 and 10. As discussed in the previous subsection, dust grains of many sizes have been detected. The distribution of sizes is dominated by the smallest grains, down to 0.1 μm in size, 'visible' in reflected sunlight. Thermal infrared measurements show that most of the grains are less than ~ 5 μm in radius, while radar observations reveal the presence of much larger (cm-sized) grains. Although the total number of such particles is much smaller than that of the μm-sized grains, most of the mass is contained in the larger particles. Submicrometer-sized grains almost reach the speed of the escaping gases (~ 1 km s^{-1}), but larger particles barely attain the gravitational escape velocity of the nucleus (~ 1 m s^{-1}). The dust decouples from the gas and becomes subject to solar radiation forces at a distance of a few tens of nuclear radii.

As discussed in §2.7.1, solar radiation pressure on (sub)micrometer-sized dust grains 'blows' the particles

outwards from the Sun relative to the trajectory of the nucleus. Particles of different sizes and with different release times become spatially separated in the dust tail. A schematic representation of tail formation is shown in Figure 10.13. The dashed lines in Figure 10.13a are the trajectories of the dust grains, if emitted at zero velocity. The grains usually have a nonzero initial velocity, and thus a slightly different trajectory than indicated. The exact path also depends on the ratio, β, between the solar radiation force and the gravitational force from the Sun (eq. 2.47a). Thus the dust orbits depend on size, shape, composition, and initial velocity. The ensemble of particles released at different times together form a curved tail, as depicted by the shaded region.

Lines connecting particles of the same β at a given instant of time are called *syndynes* (tail in Fig. 10.13a, dashed lines in Fig. 10.13b,c). The width of the tail is equal to $2v_d t$, with v_d the velocity of the dust particles in the comet's frame of reference upon release from the surface and t the time since release. The dust particles can also be sorted according to release time, regardless of their size or β. The locus of such points is called a *synchrone*; these are indicated by the solid lines in Figure 10.13b,c. The numbers indicate the time in days since the dust was released.

The simultaneous ejection of dust grains of different sizes gives rise to inhomogeneities in the dust tail, in particular the *dust jets*: highly collimated structures, which at larger distances become *streamers* – straight or slightly curved bands that converge at the nucleus. Observed infrequently are *striae*, which are parallel narrow bands at large distances from the nucleus that do not converge at the

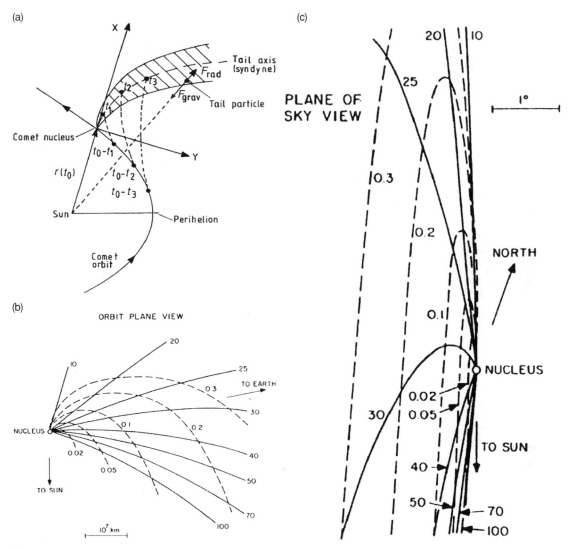

Figure 10.13 (a) A schematic presentation of the formation of a dust tail. Trajectories of dust particles that were released from the nucleus at different times t_1, t_2, and t_3 prior to the time of the observations at t_0 are shown as dashed curves. (Adapted from Finson and Probstein 1968) (b, c) Synchrones (solid lines with ages in days) and syndynes (dashed lines with β values) for C/Arend–Roland (C/1957 III) on 28 April 1957, in the orbit plane of view (b), and as viewed on the plane of the sky (c). This latter perspective explains the formation of anti-tails. (Sekanina 1976)

nucleus (Fig. 10.14). They usually intersect the comet–Sun line on the sunward side, and most likely originate from the instantaneous destruction of large particles. Striae are thus synchrones originating at the location of the parent body at the time of destruction. Images of Comet C/McNaught (C/2006 P1) show a striking display of striae (Fig. 10.14a).

Although dust tails always point away from the Sun, on some rare occasions the tail appears to point towards the Sun, as shown in an image of C/Hale–Bopp (Fig. 10.14c). Such *anti-tails* are caused by a particular viewing geometry between the Sun, the observer, and the comet, such

that its normal dust tail becomes visible in the antisolar direction. Anti-tails have also been observed for C/Arend–Roland (C/1957 III) and C/Kohoutek; the synchrones and syndynes in Figure 10.13b (in the orbit plane) and 10.13c (projected onto the plane of the sky) were calculated for C/Arend–Roland at the time the anti-tail was visible. Since the photograph in Figure 10.14c was taken near the time that the Earth was crossing Comet C/Hale–Bopp's orbital plane, this image also shows the *neck-line structure*, a narrow and very straight feature along the Sun–comet line, which is caused by sunlight reflected off dust grains along

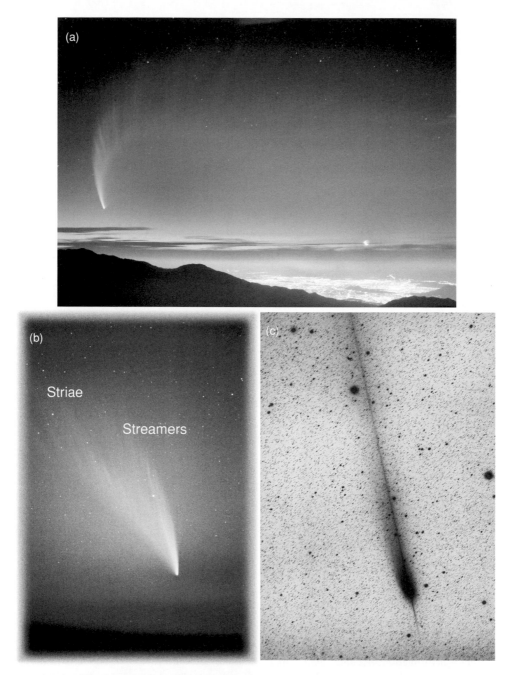

Figure 10.14 (a) Comet C/McNaught (C/2006 P1) was perhaps the most photogenic recent comet, with an extremely long and unusual dust tail. This image was taken from the Andes Mountains in Chile, looking down on the city lights of Santiago. In the lower right the crescent moon is visible. (Courtesy Stéphane Guisard, ESO PR Photo 05h/07) (b) C/West in March 1976 showing both streamers and striae. About a week after this picture was taken, the nucleus broke into four pieces. (Courtesy Akira Fujii) (c) This negative print of C/Hale–Bopp, photographed with the ESO 1-m Schmidt Telescope on 5 January 1998, reveals an anti-tail. The anti-tail is the narrow spike near the bottom of the photograph, and extends over more than 4° to the edge of the field, in the direction of the Sun. During this 1 hour exposure, an artificial satellite crossed the field – its trail may be seen as a very thin line to the right of the comet. (Courtesy Guido Pizarro, ESO Press Photo 05a/98)

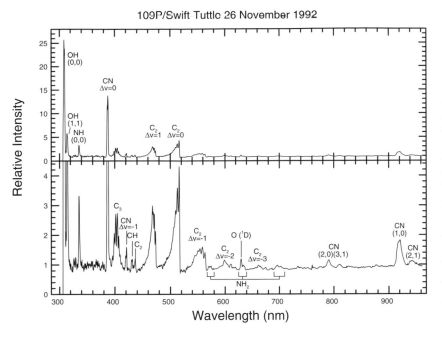

109P/Swift Tuttle 26 November 1992

Figure 10.15 Optical/near-IR CCD spectrum of Comet 109P/Swift–Tuttle obtained 26 November 1992, when the comet was at $r_\odot = 1$ AU, and $r_\Delta = 1.32$ AU. The spectrum has been divided by a solar spectrum to show the cometary emissions. The 2-pixel resolution was $\sim$0.7 nm for the blue, and $\sim$1.4 nm for the red half. The upper panel shows the spectrum scaled by the strongest feature, while the y-axis on the lower panel has been expanded to highlight the weaker features. (Adapted from Feldman *et al.* 2004)

the comet's orbit, that is viewed 'edge-on' (compare, e.g., edge-on viewing of planetary rings in §11.3.1). The true length of this feature depends on the exact viewing geometry and may reach lengths $\gtrsim 1$ AU. Since these grains are confined to the comet's orbit, their orbits have not evolved outwards due to radiation pressure and hence these grains must be relatively large (>100 μm in size).

10.4 Composition

Since a cometary nucleus is often shrouded by a coma of gas and dust, it is difficult to determine its composition directly from remote observations. Instead, spectral line measurements of molecules, radicals (Fig. 10.15), and dust grains, are used together with detailed outflow models (§10.3.3) to indirectly deduce the composition of a comet's nucleus. Information about dynamically new comets relies entirely on such remote sensing techniques.

Complementary techniques are *in situ* measurements, which have been obtained for a few periodic comets. The Giotto and Vega spacecraft measured the composition of gases and dust of 1P/Halley using mass spectrometers (§E.10). Unfortunately, identification in mass spectrometer data is somewhat hampered by the fact that masses of different species may overlap. For example, a peak at $\mu_{amu} = 28$ may contain CO, N_2, and C_2H_4 as possible main constituents. Such data, therefore, still depend to some extent on remote sensing measurements and/or theoretical arguments. A more direct method was utilized by

the Stardust mission, which captured grains from the coma of 81P/Wild 2 and returned them to Earth (§10.4.4).

Remote sensing and *in situ* sampling of material that has evolved off the surface of a comet both provide information on a comet's outermost layers, which may have undergone some chemical processing. To sample primitive materials one would wish, ideally, to break or excavate a comet. This is exactly what the Deep Impact mission did (Fig. 10.38b). Deep Impact consisted of two spacecraft: an impactor, which was made of 49% of copper to minimize chemical reactions with the comet's main constituent (water), and a flyby spacecraft. The impactor, with a mass of 370 kg and equipped with an autonavigation system and a camera, hit Comet 9P/Tempel 1 at 10.34 km s^{-1}. This excavated or sublimated a great deal of material from below the surface, the emissions of which were observed remotely, from the flyby and Rosetta spacecraft, HST, Spitzer, and ground-based telescopes (§10.4.3.2). The impact created a crater $\sim$200 m wide and 30–50 m deep. The Stardust mission flew by Comet 9P/Tempel 1 in 2011 and imaged the crater produced by the impact.

Sometimes nature helps – when comets split, their interior becomes exposed, enabling remote observations of a comet's bulk composition. Detailed observations have been obtained of individual fragments of Comet 73P/Schwassmann–Wachmann 3 (§§10.4.3.2, 10.6.4).

In the next subsections we discuss techniques to extract production rates and abundances from observations. A summary of the gaseous composition of comets is given in §10.4.3, and in §10.4.4 the composition of dust grains is discussed. It appears that, on average, the overall cometary composition (gas + dust) is, within a factor of two, similar to solar values, except for noble gases, hydrogen (deficient by a factor of ~700), and nitrogen (deficient by a factor of ~3). As all known meteorites are strongly deficient in all volatile (CHON) materials compared to solar values (§8.1), cometary dust can be considered as the most 'primitive' early Solar System material ever sampled.

10.4.1 Haser Model

To derive gas production rates from cometary observations, one needs models of the outflow of the gas and the temporal evolution (e.g., place of origin, dissociation, ionization) of the individual molecular species. The simplest and most widely used method is the *Haser* model. This model assumes isotropic radial outflow and a finite lifetime of the molecules. The density distribution for *parent* molecules, which evolve directly off the nucleus, can be written as

$$N_p(r) = \frac{Q_p}{4\pi r^2 v_p} e^{-r/\mathcal{R}_p}, \tag{10.12a}$$

where N is the number density, and $\mathcal{R}_p = v_p t_{\ell p}$ is the scale length, with $t_{\ell p}$ the lifetime of the parent molecules. The subscript p stands for parent molecules. If the *daughter* molecules, i.e., the molecules produced upon dissociation (subscript d), continue to move radially outwards like their parents with $v_d = v_p$, then the distribution for daughter molecules is

$$N_d(r) = \frac{Q_p}{4\pi r^2 v_d} \frac{\mathcal{R}_d}{\mathcal{R}_d - \mathcal{R}_p} \left(e^{-r/\mathcal{R}_d} - e^{-r/\mathcal{R}_p} \right). \tag{10.12b}$$

It is more likely, however, that dissociation products from parent molecules are ejected isotropically in the parent's frame of reference. As a result, not all molecules move radially outwards; some radicals actually move inwards, towards the nucleus. Thus, the resulting velocity distribution of daughter molecules may differ substantially from that predicted by the Haser model. The Haser model is generally a good approximation if most of the radicals are formed inside the collisional regime (short parent lifetime, high total outgassing). However, if a large fraction of the radicals is formed outside the collisional regime (low total outgassing, long parent lifetime), then the velocity distribution of the daughter molecules noticeably influences the shape of 'observed' spectral lines. This situation can be better modeled by a random walk (Monte Carlo) model.

Figure 10.16 shows observed line profiles for a parent and a daughter molecule: HCN at a wavelength of

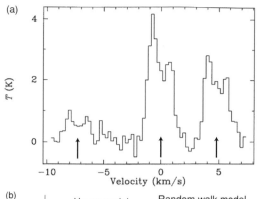

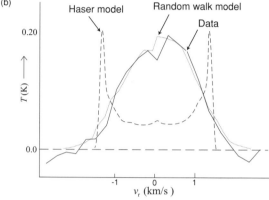

Figure 10.16 (a) HCN spectrum of C/Hale–Bopp, obtained with the BIMA array at a frequency of 89 GHz. The spectrum was taken in an 11″ × 8″ beam at the peak of the HCN emission on 3 April 1997. Note that each hyperfine line (indicated by arrows) is split, as expected from the Haser model. The stronger blueshifted component in each line suggests a higher gas production rate from the side facing the observer, which is also the side facing the Sun. (Wright *et al.* 1998) (b) Observed OH line profile (18 cm) of C/Austin (1982g) (solid line), with superposed calculated profiles based upon a Haser model (dashed line) and a random walk model (dotted line) for cometary outflow. (Adapted from Bockelee-Morvan and Gerard 1984)

3 mm and OH at a wavelength of 18 cm. On the OH line, profiles based upon a Haser and a random walk model are superposed. It is clear that the depression in OH emission predicted by the Haser outflow of gas is not observed; the observed emission typically increases towards the line center, as expected for a random walk model, rather than towards the edges of the line. In contrast, the HCN profile agrees well with a Haser model.

10.4.2 Excitation and Emission

Most atoms and molecules in a comet's coma are in the electronic ground state. When an atom or molecule is excited, it usually falls back to a lower energy state through the emission of a photon. The main emission mechanism is *fluorescence*: Absorption of a solar photon excites an atom or molecule, which is followed by spontaneous emission in a single- or multi-step decay process. Other emission mechanisms are collisional excitation (with neutrals, ions, or electrons), and dissociation (radiative or collisional) of (parent) molecules which leaves the observed radicals in an excited state.

10.4.2.1 *Fluorescence*

The strongest lines in cometary spectra are *resonance transitions*, i.e., transitions between the ground state and first excited level of an atom/molecule. The brightness of an emission line is determined by the number of molecules multiplied by the *g-factor* or emission rate per molecule. If most or all molecules are in the electronic ground state and excitation is caused by simple fluorescence, one can calculate the absorption rate of solar photons which excite the molecule, g_a, and multiply this by the appropriate branching ratio for the particular transition at wavelength λ_{ul} to determine g:

$$g = g_a \frac{A_{ul}}{\sum_{j \leq u} A_{jl}}, \tag{10.13a}$$

$$g_a = \frac{B_{lu} \mathcal{F}_\odot(\lambda_{lu})}{r^2_{\odot AU}}, \tag{10.13b}$$

where B_{lu} and A_{jl} are the Einstein coefficients (see §3.2.3.3) for transitions from the ground states and excited states, respectively, with u the upper, l the ground level, and $l < j \leq u$. The solar flux density $\mathcal{F}_\odot(\lambda_{lu})$ is measured at Earth's orbit at the wavelength of absorption λ_{lu}. Note that in the case of pure resonance, $g = g_a$; in the case of resonance fluorescence, the de-excitation of the upper level may occur via more than one transition, and $g < g_a$. For many molecular species, several electronic levels are populated and calculation of the g-factors is quite complicated. Once the g-factor or emission rate for a given transition is known, the gas column density of the emitting molecules, N_c, can be calculated from the observed brightness:

$$N_c = \frac{B_\nu}{g} \frac{4\pi}{\Omega_s}, \tag{10.14a}$$

with B_ν the brightness of the comet (photons s^{-1} cm^{-2}) and Ω_s the solid angle over which radiation is received. If excitation is by simple resonance fluorescence, the gas production rate of the molecular species is simply related to the brightness:

$$Q = \frac{4\pi r^2_\Delta B_\nu}{g t_\ell}, \tag{10.14b}$$

with t_ℓ the lifetime of the emitting molecules. Note that the product $g t_\ell$ does not depend on heliocentric distance. If the excitation is not simply resonance fluorescence, the situation is more complex since many energy levels may be populated. In this case every excitation/de-excitation process must be included in the calculations.

10.4.2.2 *Collisional Excitation*

Since the densities in the coma are usually very low, collisions between neutrals are generally not important, except very close to the nucleus. A typical cross-section, σ_x, for a neutral–neutral collision is proportional to a molecule's size, R: $\pi(R_1 + R_2)^2 \approx 10^{-15}$ cm^2. The timescale between collisions, t_c, is then given by

$$t_c = \frac{1}{N \sigma_x v_o}, \tag{10.15}$$

with v_o the thermal velocity of the molecules and N the number density. Typical collision rates for molecules near the nucleus are of the order of 10 s^{-1} (Problem 10.16); collision rates are a factor of $\sim 10^4$ less at a cometocentric distance of 10^3 km. At 1 AU, a typical absorption rate of solar photons by water out of the electronic ground state is $\sim 10^{-3}$ s^{-1}. Hence, collisional excitation by neutrals is generally not important, except very close to the nucleus. Collisions with ions or electrons are more significant, because the effective cross-sections are $\gtrsim 10^3$ larger than for neutrals and the velocity of the electrons is much higher than that of neutrals or ions. Thus, collisional excitation in the inner coma, $\lesssim 10^3$ km from the nucleus, can be important. The energy available in a collision is the kinetic energy of the molecule. With a thermal velocity of 0.1 km s^{-1}, a water molecule may transfer 1.5×10^{-15} erg, or less than 0.001 eV. This amount of energy is not enough to excite vibrational or electronic transitions, but it can excite rotational transitions, which can be observed at radio wavelengths.

10.4.2.3 *Radiation Pressure*

Radiation pressure from the Sun acts on molecules via absorption and re-emission of solar photons. The net effect is a tailward displacement of the coma. The acceleration of an atom/molecule caused by radiation pressure is

$$\frac{dr^2_\odot}{dt^2} = \frac{h}{\mu_a m_{amu} r^2_\odot} \sum_i \frac{g_i}{\lambda_i}, \tag{10.16}$$

where h is Planck's constant, $\mu_a m_{amu}$ the molecular mass, g_i the g-factor for transition i, and λ_i the wavelength of absorbed photons. For atomic hydrogen, Lyman α is the most important transition, which leads to an offset in the location of the hydrogen cloud with respect to the nucleus

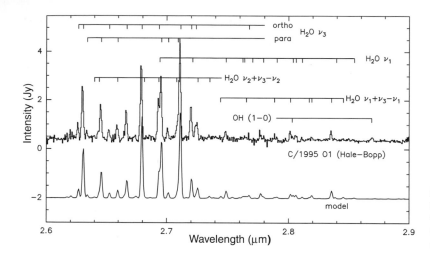

Figure 10.17 The ν_3 emission band spectrum near 2.7 μm as observed with the Infrared Space Observatory (ISO) in Comet C/Hale–Bopp. (Bockelee-Morvan *et al.* 2004)

by $\sim 10^6$ km. In the case of an isotropic uniform outflow of gas, a displacement of this magnitude should be visible in the data. However, comets generally display large anisotropies in their gas outflows, which makes it difficult to attribute any displacement to a particular process.

10.4.2.4 *Swings Effect*

Although the timescales involved in the various excitation processes strongly argue for fluorescence as the main emission mechanism, even stronger evidence for fluorescence is given by the *Swings effect*. As the comet orbits the Sun, its heliocentric velocity changes. The solar spectrum consists of numerous *Fraunhofer absorption lines*, created in the cooler atmospheric layers overlying the photosphere. In the comet's frame of reference, the solar spectrum is Doppler shifted, and the Fraunhofer absorption lines move in and out of the excitation frequencies. Hence, the *g*-factor (eq. 10.13) depends upon the radial components of a comet's heliocentric velocity. There is an excellent correlation between the relative strength of cometary OH lines at UV wavelengths and the Fraunhofer absorption-line spectrum Doppler shifted to the reference frame of the comet. A cometary line is strong if at the appropriate Doppler-shifted frequency the solar radiation is not weakened by absorption effects in the Sun's own atmosphere. The cometary line is weak or absent if the solar intensity at the exciting frequency is weak or zero.

The atoms/molecules within a comet's coma have certain velocities with respect to the nucleus; hence their heliocentric radial velocity is slightly different from that of the nucleus. Thus, in the reference frame of each molecule the solar spectrum is Doppler shifted by slightly different amounts. This leads to a modification of the Swings effect,

generally referred to as the *Greenstein effect*. The Greenstein effect may cause a certain side of a comet's coma to be brighter, which may manifest itself as an asymmetry in a spectral line profile (Problem 10.17).

10.4.3 Composition of the Gas

The basic composition of a comet as deduced from its gaseous coma can be sorted according to the following constituents: water and its products; carbon, nitrogen, and sulfur compounds; and the heavier elements, such as alkalies and metals. The noble gas abundances, ortho-to-para ratio, and isotope ratios further constrain a comet's place of formation. Below we summarize the constraints on a comet's composition as derived from gases in its coma. Measurements of the composition of dust grains as derived from spectroscopy and *in situ* data are presented in §10.4.4.

10.4.3.1 *Water Products*

Water. Although it was long surmised that water-ice dominates a comet's composition, H_2O itself was not detected until December 1985, when 1P/Halley was observed at infrared wavelengths from the Kuiper Airborne Observatory (KAO). Ten emission lines of the fundamental ν_3 band of vibration were detected near 2.7 μm, a wavelength region not observable from the ground. The spectrum agreed with models of solar infrared fluorescence from rotationally relaxed water vapor, and the water production rate appeared to vary from day to day, in concert with visible lightcurves. The same vibrational bands were later observed with the Infrared Space Observatory (ISO) in Comets C/Hale–Bopp (Fig. 10.17) and 103P/Hartley 2. High spectral resolution data in the 2–5 μm wavelength range from the Keck telescope revealed nonresonance fluorescence emission bands (hot bands) in several comets.

In addition to measuring the water production rate, such data also constrain the rotational temperature and spatial distribution of H_2O.

Most of the water vapor has a low rotational temperature throughout the coma, so radio frequencies are particularly suitable for comet observations. Tentative detections of the rotational line at 22 GHz have been reported in C/IRAS–Araki–Alcock and C/Hale–Bopp from ground-based observations. Since our own atmosphere contains water vapor, observations from above the atmosphere are more successful: the 557 GHz line has been observed in several comets with the Submillimeter Wave Astronomy Satellite (SWAS), and the combined Astronomy/Aeronomy Odin satellite. The Rosetta spacecraft, which reached 67P/Churyumov–Gerasimenko in 2014, is equipped with a 190 and 562 GHz radiometer to determine the changes in the surface and near-surface temperature and the evolution of H_2O, CO, NH_3, and CH_3OH as they sublimate off the comet, while the spacecraft joins the comet on its sunward journey.

The ions H_2O^+ and H_3O^+ have been detected in many comets. Since H_2O^+ emits in the red part of the spectrum, occasionally one may see a reddish rather than blue ion tail.

Molecular Hydrogen. Upon dissociation, a small fraction of H_2O leaves O in the excited ^1D or ^1S state,[4] or as an O^+ ion. This process simultaneously produces molecular hydrogen. The Far-UV Spectroscopy Spacecraft (FUSE) has been used to observe three lines of the H_2 Lyman series in C/2001 A_2 (Linear) to derive a column abundance that is consistent with the value expected from H_2O dissociation models.

Hydrogen. Lyman α ($\lambda = 121.6$ nm) images of comets reveal a huge atomic hydrogen cloud. The H cloud produced by C/West (C/1976 VI) is shown in Figure 10.18. The hydrogen cloud is roughly 10^7 km in size. Its asymmetric shape may result from solar radiation pressure. Models of the outflow suggest two peaks in the velocity distribution, one with $v \approx 7$ km s^{-1} and the other with $v \approx 18$ km s^{-1}, in agreement with the excess velocities expected from the dissociation of OH and H_2O (§10.3.3). The extent of the cloud suggests that the lifetime of H atoms is determined by charge exchange reactions with solar wind protons.

Oxygen. Several transitions of atomic oxygen have been detected, including the red doublet (^1D–^3P; 630.0 nm, 636.4 nm), and green (^1S–^1D; 557.7 nm) *forbidden lines*.[5] About 5% of the H_2O (and OH) molecules produce

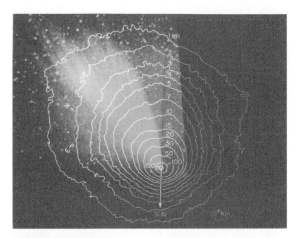

Figure 10.18 Isodensity contours of the hydrogen cloud (in Ly α) of C/West superposed on a photograph of the comet. The Lyman α emission was observed by Opal and Carruthers (1977), while the visible picture of the comet was taken by S. Koutchmy on the same day. (Courtesy Koutchmy; figure from Fernandez and Jockers 1983)

oxygen atoms in an excited (^1D or ^1S) state. About 95% of the time, atoms in the ^1S state (lifetime <1 s) decay to the ground ^3P state via the ^1D state (lifetime ~130 s), producing both the green and red lines. The remaining 5% decays directly to the ground state, emitting UV lines (297.7 nm, 295.8 nm). Although in principle CO and CO_2 could be parents to the forbidden oxygen lines, it is believed that water is the dominant, if not sole, parent molecule in the inner $\sim10^5$ km from the nucleus. In contrast to fluorescence emission, this prompt emission occurs only once for each of the O atoms; hence these lines are excellent tracers of both the abundance and distribution (atoms cannot travel far in their excited state before decaying) of the atom's parent molecule.

Hydroxyl. The OH radical has been observed extensively in the UV (~300 nm) and radio (1665 and 1667 MHz) frequencies. Prompt OH emission has been detected near 3 μm, corresponding to energy transitions from highly excited rotational levels. This most likely corresponds to OH radicals left in an excited state upon dissociation of H_2O. Solar photons excite the OH molecules, which upon de-excitation exhibit a number of UV lines around 300 nm. The relative strengths of the various lines clearly exhibit the Swings and Greenstein effects. The rotational level of the ground state is split into two levels, each of which is split again by hyperfine structure

[4] See §3.2.3, and the Further Reading section in Chapter 3.

[5] Atoms may be excited to metastable states. Decay to lower energy levels from such states is said to be 'forbidden', because on Earth

such atoms are collisionally de-excited. The spontaneous decay times for 'forbidden' transitions are very long: seconds to thousands of years. In the interplanetary medium where collisions are rare, atoms eventually decay spontaneously, resulting in forbidden line emissions.

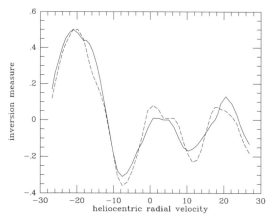

Figure 10.19 Population inversion, $i = (n_u - n_l)/(n_u + n_l)$, for the Λ-levels of the OH ground state as a function of a comet's heliocentric velocity (in km s^{-1}). When the inversion measure is positive, OH is seen in emission (maser); when the inversion measure is negative, OH is observed in absorption (see Figs. 10.20 and 10.21). The result for two pumping models are shown: solid line, from Despois *et al.* (1981) and dashed line, from Schleicher (1983). (Adapted from de Pater *et al.* 1991b)

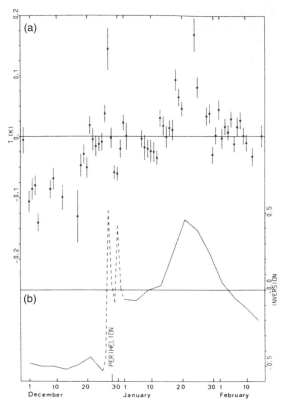

Figure 10.20 (a) A comparison between radio OH data of C/Kohoutek in 1973–1974, with (b) the inversion measure predicted by the ultraviolet pumping model as a function of time. (Biraud *et al.* 1974)

effects (§3.2.3.2). Four lines can be observed at radio wavelengths, near 18 cm, the most intense of which are the 1665 and 1667 MHz transitions. The intensity of the lines depends strongly on the heliocentric radial velocity of the comet (Swings and Greenstein effects). Usually these ground-state levels would be populated in proportion to their statistical weights (§3.2.3). Since the transition between the lines is highly forbidden, a strong *population inversion* can build up, i.e., the upper (excited) level is much more populated than the lower level relative to thermochemical equilibrium conditions (Fig. 10.19). The 3 K cosmic microwave background radiation can then induce a maser action, i.e., stimulate emission. The OH molecule is not excited when Fraunhofer absorption lines are Doppler shifted into the excitation frequency. In this case, the ground-state levels of OH become anti-inverted, and the molecules absorb photons from the 3 K background radiation. Thus, depending on the heliocentric velocity, OH is seen either in emission (maser) or in absorption against the galactic background. A comparison of the population inversion expected from the solar Fraunhofer line spectrum with OH observations is shown in Figure 10.20. The agreement is striking, though not perfect. Deviations from the calculated values can be explained by insufficient knowledge of the Fraunhofer absorption spectrum, Greenstein effects, asymmetric outgassing of the comet, and fluctuations in the gas production rate.

The population of the ground-state energy levels is also sensitive to collisions. In the inner coma, inside the collisional radius, the transitions are 'quenched': the OH

molecules are thermalized via collisions, and the maser activity ceases. This effect should show up as a 'hole' in the OH distribution around the coma. This hole may have been observed in high-resolution radio images of 1P/Halley (Fig. 10.21).

10.4.3.2 *Carbon Compounds*

At visible and UV wavelengths numerous emission lines have been observed from carbon species (e.g., C, C_2, C_3, CH, CH_2, CN, CO, and associated ions). All of these species are ultimately derived from ices in the comet's nucleus. The identification and characterization of these ices provide information on the conditions under which cometary ices formed.

Likely candidates for parent materials of the observed carbon species, in addition to dust grains, are carbon dioxide (CO_2), hydrogen cyanide (HCN), methane (CH_4), and more complex molecules such as formaldehyde (H_2CO) and methanol (CH_3OH), all of which have been detected in at least several comets. The CO_2 production rate is a few % that of water, while the CO production rate varies

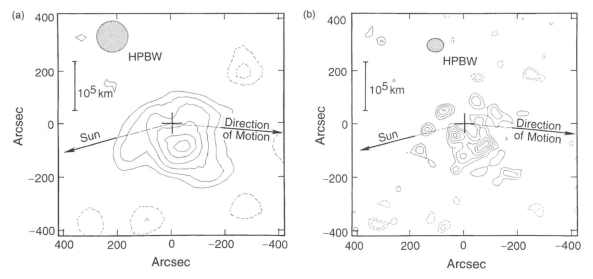

Figure 10.21 Radio image of the OH emission of 1P/Halley, taken at the peak flux density of the line (0.0 km s^{-1} in the reference frame of the comet). The cross in the center indicates the position of the nucleus. The shaded ellipse represents the resolution of the beam. Contour levels for the low-resolution image (a) are 4.9, 7.8, 10.8, 13.7, 16.7, and 18.6 mJy/beam; for the high-resolution image (b) they are: 4.4, 6.0, 7.7, 9.3, and 10.4 mJy/beam. Dashed contours signify negative values; these suggest that the emission region is much larger than that shown here (known as the 'missing short spacing problem', as discussed in the source article). (de Pater *et al.* 1986)

from $\lesssim 1\%$ up to over 30% in different comets. *In situ* measurements of Comet 1P/Halley by the Giotto spacecraft revealed that $\sim 1/3$ of the CO was released directly from the nucleus (referred to as the 'native' source), while the remaining molecules come from an 'extended' source region, which could be dust grains or complex molecules in the coma. Formaldehyde also has a native and an extended source.

The lifetimes of CO and CO_2 are roughly an order of magnitude larger than that of water, so most of these molecules enter the region of free molecular flow. A large fraction of CO and CO_2 becomes ionized; the ions are accelerated and swept away by the solar wind. Both CO^+ and CO_2^+ ions have been observed in many comets. Emission from the CO^+ ions shows up as the prominent bright-blue ion tail.

The abundances of the carbon chain molecules C_2 and C_3 relative to CN vary by less than a factor of ~ 2 for most comets, but some comets are depleted by about an order of magnitude in C_2 and C_3. Comets that show such a depletion are likely low in these compounds throughout their interiors, as evidenced by observations of individual fragments of Comet 73P/Schwassmann–Wachmann 3, a Jupiter family comet with low C_2 and C_3 abundances that split during its 1995 perihelion passage (§10.6.4). Near-infrared observations show that in its B and C fragments, the abundances of several hydrocarbons, formaldehyde and methanol, as well as C_2 and C_3, are all much (factor of ~ 5) lower than those seen in 'normal' comets.

Nearly all of the carbon chain depleted comets are members of the Jupiter family, and thus presumably originated in the Kuiper belt (§10.2.3), whereas normal carbon chain abundances are found among all dynamical groupings of comets. Hence the conditions must have been right for forming the parents of C_2 and C_3 in the region where Oort cloud comets and some, but not all, of the Kuiper belt comets formed. If C_2 and C_3 are produced from large hydrocarbons, perhaps many of the Jupiter family comets originated in a region where the temperatures were too cold for the chemistry to proceed to make these large molecules.

The (relative) abundances of CH_4, CO, C_2H_6 (ethane), and that of H_2CO and CH_3OH help characterize the ices in a comet's nucleus, and hence provide key information on the conditions of the early solar nebula. Methanol is typically a few tens of times as abundant as formaldehyde in the interstellar medium (ISM). Both species have been detected in a growing number of comets, with abundances typically up to a few percent that of H_2O (Fig. 10.22), and a CH_3OH/H_2CO ratio equal to a factor of a few at most, i.e., much less than that measured in the ISM.

The CH_4/CO abundance ratio in the condensed phase of the interstellar medium is typically around unity, with variations up to a factor of ~ 10, while the abundance

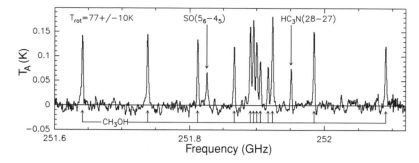

Figure 10.22 Spectrum of Comet C/Hale–Bopp observed on 21 February 1997 with the Caltech Submillimeter Observatory (CSO). Twelve J_3–J_2 A lines of CH_3OH are shown, as well as the 5_6–4_5 line of SO and, in the image sideband at 254.7 GHz, the J(28–27) line of HC_3N. (Lis *et al.* 1997)

ratio in the gas phase is much smaller, ~ 0.001–0.01. The CH_4/CO ratio has been measured in a handful of comets (mostly dynamically new comets, and the Jupiter family Comet 9P/Tempel 1) at ratios of ~ 0.1, i.e., intermediate between the condensed and gas phase ratios in the ISM. Ethane (C_2H_6) and acetylene (C_2H_2) reveal abundances of $\sim 0.6\%$ and ~ 0.2–0.4% that of water, respectively, comparable to the abundance of methane gas. Similar ethane and methane abundances, however, are inconsistent with production mechanisms in a primitive solar (or giant planet sub) nebula that is in thermochemical equilibrium; such scenarios would lead to $C_2H_6/CH_4 \lesssim 10^{-3}$.

Hence the CH_3OH/H_2CO ratio in comets is low (or H_2CO is high) compared to that in the ISM, while the C_2H_6/CH_4 ratio in comets is high. The latter has been explained via hydrogenation processes (chemical reactions where H atoms are added to molecular species) of C_2H_2 on icy grains prior to incorporation into the cometary nucleus. If such processes were common, CO should have been hydrogenated as well. Laboratory experiments involving H^+ irradiation (radiolysis) show that CO in an H_2O/CO ice mixture can be converted into the formyl radical HCO. This species is highly reactive, and forms H_2CO and CH_3OH in proportions which depend on the hydrogen density and the temperature in the environment. Formic acid (HCOOH) and longer chain molecules can also be produced this way. Both the low CH_3OH/H_2CO and high C_2H_6/CH_4 ratios in comets, as well as the detection of larger molecules, are suggestive of hydrogenation chemistry on icy grains prior to incorporation into the cometary nucleus. Large molecules such as acetaldehyde (CH_3CHO), methyl formate ($HCOOCH_3$), formic acid (HCOOH), and ethylene glycol ($HOCH_2CH_2OH$) have been detected at (sub)millimeter wavelengths in Comet Hale–Bopp, while the ion mass spectrometer on board the Giotto spacecraft detected molecules *in situ* with molecular weights up to $\mu_a = 120$. These molecules could originate from dust grains or polymerized formaldehyde (POM, or PolyOxyMethylene $(CH_2O)_n$).

10.4.3.3 *Nitrogen Compounds*

The major potential reservoirs for volatile nitrogen are N_2, HCN, and NH_3. Molecular nitrogen (N_2) has no allowed vibrational or rotational transitions in the observable spectral range, but can fluoresce at 95.9 nm; however, only upper limits (0.2% of H_2O) were derived from FUSE measurements. Although detections of N_2^+ near 391 nm are suggestive of N_2 abundances of $\sim 0.02\%$ relative to H_2O, these measurements have been called into question.

Ammonia gas readily dissociates: $NH_3 \rightarrow NH_2 + H \rightarrow NH + H + H$. Both NH and NH_2 have been detected in a number of comets, and suggest NH_3 production rates 0.3–1% that of H_2O. Ammonia gas has been observed directly in C/IRAS–Araki–Alcock, C/Hyakutake, and C/Hale–Bopp, with typical abundances (relative to water) varying from 0.3% for C/Hyakutake, to $\sim 1.5\%$ for C/Hale–Bopp, and 6% for C/IRAS–Araki–Alcock. These numbers suggest that the NH_3 production rate may vary substantially from comet to comet.

Hydrogen cyanide has been observed and imaged directly in many comets. It seems to originate both on the nucleus and as a distributed source, i.e., part of it may originate from dust grains. Its production rate is 0.3–0.6% that of water. HNC has been observed in e.g., C/Hyakutake and C/Hale–Bopp, at levels roughly 5–6 times less abundant than HCN, very similar to the ratio seen in warm molecular clouds. Based upon the relative increase in the ratio HNC/HCN in C/Hale–Bopp with decreasing heliocentric distance, it has been suggested that a large fraction of HNC is produced in the coma through dissociative electron recombination of the ion $HCNH^+$.

Nitrogen is also present in dust particles (CHON, see §10.4.4) and in the form of CN. Although a possible progenitor for CN is HCN, the production rate of HCN appears to be less than half that of CN. Comet-to-comet variations in the production rate of CN are strongly correlated with the dust-to-gas ratio in comets, and the CN emissions have revealed the presence of jets and shells in more than one comet. About 20–50% of the CN is therefore thought to come directly from the dust grains.

After adding up all the available nitrogen, the N/C and N/O abundance ratios appear to be depleted relative to the solar ratios by a factor of 2–3.

10.4.3.4 *Sulfur Compounds*

No firm measurements on the overall sulfur abundance in comets exists, although estimates for 1P/Halley and C/Hale–Bopp suggest that the overall S/O ratio is slightly above the solar value. While UV transitions of the radicals S and CS have been observed in many comets, much less data exist on possible parent molecules. Carbon disulfide (CS_2), with a lifetime of several minutes at 1 AU, is most likely the parent to CS; the abundance of each species has been measured at $\sim$0.1–0.2% that of H_2O. Since S and O are chemically quite similar, one might expect that a relatively large fraction of cometary sulfur is present in the form of hydrogen sulfide (H_2S). This molecule has been detected at millimeter wavelengths in several bright comets with an abundance varying from $\sim$0.1 up to 1.5% that of water. Other sulfur-bearing molecules (e.g., SO, SO_2, OCS, H_2CS, NS) show abundances <1% relative to H_2O. Since disulfur (S_2) is a very low temperature condensate, detections of this molecule may help constrain comet formation theories (§10.7.1). However, because of its extremely short lifetime ($\sim$450 s), observations are challenging. The first detection of S_2 was obtained in 1983 during an outburst in Comet C/IRAS–Araki–Alcock, a comet that came very close to the Earth ($r_\Delta \sim 0.03$ AU). Thirteen years later, S_2 was observed with HST in C/Hyakutake, another comet which came exceptionally close ($r_\Delta \sim 0.06$ AU) to Earth. Since then, using extremely sensitive detectors, S_2 has been measured in several other comets, with typical abundances 0.001–0.005% that of H_2O. It is probably present in most if not all long-period comets.

10.4.3.5 *Alkalies and Metals*

Sodium emission has been seen in several long-period comets that came to within $\sim$1.4 AU of the Sun. The emissions reveal a tail of sodium atoms, which can be quite extensive (Fig. 10.23). Since the boiling point for sodium is $\sim$1150 K, the observed atoms might be bound in complex molecules rather than as solid sodium. Models of C/Hale–Bopp imply that the emitting sodium atoms originate in the vicinity of the nucleus, and that the production rate is less than 0.3% of that expected based upon solar Na/O abundances.

Emission lines of calcium, potassium, and metals have only been observed in *Sun-grazing comets*, comets which

Figure 10.23 The thin straight sodium tail of C/Hale–Bopp stands out in the left image, which records the fluorescence (D-line) emission from sodium atoms. A traditional image of the plasma and dust tails is shown on the right. (Cremonese *et al.* 1997)

come to within 0.1 AU of the Sun. At $r_\odot < 0.1$–0.2 AU, potassium lines show up, and at even smaller heliocentric distances, emission lines of Fe, Ni, Co, Mn, V, Cr, Cu, Si, Mg, Al, Ti, Ca, and Ca$^+$ have been seen. The detection of these lines support the hypothesis that dust grains evaporate when close enough to the Sun. In one Sun-grazing comet, C/Ikeya–Seki (1965 VIII), the relative elemental abundances have been measured and found to be similar to those seen in carbonaceous chondritic meteorites.

10.4.3.6 *Isotopes and Ortho-to-Para Ratios*

In order to better understand how our Solar System formed and evolved, we want to know how much interstellar matter in the form of grains and ices was incorporated and retained as such in the forming bodies. Using isotope abundance ratios of various species (e.g., §8.7.2) and ortho-to-para ratios in hydrogen may help address the question of whether cometary ices formed directly, via simple condensation of its constituent gases within the primitive solar nebula, or whether the ices originated in the interstellar medium and were incorporated as such in the solar nebula.

The ortho-to-para ratio (OPR) of the nuclear spins for hydrogen molecules depends upon the rotational distribution of the molecules. At high temperatures, the population is in equilibrium with their statistical weights, $2I + 1$, with the spin $I = 1$ for ortho and $I = 0$ for para hydrogen; i.e., the OPR for water at high temperatures is 3. Typical spin temperatures derived from observed OPR values are $\sim$30 K (Fig. 10.24), which correspond to the equilibrium temperature at $r_\odot \sim 100$ AU under present Solar System conditions. Since formation of H_2O through gas phase reactions proceeds only under high temperatures, the low values for the spin temperature suggest that water molecules formed on grains, where H_2O would equilibrate with the grain temperature.

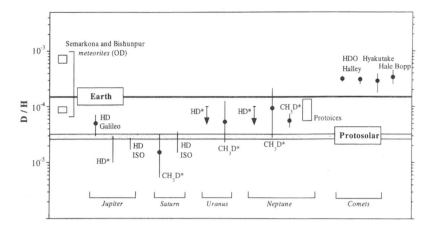

Figure 10.24 The D/H ratios of various bodies in our Solar System. The Earth and protosolar values are shown as horizontal lines. Asterisks (e.g., HD*) indicate ground-based observations. (Adapted from Bockelee-Morvan *et al.* 1998)

Isotopic abundance ratios yield information on fractionation effects from the time and the region where the comets or their constituent grains formed. In particular the D/H ratio as measured from HDO/H_2O provides important cosmogenic information. As discussed above, if cometary water formed at low temperatures, any HDO/H_2O fractionation must also have occurred at these low temperatures. In cold interstellar clouds, deuterium fractionation occurs through ion–molecule reactions and grain-surface chemistry. As observed in such clouds, these processes lead to a considerable (orders of magnitude) enhancement in the D/H ratio in some species compared to that in molecular hydrogen. In contrast, deuterium fractionation in the primitive solar nebula would have occurred via reactions between neutrals, which depend upon temperature, pressure, and time. At the low temperatures and pressures in the outer reaches of the Solar System, the rate of deuterium fractionation would have been much too low to reach an equilibrium state. Calculations suggest that under such circumstances the D/H ratio should not be much larger than three times the protosolar value.

The D/H ratio in cometary water has been determined accurately for 1P/Halley from measurements of the ion and neutral mass spectrometers aboard Giotto. Its value was measured to be $(3.2 \pm 3) \times 10^{-4}$. Similar values were derived from ground-based observations of C/Hyakutake and C/Hale–Bopp (Fig. 10.24). These ratios are almost an order of magnitude higher than the protosolar value; they are similar to that measured in hot cores of molecular clouds in which stars much more massive than our Sun form. Perhaps water-ice in comets is pristine interstellar water, incorporated as such (without volatilization) into comets. This scenario would be consistent with the OPR measurements. It would also lead to high D/H enrichments in other species. A factor of ten enrichment has

been measured in hydrogen cyanide in C/Hale–Bopp, but it has not been seen in, e.g., formaldehyde. The D/H ratio in interplanetary dust particles (IDPs), which presumably come from comets, is typically enriched by factors of 250–500. On the other hand, the D/H ratio (as associated with carbon) measured in several stardust grains (§10.4.4) is similar to or enriched by at most a factor of $\sim$3 compared to terrestrial values, i.e., a much lower enrichment than that seen in IDPs. Both the D/H ratios and OPR measurements can be reconciled with physical mixing and limited reprocessing of interstellar water in the solar nebula. Such a mixing model would also be consistent with observations of isotopic ratios of $^{12}C/^{13}C$, $^{14}N/^{15}N$, $^{32}S/^{34}S$, and $^{16}O/^{18}O$. These have been determined for a number of comets, including e.g., 1P/Halley, C/Hale–Bopp, and C/Ikeya (C/1963 A1), both from *in situ* measurements and ground-based observations of rare isotopes in HCN and CS. Results differ by a factor of $\lesssim$2 from comet to comet, and are consistent with the terrestrial value.

10.4.3.7 *Noble Gases*

Noble gases, if present and detectable in comets, would yield valuable information on the environment in which cometesimals formed. Several refractory grains brought back by the Stardust mission (§10.4.4) were heated to over 1500 K to degas He and Ne, if present. Both gases were detected. The isotope $^{20}Ne/^{22}Ne$ ratio was measured to range from 9 to 10.7, similar to our atmospheric value of 9.8, and similar to that measured in primitive carbonaceous meteorites (which ranges from 10.1 to 10.7). It is below the solar wind values (13.9 ± 0.8) as measured in samples brought back with the Genesis spacecraft. In contrast, the $^{3}He/^{4}He$ in Stardust samples is about twice that measured in meteorites ($2.7 \pm 0.2 \times 10^{-4}$ versus $1.45 \pm 0.15 \times 10^{-4}$) and Jupiter ($1.66 \pm 0.05 \times 10^{-4}$). The high $^{3}He/^{4}He$ could indicate a later He addition from the solar wind. It seems plausible that the Stardust grains and the carbonaceous carrier that contains the noble gases in meteorites originated in the same environment, a medium that must have been hot and contained high ion fluxes which implanted the noble gases in the grains.

10.4.4 Dust Composition

The structure and mineralogy of cometary dust grains is determined by the chemical and physical processes prevalent at the time when they formed, and hence analysis of these grains may reveal the conditions in our solar nebula at the time of planet formation. Our knowledge of cometary and interplanetary dust grains has advanced considerably in the past decade, with state-of-the-art mid-infrared detectors on ground-based and space telescopes, *in situ* sampling, and laboratory analysis of captured IDPs and cometary grains brought back by the Stardust mission.

The composition of dust grains has been measured *in situ* by impact-ionization mass analyzers on the Giotto and Vega spacecraft. Many grains were composed primarily of the light elements C, H, O, N, collectively referred to as *CHON* particles. Silicate grains were present in roughly equal amounts. About half the total population sampled had a ratio of C to rock-forming elements between 0.1 and 10, while the other half was equally divided between CHON and silicate grains.

Silicate and carbon emissions have been detected remotely in a growing number of comets. The 8–13 μm spectra of C/Hale–Bopp and 9P/Tempel 1 are shown in Figure 10.25a, with 5–35 μm spectra in Figure 10.25b. The featureless 10 μm continuum emission results from amorphous carbon grains. The 11.2 μm feature and the 11.8 μm shoulder are indicative of Mg-rich crystalline olivine, which implies that at least a small fraction of the silicate material must be in crystalline form. The broad 10 μm feature is characteristic of amorphous olivine, while the 9.3 μm and 10.5 μm peaks match spectra of pyroxene. The emission peaks in the 5–35 μm spectra are due to crystalline forsterite (Mg-rich olivine), and some of the spectral structure hints at crystalline enstatite (Mg-rich pyroxene). The dust is seemingly rich in magnesium, which is similar to the *in situ* findings at Comet 1P/Halley. The emissions are strong, which means that the grains that produce them have radii <1 μm, or, alternatively, the grains may be porous aggregates with small crystals embedded in them.

The crystal structure has puzzled researchers. Crystals could have condensed out directly from the solar nebula, but require temperatures of 1200–1400 K. Alternatively, annealing of amorphous grains could have turned them

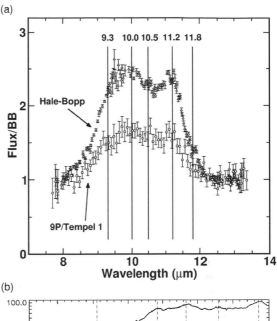

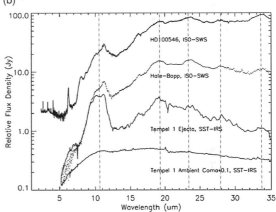

Figure 10.25 (a) Crystalline silicate features in the 8–13 μm spectra of Comet C/Hale–Bopp and of the impact-induced ejecta from Comet 9P/Tempel 1, about one hour after impact. Both spectra were divided by a best-fit blackbody spectrum. A spectrum of 1P/Halley is very similar. (Harker *et al.* 2005) (b) Comparison of the Spitzer 5–35 μm spectra of Comet 9P/Tempel 1 before and one hour after impact, and ISO spectra of Comets C/Hale–Bopp and the young stellar object HD100546. In this presentation, the spectra have been divided by factors of 10, 1, 2.7, and 10 (from bottom to top). Note the logarithmic scale. The approximate locations of crystalline forsterite are indicated by dashed lines. The spectra are normalized by their blackbody spectra. (Adapted from Lisse *et al.* 2007)

into crystalline grains, but this process still requires temperatures in excess of 1000 K.

The spectra as shown in Figure 10.25 have been seen for only a few comets, mostly long-period or Oort cloud comets. Comet 9P/Tempel 1 presents a particular case. This is a Jupiter family comet that was targeted by the Deep Impact mission. Before impact, the mid-infrared emission from the comet was dominated by thermal emission from its coma (Fig. 10.25b). The absence of 10 μm emission features from this spectrum and its ∼235 K color temperature indicates a preponderance of large dust grains in the coma (§10.3.6.1). Clear emission features were seen about one hour after impact, at which time there must have been small dust grains, of order $R \sim 0.2$ μm, composed of amorphous olivine, pyroxene, and carbon, as well as crystalline olivine. The relative abundance ratios of the individual elements are consistent with those in carbonaceous chondrites.

The cometary spectra have been compared to spectra from IDPs, 1–10 μm porous aggregates dominated by amorphous silicates and carbon (§8.5), but also containing tiny (submicron) crystals that produce the crystalline features in their spectra. When these crystals are coated by amorphous material, the emissions are absent. One may therefore wonder whether the Deep Impact fragmented such aggregates, or whether pristine subsurface crystals were excavated from the comet's subsurface layers. Evidence is accumulating in favor of the latter hypothesis.

On 15 January 2006, the Stardust mission returned cometary materials to Earth. This mission, launched in 1999, encountered Comet 81P/Wild 2 on 2 January 2004. While moving through 81P/Wild 2's coma, Stardust captured thousands of particles, 5–300 μm across. The Stardust particles were trapped in *aerogel*, a highly porous silica 'foam' with a density comparable to that of air (Fig. 10.26). Grains that impact the aerogel are gradually slowed down, without suffering substantial melting or vaporization (Fig. 10.26b,c). The recovered grains are assemblages of different minerals, in particular the crystalline silicate minerals olivine and pyroxene, as well as some troilite (FeS). Isotopic analysis shows that most minerals are similar to those found in the inner Solar System, and only a few appear to be anomalous presolar grains.

The isotopic compositions of H, C, N, O, and Ne in the Stardust grains analyzed as of November 2007 show that the cometary grains are unequilibrated aggregates, composed of materials that originated in different reservoirs. One Stardust grain is composed of high-temperature minerals found in meteoritic CAIs (calcium–aluminum inclusions; §8.5). Many other fragments are also mineralogically and isotopically linked to CAIs. In particular, graphs of isotope abundance ratios for oxygen, $^{17}O/^{16}O$ versus $^{18}O/^{16}O$ (Fig. 8.16), show that these ratios follow the CAI mixing line rather than that expected from mass-dependent fractionation. As discussed in §8.6, CAIs condense at high temperatures, $T > 1400$ K, i.e., likely close to the Sun. The presence of this grain, coupled to the

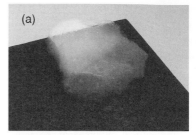

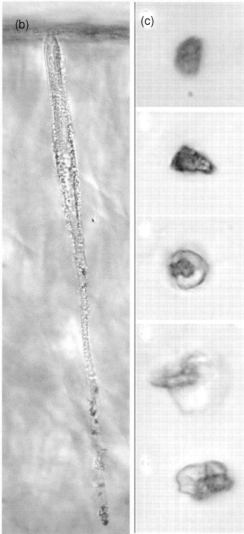

Figure 10.26 (a) A piece of aerogel, similar to that used in the Stardust mission. (Courtesy NASA Photographer Maria Garcia 1997) (b) Track of a particle from Comet 81P/Wild 2 as captured in the aerogel of the Stardust mission. The track is about 1 mm long. The particle entered at the top. The force of the impact broke up the tiny rock, and pieces can be seen all along the track (all the black dots). These particles are (sub)μm-sized. (c) A close-up of several grains captured in and recovered from the aerogel. (Adapted from Brownlee *et al.* 2006)

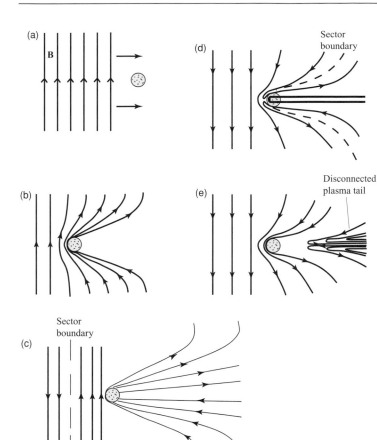

Figure 10.27 The field draping model of Alfvén, in which interplanetary magnetic field lines are deformed by a comet's ionosphere. The sequence from (a) to (c) shows the gradual draping of interplanetary magnetic field lines around the comet into a magnetic tail. When the comet encounters a sector boundary in the interplanetary magnetic field (where the magnetic field reverses direction; dashed line), the tail becomes disconnected, as depicted in the sequence from (c) to (e).

oxygen isotopic and (high-temperature) mineralogic data provide convincing evidence of strong radial mixing in the early solar nebula. Grains that formed near the Sun must have been transported radially outwards to beyond Neptune's orbit, where they were assembled into cometesimals. Although in principle turbulent mixing in the midplane could lead to such transport, the X-wind model (§8.7) where CAIs and chondrules are flung out on ballistic trajectories is perhaps a more straightforward way to explain such large-scale radial mixing.

10.5 Magnetosphere

Although sublimating gases are neutral when released by a comet, ultimately all atoms and molecules get ionized. The dominant ionization processes are photoionization and charge exchange with solar wind protons. Cometary ions and electrons interact with the interplanetary magnetic field, and 'drape' the field lines around the comet, as depicted in Figure 10.27. This process induces a magnetic field similar to the field around Venus. In the following, we describe the magnetic field morphology for an active

comet with a well-developed atmosphere; i.e., for a comet that is close to the Sun (within 1–2 AU). A rough sketch is shown in Figure 10.28.

10.5.1 Morphology

Since a comet's gravitational field is very weak, neutral gas expands supersonically outward from the nucleus (§10.3.2). When the neutrals get ionized, they are accelerated by the strong electric fields in the solar wind (§7.1.3):

$$\mathbf{E}_{sw} = \frac{\mathbf{v}_{sw} \times \mathbf{B}_{sw}}{c}. \tag{10.17}$$

The solar wind velocity, v_{sw}, is typically $\sim$400 km s^{-1}, and the interplanetary magnetic field $B_{sw} \approx 5 \times 10^{-5}$ G at $r_\odot \approx 1$ AU. Cometary ions are accelerated to the solar wind speed, either directly (if $\mathbf{B}_{sw} \perp \mathbf{v}_{sw}$), or indirectly via plasma instabilities (if $\mathbf{B}_{sw} \parallel \mathbf{v}_{sw}$). The ions gyrate around the local magnetic field lines. The pickup of these heavy ions by the solar wind leads to a mass loading of the solar wind. Based upon conservation of momentum, the solar wind is (temporarily) slowed down. As a result of the

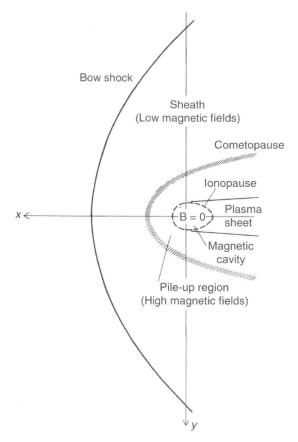

Figure 10.28 A cartoon sketch of the magnetic field morphology for a comet with a well-developed atmosphere. The Sun is located to the left, along the x-axis. (Adapted from Neubauer 1991)

with v_o the outflow velocity of the neutrals, α the ionization rate, γ the ratio of specific heats ($\gamma = 2$ for a magnetized flow), and ρ_{sw} and v_{sw} the solar wind density and velocity far from the comet. For 1P/Halley, with $\mathcal{Q} \approx 10^{30}$ molecules s^{-1} at $r_{\odot} \approx 0.9$ AU, $\mathcal{R}_{bs} \approx 10^6$ km (Problem 10.23).

The outflowing cometary neutrals collide with the incoming solar wind, which is consequently slowed down. The *cometopause* or *collisionopause* separates the collisionless solar wind plasma flow from the cometary gas, a flow in which collisions dominate. At the cometopause, significant momentum is transferred through collisions between the outflowing cometary neutrals and the solar wind ions. Although this boundary layer is not completely understood, there is general agreement that the region inside the cometopause is dominated by cometary ions (in particular H_2O^+ and H_3O^+) and compressed interplanetary magnetic field; in contrast, solar wind plasma, loaded with cometary ions created at large distances from the nucleus, dominates the region outside the cometopause. The cometopause was encountered by the Vega spacecraft $\sim 10^5$ km upstream from 1P/Halley, when the comet was at $r_{\odot} \approx 0.8$ AU. Its location can be approximated by

$$\mathcal{R}_{cp} \approx \frac{\sigma_x \mathcal{Q}}{4\pi v_o}, \tag{10.19}$$

with σ_x the ion–ion collision cross-section.

Inside the cometopause, the magnetic field is highly compressed, such that there is approximate pressure equilibrium between the magnetic pressure at the inside and the supersonic solar wind ram pressure at the outside:

$$\frac{B_c^2}{8\pi} = \rho_{sw} v_{sw}^2, \tag{10.20}$$

with B_c the magnetic field strength at the inside of the cometopause.

The magnetometer experiment on board the Giotto spacecraft discovered a well-defined boundary inside 1P/Halley's cometopause, at a cometocentric distance of 4700 km, where the magnetic field suddenly dropped to zero. This boundary has been called the *ionopause*, a tangential contact discontinuity which separates the cometary plasma from the contaminated solar wind plasma. There is no magnetic field inside the ionopause: The nucleus is surrounded by a magnetic cavity. There is approximate pressure equilibrium between the magnetic field outside of the cavity and the thermal pressure inside.

Interior to the ionopause, models predict the existence of an inner shock, which decelerates the supersonically outward flowing cometary ions and diverts them into the tail; this shock, however, has not been observed by any spacecraft. As in planetary magnetotails, a neutral sheet

mass-loading process, when the solar wind has accumulated an admixture of $\sim 1\%$ (by number) cometary ions, a collisionless reverse shock forms: the *bow shock*. Note the difference between the cometary bow shock – caused by mass loading – and the bow shocks upstream from planets, which form to divert the solar wind flow around the impenetrable boundaries of the planets or their magnetospheres (§7.1). Downstream of the cometary bow shock, the solar wind is subsonic, as in a planetary magnetosheath, and continues to interact with the cometary atmosphere. The approximate location of the bow shock, $\mathcal{R}_{bs}$, depends on the gas production rate, $\mathcal{Q}$, and is inversely proportional to the solar wind momentum, $\rho_{sw} v_{sw}$. Using the equations for mass, momentum, and energy for a plane-parallel supersonic flow in front of the bow shock and mass loading by the comet, the following formula has been derived for the standoff distance of a cometary bow shock:

$$\mathcal{R}_{bs} = \frac{\mathcal{Q} \mu_a m_{amu} \alpha (\gamma^2 - 1)}{4\pi v_o \rho_{sw} v_{sw}}, \tag{10.18}$$

Figure 10.29 This image of C/Hyakutake was obtained with a 30 cm reflector, during a 60-second exposure on 8 April 1996 UT. The image was processed to bring out the ray structure and jets emanating from the nucleus. The raw image is shown in the insert. (Courtesy Tim Puckett)

separates the two 'lobes' of the magnetotail, formed by the folding of the interplanetary magnetic field lines. When the ICE spacecraft crossed the neutral sheet in the tail of 10P/Giacobini–Zinner (1984e), the polarity of the field reversed, as expected when crossing a neutral sheet in a planetary magnetosphere.

10.5.2 Plasma Tail

Cometary ions form an ion or plasma tail in the antisolar direction. The length of this tail often exceeds 10^7 km, and the width of the main tail is roughly 10^5 km in diameter. The tail usually looks blue, as a result of fluorescent transitions of the abundant, long-lived CO^+ ions, though occasionally a reddish ion tail has been seen due to emissions by H_2O^+. The tail often consists of filaments, rays, and bright knots, and its structure changes on timescales of minutes to hours. The knots are caused by enhanced densities, and their motion can be followed down the tail. Typical speeds are $\lesssim 100$ km s^{-1}; this is clearly much more than the cometary outflow speed, but less than the solar wind velocity. Closer to the nucleus, in the cometary head, Doppler shifts of the H_2O^+ ion indicate speeds of 20–40 km s^{-1}; the plasma is clearly accelerated down the tail. Solar wind speeds are reached at distances $> 10^7$ km from the nucleus. Density concentrations in the main tail are often accompanied by enhancements in adjacent filaments, at the same cometocentric distance.

In gas-rich comets, one can often see the tail rays from the plasma envelope. Figure 10.29 presents a processed image of C/Hyakutake which shows the rays 'draped' around the coma (compare the field line draping

depicted in Fig. 10.27). Individual rays also show up in Figure 10.30.

The precise structure of a cometary plasma tail depends on the interplanetary medium and its magnetic field. One can often see large disturbances in a plasma tail (Fig. 10.30), and sometimes the comet appears to lose its tail, and starts forming a new one. Such events have been attributed to 'disconnection' or magnetic reconnection events, caused by a sudden reversal of the interplanetary magnetic field. Such reversals take place when the comet meets an interplanetary sector boundary, or when it crosses the heliospheric current sheet (Fig. 10.27).

10.5.3 X-Ray Emissions

The interaction between the solar wind and a comet's atmosphere is visible through the emissions of soft X-ray ($E < 1$ keV) and EUV photons. The emissions were discovered in 1996 when ROSAT and EUVE satellites observed Comet C/Hyakutake (Fig. 10.31); since that time such emissions have been observed from many comets. The emissions are typically displaced from the nucleus in the direction of the Sun. For C/Hyakutake, the peak emission was displaced by $\sim 20\,000$ km, but for the most active comets this distance can exceed 10^6 km. The X-rays are produced via charge exchange interactions of highly charged heavy solar wind ions ($C^{+q}, O^{+q}, N^{+q}, Si^{+q}, Ne^{+q}$, with $q = 4, 5, 6, 7, \ldots$) with cometary neutrals:

$$O^{+6} + H_2O \rightarrow O^{+5} + H_2O^+ + h\nu. \tag{10.21}$$

Reaction (10.21) leaves the heavy solar wind ion in an excited state, while transforming the cometary neutral (H_2O, OH, O, H, $\ldots$) into an ion.

10.6 Nucleus

It is difficult to observe cometary nuclei directly (except at close range from interplanetary spacecraft), since a comet is usually not visible until it has developed a coma surrounding its nucleus. Hence most observations, even at relatively large heliocentric distances, are to some extent contaminated by the comet's coma (Problem 10.27). Although models of the brightness (distribution) of a coma can be used to subtract the coma from the data, this can introduce significant errors. Despite these problems, both size and albedo estimates of cometary nuclei have been obtained from Earth-based observations. The albedo of most comets is very low; at visible wavelengths it ranges from $\lesssim 2\%$ up to 6%. Reflection spectra at visible–IR wavelengths (380–850 nm) usually show a smooth slightly upward slope towards longer wavelengths.

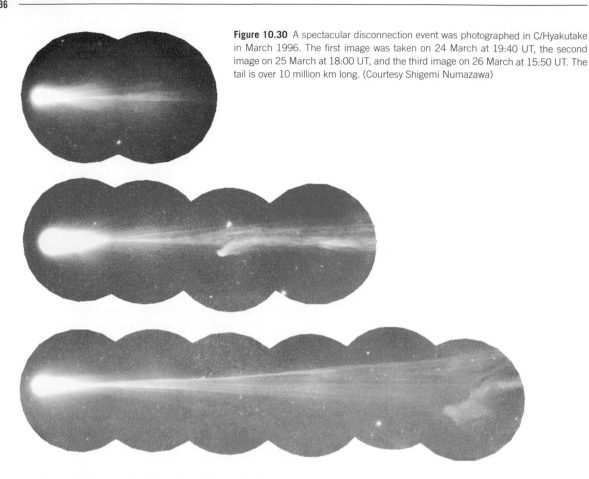

Figure 10.30 A spectacular disconnection event was photographed in C/Hyakutake in March 1996. The first image was taken on 24 March at 19:40 UT, the second image on 25 March at 18:00 UT, and the third image on 26 March at 15:50 UT. The tail is over 10 million km long. (Courtesy Shigemi Numazawa)

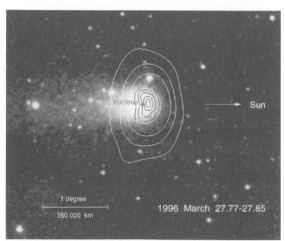

Figure 10.31 Image of the X-ray emission from C/Hyakutake, as observed with the ROSAT (Röntgen) satellite. The image shows the measured intensity of both the high (0.1–2.0 keV inner contours) and low (0.09–0.2 keV outer contours) energy bands as contour lines overlayed on an optical image taken with a common camera during the ROSAT observation. (Courtesy Max-Planck-Institut für Extraterrestrische Physik)

10.6.1 Size, Shape, and Rotation

The size distributions as derived for 65 ecliptic comets (EC, §10.2.4) and, separately, 11 comets with a more isotropic distribution in inclination angle (essentially long-period comets) are shown in Figure 10.32. Typical cometary radii range from $\lesssim 1$–10 km; however, this may be primarily an observational effect, with smaller bodies being more numerous, but difficult to detect, and larger bodies rarer, but still dominating the distribution by mass. The EC population is best fit by a power law (eq. 9.3) with $\zeta = 2.9 \pm 0.3$ at radii $R > 1.6$ km, which may be slightly flatter than the size distribution for asteroids and TNOs (§9.4.1), and than that expected for a collisionally evolved population of bodies ($\zeta = 3.5$). As discussed in detail in §9.4.1, in a steady-state situation the number of bodies destroyed in a certain mass bin is equal to the number of objects created in that mass bin. A steeper distribution is expected during the accretion phase, while a shallower slope suggests a relatively larger loss of bodies than expected based upon collisional processes. Indeed, one might expect a shallower slope for comets since (*i*) noncollisional fragmentation, such as splitting (§10.6.4),

Table 10.2 First four comets visited close-up by spacecraft.

Comet	Spacecraft	Year	Image resolution	q (AU)	P_{orbit} (yr)	$R_1 \times R_2 \times R_3$ (km)	A_v	P_{rot} (hr)
1P/Halley	Giotto[a]	1986	45 m/pixel	0.59	75.3	$7.21 \times 3.70 \times 3.70$	0.04	177.6
9P/Tempel 1	Deep Impact	2005	5 m/pixel	1.51	5.5	$8.2 \times 4.9 \times 3.5$	0.04	40.83
19P/Borrelly	Deep Space 1	2001	47 m/pixel	1.35	6.4	$4.0 \times 1.60 \times 1.60$	0.03	25.0
81P/Wild 2	Stardust	2004	20 m/pixel	1.60	6.4	$2.75 \times 2.0 \times 1.65$	0.03	12.3 or 25

[a] Four other spacecraft flew past 1P/Halley in 1986: the large probes Vega 1 and 2, and the smaller Suisei and Sakigaki spacecraft (Table F.2). Closest approach for these spacecraft varied from 3000 to 200 000 km. The first spacecraft encounter with a comet took place in September 1985, when the International Cometary Explorer (ICE) flew through the tail of Comet 21P/Giacobini–Zinner.

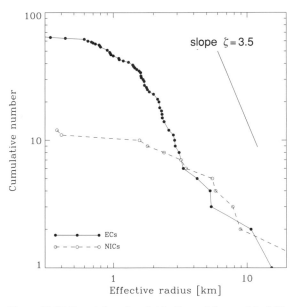

Figure 10.32 Cumulative size distributions of the nuclei of 65 ecliptic comets (upper curve) and 11 near-isotropic comets (lower curve). A power-law slope as expected for a collisionally evolved population of bodies, $\zeta = 3.5$, is indicated on the graph. (Adapted from Lamy *et al.* 2004; courtesy I. Toth)

is common amongst comets, and (*ii*) the nuclei are eroded through sublimation processes during each perihelion passage. A typical EC comet may lose 400 m in radius over half its lifetime. In particular for small comets, this would lower ζ substantially. The paucity of small long-period comets may be the result of few small bodies reaching the Oort cloud (§10.7.2) because of the optical depth of the protoplanetary disk to scattering.

Since comet nuclei are small, they are most likely not spherical (§6.1). Indeed, the four comets encountered by spacecraft show irregular, in most cases quite elongated, shapes (Table 10.2).

Measuring the rotational properties of comets is nontrivial, because a comet's coma generally dominates the

body's reflected light, except at large heliocentric distances where most comets are very faint and difficult to detect. As active regions on a comet are not uniformly distributed across its surface, rotation of these spots can produce periodic variations in coma brightness that can be used to estimate the rotation period. Radar measurements can probe through the coma, but the comet has to come very close to Earth to be observed. Most comets have rotation periods between a few hours and several days. These periods are typically somewhat longer than for main belt asteroids (§9.4.6).

The physics of cometary rotation is analogous to that of asteroid rotation (discussed in §9.4.6), with the added complication that outgassing of comets can result in torques which alter the rotational angular momentum vector. Comets that are (or were in the geologically recent past) active are thus more likely to exhibit complex (nonprincipal axis) rotation states than are asteroids. Moreover, outgassing-induced spin-up of cometary nuclei may cause some comets to split into two or more pieces (§10.6.4), which can instantaneously alter the rotation rate.

Close-up images of the nucleus of 1P/Halley obtained by three spacecraft in 1986, together with studies of coma brightness variations observed from the ground, exclude pure rotation about either the short or the long principal axis. Thus, 1P/Halley is in a complex rotation state, with its instantaneous axis of rotation precessing about its rotational angular momentum vector. The particulars of 1P/Halley's spin are not fully constrained, but the most probable rotational state is one in which the long axis executes precessional motion about the angular momentum vector with a period ∼3.7 days combined with rotation around the long axis at a period of ∼7.3 days.

10.6.2 Processing far from the Sun

Since comets are rather small and rich in volatiles, they cannot have undergone much thermal evolution, and are therefore regarded as the most pristine objects observed in our Solar System. The outer layers of a comet, however, may have undergone significant processing while in the Kuiper belt or Oort cloud. Bombardment by energetic charged particles breaks up molecules on a comet's surface. The light hydrogen atoms may escape into interplanetary space, migrate through the ice matrix to form H_2, and/or initiate spin conversion in water and other symmetric molecules through exchange reactions. The heavier atoms/molecules stay on the surface, and may form new carbon-rich materials (e.g., CHON particles, hydrocarbon chains), which are usually dark and red. The penetration depth of a charged particle depends upon its energy; low-energy protons (1–300 keV) penetrate $\sim$10 μm into the crust, while GeV particles penetrate 1–2 m (if the comet's density is 1 g cm^{-3}). The resulting refractory surface layer, referred to as the *irradiation mantle*, is probably about a meter thick. This is comparable in thickness to the average amount of material lost in every perihelion passage once the perihelion distance gets below $\sim$1.5 AU (§10.6.3). Hence, for many Jupiter family comets the loss of material from the surface is negligible relative to the thickness of the irradiation mantle, but for some it is highly relevant, e.g., for 81P/Wild 2 that recently had its perihelion lowered considerably. The visible effect of irradiation by energetic photons (UV and X-ray) on a cometary surface is similar to that induced by charged particles (darkening and reddening of the material), but pertains only to the upper few microns of the surface.

In addition to the radiation damage described above, a comet's surface is also modified by erosion and/or gardening by Solar System debris and interstellar grains. Heating by passing stars and supernova explosions may affect comets in the Oort cloud.

10.6.3 Sublimation of Ices

Dynamically new comets, which have never entered the inner Solar System before, are likely to have highly volatile ices on their surfaces. This volatile material sublimates and evolves off the surface as soon as its sublimation temperature is reached. This, together with the explosive release of unstable species created by 4.5 Gyr of irradiation by galactic cosmic rays outside the heliosphere, explains the relatively high activity in new comets at large heliocentric distances. In contrast, an old, periodic comet has lost much or all of its volatile surface material. Such a comet is covered by a *dust crust*. This crust is built up by dust grains too heavy to be dragged off the surface by the sublimating gases. Such a dust crust usually forms when the comet recedes from the Sun, and sublimation gradually ceases. On the comet's return to the

inner Solar System, gas pressure builds up in cavities as soon as subsurface ices reach their sublimation temperature. When the gas pressure is high enough, a portion of the dust crust is blown off, exposing fresh ice. This effect produces the sudden increase in activity as seen for many periodic comets when they approach perihelion. During each perihelion passage a typical comet sublimates away a layer of ice only $\sim$1 m thick (Problems 10.14, 10.15), which is small compared to the size of a comet. However, periodic comets can develop thick dust crusts, and their activity can drop so low that no coma can be observed, even near perihelion. Such extinct comets are classified as asteroids.

One might expect the most volatile near-surface material in a comet to sublimate away long before perihelion is reached. Such activity has been invoked to explain cometary brightenings at large heliocentric distances (e.g., Fig. 10.7). Nonetheless, the production rate of highly volatile gases is still large when a comet reaches perihelion. Perhaps some of these molecules were 'trapped' in the form of solid clathrate hydrates, where a guest molecule occupies a cage in the water-ice lattice. Initially, water-ice in a cometary nucleus is presumably amorphous rather than crystalline, since comets formed at large heliocentric distances where the temperature is very low (§10.7). Amorphous ice can easily trap large quantities of guest molecules. Timescales for crystallization of water-ice are large, and depend exponentially on temperature. At a temperature of 140 K, amorphous ice converts to a crystalline structure within about an hour, but at temperatures of $\sim$75 K this process takes about 40 Gyrs. The reaction is exothermic and irreversible. When a new comet enters the inner Solar System, amorphous ice converts to a crystalline structure at $\sim$5 AU. Because the reaction is exothermic, a *heat front* propagates into the nucleus, so that 10–15 m of ice may be converted at once. This transition may supply enough energy to 'blow off' the original crust from a new comet, which could explain sudden brightenings at large heliocentric distances. The guest molecules trapped in the amorphous ice are set free. Those that have low sublimation temperatures diffuse through the crystalline ice and escape into space. Other molecules may recondense near the surface of the nucleus. These pockets of ice may sublimate away closer to the Sun, and cause *flare-ups* at smaller heliocentric distances.

10.6.4 Splitting and Disruption

Comets sometimes possess multiple nuclei that are spatially separated. The gravitational fields of comets are too weak for these nuclei to be bound binary or multiple systems. Rather, they were presumably formed by recent

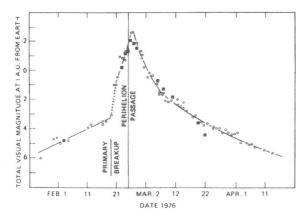

Figure 10.33 The lightcurve of C/West, which shows evidence of a splitting event (indicated in the figure). (Sekanina and Farrell 1978)

breakup (splitting) events. Over 40 split comets have been observed over the past 170 years, with over 100 splitting events. The first clear case was 3D/Biela in 1845/1846 (3D/1846 II). After the comet broke up, the brightest fragment was left with a large companion, which evolved as a separate comet. On its next return, in 1852, 3D/Biela appeared as a double comet. Neither piece has been seen since then, but in 1872 an intense meteor shower was seen when the Earth crossed the orbit of D/Biela, indicative of the 'death' of this comet. This shower gradually diminished in intensity over the next century.

In many split comets, only tiny fragments separate from the principal nucleus; such small pieces last for at most a few weeks. Since the splitting of a comet releases a substantial amount of dust and exposes fresh ice, it is typically accompanied by a 'flare-up' in the comet's brightness and a (temporary) increase in dust emission. Indeed, a sudden increase in the observed brightness of a comet is often taken as evidence of a 'splitting' event. A clear example is seen in visual lightcurves of C/West (Fig. 10.33), where a sudden increase in brightness before perihelion passage is attributed to a partial fragmentation of the primary nucleus. Four fragments (3 faint and 1 bright one) survived for many months. Near-simultaneous images of C/West's dust tail showed a substantial increase in the dust emission, in the form of broad streamers. The presence of striae (§10.3.7) also hints at fragmentation, since the particles that form the striae originated from the disintegration of large fragments that broke off the comet at an earlier time. The most extraordinary flare-up ever observed was an overnight brightening of Comet 17P/Holmes in 2007, from a magnitude of ∼17 to 2.8, i.e., by a factor of almost a million. This same comet also showed a major outburst in 1892, when Edwin Holmes

discovered the comet (hence the comet's name). Most splitting events occur when the comet comes close to the Sun. Some comets, like 73P/Schwassmann–Wachmann 1, show (frequent) outbursts even beyond Jupiter's orbit.

The best understood cause for the breakup of a cometary nucleus is tidal disruption during a close encounter with the Sun or a planet. Cometary nuclei may also split as a result of the centrifugal strain of rapid rotation, which can happen anywhere in the Solar System. Thermal stress, such as caused by the propagation of a heat wave, due e.g., to the conversion of amorphous into crystalline ice (§10.6.3), could cause a splitting or partial fragmentation of a comet even at a few tens of AU from the Sun. When ices sublimate in subsurface pockets, 'geyser' eruptions may result if the gas pressure exceeds the tensile strength of the nucleus material. Although this certainly leads to a localized blowoff of material, the nucleus may also be completely disrupted in the process. Another possible cause for a disruption or splitting event is a collision with interplanetary boulders. Although such events would occur most often in the asteroid belt, comets have not been observed to split while crossing this region.

An extraordinary example of a split comet was D/Shoemaker–Levy 9 (D/1993 F2), which was initially found to orbit and later to crash into Jupiter (July 1994; §5.4.5). Ground-based and HST images revealed more than 20 cometary nuclei, strung out like pearls on a string (see Fig. 5.41). Orbital calculations show that the comet was captured by Jupiter around 1930, and must have completed dozens of (chaotic) orbits about the planet before coming so close (1.3 $R_{\mathcal{Z}}$) that tidal forces disrupted the comet just after its close approach in July 1992. Simulations of the effect suggest the parent body to have had a low material strength, with a bulk density between 0.3 and 0.7 g cm^{-3}. The individual subkilometer-sized clumps must have been composed of loose agglomerates of material. Tidal breakup models predict that the largest fragments come from the center of the original nucleus, and should end up near the center of the chain of fragments. This pattern of size versus location is observed in D/Shoemaker–Levy 9 and other comets, as well as in crater chains on satellites (Fig. 5.85).

During its perihelion passage in 1995, Comet 73P/Schwassmann–Wachmann 3 had a huge outburst in activity when it split into at least five pieces. Fragments B and C were recovered during its next apparition 5.4 years later. In 2006, the viewing geometry was ideal, and a chain of more than three dozen individual fragments were seen, the largest of which were hundreds of meters in diameter. Some of these fragmented further during the 2006

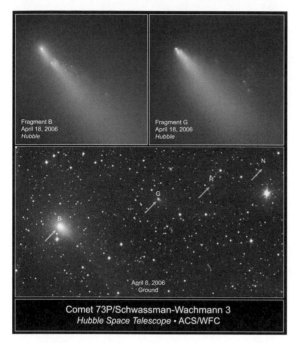

Comet 73P/Schwassman-Wachmann 3
Hubble Space Telescope • ACS/WFC

Figure 10.34 Breakup of Comet 73P/Schwassmann–Wachmann 3. The top frames show the 'second generation' fragmentation of fragments B and G shortly after large outbursts in activity. The original fragments were created during a splitting event in 1995. The bottom panel displays a wider field of view, showing several of the original fragments. (Courtesy Hal Weaver and NASA/HST)

apparition (Fig. 10.34); it remains to be seen whether any fragment will survive to the next apparition.

Many comets disappear from view, and for most comets the reason is unknown. Do they usually disintegrate completely, as did C/1999 S4 (Linear)? This comet broke up when it approached perihelion, and all fragments, over 20 initially, disappeared within a few weeks. It is not known whether these fragmented further, or became invisible because all ices had evaporated. Comets may also simply fade into obscurity by suddenly becoming dormant. Active areas may suddenly stop sublimating when they no longer receive sunlight, as when they move into a shadow. Based upon analysis of the orbital characteristics of NEOs, there probably are numerous dormant comets in this population.

Sun-grazing comets have a perihelion distance <2.5 $R_\odot$. Amateur astronomers have discovered 1685 Sun-grazing comets on SOHO images as of August 2009. The majority of these have similar orbital properties and belong to the *Kreutz* Sun-grazing family of comets. This comet family is named after Heinrich Kreutz, who made the first extensive observations of Sun-grazing comets in the nineteenth century. He suggested that these comets are

fragments of a single object that broke up about two thousand years ago. In addition to the Kreutz group, several other families have now been identified among the SOHO Sun-grazing comets (e.g., the Kracht, Marsden, and Meyer groups). Most Sun-grazing comets do not survive perihelion passage – they evaporate or disrupt completely or collide with the Sun (Fig. 10.35). The SOHO comets are not distributed randomly in space/time. In particular, many pairs of comets have been discovered with a perihelion passage within a day of each other. Such pairs and the apparent clumping of Sun-grazing comet apparitions are indicative of fragmentation events of family members at a variety of positions along their orbit, i.e., not confined to perihelion passages.

10.6.5 Structure of the Nucleus

The composition and structure of cometary nuclei were initially determined from observations of the material that they release into their comas, plasma, and dust tails, and from dynamical observations, in particular nongravitational forces and breakup of nuclei. Such observations led Whipple to propose his *dirty snowball* theory in 1950. In his model, a cometary nucleus is a loosely bound agglomeration of frozen volatile material interspersed with meteoritic dust. These conglomerates may be welded into a single solid body by thermal processing and *sintering*. Sintering is a process by which weak chemical bonds form along grain contacts of a particulate material that is near but below its melting temperature. Observations of the tidally disrupted Comet D/Shoemaker–Levy 9, however, suggested that the original body must have had a tensile strength less than that of dry snow, $\lesssim 1000$ dynes cm^{-2} for a 1–2 km sized body with a density of 0.6 g cm^{-3}. The comet nucleus must thus have retained its rubble-pile makeup, and not transitioned into a single coherent body.

Four comets have been imaged at close range by spacecraft (Table 10.2; Figs. 10.8, 10.36–10.38), one at a spatial resolution as high as 5 m/pixel. These images have led to a deeper understanding of both the surface and the interior structure of cometary nuclei. The first comet that was imaged in detail was 1P/Halley (Fig. 10.8). Giotto images revealed Halley's extremely low albedo, $A_v = 0.04$, which now appears to be a ubiquitous (or at least typical) characteristic of comets. Halley is an elongated potato-shaped object, covered with craters, valleys, and hills.

Deep Space 1 images of Comet Borrelly reveal a complex surface, with a variety of morphological features (Fig. 10.36). There are several dark spots that have a typical surface albedo of only 0.012–0.015, less than one-third that of the brightest areas at 0.045. Most notable are the flat-topped mesas on this comet, which appear to be

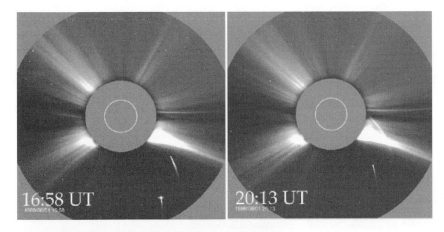

Figure 10.35 The LASCO coronagraph on the Solar and Heliospheric Observatory (SOHO) spacecraft observed two comets plunging into the Sun's atmosphere in close succession, on 1 and 2 June 1998. The inner few solar radii are blocked by a coronagraph. The circle is drawn in to represent the size and location of the solar disk. (Courtesy SOHO/LASCO consortium; ESA and NASA)

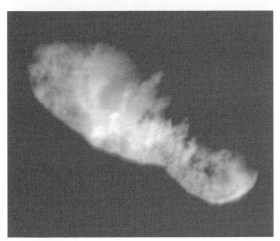

Figure 10.36 Comet 19P/Borrelly imaged by Deep Space 1 on 22 September 2001. The image reveals a variety of terrains and surface textures, including smooth, rolling plains that appear to be the source of dust jets seen in the coma. The rough terrain near the edges contains very dark patches, which are elevated compared to surrounding areas. Sunlight is coming from the bottom of the frame. (NASA/JPL, PIA03500)

associated with some of the active jets. Several ridges, oriented normal to the long axis of the comet, may result from compressional shortening, in which case the nucleus should have some tensile strength. No impact craters were seen.

Images of Comets 81P/Wild 2 by the Stardust mission (Fig. 10.37) and 9P/Tempel 1 by the Deep Impact mission (Fig. 10.38) reveal structure on the surfaces of these comets that was never seen or anticipated before. Although both comets belong to the Jupiter family, 81P/Wild 2 was thrown into its current small a, small q, orbit only in the 1970s, and hence has had only a handful of apparitions, whereas 9P/Tempel 1 has been in the Jupiter family at least since 1867. 81P/Wild 2's roundish shape suggests that this comet is not a collisional fragment, in contrast to the three other cometary nuclei imaged by spacecraft, which are more potato-shaped. Both Comets 81P/Wild 2 and 9P/Tempel 1 show evidence of impact craters.

The nucleus of Comet 81P/Wild 2 is dominated by depressions (Fig. 10.37). Some of these are characterized as pit-halo craters, which have a rounded central pit surrounded by 'ejecta', whereas others are characterized as flat-floored craters, surrounded by steep cliffs. Both types

(a)   (b)

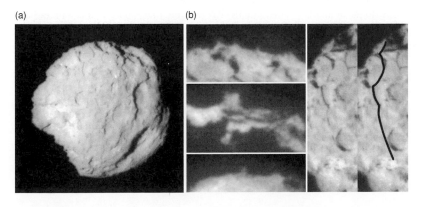

Figure 10.37 Images of Comet 81P/Wild 2 taken by the Stardust spacecraft highlight the diverse features that make up its surface. (a) A full view of the comet shows the numerous depressions. (b) These higher resolution images show a variety of small pinnacles and mesas on the limb of the comet (left side), and a 2 kilometer long scarp on the right (outlined by the black line on the rightmost image). (NASA/JPL-Caltech, PIA06285, PIA06284)

(a)   (b)

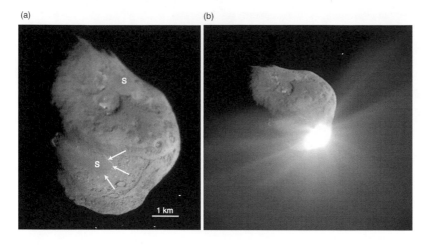

Figure 10.38 Deep Impact imaged Comet 9P/Tempel 1 before (a) and after (b) impact. (a) This composite image was constructed by scaling all images to 5 m/pixel, and aligning images to fixed points. The impact site has the highest resolution because images were acquired by the impactor spacecraft until about 4 seconds before impact. Smooth areas are indicated with the letter S, and the arrows highlight a bright (due to viewing geometry) scarp, which shows that the smooth area is elevated above the rough terrain. (NASA/JPL/UMD, PIA02142, PIA02137)

of craters can be reproduced by laboratory cratering experiments, but both kinds require the target to have substantial strength, i.e., this comet must be a cohesive body, not a rubble pile. Features reminiscent of *pinnacles, spires*, or *hoodoos* (like a pinnacle, but with a variable thickness, like a totem pole), tens of meters to over 100 m high, are visible on the limb (Fig. 10.37). These features might be erosional remnants, where the environment has been eroded away through, e.g., sublimation. A similar process may have led to the formation of mesas on 19P/Borrelly, where activity clearly seems associated with some of the mesas. The pinnacles could also be fumarole conduits, in analogy to volcanic processes. On comets, however, they must be formed by cryogenic geyser processes, where the escaping vapors 'line' and thereby harden the conduit with dust and less-volatile materials.

Images of 9P/Tempel 1 look quite different. Several dozen circular features, 40–400 m in diameter, cover the surface. These show a size distribution consistent with impact crater populations. More intriguing are the smooth areas, some bounded by scarps tens of meters high. One such area appears to be eaten away at the edges, revealing another, older, layer underneath. The nucleus of 9P/Tempel 1 seems to be layered with geologic strata of uncertain origin. Impacts, sublimation, mass wasting, and ablation are all important in shaping the morphological features that are so prominent on this and other comets.

Before the encounters with Comets 9P/Tempel 1 and 81P/Wild 2, the structural model of a cometary nucleus, though in essence still a 'dirty snowball', was converging to that of a collisionally processed or primordial rubble pile of icy planetesimals (Fig. 10.39). Now modelers envision comets as layered structures, known as the *talps* (Thin Active Layers on a Passive Substrate), composed of a core

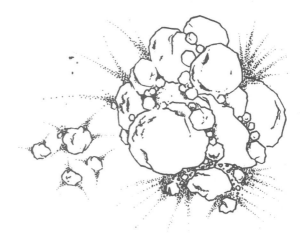

Figure 10.39 A schematic representation of a cometary nucleus according to the rubble-pile model, in which the individual fragments are lightly bonded by thermal processing or sintering. (Weissman 1986)

and piles of randomly stacked layers. Perhaps a nucleus is composed of some kind of combination of these models. The spacecraft images also make it abundantly clear that comets are a diverse population of objects, some of which have practically zero internal strength, and others have considerable cohesiveness.

10.7 Comet Formation

The failure to observe comets on significantly hyperbolic orbits (Fig. 10.3), unless they have recently been perturbed onto escape orbits by the gravitational pull of the planets, presents a convincing argument that comets are bona fide members of and originated within the Solar System. The volatile nature of comets implies that they could not have formed close to the Sun. Although comets are regarded as the most pristine objects in our Solar System, they do change on timescales of eons (§10.6.2), which makes it challenging to extract information from observations of comets about conditions in the primitve solar nebula.

10.7.1 Constraints from Chemistry

Comets are a compositionally diverse group of bodies, as evidenced by large variations in dust-to-gas ratios, the relative abundances of different molecular species (§10.4) and apparent changes in brightness and composition over time. Comets have, however, one thing in common: they are all composed of ices (water-ice and more volatile species) and dust. Among cometary dust grains, we find silicates, the much more volatile CHON particles, and extremely refractory minerals, such as have been found in CAIs embedded in carbonaceous chondrites.

The most volatile species detected in comets are S_2, N_2^+, and CO. The maximum ambient temperature in the region where the comets formed can be estimated from the condensation temperatures of these gases, which are 20 K for S_2, 22 K for N_2 (the parent molecule of N_2^+), and 25 K for CO, if these gases condensed directly. These values, together with the low spin temperature for water ($\sim$30 K, §10.4.3.6), suggest that comets formed in regions where the temperature was $\lesssim$30 K, which corresponds to a heliocentric distance in the solar nebula $r_\odot > 20$ AU, i.e., beyond the orbit of Uranus.

The (relative) abundances of hydrocarbons, formaldehyde, and methanol, as well as many isotope ratios and the noble gas abundances (§10.4.3), suggest that cometary ices formed in the outer regions in the primitive solar nebula. However, comets are clearly composed of material that condensed in many different regions within the solar nebula, plus some interstellar grains (§10.4.4). Clearly, radial mixing must have been efficient, and some interstellar grains must have entered the solar nebula without being vaporized. The various isotope ratios, including the D/H ratio and oxygen isotopes, as well as the high-temperature minerals can probably be reconciled with a formation model that includes physical radial mixing and limited reprocessing of interstellar water in the solar nebula.

10.7.2 Dynamical Constraints

There is broad agreement that two comet reservoirs exist: the Kuiper belt (including the scattered disk) and the Oort cloud. The composition of comets places their origin in the outer regions of the planet-forming disk. Kuiper belt objects on nearly circular orbits likely formed close to

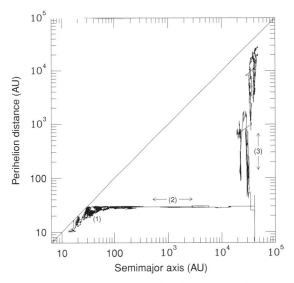

Figure 10.40 The dynamical evolution of an object as it evolves into the Oort cloud. The object began on a nearly circular orbit between the giant planets. In the initial phase of the evolution (1), the object remains in a moderate eccentricity orbit in the giant planet region. Neptune eventually scatters it outward, after which it undergoes a random walk in inverse semimajor axis (2). When the orbit becomes almost parabolic, the galactic tidal force can raise its perihelion above the planetary region (3). (Levison and Dones 2007)

their present locations. Since models of star formation imply that the density of gas and dust in the Oort cloud were much too small for planetesimals to form, Oort cloud comets likely formed in or near the region now inhabited by the giant planets. Objects now in the Oort cloud and the scattered disk formed on near-circular orbits, which were subsequently perturbed by the giant planets.

Dynamical simulations show that as long as the giant planets provide the dominant perturbations, the small body's perihelion remains within the planetary region and the inclination of its orbit does not change much, i.e., the body stays near the ecliptic plane. When the body reaches distances of over 10 000 AU, perturbations from the tidal pull of the galaxy can lift its perihelion out of the planetary region, and the body can thus be 'stored' in the Oort cloud (Fig. 10.40). Stellar perturbations can also lift perihelia. At present, the galactic field is the primary perturber of comets within the Oort cloud (§10.2.2), but if the giant planets formed when the Solar System was embedded in a dense star cluster (a common environment for young stars), then perturbations from nearby stars would have been much larger. Such enhanced stellar perturbations could have detached the orbits of bodies with smaller aphelia from planetary perturbations, forming the inner Oort cloud and explaining the anomalous orbit of Sedna (§9.1), with $a = 468$ AU and $q = 76$ AU.

The Oort cloud and Kuiper belt have survived as comet reservoirs for over four billion years. Icy bodies in these reservoirs occasionally are perturbed into orbits that bring them into the planetary region. The galactic tide, passing stars, and giant molecular clouds provide the primary perturbations for bodies in the Oort cloud, whereas resonant perturbations of the outer planets (sometimes aided by orbit-altering collisions) dominate for Kuiper belt objects, or more precisely for KBOs in the scattered disk, SDOs (§9.1).

When these icy bodies approach the Sun, their most volatile constituents sublimate and evolve off the nucleus, taking along with them more refractory dust and producing comets which may be spectacular in their visual appearance. Most comets are quickly ejected from the Solar System as a result of gravitational perturbations by the planets. Some comets crash into the Sun, while others end their active lives releasing all of their volatiles or completely disintegrating. A small minority of comets collide with planets. The source regions of comets are gradually being depleted, but the average rate at which comets are being supplied to the planetary region will probably drop by at most a factor of a few between the present epoch and the end of the Sun's main sequence (hydrogen burning) lifetime six billion years hence.

Further Reading

We recommend the following books on comets:

Krishna Swamy, K.S., 1986. *Physics of Comets*. World Scientific Publishing Co. Pte. Ltd., Singapore. 273pp.

Huebner, W.F., Ed., 1990. *Physics and Chemistry of Comets*. Springer-Verlag, Berlin. 376pp.

Festou, M.C., H.U. Keller, and H.A. Weaver, 2004. *Comets II*. University of Arizona Press, Tucson, 733pp.

We recommend the review papers by Brandt, and by Levison and Dones in McFadden, L., P. R. Weissman, and T.V. Johnson, Eds., 2007. *Encyclopedia of the Solar System*, 2nd Edition. Academic Press, San Diego. 982pp.

The University of Arizona Press series on *Protostars and Planets* has several good reviews, including the article:

Wooden, D., S. Desch, D. Harker, H.-P. Gail, and L. Keller, 2007. Comet grains and implications for heating and radial mixing in the protoplanetary disk. In *Protostars and Planets V*. Eds. B. Reipurth, D. Jewitt, and K. Keil. University of Arizona Press, Tucson, pp. 815–830.

Special issues presenting results of spacecraft encounters with comets have been published in *Science* and *Icarus*:

Deep Impact: *Science*, **310** #5746 (2005), and *Icarus*, **187** #2 and **191** #1 (2007).

Stardust. I: Encounter with C/Wild 2. *Science*, **304** #5678 (2004), and II: Stardust samples. *Science*, **314** #5806 (2006).

The most recent catalog of cometary orbits is published by:

Marsden, B.G., and G.V. Williams, 2008. *Catalogue of Cometary Orbits*, 17th Edition. The International Astronomical Union, Minor Planet Center and Smithsonian Astrophysical Observatory, Cambridge, MA.

Problems

10.1.E Estimate the largest negative value of $1/a_0$ that nongravitational forces are reasonably likely to produce for a comet that is initially in heliocentric orbit and is not subjected to significant planetary perturbations. You may assume that the comet's initial orbit is so eccentric that it may be taken as parabolic and that the comet outgasses 0.1% of its mass, with a 10% asymmetry and that this outgassing occurs in a single burst at perihelion, which is at 0.2 AU.

10.2.I Draw a theoretical histogram of $1/a_0$ for interstellar comets. You may assume that the velocity distribution (or, more precisely, the distribution of speed, since direction does not matter) of interstellar comets relative to the Sun at 'infinity' is comparable to that of neighboring stars, which can be approximated by a Maxwellian of mean velocity 30 km s^{-1}, i.e.,

$$N(v) \propto \frac{v}{v_*} e^{-(v/v_*)^2}, \tag{10.22}$$

where $v_* = 30$ km s^{-1}.

10.3.E Draw a histogram of $1/a$ for short-period comets, using data in a recent edition of Marsden and Williams's comet catalog, or on the web, via: http://cfa-www.harvard.edu/iau/ or http://pdssbn.astro.umd.edu/.

10.4.I Estimate the frequency at which we would expect to observe interstellar comets. Assume that the Oort cloud contains 10^{12} comets that are large enough to be observable if they were to reach the inner Solar System, and that five times as many comets have been ejected from the Solar System over time. Postulate that the number of comets produced and released by a planetary system is proportional to the star's mass, and that our Solar System is average in this regard. Furthermore, assume that the density of stellar mass in the solar neighborhood is 0.065 M$_\odot$ parsec^{-3} (1 parsec = 3.1×10^{18} cm), the typical velocity of a comet relative to the Sun 'at infinity' is comparable to

that of neighboring stars, 30 km s^{-1}, and a comet must come within 2 AU of the Sun to be visible.
(a) Neglecting gravitational focusing by the Sun, i.e., approximating the trajectories of the comets by straight lines.
(b) Including gravitational focusing by the Sun, i.e., considering the comets to be traveling on hyperbolic orbits about the Sun.
(c) Repeat parts (a) and (b) assuming that we can observe all comets that approach to within 5 AU of the Sun.

10.5.E The magnitude (brightness) of a body is related to its flux density via the first part of equation (10.4). Calculate the apparent magnitude of a Kuiper belt object at a heliocentric distance of 40, 70, and 150 AU, assuming a visual albedo of 0.04 and a radius of 150 km. (Hint: What is the solar flux at these distances?)

10.6.I Use diagrams and/or equations to show that under the assumptions of the standard (symmetric) model of nongravitational forces, the tangential term of the acceleration resulting from cometary outgassing can produce a secular change in a comet's orbital period, but the radial term cannot.

10.7.E Show that if seasonal variations break the symmetry about perihelion of a comet's outgassing, then the radial component of the nongravitational force can secularly alter the comet's orbital period.

10.8.E Suppose a comet has a velocity of 40 km s^{-1} at perihelion. The perihelion distance is 1 AU. Calculate the aphelion distance, the velocity of the comet at aphelion, and the orbital period of the comet.

10.9.E An asteroid and a comet have the same apparent brightness while both are at $r_\Delta = 2$ AU and $r_\odot = 3$ AU. At a later time they are both observed at $r_\Delta = 2$ AU and $r_\odot = 2$ AU. Which object is brighter, and by approximately how much?

10.10.E The apparent visual magnitude of Comet X is $m_v = 21.0$ when it is discovered. At this time $r_\odot = r_\Delta = 10$ AU.
(a) Calculate the absolute magnitude of the comet, H_{10}.
(b) Estimate m_v when the comet reaches perihelion at $r_\odot = 0.3$ AU; at this time $r_\Delta = 1$ AU.

10.11.E Calculate the surface temperature of a comet with Bond albedo $A_b = 0.1$ at a heliocentric distance of 15 AU.

10.12.I At a distance $r_\odot = 1$ AU, outgassing is dominated by H_2O, and a comet's surface temperature is 190 K. The latent heat of sublimation $L_s \approx 5 \times 10^{11}$ erg mole^{-1}. Assume the Bond albedo $A_b = 0.1$ and the infrared emissivity $\epsilon_{ir} = 0.9$.
(a) Derive equation (10.6), and use it to determine the gas production rate of this comet.
(b) Calculate the thermal expansion velocity of the gas, and compare this to the escape velocity of a typical comet (radius 1–10 km).

10.13.E (a) Perform the integration in equation (10.8) and solve for the collisional radius $\mathcal{R}_c$.
(b) Use your result to calculate the collision radius of 1P/Halley's comet at 1 AU from the Sun. Assume the comet outgasses $\sim 10^{30}$ molecules s^{-1}, the terminal expansion velocity $v = 1$ km s^{-1} and a typical molecular radius is 0.15 nm.

10.14.E A comet's perihelion distance is 1 AU, and its aphelion distance is 15 AU. In the following we make a very, very crude calculation of the average rate of shrinkage of the comet.
(a) Calculate the comet's orbital period.
(b) Calculate how many meters of ice the comet will lose each time it orbits the Sun. (Hint: In order to simplify the calculations, you may assume that ice sublimates off the comet's surface during 1/10 its orbital period, that the average cometary distance over that period is 1.5 AU and that the density of the cometary ice is 0.6 g cm^{-3}.)

10.15.I Assume the comet from Problem 10.14 is active inside 3 AU and inert at large heliocentric distances. By making use of Kepler's second and third laws, together with the gas production rate as a function of heliocentric distance (eq. 10.6), calculate how many meters of ice the comet loses each orbit.

10.16.E A comet consists primarily of water-ice; when the ice sublimates, the water molecules flow off the surface at the thermal expansion velocity. Assume the comet to be at $r_\odot = 0.6$ AU, and the sublimation temperature of water to be 200 K.
(a) Calculate the thermal expansion velocity.
(b) The typical lifetime of H_2O molecules is 6×10^4 s, for OH it is 2×10^5 s, and for H it is 10^6 s. Assume the outflow velocity of OH to be equal to that of H_2O, and for H it is on average 12 km s^{-1}. Calculate the typical sizes of the H_2O, OH, and H comae.

(c) Suppose the comet is perfectly spherical and homogeneous, with a radius of 10 km. Calculate the collisional radius, $\mathcal{R}_c$.

(d) When the particle density exceeds 1×10^6 molecules cm^{-3}, collisions with OH molecules are so numerous that these molecules are quickly thermalized. Calculate the radius inside of which the coma density exceeds 1×10^6 molecules cm^{-3}.

(e) Calculate the time between collisions at a distance of 20 km from the center of the comet; repeat your calculations at distances of 100 km, 1000 km, and 10^4 km.

(f) Compare your answers from (c) and (d), together with (e), and comment on your results.

10.17.I A comet at a heliocentric distance of 1 AU shows a gas production rate $\mathcal{Q} = 10^{29}$ molecules s^{-1}. The outgassing is dominated by H_2O.

(a) Plot the H_2O number density as a function of distance from the comet. Assume an outflow velocity of 1 km s^{-1} and a lifetime for H_2O molecules of 6×10^4 s.

(b) Assume that all H_2O molecules dissociate into OH and H, and OH has a lifetime of 1.7×10^5 s. Plot, according to the Haser model, the OH number density as a function of distance from the comet.

(c) How, qualitatively, does $N_d(r)$ differ in the vectorial model from that given by the Haser model? Describe approximately using a graph and associated text.

10.18.I Assume the OH brightness of a comet is proportional to N_d, the number of OH molecules, and that the comet under consideration is not resolved by the telescope. We receive the OH emission at a frequency of 1667.0 MHz. There are 15 frequency channels, centered at 1667.0 MHz, which is the rest frequency of the comet (i.e., the comet's velocity is zero at 1667.0 MHz). Each channel is 1.11 kHz wide. Calculate the line profile (in relative numbers, as a function of frequency and gas flow velocity) you expect for the comet from Problem 10.16, assuming the Haser model represents the gas outflow accurately.

10.19.I Using the assumptions of the Haser model, derive equation (10.12b) from equation (10.12a). (Hint: Follow the total flux of molecules moving through concentric spheres centered on the comet.)

10.20.E Comet X is observed in the CO (151.0 nm), CI (165.7 nm), and OH (309.0 nm) transitions,

while at a geocentric distance of $r_\Delta = 0.8$ AU and heliocentric distance of $r_\odot = 0.4$ AU. The g-factors at $r_\odot = 1$ AU for the three molecules are 2.2×10^{-7}, 2.5×10^{-5}, and 1.2×10^{-3}, respectively. The lifetimes of the various molecules (at $r_\odot = 1$ AU) are approximately: 1×10^6, 2.5×10^5, and 1.6×10^5 seconds, respectively. The observed brightnesses are 30, 770, and 105 000 photons s^{-1} cm^{-2} for CO, CI, and OH, respectively. Assume that the comet is unresolved in the telescope beam, and that it subtends an angle of 10 arcminutes on the sky.

(a) Calculate the column density for each of the three species.

(b) Assume that the comet is spherically symmetric and outgassing equally in all directions. Determine the gas production rate, $\mathcal{Q}$, for all three compounds. Compare the results and comment on the similarities/differences.

10.21.E Consider a comet in orbit around the Sun. The molecules evolve off the nucleus at 1 km s^{-1}.

(a) Sketch the OH line profiles at 1667 MHz that you would expect to observe when the radial component of the comet's heliocentric velocity $v_r = -20$ km s^{-1}. (Hint: Use the graph for the inversion measure shown in Fig. 10.19.)

(b) Repeat for $v_r = -8$ km s^{-1}.

(c) Repeat for $v_r = -14$ km s^{-1}.

10.22.E Derive equation (10.11). (Hint: Assume that the grain is spherical, and that it radiates uniformly in all directions.)

10.23.E Calculate approximately the standoff distance of the bow shock of Halley's comet at $r_\odot = 0.9$ AU. Assume an icy comet with a production rate of 10^{30} molecules s^{-1}, a solar wind density of 5 protons cm^{-3}, and velocity of 400 km s^{-1}. Estimate the ionization rate from molecular lifetimes.

10.24.I Comet X is on a nearly parabolic orbit with a perihelion at 0.5 AU. Calculate the (generalized) eccentricity of grains released near perihelion as a function of β. Give a formula for the separation between the grains and the nucleus as a function of β and t that is valid for the first few days following release.

10.25.I Derive equation (10.10a). (Hint: Set the upward force from gas pressure equal to the downward force from gravity and solve for $R_{d,max}$.)

10.26.E Calculate the size of the largest icy particle that can be released by gas drag from the surface of an icy comet with a radius of 1 km at heliocentric distances of 3, 1, and 0.3 AU. List and explain all of your assumptions.

10.27.E Make a very crude estimate of the relative brightness of a comet's coma and nucleus by assuming that the coma consists of 10 μm grains which emanate from a uniform 10 cm thick layer on the comet's surface, the composition of the comet is 50% ice and 50% dust and the albedo of the grains is the same as that of the nucleus. Using your result, comment on the possibility of observing the nucleus of an active comet from the ground.

11 Planetary Rings

It (Saturn) is surrounded by a thin flat ring, nowhere touching, and inclined to the ecliptic.
Christiaan Huygens, published in Latin in anagram form in 1656

Each of the four giant planets in our Solar System is surrounded by flat, annular features known as *planetary rings*. Planetary rings are composed of vast numbers of small satellites, which are unable to accrete into large moons because of their proximity to the planet.

When Galileo Galilei first observed Saturn's rings in 1610, he believed them to be two giant moons in orbit about the planet. However, these 'moons' appeared fixed in position, unlike the four satellites of Jupiter which he had previously observed. Moreover, Saturn's 'moons' had disappeared completely by the time Galileo resumed his observations of the planet in 1612. Many explanations were put forth to explain Saturn's 'strange appendages', which grew, shrank, and disappeared every 15 years (Fig. 11.1a). In 1656, Christiaan Huygens finally deduced the correct explanation, that Saturn's strange appendages are a flattened disk of material in Saturn's equatorial plane, which appear to vanish when the Earth passes through the plane of the disk (Fig. 11.1b).

For more than three centuries, Saturn was the only planet known to possess rings. Although Saturn's rings are quite broad, little structure within the ring system was detected from Earth (Fig. 11.2). Observational and theoretical progress towards understanding the physics of planetary rings was slow. But then, in March of 1977, an occultation of the star SAO 158687 revealed the narrow opaque rings of Uranus (Fig. 11.3) and launched a golden age of planetary ring exploration. The Voyager spacecraft first imaged and studied the broad but tenuous ring system of Jupiter in 1979 (§11.3.1). Pioneer 11 and the two Voyagers obtained close-up images of Saturn's spectacular ring system in 1979, 1980, and 1981 (Fig. 11.4; §11.3.2). Neptune's rings, whose most prominent features are azimuthally incomplete arcs, were discovered by stellar occultation in 1984. Voyager 2 obtained high-resolution images of the rings of Uranus in 1986 (§11.3.3)

and the rings of Neptune in 1989 (§11.3.4). Technological advances allowed ground-based and Earth-orbiting studies of planetary rings over a far larger range of wavelengths and with much higher precision than previously attained. The Galileo spacecraft obtained high-resolution images of Jupiter's rings from jovian orbit in the late 1990s, and the New Horizons spacecraft obtained valuable images from a wide variety of phase angles as it swung by Jupiter in 2007, en route to Pluto. The Cassini spacecraft began an intensive and multifaceted study of Saturn's rings from saturnian orbit in 2004. HST observations and advanced technology telescopes on the ground have provided new information on all four planetary ring systems. Finally, our theoretical understanding of rings advanced by leaps and bounds. Despite these advances, we now have more outstanding questions concerning rings than researchers had in the mid 1970s!

In this chapter, we summarize our current observational and theoretical understanding of planetary rings. We begin with an explanation of why rings exist, and why they are generally located much closer to planets than are large moons (Fig. 11.5). A more detailed observational summary is then presented, followed by theoretical models for some of the features observed. We conclude with a discussion of the evolution of planetary ring systems and models of planetary ring formation.

11.1 Tidal Forces and Roche's Limit

The strong tidal forces close to a planet lead orbital debris to form a planetary ring rather than a moon. The closer a moon is to a planet, the stronger the tidal forces that it is subjected to. If it is too close, then the difference between the gravitational force exerted by the planet on the point of the moon nearest to (and furthest from) the planet from that exerted on the center of the moon is stronger than the

(a)

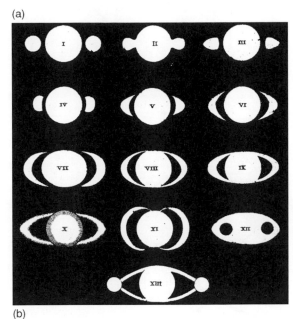

(b)

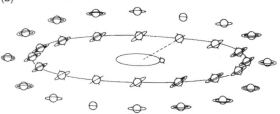

Figure 11.1 (a) Seventeenth-century drawings of Saturn and its rings. I: Galileo, 1610; II: Scheiner, 1614; III: Riccioli, 1641 and 1643; IV–VII: Havel, theoretical forms; VIII, IX: Riccioli, 1648–50; X: Divini, 1646–48; XI: Fontana, 1636; XII: Biancani, 1616; Gassendi, 1638–39; XIII: Fontana and others at Rome, 1644–45. (Huygens 1659) (b) Cartoon views of Saturn and its rings over one saturnian orbit according to Huygens's model.

Figure 11.2 Saturn and its rings over one-half of a saturnian orbit, as seen in ground-based photographs taken in the middle of the 20th century.

moon's self-gravity. Under such circumstances, the moon is ripped apart, unless it is held together by mechanical strength, and a planetary ring results.

In order to understand tidal disruption more quantitatively, we make the following assumptions:

(1)   The system consists of one large primary body (the planet) and one small secondary body (the moon).

(2)   The orbit is circular, the rotational period of the moon is equal to its orbital period, and the moon's obliquity is nil. (These assumptions make the analysis far simpler, because the problem becomes stationary in a rotating frame.)

(3)   The moon is spherical, and the planet can be treated as a point mass.

(4)   The moon is held together by gravitational forces only.

The 'external' forces per unit mass on material in orbit about a planet of mass M_p are gravity:

$$\mathbf{g}_\rho = -\frac{GM_p}{r^2}\hat{\mathbf{r}}, \tag{11.1}$$

and centrifugal force:

$$\mathbf{g}_n = n^2 r\hat{\mathbf{r}}, \tag{11.2}$$

where the origin is at the center of the planet and n is the angular velocity of the system. Steady state in the frame rotating with the system gives

$$n^2 r\hat{\mathbf{r}} - \frac{GM_p\hat{\mathbf{r}}}{r^2} = 0, \tag{11.3a}$$

therefore:

$$n^2 = \frac{GM_p}{r^3}. \tag{11.3b}$$

Note that equation (11.3b) implies Kepler's third law for the case of circular orbits.

The sum of the gravitational force and the effect of the rotating frame of reference ('centrifugal force') is referred to as the *effective gravity*; the local effective gravity vector points normal to the *equipotential surface* in the rotating frame. The effective gravity, $\mathbf{g}_{\text{eff}}$, felt by an object that is

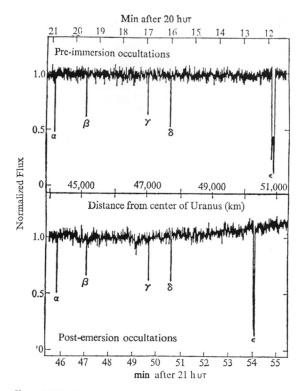

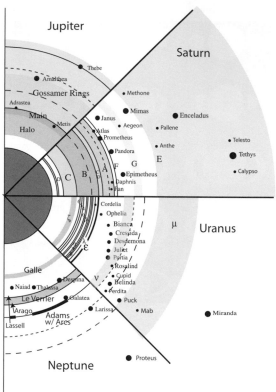

Figure 11.3 Lightcurve of the star SAO 158687 as it was observed to pass behind Uranus and its rings. Dips in the lightcurves corresponding to the occultation of the star by five rings are clearly seen both prior to immersion and following emersion of the star from behind the planet. Four of these pairs of features are symmetric about the planet, but the location, depth and duration of the outermost pair imply that the ϵ ring is both noncircular and nonuniform. (Adapted from Elliot *et al.* 1977)

Figure 11.5 Diagram of the rings and inner moons of the four giant planets. The systems have been scaled according to planetary equatorial radius. The long-dashed curves denote the radius at which orbital motion is synchronous with planetary rotation. The short-dashed curves show the location of Roche's limit for particles of density 1 g cm^{-3}. (Courtesy Judith K. Burns)

Figure 11.4 COLOR PLATE This approximately natural-color image shows Saturn and its rings as seen by the Cassini spacecraft in October 2004. The resolution is 38 km/pixel and the phase angle is 72°. A total of 126 images taken through red, green, and blue filters were used to produce this mosaic. The pronounced concentric gap in the rings, the Cassini division, is a 3500 km wide region that is much less populated with ring particles than are the brighter B and A rings to either side. Note the subtle color variations across the rings and the thread-like F ring. The shadows of the rings darken parts of Saturn's cold, blue northern hemisphere, and Saturn shadows part of the rings. (NASA/JPL/CICLOPS, PIA06193)

at a distance r from the planet's center and traveling on a circular orbit at semimajor axis a is

$$\mathbf{g}_{\text{eff}} = GM_{\text{p}} \left(\frac{r}{a^3} - \frac{1}{r^2} \right) \hat{\mathbf{r}}. \tag{11.4}$$

The (effective) tidal force upon such a body is

$$\frac{d\mathbf{g}_{\text{eff}}}{dr} = GM_{\text{p}} \left(\frac{1}{a^3} + \frac{2}{r^3} \right) \hat{\mathbf{r}} \approx \frac{3GM_{\text{p}}}{a^3} \hat{\mathbf{r}}, \tag{11.5}$$

where the approximation $r^3 \approx a^3$ has been used in the last step.[1] Note that equation (11.5) differs from equation (2.40) because it also includes a contribution from the

[1] The approximation $r \approx a$ is valid at this stage in the derivation as long as the size of the moon is much smaller than that of its orbit, $R_s \ll a$. It could not have been used prior to taking the derivative, as this would have omitted the gradients that are the essence of the tidal force.

centrifugal force. The moon's self-gravity just balances the tidal force at the surface of the moon when

$$\frac{GM_s}{R_s^2} = \frac{3GM_pR_s}{a^3}, \tag{11.6}$$

where the subscript s refers to the satellite (moon). This occurs at a planetocentric distance of

$$\frac{a}{R_p} = 3^{\frac{1}{3}} \left(\frac{\rho_p}{\rho_s}\right)^{\frac{1}{3}} = 1.44 \left(\frac{\rho_p}{\rho_s}\right)^{\frac{1}{3}}. \tag{11.7}$$

In the above derivation, we made a number of simplifying assumptions. Let us now review the accuracy of these assumptions in order to assess the applicability of our calculations:

(1)   We used the small moon/large planet approximation in order to neglect the influence of the moon on the planet and to neglect terms containing higher powers of the ratio of the radius of the moon to its orbital semimajor axis. This assumption is thus very accurate for bodies within our Solar System.

(2)   All known inner moons have low eccentricities, so the approximation of circular orbits is very good. All moons near planets for which rotation rates have been measured are in synchronous rotation and have low obliquity. Young moons that have not had time to be tidally despun, and thus rotate rapidly, would be less stable.

(3)   Although the giant planets are noticeably oblate, the departures of their gravitational potentials from those of point masses have only an $\mathcal{O}(1\%)$ effect on these tidal stability calculations. A much larger effect results from moons being stretched out along the planet–moon line due to the planet's gravitational tug (Fig. 2.20). This stretching brings the tips of the moon further from its center, which both decreases the magnitude of self-gravity and increases the tidal force. In 1847, Roche performed a self-consistent analysis for a liquid (fully deformable) moon and obtained

$$\frac{a_R}{R_p} = 2.456 \left(\frac{\rho_p}{\rho_s}\right)^{\frac{1}{3}}. \tag{11.8}$$

Such a marginally gravitationally bound fluid moon would fill its entire Roche lobe, extending to its inner Lagrangian point, L_1, a distance of one Hill radius, R_H, from the moon's center (§2.2.3). The shape of such a moon would be intermediate between that of an almond and a sphere, with a volume equal to about one-third that of a sphere of radius R_H. The location a_R is known as *Roche's limit* for tidal disruption.

(4)   Most small bodies have significant internal coherence, e.g., small moons are not always spherical. Internal friction and/or tensile strength of small bodies allows

moons smaller than $\sim$100 km radius to be stable somewhat inside Roche's limit. Ring particles, which typically are so small that internal strength exceeds self-gravity by orders of magnitude provided the particles are not loose aggregates, can remain coherent well inside Roche's limit.

The concept of Roche's limit explains in a semi-quantitative manner why we observe rings near giant planets, small moons a bit further away, and large moons only at greater distances (Fig. 11.5). However, the interspersing of some rings and moons implies that other factors are important in determining the precise configuration of a planet's satellite system. We will return to theories on the origin and evolution of ring/moon systems in §11.7.

11.2 Flattening and Spreading of Rings

A ring particle orbiting a planet passes through the planet's equatorial plane twice each orbit, unless its trajectory is diverted by a collision with another particle or by the ring's self-gravity. The average number of collisions that a particle experiences during each vertical oscillation is a few times as large as the optical depth of the rings, τ (Problem 11.6; collisions can be much more frequent in physically thin but optically thick self-gravitating rings). Typical orbital periods for particles in planetary rings are 6–15 hours. As τ is $\mathcal{O}(1)$ in the most prominent rings of Saturn (and Uranus), collisions are very frequent. Collisions dissipate energy but conserve angular momentum. Thus, the particles settle into a thin disk on a timescale

$$t_{flat} = \frac{\tau}{\mu}, \tag{11.9}$$

where μ is the frequency of the particles' vertical oscillations. An oblate planet exerts torques which alter the orbital angular momenta of orbiting particles. These torques cause inclined orbits about oblate planets to precess (see §2.5.2). When coupled with collisions among ring particles, a secular transfer of angular momentum between the planet and the ring can result. Only the component of the ring's angular momentum that lies along the planet's spin axis is conserved. Any net angular momentum of the disk parallel to the planet's equator is quickly dissipated (i.e., returned to the planet via the torque between the planet's equatorial bulge and the ring). As collisions at high speeds damp relative motions rapidly, the ring settles into the planet's equatorial plane on a timescale of a few orbits (or $\sim\tau^{-1}$ orbits, if $\tau \ll 1$).

Several mechanisms act to maintain a nonzero thickness of the disk. Finite particle size implies that even particles on circular orbits in a planet's equatorial plane collide with a finite velocity, when an inner particle catches up with a particle further out that moves less rapidly. Unless the collision is completely inelastic, i.e., unless the two particles stick, some of the energy involved goes into random particle motions. The ultimate consequence of these collisions is a spreading of the disk. Viewed in another way, spreading of the disk is the source of the energy required to maintain particle velocity dispersion in the presence of inelastic collisions. Gravitational scatterings between slowly moving particles is another process that converts energy from ordered circular motions to random velocities, in this case without the losses resulting from the inelasticity of physical collisions. External energy sources may also contribute to maintaining random velocities of particles against energy loss from inelastic collisions, especially near strong orbital resonances with moons (§11.4).

As a result of continuing collisions, rings spread in the radial direction. Since diffusion is a random walk process, the diffusion timescale is

$$t_d = \frac{\ell^2}{\nu_v}, \tag{11.10}$$

where ℓ is the radial length scale (the width of the ring, or of a particular ringlet). The viscosity, ν_v, depends on the spread in particle velocities, c_v, and the local optical depth approximately as

$$\nu_v \approx \frac{c_v^2}{2\mu}\left(\frac{\tau}{1+\tau^2}\right) \approx \frac{c_v^2}{2n}\left(\frac{\tau}{1+\tau^2}\right). \tag{11.11}$$

Equation (11.11) was derived assuming that ring particles behave like a diffuse gas, and so is valid only if particles move several times their diameters relative to one another between collisions. If the filling factor is large, i.e., if typical distances between particles are not much larger than particle sizes, then this approximation is not valid. The viscosity of dense, high τ, rings depends on other factors, including particle size and the *coefficient of restitution* (ratio of the relative speed of two particles immediately after a collision to that just prior to impact) for inelastic collisions.

Equation (11.10) with $\ell = 6 \times 10^9$ cm (the approximate radial extent of Saturn's main rings) and a viscosity of $\nu_v = 100$ cm^2 s^{-1}, yields a diffusion timescale comparable to the age of the Solar System. For smaller ℓ, timescales are much shorter (Problem 11.9). Even in regions of planetary rings where the viscosity is substantially lower than the value quoted above, viscous diffusion should be able to rapidly smooth out any fine-scale density variations unless other processes counteract viscosity. Most structure in optically thick planetary ring systems must therefore be actively maintained, except on the largest length scales, where it may have resulted from 'initial' conditions.

Although under most circumstances viscosity acts to wipe out structure, it is possible that in some regions of planetary rings a *viscous instability* may occur. A ring is unstable to clumping in the radial direction if the viscous torque, $\nu_v\sigma_\rho$, is a decreasing function of surface density:

$$\frac{d}{d\sigma_\rho}\left(\nu_v\sigma_\rho\right) < 0 \rightarrow \text{instability}. \tag{11.12a}$$

If the surface density is proportional to optical depth, then equations (11.11) and (11.12a) may be combined to yield the stability condition:

$$\frac{\tau}{c_v}\frac{dc_v}{d\tau} + \frac{1}{\tau^2 + 1} < 0 \rightarrow \text{instability}. \tag{11.12b}$$

As c_v and τ are positive, equation (11.12b) implies that a ring is viscously unstable if the velocity dispersion of the ring particles decreases sufficiently rapidly as optical depth is increased. Experimental results for low-velocity impacts of icy particles suggest that c_v does in fact decrease as τ increases. A steep-enough decrease would imply that more particles are able to diffuse from regions of lower optical depth into regions of higher optical depth than vice versa. A density perturbation would thus be amplified as a result of diffusion and 'ringlets' may be formed. If the divergence in optical depth tends to overshoot, rather than to approach, a nonuniform equilibrium configuration, then the instability is referred to as a *viscous overstability*. The viscous overstability occurs when viscous stresses vary with surface density in such a way that energy from the Keplerian shear of orbital motion can be directed into growing oscillations.

Since the second term in equation (11.12b) is always positive and decreases as τ increases, viscous instability or overstability is most likely to occur in regions of high optical depth, such as Saturn's B ring. As noted above, the viscosity in high τ regions may not be well approximated by equation (11.11), so the theory is not yet predictive.

Collective gravitational effects are important when the velocity dispersion of the particles is so small that *Toomre's stability parameter*,

$$Q_T \equiv \frac{\kappa c_v}{\pi G \sigma_\rho}, \tag{11.13}$$

is less than unity. Here κ is the particles' epicyclic (radial) frequency (for orbits near the equatorial plane of an oblate planet, κ is slightly smaller than n, see eqs. 2.34 and 2.35), and σ_ρ is the surface mass density of the rings. The dispersion velocity of the ring particles is related to the Gaussian scale height of the rings, H_z, via the formula

$$c_v = H_z\mu. \tag{11.14}$$

When $Q_T < 1$, the disk would be unstable to axisymmetric clumping of wavelength:

$$\lambda = \frac{4\pi G\sigma_\rho}{\kappa^2}. \qquad (11.15)$$

For parameters typical of Saturn's rings, λ is on the order of 10–100 m. However, numerical simulations show that clumping occurs for $Q_T \sim< 2$, and that these clumps stir particle velocities and thereby keep Q_T above unity.

11.3 Observations

Although the same basic physical processes govern the particles in all planetary ring systems, each system has its own distinctive character. Most rings lie inside or near the Roche limit. However, tenuous rings of ephemeral dust particles are also observed in less tidally hostile environments. The differences in dynamical structure of the planetary ring systems are apparent in casual inspection of Figures 11.6–11.10, and 11.17–11.19.[2] Saturn's ring particles have high albedo, whereas particles in other ring systems are generally quite dark. Particle sizes range from submicrometer dust to bodies large enough to be considered moons. Indeed, there is no fundamental demarcation line between large ring particles and small moons. An operational definition useful at present is that bodies viewed as individual objects are defined to be moons and given names, whereas (smaller) bodies detected only as a collective ensemble are known as ring particles. This definition will become obsolete when extremely high resolution images of rings revealing small bodies as distinct entities become available; at such time a new definition, probably based on the ability of a body to gravitationally clear a gap around itself (§11.4.3), will be needed.

Our knowledge of ring properties has been obtained almost exclusively from photons of various wavelengths that have been scattered, reflected, absorbed, or emitted by ring particles, although a small amount of data is available from ring absorption of charged particles and impacts of microscopic particles on spacecraft passing through very tenuous regions of planetary ring systems.

Observations of starlight and (spacecraft) radio signals that have passed through partially transparent rings, i.e., *stellar occultations* and *radio occultations* (§E.5), provide direct measurements of the optical depth of the rings along the observed line of sight. In the simplest model, rings can be approximated by a uniform disk of normal optical depth, τ, where particles are widely separated (compared to their sizes). For such a ring, the observed line of sight optical depth, τ_{sl}, is given by (§3.2.3.4)

$$\tau_{sl} = \ln(I_0/I) = \tau/\sin B_{oc} = \tau/\mu_\theta, \qquad (11.16)$$

where I_0 is the unoccluded star (or radio signal) intensity, I is the intensity of the observed starlight, and $\sin B_{oc}$, with B_{oc} the angle between the direction to the star and the ring plane, is a projection factor to convert from line-of-sight optical depth, τ_{sl}, to normal-incidence-angle optical depth, τ. The last equality in equation (11.16) uses the language of planetary atmospheres, $\mu_\theta \equiv \cos\theta$, with $\theta = 90° - B_{oc}$ the angle between the line of sight (i.e., towards the occulted star) and the normal to the ring plane (see Fig. 11.28a). Equation (11.16) is based on a model of the rings as a planar, homogeneous medium with nonzero and finite transparency, analogous to a cloud. For a uniform distribution of particles that are not in a monolayer, the factor $\sin B_{oc}$ corrects the optical depth for the length of the observed line of sight, which is always greater than the vertical thickness of the ring. If the positions of the ring particles are uncorrelated and the rings are azimuthally symmetric, and if individual particles are small compared to the area of the ring sampled in a single occultation integration period, then equation (11.16) gives the same value of the ring normal optical depth for all B_{oc}. However, deviations in the distribution of particles from a homogeneous axisymmetric distribution can result in computed values of τ being dependent upon viewing geometries and resolution. Low optical depth rings generally obey equation (11.16), whereas observed deviations from this simple behavior imply that high τ rings tend to be clumpy.

In this section, we summarize the properties of planetary ring particles and structure in planetary rings. Theoretical explanations for some of this structure are presented in §§11.4–11.6; the mechanisms responsible for creating/maintaining the particle size distribution are not well understood theoretically, but a few general principles are discussed together with the observations.

11.3.1 Jupiter's Rings

The jovian ring system is extremely tenuous, so many of the best images are taken when the rings are edge-on, when all particles merge into a single line in the camera. Under such viewing geometries, optically thin rings are

[2] Note that the differences between ring systems are generally substantially larger than they appear to be in most processed images. Camera exposures and a variety of image processing techniques are usually selected to compensate for variations in overall brightness and are often used to stretch and/or filter these data in order to make structure more apparent to the eye. Thus, processed images usually display rings as fairly prominent features with internal brightness variations of order unity, even though many actual rings are very tenuous and/or nearly uniform in brightness.

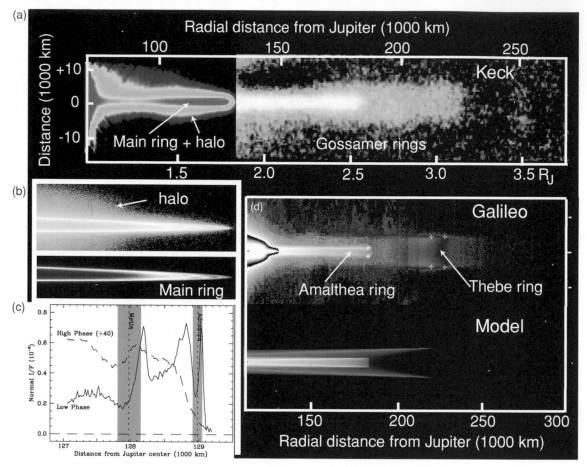

Figure 11.6 Images of Jupiter's ring system. (a) The entire ring system as imaged at 2.27 μm with the 10 m Keck telescope in 2002–2003 when the rings were edge-on. The 0.6″ seeing (full width at half maximum, FWHM) corresponds to a resolution of ~1800 km. The brightness contrast used for the portion of the image showing the gossamer rings (right side) is enhanced relative to that for the main ring + halo (left side). (Adapted from de Pater *et al.* 2008) (b) Galileo images of Jupiter's ring system processed to emphasize the ring halo (upper panel) and the main ring (lower panel). (NASA/Galileo, PIA01622) (c) A radial scan through Galileo images of Jupiter's main ring, in backscattered and forward scattered light. The gray bands indicate zones of width about 3 times the Hill radius (eq. 2.22) around the orbits of the embedded moons Metis and Adrastea. (Burns *et al.* 2004) (d) Upper panel: A mosaic of Jupiter's gossamer rings made from four Galileo images. Images were obtained through the clear filter (central wavelength = 0.611 μm, passband = 0.440 μm) from within Jupiter's shadow, in near-forward-scattered light (phase angle of 177–179° at an elevation of 0.15° – i.e., almost edge-on). The top and bottom edges of both gossamer rings are approximately twice as bright as their central cores, although this difference is subdued by the logarithmic scale used in displaying the image. The two gossamer rings have crosses showing the four extremes of the eccentric and inclined motions of Amalthea and Thebe. Lower panel: A model of debris rings formed from Amalthea and Thebe ejecta. Each ring is composed of material created continually at its source moon and decaying inward at a uniform rate, retaining its initial inclination but having nodes randomized. (Adapted from Burns *et al.* 1999)

much brighter compared to conditions when the rings are partially open. Jupiter's ring system consists of four principal components: the main ring, the halo, and two gossamer rings (Fig. 11.6 and Table 11.1). As Jupiter's rings appear much brighter in forward scattered light compared

to backscattered light, the surface area and optical depth are dominated by dust, even if most of the mass is in larger bodies. (See equations (3.60)–(3.62) for a discussion of forward versus backscattered light.) The *main ring*, with a normal optical depth $\tau \approx$ few $\times 10^{-6}$, is the most

Table 11.1 Properties of Jupiter's ring system [a]

	Halo[b]	Main ring	Amalthea ring	Thebe ring	Thebe extension
Radial location ($R_{\jmath}$)	1.4–1.71	1.72–1.806	1.8–2.55	1.8–3.10	3.1–3.8
Radial location (km)	100 000–122 400	122 400–129 100	122 400–181 350	122 400–221 900	221 900–270 000
Vertical thickness	$\sim 5 \times 10^4$ km	30–100 km	~ 2300 km	~ 8500 km	~ 9000 km
Normal optical depth	few $\times 10^{-6}$	few $\times 10^{-6}$	$\sim 10^{-7}$	$\sim 10^{-8}$	$\sim 10^{-9}$
Particle size	(sub)µm	broad distribution	broad distribution	broad distribution	

[a] Data from Ockert-Bell *et al.* (1999) and de Pater *et al.* (1999, 2008).

[b] Numbers quoted are based upon the Galileo data (visible light data, in forward scattered light). Relative to the main ring, the halo is much less bright and more spatially confined at longer wavelengths and in backscattered light.

prominent component, especially in backscattered light, the latter being indicative of a greater fraction of macroscopic material in this component of the rings.

Spacecraft and ground-based images reveal variations in the main ring's radial structure, and several Galileo and New Horizon images show intriguing azimuthal arc-like features. The main part of the main ring, the *main ring annulus*, is ~ 800 km wide, and is located between the orbits of Adrastea and Metis. In fact, it extends ~ 100 km beyond Adrastea, where the tiny moon itself appears to clear a gap (Fig. 11.6c). The ring is likely composed of a power-law differential particle size distribution (eq. 9.3) with $\zeta \approx 2$ for particles with radii <15 µm, and steeper for larger particles. 'Parent-sized bodies' (radii over ~ 5 cm), including Metis and Adrastea, are confined to this main ring annulus and make up $\sim 15\%$ of its optical depth. Most of the main ring's mass is contained in such large objects. The red color of the rings, very similar to that of the small satellites, can only be partially explained by light scattering off the ring's dust population. The larger particles must be distinctly red, like Metis, to fully explain the ring's color in backscattered light.

Radial profiles in forward and backscattered light are very different, as exemplified in Fig. 11.6. While backscattered-light profiles highlight the macroscopic bodies, profiles in forward scattered light reveal the distribution of tiny dust grains, presumably generated via collisions between, and micrometeorite impacts on, the macroscopic bodies that dominate the signal seen in backscattered light. The abundance of micron-sized dust appears to increase inside the orbit of Adrastea, throughout the main portion of the main ring, as expected for dust created in this region. The apparent vertical thickness of the main ring varies from $\lesssim 30$ km up to 100 km, depending on solar phase angle.

Interior to the main ring annulus is a ~ 4000 km broad extension, composed of micron-sized dust transported

inwards by Poynting–Robertson drag. The physically thin main ring stops at 1.71 $R_{\jmath}$, interior of which is the *halo*, which extends inwards to 1.4 $R_{\jmath}$. Its normal optical depth is very similar to that of the main ring, $\tau \approx$ few $\times 10^{-6}$. Although most halo particles are within a few thousand km of the ring plane, the halo's full extent is close to 40 000 km. The locations of the inner and outer boundaries of the halo coincide with *Lorentz resonances*. Lorentz resonances arise from electromagnetic forces on charged particles, and are discussed in §11.5.2. The particles within the halo probably had their inclinations increased by interactions with the planet's magnetic field at Jupiter's 3:2 Lorentz resonance, located at 1.712 $R_{\jmath}$. A second interaction with the 2:1 Lorentz resonance, at 1.407 $R_{\jmath}$, can perturb particle orbits into Jupiter's atmosphere. Inside of 1.41 $R_{\jmath}$ the particle density is too low to be detectable from the ground. Galileo images have revealed a faint extended halo closer to the planet.

The much fainter gossamer rings ($\tau \sim 10^{-7}$) consist of several parts (Fig. 11.6a,d): The *Amalthea ring* lies immediately interior to the orbit of Amalthea; seen edge-on, this ring is almost uniform in brightness in both forward and backscattered light. Interior to Thebe is the *Thebe ring*, fainter and thicker than the Amalthea ring. Exterior to Thebe one can, albeit barely, distinguish material out to $\gtrsim 3.8$ $R_{\jmath}$, with an intensity $\sim 10\%$ of that of the Thebe ring (Thebe is at 3.11 $R_{\jmath}$, Amalthea at 2.54 $R_{\jmath}$). In a high-resolution Galileo image (Fig. 11.6d), the upper and lower edges of the gossamer rings are much brighter than their central cores. The vertical location of the peak brightness of each of the gossamer rings, as well as the vertical extent seen in backscattered light, is proportional to distance from Jupiter's center. These characteristics imply that the particles originate from the bounding satellites (§11.6.3). Some 'clumping' of material just interior to the orbits of Amalthea and Thebe is indicative of the presence of larger particles, which also serve as source for ring material.

Table 11.2 Properties of Saturn's ring system.[a]

			Main rings					
	D ring	C ring	B ring	Cassini division	A ring	F ring	G ring	E ring
Radial location (R_h)	1.08–1.23	1.23–1.53	1.53–1.95	1.95–2.03	2.03–2.27	2.32	2.73–2.90	3.7–11.6
Radial location (km)	65 000–74 500	74 500–91 975	91 975–117 507	117 507–122 340	122 340–136 780	140 219	166 000–173 200	180 000–700 000
Vertical thickness		<4 m	<100 m	<50 m	<100 m			10^3–2×10^4 km (increases with radial location)
Normal optical depth	$\sim 10^{-4}$–10^{-3}	0.05–0.2	1–10	0.1–0.15	0.4–1	1	10^{-6}	10^{-7}–10^{-5}
Particle size	μm–100 μm	mm–m	cm–10 m	1–10 cm	cm–10 m	μm–cm	μm–cm	~1 μm

[a] Data for main rings primarily from Cuzzi et al. (1984); data for ethereal rings primarily from Burns et al. (1984), de Pater et al. (2004b), and Horányi et al. (2009); data for the D ring from Showalter (1996). See text for the few known properties of the (recently-discovered) Phoebe ring.

The mass of Jupiter's rings is poorly constrained. The dust component of the ring system is very low in mass. The macroscopic particles observed in backscattered radiation clearly provide a substantially larger contribution, but one that is uncertain by several orders of magnitude (Problem 11.11). At Jupiter's distance from the Sun, icy ring particles would evaporate rapidly, so Jupiter's ring particles must be composed of more refractory materials. But the lifetime of such grains is short as well. Sputtering by energetic ions and (micro)meteorite impacts limit the lifetimes of μm-sized particles to $\lesssim 10^3$ years; orbital evolution of (sub)micron-sized particles via processes discussed in §11.5 is also very rapid. An ongoing source of particles is thus required, unless we are observing the rings at a very special time. The primary formation mechanism for the dust grains in the main and gossamer rings is thought to be erosion (probably by micrometeorites) from the small moons bounding the rings, a model based upon the morphology and vertical structure of the rings (§11.6.3). The red color of the jovian rings, discussed above, is also consistent with the satellites being the source of the ring particles.

11.3.2 Saturn's Rings

Saturn's ring system is the most massive, the largest, the brightest, and the most diverse in our Solar System (Figs. 11.4, 11.8, and 11.10). Most ring phenomena observed in other systems are present in Saturn's rings as well. The large-scale structure and bulk properties of Saturn's rings are listed in Table 11.2. A schematic illustration of Saturn's rings and inner moons is shown in Figure 11.5.

11.3.2.1 *Radial Structure of Saturn's Rings*

As seen through a small- to moderate-size telescope on Earth, Saturn appears to be surrounded by two rings (Fig. 11.2). The inner and brighter of the two is called the *B ring* (or Ring B) and the outer one is known as the *A ring*. The dark region separating these two bright annuli is named the *Cassini division* after Giovanni Cassini, who discovered it in the 1670s. The Cassini division is not a true gap, rather it is a region in which the optical depth of the rings is only about 10% of that of the surrounding A and B rings. A larger telescope with good seeing can detect the faint *C ring*, which lies interior to the B ring. The *Encke gap*, a nearly empty annulus in the outer part of the A ring, can also be detected from the ground under good observing conditions. Rings A, B, and C and the Cassini division are known collectively as *Saturn's main rings* or *Saturn's classical ring system*. Interior to the C ring lies the extremely tenuous *D ring*, which was imaged by the Voyager and Cassini spacecraft but has not

(yet) been detected from the ground. The narrow, multi-stranded, kinky *F ring* is 3000 km exterior to the outer edge of the A ring, with the region between the A and F rings known as the *Roche division*. Several tenuous dust rings lie well beyond Saturn's Roche limit (assuming a particle density equal to that of nonporous water-ice); the most prominent by far being the fairly narrow *G ring* and the extremely broad *E ring*.

The largest known ring of Saturn is associated with the planet's largest irregular moon, Phoebe. The *Phoebe ring* extends at least over the range 125–207 R_h. The vertical extent of this ring is 40 R_h, which matches the vertical motion of Phoebe along its orbit, and the ring midplane matches the plane of Saturn's orbit around the Sun rather than the planet's equatorial plane. The ring's normal optical depth of $\sim 2 \times 10^{-8}$ is similar to that of Jupiter's Thebe ring, although the particle density in the much thicker Phoebe ring is several hundred times smaller. Impacts on Phoebe presumably eject the particles that make up this enormous ring.

The classically known components of Saturn's ring system are quite inhomogeneous upon close examination, displaying both radial and azimuthal variations (Figs. 11.7–11.9, 11.11, 11.22, 11.24, 11.30, 11.32, 11.34). The character of this structure is correlated with the overall optical depth of the region in which it exists. The A ring, with its moderate optical depth, $\tau \approx 1/2$, has many regions that are relatively uniform in appearance (Fig. 11.9a). The observed features in the A ring are better understood than the majority of structure elsewhere in Saturn's ring system. Most of the A ring's structure results from resonant perturbations by external moons (§11.4.2). The Encke gap is maintained by the embedded moonlet Pan, and the tiny moonlet Daphnis clears out the narrow *Keeler gap* near the outer edge of the A ring; the theory of gap clearing by embedded moonlets is discussed in §11.4.3. The outer edges of the B and A rings are maintained by the Mimas 2:1 and Janus 7:6 resonances, which are the strongest resonances within the ring system (§11.4.1). The optically thick B ring (as well as a region of high optical depth in the inner portion of the A ring) displays irregular structure in the radial direction (Fig. 11.9b); the cause of this structure is still unknown. The optically thin C ring and Cassini division contain several gaps (Fig. 11.9c) that may be produced by embedded moonlets. The causes of the large-scale optical depth variations observed in the C ring and the Cassini division are not known.

The D ring is very faint and consists of many ringlets (Fig. 11.9d). Some of these ringlets changed substantially in appearance in the 25 years between Voyager and Cassini

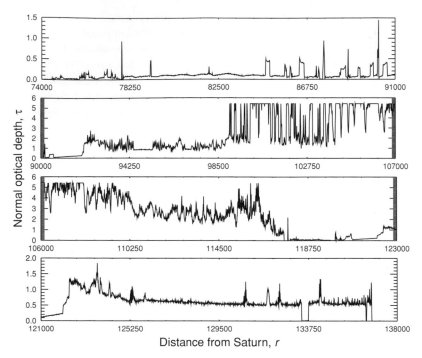

Figure 11.7 Optical-depth profile of Saturn's main rings obtained by observing the star α Arae through the rings with the Cassini Ultraviolet Imaging Spectrograph (UVIS) during November 2006. The angle between the direction of starlight and the plane of the rings is $B_{oc} = 54.43°$. The plot shows normal optical depth, τ, computed by multiplying the directly measured slant optical depth averaged at 10 km resolution by $\sin B_{oc}$. The observed starlight is in the wavelength range 110–190 nm. Note that the optical depth scale differs among the panels, radial ranges of the panels overlap by a small amount, and the regions in the B ring shown with $\tau = 5.5$ do not allow enough light to pass to allow for anything more than a lower bound on the optical depth. (Josh Colwell and the Cassini UVIS team)

observations. Particularly interesting is a regular, periodic structure with a wavelength decreasing from ∼60 km to about half this value a decade later. This structure, extending between orbital radii of 73 200 and 74 000 km, appears to be a vertical corrugation produced by differential nodal regression (eqs. 2.31 and 2.33) of an initially inclined ring that may have formed by a heliocentric impactor striking the D ring in early 1984.

The G ring has a sharp inner boundary and a faint diffuse outward extension. Within the G ring, the Cassini spacecraft discovered a bright (relative to the rest of the ring) arc. The longitudinally confined material in this arc is trapped in a 7:6 corotation eccentricity resonance with the satellite Mimas. The extremely small moon Aegaeon ($R \sim 250$ m) orbits within the G ring and is likely to be the source of many of its particles. Other faint narrow rings and arcs, including features associated with small moons Pallene, Anthe, and Methone, have been observed exterior to the main rings.

Although the E ring is also quite ethereal, it is so broad that it can readily be observed from Earth when the ring system appears almost edge-on (Fig. 11.10). The inner boundary of the E ring is fairly abrupt, just ∼12 000 km inside the orbit of Enceladus. The peak intensity of the E ring is located ∼10 000 km exterior to Enceladus's orbit. The density of the E ring drops gradually outside of this location, until it disappears into the sky background near 8.0 R_h. Cassini has encountered dust in Saturn's ring plane

up to distances of 18 R_h. Geysers at Enceladus's south pole (§5.5.6.2) provide the bulk of the E ring's material (see §§11.5, 11.6).

11.3.2.2 Azimuthal Variations

To a first approximation, Saturn's rings are uniform in longitude, i.e., the character of the rings varies much more substantially with distance from the planet than with longitude. This is, presumably, a consequence of the much shorter timescale for wiping out azimuthal structure via Kepler shear compared to radial diffusion times (Problem 11.7). However, various types of significant azimuthal structure have been observed in Saturn's rings. The most spectacular longitudinal structures seen in Saturn's rings are the nearly radial features known as *spokes* (Fig. 11.34), which are described in detail in §11.5.3.

Several narrow rings and ring edges are eccentric. Some of these features are well modeled by Keplerian ellipses that precess slowly as a result of the planet's quadrupole (and higher order) gravitational moments (eq. 2.37). However, a few features, such as the outer edges of rings B (Fig. 11.11) and A and the edges of Encke's gap (Fig. 11.30), are multi-lobed patterns which are controlled by satellite resonances; the dynamical mechanisms responsible for such features are described in §11.4.3.

Saturn's narrow F ring (Fig. 11.32) exhibits several types of unusual features that vary on timescales of hours

(a)

(b)

Figure 11.8 Clear-filter images of Saturn's rings taken by the Cassini spacecraft, at a distance of ~900 000 km from Saturn with 48 km/pixel resolution. (a) The optically thick B and A rings appear brightest in this view of the sunlit face of the rings from 9° south of the ring plane. (b) Regions of moderate optical depth, such as the C ring and Cassini division, are most prominent in this perspective of the unlit face of the rings taken 8° north of the ring plane. In this geometry, the rings are illuminated primarily via *diffuse transmission* of sunlight, so both optically thick parts of the rings (which do not allow sunlight to pass through) and very optically thin regions (which do not scatter sunlight) appear dark. (Movie) The movie consists of 34 images, beginning with the one shown in panel (a) and ending with the one in panel (b), that were taken over the course of 12 hours as Cassini pierced the ring plane. Additional frames were inserted between the spacecraft images in order to smooth the motion in the sequence. Six moons move through the field of view during the sequence. The first large one is Enceladus, which moves from the upper left to the center right. The second large one, seen in the second half of the movie, is Mimas, going from right to left. (Cassini Imaging Team and NASA/JPL/CICLOPS; PIA08356)

to years. The ring consists of a relatively optically thick central core surrounded by a fairly diffuse multistranded structure, a variety of clumps, and a regular series of longitudinal channels associated with the nearby moon Prometheus. The F ring lies near the Roche limit for moderately porous ice, and it appears that a wide range of accretion, disruption, and ring–moon interactions are occurring within it. Key factors in shaping the structure of the F ring include gravitational perturbations by Prometheus plus collisions with a family of substantially smaller moons or clumps that cross the ring and impact at high speeds. The ringlet in Encke's gap, which orbits at the same radius as the moonlet Pan, shares some characteristics of longitudinal variability with the F ring.

Azimuthal variations in brightness associated with the Sun/ring/spacecraft geometry have been observed at ring locations that are in vertical resonance with Saturn's moons; these variations are produced by spiral bending waves (§11.4.2). Subtle azimuthal variations in resonantly excited spiral density waves have also been detected.

The reflectivity of the A ring exhibits an intrinsic longitudinal variation known as the *azimuthal asymmetry*. The pattern is not symmetric about the *ansae* (the portions of the rings that appear furthest from the disk of the planet), rather the minimum brightness observed at low phase angles (e.g., from Earth) occurs ~24° before each ansa, measured in the direction of the particle orbits. The amplitude of the azimuthal asymmetry is largest in the middle of the A ring, and greatest at low ring tilt angles, where the peak brightness is ~40% larger than the brightness minima. These asymmetries are also seen at radio wavelengths, in images obtained with the Very Large Array (VLA) (Fig. 11.12) and with the radiometer on board the Cassini spacecraft, where radio waves emitted by Saturn (thermal radiation) are scattered off the ring particles. Radar experiments also reveal these structures, as does comparison between stellar occultation profiles obtained with differing geometries.

11.3.2.3 *Local Structure*
The azimuthal asymmetry described in the previous paragraph is caused by ephemeral *self-gravitating clumps* of material that are sometimes referred to as *self-gravity wakes* or, even more misleadingly, as *density wakes*. Self-gravitating clumps are elongated (by Keplerian shear), temporary, optically thick groupings of ring particles that form near large particles or clusters of particles as a result of local gravitational forces (Fig. 11.13). Kepler shear causes such wakes to trail at an angle of ~23° in planetary rings. Self-gravitating clumps are also present within the B ring.

Clumps caused by viscous overstabilities (§11.2) are observed by comparing occultation measurements in high τ regions taken in a variety of geometries. The optical depth varies less rapidly with observation angle than predicted by equation (11.16), and it also depends on the

(a)

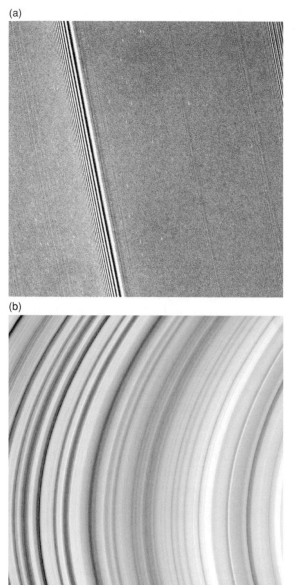

(b)

(c)

(d)

Figure 11.9 (*cont.*) (c) The characteristic plateau and oscillating structure of Saturn's inner C ring is shown at a resolution of 4.7 km/pixel. The dark feature in the middle of the image is the Colombo gap, which contains the bright narrow Colombo ringlet that is in resonance with Saturn's largest moon, Titan. This Cassini image views the lit face of the rings from an elevation of 9° above the ring plane. (NASA/JPL/CICLOPS, PIA06537) (d) Cassini image of Saturn's D ring, taken from a distance of 272 000 km with resolution of 13 km/pixel. The inner edge of the C ring is seen as the bright area in the lower right corner of the image. (NASA/JPL/CICLOPS, Portion of PIA07714)

Figure 11.9 Close-up images of portions of Saturn's A, B, C, and D rings taken by the Cassini spacecraft are shown in panels (a)–(d) respectively. (a) Dozens of propeller-shaped features (§11.3.2.5) can be discerned in this substantially stretched (contrast-enhanced) image of the lit face of a quite bland ∼800 km wide region in Saturn's A ring. The prominent density wave is excited at the Prometheus 9:8 ILR (§11.4), which is located 128 946 km from the center of Saturn. This wave propagates outwards, away from the planet. A less stretched, broader high-resolution image of structure within the A ring is shown in Fig. 11.24. (Courtesy Jeff Cuzzi and NASA) (b) This image (at $\lambda = 0.75\,\mu$m) shows the lit face of the mid to outer B ring 107 200–115 700 km from Saturn at a resolution of 6 km/pixel. (NASA/JPL/CICLOPS, PIA07610)

direction of the observation path projected into the ring plane.

11.3.2.4 *Thickness*

Saturn's rings are extremely thin relative to their radial extent. Upper limits to the local thickness of the rings of ∼150 m at several ring edges were obtained by the abruptness of some of the ring boundaries detected in stellar occultations by the rings observed from Voyager 2 and from diffraction patterns in the radio signal transmitted through the rings by Voyager 1. Estimates of ring thickness from the viscous damping of spiral bending waves and density waves, models of self-gravitating clumps, and

(a)
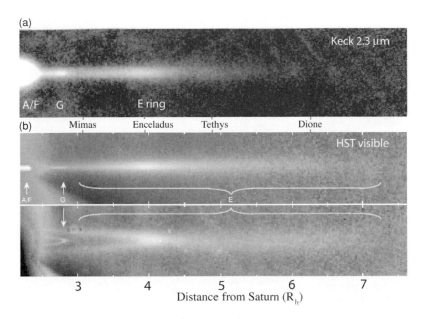

(b)

Figure 11.10 Images of Saturn's tenuous E and G rings. Saturn is off to the left. (a) Infrared photograph ($\lambda = 2.3$ μm) taken on 8–10 August 1995 with the Keck telescope. Even in this view of the dark face of the rings, seen nearly edge-on, the main rings still appear substantially brighter than the E and G rings. (de Pater *et al.* 2004b) (b) HST images of Saturn's G and E rings at visible wavelengths, seen edge-on in August 1995 (top panel) and the unlit face open by 2.5° in November 1995 (bottom panel). The G ring is the relatively bright and narrow annulus whose ansa appears in the leftmost portion of the image. The E ring is much broader and more diffuse. (Courtesy J.A. Burns, D.P. Hamilton, and M.R. Showalter)

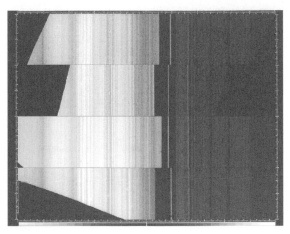

Figure 11.11 The eccentric outer edge of Saturn's B ring as imaged at four different longitudes by Voyager 2. The lit face of the rings is seen and Saturn is off to the left in all images. The left portion of the images show the outermost region of the B ring, and the right part shows the inner part of the Cassini division. The middle two slices were taken from high-resolution (< 8 km pixel^{-1}) images of the east ansa and the outer two slices are from the west ansa. The width of the gap separating the B ring from the Cassini division varies by up to 140 km. These variations are caused by perturbations exerted by Mimas near its 2:1 inner Lindblad resonance. Also visible are variations in fine structure in the B ring and the eccentric Huygens ringlet within the variable width gap. All slices were taken within about 7 hours. (Smith *et al.* 1982)

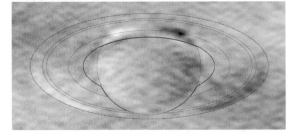

Figure 11.12 Residual asymmetry map of the 3.6 cm VLA image of Saturn shown in Figure 4.37d. At radio wavelengths the rings are visible because they reflect thermal radio emission emanating from the planet. This particular map has been produced by subtracting the transpose of the data about the polar axis from the original image. (Adapted from Dunn *et al.* 2007)

from the characteristics of the outer edge of the B ring yield values ranging from one meter to tens of meters.

The amount of light reflected from the rings when they appear edge-on from Earth is equal to that of a 1 km thick slab with reflectivity comparable to that of the lit face of

the rings. Such a thickness is far too small to be resolvable from Earth-based telescopes. This effective thickness is probably dominated by the F ring, which is quite bright near ring plane crossing and is likely warped and/or slightly inclined with respect to Saturn's equatorial plane. Additional contributions to the rings' edge-on brightness come from the actual thickness of the main rings, local corrugations of the ring plane in bending waves (§11.4.2), more gradual warping of the ring plane on larger scales by Saturn's moons, and the dusty outer rings, primarily the E ring.

11.3.2.5 *Particle Properties*

The particles in Saturn's main ring system are better characterized than are particles in other planetary rings. Spectra of infrared light reflected off Saturn's main rings are

(a)   (b)

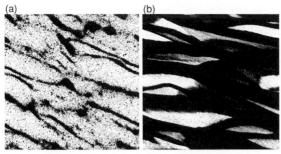

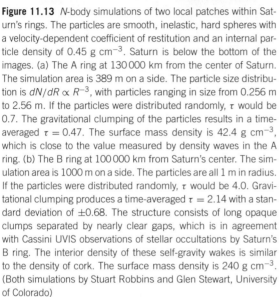

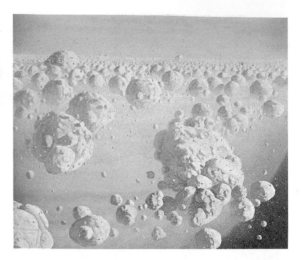

Figure 11.13 *N*-body simulations of two local patches within Saturn's rings. The particles are smooth, inelastic, hard spheres with a velocity-dependent coefficient of restitution and an internal particle density of 0.45 g cm^{-3}. Saturn is below the bottom of the images. (a) The A ring at 130 000 km from the center of Saturn. The simulation area is 389 m on a side. The particle size distribution is $dN/dR \propto R^{-3}$, with particles ranging in size from 0.256 m to 2.56 m. If the particles were distributed randomly, τ would be 0.7. The gravitational clumping of the particles results in a time-averaged $\tau = 0.47$. The surface mass density is 42.4 g cm^{-3}, which is close to the value measured by density waves in the A ring. (b) The B ring at 100 000 km from Saturn's center. The simulation area is 1000 m on a side. The particles are all 1 m in radius. If the particles were distributed randomly, τ would be 4.0. Gravitational clumping produces a time-averaged $\tau = 2.14$ with a standard deviation of ± 0.68. The structure consists of long opaque clumps separated by nearly clear gaps, which is in agreement with Cassini UVIS observations of stellar occultations by Saturn's B ring. The interior density of these self-gravity wakes is similar to the density of cork. The surface mass density is 240 g cm^{-3}. (Both simulations by Stuart Robbins and Glen Stewart, University of Colorado)

similar to that of water-ice, implying that water-ice is a major constituent. The high albedo of Saturn's rings suggests that impurities are few and/or not well mixed at the microscopic level.

The frequent collisions among ring particles cause particle aggregation as well as erosion. Scale-independent processes of accretion and fragmentation lead to power-law distributions of particle number vs. particle size. Such power-law distributions are observed over broad ranges of radii in the asteroid belt and in most of those planetary rings for which adequate particle size information is available. Data on particle sizes are thus often fit to a distribution of the form (see eq. 9.3)

$$N(R)dR = \frac{N_{\mathrm{o}}}{R_{\mathrm{o}}}\left(\frac{R}{R_{\mathrm{o}}}\right)^{-\zeta} dR \quad (R_{\mathrm{min}} < R < R_{\mathrm{max}}) \quad (11.17)$$

and zero otherwise, where $N(R)dR$ is the number of particles with radii between R and $R + dR$ and N_{o} and R_{o} are normalization constants. The distribution is characterized by the values of its power-law index, ζ, and the minimum and maximum particle sizes, R_{min} and R_{max}, respectively.

Figure 11.14 Hypothetical view from within Saturn's rings, showing macroscopic particles as loosely bound agglomerates. (Painted by W.K. Hartmann)

A uniform power law over all radii implies infinite mass in either large or small radii particles (Problem 11.10); thus, such a power law must be truncated at large and/or small radius. Note that the value of the upper size limit is not very important for the total mass or surface area of the system if the distribution is sufficiently steep (ζ significantly larger than 4) and that the lower limit is not a major factor provided the distribution is shallow enough (ζ significantly smaller than 3). No images showing individual ring particles have yet been obtained, but imagined views from within Saturn's rings are shown in Figure 11.14.

Radar signals, with wavelengths of several centimeters, have been bounced off Saturn's rings. The high radar reflectivity of the rings implies that a significant fraction of their surface area consists of particles with diameters of at least several centimeters. Radio signals sent through rings by Voyager 1 and Cassini give information on particle sizes from a comparison of optical depths at two wavelengths and from diffraction patterns of the signal. The combination of these data implies a broad range of sizes from $\sim$5 cm to 5–10 m in the optically thick B ring and inner A ring, with approximately equal areas in equal logarithmic size intervals (i.e., $\zeta \approx 3$) and most of the mass being in the largest particles. The power-law index in the size distribution is $2.8 < \zeta < 3.4$ for 5 cm $< R <$ 5 m and $\zeta > 5$ for $R > 10$ m. The C ring and outer A ring also contain a substantial quantity of particles somewhat smaller than 5 cm. Attempts to derive characteristic particle sizes from first principles (using Hertz's theory of elastic solids) have not been successful.

Micrometer-sized dust particles are comparable in size to the wavelength of visible light, so they preferentially scatter in the forward direction. Micrometer-sized particles are most common in the dusty outer rings and the F ring, but also dominate the spokes in the B ring (see Fig. 11.34) and are apparent in the very outer part of the A ring, exterior to the Keeler gap. It is likely that the μm-sized dust seen in the spokes is rapidly reaccumulated by larger ring particles, causing the spoke to vanish.

Moonlets several km in radius clear gaps in the rings, with gap width several times the moonlet's own radius (§11.4.3). The embedded moonlets Daphnis and Pan (Fig. 11.30), with mean radii of 4 and 14 km, respectively, have very low densities, $\lesssim 0.5$ g cm^{-3}, implying high porosities. The gravitational effect by smaller moonlets is not sufficient to clear a gap. Instead, very small bodies are surrounded by a region of low density, which is flanked by density enhancements a few km in extent.

Propeller-shaped features (Figs. 11.9a and 11.15a) provide indirect evidence of many bodies of radius ~ 20–250 m, assuming densities comparable to that of water-ice. These bodies are intermediate in size between the ring particles at the upper limits of the power-law size distribution and moonlets that are able to clear gaps around their orbits. Although the propeller-producing objects are too small to be detected directly, their gravitational effect on the surrounding ring material betrays their presence by perturbing particle trajectories in a manner analogous to that shown in Figure 2.7. Collisions among the perturbed swarm of ring particles complicate the situation, and produce the observed features, which appear bright on both the lit and unlit faces of the rings. Shadows of about a dozen propellers were imaged by the Cassini spacecraft in 2009, near the time of the saturnian equinox. Propellers have only been observed within Saturn's A ring, and the vast majority have been seen within three ~ 1000 km zones in the middle portion of the A ring. The objects producing the propellers are sufficiently rare that they make up only a tiny fraction of the mass within these regions of the ring. They may have formed by the disruption of a Pan-sized moonlet.

A single large ring particle together with its shadow were photographed by Cassini close to 2009 equinox (Fig. 11.15b). This body orbits within the outermost portion of the B ring, and the length of its shadow implies that it sticks out 200 meters above the ring plane. No features within nearby ring material have been associated with this giant particle.

Spectra of Saturn's rings depend upon both the composition and the size distribution of grains, including that

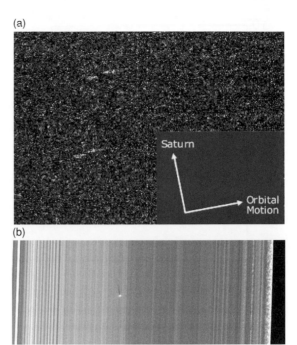

Figure 11.15 (a) Propeller-shaped features are seen on this highly stretched Cassini image of Saturn's A ring. These propellers measure 5 km tip to tip and are produced by the gravitational perturbations of ring particles that are too small to be seen directly in this image, which has a resolution of 52 m/pixel. (Courtesy Mark Showalter and NASA/JPL) (b) The Cassini spacecraft captured this image of a small object in the outer portion of Saturn's B ring casting a shadow on the rings as Saturn approached its August 2009 equinox. This new object, located ~480 km inward from the outer edge of the B ring, was found by detecting its shadow across the rings. The shadow length of 41 km implies the body is protruding about 200 m above the ring plane. If the object is orbiting in the same plane as the ring material surrounding it, which is likely, it must be about 400 m across. This view was obtained at a phase angle of 120°. Image scale is 1 km/pixel. (NASA/JPL/Space Science Institute)

of ring particle regolith, i.e., the grains and ice crystals that cover the larger (cm–m sized) boulders in the rings. The depth of water-ice absorption bands in the infrared show that the ice in the rings is quite pure. The deepest absorptions are seen in the A ring, implying the purest ice and/or the largest grain sizes; absorption is deeper in the B ring than in the Cassini division and the C ring. Small but significant color variations are seen from ring to ring. Particles within regions that are optically thin, especially the Cassini division and the C ring, appear to be much 'dirtier' and more neutral in color (less red). These differences may result from the more rapid contamination of particles in low τ regions by impacting micrometeoroids (particles in regions of high optical depth partially shield one another from such bombardment) and the higher impact flux closer

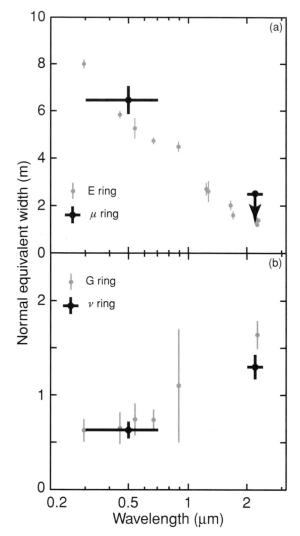

Figure 11.16 Very low resolution spectra of Saturn's E ring and Uranus's μ ring (a), and Saturn's G ring together with Uranus's ν ring (b). Measurements of Uranus's rings are shown in heavy black, Saturn's in gray. Horizontal bars indicate the range of wavelengths averaged; vertical bars indicate $\pm 1\sigma$ uncertainties. The vertical axis plots the brightness in terms of the equivalent width, which is the width of a hypothetical ring of the same total brightness and with a reflectivity $I/F = 1$ (§3.1.2.1). The E ring values are scaled downwards in intensity to match the μ ring values in the visible. Ring ν and the G ring are both radial integrals and are plotted at the same scale. (de Pater *et al.* 2006b)

to the planet due to gravitational focusing of heliocentric debris (§13.5.3.2).

The E ring is distinctly blue in color (Fig. 11.16a), in contrast to, e.g., Jupiter's dusty rings and Saturn's G ring, which are red (Fig. 11.16b). The blue color indicates that the particle size distribution is dominated by tiny grains. In Saturn's E ring, the particle radii cluster near 1 μm;

hence, this population is not collisionally evolved. The E ring particles are ice crystals formed in the plumes of water-ice volcanoes or geysers on the moon Enceladus (§5.5.6). At Enceladus's distance from Saturn, radiation pressure (from sunlight) and magnetic forces on charged grains are much more efficient at dispersing ~ 1 μm grains over a broad annulus than they are at spreading out grains of larger or smaller size (§11.5.1). If a broad particle size distribution were created by water ejected from Enceladus, grains of other sizes would remain clustered near the orbit of their parent moon, and most would soon either re-impact the moon or be ground down via mutual collisions. Thus, the narrowness of the E ring's size distribution may be a consequence of survival advantages for 1 μm grains, rather than a monodisperse particle formation mechanism. Moreover, numerical models of particles with ~ 1–1.5 μm radii ejected by Enceladus show that the surrounding plasma drags the particles outwards, in accord with the observed slight outward displacement of the peak density of the E ring.

11.3.2.6 *Mass*

The mass of Saturn's rings is much larger than that of any other ring system within the Solar System, but is too low to have been measured by its gravitational effects on moons or spacecraft. Thus, it must be deduced from more circuitous theoretical arguments. Several different techniques have been used, and all give similar answers, lending confidence to the results.

The wavelengths of spiral density waves and spiral bending waves are proportional to the local surface mass density of the rings; therefore, we can deduce the mass where we see waves. The theory behind this analysis is presented in §11.4, and a sample exercise in this technique is given in Problem 11.14. The surface density, σ_ρ, at the two wave locations observed in the B ring is ~ 50–80 g cm^{-2}. Analysis of dozens of waves within the A ring reveals surface densities of ~ 50 g cm^{-2} in the inner to middle A ring, dropping to $\lesssim 20$ g cm^{-2} near this ring's outer edge. Measured values in the optically thin C ring and Cassini division are ~ 1 g cm^{-2}. Since observable spiral waves cover only a small fraction of the area of Saturn's rings, we assume that the *opacity*, σ_ρ/τ, is constant within a given region of the rings in order to estimate the mass of the ring system. This approximation is fairly good wherever there are several waves near one another (so that it can be checked). It must be noted, though, that the energy input by the moons into the waves may make regions in which strong waves propagate somewhat anomalous. The large amplitude of the eccentricity of the outer edge of the B ring is likely caused by a combination of resonant

Table 11.3 Properties of Uranus's ring system.[a]

	Ring ζ	Rings 6, 5, 4, $\alpha, \beta, \gamma, \eta, \delta$	Ring λ	Ring ϵ	Ring ν	Ring μ
Radial location (R_δ)	1.55	~1.64–1.90	1.96	2.01	2.63	3.82
Radial location (km)	~39 600	~41 837–48 300	50 024	51 149	67 300	97 700
Radial width (most narrow rings vary with azimuth)	3500 km[b]	1–10 km	~2 km	20–96 km	3800 km	17 000 km
Normal optical depth	~10^{-6}–10^{-3}	~0.3–0.5	0.1	0.5–2.3	~6×10^{-6}	~8×10^{-6}
Particle size	(sub)µm	~10 cm – 10 m	(sub)µm	~10 cm – 10 m	µm	(sub)µm

[a] Data from French *et al.* (1991), Esposito *et al.* (1991), de Pater *et al.* (2006a,b), Showalter and Lissauer (2006).

[b] A fainter component of the ζ ring extends $\gtrsim$ 5000 km radially, towards Uranus.

forcing by the moon Mimas and ring self-gravity; calculations based on this model suggest that the surface density near the (optically thick) outer edge of the B ring is ~100 g cm^{-2}, consistent with the density wave results. The particle flux produced from cosmic ray interactions with the ring system suggests an average surface density of 100–200 g cm^{-2}, a few times as large as the average from density wave estimates. These values can be reconciled if the surface density in the optically thick parts of the B ring (Fig. 11.7), where no density waves have been observed, is significantly higher than the surface density in the wave regions.

The total mass of Saturn's ring system is estimated (using the local surface density deduced from spiral waves and the assumption of surface density being proportional to optical depth) to be: $M_{rings} \sim 5 \times 10^{-8} M_h \sim M_{Mimas}$, where Mimas, the innermost and smallest of Saturn's nearly spherical moons, has a mean radius of 196 km (Table 1.5). The particle size distribution derived from Voyager radio occultation measurements suggests a ring mass 40% less than the value quoted above, assuming particle densities equal to that of solid water-ice. Uncertainties in these estimates are large, and thus do not tightly constrain the porosity of Saturn's ring particles, which are composed of nearly pure water-ice. Moreover, the mass of the rings might be substantially larger than suggested by either of the above measurements, because the ring's clumpiness on a local scale breaks down the simple relationship between optical depth and local mass density. Thus, significant amounts of mass might be 'hiding' in regions of the rings that are composed of optically thick (and relatively massive) clumps with open space between them.

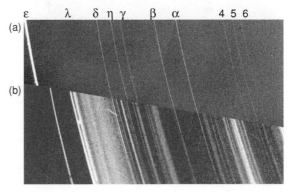

Figure 11.17 The main rings of Uranus, as imaged by Voyager 2 in 1986. The region shown ranges from ~40 000 km from the planet's center (right) to ~50 000 km away (left). (a) Mosaic of two low phase angle (21°), high-resolution (10 km/pixel) images. The planet's nine narrow optically thick rings are clearly visible, and the very narrow moderate optical depth λ ring is marginally detectable. (NASA/Voyager 2, PIA00035) (b) High phase angle (172°) view. The forward scattering geometry dramatically enhances the visibility of the micrometer-sized dust particles. The streaks are trailed star images in this 96-second exposure. (PIA00142)

11.3.3 Uranus's Rings

Most of the material in the uranian ring system is confined to nine narrow annuli whose orbits lie between 1.64 and 2.01 R_δ from the planet's center (Table 11.3, Figs. 11.3, 11.17, and 11.18). These nine optically thick rings were discovered from Earth-based observations of stars whose light was seen to diminish as they were occulted by Uranus's rings in the late 1970s. Most are 1–10 km wide and have eccentricities of order 10^{-3} and inclinations of $\lesssim 0.06°$. The outermost annulus, *Ring ϵ*, is the widest and most eccentric, with $e = 8 \times 10^{-3}$ and width ranging from

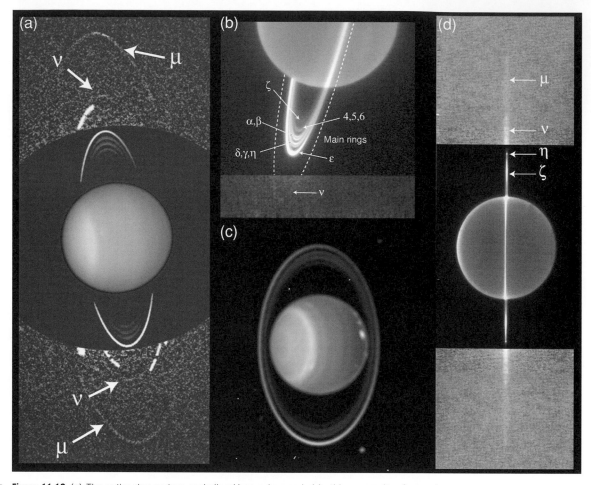

Figure 11.18 (a) The entire ring system encircling Uranus is revealed in this composite of several HST images taken at visible wavelengths in 2003. The regions furthest from the planet have been processed differently in order to display very faint features, in particular, the newly discovered μ and ν rings, as indicated. Because of the long exposures, the moons are smeared out and appear as arcs within the ring system. (Courtesy M. Showalter, NASA/ESA) (b) Composite image at 2.3 μm taken with the adaptive optics system on the Keck telescope in August 2005. Only the south side of the rings is shown here to emphasize the main ring system, the inner ζ and outer ν rings. (Adapted from de Pater *et al.* 2006b) (c) This HST image of Uranus, composed of images at 0.9, 1.1, and 1.7 μm (in blue, green, and red on the web color image), was taken on 8 August 1998, with Hubble's Near Infrared Camera and Multi-Object Spectrometer. The asymmetry of the ϵ ring is clearly visible, as well as a moon in the lower right and numerous clouds in the northern (on the right) hemisphere. (Courtesy Erich Karkoschka and NASA/ESA) (d) Composite image at 2.3 μm taken with the adaptive optics system on the Keck telescope on UT 26–27 July 2007, when the rings are seen (almost) edge-on (ring plane crossing). The 'dark' side of the rings are viewed, i.e., the Sun and Earth are on opposite sides of the rings. Most main rings are very dark (invisible) in this geometry, while dust in the rings is bright. The bright tip of the main rings coincides with ring η; the ζ ring is very bright, and both the ν and μ rings are visible. (I. de Pater, H.B. Hammel, and M. Showalter; see de Pater *et al.* 2007 for details)

20 km at periapse to 96 km at apoapse. As the ϵ ring is optically thick, this difference in width leads to a pronounced asymmetry in ring brightness, with the ring at apoapse more than twice as bright as at periapse (Fig. 11.18c). The majority of ring edges, including both the inner and outer boundaries of the ϵ ring, are quite sharp compared to ring width, but in a few cases a more gradual dropoff in optical depth is observed. The main uranian ring system also includes the narrow, moderate optical depth, dusty λ *ring*. Wide, radially variable, low optical depth dust

sheets are interspersed with the optically thick uranian rings (Fig. 11.17). Interior to the main rings lies the broad, tenuous ζ ring.

Within each of the uranian rings that possesses a nonzero measured eccentricity, including the ϵ ring, e increases with distance from the planet, with the *eccentricity gradient*, $a(de/da) \sim 0.5$. Two of the uranian rings are not well modeled by precessing Keplerian ellipses. The δ *ring* looks like an ellipse *centered* on Uranus, whereas the γ *ring* combines a standard eccentric ellipse pattern with temporally coherent radial motions of the particles (i.e., periapse times are correlated for all longitudes, so that the ring appears to 'breathe'). The λ ring varies azimuthally with a periodicity of 72° (five-lobe symmetry).

The mass of the ϵ ring estimated from the particle size distribution and an assumed particle density of 1 g cm^{-3} is 1–5×10^{19} g. Dynamical models for the maintenance of the ϵ ring's eccentricity by ring self-gravity yield a mass estimate of $\sim 5 \times 10^{18}$ g, but this estimate may not be very accurate as these models do not adequately reproduce certain aspects of ring structure. The combined mass of all of Uranus's other rings is probably a factor of a few smaller than that of ring ϵ.

The particles in the nine optically thick uranian rings have a similar size distribution to those in Saturn's main ring system, except that the lower limit is closer to ~ 10 cm, roughly an order of magnitude larger than in Saturn's rings. Particle sizes thus range from ~ 10 cm to ~ 10 m. Uranus's ring particles are extremely dark, with ring particle reflectivity at visible–near-infrared wavelengths being ~ 0.04. They appear as dark as the darkest asteroids and carbonaceous chondrite meteorites. However, they probably consist of *radiation-darkened ice*, which is a mixture of complex hydrocarbons embedded in ice that includes CH_4, CO, and/or CO_2 produced as a result of the removal of H atoms via sputtering processes. Such radiation-darkened ice may also account for the low albedos of cometary nuclei (§10.6.1).

Two broad, low optical depth, rings of Uranus located well exterior to the planet's nine main rings were discovered by HST (Fig. 11.18). The outermost, the μ *ring*, is more than 15 000 km wide. Like Saturn's E ring, the μ ring is distinctly blue (Fig. 11.16), indicative of a particle size distribution dominated by submicrometer-sized material. Its peak intensity coincides with the orbit of the tiny moon Mab, which presumably is the source of its particles. The other ring, *ring ν*, is less than 4000 km wide. It has a 'normal' red color, suggestive of a much larger fraction of micrometer and larger particles. It does not coincide with any known moons. Most of the surface area of the tenuous material interspersed between the main rings is covered by submicrometer- and micrometer-sized particles (Fig. 11.17b).

11.3.4 Neptune's Rings

Neptune's ring system is quite diverse, showing structure in both radius and longitude (Figs. 11.19, 11.20). The most prominent feature of the ring system is a set of *arcs* of optical depth $\tau \approx 0.1$ within the *Adams ring*. The arcs vary in extent from $\sim 1°$ to $\sim 10°$, and are grouped in a 40° range in longitude; they are about 15 km wide. At other longitudes within the Adams ring, $\tau \approx 0.003$; a comparable optical depth has been measured for the *Le Verrier ring*. Neptune's other rings are even more tenuous. Several moderately large moons orbit within Neptune's rings; these satellites are believed to be responsible for much of the radial and longitudinal ring structure that has been observed. Figure 11.20 is a cartoon sketch of the known components of Neptune's ring system; ring locations and optical depths are listed in Table 11.4.

Images obtained with HST and the Keck telescope have revealed that the arrangement of arcs is changing over time. The trailing arc, Fraternité, appears to be stable and tracks at a rate in resonance with the moon Galatea (§11.4.4). The leading arcs are shifting in location and decreased in brightness since 1989 (Voyager flyby); they had almost disappeared in 2003.

The particles in Neptune's rings are very dark and (at least in the arcs) red. They may be as dark as the particles in the uranian rings, but the properties of Neptune's ring particles are less well constrained by data currently available. The fraction of optical depth due to micrometer-sized dust is very high, $\sim 50\%$, and appears to vary from ring to ring. The limited data available are not sufficient to make even an order of magnitude estimate of the mass of Neptune's rings, although they suggest that the rings are significantly less massive than the rings of Uranus, unless they contain a substantial population of undetected large ($\gtrsim 10$ m) particles, which is unlikely given the paucity of smaller macroscopic ring particles.

11.4 Ring–Moon Interactions

Observations of planetary rings reveal a complex and diverse variety of structure, mostly in the radial direction, over a broad range of length scales, in contrast to naive theoretical expectations of smooth, structureless rings (§11.2). The processes responsible for some types of ring structure are well understood. Partial or speculative explanations are available for other features, but the causes of many structures remain elusive. The agreement between theory and observations is best for ring features

(a)

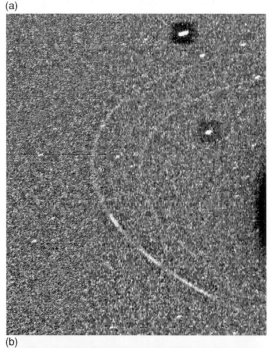

(b)

Figure 11.19 Neptune's two most prominent rings, Adams (which includes higher optical depth arcs) and Le Verrier, as seen by Voyager 2. (a) The rings appear faint in this image, FDS 11350.23 (FDS numbers refer to the Voyager flight data system timeline. Each image has a unique FDS number), taken in backscattered light (phase angle 15.5°) with a resolution of 19 km per pixel. The moon Larissa at the top of the image appears streaked as a result of its orbital motion. The other bright object in the field is a star. The image appears very noisy because a long exposure (111 s) and large stretch were needed to show the faintly illuminated, low optical depth dark rings of Neptune. (NASA/Voyager 2, PIA00053) (b) This forward scattered light (phase angle 134°) image, FDS 11412.51, was obtained using a 111 second exposure with a resolution of 80 km per pixel. The rings are much brighter in forward scattered light than in backscatter, indicating that a substantial fraction of the ring optical depth consists of micrometer-size dust. (NASA/Voyager 2, PIA01493)

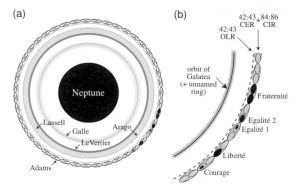

Figure 11.20 Cartoon sketch of Neptune's rings and associated moons as viewed from the south pole showing: (a) the location and names of the main rings; (b) a model for the arcs in the Adams ring (as it appeared when Voyager 2 encountered Neptune in 1989); the arcs are the filled centers of libration of the Galatea's 42:43 corotation eccentricity resonance (CER) and 84:86 corotation inclination resonance (CIR). The location of the two resonances (CER + CIR) as shown in the figure is not that of the nominal CIR but that of the observed arc mean motion. (Courtesy Carl Murray)

thought to be produced by gravitational perturbations from known moons, and we examine such models first.

11.4.1 Resonances

A common process in many areas of physics is resonance excitation: When an oscillator is excited by a varying force whose period is very nearly equal to the oscillator's natural frequency, the response can be quite large even if the amplitude of the force is small (eq. 2.26). In the planetary ring context, the perturbing force is the gravity of one of the planet's moons, which is generally much smaller than the gravitational force of the planet itself.

Resonances occur where the radial (or vertical) frequency of the ring particles is equal to the frequency of a component of a satellite's horizontal (or vertical) forcing, as sensed in the frame rotating at the frequency of the particle's orbit. In this case, the resonating particle is repeatedly near the same phase in its radial (vertical) oscillation when it experiences a particular phase of the satellite's forcing. This situation enables continued coherent 'kicks' from the satellite to build up the particle's radial (vertical) motion, and significant forced oscillations may thereby result. Particles nearest resonance have the largest eccentricities (inclinations), as they receive the most coherent kicks; the forced eccentricity (inclination) is inversely proportional to the distance from resonance for noninteracting particles in the linear regime. Collisions among ring particles and the self-gravity of the rings complicate the situation, and resonant forcing of planetary rings can produce a variety of features, including gaps and

Table 11.4 Properties of Neptune's ring system.[a]

	Galle ring	Le Verrier ring	Lassell ring	Arago ring	unnamed ring	Adams ring
Radial location (R_Ψ)	1.7	2.15	2.23	2.31	2.50	2.54
Radial location (km)	42 000	53 200	55 200	57 200	61 953	62 933
Radial width (km)	2000	~100	4 000			15 (in arcs)
Normal optical depth	~10^{-4} (of dust)	~0.003	~10^{-4}			0.1 in arcs 0.003 elsewhere
Dust fraction	? (large particles not detected)	~50%				~50% in arcs ~30% elsewhere

[a] Data from Porco *et al.* (1995).

spiral waves, which are discussed in detail in the following subsections.

The locations and strengths of resonances with any given moon can be calulated by decomposing the gravitational potential of the moon into its Fourier components. The *disturbance* (forcing) *frequency*, ω_f, can be written as the sum of integer multiples of the satellite's angular, vertical, and radial frequencies:

$$\omega_f = m_\theta n_s + m_z \mu_s + m_r \kappa_s, \qquad (11.18a)$$

where the azimuthal symmetry number, m_θ, is a nonnegative integer, and m_z and m_r are integers, with m_z being even for horizontal forcing and odd for vertical forcing. The subscript s refers to the satellite (moon).

A particle placed at a distance $r = r_L$ from the planet is in horizontal (Lindblad) resonance if r_L satisfies

$$\omega_f - m_\theta n(r_L) = \pm \kappa(r_L). \qquad (11.18b)$$

Vertical resonance occurs if its radial position r_v satisfies

$$\omega_f - m_\theta n(r_v) = \pm \mu(r_v). \qquad (11.18c)$$

When equation (11.18b) is valid for the lower (upper) sign, we refer to r_L as the inner (outer) Lindblad or horizontal resonance, which are frequently abbreviated as ILR and OLR, respectively. The radius r_v is called an inner (outer) vertical resonance, IVR (OVR), if equation (11.18c) is valid for the lower (upper) sign. Since all of Saturn's large satellites orbit the planet well outside the main ring system, the moon's angular frequency, n_s, is less than the angular frequency of the particle, and inner resonances are more important than outer ones. The differences between the orbital, radial, and vertical[3] frequencies are at most a few

percent within Saturn's rings. Thus, when $m_\theta \neq 1$, the approximation $\mu \approx n \approx \kappa$ may be used to obtain the ratio

$$\frac{n(r_{L,v})}{n_s} \approx \frac{m_\theta + m_z + m_r}{m_\theta - 1}. \qquad (11.19)$$

The notation $\frac{(m_\theta + m_z + m_r)}{(m_\theta - 1)}$ or $(m_\theta + m_z + m_r):(m_\theta - 1)$ is commonly used to identify a given resonance. If $n = \mu = \kappa$, the inner horizontal and vertical resonances would coincide: $r_L = r_v$. Since, due to Saturn's oblateness, $\mu > n > \kappa$ (eqs. 2.31–2.33), the positions r_L and r_v do not coincide, and $r_v < r_L$.

The strength of the forcing by the satellite depends, to lowest order, on the satellite's mass, M_s, eccentricity, e, and inclination, i, as $M_s e^{|m_r|} \sin^{|m_z|} i$. The strongest horizontal resonances have $m_z = m_r = 0$, and are of the form m_θ: $(m_\theta - 1)$. The strongest vertical resonances have $m_z = 1$, $m_r = 0$, and are of the form $(m_\theta + 1):(m_\theta - 1)$. The location and strengths of such orbital resonances can be calculated from known satellite masses and orbital parameters and Saturn's gravity field. By far the lion's share of the strong resonances in Saturn's ring system lie within the outer A ring (Figs. 11.21 and 11.22), near the orbits of the moons that excite them.

Resonant forcing leads to a secular transfer of orbital angular momentum from Saturn's rings to its moons. These torques produce two classes of structure in Saturn's rings: gaps/ring boundaries and spiral density and bending waves. The outer edges of Saturn's two major rings are maintained by the two strongest resonances in the ring system. The outer edge of the B ring is located at Mimas's 2:1 ILR, and is shaped like a two-lobed oval *centered* on Saturn (Fig. 11.23). The A ring's outer edge is coincident with the 7:6 resonance of the coorbital moons Janus and Epimetheus, and has a seven-lobed pattern, consistent with theoretical expectations. In order for a resonance to clear a gap or maintain a sharp ring edge, it must exert enough

[3] Disk self-gravity can significantly increase the vertical frequencies of particles in a vertically thin ring. However, self-gravity does not alter the vertical frequency of the local disk midplane, which is the relevant quantity for vertical resonances.

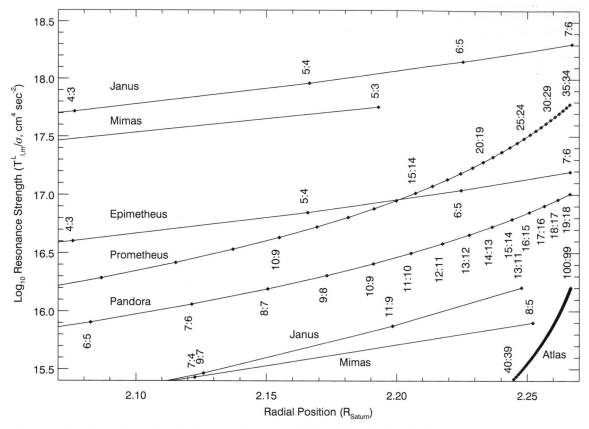

Figure 11.21 Locations and strengths of major Lindblad resonances in Saturn's A ring for Saturn's five largest 'ring moons' whose orbits lie interior to the planet's more massive spherical moons: Janus, Epimetheus, Pandora, Prometheus, and Atlas, as well as those of the innermost spherical moon, Mimas. The moons that orbit nearer to the A ring have more closely spaced resonances with strength increasing outward more rapidly. (Courtesy Matt Tiscareno)

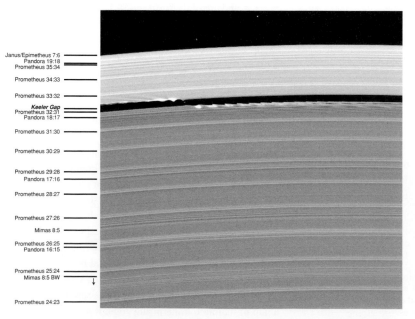

Figure 11.22 Cassini image of the lit face of the outer portion of Saturn's A ring, with the locations of strong satellite resonances and the Keeler gap marked on the left. The Janus/Epimetheus 7:6 ILR confines the outer edge of the A ring. The first few crests and troughs of the inwardly propagating Mimas 8:5 bending wave and the outwardly propagating Mimas 8:5 density wave can be seen on this image, whereas the wavelengths of the higher m_θ density waves at the resonances of the nearby small moons are shorter (eq. 11.22) and not clearly discernible in this version. The tiny moon Daphnis is to the left of center and its effects on the edges of the Keeler gap are quite prominent. (Image PIA07809 from NASA/JPL/CICLOPS, annotated by Matt Tiscareno)

torque to counterbalance the ring's viscous spreading. In the low optical depth C ring, resonances with moderate torques create gaps, but in the higher optical depth A and B rings, resonances with similar strengths excite spiral density waves.

Alternating bright and dark features are observed in high phase angle images of Saturn's low optical depth, dusty G ring. These features, which are slightly canted relative to the azimuthal direction, appear to be generated by the Mimas 8:7 ILR. Models require gradual damping of the resonantly excited particle eccentricities in order to match the observations. Similar structures are seen in the D ring and Roche division. No moon orbits at the correct location to account for these features, but the excitation frequencies correspond to clouds in Saturn's atmosphere and asymmetries in Saturn's magnetic field; Lorentz resonances (§11.5.2) acting on charged grains may be responsible for some of these structures.

Nearly empty gaps with embedded optically thick ringlets have been observed at strong resonances located in optically thin regions of the rings. Just exterior to the B ring lies a small gap with an opaque eccentric ringlet (Fig. 11.11). Similar ringlets are also observed in gaps at the strongest resonances in Saturn's C ring. Qualitatively similar features have been reproduced in numerical simulations of resonantly forced particulate rings. Note, however, that gaps with embedded ringlets have also been observed at nonresonant locations.

11.4.2 Spiral Waves

Spiral density waves generated by gravitational perturbations of external moons have been observed at many dozen locations within Saturn's rings and have tentatively been detected within the rings of Uranus. Analogous bending waves have been detected at several places within Saturn's rings. Spiral waves are among the best understood forms of structure within planetary rings, and have been very useful as diagnostics of ring properties such as surface mass density, σ_ρ, and local thickness. However, the angular momentum transfer associated with the excitation of density waves within Saturn's rings leads to characteristic orbital evolution timescales of Saturn's A ring and inner moons which are much shorter than the age of the Solar System, creating a major theoretical puzzle.

Spiral density waves are horizontal density oscillations that result from the bunching of streamlines of particles on eccentric orbits (Fig. 11.23a,b). *Spiral bending waves*, in contrast, are vertical corrugations of the ring plane resulting from the inclinations of particle orbits (Fig. 11.23c). Both types of spiral waves are excited at resonances with moons and propagate as a result of the collective

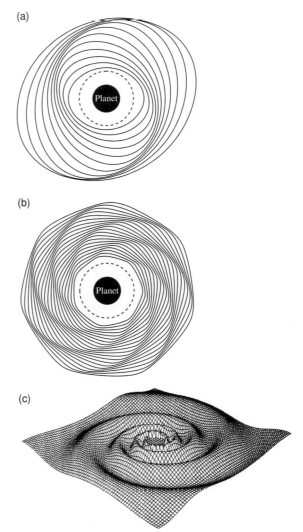

Figure 11.23 Schematic diagrams of the coplanar particle orbits that give rise to trailing spiral density waves near a resonance with an exterior satellite are shown in panels (a) and (b). (a) The two-armed spiral density wave associated with the 2:1 ($m = 2$) inner Lindblad resonance. (b) The seven-armed density wave associated with the 7:6 ($m = 7$) inner Lindblad resonances. The pattern rotates with the angular velocity of the satellite and propagates outwards from the exact resonance (denoted by a dashed circle). (Murray and Dermott 1999) (c) Schematic of an inward-propagating spiral bending wave showing variation of vertical displacement with angle and radius for a two-armed spiral. Spiral waves observed in Saturn's rings are much more tightly wound. (Shu *et al.* 1983)

self-gravity of the particles within the ring disk (Fig. 11.24). Ring particles move along paths that are very nearly Keplerian ellipses with one focus at the center of Saturn. However, small perturbations caused by the wave force a coherent relationship between particle

Figure 11.24 A portion of the lit face of Saturn's A ring is seen in this mosaic of Cassini images which were taken with resolution of 940 m/pixel just after the spacecraft entered orbit about Saturn. Saturn is off to the left. Various features produced by gravitational perturbations of moons are seen against an otherwise uniform background. The most prominent features in the above panel are the Mimas 5:3 bending wave, which propagates inwards towards the planet, and the Mimas 5:3 density wave, which propagates away from Saturn. The separation between the locations of the two waves results from the nonclosure of orbits caused by Saturn's oblateness. The other density waves are excited by the moons Janus/Epimetheus, Pandora, and Prometheus. The dark region in the opposite panel is Encke's gap; the scalloped inner edge of the Encke gap and the associated satellite wake to the interior are produced by the moonlet Pan. (Lovett *et al.* 2006 and NASA/JPL)

eccentricities/periapses (in the case of density waves) or inclinations/nodes (in the case of bending waves), which produces the observed spiral pattern.

11.4.2.1 *Theory of Bending Waves*

A moon on an orbit inclined with respect to the plane of the rings excites motion of the ring particles in the direction perpendicular to the mean ring plane. The vertical excursions of the particles are generally quite small (up to ~400 m in Saturn's rings, resulting primarily from perturbations by Titan and the Sun) and vary coherently over scales of tens of thousands of kilometers, warping the rings like the brim of a hat. However, at vertical resonances, the natural vertical oscillation frequency of a particle, $\mu(r)$, is equal to the frequency at which a moon tugs the particle perpendicular to the ring midplane. Such coherent vertical perturbations can produce significant out-of-plane

motions (§2.3.2). Self-gravity of the ring disk supplies a restoring force that distributes the torque exerted by the moon upon the ring at the resonance to nearby (but not resonant) regions of the ring. This process enables bending waves to propagate away from resonance, creating a corrugated spiral pattern. The number of spiral arms is equal to the value of m_θ. For $m_\theta > 1$, bending waves propagate toward Saturn; *nodal bending waves* ($m_\theta = 1$) propagate away from the planet.

In the inviscid linear theory, where viscous damping is ignored and the slope of the bent ring midplane is assumed to be small, the height of the local ring midplane relative to the Laplacian plane is given by a Fresnel integral,[4] which is evaluated and plotted in Figure 11.25. In the asymptotic

---

[4] Fresnel integrals are described in various applied mathematics texts.

Figure 11.24 (*cont.*)

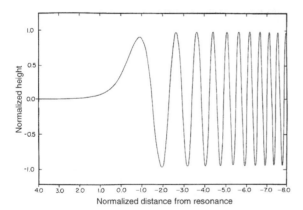

Figure 11.25 Theoretical height profile of an undamped linear spiral bending wave as a function of distance from resonance. The length and height scales are in arbitrary units. (Adapted from Shu *et al.* 1983)

far-field approximation, the oscillations remote from resonance have a wavelength

$$\lambda = \frac{4\pi^2 G \sigma_\rho}{m_\theta^2 [\omega_f - n(r)] - \mu^2(r)}. \tag{11.20}$$

Equation (11.20) can be simplified by approximating the orbits of the ring particles as Keplerian, $n(r) \approx (GM_p/r^3)^{\frac{1}{2}} \approx \mu(r)$, for the $m_\theta > 1$ case and approximating the departure from Keplerian behavior to be due

exclusively to the quadrupole term of Saturn's gravitational potential for the $m_\theta = 1$ case. The resulting formulae are

$$\lambda(r) \approx 3.08 \left(\frac{r_v}{R_h} \right)^4 \frac{\sigma_\rho}{m_\theta - 1} \frac{1}{r_v - r} \quad (m_\theta > 1) \text{ and}$$

$$\tag{11.21a}$$

$$\lambda(r) \approx 54.1 \left(\frac{r_v}{R_h} \right)^6 \sigma_\rho \frac{1}{r_v - r} \quad (m_\theta = 1), \tag{11.21b}$$

where λ, r, and r_v are measured in kilometers, and σ_ρ is in g cm^{-2}. Equations (11.21) afford a means of deducing the surface density from measured wavelengths (Problem 11.14b).

Inelastic collisions between ring particles act to damp bending waves. Larger velocities lead to more rapid damping. The damping rate of bending waves can be used to estimate the ring viscosity, which can be converted into an estimate of the ring thickness using equations (11.11) and (11.14).

11.4.2.2 *Theory of Density Waves*

The gravity of a moon on an arbitrary orbit about Saturn has a component which produces epicyclic (radial and azimuthal) motions of ring particles. However, as in the case of vertical excursions induced by moons on inclined orbits, the epicyclic excursions are generally extremely small. An exception occurs near Lindblad (horizontal) resonances (eq. 11.18b), where coherent perturbations are able to excite significant epicyclic motions. In a manner analogous to the situation at vertical resonances, self-gravity of the ring disk supplies a restoring force that enables density waves to propagate away from Lindblad resonance. Nearly all of the density waves identified within Saturn's rings are excited at inner Lindblad resonances and propagate outward, away from the planet, but some wave-trains in the outer portion of the A ring are excited at outer Lindblad resonances of the moonlet Pan (which orbits within Encke's gap) and therefore propagate towards Saturn.

The theory of spiral density waves is analogous to that of spiral bending waves, with the fractional perturbation in surface mass density, $\Delta\sigma_\rho/\sigma_\rho$, replacing the slope of the disk, dZ/dr. The relationship in the linear theory $\left(\Delta\sigma_\rho/\sigma_\rho \ll 1\right)$ that is analogous to equation (11.20) is

$$\lambda = \frac{4\pi^2 G\sigma_\rho}{m_\theta^2[\omega_f - n(r)] - \kappa^2(r)}. \quad (11.22)$$

The approximations given by equations (11.21) are also valid for density waves, provided r_v is replaced by r_L. Spiral density waves with values of $m_\theta = \mathcal{O}(100)$ have been detected in Saturn's rings, although the strongest ones have much smaller m_θ. In contrast, most spiral galaxies have 2–4 spiral arms.

The most prominent density waves observed in Saturn's rings have $\Delta\sigma_\rho/\sigma_\rho \approx 1$. At such large amplitudes the linear theory breaks down, and a nonlinear model is required. The principal results of the nonlinear theory are as follows: (*i*) Nonlinear density waves depart from the smooth sinusoidal pattern predicted by the linear model, and become highly peaked (Fig. 11.26). (*ii*) The theoretical wave profiles have broad, shallow troughs with surface density never dropping below half of the ambient value, and are qualitatively similar to observed waves (compare Figs. 11.26 and 11.29). (*iii*) The nonlinear torques exerted by Saturn's moons are similar to those calculated using linear theory.

11.4.2.3 *Spiral Waves in Saturn's Rings*

Spiral waves in planetary rings are extremely tightly wound, with typical *winding angles* (departures from circularity) being 10^{-5}–10^{-4} radian (compared to $\gtrsim 10^{-1}$ radian in most spiral galaxies). Such waves have very short wavelengths, of the order of 10 km. Brightness contrasts between crests and troughs of density waves, both in reflected light on the sunlit face of the rings (Fig. 11.24) and in diffuse transmission of sunlight on the dark side

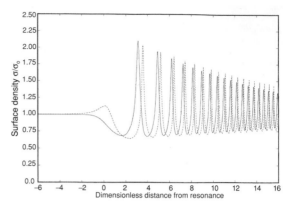

Figure 11.26 Theoretical surface density profile of a damped non-linear spiral density wave. The solid and dashed lines represent two profiles of the same wave plotted at different azimuths. (Adapted from Shu *et al.* 1985)

of the ring plane, can be seen in Voyager and Cassini images. Bending waves are visible on spacecraft images of the lit face of the rings (Fig. 11.24) as a result of the dependence of brightness on local solar elevation angle. Bending waves appear on images of the unlit face of the rings because the slant optical depth, through which sunlight diffused, depends on local ring slope.

Both density waves and bending waves were detected in Voyager and Cassini radio occultation data (Fig. 11.27). Density waves are observable by occultation experiments because bunching of particle streamlines increases the optical depth, τ, of crests. The oscillations of bending waves leave the optical depth normal to the ring plane unchanged, but can be detected because the tilt of the ring plane causes variations in the observed slant optical depth (Fig. 11.28). Similarly, spacecraft have observed the diminution of light as certain targeted stars passed behind the rings (Fig. 11.29).

Five bending waves and ~100 density waves in Saturn's rings have thus far been identified with the resonances responsible for exciting them and have been analyzed to determine the local surface mass density of the rings. The surface density at most wave locations in the optically thick A and B rings is ~30–60 g cm^{-2}. Measured values in the optically thin C ring and Cassini division are ~1 g cm^{-2}. As the ratio of optical depths are ~ 10, the larger magnitude of the difference in surface densities implies that the average particle size in the C ring and Cassini division is smaller than that in the B and A rings.

The damping behavior of spiral density and bending waves has also been analyzed to place upper bounds on the viscosity and local thickness of the rings. The A ring

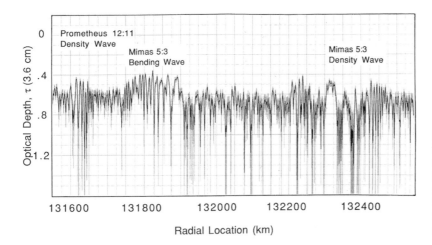

Figure 11.27 Examples of wave features observed in the radio occultation data of Saturn's A ring. The solid curve is measured normal optical depth, $\tau(r)$, plotted to increase downward. The gray shaded region represents the 70% confidence bounds on the measurement. (Rosen 1989)

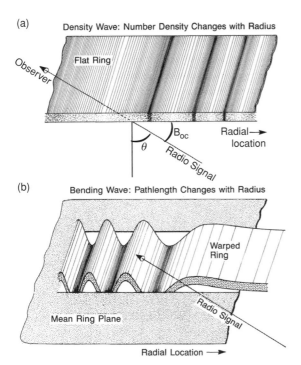

Figure 11.28 Schematics of a radio occultation of (a) a spiral density wave and (b) a bending wave. (Adapted from Rosen *et al.* 1991)

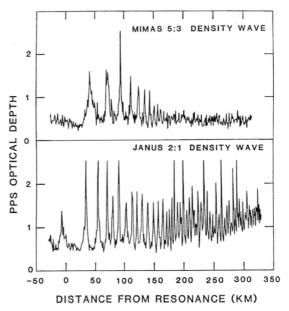

Figure 11.29 The Mimas 5:3 and Janus 2:1 density waves as viewed from the Voyager 2 PPS stellar occultation of the star δ Sco with $B_{oc} = 28.7°$, plotted so that $\tau(r)$ increases upwards. Note the sharp peaks and broad flat troughs caused by nonlinearities (Fig. 11.26). See the caption to Fig. 11.7 for information about interpreting optical depths measured by occultation. (Esposito 1993)

appears to have a local thickness of a few tens of meters or less; the thickness of the C ring is $\lesssim 5$ m. Complications resulting from wave nonlinearities and nonuniformities in disk properties make measurements of viscosity from the damping of spiral waves less reliable than corresponding techniques used to estimate ring surface mass density.

In summary, resonantly excited spiral density waves and bending waves are among the best understood features in planetary rings. Waves are seen to propagate from all of the strong satellite resonances, except those resonances that produce gaps. The locations of these waves agree with predicted values to within the observational uncertainties, which in some cases are less than 1 part in 10^5. The wavelength behavior also agrees with the theory, and wavelength analysis has been used to obtain the best available estimates of the surface mass density of the rings. The damping behavior of spiral waves appears to be very complex, and theoretical studies suggest that damping rates may be very sensitive to particle collision properties.

11.4.2.4 Angular Momentum Transport

The Voyager and Cassini spacecraft found a multitude of density waves excited by small satellites orbiting near the rings. The torque at these individual resonances is substantially less than that at Mimas's 2:1 resonance, but the sum of their torques is of comparable magnitude. These waves are observed with amplitudes that agree with theoretical predictions to within a factor of order unity.

The back torque that the rings exert upon the inner moons causes these moons to recede on a timescale short compared to the age of the Solar System; current estimates suggest that Atlas, Pandora, Prometheus, and Janus/Epimetheus should all have been at the outer edge of the A ring within the past ~2×10^8 years, with the journey of Prometheus, a relatively large moon located quite close to the A ring, occurring on a timescale of $\lesssim 20$ million years. Resonance locking to outer, more massive, moons could slow the outward recession of the small inner moons; however, angular momentum removed from the ring particles should force the entire A ring into the B ring in $<10^9$ yr. If the calculations of torques are correct, and if no currently unknown force counterbalances them, then the small inner moons and/or the rings must be 'new', i.e., much younger than the age of the Solar System. However, a 'recent' origin of Saturn's rings appears to be a priori highly unlikely. We return to this problem when we discuss the origins of planetary ring systems in §11.7.

11.4.3 Shepherding

Moons and rings repel one another through resonant transport of angular momentum via density waves. As is the case for viscous spreading of a ring, the majority of the angular momentum is transferred outwards and most of the mass inwards (in this case, the ring and moon are considered together as parts of the same total reservoir of energy and angular momentum). This is a general result for dissipative astrophysical disk systems, as a spread-out disk of material on circular (nearly) Keplerian orbits has a lower energy state for fixed total angular momentum than does more radially concentrated material. Analogously, resonant transfer of energy from an inner moon to a moon on an orbit further out frees up energy for tidal heating (Problem 2.29), whereas transfer in the opposite direction is almost always unstable (i.e., the resonance lock is only temporary).

We now consider the process of *shepherding*, by which a moon repels ring material on nearby orbits. The essence of the interaction is that ring particles are gravitationally perturbed into eccentric orbits by a nearby moon and collisions among ring particles damp these eccentricities. The net result is a secular repulsion between the ring and the moon. Details of the interactions depend on whether a single resonance dominates the angular momentum transport or the moon and ring are so close that individual resonances don't matter, either because the resonances overlap or because the synodic period between the ring and the moon is so long that collisions damp out perturbations between successive close approaches. However, many aspects of the basic qualitative picture are the same in both cases.

A moon exerts its strongest torques on a ring particle when they are near one another. Most of the interaction during a single encounter thus occurs near conjunction. If both the ring particles and the moon are initially on circular orbits, the perturbations received by each ring particle can only depend on its semimajor axis. By conservation of the Jacobi parameter (eq. 2.21), the amplitude of the eccentric excursions induced in the ring particles, ae, is much larger than the change in their semimajor axes. Moreover, for given semimajor axes of the moon and particle, any induced eccentricity can only depend on the longitude of the particle's orbit relative to where it passed the moon. Thus, if we may approximate the encounter by an impulse and the moon initially sends ring particles inwards, the particles will execute epicyclic motion, and be at periapse in $\frac{1}{4}$ orbit, at apoapse $\frac{3}{4}$ orbit after encounter, periapse at $\frac{5}{4}$ orbit after encounter, etc. A plot of radius vs. time would be a sinusoid. In fact, the shape of a ring's edge represents just such a plot. The motion of the particles relative to the moon is nearly constant, as long as the difference in their semimajor axes is large compared to the size of the induced epicyclic motion. Thus, distance along the edge is essentially a measure of time, with the normalization being the synodic velocity. The wavelength of the induced oscillation is the relative motion of the ring and moon during one epicyclic period:

$$\lambda_{\text{edge}} = 3\pi \ |\Delta a| \ \frac{n}{\kappa} \approx 3\pi \ |\Delta a|, \tag{11.23}$$

where $\Delta a \equiv a_r - a_s$ is the separation in the orbital semimajor axes of the ring and moon (Fig. 11.30). Calculation of the amplitude of the oscillation is more complicated. Perturbations before and after encounter almost cancel out, except for a second-order (in torque) term that results from the particles having been pulled slightly closer to the moon by the encounter, and thus the torque being slightly larger after closest approach than before. Particles initially on circular orbits about a planet of mass M_p are excited by

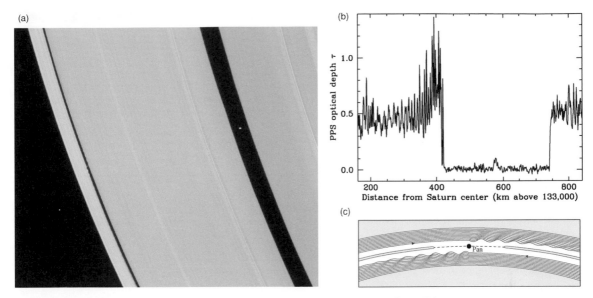

Figure 11.30 (a) This Cassini image of Saturn's A ring shows the gap-forming moons Pan within the Encke gap and Daphnis in the Keeler gap. Pan, with a radius of 14 km, can easily be discerned here. Daphnis, only 4 km in radius, is a mere speck, although its presence is made obvious by the edge waves it creates in the surrounding ring material. Pan also raises waves in the edges of the Encke gap (see Fig. 11.24). However, even though Pan is more massive than Daphnis, Pan is further from the edges of its gap than is the smaller moon. This causes Pan's edge waves to have a much longer wavelength (eq. 11.23) and a smaller amplitude, making them more difficult to see, except when the azimuthal direction appears foreshortened. This view looks toward the sunlit side of the rings from about 24° south of the ring plane. Resolution is 5 km per pixel, and the phase angle is 21°. (NASA/JPL/CICLOPS, PIA08926) (b) Optical depth obtained from the Voyager stellar occultation profile of the region of the Encke gap and environs. The regular pattern of oscillations in the region of the A ring immediately interior to the gap represents a cross-section of optical depth variations resulting from Pan's satellite wake. These data are from a different segment of the same stellar occultation used to produce Fig. 11.29. (Showalter 1991) (c) Schematic illustration of the pattern that the moonlet Pan creates on and near the edges of the Encke gap in Saturn's rings. This pattern remains stationary in the frame rotating with the moonlet's orbital frequency. The radial scale is greatly exaggerated relative to the angular scale for clarity.

the gravitational tug of a moon of mass M_s to eccentricities of

$$e \approx 2.24 \frac{M_s}{M_p} \left(\frac{a}{\Delta a} \right)^2. \tag{11.24}$$

The shape and amplitude differ if the ring or moon has an initial eccentricity, or if ae is not small compared to $|\Delta a|$, but the wavelength remains the same as long as the potential is Keplerian and $|\Delta a| \ll a$. Wavy edges of Encke's gap excited by the moonlet Pan are visible in Voyager and Cassini images (Fig. 11.24). This pattern satisfies the wavelength relationship given by equation (11.23), and the mass of Pan has been estimated from the observed wave amplitudes via equation (11.24). Likewise, the edge waves excited by Daphnis have been observed in the Keeler gap (Fig. 11.22). The characteristic length scales of longitudinal variations in Saturn's

F ring are comparable to wavelengths predicted from forcing by the nearby moon Prometheus according to equation (11.23), but the situation is more cluttered and complicated than for the edges of the Encke and Keeler gaps.

The observable manifestations of Pan's interactions with Saturn's A ring extend beyond the wavy edges of Encke's gap. Pan also excites eccentricities of ring particles more distant than the gap edges, producing a pattern that decreases in amplitude and increases in wavelength with increasing distance from the moonlet. The variations in particle response with semimajor axis produce a *satellite wake* which is observable as radial variations in ring optical depth (Fig. 11.24). The wake can be used to obtain an estimate of Pan's mass which is more accurate than that provided by available measurements of the amplitude of the wavy edges of Encke's gap.

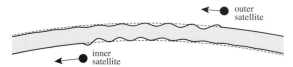

Figure 11.31 Schematic illustration of the shepherding of a planetary ring by two moons. Ring particles' eccentricities are excited as the particles pass by a moon. Interparticle collisions subsequently damp eccentricities and leave the particles orbiting further from the moon than prior to the encounter. Particles orbiting closest to the moon are affected most strongly. (Murray and Dermott 1999)

In order to study the shepherding problem in more detail, let us move to the frame in which the moon's position is fixed (the frame rotating at the moon's orbital frequency). For our analysis, we assume that the moon's orbit is circular and exterior to that of the ring. We further assume that the separation in the orbital semimajor axes of the ring and moon is small, $|\Delta a| \ll a$, but that they are far enough apart that the moon's perturbations alter the radius of the ring's orbit by an amount $ae \ll |\Delta a|$. Interactions among the ring particles are ignored for the brief interval near conjunction with the moon, during which essentially all of the angular momentum transport occurs, but the orbits of the ring particles are assumed to be circularized between successive encounters with the moons as a result of interparticle collisions. Note that the eccentricity of the ring particles must be damped out by collisions, at least partially, before the next encounter with the moon, or the reverse angular momentum transfer may occur, and over long periods of time the net torque is zero. The system is sketched in Figure 11.31.

As the ring particles approach a moon that orbits further from the planet than they do, they are pulled in the forward direction by its gravity, thereby increasing the particles' energy (and angular momentum) and causing their orbits to move closer to that of the moon. The magnitude of the force is proportional to $M_s/\Delta a^2$. However, after the ring particle passes the moon, the tug is in the opposite direction, and removes energy and angular momentum. To the lowest order, the forces cancel, but a small asymmetry exists because the particle's orbit is slightly closer to that of the moon on the outbound journey as a result of perturbations prior to conjunction. The net torque is thus second order in the forcing. A more detailed analysis gives the magnitude of the torque to be

$$T_g \approx 0.40 \frac{\Delta a}{|\Delta a|} \sigma_\rho \left[\frac{GM_s}{n(\Delta a)^2} \right]^2. \qquad (11.25)$$

A moon thus pushes material away from it (on both sides). Ring material between two moons is forced into a narrow annulus between them. The annulus should be

Figure 11.32 The sunlit face of Saturn's multistranded F ring and its companion moons, as imaged by the Cassini spacecraft from 6–7° above the ring plane. Prometheus, the inner moon, is more massive than the outer moon, Pandora, and orbits closer to the ring, so it exerts stronger perturbations. Note that the long axes of these decidedly nonspherical moons are aligned with the direction to the planet, which is expected because this is the lowest energy state for a synchronously rotating moon. (NASA/JPL/CICLOPS, PIA07712)

located nearer the smaller moon, so that the torques balance. Viscous diffusion maintains a finite width of this annulus. Saturn's F ring is confined between the moons Prometheus and Pandora, but additional processes are also clearly operating to create the complicated features observed (Fig. 11.32). The shepherding moons Cordelia and Ophelia confine the uranian ϵ ring, although in this case the ring edges are maintained by individual 'isolated' resonances of these moons. Uranus's other narrow rings are also believed to be confined by shepherding torques; a few ring edges may be maintained by resonances with Cordelia and Ophelia, but the small moons that this model requires to hold most of Uranus's narrow rings in place have yet to be observed.

What about a small moon embedded in the middle of a ring plane? Material is cleared on both sides, so a gap is formed around it. Diffusion acts to fill in the gap and blur the edges, but optical depth gradients can occur on length scales comparable to typical particle collision distances, allowing sharp edges to be produced. If a moon is too small, then it can't clear a gap larger than its own size, so no gap is formed. As discussed above, Encke's gap is cleared by the moonlet Pan, and the smaller moonlet Daphnis maintains the Keeler gap.

11.4.4 Longitudinal Confinement

Neptune's Adams ring contains prominent arc-shaped features that orbit at the Keplerian rate (Figs. 11.19 and 11.20). The arcs have been observed to persist for at least several years, which is more than their lifetime against Kepler shear (Problem 11.7). Thus, there must be some mechanism that confines the ring particles. A combination of Lindblad and corotation resonances of a moon or moons is capable of confining rings in both radius and longitude.

The most prominent example of a corotation resonance is the 1:1 commensurability. Jupiter's 1:1 corotation resonance is responsible for the confinement of the Trojan asteroids (§2.3.2.2 and §9.1), which librate in 'tadpole' orbits about Jupiter's triangular Lagrangian points. A similar form of confinement is not stable for a dissipative collisional system such as a planetary ring. The triangular Lagrangian points, L_4 and L_5, are potential energy maxima, and thus are unstable to most forms of dissipation, including interparticle collisions. A ring would gradually spread in both radius and longitude as a result of such collisions. An arc ring could be confined about one of the triangular Lagrangian points of a moon if a second moon on a nearby or nearly resonant orbit exerted a shepherding torque on the ring at a Lindblad resonance.

Moons on eccentric or inclined orbits have corotation torques with various pattern speeds, and these torques can provide azimuthal confinement at orbital radii different from that of the moon. Although such corotation resonances are usually much weaker than the 1:1 resonance, in a nearly Keplerian potential (such as those of all four ringed planets in the Solar System) these other corotation resonances are associated with a nearby Lindblad resonance, which may provide the torque required to counteract dissipation. Thus, an arc ring could be confined by the combination of a corotation resonance and a Lindblad resonance of a single moon. Indeed, dynamical models imply that the arcs in Neptune's Adams ring are confined in whole or in part by the nearby moon Galatea at its 42:43 corotation and Lindblad resonances. Large particles orbiting within the Adams ring may have a role in producing the detailed structure of the arcs.

11.5 Physics of Dust Rings

Thus far, we have analyzed the motions of ring particles by considering only the forces of gravity and physical collisions. While these are the dominant forces on ring particles that are $\gtrsim 1$ mm in size, micrometer-size dust is significantly affected by electromagnetic forces. Radiation forces, primarily Poynting–Robertson drag (§2.7.2), affect all long-lived small particles in planetary rings. A planet's magnetic field can be an important influence on the motion of charged dust particles, with substantial effects occurring on both orbital and secular timescales.

11.5.1 Radiation Forces

A general introduction to the effects of radiation forces on the motion of Solar System particles was presented in §2.7. In Chapter 2, we concentrated on particles in heliocentric orbits. Here, we focus our discussion on those aspects of radiation forces pertinent to particles in planetocentric orbits. In most cases, radiation directly from the Sun dominates over radiation reflected from and emitted by the planet.

Radiation pressure has little effect on the orbits of most planetocentric grains. Even if the ratio of the solar radiation force to solar gravity, $\beta \equiv F_{\rm r}/F_{\rm g} \gtrsim 1$, the planet's gravity is usually far greater than the Sun's, so the perturbation is small. The shapes of the orbits of small grains are altered slightly, and forces on particles change when they enter and exit the shadows of planets. However, in most circumstances the net effects of these forces cancel out over time and they do not produce any secular evolution of particle orbits. A notable exception occurs for particles of radius $\sim 1 \pm 0.3$ μm in Saturn's E ring. The periapse precession rate induced by the Lorentz force (§11.5.2) on charged 1 μm grains at Enceladus's orbit is opposite in sign and approximately equal in magnitude to the precession resulting from Saturn's oblateness (§2.5.2). The very slow rate of periapse precession for particles within this narrow size interval that results from cancelation of precession rates allows perturbations to build up over many orbits, so that radiation pressure can alter almost circular orbits of micrometer-size dust into highly eccentric trajectories. Eruptions from the moon Enceladus (§5.5.6) are the source of the particles in the E ring, and this model explains both the large radial extent of the E ring (produced by particles on eccentric orbits) and the unusual narrowness of the particle size distribution.

Poynting–Robertson drag, in contrast, leads to substantial evolution of tiny particles within planetary rings over a much broader range of parameter space. The secular rates of change of orbital semimajor axis and eccentricity are given by

$$\frac{da}{dt} = -\frac{a}{t_{\rm pr}} \frac{5 + \cos^2 i_*}{6} \tag{11.26}$$

and

$$\frac{de}{dt} = 0, \tag{11.27}$$

respectively. In equation (11.26), i_* represents the inclination of the particle's orbit *to the planet's orbital plane* *about the Sun*. The characteristic decay time, t_{pr}, is approximately the time that it takes a particle to absorb the equivalent of its own mass in solar radiation and is given by

$$t_{pr} = \frac{1}{3\beta} \frac{r_\odot}{c} \frac{r_\odot}{GM/c^2} \approx 530 \frac{r_{\odot AU}^2}{\beta} \text{ yr}, \qquad (11.28)$$

where $r_\odot$ is the planet–Sun distance, and $r_{\odot AU}$ is this distance in astronomical units. The orbital eccentricity is thus constant, apart from minor short-period variations, and the particle's semimajor axis decreases in an exponential fashion. This contrasts to particles in heliocentric orbits, whose eccentricities are reduced by Poynting–Robertson drag (eq. 2.50*b*) and whose semimajor axes decrease more rapidly as they approach the Sun (eq. 2.50*a*). Poynting–Robertson decay times for microscopic grains given by equation (11.28) are short compared to the age of the Solar System, even for particles in orbit about Neptune. Jupiter's ring particles are thought to be ejecta from jovian moons, and subsequently to spiral inwards through Poynting–Robertson drag, thereby forming very broad rings.

11.5.2 Charged Grains

Ring particles orbit close to their planets, in environments characterized by high densities of energetic charged particles trapped by strong planetary magnetic fields. Uncharged dust grains are impacted by electrons more frequently than by ions because the thermal speeds of electrons are much larger. Grains thus acquire sufficient negative charge to achieve a balance between the rates at which they accumulate additional electrons and ions via electrostatic attraction and repulsion. For parameters typical of micrometer-size grains in planetary rings, equilibrium is achieved in less than an orbital period. In Jupiter's rings, grains reach equilibrium at a potential $\Phi_V \approx -10$ volts. The charge that a grain can accumulate depends upon its proximity to other grains as well as the charged particle environment; if other grains are within the plasma's Debye shielding length (the characteristic distance beyond which the electric field of the grain is counterbalanced by particles in the plasma having opposite charge, §7.1.2.1), then they contribute to the repulsion of additional electrons, and a given potential can be maintained with less charge per grain. For an isolated grain of radius R, the potential, Φ_V, and charge, q, are related according to

$$\Phi_V = -\frac{q}{R}. \qquad (11.29)$$

Other charging mechanisms, such as photoelectron currents, can perturb the equilibrium value of q. Stochastic variations in particle charge have minor effects on particle motions; however, the systematic variations in charge experienced by grains moving into and out of

Saturn's shadow or on highly eccentric orbits can significantly affect particle trajectories.

The motion of charged grains within planetary magnetospheres is influenced by trapped plasma, as well as by the planetary magnetic field itself. Electric and magnetic forces are most important for very small particles, because mass increases much faster than mean charge as particle radius grows. The collisions of charged particles with dust grains result in an exchange of angular momentum between the plasma and dust, a process known as *plasma drag*. The drag force depends upon the velocity of the grain relative to the plasma. The plasma rotates with the planet's magnetic field, so grains orbiting at the corotation radius, r_c, don't move relative to the plasma and thus feel no drag. Grains orbiting interior to r_c lose energy and angular momentum to the plasma and spiral inwards, whereas grains at $r > r_c$ gain energy and spiral outwards (unless energy losses from Poynting–Robertson drag exceed the energy gains from plasma drag, as is the case for Jupiter's rings). Jupiter's corotation radius is located at $r_c = 2.24$ R₂, within the gossamer rings. Although most of Jupiter's rings are formed from particle ejecta off Jupiter's moons brought inwards by Poynting–Robertson drag (§11.6.3), the feeble outward extension of the gossamer Thebe ring may have formed from a shadow resonance that arises from the abrupt shutoff of photoelectric charging when a dust particle enters Jupiter's shadow.

Charged dust grains are acted upon by the planet's magnetic field via the Lorentz force (eq. 7.44):

$$\mathbf{F}_L = \frac{q}{c} \mathbf{v} \times \mathbf{B}. \qquad (11.30)$$

The Lorentz force couples charged dust grains to the magnetic field. The Lorentz force is especially important for Jupiter's rings, because they have many small particles, are located close to the planet, and Jupiter's magnetic field is very strong and inclined with respect to its rotation axis. The jovian Lorentz force is approximately 1% of Jupiter's gravitational force for typical ring particles (Problem 11.18). The tilt of the field with respect to the rotation axis implies that particles not orbiting at r_c experience a time-varying force as they move relative to the field (Fig. 11.33). At certain locations, the frequency at which a particle experiences variations in the Lorentz force is commensurate with the particle's epicyclic or vertical frequency, leading to *Lorentz resonances*. Lorentz resonances have many things in common with the gravitational resonances discussed in §11.4.1, but there are also several differences. The Lorentz force affects particle mean motions, so orbital, radial, and vertical frequencies are functions of charge-to-mass ratio. Thus, the locations of Lorentz

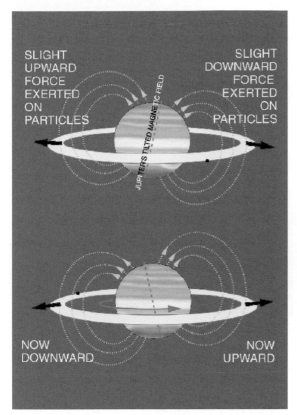

 Figure 11.33 This graphic illustrates the effect of the magnetic force (shown as arrows) on charged ring particles. Since Jupiter's dipolar magnetic field (shown dotted) is tilted about 10° from the planet's spin axis (vertical in these sketches), the direction of this out-of-plane force depends on where the particle is in the ring plane and on Jupiter's rotational orientation. With Jupiter's orientation as shown in the top panel, the magnetic force on a charged ring particle has a slightly upward (downward) force on particles at the left (right) of Jupiter. This situation is reversed in direction five hours later, after Jupiter has rotated 180° (into the orientation depicted in the bottom panel). Thus, every charged ring grain experiences an oscillating vertical force; the period of these forces depends on the orbital radius, so that at certain locations the periods become multiples of the particles' orbital periods, which leads to Lorentz resonances. (Courtesy J.A. Burns)

resonances vary (slightly) with particle size, in contrast to the sharply defined gravitational commensurabilities. The consequences of forcing at Lorentz resonances in Jupiter's rings also differ substantially from those at vertical and Lindblad resonances in Saturn's rings. Self-gravity and collisions are important in Saturn's rings, whereas self-gravity is negligible for Jupiter's ring system and collisions are very infrequent. The orbits of (sub)micrometer-sized inward-migrating particles are perturbed significantly in the direction perpendicular to the ring plane by the strong Lorentz 3:2 resonance located at 1.71 R_J. These orbits become much more inclined, thus forming the halo. At the location of the Lorentz 2:1 resonance, 1.40 R_J, the particle orbits are perturbed again, this time onto orbits which impact the planet, leading to a loss of halo particles. The normal optical depth of the halo is substantially reduced inwards of 1.40 R_J.

11.5.3 Spokes in Saturn's Rings

Electric and/or magnetic effects are responsible for *spokes*, the only known planetary ring features that are predominantly radial in shape. They are centered within the B ring, where particles orbit synchronously with Saturn's magnetic field. Spokes appear darker than their surroundings in backscatter, but brighter in forward scatter (Fig. 11.34). The strongly forward-scattered appearance of spokes implies they contain a significant component of micrometer- and submicrometer-sized dust grains. Spokes exhibit the greatest contrast in backscatter when the tilt angle of the ring plane to the Sun is small; this enhanced visibility at low tilt angle implies that the vertical thickness of the dust is greater than that of the macroscopic particle layer.

Spokes form rapidly, on timescales of minutes to tens of minutes, and initially appear as linear features pointing towards the center of Saturn. One edge remains radial (and evolves with the period of Saturn's magnetic field) as long as new material is being added to the spoke. Dust in the spokes orbits at essentially the Keplerian rate, so spokes get smeared out as they age. The spoke fades gradually as dust is reaccumulated by the larger ring particles, and disappears from view in about one-fourth to one-third of an orbital period. Spokes occur at all azimuths, but not with equal frequency. The formation of spokes occurs most frequently soon after the ring emerges from Saturn's shadow. Spoke formation is also strongly correlated with magnetic field longitude, and is weakly correlated with the longitude of certain cloud features near Saturn's equator, where strong winds cause an ~5% reduction in rotation period relative to the rotation period of Saturn's magnetic field. A much larger number of spokes have been observed near saturnian equinox than when the rings are more open to solar illumination. No consensus exists concerning the physical mechanism responsible for spoke formation.

11.6 Meteoroid Bombardment of Rings

Planetary rings have very large surface area to mass ratios, and thus are heavily bombarded with and significantly affected by the flux of small stray debris that is present throughout our Solar System. Such debris can change the orbits and composition of planetary rings, and may be responsible for the formation and destruction of planetary ring systems.

(a)   (b)

Figure 11.34 Two Voyager images of spokes in Saturn's B ring. Both images show the lit face of the rings, but the spokes look quite different because of the differing phase angles of the observations. The spokes appear dark in frame (a), which was taken in backscattered light, whereas they are brighter than the surrounding ring material in frame (b), which was imaged in forward scattered light with a resolution of ~80 km/pixel. The black dots in the images are reseau markings. (Within the optics of the Voyager cameras was a grid of black dots called 'reseau markings'. These appear superimposed upon each image sent back from Voyager, and are used to correct for geometric distortions in the camera. They are often suppressed in published images by replacing them by the local average brightness. Nevertheless, you can almost always find them if you look closely. Reseau marks were used for most early spacecraft images, but they are not needed for CCD cameras, which are intrinsically more stable.) (NASA; a: PIA02275; b: Voyager 1, FDS 34956.55)

When hypervelocity dust strikes a ring particle, it can excavate ~10^4–10^5 times its mass in impact ejecta. Some of the debris from the impact is permanently lost to the rings, either by flying off on orbits which are unbound or intersect the planet, or by being vaporized/ionized and subsequently escaping. However, under most circumstances, the bulk of the ejecta is re-accreted by the ring system, although not necessarily at the same distance from the planet from which it originated. In this section, we first consider the changes in mass, composition, and total ring angular momentum resulting from meteoroid impacts, and then discuss ballistic transport of ejecta and its possible role in creating ring structure. Finally, we consider impact ejecta from moons as a source of ring material.

11.6.1 Accretion of Interplanetary Debris

Interplanetary impactors add material to ring systems, but they also remove matter from the rings. The net effect is uncertain; either a gain or loss may result, depending upon many factors including impact speed, proximity to the planet's atmosphere, depth within the planet's gravitational potential well, and the fragility and volatility of the materials involved. The impactor flux is also quite uncertain, but is consistent with Saturn's rings being bombarded with their own mass of material over the age of the Solar System, so material losses and gains from impacts may be a significant factor in the evolution of planetary ring systems.

Planetary rings possess substantial net orbital angular momentum, as all particles orbit in the same direction. Impacting debris from heliocentric orbits provides essentially no net angular momentum to an optically thick ring system. (Optically thin rings accrete a net negative angular momentum flux because bodies traveling in the retrograde direction have larger impact probabilities (Problem 11.20).) Ejecta lost from the ring system generally take with them 'positive' angular momentum. Thus, in most circumstances, the net effect of interplanetary impactors is to cause a ring system to lose (specific) angular momentum, and thus to decay slowly inwards.

Impacting debris can change the mineralogical composition of planetary rings in three ways: New material brought to the ring system is, on average, roughly solar in composition (apart from a substantial depletion in volatiles). The more volatile and fragile components of the rings are preferentially lost from the ring system. Some of these volatiles can remain temporarily near the ring in a gaseous state, producing a *ring atmosphere*. High pressures and temperatures during the impacts produce chemical and mineralogical changes (§5.4.2).

The most interesting observable chemical consequences of interplanetary debris on planetary rings may well be the 'pollution' of Saturn's rings. Observations suggest that Saturn's rings are almost pure water-ice. The largest fraction of other material is observed in regions of low optical depth, which should receive the largest fractional contamination by interplanetary debris. The age of Saturn's rings estimated by attributing all of the observed darkening to interplanetary debris is on the order of 10^8 years. Although the uncertainties in the meteoroid flux and vaporization/loss fractions are quite high, it is interesting to note that this timescale is consistent with age estimates from density wave torques, and suggests that Saturn's ring system is not primordial.

11.6.2 Ballistic Transport

Most of the ejecta from hypervelocity impacts leave ring particles at speeds that are much smaller than the particles' orbital speed. This debris thus travels on orbits of low eccentricity and inclination, and unless the ring optical depth is very small, most ejecta re-impact ring particles within a few orbital periods. The rates at which a given region of a ring gains and loses material by this process depends on its optical depth and the optical depth of neighboring regions, within a 'throw distance' of the ejecta. Structure can be formed by this *ballistic transport*, especially near abrupt boundaries in ring optical depth. Numerical simulations of ballistic transport have successfully reproduced the ramp-like structure seen at the boundaries between the C ring and B ring and between the Cassini division and the A ring (Fig. 11.35), and provide the most plausible explanation for such features.

11.6.3 Mass Supplied by Satellite Ejecta

The mass of a planetary ring, M_{ring}, can increase as a consequence of impacts on a moon:

$$\frac{dM_{ring}}{dt} = f_i Y_e Y_i \pi R^2, \tag{11.31}$$

where R is the radius of the moon, f_i is the mass flux density of hypervelocity impactors, and Y_i is the impact yield or the

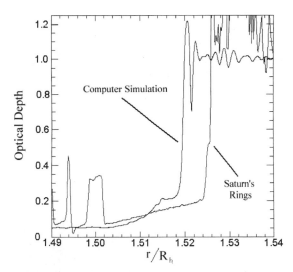

Figure 11.35 Comparison between observations of the transition region from Saturn's optically thin C ring to the optically thick B ring and a numerical simulation of the pattern produced by ballistic transport resulting from a few million years of meteoroid bombardment. The observed optical depth profile was obtained from Voyager 1 images of the ring taken against the planet's limb. The initial profile in the simulation was assumed to be sharp-edged, but the tendency for viscosity to spread such an edge was also included. Note the similarity between the sharpness of the inner B ring edge as well as the gradual ramp located just interior to it in both curves. (Courtesy Richard Durisen; for details about the simulations, see Durisen *et al.* 1996)

ratio of ejected mass to projectile mass, which is typically of the order of $40v_i^2$ (i.e., roughly proportional to kinetic energy), with the impact velocity v_i in km s^{-1}. Based upon empirical fits to hypervelocity cratering experiments, the fraction of ejecta that escapes the moon, Y_e, can be approximated by

$$Y_e \approx \left(\frac{v_{min}}{v_e}\right)^{9/4}, \tag{11.32}$$

with v_{min} the minimum speed at which ejecta are launched (typically 10–100 m s^{-1}) and v_e is the escape velocity. For an isolated moon, $v_e \propto R$, and is $\sim$10–100 m s^{-1} for spheres with sizes similar to Jupiter's small inner moons. Using these approximations one can show that $Y_e \propto R^{-9/4}$, and therefore

$$\frac{dM_{rings}}{dt} \propto R^{-1/4} \quad (v_e > v_{min}), \tag{11.33a}$$

$$\frac{dM_{rings}}{dt} \propto R^2 \quad (v_e < v_{min}). \tag{11.33b}$$

Thus smaller moons, even though they have less surface area, provide more material than larger ones because their gravitational potential well is less deep. If the moons reside near or interior to the Roche limit, the ejecta losses by impacts are even larger. The optimum source for ring material is a satellite for which $v_e \approx v_{min}$. If the moon has a density of 2 g cm^{-3} and is covered by a soft regolith, it would be an optimum source for ring material if its radius $R \approx$ 5–10 km, roughly the size of Adrastea and of Uranus's moon Mab.

Ejecta from impacts on the jovian satellites Thebe, Amalthea, Metis, and Adrastea migrate inwards via Poynting–Robertson drag, producing the morphology of the jovian ring system. The ejecta start on orbits that resemble the orbit of the parent satellite. The orbits precess quickly (4 and 8 months at 2.5 and 3.1 R$_2$, respectively), while the particles spiral slowly inward due to Poynting–Robertson drag; meanwhile, the inclinations of the orbits are preserved. Within a relatively short period of time, the ensemble of particle orbits resembles a donut with a vertical thickness determined by the inclination of the parent satellite. The top and bottom edges are brightest, since a particle spends most of its time near its turning points. The inclination of the orbits of Thebe, Amalthea, and Metis/Adrastea are 1.1°, 0.4°, and 0.0°, respectively. The vertical extent of the Thebe ring is thus largest, as shown on the images (Fig. 11.6), while Jupiter's main ring lies in the equatorial plane. Although the Thebe and Amalthea gossamer rings extend inwards beyond Metis, they are outshone by the main ring.

11.7 Origins of Planetary Rings

Are ring systems primordial structures (dating from the epoch of planetary formation, $\sim 4.5 \times 10^9$ years ago) that are remnants of protosatellite accretion disks, or did they form more recently, as the result of the disruption of larger moons or interplanetary debris? Various evolutionary processes that occur on timescales far shorter than the age of the Solar System imply that tenuous, dust-dominated, and/or narrow rings must be geologically very young. However, such a recent origin of Saturn's main rings is a priori quite unlikely.

Micrometer-size dust is removed from rings quite rapidly. Some loss mechanisms lead to permanent removal of grains from ring systems. Processes leading to permanent loss of grains include: sputtering, which is dominant for Jupiter's dust; gas drag, which is important for particles orbiting Uranus because of that planet's hot extended atmosphere; and Poynting–Robertson drag. Other dust removal mechanisms, such as re-accretion by large particles, which dominates for the dust comprising Saturn's spokes, allow for recycling. In all ring systems, the dust requires continual replenishment, which means macroscopic parent particles. The dust mass is so small that in

most cases a quasi-steady state could exist over geologic time.

If rings don't last long, why do we observe them around all four giant planets? This question is most relevant to the case of the saturnian ring system, as it is far more massive than the rings of other planets. In the cases of Jupiter, Uranus, and Neptune, it is easier to envisage the current ring/moon systems as remnants of eons of disruption, re-accretion (to the extent possible so near a planet), and gradual net losses of material. In the previous section we showed that Jupiter's ring system is formed from impact ejecta from small moons. That theory, in fact, suggests that all small moons orbiting near planets (including the martian satellites Phobos and Deimos) should cause the formation of a ring.

Let us examine some specific origin scenarios for the macroscopic particles present within planetary rings. A stray body passing close to a planet may be tidally disrupted (e.g., Comet D/Shoemaker–Levy 9, see §10.6.4). However, under most circumstances, the vast majority of the pieces escape from or collide with the planet. Moreover, such an origin scenario does not explain why all four planetary ring systems orbit in the prograde direction. And this mechanism doesn't easily solve the 'short timescale' problems because the flux of interplanetary debris has decreased substantially over geologic time (at least in the inner Solar System, where we have 'ground truth' from radioisotope dating of lunar craters; see §5.4.4.1). Ring particles are thus most likely (possibly second or later generation) products of circumplanetary disks.

Planetary rings may be the debris from the disruption of moons that got too close to their planets and were broken apart by tidal stress, or that were destroyed by impacts and did not re-accrete because of tidal forces. There are two difficulties with the hypothesis that the rings that we observe are the products of recent disruptions of moons: Why did the rings form recently, i.e., why is now special? Can moons form at or move inwards to the radii where planetary rings are seen? It is more difficult for a body to accrete than to remain held together (Problem 11.2). A ring parent moon must form beyond the planet's Roche limit and subsequently drift inwards. Tidal decay towards a planet only occurs for moons inside the orbit that is synchronous with the planet's spin period (unless the moon's orbit is retrograde), but Roche's limit is outside the synchronous orbit for Jupiter and Saturn. However, the presence of moons interspersed with ring particles in all four planetary ring systems (Problems 11.3 and 11.4) argues strongly for the viability of this model, especially for the rings of Jupiter, Uranus, and Neptune, which are less massive than (or, in the case of Uranus possibly of comparable mass to) the nearby moons.

Thus, a major outstanding issue in planetary rings is the origin and age of Saturn's main ring system. The strongest evidence for a geologically recent origin of Saturn's rings is orbital evolution of rings and nearby moons resulting from resonant torques (responsible for density wave excitation) and ring pollution by accretion of interplanetary debris. Testing the validity of satellite torques is especially important, because models also suggest that angular momentum transport via resonant torques and density waves is a significant factor in the evolution of other, less well observed, astrophysical disk systems, such as protoplanetary disks and accretion disks in binary star systems. For example, density wave torques may lead to significant orbital evolution of young planets within the protoplanetary disks on timescales of $\sim 10^5$–10^6 years (§13.8).

11.8 Summary

Planetary rings are a diverse lot. Jupiter's rings are broad but extremely tenuous. Most of the light observed from the jovian ring system has been scattered off micrometer-size silicate dust. Saturn's main rings are broad and optically thick, with most of the area covered by centimeter–meter bodies composed primarily of water-ice; Saturn's broad outer rings are tenuous and consist of micrometer-size, ice-rich particles. Most of the mass in the uranian rings is confined to narrow rings of particles similar in size to the bodies in Saturn's main ring system, but much darker in color. Neptune's brightest ring is narrow and highly variable in longitude; more tenuous, broader rings have also been observed around Neptune. The fraction of micrometer-size dust in Neptune's rings is larger than in Saturn's rings and Uranus's rings, but smaller than in Jupiter's. The particles in Neptune's rings have very low albedos.

The reason(s) why Saturn's main ring system is divided into four major components, and the causes of the major ring boundaries, remain for the most part unexplained, although there are two major exceptions. The classical identification of the outer edge of the B ring with the Mimas 2:1 inner Lindblad resonance was confirmed by Voyager, and the unexpected sharpness and noncircularity of this edge have been explained theoretically. The outer edge of the A ring has been identified with the location of the 7:6 resonance of the coorbital moons Janus and Epimetheus. The two strongest resonances in the ring system are thus responsible for the two major ring outer boundaries.

Attempts to explain inner boundaries of the rings have been less successful. In order to maintain an inner boundary, positive angular momentum must be transferred to the ring particles. If angular momentum is transferred inwards from an outer moon, excess energy, above that which is required for circular orbits, must be supplied. As this energy loss would rapidly damp the perturbing moon's eccentricity (or inclination), the mechanism is not effective over long periods of time, and some other process must be responsible for the maintenance of inner edges. This problem is absent for outer ring boundaries; the transfer of angular momentum outwards in this case leads to increase in noncircular orbital energy which can be dissipated by interparticle collisions. (An analogous situation exists for the stability of orbital resonances between moons to tidal evolution. If the inner moon of a resonant pair is tidally forced outwards faster, excess energy is available, and can lead to tidal heating (Problem 2.29), whereas if the outer moon is tidally receding from the planet more rapidly, the lock is unstable.) Ballistic transport resulting from cratering and disruptive impacts by hypervelocity micrometeoroids on ring particles has the potential of reproducing the morphology of the inner edges of rings A and B, but it requires special initial conditions, and does not explain why the edges are positioned as observed.

Moons produce rings (jovian ring system, Saturn's E ring, Uranus's μ ring), and moons cause much of the identified ring structure. Electromagnetic forces influence the motion of small grains. Some structure is produced by internal instabilities within rings, and ballistic transport of ring material might also play a significant role. However, these processes do not explain all of the diversity observed in planetary ring systems, and the identification of other structure-producing mechanisms in planetary rings remains an active research area.

Further Reading

The following review book provides a comprehensive overview of our knowledge of planetary rings at that date:

Greenberg, R., and A. Brahic, Eds., 1984. *Planetary Rings*. University of Arizona Press, Tucson. Particular attention should be given to the articles by Cuzzi *et al.* (Saturn's rings), Burns *et al.* (ethereal rings), and Shu (spiral waves).

More recent reviews are available for the individual ring systems:

Grün, E., I. de Pater, M. Showalter, F. Spahn, and R. Srama, 2006. Physics of dusty rings: History and perspective. *Planetary and Space Science*, **54**, 837–843.

Burns, J.A., D.P. Simonelli, M.R. Showalter, D.P. Hamilton, C.C. Porco, H. Throop, and L.W. Esposito, 2004. Jupiter's ring–moon system. In *Jupiter: Planet, Satellites and*

Magnetosphere. Eds. F. Bagenal, T. E. Dowling, and W. McKinnon. Cambridge University Press, Cambridge. pp. 241–262.

Cuzzi, J.N., *et al.*, 2002. Saturn's rings: Pre-Cassini status and mission goals. *Space Science Reviews*, **118**, 209–251.

French, R.G., P.D. Nicholson, C.C. Porco, and E.A. Marouf, 1991. Dynamics and structure of the uranian rings. In *Uranus*. Eds. J.T. Bergstrahl, E.D. Miner, and M.S. Matthews. University of Arizona Press, Tucson, pp. 327–409.

de Pater, I., M. Showalter, and B. Macintosh, 2008. Structure of the jovian ring from Keck observations during RPX 2002–2003. *Icarus*, **195**, 348–360.

Showalter, M.R., I. de Pater, G. Verbanac, D.P. Hamilton, and J.A. Burns, 2008. Properties and dynamics of Jupiter's gossamer rings from Galileo, Voyager, Hubble and Keck images. *Icarus*, **195**, 361–377.

Porco, C.C., P.D. Nicholson, J.N. Cuzzi, J.J. Lissauer, and L.W. Esposito, 1995. Neptune's ring system. In *Neptune*. Ed. D.P. Cruikshank. University of Arizona Press, Tucson, pp. 703–804.

Problems

11.1.E (a) Derive equation (11.7) from equation (11.6). (Hint: Recall that the analysis assumes that the planet is spherical.)

(b) Derive equation (11.7) from equation (2.22).

11.2.I In §11.1, we estimated the limits for tidal stability of a spherical satellite orbiting near a planet by equating the self-gravity of the satellite to the tidal force of the planet at a point on the satellite's surface which lies along the line connecting the planet's center to the satellite's center. The results of this study are directly applicable to the ability of a spherical moon to accrete a much smaller particle that lands on the appropriate point on its surface. Perform a similar analysis for the mutual attraction of two spherical bodies of equal size and mass whose centers lie along a line which passes through the planet's center. The result is known as the *accretion radius*. Comment on the qualitative similarities and quantitative differences between the accretion radius, the tidal disruption radius estimated for a single spherical body (eq. 11.7), and Roche's tidal limit (eq. 11.8), which was calculated for a deformable body.

11.3.E Calculate the Roche limit of each of the giant planets as a function of satellite density. Compare your results with the observed positions and sizes of some of these planets' inner satellites and rings, and comment.

11.4.E Calculate the densities required for Neptune's six inner satellites assuming each is located just exterior to the Roche limit for its density. Are these densities realistic? What holds these moons together?

11.5.E Why are planetary rings so flat? Why do they always orbit in a planet's equatorial plane?

11.6.E If a ring has a normal optical depth τ, then using the geometrical optics approximation, a photon traveling perpendicular to the ring plane has a probability of $e^{-\tau}$ of passing through the rings without colliding with a ring particle. For this problem, you may assume that the ring particles are well separated, and that the positions of their centers are uncorrelated.

(a) What fraction of light approaching the rings at an angle θ to the ring plane will pass through the rings without colliding with a particle?

(b) The cross-section of a ring particle to collisions with other ring particles is larger than its cross-section for photons. Assuming all ring particles are of equal size, what is the probability that a ring particle moving normal to the ring plane would pass through the rings without a collision?

(c) On average, how many ring particles does a line perpendicular to the ring plane pass through?

(d) On average, how many ring particles does a particle passing normal to the ring plane collide with? (As ring particles on Keplerian orbits pass through the ring plane twice per orbit, two times your result is a good estimate for the number of collisions per orbit of a particle in a sparse ring. The in-plane component of the particles' random velocities increases the collision frequency by a small factor, the value of which depends upon the ratio of horizontal to vertical velocity dispersion. For a ring that is thin and massive enough, local self-gravity can increase the vertical frequency of particle orbits, and combined with the gravitational pull of individual particles, greatly increase particle collision frequencies.)

11.7.E (a) Calculate the time necessary for Kepler shear to spread rings with the following parameters over 360° in longitude:

(i)   Width 1 km, orbit 80 000 km from Saturn.

(ii) Width 100 km, orbit 80 000 km from Saturn.

(iii) Width 1 km, orbit 120 000 km from Saturn.

(iv) Width 2 km, orbit 63 000 km from Neptune.

(b) Calculate the radial diffusion time for the doubling of the widths of the rings in part (a) assuming a viscosity $\nu_v = 100$ cm^2 s^{-1}. How do these times vary with viscosity (give the functional form)?

(c) Compare your results in parts (a) and (b), and comment on the observation that rings are generally observed to vary much more substantially with radius than with longitude.

11.8.I Consider the situation in which a small moon on an eccentric and inclined orbit near Roche's limit gets disrupted by a hypervelocity impact. The debris is initially quite localized, and has a velocity dispersion small compared to its orbital velocity. Using the results of the previous problem, describe the subsequent evolution of the debris swarm.

11.9.I Assume that an isolated ringlet of width $\Delta r = 20$ km, full thickness $2H = 2c_v/\mu = 20$ m, and optical depth $\tau = 1$ is in orbit 100 000 km from Saturn's center at time $t = 0$. The symbols c_v and μ represent the velocity dispersion and vertical frequency, respectively, and $\mu \approx n$. What is the approximate width of the ring at $t = 10^3$ years? At $t = 10^8$ years?

You should use the viscosity formula given by equation (11.11), and the diffusion relationship:

$$(\Delta r(t))^2 = (\Delta r(0))^2 + \nu_v t. \qquad (11.34)$$

You may make any plausible assumption about the time evolution of H and τ. (Choose something convenient and justify it.) Quote your answer in simple units.

11.10.E Assume that the differential size distribution of particles in a planetary ring is given by a power law over a finite range (eq. 11.17).

(a) Compute the critical values of ζ for which equal amounts of (i) mass and (ii) surface area are presented by particles in each factor of two interval in radius. For larger ζ (steeper distributions), most of the mass or surface area is contained in small particles, whereas for smaller ζ, most is in the largest bodies in the distribution. (Hint: This problem requires you to integrate over the size distribution.)

(b) Prove that it is impossible to have both $R_{min} = 0$ and $R_{max} = \infty$ for any value of ζ.

11.11.E Estimate (very crudely) the mass of Jupiter's ring system and associated inner moons. In parts (a)–(c) of this problem, you may assume that ring particles have a density of 1 g cm^{-3}.

(a) Estimate the mass of the dust in each of the three parts of the ring system, assuming particle

radii of 0.5 μm in the halo and 1 μm in the other regions. Use the optical depths given in Table 11.1.

(b) Estimate the mass of the macroscopic particles in the main ring, assuming particle radii are

(i) all 5 cm;

(ii) all 5 m;

(iii) distributed as a power law with $N(R) \propto R^{-3}$ from 5 cm to 5 m;

(iv) distributed as a power law with $N(R) \propto R^{-2}$ from 5 cm to 5 m;

(v) distributed as a power law with $N(R) \propto R^{-3}$ from 1 cm to 500 m;

(vi) distributed as a power law with $N(R) \propto R^{-2}$ from 1 cm to 500 m.

(c) Estimate the mass of the moon Metis, using the size given in Table 1.5.

(d) Compare the uncertainties introduced by the assumed density to those resulting from uncertainties in the particle size distribution.

11.12.I The moonlet Pan orbits within the Encke gap in Saturn's rings.

(a) Calculate the location of Pan's 2:1 inner Lindblad resonance using the Keplerian approximation for orbits.

(b) Calculate the location of Pan's 2:1 inner Lindblad resonance more precisely.

(c) Is this resonance within Saturn's rings? If yes, state which ring.

(d) What type of wave is such a resonance capable of exciting? How many spiral arms would it have?

11.13.E Saturn's equatorial radius $R_h = 60\,330$ km, and its gravitational moments are $J_2 = 1.63 \times 10^{-2}$, $J_4 = -9.17 \times 10^{-4}$.

(a) Calculate the location of the Enceladus 3:1 inner Lindblad (horizontal) resonance using the Keplerian approximation for orbits.

(b) Is this resonance located within Saturn's ring system? If so, in which ring?

11.14.D (a) Calculate the location of the Enceladus 3:1 inner Lindblad resonance more accurately than you did in the previous problem. State whatever approximations you use.

(b) Calculate the location of the expected crests of the density wave excited by this resonance if the surface density of the ring is 50 g cm^{-2}. You may assume that a crest occurs at exact resonance. Is this a reasonable assumption? Comment.

(c) Calculate the location of the Enceladus 3:1 inner vertical resonance using the Keplerian approximation for orbits.

(d) Calculate the location of the Enceladus 3:1 inner vertical resonance more accurately. State whatever approximations you use.

11.15.E The moonlet Pan orbits in the middle of the 325 km wide Encke gap. Calculate the wavelength and amplitude of the ripples that Pan excites in Saturn's A ring at the edges of Encke's gap.

11.16.E Imagine you were to construct a scale model of Saturn's rings 5 km in radius (about the size of San Francisco). How thick would this model be locally? What would be the height of the corrugations corresponding to the highest bending waves? How wide would the A ring be? What would be the width of the F ring?

11.17.E Briefly explain how shepherding of planetary rings works and why shepherding does not occur in the asteroid belt.

11.18.I Calculate the ratio of the Lorentz force to the gravitational force on an unshielded charged grain with a potential of -10 volts in Jupiter's ring as a function of grain size and density and of distance from Jupiter. Assume Jupiter's

magnetic field can be approximated by a dipole with a surface magnetic field strength of 4 G.

11.19.E Assume that a spoke in Saturn's rings forms radially and stretches out from $r = 1.6 \, R_h$ to $r = 1.9 \, R_h$. Calculate the orbital period for particles at the two ends of the spoke. Sketch the spoke after the particles at $r = 1.6 \, R_h$ have completed one-fourth of an orbit.

11.20.I The probability that a photon traveling towards a planetary ring will interact with a ring particle depends upon the angle between the photon's path and the ring plane (Problem 11.6), but it does not depend (significantly) upon whether the in-plane component of the photon's velocity is in the direction of particle orbits or opposite to them, because photons move much more rapidly than do ring particles. However, the velocity of interplanetary debris is only slightly larger than that of ring particles, so particle motion during the passage of this debris through the ring disk cannot be neglected in determining collision probabilities. By decomposing the velocity of the debris into cylindrical polar coordinates in the planetocentric system, derive a formula for the impact probability as a function of speed (relative to the orbital velocity of the ring particles), direction, and ring normal optical depth, τ.

12 Extrasolar Planets

Since one of the most wondrous and noble questions in Nature is whether there is one world or many, a question that the human mind desires to understand, it seems desirable for us to inquire about it.

Albertus Magnus, 13th century

The first eleven chapters of this book covered general aspects of planetary properties and processes, and described specific objects within our Solar System. We now turn our attention to far more distant planets. What are the characteristics of planetary systems around stars other than the Sun? How many planets are typical? What are their masses and compositions? What are the orbital parameters of individual planets, and how are the paths of planets orbiting the same star(s) related to one another? What are the relationships between stellar properties such as mass, composition, and multiplicity and the properties of the planetary systems that orbit them? These questions are hard to answer because extrasolar planets, often referred to as *exoplanets*, are far more difficult to observe than are planets within our Solar System.

Just as the discoveries of small bodies orbiting the Sun have forced astronomers to decide how small an object can be and still be worthy of being classified as a planet (Chapter 9), detections of substellar objects orbiting other stars have raised the question of an upper size limit to planethood. We adopt the following definitions, which are consistent with current International Astronomical Union (IAU) nomenclature:

• *Star*: self-sustaining fusion is sufficient for thermal pressure to balance gravity ($\gtrsim 0.075$ M$_\odot \approx 80$ M$_2$ for solar composition; the minimum mass for an object to be a star is often referred to as the *hydrogen burning limit*).
• *Stellar remnant*: dead star – no more fusion (or so little that the object is no longer supported primarily by thermal pressure).
• *Brown dwarf*: substellar object with substantial deuterium fusion – more than half of the object's original inventory of deuterium is ultimately destroyed by fusion.
• *Planet*: negligible fusion ($\lesssim 0.012$ M$_\odot \approx 13$ M$_2$, with the precise value again depending upon composition), plus it orbits one or more stars and/or stellar remnants.

Gravitational contraction is a major source of the energy radiated by giant planets and brown dwarfs. These objects shrink and (after some initial warming) cool as they age (Fig. 12.1), so there is not a unique relationship between luminosity and mass.

Radial velocity surveys in the 1990s first demonstrated that planets orbit many stars other than our Sun. Most of these planets have combinations of mass and orbit quite different from those within our own Solar System. All of the extrasolar planets orbiting nearby stars discovered as of 2009 induce variations in stellar reflex motion larger in amplitude and/or of shorter period than would a planetary system like our own, and surveys accomplished to date are strongly biased against detecting low-mass and long-period planets.

12.1 Physics and Sizes of Planets, Brown Dwarfs, and Low-Mass Stars

Nuclear reactions maintain the temperature in the cores of low-mass stars close to $T_{\text{nucl}} \approx 3 \times 10^6$ K (§13.2.2) because the fusion rate is roughly proportional to T^{10} near T_{nucl}. The virial theorem (eq. 2.61) can be used to show that the radii of such stars must be roughly proportional to mass. In equilibrium, the thermal energy and the gravitational potential energy are in balance:

$$\frac{GM_\star^2}{R_\star} \sim \frac{M_\star k T_{\text{nucl}}}{m_{\text{amu}}}. \tag{12.1a}$$

Therefore

$$R_\star \propto M_\star, \tag{12.1b}$$

and the star's mean density

$$\rho_\star \propto M_\star^{-2}. \tag{12.1c}$$

At low densities, the hydrostatic structure of a star is determined primarily by a balance between gravity and

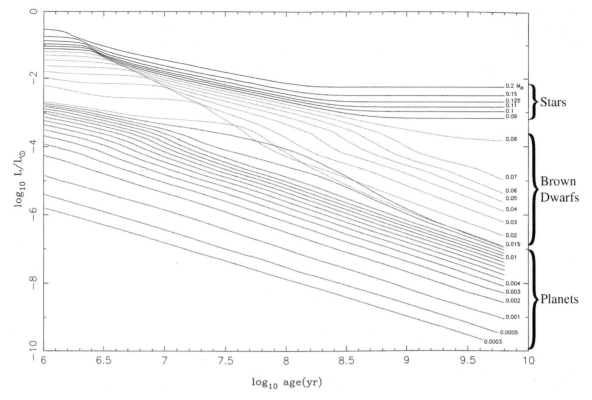

Figure 12.1 Evolution of the luminosity (in $\mathcal{L}_\odot$) of initially hot and distended isolated solar-metallicity very low mass (M and L dwarf) stars and substellar objects plotted as functions of time (in years) after formation. The stars, brown dwarfs, and planets are shown as the upper, middle, and lower sets of curves, respectively. Most of the curves are labeled by the object's mass in units of $M_\odot$; the three lowest curves correspond to the mass of Saturn, half the mass of Jupiter, and the mass of Jupiter. All of the substellar objects become less luminous as they radiate away the energy released by their gravitational contraction from large objects to much more compact bodies with sizes of order R. Objects with $M \gtrsim 0.012\ M_\odot$ exhibit plateaus between 10^6 and 10^8 years as a result of deuterium burning (the initial deuterium mass fraction was assumed to be 2×10^{-5}). Deuterium burning occurs earliest and is quickest in the most massive objects. Stars ultimately level off in luminosity when they reach the hydrogen burning main sequence, and some of the curves just above the deuterium burning limits cross twice because lower mass objects reach sufficient temperature for deuterium burning at later times and take longer to exhaust their deuterium supply. In contrast, the luminosities of brown dwarfs and planets decline indefinitely. The luminosities of the younger objects, especially giant planets $\lesssim 10^8$ years old, may be substantially smaller than the values shown above if they radiate away substantial portions of their accretion energy while they are growing. (Burrows *et al.* 1997)

thermal pressure. At sufficiently high densities, another source of pressure becomes significant. Electrons, because they have half-integer spins, must obey the *Pauli exclusion principle* and are accordingly forbidden from occupying identical quantum states. The electrons thus successively fill up the lowest available energy states. Those electrons that are forced into higher energy levels contribute to *degeneracy pressure*. The degeneracy pressure scales as $\rho^{5/3}$ and is important when it is comparable in magnitude to or larger than the ideal gas pressure (which scales

as ρT; see eq. 3.27). Near $T_{\rm nucl}$, the degeneracy pressure dominates when densities exceed a few hundred grams per cubic centimeter.

Bodies supported primarily by degeneracy pressure are referred to as *compact objects*. In compact objects, the virial theorem implies that the energy of the degenerate particles (electrons in the case of brown dwarfs) is comparable to the gravitational potential energy:

$$\rho^{5/3}R^3 \sim \frac{GM^2}{R}. \qquad (12.2a)$$

Therefore,

$$R \propto M^{-1/3}, \qquad (12.2b)$$

(Problem 12.1), and compact objects shrink if more mass is added to them. The most massive cool brown dwarfs are indeed expected to have slightly smaller radii than their lower mass brethren. Young brown dwarfs can be hot and distended, depending on their age and formation circumstances; see Figure 12.1.

For lower mass objects, *Coulomb pressure*, provided by the electromagnetic repulsion of electrons in one molecule from those in another, plays a larger role relative to degeneracy pressure. Coulomb pressure is characterized by constant density, which implies that the radius of such bodies scale as:

$$R \propto M^{1/3}. \qquad (12.3)$$

The combination of Coulomb and degeneracy pressures result in radii similar to that of Jupiter for all cool brown dwarfs and giant planets of solar composition, as well as for the very lowest mass stars (Fig. 6.25). The largest size cool planets are expected to have $M_p \approx 4\ M_{\odot}$.

Planets of order 1 $M_{\oplus}$ that are composed primarily of silicates, iron, and H_2O are subject to pressure-induced compression that increases their average densities by tens of percent. For a given composition, planetary radius varies with mass roughly as $R \propto M_p^{0.3}$ for 1 $M_{\mathbb{C}} < M_p < 1\ M_{\oplus}$, and $R \propto M_p^{0.27}$ for 1 $M_{\oplus} < M_p < 10\ M_{\oplus}$. Throughout these ranges, the radius of a planet composed of an Earth-like mixture is expected to be ~20% less than that of a planet of the same mass containing equal amounts (by mass) of Earth-like mixture and H_2O (Figs. 12.2 and 12.3). Many different mixtures for a given mass share the same radius, and hence, the composition of a super-Earth cannot be uniquely determined from mass and radius measurements. The maximum radius for a rocky planet of a specified mass corresponds to the 'silicate mantle' vertex in Figure 12.3. A planet with a larger radius necessarily implies that it has H_2O or other light components.

Planets orbiting very close to stars are subjected to intense stellar heating. This heating retards convection in the upper envelope of gas-rich inner planets. The excess entropy retained throughout jovian mass hydrogen-rich close-in planets produces radii that can be tens of percent larger than colder planets of the same mass, composition, and age. This effect is greater for lower mass planets. Note that if a planet migrates close to a star after radiating away much of its initial accretion energy and shrinking to near 1 $R_{\odot}$ (a process which requires tens of millions of years), its radius would likely grow only slightly larger

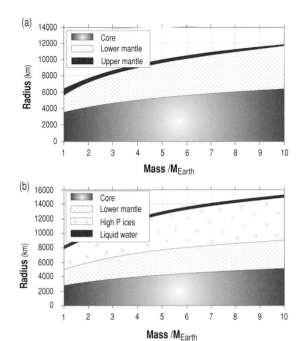

Figure 12.2 Radii of (a) Earth-composition planets and (b) ocean planets that are equal mixtures (by mass) of H_2O and Earth composition. The inset keys show the various layers of the planets. The crust is included in the upper mantle portion of the Earth-composition planets, and the amount of water in Earth is insignificant on the scale of these plots. For hot H_2O-rich planets, the upper part of the region shown as high-pressure ice is replaced by liquid water, and for cold H_2O-rich planets, the upper part of the region shown as liquid water is replaced by ice. (Courtesy C. Sotin; see Sotin *et al.* 2007 for details)

than when the planet was in colder environs, as the star's heating would only affect and expand the outer part of the planet's atmosphere that had cooled below the planet's new T_{eff}.

The dynamics of atmospheres of planets that orbit very close to stars also differ from those of the planets within our Solar System. To illustrate the very different dynamical regimes present in the atmospheres of hot and cold giant planets, it is useful to introduce the concept of the *radiative time constant*. This is the timescale over which a given pressure level in the atmosphere can cool substantially. It depends on both the local temperature and the overlying optical depth. Radiative timescales in the Solar System's giant planets approach the orbital timescale, and these planets complete $> 10^4$ rotations per orbit, so the atmospheres of these planets show no diurnal variations and change very slowly, if at all, over seasons. In giant planets orbiting only ~0.05 AU from their stars, radiative timescales can be much shorter (because energy content

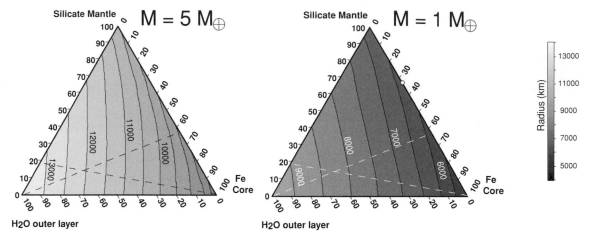

Figure 12.3 Ternary diagrams showing the relationship between composition, mass, and radius for 1 $M_\oplus$ and 5 $M_\oplus$ planets. Lines representing constant radius are shown in black at increments of 500 km. Different mixtures of the three most likely end-member components (iron cores, silicate mantles, and H_2O outer regions) yield planets of different sizes. Each point in the ternary diagrams depicts a unique composition with a corresponding radius shown by the shade of gray color. The three vertices correspond to pure compositions of H_2O, iron, or silicates, and the opposite sides of the triangle correspond to 0% of that end member. Thus, the side that connects Fe core and silicate mantle represents waterless planets. Earth's composition is shown by a circle essentially on this line in the 1 $M_\oplus$ diagram. Planets that formed in disks of solar nebula composition that are composed of all substances that condense above any specified temperature (Fig. 13.13), or mixtures of materials that condense at a range of temperatures, lie above both of the dashed lines. (Courtesy Diana Valencia)

is proportional to T and radiation is emitted in proportion to T^4; see eqs. 3.1 and 3.8 and Problem 12.2), on the order of hours, comparable to or less than the orbital period. Thus, dynamical processes and radiation compete in these atmospheres to redistribute heat.

12.2 Detecting Extrasolar Planets

Prior to the 1990s, our ability to understand how planets form was constrained because we had observed only one planetary system, our own Solar System. Hundreds of extrasolar planets have been discovered within the past fifteen years, and far larger numbers are likely to be found in the upcoming decades. Various methods for detecting planets around other stars are being used or studied for possible future use. As distant planets are extremely faint, most methods are indirect, in the sense that the planet is detected through its influence on the star that it orbits. The methods are sensitive to different classes of planets, and provide us with complementary information about the planets they do find, so most or all of them are likely to provide valuable contributions to our understanding of the diversity of planetary system characteristics. A brief review of detection techniques is presented in this section.

12.2.1 Timing Pulsars and Pulsating Stars

The first confirmed detection of extrasolar planets was provided by *pulsar timing*. Pulsars, which are magnetized rotating neutron stars,[1] emit radio waves that appear as periodic pulses to an observer on Earth. The pulse period can be determined very precisely, and the most stable pulsars rank among the best clocks known. The mean time of pulse arrival can be measured especially accurately for the rapidly rotating millisecond pulsars, whose frequent pulses provide an abundance of data. Even though pulses are emitted periodically, the times at which they reach the receiver are not equally spaced if the distance between the pulsar and the telescope varies in a nonlinear fashion. Earth's motion around the Sun and Earth's rotation cause such variations, which can be calculated and subtracted from the data. If periodic variations are present in these reduced data, they may indicate the presence of companions orbiting the pulsar.

Pulsar timing effectively measures variations in the distance to the pulsar, relative to a trajectory with constant

[1] 'Neutron stars', white dwarfs, and some black holes are stellar remnants, not true stars. As is the case for cool brown dwarfs, degeneracy pressure balances gravity in neutron stars and white dwarfs. Nothing balances gravity in black holes, which are singularities in the space–time continuum.

velocity with respect to the barycenter of our Solar System. It thus reveals only one dimension of the pulsar's motion. The easiest planets to detect via pulsar timing are massive planets whose orbital planes lie close to the line of sight and with orbital periods comparable to or somewhat less than the length of the interval over which timing measurements are available.

Some variable stars pulsate with very regular periods. Such pulsations produce periodic variations in stellar brightness. The time intervals at which these oscillations are observed at Earth vary as a pulsating star moves in response to the gravitational tugs of orbiting planets. Pulsation times can be measured to deduce the presence of planets using the same principles as pulsar timing. However, the precision of timing stellar pulsations is not nearly as good as that of pulsar timing, so the minimum detectable planet masses are substantially larger.

12.2.2 Radial Velocity

Radial velocity surveys have been the most successful method for detecting planets around main sequence stars. By fitting the Doppler shift of a large number of features within a star's spectrum, the velocity at which the star is moving towards or away from the observer can be precisely measured. After removing the motion of the observer relative to the barycenter of the Solar System and other known motions, radial motions of the target star resulting from planets that are orbiting the star remain.

The amplitude, K, of the radial velocity variations of a star of mass $M_\star$ that are induced by an orbiting planet of mass M_p is

$$K = \left(\frac{2\pi G}{P_{orb}}\right)^{1/3} \frac{M_p \sin i}{(M_\star + M_p)^{2/3}} \frac{1}{\sqrt{1 - e^2}}, \tag{12.4}$$

where P_{orb} is the orbital period, i is the angle between the normal to the orbital plane and the line of sight, and e is the orbit's eccentricity. As in the case of pulsar timing, radial velocity measurements yield the product of the planet's mass (divided by $M_\star^{2/3}$; the star's mass can usually be estimated to an accuracy of ~10% from its spectral characteristics) and the sine of the angle between the orbital plane and the plane of the sky, as well as the period and the eccentricity of the orbit. This technique is most sensitive to massive planets and to planets in short-period orbits (Fig. 12.4). The best observers are now achieving a precision of 1 m s^{-1} (representing a Doppler shift of three parts in 10^9) on spectrally stable stars. With this precision, Jupiter-like planets orbiting Sun-like stars are detectable, although these detections require a long baseline of observations (comparable to the planet's orbital period). Planets as small as a few M$_\oplus$ orbiting very close to stars also can

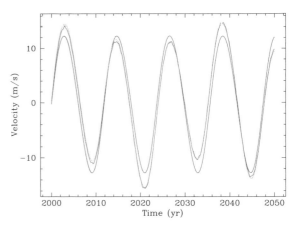

Figure 12.4 COLOR PLATE Velocity variations of the Sun in response to Jupiter (nearly sinusoidal narrow blue curve), Jupiter plus Saturn (faint green curve), and all eight planets plus Pluto (thick red curve). Jupiter's tug dominates the variations, with Saturn having much less influence than Jupiter but still far more than all of the remaining planets combined. The pull of Earth and Venus is evident in the short-period variations seen in the thick red curve. (Courtesy Elisa V. Quintana)

be detected; however, Earth-like planets orbiting at 1 AU are beyond the capabilities currently envisioned for this technique. Precise radial velocity measurements require a large number of spectral lines, and thus cannot be achieved for the hottest stars (spectral types A, B, and O), which have far fewer features in their spectra than do cooler stars like the Sun. Stellar rotation and intrinsic variability (including starspots) represent major sources of noise for radial velocity measurements.

12.2.3 Astrometry

Planets may be detected via the wobble that they induce in the motion of their stars projected onto the plane of the sky. This *astrometric* technique is most sensitive to massive planets orbiting about stars that are relatively close to Earth. The amplitude of the wobble, $\Delta\theta$, is given by the formula

$$\Delta\theta \leq \frac{M_p}{M_\star} \frac{a}{r_\odot}, \tag{12.5}$$

where $r_\odot$ is the distance of the star from our Solar System, and a is the semimajor axis of the orbit. If $r_\odot$ and a are measured in the same units, then the value of $\Delta\theta$ in equation (12.5) is in radians; if $r_\odot$ is measured in parsecs and a in AU, then the units of $\Delta\theta$ are arcseconds. For example, a 1 M$_\lambda$ planet orbiting 5 AU from a 1 M$_\odot$ star located 10 parsecs (1 parsec = 3.26 light-years = 2.06 × 10^5 AU = 3.0857 × 10^{18} cm) from Earth would produce an astrometric wobble 0.5 milliarcseconds (mas) in amplitude. The path of the star on the plane of the sky

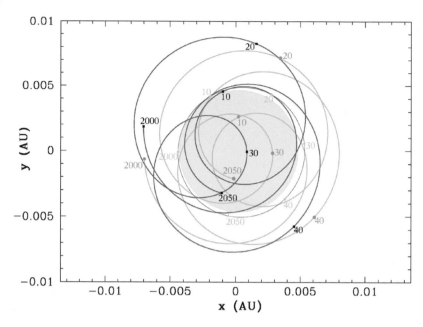

Figure 12.5 COLOR PLATE Motion of the Sun during the first half of the twenty-first century in response to Jupiter (narrow blue ellipse, faint dates), Jupiter plus Saturn (light green curve), and all eight planets plus Pluto (thick dark red curve, dark dates). The solar disk (shaded yellow) is shown for comparison. The Sun moves counterclockwise in this perspective, completing slightly less than one trip around the elliptical curve per decade. Jupiter's tug dominates the variations on short timescales, but Saturn, Uranus, and Neptune have more significant influence on the Sun's position than they do on the Sun's velocity (Fig. 12.4). Solar motion induced by the terrestrial planets is very small. (Courtesy E. V. Quintana)

depends on all of the planet's orbital elements (§2.1.3), but the equality in equation (12.5) holds for orbits that are circular and/or lie in the plane of the sky. Because the star's motion is detectable in two dimensions, the plane of the planet's orbit can be measured, so there is no sin i ambiguity analogous to that in equation (12.4), and thus a better estimate of the planet's mass can be obtained astrometrically than using radial velocities.

Planets on more distant orbits are ultimately easier to detect using astrometry (Fig. 12.5) because the amplitude of the star's motion is larger, but finding these planets requires a longer baseline of observations due to their greater orbital periods. Astrometric systems require considerable stability over long times to reduce the noise that can lead to false detections. The best long-term precision demonstrated by single ground-based telescopes employing adaptive optics is better than 1 mas. The Gaia space telescope launched by ESA in December 2013 is expected to achieve two orders of magnitude better astrometric precision. No astrometric claim of detecting an extrasolar planet has yet been confirmed, but data obtained by the Hipparcos satellite in the early 1990s have shown that several candidate brown dwarfs observed in radial velocity surveys are actually low-mass stellar companions whose orbits are viewed almost face-on.

12.2.4 Transit Photometry

If Earth lies in or near the orbital plane of an extrasolar planet, that planet passes in front of the disk of its star once each orbit as viewed from Earth. Precise *photometry* can reveal such transits, which can be distinguished from rotationally modulated starspots and intrinsic stellar variability by their periodicity, approximately square-well shapes, and relative spectral neutrality. Transit observations provide the size and orbital period of the detected planet. Although geometrical considerations limit the fraction of planets detectable by this technique, thousands of stars can be surveyed within the field of view of one telescope, so surveys using transit photometry can be quite efficient.

Neglecting variations of brightness across the stellar disk (resulting from limb darkening, starspots, etc.), the depth of a transit, i.e., the fractional decrease in the star's apparent luminosity, is given by

$$\frac{\Delta \mathcal{L}}{\mathcal{L}} = \left(\frac{R_p}{R_\star} \right)^2. \tag{12.6}$$

For a transit to be observed, the orbit normal must be nearly 90° from the line of sight,

$$\cos i < \frac{R_\star + R_p}{r}, \tag{12.7}$$

where $R_\star$ and R_p are the stellar and planetary radii, respectively, and r the distance between the two bodies when the planet is nearest the observer. The probability of observing a transit of a randomly-oriented planet, $\mathcal{P}_{tr}$, is given by

$$\mathcal{P}_{tr} = \frac{R_\star + R_p}{a(1 - e^2)}. \tag{12.8}$$

The duration of a *central transit*, wherein the center of the planet blocks light from the center of the stellar disk, is

$$T_{tr} = \frac{R_\star + R_p}{\pi a} \frac{1 - e^2}{1 + e \cos \varpi}, \tag{12.9}$$

where the longitude of periapse, ϖ, is measured relative to the line of sight. Neglecting transits with a duration less than half that of a central transit (sometimes misleadingly referred to as grazing transits), which are more difficult to detect, the probability of a planet orbiting 1 AU from a 1 $R_\odot$ star transiting across the stellar disk is 0.4%, whereas the transit probability of a planet at 0.05 AU from the same star is 8%.

Scintillation in and variability of Earth's atmosphere limit photometric precision to roughly one-thousandth of a magnitude (1 *millimagnitude* or *mmag*), allowing detection of transits by Jupiter-sized planets (but not by Earth-sized planets) from the ground. Far greater precision is achievable above the atmosphere, with planets as small as Earth likely to be detectable.

One major advantage of the transit technique is that many planets detected in this manner are observable via the radial velocity method as well, yielding a mass (as the inclination is known from the transits). Such combined and complementary measurements provide the density of the planet, an especially valuable datum for formation studies.

A space-based photometric telescope could also detect the sinusoidal phase modulation of light reflected by an inner giant planet as it orbits its star, provided the planet does not induce variations in the appearance of the star's photosphere that track its orbit. Observations of an exoplanet in both transit photometry and reflected light photometry could yield its albedo and phase function. Analogous observations in the thermal infrared can reveal variations in the planet's temperature with longitude.

12.2.5 Transit Timing Variations

For a planet traveling on a Keplerian orbit, the time interval between successive transits remains constant. However, planets which are perturbed by additional planets (or by a companion star or brown dwarf) do not travel on purely Keplerian orbits. Transits of such perturbed planets are not strictly periodic. The amplitude of the variations depends on the mass of the perturber and relative orbits of the two bodies. Resonant orbits (§2.3.2) can produce especially large variations in transit time series.

Observing irregularities in transit times can betray the existence of unseen planets. Precise timing of eclipses of eclipsing binary stars or transits of planets has the potential of revealing the masses and orbits of unseen companions.

However, uniquely determining the masses and orbital properties of these unseen planets is much more difficult. If two or more planets within a system are observed to transit, then measuring transit timing variations (TTVs) has the potential of providing good estimates of planetary masses and bounds on orbital eccentricities.

12.2.6 Microlensing

According to Einstein's general theory of relativity, the path of the light from a distant star that passes by a massive object (lens) between the source and the observer is bent. The bending angle is typically very small, and the effect is known as *microlensing*. The lens magnifies the light from the source by a substantial factor when it passes closer to the line of sight than the radius of the *Einstein ring*, R_E, which is given by

$$R_E = \sqrt{\frac{4GM_L r_{\Delta L}}{c^2}} \left(1 - \frac{r_{\Delta L}}{r_{\Delta S}}\right)^{1/2}, \tag{12.10}$$

where M_L is the mass of the lens, c is the speed of light, and $r_{\Delta L}$ and $r_{\Delta S}$ are the distances from the Earth to the lens and the source, respectively.

Microlensing is used to investigate the distribution of faint stellar and substellar mass bodies within our galaxy. The brightness of the source can increase several fold for a period of weeks during a microlensing event, and the pattern of brightening can be used to determine (in a probabilistic manner) properties of the lens. If the lensing star has planetary companions, then these less-massive bodies can produce characteristic blips on the observed lightcurve provided the line of sight passes within the planet's (much smaller, eq. 12.10) Einstein ring (Fig. 12.6). Under favorable circumstances, planets as small as Earth can be detected.

Microlensing provides information on the mass ratio and projected separation of the planet and star. This technique is capable of detecting systems with multiple planets and/or more than one star. The properties of individual microlensing planets (especially orbital eccentricities and inclinations) can often only be estimated in a statistical sense because of the many parameters that influence a microlensing lightcurve. However, additional information about the planets may be deduced under some special circumstances, such as very high magnification microlensing events. Follow-up observations of planets detected via microlensing (using other techniques, because a given system's chance of producing a second observable microlensing event is extremely small) is very difficult due to the faintness of these distant systems. But when light from the lensing star can be distinguished from that of the source star, the mass of the planet's host star can be determined.

(a)   (b)

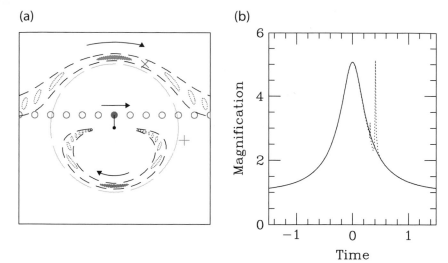

Figure 12.6 Cartoon sketch of a microlensing event illustrating the effect of the light from a distant source being bent by a lensing star that possesses a planetary companion. (a) The images (dotted ovals) are shown for several different positions of the source (solid circles), along with the primary lens (dot) and Einstein ring (long-dashed circle). The source is moving from left to right relative to the lens, and the images of its bent light move in a clockwise sense, as indicated by the arrows. The filled ovals correspond to the images of the source when it is at the position of the filled circle. If the primary lens has a planet near the path of one of the images, i.e., within the short-dashed lines, then the planet will perturb the light from the source, creating a deviation to the single lens lightcurve. (b) The observed amplification of the amount of light from the source received at the telescope as a function of time is shown for the case of a single stellar-mass lens (solid line) and a star with an accompanying planet located at the position of the × (dotted line). If the planet was located at the + instead, then there would be no detectable perturbation, and the resulting lightcurve would be essentially identical to the solid curve. The units of the time are R_E/v, where v is the velocity of source's light relative to the lens on the plane of the sky. (Courtesy Scott Gaudi)

Careful monitoring of many microlensing events could provide a very useful data set on the distribution of planets within our galaxy.

12.2.7 Imaging

Extrasolar planets are very faint objects which are located near much brighter objects (the star or stars that they orbit), making them extremely difficult to image. The reflected starlight from planets with orbits and sizes like those in our Solar System is roughly one-billionth as large as the stellar brightness, although the contrast is ∼3 orders of magnitude more favorable in the thermal infrared (Fig. 12.7). Diffraction of light by telescope optics and atmospheric variability add to the difficulty of *direct detection* of extrasolar planets. However, an ∼30 M₂ brown dwarf companion at a projected distance of ∼30 AU from the 0.6 M$_\odot$ star Gliese 229 was first imaged in the thermal infrared in 1994.

Substellar objects that are closer to their stars and/or significantly smaller in mass have subsequently been imaged in the thermal infrared (Figs. 12.8a,b). Many of these substellar companions have been studied spectroscopically as well. Technological advances in interferometry (§E.6.3) and nulling (i.e., by choosing the right parameters, the intensity of the star's image can be decreased to near-zero levels) should eventually allow for imaging and spectroscopic studies of planets resembling those within our Solar System that are in orbit about nearby stars.

Many substellar objects that don't orbit stars have been imaged in the infrared. This newly discovered class of objects may have members less massive than the deuterium burning limit. The term *free-floating giant planets* has been used to describe these objects, even though they appear to be in many ways more akin to low-mass stars and brown dwarfs than to the planets within our Solar System.

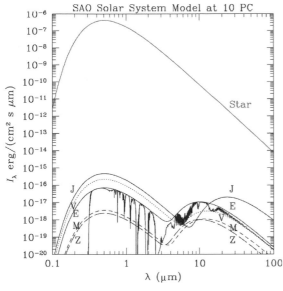

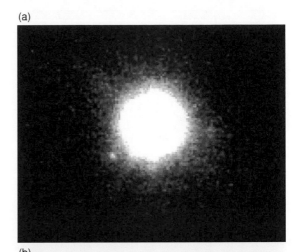

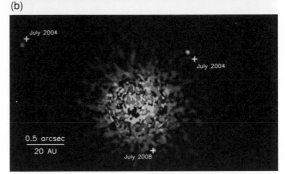

Figure **12.7** Spectral energy distribution of the Sun, Jupiter, Venus, Earth, Mars, and the zodiacal cloud. The bodies are approximated by blackbodies of uniform albedo, with an additional curve showing Earth's atmospheric absorption features. (Des Marais *et al.* 2002)

12.2.8 Other Techniques

Several other methods can be used to detect and study extrasolar planets. Planets transiting nearby stars could be detected as dark dots moving across high-resolution images of stellar disks that will be obtainable using interferometry. Radio emissions similar to those detected from Jupiter (§7.5) could reveal the presence of extrasolar planets. Very high spectral resolution and SNR spectra that include light from both the star and the planet could be used to identify gases that would be stable in planetary atmospheres but not in stars, and Doppler variations of such signals could yield planetary orbital parameters. Finally, *artificial signals* (§12.6) from an alien civilization could betray the presence of the planets on which they live (and the aliens might be willing to provide us with substantially more information!).

12.2.9 Exoplanet Characterization

Short of making contact with an alien civilization, detailed studies of exoplanets, especially small ones resembling Earth, will require technological advances. A planet's density can be computed if it is detected both by transit photometry and by radial velocity variations. The tenuous upper reaches of planetary atmospheres transmit continuum radiation from the star, yet absorb it in some spectral

Figure **12.8** (a) This image shows the young (5–10 Myr old) system 2MASS 1207-3932 AB viewed in the thermal infrared at 1.2 μm. The brighter object is clearly a brown dwarf. The fainter object at 8 o'clock, ~600 times less luminous at this wavelength and ~40 AU distant on the plane of the sky, may well be below the 13 $M_\gimel$, planet/brown dwarf boundary. (Mohanty *et al.* 2007) (b) COLOR PLATE A near-infrared color image (1.2, 1.6, and 2.2 microns) of the HR 8799 system made using Keck 2 telescope images acquired during the summer of 2008. Companion 'b' is 68 AU NE, 'c' at 38 AU NW, and 'd' at 24 AU SW from the star (north is up and east is left). All three companions are substantially substellar, with masses ~7–10 $M_\gimel$, radii ~1.2–1.3 $R_\gimel$, and T_{eff} ~900 K. Observations at earlier epochs, shown by +s on the figure, show counterclockwise Keplerian orbital motion for all three companions. The orbits appear to have small eccentricities and to be viewed nearly face-on. (Courtesy NRC Canada/C. Marois)

bands, producing deeper overall transits at these wavelengths. Comparison of spectra taken during transit with those taken outside transit yields information about the composition and temperature of the planet's atmosphere.

Most planets which transit in front of their stars also pass behind their stars; as planets are usually much smaller than their stars, their disks are generally completely occulted, in an event known as an *occultation* (but

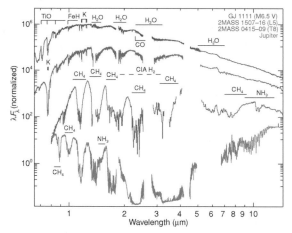

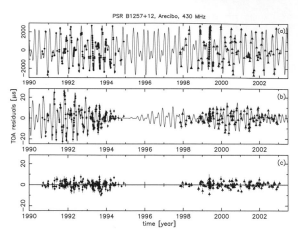

Figure 12.9 This set of observed spectra of an M dwarf star, brown dwarfs with two different types of spectra (L and T), and Jupiter illustrates the changes in atmospheric chemistry and spectra that are seen as low-mass hydrogen–helium dominated objects (like giant planets) cool over time. Water and refractory diatomic species dominate the M dwarf spectrum. In L and T dwarfs, the water absorption bands get progressively deeper and methane appears, while the refractory gases disappear as they condense into solid grains. At still cooler temperatures, ammonia appears and all water is condensed into clouds at Jupiter. The effective temperatures are about 2900 K in the M dwarf, 1600 K in the L5 brown dwarf, 700 K in the T8 brown dwarf, and 130 K for Jupiter. Jupiter's spectrum shortwards of 3.5 microns is entirely reflected sunlight. (Courtesy Mark Marley and Mike Cushing)

Figure 12.10 The best-fit residuals from modeling the times of arrival (TOAs) of pulses from PSR B1257+12 measured with the Arecibo radio telescope at 430 MHz. (a) The points represent the residuals for a standard pulsar timing model without planets, and the curve shows the best-fit three-planet Keplerian model to these data. Note that the vertical scale on this panel is 100 times as large as are those on the two panels below it. (b) The points represent the residuals for the best-fit three-planet Keplerian model to these data, and the curve shows the changes to the residuals with the addition of gravitational perturbations between the two larger planets. (c) Residuals to the best-fit three-planet model with perturbations included. (Konacki and Wolszczan 2003)

often referred to less precisely as a *secondary eclipse*). The surface brightness of a planet is typically far less than that of a star, so secondary eclipses are much less deep than are transits. The ratio of depths is smallest in the infrared, where the planet's brightness is a larger fraction of that of the star (Fig. 12.7).

Within the next few decades, we should be able to take images of Earth-like exoplanets in both reflected starlight (using coronagraphy) and the thermal infrared (using interferometry). We should also be able to determine atmospheric and surface properties spectroscopically, as is currently being done for the atmospheres of brown dwarfs (Fig. 12.9). These advances are most likely to be achieved from space, above the interference of Earth's atmosphere.

12.3 Observations of Extrasolar Planets

First one, and then two jovian-mass planets in orbit about Barnard's star were announced with great fanfare in the 1960s. Only 6 light-years away, but still impossible to see with the unaided eye because of its faintness, Barnard's star is Sun's nearest isolated neighbor (only the α Centauri triple-star system is closer). However, the astrometric evidence for Barnard's star's purported planets

was discredited in the 1970s. Subsequent claims for the discovery of the first extrasolar planet via astrometry continued to capture newspaper headlines, but failed to stand up to further analysis or additional data.

12.3.1 Pulsar Planets

The first extrasolar planets were discovered in the early 1990s by Alexander Wolszczan and Dale Frail. Wolszczan and Frail found periodic variations in the arrival time of pulses from pulsar PSR B1257+12 (which has a 6 millisecond rotational/pulse period) that remained after the motion of the telescope about the barycenter of the Solar System had been accounted for, and they attributed these variations to two companions of the pulsar. One companion has an orbital period of 66.54 days and the product of its mass and orbital tilt to the plane of the sky is $M_p \sin i = 3.4\ M_\oplus$; the other planet has a period of 98.21 days and $M_p \sin i = 2.8\ M_\oplus$ (these masses assume that the pulsar is 1.4 times as massive as the Sun). Both planets have orbital eccentricities of ~ 0.02. Subsequent observations showed the effects of mutual perturbations of these two bodies on their orbits, thereby confirming the planet hypothesis and implying that both planets have $i \approx 50°$ (Fig. 12.10). Additionally, the data imply that there is a

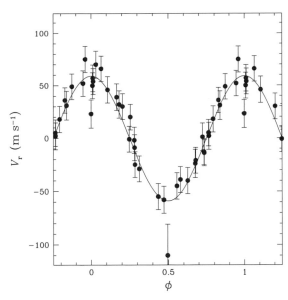

Figure 12.11 Radial velocity measurements of the star 51 Pegasi (points with error bars) as a function of phase of the orbital fit (solid line). These data were used to discover the first planet found around a main sequence star other than the Sun. One and a half cycles are shown for clarity. (Adapted from Mayor and Queloz 1995; courtesy Didier Queloz)

lunar mass object with period of 25 days orbiting interior to the two near-resonant planets.

The fourth pulsar planet to be detected is a $\sim$2.5 M$_{2\!\!\!/}$ object orbiting $\sim$23 AU from the close (191.4 day period) pulsar/white dwarf binary PSR B1620-26. This system lies within the low-metallicity globular cluster Messier 4 (M4).

Pulsar timing has been demonstrated to be a very sensitive detector of planetary objects, but it only works for planets in orbit about a rare and distinctly nonsolar class of stellar remnants. The paucity of known pulsar planets is due to the combination of a small number of available search targets and a low frequency of planets orbiting pulsars.

12.3.2 Radial Velocity Detections

The first planet known to orbit a main sequence star other than the Sun is the $M_p \sin i = 0.47$ M$_{2\!\!\!/}$, $P_{orb} = 4.23$ days companion discovered to orbit the star 51 Pegasi by Michel Mayor and Didier Queloz in 1995 (Fig. 12.11). Over the next fourteen years, radial velocity surveys identified more than 300 objects with $M_p \sin i < 13$ M$_{2\!\!\!/}$ in orbit about main sequence stars other than the Sun. The vast majority of these planets share at least two of the following three characteristics, each of which acts to increase their detectability: Their masses exceed that of Saturn, their orbital semimajor axes

are $\lesssim$3 AU, and they dominate the radial velocity variations of their parent stars over a broad range of timescales (thus, the most massive planet near these stars surpasses the second most massive planet by a factor larger than the ratio of the mass of Jupiter to that of Saturn).

Exoplanets have not yet been assigned official names. They are generally referred to using a convention that is an extension of the system used for multiple star systems. Many different algorithms have been used to name individual stars; most are based upon catalog identifier and number, a few on discoverer. For the brightest stars, classical names or constellation name with a Greek letter prefix are often used. However, it is standard to designate the primary star within a bound multiple star system with an 'A' following its name, the secondary with a 'B', etc. Exoplanets are designated analogously, using lower case letters beginning with 'b' and assigned in the order in which the planets are detected. Thus, 51 Pegasi's planet is known as 51 Peg b.

Many of these planets, henceforth referred to as *vulcans*, after the hypothetical planet once believed to travel about the Sun interior to the orbit of Mercury, have periods less than one week. (Jupiter-sized vulcan planets are often referred to as *hot Jupiters*, a term which may well overemphasize the similarities of these planets with our Solar System's largest planet.) The orbits of most of the vulcans are nearly circular, as expected because eccentric orbits this close to a star should be damped relatively rapidly by tidal forces (§2.6). A few vulcans have orbits with substantial eccentricity, probably caused by a third body (another planet, a brown dwarf, or a stellar companion, in some cases not yet observed) within the system.

12.3.3 Transiting Planets

Many vulcan planets, as well as a few more distant planets, have been observed both by radial velocity, giving $M_p \sin i$, and in transit, yielding i and R_p.[2] The densities of these planets can be calculated from these data, and educated guesses can be made about their compositions.

The giant vulcan HD 209458 b was the first exoplanet to be observed in transit. Transits of HD 209458 b have now been observed from space at many wavelengths (Fig. 12.12). This planet's orbital period is 3.525 days, its mass is 0.63 M$_{2\!\!\!/}$ and its radius is 1.35 ± 0.05 R$_{2\!\!\!/}$, implying that it is composed primarily of H_2 and He and that its

---

[2] Because even very tenuous regions of an atmosphere have high optical depths when viewed at highly oblique angles at the limb of a planet, transit radii for giant vulcans are estimated to be $\sim$1–4% larger than the 1 bar radii that are conventionally used to define the sizes of those planets within the Solar System which lack solid surfaces.

(a)

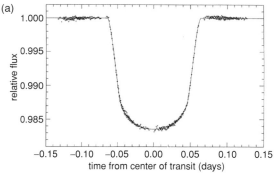

(b)

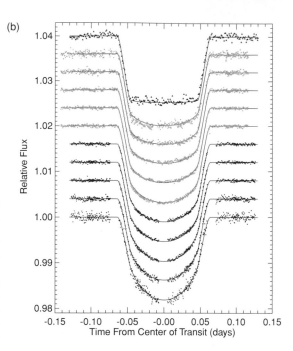

Figure 12.12 Data obtained from spaceborne observations of a total of nine transits of the planet HD 209458 b. Note that HST cannot observe complete transits because the Earth interferes during a portion of HST's orbit. (a) Superposed lightcurve of the portions of four transits of HD 209458 b observed at an average wavelength of 610 nm by the Hubble Space Telescope from 25 April through 12 May 2000. Plotted points represent individual measurements, while the solid line is the fit to a model of a circular planet passing in front of a limb-darkened star. There is one time sample per minute, each with precision of about 0.01%. Fits to the shape of the curve yield estimates of the diameters of both the star and the planet, the inclination of the orbit, and one parameter describing the star's limb darkening. (Courtesy Tim Brown) (b) Lightcurves taken in 11 different bandpasses, vertically offset from one another for clarity. The uppermost lightcurve (black triangles) was observed using the Spitzer Space Telescope on 23 December 2007 at an average wavelength of 8 μm. Moving downwards, the next five curves are composites of two transits observed using HST on 31 May 2003 (gray 'x' symbols) and 5 July 2003 (gray circles) at average wavelengths of (from top to bottom) 971, 873, 775, 677, and 581 nm, respectively. The lowest five curves are composites of two transits observed using HST on 3 May 2003 (black inverted triangles) and 25 June 2003 (black '+' symbols) at average wavelengths of (from top to bottom) 540, 485, 430, 375, and 320 nm, respectively. The shape of each transit curve is determined by the limb darkening of the star, which is more pronounced at shorter wavelengths (see text). (Courtesy Heather Knutson)

envelope is bloated because the intense stellar radiation that it absorbs has prevented it from contracting. Differences between the observed and model lightcurves would allow detection of a planetary ring system similar in size to Saturn's, or of satellites as small as 1.5 $R_\oplus$, but no evidence for either rings or moons has been found.

The shape of the transit curve is affected by the limb darkening of the star in each bandpass. Limb darkening causes the edges of the star to appear fainter than the central region. Thus the planet blocks a relatively small fraction of the star's light when it is near the edge of the star. As the planet moves towards the central region of the star where the intensity is greatest, it occults an increasingly large fraction of the star's light. This means that the center of the transit is always deeper than the edges, even though the geometric area occulted by the planet remains the same for most of the event. If the star had a uniform brightness, the transit would appear to be flat from the end of ingress to the beginning of egress. The amount of limb

darkening varies with wavelength; this is because at the edge of the star we are looking into the stellar atmosphere along a slant path, and the point where the optical depth equals one is higher up in the stellar atmosphere where the temperatures are cooler. When we look at the center of the star we see deeper into the stellar atmosphere where the temperatures are higher. The difference in brightness between the edge of the star and its center can be approximated as the difference between two Planck functions. At short wavelengths a small change in temperature produces a large change in brightness, and the edges of the star appear to be significantly fainter than the center of the star. At these wavelengths, limb darkening causes the transit to have a smoothly curved shape. Moving towards longer wavelengths (upwards in Fig. 12.12b), observations shift onto the Rayleigh–Jeans tail and the difference between the two blackbodies becomes smaller. As the amount of limb darkening diminishes, the transit has an increasingly angular, box-like shape.

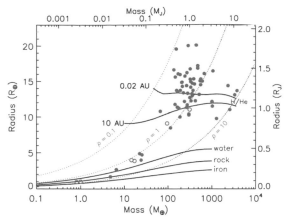

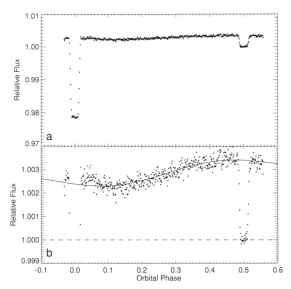

Figure **12.13** Mass–radius relationships for theoretical planets and observed planets. The planets of the Solar System are shown as open points. Transiting exoplanets are represented by solid points. The solid curves represent models of 4.5 Gyr old planets of the specified compositions. Two curves are given for H/He planets to illustrate the effects of stellar radiation; they are labeled by distance from their star, which is assumed to be 1 $M_\odot$. (Courtesy Jonathan Fortney)

Figure **12.14** Photometric observation of the combined 8 μm radiation from the star HD 189733 and its transiting giant vulcan planet. The orbital phase is measured relative to the midpoint of the transit. The observed flux is normalized to that of the star alone, with the range in panel (a) being large enough to show the full depth of the transit and panel (b) showing a magnified view that emphasizes the smaller variations from the occultation of the planet (centered at orbital phase 0.5) and the increase in radiation emitted from the planet as the hemisphere facing the star comes into view. Note that the transit lightcurve is nearly flat-bottomed, because the star is barely limb darkened at 8 μm. (Knutson *et al.* 2007)

The radii of some ∼1 M_2 vulcans, including HD 209458 b, are significantly larger than 1 R_2 (Fig. 12.13), which is near the maximum possible radius for any 'cool' cosmic composition (i.e., H–He dominated, Table 8.1) body (Fig. 6.25). These planets must thus be warm. Moreover, assuming that they have been heated (post-formation) from their external layers, they must have remained warm for most of their lifetimes (§12.1). The higher equilibrium temperatures of these planets resulting from the greater stellar fluxes impinging upon them can explain the radii of some of the observed giant vulcans, but others are so large that they must be heated by additional processes (Fig. 12.13). Three types of mechanisms have been proposed: large atmospheric opacity to outgoing radiation combined with a substantial source of energy many optical depths below the surface, downward pumping of waves created from stellar heating and/or tidal dissipation, and internal structure that features gradients in the distribution of heavy elements that suppress convective cooling.

The vulcan planet HD 149026 b has a mass of 0.36 M_2 and its radius is 0.725 ± 0.03 R_2, implying that more than half of its mass consists of elements heavier than helium. More quantitative statements about the compositions of the giant vulcan planets will not be possible until the physical process that keeps HD 209458 b and some of its brethren from shrinking is known, and it can be determined whether, and if so to what extent, this process affects other transiting vulcan planets. Three of the

five least massive and smallest known (as of early 2010) transiting planets, Glicsc 436 b, HAT-P-11 b, and Kepler-4 b, are slightly larger and more massive than Neptune. These three exoplanets must thus be composed primarily of heavy elements (by mass), but also have a substantial H_2/He component, occupying more than half of each planet's volume. Gliese 1214 b is a bit smaller and less massive, and is subject to the same compositional constraints. CoRoT-7 b is smaller still, and may be composed primarily of rock or of a mixture of rock and lighter constituents.

The combined luminosity of the star HD 189733 and its transiting giant vulcan planet have been observed in the thermal infrared (i.e., at a wavelength where the luminosity of the planet is primarily its own thermal radiation rather than reflected starlight) for more than half an orbital period of the planet, including both the transit and the occultation (Figs. 12.14 and 12.15). These data provide information on the temperature of the planet as a function of longitude relative to the substellar point. In some transiting planets, atmospheric winds seem to redistribute heat rapidly, while in other cases there is a large day–night contrast. In these

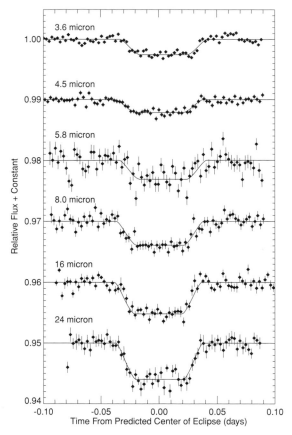

Figure 12.15 Photometric observation of the planet HD 189733 b before, during, and after the epoch when it passes behind its star at various wavelengths in the infrared. The eclipse depth is greater at longer wavelengths, where the planet's radiation is a greater fraction of the output of the star (see Fig. 12.7). Note that the eclipse lightcurves are all nearly flat-bottomed. (Charbonneau *et al.* 2008)

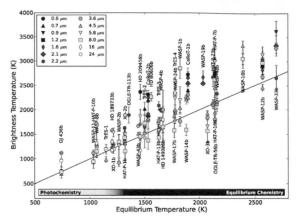

Figure 12.16 Comparison of measured brightness temperatures, T_b (§3.1.2), vs. predicted equilibrium temperatures, T_{eq} (§3.1.2.2), for close-in exoplanets that are observed to pass behind their stars. The T_b numbers apply to the substellar hemisphere at the specified wavelength, and were obtained from Spitzer Space Telescope occultation measurements. The T_{eq} values assume a Bond albedo (wavelength-independent reflectivity) $A_b = 0$ (which leads to an overestimate of T_{eq}, albeit a small one if the planet's albedo is low), uniform emission across the planet's surface (eq. 3.17; the actual temperature on the star-facing hemisphere is likely higher than this planet-wide equilibrium average value by up to a factor of $\sim 2^{1/4}$), and emissivity $\epsilon = 1$. The line indicates $T_b = T_{eq}$. The hotter planets tend to be well above this line, possibly due to formation of a dark absorber (e.g., TiO) in their upper atmospheres. (Adapted and updated from Harrington *et al.* 2007 by J. Harrington in January 2014)

latter cases, the atmosphere cools so rapidly (§12.1) that winds cannot redistribute energy efficiently around the planet and air parcels cool substantially after they rotate from underneath the substellar point. The infrared luminosities of dozens of planets have been deduced by differencing measurements taken near and during occultation (Fig. 12.16).

Variations in the depth of the transit of HD 209458 b with wavelength reveal the presence of sodium in the upper atmosphere of this giant vulcan. The very large depth of the transit at the wavelength of the Lyman α line (Fig. 3.7) shows that hydrogen associated with the planet extends over an area larger than the size of the planet's Hill sphere (eq. 2.22); this implies that hydrogen is escaping the planet at a considerable rate, albeit not so rapidly as to have removed a substantial fraction of

the planet's mass over its lifetime to date. Infrared spectra of this system during and surrounding the secondary eclipse suggest the presence of silicates in the planet's atmosphere. Near-infrared transmission spectra of HD 189733 b reveal the presence of methane in this planet's atmosphere.

As stars rotate, gas in half of the stellar disk moves towards us and gas in the other half moves away. Thus, planetary transits can affect the apparent radial velocity of the star by blocking light from either the rotationally blueshifted or redshifted half of the stellar disk. This is known as the *Rossiter–McLaughlin effect*. Planets orbiting in the prograde direction initially block a blueshifted portion of the stellar disk, leading to an apparent redshift, and then the opposite occurs (Fig. 12.17). For noncentral transits, the inclination of the planet's orbit relative to the star's equator can be determined by the asymmetry of the Rossiter–McLaughlin effect. The Rossiter–McLaughlin effect has been detected for more than a dozen planets as of late 2009. The vast majority of these have been found to orbit in the prograde direction near their star's equatorial plane, as in our Solar System, but a few

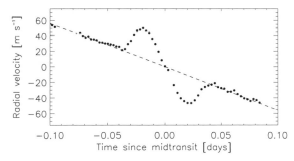

Figure 12.17 The apparent radial velocity variation of the star HD 189733 during and immediately surrounding a transit of its giant vulcan planet. The dashed line is a Keplerian fit to the data outside of transit, for which the observed radial velocity variation is due entirely to the star's orbital motion. The velocity anomaly, first high then low, is the Rossiter–McLaughlin effect, the apparent Doppler shift due to the partial eclipse of the rotating stellar surface. The symmetry of the velocity variations shows that the stellar spin angular momentum and the orbital angular momentum (both projected onto the plane of the sky) are nearly aligned. (Courtesy Josh Winn)

travel on substantially inclined (perhaps even retrograde) paths.

12.3.4 Planets Orbiting Pulsating Stars

The first planet to have been observed around a pulsating star is V391 Pegasi b. This planet has $M_p \sin i \approx 3.2$ M$_{\mathrm{2}}$, $a \sim 1.7$ AU, and small eccentricity. The stellar host is a post-red-giant helium-burning star with a current mass of ~ 0.5 M$_\odot$; models suggest that the star had a mass of ~ 0.85 M$_\odot$ when it was on the main sequence, implying that the planet used to have a significantly smaller orbit (§2.8; Problem 12.9).

12.3.5 Microlensing Detections

The first planet to be detected via microlensing is a ~ 2.6 M$_{\mathrm{2}}$ object seen ~ 4.3 AU away (on the plane of the sky) from an ~ 0.63 M$_\odot$ star (Fig. 12.18). More than twenty other planets, some only a few times as massive as Earth, have been detected via microlensing as of 2014. The OGLE-06-109L system is especially interesting, as it has two planets and resembles a somewhat smaller version of our Solar System: The star and its inner and outer planets are a bit less massive than the Sun, Jupiter, and Saturn, and their separations (on the plane of the sky) are just under half those of the three largest members of our Solar System.

12.3.6 Multiple Planet Systems

Almost three dozen nearby stars are now known to possess two or more planets, and most of these planets are roughly of jovian mass (Fig. 12.19). Three jovian-mass planets

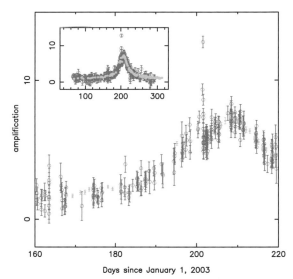

Figure 12.18 Lightcurve of the OGLE 2003-BLG-235/MOA 2003-BLG-53 microlensing event. The OGLE and MOA measurements are shown as small filled and large open circles, respectively. The main panel presents the complete data set during August 2003, and the insert shows the data during all of 2003. The overall rise in brightness is primarily due to light being bent by the lensing star, but the high narrow peak one week prior to the midpoint of the event is caused by the combined gravitational tugs of the star and its 2–3 M$_{\mathrm{2}}$ planet. (Courtesy Ian Bond)

have been detected in orbit about the star υ Andromedae (Fig. 12.20). The innermost of these objects is a vulcan planet; the other two planets are far more distant from the star and travel on eccentric orbits. Dynamical calculations imply that at least the outermost two planets must have $\sin i > 1/5$ for the system to remain stable for the star's 2.5×10^9 years age.

Some exoplanets are dynamically isolated from other known planets orbiting their star, as is the inner planet of υ Andromedae. However, other giant planets have more similar periods, and some pairs of giant planets are in low-order mean motion resonances with one another. The two giant planets orbiting the star Gliese 876 are locked in a 2:1 orbital mean motion resonance and have particularly strong interactions because of their large masses relative to that of their one-third M$_\odot$ star. Fits to the radial velocity data which account for these perturbations (Fig. 12.21) are far superior to those in which the stellar response to Keplerian motions of the planets are superposed. Moreover, the orbital periods of the two planets are only one and two months (Table 12.1), so their motion has been followed for many orbits. Mutual perturbations of these planets have been detected, and these perturbations allow the determination of the inclinations of the planets' orbits.

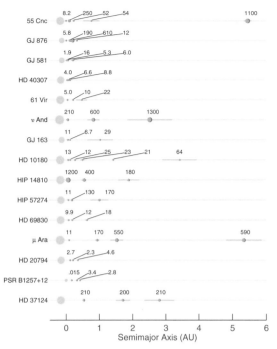

 Figure 12.19 Chart of planetary semimajor axes and masses for the 14 stars with three or more planets found by radial velocity surveys (as of March 2014) as well as the one multi-planet pulsar system. The depicted planetary radii are proportional to $(M\sin i)^{1/3}$, and numerical values are in quoted terms of $M_\oplus$. The periapse to apoapse excursion of each planet is shown by a horizontal line centered at the planet's semimajor axis. The radii depicted for the stars are proportional to $M_\star^{1/3}$ (with a different constant of proportionality from that used for the planets). Note the large number of planets with masses smaller than 0.1 $M_{\jupiter}$, in contrast to their scarcity among the overall radial velocity sample (Figure 12.23). This implies that smaller planets are more likely to be found in multiple planet systems than are giant planets. (Courtesy Jason Wright)

A small ($M_p \sin i \approx 6\ M_\oplus$) planet Gliese 876 d, orbiting very close to the star, has also been detected.

12.4 Exoplanet Statistics

Extrasolar planet discoveries have expanded our database by increasing the number of known planets by well over an order of magnitude. The distribution of known extrasolar planets is highly biased towards those planets that are most easily detectable using the Doppler radial velocity technique (Fig. 12.4), which has been by far the most effective method of discovering exoplanets. These extrasolar planetary systems are quite different from our Solar System; however, it is not yet known whether our planetary

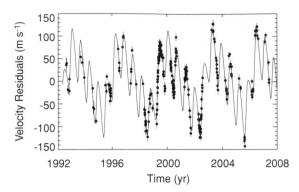

Figure 12.20 The variations in the radial velocity of the star υ Andromedae after subtracting off the star's motion due to its inner planet (υ And b, which has an orbital period of only 4.6 days) are shown as a function of time. Uncertainties of individual measurements are indicated. The solid curve represents the model response of the star υ Andromedae to two additional giant planets with much longer orbital periods, υ And c and υ And d. (Courtesy Debra Fischer)

system is the norm, quite atypical, or somewhere in between.

Nonetheless, some unbiased statistical information can be distilled from available exoplanet data: Roughly 0.7% of Sun-like stars (late F, G, and early K spectral class single main sequence stars that are chromospherically quiet, i.e., have inactive photospheres) have planets more massive than Saturn within 0.1 AU. Approximately 7% of Sun-like stars have planets more massive than Jupiter within 3 AU. Only about 1% of low-mass stars (M dwarfs with masses 0.3–0.5 $M_\odot$) are orbited by giant planets within 2 AU. Most planets orbiting interior to $\sim$0.1 AU, a region where tidal circularization timescales are less than stellar ages, have small orbital eccentricities. The median e of giant planets with $0.1 < a < 3$ AU is $\sim$0.26, and some of these planets travel on very eccentric orbits (Fig. 12.22). Within 5 AU of Sun-like stars, Jupiter-mass planets are more common than planets of several Jupiter masses, and substellar companions of mass $\gtrsim$10 $M_{\jupiter}$ are rare (Fig. 12.23). The paucity of objects of mass $\sim$10−60 $M_{\jupiter}$ near Sun-like stars is known as the *brown dwarf desert*.

Stars with higher metallicity are much more likely to host giant planets within a few AU than are metal-poor stars, with the probability of hosting such a planet varying roughly as the square of stellar metallicity (Fig. 12.24). The Sun itself has a higher metallicity than do most $\sim$1 $M_\odot$ stars in the solar neighborhood. At least over the range 0.3–1.5 $M_\odot$, more massive stars also appear more likely to host giant planets orbiting within a few AU.

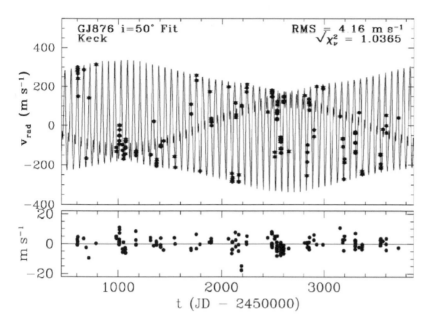

Figure 12.21 The variations in the radial velocity of the faint M dwarf star Gliese 876 plotted as a function of time. The curve passing through most of the points was calculated by varying the parameters of three mutually interacting planets in orbit about the star to best fit the data (the values of which are similar to the more recent results that are listed in Table 12.1). Error bars for the points represent intrinsic uncertainties of the measured velocities. The points below the main curve show the departures of the data from the best-fit model. (Courtesy Eugenio J. Rivera)

Table 12.1 Gliese 876's planetary system (Rivera *et al.*, 2010).

	Period (d)	a (AU)	e	ω (°)	M_p^a
d	1.938	0.021	0.21	234	$6.8\,M_\oplus$
c	30.09	0.130	0.260	49	$0.71\,M_{\tiny 2}$
b	61.12	0.208	0.032	50	$2.28\,M_{\tiny 2}$
e	124.26	0.33	0.055	239	$14.6\,M_\oplus$

[a] Actual masses shown assuming a planar system with best-fit inclination of $i = 59.5°$.

Multiple planet systems are more common than if detectable giant planets were randomly distributed among stars (i.e., than if the presence of a detectable planet around a given star was not correlated with the presence of other planets around that same star). Nonetheless, far more stars have just one identified planet, and in most cases these 'single' planets must be much more massive ($M\sin i$) than any other companions with periods of a few years or less that the star may possess. This is in contrast to our Solar System, where planets of comparable size have orbital periods within a factor of two or three of their neighbors. These extrasolar planetary systems thus appear to be more overstable than is our own, which may indicate that different mechanisms are important in the formation process.

Most transiting extrasolar giant planets are predominantly hydrogen, as are Jupiter and Saturn. However, HD 149026 b, which is slightly more massive than Saturn, appears to have comparable amounts of hydrogen + helium versus heavy elements. So the bulk composition

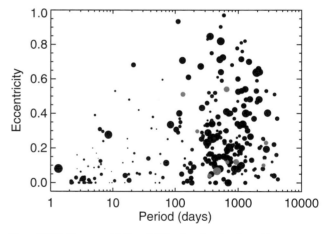

Figure 12.22 The eccentricities of 438 extrasolar planets and low-mass brown dwarf companions to main sequence stars discovered by the radial velocity method that have well-determined orbits (as of March 2014) are plotted against orbital period. The dot size is proportional to $(M_p \sin i)^{1/3}$, and the gray points represent planets in systems known to possess more than one planet. The eccentricities of almost all of those planets with periods of less than a week are quite small (consistent with 0), presumably as a result of tidal damping, whereas the eccentricities of planets with longer orbital periods are generally much larger than those of the giant planets within our Solar System. (Courtesy Jason Wright, data from exoplanets.org)

of HD 149026 b is intermediate between those of Saturn and Uranus, and HD 149026 b is more richly endowed in terms of total amount of 'metals' than is any planet in our Solar System.

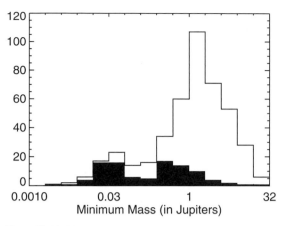

Figure 12.23 Histogram of the number of planets and low-mass brown dwarfs observed by radial velocity surveys as a function of minimum planet mass, $M_p \sin i$, obtained from the same data set used to produce Figure 12.22. Each bin encompasses a factor of two in minimum mass. The black region at the bottom represents planets with orbital periods of less than 30 days, and the white region shows the number of planets with longer periods. As more massive planets are easier to detect, the tail off in the distribution above 1–2 $M_{\textrm{\tiny 2}}$ is real, whereas the dropoff at smaller masses is a consequence of observational selection effects. Likewise, the shift in the distribution to smaller masses at shorter periods results from the larger radial velocity perturbations by planets orbiting closer to the star (eq. 12.4) and the greater number of orbits that these planets have covered since the highest precision radial velocity surveys have been underway. (Courtesy Jason Wright, data from exoplanets.org)

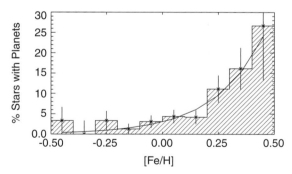

Figure 12.24 The fraction of Sun-like stars possessing giant planets with orbital periods of less than four years is shown as a function of stellar metallicity. Metallicity is measured on a logarithmic scale, with zero corresponding to the solar value. (Fischer and Valenti 2005)

Transit observations have also yielded an important negative result: Hubble Space Telescope photometry of a large number of stars in the globular cluster 47 Tucanae failed to detect any transiting inner giant planets, even though ~17 such transiting objects would be expected were the frequency of such planets the same as that for

Sun-like stars in the solar neighborhood of our galaxy. A subsequent ground-based search for short-period transiting planets in the outer reaches (halo) of 47 Tucanae also failed to yield any planets; the expected yield of this search would have been ~7 for occurrence frequency of giant vulcan planets equal to that around stars in the vicinity of our Solar System. The paucity of transiting planets in 47 Tucanae is best understood as a result of the very low metallicity (heavy element content) of the stars in this globular cluster.

12.5 Planets and Life

One of the most basic questions that has been pondered by people since antiquity concerns humanity's place in the Universe: Are we alone? This question has been approached from a wide variety of viewpoints, and similar reasoning has led to widely diverse answers. Aristotle believed that Earth, the densest of the four 'elements' of ancient Greek cosmology, fell towards the center of the Universe, so no other worlds could possibly exist; in contrast, Democritus and other early atomists surmised that the ubiquity of physical laws implies innumerable Earth-like planets must exist in the heavens.

A major scientific debate concerning the possibility of life, advanced or otherwise, on Mars was ongoing at the beginning of the twentieth century. As we learned more about the current surface conditions on the planet whose climate is more Earth-like than any of our other neighbors, the chances of life seemed to be far more remote. However, recent theoretical and observational results suggest that early Mars may have been as hospitable to life as was early Earth (§4.9), and that the descendants of such life may survive deep under Mars's surface. Martian microbes may even have traveled to Earth within meteorites and might be our very distant ancestors!

The conventional, very anthropocentric, picture is that to be habitable, a planet must have liquid water on its surface for a very long time. This single factor is unlikely to be either necessary or sufficient. Nonetheless, it provides a useful guide for habitability by life similar to that on Earth. The main sequence phase of low- to moderate-mass stars provides long-lived regions of orbital space where planets may maintain liquid water on their surfaces; these regions are referred to as *continuously habitable zones*. The lifetime of continuously habitable zones is shorter for more massive stars; also, although the quantity of radiation received by planets in habitable zones is the same regardless of stellar type, the 'quality' differs, as more-massive stars radiate a larger fraction of their energy at shorter wavelengths. Shorter-lived habitable zones occur at greater distances from post-main-sequence stars that burn helium in their cores and shine brightly as red giants.

The greater flux of ultraviolet radiation could conceivably speed up biological evolution enough to compensate for a few $M_\odot$ star's shorter lifetime. At the other end of the spectrum, the smallest, faintest stars can live for trillions of years, but they emit almost all of their luminosity at infrared wavelengths and their luminosity varies because they emit large flares. Also, rocky habitable-zone planets orbit so close to these faint stars that their rotation would be tidally synchronized unless their orbits are quite eccentric (§2.6.2). Tidally synchronized planets have no day–night cycle, and if their atmosphere is thin it freezes out on the planet's cold, perpetually dark, hemisphere.

The minimum separation of ~ 1 $M_\oplus$ planets on low-eccentricity orbits required for the system to be stable for long periods of time is comparable to the width of a star's continuously habitable zone. Thus, orbital stability arguments support the possibility that most stars could have one or even two planets with liquid water on their surfaces, but unless greenhouse effects conspire to substantially compensate for increasing distance from the star, larger numbers of habitable planets around an individual star are unlikely (although it is conceivable that a giant planet orbiting at the appropriate distance from a star could possess several ~ 1 $M_\oplus$ moons).

Because of the destruction that impacts may produce, impact frequency is an important factor in planetary habitability. The impact rate on the terrestrial planets of our Solar System was orders of magnitude larger 4 billion years ago than it is at present (§5.4.4.1). In another planetary system, large impact fluxes could continue, making planets with Earth-like compositions and radiation fluxes hostile abodes for living organisms. Life on Earth has thrived thanks to billions of years of benign climate. Mars appears to have had a climate sufficiently mild for liquid water to have flowed on its surface when the Solar System was roughly one-tenth its current age (§4.9.2), but at the present epoch, the low atmospheric pressure and usually low temperature mean that liquid water would not be stable on the martian surface. Venus is too hot, with a massive carbon-dioxide dominated atmosphere; we cannot say whether or not young Venus had a mild Earth-like climate. Indeed, as stellar evolution models predict that the young Sun was about 25% less luminous than at present (§1.3), we don't understand why Earth, much less Mars, was warm enough to be covered by liquid oceans 4 billion years ago (§4.9.1).

Carbon dioxide on our planet cycles between the atmosphere, the oceans, life, fossil fuels, and carbonate rocks on a wide range of timescales (§4.9). The carbonate rocks form the largest reservoir; they are produced by reactions involving water, in some cases living organisms act as catalysts, in other cases not. Carbon dioxide is recycled from carbonates back into the atmosphere as plates are subducted and heated within the Earth's mantle. Carbonates are not readily recycled on a geologically inactive planet such as Mars; in contrast they are not formed on planets like Venus, which lack surface water. Larger planets of a given composition remain geologically active for longer, as they have smaller surface-area-to-mass ratios, enabling them to retain heat from accretion and radioactive decay longer. The number of variables involved in determining a planet's habitability precludes a complete discussion, but some of the major issues are summarized in Figure 12.25.

12.6 SETI

The Search for Extra-Terrestrial Intelligence (*SETI*) is an endeavor to detect signals from alien life-forms. A clear detection of such a signal would likely change humanity's worldview as much as any other scientific discovery in history. As our society is in its technological infancy, another civilization capable of communicating over interstellar distances is likely to be enormously advanced compared to our own – compare our technology to that of a mere millennium ago and then extrapolate millions or billions of years into the future! Thus, a dialog with extraterrestrials could alter our society in unimaginable ways.

The primary instrument used by SETI is the radio telescope. Most radio waves propagate with little loss through the interstellar medium, and many wavelengths also easily pass through Earth's atmosphere. They are easy to generate and to detect. Radio thus appears to be an excellent means of interstellar communication, whether data are being exchanged between a community of civilizations around different stars or are being broadcast to the galaxy in order to reach unknown societies in their technological infancy. Signals used for local purposes, such as radar and TV on Earth, also escape and can be detected at great distances.

The first deliberate SETI radio telescope observations were performed by Frank Drake in 1960. Since that time, improvements in receivers, data processing capabilities and radio telescopes have doubled the capacity of SETI searches roughly once per year. While a betterment by a factor of $\sim 10^{14}$ is quite impressive, only a minuscule fraction of directions and frequencies have been searched, so SETI proponents are not discouraged at the lack of success to date.

Figure 12.25 COLOR PLATE Theoretical comparison of planets of different sizes with the same composition as Earth. (Left) A smaller planet would be less dense, because the pressure in the interior would be lower. Such a planet would have a larger ratio of surface area to mass, so its interior would cool faster. Its lower surface gravity and more rigid crust would allow for higher mountains and deeper valleys than are seen on Earth. Most important to life is that the atmosphere of the mini-Earth would have a much smaller surface pressure as a result of four factors: larger surface area to mass, lower surface gravity, more volatiles sequestered in the crust because there would be less crustal recycling, and more atmospheric volatiles escaping to space. This would imply, among other things, lower surface temperature, resulting from less greenhouse gas in the atmosphere. Some remedial measures which could improve the habitability of such a mass-deprived planet are: (1) Move it closer to the star, so less greenhouse effect would be needed to keep surface temperatures comfortable. (2) Add extra atmospheric volatiles. (3) Include a larger fraction of long-lived radioactive nuclei than on Earth, to maintain crustal recycling. (Center) Earth; home sweet home. (Right) A larger planet made of the same material as Earth would be denser and have a hotter interior. Its higher surface gravity and more ductile crust would lead to muted topography. It would have a much greater atmospheric pressure, and, unless its greenhouse was strong enough to boil away the planet's water, much thicker oceans, probably covering the planet's entire surface. Some remedial measures which could improve the habitability of such a mass-gifted planet are: (1) Move it further from the star. (2) Include a smaller fraction of atmospheric volatiles. It is not clear that more active crustal recycling would be a problem, within limits, but crustal activity would be lessened if the planet had a smaller inventory of radioactive isotopes. (3) Give it a wide, optically thick ring. Provided the planet has a moderate to large obliquity, such a ring would shadow a significant portion of the planet for much of its 'year' (Figs. 11.2 and 11.4). (Lissauer 1999)

12.7 Conclusions

Prior to the discovery of extrasolar planets, models of planetary growth suggested that most single solar-type stars possess planetary systems that are grossly similar to our Solar System. Observations have subsequently demonstrated that nature is more creative than the human imagination. It was realized that stochastic factors are important in planetary growth, so that the number of terrestrial planets (as well as the presence or lack of an asteroid belt) would vary from star to star, even if their protoplanetary disks were initially very similar. The difficulty in accreting giant planet atmospheres prior to dispersal of circumstellar gas suggested that many systems might lack gas giants. The low eccentricities of the giant planets in our Solar System (especially Neptune) are difficult to account for, so systems with planets on highly eccentric orbits were viewed as possibilities, although researchers did not hazard to estimate the detailed characteristics of such systems. A maximum planetary mass similar to that of Jupiter was suggested as a possibility if Jupiter's mass was determined by a balance between a planet's gap-clearing ability and viscous inflows (§13.7.2), although it was noted that the value of the viscosity could well vary from disk to disk. Orbital migration of some giant planets towards their parent star (§13.8) was also envisioned, but since migration rates increased as the planet approached the star, such planets were expected to be accreted by their star, and the existence of numerous giant planets with orbital periods ranging from a few days to several weeks was not predicted.

It must thus be admitted that theoretical models based upon observations within our Solar System failed to predict the types of planets that have been detected by radial velocity surveys. But such planets are fairly scarce, occurring in a minority of the systems. Radial velocity surveys are biased in favor of detecting massive planets orbiting close to stars, and planets similar to those in our own Solar System would not yet have been detected. Thus, it is possible that the majority of single Sun-like stars possess planetary systems quite similar to our own. Alternatively, although theoretical considerations suggest that terrestrial planets are likely to grow around most Sun-like stars, they may typically be lost if most systems also contain giant planets which migrate into the central star.

Discovering an exoplanet with size, mass, star, and orbit like that of our own is beyond the present capabilities of any planet-finding technique other than photometry with the Kepler spacecraft. While an Earth analog at a distance of 10 parsecs would be brighter than the faintest objects observed by HST, the adjacent massive, huge, and overwhelmingly bright Sun-like star makes detecting such a planet exceedingly challenging. The Sun's radius is 100 times that of Earth, its mass 300 000 times as large, and its brightness is $10^6 - 10^{10}$ that of our home planet (Fig. 12.7).

We still do not know whether terrestrial planets on which liquid water flows are rare, are the norm for solar-type stars or have intermediate abundances. Nonetheless, even if planetary migration destroys some promising systems, planets qualifying as continuously habitable for long periods of time by the liquid-water criterion are expected to be sufficiently common that if we are the only advanced life-form in our sector of the galaxy, biological and/or local planetary factors are much more likely to be the principal limiting factor than are astronomical causes.

Further Reading

Good overviews of planet detection techniques and early results, concentrating on the radial velocity method, are given by:

Marcy, G.W., R.P. Butler, D. Fischer, S. Vogt, J.T. Wright, C.G. Tinny and H.R.A. Jones, 2005. Observed properties of exoplanets: Masses, orbits and metallicities. *Prog. Theor. Phys. Supp.*, **158**, 24–42.

Udry, S., D. Fischer and D. Queloz, 2007. A decade of radial-velocity discoveries in the exoplanet domain. In *Protostars and Planets V*. Eds. B. Reipurth, D. Jewitt and K. Keil. University of Arizona Press, Tucson, pp. 685–699.

A general overview of extrasolar planet research with frequently updated news on planetary discoveries is given by the Extrasolar Planet Encyclopedia: http://exoplanet.eu.

The websites of the two leading radial velocity planet search teams are: http://exoplanets.org and http://exoplanets.eu.

The relationship between life and the planet(s) that host it is discussed in greater depth in:

Lissauer, J.J. and I. de Pater, 2013. *Fundamental Planetary Science: Physics, Chemistry and Habitability*. Cambridge University Press, Cambridge.

Problems

12.1.**I** For small bodies the relationship between mass and size is given by equation (12.3). When more mass is added to a planet, the material will be compressed. When the internal pressure becomes very large, the matter becomes degenerate, as is the case for white dwarf stars. Consider the central pressure, P_c, of a white dwarf star, which can be calculated from equation (6.1), as you did in Problem (6.3). The polytropic constant n in the equation of state (eq. 6.5) is 2/3 in the limit of high pressure. Show that for a white dwarf $M \propto R^{-3}$.

12.2.**E** Estimate the ratio of the radiative timescale of the atmosphere (the time it would take for the thermal energy to fall by half) of a planet whose temperature is 1500 K to that of a planet whose atmosphere has $T = 100$ K at a comparable pressure level. You may assume that the heat capacities and radiative efficiencies are the same, so you only need to consider the thermal energy contents above the 1 bar pressure level and the blackbody luminosities of the planets.

12.3.**E** A planet of mass $M_p = 2\ M_{2}$ travels on a circular orbit of radius 4 AU about a 1 $M_{\odot}$ star. The Solar System lies in the plane of the orbit. Write the equation for the star's radial velocity variations caused by the planet and sketch the resulting curve.

12.4.**E** A planet of mass $M_p = 2\ M_{2}$ travels on a circular orbit of radius 4 AU about a 1 $M_{\odot}$ star. The Solar System lies 60° from the plane of the orbit. Write the equation for the star's radial velocity variations caused by the planet and sketch the resulting curve.

12.5.**I** A planet of mass $M_p = 2\ M_{2}$ travels on an orbit with $a = 4$ AU and $e = 0.5$ about a 1 $M_{\odot}$ star. The Solar System lies 60° from the plane of the orbit, and the major axis of the orbit ellipse is oriented perpendicular to the line of sight. Compute the extrema of the star's radial velocity, estimate a few other points, and sketch the resulting curve.

12.6.**E** What is the amplitude of the astrometric wobble induced by a planet of mass $M_p = 2\ M_{2}$ that travels on a circular orbit of 4 AU radius about a 1 $M_{\odot}$ star located 4 parsecs from the Sun?

12.7.**E** Calculate the probability of transits of the planets Venus and Jupiter being observable from another (randomly positioned) planetary system.

12.8.**I** (a) Calculate the ratio of the light reflected by Earth at 0.5 μm to that emitted by the Sun at the same wavelength.

(b) Calculate the ratio of the thermal radiation emitted by Earth at 20 μm to that emitted by the Sun at the same wavelength.

(c) Repeat the above calculations for Jupiter.

12.9.**E** Estimate the semimajor axis of the planet V391 Pegasi b (§12.3.4) when its star was on the main sequence.

12.10.**E** Consider a Sun-like star with radius of 1 $R_{\odot}$ and effective temperature of 6000 K that has a

close-in giant planet with Jupiter's radius and effective temperature of 1500 K.

(a) Calculate the ratio of the (bolometric) luminosity of the planet to that of its star using the wavelength-integrated blackbody radiation formula (eq. 3.9).

(b) Calculate the ratio of the infrared flux emitted by the two bodies at a wavelength of 24 microns. You may use the Rayleigh–Jeans approximation to the blackbody radiation formula, which is given in equation (3.4a).

(c) Imagine that you measure, with a space telescope, the combined flux emitted by both the planet and the star. The planet then passes behind the star so that you no longer see its contribution to the total system flux. What is the percentage drop in the total flux that you would measure? Modern infrared detectors can measure changes on the order of 0.1%. Comment if you think the passage of such a giant vulcan planet behind its star would be detectable.

(d) Re-do part (b) but for wavelengths of 5 and 0.5 microns. Note that at these shorter wavelengths, the full blackbody formula (eq. 3.3) must be used. Are all wavelengths equally good for detecting the planet passing behind the star? Why or why not?

12.11.**E** Imagine a terrestrial planet with Earth's radius orbiting 0.03 AU away from a cool M-type star with luminosity $\mathcal{L} = 10^{-3}\mathcal{L}_{\odot}$.

(a) Calculate the equilibrium temperature for this planet. Does it lie in the habitable zone?

(b) Planets orbiting this close to their star are likely to become 'tidally locked' and keep one side always facing the star (like the Moon keeps one side facing Earth). The equilibrium temperature computed in (a) is really a planet-wide average. Assuming that this planet does not possess an atmosphere, describe qualitatively how the surface temperature varies with location on the globe. Discuss which locations on the planet might be 'habitable' and 'uninhabitable'.

12.12.**E** As demonstrated in Problems 2.14 and 2.15, Newton's theory of gravity is quite accurate for most Solar System situations, and the most easily observable effect of general relativity is the precession of orbits. Those extrasolar planets that orbit much closer to their star than Mercury's distance from the Sun travel much faster, and they are also subjected to a greater gravitational field, so relativistic deviations from the trajectories predicted by Newton's laws should be larger. The first-order (weak field) general relativistic corrections to Newtonian gravity imply the periapse precession at the rate given by equation (2.63). Calculate the general relativistic precession of the periapse of:

(a) The transiting planet Gliese 436 b, for which $M_\star = 0.45$ $M_\odot$, $M_p = 0.0692$ M_2, $e = 0.16$, $P_{orb} = 2.644$ days.

(b) The highly eccentric planet HD 80606 b, for which $M_\star = 0.9$ $M_\odot$, $M_p \sin i = 4.3$ M_2, $e = 0.93$, $P_{orb} = 111.5$ days.

12.13.E Estimate the equilibrium temperature of Kepler-11 g using the data given in Table G.1:

(a) Assuming $A_b = 0.5$ and efficient global redistribution of heat.

(b) Assuming $A_b = 0.05$ and efficient global redistribution of heat.

(c) At the subsolar point assuming $A_h = 0.5$ and no redistribution of heat.

(d) Suppose that the James Webb Space Telescope measured the brightness temperature of this planet in a bandpass near 3 μm to be 400 K. Discuss what this information tells you about the planet.

12.14.I Radii and masses measured for five hypothetical transiting planets are listed below. State the ranges in composition consistent with these measurements. In some cases, the ranges allowed by the error bars include some unlikely or unphysical compositions. Which cases are these, and why?

(a) $R_p = 3 \pm 1$ $R_\oplus$, $M_p = 3 \pm 1$ $M_\oplus$

(b) $R_p = 1 \pm 0.5$ $R_\oplus$, $M_p = 3 \pm 1$ $M_\oplus$

(c) $R_p = 12 \pm 2$ $R_\oplus$, $M_p = 300 \pm 100$ $M_\oplus$

(d) $R_p = 3 \pm 1$ $R_\oplus$, $M_p = 30 \pm 10$ $M_\oplus$

(e) $R_p = 2 \pm 0.2$ $R_\oplus$, $M_p = 10 \pm 1$ $M_\oplus$

13 Planet Formation

From a consideration of the planetary motions, we are therefore brought to the conclusion, that in consequence of an excessive heat, the solar atmosphere originally extended beyond the orbits of all the planets, and that it has successively contracted itself within its present limits.

Pierre Simon de Laplace, *The System of the World*, 1796

The origin of the Solar System is one of the most fundamental problems of science. Together with the origin of the Universe, galaxy formation, and the origin and evolution of life, it is a crucial piece in understanding where we come from. Because planets are difficult to detect and study at interstellar distances, we have detailed knowledge of only one planetary system, the Solar System. Data from other planetary systems are now beginning to provide further constraints (Chapter 12). But even though more than 99% of known planets orbit stars other than the Sun, the bulk of the data available to guide modelers of planet formation is from objects within our Solar System. Models of planetary formation are developed using the detailed information we have of our own Solar System, supplemented by astrophysical observations of extrasolar planets, circumstellar disks, and star-forming regions. These models are used together with observations to estimate the abundance and diversity of planetary systems in our galaxy, including those planets which may harbor conditions conducive to the formation and evolution of life (§12.5).

13.1 Solar System Constraints

Any theory of the origin of our Solar System must explain the following observations:

Orbital Motions, Spacings, and Planetary Rotation: The orbits of most planets and asteroids are nearly coplanar, and this plane is near that of the Sun's rotational equator. The planets orbit the Sun in a prograde direction (the same sense as the Sun rotates), and travel on nearly circular trajectories. Most planets rotate around their axis in the same direction in which they revolve around the Sun, and have obliquities of $<30°$. Venus and Uranus are exceptions to this rule (see Tables 1.2 and 1.3). Major planets are confined to heliocentric distances $\lesssim 30$ AU, and the separation between orbits increases with distance from the Sun

(Table 1.1). Most of the smaller bodies orbiting the Sun (asteroids, Kuiper belt objects, etc.) move on somewhat more eccentric and inclined paths (Tables 9.1 and 9.2), and their rotation axes are more randomly oriented. Aside from the asteroid belt between 2.1 and 3.3 AU and the regions centered on Jupiter's stable triangular Lagrangian points, interplanetary space contains very little stray matter. But this does not provide a significant constraint on the planet formation epoch, since most orbits within these empty regions are unstable to perturbations by the planets on timescales short compared to the age of the Solar System (Fig. 2.17). Thus bodies initially in orbits traversing these regions would likely collide with a planet or the Sun or be ejected from the Solar System. Thus, in a sense, the planets are about as closely spaced as they could possibly be.

Angular Momentum Distribution: Although the planets contain $\lesssim 0.2\%$ of the Solar System's mass, more than 98% of the angular momentum in the Solar System resides in the orbital motions of the giant planets. In contrast, the orbital angular momenta of the satellite systems of the giant planets are far less than the spin angular momenta of the planets themselves.

Age: Radioisotope ^{207}Pb/^{206}Pb dating of refractory inclusions (CAIs) found within chondritic meteorites, the oldest Solar System solids known, yields an age of 4.568 Gyr. Dating with other isotope systems yields similar ages. Chondrules, as well as most differentiated meteorites that originated within small bodies, solidified only a few million years later (§8.7). Rocks formed on the Moon and Earth are younger: lunar rocks are typically between 3 and 4.4 Gyr old, and terrestrial rocks are $\lesssim 4$ Gyr old, although terrestrial mineral grains as old as 4.4 Gyr have been found.

Sizes and Densities of the Planets: The relatively small terrestrial planets and the asteroids, which are mainly composed of rocky material, lie closest to the Sun. The uncompressed (zero pressure) density (§6.1) decreases with heliocentric distance, which suggests a larger fraction of heavier elements, like metals and other refractory (high condensation temperature) material, in planets closer to the Sun. At larger distances we find the giants Jupiter and Saturn, and further out the somewhat smaller Uranus and Neptune. The low densities of these planets imply lightweight material. Jupiter and Saturn are primarily composed of the two lightest elements, hydrogen and helium (Jupiter has ∼90% H and He by mass, Saturn ∼80%), while Uranus and Neptune contain relatively large amounts of ices and rock (they contain ∼10–15% H and He by mass).

Shapes and Densities of Small Bodies: Smaller bodies tend to be more irregularly shaped. This is a consequence of their weaker gravity, but it also implies that either they were never molten or have suffered disruptive collisions subsequent to their resolidification. Bodies of radius $R \lesssim 100$ km tend to be of lower density (for a given surface composition), implying substantial porosity on microscopic and/or macroscopic scales.

Asteroid Belt: Between the orbits of Mars and Jupiter are countless minor planets. The total mass of this material is ∼1/20 the mass of the Moon. Except for the largest asteroids ($R \gtrsim 100$ km), the size distribution of these objects is similar to that expected from a collisionally evolved population of bodies (Chapter 9).

Kuiper Belt: Most small bodies within the Solar System orbit beyond Neptune. The greatest concentration of such bodies is within a flattened disk at heliocentric distances between 35 and 50 AU.

Comets: There is a 'swarm' of ice-rich solid bodies orbiting the Sun at $\gtrsim 10^4$ AU, commonly referred to as the Oort cloud. There are roughly 10^{12}–10^{13} objects larger than one kilometer in this 'cloud'. The bodies are isotropically distributed around the Sun, aside for a slight flattening produced by galactic tidal forces. The Kuiper belt and the scattered disk represent a second comet reservoir, and provide most of the Jupiter family comets.

Moons: Most planets, including all giant planets, have natural satellites. Almost all close-in satellites orbit in a prograde sense, in a plane closely aligned with the planet's rotational equator. They are locked in synchronous rotation, so their orbital periods are equal to their rotation periods. Most of the smaller, distant satellites (as well as Triton, Neptune's large and not so distant moon) orbit the planet in a retrograde sense, and/or on orbits with high eccentricity and inclination (Table 1.4). Planetary satellites are primarily composed of a mixture of rock and ice in varying proportions. Jupiter's Galilean moons imitate a miniature planetary system, with the density of the satellites decreasing with distance from their planet.

Planetary Rings: All four giant planets have ring systems orbiting in their equatorial planes. Ring particles travel on prograde trajectories, and most rings lie interior to most sizable moons.

Satellites of Minor Planets: Numerous small bodies orbiting the Sun have satellites. In some cases, the primary and secondary are similarly sized, whereas in others, one body dominates. The distribution of satellite sizes and orbits within the Kuiper belt is different from that in the asteroid belt (§9.4.4).

Meteorites: Meteorites display a great deal of spectral and mineralogical diversity. The crystalline structure of many inclusions within primitive meteorites indicates rapid heating and cooling events. Interstellar grains contained in chondrites imply that some parts of the protoplanetary disk remained cool, whereas high-temperature inclusions that clearly formed in the Solar System show that other parts were subjected to far hotter conditions. The intermingling of grains with different thermal histories within individual meteorites indicates substantial mixing of solid material within the disk. The small spread in ages among most meteorites indicates that the accretion epoch was brief, and the presence of decay products of various short-lived nuclei in chondritic meteorites demonstrates that solid material accreted rapidly. There is also evidence for (local) magnetic fields of the order of 1 G during the planet formation epoch.

Isotopic Composition: Although elemental abundances vary substantially among Solar System bodies, isotopic ratios are remarkably uniform. This is true even for bulk meteorite samples. Most isotopic variations that have been observed can be explained by mass fractionation (§8.4), or as products of radioactive decay. Some of these decay products imply that short-lived radionuclides were present when the material in which they are now situated solidified. The similarity of the isotopic ratios suggests a well-mixed environment. However, small-scale variations in the isotopic ratios of oxygen and a few trace elements in some primitive meteorites imply that the protoplanetary nebula was not completely mixed on the molecular level, i.e., that some presolar grains did not vaporize.

Differentiation and Melting: The interiors of all of the major planets, many asteroids, and most if not all large moons are differentiated, with most of the heavy material confined to their cores. This implies that each of these bodies was warm at some time in the past.

Composition of Planetary Atmospheres: The elements which make up the bulk of the atmospheres of both the terrestrial planets and planetary satellites can form compounds that are condensable at temperatures which prevail on Solar System bodies; hydrogen and noble gases are present in far less than solar abundances. Giant-planet atmospheres consist primarily of H_2 and He, but have enhanced abundances of most if not all ice-forming elements; this enhancement increases from Jupiter to Saturn to Uranus/Neptune.

Surface Structure: Most planets and satellites show many impact craters, as well as past evidence of tectonic and/or volcanic activity. A few bodies show signs of volcanism at the present time. Other surfaces appear to be saturated with impact craters. At current impact rates, such a high density of craters could not have been produced over the age of the Solar System.

13.2 Nucleosynthesis: A Concise Summary

The nuclei of the atoms that compose stars, planets, life, etc. formed in a variety of astrophysical environments. Models of *nucleosynthesis*, together with observational data from meteorites and other bodies, yield clues about the history of the material which was eventually incorporated into our Solar System. The two most important environments for nucleosynthesis are the very early Universe and the interiors of stars. However, other environments are important for some isotopes; e.g., energetic cosmic rays can split nuclei with which they collide, and this *spallation* process is a major source of some rare odd-number light isotopes. Also, many isotopes form via radioactive decay (§8.6.1).

13.2.1 Primordial Nucleosynthesis

The Universe began in an extremely energetic *hot big bang* roughly 13.7 billion years ago. The very young Universe was filled with rapidly moving particles. There were untold numbers of protons (ordinary hydrogen nuclei, ^{1}H, or more precisely ^{1}p$^+$) and neutrons (^{1}n). Free neutrons are unstable with a half-life of 10.3 minutes and decay via the reaction

$$^1\text{n} \xrightarrow[t_{1/2}=10.3\text{min}]{} {}^1\text{p}^+ + e^- + \bar{\nu}_e, \tag{13.1}$$

where e^- represents an electron, and $\bar{\nu}_e$ an (electron) antineutrino. Protons and neutrons collided and sometimes fused together to form deuterium (^{2}H) nuclei, but during the first few minutes, the cosmic background (blackbody) radiation field was so energetic that the latter nuclei were photodissociated very soon after they formed. After about three minutes, the temperature cooled to the point that deuterium was stable for long enough to merge with protons, neutrons, and other deuterium nuclei. Within the next few minutes, about one-fourth of the baryonic matter (nucleons) in the Universe agglomerated into alpha particles (^{4}He); most of the baryonic matter remained as protons, with small amounts forming deuterium, light helium (^{3}He), and tritium (^{3}H, which decays into ^{3}He with a half-life of 12 years), as well as very small but astrophysically significant amounts of the rare light elements lithium, beryllium, and boron,[1] and minute amounts of heavier elements. Big bang nucleosynthesis did not proceed much beyond helium, because by the time the blackbody radiation was cool enough for nuclei to be stable, the density of the Universe had dropped too low for fusion to continue to form heavier nuclei. After about 700 000 years, the blackbody radiation had cooled sufficiently for electrons to join the remaining protons and the larger nuclei that had formed in the early minutes of the Universe, producing atoms.

13.2.2 Stellar Nucleosynthesis

Most nuclei heavier than boron, as well as a small but still significant fraction of the helium nuclei, were produced in stellar interiors. Main sequence stars, such as the Sun, convert matter into energy via nuclear reactions that ultimately transform hydrogen nuclei into alpha particles. In normal (nondegenerate) stars, thermal pressure acts to counter gravitational compression. Protostars and young stars contract as they radiate away their thermal energy, and this contraction leads to an increase in pressure and density in the stellar core. Contraction continues until the core becomes hot enough to generate energy from *thermonuclear fusion*. The rates of fusion reactions increase steeply with temperature because only the tiny minority of nuclei in the high-velocity tail of the Maxwell–Boltzmann distribution (eq. 4.14) have enough kinetic energy to have a noninfinitesimal probability of quantum-mechanically tunneling through the barrier that is produced by Coulomb repulsion. If fusion proceeds too rapidly, the core expands and cools; if not enough energy is supplied by fusion, the core shrinks and heats up; in this fashion, equilibrium can be maintained (§1.3). Deuterium fusion requires a lower temperature than fusion of ordinary hydrogen, so it occurs first and stars rapidly deplete their supply of deuterium, although a significant amount of deuterium can remain in the outer (cooler) portion of a star if it is not convectively mixed with the lower hot regions. The cores of very low mass objects (brown dwarfs, §12.1) get so dense that they are stopped from collapse by degenerate electron pressure before they reach a temperature high enough for fusion to occur at a significant rate.

---

[1] Cosmic ray nucleosynthesis is another major producer of lithium, beryllium, and boron.

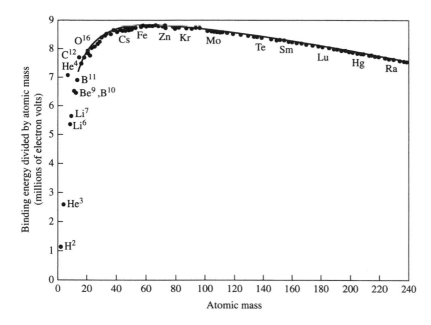

Figure 13.1 The nuclear binding energy per nucleon is shown as a function of atomic weight. For most elements, only the most stable isotope is plotted. Note that ^{4}He, ^{12}C, and to a lesser extent somewhat heavier α-particle multiples, lie above the general curve, indicating greater stability. The peak occurs at ^{56}Fe, indicating that iron is the most stable element. (Lunine 2005)

In main sequence stars of solar mass and smaller, the primary reaction sequence is the *pp-chain*. The principal branch of the pp-chain occurs as follows:

$$2(^1\text{H} + {}^1\text{H} \rightarrow {}^2\text{H} + \text{e}^+ + \nu_e), \qquad (13.2a)$$

$$2(^2\text{H} + {}^1\text{H} \rightarrow {}^3\text{He} + \gamma), \qquad (13.2b)$$

$$^3\text{He} + {}^3\text{He} \rightarrow {}^4\text{He} + 2\,{}^1\text{H} + 2\gamma, \qquad (13.2c)$$

where e^+ represents a positron, ν_e an (electron) neutrino, and γ a photon. The reaction rate for the pp-chain becomes significant near $T_{\text{nucl}} = 3 \times 10^6$ K (§12.1). At temperatures close to that of the Sun's core, $T \approx 15 \times 10^6$ K, the fusion rate is roughly proportional to T^4. While the rate of fusion in a 15 million degree plasma is not as sensitive to temperature change as the T^{10} dependence near T_{nucl}, energy generation in the solar core still varies steeply with temperature. This steep temperature dependence implies that fusion acts as an effective thermostat: If the core gets too hot, it expands, cools, and energy production drops; if the core is too cold, it shrinks until adiabatic compression heats it enough for fusion rates to generate enough energy to balance the energy transported outwards. Moreover, the steep temperature dependence of fusion rates implies that more massive main sequence stars only require a slightly higher core temperature in order to generate a substantially higher luminosity than their smaller brethren.

In main sequence stars more massive than the Sun, the core temperature is somewhat higher, and the even more temperature-sensitive catalytic *CNO cycle* predominates. The principal branch of the CNO cycle is

$$^{12}\text{C} + {}^1\text{H} \rightarrow {}^{13}\text{N} + \gamma, \qquad (13.3a)$$

$$^{13}\text{N} \xrightarrow[t_{1/2}=10\,\text{min}]{} {}^{13}\text{C} + \text{e}^+ + \nu_e, \qquad (13.3b)$$

$$^{13}\text{C} + {}^1\text{H} \rightarrow {}^{14}\text{N} + \gamma, \qquad (13.3c)$$

$$^{14}\text{N} + {}^1\text{H} \rightarrow {}^{15}\text{O} + \gamma, \qquad (13.3d)$$

$$^{15}\text{O} \xrightarrow[t_{1/2}=2\,\text{min}]{} {}^{15}\text{N} + \text{e}^+ + \nu_e, \qquad (13.3e)$$

$$^{15}\text{N} + {}^1\text{H} \rightarrow {}^{12}\text{C} + {}^4\text{He}. \qquad (13.3f)$$

Note that while the half-lives are given for both of the inverse beta decays in equation (13.3), the timescales for the four fusion reactions within the CNO cycle depend upon the temperature and the abundances (densities) of the nuclei involved.

The most stable nucleus is ^{56}Fe (Figure 13.1), so fusion up to this mass can release energy. However, fusion of alpha particles (helium nuclei) into heavier nuclei (up to $Z = 28$) requires higher temperatures in order to overcome the *Coulomb barrier* (electromagnetic repulsion between nuclei dominates the strong nuclear force unless the nuclei are very close, Figure 13.2). Moreover, no nuclide with atomic mass 5 or 8 is stable, so to produce carbon from helium requires two fusions in immediate succession: first a pair of alpha particles combine to produce a (highly unstable, $t_{1/2} = 2 \times 10^{-16}$ s) beryllium 8 nucleus, and then

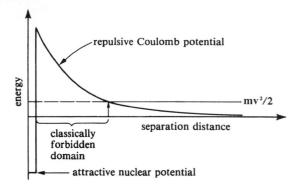

Figure 13.2 The electromagnetic repulsion between nuclei domi-nates the strong (but short-ranged) nuclear force unless the nuclei are very close. (Adapted from Shu 1982)

another alpha particle is added before this nucleus decays:

$$^4\text{He} + {}^4\text{He} \leftrightarrow {}^8\text{Be} \tag{13.4a}$$

followed immediately by

$$^8\text{Be} + {}^4\text{He} \rightarrow {}^{12}\text{C} + \gamma. \tag{13.4b}$$

This *triple alpha process* requires much higher densities than do the pp-chain and CNO process described above. Helium fusion occurs when a sufficiently massive star ($\gtrsim 0.25\ \text{M}_\odot$) has exhausted the supply of hydrogen in its core, so the thermostat that maintained equilibrium during the star's main sequence phase is no longer active. Hydro-gen fusion occurs in a shell surrounding the hydrogen-depleted core, and total stellar energy production greatly exceeds that during the star's main sequence phase, so its outer layers expand and cool, and the star becomes a *red giant*.

Nuclear growth beyond carbon does not require two reactions in immediate succession (as does the triple alpha process), and thus could occur in a lower density envi-ronment, but the increased Coulomb barrier implies that even higher temperatures and thus larger stellar masses are required. Growth can proceed by successive addition of alpha particles:

$$^{12}\text{C} + {}^4\text{He} \rightarrow {}^{16}\text{O} + \gamma, \tag{13.5a}$$

$$^{16}\text{O} + {}^4\text{He} \rightarrow {}^{20}\text{Ne} + \gamma, \tag{13.5b}$$

$$^{20}\text{Ne} + {}^4\text{He} \rightarrow {}^{24}\text{Mg} + \gamma, \tag{13.5c}$$

or at somewhat higher temperatures by reactions such as

$$^{12}\text{C} + {}^{12}\text{C} \rightarrow {}^{24}\text{Mg} + \gamma. \tag{13.5d}$$

Nuclei composed of 3–10 alpha particles are quite sta-ble and easy to produce, so they are relatively abundant. Larger nuclei of this form are too proton-rich, and rapidly inverse β-decay (emit positrons), thereby transforming

themselves into more neutron-rich nuclei. Nonetheless, heavy elements with even atomic number tend to be more abundant than odd-numbered elements (Table 8.1). The proton and neutron numbers of all stable nuclides are shown in Figure 13.3.

Large quantities of elements up to the iron binding-energy peak can be produced by reactions of the type discussed above, but the Coulomb barrier (Fig. 13.2) is too great for significant quantities of substantially more massive elements such as lead and uranium to be generated in this manner. Such massive nuclei are produced primarily by the addition of free neutrons, which are uncharged and thus do not need to overcome electrical repulsion. Free neutrons are released by reactions such as

$$^4\text{He} + {}^{13}\text{C} \rightarrow {}^{16}\text{O} + {}^1\text{n}, \tag{13.6a}$$

and

$$^{16}\text{O} + {}^{16}\text{O} \rightarrow {}^{31}\text{S} + {}^1\text{n}. \tag{13.6b}$$

Neutron addition does not produce a new element directly, but if enough neutrons are added, nuclei can become unsta-ble and β-decay into elements of higher atomic number. The mix of nuclides produced by neutron addition depends upon the flux of neutrons. When the time between succes-sive neutron absorptions is long enough for most unstable nuclei to decay, the mixture of nuclides produced lies deep within the *valley of nuclear stability*, where the mixture of neutrons and protons leads to the greatest binding energy for a nucleus with a given total number of nucleons; this 'slow' type of heavy element nucleosynthesis is referred to as the *s-process*. Nuclei with atomic masses as large as 209 may be formed via the s-process. The rapid *r-process* chain of nuclear reactions occurs during explosive nucleo-synthesis (such as core-collapse supernovae and mergers of neutron stars with other neutron stars and with black holes), when there is a very high flux of neutrons, and pro-duces a more neutron-rich distribution of elements. Ura-nium and other very heavy naturally occurring elements are produced via r-process nucleosynthesis. Rare proton-rich heavy nuclei are produced by *p-process nucleosyn-thesis*. Detailed nuclear physics calculations suggest that the most likely explanation of p-process nucleosynthesis is removal of neutrons through partial nuclear photodis-sociation in a high-temperature (10^9 K) environment. An alternative model is β-decay induced by a high neutrino flux.

Note that most of the elements produced via stellar nucleosynthesis are never released from their parent stars; only material ejected by stellar winds, nova outbursts, and supernova explosions is available to enrich the interstellar medium and to form subsequent generations of stars and

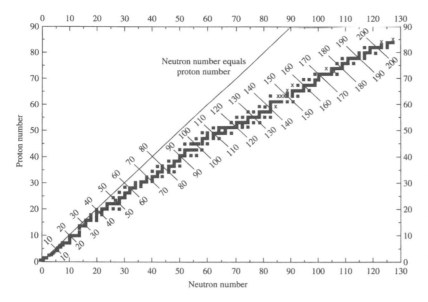

Figure 13.3 Distribution of stable nuclei, plotted as atomic number vs. number of neutrons. The diagonal line representing equal numbers of protons and neutrons is plotted for reference. The short lines perpendicular to this diagonal represent nuclides with the same atomic weight. Long-lived but unstable isotopes are represented by crosses. (Adapted from Lunine 2005)

planets. The distributions of elements and isotopes found in individual interstellar grains and in the Solar System as a whole are indicative of the various environments in which stellar nucleosynthesis occurs, and the conditions under which material is released from stars.

13.3 Star Formation: A Brief Overview

In analogy with current theories on star formation, it is generally thought that our Solar System was 'born' in a dense (by the standards of interstellar space) molecular cloud, as the result of gravitational collapse. In the remainder of this chapter we review current ideas on star formation, the formation of a disk around a (proto)star, and, finally, the evolution of such a disk and the accretion (growth) of planets.

13.3.1 Molecular Cloud Cores

Our Milky Way galaxy contains a large number of cold, dense molecular clouds (Fig. 13.4), varying in size from giant systems with masses of $\sim 10^5$–10^6 $M_\odot$ to small ~ 0.1–10 $M_\odot$ cores. The small cores are usually embedded in the larger complexes, and are observed at radio wavelengths in molecular line transitions such as CO, NH_3, HCN, CS, or H_2CO. Molecular clouds have typical temperatures of ~ 10–30 K and densities of a few thousand molecules cm^{-3}. The cores from which stars form may have densities 10–1000 times larger than this and temperatures of only $\lesssim 10$ K. Molecular clouds consist mostly of H_2 and presumably He (cold helium is extremely difficult to detect remotely because this noble gas is chemically inert and holds on to its electrons quite tightly). Many

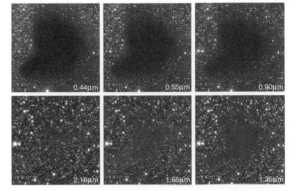

Figure 13.4 The sky area of the globule Barnard 68 in the Ophiuchus star-forming region, imaged in six different wavebands, clockwise from the blue to the near-infrared spectral region. The three frames at the top show images at wavelengths of 0.44 μm, 0.55 μm, and 0.90 μm that were obtained with the VLT (Very Large Telescope); the three longer wavelength images at the bottom were taken through near-infrared filters (at 1.25 μm, 1.65 μm, and 2.16 μm) with the NTT (New Technology Telescope). The obscuration caused by the cloud diminishes dramatically with increasing wavelength, implying that most of the dust is in the form of sub-μm grains. Since the outer regions of the cloud are less dense than the inner ones, the apparent size of the cloud also decreases as wavelength increases, with more background stars shining through the outer parts. (European Southern Observatory PR Photo 29b/99)

other molecules are present, including CO, CN, CS, SiO, OH, H_2O, HCN, SO_2, H_2S, NH_3, H_2CO, and numerous other combinations of H, C, N, and O, some containing more than a dozen atoms in a molecule. All of these more massive molecules combined, however, make up only a small fraction of the total mass of the cloud.

The typical interstellar cloud is stable against collapse. Its internal pressure (ordinary gas pressure augmented by magnetic fields, turbulent motions, and rotation), is more than sufficient to balance the inward pull of self-gravity. This excess pressure would cause the cloud to expand, were it not for the counterbalancing pressure of surrounding gas of higher temperature ($\sim 10^4$ K) and lower density (~ 0.1 atoms cm^{-3}).

In equilibrium systems where magnetic pressure and external pressure can be ignored, the *virial theorem* states that the gravitational potential energy, E_G, is equal to negative twice the kinetic energy, E_K (Problem 2.5). The kinetic energy of a gas cloud is primarily thermal energy, unless the cloud is highly turbulent or rapidly rotating. When $|E_G| > 2E_K$, the cloud may collapse under its own self-gravity. One can solve for the minimum mass of such a cloud, the *Jeans mass*, M_J (Problem 13.1):

$$M_J \approx \left(\frac{kT}{G\mu_a m_{amu}} \right)^{3/2} \frac{1}{\sqrt{\rho}}. \tag{13.7}$$

A cloud with $M > M_J$ will collapse if its only means of support is thermal pressure. Note that the critical mass, M_J, decreases if the density in the cloud increases. Low-density but massive clouds may collapse into galaxies; less massive clouds with a higher density may collapse into clusters of stars or a single star. Observed cores within molecular clouds appear to be dense enough to collapse gravitationally into objects of stellar masses. However, the density in a small, cold (10 K) cloud would need to exceed $\sim 10^{-11}$ g cm^{-3} ($\sim 10^{13}$ atoms cm^{-3}) to form Jupiter-mass objects from gravitational collapse. This is much larger than the observed densities of interstellar clouds.

When a marginally stable molecular cloud passes through a spiral arm of a galaxy it is compressed; such compression may be sufficient to trigger collapse. Clouds pass through spiral arms since the clouds and stars in a galaxy orbit faster than the pattern of spiral density waves rotates. Other phenomena that may trigger gravitational collapse are (super)nova explosions, where shells of gas and dust are thrown into space, and stellar outflows or expanding HII (ionized hydrogen) regions. Collapse converts gravitational potential energy into kinetic energy of the collapsing material. If this energy is retained, either in ordered motion or random thermal motions, virial equilibrium may be achieved and the collapse ceases. However, if this energy is lost, for example via radiation, then the cloud becomes even more unstable. Once a gravitational collapse begins, densities increase, causing collapse to proceed faster.

Observations of molecular cloud cores show them to be far more complicated than the simplistic picture described above. Many cores are far from spherical. Magnetic fields, turbulence, and to a lesser extent rotation, all oppose collapse. The typical lifetime of a core is only a few hundred thousand years, which is more than two orders of magnitude less than the time interstellar gas in the solar neighborhood takes to pass from one spiral arm of the galaxy to the next.

13.3.2 Collapse of Molecular Cloud Cores

For a molecular cloud that is in equilibrium, pressure gradients (and sometimes magnetic forces) balance gravitational forces. In contrast, when the pressure gradient and other forces in a cloud are small enough relative to the gravity that they can be ignored, the cloud (or core) collapses on a *free-fall timescale*, t_{ff} (Problem 13.2):

$$t_{ff} = \left(\frac{3\pi}{32 G\rho} \right)^{1/2}. \tag{13.8}$$

The free-fall solution does not directly apply to star formation because pressure is important, at least initially, but it yields a lower bound on the timescale of collapse that can be a good approximation under some circumstances. Equation (13.8) implies that the denser clumps collapse more rapidly, which may lead to separation and fragmentation. As cores are observed to be densest near their centers, the interiors cave in most quickly, producing an inside-out collapse. The collapse of a molecular cloud core can be halted if the gas temperature increases to the point that thermal pressure balances gravity; numerical simulations indicate that this only happens at sizes much smaller than those of observed molecular cloud cores. Rotation can also prevent material from continuing its collapse, leading to the formation of a disk around the *protostar* growing at the center and/or to fragmentation.

Rotation becomes a dominant effect when the centrifugal force balances the gravitational force. Unless angular momentum is redistributed during collapse, material with initial specific angular momentum L_s joins the disk at a radius $r_c \approx L_s^2/GM$, where M is the mass interior to the radius under question (the relationship would be exact if this mass were distributed in a spherically symmetric manner). If the core rotates rapidly, it may break up into two or more subclouds, where the angular momentum is taken up by the individual fragments orbiting one another. Each of the subclouds may collapse into a star, forming a binary or multiple star system. The majority of stars are observed to be in such binary or multiple systems. Cores with less angular momentum may form only a single star. Since a core must contract by orders of magnitude to form a star, even an initially very slowly rotating clump contains much more angular momentum than the final star can take without breaking up. We expect, therefore, that virtually all single stars, and probably many binary/multiple systems, are surrounded by a flat disk of material at some stage during their formation. Although the star may contain most of the core's initial mass, most of the angular momentum is in the disk. Recall that in the Solar System today, 99.8% of the mass is in the Sun, and over 98% of the angular momentum resides in planetary orbits.

While a dense core is collapsing, its temperature rises as a result of the conversion of gravitational energy into kinetic energy. If the cloud core is sufficiently transparent at infrared wavelengths, then most of the thermal energy is radiated away, and the core stays relatively cool. The increasing density eventually makes the core opaque, so thermal energy can no longer escape. Released gravitational energy then heats the protostar growing at the center of the core, thereby building up the internal pressure, until hydrostatic equilibrium (balance between gravity and the pressure gradient; eq. 3.26) is reached. When the temperature inside the protostar gets hot enough ($\sim 10^6$ K), nuclear reactions (the conversion of deuterium into helium; eq. 13.2b) start. The energy generated by this process is sufficient to temporarily forestall further contraction. When the supply of deuterium becomes exhausted, the star shrinks and heats up until the central temperature reaches the $\sim 10^7$ K value required for ^{1}H fusion (eqs. 13.2a,b) at a rate sufficient to prevent further collapse.

During these accretion phases, the protostar is blocked from view by dust in the outer layers of the cloud. In the early phases of gravitational collapse, the dust stays relatively cool, ~ 30 K, and thus emits infrared radiation whose distribution peaks near 100 μm. When the protostar forms, the inner layers of dust heat up dramatically, as is observed at shorter infrared wavelengths.

13.3.3 Observations of Star Formation

Numerous young stars have been observed within several molecular clouds. The ages of these stars are estimated in several ways: kinematic ages of groups of stars (the size of the region divided by the relative velocities of the stars), the ages of individual stars on the Hertzsprung–Russell (H–R) diagram (Fig. 1.7), and the presence and intensity of Li absorption lines in the stellar spectra. (Lithium is convectively transported downwards to depths at which it can be destroyed by thermonuclear reactions in stars of mass $\lesssim 1$ M$_\odot$.) These methods all yield ages of the youngest of the well-studied star-forming regions of $\lesssim 10^7$ years. Moreover, observations imply that the same process which forms stars also produces objects with masses significantly lower than the minimum required for hydrogen fusion (§12.1).

Young stars that are still contracting towards the main sequence are called *pre-main-sequence stars*. Among these stars we find *T Tauri stars*, named after the first such star discovered, the variable star T in the constellation Taurus. T Tauri stars are usually found within dense patches of gas and dust. The luminosity of many T Tauri stars varies considerably in an irregular manner, over timescales as short as a few hours. The spectral energy distributions of T Tauri stars are much broader than blackbody spectra, are dominated by intense emission lines, and show the presence of strong stellar winds. Most T Tauri stars have large starspots, which modulate their lightcurves and allow their rotation periods to be measured. Typical observed rotation periods for deeply embedded T Tauri stars are a few days, much shorter than the Sun's current 27 day period, yet several times as long as that required for rotational breakup (Problem 9.11). T Tauri stars also emit substantially more X-rays than do older stars of similar mass; this implies that very energetic nonequilibrium processes occur during star formation.

The formation and early evolution of stars can be quite erratic and violent. *Bipolar outflows* of gas are ejected perpendicular to disks around accreting stars at ~ 100 km s^{-1}. Interaction of this gas with the surrounding interstellar medium produces shocks which are observed as the bright emission nebulae known as *Herbig–Haro (HH) objects*. Outbursts, presumably related to enhancements in the stellar accretion rate by a factor $\gtrsim 100$ for tens of years, are seen in *FU Orionis stars*.

13.3.4 Observations of Circumstellar Disks

Excess emission at infrared wavelengths, indicative of circumstellar material extending out to tens or hundreds of AU from the star (Fig. 13.5), is observed in 25–50% of pre-main-sequence solar mass stars, including T Tauri stars. The lack of near-infrared radiation from some disks suggests that gaps exist in the inner parts of these disks. This absence of near-infrared radiation is observed in a larger fraction of the somewhat older objects. Resolved images of circumstellar dust disks – the most direct evidence of such disks – has come from HST and millimeter observations. Millimeter data have indicated disk-like structures with masses between ~ 0.001 and 0.1 M$_\odot$ (assuming that the observed dust is mixed with gas at the interstellar abundance ratio) around several protostars. Such massive disks are also observed around some young stars, but only much less massive disks have been seen around older stars (Fig. 13.6). HST images have revealed disk-like structures of order 100 AU in radius around young stars; these structures are referred to as *proplyds* (Fig. 13.7). The disks around the stars HD 141569 and HR 4796A as imaged

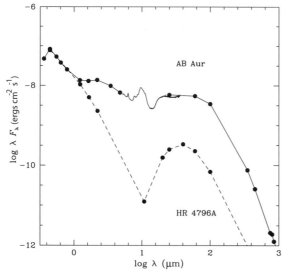

Figure 13.5 Broadband spectra, normalized to the stellar photospheres, of AB Aurigae, a ~2 Myr old Herbig Ae/Be star (the equivalent of the T-Tauri phase for stars a few times as massive as the Sun) and the ~8 Myr old star HR 4796A (an image of which is shown in Fig. 13.8b). The disk around HR 4796A is thought to be composed of debris from collisions between 'large' bodies, and is the brightest such disk known; the absence of excess emission at wavelengths $\lesssim$10 μm indicates a strong evacuation of dust within 40 AU of the star. (Furlan *et al.* 2006)

by HST (Fig. 13.8) appear to contain inner holes. The former disk, ~750 AU wide, shows a dark band near the center, i.e., absence of material. This material may have accreted into a planet or been pushed away by a planet's shepherding torque. The narrowness of the ring of dust

around HR 4796A might be explained via shepherding (§11.4.3) and/or an unseen planet accreting or scattering dust particles whose orbits decay as a result of Poynting–Robertson drag (§2.7.2).

Cool debris disks of solid particles have been observed around the nearby star Vega and many other young main sequence stars, primarily via photometry in the far-IR. These circumstellar disks typically extend a few hundred AU from the star, but their optical depths are small, and the observed particles may contain as little as ~ one lunar mass of material. Although small dust dominates the radiating area of these disks, such small particles could not have survived for the lifetimes of the stars, so larger (source) particles must also be present. These disks are typically more prominent around younger main sequence stars, but some older stars have fairly bright disks. The first image of a circumstellar dust disk around a (main sequence) star, β Pictoris, was obtained with a CCD camera from the ground. HST images of this 1500 AU wide disk (Fig. 13.9) show that the inner part of the disk is warped, a feature which may be produced by the gravitational pull of one or more nearby planets or a brown dwarf on an inclined orbit.

13.4 Evolution of the Protoplanetary Disk

Based on observations of star formation in our galaxy at the present epoch, we assume that our Sun and planetary system formed in a molecular cloud. The growing Sun together with its surrounding disk are referred to as the *primitive solar nebula*; the planetary system formed from

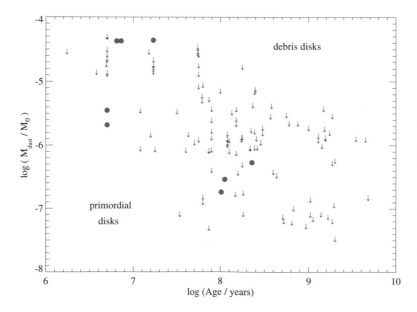

Figure 13.6 Dust masses in circumstellar disks around ~1 $M_\odot$ stars plotted against stellar age. Filled circles represent detections, and downward-pointing arrows signify (3σ) upper limits. Note that dust disks more massive than Earth have only been observed around stars less than 20 Myr old; these massive disks may well be primordial, and contain an amount of gas up to two orders of magnitude more than the observed dust. Older debris disks are likely to be gas-poor. Disk masses were calculated based upon observations made at 1.2 mm, 2.7 mm, and 3 mm. (Courtesy Veronica Roccatagliata)

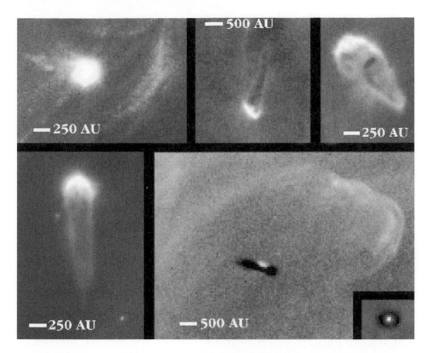

Figure 13.7 Young stars with disks in the Orion Nebula. The top row and the image at the lower left show disks of gas and dust which are being photoevaporated by ultraviolet radiation from nearby massive stars. The other two images in the bottom row show silhouettes where disks associated with young stars obscure light from background hot gas. Note that the sizes of these disks are considerably larger than the planetary region of our Solar System. (NASA/HST images by J. Bally, D. Devine, and R. Sutherland)

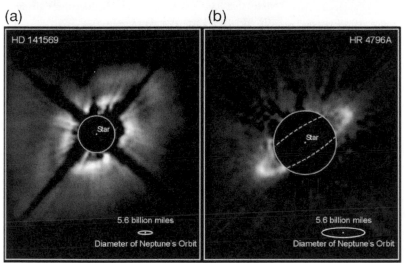

(a) (b)

Figure 13.8 Images of disks around young stars. In both cases a coronagraph was used to block off the light from the star. (a) A near-infrared image of a disk around the star HD 141569, located about 320 light-years away in the constellation Libra. A dark band separates a bright inner region of the $\sim$750 AU wide disk from a fainter outer region. This band may be the result of the formation of a planet in the disk. (Courtesy B. Smith and G. Schneider, HST/NASA) (b) A near-infrared image of a dust ring around the young ($\lesssim 10^7$ yr old) star HR 4796A. This ring, at a distance of $\sim$40 AU from the central star, is <17 AU wide. The confinement of this ring suggests the presence of unseen planets in orbit about the star. (Courtesy E. Becklin and A. Weinberger, HST/NASA)

the *protoplanetary disk* within this nebula. A *minimum mass* of $\sim$0.02 M$_\odot$ for the protoplanetary disk can be derived from the present abundance of refractory elements in the planets and the assumption that the abundances of the elements throughout the nebula were solar. The actual mass was probably significantly larger, since some (perhaps most) of the refractory component of this mixture was not ultimately incorporated into planets. The history of our solar nebula can be divided into three stages: infall, internal evolution, and clearing.

13.4.1 Infall Stage

When a molecular cloud core becomes dense enough that its self-gravity exceeds thermal, turbulent, and magnetic support, it starts to collapse. Collapse proceeds from the inside out, and continues until the reservoir of cloud material is exhausted, or until a strong stellar wind reverses the flow. The duration for the infall stage is comparable to the free-fall collapse time of the core, $\sim 10^5$–10^6 yr.

Initially, gas and dust with low specific angular momentum relative to the center of the core falls towards

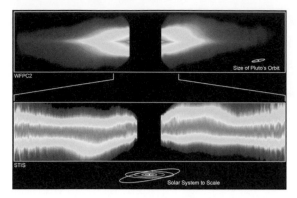

 Figure 13.9 HST images of the inner portion of the dust disk around the star β Pictoris. The bright glare of the central star is blocked by a coronagraph. The warps in the disk might be caused by the gravitational pull of one or more unseen (planetary?) companions. (Courtesy Al Schultz, HST/NASA)

the center, forming a protostar. Eventually, matter with high specific angular momentum falls towards the protostar, but cannot reach it due to centrifugal forces. Essentially, the material is on orbits that do not intersect the central, pressure-supported star. However, as the gas and dust mixture falls to the equatorial plane of the system, it is met by material falling from the other direction, and motions perpendicular to the plane cancel. The energy in this motion is dissipated as heat in the forming disk. Significant heating can occur, especially in the inner portion of the disk, where the material has fallen deep into the potential well. The equatorial plane of the resultant disk is roughly perpendicular to the rotation axis of the initial collapsing molecular cloud core. The direction of the core's angular momentum determines the plane of the disk, whereas the magnitude of the angular momentum governs how the material is divided between the protostar and its disk.

Let us consider a parcel of gas which falls from infinity to a circular orbit at $r_\odot$. Half of the gravitational energy per unit mass is converted to orbital kinetic energy:

$$\frac{GM_{protostar}}{2r_\odot} = \frac{v_c^2}{2};$$ (13.9)

the other half is available for heat. At 1 AU, the circular velocity $v_c = 30$ km s^{-1} if $M_{protostar} = 1$ M$_\odot$. If no energy escapes the system, it follows that the temperature in a hydrogen gas would be ∼7×10^4 K (Problem 13.3). However, this very high temperature is never actually attained, as the timescale for radiative cooling is much shorter than the heating time.

Gas reaches supersonic velocities as it descends towards the midplane of the nebula. The gas slows abruptly when it passes through a *shock front* as it is accreted onto

the disk. The highest temperature that is attained in the nebula depends upon the structure of the shock through which material passes. Models of protoplanetary disk formation suggest that typical post-shock temperatures for the protoplanetary disk are ∼1500 K at 1 AU and ∼100 K at 10 AU. Equilibrium is reached when all forces balance, i.e., the gravitational force towards the center balances with the centrifugal force outward and the gravitational force toward the midplane balances with the pressure gradient outwards. Provided the star's gravity dominates that of the disk and temperature variations in the z-direction can be ignored, the gas density and pressure variations in the vertical direction are given by

$$\rho_{g_z} = \rho_{g_{z_0}} e^{-z^2/H_z^2},$$ (13.10a)

$$P_z = P_{z_0} e^{-z^2/H_z^2},$$ (13.10b)

where the Gaussian scale height, H_z, is given by:

$$H_z = \sqrt{\frac{2kTr_\odot^3}{\mu_a m_{amu} GM_\odot}}.$$ (13.11)

Note that H_z increases with heliocentric distance $(dH_z/dr_\odot > 0)$ provided $d \ln T/d \ln r_\odot < 3$.

13.4.2 Disk Dynamical Evolution

Unless the collapsing cloud has negligible rotation, a significant amount of material lands within the disk. Redistribution of angular momentum within the disk can then provide additional mass to the star (Fig. 13.10). The structure and evolution of the disk are primarily determined by the efficiency of the transport of angular momentum and heat. Angular momentum and mass can be transported in the following ways:

Magnetic Torques: If magnetic field lines from the star thread through the disk, then there is a tendency towards corotation, i.e., material orbiting more rapidly than the star's spin period loses angular momentum and that orbiting less rapidly gains angular momentum. The field lines couple the star to the disk if the gas in the disk is sufficiently ionized, which it tends to be in the innermost parts of the solar nebula, where temperatures are high. Thus, the spin rate of a rapidly rotating star is slowed by *magnetic braking* torques from the inner parts of the disk, where ionized gas couples to the stellar magnetic field but is slowed by frequent collisions with neutral gas orbiting at the Keplerian rate. Angular momentum is transferred outwards, from the star to the disk, but transfer to larger radii is inhibited by the lack of ionized gas in the disk and the weakness of stellar magnetic field lines at greater distances.

Observations and detailed modeling suggest that protoplanetary disks do not extend all of the way to the star's

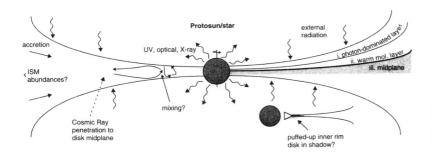

Figure 13.10 Schematic diagram of a protoplanetary disk that is subject to radiation both from its central star and more distant but brighter stars. (Pudritz *et al.* 2007)

surface. Rather, magnetic interactions between the star and the disk near the *corotation point* (where the Keplerian orbital angular velocity in the disk equals the star's rotational angular velocity) funnel some of the disk's gas onto the star and expel other gas in a rapid centrifugally driven bipolar outflow which carries with it a substantial amount of angular momentum. (Although this *bipolar wind* is concentrated near the star's poles, the gas does not move exactly parallel to the star's rotation axis, so it is able to carry away angular momentum.) This loss of angular momentum explains why protostars are observed to rotate substantially less rapidly than breakup speed (§13.3.3). The solid particles that are expelled along with the gas are subjected to brief but intense heating by bright starlight; this process may produce the chondrules and CAIs that are found in most primitive meteorites (§8.7.2).

Gravitational Torques: Local or global gravitational instabilities may lead to rapid transport of material within the protoplanetary disk. As discussed in §11.2, a thin rotating disk is unstable to local axisymmetric perturbations if Toomre's parameter, $Q_T < 1$ (eq. 11.13; note that for a gaseous disk, the sound speed replaces the velocity dispersion). Nonaxisymmetric local instabilities also occur when $Q_T \lesssim 1$. These instabilities can produce spiral density waves, which transport mass and angular momentum on a dynamical timescale until a stable configuration is once again reached. This limits the disk mass to a value less than or comparable to the protostar's mass. Under most circumstances, a global one-armed spiral instability limits the disk mass to less than 1/3 that of the star.

Large protoplanets may clear annular gaps surrounding their orbits and excite density waves at resonant locations within the protoplanetary disk (see Fig. 13.23). These density waves transfer angular momentum outwards. Such processes have been observed in Saturn's rings (§11.4), albeit on a much smaller scale. If the protostar had enough angular momentum, its lowest energy configuration would be triaxial (i.e., ellipsoidal with three unequal principal axes, and thus not cylindrically symmetric; see §6.1.4.2), which would present an asymmetric and rotating gravitational potential to the disk, much like that due to

a protoplanet. This would trigger resonances and density waves, which transport angular momentum from the protostar to the disk. However, observations imply that ~ 1 $M_\odot$ protostars generally rotate far too slowly to become triaxial.

Viscous Torques: Since molecules revolve around the protosun in roughly Keplerian orbits, those closer to the center move faster than those further away. Collisions among the gas molecules speed up the outer molecules, hence driving them outwards, and slow down the inner molecules, which then fall towards the center. The net effect is that most of the matter diffuses inwards, angular momentum is transferred outwards, and the disk as a whole spreads.

The disk evolves on a diffusion timescale, given by equation (11.10), where the length scale, ℓ, is equal to the radius of the disk (or that portion of the disk under consideration). Viscous diffusion is thus more rapid in the inner portions of protoplanetary disks, provided the kinematic viscosity is roughly constant. Unfortunately, the magnitude of the viscosity in protoplanetary disks is uncertain by several orders of magnitude. Thermal motions of the gas in the disk produce a *molecular viscosity*, ν_m, of order

$$\nu_m \sim \ell_{fp} c_s, \tag{13.12}$$

where ℓ_{fp} is the mean free path of the molecules and c_s is the sound speed. This molecular viscosity is far too small to produce significant viscous evolution over the lifetime of a protoplanetary disk (Problem 13.5). However, if turbulence, caused for example by convection in the nebula, plays a role, it may be able to transport substantial mass and angular momentum within the protoplanetary disk. The physics of turbulence is extremely complicated and poorly understood. Because turbulent velocities are unlikely to exceed the sound speed, and eddy sizes are probably no larger than the scale height of the disk, the turbulent viscosity of accretion disks is often parameterized as

$$\nu_v = \frac{2}{3}\alpha_v c_s H_z, \tag{13.13}$$

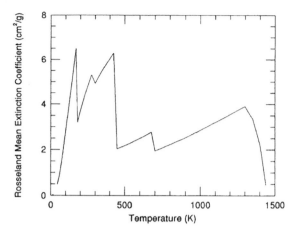

Figure 13.11 The Rosseland mean opacity of a solar composition mixture of gas and dust as a function of temperature. The size spectrum of the dust is assumed to resemble that of interstellar grains; the opacity is lower if the condensed material is contained in macroscopic bodies. The contribution of gas to the opacity is not included in these calculations, but for the low pressures occurring in protoplanetary disks, gas opacity is small compared to dust opacity at temperatures below 1400 K. (Adapted from Pollack *et al.* 1994)

where the dimensionless viscosity parameter $\alpha_v \lesssim 1$. Theoretical estimates based on convection in an optically thick disk yield values of α_v between 10^{-4} and 10^{-2}, which imply significant viscous evolution of (at least the inner portions of) protoplanetary disks (Problem 13.7).

The rate at which the disk evolves may thus depend on its optical depth perpendicular to the midplane. Very near the protostar, well inside the distance of Mercury's orbit, the disk is too hot for grains to condense out, and interstellar grains all have evaporated. The dominant sources of opacity in this region are due to molecular transitions in, for example, H_2O and CO molecules and atomic hydrogen ionization processes. At larger distances from the star the temperature in the nebula is well below 2000 K, and micrometer-sized dust provides the dominant source of opacity. The magnitude of this *Rosseland mean* (energy-averaged) *opacity* varies with temperature roughly as T^2, aside from occasional sharp drops when abundant species evaporate (Fig. 13.11). The disk cools by radiating energy from its faces. Portions of the disk may become unstable against convection perpendicular to the disk's midplane, allowing heat to be convectively transported from the hot midplane of the disk to the faces, where it is radiated away into space. Turbulence, induced by this convection, may also mix material in the radial direction over a distance comparable to the size of the largest convective eddies. Under most circumstances, the temperature in the nebula decreases with increasing distance from the

Sun, and, for $r_\odot \lesssim 100$ AU, with distance from the nebula midplane.

Partially ionized regions of protoplanetary disks are subject to *magnetorotational instabilities*, which can lead to magnetohydrodynamic turbulence resulting in substantial viscosity. Most regions within protoplanetary disks are predicted to be sufficiently ionized for these instabilities to be effective. Photons emitted by the central star ionize the inner regions of the disks, and the faces of the disk and its thin distant regions are ionized by photons from other stars and cosmic rays. However, material near the midplane of the disk a few AU from the star probably is sufficiently shielded from these sources of radiation that it does not have a large enough ionization fraction for the magnetorotational instability to be effective. Regions of a disk in which magnetorotational instabilities are active may thus viscously evolve relatively rapidly, whereas material piles up in low-viscosity un-ionized *dead zones*.

13.4.3 Chemistry in the Disk

The chemical composition of protoplanetary disks is important because it determines what raw materials are available for planetesimal formation. Protoplanetary disks are regions in which interstellar matter is made into planets. Disks are dynamically evolving objects: the physical conditions within them, and thus their chemical composition, change over time as disk material accretes onto the central star, the disk cools, and planetesimals form, as planets form, and as the disk eventually disperses, leaving behind a planetary system such as the Solar System. Comets and chondritic meteorites are relics from the planetesimal-forming era of the Solar System's protoplanetary disk.

The initial chemical state of the disk depends upon the composition of the gas and dust in the interstellar medium and subsequent chemical processing during the collapse phase. The chemical composition varies both with time and with distance from the central (proto)star. Because the Sun formed from the same raw materials as its protoplanetary disk, the abundances of the elements in the Sun tell us what the original *elemental* composition of the disk was. But as the Sun is too hot for molecules to be stable, it does not provide information about the chemical compounds within which these elements resided while in the disk.

Even though there still are many uncertainties in the chemical evolution of the cooling and evolving protoplanetary disk, it is clear that silicates and metal-rich condensates exist throughout almost the entire disk, but ices only in the outer parts. Because ice-forming elements are more abundant than refractory elements in the Sun and the interstellar medium (Table 8.1), the outer parts of the primitive solar nebula contained much more material than the inner regions. Well inside Mercury's orbit, the temperature was too high for solids to exist. However, the presence of complex disequilibrium compounds and both refractory and volatile grains of presolar origin (§§8.7.2, 10.4.4, 10.7.1) implies that the basic equilibrium condensation models are too simplistic.

13.4.3.1 Equilibrium Condensation

The chemical evolution of interstellar matter as it is incorporated into planetesimals is fundamental to our understanding of planet formation. Gas cools after passing through the shock front that it encounters while entering the protoplanetary disk, but can subsequently be heated as material is added above it and/or if dynamical evolution of the disk brings it closer to the protostar. The chemical composition can be calculated within those regions in which nebular material has experienced temperatures high enough to completely evaporate and dissociate all incoming interstellar gas and dust (>2000 K). At such high temperatures, the chemistry can be assumed to be in thermodynamic equilibrium, since the chemical reaction rates are rapid compared to the cooling rate of the disk. This situation is likely to occur close to the protostar.

When the nebula cools below a temperature at which the chemical reaction times become comparable to the timescale of cooling, the chemistry becomes more complicated. This *freeze-out temperature* is different for different species. For example, the CO/CH_4 and N_2/NH_3 ratios are sensitive functions of the temperature and pressure in the nebula. At the low pressures given by models of the solar nebula, carbon is thermodynamically most stable in the form of CO at $T \gtrsim 700$ K, and in the form of CH_4 at lower temperatures. Nitrogen is most stable as N_2 at $T \gtrsim 300$ K, and as NH_3 at lower temperatures. Thus, were the protoplanetary disk in thermodynamic equilibrium, CO and N_2 would dominate in the warm inner nebula, while in the cold outer nebula, CH_4 and NH_3 would be the favored forms of C and N, respectively (Fig. 13.12). Several lines of evidence imply that the solar nebula was not in equilibrium. The existence of N_2 and CO ices on Pluto and Triton, for example, suggests that the outer solar nebula did not have enough time to equilibrate chemically. The depletion of the N/C ratio in comets relative to the solar abundances of these elements (§10.4.3), despite NH_3 having a significantly higher condensation temperature than CH_4, also indicates that these substances did not achieve chemical equilibrium.

As a protoplanetary disk cools, elements condense out of the gas and undergo chemical reactions at different

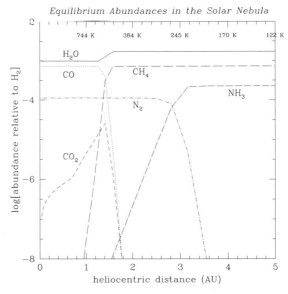

Figure 13.12 Calculations of the thermodynamically stable forms of C, O, and N at the time that icy condensates began to form at 5 AU. The mass accretion rate of the protoplanetary disk for this calculation is 10^{-7} $M_\odot$ per year. Note that the abundances of ices observed to emanate from comets (§10.4.3) are substantially different from the values expected in chemical equilibrium, implying that disequilibrium processes played an important role in the region of the protoplanetary disk where comets formed. In addition, although very warm (~1000 K) CO and H_2O have been observed by Spitzer in planetary disks, as shown in the diagram, this does not necessarily imply chemical equilibrium. Several other complex molecules have also been seen in these disks, such as NH_3. (Courtesy Monika Kress)

temperatures (Figs. 13.13 and 13.14). Refractory minerals such as REE (Rare Earth Elements) and oxides of aluminum, calcium, and titanium (e.g., corundum, Al_2O_3, and perovskite, $CaTiO_3$) condense at a temperature of ~1700 K. At $T \sim 1400$ K, iron and nickel condense to form an alloy; at slightly lower temperatures magnesium silicates appear, including forsterite (Mg_2SiO_4) and enstatite ($MgSiO_3$). Upon further cooling, at $T \lesssim 1200$ K the first feldspars appear; initially the more refractory compounds such as plagioclase anorthite ($CaAl_2Si_2O_8$), and later ($T \sim 1100$ K) sodium and potassium feldspars ((Na,K)$AlSi_3O_8$). Note that, as essentially all of the aluminum condenses out of the gas at much higher temperatures, aluminum is available for inclusion in feldspars only if it is contained in small grains, which can reach equilibrium with the surrounding gas. If grain growth is rapid compared to cooling, different minerals are formed. If chemical equilibrium is maintained, chemical reactions in the gas and with the dust take place as the temperature drops, such as the reactions of iron with H_2S to form troilite (FeS) at ~700 K, and with water to form iron

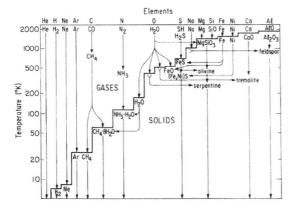

Figure 13.13 Flow chart of major reactions during fully equilibrated cooling of solar nebula material from 2000 to 5 K. The 15 most abundant elements are listed across the top, and directly beneath are the dominant gas species of each element at 2000 K. The staircase curve separates gases from condensed phases. (Barshay and Lewis 1976)

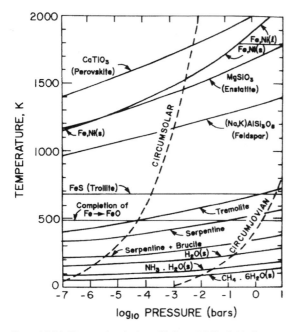

Figure 13.14 Thermochemical equilibrium stability fields for condensed material in a solar composition medium. The species diagrammed are in primarily solid form below the lines and primarily gaseous at higher temperatures. The dashed lines are estimated temperature–pressure profiles for the circumsolar and circumjovian disks. (Prinn 1993)

oxide ($Fe + H_2O \rightarrow FeO + H_2$) at ~500 K. Reactions between FeO and compounds like enstatite and forsterite form olivines and pyroxenes of intermediate iron content (e.g., $(Mg, Fe)_2SiO_4$; $(Mg, Fe)SiO_3$).

Water plays an extremely important role below 500 K, and condenses as pure water-ice at temperatures below

200 K. Assuming equilibrium is maintained, ammonia and methane gas condense as hydrates and clathrates, respectively ($NH_3 \cdot H_2O$, $CH_4 \cdot 6H_2O$), at temperatures somewhat lower than that at which ice condenses. At temperatures of below ~40 K, CH_4- and Ar-ices form.

13.4.3.2 Disequilibrium Processes

Reactions such as $CO + 3H_2 \rightarrow CH_4 + H_2O$, $N_2 + 3H_2 \rightarrow 2NH_3$, and the formation of hydrated silicates are thermodynamically favored at low temperatures, but they have high activation energies. Such reactions are *kinetically inhibited* because they take a very long time to reach equilibrium, longer than nebular evolution allows. (These reactions are more likely to proceed to equilibrium at the higher densities believed to characterize circumplanetary nebulae.) As equilibrium cannot be assumed, the condensation sequence at low temperatures is quite uncertain. If equilibrium is maintained, water vapor reacts with olivines and pyroxenes to form hydrated silicates (e.g., serpentine $Mg_6Si_4O_{10}(OH)_8$, talc $Mg_3Si_4O_{10}(OH)_2$) and hydroxides (e.g., brucite $Mg(OH)_2$). Although hydrated silicates are plentiful in some meteorites (§8.5), most of the hydrated silicates seen in meteorites appear to have formed by ice-bearing asteroids after accretion, rather than by hydration reactions within the protoplanetary disk. The 'nonequilibrium' species CO and N_2 can form clathrates with water-ice below ~60 K, and they can be physically 'trapped' in water-ice if it is cold enough. At $T \lesssim 25$ K, CO and N_2 condense into ice.

The time required to reach equilibrium at the low gas densities ($\rho \lesssim 10^{-9}$ g cm^{-3}) and low temperatures of outer regions characteristic of protoplanetary disks is likely longer than cooling and condensation times, and may even exceed the lifetime of the disk. Thus, many chemical reactions may be kinetically inhibited from reaching (or even closely approaching) equilibrium, and the chemistry in the outer nebula depends on the kinetics of the reactions involving interstellar medium constituents. At the onset of gravitational collapse, roughly 40% of the carbon in the interstellar medium is in the form of dust and ~10% is contained in PAHs (polycyclic aromatic hydrocarbons), while most of the gas-phase C is in the form of CO molecules. The interstellar nitrogen is expected to be gaseous N_2, but a significant fraction of the nitrogen is also present as NH_3. It is feasible that CO and N_2 in the cold outer regions of the protoplanetary disk never converted to CH_4 and NH_3, and similarly it may be possible that interstellar grains never evaporated. Thus, NH_3 and CH_4 in cometary ices might be of interstellar origin.

We have seen in §8.7, §10.3, and §10.4 that the D/H ratio in meteorites and comets is much higher than the protosolar value, as is the $^{15}N/^{14}N$ ratio in meteorites. The observed nuclear spin temperature for water in comets is ~30 K, indicative of formation on cold grains rather than direct condensation from the gas phase (§10.4.3.6). Both the high fractionation and low spin temperatures can be explained if some of the material formed in cold interstellar clouds. The (relative) abundances of CH_4:CO:C_2H_6 and those of H_2CO:CH_3OH, as well as the presence of some large molecules have been explained by hydrogenation processes of species on icy grains prior to incorporation in comets (§10.4.3.2). The Galileo probe discovered enhancements of ~3–6 times solar in the moderately volatile elements C, N, S, as well as the noble gases Ar, Kr, and Xe in Jupiter's atmosphere (at a pressure of ~10 bar). As these elements have a broad range of condensation temperatures, the small range in enhancements suggests that these elements were brought in by planetesimals that condensed at temperatures low enough for these elements to either be trapped within H_2O-ice or to be stable as solids.

Some models of protoplanetary disks account for the evolution of molecular abundances while the matter accretes to the central star. In such models, CO, N_2, and other gases as well as interstellar grains are brought in from the interstellar medium. When they are close enough to the star, where the temperature is high, equilibrium chemistry becomes important, but at larger distances ($\gtrsim$ several AU from the Sun) disequilibrium chemistry is the norm. Cosmic-ray and X-ray ionization may play an important role. The ions formed in the disk by cosmic rays and X-rays influence the evolution of molecular species. For example, CO is transformed through such reactions into CO_2, H_2CO, and CH_4, while N_2 is transformed to NH_3 and HCN. So over time, depending on the ionization rate, the CO and N_2 abundances decrease even though the equilibrium reactions (CO $\rightarrow$ CH_4, N_2 $\rightarrow$ NH_3) are kinetically inhibited. If the temperature is low enough for gases to freeze out, such gases may be adsorbed onto grains. The ices may sublimate again when the grains migrate inwards towards the star. Such more advanced models can thus explain the coexistence of certain ices (i.e., CH_4, CO, and CO_2) in comets, and the survival of interstellar grains in the protoplanetary disk (a significant fraction of the water molecules in these grains are of interstellar origin). The chemistry is extremely complex, and detailed models of the formation of our Solar System are becoming increasingly more sophisticated.

13.4.4 Clearing Stage

As there is no gas left between the planets, the gas must have been cleared away at some stage during the evolution process. The gas may have been cleared out via ablation from the faces of the protoplanetary disk by ultraviolet radiation emanating from nearby hot stars (Fig. 13.7) and/or from the early Sun during its active T-Tauri phase, a process called *photoevaporation*. The pre-main-sequence Sun went through its T-Tauri phase of stellar evolution approximately 10^6–10^7 yr after the protosun formed. The timing of the gas loss is a crucial issue concerning the growth of giant planets, but is not directly constrained; it is therefore usually assumed that the lifetime of the gaseous protoplanetary disk was the same as that typical of massive dust disks around young stars, $\lesssim 10^7$ yr. Note that all four giant planets in our Solar System have roughly the same mass of rock- and ice-forming elements, but their H and He abundances vary by a factor of 100. This may be a consequence of the time at which the gas within the protoplanetary disk was dissipated.

13.5 Growth of Solid Bodies

13.5.1 Timescale Constraints

Chondritic meteorites contain the oldest rocks known in our Solar System. As discussed in Chapter 8, the age of most chondrites (primitive meteorites) is 4.56 Gyr, and they formed within a period of $\lesssim 5$ Myr at the beginning of Solar System history. Evidence for extinct radionuclides in meteorites implies that the material which formed the Solar System contained an admixture of recently nucleosynthesized isotopes. In particular, live ^{26}Al ($t_{1/2} = 0.72$ Myr) in the protoplanetary disk suggests that the first solid planetary material formed at most a few million years after the last injection of freshly nucleosynthesized matter, a timescale which is similar to that required for the collapse of a molecular cloud core. The short-lived isotopes evidenced in meteorites could have been produced in an asymptotic giant branch (AGB) star. It is possible that the gravitational collapse of the solar nebula was triggered by a strong wind from a nearby AGB star. However, alternative models suggest that ^{26}Al and some of the other short-lived isotopes were produced in the early Solar System through particle irradiation of solid particles within the inner portions of the protoplanetary disk by the active protosun (§8.6.4).

The presence of nearly pure ^{22}Ne in some meteorites (presumably from ^{22}Na decay, with $t_{1/2} = 2.6$ yr; §8.6.4.2), as well as the isotopic ratios measured in several different elements (i.e., D/H, ^{15}N/^{14}N; §8.7) implies that some interstellar grains survived their journey within the solar nebula and were included into meteoritic material. As interstellar grains heat up considerably when falling to the midplane and passing through the accretion shock, most of the interstellar grains may evaporate completely before arrival in (at least the inner part of) the protoplanetary disk.

13.5.2 Planetesimal Formation

As a disk of gaseous matter cools, various compounds condense into microscopic grains. For a disk of solar composition, the first substantial condensates are silicates and iron compounds. At lower temperatures, characteristic of the outer region of our planetary system, large quantities of water-ice and other ices can condense (§13.4.3). In these regions there also may have been a significant fraction of pre-existing condensates from the interstellar medium and stellar atmospheres. Growth of solid particles then proceeds primarily by mutual collisions.

The microphysics of the growth of subcentimeter-sized grains is quite different from the dynamical processes important to later stages of planetary accretion. The mechanical and chemical processes related to grain agglomeration are poorly understood. Data from smokestack studies and numerical models suggest that loosely packed fractal structures which are held together by van der Waals forces may be formed (Fig. 13.15). However, most primitive meteorites differ from the subject of these studies as they contain chondrules, which are small igneous inclusions ∼1 mm in size (§8.1). The large abundance of chondrules implies that a significant fraction of the hypothesized fluffy (very porous) aggregates were rapidly heated and cooled prior to being incorporated into larger bodies. Various models of chondrule formation exist, but no consensus has yet been reached (§8.7.2).

The motions of small grains in a protoplanetary disk are strongly coupled to the gas. For the parameter regime thought have existed in the Solar System's protoplanetary disk, the coupling between the gas and solid particles smaller than 1 cm is well described by Epstein's drag law (eq. 2.55). When grains condense, the vertical component of the star's gravity causes the dust to sediment out towards the midplane of the disk. The acceleration of a grain is given by

$$\frac{dv_z}{dt} = -\frac{\rho_g c_s}{R\rho}v_z - n^2 z, \tag{13.14}$$

where v_z is the grain's velocity in the z-direction (perpendicular to the midplane of the disk), $\rho_g(z)$ the gas density, ρ the grain's density, R the radius of the grain, c_s the local speed of sound, which is equal to the thermal gas velocity (and thus $c_s \propto T^{1/2}$), and the Keplerian orbital angular velocity $n = \sqrt{GM_\odot/r_\odot^3}$. Equation (13.14) describes the behavior of a damped oscillator. The vertical motion of a large solid object is nearly sinusoidal, with the ρ_g term providing a slow damping of the amplitude. In contrast, the small grains considered here are highly overdamped, and they move very slowly relative to the gas – the

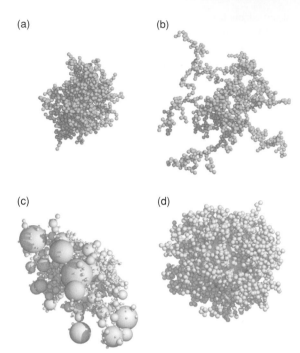

(a)   (b)

(c)   (d)

Figure 13.15 Examples of fractal aggregates produced by numerical simulations of the collisional agglomeration of dust grains. In a ballistic particle–cluster agglomeration (BPCA) process, a seed particle grows by the accumulation of single particles that collide with random impact parameters and from random directions on linear trajectories (hit-and-stick process). A ballistic cluster–cluster agglomeration (BCCA) process proceeds through the coagulation of equal-mass aggregates (again on linear trajectories with random impact parameters and from random directions). (a) BPCA with 1024 monodisperse spherical particles; the simple BPCA process leads to aggregates with a fractal dimension of 3.0. (b) BCCA with 1024 monodisperse spherical particles; these aggregates have a fractal dimension of 1.9. (c) BPCA with 2001 spherical constituent particles, following a power-law size distribution with an exponent of −3.15. (d) BPCA with 2000 monodisperse spherical particles, aggregated onto a large spherical core. (Blum *et al.* 1994)

two terms on the right-hand side are both much larger in magnitude than the acceleration term on the left. Neglecting this small acceleration, the equilibrium settling speed is

$$v_z = -\frac{n^2 z \rho R}{\rho_g c_s}. \tag{13.15}$$

Note that for particles of a given density, the settling rate is proportional to particle radius, and particles far from the midplane, where ρ_g is small (eq. 13.10a), settle much faster than do those close to the midplane.

At a heliocentric distance of 1 AU, the temperature of the disk is approximately 500–800 K, and the gas density near the midplane $\rho_g \approx 10^{-9}$ g cm^{-3}. The thermal velocity $c_s \approx 2.5 \times 10^5$ cm s^{-1} for an H$_2$-dominated nebula. For 1 µm grains with a density of 1 g cm^{-3}, $v_z \approx 0.03(z/H_z)$ cm s^{-1}. With this sedimentation rate, it would take a 1 µm sized particle within about one gas scale height of the midplane $\sim 10^6$ yr to fall halfway towards the disk midplane, or about 10^7 yr for 99.9% of the distance. Such a long settling time is inconsistent with timescales of grain condensation and growth into planetesimals based upon dating of meteorites. Additional processes must, therefore, be at work.

Collisional growth of grains during their descent to the midplane of the disk increases R in equation (13.15), thereby shortening sedimentation times by as much as several orders of magnitude, and differential settling velocities increase the collision rates between particles of differing sizes. For fluffy fractal aggregates, the settling rate increases much more gradually with increasing particle size, but the larger collision cross-sections of these low-density agglomerates compensates for the slower settling rate. Current models suggest that (for the estimated parameters of the terrestrial planet region of the solar nebula) the bulk of the solid material agglomerated into bodies of macroscopic size within $\lesssim 10^4$ years at 1 AU. Most of these bodies were confined to a relatively thin region about the midplane of the disk in which the density of condensed material was comparable to, or exceeded, that of the gas.

At a larger scale, growth from centimeter-sized particles to kilometer-sized planetesimals depends primarily on the relative motions between the various bodies. The motions of (sub)centimeter-sized material in the proto-planetary disk are strongly coupled to the gas (Fig. 13.16). The gas in the protoplanetary disk is partially supported against stellar gravity by a pressure gradient in the radial direction, so gas circles the star slightly less rapidly than the Keplerian rate. The 'effective' gravity felt by the gas is (eq. 2.56)

$$g_{\text{eff}} = -\frac{GM_\odot}{r_\odot^2} - \frac{1}{\rho_g}\frac{dP}{dr_\odot}. \tag{13.16}$$

The second term on the right-hand side of equation (13.16) is the acceleration produced by the pressure gradient. For circular orbits, the effective gravity must be balanced by centrifugal acceleration, $g_{\text{eff}} = -r_\odot n^2$. Since the pressure gradient is much smaller than the gravity, we can approximate the angular velocity of the gas, n_{gas}, as

$$n_{\text{gas}} \approx \sqrt{\frac{GM_\odot}{r_\odot^3}}(1-\eta), \tag{13.17}$$

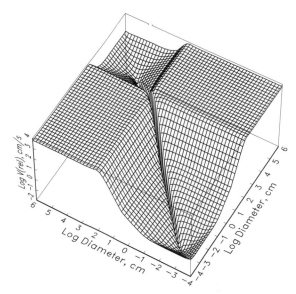

Figure 13.16 Contours of constant relative velocity (in cm s^{-1}) between pairs of particles of density 2 g cm^{-3} orbiting within a partially pressure-supported gaseous protoplanetary disk are displayed as a function of particle size. Sizes from 1 µm to 10 km are shown; relative velocities are due to thermal motions (dominant at sizes <10 µm), as well as radial and transverse velocities induced by gas drag. Disk parameters are for the midplane at 1 AU in a nonturbulent minimum mass solar nebula: gas density $\rho_g = 3.4 \times 10^{-9}$ g cm^{-3}, $T = 320$ K, $\Delta v = 61.7$ m s^{-1}. The narrow 'valley' in the contour plot results from the fact that equal-sized bodies have identical velocities. (Courtesy Stuart J. Weidenschilling)

where

$$\eta \equiv \frac{-r_\odot^2}{2GM_\odot\rho_g}\frac{dP}{dr_\odot} \approx 5 \times 10^{-3}. \tag{13.18}$$

For estimated protoplanetary disk parameters, the gas rotates $\sim 0.5\%$ slower than the Keplerian speed.

Large particles moving at (nearly) the Keplerian speed thus encounter a headwind that removes part of their orbital angular momentum and causes them to spiral inwards towards the star. Small grains drift less, as they are so strongly coupled to the gas that the headwind they encounter is very slow. Kilometer-sized planetesimals also drift inwards very slowly, because their surface area to mass ratio is small. Peak rates of inward drift occur for particles that collide with roughly their own mass of gas in one orbital period (Figure 13.17). Meter-sized bodies in the terrestrial planet region of the solar nebula drift inwards at the fastest rate (Problem 13.9), up to $\sim 10^6$ km yr^{-1}. Thus, a meter-sized body at 1 AU would spiral inwards approaching the Sun in ~ 100 years! As a consequence of the difference in (both radial and azimuthal)

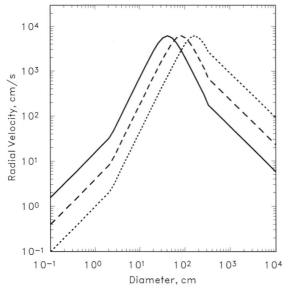

Figure 13.17 The inward radial drift rates of solid particles in a protoplanetary disk as a function of size for three values of density: 0.5 (dotted line), 2.0 (dashed line), and 7.9 (solid line) g cm^{-3}. Gas parameters are the same as for Figure 13.16. Tiny particles, with small mass/surface area ratios, are strongly coupled to the gas and compelled to move with (nearly) its angular velocity. As the gas velocity is less than the Keplerian orbital rate, the solid particles feel a residual component of the (proto)Sun's gravity, and settle inward at a terminal velocity at which gas drag balances this radial acceleration. Thus, larger and/or denser particles drift more rapidly in this regime. Bodies with large mass/surface area ratios travel in (nearly) Keplerian orbits, moving faster than the gas. They experience a 'headwind' that causes their orbits to decay; larger and/or denser bodies are less affected by this drag, so the decay rate decreases with increasing particle radius. The radial velocity reaches a peak at the transition between these regimes, at sizes of about a meter. The abrupt changes in slope result from transitions between drag laws for different Knudsen and Reynolds numbers. (Courtesy Stuart J. Weidenschilling)

velocities, small (sub)centimeter grains can be swept up by the larger bodies, while gas drag on the meter-sized planetesimals may induce considerable radial motions. This radial migration can remove solids from the planetary region, or bring particles of various sizes together to enhance accretion rates. Thus, the material that survives to form planets must complete the transition from centimeter to kilometer size rather quickly, unless it is confined to a thin dust-dominated subdisk in which the gas is dragged along at essentially Keplerian velocity.

Two alternative hypotheses describe the growth through this size range. First, if the nebula is quiescent, the dust and small particles settle into a layer thin enough to be gravitationally unstable to clumping, and planetesimals presumably are formed as a result of this instability. The

planetesimals produced by this mechanism have masses of order

$$M_{\text{planetesimal}} \sim \frac{16\pi^2 G^2 \sigma_\rho^3}{n^4}, \tag{13.19}$$

where σ_ρ is the surface mass density of the particle layer at the time the instability occurs. Planetesimals formed in the inner regions of the solar nebula would have been ~ 1 km in radius, with larger planetesimals forming further from the Sun (Problem 13.10). However, current models suggest that interactions between the gas and solids produced sufficient turbulence to prevent the particulate layer from becoming thin enough to be gravitationally unstable, at least in the terrestrial region of the solar nebula. According to the second scenario, the dust subdisk never becomes gravitationally unstable. Rather, growth in a turbulent nebula continues via simple two-body collisions. Under these circumstances, there is no fine line between planetesimal formation and accretion from planetesimals to planets. Molecular forces can lead to ~ 1 km sized planetesimals by coagulation, since the van der Waals (chemical) binding energies of $\sim 10^3$ erg g^{-1} are comparable to the gravitational binding energy of a 1 km body. When the planetesimals reach sizes of ~ 1 km, their mutual gravitational perturbations become important.

The growth of solid bodies from millimeter size to kilometer size within a turbulent protoplanetary disk presents particular problems. The physics of interparticle collisions in this size range is not well understood. Furthermore, the high rate of orbital decay due to gas drag for meter-size particles implies that growth through this size range must occur very rapidly. Solid particles may be concentrated within vortexes (which act as temporary nodes in the turbulent flow), leading to more rapid growth. A small fraction of the grains may grow into solid planetesimals via fortuitous circumstances, and these planetesimals may subsequently sweep up many times their mass in small particles. Differing models of planetesimal formation yield a wide variety of size distributions for the initial population of planetesimals. In some scenarios, planetesimals much larger than 1 km are expected.

The large radial motions of planetesimals may provide an explanation for some of the anomalies observed in meteorite composition, where isotopically distinct components of a single meteorite must have condensed separately, at different heliocentric distances, and been brought together as solid bodies. However, note that current theories of planetesimal formation are clearly oversimplified. For example, planetesimal growth models do not account for chondrule formation and other violent and disruptive events which may occur in a turbulent nebula.

13.5.3 From Planetesimals to Planetary Embryos

The primary factors controlling the growth of planetesimals into planets differ from those responsible for the accumulation of dust into planetesimals. Solid bodies larger than $\sim$1 km in size face a headwind only slightly faster than that experienced by 10 m objects (for parameters thought to be representative of the terrestrial region of the solar nebula), and because of their much greater mass-to-surface-area ratio they suffer far less orbital decay from interactions with the gas in their path (Fig. 13.17). The primary perturbations on the Keplerian orbits of kilometer-sized and larger bodies in protoplanetary disks are mutual gravitational interactions and physical collisions. These interactions lead to accretion (and in some cases erosion and fragmentation) of planetesimals. Gravitational encounters are able to stir planetesimal random velocities up to the escape speed from the largest common planetesimals in the swarm. The most massive planetesimals have the largest gravitationally enhanced collision cross-sections, and accrete almost everything with which they collide. If the random velocities of most planetesimals remain much smaller than the escape speed from the largest bodies, then these large *planetary embryos* (also referred to as *protoplanets*) grow extremely rapidly. The size distribution of solid bodies becomes quite skewed, with a few large bodies growing much faster than the rest of the swarm in a process known as *runaway accretion*. Eventually planetary embryos accrete most of the (slowly moving) solids within their gravitational reach, and the runaway growth phase ends. We examine the growth process from kilometer-sized planetesimals to 10^3–10^4 kilometer-sized planetary embryos in detail below.

13.5.3.1 *Planetesimal Velocities*

The distribution of planetesimal velocities is one of the key factors that controls the rate of planetary growth. Planetesimal velocities are modified by mutual gravitational interactions, physical collisions (which can be partially elastic, leading to rebound or fragmentation, or completely inelastic, leading to accretion), and gas drag. Gravitational scatterings and (nearly) elastic collisions convert energy present in the ordered relative motions of orbiting particles (Keplerian shear) into random motions. These interactions also tend to reduce the *random velocities* (velocities relative to that of a circular orbit in the disk midplane) of the largest bodies in a swarm and increase those of smaller

bodies, a process referred to as *dynamical friction*. Inelastic collisions and gas drag damp eccentricities and inclinations, especially of small planetesimals.

A statistical approach (referred to as the *particle-in-a-box* approximation) uses the methods of the kinetic theory of gases for calculating the evolution of planetesimals. The particle-in-a-box approximation ignores the details of individual planetesimal orbits and instead follows the evolution of the mean squared speeds of planetesimals (which depend on planetesimal size). A probability density function is used to describe the distribution of orbital elements in the planetesimal population. During the final stages of planetesimal accumulation, the number of planetesimals eventually becomes small enough that direct N-body numerical integrations of individual planetesimal orbits is feasible.

13.5.3.2 *Collisions and Accretion*

The size distibution of planetesimals evolves principally via physical collisions among its members. Physical collisions between solid bodies can lead to accretion, fragmentation, or inelastic rebound of relatively intact bodies; intermediate outcomes are possible as well. The outcome of a collision depends upon the internal strength of the planetesimals, the coefficient of restitution of the bodies, and most sensitively on the kinetic energy of the collision. The speed at which two bodies of radii R_1 and R_2 and masses m_1 and m_2 collide is given by

$$v_i = \sqrt{v^2 + v_e^2}, \tag{13.20}$$

where v is the speed of m_2 relative to m_1 far from encounter, and v_e is their mutual escape velocity from the point of contact:

$$v_e = \left(\frac{2G(m_1 + m_2)}{R_1 + R_2} \right)^{1/2}. \tag{13.21}$$

The impact speed is thus at least as large as the escape velocity, which for a rocky 10 km sized object is $\sim$6 m s^{-1}. The rebound speed is equal to ϵv_i, where the coefficient of restitution $\epsilon \leq 1$. If $\epsilon v_i < v_e$, then the bodies remain bound gravitationally and soon recollide and accrete. Net disruption requires both fragmentation, which depends upon the internal strength of the bodies, and post-rebound velocities greater than the escape speed. Since relative velocities of planetesimals are generally less than the escape velocity from the largest common bodies in the swarm, the largest members of the swarm are likely to accrete the overwhelming bulk of the material with which they collide, unless ϵ is very close to unity. Very small planetesimals are most susceptible to fragmentation. The largest bodies in the

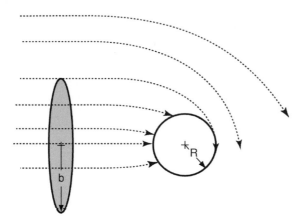

Figure 13.18 Schematic diagram of the gravitational focusing of planetesimal trajectories by an accreting planet. The critical trajectory which collides tangentially with the planet has an unperturbed impact parameter larger than the radius of the planet, $b > R$. (Adapted from Brownlee and Kress 2007)

swarm accrete at a rate essentially identical to the collision rate. Subcentimeter-sized grains corotating with the gas may impact kilometer-sized planetesimals at speeds well above the escape velocities of these planetesimals. This process could lead to erosion of planetesimals via 'sandblasting'.

The simplest model for computing the collision rate of planetesimals ignores their motion around the Sun completely. A collision occurs when the separation between the centers of two particles equals the sum of their radii. The mean rate of growth of a planetary embryo's mass, M, is:

$$\frac{dM}{dt} = \rho_s v \pi R^2 \mathcal{F}_g, \qquad (13.22)$$

where v is the average relative velocity between the large and small bodies, ρ_s the volume mass density of the swarm of planetesimals, and the planetary embryo's radius, R, is assumed to be much larger than the radii of the planetesimals. The last term in equation (13.22) is the gravitational enhancement factor, which in the 2 + 2-body approximation is given by:

$$\mathcal{F}_g = 1 + (v_e/v)^2. \qquad (13.23)$$

The gravitational enhancement factor arises from the ratio of the distance of close approach to the asymptotic unperturbed impact parameter, b, in a two-body hyperbolic encounter (Fig. 13.18), and can be derived using conservation of angular momentum and energy of the planetesimal relative to the planetary embryo (Problem 13.11). In the 2 + 2-body (patched-conics) approximation, often abbreviated as the *two-body approximation*, one ignores the influence of planetesimals/planetary embryos on one another except during close encounters, and during such

close encounters the influence of the Sun upon the bodies is neglected. Thus, the analysis reduces the problem to a pair of two-body calculations for the planetesimal.

It is often convenient to state the growth rate of the planet in terms of the surface density of the planetesimals in the disk rather than the volume density of the swarm. If the protosun's gravity is the dominant force in the vertical direction and if the relative velocity between planetesimals is isotropic, then the vertical Gaussian scale height H_z (eq. 13.11) of the planetesimal disk is

$$H_z = \frac{1}{\sqrt{3}} \frac{v}{n}. \qquad (13.24)$$

The surface mass density, also referred to as the column density (g cm^{-2}) of solids in the disk, σ_ρ, can be written as

$$\sigma_\rho = \sqrt{\pi} \rho_s H_z = \sqrt{\frac{\pi}{3} \frac{\rho_s v}{n}}. \qquad (13.25)$$

Equations (13.22)–(13.25) can be used to express the rate of growth in the planetary embryo's radius:

$$\frac{dR}{dt} = \frac{\frac{dM}{dt}}{4\pi \rho_p R^2} = \sqrt{\frac{3}{\pi}} \frac{\sigma_\rho n}{4\rho_p} \mathcal{F}_g, \qquad (13.26)$$

where ρ_p is the density of the planetary embryo. The planetary embryo's radius thus grows at a constant rate if $\mathcal{F}_g$ remains constant.

Random velocities of the planetesimals are determined by a balance between gravitational stirring and damping via inelastic collisions. If most of the mass is contained in the largest bodies, the equilibrium velocity dispersion is comparable to the escape speed of the largest bodies, implying $\mathcal{F}_g < 10$. Let us assume $\mathcal{F}_g = 7$ for the proto-Earth. In the minimum mass model, at 1 AU, the surface mass density $\sigma_\rho = 10$ g cm^{-2}, $n = 2 \times 10^{-7}$ s^{-1}, and $\rho_p = 4.5$ g cm^{-3}, which implies a growth time for the Earth of 2×10^7 yr. More detailed calculations yield times closer to 10^8 years, since the accretion rate drops during the later stages of planetary growth because a substantial fraction of the planetesimals have already been accreted.

For the giant planets, however, growth times computed in this manner are much larger. For a minimum mass nebula, the surface density drops with heliocentric distance, approximately as $r^{-3/2}$, except for a jump by a factor of about three at ~4 AU resulting from the condensation of water-ice. At Jupiter's distance from the Sun, the surface mass density in a minimum mass nebula would be $\sigma_\rho \approx 3$ g cm^{-2}. Jupiter's heavy element mass is approximately 15–25 M$_\oplus$, which results in a growth time of over 10^8 yr. A similar calculation for Neptune yields an estimate many times the age of the Solar System (Problem 13.13). Since at least Jupiter and Saturn must have formed within ~10^7 years, before the gas in the solar nebula was swept away, additional factors must be involved in the growth of giant planets.

13.5.3.3 Runaway Growth

When the relative velocity between planetesimals is comparable to or larger than the escape velocity, $v \gtrsim v_e$, the growth rate is approximately proportional to R^2, and the evolutionary path of the planetesimals exhibits an orderly growth of the entire size distribution. When the relative velocity is small, $v \ll v_e$, one can show, by rewriting the escape velocity in terms of the protoplanet's radius, that the growth rate is proportional to R^4 (eqs. 13.21–13.23). In this situation, the planetary embryo rapidly grows larger than any other planetesimal, which can lead to *runaway growth* (see Fig. 13.19 and Problem 13.15). The growth rate of an embryo of mass M in a disk whose velocity dispersion is governed by planetesimals of mass m varies as $\dot{M} \propto M^{4/3} m^{-2/3}$. If embryos dominate the stirring, then $\dot{M} \propto M^{2/3}$; in this circumstance, if individual embryos control the velocity in their own zones, larger embryos take longer to double in mass than do smaller ones, although embryos of all masses continue their runaway growth relative to surrounding planetesimals; this phase of rapid accretion of planetary embryos is known as *oligarchic growth*. A runaway embryo can grow so much larger than the surrounding planetesimals that its $\mathcal{F}_g$ can exceed 1000; however, three-body stirring by the embryo prevents $\mathcal{F}_g$ from growing much larger than this.

Runaway accretion requires low random velocities, and thus small radial excursions of planetesimals. The planetary embryo's feeding zone is therefore limited to the annulus of planetesimals that it can gravitationally perturb into intersecting orbits. Thus, rapid growth ceases when a planetary embryo has consumed most of the planetesimals within its gravitational reach. Planetesimals within ∼4 times the planetary embryo's Hill sphere eventually will come close enough to the planetary embryo during one of their orbits that they may be accreted (unless their semimajor axis is very similar to that of the embryo, in which case they may be locked in tadpole or horseshoe orbits that avoid close approaches, §2.2.2). The mass of a planetary embryo that has accreted all of the planetesimals within an annulus of width $2\Delta r_\odot$ is:

$$M = \int_{r_\odot - \Delta r_\odot}^{r_\odot + \Delta r_\odot} 2\pi r' \sigma_\rho(r') dr' \approx 4\pi r_\odot \Delta r_\odot \sigma_\rho(r_\odot).$$

$$(13.27)$$

Setting $\Delta r_\odot = 4\,R_H$ (eq. 2.28) and generalizing to a star of any mass, $M_\star$, we obtain the *isolation mass*, M_i (in grams),

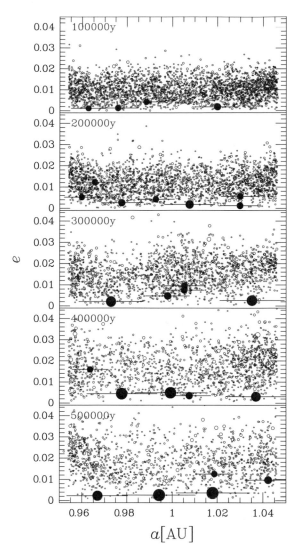

Figure 13.19 Snapshots of a planetesimal system on the a–e plane. The circles represent planetesimals and their radii are proportional to the radii of planetesimals. The system initially consists of 4000 planetesimals whose total mass is 1.3×10^{27} g. The initial mass distribution is a power law with index $\zeta = -2.5$ over the mass range 2×10^{23} g $\leq m \leq 4 \times 10^{24}$ g. The system is followed using an N-body integrator, and physical collisions are assumed to always result in accretion. The numbers of planetesimals are 2712 ($t = 100\,000$ yr), 2200 ($t = 200\,000$ yr), 1784 ($t = 300\,000$ yr), 1488 ($t = 400\,000$ yr), and 1257 ($t = 500\,000$ yr). The filled circles represent planetary embryos with mass larger than 2×10^{25} g, and lines from the center of each planetary embryo extend 5 R_H outwards and 5 R_H inwards. (Kokubo and Ida 1999)

which is the largest mass to which a planetary embryo orbiting can grow by runaway accretion:

$$M_i \approx 1.6 \times 10^{25} (r_{AU}^2 \sigma_\rho)^{3/2} \left(\frac{M_\odot}{M_\star}\right)^{(1/2)}$$

$$(13.28)$$

where σ_ρ is in g cm^{-2}. For a minimum-mass solar nebula, the mass at which runaway must have ceased in Earth's accretion zone would have been $\sim$6 M$_\oplus$, and in Jupiter's accretion zone $\sim$1 M$_\oplus$.

Runaway growth can persist beyond the isolation mass given by equation (13.28) only if additional mass can diffuse into the planet's accretion zone. Three plausible mechanisms for such diffusion are scattering between planetesimals, perturbations by planetary embryos in neighboring accretion zones, and gas drag. Alternatively, radial motion of the planetary embryo may bring it into zones not depleted of planetesimals. Gravitational torques resulting from the excitation of spiral density waves in the gaseous component of the protoplanetary disk have the potential of inducing rapid radial migration of planets (§13.8). Gravitational focusing of gas could also vastly increase the rate of inward drift of planetary embryos.

The limits of runaway growth are less severe in the outer Solar System than in the terrestrial planet zone. If the surface density of condensed material at 5 AU was $\gtrsim$10 g cm^{-2}, it is possible that runaway growth of Jupiter's core continued until it attained the mass necessary to rapidly capture its massive gas envelope. The 'excess' solid material in the outer Solar System could have been subsequently ejected to the Oort cloud or to interstellar space via gravitational scattering by the giant planets.

In contrast, the small terrestrial planets, orbiting deep within the Sun's gravitational potential well, could not have ejected substantial amounts of material. Thus, unless external factors such as one or more giant planets migrating sunwards through the terrestrial planet zone (§12.8.1) altered the evolution, the total mass of solids in the terrestrial planet zone during the runaway accretion epoch was probably not substantially larger than the current mass of the terrestrial planets. This implies that a high-velocity growth phase subsequent to runaway accretion was required to yield the present configuration of terrestrial planets.

13.6 Formation of the Terrestrial Planets

13.6.1 Dynamics of the Final Stages of Planetary Accumulation

The self-limiting nature of runaway and oligarchic growth implies that massive planetary embryos form at regular intervals in semimajor axis. The agglomeration of these embryos into a small number of widely spaced terrestrial planets necessarily requires a stage characterized by large orbital eccentricities, significant radial mixing, and giant impacts. At the end of the rapid-growth phase,

most of the original mass is contained in the large bodies, so their random velocities are no longer strongly damped by energy equipartition with the smaller planetesimals. Mutual gravitational scattering can pump up the relative velocities of the planetary embryos to values comparable to the surface escape velocity of the largest embryos, which is sufficient to ensure their mutual accumulation into planets. The large velocities imply small collision cross-sections and hence long accretion times (eq. 13.26).

Once the planetary embryos have perturbed one another into crossing orbits, their subsequent orbital evolution is governed by close gravitational encounters and violent, highly inelastic collisions. This process has been studied using N-body integrations of planetary embryo orbits, which include the gravitational effects of the giant planets, but neglect the population of numerous small bodies that must also have been present in the terrestrial zone; physical collisions are assumed to always lead to accretion (i.e., fragmentation is not considered). This approximation is almost certainly not justified, because the rotation of many objects exceeds that required for breakup, but it is better for the largest bodies than it is for smaller objects. Few bodies initially in the terrestrial planet zone are lost; in contrast, most planetary embryos in the asteroid region are ejected from the system by a combination of jovian perturbations and mutual gravitational scatterings. As the simulations endeavor to reproduce our Solar System, they generally begin with about 2 M$_\oplus$ of material in the terrestrial planet zone, typically divided (not necessarily equally) among hundreds of bodies. The end result is the formation of 2–5 terrestrial planets on a timescale of about 10^8 years (Fig. 13.20). Some of these systems look quite similar to our Solar System, but most have fewer terrestrial planets, and these planets travel on more eccentric orbits. It is possible that the Solar System is by chance near the quiescent end of the distribution of terrestrial planets. Alternatively, processes such as fragmentation and gravitational interactions with a remaining population of small debris, thus far omitted from the calculations because of computational limitations, may lower the characteristic eccentricities and inclinations of the ensemble of terrestrial planets.

An important result of these N-body simulations is that planetary embryo orbits execute a random walk in semimajor axis as a consequence of successive close encounters. The resulting widespread mixing of material throughout the terrestrial planet region diminishes any chemical gradients that may have existed when planetesimals formed, although some correlations between the final heliocentric distance of a planet and the region where most of its constituents

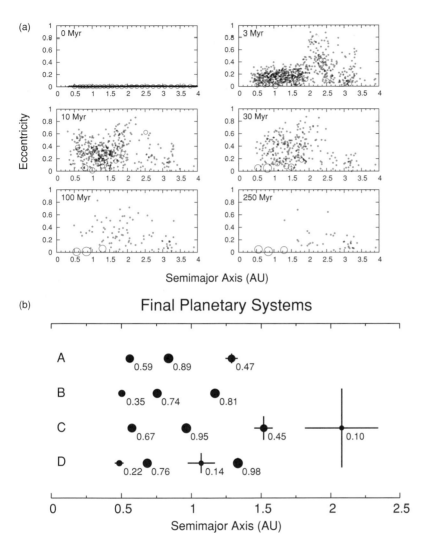

Figure 13.20 (a) Simulation of the final stages of terrestrial planet growth in our Solar System using an *N*-body code that assumes all physical collisions lead to mergers. The simulation begins with 25 planetary embryos as massive as Mars, ~1000 planetesimals each of mass 0.04 M_{σ}, and Jupiter and Saturn on their current orbits. The planetary embryos and planetesimals are represented as circles whose radii are proportional to the body's radius and whose locations are displayed in *a–e* phase space at the times indicated. (b) Synthetic terrestrial planet systems produced by four different *N*-body simulations of the final stages of planetary accretion. The final planets are indicated by filled circles centered at the planet's semimajor axis. The horizontal line through each circle extends from the planet's perihelion to its aphelion; the length of the vertical line extending upward and downward from a planet's center represents its excursions perpendicular to the invariant plane at the same scale. The numbers to the lower right of each circle represent the planet's final mass in $M_{\oplus}$. For example, the outermost planet in simulation A has $a = 1.29$ AU, $e = 0.035$, $i = 1.55°$, and $M_{p} = 0.47$ $M_{\oplus}$. The results of the simulation shown in part (a) are presented in row A. The initial disks used for these four simulations are very similar, and the different outcomes arise from stochastic variations of accretion dynamics. See O'Brien *et al.* (2006) for particulars of the calculations. (Courtesy David O'Brien)

originated are preserved in the simulations. Nonetheless, these dynamical studies imply that Mercury's high iron abundance is unlikely to have arisen from chemical fractionation in the solar nebula.

The mutual accumulation of numerous planetary embryos into a small number of planets must have entailed many collisions between protoplanets of comparable size. Mercury's silicate mantle was probably partially stripped off by one or more of such giant impacts, leaving behind an iron-rich core. Accretion simulations also lend support to the giant impact hypothesis for the origin of the Earth's Moon (§13.11.2); during the final stage of accumulation, an Earth-size planet is typically found to collide with several objects as large as the Moon and frequently one body as massive as Mars. The obliquities of the rotation axes

of the giant planets provide independent evidence of the occurrence of giant impacts during the accretionary epoch.

13.6.2 Accretional Heating and Planetary Differentiation

Impacting planetesimals provide a planet with energy as well as mass. This energy heats a growing planet. Decay of radioactive elements (§8.6.1) also heats planetary bodies, with short-lived nuclides such as ^{26}Al being the most important for growing bodies during the first few million years of Solar System history (Problem 13.26), and potassium, thorium, and uranium dominating over billion year timescales (Problem 13.24). A heated planet, or even a small planetesimal if it contains a sufficient quantity of short-lived radioisotopes, may become warm enough that portions melt, allowing denser material to sink and the planet to differentiate.

The nonradiogenic energy available to a growing planet is supplied by accreted planetesimals (which contribute both their kinetic energy 'at infinity' and the potential energy released as the planetesimal falls onto the planet's surface), gravitational potential energy released as the planet contracts (in response to increased pressure) or differentiates, decay, and exothermic chemical processes. The primary energy-loss mechanism is radiation to space, although the planet may also cool via endothermic reactions or give up gravitational energy as a result of expansion if it heats up significantly or if water freezes. Energy may be transported within the planet via conduction, or if the planet is (partially or fully) molten, via convection. Energy transport within a planet is important to the global as well as local heat budgets, as radiative losses can only occur from the planet's surface or atmosphere.

Conduction is rather slow over planetary distances, and convection operates only in regions that are sufficiently molten to allow fluid motions to occur (§6.1.5.2). Thus, to a first approximation, a growing solid planet's temperature is given by a balance between accretion energy deposited at the planet's surface, radioactive decay, and radiative losses from the surface. The temperature of a given region changes slowly once it becomes buried deep below the surface (unless short-lived radionuclides are sufficiently abundant or large-scale melting and differentiation occur). For gradual accretion, the temperature at a given radius can thus be approximated by balancing the accretion energy source with radiative losses at the time when the material was accreted. For the 10^8 year accretion times estimated for the terrestrial planets, such an estimate implies far less heating than would be required to melt and differentiate a $\lesssim 1\,M_\oplus$ planet (Problem 13.19).

However, modern theories of planetary growth imply that terrestrial planets accumulate most of their mass in planetesimals of radius 100 km and larger. Impactors deposit $\sim$70% of their kinetic energy as heat in the target rocks directly beneath the impact site, with the remaining $\sim$30% being carried off with the ejecta. If an impactor is large, it may raise deeply buried heat to near the surface, where energy may be radiated away. A more important effect is that heat may become buried by deep ejecta blankets. The ejecta blankets produced by such large impactors are thick enough that most of the heat from the impacts remains buried. Planets can thus become quite warm, with temperature increasing rapidly with radius (Problem 13.19). Accretion energy can lead to the differentiation of planetary (but not asteroidal) sized bodies.

A planetary embryo can form a proto-atmosphere as it accretes solid bodies. When the mass of a growing planet reaches $\sim$0.01 $M_\oplus$, impacts are energetic enough for water to evaporate, while impact devolatilization of NH_3 and CO_2 may occur a little sooner. Complete degassing of accreting planetesimals occurs when the radius of the protoplanet reaches about 0.3 $R_\oplus$. A massive proto-atmosphere that is optically thick to outgoing radiation can trap energy provided by the impacting planetesimals. This process, known as the *blanketing effect*, is capable of increasing the surface temperature of the protoplanet by even more than the greenhouse effect. Solar radiation determines the temperature at the top of the atmosphere, and is scattered and absorbed at lower altitudes. The atmosphere provides a partially insulating blanket to the heat released from impacting planetesimals, so the surface becomes quite hot.

Calculations show that the proto-atmosphere's blanketing effect becomes important when the growing planet's mass exceeds 0.1 $M_\oplus$. The surface temperature exceeds $\sim$1600 K, the melting temperature for most planetary materials, when the planet's mass is 0.2 $M_\oplus$. As a result, the surface melts and newly accreting planetesimals on the molten surface will also melt. Heavy material migrates downwards, while lighter elements float on top. This process of differentiation liberates a large amount of gravitational energy in the planet's interior; together with adiabatic compression that results from the increase in the planet's mass, enough energy can be released to cause melting of a large fraction of the planet's interior, allowing the planet to differentiate throughout.

13.6.3 Accumulation (and Loss) of Atmospheric Volatiles

Atmospheric gases form a tenuous veneer surrounding many of the smaller planets and moons in the Solar System, amounting to far less than 1% of the mass of each body. These atmospheres consist primarily of high-Z ($Z \geq 3$) elements. The atmospheres of the terrestrial planets and other small bodies were probably outgassed from material accreted as solid planetesimals. The problem of the origin of terrestrial planet atmospheres is not simply bringing the required volatiles to the planets, as losses were also important. Impacting planetesimals on a growing planet surrounded by a proto-atmosphere may lead to the following phenomena (§5.4.3):

(1) If the planetesimals are small enough to be stopped by atmospheric drag or disrupted by ram pressure, all of their kinetic energy is deposited in the atmosphere. Most rocky objects smaller than a few dozen meters in radius are

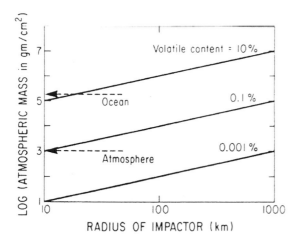

Figure 13.21 The mass per unit surface area of an atmosphere of a 1 $M_\oplus$ planet that is in equilibrium between the rate of addition of volatiles to the planet by accretion of material with the indicated volatile content and impact erosion of the atmosphere by impactors of the indicated radii. The arrows show the mass per unit surface area of the present terrestrial ocean and atmosphere. (Hunten *et al.* 1989)

stopped in an atmosphere like that of Earth at the present time, and deposit all of their energy in the atmosphere.

(2) Ejecta excavated by larger impacting planetesimals are slowed down by the atmosphere, and transfer kinetic energy to it. The interaction is quite complicated. But note that an atmosphere has a large compressibility, in contrast to a solid surface, and that the gas can be raised to very high temperatures and pressures. Additionally, the energy from atmospheric impacts is released over an extended area and over an interval of tens of seconds.

(3) If the impactor is large, the energy transferred to the atmosphere may be sufficient to blow off part of the atmosphere via hydrodynamic escape (§4.8.3.1). If the size of the impactor is comparable to or larger than the atmospheric scale height, impact erosion blows off a large portion of the atmosphere, i.e., an atmospheric mass equal to the mass intercepted by the impactor (§4.8.3.2). The same impactor may also add volatiles to the accreting planet. Whether this mass is more than that blown off from the atmosphere depends upon the size of the impactor, its volatile content, and the density of the atmosphere. Impactors with radii of ~100 km and volatile content of 1% would yield a balance between impact erosion and accretion of volatiles for an atmospheric mass per unit area similar to that of the terrestrial ocean. A similarly sized impactor population with a volatile content of 0.01% would keep a present-day Earth's atmosphere in equilibrium. Figure 13.21 graphs the

mass per unit area of an atmosphere in equilibrium between impact erosion and addition of volatiles as a function of impactor radius and volatile content. Atmospheric blowoff is more likely to occur on smaller planets, such as Mars. A growing planet may lose its atmosphere several times during the accretion period, since impacts with large planetesimals are quite common.

In addition to impact erosion, atmospheric gases may be lost via Jeans escape (§4.8.1). In particular, light elements such as H and He easily escape from the top of a terrestrial atmosphere, while heavier gases may have escaped this way in the early hot proto-atmospheres. The present-day terrestrial planet atmospheres were probably formed towards the end of the accretion epoch, by outgassing of the hot planet and impacts by small planetesimals.

13.7 Formation of the Giant Planets

The large amounts of H_2 and He contained in Jupiter and Saturn imply that these planets formed within ~10^7 years, before the gas in the protoplanetary disk was swept away. Any formation theory of the giant planets must account for these timescales. In addition, formation theories should explain the elemental and isotopic composition of these planets and variations therein from planet to planet, their presence and/or absence of internal heat fluxes, their axial tilts, and the orbital and compositional characteristics of their ring and satellite systems. In this section, we discuss the formation of the planets themselves; the formation of their satellites and ring systems is addressed in §13.11.1.

Elements heavier than helium constitute <2% of the mass of a solar composition mixture. The giant planets, however, are enriched in heavy elements relative to the solar value by roughly 5, 15, and 300 times for Jupiter, Saturn, and Uranus/Neptune, respectively. Thus, all four giant planets accreted solid material much more effectively than gas from the surrounding nebula. Moreover, the total mass in heavy elements varies by only a factor of a few between the four planets, while the mass of H and He varies by about two orders of magnitude between Jupiter and Uranus/Neptune.

Table 4.6 shows the composition of the giant-planet atmospheres. The enhancement in heavy elements increases from Jupiter to Neptune.[2] This gradual, nearly

---

[2] The nitrogen mixing ratio derived from the NH_3 abundance may form an exception to this rule. But if nitrogen is present as N_2 rather than NH_3, its abundance could scale similarly to the other heavy elements. We also note that sulfur, presumably present in the form of H_2S, has only been detected directly in Jupiter by the Galileo probe.

monotonic relationship between mass and composition argues for a unified formation scenario for all of the planets and smaller bodies. Moreover, the continuum of observed extrasolar planetary properties, which stretches to systems not very dissimilar to our own, suggests that extrasolar planets formed in a similar way to the planets within our Solar System.

The D/H ratio in the giant-planet atmospheres may provide important clues to the formation history of these planets. The D/H ratios in Jupiter and Saturn, measured from monodeuterated methane gas, are equal to the interstellar D/H ratio of 2×10^{-5}. Since 90% of Jupiter's mass and 75% of Saturn's mass consist of H and He, one would indeed expect the D/H ratios to agree with the interstellar value. Uranus and Neptune are only about 10% H and He by mass. The observed D/H values on Uranus and Neptune (Fig. 10.24) are higher than the interstellar value, which can be attributed to exchange of deuterium with an icy reservoir.

Various classes of models have been proposed to explain the formation of giant planets and brown dwarfs. The mass function (abundance of objects as a function of mass) of young compact objects in star-forming regions extends down through the brown dwarf mass range to below the deuterium-burning limit. This observation, together with the lack of any convincing theoretical reason to think that the collapse process that leads to stars cannot also produce substellar objects, strongly implies that most isolated (or distant companion) brown dwarfs and isolated high planetary mass objects form via the same collapse process as do stars.

By similar reasoning, the brown dwarf desert, a profound dip over the range $\sim$10–50 M_{J} in the mass function of companions orbiting within several AU of Sun-like stars (§12.4), strongly suggests that the vast majority of extrasolar giant planets formed via a mechanism different from that of stars. Within our Solar System, bodies up to the mass of Earth consist almost entirely of condensable material, and even bodies of mass $\sim$15 $M_{\oplus}$ consist mostly of condensable material.[3]

The theory of giant-planet formation favored by most researchers is the *core nucleated accretion model*, in which the planet's initial phase of growth resembles that of a terrestrial planet, but when the planet becomes sufficiently

massive (several $M_{\oplus}$) it is able to accumulate substantial amounts of gas from the surrounding protoplanetary disk. Aside from core nucleated accretion, the only giant-planet formation scenario receiving significant attention is the *disk instability hypothesis*, in which a giant gaseous protoplanet forms directly from the contraction of a clump that was produced via a gravitational instability in the protoplanetary disk. Star-like direct quasi-spherical collapse is not considered viable, both because of the observed brown dwarf desert and theoretical arguments against the formation of Jupiter-mass objects via fragmentation.

13.7.1 Disk Instability Hypothesis

Numerical calculations show that 1 M_{J} clumps can form in sufficiently gravitationally unstable disks ($Q_{\mathrm{T}} \lesssim 1$, see eq. 11.13). However, weak gravitational instabilities excite spiral density waves; density waves transport angular momentum that leads to spreading of a disk, lowering its surface density, and making it more gravitationally stable. Rapid cooling and/or mass accretion is required to make a disk highly unstable. Long-lived clumps capable of shrinking into bodies of planetary dimensions can thus only be produced in protoplanetary disks with highly atypical physical properties. Additionally, gas instabilities would yield massive stellar-composition planets, requiring a separate process to explain the smaller bodies in our Solar System and the heavy-element enhancements in Jupiter and Saturn. The existence of compositionally intermediate objects like Uranus, Neptune, and dense exoplanets such as HD 149026 b is particularly difficult to account for in such a scenario.

Metal-rich stars are more likely to host giant planets within a few AU than are metal-poor stars (Fig. 12.24); this trend is consistent with the need of having sufficient condensables to form a massive core, but runs contrary to the requirement of rapid disk cooling needed to form long-lived clumps via gravitational instabilities.

The above listed points do not preclude a modified version of giant gaseous protoplanets in which Jupiter (and possibly Saturn) formed via gravitational instability and subsequently accumulated their heavy-element excesses by accreting planetesimals. However, the possibility of forming Jupiter- and Saturn-mass objects via disk instabilities has not been convincingly demonstrated, as gravitational instabilities might produce density waves that redistribute mass and stabilize the disk rather than bound planetary mass blobs. Moreover, the gradual progression of masses and composition discussed above argues for a single formation scenario for at least the four giant planets in our Solar System. Nonetheless, it is possible that

---

[3] The definition of 'condensable' is best thought of as the value of the specific entropy of the constituent relative to that for which the material can form a liquid or solid. Hydrogen and helium within protoplanetary disks have entropies far in excess of that required for condensation, even if they are compressed isothermally to pressures of order one bar, even for a temperature of only a few tens of degrees. Thus, H_2 and He remain in a gaseous state.

some giant planets form via disk instability, most likely in the regions of protoplanetary disks distant from the central star, where Keplerian shear is small and orbital timescales are long.

13.7.2 Core Nucleated Accretion

The core nucleated accretion model relies on a combination of planetesimal agglomeration and gravitational accumulation of gas. According to this scenario, the initial stages of growth of a gas giant planet are identical to those of a terrestrial planet. Dust settles towards the midplane of the protoplanetary disk, agglomerates into (at least) kilometer-sized planetesimals, which continue to grow into larger solid bodies via pairwise inelastic collisions. As the (proto)planet grows, its gravitational potential well deepens, and when its escape speed exceeds the thermal velocity of gas in the surrounding disk, it begins to accumulate a gaseous envelope. The gaseous envelope is initially optically thin and isothermal with the surrounding protoplanetary disk, but as it gains mass it becomes optically thick and hotter with increasing depth. While the planet's gravity pulls gas from the surrounding disk towards it, thermal pressure from the existing envelope limits accretion. For much of the planet's growth epoch, the primary limit on its accumulation of gas is the planet's ability to radiate away the gravitational energy provided by accretion of planetesimals and envelope contraction; this energy loss is necessary for the envelope to further contract and allow more gas to reach the region in which the planet's gravity dominates. The size of the planet's gaseous envelope is typically a few tens of percent of the planet's Hill sphere radius, R_H, whose size is given by equation (2.22). Eventually, increases in the planet's mass and radiation of energy allow the envelope to shrink rapidly. At this point, the factor limiting the planet's growth rate becomes the flow of gas from the surrounding protoplanetary disk.

The rate and manner in which a forming giant planet accretes solids substantially affect the planet's ability to attract gas. Initially accreted solids form the planet's core, around which gas is able to accumulate. Calculated gas accretion rates are very strongly increasing functions of the total mass of the planet, implying that rapid growth of the core is a key factor in enabling a planet to accumulate substantial quantities of gas prior to dissipation of the protoplanetary disk. Continued accretion of solids acts to reduce the planet's growth time by increasing the depth of its gravitational potential well, but has counteracting effects by providing additional thermal energy to the envelope (from solids that sink to or near the core) and increased atmospheric opacity from grains that are released in the upper parts of the envelope. Major

questions remain to be answered regarding solid-body accretion in the giant-planet region of a protoplanetary disk, with state-of-the-art models providing a diverse set of predictions.

A $\sim 1 \, M_\oplus$ planet is able to capture an atmosphere from the protoplanetary disk because the escape speed from its surface is large compared to the thermal velocity of gas in the disk. However, such an atmosphere is very tenuous and distended, with thermal pressure pushing outwards to the limits of the planet's gravitational reach and thereby limiting further accretion of gas. The key factor governing the planet's evolution at this stage is its ability to radiate energy so that its envelope can shrink and allow more gas to enter the planet's gravitational domain.

During the runaway planetesimal accretion epoch (§13.5.3), the (proto)planet's mass increases rapidly (Fig. 13.22). The internal temperature and thermal pressure increase as well, preventing nebular gas from falling onto the protoplanet. When the feeding zone is depleted, the planetesimal accretion rate, and therefore the temperature and thermal pressure, decrease. This allows gas to fall onto the planet much more rapidly. Gas accumulates at a gradually increasing rate until the mass of gas contained in the planet is comparable to the mass of solid material. The rate of gas accretion then accelerates more rapidly, and runaway gas accretion occurs.

Once a planet has a mass large enough for its self-gravity to compress the envelope substantially, its ability to accrete additional gas is limited only by the amount of gas available. Hydrodynamic limits allow quite rapid gas flow on a planet of mass $10 \, M_\oplus \lesssim M_p \lesssim 1 \, M_2$. As the planet grows, it alters the disk by accreting material from it and by exerting gravitational torques on it. These processes can lead to gap formation and, eventually, to isolation of the planet from the surrounding gas (Fig. 13.23). Gaps that small moons clear via a similar process are observed within Saturn's rings (Figs. 11.24, 11.30).

The planet starts to contract when the factor limiting the rate at which the planet accumulates gas transitions from internal thermal pressure to the disk's ability to provide gas. Initially, contraction takes place rapidly on a Kelvin–Helmholtz timescale, t_{KH}, which is the ratio of the planet's gravitational potential energy, E_G, to its luminosity, $\mathcal{L}$:

$$t_{KH} \equiv \frac{E_G}{\mathcal{L}} \sim \frac{GM^2}{R\mathcal{L}}. \qquad (13.29)$$

The temperatures in the envelope increase rapidly, so that the protoplanet's luminosity stays approximately constant, despite the planet's decrease in size. Vigorous convection

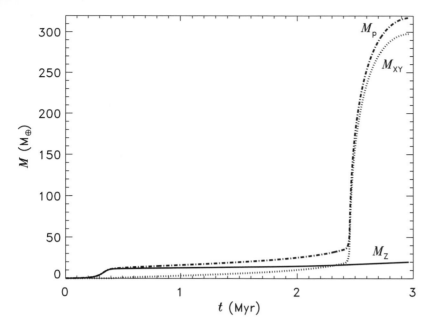

Figure 13.22 The mass of a giant planet that grows to 1 $M_{2\!\!\!\!\!\backslash}$ is shown as a function of time according to one particular simulation based upon the core nucleated accretion model. The planet's total mass is represented by the dot-dashed curve, the mass of the solid component is given by the solid curve, and the dotted curve represents the gas mass. The solid core grows rapidly by runaway accretion in the first 4×10^5 years. The rate of solid body accumulation decreases once the planet has accreted nearly all of the condensed material within its gravitational reach. The envelope accumulates gradually, with its settling rate determined by its ability to radiate away the energy of accretion. Eventually, the planet becomes sufficiently cool and massive that gas can be accreted rapidly. This simulation is for growth at 5.2 AU from a 1 $M_\odot$ star, with a local surface mass density of solids equal to 10 g/cm^2. (Lissauer *et al.* 2009)

mixes the envelope during this period, homogenizing the distribution of heavy elements. After a few thousand years (for Jupiter/Saturn), contraction slowed down due to the increasing incompressibility of the fluid envelope, and the temperature and luminosity decreased with time. The slow cooling of the envelope is a major source of the excess thermal energy emitted into space by the giant planets.

The fact that Uranus and Neptune contain less H_2 and He than do Jupiter and Saturn suggests that our Solar System's two outermost planets never quite reached runaway gas accretion conditions, possibly due to a slower accretion of planetesimals. As per equation (13.26), the rate of accretion of solids depends upon the surface density of condensates and the orbital frequency, both of which decrease with heliocentric distance.

The current composition of the atmospheres of the giant planets is largely determined by how much heavy material was mixed with the lightweight material in the planets' envelopes. Once the core mass exceeds about 0.01 $M_\oplus$, the temperature becomes high enough for water to evaporate into the protoplanet's envelope. While accretion continues, the envelope becomes more massive, and late-accreting planetesimals have an increasing difficulty penetrating through the growing envelope. These bodies sublimate in the envelopes of the giant planets, thereby enhancing the heavy-element content of the outer regions considerably.

13.8 Planetary Migration

13.8.1 Torques from Protoplanetary Disks

Planetary orbits can *migrate* towards (or in some circumstances away from) their star as a consequence of angular momentum exchange between the protoplanetary disk and the planet. As is the case for moons near planetary rings (§11.4), protoplanets drift away from the disk material with which they interact. Planets beyond a disk's edge are pushed in only one direction, but others are subjected to partially offsetting torques. For conditions thought to exist in most protoplanetary disks, the net torque is negative, and therefore planets lose angular momentum to the disk and drift towards the central star. If a planet's mass is small enough that the disk is weakly perturbed, the planet's migration timescale, $a/|\dot{a}|$, is inversely proportional to the planet's mass and the local disk mass, and is directly proportional to the square of the disk aspect ratio, $(H_z/r)^2$. Radial motion within this linear regime is referred to as *Type 1 migration*. Type 1 migration is rapid for cores in excess of a few $M_\oplus$. A 10 $M_\oplus$ planet at 5 AU in a minimum-mass solar nebula would drift towards the star on a timescale of $\sim 10^5$ years, which is at least one order of magnitude shorter than both the formation timescale of giant planets and the disk lifetime. Therefore, Type 1 migration as predicted by the linear theory poses difficulties for giant-planet formation via the core nucleated accretion scenario, since it suggests

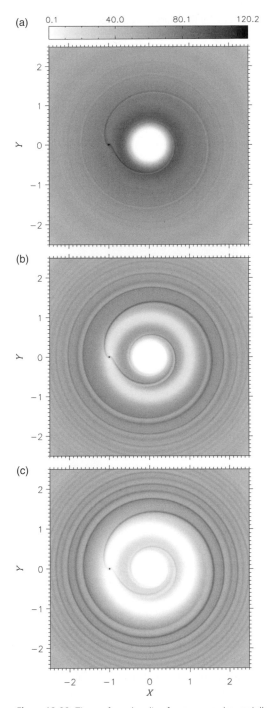

Figure 13.23 The surface density of a gaseous circumstellar disk containing an embedded planet on a circular orbit located 5.2 AU from a 1 $M_\odot$ star. The ratio of the scale height of the disk to the distance from the star is $H_z/r = 1/20$, and the viscosity is $\nu_v = 1 \times 10^{15}$ cm^2/s. The distance scale is in units of the planet's orbital distance, and the scale bar gives the surface density in units of g/cm^2. The planet is located at $(-1, 0)$ and the star at $(0, 0)$. (a) $M_p = 10$ M$_\oplus$. (b) $M_p = 0.3$ M$_2$. (c) $M_p = 1$ M$_2$. Details of the calculations can be found in D'Angelo *et al.* (2003). (Courtesy Gennaro D'Angelo)

that most planets in the Neptune mass range should be consumed by their star.

However, three-dimensional numerical hydrodynamic calculations show that the disk response to the perturbation induced by the planet becomes nonlinear when $M_p \approx$ 5–20 M$_\oplus$ (depending on the disk thickness, H_z, where H_z is given by eq. 13.11). This nonlinearity results in much longer migration timescales than those predicted in the linear regime. A planet may thus be able to begin its growth by runaway accretion of solids, augmented by inward migration as its mass increases up to several M$_\oplus$, then slow down as a result of nonlinear interaction with the disk, and cease to migrate rapidly when its mass reaches about ∼10 M$_\oplus$ (depending on the local conditions of the nebula). Such a stalled core could then accumulate gas in a relatively benign environment.

Once the mass of a (proto)planet has reached ∼1 M$_2$, i.e., when $R_H/H_z \gtrsim 1$, the disk–planet interaction is strongly nonlinear. This also corresponds to the thermal condition for gap formation (the size of the region strongly influenced by the planet's gravity being comparable to the thickness of the disk). Orbital migration of a planet becomes unavoidable after it has opened up a gap in the disk. *Type 2 migration* is generally slower than Type 1 migration, and its speed does not vary with planetary mass unless the planet's mass is comparable to or larger than the mass of the local disk. Once the planet's mass becomes similar to that of the protoplanetary disk, inertial effects become important and the rate of migration slows.

Observations of classical T Tauri stars indicate that gas in the inner region of protostellar disks is being continually depleted by accretion onto the young stellar objects. Without a mass supply, the surface density and the tidal angular momentum transfer rate interior to the planet's orbit decrease. In the outer disk, these quantities maintain their value as gas is prevented from viscously diffusing inwards by the planet's tidal torque. The imbalance between the inner and outer disk leads to the inward orbital migration of the planet. If the planet's mass is less than that of the disk, its orbital migration is coupled to the viscous evolution of the disk (§13.4.2), which is estimated to occur on the timescale of ∼10^6 yr.

These considerations have led to the speculation that tidal evolution may cause some first-born protogiant planets to migrate towards and eventually merge with their host stars. This 'infant mortality' would continue until the nebula mass is depleted to such an extent that the residual gas can no longer induce any significant evolution of the protogiant planets' orbits.

Protoplanetary disk-induced migration is considered to be the most likely explanation for giant exoplanets with orbital periods of a few years or less, especially the giant vulcan planets that are very close to their stars (§13.12). In contrast, the giant planets within our Solar System do not show evidence for similar inward wanderings. Planet-formation models were developed to explain the configuration of the Sun's planetary system, so perhaps there are more fundamental problems with the models that result from prejudices created by our particular vantage point. Our own Solar System may indeed represent a biased sample (in addition to poor statistics), because it contains a planet with conditions suitable for life to evolve to the point of being able to ask questions about other planetary systems!

13.8.2 Scattering of Planetesimals

Planets can also migrate as a back-reaction to clearing large amounts of planetesimals from the regions within their gravitational reach. The distribution of orbits within the Kuiper belt and the existence of the Oort cloud provide strong evidence for *planetesimal-induced migration* of the four giant planets within our Solar System. This process is distinct from the interactions with the gaseous disk responsible for the types of migration discussed in §13.8.1. The mechanism operates as follows: Gravitational stirring by Uranus and Neptune excites high eccentricities in the surrounding planetesimals. Those which acquire sufficiently small perihelia can be 'handed off' to the next-innermost planet, with a resultant gain in angular momentum for the first planet. In this way, planetesimals get passed inward from Neptune to Uranus to Saturn and finally to Jupiter, which is massive enough to readily eject them from the Solar System or onto nearly parabolic paths about the Sun. When bodies on nearly parabolic paths reach distances of $\sim 10^4$ AU from the Sun, their heliocentric velocities are so slow that the tidal forces of our galaxy and the tugs of nearby stars can raise their perihelia out of the planetary region, placing them into the Oort cloud. The other giant planets are also massive enough to eject planetesimals from the Solar System or to the Oort cloud; however, the characteristic timescales for direct ejection are longer than for passing the planetesimals inwards to the control of Jupiter.

For a planet on a circular orbit, change in angular momentum L is related to change in semimajor axis a according to

$$\Delta L = \frac{1}{2} M_{\rm p} \sqrt{\frac{GM_\odot}{a}} \Delta a. \qquad (13.30)$$

Thus Jupiter, being the innermost and most massive giant planet, migrates the shortest distance, while Neptune migrates furthest.

As Neptune migrated outwards, it would have pumped the eccentricities of objects carried along in its outer mean motion resonances. The orbital distribution of Kuiper belt objects, especially the many plutinos (including Pluto) with $e \sim 0.3$, provides strong evidence for the outwards migration of Neptune by several AU. This amount of migration requires a planetesimal disk of a few dozen $M_\oplus$.

Expelling small bodies from the outer Solar System to the Oort cloud and to interstellar space caused Jupiter to migrate inwards, probably by a few tenths of an AU. In contrast, the other giant planets, which perturbed more planetesimals inwards to Jupiter-crossing orbits than directly outwards to the Oort cloud and beyond, are thought to have migrated away from the Sun.

13.9 Small Bodies Orbiting the Sun

13.9.1 Asteroid Belt

Thousands of minor planets of radii > 10 km orbit between Mars and Jupiter (Chapter 9), yet the total mass of these bodies is $< 10^{-3}$ $M_\oplus$. This is orders of magnitude less than would be expected for a planet accreting at ~ 3 AU within a smoothly varying protoplanetary disk. Why is there so little mass remaining in the asteroid region? Why is this mass spread among so many bodies? Why are the orbits of most asteroids more eccentric and inclined to the invariable plane of the Solar System than are those of the major planets? Why are the asteroids so diverse in composition, as indicated by their spectra and the wide variety of meteorites found on Earth?

Many small asteroids are differentiated. Accretional heating and long-lived radionuclides could not have supplied sufficient energy to cause the melting required for differentiation. Proposed energy sources are electromagnetic induction heating (§8.7.1) and the decay of short-lived radionuclides, especially ^{26}Al. The observed segregation of asteroidal spectral types by semimajor axis (Fig. 9.10) places an upper limit on the amount of planetesimal mixing that could have occurred within the asteroid belt.

Proximity to Jupiter is almost certainly responsible for the mass depletion in the asteroid belt, as well as for the orbital properties of asteroids. Large planetary embryos scattered into the asteroid zone by Jupiter, and/or direct resonant perturbations of Jupiter, are capable of exciting eccentricities and inclinations of asteroid zone planetesimals and planetary embryos. Much of the material once contained in small bodies orbiting between Mars and Jupiter could thereby have been scattered into Jupiter-crossing orbits, from which it would have been ejected from the Solar System or accreted by Jupiter. Other planetesimals could have been ground to dust or even partially vaporized by high-velocity collisions. Planetary embryos that formed within the present asteroid belt near resonances with Jupiter may have been resonantly pumped to high eccentricities and perturbed their nonresonant neighbors; orbital migration, as well as the dispersal of the gaseous component of the protoplanetary disk, could have enhanced these perturbations by sweeping resonance locations over a large portion of the asteroid region.

Changes in Jupiter's orbit caused by gravitational interactions with the remnant planetesimal disk and/or other planets played a major role in clearing some parts of the asteroid belt. Such variations in Jupiter's orbit may have been instrumental in moving some asteroids (perhaps bodies originating exterior to Jupiter's orbit) into the Trojan regions, 60° ahead of and behind Jupiter in its orbit.

13.9.2 Comet Reservoirs

Current theories of Oort cloud formation imply that substantial quantities of small planetesimals which formed between ~3 and 30 AU from the Sun were ejected from the planetary region by gravitational perturbations from the giant planets. Accounting for the inefficiency in transporting bodies from the planetary region into bound Oort cloud orbits and for losses over the age of the Solar System, the mass of solid material ejected from the planetary region could have been 10–1000 $M_{\oplus}$. This implies that the minimum-mass protoplanetary disk (§13.5) must be a moderate to severe underestimate of the actual mass of the disk out of which our planetary system formed.

The Kuiper belt requires that planetesimals existed beyond the orbit of Neptune. Thus, the abrupt cutoff of observed massive planets beyond the orbit of Neptune (the masses of Pluto and Eris are each less than 2×10^{-4} times that of Neptune, Tables 1.3 and 9.5) cannot be explained solely by the lack of material in this region of the Solar System. Better observational estimates of the total mass, size distribution, and orbital characteristics of bodies in this region may provide helpful constraints on the dynamics of the accretionary process in the outer, loosely bound, regions of a protoplanetary disk.

13.10 Planetary Rotation

The origin of planetary rotation is one of the most basic questions of cosmogony. It has also proven to be one of

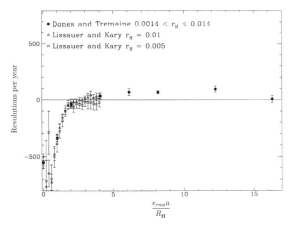

Figure 13.24 The number of rotations per orbit for an Earth-like planet which accreted within a uniform surface density 2-D disk of planetesimals is displayed as a function of the root mean square planetesimal eccentricity, e_{rms}, normalized by the size of the planet's Hill sphere. Negative values indicate rotation in the retrograde direction. The different symbols indicate different sets of numerical experiments, and the error bars result from statistical uncertainties. Note that the rapid prograde rotation observed for Earth and Mars cannot be produced by accretion of small planetesimals from a uniform disk, regardless of the value of planetesimal rms eccentricity. (Courtesy Luke Dones)

the most difficult to answer. Planets accumulate rotational angular momentum from the relative motions of accreted material (Problem 13.18). The stochastic nature of planetary accretion from planetesimals allows for a random component to the net spin angular momentum of a planet in any direction. As planets might accumulate a significant fraction of their mass and spin angular momentum from only a very few impacts, stochastic effects may be very important in determining planetary rotation. From the observed rotational properties of the planets, the size of the largest bodies to impact each planet during the accretionary epoch has been estimated to be 1–10% of the planet's final mass.

Very little net spin angular momentum is accumulated by a planet that accretes while on a circular orbit within a uniform surface density disk of small planetesimals (Fig. 13.24). A planet that partially clears a gap in the disk, and thus accretes a larger fraction of material from the edges of its accretion zone, may accumulate sufficient prograde angular momentum to explain the planetary rotation rates observed in our Solar System. Rapid prograde rotation can also result if planetesimal orbits decay slowly towards the protoplanet as a result of gas drag. Alternatively, stochastic impacts of large bodies may be the primary source of the rotational angular momentum of the terrestrial planets, with the observed preference of low obliquities being a chance occurrence (and, in the case of Mercury, tidal torques exerted by the Sun, §2.6.2).

Jupiter and Saturn are predominantly composed of hydrogen and helium, which they must have accreted hydrodynamically, in flows quite different from those that govern the dynamics of planetesimals. Such flows lead to prograde rotation. The nonzero obliquities of the giant planets might have been produced by giant impacts. Spin-orbit resonances also might have tilted the rotation axes of some or all of the giant planets. The precession periods of Saturn's spin axis and Neptune's orbital plane are similar, and may well have passed through a resonance when the Kuiper belt was more massive. Passage through this resonance could have tilted Saturn's rotational axis, producing the current obliquity of 27°.

13.11 Satellites of Planets and Minor Planets

13.11.1 Giant Planet Satellites

The moons and rings of the giant planets are analogous to miniature planetary systems in many respects. In each case, multiple secondaries orbit their primary, with most of the larger bodies traveling on nearly circular, coplanar prograde orbits having a certain regularity to their spacings.

Satellites orbiting closest to giant planets (near the Roche limit) are generally small. Planetary rings dominate where tidal forces from the planet are sufficient to tear apart a moon held together solely by its own gravity. Larger moons orbit at distances ranging from a few planetary radii to several dozen planetary radii. The outer regions of the satellite systems of all four giant planets contain small bodies on highly eccentric and inclined orbits. The diversity of planetary satellites suggests that they are formed by more than a single mechanism.

The satellite systems of the giant planets consist of *regular* and *irregular* satellites. Regular satellites move on low-eccentricity prograde orbits near the equatorial plane of their planet. They orbit close to the planet, well within the bounds of the planet's Hill sphere. These properties imply that regular satellites formed within a disk orbiting in the planet's equatorial plane. Irregular satellites generally travel on high-eccentricity, high-inclination orbits lying well exterior to a planet's regular satellite system; most irregular satellites are quite small. Most, if not all, of them were captured from heliocentric orbits.

Several models for the capture of irregular satellites have been proposed. One possibility is a slowdown of nearby bodies due to gas drag in the protoplanet's envelope or incorporation into the accretion disk. Most of such planetesimals would then have ended up in the planet itself, but some survived and became captured satellites. Gas drag is a viable capture mechanism only during the planet-formation epoch, but two other proposed mechanisms, *tidal disruption of binaries* and collisions with regular satellites, are viable at later times as well. Satellites that are captured via tidal disruption and collisions typically have very eccentric orbits about their planet. However, Neptune's large moon Triton, whose orbit is retrograde and highly inclined but nonetheless has a very low eccentricity, was probably captured by one of these mechanisms; Triton's orbit could subsequently have been tidally circularized by virtue of its large mass and proximity to Neptune (see §2.6).

The regular satellites of the giant planets likely formed by a solid body accretion process in a gas/dust disk surrounding the planet. Such disks may consist of material from the outer portions of the protoplanet's envelope, or matter that was directly captured from the protoplanetary disk. Solid-body accretion rates within a minimum mass 'subnebula' disk surrounding a giant planet are very rapid (Problem 13.30). The density of gas within the giant planets' circumplanetary disks exceeded that of the nearby protoplanetary disk and temperatures were also higher. Thus, chemical reactions proceeded further towards equilibrium. This explains the gross compositional differences between the regular and irregular satellites around Jupiter and Saturn. When youthful Jupiter's high luminosity is included, the model naturally accounts for the decrease in density of the Galilean satellites with increasing distance from Jupiter. The densities of the moons around Saturn and Uranus do not vary in such a systematic manner with distance from the planet; however, these lower mass planets were never as luminous as was young Jupiter. Since tidal forces prevent material from accreting within a planet's Roche limit, rings formed around the giant planets. Note, however, that most if not all of the ring systems that we see at present are not primordial (§11.7).

The rock/(rock + ice) mass fraction of the icy satellites and Pluto yield clues to the place of formation of these bodies. Figure 13.25 shows the estimated rock fraction of various satellites and Pluto, together with the range in expected mass fractions for the protoplanetary disk, where CO was much more abundant than CH_4 (§13.4.3), and the circumplanetary disks, where CH_4 may have been dominant. The range in mass fractions is caused by the uncertainty in the solar C/O ratio, which lies between 0.43 and 0.60. The larger planetary satellites (Ganymede, Callisto, Titan) contain more rock than expected for a body accreted in a circumplanetary disk; however, vaporization and escape of volatile material during accretion could have

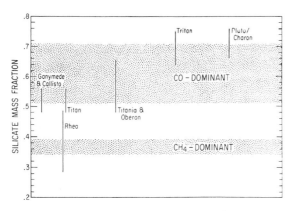

Figure 13.25 The mass ratio of rock (silicates, metals, and other refractory compounds) to rock + ice in outer Solar System objects. The ranges for individual objects represent uncertainties in densities and assumptions regarding interior models. The shaded areas represent expected compositions for a solar composition mixture at outer Solar System temperatures, assuming that either carbon monoxide or methane is the dominant carbon-bearing species. (Lunine and Tittemore 1993)

depleted the ice. The uranian satellites also contain a substantial rock component. Io and Europa (not shown) are composed primarily of rock. Since these moons formed very close to Jupiter at high temperatures, there was less water-ice available for accretion. Triton and Pluto/Charon have very similar high rock-mass fractions, consistent with that expected in a CO-rich solar nebula. One also would expect a large loss of volatile material from these small bodies if they were hit by sizable impactors subsequent to being differentiated.

The wide variety of properties exhibited by the satellite systems of the four giant planets in our Solar System suggests that stochastic processes may be even more important for satellite formation than current models suggest them to be in planetary growth. A possible explanation for this difference is that satellite systems are subjected to a very heavy bombardment of planetesimals on heliocentric orbits, which may fragment moons and also produce them. Deterministic models of satellite formation must thus be interpreted with caution.

Terrestrial planets and smaller objects presumably never possessed gas-rich circumplanetary disks; thus other explanations are required for the origins of the moons of Mars, Earth, asteroids, and Kuiper belt objects.

13.11.2 Formation of the Moon

The Earth's Moon is a very peculiar object. The Moon/Earth mass ratio greatly exceeds that of any other satellite/planet (although Charon/Pluto and various other satellite/minor

planet ratios are larger, see Table 9.5), raising the question of how this much material was placed into orbit about Earth. The density of the Moon is ~25% smaller than the uncompressed density of the Earth (Table 6.1). Yet the Moon is severely depleted in volatiles (which tend to have low densities), having less than half the potassium abundance of Earth and very little water. The combination of low mean density and lack of volatiles implies that the Moon is not simply an amalgam of that solar composition material which is able to condense above a certain temperature. Rather, the Moon's bulk composition resembles Earth's mantle, albeit depleted in volatiles. The bulk composition of the lunar crust and mantle could be understood if the Moon equilibrated with a large iron core, but the Moon's core is quite small (§6.3.1). Capture, coaccretion, and fission models of lunar origin have all been studied in great detail, but none satisfies both the dynamical and chemical constraints in a straightforward manner.

The favored hypothesis is the *giant impact model*, in which a collision between the Earth and a Mars-sized or larger planetary embryo ejects a lunar mass (or more) of material into Earth's orbit (Fig. 13.26). Assuming both bodies were differentiated prior to the impact, this model would explain the apparent similarities between the lunar composition and that of the Earth's mantle and at the same time the lack of volatile material on the Moon. Volatile material would have been vaporized completely by the impact, and it would have remained in a gaseous state within the circumterrestrial disk, allowing most of the volatiles to escape into interplanetary space. A range of impactor and impact parameters can place roughly one lunar mass of material primarily from the mantles of the impactor and/or Earth into terrestrial orbit beyond Roche's limit. Once this material is cool enough to form condensed bodies, it can quickly accumulate into a single large moon (Fig. 13.27). A giant impact origin of Pluto's satellite system also appears likely.

13.11.3 Satellites of Small Bodies

Mars's moons Phobos and Deimos are similar in composition to C-class asteroids (§5.5.4.6). Since these satellites orbit in the plane of the martian equator, they most likely accreted from a small disk formed by the capture of one or more planetesimals. As in the cases of irregular moons of giant planets and satellites of small bodies, capture of the material for the circummartian disk may have occurred via binary exchange reactions or collisions between two planetesimals passing close to Mars. Tidal disruption of one or more planetesimals also presents a viable mechanism to form such a disk. An alternative possibility is that they were

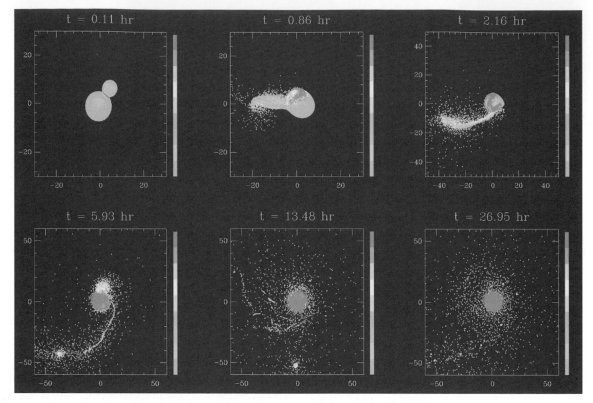

Figure 13.26 COLOR PLATE These computer-generated images illustrate the first day after a Mars-sized protoplanet and the proto-Earth collide with a velocity upon contact of 9 km s^{-1}. This contact velocity corresponds to a nil relative approach velocity, $v_\infty = 0$, and is less than the escape velocity from Earth because the collision occurs when the centers of the two bodies were separated by almost 1.5 $R_\oplus$. The angular momentum of the collision was 73% that of a grazing impact. Both bodies were differentiated prior to the impact. This collision produced a circumterrestrial disk 1.62 times as massive as the Moon; only 5% of the mass of the disk was metallic iron. The simulations were performed using a smooth particle hydrodynamics (SPH) code with a total of 60 000 particles. The particles are color-coded by temperature, and time is indicated within each panel. See Canup (2004) for details on the calculation. (Courtesy Robin Canup)

captured more or less intact and eons of subsequent interactions with the tori of debris produced by impacts onto the surfaces of these moons has damped their eccentricities and inclinations.

The relative sizes and orbits of asteroidal and Kuiper belt binaries imply that some of these pairs formed as the result of collisions, whereas others probably originated when three or more objects came into close proximity and the binary pair was able to gravitationally transfer mechanical energy to one or more objects.

13.12 Exoplanet Formation Models

The orbits of most of the extrasolar giant planets thus far observed are quite different from those of Jupiter, Saturn, Uranus, and Neptune (§12.4), and new models have been proposed to explain them. It has been suggested that most of these planets, and particularly the giant vulcans, formed substantially further from the star and subsequently migrated inwards to their current short-period orbits. Planetary orbital decay had been studied prior to the discovery of extrasolar planets, but no one predicted giant planets near stars because migration speeds were expected to increase as the planet approached the star, so the chance that a planet moved substantially inwards and was not subsequently lost was thought to be small (§13.8). Two possible mechanisms have been proposed for stopping a planet less than one-tenth of an AU from the star: tidal torques from the star counteracting disk torques or a substantial reduction in disk torque once the planet was well within a nearly empty zone close to the star. However, the substantially larger abundance of giant planets with orbital periods ranging from 15 days to 3 years, which feel negligible tidal torque from their star and are exterior to the region of the disk expected to be

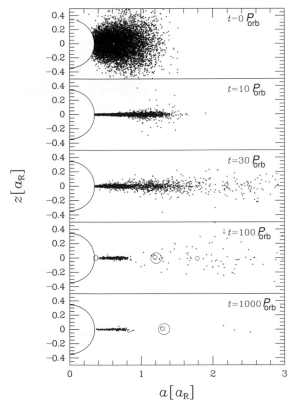

Figure 13.27 Snapshots of the protolunar disk in the r–z plane at $t = 0, 10, 30, 100, 1000\ P_{orb}$, where P_{orb} is the Keplerian orbital period at the Roche limit. The initial number of disk particles is 10 000, and the disk mass is four times the present lunar mass. The semicircle centered at the coordinate origin stands for the Earth. Circles represent disk particles and their sizes are proportional to the physical sizes of the disk particles. The horizontal scale shows the semimajor axis of disk particles in units of the Roche limit radius, a_R (see eq. 11.8). Note the very massive transient ring around the Earth. (Kokubo *et al.* 2000)

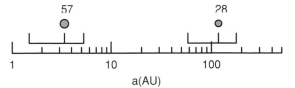

Figure 13.28 Orbits of the two remaining planets from a synthetic system of giant planets that was disrupted by chaotic scattering which sent most of the planets off into interstellar space. The planets' masses are given in units of $M_\oplus$ and the periastra, semimajor axes, and apoastra are shown below. The inner planet in this system, which is twice as massive as the outer planet, would dominate the radial velocity signature of the star, especially for observations over a time span of a few decades or less. (Adapted from Levison *et al.* 1998)

The sample of known extrasolar planets contains strong biases. Most solar-type stars could well have planetary systems which closely resemble our own. Nonetheless, if giant planets (even of relatively modest Uranus masses) orbiting near or migrating through 1 AU are the norm, then terrestrial planets in habitable zones may be scarcer than they were previously believed to be. However, giant planets could have large moons which themselves might be habitable.

13.13 Confronting Theory with Observations

The current theory of planetary growth via planetesimal accretion within a circumstellar disk provides excellent explanations of the causes of many of the observed Solar System and exoplanet properties, but less complete or less satisfactory explanations for several others.

13.13.1 Solar System's Dynamical State

Dynamical models of planetary accretion within a flattened disk of planetesimals produce moderately low eccentricity, almost coplanar orbits of planets, except at the outer fringes of the Solar System. The ultimate sizes and spacings of solid planets are determined by their ability to gravitationally perturb one another into crossing orbits. Such perturbations are often caused by weak resonant forcing and occur on timescales much longer than the bulk of planetesimal interactions discussed in §13.5.3. A more massive protoplanetary disk probably produces larger, but fewer planets. Stochastic processes are important in planetary accretion, so nearly identical initial conditions could lead to quite different outcomes; for example, the fact that there are four terrestrial planets in our Solar System as opposed to three or five is probably just the luck of the draw.

Jupiter played a major role in preventing the formation of a planet in the asteroid zone. Jovian resonances

cleared by magnetic accretion onto the star (§13.4.2), are more difficult to explain by this model. Perhaps (at least the inner few AU of) protoplanetary disks are cleared from the inside outwards, leaving migrating planets stranded.

Some giant exoplanets move on quite eccentric orbits (Fig. 12.22). These eccentric orbits may be the result of stochastic gravitational scatterings among massive planets (which have subsequently merged or been ejected to interstellar space, see Fig. 13.28), perturbations from a stellar binary companion (which might no longer be present if the now-single stars were once members of unstable multiple star systems), or from the complex and currently ill-constrained interactions between the planets and the protoplanetary disk (§13.8.1).

could have directly stirred planetesimals in the asteroid zone, or Jupiter could have scattered large failed planetary embryos inwards from 5 AU, yielding the same effect. The resulting stirring could have prevented further planetary growth and/or ejected an already-formed planet from the Solar System.

The giant planets ejected a substantial mass of solid bodies from the planetary region. The majority of these planetesimals escaped from the Solar System, but $\gtrsim 10\%$ ended up in the Oort cloud. Perturbations from the galactic tide, passing stars, and giant molecular clouds have randomized the orbits of Oort cloud comets over the past 4.5×10^9 years. Aside from a small flattening caused by the tidal potential of the galaxy, the Oort cloud is nearly spherical with prograde as well as retrograde objects. The Kuiper belt likely formed *in situ*, from planetesimals orbiting exterior to Neptune's orbit. The dynamical structure of the Kuiper belt suggests that Neptune slowly migrated outwards by several AU during the final stages of planetary formation. During this gradual migration, Neptune could have trapped objects in resonance and excited their eccentricities, thereby producing the observed populations of plutinos in the 2:3 resonance as well as resonant KBOs (Fig. 9.3).

The prograde rotation of Jupiter and Saturn can be explained as a deterministic result of gas accretion, while the excess of prograde rotation among the other planets may have been produced in a systematic way via expansion of their accretion zones, or may just be a chance result. Planetary obliquities result from stochastic impacts of large bodies and/or spin-orbit resonances between planets.

The gross features of the regular satellite systems of the giant planets can be understood if these planets had disks orbiting them; various models for the formation of such disks exist. The angular momentum distribution of the Solar System resulted from outward transport of mass and angular momentum (via poorly characterized viscous, gravitational, and/or magnetic torques) within the protosun/protoplanetary disk, plus a subsequent removal of most of the Sun's spin angular momentum by the solar wind.

The high cratering rate in the early Solar System and the lower rate at the current epoch are a consequence of the sweep-up of debris from planetary formation. The early high bombardment rate caused ancient surfaces on planets, asteroids, and satellites to be covered with craters. Huge impacts led to the formation of the Moon, stripped off the outer layers of Mercury, and may have changed the spin orientation of Uranus. Some large planetesimals probably were captured into planetocentric orbits, e.g., Triton about Neptune.

13.13.2 Composition of Planetary Bodies

The masses and bulk compositions of the planets can be understood in a gross sense as resulting from planetary growth within a disk whose temperature and surface density decreased with distance from the growing Sun. The terrestrial planets are rocky because the more volatile elements could not condense (or survive in solid form) so close to the Sun, whereas comets and the moons of the giant planets retain ices because they grew in a colder environment. The condensation of water-ice beyond $\sim$4 AU provided the outer planets with enough mass to gravitationally trap substantial amounts of H_2 and He from the solar nebula. Longer accretion times at greater heliocentric distances together with the timely disappearance of gas may account for the decrease in the gas fractions of the giant planets with increasing semimajor axis.

Some solids were transported over significant radial distances within the protoplanetary disk, leading to mixing of material that condensed in different regions of the solar nebula and/or had survived from the presolar era. This mixing helps explain the bulk compositions of those chondritic meteorites that contain both refractory inclusions and volatile-rich grains, as well as the very refractory Solar System condensates included among the samples of Comet P/Wild 2 returned by Stardust (§10.4.4). Radial and vertical mixing could have brought together material for meteorites that are progressively more depleted in volatiles over too large a range in condensation temperature to be explained by equilibrium solidification at any one time and place. Explanations of many detailed characteristics of meteorites (especially the formation of chondrules and remanent magnetism) remain controversial. The (well-established) planetesimal hypothesis explains the similarity in ages among primitive meteorites and the fact that all Solar System rocks, whose components presumably at one point passed through a stage similar to primitive meteorites, are the same age or younger.

All planets were hot during the accretionary epoch. The present terrestrial planets show evidence of this early hot era in the form of extensive tectonic and/or volcanic activity. Jupiter, Saturn, and Neptune have excess thermal emissions resulting from accretional and differentiation heating. Outgassing of the hot newly formed planets, combined with late accretionary veneers from the asteroid belt and comet reservoirs, led to the formation of atmospheres on the terrestrial planets.

13.13.3 Extrasolar Planets

Although many more planets are now known outside of our Solar System than within it, we have far less data on these planets, detection statistics are highly biased, and the data in hand are relatively new and have not been used to constrain many planet formation models. The orbits of exoplanets provide strong evidence that radial migration in protoplanetary disks is an important process. Gravitational interactions between planets and the disks in which they formed can provide the torques required for these orbital changes. Disk–planet interactions can indeed be so powerful that it is difficult to explain why so many giant planets have not migrated all of the way inwards and been consumed by their stars. The high eccentricities of many exoplanets came as a surprise, and in general the discoveries of exoplanets to date have shown that Nature is more creative than theorists, making predictions quite difficult!

13.13.4 Conclusions

The planetesimal hypothesis provides a viable theory of the growth of the terrestrial planets, the cores of the giant planets, and the smaller bodies present in the Solar System. The formation of solid bodies of planetary size should be a common event, at least around young stars that do not have binary companions orbiting at planetary distances. Planets could form by similar mechanisms within circumpulsar disks if such disks have adequate dimensions and masses. The formation of giant planets, which contain large quantities of H_2 and He, requires rapid growth of planetary cores, so that gravitational trapping of gas can occur prior to the dispersal of the gas from the protoplanetary region. According to the scenario outlined in this chapter, the largest body in any given zone is the most efficient accreter, and its mass 'runs away' from the mass distribution of nearby bodies in the sense that it doubles in mass faster than typical bodies. Such rapid accretion of a few large solid protoplanets can lead to giant planet core formation in $\sim 10^6$ years, provided disk masses are a few times as large as those given by 'minimum mass' models of the solar nebula. Thus, we appear to have a basic understanding of giant-planet formation, although our models of the origin of giant planets must be regarded as somewhat more uncertain than those of terrestrial planet accretion (because a wider variety of physical processes need to be considered in order to account for both the massive gaseous components and the solids enrichments of giant planets).

Further Reading

The series of *Protostars and Planets* books contain review papers on molecular clouds and star and planet formation; the latest volume in the series is:

Reipurth, B., D. Jewitt, and K. Keil, Eds., 2007. *Protostars and Planets V*. University of Arizona Press, Tucson. 951pp.

A comprehensive textbook on stellar formation is:

Stahler, S.W., and F. Palla, 2005. *The Formation of Stars*. Wiley-VCH, Weinheim, Germany. 865pp.

Several papers related to different aspects of the formation of our Solar System are given by:

Lin, D.N.C., 1986. The nebular origin of the Solar System. In *The Solar System: Observations and Interpretations*. Ed. M.G. Kivelson. Rubey Vol. IV. Prentice Hall, Englewood Cliffs, NJ, pp. 28–87.

Lissauer, J.J., 1993. Planet formation. *Annu. Rev. Astron. Astrophys.*, **31**, 129–174.

Lissauer, J.J., 1995. Urey Prize lecture: On the diversity of plausible planetary systems. *Icarus*, **114**, 217–236.

Lissauer, J.J., O. Hubickyj, G. D'Angelo, and P. Bodenheimer, 2009. Models of Jupiter's growth incorporating thermal and hydrodynamics constraints. *Icarus*, **199**, 338–350.

A good paper on the formation of terrestrial atmospheres is:

Ahrens, T.J., J.D. O'Keefe, and M.A. Lange, 1989. Formation of atmospheres during accretion of the terrestrial planets. In *Origin and Evolution of Planetary and Satellite Atmospheres*. Eds. S.K. Atreya, J.B. Pollack, and M.S. Matthews. University of Arizona Press, Tucson, pp. 328–385.

Equilibrium chemistry in the solar nebula is described in:

Prinn, R.G., and B. Fegley, Jr., 1989. Solar nebula chemistry: Origin of planetary, satellite and cometary volatiles. In *Origin and Evolution of Planetary and Satellite Atmospheres*. Eds. S.K. Atreya, J.B. Pollack, and M.S. Matthews. University of Arizona Press, Tucson, pp. 78–136.

A rather detailed chemical model of the evolution of our protoplanetary disk is provided in:

Aikawa, Y., T. Umbebayashi, T. Nakano, and S.M. Miyama, 1999. Evolution of molecular abundances in protoplanetary disks with accretion flow. *Astrophys. J.*, **519**, 705–725.

A (mostly still current) thorough review of nucleosynthesis within stars is given in:

Clayton, D.D., 1983. *Principles of Stellar Evolution and Nucleosynthesis*. University of Chicago Press. 612pp.

An excellent popular account of big bang nucleosynthesis is provided by:

Weinberg, S., 1988. *The First Three Minutes*. Basic Books, New York. 198pp.

Problems

13.1.**E**   (a) Calculate the gravitational potential energy of a uniform spherical cloud of density ρ and radius R.

(b) Determine the Jeans mass, M_J, of an interstellar cloud of solar composition with density ρ and temperature T. (Hint: Set the gravitational potential energy equal to negative twice the cloud's kinetic energy and solve for the radius of the cloud.)

(c) Show that if the cloud collapses isothermally, it becomes more *unstable* as it shrinks.

(d) Show that if the cloud retains the gravitational energy of its collapse as heat, it becomes more *stable* as it shrinks.

13.2.E Derive the formula for the free-fall gravitational collapse time of a uniform spherical cloud of density ρ (eq. 13.8). (Hint: The trajectory of a gas parcel initially at rest at a distance r from the center of the cloud can be approximated as a very eccentric ellipse with semimajor axis $r/2$.)

13.3.E Consider an H_2 molecule that falls from ∞ to a circular orbit at 1 AU from a 1 $M_\odot$ star.

(a) Calculate the circular velocity at 1 AU, and determine the total mechanical (kinetic + potential) energy of a molecule on a circular orbit at 1 AU. Note that the total energy of the molecule at rest at infinity is zero.

(b) Calculate the temperature increase of the hydrogen gas assuming it has not suffered radiative losses.

13.4.I Show that the density of gas in a thin isothermal circumstellar disk varies in the direction perpendicular to the midplane of the disk as

$$\rho = \rho_0 e^{-z^2/H_z^2}, \qquad (13.31)$$

where the Gaussian scale height is given by equation (13.11). (Hint: Consider a balance between pressure and the component of the star's gravity perpendicular to the midplane of the disk.)

13.5.E (a) Compute the molecular viscosity in a protoplanetary disk at a radius of 10^{14} cm, where the mean free path is $\ell_{\rm fp} = 10$ cm and the sound speed $c_s = 1$ km s^{-1}.

(b) What is the viscous accretion timescale of such a disk?

13.6.E Compute the value of α_v necessary for a turbulent disk with the parameters listed in Problem 13.5 to have a viscous evolution time of 10^6 years.

13.7.I (a) Diffusion inwards within a viscous circumstellar disk leads to accretion. The timescale for this accretion is equivalent to the diffusion timescale (eq. 11.10) between the radius in question and the radius of the star. Derive a formula for the viscous accretion timescale (in years) of a protoplanetary disk with scale height $H_z = 0.1\,r$ and sound speed $c_s = 10^5\,r_{\rm AU}^{-1/2}$ cm s^{-1} as a

function of the viscosity parameter α_v and radius, $r_{\rm AU}$.

(b) Evaluate your formula at 1 AU and 5 AU for $\alpha_v = 0.01$.

13.8.I As discussed in the text, the ice/rock ratio in a protoplanetary disk depends both on elemental composition and chemical state. In this problem, you will calculate the ice/rock ratio under a variety of assumptions.

(a) Assume anhydrous rock to consist of SiO_2, MgO, FeO, and FeS, using up the entire inventory of all of these elements other than oxygen, which is the most abundant. Referring to the elemental abundances listed in Table 8.1, calculate the amount of oxygen available to combine with lighter elements, forming compounds such as CO and H_2O. Express your answer in atoms of available oxygen per 10^6 silicon atoms.

(b) Calculate the mass of rock in amu/silicon atom. Augment your result by 10% to approximately account for the less abundant rock-forming elements not included in your calculation.

(c) Assuming that all of the carbon is in CO, calculate the mass of H_2O-ice (in amu/silicon atom), the H_2O ice/rock (mass) ratio, the $H_2O + CO$ ice/rock ratio, and the $H_2O + CO + N_2$ ice/rock ratio.

(d) Assuming that all of the carbon is in CH_4, calculate the mass of H_2O-ice, the H_2O ice/rock ratio, the $H_2O + CH_4$ ice/rock ratio, and the $H_2O + CH_4 + NH_3$ ice/rock ratio.

(e) Repeat the above calculations assuming that the abundance of O is 10% greater than listed in Table 8.1. (This is within the uncertainty to which Solar System abundances are known, and probably significantly less than the differences between various protoplanetary disks.)

13.9.E (a) Calculate the amount of gas that a particle R cm in radius orbiting at 1 AU from a 1 $M_\odot$ star passes through (collides with) during one year. You may assume that the density of the protoplanetary disk is 10^{-9} g cm^{-3} and $\eta = 5 \times 10^{-3}$.

(b) Assuming a particle density of 3 g cm^{-3}, calculate the radius of a particle that passes through its own mass of gas during one orbit.

13.10.E Estimate the masses and radii of planetesimals formed via gravitational instabilities in a quiescent (nonturbulent) protoplanetary disk orbiting a 1 $M_\odot$ star. Assume that the surface mass density of solid material in the disk varies as $\sigma_\rho = 10\, r_{AU}^{-1}$ g cm^{-2}. Perform your calculations at:

(a) 1 AU.

(b) 5 AU.

(c) Repeat your calculation at 5 AU with a surface density twice as large (to account for the condensation of water-ice).

13.11.**E** Derive the two-body gravitational accretion cross-section of a planetary embryo of radius R and mass M for small planetesimals whose velocity at ∞ relative to the embryo is v. (Hint: First determine the maximum unperturbed impact parameter of a planetesimal which collides with a planetary embryo. Note that the periapsis of this planetesimal's orbit is at a distance R from the planetary embryo's center. Use conservation of angular momentum and energy.)

13.12.**E** (a) Calculate the rate of growth, dR/dt, of a planetary embryo of radius $R = 4000$ km and mass $M = 10^{27}$ g, in a planetesimal disk of surface density $\sigma_\rho = 10$ g cm^{-2}, temperature $T = 300$ K, and velocity dispersion $v = 1$ km s^{-1}, at a distance of 2 AU from a star of mass 3 M$_\odot$. You may use the two-body approximation for planetesimal/planetary embryo encounters.

(b) What will halt (or at least severely slow down) the accretion of such a planetary embryo? What will its mass be at this point? (Hint: See Problem 13.16.)

13.13.**E** Calculate the growth time for Neptune assuming *in situ* ordered growth (i.e., not runaway accretion; use $\mathcal{F}_g = 10$) in a minimum-mass nebula. (Hint: Determine the surface density by spreading Neptune's mass over an annulus from 25 to 35 AU.) Is this model realistic? Why, or why not?

13.14.**I** Two asteroids, each of mass 10^{21} g, collide and accrete (i.e., the collision may be assumed to be completely inelastic). Their initial orbits were $a_1 = 2.75$ AU, $e_1 = 0.1$, $i_1 = 10°$; $a_2 = 3.0$ AU, $e_2 = 0$, $i_2 = 0°$.

(a) Calculate the orbital elements of the single body after accretion and the energy dissipated by the collision. (Hint: Convert to Cartesian coordinates and use conservation of momentum.)

(b) In reality, is such a collision likely to result in accretion or disruption? Why?

13.15.**I** Consider a few relatively large planetesimals in a swarm of much smaller bodies. Assume that the densities of all bodies are the same and that the velocity dispersion is comparable to the escape speed of the *small* bodies.

(a) Show that the cross-sections (and the accretion rates) of the large planetesimals are proportional to the fourth power of their radii.

(b) Use this result to demonstrate that the largest planetesimal doubles in mass the fastest and thus 'runs away' from the rest of the distribution of bodies.

13.16.**E** Equation (13.28) gives the isolation mass for runaway growth of a planet around a 1 M$_\odot$ star as $M_i \approx 10^{24} \left(r_{AU}^2 \sigma_\rho\right)^{3/2}$ grams. Generalize this formula to stars of arbitrary mass.

13.17.**I** Compare a hypothetical planetary system that formed in a disk with the same size as the solar nebula but only half the surface mass density to our own Solar System. Assume that the star's mass is 1 M$_\odot$ and that it does not have any stellar companions. Concentrate on the final number, sizes, and spacings of the planets. Explain your reasoning. Quote formulae and be quantitative where possible.

13.18.**I** A planet of mass M and radius R initially spins in the prograde direction with zero obliquity and rotation period P_{rot}. It is impacted nearly tangentially at its north pole by a body of mass m, whose velocity prior to encounter was small compared to the escape speed from the planet's surface.

(a) Derive an expression for the planet's spin period and obliquity after the impact. You may assume that the projectile was entirely absorbed.

(b) Numerically evaluate your result for $M = $ M$_\oplus$, $R = $ R$_\oplus$, $P_{rot} = 10^5$ sceonds, $m = 0.02$ M$_\oplus$. (Of course, a truly tangential impactor is likely to 'skip off' rather than be absorbed, but even for a trajectory only $\sim 10°$ from the horizontal, most ejecta can be captured at the velocities considered here. The case of a polar impactor is also a singular extremum, but both of these effects together only add a factor of a few to the angular momentum provided by a given mass impacting with random geometry, and they make the algebra much easier.)

13.19.**E** Consider a simple model for the accretion of the Earth. Suppose the radius of proto-Earth increases linearly with time from some starting point until accretion ends t_{acc} years later, i.e., $R(t) = (t/t_{acc})$R$_\oplus$. Ignore the insulating effects of large impacts (which bury hot ejecta) and any possible atmosphere (which can prevent the surface from radiating freely to space). For further simplicity neglect compressibility, heat conduction, and internal heat sources, and set the emissivity $\epsilon = 1$. The energy balance (per unit area) at the surface of the growing Earth is then quantified by equation (6.27).

(a) Describe physically each of the three terms in equation (6.27).

(b) Assume the accreting material has an initial temperature $T_0 = 300$ K, and that the solid particles have an average density $\rho = 4.5$ g cm^{-3} and heat capacity $c_p = 10^7$ erg g^{-1} K^{-1}. Find the approximate temperature of the planet as a function of radius assuming $dR/dt = $ constant, for $t_{acc} = 10^8$ yr and for $t_{acc} = 10^6$ yr.

(c) Find t_{acc} so that $T(R_\oplus) = 2000$ K at the end of accretion.

13.20.I A serious omission from equation (6.27) is the lack of a term involving heat transport within the planet. Convection is suppressed, as the outer layers are warmer than the inner ones in our simple model. (If we consider the possibility of variable accretion rates, such as the impact of a single large body allowing heat to be buried below the surface, then temperature may decrease sufficiently rapidly with radius to allow convection at some locations. However, rapid convection can only occur in a fluid, and if a planet melts, the thermal effects of differentiation must also be included. Solid-state convection can occur, albeit at a slow rate, in material slightly below its melting point, bringing heat out sufficiently rapidly to prevent melting, especially in icy satellites. In any case, we shall ignore convection here.) Radiation within a solid planet is negligible compared to conduction, as the mean free path of a photon is extremely small. Conduction can be included by the addition of a conductivity term to the right-hand side of equation (6.27). However, this requires knowing the temperature as a function of both position and time, because conduction changes temperatures below the surface. A simpler approach is to examine the infinite conductivity limit instead. In this limit, the body is isothermal, so, as before, temperature is only a function of time. The last term in equation (6.27) is thus replaced by

$$\frac{1}{4\pi R^2}\frac{d}{dt}\left(\frac{4\pi}{3}\rho R^3 c_p (T - T_n)\right). \quad (13.32)$$

(a) By taking the derivative specified in expression (13.32), derive an equation identical to equation (6.27) except for one extra term on the right-hand side.

(b) Describe, qualitatively, the effects of this new term on the planet's surface temperature.

(c) Find, approximately, the planet's temperature at the end of accretion if $t_{acc} = 10^8$ yr and if $t_{acc} = 10^6$ yr, assuming infinite conductivity.

13.21.I Find the initial temperature profile of the Earth, assuming that it was homogeneous and it accreted so fast (or that large impacts buried the heat so deep) that radiation losses were negligible. Do this both for the zero and infinite conductivity cases.

13.22.E Calculate the rise in temperature if the Earth differentiated from an initially homogeneous density distribution to a configuration in which one third of the planet's mass was contained in a core whose density was twice that of the surrounding mantle. You may assume infinite conductivity.

13.23.I Repeat the four previous problems for asteroids of radius 50 km and 500 km.

13.24.I Most of the heating by radioactive decay in our planetary system at the present epoch is due to decay of four isotopes, one of potassium, one of thorium, and two of uranium. Chondritic elemental abundances are listed in Table 8.1. Isotopic fractions and decay properties are given in the *CRC Handbook*.

(a) What are these four major energy-producing isotopes? What energy is released per atom decayed? What energy is released per gram decayed? What is the *rate* of energy produced by one gram of the pure isotope? What is the rate released per gram of the element in its naturally occurring isotopic ratio? Note: Some of these isotopes decay into other isotopes with short half-lives (e.g., radon). The decay chain must be followed until a stable (or very long-lived) isotope is reached, adding the energy contribution of each decay along the path.

(b) What is the rate of heat production per gram of chondritic meteorite (or, equivalently, per gram of the Earth as a whole, neglecting the fact that the volatile element potassium is less abundant in the Earth than in CI chondrules) from each of these sources?

(c) What was the heat production rate from each of these sources 4.56×10^9 years ago?

(d) There are very many radioactive isotopes known. What characteristics do these isotopes share which make them by far the most important? (Hint: There are two very important characteristics shared by all four, and one other by three of the four.)

13.25.E (a) How long would it take radioactive decay at the early Solar System rate calculated in the previous problem to generate enough heat to melt a rock of chondritic composition, assuming an initial temperature of 300 K and no loss of energy from the system?

(b) How long would it take radioactive decay to generate as much energy as the gravitational potential energy obtained from accretion for an asteroid of radius 500 km? How long for an asteroid 50 km in radius?

13.26.I The radioactive isotope ^{26}Al is formed by a variety of nucleosynthetic processes and has been observed in interstellar space. It has a half-life of 7.2×10^5 years; thus none remains on Earth from the origin of the Solar System.

(a) How many *grams* of pure ^{26}Al would have had to have been present 4.56×10^9 years ago in order for one *atom* to likely be present today? Evidence for extinct ^{26}Al exists in some primitive meteorites. As isotopes are chemically (almost) indistinguishable, isotopic ratios are nearly uniform throughout the Solar System. The major exceptions to this rule are deuterium, which, being twice as massive as hydrogen, tends to preferentially occupy positions within heavier molecules, and atmospheric gases, for which the lighter isotopes escape more easily. Other isotopic variations are caused by radioactive decay. The decay product of ^{26}Al is ^{26}Mg. Excess amounts of ^{26}Mg (compared to other Mg isotopes) have been found in some aluminum-rich meteoritic inclusions; these detections imply that the inclusions condensed with 'live' ^{26}Al.

(b) What ratio of ^{26}Al/^{27}Al would have required for the heat produced by ^{26}Al decay to have equaled that produced by the four radioactive isotopes mentioned in Problem 13.24 during the epoch of planetary formation 4.56×10^9 years ago?

(c) What ratio of ^{26}Al/^{27}Al would have been required for the heat produced by ^{26}Al decay to have been enough to melt chondritic rock (assuming no loss of heat)? How long would it have taken for 90% of this energy to have been released?

13.27.I (a) Estimate the radius of the smallest icy satellite which could have melted as a result of accretional heating. For your calculation, you may assume (1) this moon is pure water-ice, (2) the specific heat of ice is 2×10^7 erg g^{-1} K^{-1} and the latent heat is 10^9 erg g^{-1}, and (3) accretion was rapid (or large impacts buried the heat released) and random velocities of accreting bodies were small.

(b) How well do your results agree with the observed 'boundary' between spherical and nonspherical moons? What other heat sources may have been important? How could the energy requirement be substantially reduced?

13.28.E Describe and sketch the temperature profiles you would expect after the accretion of Mars due to:
(a) Accretion heating only.
(b) Radioactive heating only.
(c) Suppose Mars had accreted very slowly ($>10^8$ years) from tiny planetesimals devoid of radioactive material. Would you expect Mars to be differentiated? Why or why not?

13.29.I From the results of the calculations described in the previous ten problems, what can you conclude regarding heating, melting, and differentiation of asteroids and terrestrial planets?

13.30.E (a) Compute the surface density of solids in a minimum-mass circumjovian protosatellite disk by spreading the masses of Jupiter's four large moons over a region comparable to their current orbits.

(b) Determine the growth time of Io and Callisto using equation (13.26) with $\mathcal{F}_g = 2$.

Appendix A: List of Symbols Used

a	semimajor axis of an orbit
a_{AU}	semimajor axis of planetary orbit in AU
A	surface area
$\mathcal{A}$	area enclosed by orbit; cross-sectional area
A_0	geometric albedo (head-on reflectance)
A_b	Bond albedo
A_v	albedo at visual wavelengths
A_ν	albedo at frequency ν
$A_{0,\nu}$	geometric albedo at frequency ν
A_{rdr}	radar albedo
b	impact parameter
b_m	semiminor axis of an orbit
B	ring opening angle
$B, \mathbf{B}$	magnetic field strength and vector
B, B_ν	brightness, at frequency ν
B_e	magnetic field strength at the magnetic equator
B_o	surface magnetic field strength
B_{oc}	angle between occulted signal and ring plane
B_{sw}	magnetic field strength in solar wind
c	speed of light in vacuum
c_s	speed of sound
c_v	velocity dispersion
c_A	Alfvén velocity
c_P	specific heat at constant pressure
c_V	specific heat at constant volume
C	a constant
C_D	drag coefficient
C_H	heat transfer coefficient
C_J	Jacobi's constant
C_P	thermal heat capacity (or molecular heat) at constant pressure
C_T	Tisserand parameter
C_V	thermal heat capacity (or molecular heat) at constant volume
C_{mn}	harmonic coefficients
D	diameter

D_i	molecular diffusion coefficient of species i
D_{st}	worldwide average of geomagnetic disturbances
$D_{\mathcal{LL}}$	radial diffusion coefficient
e	(generalized) eccentricity
e, e^-	electron
e^+	positron
E	total energy
$E, \mathbf{E}$	electric field strength and vector
E_G	gravitational potential energy
E_K	kinetic energy
E_{rot}	kinetic energy of rotation
E_{sw}	electric field due to $\mathbf{v}_{sw} \times \mathbf{B}_{sw}$ in solar wind
$\varepsilon_e, \varepsilon_v$	enhancement factor; evaporative; loading parameter
EW	equivalent width
f	true anomaly (angle between planet's periapse and instantaneous position)
f_p	phase space density
f_C	Coriolis parameter
f_{osc}	oscillator strength
$\mathbf{F}$	force
$\mathbf{F}_c$	centripetal force
F_d	amplitude of the driving force
F_f	amplitude of driving force
$\mathbf{F}_g$	force of gravity
$\mathbf{F}_{g,eff}$	effective gravitational force
$\mathbf{F}_D$	drag force
$\mathbf{F}_L$	Lorentz force
$\mathbf{F}_T$	tidal force
$\mathbf{F}_Y$	Yarkovsky force
$\mathbf{F}_{rad}$	radiation force
$\mathcal{F}$	flux
$\mathcal{F}_e$	enhancement factor
$\mathcal{F}_g$	gravitational enhancement factor
$\mathcal{F}_\nu$	flux density at frequency ν
$\mathcal{F}_\odot$	solar constant

g, g_i	g-factor or emission rate	l_D	Rossby deformation radius
g_B	Bouguer anomaly	L_e	electrical skin depth
g_n	acceleration due to centripetal force	L_s	latent heat of sublimation/condensation
g_p	gravitational acceleration	L_s	Mars solar longitude
g_r	radar backscatter gain	L_T	thermal skin depth
g_{fa}	free-air gravity anomaly	$\mathcal{L}_\odot$	solar luminosity
g_{hg}	asymmetry parameter in Henyey–Greenstein phase function	$\mathcal{L}_\star$	stellar luminosity
		$\mathcal{L}$	luminosity
g_{eff}	effective gravitational acceleration	L	McIlwain's parameter
G	gravitational constant	m	mass
G_t	gain radar transmitter	m_v	(visual) apparent magnitude
h	Planck's constant	m_H	mass of hydrogen atom
$\hbar$	normalized Planck's constant, $h/(2\pi)$	m_{amu}	mass of an atomic mass unit
h	vertical scale length	m_{gm}	mass of one gram-mole
h_i	spacing between isentropes	M	mass
h_p	polarimetric slope	M_p	mass of planet
h_{cp}	height of crater peak	M_s	total mass of bodies
H	scale height	M_v	absolute magnitude (visual wavelengths)
H	enthalpy	M_J	Jeans mass
H_{10}	absolute magnitude	$M_\star$	mass of star
H_z	Gaussian scale height	$M_\odot$	solar mass
$\mathcal{H}$	heating rate	$M_\oplus$	Earth mass
$\mathcal{H}_\nu$	Eddington flux	$\mathcal{M}_o$	magnetosonic Mach number
i	inclination angle	$\mathcal{M}_B$	magnetic dipole moment
I, I_{jk}	moment of inertia (along axes j, k)	$\mathcal{M}_R$	Richter magnitude
I_B	integral of $J_B/(2mv)$	n	neutron
I_ν	specific intensity	n	energy level of an electron
j	differential energy flux of particles	n	index of refraction
j_ν	mass emission coefficient	n	mean angular velocity of body in orbit
j_{ν_0}	mass emission coefficient at line center	n_o	Loschmidt's number (Table C.3)
J	photodissociation rate	n_{po}	polytropic index
$\mathbf{J}$	electric current	n_s	number of structural sites in a mineral
J, J_ν	mean intensity, at frequency ν	N	number density of particles (cm^{-3})
J_i	action variable	N	Brunt–Väisälä (buoyancy) frequency
J_n	gravitational moments	N_c	column density (cm^{-2})
$\mathbf{J}_u$	energy flux vector	N_A	Avogadro's number (Table C.3)
J_B	second adiabatic invariant	$\mathcal{N}_u$	Nusselt number
k	Boltzmann constant	$\mathcal{O}$	order
$\mathbf{k}$	wave vector	$\mathbf{p}$	momentum
k_d	diffusivity	p	as subscript: polarization, planet, particle
k_T	tidal Love number	p, p^+	proton
k_{ri}	chemical reaction rate for reaction i	p_r	momentum due to radiation pressure
K_m	incompressibility modulus	p_{sw}	momentum of solar wind motion
K_{po}	polytropic constant	P	pressure
K_T	thermal conductivity	P_c	circular polarization
$\mathcal{K}$	eddy diffusion coefficient	P_n	Legendre polynomials
$\mathcal{K}_\nu$	the $\mathcal{K}$-integral	P_L	linear polarization
ℓ	characteristic length or depth scale	P_{orb}	orbital period
ℓ_{fp}	mean free path	P_{rot}	rotation period
$L, \mathbf{L}$	angular momentum magnitude and vector	P_{yr}	orbital period in years
$L_1–L_5$	Lagrangian (equilibrium) points	$\mathcal{P}$	total power (flux integrated over surface area)

q	electric charge	$\mathcal{R}_{cp}$	distance of cometopause from nucleus
q	pericentric separation	$\mathcal{R}_{mp}$	distance of magnetopause
q_r	rotation parameter	$\mathfrak{R}_a$	Rayleigh number
q_T	tidal coefficient	$\mathfrak{R}_e$	Reynolds number
$q_{ph,\lambda}$	phase integral at wavelength λ	$\mathfrak{R}_m$	magnetic Reynolds number
Q	amount of heat	$\mathfrak{R}_o$	Rossby number
Q_J	Joule heating	S	entropy
Q_T	Toomre's stability parameter	S_ν	source function
Q_{pr}	radiation pressure coefficient	t	time
$\mathcal{Q}$	gas production rate	t_c	time between collisions
$r, \mathbf{r}$	distance, separation	t_d	diffusion timescale
r_c	corotational radius	t_d	Ohmic dissipation timescale (§7.1)
r_e	equatorial crossing distance of field line	t_m	mean lifetime
r_g	guiding center	t_ℓ	lifetime
r_v	location of vertical resonance	$t_{1/2}$	half-life of nuclide
r_L	location of Lindblad (horizontal) resonance	t_{cf}	crater formation time
r_{AU}	distance in AU	t_{damp}	damping timescale
r_{Bohr}	Bohr radius	t_{ff}	free-fall timescale
r_Δ	distance from observer (or Earth)	t_{KH}	Kelvin–Helmholtz timescale
$r_\odot$	heliocentric distance	t_{pr}	decay time due to Poynting–Robertson drag
$r_{\odot AU}$	heliocentric distance (AU)	t_{rx}	relaxation time
$r_{\Delta AU}$	distance from observer (or Earth) (AU)	t_ϖ	time of periapse passage
r_{CM}	distance from center of mass	$\tan \Delta$	loss tangent, ϵ_i/ϵ_r
r_{cr}	critical radius	T	temperature
R	radius of object	T_a	atmospheric temperature
R_λ	spectral resolution, $\lambda/\Delta\lambda$	T_b	brightness temperature
R_0	Fresnel reflection coefficient	T_e	effective temperature
R_c	field line curvature radius	T_m	melting temperature
R_d	dust grain radius	T_s	surface temperature
R_e	equatorial radius	T_g	magnitude torque
R_p	polar radius	T_{cr}	critical temperature
R_s	distance of closest approach	T_{eq}	equilibrium temperature
R_{CF}	radius of body from center of figure	T_{tr}	triple point
R_{CM}	radius of body from center of mass	u_a	annihilation rate
R_E	radius of Einstein ring	u_ν	radiation density
R_H	radius of Hill sphere	u_{ij}	deformation or displacement component
R_L	gyro, cyclotron, Larmor radius	U	total energy
$R_\oplus$	radius of Earth	$v, \mathbf{v}$	velocity magnitude and vector
$R_\star$	radius of star	v'	velocity in rotating coordinate frame
R_{gas}	universal gas constant	v_c	circular orbit velocity
R_{Sch}	Schwarzschild radius	v_e	escape velocity
$R_\parallel$	Fresnel reflection coefficient linearly polarized in the plane of incidence	v_g	wave group velocity
		v_h	horizontal wind speed
$R_\perp$	Fresnel reflection coefficient linearly polarized normal to the plane of incidence	v_i	impact velocity
		v_o	thermal velocity
$\mathcal{R}$	disturbing function	v_r	radial component of the velocity
$\mathcal{R}$	Rydberg's constant	v_P	P wave velocity
$\mathcal{R}_c$	collisional radius	v_S	S wave velocity
$\mathcal{R}_d$	scale length of daughter molecules	v_θ	tangential component of the velocity
$\mathcal{R}_p$	scale length of parent molecules	v_{ph}	wave phase velocity
$\mathcal{R}_{bs}$	distance of bow shock from planet/comet	v_{sw}	solar wind velocity

v_∞	terminal velocity	θ	angle between line of sight and normal to surface
V	volume		
$V(1,0)$	visual equivalent magnitude at 1 AU and zero phase angle	θ	colatitude
		Θ	potential temperature
X_i	fractional concentration of constituent i	θ_B	magnetic axis tilt angle
Y_e	fraction of ejecta escaping	κ	epicyclic (radial) frequency
Y_i	ratio of ejecta mass to impactor mass	κ_ν	mass absorption coefficient
z_{ex}	altitude exobase	κ_{ν_0}	mass absorption coefficient at line center
Z	atomic number		
Z_p	partition function	λ	wavelength
		λ_D	Debye length
x, y, z	Cartesian coordinate axes	λ_m	mean longitude
r, ϕ, θ	spherical coordinate system	λ_T	threshold wavelength for photodestruction
u, v, w	wind velocities along the x, y, z axes	λ_{esc}	escape parameter
		λ_{np}	longitude of magnetic north pole
α	pitch angle of a particle	λ_{III}	longitude based on the rotation period of Jupiter's magnetic field
α_e	particle's pitch angle at the magnetic equator		
		Λ_2	response coefficient
α_i	thermal diffusion parameter for constituent i	μ	frequency of vertical oscillation
α_l	particle's loss cone	μ_a	molecular mass in amu
α_R	Rosseland mean absorption coefficient	μ_b	magnetic moment, μ_B/γ_r
α_v	viscosity parameter	μ_c	circular polarization ratio
α_ν	mass extinction coefficient	μ_i	chemical potential of species i, or partial mole free energy
β	ratio of radiation to gravitational force, F_{rad}/F_g		
		μ_r	reduced mass
β_{cp}	ratio of corpuscular to radiation drag	μ_B	first adiabatic invariant
δ_{jk}	Kronecker delta (along axes j, k)	μ_θ	$\equiv \cos\theta$
γ	photon	μ_{rg}	rigidity modulus
γ	ratio of specific heats, C_P/C_V	ν	frequency
γ_c	Lyapunov exponent	ν_c	critical frequency
γ_c^{-1}	Lyapunov timescale	ν_e	collisional frequency for electrons
γ_r	relativistic correction factor, $\left(1 - (v^2/c^2)\right)^{-1/2}$	ν_i	collisional frequency for ions
		ν_e	electron neutrino
γ_T	thermal inertia	$\overline{\nu}_e$	electron antineutrino
δg_T	terrain correction factor	ν_m	molecular viscosity
Δ	bow shock thickness	ν_v	kinematic viscosity
Δh_g	geoid height anomaly	$\nu_v \rho$	dynamic viscosity
$\Delta\Phi_d$	deformation potential	ρ	density
ϵ	flattening (geometric oblateness) $((R_e - R_p)/R_e)$	ρ_d	density of dust grain
		ρ_g	gas density
ϵ, ϵ_ν	emissivity, at frequency ν	ρ_p	density of (proto)planet
ϵ_i	imaginary part of the dielectric constant	ρ_s	volume mass density of swarm planetesimals
ϵ_r	real part of the dielectric constant	$\rho_\star$	density of star
ϵ_{ij}	strain	σ	Stefan–Boltzmann constant
ϵ_{ir}	infrared emissivity	σ_c	Cowling conductivity
ζ	exponent in power-law distribution	σ_h	Hall conductivity
$\eta(r), \eta_i$	nongravitational force parameters	σ_o	electrical conductivity
$\eta_\nu(\phi)$	beaming factor at frequency ν, phase angle ϕ	σ_p	Pederson conductivity
		σ_x	(molecular) cross-section

σ_ν	mass scattering coefficient	ω_{pi}	ion plasma frequency
σ_ρ	surface mass density	ω_B	gyro, cyclotron, Larmor frequency
σ_{xr}	radar cross-section	ω_{Be}	electron cyclotron frequency
τ	optical depth	ω_{Bi}	ion cyclotron frequency
τ_{sl}	optical depth along a path slanted relative to ring normal	ω_{rot}	spin angular velocity
		ω_{LHR}	lower hybrid resonance frequency
ϕ	solar phase (Sun–target–observer) angle	ω_{UHR}	upper hybrid resonance frequency
ϕ	longitude, azimuth	Ω	longitude of ascending node
ϕ_{sc}	scattering angle (seen from photon; $\phi = 180° - \phi_{sc}$)	Ω_s	solid angle
		ϖ	longitude of periapse
Φ_c	centrifugal potential	ϖ_v	vorticity
Φ_g	gravitational potential	ϖ_ν	single scattering albedo at frequency ν
Φ_i	upward particle flux in an atmosphere	ϖ_{pv}	potential vorticity
Φ_ℓ	limiting flux		
Φ_B	magnetic flux	☉	Sun
Φ_J	Jeans escape rate	☿	Mercury
Φ_T	tidal potential	♀	Venus
Φ_V	electric potential	⊕	Earth
Φ_ν	line shape	☽	Moon
ψ	obliquity of a body (angle between rotation axis and orbit pole)	♂	Mars
		♃	Jupiter
ω	argument of periapse	♄	Saturn
ω_e	cyclotron frequency (electrons)	♅	Uranus
ω_f	forcing frequency	♆	Neptune
ω_i	cyclotron frequency (ions)	♇	Pluto
ω_o	frequency of oscillator, wave frequency		
ω_p	plasma frequency		movie
ω_{pe}	electron plasma frequency		WebColor

Appendix B: Acronyms Used

ACR	Anomalous Cosmic Rays	DISR	Descent Imager/Spectral Radiometer
AGU	American Geophysical Union	DS2	Dark Spot 2 (on Neptune)
AKR	Auroral Kilometric Radiation	EC	Ecliptic Comets
ALH	ALlen Hills, meteorite recovery area in Antarctica	EH, EL	Enstatite chondrite meteorites (with High and Low iron abundances)
AOA	Amoeboid Olivine Aggregate	EL	Equilibrium Level
AU	Astronomical Unit	ENA	Energetic Neutral Atom
BCCA	Ballistic Cluster–Cluster Agglomeration	ESA	European Space Agency
		ESO	European Southern Observatory
BIF	Banded Iron Formations	EUV	Extreme UltraViolet wavelengths
bKOM	broadband KiloMetric radiation	EUVE	Extreme UltraViolet Explorer
BLG	BuLGe (thick central portion of the Milky Way galaxy)	FDS	Flight Data System
BPCA	Ballistic Particle–Cluster Agglomeration	FUSE	Far Utraviolet Spectroscopic Explorer
		FWHM	Full Width at Half Maximum
CA	Closest Approach	GCM	General Circulation Model
CAI	Calcium–Aluminum Inclusion (found in chondritic meteorites)	GCMS	Gas Chromatograph Mass Spectrometer (on the Huygens probe)
CAPE	Convective Available Potential Energy	GDS	Great Dark Spot (on Neptune)
CCD	Charge Coupling Device	GEMS	Glass with Embedded Metal and Sulfides
CER	Corotation Eccentricity Resonance	GRS	Great Red Spot (on Jupiter)
CI, CM, CO, CV, CR, CH, CB, CK	types of Carbonaceous chondrite meteorites	H–H	Herbig–Haro objects
CICLOPS	Cassini Imaging Central Laboratory for OPerationS	H–R	Hertzsprung–Russell (color–luminosity diagram for stars)
CIRS	Composite InfraRed Spectrometer	HD	Henry Draper (who assembled an important star catalog)
CME	Coronal Mass Ejection	HED	Howardite–Eucrite–Diogenite (achondrite meteorite types from asteroid 4 Vesta)
CNO	Carbon Nitrogen Oxygen (hydrogen fusion catalyst) cycle		
CP	Chondritic Porous (type of interplanetary dust particle)	HF	Higher Frequency emissions
		HFC	Halley Family Comets
CSO	Caltech Submillimeter Observatory	HiRISE	High Resolution Imaging Science Experiment (on MRO)
DAM	DecAMetric radiation		
DIM	DecIMetric radiation		

HOM	HectOMetric radiation
HRSC	High Resolution Stereo Camera (on Mars Express)
HST	Hubble Space Telescope
IAU	International Astronomical Union
ICE	International Cometary Explorer
ICME	Interplanetary Coronal Mass Ejection
IDP	Interplanetary Dust Particle
ILR	Inner Lindblad Resonance
IMF	Interplanetary Magnetic Field
IMP-1	Interplanetary Monitoring Platform
INMS	Ion and Neutral Mass Spectrometer (on Cassini)
IR	InfraRed
IRAS	InfraRed Astronomical Satellite
IRTF	InfraRed Telescope Facility
ISM	InterStellar Medium
ISO	Infrared Space Observatory
ISRO	Indian Space Research Organisation
ITCZ	InterTropical Convergence Zone
IVR	Inner Vertical Resonance
JAXA	Japan Aerospace eXploration Agency
JFC	Jupiter Family Comets
JPL	Jet Propulsion Laboratory
K–T	Cretacious–Tertiary
KAO	Kuiper Airborne Observatory
KBO	Kuiper Belt Object
KREEP	Potassium (K), Rare Earth Elements, Phosphorus (P)
LCL	Lifting Condensation Level
LCROSS	Lunar CRater Observation and Sensing Satellite
LFC	Level of Free Convection
LH	Left-Hand sense (of circular polarization)
LHR	Lower Hybrid Resonance
LINEAR	LIncoln laboratory Near-Earth Asteroid Research project
LL	Low-iron, Low-metal chondrite meteorites
LORRI	LOng Range Reconnaissance Imager
LP	Long-Period comets ($P_{orb} > 200$ yr)
LRO	Lunar Reconnaissance Orbiter
LT	Local Time (on planet)
LTE	Local Thermodynamic Equilibrium
LWS	Long Wavelength Spectrometer (on Keck)
M^3	Moon Mineralogy Mapper
2MASS	2 Micron All-Sky Survey

MBA	Main Belt Asteroid
MC	Mars Crossing
MER	Mars Exploration Rovers
MESSENGER	MErcury Surface, Space ENvironment, GEochemistry, and Ranging spacecraft
MGS	Mars Global Surveyor
MIMI	Magnetospheric IMaging Instrument (on the Cassini spacecraft)
mini-TES	mini Thermal Emission Spectrometer
MOA	Microlensing Observations in Astrophysics
MOC	Mars Orbiter Camera
MOLA	Mars Orbiter Laser Altimetry
MRO	Mars Reconnaissance Orbiter
NASA	National Aeronautics and Space Administration (of the United States)
NEA	Near-Earth Asteroid
NEAR	Near-Earth Asteroid Rendezvous spacecraft
NEAT	Near-Earth Asteroid Tracking project
NEO	Near-Earth Object
NIMS	Near Infrared Mapping Spectrometer (on the Galileo spacecraft)
nKOM	narrowband KilOMetric radiation
NKR	Neptune Kilometric Radiation
NTC	NonThermal Continuum
NTT	New Technology Telescope
OC	radar echo with 'Opposite sense' Circular polarization
OC	Oort Cloud comets
OGLE	Optical Gravitational Lensing Experiment
OLR	Outer Lindblad Resonance
O mode	Ordinary mode of propagation
OPR	Ortho-to-Para Ratio
OTD	Offset Tilted Dipole
OVR	Outer Vertical Resonance
PA	Position Angle
PAH	Polycyclic Aromatic Hydrocarbons
PIA	Photo Identification Access (JPL)
POM	PolyOxyMethylene (formaldehyde polymer: (-CH_2-O-)n)
PPS	PhotoPolarimeter Subsystem (instrument on the Voyager spacecraft)

pp-	proton–proton (hydrogen fusion) chain
PREM	Preliminary Reference Earth Model
PSR	PulSaR
P waves	Primary, push, or pressure seismic waves
QP	Quasi-Periodic bursts
REE	Rare Earth Elements (elements with atomic numbers 57–70)
RH	Right-Hand sense (of circular polarization)
RKBO	Resonant KBO
ROSAT	Röntgen SATellite
RPWS	Radio and Plasma Wave Science
SAMPEX	Solar Anomalous and Magnetospheric Particle EXplorer
SAO	Smithsonian Astrophysical Observatory
SC	radar echo with 'Same sense' Circular polarization
SCET	SpaceCraft Event Time
SDO	Scattered Disk Object
SED	Saturn Electrostatic Discharges
SETI	Search for ExtraTerrestrial Intelligence
SKR	Saturn's Kilometric Radiation
SL9	Comet D/Shoemaker–Levy 9
SMM	Solar Maximum Mission
SMOW	Standard Mean Ocean Water
SOHO	SOlar and Heliospheric Observatory
SP	Short-Period comets ($P_{orb} < 200$ yr)
STIS	Hubble Space Telescope Imaging Spectrograph
SWAS	Submillimeter Wave Astronomy Satellite
S waves	Secondary, shake, or shear seismic waves

TAOS	Taiwan–America Occultation Survey
TDV	Transit Duration Variation
TES	Thermal Emission Spectrometer
THEMIS	Time History of Events and Macroscale Interactions during Substorms in Earth's magnetosphere (5 spacecraft)
TNO	Trans-Neptunian Object
TRACE	Transition Region And Coronal Explorer
TTV	transit timing variation
UED	Uranus Electrostatic Discharges
UFO	Unidentified Flying Object
UHR	Upper Hybrid Resonance
UKR	Uranus Kilometric Radiation
UV	UltraViolet wavelengths
UVIS	UltraViolet Imaging Spectrograph (on the Cassini spacecraft)
VIMS	Visual and Infrared Mapping Spectrometer (on the Cassini spacecraft)
VIRTIS	Visible and InfraRed Thermal Imaging Spectrometer (on Venus Express)
VLA	Very Large Array radio telescope
VLBI	Very Long Baseline Interferometry
VLF	Very Low Frequency emissions
VLT	Very Large Telescope
WMAP	Wilkinson Microwave Anisotropy Probe satellite
X mode	extra-ordinary mode of propagation
YORP	Yarkovsky–O'Keefe–Radzievskii–Paddack effect

Appendix C: Units and Constants

Table C.1 Prefixes.

Prefix	Value in SI units
y (yocto-)	10^{-24}
z (zepto-)	10^{-21}
a (atto-)	10^{-18}
f (femto-)	10^{-15}
p (pico-)	10^{-12}
n (nano-)	10^{-9}
μ (micro-)	10^{-6}
m (milli-)	10^{-3}
c (centi-)	10^{-2}
d (deci-)	10^{-1}
da (deca-)	10
h (hecto-)	10^{2}
k (kilo-)	10^{3}
M (mega- or million)	10^{6}
G (giga- or billion)	10^{9}
T (tera-)	10^{12}
P (peta-)	10^{15}
E (exa-)	10^{18}
Z (zetta-)	10^{21}
Y (yotta-)	10^{24}

Table C.2 Units.

Symbol	Value in cgs units
Å (angstrom)	10^{-8} cm
μm (micrometer)	10^{-4} cm
m (meter)	100 cm
km (kilometer)	10^{5} cm
L (liter)	10^{3} cm^3
kg (kilogram)	10^{3} g
t (tonne)	10^{6} g
J (joule)	10^{7} erg
eV (electron volt)	1.602×10^{-12} erg
W (watt)	10^{7} erg s^{-1}
N (newton)	10^{5} dyne
atm (atmosphere)	1.013 25 bar
Pa (pascal)	10 dyne cm^{-2}
bar	10^{6} dyne cm^{-2}
Hz (hertz)	1 cycle s^{-1}
Ω (ohm)	1.1126×10^{-12} esu
mho (ohm^{-1})	8.988×10^{11} esu
A (ampere)	2.998×10^{9} esu
γ (gamma)	10^{-5} gauss
T (tesla)	10^{4} gauss
Jy (jansky)	10^{-23} erg cm^{-2} Hz^{-1} s^{-1}

Table C.3 Physical constants.

Symbol	Value in cgs units	Value in SI units	Quantity
c	$2.997\,925 \times 10^{10}$ cm s^{-1}	$2.997\,925 \times 10^{8}$ m s^{-1}	Velocity of light
G	6.674×10^{-8} dyn cm^2 g^{-2}	6.674×10^{-11} m^3 kg^{-1} s^{-2}	Gravitational constant
h	$6.626\,069 \times 10^{-27}$ erg s	$6.626\,069 \times 10^{-34}$ J s	Planck's constant
k	$1.380\,650 \times 10^{-16}$ erg deg^{-1}	$1.380\,650 \times 10^{-23}$ J deg^{-1}	Boltzmann's constant
m_e	$9.109\,382 \times 10^{-28}$ g	$9.109\,382 \times 10^{-31}$ kg	Electron mass
m_p	$1.672\,622 \times 10^{-24}$ g	$1.672\,622 \times 10^{-27}$ kg	Proton mass
m_{amu}	$1.660\,539 \times 10^{-24}$ g	$1.660\,539 \times 10^{-27}$ kg	Atomic mass unit
n_o	2.686×10^{19} cm^{-3}	2.686×10^{25} m^{-3}	Loschmidt's number
N_A	$6.022\,142 \times 10^{23}$ mole^{-1}	$6.022\,142 \times 10^{23}$ mole^{-1}	Avogadro's number
r_{Bohr}	$5.291\,77 \times 10^{-9}$ cm	$5.291\,77 \times 10^{-11}$ m	Bohr radius or atomic unit
R_{gas}	8.3145×10^{7} erg deg^{-1} mole^{-1}	8.3145 J deg^{-1} mole^{-1}	Universal gas constant
$\mathcal{R}$	$1.097\,373 \times 10^{5}$ cm^{-1}	$1.097\,373 \times 10^{7}$ m^{-1}	Rydberg constant
q	4.803×10^{-10} esu	$1.602\,176 \times 10^{-19}$ C	Electron charge
σ	5.6704×10^{-5} erg cm^{-2} deg^{-4} s^{-1}	5.6704×10^{-8} W m^{-2} deg^{-4}	Stefan–Boltzmann constant

Table C.4 Material properties.

Symbol	Value in cgs units	Value in SI units	Quantity
ρ	1.293×10^{-3} g cm^{-3}	1.293 kg m^{-3}	Density of air at STP^a
ν_v	0.134 cm^2 s^{-1}	1.34×10^{-5} m^2 s^{-1}	Kinematic viscosity of air at STP^a
c_P	1.0×10^{7} erg g^{-1} deg^{-1}	1.0×10^{3} J kg^{-1} deg^{-1}	Isobaric specific heat capacity of air at STP^a
c_V	7.19×10^{6} erg g^{-1} deg^{-1}	7.19×10^{2} J kg^{-1} deg^{-1}	Isochoric specific heat capacity of air at STP^a
c_P	1.2×10^{7} erg g^{-1} deg^{-1}	1.2×10^{3} J kg^{-1} deg^{-1}	Typical value for the specific heat of rock
L_v	2.50×10^{10} crg g^{-1}	2.50×10^{6} J kg^{-1}	Specific latent heat of vaporization for water
L_s	2.83×10^{10} erg g^{-1}	2.83×10^{6} J kg^{-1}	Specific latent heat of sublimation for water-ice

^{a} STP: Standard Temperature (273 K) and Pressure (1 bar).

Table C.5 Astronomical constants.

Symbol	Value in cgs units	Value in SI units	Quantity
AU	1.496×10^{13} cm	1.496×10^{11} m	Astronomical unit of distance
ly	9.4605×10^{17} cm	9.4605×10^{15} m	Light-year
pc	3.086×10^{18} cm	3.086×10^{16} m	Parsec
$M_\odot$	1.989×10^{33} g	1.989×10^{30} kg	Solar mass
$R_\odot$	6.96×10^{10} cm	6.96×10^{8} m	Solar radius
$\mathcal{L}_\odot$	3.827×10^{33} erg s^{-1}	3.827×10^{26} J s^{-1}	Solar luminosity
$F_\odot$	1.37×10^{6} erg cm^{-2} s^{-1}	1.37×10^{3} J m^{-2} s^{-1}	Solar constant
$M_\oplus$	5.976×10^{27} g	5.976×10^{24} kg	Earth's mass
$R_\oplus$	6.378×10^{8} cm	6.378×10^{6} m	Earth's equatorial radius
$g_p(eq)$	978 cm s^{-2}	9.78 m s^{-2}	Gravity at sea level on Earth's equator
$g_p(pole)$	983 cm s^{-2}	9.83 m s^{-2}	Gravity at sea level at Earth's poles

Appendix D: Periodic Table of Elements

Key to chart

Atomic number → 97
Atomic mass → (247)
Oxidation states → +3 +4
Symbol → Bk
Name → Berkelium

— Metals
— Non-metals
— Semi-metals
— Artificially prepared elements

* Lanthanide

◆ Actinide

Numbers in parentheses are mass numbers of most stable known isotope of radioactive elements that are rare or not found in nature.

Appendix E: Observing Techniques

In this appendix we briefly summarize aspects of observational planetary science. Techniques specific to extrasolar planets are discussed in §12.2. References to more extensive treatments are provided in the Further Reading section at the end.

E.1 Photometry

Photometry is a photon-counting technique, wherein the brightness of an object is measured. Time series of brightness measurements can be combined into a *photometric lightcurve* that shows, e.g., the variations in a body's brightness as the object rotates around its axis (Fig. 9.4).

Graphs of an asteroid's brightness in reflected sunlight as a function of phase angle, ϕ, usually show an abrupt increase in intensity at $\phi \lesssim 2°$, referred to as the *opposition effect*. The opposition effect for the Moon, shown in Figure E.1, is very large; the intensity increases by ~20% from $\phi \sim 2°$ down to $\phi = 0°$. This is why a 'full Moon' can appear to be much brighter than a nearly full gibbous Moon. Part of the opposition effect can be attributed to the hiding of shadows when the Sun and observer are located in the same direction as seen from the object. Laboratory simulations of this phase angle effect, however, show that this is not the complete story. In any particulate material, multiple reflections diffusely scatter the incoming waves in all directions. At zero phase angle the waves interfere constructively, and the reflected intensity can be amplified considerably. This process is known as the *coherent-backscatter effect*. Both the shadow-hiding and coherent-backscatter contribute to the opposition effect. For the Moon, the coherent-backscatter effect results in the narrow peak near opposition at phase angles $\phi < 2°$, whereas the broader component at $\phi < 20°$ can be explained by the shadow-hiding theory.

E.2 Spectroscopy

Spectroscopy pertains to the dispersion of light as a function of wavelength. Spectra can be used to derive, e.g., the composition of gaseous and solid objects, the temperature and pressure in an atmosphere (§4.3), and the radial velocity of an object via the Doppler shift (§12.2.2). A spectral resolution as low as $R_\lambda \equiv \lambda/\Delta\lambda \approx$ few $\times$ 100 usually suffices for surface reflectance spectroscopy (λ is the wavelength, and $\Delta\lambda$ the spectral resolution), but for atmospheres a minimum of $R_\lambda \approx$ few $\times$ 1000 and often times $R_\lambda \approx$ few $\times$ 10^4–10^5 is desired because spectral lines in gases are generally much narrower than those in solids. Broadband spectra that cover many wavelengths consist of individual data points at different wavelengths; such data are often obtained with standard broadband filters (Table E.1), or narrowband filters at specific wavelengths.

E.3 Radiometry

Irregularities in a body's figure and variations in surface albedo both influence the shapes of lightcurves. The brightness of an object in reflected sunlight is proportional to the product of the body's projected area, A, and its average albedo at frequency v, A_v. The thermal emission from a body varies as $A(1-A_b)$, with A_b the Bond albedo (§3.1.2.1). If the relationship between A_v and A_b is known, a combination of lightcurves at visible wavelengths, where reflected light is measured, and at wavelengths sensitive to a body's thermal emission, enables one to: (1) determine the size and albedo of the body, a technique referred to as *radiometry*, and (2) distinguish amplitude variations caused by the object's figure from those resulting from albedo variations on its surface. Irregularly shaped objects show lightcurves with two maxima and two minima per rotation, whereas albedo variations on a body alone may produce lightcurves that are single-peaked. Lightcurves

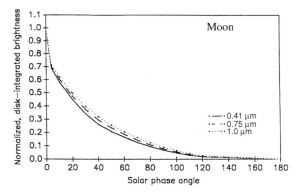

Figure E.1 Brightness of the Moon as a function of solar phase angle, at three different wavelengths. The data at small phase angles ($\phi < 5°$) were derived from Clementine observations. (Buratti *et al.* 1996)

obtained at different viewing angles are used to determine a body's pole position and its sense of rotation (Fig. 9.4). It may take, however, many years to gather data over enough different viewing aspects to determine the spin orientation of a minor planet.

E.4 Polarimetry

The linear polarization of light reflected by a solid surface depends upon the scattering geometry, refractive index, and texture. When unpolarized light, e.g., sunlight, is reflected or scattered off a rough surface, it becomes (partially) linearly polarized. The degree of polarization is given by

$$P_L = \frac{I_\perp - I_\parallel}{I_\perp + I_\parallel}, \tag{E.1}$$

where $I_\perp$ and $I_\parallel$ are the components of the intensity measured perpendicular and parallel to the plane of scattering. The polarization of the Moon (and other bodies covered by relatively dark particulate materials, such as asteroids), is almost constant over the disk, and depends only on phase angle, ϕ, and frequency, ν.

E.5 Occultation Techniques

When one body passes in front of another, the total light received at the telescope is reduced. If the intervening body blocks all of the light from the distant body, the event is referred to as an *occultation*; if only part of the light is blocked, it is called a *transit*. Thus, a total solar eclipse is an occultation, whereas an annular eclipse would be classified as a transit. Lightcurves of stars can reveal occultations and transits. Such events can be used to determine the size of the occulting object if its speed and distance to the telescope are known. The sizes of several asteroids

Table E.1 Filter nomenclature in astronomy.

Band	Central wavelength	FWHM[a]
Visible/IR		
U	357 nm	65 nm
B	436 nm	100 nm
V	537 nm	94 nm
R	644 nm	151 nm
I	805 nm	150 nm
Z	1.0 μm	0.2 μm
J	1.21 μm	0.3 μm
H	1.65 μm	0.3 μm
K	2.2 μm	0.4 μm
L	3.5 μm	0.6 μm
M	4.7 μm	0.5 μm
N	10.5 μm	5.2 μm
Q	20.1 μm	7.8 μm
Radio		
W	0.4 cm	∼0.1 cm
V	0.6 cm	∼0.1 cm
Q	0.7 cm	∼0.1 cm
Ka	1 cm	0.1 cm
K	1.3 cm	∼0.1 cm
Ku	2 cm	∼0.1 cm
X	3.6 cm	∼0.2 cm
C	6.3 cm	∼0.4 cm
S	13 cm	∼3 cm
L	21 cm	∼4 cm
P	90 cm	∼15 cm

[a] FWHM, the full width at half maximum, gives the approximate width of each filter. At radio wavelengths, the FWHM presents the approximate extent of the wavelength range as used by radio observatories.

have been measured this way. Further, this technique led to, e.g., the discovery of rings around Uranus and Neptune (§11.3), and the presence of an atmosphere around Pluto (§4.3.3.3). A variant of this technique, where the object of interest is occulted by another body, has enabled precise astrometry of hot spots on Io (§5.5.5.1). When two objects in a system (two satellites, or the primary and secondary in a binary system) pass in front of and behind each other, the occultations/transits are referred to as *mutual events*. Such mutual events have been used to characterize the Pluto–Charon system (§9.5). Transits of extrasolar planets in front of their host star can be used to derive the size and orbital plane of such a planet. Spectroscopic measurements during such events have led to the detection of atmospheric constituents in some exoplanets (Chapter 12).

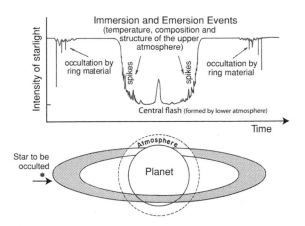

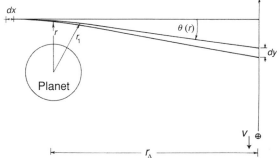

Figure E.2 Lightcurve for a central occultation by a ringed planet. The upper portion shows the light intensity from the star as observed from Earth when the star follows the path as indicated in the lower panel. (Adapted from Elliot 1979)

Figure E.3 Geometry for an occultation of a star or radio source by a planet with an extended atmosphere. The refractive bending angle θ of a ray decreases with increasing distance from the planet. The Earth is moving along the y-axis, as indicated, and the divergence of rays is given by dy. (Adapted from Hunten and Veverka 1976)

Occultations of bright stars by planets, satellites, asteroids, and KBOs are relatively rare, but when these occur one can deduce important information from the light curves, such as (with more than one chord) the size and shape of the occulting body, the optical depth and precise locations of rings surrounding a planet, the presence and extent (scale height) of an atmosphere, as well as its temperature, pressure, and density profile. Stellar occultations have been carried out at UV, visible, and IR wavelengths. Radio occultations have been conducted by transmitting radio signals from spacecraft (such as the Voyagers, Cassini) through, e.g., Saturn's rings, and the atmospheres of the giant planets, Venus, Mars, and Titan.

Figure E.2 shows a schematic for a central occultation by a ringed planet that possesses an atmosphere. Starlight is first dimmed when it passes through the rings. Analysis of these signals yields the optical depth and extent of rings. When the rays traverse the atmosphere, they are refracted as sketched in Figure E.3. The refractive bending angle, θ, is increased for rays passing closer to the planet. Therefore, the rays diverge at Earth. In a stellar occultation, the intensity of starlight is thus reduced both through absorption by atmospheric gases and by refraction of the starlight through the atmosphere. These two effects cannot be separated. In a radio occultation the atmospheric refraction introduces a delay (or shift) in the radio signal's phase (or frequency), which can be measured with high accuracy. Hence the refractive bending angle can be determined independently of the change in intensity. Using this information, the decrease in intensity due to refraction and to absorption by atmospheric gases can be determined separately. The dynamic range of radio occultation measurements is typically of the order of

10^3–10^4:1, in contrast to perhaps 10:1 for stellar occultation measurements.

The bending angle, θ, depends upon the gradient in refractivity normal to the ray's path through the atmosphere, which is a consequence of the gradient in atmospheric density. The intensity received on Earth is decreased by a factor of 2 when the refractive bending angle, θ, is equal to H/r_Δ, with H the scale height (eq. 4.2) and r_Δ the geocentric distance. In an isothermal atmosphere, or if the thermal profile is known (e.g., through inversion of thermal IR spectra), analysis of an occultation lightcurve can thus be used to determine the mean molecular weight of the atmosphere as a function of altitude. Since the mean molecular weight in the atmospheres of the giant planets is mostly determined by H_2 and He, analysis of thermal infrared spectra and radio occultation data have been used to estimate the He abundance in these planets' atmospheres. For radio occultations, the thermal and density profiles can be determined if the composition is known.

When a star passes exactly behind the center of the object, there is a brief *central flash*, formed by the convergence of rays from the entire circumference of the planet's limb. For oblate planets, the detailed structure of the flash provides information on the shape of the atmosphere. Although extinction of starlight in an atmosphere can usually be ignored in occultation experiments, the amplitude of the central flash is reduced by atmospheric extinction. The signals can therefore be used to determine the extinction in the atmosphere.

In stellar occultations and some long-distance radio observations, density variations in the atmosphere, even if only a few percent, give rise to *spikes*, as indicated in

(a)

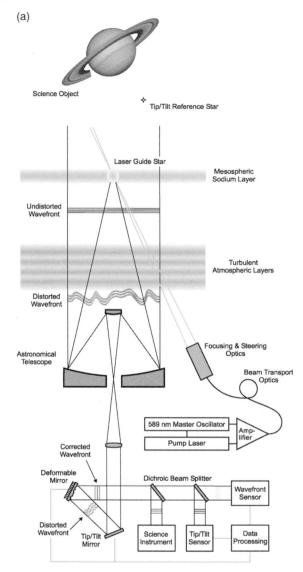

(b)

Figure E.4 (*cont.*)

Figure E.4 (a) A schematic representation of the adaptive optics (AO) technique. The incoming undisturbed wavefront is distorted by atmospheric turbulence. The distortions can be measured by observing a bright guide star near the object of interest. The guide star can be either a natural star or a laser beacon. The laser is capable of producing a guide star over most locations in the sky by inducing a fluorescent glow in sodium, which is naturally present in a thin layer in the mesosphere, at an altitude of ∼90 km. (Courtesy Wolfgang Hackenberg and Andreas Quirrenbach) (b) The distorted wavefront from the guide star, as measured with the wavefront sensor, can be compensated by using a deformable mirror that, at each instant of time, has the same deformation as the incoming wavefront but with only half the amplitude.

Figure E.2. Detection of such spikes can thus be used to analyze the atmospheric structure.

Finally, Fresnel diffraction must be taken into account when analyzing occultation signals caused by sharp edges, such as when the limb of a solid body, narrow rings, or tiny (∼km-sized) bodies (as, e.g., TNOs) pass in front of a star.

E.6 High-Resolution Imaging

Two factors limit the spatial resolution of telescopes on the ground. The first is optical *diffraction*, which limits the resolution (FWHM: full width at half maximum) at a wavelength λ to $1.22 \times \lambda/D$ radians, with D the telescope diameter. For a 10 m telescope at a wavelength of 1 μm, this diffraction limit is $0.026''$. The second limit is *seeing*, the blurring of images caused by turbulence in the Earth's atmosphere. For even the largest ground-based telescopes at the best sites, seeing limits the resolution to $0.4–1''$. The Hubble Space Telescope, 2.4 m in diameter, was launched largely to overcome this limitation, in addition to observing ultraviolet light to which the atmosphere is opaque, and do precise photometry.

Before entering the atmosphere, light from a distant source forms a plane wave. The speed of light varies as the inverse of the refractive index (§7.4.2), and fluctuations in this quantity are essentially proportional to fluctuations in atmospheric temperature. Such fluctuations are common at the interface between different atmospheric layers, where winds produce turbulence. Hence, light passing through different parts of the atmosphere travels at different speeds, which produces a deformation in the originally plane wavefront (Fig. E.4). The wavefront phase fluctuations also depend on wavelength, since the wave vector is inversely proportional to wavelength, $|\mathbf{k}| = 2\pi/\lambda$. The wavefront perturbations are therefore smaller at longer wavelengths, and less detrimental to image quality. The development of speckle imaging and adaptive optics techniques has made it possible to overcome this atmospheric seeing.

E.6.1 Speckle Imaging

Speckle imaging corrects for the effects of atmospheric turbulence in software, via post-processing. At the telescope one takes many exposures, each short enough ($\lesssim$200–300 ms) that atmospheric turbulence does not vary substantially. These images are subsequently coadded after careful alignment.

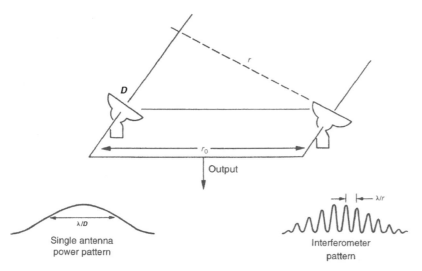

Figure E.5 Top: Geometry of a two-element interferometer. Bottom: Antenna response for a single element of the interferometer (left) and the response of the interferometer (right) to an unresolved radio source. (Adapted from Gulkis and de Pater 2002)

E.6.2 Adaptive Optics

With *Adaptive Optics (AO)*, the effects of seeing are overcome by monitoring the atmospheric distortions in real time and compensating for wavefront errors in the incident beam by means of a deformable mirror (Fig. E.4). This technique makes the wavefront planar again, and thus overcomes the effect of seeing in the atmosphere, so that one can image at the diffraction limit of the telescope. Many images in this book were obtained using AO techniques (e.g., Figs. 4.36b, 5.78, 9.18, 11.18b,d).

E.6.3 Interferometry

The resolution of a telescope can be improved by connecting the outputs of two antennas, separated by a distance r_0, at the input of a receiver. Such a system is called an *interferometer*. Interferometers are common in radio astronomy (e.g., the Very Large Array, VLA), and are sometimes used at infrared wavelengths. The response of an interferometer to an unresolved source is an interference pattern, as sketched in Figure E.5, where the maxima are separated by an angle λ/r radians, with r the baseline length as projected on the sky. As shown in Figure E.5, the single antenna output is essentially modulated on a scale λ/r. This angle is the *resolving power* of the interferometer in the direction of the projected baseline r. If the interferometer is built along the east–west direction, the Earth's rotation causes the baseline as projected onto the sky to trace out an ellipse during the course of one day. The coordinates of this ellipse are generally referred to as the u (east–west on the sky) and v (north–south on the sky) coordinates in the (u, v) plane. The size/shape of the ellipse determines the angular resolution on the source (the larger the size, the higher the resolution).

The VLA consists of a Y-shaped track, with nine antennas along each of the arms. The antennas operate at centimeter wavelengths, and each antenna is connected with each of the others forming a set of interferometers. One can thus gather data from $\frac{27 \times 26}{2} = 351$ individual interferometer pairs, which each trace out their own unique ellipse in the (u, v) plane; in other words, each interferometric pair has its own instantaneous resolution along its projected baseline r. Such an array of antennas can be used to build up an image that shows both the large- and small-scale structure of a radio source. At short spacings the entire object can be 'seen', but details on the planet are washed out due to the low resolution of such baselines. At longer baselines, details on the planet can be distinguished, but the large-scale structure of the object is 'overresolved', and hence is invisible on the image unless short spacing data are included as well. To recover the spatial brightness distribution of an object, one needs to measure both the amplitude and phase of fringes received (called the *complex visibilities*) by many interferometer pairs. This measurement forms the basis of mapping by Fourier synthesis in (radio) astronomy. The response of all the individual interferometer pairs, the visibility data, is then gridded into cells having uniform intervals in the (u, v) plane. This grid of data is then Fourier transformed to give a map of the brightness distribution on the sky. Examples of radio interferometric images are shown in Figures 4.35b,c, 4.37d, and 7.37.

E.7 Radar Observations

In a typical radar experiment, a signal with known properties (intensity, polarization, time/frequency spectrum) is

Appendix E: Observing Techniques

transmitted, usually for a duration equal to the round-trip propagation time, and received by the same antenna for a comparable length of time. In contrast to these *monostatic radar experiments*, there are *bistatic experiments* wherein the signal is transmitted continuously by one telescope and received by a different telescope, or by an array of telescopes, such as the VLA. Bistatic experiments are also carried out by substituting a spacecraft transmitter or receiver for one end of the path.

The transmission frequency is adjusted continuously so that if the ephemeris of the object is well known and the target is a point source, the echo received will be a spike at a particular frequency. Any frequency spreading of the signal can then be attributed to the object's rotation and geometry.

The radar power incident on an object per unit area (erg cm^{-2} s^{-1}) is equal to

$$\mathcal{P}_i = \frac{\mathcal{P}_t G_t}{4\pi r_t^2}, \tag{E.2}$$

with $\mathcal{P}_t$ the transmission power, G_t the *gain* or 'effectiveness' of the transmitting antenna in the direction of interest as compared with that of an isotropic radiator, and r_t the distance from the transmitter to the object. For transmission and reception on the Earth, the total power received from the object (i.e., integrated over frequency), $\mathcal{P}_r$, can be written

$$\mathcal{P}_r = \frac{\mathcal{P}_i \sigma_{xr} A_r}{4\pi r_\Delta^2} = \frac{\sigma_{xr} A_r \mathcal{P}_t G_t}{16\pi^2 r_\Delta^4}, \tag{E.3a}$$

with A_r the receiving antenna's effective aperture, r_Δ the geocentric distance, and σ_{xr} the radar cross-section of the object:

$$\sigma_{xr} = A_{rdr}\pi R^2, \tag{E.3b}$$

where A_{rdr} is the radar albedo and πR^2 the target's projected surface area. To first order, for solid bodies the radar cross-section is equal to the area of an equivalently backscattering dielectric sphere, and $\sigma_{xr}/(4\pi)$ is the backscattered power per steradian per unit of flux incident at the target. Note that for ground-based measurements, the signal travels from Earth to the object, and then from the object to Earth, so that the received power is diminished by the fourth power of the distance. It is therefore important to make radar observations as close to the object as possible, i.e., for ground-based observations when the object is in opposition[1] or inferior conjunction, or for spacecraft during close encounters with the object.

RADAR PROPERTIES

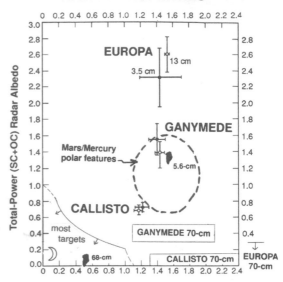

Figure E.6 Radar properties of Europa, Ganymede, and Callisto compared to those of some other targets. The icy Galilean satellites' total power radar albedos do not depend on wavelength between 3.5 and 13 cm, but are much lower at 70 cm. Solid symbols shaped like Greenland indicate properties of that island's percolation zone at 5.6 and 68 cm. The domain of most of the bright features on Mars and Mercury is outlined. (Ostro 2007)

The geometric albedo $A_{0,\nu}$ (at radio frequencies) is related to the radar albedo:

$$A_{0,\nu} = \frac{A_{rdr}}{4}. \tag{E.4}$$

The radar cross-section is thus a measure of the backscatter efficiency of the target and its size. Typical radar cross-sections of solid surfaces are of the order of 10% of the body's projected surface area, although sometimes, as, e.g., for the Galilean satellites, it can exceed unity (Fig. E.6).

The radar albedo is equal to the product of the effective Fresnel coefficient at normal incidence, R_0 (eq. 3.88), and the radar backscatter gain, g_r:

$$A_{rdr} = g_r R_0^2. \tag{E.5}$$

The radar backscatter gain is usually close to unity, in which case the radar albedo can be related directly to the dielectric constant, ϵ_r, of the material probed, and hence to the composition and compactness of the surface layers.

[1] Astronomers refer to a body being at *opposition* when the Sun, Earth, and a body outside Earth's orbit line up at the same side of the Sun. It is called *conjunction* when the body is along the Sun–Earth line on the opposite side of the Sun. For a body in

an orbit around the Sun that is inside the Earth's orbit, the body is in *inferior conjunction* when between the Sun and Earth, and *superior conjunction* at the other side of the Sun.

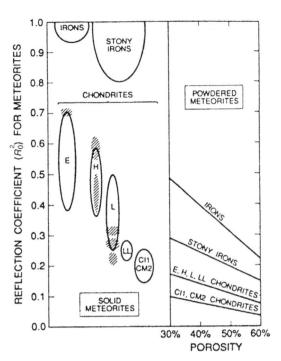

Figure E.7 Reflection coefficient for meteorite types (left) and powdered meteorites (right) as a function of porosity. (Ostro *et al.* 1991)

Laboratory measurements show that most lunar and terrestrial rock powders follow the *Rayleigh mixing formula*:

$$\frac{1}{\rho} \left(\frac{\epsilon_r - 1}{\epsilon_r + 2} \right) = \frac{1}{\rho_o} \left(\frac{\epsilon_{ro} - 1}{\epsilon_{ro} + 2} \right), \tag{E.6}$$

where ρ and ϵ_r are the density and dielectric constant of the rock powder, and ρ_o and ϵ_{ro} those of the parent rock. Typical values for terrestrial and lunar rocks are $\rho_o = 2.8$ g cm^{-3} and $\epsilon_{ro} = 6-8$. Such measurements for solid and powdered meteorites (Fig. E.7) can be used in the analysis of radar echoes from asteroids (§9.2).

If the transmitted radar signal is circularly polarized and the object is a smooth sphere, the direction of rotation is reversed upon reflection, and the received signal is called an *OC* (opposite circularly polarized) signal. Power in the same sense of polarization, *SC*, is generally indicative of multiple reflections, and hence the ratio SC/OC is a measure of the near-surface roughness at scales comparable with the observing wavelength. Main belt asteroids usually show $SC/OC < 0.2$, whereas near-Earth objects often show higher numbers, $0.2 \lesssim SC/OC \lesssim 0.5$, indicative of a rougher surface. The circular polarization ratio can exceed unity for icy bodies (Fig. E.6). Such data revealed, e.g., that Mercury's poles contain ice (§5.5.2).

Radar echoes are Doppler broadened as a result of the rotation of the body. The full or maximum bandwidth, B_{max} (in Hz) depends upon the size of the object (below we assume the body to be spatially unresolved), its rotation period, P_{rot}, and the viewing geometry from the radar:

$$B_{max} = \frac{8\pi R}{\lambda P_{rot}} \sin \theta, \tag{E.7}$$

with λ the radar wavelength and θ the aspect angle, which is the angle between the spin vector and the line of sight.

OC echoes are dominated by specular reflections from surface slopes that are tilted in the direction of the radar system. For near-spherical objects, such as the Moon and Mercury, these slopes are concentrated close to the center of the projected disk, where Doppler shifts caused by the object's rotation are relatively small. *OC* echoes from near-spherical objects therefore display a 'specular spike' at small Doppler shifts and relatively weak wings (referred to as diffuse echo power, in contrast to specular reflections) that extend out to B_{max}. Consequently, the shape of a radar echo spectrum can be used to derive information on the body's size, shape, spin state, and surface characteristics (Fig. E.8).

In a bistatic radar experiment, when the radar echo is received by an interferometer such as the VLA, the signal can be imaged in each frequency channel. Although these observations do not produce images at a higher spatial resolution, the measurements yield an unambiguous result regarding the sense of rotation of the object, since the Doppler red- and blueshifted signals are displaced from the center of the object in opposite directions (Fig. E.9).

Radar echoes can be used to extract the three-dimensional shape of a body (§9.2); radar images are shown in Figs. 9.5, 9.17. Radar echoes from spacecraft during close encounters can be inverted to provide high-angular-resolution maps of (portions of) a body, such as Venus (e.g., Fig. 5.54) and Titan (e.g., Figs. 5.88, 5.89).

E.8 Active Fluorescence Spectroscopy

The elemental composition of a body's surface can be determined using *X-ray* and/or *γ-ray fluorescence spectroscopy*. With this technique, the target material is excited through bombardment with α particles, high-energy X- or γ-rays. Upon de-excitation, X-rays and/or γ-rays are emitted and observed. Depending on which emissions are targeted to be detected, one refers to the technique as either X- or γ-ray fluorescence spectroscopy. In most cases the innermost electron shells of an element are involved in the process. If alpha particles are used for the excitation, the energy distribution of these particles, recoiling from

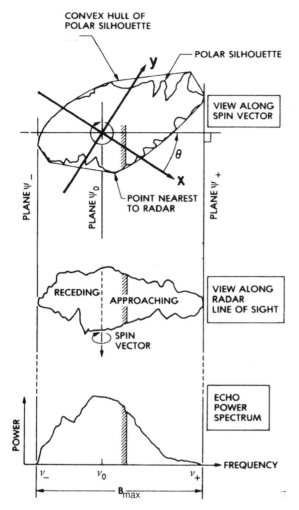

Figure E.8 A cartoon illustrating the geometric relationship between echo power and an asteroid's shape. The upper picture shows the convex hull of the polar silhouette, i.e., the asteroid shape as viewed from the pole. The middle panel shows a view along the radar line of sight. The actual radar echo is shown at the bottom. The plane Ψ_0 contains the line of sight and the asteroid's spin vector. The radar echo from any part of the asteroid that intersects Ψ_0 has a Doppler frequency ν_0. The cross-hatched strip of power in the spectrum corresponds to echoes from the cross-hatched strip on the asteroid. The asteroid's polar silhouette can be estimated from echo spectra that are adequately distributed in rotational phase. (Adapted from Ostro 1989)

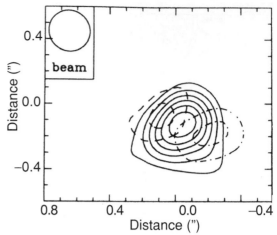

Figure E.9 Radio images of 324 Bamberga at 8510 MHz, obtained with a bistatic radar system wherein the Goldstone antenna was used to transmit the signal and the VLA to receive and image the radar echo. Radar echoes are shown for 13 September 1991, in the center channel (solid contours), and channels at a Doppler frequency of −381 Hz (dashed contours; redshifted) and +381 Hz (dot-dashed contours; blueshifted). Contour levels are from 3 to 19 standard deviations, where one standard deviation is 5.5 mJy per beam. (de Pater *et al.* 1994)

conservation of momentum upon absorption or emission of a photon. This recoil energy is very small in the case of electron (de-)excitation in an atom, so fluorescence is readily achieved. For nuclei, however, this recoil energy is much larger, and therefore fluorescence normally does not occur. However, in the case of solid material, the recoil energy could be taken up by the entire crystal, rather than a single atom, in which case recoilless emission and absorption of gamma rays is possible, and nuclei can fluoresce. This is referred to as the *Mössbauer effect*. In Mössbauer spectroscopy one investigates hyperfine splitting of nuclear lines by vibrationally modulating (i.e., Doppler shifting) the energy of the excitation source. This method is very effective to study iron-bearing minerals, and has been used by, e.g., the Mars rover Opportunity to analyze the hematite 'blueberries' (§5.5.4).

E.9 Nuclear Spectroscopy

The elemental composition of a planet's (sub)surface and atmosphere can also be ascertained by using *nuclear spectroscopy*, which involves obtaining spectra of γ-rays and neutrons. Because the signals are weak, one needs to be within about 0.1 of the planet's radius above its surface to get decent counting rates and spatial resolution. These neutron and gamma-ray measurements provide information about the composition down to a few tens of cm below the surface.

the target, as well as the X-rays and γ-rays (whichever are measured), provide information on the elemental composition of a rock, i.e., it provides the amounts of the different chemical elements that make up the minerals within a rock. These methods have been used, e.g., on the Mars rovers Spirit and Opportunity.

    Nuclei can fluoresce in a similar way to atoms. The difference, however, lies in the recoil energy, i.e., the

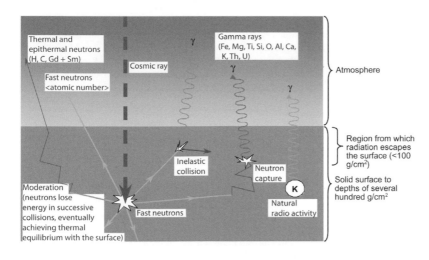

Figure E.10 A graphical presentation of the production of γ-rays and neutrons through the interaction of cosmic rays with a body's (sub)surface and radioactive decay. A nuclear spectrometer measures the flux and spectra of the γ-rays and neutrons that escape into space. (Adapted from Prettyman 2007)

Interplanetary space contains copious amounts of energetic particles and cosmic rays, which interact with planetary surfaces and atmospheres (§7.1). Galactic cosmic rays are the most energetic of these (protons have energies $\sim$0.01–10 GeV) and, therefore, penetrate deepest into a body's surface, up to several hundred g cm^{-2} (i.e., typically a few meters). The flux of galactic cosmic rays is anticorrelated with solar activity. The interaction of these with matter in a body's surface produces fast neutrons, a subset of which undergoes a variety of processes before they surface into space (Fig. E.10). These interactions alter the overall energy of the neutrons that escape into space, and produce γ-rays of specific energies, depending on the atoms they interact with. In addition, decay of radioactive elements, such as potassium, thorium, and uranium, also produce γ-rays that can be detected in space. The γ-ray spectrum thus provides a fingerprint of the elemental composition of a body's (sub)surface (or atmosphere).

Neutrons lose some or all of their energy through successive collisions and/or absorption with nuclei. The neutrons that escape into space are categorized as: (*i*) *Thermal neutrons*, which have undergone many collisions and consequently have low energies, <0.1 eV. (*ii*) *Epithermal neutrons*, which have energies between $\sim$0.1 eV and a few $\times 10^5$ eV. (*iii*) *Fast neutrons*, which have lost little, if any, energy. Elastic scattering is the most important loss mechanism in planetary environments. The maximum energy that a neutron can lose in a collision with a nucleus of mass μ_a (in amu) is given by

$$\Delta E = \left[1 - \left(\frac{\mu_a - 1}{\mu_a + 1} \right)^2 \right] E. \tag{E.8}$$

A neutron can thus lose all of its energy when colliding with hydrogen ($\mu_a = 1$), while at most 28% is lost when colliding with carbon. Neutrons thus lose much energy when the surface of a planet contains water. However, rather than measuring an excess of thermal neutrons above a surface containing water-ice, the thermal flux itself is low because elements such as H, Cl, Fe, and Ti have high absorption cross-sections for thermal neutrons. The most sensitive parameter to give the hydrogen abundance is the flux of epithermal neutrons, because it depends only on the fractional energy loss given in equation (E.8), regardless of the composition of the regolith. Hence, the energy spectrum of neutrons yields information on the abundance of hydrogen, and hence indirectly of water-ice, near a planet's surface. The γ-rays provide additional information on surface composition. These techniques have been used to infer the presence and spatial distribution of water-ice on the Moon (§5.5.1) and Mars (§5.5.4).

E.10 Mass Spectrometry

The composition of a gas can be measured *in situ* with a *mass spectrometer*, an instrument in which ions are sorted and separated according to their mass-to-charge ratios with help of electric and magnetic fields. Ions are accelerated by an electric field, and their paths can be deflected into an arc by applying a magnetic field perpendicular to the direction of motion of the ion beam. Since the radius of curvature of the ions' path is inversely proportional to their mass-to-charge ratio, lighter ions are deflected more than heavier ions. This way the number of ions in each mass 'bin' can be counted, providing a mass spectrum (e.g., Fig. 4.10b). This method gives just one example of how a mass spectrometer works; several different types of mass spectrometers have been developed over the years.

For the sampling of neutral gases, a mass spectrometer carries an ionization source. The ions are then extracted from this source (using electrostatic potentials), and focused into the mass spectrometer, where they can be sorted and counted. In low-density environments such as in an upper atmosphere of a planet, mass spectrometers measure either ambient ions or ions created in the ion source of the instrument from ambient neutrals. Some instruments such as the mass spectrometer on the Cassini Orbiter (the Ion and Neutral Mass Spectrometer – INMS) measure both ambient ions and neutrals sequentially. These instruments have therefore been referred to as 'ion mass spectrometers' and 'neutral mass spectrometers'. For planetary entry probes (at Venus, Mars, Titan, and Jupiter), neutral mass spectrometers are used.

The one drawback of the mass spectrometry technique is that a molecular mass does not provide unique information on the composition of the molecule. In particular for the higher mass molecules, there are ambiguities. For example, CO_2 and C_3H_8 each have a mass of 44 amu. This ambiguity can be overcome with very high mass resolution. Although most mass spectrometers on spacecraft have been low-resolution instruments, the Rosetta mass spectrometer has sufficient resolution to separate CO and N_2.

Further Reading

The *Encyclopedia of the Solar System* (2nd Edition, 2007) contains a number of papers describing the Solar System at specific wavelengths, or using specific techniques. Eds. L. McFadden, P. Weissman, and T.V. Johnson. Academic Press, San Diego, 982pp:

Bhardwaj, A., and C.M. Lisse: X-rays in the Solar System, pp. 637–658.

de Pater, I., and W.S. Kurth: The Solar System at radio wavelengths, pp. 695–718.

Hendrix, A.R., R.M. Nelson, D.L. Domingue: The Solar System at ultraviolet wavelengths, pp. 659–680.

Ostro, S.J.: Planetary radar, pp. 735–764.

Prettyman, T.H.: Remote chemical sensing using nuclear spectroscopy, pp. 765–786.

Tokunaga, A.T., and R. Jedicke: New generation ground-based optical/infrared telescopes, pp. 719–734.

A review on mass spectrometers is given by:

Mahaffy, P.R., 1998. Mass spectrometers developed for planetary missions. In *Laboratory Astrophysics and Space Research*. Eds. P. Ehrenfreund, H. Kochan, K. Krafft, V. Pirronello. Kluwer Academic Publishers, Dordrecht, pp. 355–376.

An excellent review on stellar occultation experiments is given by:

Elliot, J.L., 1979. Stellar occultation studies of the Solar System. *Annu. Rev. Astron. Astrophys.*, **17**, 445–475.

The radio occultation technique is reviewed in:

Tyler, G.L., 1987. Radio propagation experiments in the outer Solar System with Voyager. *Proc. IEEE*, **75**, 1404–1431.

Tyler, G.L., I.R. Linscott, M.K. Bird, D.P. Hinson, D.F. Strobel, M. Pätzold, M.E. Summers, and K. Sivaramakrishan, 2008. The New Horizon radio science experiment (REX). *Space Sci. Rev.*, **140**, 217–259.

Books on remote sensing and image processing:

Cox, A.N., Ed., 2000. *Allen's Astrophysical Quantities*, 4th Edition. Springer-Verlag, New York, Inc. 719pp.

Hanel, R.A., B.J. Conrath, D.E. Jennings, and R.E. Samuelson, 1992. *Exploration of the Solar System by Infrared Remote Sensing*. Cambridge University Press, Cambridge. 458pp.

Hardy, J.W., 1998. *Adaptive Optics for Astronomical Telescopes*. Oxford University Press, New York. 438pp.

Kraus, J.D., 1986. *Radio Astronomy*, 2nd Edition. Cygnus Quasar Books, Powell, Ohio.

Perley, R.A., F.R. Schwab, A.H. Bridle, 1989. *Synthesis Imaging in Radio Astronomy*, NRAO Workshop No. 21, Astronomical Society of the Pacific. 509pp.

Schowengerdt, R. A., 2007. *Remote Sensing, Models, and Methods for Image Processing*. 3rd Edition. Elsevier, Academic Press, Burlington, MA. 515pp.

Thompson, A.R., J.M., Moran, G.W. Swenson, Jr., 2001. *Interferometry and Synthesis in Radio Astronomy*, 2nd Edition. John Wiley and Sons, New York. 692pp.

A thorough overview of the shadow-hiding and coherent-backscatter effects near opposition are presented in a series of papers by Hapke, B., *et al.*, and Helfenstein, P., *et al.*, which include:

Hapke, B., R. Nelson, and W. Smythe, 1998. The opposition effect of the Moon: Coherent backscatter and shadow hiding. *Icarus*, **133**, 89–97.

Helfenstein, P., J. Veverka, and J. Hiller, 1997. The lunar opposition effect: A test of alternative models. *Icarus*, **128**, 2–14.

Appendix F: Interplanetary Spacecraft

Figure F.1 Artist impression of the Luna 3 spacecraft. The probe's mass was 278 kg; it was 130 cm in length and had a maximum diameter of 120 cm. The interior of the spacecraft held the cameras and film processing system, radio equipment, propulsion systems, batteries, gyroscopic units for attitude control, and circulating fans for temperature control. The spacecraft was spin stabilized and was directly radio controlled from Earth. Solar cells were mounted along the outside of the cylinder and provided power to the chemical batteries stored inside the spacecraft. The spacecraft had no rockets for course adjustment.

A substantial fraction of our data on many Solar System objects has been obtained by close-up studies conducted by spacecraft. This appendix starts with a short section on *rocketry* (how a rocket works). Section F.2 contains tables listing many of the most significant lunar and interplanetary spacecraft and astronomical observations in space. This appendix further includes diagrams of two historically significant spacecraft (Figs. F.1 and F.5), and two historic images (Figs. F.2 and F.6).

F.1 Rocketry

The principles of 'rocket science' are actually quite simple, although many practical aspects of 'rocket engineering' are far more complicated. A rocket accelerates by

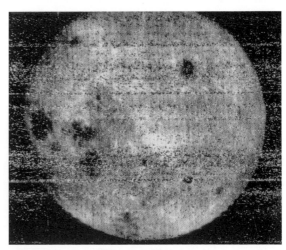

Figure F.2 This fuzzy image taken on 7 October 1959 provided humanity with its first view of the lunar far side.

expelling gas (or plasma) at high velocity. Conservation of momentum implies that the velocity, v, of the rocket of mass M (which includes propellent), expelling gas at velocity $v_{\exp}$ and rate dM/dt satisfies:

$$M\frac{d\mathbf{v}}{dt} = -\mathbf{v}_{\exp}\frac{dM}{dt} + \mathbf{F}_{\text{ext}}, \qquad (\text{F.1})$$

where $\mathbf{F}_{\text{ext}}$ accounts for all external forces on the rocket. Equation (F.1) is known as the fundamental *rocket equation*.

In a uniform gravitational field that induces an acceleration $\mathbf{g}_{\text{p}}$ with no other external forces, the rocket equation reduces to

$$\frac{d\mathbf{v}}{dt} = -\frac{\mathbf{v}_{\exp}}{M}\frac{dM}{dt} + \mathbf{g}_{\text{p}}. \qquad (\text{F.2})$$

Integrating equation (F.2) and setting $v = 0$ at $t = 0$ gives

$$\mathbf{v} = -\mathbf{v}_{\exp}\ln\frac{M_0}{M} - \mathbf{g}_{\text{p}}t, \qquad (\text{F.3})$$

where M_0 is the mass at $t = 0$, and there is a minus sign in front of the last term in equation (F.3) because the gravitational force is directed downwards. Note that there is a premium to burning fuel rapidly – the shorter the burn time, the greater the velocity for given ejection speed and mass. This is why high-thrust rocket engines are used to attain escape velocity from Earth; at present, such large thrusts can only be obtained using *chemical propulsion* (Fig. F.3).

For acceleration in free space, the final velocity of the rocket does not depend upon the ejection rate. Thus, *electric propulsion* (Fig. F.4), which can achieve higher expulsion speeds (specific impulse) than chemical rockets, can be very efficient at modifying trajectories of orbiting bodies. Trajectories of interplanetary spacecraft are discussed in §2.1.5.

F.2 Tabulations

We list significant lunar missions in Table F.1, and interplanetary spacecraft in Table F.2. Table F.3 lists several

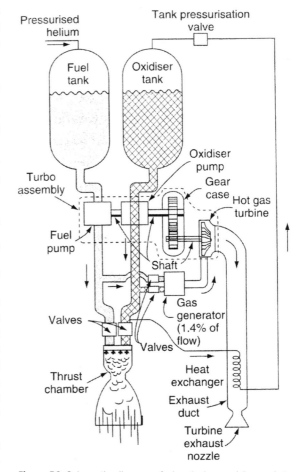

Figure F.3 Schematic diagram of chemical propulsion rockets. (Sutton and Biblarz 2001)

spacecraft that conducted observations of Earth's magnetosphere and aurora, and of the Sun and the heliosphere. Table F.4 provides a list of space observatories. A more complete listing of spacecraft observation of Solar System objects is provided in the Appendix of the *Encyclopedia of the Solar System* (2007).

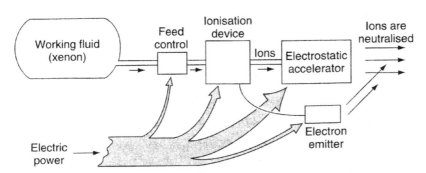

Figure F.4 Schematic diagram of electric propulsion rockets. (Sutton and Biblarz 2001)

Table F.1 Selected lunar spacecraft.

Spacecraft	Sender	Launch date	Type	Remarks
Luna 3	USSR	1959 Oct. 4	flyby	Photographed far side of the Moon
Ranger 7	USA	1964 July 28	impact	Returned 4308 photos
Ranger 8	USA	1965 Feb. 17	impact	Returned 7137 photos
Ranger 9	USA	1965 Mar. 21	impact	Returned 5814 photos
Luna 9	USSR	1966 Jan. 31	lander	Soft landing, returned photos
Luna 10	USSR	1966 Mar. 31	orbiter	First lunar orbiter
Surveyor 1	USA	1966 May 30	lander	Soft landing, returned 11 150 photos
Lunar Orbiter 1	USA	1966 Aug. 10	orbiter	Photographic mapping
Luna 11	USSR	1966 Aug. 24	orbiter	Science return
Luna 12	USSR	1966 Oct. 22	orbiter	Photographic mapping
Lunar Orbiter 2	USA	1966 Nov. 6	orbiter	Photographic mapping
Luna 13	USSR	1966 Dec. 21	lander	Surface science
Lunar Orbiter 3	USA	1967 Feb. 4	orbiter	Photographic mapping
Surveyor 3	USA	1967 Apr. 17	lander	Surface science
Lunar Orbiter 4	USA	1967 May 4	orbiter	Photographic mapping
Lunar Orbiter 5	USA	1967 Aug. 1	orbiter	Photographic mapping
Surveyor 5	USA	1967 Sep. 8	lander	Surface science
Surveyor 6	USA	1967 Nov. 7	lander	Surface science
Surveyor 7	USA	1968 Jan. 8	lander	Surface science
Luna 14	USSR	1968 Apr. 7	orbiter	Mapped gravity field
Apollo 8	USA	1968 Dec. 21	manned orbiter	First humans in deep space
Apollo 10	USA	1969 May 18	manned orbiter	2 spacecraft; undocking, docking
Apollo 11	USA	1969 July 16	manned lander	First humans on Moon; 22 kg sample
Apollo 12	USA	1969 Nov. 14	manned lander	34 kg sample return
Luna 16	USSR	1970 Sep. 12	sample return	First robotic sample return (101 g)
Luna 17	USSR	1970 Nov. 10	rover	First lunar rover
Apollo 14	USA	1971 Jan. 31	manned lander	42 kg sample return
Apollo 15	USA	1971 July 26	manned rover	77 kg sample return
Luna 19	USSR	1971 Sep. 28	orbiter	Photographic mapping
Luna 20	USSR	1972 Feb. 12	sample return	Returned 55 g sample
Apollo 16	USA	1972 Apr. 16	manned rover	95 kg sample return
Apollo 17	USA	1972 Dec. 10	manned rover	111 kg sample return
Luna 21	USSR	1973 Jan. 8	rover	Lunokhod 2; traversed 39 km
Luna 22	USSR	1974 May 29	orbiter	Photographic mapping
Luna 24	USSR	1976 Aug. 9	sample return	Returned 170 g sample
Clementine	USA	1994 Jan. 25	orbiter	Photographic mapping
Lunar Prospector	USA	1998 Jan. 6	orbiter	Photographic mapping (plus impact)
SMART 1	ESA	2003 Sep. 27	orbiter	Photographic mapping (plus impact)
SELENE/Kaguya	Japan	2007 Sep. 14	orbiter	Photographic mapping
Chang'e 1	China	2007 Oct. 24	orbiter	Photographic mapping
Chandrayaan 1	India	2008 Oct. 22	orbiter	Photographic mapping, radar
LRO	USA	2009 June 18	orbiter	Photographic mapping
LCROSS	USA	2009 June 18	impactor	Launched with LRO
Chang'e 2	China	2010 Oct. 1	orbiter	Photographic mapping
GRAIL	USA	2011 Sep. 10	two orbiters	Gravity mapping
LADEE	USA	2013 Sep. 7	orbiter	Atmosphere and dust
Chang'e 3	China	2013 Dec. 1	lander	Also a small rover

Table F.2 Selected interplanetary spacecraft.

Spacecraft	Sender	Launch date	Target	Type	Remarks
Mariner 2	USA	1962 Aug. 27	Venus	flyby	First close-up data from Venus
Mariner 4	USA	1964 Nov. 28	Mars	flyby	First 21 close-up photos of Mars
Venera 4	USSR	1967 June 12	Venus	probe	Atmospheric measurements
Mariner 6	USA	1969 Feb. 24	Mars	flyby	Returned 75 photos
Mariner 7	USA	1969 Mar. 27	Mars	flyby	Returned 125 photos
Venera 7	USSR	1970 Aug. 17	Venus	lander	First soft landing
Mars 2	USSR	1971 May 19	Mars	orbiter+lander	Orbiter succeeded, lander failed
Mars 3	USSR	1971 May 28	Mars	orbiter+lander	Lander failed after 20 seconds
Mariner 9	USA	1971 May 30	Mars	orbiter	Returned many photos
Pioneer 10	USA	1972 Mar. 3	Jupiter	flyby	First Jupiter flyby, 1973 Dec. 3
Venera 8	USSR	1972 Mar. 27	Venus	lander	Landed 1972 July 22
Pioneer 11	USA	1973 Apr. 6	Jupiter	flyby	Closest approach 1974 Dec. 4
			Saturn	flyby	First Saturn flyby, 1979 Sep. 1
Mars 5	USSR	1973 July 25	Mars	orbiter	Orbited 1974 Feb. 12
Mariner 10	USA	1973 Nov. 3	Venus	flyby	Closest approach 1974 Feb. 5
			Mercury	flyby	Three flybys in 1974–1975
Venera 9	USSR	1975 June 8	Venus	orbiter+lander	First images of surface
Venera 10	USSR	1975 June 14	Venus	orbiter+lander	Landed 1975 Oct. 25
Viking 1	USA	1975 Aug. 20	Mars	orbiter+lander	First long-term surface science
Viking 2	USA	1975 Sep. 9	Mars	orbiter+lander	Landed 1976 Sep. 3
Voyager 2	USA	1977 Aug. 20	Jupiter	flyby	Closest approach 1979 July 9
			Saturn	flyby	Closest approach 1981 Aug. 26
			Uranus	flyby	First Uranus flyby, 1986 Jan. 24
			Neptune	flyby	First Neptune flyby, 1989 Aug. 24
Voyager 1	USA	1977 Sep. 5	Jupiter	flyby	Closest approach 1979 Mar. 5
			Saturn	flyby	Closest approach 1980 Nov. 12
Pioneer 12	USA	1978 May 20	Venus	orbiter	Entered orbit 1978 Dec. 8
Pioneer 13	USA	1978 Aug. 8	Venus	probe	4 probes entered atmosphere 1978 Dec. 9
Venera 11	USSR	1978 Sep. 9	Venus	lander	Landed 1978 Dec. 25
Venera 12	USSR	1978 Sep. 14	Venus	lander	Landed 1978 Dec. 21
Venera 13	USSR	1981 Oct. 30	Venus	lander	Landed 1982 Feb. 27
Venera 14	USSR	1981 Nov. 4	Venus	lander	Landed 1982 Mar. 5
Vega 1	USSR	1984 Dec. 15	Venus	balloon+lander	First Venus balloon
			1P/Halley	flyby	First comet nucleus images
Vega 2	USSR	1984 Dec. 21	Venus	balloon+lander	Landed 1985 June 15
			1P/Halley	flyby	Closest approach 1986 Mar. 9
Sakigake	Japan	1985 Jan. 8	1P/Halley	flyby	Closest approach 1986 Mar. 11
Suisei	Japan	1985 Aug. 18	1P/Halley	flyby	Closest approach, 151 000 km, 1986 Mar. 8
Giotto	ESA	1985 July 2	1P/Halley	flyby	Closest approach, 596 km, 1986 Mar. 14
Phobos 2	USSR	1988 July 12	Phobos	lander	Mars + Phobos images; failed prior to landing
Magellan	USA	1989 May 5	Venus	orbiter	Global radar mapper
Galileo	USA	1989 Oct. 18	951 Gaspra	flyby	First asteroid flyby, 1991 Oct. 29
			243 Ida	flyby	Discovered Dactyl, first asteroid moon
			Jupiter	orbiter+probe	First Jupiter probe, arrived 1995 Dec. 7
NEAR	USA	1996 Feb. 17	Mathilde	flyby	Closest approach, 1997 June 27
			Eros	orbiter	First asteroid orbiter; orbited 2000 Feb. 14
MGS	USA	1996 Nov. 7	Mars	orbiter	Entered orbit 1997 Sep. 12
Mars Pathfinder	USA	1996 Dec. 2	Mars	lander+rover	First Mars rover
Cassini	USA	1997 Oct. 15	Jupiter	flyby	Closest approach, 2000 Dec. 30
			Saturn	orbiter	Entered orbit 2004 July 1

Table F.2 (*cont.*)

Spacecraft	Sender	Launch date	Target	Type	Remarks
Cassini	USA	1997 Oct. 15	Jupiter	flyby	Closest approach, 2000 Dec. 30
			Saturn	orbiter	Entered orbit 2004 July 1
Huygens	ESA	1997 Oct. 15	Titan	probe/lander	Travelled with Cassini; first Titan probe
Deep Space 1	USA	1998 Oct. 24	9969 Braille	flyby	Closest approach 1999 July 29
			19P/Borrelly	flyby	Closest approach 2001 Sep. 22
Stardust	USA	1999 Feb. 6	81P/Wild 2	sample return	Closest approach/flyby science 2004 Jan. 2
			9P/Temple 1	flyby	Closest approach 2011 Feb. 14
Mars Odyssey	USA	2001 Apr. 7	Mars	orbiter	Entered orbit 2001 Oct. 23
Hayabusa	Japan	2003 May 9	25143 Itokawa	orbiter	Also returned a small sample
Mars Express	ESA	2003 June 2	Mars	orbiter	Entered orbit 2003 Dec. 25
Spirit	USA	2003 June 10	Mars	orbiter	Mars Exploration Rover I
Opportunity	USA	2003 July 7	Mars	orbiter	Mars Exploration Rover II
Rosetta	ESA	2004 Mar. 2	2867 Steins	flyby	closest approach 2008 Sep. 5
			21 Lutetia	flyby	closest approach 2010 July 10
			67P/Churyumov-Gerasimenko	rendezvous and lander	arrived 2014 Aug. 6
MESSENGER	USA	2004 Aug. 3	Mercury	orbiter	Entered orbit 2011 Mar. 17
Deep Impact	USA	2005 Jan. 12	9P/Temple 1	flyby+impactor	Impacted 2005 July 4
			103P/Hartley 2	flyby	Closest approach 2010 Nov. 4; renamed EPOXI
MRO	USA	2005 Aug. 12	Mars	orbiter	Entered orbit 2006 Mar. 10
Venus Express	ESA	2005 Nov. 9	Venus	orbiter	Entered orbit 2006 Apr. 11
New Horizons	USA	2006 Jan. 19	Jupiter	flyby	Perijove 2007 Feb. 28; en route to Pluto
Mars Phoenix	USA	2007 Aug. 4	Mars	lander	Explored polar region
Dawn	USA	2007 Sep. 27	4 Vesta	orbiter	Orbited 2011-2012; en route to 1 Ceres
Curiosity	USA	2011 Nov. 26	Mars	rover	Mars Science Laboratory
Mars Orbiter	India	2013 Nov. 5	Mars	orbiter	very low cost mission
MAVEN	USA	2013 Nov. 18	Mars	orbiter	Mars Atmosphere & Volatile Evolution

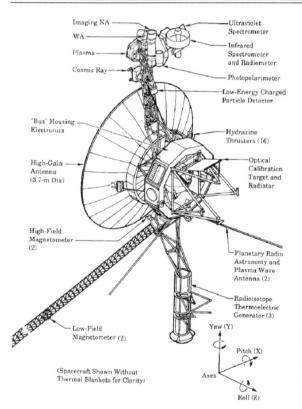

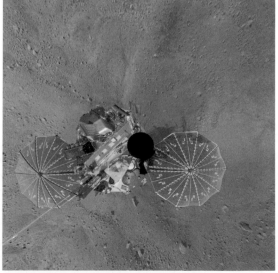

Figure F.5 Schematic diagram of Voyager spacecraft. Voyager 1 sent back the first high-resolution images of Jupiter, Saturn, and their moons. Voyager 2 was the first, and thus far only, spacecraft to visit Uranus and Neptune. Many of the photographs of the outer planets and their moons that appear in this book were taken by the Voyager Narrow Angle (NA) and Wide Angle (WA) cameras illustrated at the top of this diagram.

Figure F.6 This view is a vertical projection that combines hundreds of exposures taken by the Surface Stereo Imager camera on NASA's Mars Phoenix Lander and projects them as if looking down from above. The black circle is where the camera itself is mounted. (NASA/JPL-Caltech/University of Arizona/Texas A&M, PIA11719)

Table F.3 Selected magnetosphere, solar, and heliosphere spacecraft.

Spacecraft	Sender	Launch date	Target	Remarks
Explorer 1	USA	1958 Jan. 31	Earth orbit	Discovered Van Allen Belts
IMP-8	USA	1973 Oct. 26	solar wind	Interplanetary Monitoring Platform-8
Helios 1	W. Germany	1974 Dec. 10	solar wind	Monitored solar wind and dust
Helios 2	W. Germany	1976 Jan. 15	solar wind	Monitored solar wind and dust
ISEE 3/ICE	USA	1978 Aug. 12	solar wind	Monitored solar wind. Then flew through tail of 21P/Giacobini–Zinner
SMM	USA	1980 Feb. 14	Sun	Observatory: Solar Maximum Mission
Ulysses	ESA/USA	1990 Oct. 6	Sun	Solar polar orbit
Yokhoh	Japan	1991 Aug. 30	Sun	Solar observatory in Earth's orbit
SAMPEX	USA	1992 Jul. 3	energetic particles	Magnetospheric particle explorer
WIND	USA	1994 Nov. 1	solar wind/magnetosphere	Magnetosphere and solar wind
SOHO	ESA	1995 Dec. 2	Sun	Discovered many Sun-grazing comets
POLAR	USA	1996 Feb. 24	Earth aurora	Polar orbit
FAST	USA	1996 Aug. 21	Earth aurora	Polar orbit
ACE	USA	1997 Aug. 25	solar wind	Advanced Composition Explorer
TRACE	USA	1998 Apr. 2	Sun	Sun-synchronous orbit about Earth
IMAGE	USA	2000 Mar. 25	Earth's magnetosphere	Earth's orbit
CLUSTER II	ESA	2000 Jul. 16	Earth's magnetosphere	4 spacecraft to conduct joint studies
		2000 Aug. 9	Earth's magnetosphere	
Genesis	USA	2001 Aug. 8	solar wind	Sample returned to Earth 2004 Sep. 8
RHESSI	USA	2002 Feb. 5	solar flares	Observatory in Earth's orbit
Hinode	Japan/USA/UK	2006 Sep. 23	Sun	In polar orbit about Earth
STEREO	USA	2006 Oct. 25	Sun, CMEs	Two spacecraft
THEMIS	USA	2007 Feb. 17	Earth aurora	5 probes in Earth's orbit
		2010	Moon, solar wind	2 probes renamed ARTEMIS
SDO	USA	2010 Feb. 11	Sun	Solar Dynamics Observatory
Van Allen Probes	USA	2012 Aug. 30	Earth's radiation belts	2 spacecraft

Table F.4 Selected space observatories.

Spacecraft	Sender	Launch date	Orbit	Remarks
IUE	USA/ESA	1978 Jan. 26	low Earth	International Ultraviolet Explorer
IRAS	USA/UK/NL	1983 Jan. 25	low Earth	Infrared Astronomical Satellite
HST	USA/ESA	1990 Apr. 24	low Earth	Hubble Space Telescope
ROSAT	Germany	1990 June 1	low Earth	X-ray observatory
EUVE	USA	1992 June 7	low Earth	Extreme Ultraviolet Explorer
ISO	ESA	1995 Nov. 17	low Earth	Infrared Space Observatory
Spitzer	USA	2003 Aug. 25	heliocentric	Infrared telescope
CoRoT	France/ESA	2006 Dec. 27	Earth polar	Search for extrasolar planets
Kepler	USA	2009 Mar. 6	heliocentric	Search for extrasolar planets
Herschel	ESA	2009 May 14	Sun–Earth L_2	Far-IR and sub-mm

Appendix G: Recent Developments in Planetary Sciences

G.1 Introduction

This appendix presents new material to update the book. We focus on discoveries from the past few years and also include some material that was omitted from the first printings. We arrange this material by the chapter in the main text to which it is associated.

The first printings of the Second Edition of this book included as Appendix G a selection of some of the most spectacular images sent back by NASA spacecraft that were released to the public during 2009 and early 2010. That version of Appendix G is now provided on the website of this book, www.cambridge.org/depater.

Good web sources to view recent planetary images include:

www.nineplanets.com/,
www.nasa.gov/topics/solarsystem/index.html,
saturn.jpl.nasa.gov/index.cfm,
hubblesite.org/gallery/album/solar_system,
photojournal.jpl.nasa.gov/.

G.2 Dynamics

Quasi-satellites of Jupiter

Several asteroids and comets have recently been discovered to be quasi-satellites (§2.2.4) of Jupiter. A few of the asteroids will remain as quasi-satellites for at least a few thousand years. Some others will soon transition to horseshoe orbits or large-amplitude tadpole orbits. The trajectories of the comets are more difficult to predict, as small nongravitational forces (§10.2.1) on these bodies may alter their velocities by the tiny amounts needed to cause them to transition between these various types of co-orbital configurations with Jupiter.

ν_6 Secular Resonance

The ν_6 secular resonance occurs where the perihelion angle of an asteroid precesses at the rate of the sixth secular frequency of our Solar System, which is essentially the same as the precession rate of Saturn's periapse. Perturbations resulting from the ν_6 resonance can excite asteroidal eccentricities to such high values that they can collide with Mars or even with Earth. The ν_6 resonance is largely responsible for the inner edge of the asteroid belt near 2.1 AU.

Lidov–Kozai Mechanism

In the restricted circular three-body problem, secular perturbations can change the small body's eccentricity and its inclination relative to the orbit of the two massive bodies, but the quantity $\sqrt{1-e^2}\cos i$ remains constant. Orbital inclination can thus be traded for eccentricity. For high values of inclination, $\cos^2 i < 3/5$, the *Lidov–Kozai mechanism* (sometimes referred to as simply the *Kozai mechanism*) forces the argument of periapse to remain fixed, and large periodic variations in eccentricity and inclination are produced. The Lidov–Kozai mechanism causes some asteroids and comets to approach closely to and even collide with the Sun (§10.6.4) and highly inclined irregular satellites to collide with their planets. It is also likely to be one of the mechanisms responsible for the high observed eccentricities of some extrasolar planets (Fig. 12.22) and the high inclinations some of the hot Jupiters' orbits have relative to the plane of their star's equator (§§12.3.3 and G.12.5).

Chaotic Orbits of the Planets

As noted in §2.4.2, numerical integrations show that the orbits of the eight major planets in our Solar System are chaotic with a timescale for exponential divergence of $\sim$5 million years. The effect is most apparent in the orbits of the inner planets. Despite this chaos, gross changes in planetary orbits are unlikely on astrophysically important timescales. But in $\sim$1% of systems integrated forward for 5 billion years, Mercury and Venus suffer a close approach, which can lead to Mercury colliding with another planet or the Sun or being ejected from the Solar System.

The chaotic divergence seen in all long-term integrations implies that the accuracy of the deterministic equations of celestial mechanics to predict the future positions of the planets will always be limited by the accuracy with which their orbits can be measured. For example, if the position of Earth along its orbit is uncertain by 1 cm today, then the exponential propagation of errors that is characteristic of chaotic motion implies that, even within the framework of purely Newtonian interactions among the Sun and the eight planets, we have no knowledge of Earth's orbital longitude 200 million years in the future. The situation is even less predictable when the gravitational influence of smaller bodies is taken into account. Asteroids exert small perturbations on the orbits of the major planets. These perturbations can be accounted for and don't adversely affect the precision to which planetary orbits can be simulated on timescales of tens of millions of years. However, unlike the major planets, asteroids suffer close approaches to one another. Close approaches between the two largest and most massive asteroids, 1 Ceres and 4 Vesta, lead to exponential growth in uncertainty for backwards integrations of planetary orbits with doubling times of $<10^6$ years prior to $50-60$ million years ago.

Orbital Decay of Phobos

The orbital motion of Mars's inner moon Phobos is observed to be accelerating at a rate of 1.27×10^{-3} degree/yr^2. This acceleration is caused by the lag in the tidal bulge that Phobos raises on Mars (§2.6.2), and it implies that $Q_{\mars} = 83$.

Tidal Heating

The rate at which tides on a moon convert orbital energy to heat depends on a complicated combination of orbital and physical properties (e.g., rigidity) of that moon. For the Galilean moons, the key factors determining average energy dissipation rates are those that control the amount of energy that goes into the moon's orbital eccentricity. However, for nonequilibrium situations such as those that occurred in the distant past for Neptune's moon Triton, which was subjected to a major infusion of tidal heat after being captured into a highly eccentric orbit about Neptune, the strong dependences of tidal forces on eccentricity and periapse location dominated, and most of the heating occurred quite rapidly.

G.3 Solar Heating and Energy Transport

Chapter 3 presents basic physics that has not changed. For students desiring additional background in basic thermodynamics, we recommend §3.1 of our new advanced undergraduate level textbook *Fundamental Planetary Science: Physics, Chemistry and Habitability* by J.J. Lissauer and I. de Pater, Cambridge University Press, 2013. A more pedagogical explanation of the greenhouse effect is provided in §4.6 of the above mentioned text.

G.4 Planetary Atmospheres

Earth

The banded iron formations (§4.9.2.1) shown in Figure G.1 provide striking evidence for the low oxygen abundance in Earth's atmosphere billions of years ago.

Figure G.2 shows spectra of light reflected by Earth over various ranges of wavelength. It illustrates what can be learned about a planet with an atmosphere similar to that of Earth (i.e., an inhabited planet) using remote observations.

2 cm

Figure G.1 Photograph of a 3.8-Gyr-old rock from the Isua Formation in Greenland showing a banded iron formation (BIF). The minerals comprising this rock cannot form in equilibrium with Earth's current oxygen-rich atmosphere, but banded iron formations are abundant in the geologic record of rocks more than 2.4 Gyr old. (Courtesy Minik Rosing)

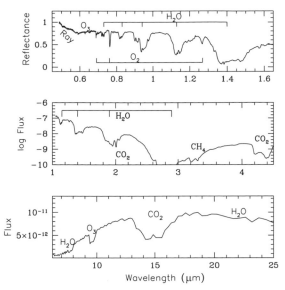

Figure G.4 NASA's Curiosity rover and its parachute were photographed by HiRISE on MRO as Curiosity descended to the martian surface. The parachute and rover are seen in the center of the white box. The inset image is a cutout of the rover stretched to avoid saturation. (NASA/JPL-Caltech/Univ. of Arizona, PIA15978)

Figure G.2 Observed disk-integrated reflection spectra of Earth over a wide range of wavelengths. The top panel shows the spectrum at visible wavelengths measured from earthshine reflected off the dark side of the Moon. The brightening at short wavelengths is caused by Rayleigh scattering (§3.2.3.4). The middle panel shows Earth in the near-infrared (IR) as measured by NASA's Deep Impact spacecraft, with flux in units of W m^{-2} μm^{-1}. The bottom panel displays a mid-IR spectrum from NASA's Mars Global Surveyor en route to Mars, with flux in units of W m^{-2} Hz^{-1}. Major molecular features are noted on the plots. (From Meadows and Seager 2010)

The overarching science goal of this mission is to assess whether the landing area has ever had or still has environmental conditions favorable to microbial life, both its habitability and its preservation.

The mass spectrometer of the Sample Analysis at Mars (SAM) instrument on Curiosity rover measured the argon isotope ratios $^{36}Ar/^{38}Ar = 4.2 \pm 0.1$ and $^{40}Ar/^{36}Ar = (1.9 \pm 0.3) \times 10^3$. The relatively low (compared to Earth) $^{14}N/^{15}N$ ratio was also confirmed, at a value of 173 ± 9. These isotope ratios, in particular that of the nonradiogenic $^{36}Ar/^{38}Ar$, are highly suggestive of fractionation due to a substantial loss of mass from the martian atmosphere. The nonradiogenic argon numbers also provide evidence that the SNC and related meteorites (§8.2) indeed originated on Mars.

A main goal of SAM was to measure the precise abundance of methane gas (see §4.3.3.1). No sign of CH_4 has been found, implying an upper limit to its abundance of ~1 ppb.

Figure G.3 This MRO image shows a martian dust devil roughly 20 kilometers high winding its way along the Amazonis Planitia region of northern Mars. Despite its height, the plume is only 70 meters wide. (NASA/JPL-Caltech/Univ. of Arizona)

Mars

Figure G.3 shows a spectacular image of a dust devil (§4.5.5.3) obtained from martian orbit.

On 6 August 2012, the rover Curiosity landed on Mars at Gale Crater. The HiRISE camera on MRO captured the image of Curiosity and its parachute shown in Figure G.4.

Saturn

Saturn displays a large variety of atmospheric phenomena, although such features are usually not as prominent as those on Jupiter. A spectacular photograph taken by the Cassini spacecraft is shown in Figure G.5. Clear zonal bands can be discerned, together with a highly unusual storm in the planet's northern hemisphere. Although large storms on Saturn typically appear once every few decades, a storm as dramatic as this one has never before been seen. The shadow cast by Saturn's rings has a strong seasonal effect, and the powerful storm in the northern hemisphere in 2011 may have been related to the change of seasons after the planet's August 2009 equinox.

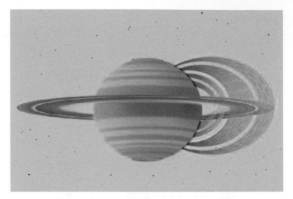

Figure G.6 This artist's concept illustrates how charged water particles flow into the Saturnian atmosphere from the planet's rings, causing a reduction in atmospheric brightness. This 'rain' of charged water particles into the atmosphere of Saturn explains why electron densities and the H_3^+ abundance is unusually low at some latitudes at Saturn. (PIA16842, Keck Observatory and NASA/JPL-Caltech/Space Science Institute/University of Leicester)

Figure G.5 Image of Saturn taken by the Cassini spacecraft on 25 February 2011 about 12 weeks after a powerful storm was first detected in Saturn's northern hemisphere. This storm is seen overtaking itself as it encircles the entire planet. (NASA/JPL/Space Science Institute, PIA12826)

The Voyager 2 spacecraft observed latitudinal variations in the electron density in Saturn's ionosphere, which were interpreted to be caused by the influx of water from Saturn's rings. Similar latitudinal variations have been mapped in the intensity of the H_3^+ rotational–vibrational emission lines. As illustrated in Figure G.6, the electron and H_3^+ densities are lower at latitudes where the magnetic field lines connect the ionosphere to Saturn's rings than they are at latitudes where the field lines thread through gaps in the rings. The rings are surrounded by an atmosphere composed of water-products, which are partially ionized. Water-related ions and electrons move along the field lines into the ionosphere, reducing the local electron density through rapid chemical recombination. Charged water-derived particles also deplete H_3^+ through charge-exchange. Hence, this 'rain' of charged water particles into the atmosphere of Saturn leads to a series of light and dark bands in the H_3^+ abundance with a pattern mimicking the planet's rings.

Uranus

Since Uranus's 2007 equinox, it has become much easier to view the planet's north polar region. Thanks to the development of new image processing techniques, stunning images of discrete cloud features in this region have been obtained, as shown in Figure G.7. Images were taken in two different near-infrared filters to determine the altitude of clouds and hazes. The broad H filter centered

near 1.6 μm (blue and green colors) samples both weak and strong absorptions by methane gas, while the narrow Hcont filter (red color) is near the same wavelength, but samples only regions of weak methane absorption. Most of the features in these images are quite subtle and require long exposures to be detectable above the background noise. But during long exposures, the features are smeared out by planetary rotation and zonal winds. To deal with that, many short-exposure images were taken and the effects of rotation and winds were removed before averaging. These polar features are visible while it is spring at the north pole; no such features were visible at the south pole during its summer and fall. This technique also revealed a never-before-seen scalloped wave pattern just south of the equator, similar to instabilities that develop in regions of horizontal wind shear.

G.5 Planetary Surfaces

Small Impacts on Jupiter

Figure G.8 shows HST's view of the atmospheric debris produced by the July 2009 impact on Jupiter. The size of the impactor is estimated to be a few hundred meters.

Modern digital imaging systems now enable amateur astronomers to spot occasional flashes from small ($R \sim 10$ m) bolides entering Jupiter; an example is shown in Figure G.9. Because subkilometer objects near and beyond Jupiter's orbit cannot be detected directly, analysis of such bolides may lead to more reliable estimates of the population of small bodies in the outer regions of our Solar System.

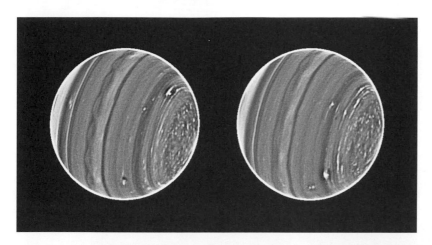

Figure G.7 These two images of Uranus are composites of 117 images from 25 July 2012 (left) and 118 images from 26 July 2012 (right), all obtained with the near-infrared NIRC2 camera coupled to the adaptive optics system on the Keck II telescope. In each image, the north pole is on the right. The white features are high altitude clouds like Earth's cumulus clouds, while the bright blue-green features are thinner high-altitude clouds akin to cirrus clouds. Reddish tints indicate deeper cloud layers. (Lawrence Sromovsky, Pat Fry, Heidi Hammel, Imke de Pater, and the Keck Observatory)

Jupiter • July 23, 2009
Hubble Space Telescope
Wide Field Camera 3

Figure G.8 The atmospheric debris from a comet or asteroid that collided with Jupiter on 19 July 2009 appears dark in this visible-light Hubble image taken four days after the impact. (NASA, ESA, H. Hammel, and the Jupiter Impact Team)

Chelyabinsk Meteor Explosion

A meteor that was about 10 meters in radius when it entered Earth's atmosphere exploded at an altitude of ∼20 km near the Russian city of Chelyabinsk in early 2013 (Fig. G.10). This explosion, with an energy release very roughly equivalent to that of 0.5 Mt of TNT, was less than one-tenth as energetic as that over Tunguska (§5.4.6), but because it occurred in a much more densely populated region, more than 1000 people were injured seriously enough to seek medical attention.

The Moon

Some locations within craters near the lunar poles are in permanent shadow, as illustrated in Figure G.11. These regions may remain as cold as 40 K. Based on data by several spacecraft (especially the impact by the Lunar CRater Observation and Sensing Satellite, LCROSS), some of the shadowed craters contain tiny ice crystals mixed in with the soil at a concentration of a few percent by weight. This

Figure G.9 Color composite of images taken by amateur astronomer Anthony Wesley, revealing the flash (at 4 o'clock) from a bolide entering Jupiter's atmosphere in June 2010. (Adapted from Hueso *et al.* 2010)

ice must have been brought in by comets and asteroids well after the Moon had formed because the Moon, in general, is very dry.

The multi-year, high-resolution, imaging done by the Lunar Reconnaissance Orbiter, LRO, has obtained before and after pictures of recent cratering events. In some cases, such as the pair of images shown in Figure G.12, the new crater can be associated with an impact flash observed from Earth.

Mercury

Images of Mercury's surface, such as shown in Figure G.13, resemble those of the Moon because craters are the dominant landform on both bodies.

Figure G.10 A meteorite contrail is seen over Chelyabinsk on 15 February 2013. (AP Photo/Chelyabinsk.ru)

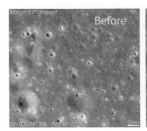 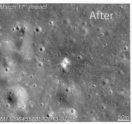

Figure G.12 Before and after images of an 18-meter-diameter lunar crater that formed on 17 March 2013. Both the crater and its rays are much brighter than the surrounding terrain in Mare Imbrium (NASA/ASU/LORC). The associated movie shows the flash of the impact of the ~40 kg meteoroid that produced the crater.

Figure G.11 Mosaic of the Moon's north pole, constructed from images taken at various Sun angles with the wide-angle camera on LCROSS. A polar stereographic projection between latitudes of 60°N to 90°N is shown, including permanently shadowed craters. (PIA14024; NASA GSFC/ASU)

Figure G.13 Image of Mercury taken with the MESSENGER spacecraft. Most striking on this image are the large rays that appear to emanate from a relatively young crater in the far north. (NASA/JHU/CIW, PIA11245)

Mercury's surface composition appears to be very different from that of the other terrestrial planets and the Moon. Both ground-based microwave data and X-ray spectroscopy from the MESSENGER spacecraft reveal a surface that is depleted in Fe and Ti, by roughly an order of magnitude. The Mg/Si ratio at Mercury's surface is ~2–3 times as high as the ratios seen in terrestrial ocean and lunar mare basalts, whereas the Al/Si and Ca/Si are lower by a factor of two; the discrepancy with lunar highland rocks is about two times larger still for these three ratios, and hence rules out a lunar-like feldspar-rich crust. These ratios were inferred from ground-based infrared spectral data, and confirmed by X-ray spectrometry from MESSENGER. Perhaps even more intriguing is

the $\gtrsim$ order of magnitude enhancement in the abundance of sulfur, a rather volatile element, compared to that of the silicate portion of Earth, the Moon, Mars, asteroids, and stony meteorites. This surface composition, in particular the high volatile content, appears to be at odds with the conventional formation scenario of Mercury, in which a large impact 'blasted' Mercury's mantle away (§13.6.1). The surface composition suggests that Mercury probably formed from highly reduced, but not strongly volatile depleted planetesimals, perhaps materials similar to comet dust.

Abundances of the radiogenic elements potassium, thorium, and uranium, measured with the gamma-ray spectrometer on board the MESSENGER spacecraft, revealed that the K/Th ratio is similar to that measured on other terrestrial planets. (It is an order of magnitude lower on the Moon, indicative of the depletion of lunar volatiles compared to Earth.) The measured values of the radiogenic elements, which are the primary long-lived source of internal heat generation, indicate that heat production was about four times as high 4.5 Gyr ago as it is today. Calculations indicate that the internal heat production declined substantially since Mercury's formation, which is consistent with widespread volcanism shortly after the end of the late heavy bombardment 3.8 Gyr ago.

Volcanism appears to be widespread on Mercury, as evidenced from, e.g., dome-like features, pyroclastic vents, and pit craters. Smooth plains seem to fill craters and embay crater rims. At places, volcanic plains are more than 1 km thick and appear to have been formed in multiple phases of emplacement in a flood-basalt style consistent with the measured surface compositions being more refractory than those of basalts. Most intriguing are irregular shallow depressions, referred to as *hollows*, shown in Figure G.14. Many hollows are associated with impact craters. These hollows likely involve a recent loss of volatiles through (some combination of) sublimation, outgassing, volcanic venting, or space weathering.

Confirmation of the presence of water-ice in permanently shaded craters near the poles is shown in Figure G.15, which displays a composite image with radar reflections obtained with the Arecibo Observatory superposed on a MESSENGER image of Mercury's south pole.

Mars

The variable ice caps, other seasonal changes and 'canals' on Mars convinced some scientists that life existed on Mars. Although we now know that the surface of the red planet is not currently inhabited, the questions of whether there is life underground or there has been life in the past are central to Mars exploration programs today, in particular in sending the Mars Exploration Rovers (MER) Spirit and Opportunity, and the much larger rover Curiosity (Fig. G.4).

Spirit's mission ended in March 2010, but Opportunity is still driving around (after 10 years!) and investigating the martian surface. After Victoria crater, it set course to the 22-km-diameter Endeavour crater, where it arrived in the summer of 2011. Layers of bedrock exposed at Victoria crater and other locations revealed a sulfate-rich composition indicative of an ancient era when acidic water was present. After arriving at the rim of Endeavour crater, the rover stumbled upon a vein, shown in Figure G.16, rich in calcium and sulfur, possibly made of the calcium-sulfate mineral gypsum. This vein shows that water must

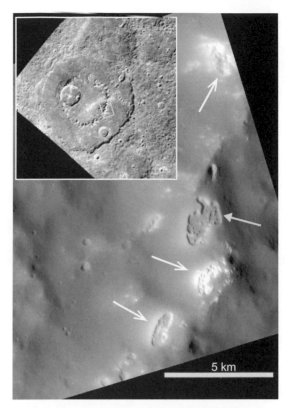

Figure G.14 The MESSENGER spacecraft imaged an unexpected class of shallow, irregular depressions (arrows), referred to as hollows. Some hollows have bright interiors and halos. This image shows hollows on the peak-ring mountains of an unnamed 170-km-diameter impact basin (inset). (Blewett *et al.* 2011)

have flowed through underground fractures in the rock, forming the chemical deposit gypsum ($CaSO_4 \cdot 2H_2O$).

The Curiosity rover landed at Gale crater in 2012. Its goal was to test whether ancient aqueous environments could also have been habitable. Data from Curiosity revealed that water once flowed from the rim of Gale crater and pooled at the base of its central mountain to form a lake-stream-groundwater system that may have existed for millions of years. Clay minerals and lake deposits show evidence for a moderate to neutral pH in the lake, and a very low concentration of salt. The data further indicate that minimal weathering of the crater rim occurred, suggesting that a colder and/or drier climate was prevalent. Taken together, the Curiosity results strongly suggest that early Mars was habitable, but whether life did exist is still a largely open question.

One important question is the effect of radiation on the preservation of chemical signatures of past life. Extrapolating measurements over the first year of Curiosity's operations predict a 1000-fold decrease in organic

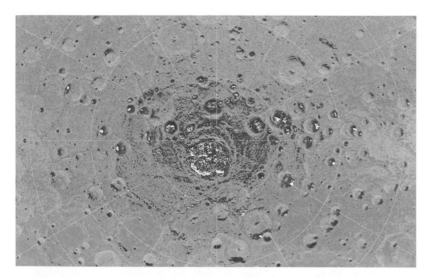

Figure G.15 The highest-resolution radar image of Mercury's south polar region made from the Arecibo Observatory is shown in white on an image obtained with the MESSEN-GER spacecraft. The image is col-orized by the fraction of time the surface is illuminated. Areas in per-manent shadow are black. Radar-bright features in the Arecibo image all co-locate with craters that are in permanent shadow. This image is shown in a polar stereographic projection with every 5° of latitude and 30° of longitude indicated and with 0° longitude at the top. The large crater near Mercury's south pole, Chao Meng-Fu, has a diameter of 180 km. (PIA15533, NASA/JHU APL/CIW)

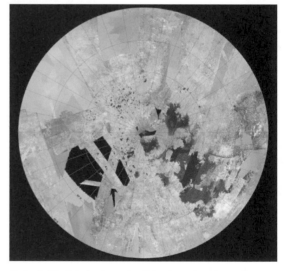

Figure G.16 A false-color view of a mineral vein imaged with the panoramic camera (Pancam) on NASA's Mars Exploration Rover Opportunity. The vein is about 2 cm wide and 45 cm long. Opportu-nity found it to be rich in calcium and sulfur, possibly the calcium-sulfate mineral gypsum. (NASA/JPL/Cornell, PIA15034)

molecules of over 100-atomic mass units in ~650 mil-lion years. Calculations suggest that preservation of any organics that accumulated in the primary environment is possible, although the signal may be substantially reduced.

Europa

Hubble Space Telescope (HST) images from Decem-ber 2012 revealed evidence of water plume activity over Europa's south pole. The data are consistent with two 200-km-high plumes of water vapor with line-of-sight column densities of about 0.1 cm^{-2}. The fact that no plumes were detected at other times (November 2012, and in 1999), suggests variations in plume activity that might depend on the variations of surface stresses with Europa's orbital phases. The plume was present when Europa was near apojove and was not detected close to its perijove, in agree-ment with tidal modeling predictions.

Figure G.17 Colorized mosaic of Titan's northern land of lakes and seas. The data were obtained by Cassini's radar instrument from 2004 to 2013. In this projection, the north pole is at the center. The view extends down to 50° N latitude. In this color scheme, liquids appear blue and black depending on the way the radar bounced off the surface. Land areas appear yellow to white. A haze was added to simulate the Titan atmosphere. The area above and to the left of the north pole is dotted with smaller lakes. Lakes in this area are about 50 km across or less. (PIA17655; NASA/JPL-Caltech/ASI/USGS)

Titan

Figure G.17 shows a composite image of Titan's northern lakes and seas. The liquid in Titan's lakes and seas is mostly methane and ethane. Most of the bodies of liquid on Titan lie in the northern hemisphere. In fact nearly all the lakes and seas on Titan fall into a rectangle covering

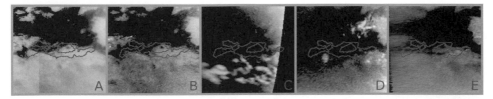

Figure G.18 (a) This huge arrow-shaped storm that blew across the equatorial region of Titan on 27 September 2010 created large effects on Titan's surface in the form of dark – likely wet – areas. After this storm dissipated, Cassini observed significant changes on Titan's surface at the southern boundary of the dune field named Belet, shown in panels b. (b) This series of images shows the changes on Titan's surface, attributed to the rainstorm in panel a. These changes covered an area of 500 000 km^2. Image A in this montage was taken on 22 October 2007, and shows how this region had appeared before the storms. In image B, taken on 27 September 2010, the huge arrow-shaped cloud is on the left, just out-of-frame. The arrow-shaped cloud was quickly followed by extensive changes on the surface that can be seen in image C (14 October 2010) and image D (29 October 2010). By 15 January 2011 (image E), the area mostly appears dry and bright, with a much smaller area still dark, i.e., wet. The brightest spots in these images are methane clouds in the troposphere, the lowest part of the atmosphere, which are most visible on the left of image B, the lower half of image C, and the right of image D. (NASA/JPL/SSI)

about 900 by 1800 km. Only three percent of the liquid 'rain' at Titan falls outside of this area.

Images taken in 2006, when it was winter in the northern hemisphere, and 2012 when it was spring, show no apparent change in the lakes, consistent with climate models that predict stability of liquid lakes over several years. This shows that the northern lakes are not transient weather events, in contrast to the temporary darkening of parts of the equator after a rainstorm in 2010 (Fig. G.18).

G.6 Planetary Interiors

Moon

The Moon's gravitational field has been determined to spherical harmonic degree and order 420 (which corresponds to 13-km-sized blocks) using tracking data from the Gravity Recovery And Interior Laboratory (GRAIL). The GRAIL mission consisted of two spacecraft, each equipped with a gravity ranging system, which measured the change in distance by intersatellite ranging as the spacecraft flew above the lunar surface. In contrast to previous results at low harmonic degrees (§6.3.1), GRAIL revealed that over 98% of the gravitational signature between degree 80 (68 km) and 320 (17 km) is associated with topography; this has not been observed for any major planet, although Mercury might show a similar correlation once global gravity maps at high enough degrees are obtained. Indeed, one might expect gravity and topography to better correlate with increasing degree, because the lithosphere is increasingly able to support topographic loads at shorter wavelengths without compensating masses at depth. Most of the correlation at the 30–130 km scales is related to impact craters.

A map of the Bouguer anomaly (§6.1.4.4), obtained after subtracting the expected signal of surface topography

from the free-air gravity map, reveals the gravitational structure of the subsurface. For a perfect correlation between gravity and topography, the Bouguer map would be zero everywhere. Positive and negative anomalies in the Bouguer map can be explained either by lateral variations in the subsurface density or by variations in crust thickness.

Using the GRAIL data, the average density of the highlands crust was found to be 2.55 ± 0.02 g cm^{-3}, which is much lower than the hitherto assumed density of 2.8–2.9 g cm^{-3} that is typical for the anorthositic crustal materials on the Moon. Lateral variations have been identified of up to ± 0.25 g cm^{-3}. For example, the South Pole Aitken basin shows a density of 2.80 g cm^{-3}, while regions with lower-than-average densities are seen around the impact basins Orientale and Moscoviense, two of the largest young impact basins on the Moon. The overall lower density of 2.55 g cm^{-3} has been attributed to impacts fracturing the crust. Typical crustal porosities of a few up to ∼20% can explain the relative low densities compared to the expected 2.8–2.9 g cm^{-3}.

With GRAIL's high spatial resolution, distinctive gravitational signatures can be recognized, such as impact basin rings, central peaks of complex craters, volcanic landforms, and smaller simple bowl-shaped craters. The free-air gravity field over lunar mascons (§6.3.1) has been characterized in detail, and reveals a bull's-eye pattern, with a central positive anomaly, i.e., the mascon, surrounded by a negative collar, and a positive outer annulus. Numerical models show that this pattern is a natural consequence of excavation from the impact crater, followed by post-impact isostatic adjustment and cooling and contraction of a voluminous melt pool. GRAIL further showed that the central peak ring in the 417-km-diameter Korolev basin encloses a Bouguer high, while the surrounding low

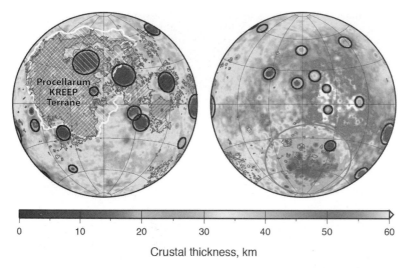

0   10   20   30   40   50   60

Crustal thickness, km

Figure G.19 Global map of crustal thickness of the Moon derived from gravity data obtained by NASA's GRAIL spacecraft. The lunar near side is shown on the left; the far side on the right. On the left, outlined in white, is the Procellarum KREEP Terrane, which contains high abundances of potassium, rare earth elements, and phosphorus. In addition to the South Pole Aitken basin (the gray circle on the right), there are 12 impact basins with crustal thinning that have diameters over 200 km on each hemisphere, marked with black circles. (PIA17674; NASA/JPL-Caltech/S. Miljkovic)

Bouguer anomaly resides on the crater floor and is not co-located with the crater walls. This suggests that there is a high density at the center of the crater, and a density deficit under the floor, caused perhaps by less dense, possibly brecciated, material in this area.

Assuming everywhere a crustal porosity of 12% and a mantle density of 3.22 g cm^{-3}, the data can be used to derive the map of the Moon's crustal thickness displayed in Figure G.19. The minimum crustal thickness of <1 km is in the interior of the far-side basin Moscoviense; the thickness at the Apollo 12 and 14 landing sites is 30 km.

Enceladus

As discussed in §§4.3.3.3, 5.5.6.2, and 6.3.5.2, the Cassini spacecraft imaged active geysers on the south pole of Enceladus, the origin of which has been much debated. Two leading theories include the presence of a liquid ocean under the south pole driving the jets, or the presence of diapirs driving the geysers. Using precise measurements of the orbit of the Cassini spacecraft through Doppler tracking, the satellite's quadrupole gravity field and harmonic coefficient J_3 have been determined. The ratio J_2/C_{22} differs slightly from the value of 10/3 required for hydrostatic equilibrium, which suggests that the satellite is not in a fully relaxed shape. The moment of inertia is $0.335MR^2$, and the value of the J_3 coefficient implies a negative mass anomaly in the south-polar region, which is largely compensated by a positive subsurface anomaly compatible with a subsurface ocean 30–40 km below the surface and extending over southern latitudes from ∼50° to the south pole.

An image of Enceladus in Saturn's E ring is displayed in Figure G.20.

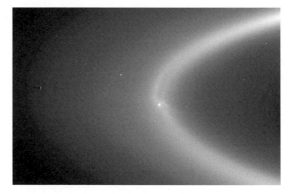

Figure G.20 Wispy streaks of bright, icy material reach tens of thousands of km outward from Saturn's moon Enceladus into the E ring, and the moon's active south polar jets continue to eject material. The Sun is almost directly behind the Saturn system from Cassini's vantage point, so small particles are 'lit up', but Enceladus itself is dark. Tethys is visible to the left of Enceladus. The image was taken in visible light when Cassini was at ∼2.1 million km from Enceladus. (NASA/JPL/SSI, PIA08321)

G.7 Magnetic Fields and Plasmas

Heliopause

Voyager 1 entered the Interstellar Medium (ISM) on 25 August 2012, when it was at a heliocentric distance of 121 AU. At that time, an abrupt decrease was detected in the charged particle and anomalous cosmic rays (ACRs) intensities, with a simultaneous increase in the galactic cosmic ray (GCR) intensity. In April 2013, the Voyager 1 plasma wave instrument started to detect locally generated electron plasma oscillations at a frequency of about 2.6 kHz. This oscillation frequency corresponds to an electron density of about 0.08 cm^{-3}, very close to the value expected in the ISM. Voyager 1 is now in a region referred to as the *disturbed interstellar medium.*

Mercury

Mercury's magnetic field is very weak compared to that of Earth, with a surface field strength of 195 nT, a bit less than 1% of that on Earth. A sketch of Mercury's magnetosphere and plasma population based on the MESSENGER findings is shown in Figure G.21.

Saturn

The auroral footprint of Enceladus has been seen in images of Saturn's aurora (Fig. G.22). It resembles the footprints of Io, Europa of and Ganymede on Jupiter (Fig. 7.39).

G.8 Meteorites

Two CM carbonaceous chondrite meteorite falls, Maribo in 2009 and Sutter's Mill in 2012, had encounter velocities with the Earth of $\sim$28 km/s, much higher than any other meteorites with known pre-impact orbits. Both had orbits with Tisserand parameter (§10.2.4) $C_T \approx 3$ and aphelia $\sim$4–5 AU. Such orbits are similar to those of some Jupiter family comets. The Orgueil meteorite is a (very primitive) CI chondrite that fell in France in 1864. Visual observations of Orgueil's fireball have been analyzed to estimate the pre-impact orbit. The derived orbit also has an aphelion near Jupiter, and is similar to the orbits of Jupiter family comets. These dynamical results support arguments based on compositional grounds that comets are the source of some chondritic meteorites.

G.9 Minor Planets

Figure G.23 shows close-up views of those comets and asteroids that had been imaged by interplanetary spacecraft as of 2010.

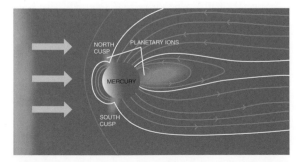

Figure G.21 Sketch of Mercury's magnetic field and plasma population, as derived from *MESSENGER* data. Maxima in heavy ion fluxes are indicated. (Adapted from Zurbuchen *et al.* 2011)

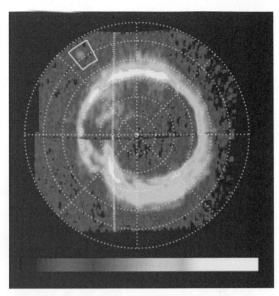

Figure G.22 Cassini UVIS EUV polar-projected image of Saturn's northern aurora, including the Enceladus auroral footprint (inside white box). The color bar shows EUV emission per pixel. The north pole is at the center; the latitude circles are 5° apart, and the hashed white line indicates the day/night terminator. The Sun is to the left. (PIA13764; NASA/JPL/University of Colorado/Central Arizona College)

Near-Earth objects (§9.1.1) whose orbits come within 8 million km of Earth and that are large enough to cause significant damage if they hit Earth are referred to as *potentially hazardous asteroids*, or PHAs. The number of PHAs with $R > 50$ meters has been estimated at 4700 ± 1500.

21 Lutetia

The largest asteroid in Figure G.23 is 21 Lutetia, which was photographed by ESA's Rosetta spacecraft. Although classified as an M-type asteroid, Lutetia does not display much evidence of metals on its surface. Moreover, its spectrum resembles that of carbonaceous chondrites and C-type asteroids and is not at all like that of metallic meteorites.

Lutetia has a complex and morphologically diverse surface. There are several large (tens of kilometers) craters and many smaller craters; it clearly is a very old object. Lutetia is covered by a $\sim$600 m thick layer of regolith, which is revealed by unique landslide structures along the walls of some craters. Many other structures can be recognized, such as pits, crater chains, ridges, scarps, and grooves, some radially aligned and others concentric around relatively young craters.

Figure G.23 Views of the first four comets (lower right) and nine asteroid systems that were imaged close-up by interplanetary spacecraft, shown at the same scale. The object name and dimensions ('diameters' along two or three axes), as well as the name of the imaging spacecraft and the year of the encounter, are listed below each image. Note the wide range of sizes. Dactyl is a moon of Ida.

4 Vesta

The Dawn spacecraft orbited 4 Vesta for more than one Earth year before continuing its journey to 1 Ceres. A full-disk image of the asteroid taken by Dawn is shown in Figure G.24. Vesta's mean radius is 265 km, which makes it the largest asteroid apart from Ceres. Vesta, a V-type asteroid, is unique among large objects because it has a basaltic surface. Two prominent pyroxene absorption bands dominate its spectrum, as is the case for HED achondrite meteorites (§8.7.1).

Vesta's surface topography is characterized by a 460-km-diameter crater at the south pole. The Dawn spacecraft identified a pronounced central peak, ~100 km across, that rises 20–25 km above the relatively flat crater floor. A set of circumferential troughs associated with Vesta's south pole crater are seen near the equator. An older basin is offset from the south pole crater; this ancient basin has its own set of troughs. Numerous other depressions on Vesta's surface may also be remains of large impact basins. Figure G.25a shows a more detailed view of part of the surface, revealing a scarp with landslides and craters in the scarp wall. As

shown, these smaller craters have a simple bowl-shaped morphology; some craters have central peaks.

Vesta has a higher albedo than do most asteroids, with many brighter and extremely dark spots across its surface, some of which are shown in Figures G.25b and c. The bright areas occur mostly in and around craters, but the dark materials also seem to be related to impacts and include carbon-rich compounds.

Several small V-type, or *vestoid*, asteroids have been detected in orbits similar to that of Vesta. These asteroids and the parent bodies of the HED meteorites may have been ejected in the impact that produced the south pole crater. Assuming this to be the case, these HED meteorites were used to develop a model of the origin and interior structure of Vesta. This model suggests that Vesta accreted within the first ~2 Myr of the formation of our Solar System, trapping short-lived radionuclides in its interior, the heat of which led to melting, fractionation, and the formation of an iron core. Dawn's gravity data (Fig. G.26) indeed suggest the presence of a ~110-km-radius iron core. This, together with the mineral composition and

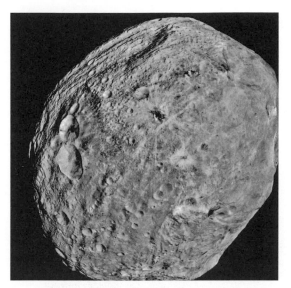

Figure G.24 This full view of the giant asteroid Vesta was taken by NASA's Dawn spacecraft from a distance of ~5000 km. The northern hemisphere (upper left) is heavily cratered, in contrast to the south. The cause of this contrast, as well as the origin of grooves that circle the asteroid near its equator, is unknown. The movie best illustrates the equatorial grooves and also shows the ~460-km-diameter crater near Vesta's south pole. The resolution of this image is about 500 m per pixel. (NASA/JPL-Caltech/UCLA/MPS/DLR/IDA, PIA14894)

crater chronology of Vesta's surface, is consistent with the aforementioned formation scenario.

1 Ceres

Water vapor has been observed to be escaping from 1 Ceres. The sources are localized and time-variable. The water appears to be coming from dark regions near the equator.

136472 Makemake

Makemake, the third largest TNO known, is an icy dwarf planet with a spectrum similar to those of Pluto and Eris, although its methane-ice absorption bands are even stronger than for either Pluto or Eris. Occultation measurements reveal a body with radii of 717×710 km. Makemake's rotation period is 7.77 hr. Its visible albedo, $A_0 \approx 0.8$, is in between those of Pluto and Eris, and points at an icy surface. However, there is also a small, much warmer region on its surface, characterized by an albedo $A_0 \sim 0.12$.

Stellar occultation measurements, when Makemake was at a heliocentric distance of 52.2 AU, revealed an upper limit of ~ 100 nbar to the surface pressure produced by an atmosphere. The absence of an atmosphere suggests a lack of nitrogen-ice on the body's surface, as the vapor pressure of N_2-ice is at microbar levels, assuming Makemake's temperature is equal to its equilibrium temperature.

136199 Eris

The dwarf planet and TNO Eris was discovered in 2005. Eris's orbit about the Sun is quite eccentric, with $e = 0.44$ and $i = 44°$. The orbit of its satellite, Dysnomia, indicates a mass of Eris that is $\sim 27\%$ larger than that of Pluto (Table 9.2). Eris's size has been determined via a multi-chord stellar occultation experiment, which showed the dwarf planet to be quite spherical with a radius of 1163 km, so that its density is ~ 2.52 g cm^{-3}.

Eris has a very high visible geometric albedo, $A_0 = 0.96 \pm 0.07$. At present, Eris is close to its aphelion, about 97 AU from the Sun. At this distance, one would not expect Eris to have an atmosphere. However, one might expect

Figure G.25 Views of asteroid 4 Vesta as imaged by NASA's Dawn spacecraft. (a) This image was taken through the camera's clear filter and shows a steep scarp with landslides and vertical craters in the scarp wall. (b) Image of bright material that extends out from the crater Canuleia on Vesta. The bright material appears to have been thrown out of the crater during the impact that produced it, and extends 20–30 km beyond the crater's rim. (c) Image of a dark-rayed impact crater and several dark spots on Vesta. (NASA/JPL-Caltech/UCLA/MPS/DLR/IDA, a: PIA14716, b: PIA15235, c: PIA15239)

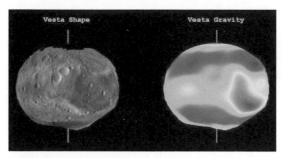

Figure G.26 The gravity field of Vesta closely matches its surface topography. The movie shows shaded topography with troughs and craters on the left, and color-contoured data from Dawn's gravity experiment on the right. Red denotes the areas with a higher than average gravity field and blue-purple denotes the areas where the field is weaker on average. The highest topography, on the rim of the Rheasilvia basin deep in the southern hemisphere, shows a particularly strong gravity field. The dashed line indicates the north–south axis. The topography model is derived from images taken from a high altitude (680 km above the surface), and the gravity data come from the low-altitude orbits (210 km above the surface). (PIA15602; NASA/JPL-Caltech/UCLA/MPS/DLR/IDA)

ices to sublime off its surface when Eris approaches its perihelion, which is only 37.8 AU from the Sun.

2012 VP₁₁₃

The TNO 2012 VP_{113}, with a perihelion of 80 AU and semimajor axis of 266 AU (i.e., aphelion distance of 452 AU), is the second body discovered to date (April 2014) that is likely to reside in the inner Oort Cloud. The first such body was the TNO 90377 Sedna (§9.1.2). As Sedna's perihelion distance is 4 AU smaller than that of 2012 VP_{113}, the newly discovered object has the largest perihelion distance yet discovered. Based on this new observation and new simulations, the total mass of the inner Oort Cloud is expected to be ~0.013 $M_{\oplus}$, very similar to the total mass of KBOs (0.01 $M_{\oplus}$).

G.10 Comets

Figure G.27 shows a small comet approaching the Sun; in the associated movie, one can see the comet completely vaporize because of the heat from the Sun.

Main Belt Comets

As of March 2014, about ten main belt comets have been detected and, in addition, mass loss has been reported from three highly eccentric bodies that have aphelia well interior to Jupiter's orbit (3200 Phaethon, 2201 Oljato, and 107P/Wilson-Harrington). The loss of mass from these bodies has been attributed to a variety of causes: rotational

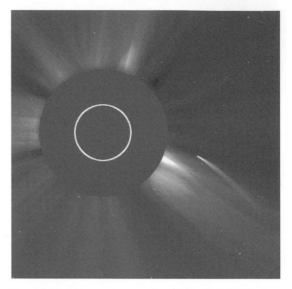

Figure G.27 A Sun-grazing comet caught by the LASCO coronagraph on the SOHO spacecraft as it moved toward the Sun on 5–6 July 2011. The inner few solar radii are blocked by a coronagraph. The circle is drawn in to represent the size and location of the solar disk. (Courtesy SOHO/LASCO consortium; ESA and NASA). The associated movie taken by the Solar Dynamics Observatory (SDO) shows the comet to evaporate completely.

instability, impact ejection, electrostatic repulsion, radiation pressure sweeping, dehydration stresses, and thermal fracture, in addition to the sublimation of ice that was mentioned in §10.2.4. The repetitive nature of the observed activity of 133P/Elst-Pizarro and 238P/Read is suggestive of ice sublimation, while a recent impact is a most likely cause for the activity in (596) Scheila and P/2010 A2, although the activity in the latter object may also be explained as having shed mass after reaching rotational instability. Rotational instability is the most likely cause for the shedding of mass from P/2013 P5 (Fig. G.28) and P/2013 R3.

67P/Churyumov–Gerasimenko

On 6 August 2014, the Rosetta spacecraft arrived at Comet 67P/Churyumov–Gerasimenko. Churyumov–Gerasimenko is a Jupiter-family comet, with an aphelion of 5.68 AU and perihelion at 1.24 AU. Rosetta is studying the comet with a combination of remote sensing and *in situ* measurements to characterize the environment and the comet nucleus. An image taken by Rosetta just prior to orbit insertion is shown in Figure G.29.

103P/Hartley 2

The Deep Impact spacecraft, whose prime mission was to produce and observe an impact crater at Comet 9P/Tempel

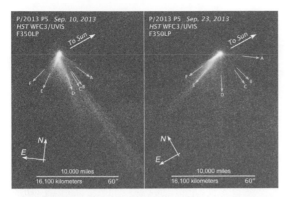

Figure G.28 Composite images of P/2013 P5 at UT 2013 September 10 (left) and 23 (right). This main belt comet reveals an extraordinary system of 6 tails, indicated by letters. HST observations taken 2 weeks apart show dramatic morphological changes in the tails. Each tail is associated with a unique ejection date, indicative of a continued, though episodic, ejection of mass. (Jewitt *et al.*, 2013)

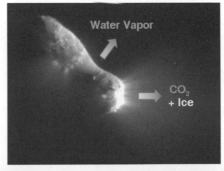

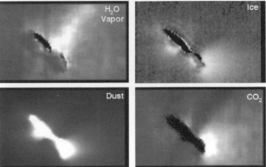

Figure G.30 Infrared scans of Comet Hartley 2 by NASA's EPOXI mission spacecraft on 4 November 2010 show carbon dioxide, dust, and ice being distributed in a similar way and emanating from apparently the same locations on the nucleus. Water vapor, however, has a different distribution. (PIA13628; NASA/JPL-Caltech/UMD)

Figure G.29 Comet 67P/Churyumov Gerasimenko imaged by the Rosetta spacecraft on 3 August 2014, as the spacecraft approached the comet. As shown, the comet has a double-lobed structure. Whether it is a 'contact binary' (two comets that joined together) or was a single comet whose shape has been sculpted by subliming ices, is yet to be determined. (Rosetta/ESA)

1 (§10.6.5), was later renamed EPOXI and flew past Comet 103P/Hartley 2 in November of 2010. High-resolution images of Hartley 2 (Fig. G.30) revealed that the spatial distribution of carbon dioxide, dust, and ice is very similar, but that water vapor has a different distribution, which implies a different source region and process.

G.11 Planetary Rings

NASA's Cassini spacecraft has continued to return spectacular data illuminating the properties of Saturn's rings.

Among the most interesting findings are signatures of impacts onto the rings. Figure G.31 shows images of a corrugation pattern in the C ring that were taken near the 2009 equinox, when the rings were illuminated very obliquely. The winding angle of the pattern suggests that it was formed by an impact that tilted a portion of the rings around 1983, very likely the same impact that was responsible for the corrugation pattern observed in the neighboring D ring (§11.3.2.1). Modeling suggests that the cloud shown in Figure G.32 was produced by a stream of debris derived from the breakup of a 1–10-meter object that impacted Saturn's A ring about 24 hours prior to the image being taken.

Analysis of images of Jupiter's main ring taken by the Galileo and New Horizons spacecrafts reveals vertical corrugations similar to those found in the low optical depth inner rings of Saturn. The most prominent pattern appears to have originated in the third quarter of 1994, and so is attributed to Comet D/Shoemaker-Levy 9, which impacted Jupiter in July of 1994 (§5.4.5). Impacts by comets or their dust streams are regular occurrences in planetary

Radius (1000 km)

90   85   80   75

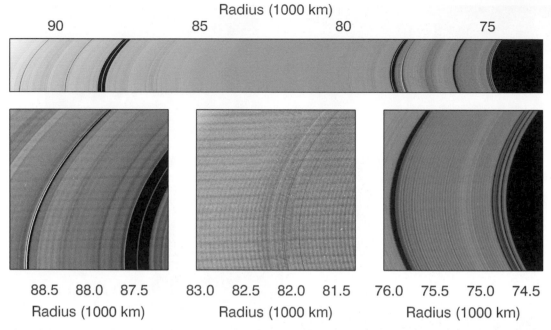

88.5 88.0 87.5   83.0 82.5 82.0 81.5   76.0 75.5 75.0 74.5
Radius (1000 km)   Radius (1000 km)   Radius (1000 km)

Figure G.31 Mosaic of images of Saturn's C ring taken during Cassini's orbit 117, together with close-ups of selected radial regions showing the periodic bright and dark bands that extend across the entire C ring. The contrast has been adjusted in each close-up image to enhance the appearance of the banded structure. Horizontal lines within the close-ups are camera artifacts. (Hedman *et al.* 2011)

Figure G.32 An ejecta cloud from a small impact on Saturn's A ring is apparent as a canted feature several thousand km long in this Cassini image taken near the time of the saturnian equinox. (Tiscareno *et al.* 2013)

rings, altering them in ways that remain detectable for decades.

Figure G.33 shows a propeller-shaped disturbance in Saturn's A ring produced by a large ring particle/small moon that is probably ~500 m in radius.

The entire 360° extent of the F ring is visible in the projected mosaic of Cassini images shown in Figure G.34. The basics of confinement are explained in §11.4.3, but additional processes are also clearly operating to create the complicated features observed.

Rings of 10199 Chariklo

When the Centaur 10199 Chariklo occulted a star, the occultation profile revealed the presence of two rings (Fig. G.35). Chariklo has an equivalent radius of 124 ± 9 km and orbits between Uranus and Saturn, with aphelion near Uranus's orbit. The rings, with respective widths of about 7 and 3 km, have (normal) optical depths of 0.4 and 0.06, and mean orbital radii of 391 and 405 km. The present orientation of the ring is consistent with an edge-on geometry in 2008, which provides a simple explanation for the dimming of the Chariklo system between 1997 and 2008, and for the gradual disappearance of ice and other absorption features in its spectrum over the same period. This implies that the rings are partly composed of water-ice. Note that for the rings to be located interior to Roche's limit (eq. 11.8), the ring particles must be significantly less dense than is Chariklo.

Figure G.33 NASA's Cassini spacecraft captured a propeller-shaped disturbance in Saturn's A ring produced by a large ring particle/small moon that is too small to be seen here. The body, likely ~500 m in radius, is at the center of the image. It has cleared ring material from the dark wing-like structures to its left and right in the image. Disturbed ring material closer to the 500 m object reflects sunlight brightly and appears like a white airplane propeller. The propeller structure is 5 km in the radial dimension. The dark wings appear 1100 km in the azimuthal direction, and the central propeller structure is 110 km long. This image has been reprojected so that orbiting material moves to the right and Saturn is down. This view looks at the sunlit side of the rings, and resolution is ~1 km/pixel. (NASA PIA 12789)

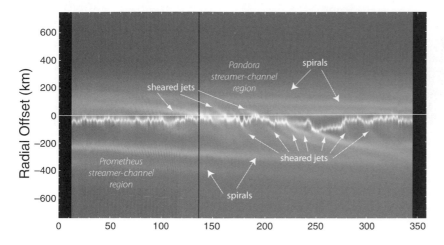

Figure G.34 Mosaic of *Cassini* spacecraft images of Saturn's multistranded F ring, annotated to show the prominent jets, spirals, and channels that are produced by the nearby moons Prometheus and Pandora. (Colwell *et al.* 2009)

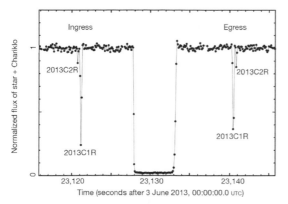

Figure G.35 Lightcurve of the occultation by the Chariklo system. The data were taken with the Danish 1.54 m telescope (La Silla) on 3 June 2013, at a rate of almost 10 Hz. The sum of the stellar and Chariklo fluxes has been normalized to unity outside the occultation. The central drop is caused by Chariklo, and two secondary events, 2013C1R and 2013C2R, are observed, first at ingress (before the main Chariklo occultation) and then at egress (after the main occultation). (Braga-Ribas *et al.* 2014)

G.12 Exoplanets

G.12.1 Images and Spectra of Exoplanets

Massive young exoplanets and brown dwarfs located several AU or further from the star that they orbit have been imaged in the thermal infrared (e.g., Figure G.36). In most

cases, only one substellar object has been imaged about a given star. But the HR 8799 system has four planets (the outer three are shown in Fig. 12.8b), each of mass ~5–10 M_{2j} based on the age of the star and the thermal luminosities of the planets (Fig. 12.1). At least the middle pair of planets, HR 8799 c and d, must travel on nearly circular orbits for the system to have survived for the millions of years since the star has formed. Dynamically, this is a closely packed group of planets with orbits that are likely to be nearly circular and coplanar. But the planets' masses and the large distances of the outer planets from their star present substantial challenges to planet formation theories.

New instrumentation and observing techniques now allow the thermal emission spectra of extrasolar planets to be measured directly. Figure G.37 shows a spectrum at moderate resolution of HR 8799 b, the outermost of the star's four known planets. The hot ($T_{\rm eff} \approx 1000$ K), young (estimated age of tens of millions of years) planets show absorption features from H_2O, CO, and CH_4. Gaseous H_2O is seen in these hot young objects because, unlike in the case of Jupiter, the atmosphere is too warm for water clouds to form; thus, H_2O is found throughout the atmosphere. Although methane is seen in HR 8799 b, the planet's atmosphere is warm enough and atmospheric mixing is vigorous enough (§4.7) to also allow for a sizable

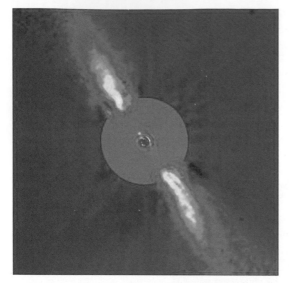

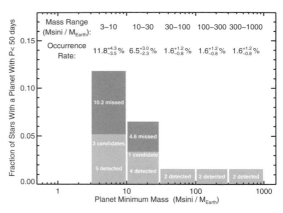

Figure G.36 This composite image represents the close environment of β Pictoris as seen in near-infrared light. This very faint structure is revealed after a very careful subtraction of the much brighter stellar halo. The outer part of the image shows the reflected light from the dust disk; the planet β Pic b in the inner part is the innermost part of the system, and is shown as observed at 3.6 microns. The planet is less than 10^{-3} times as bright as β Pictoris, aligned with the disk, at a projected distance of 8 AU. Because the planet is still very young, it is still very hot, with a temperature around 1500 K. Both parts of the image were obtained on ESO telescopes equipped with adaptive optics. (ESO/A.-M. Lagrange *et al.*)

Figure G.38 Occurrence rates of planets with orbital period $P <$ 50 days as a function of minimum mass based on radial velocity observations. The bottom (green) portions of the histogram represent confirmed planets, and the middle (yellow) and upper (blue) rectangles in the two lowest mass bins represent unconfirmed but likely candidate planets and a correction factor to account for an estimate of the fraction of planets that were not detected, respectively. (Howard *et al.* 2010)

spectroscopy. At very high resolution, molecular bands are resolved into the individual lines, allowing robust identification of molecular species. Furthermore, while the absorption by molecules in Earth's atmosphere is static in wavelength, the exoplanet's molecular lines are Doppler-shifted by amounts that vary by $\sim$100 km s^{-1}. For transiting planets, the maximum signal occurs during transit (transmission spectroscopy) and around – but not during – occultation (day-side spectroscopy). The day-side spectroscopy views the planet's thermal emission, so nontransiting planets can be studied as well, and their masses and orbital inclinations can be measured by determining the radial component of the orbital velocity. In recent years, the Very Large Telescope has detected CO absorption by several hot Jupiters in transmission and in day-side spectroscopy, and H$_2$O absorption was found near 3.2 μm in the day-side spectrum of HD 189733 b.

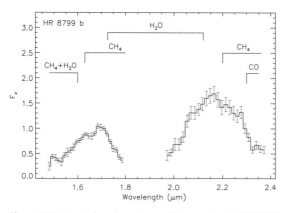

Figure G.37 Near-infrared spectrum of the exoplanet HR 8799 b plotted with 1σ uncertainties. The locations of prominent water, methane, and carbon monoxide absorption bands are indicated. (Courtesy Travis Barman)

quantity of CO to be present as well. The shape of the spectrum can be compared with models to estimate the gravity and temperature of the young planet.

The atmospheres of hot Jupiters (giant vulcans) are being studied via ground-based, high-resolution

G.12.2 Exoplanet Statistics from Ground-Based Surveys

Occurrence rates of planets with orbital periods less than 50 days detected by radial velocity are presented as a function of planetary mass in Figure G.38. The numbers of planets detected supplemented by likely candidates and the estimated fraction of such objects that would have gone undetected represent the mean number of short-period planets in the quoted mass range for Sun-like stars.

Radial velocity surveys have shown that whereas in general multiple giant planet systems are more common than if detectable giant planets were randomly distributed among stars, few additional planets have been found about stars that host hot Jupiters.

With a scant two dozen confirmed planet detections and very limited characterization of most planet and host star properties, microlensing planets do not offer nearly as extensive results for statistical characterization of exoplanets as do radial velocity planets (above and §12.4) and Kepler transiting planet candidates (§G.3). Nonetheless, because of the complementarity of the techniques, microlensing allows the best estimates of the frequencies of planets orbiting several AU from low-mass stars, which are the most common stars in our galaxy. The average number of planets (per star) with masses $5 \, M_{\oplus} < M_p < 10 \, M_{2\!\!\!/}$ at distances of 0.5–10 AU is $1.6^{+0.7}_{-0.9}$. Within these ranges, smaller planets are more common than larger ones (measured in planets per logarithmic mass range), but large planets contain a greater fraction of the total planetary mass.

G.12.3 NASA's Kepler Mission

The major effort to find exoplanets in recent years has been the Kepler mission, a spacecraft launched by NASA in 2009. Kepler's sole scientific instrument is a differential photometer with a wide (105 square degrees) field-of-view that monitored the brightness of approximately 160 000 main-sequence stars, most with a duty cycle of >85% for four years. The Kepler mission was designed to detect Earth-sized planets in the habitable zone of the host star, necessitating both a large sample size and sensitivity to a much larger range of orbital separations than ground-based transit surveys.

Its sensitivity to small planets over a wide range of separations gives Kepler the capability of discovering multiple planet systems. For closely packed planetary systems, nearly coplanar systems, or systems with a fortuitous geometric alignment, Kepler is able to detect transits of more than one planet. For systems with widely spaced planets

or large relative inclinations, not all planets transit, but some non transiting planets are still detectable based on TTVs that their gravitational perturbations produce on one or more transiting planets (§12.2.5).

Kepler's primary goal was to conduct a statistical census of transiting exoplanets. We present preliminary results of this census at the end of this subsection. Additionally, Kepler has discovered many interesting planets and planetary systems. Some highlights are presented below.

Notable Planets & Planetary Systems

Kepler's first major exoplanet discovery was the Kepler-9 planetary system, which includes two transiting near-giant planets with orbital periods of 19.2 and 38.9 days. The nearby 2:1 mean motion resonance induces TTVs of tens of minutes. Analysis of these TTVs enabled the planets to be confirmed and provided estimates of their masses. Each of these planets is slightly smaller and significantly less massive than Saturn. The star also has a substantially smaller transiting planet of unknown mass with an orbital period of 1.6 days. Kepler-18 is analogous to Kepler-9, with two Neptune-mass planets near the 2:1 orbital resonance and a smaller inner planet.

The first rocky planet found by Kepler was Kepler-10 b, which has $M_p = 4.6 \, M_{\oplus}$, $R_p = 1.4 \, R_{\oplus}$, and an orbital period of only 20 hours. This planet's high density results from compression caused by high internal pressure, and its bulk properties are consistent with an Earth-like composition.

Kepler-11 is a Sun-like star with six transiting planets that range in size from ∼1.8–4.2 $R_{\oplus}$. Orbital periods of the inner five of these planets are between 10 and 47 days, implying a very close-packed dynamical system. Transit timing variations have been used to estimate the planets' masses. The orbital and physical properties of these planets are listed in Table G.1, which also gives the flux of stellar radiation intercepted by each of the planets

Table G.1 Kepler-11's planetary system.

	Period (d)	a (AU)	e	i (°)	R_p ($R_{\oplus}$)	M_p ($M_{\oplus}$)	ρ (g/cm^3)	$\mathcal{F}$ ($\mathcal{F}_{\odot}$)
b	10.304	0.091	0.04	89.6	1.8	1.9	1.72	125
c	13.024	0.107	0.03	89.6	2.9	2.9	0.66	92
d	22.685	0.155	0.004	89.7	3.1	7.3	1.28	44
e	32.000	0.195	0.01	88.9	4.2	8.0	0.58	27
f	46.689	0.250	0.01	89.5	2.5	2.0	0.69	17
g	118.381	0.466	<0.15	89.9	3.3	<25	–	4.8

Source: Lissauer *et al.* (2013).

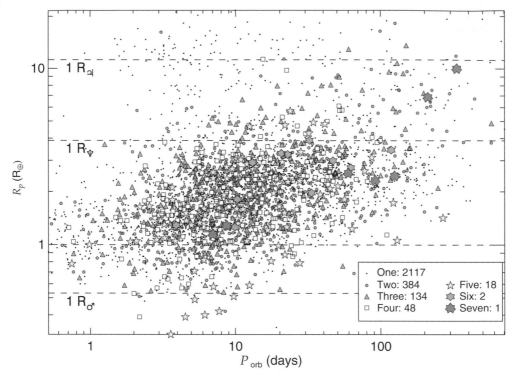

Figure G.39 Planetary candidates found by searching the first three years of Kepler data shown plotted on the orbital period - physical radius plane. Planets that are the sole candidate for their given star are represented by black dots, those in two-planet systems as green circles, those in three-planet systems as blue triangles, those in four planet candidates systems as open black squares, those in five-planet systems as yellow five-pointed stars, those in six-planet systems as orange six-pointed stars and the seven planets orbiting Kepler-90 as red seven-pointed stars. It is immediately apparent that there is a paucity of giant planets in multi-planet systems. The upward slope in the lower envelope of these points is caused by the low SNR of small transiting planets with long orbital periods (for which few transits were observed). (Courtesy Rebekah Dawson)

in units of the solar constant, $\mathcal{F}_\odot$ (the mean solar flux intercepted by Earth; see eq. 3.13), from which the planets' temperatures may be estimated (Problem 12.13). None of these planets is rocky; most if not all have a substantial fraction of their volume occupied by the light gases H_2 and He. Observations of low density sub-Neptune exoplanets such as those in the Kepler-11 system imply that H/He can dominate the volume of a planet that is only a few times as massive as the Earth.

Kepler-20 e was the first planet smaller than Earth to be verified around a main sequence star other than the Sun; its 6-day orbit means that it is far too hot to be habitable. Kepler-37 b, only slightly larger than Earth's Moon, was the first planet smaller than Mercury to be found orbiting a normal star; its orbital period is 13 days, and the stellar host is 80% as massive as the Sun.

Kepler-36 hosts two planets whose semimajor axes differ by little more than 10% but whose compositions are dramatically different. The inner, rocky, Kepler-36 b has a radius of 1.5 $R_\oplus$ and mass of ~4.4 $M_\oplus$, whereas puffy Kepler-36 c has a radius of 3.7 $R_\oplus$ and mass of 8

$M_\oplus$ implying that most of its volume is filled with H/He. Kepler-78b has the shortest period of any confirmed exoplanet, orbiting its star in 8.5 hours. This vulcan planet is slightly larger than Earth, and its mass, measured from the radial velocity variations it induces in its nearby host star, implies a rocky composition.

Circumstellar habitable zones are generally defined to be the distances from stars where planets with an atmosphere similar to that of Earth receive the right amount of stellar radiation to maintain reservoirs of liquid water on their surfaces. Kepler-62 f is the first known exoplanet whose size (1.4 $R_\oplus$) and orbital position suggests that it could well be a rocky world with stable liquid water at its surface.

Planets in Multiple Star Systems

Kepler has found several circumbinary transiting planets. Kepler-16(AB) b is a Saturn-sized ($R_p = 0.7538 \pm 0.0025$ $R_{2\!\!/}$), Saturn-mass ($M_p = 0.333 \pm 0.016$ $M_{2\!\!/}$) planet that orbits about a pair of tightly bound stars. This *circumbinary planet* travels on a nearly circular orbit of period 229 days about a 41-day-period eclipsing binary composed of one star that is about two-thirds the size and mass of the Sun and another less than one-fourth as large and massive as our Sun. The planet is observed to transit both of the stars, allowing for good size estimates. Timing of the transits and eclipses reveals the mutual gravitational interactions of the three bodies enabling the masses to be measured. Kepler-34(AB) b and Kepler-35(AB) b are analogous circumbinary planets. Kepler-47(AB) contains two stars orbiting one another with a period of 7.45 days that in turn are orbited by three known transiting planets, the outermost having an orbital period of 303 days.

Many planets have been detected on close-in orbits around one of the stars in a binary/multiple star system with binary orbital semimajor axes exceeding a few dozen AU. Radial velocity surveys suggest that giant planets within a few AU of individual stars are about as common around stars with stellar companions separated by more than 100 AU as they are around single stars, and binaries with semimajor axis $35\,\text{AU} < a_b < 100\,\text{AU}$ are slightly less likely to host such planets. Kepler-132(AB) has two stars, each a bit larger than our Sun, separated by $\sim$500 AU. One of the stars has two known transiting planets, the other has one transiting planet.

Kepler Planet Candidates

Analysis of the first three years of Kepler data has revealed more than 3500 exoplanet candidates, the vast majority of which are likely to be true exoplanets. Figure G.39 shows the radii and period of Kepler planet candidates, as well as the number of candidates in the system in which they reside. Figure G.40 shows the fraction of small Kepler stars that host planets within various ranges of size and period.

Although little is known for certain about most of Kepler's planet candidates individually, the statistical properties of the ensemble provide key information about planets orbiting within $\sim$0.5 AU of their star. Neptune-sized planets are far more common than are Jupiter-sized ones; the number of planets in a given (fractional) size bin increases with decreasing planet size down to at least $2\,\text{R}_\oplus$, below which the survey is incomplete because of an inadequate signal-to-noise ratio. The number of planets per logarithmic period bin increases with increasing period; one exception to this trend is an excess of Jupiter-sized planets with periods of $\sim$4 days, the hot Jupiter population referred to in §12.3.2; smaller planets do not have a similar concentration of periods.

Almost 1500 transiting planet candidates detected in the first three years of Kepler data are members of multiple candidate systems. There are 384 target stars with two

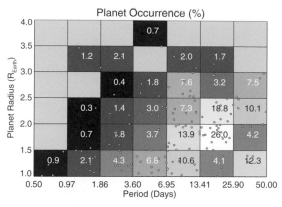

Figure G.40 Occurrence rate (in percent) of planets smaller than Neptune around small (M dwarf) stars as a function of planet radius and orbital period. The color-coding of each cell indicates the planet occurrence (mean number of planets per star) within that cell. The circles mark the radii and periods of planet candidates from the list of Kepler planet candidates as of mid 2014. The occurrence rates shown are corrected for geometric factors and the difficulty in detecting small planets transiting faint and/or noisy stars. (Courtesy Courtney Dressing and David Charbonneau)

candidate transiting planets, 134 with three, 48 with four, 17 with five, 2 with six, and 1 (Kepler-90) with seven. Figure G.41 illustrates the characteristics of the 20 systems with five or more planet candidates discovered using the first three years of Kepler data.

As shown in Figure G.42, the distribution of observed period ratios implies that the vast majority of planet pairs are neither in nor near low-order mean motion resonances. Nonetheless, there are small but statistically significant excesses of planet pairs both in resonance and spaced slightly too far apart to be in resonance, particularly near the 2:1 and 3:2 resonances. Virtually all candidate planetary systems are stable, as tested by numerical integrations that assume a nominal mass–radius relationship derived from planets within our Solar System.

Many Kepler targets with one or two observed transiting planet candidate(s) must be multi-planet systems where additional planets are present that are either not transiting and/or too small to be detected. When geometrical considerations concerning transit probabilities as well as completeness (planet detectability using currently available data) considerations are factored in, approximately 3%–5% of Kepler target stars have multiple planets in the $1.5\,\text{R}_\oplus < R_p < 6\,\text{R}_\oplus$ and $3 < P < 125$ day range. The large number of candidate multiple transiting planet systems observed by Kepler shows that many of the multi-planet systems in short-period orbits around other stars are nearly coplanar.

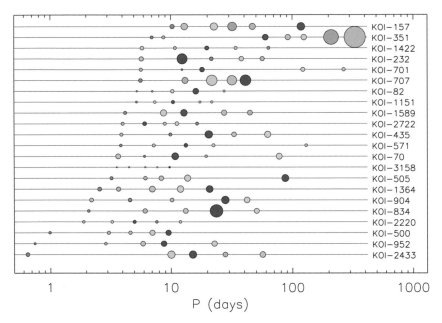

Figure G.41 Kepler candidate planetary systems of five or more planets found by searching the first three years of Kepler data. Each line corresponds to one system, as labeled on the right side. Ordering is by the orbital period of the innermost planet. Planet radii are to scale relative to one another and are colored by decreasing size within each system: red, orange, light green, light blue, dark blue, and dark green. (Courtesy Daniel Fabrycky)

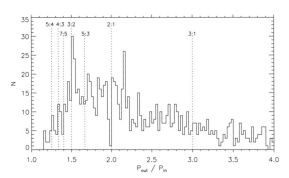

Figure G.42 Histogram of period ratios for all pairs of planets (excluding the two dynamically unstable candidate pairs) within multiple planet systems found by searching the first three years of Kepler data, out to a period ratio of 4. First-order (period ratios given in top row) and second-order (period ratios given in second row) resonances are marked by dashed lines. (Courtesy Daniel Fabrycky)

G.12.4 Mass–Radius Relationship for Small Planets

Measurements of both the mass and radius of a planet provide constraints on its composition. The planet's temperature can also affect its radius.

Figure G.43 shows the mass–radius–temperature relationship for planets with masses $M_p < 20$ $M_⊕$. The only exoplanets with mass and radius measurements that imply a rocky composition have $R_p < 1.7$ $R_⊕$, and all of them apart from Kepler-36 b have orbital periods under 1 day. All planets with known masses and radii $R_p > 3$ $R_⊕$ have a significant fraction of their volumes occupied by H_2 and/or

He. Most planets with 1.7 $R_⊕$ $< R_p < 3$ $R_⊕$ could either be rich in water (and perhaps other astrophysical 'ices') or are primarily rocky compounds by mass but with H/He dominating by volume.

G.12.5 Orbital Inclinations to Stellar Equators

The Rossiter-McLaughlin effect (§12.3.3) has been detected for a few dozen non-Kepler planets, most of which are hot Jupiters, as of early 2014. The observed values provide clues to the origin of the population of hot Jupiters. A clear majority of these, including HD 189733 as illustrated in Figure 12.17, have been found to orbit in the prograde direction near their star's equatorial plane, as in our Solar System. But some, such as those represented in Figure G.44, travel on substantially inclined (in some cases even retrograde) paths. Most of the high inclination planets orbit hotter stars that are fully radiative; indeed, the distribution of inclinations around such hot stars is consistent with random orientations. In contrast, most hot Jupiters orbiting cooler stars have low orbital inclinations with respect to their star's equatorial plane. This difference between hotter and cooler stars has been interpreted as a consequence of tidal dissipation: the cooler stars have deeper convective zones and consequently more rapid tidal dissipation, which leads to more efficient damping of orbital inclinations of close-in planets. This overall inclination distribution suggests that the mechanism that drives hot Jupiters close to their stars randomizes inclinations, which argues in favor of planet–planet scattering and Lidov–Kozai resonances (§G.2) and against orderly

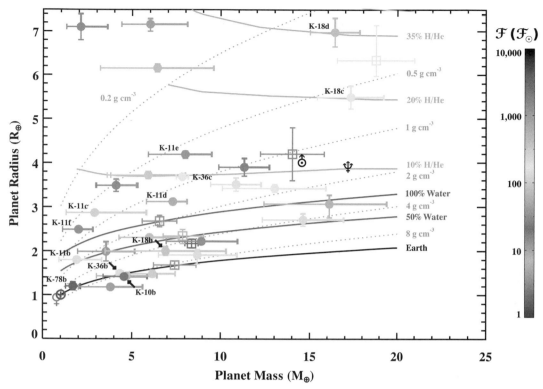

Figure G.43 Mass–radius diagram for low-mass planets. The planet symbols denote Venus, Earth, Uranus, and Neptune. Kepler planets are shown by filled circles. Other transiting exoplanets in this range are shown as open squares, with numbers and letters indicating planets referred to in the text. The colors represent the flux of stellar radiation intercepted by the planets in units of the solar constant, $\mathcal{F}_\odot$, as indicated by the scale bar on the right (Uranus and Neptune, which receive far less radiant energy than any of the other planets graphed, are shown in black). Model mass–radius curves for planets that are 5 Gyr old and subjected to a flux equal to $100\mathcal{F}_\odot$, with the various compositions specified below, are plotted for comparison. The red curve shows an Earth-like rock/iron composition. The blue curves show models that are 100% water and 50% water by mass atop an Earth-like core. The orange curves show models that are 10%, 20%, and 35% H/He by mass atop a core that has an Earth-like composition. The gray dotted curves show loci of constant density. (Courtesy Eric Lopez)

disk-induced migration because orderly migration would leave inclinations small provided the star's equator was near the plane of the protoplanetary disk.

The tilt of the rotation axis of a star relative to the plane of the sky can be estimated by using a combination of stellar mass and radius (from spectroscopy), rotation period (from spot modulation for Kepler targets), and rotational broadening of spectral lines (maximized for rotation axis in the plane of the sky, zero for a star observed pole-on). This technique, as well as estimations using the patterns of starspot crossings during transits and Rossiter–McLaughlin, has been used to measure tilts of the orbits of Kepler planets relative to the stellar equator. Most of these planets are Neptune-size and smaller and have orbital periods of weeks or months, so minimal tidal damping is expected. Nonetheless, most single planet and multi-planet Kepler systems orbit near the plane of

their star's equator, although high inclinations have been observed for both singles and multis.

G.13 Planet Formation

As shown in Figure G.45, the bulk of condensation occurs at temperatures between 1200–1300 K and below 200 K.

Figure G.46 shows the relative abundance of elements in the Solar System as well as those in galactic cosmic rays in the solar neighborhood. The interstellar medium has a composition similar to that of the Solar System, but cosmic ray abundances are strongly affected by spallation (§13.2).

New models of the tidal evolution of Earth's Moon suggest that the Earth–Moon system may have lost a substantial amount of angular momentum to the Earth's orbit around the Sun. This transfer could have occurred

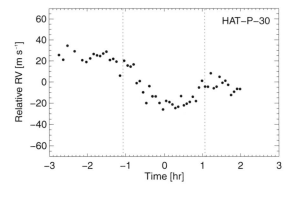

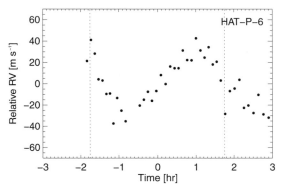

Figure G.44 The apparent radial velocity variation of the stars HAT-P-30 (top panel, data from Johnson *et al.* 2011), HAT-P-6 (bottom panel, data from Albrecht *et al.* 2012), during and immediately surrounding a transit of its giant vulcan planet. The curves represent Keplerian fits to the data outside of transit, for which the observed radial velocity variation is due entirely to the star's orbital motion and best-fit physical models (allowing arbitrary inclination of the planet's orbit to the star's equatorial plane) that account for the Rossiter–McLaughlin effect. The highly asymmetric anomaly in the top panel implies a nearly perpendicular orbit with displacement of $73.5 \pm 0.9°$, and the pattern in the bottom panel implies a nearly retrograde orbit ($165 \pm 6°$). (Courtesy Simon Albrecht and Josh Winn)

if the Moon remained trapped for a significant amount of time in the *evection resonance*, which occurs when

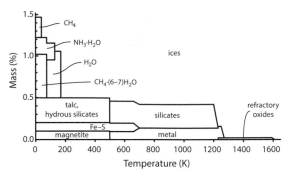

Figure G.45 Amount and composition of major condensed components formed during fully equilibrated cooling of solar nebula material. (Lodders 2010)

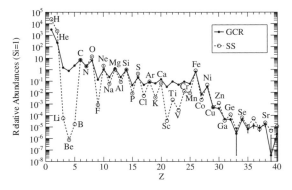

Figure G.46 Abundances of elements relative to silicon in galactic cosmic rays (GCR, solid curve) and the Solar System (SS, dashed curve). (Israel 2012; adapted from George *et al.* 2009 and Rauch *et al.* 2009)

the period of precession of the lunar perigee is equal to Earth's orbital period about the Sun (i.e., 1 year). If such angular momentum transfer occurred, it would remove the angular momentum constraint on the parameters of the giant impact that led to the Moon's formation, allowing a broader range of impactor size, collision parameters, and pre-impact rotation state of the Earth.

References

Ahrens, T.J., J.D. O'Keefe, and M.A. Lange, 1989. Formation of atmospheres during accretion of the terrestrial planets. In *Origin and Evolution of Planetary and Satellite Atmospheres*. Eds. S.K. Atreya, J.B. Pollack, and M.S. Matthews. University of Arizona Press, Tucson, pp. 328–385.

Aikawa, Y., T. Umbebayashi, T. Nakano, and S.M. Miyama, 1999. Evolution of molecular abundances in protoplanetary disks with accretion flow. *Astrophys. J.*, **519**, 705–725.

Albrecht, S., +12 co-authors, 2012. Obliquities of hot Jupiter host stars: Evidence for tidal interactions and primordial misalignments. *Astrophys. J.*, **757**, 18.

Alvarez, W., 1997. *T. Rex and the Crater of Doom*. Princeton University Press, Princeton, NJ. 185pp.

Anders, E., and N. Grevesse, 1989. Abundances of the elements: Meteoritic and solar. *Geochim. Cosmochim. Acta*, **53**, 197–214.

Anderson, J.D., and G. Schubert, 2007. Saturn's gravitational field, internal rotation, and interior structure. *Science*, **317**, 1384–1387.

Armstrong, J.C., C.B. Leovy, and T. Quinn, 2004. A 1 Gyr climate model for Mars: New orbital statistics and the importance of seasonally resolved polar processes. *Icarus*, **171**, 255–271.

Aschwanden, M.J., 2007. The Sun. In *Encyclopedia of the Solar System*, 2nd Edition. Eds. L. McFadden, P.R. Weissman, and T.V. Johnson. Academic Press, San Diego, pp. 71–98.

Atreya, S.K., 1986. *Atmospheres and Ionospheres of the Outer Planets and Their Satellites*. Springer-Verlag, Heidelberg. 224pp.

Atreya, S.K., J.B. Pollack, and M.S. Matthews, Eds., 1989. *Origin and Evolution of Planetary and Satellite Atmospheres*. University of Arizona Press, Tucson. 881pp.

Atreya, S.K., + 7 co-authors, 1999. A comparison of the atmospheres of Jupiter and Saturn: Deep atmospheric composition, cloud structure, vertical mixing, and origin. *Planet. Space Sci.*, **47**, 1243–1262.

Atreya, S.K., P.R. Mahaffy, H.B. Niemann, M.H. Wong, and T.C. Owen, 2003. Composition and origin of the atmosphere of Jupiter: An update, and implications for the extrasolar giant planets. *Planet. Space Sci.*, **51**, 105–112.

Bagenal, F., 1989. Torus–magnetosphere coupling. In *Time Variable Phenomena in the Jovian System*. Eds. M.J.S. Belton, R.A. West, and J. Rahe. Proceedings of a conference held in Flagstaff, August 25–27, 1987, pp. 196–210.

Bagenal, F., 1992. Giant planet magnetospheres. *Annu. Rev. Earth Planet. Sci.*, **22**, 289–328.

Bagenal, F., T.E. Dowling, and W.B. McKinnon, Eds., 2004. *Jupiter: The Planet, Satellites and Magnetosphere*. Cambridge University Press, Cambridge. 719pp.

Barshay, S.S., and J.S. Lewis, 1976. Chemistry of primitive solar material. *Annu. Rev. Astron. Astrophys.*, **14**, 81–94.

Barth, C.A., + 5 co-authors, 1992. Aeronomy of the current martian atmosphere. In *Mars*. Eds. H.H. Kieffer, B.M. Jakosky, C.W. Snyder, and M.S. Matthews. University of Arizona Press, Tucson, pp. 1054–1089.

Barucci, M.A., H. Boenhardt, D.P. Cruikshank, and A. Morbidelli, Eds., 2008. *The Solar System Beyond Neptune*. University of Arizona Press, Tucson. 592pp.

Beatty, J.K., C.C. Peterson, and A. Chaikin, Eds., 1999. *The New Solar System*, 4th Edition. Sky Publishing Co., Cambridge, MA and Cambridge University Press, Cambridge. 421pp.

Becker, G.E., and S.H. Autler, 1946. Water vapor absorption of electromagnetic radiation in the centimeter wavelength range. *Phys. Rev.*, **70**, 300–307.

Belcher, J.W., 1987. The Jupiter–Io connection: An Alfvénic engine in space. *Science*, **238**, 170–176.

Bell, J.F., D.R. Davis, W.K. Hartmann, and M.J. Gaffey, 1989. Asteroids: The big picture. In *Asteroids II*. Eds. R.P. Binzel, T. Gehrels, and M.S. Matthews. University of Arizona Press, Tucson, pp. 921–945.

Bernath, P.F., 2005. *Spectra of Atoms and Molecules*. Oxford University Press, Oxford. 439pp.

Bernstein, G.M., + 5 co-authors, 2004. The size distribution of transneptunian objects. *Astron. J.*, **128**, 1364–1390.

Bhardwaj, A., and C.M. Lisse, 2007. X-rays in the Solar System. *Encyclopedia of the Solar System*, 2nd Edition. Eds. L. McFadden, P.R. Weissman, and T.V. Johnson. Academic Press, San Diego, pp. 637–658.

Bida, T., T. Morgan, and R. Killen, 2000. Discovery of calcium in Mercury's atmosphere. *Nature*, **404**, 159–161.

Biraud, F., + 5 Co-authors, 1974. OH observation of Comet Kohoutek (1973f) at 18 cm wavelength. *Astron. Astrophys.*, **34**, 163–166.

Bird, M.K., O. Funke, J. Neidhöfer, and I. de Pater, 1996. Multi-frequency radio observations of Jupiter at Effelsberg during the SL–9 impact. *Icarus*, **121**, 450–456.

Bishop, J., + 5 co-authors, 1995. The middle and upper atmosphere of Neptune. In *Neptune*. Ed. D.P. Cruikshank. University of Arizona Press, Tucson, pp. 427–488.

Biver, N., + 22 co-authors, 2002. The 1995–2002 long-term monitoring of Comet C/1995 O1 (HALE – BOPP) at radio wavelength. *Earth, Moon and Planets*, **90**, 5–14.

Blewett, D.T., +17 co-authors, 2011. Hollows on Mercury: MESSENGER evidence for geologically recent volatile-related activity. *Science*, **333**, 1856–1859.

Blum, J., + 5 co-authors, 1994. Fractal growth and optical behaviour of cosmic dust. In *Fractals in the Natural and Applied Sciences*. Ed. M.M. Novak. Elsevier Science B.V. (North-Holland), pp. 47–59.

Bockelee-Morvan, D., and E. Gerard, 1984. Radio observations of the hydroxyl radical in comets with high spectral resolution. Kinematics and asymmetries of the OH coma in C/Meier (1978XXI), C/Bradfield (1979X), and C/Austin (1982g). *Astron. Astrophys.*, **131**, 111–122.

Bockelee-Morvan, D., + 11 Co-authors, 1998. Deuterated water in Comet C/1996 B2 (Hyakutake) and its implications for the origin of comets. *Icarus*, **133**, 147–162.

Bockelee-Morvan, D., J. Crovisier, M.J. Mumma, and H.A. Weaver, 2004. The composition of cometary volatiles. In *Comets II*. Eds. M.C. Festou, H.U. Keller, and H.A. Weaver. Arizona University Press, Tucson, pp. 391–423.

Bolt, B.A., 1976. *Nuclear Explosions and Earthquakes: The Partial Veil*. San Francisco, California, Freeman, Cooper and Co.

Bottke, W.F. Jr., A. Cellino, P. Paolicchi, and R.P. Binzel, Eds., 2002. *Asteroids III*. University of Arizona Press, Tucson. 785pp.

Bottke, W.F., + 6 co-authors, 2005a. The fossilized size distribution of the main asteroid belt. *Icarus*, **175**, 111–140.

Bottke, W.F., + 6 co-authors, 2005b. Linking the collisional history of the main asteroid belt to its dynamical excitation and depletion. *Icarus*, **179**, 63–94.

Bouchez, A.H., M.E. Brown, and N.M. Schneider, 2000. Eclipse spectroscopy of Io's atmosphere, *Icarus*, **148**, 316–319.

Boyd, T.J.M., and J.J. Sanderson, 2003. *The Physics of Plasmas*. Cambridge University Press, Cambridge. 532pp.

Braga-Ribas, F., + many co-authors, 2014. A ring system detected around the Centaur (10199) Chariklo. *Nature*, **508**, 72–75.

Brain, D.A., 2006. Mars Global Surveyor measurements of the martian solar wind interaction. *Space Science Reviews*, **126**, 77–112.

Brandt, J.C., 2007. Physics and chemistry of comets. *Encyclopedia of the Solar System*, 2nd Edition. Eds. L. McFadden, P. Weissman, and T.V. Johnson. Academic Press, Inc., pp. 557–588.

Britt, D.T., D. Yeomans, K. Housen, and G. Consolmagno, 2002. Asteroid density, porosity, and structure. In *Asteroids III*, Eds. W.F. Bottke Jr.,

A. Cellino, P. Paolicchi, and R.P. Binzel. University of Arizona Press, Tucson, pp. 485–500.

Brouwer, D., and G.M. Clemence, 1961. *Methods of Celestial Mechanics*. Academic Press, New York. 598pp.

Brown, G.C., and A.E. Mussett, 1981. *The Inaccessible Earth*. George Allen and Unwin, London. 235pp.

Brown, M.E., 2001. Potassium in Europa's atmosphere. *Icarus*, **151**, 190–195.

Brown, M.E., C.A. Trujillo, and D.L. Rabinowitz, 2005a. Discovery of a planetary-sized object in the scattered Kuiper belt. *Astrophys. J. Lett.*, **635**, L97–L100.

Brown, M.E., + 14 co-authors, 2005b. Keck Observatory laser guide star adaptive optics discovery and characterization of a satellite to the large Kuiper belt object 2003 EL61. *Astrophys. J. Lett.*, **632**, L45–L48.

Brown, M.E., D. Ragozzine, J. Stansberry, and W.C. Fraser, 2010. The size, density, and formation of the Orcus-Vanth system in the Kuiper belt. *Astron. J.*, **139**, 2700–2705.

Brownlee, D.E., and M.E. Kress, 2007. Formation of Earth-like habitable planets. In *Planets and Life: The Emerging Science of Astrobiology*. Eds. W.T. Sullivan III and J.A. Baross. Cambridge University Press, Cambridge, pp. 69–90.

Brownlee, D., + many co-authors, 2006. Comet 81 P/Wild 2 under a microscope. *Science*, **314**, 1711–1716.

Buratti, B.J., J.K. Hillier, and M. Wang, 1996. The lunar opposition surge: Observations by Clementine. *Icarus*, **124**, 490–499.

Burgdorf, M., + 6 co-authors, 2003. Neptune's far-infrared spectrum from the ISO long-wavelength and short-wavelength spectrometers. *Icarus*, **164**, 244–253.

Burgdorf, M., G.S. Orton, J. van Cleve, V. Meadows, and J. Houck, 2006. Detection of new hydrocarbons in Uranus' atmosphere by infrared spectroscopy. *Icarus*, **184**, 634–637.

Burlaga, L.F., + 5 co-authors, 2008. Magnetic fields at the solar wind termination shock. *Nature*, **454**, 75–77.

Burns, J.A., P.L. Lamy, and S. Soter, 1979. Radiation forces on small particles in the Solar System. *Icarus*, **40**, 1–48.

Burns, J.A., M.R. Showalter, and G.E. Morfill, 1984. The ethereal rings of Jupiter and Saturn. In *Planetary Rings*. Eds. R. Greenberg and A. Brahic, University of Arizona Press, Tucson, pp. 200–272.

Burns, J.A., + 5 co-authors, 1999. The formation of Jupiter's faint rings. *Science*, **284**, 1146–1150.

Burns, J.A., + 6 co-authors, 2004. Jupiter's ring–moon system. In *Jupiter: Planet, Satellites and Magnetosphere*, Eds. F. Bagenal, T.E. Dowling, and W. McKinnon. Cambridge University Press, Cambridge, pp. 241–262.

Burrows, A., + 8 co-authors, 1997. A non-gray theory of extrasolar giant planets and brown dwarfs. *Astrophys. J.*, **491**, 856–875.

Bus, S.J., M.F. A'Hearn, E. Bowell, and S.A. Stern, 2001. (2060) Chiron: Evidence for activity near aphelion. *Icarus*, **150**, 94–103.

Butler, B.J., and R.J. Sault, 2003. Long wavelength observations of the surface of Venus. IAUSS, 1E, 17B.

Butler, B.J., P.G. Steffes, S.H. Suleiman, M.A. Kolodner, and J.M. Jenkins, 2001. Accurate and consistent microwave observations of Venus and their implications. *Icarus*, **154**, 226–238.

Canup, R.M., 2004. Simulations of a late lunar-forming impact. *Icarus*, **168**, 433–456.

Carlson, R.W., and the Galileo NIMS team, 1991. Galileo infrared imaging spectroscopy measurements at Venus. *Science*, **253**, 1541–1548.

Carr, M.H., 1999. Mars: Surface and interior. In *Encyclopedia of the Solar System*. Eds. L. McFadden, P.R. Weissman, and T.V. Johnson. Academic Press, San Diego, pp. 291–308.

Carry, B., + 7 co-authors, 2008. Near-infrared mapping and physical properties of the dwarf-planet Ceres. *Astron. Astrophys.*, **478**, 235–244.

Carry, B., + 10 co-authors, 2010. Physical properties of (2) Pallas. *Icarus*, **205**, 460–472.

Chamberlain, J.W., and D.M. Hunten, 1987. *Theory of Planetary Atmospheres*, Academic Press, New York. 481pp.

Chandrasekhar, S., 1960. *Radiative Transfer*. Dover, New York. 392pp.

Charbonneau, D., + 7 co-authors, 2008. The broadband spectrum of the exoplanet HD 189733b. *Astropyhs. J.*, **686**, 1341–1348.

Chaussidon, M., and M. Gounelle, 2006. Irradiation processes in the early Solar System. In *Meteorites and the Early Solar System II*. Eds. D.S. Lauretta and H.Y. McSween. University of Arizona Press, Tucson, pp. 323–339.

Chesley, S.R., + 9 co-authors, 2003. Direct detection of the Yarkovsky effect via radar ranging to asteroid 6489 Golevka. *Science*, **302**, 1739–1742.

Chiang, E.I., + 5 co-authors. 2007. A brief history of transneptunian space. In *Protostars and Planets V*. Eds. B. Reipurth, D. Jewitt, and K. Keil. University of Arizona Press, Tucson, pp. 895–911.

Clancy, R.T., B.J. Sandor, and G.H. Moriarty-Schieven, 2004. *Icarus*, **168**, 116–121.

Clark, B.E., + 10 co-authors, 1999. NEAR photometry of asteroid 253 Mathilde. *Icarus*, **140**, 53–65.

Clark, R.N., F.P. Fanale, and M.J. Gaffey, 1986. Surface composition of natural satellites. In *Satellites*. Eds. J.A. Burns and M.S. Matthews. University of Arizona Press, Tucson, pp. 437–491.

Clayton, D.D., 1983. *Principles of Stellar Evolution and Nucleosynthesis*, University of Chicago Press, Chicago. 612pp.

Cole, G.H.A., and M.M. Woolfson, 2002. *Planetary Science: The Science of Planets around Stars*, Institute of Physics Publishing, Bristol and Philadelphia. 508pp.

Colina, L., R.C. Bohlin, and F. Castelli, 1996. The 0.12–2.5 micron absolute flux distribution of the Sun for comparison with solar analog star. *Astron. J.*, **112**, 307–315.

Colwell, J.E., + 5 co-authors, 2009. The structure of Saturn's rings. In *Saturn from Cassini-Huygens*. Eds. M. Dougherth, L. Esposito, and T. Krimigis. Springer, Heidelberg, pp. 375–412.

Conrath, B.J., + 15 co-authors, 1989a. Infrared observations of the Neptunian system. *Science*, **246**, 1454–1459.

Conrath, B.J., R.A. Hanel, and R.E. Samuelson, 1989b. Thermal structure and heat balance of the outer planets. In *Origin and Evolution of Planetary and Satellite Atmospheres*. Eds. S.K. Atreya, J.B. Pollack, and M.S. Matthews. University of Arizona Press, Tucson, pp. 513–538.

Coustenis, A., and R.D. Lorenz, 1999. Titan. In *Encyclopedia of the Solar System*. Eds. L. McFadden, P.R. Weissman, and T.V. Johnson. Academic Press, San Diego, pp. 377–404.

Coustenis, A., + 10 co-authors, 1998. Titan's atmosphere from ISO observations: Temperature, composition and detection of water vapor. *BAAS*, **30**, 1060.

Coustenis, A., + 24 co-authors, 2007. The composition of Titan's stratosphere from Cassini/CIRS mid-infrared spectra. *Icarus*, **189**, 35–62.

Cowley, S.W.H., 1995. The Earth's magnetosphere: A brief beginner's guide. *EOS*, **51**, 525–529.

Cox, A.N., Ed., 2000. *Allen's Astrophysical Quantities*, 4th Edition. Springer-Verlag, New York. 719pp.

Cremonese, G., + 9 co-authors, 1997. Neutral sodium from Comet Hale–Bopp: A third type of tail. *Astrophys. J. Lett.*, **490**, L199–L202.

Cruikshank, D.P., + 6 co-authors, 2007. Physical properties of transneptunian objects. In *Protostars and Planets V*. Eds. B. Reipurth, D. Jewitt, and K. Keil. University of Arizona Press, Tucson, pp. 879–893.

Cuzzi, J.N., + 6 co-authors, 1984. Saturn's rings: Properties and processes. In *Planetary Rings*. Eds. R. Greenberg and A. Brahic. University of Arizona Press, Tucson, pp. 73–199.

Cuzzi, J.N., + 10 co-authors, 2002. Saturn's rings: Pre-Cassini status and mission goals. *Space Science Reviews*, **118**, 209–251.

Danby, J.M.A., 1988. *Fundamentals of Celestial Mechanics*, 2nd Edition. Willmann-Bell, Richmond, VA. 467pp.

D'Angelo, G., W. Kley, and T. Henning, 2003. Orbital migration and mass accretion of protoplanets in three-dimensional global computations with nested grids. *Astrophys. J.*, **586**, 540–561.

Del Genio, A., + 6 co-authors, 2009. Saturn atmospheric structure and dynamics. In *Saturn from Cassini-Huygens*. Eds. M. Dougherty, L. Esposito, and T. Krimigis. Springer-Verlag, Berlin. 805pp.

de Pater, I., 1981. Radio maps of Jupiter's radiation belts and planetary disk at $\lambda = 6$ cm. *Astron. Astrophys.*, **93**, 370–381.

de Pater, I., and M.J. Klein, 1989. Time variability in Jupiter's synchrotron radiation. In *Time Variable Phenomena in the Jovian System*. Eds. M.J.S. Belton, R.A. West, and J. Rahe. Proceedings of a conference held in Flagstaff, August 25–27, 1987, pp. 139–150.

de Pater, I., and W.S. Kurth, 2007. The Solar System at radio wavelengths. *Encyclopedia of the Solar System*, 2nd Edition. Eds. L. McFadden, P. Weissman, and T.V. Johnson. Academic Press, San Diego, pp. 695–718.

de Pater, I., and D.L. Mitchell, 1993. Microwave observations of the planets: The importance of laboratory measurements. *J. Geophys. Res. Planets*, **98**, 5471–5490.

de Pater, I., and R.J. Sault, 1998. An intercomparison of 3-D reconstruction techniques using data and models of Jupiter's synchrotron radiation. *J. Geophys. Res. Planets*, **103**, 19 973–19 984.

de Pater, I., P. Palmer, and L.E. Snyder, 1986. The brightness distribution of OH around Comet Halley. *Astrophys. J. Lett.*, **304**, L33–L36.

de Pater, I., P.N. Romani, and S.K. Atreya, 1989. Uranus' deep atmosphere revealed. *Icarus*, **82**, 288–313.

de Pater, I., F.P. Schloerb, and A. Rudolph, 1991a. CO on Venus imaged with the Hat Creek Radio Interferometer. *Icarus*, **90**, 282–298.

de Pater, I., P. Palmer, and L.E. Snyder, 1991b. Review of interferometric imaging of comets. In *Comets in the Post-Halley Era*. Eds. R.L. Newburn, and J. Rahe. A book as a result from an international meeting on Comets in the Post-Halley Era, Bamberg, April 24–28, 1989, 175–207.

de Pater, I., + 5 co-authors, 1994. Radar aperture-synthesis observations of asteroids. *Icarus*, **111**, 489–502.

de Pater, I., F. van der Tak, R.G. Strom, and S.H. Brecht, 1997. The evolution of Jupiter's radiation belts after the impact of Comet D/Shoemaker–Levy 9. *Icarus*, **129**, 21–47. Erratum (Fig. reproduction): 1998, **131**, 231.

de Pater, I., + 6 co-authors, 1999. Keck infrared observations of Jupiter's ring system near Earth's 1997 ring plane crossing. *Icarus*, **138**, 214–223.

de Pater, I., D. Dunn, K. Zahnle, and P.N. Romani, 2001. Comparison of Galileo probe data with ground-based radio measurements. *Icarus*, **149**, 66–78.

de Pater, I., H.G. Roe, J.R. Graham, D.F. Strobel, and P. Bernath, 2002. Detection of the Forbidden SO $a^1\Delta \rightarrow X^3\Sigma^-$ Rovibronic Transition on Io at 1.7 μm. *Icarus*, **156**, 296–301.

de Pater, I., + 6 co-authors, 2004a. Keck AO observations of Io in and out of eclipse. *Icarus*, **169**, 250–263.

de Pater, I., S. Martin, and M.R. Showalter, 2004b. Keck near-infrared observations of Saturn's E and G rings during Earth's ring plane crossing in August 1995. *Icarus*, **172**, 446–454.

de Pater, I., + 8 co-authors, 2005. The dynamic neptunian ring arcs: Evidence for a gradual disappearance of Liberté and a resonant jump of Courage. *Icarus*, **174**, 263–272.

de Pater, I., S.G. Gibbard and H.B. Hammel, 2006a. Evolution of the dusty rings of Uranus. *Icarus*, **180**, 186–200.

de Pater, I., H.B. Hammel, S.G. Gibbard, and M.R. Showalter, 2006b. New dust belts of Uranus: One ring, two ring, red ring, blue ring. *Science*, **312**, 92–94.

de Pater, I., + 8 co-authors, 2006c. Titan imagery with Keck AO during and after probe entry, *J. Geophys. Res.*, **111**, E07S05.

de Pater, I., H.B. Hammel, M.R. Showalter, and M. van Dam, 2007. The dark side of the rings of Uranus. *Science*, **317**, 1888–1890.

de Pater, I., M. Showalter, and B. Macintosh, 2008. Keck observations of the 2002–2003 jovian ring plane crossing. *Icarus*, **195**, 348–360.

de Pater, I., + 7 co-authors, 2010. HST and Keck AO images of vortices on Jupiter. *Icarus*, **210**, 742–762.

Dermott, S.F., and C.D. Murray, 1981. The dynamics of tadpole and horseshoe orbits. I: Theory. *Icarus*, **48**, 1–11.

Descamps, P., + 19 co-authors, 2007. Figure of the double asteroid 90 Antiope from adaptive optics and lightcurve observations. *Icarus*, **187**, 482–499.

Descamps, P., + 18 co-authors, 2008. New determination of the size and bulk density of the binary asteroid 22 Kalliope from observations of mutual eclipses. *Icarus*, **196**, 578–600.

Descamps, P., + 18 co-authors, 2010. Triplicity, and physical characteristics of asteroid 216 Kleopatra, *Icarus*, **211**, 1022–1033.

Desch, M.D., + 6 co-authors, 1991. Uranus as a radio source. In *Uranus*. Eds. J.T. Bergstrahl, A.D. Miner, and M.S. Matthews. University of Arizona Press, Tucson, pp. 894–925.

Des Marais, D.-J., + 9 co-authors, 2002. Remote sensing of planetary properties and biosignatures on extrasolar terrestrial planets. *Astrobiology*, **2**, 153–181.

Despois, D., E. Gerard, J. Crovisier, and I. Kazes, 1981. The OH radical in comets: Observation and analysis of the hyperfine microwave transitions at 1667 MHz and 1665 MHz. *Astron. Astrophys.*, **99**, 320–340.

Dollfus, A., M. Wolff, J.E. Geake, D.F. Lupishko, and L.M. Dougherty, 1989. Photopolarimetry of asteroids. In *Asteroids II*. Eds. R.P. Binzel, T. Gehrels, and M.S. Matthews. University of Arizona Press, Tucson, pp. 594–616.

Domingue, D.L., + 5 co-authors, 2002. Disk-integrated photometry of 433 Eros. *Icarus*, **155**, 205–219.

Dowling, T.E., 1999. Earth as a planet: Atmosphere and oceans. In *Encyclopedia of the Solar System*. Eds. L. McFadden, P.R. Weissman, and T.V. Johnson. Academic Press, San Diego, pp. 191–208.

Duffard. R., + 6 co-authors, 2002. New activity of Chiron: Results from 5 years of photometric monitoring. *Icarus*, **160**, 44–51.

Duncan, M.J., and T. Quinn, 1993. The long-term dynamical evolution of the Solar System. *Annu. Rev. Astron. Astrophys*, **31**, 265–295.

Dunn, D.E., I. de Pater and L.A. Molnar, 2007. Examining the wake structure in Saturn's rings from microwave observations over varying ring opening angles and wavelengths. *Icarus*, **192**, 56–76.

Durisen, R.H., + 5 co-authors, 1996. Ballistic transport in planetary ring systems due to particle erosion mechanisms. III: Torques and mass loading by meteoroid impacts. *Icarus*, **124**, 220–236.

Dziewonski, A.M., and D.L. Anderson, 1981. Preliminary reference Earth model. *Phys. Earth Planet. Inter.*, **25**, 297–356.

Ebel, D.S., 2006. Condensation of rocky material in astrophysical environments. In *Meteorites and the Early Solar System II*. Eds. D. Lauretta *et al*. University of Arizona Press, Tucson, pp. 253–277, + 4 plates.

Elliot, J.L., 1979. Stellar occultation studies of the Solar System. *Annu. Rev. Astron. Astrophys.*, **17**, 445–475.

Elliot, J.L., E. Dunham, and D. Mink, 1977. The rings of Uranus. *Nature*, **267**, 328–330.

Encrenaz, Th., 2005. Neutral atmospheres of the giant planets: An overview of composition measurements. *Space Sci. Rev.*, **116**, 99–119.

Encrenaz, T., + 5 co-authors, 2004. *The Solar System*, 3rd Edition. Springer-Verlag, Berlin. ∼ 500pp.

Esposito, L.W., 1993. Understanding planetary rings. *Annu. Rev. Earth Planet. Sci.*, **21**, 487–521.

Esposito, L.W., A. Brahic, J.A. Burns, and E.A. Marouf, 1991. Particle properties and processes in Uranus' rings. In *Uranus*. Eds. J.T. Bergstrahl, E.D. Miner, and M.S. Matthews. University of Arizona Press, Tucson, pp. 410–465.

Etheridge, D.M., + 5 co-authors, 1996. Natural and anthropogenic changes in atmospheric CO_2 over the last 1000 years from air in Antarctic ice and firn. *J. Geophys. Res.*, **101**, 4115–4128.

Farrington, O., 1915. *Meteorites, Their Structure, Composition and Terrestrial Relations*. Chicago, published by the author.

Feaga, L.M., M.A. McGrath, and P.D. Feldman, 2002. The abundance of atomic sulfur in the amosphere of Io. *Astrophys. J.*, **570**, 439–446.

Feaga, L.M., M.A. McGrath, P.D. Feldman, and D.F. Strobel, 2004. Detection of atomic chlorine in Io's atmosphere with the Hubble Space Telescope GHRS. *Astrophys. J.* **610**, 1191–1198.

Fedorov, A.V., + 7 co-authors, 2006. The Pliocene paradox (mechanisms for a permanent El Nino), *Science*, **312**, 1485–1491.

Feldman, P.D., A.L. Cochran, and M.R. Combi, 2004. Spectroscopic investigations of fragment species in the coma. In *Comets II*. Eds. M.C. Festou, H.U. Keller, and H.A. Weaver. Arizona University Press, Tucson, pp. 425–447.

Fernandez, J.A., and K. Jockers, 1983. Nature and origin of comets. *Rep. Prog. Phys.*, **46**, 665–772.

Festou, M.C., H.U. Keller, and H.A. Weaver, 2004. *Comets II*. University of Arizona Press, Tucson. 733pp.

Finson, M.L., and R.F. Probstein, 1968. A theory of dust comets. 1: Model and equations. *Astrophys. J.*, **154**, 327–380.

Fischer, D.A., and J. Valenti, 2005. The planet–metallicity correlation. *Astrophys. J.*, **622**, 1102–1117.

Flasar, F.M., + 45 co-authors, 2005. Temperatures, winds, and composition in the saturnian system. *Science*, **307**, 1247–1251.

Fletcher, L. N., + 9 co-authors, 2007. Characterising Saturn's vertical temperature structure from Cassini/CIRS. *Icarus*, **189**, 457–478.

Fletcher, L.N., + 6 co-authors, 2008. Deuterium in the outer planets: New constraints and new questions from infrared spectroscopy. AGU Fall Meeting Abstracts, #P21B-04.

Forbes, J.M., F.G. Lemoine, S.L. Bruinsma, M.D. Smith, and X. Zhang, 2008. Solar flux variability of Mars' exosphere densities and temperatures. *Geophys. Res. Lett.*, **35**, L01201.

Formisano, V., S. Atreya, T. Encrenaz, N. Ignatiev, and M. Giuranna, 2004. Detection of methane in the atmosphere of Mars. *Science*, **306**, 1758–1761.

Fowler, C.M.R., 2005. *The Solid Earth: An Introduction to Global Geophysics*. 2nd Edition. Cambridge University Press, New York. 685pp.

French, R.G., P.D. Nicholson, C.C. Porco, and E.A. Marouf, 1991. Dynamics and structure of the uranian rings. In *Uranus*. Eds. J.T. Bergstrahl, E.D. Miner, and M.S. Matthews. University of Arizona Press, Tucson, pp. 327–409.

Fujiwara, A., + 21 co-authors, 2006. A rubble-pile asteroid Itokawa as observed by Hayabusa. *Science*, **312**, 1330–1334.

Fulchignoni, M., *et al.*, 2005. In situ measurements of the physical characteristics of Titan's environment. *Nature*, **438**, 785–791.

Furlan, E. *et al.*, 2006. A survey and analysis of Spitzer Infrared Spectrograph spectra of T Tauri stars in Taurus. *Astrophys, J. Supp.*, **165**, 568–605.

Gaffey, M.J., J.F. Bell, and D.P. Cruikshank, 1989. Reflectance spectroscopy and asteroid surface mineralogy. In *Asteroids II*. Eds. R.P. Binzel, T. Gehrels, and M.S. Matthews. University of Arizona Press, Tucson, pp. 98–127.

Gautier, D., and T. Owen, 1989. The composition of outer planet atmospheres. In *Origin and Evolution of Planetary and Satellite Atmospheres*. Eds. S.K. Atreya, J.B. Pollack, and M.S. Matthews. University of Arizona Press, Tucson, pp. 487–512.

Gautier, D., B.J. Conrath, T. Owen, I. de Pater, and S.K. Atreya, 1995. The troposphere of Neptune. In *Neptune and Triton*. Eds. D.P. Cruikshank and M.S. Matthews, University of Arizona Press, Tucson. pp. 547–612.

George, J.S., +15 co-authors, 2009. Elemental composition and energy spectra of galactic cosmic rays during Solar Cycle 23. *Astrophys. J.*, **698**, 1666–1681.

Ghil, M., and S. Childress, 1987. *Topics in Geophysical Fluid Dynamics: Atmospheric Dynamics, Dynamo Theory, and Climate Dynamics*. Springer-Verlag, New York. 485pp.

Gibson, J., W.J. Welch, and I. de Pater, 2005. Accurate jovian flux measurements at λ1cm show ammonia to be sub-saturated in the upper atmosphere. *Icarus*, **173**, 439–446.

Ginzburg, V., and S. Syrovatskii, 1965. Cosmic magnetobremsstrahlung (synchrotron radiation). *Annu. Rev. Astron. Astrophys.*, **3**, 297–350.

Goldstein, H., 2002. *Classical Mechanics*, 3rd Edition. Addison Wesley, MA. 638pp.

Goody, R.M., and J.C.G. Walker, 1972. *Atmospheres*. Prentice Hall, Englewood Cliffs, NJ. 160pp.

Gosling, J.T., 2007. The solar wind. In *Encyclopedia of the Solar System*. 2nd Edition. Eds. L. McFadden, P.R. Weissman, and T.V. Johnson. Academic Press, San Diego, pp. 99–116.

Gradie, J.C., C.R. Chapman, and E.F. Tedesco, 1989. Distribution of taxonomic classes and the compositional structure of the asteroid belt. In *Asteroids II*. Eds. R.P. Binzel, T. Gehrels, and M.S. Matthews. University of Arizona Press, Tucson, pp. 316–335.

Graham, J.R., I. de Pater, J.G. Jernigan, M.C. Liu, and M.E. Brown, 1995. W.M. Keck telescope observations of the Comet P/Shoemaker–Levy 9 fragment R Jupiter collision. *Science*, **267**, 1320–1323.

Greeley, R., 1994. *Planetary Landscapes*. 2nd Edition. Chapman and Hall, New York, London. 286pp.

Greeley, R., + 6 co-authors, 2004. Geology of Europa. In: *Jupiter. The Planet, Satellites and Magnetosphere*. Eds. F. Bagenal, T.E. Dowling, and W.B. McKinnon. Cambridge Planetary Science, Vol. 1. Cambridge University Press, Cambridge, pp. 329–362.

Greenberg, R., and A. Brahic, Eds., 1984. *Planetary Rings*. University of Arizona Press, Tucson. 784pp.

Grevesse, N., M. Asplund, and A.J. Sauval, 2007. The solar chemical composition. *Space Sci. Rev.*, **130**. 105–114.

Grotzinger, J., T. Jordan, F. Press, and R. Siever, 2006. *Understanding Earth*, 5th Edition. W.H. Freeman and Company, New York. 579pp.

Grün, E., I. de Pater, M. Showalter, F. Spahn, and R. Srama, 2006. Physics of dusty rings: History and perspective. *Planet. Space Sci.*, **54**, 837–843.

Grundy, W.M., + 8 co-authors, 2007. The orbit, mass, size, albedo, and density of (65489) Ceto/Phorcys: A tidally-evolved binary Centaur. *Icarus*, **191**, 286–297.

Guillot, T., 1999. Interiors of giant planets inside and outside the Solar System. *Science*, **286**, 72–77.

Guillot, T., 2005. The interiors of giant planets: Models and outstanding questions. *Annu. Rev. Earth Planet. Sci.*, **33**, pp. 493–530.

Guillot, T., G. Chabrier, D. Gautier, and P. Morel, 1995. Effect of radiative transport on the evolution of Jupiter and Saturn. *Astrophys. J.*, **450**, 463–472.

Guillot, T., D.J. Stevenson, W.B. Hubbard, and D. Saumon, 2004. The interior of Jupiter. In *Jupiter: Planet, Satellites and Magnetosphere*. Eds. F. Bagenal, T. E. Dowling, and W. McKinnon. Cambridge University Press, Cambridge, pp. 19–34.

Gulkis, S., and I. de Pater, 2002. Radio astronomy, planetary. In *Encyclopedia of Physical Science and Technology*, 3rd Edition, Vol. 13. Academic Press, Inc., Burlington, MA, pp. 687–712.

Gurnett, D.A., W.S. Kurth, A. Roux, S.J. Bolton, and C.F. Kennel, 1996. Evidence of a magnetosphere at Ganymede from Galileo plasma wave observations. *Nature*, **384**, 535–537.

Gurnett, D.A., + 6 co-authors, 2007. The variable rotation period of the inner region of Saturn's plasma disk. *Science*, **316**, 442–445.

Hamblin, W.K., and E.H. Christiansen, 1990. *Exploring the Planets*. Macmillan Publishing Company, New York. 451pp.

Hamilton, D.P., 1993. Motion of dust in a planetary magnetosphere: Orbit-averaged equations for oblateness, electromagnetic, and radiation forces with application to Saturn's E ring. *Icarus*, **101**, 244–264.

Hammel, H.B., + 9 co-authors, 1995. HST imaging of atmospheric phenomena created by the impact of Comet Shoemaker–Levy 9. *Science*, **267**, 1288–1295.

Hanel, R.A., B.J. Conrath, D.E. Jennings, and R.E. Samuelson, 1992. *Exploration of the Solar System by Infrared Remote Sensing*. Cambridge University Press, Cambridge. 458pp.

Hapke, B., R. Nelson, and W. Smythe, 1998. The opposition effect of the Moon: Coherent backscatter and shadow hiding. *Icarus*, **133**, 89–97.

Hardy, J.W., 1998. *Adaptive Optics for Astronomical Telescopes*. Oxford University Press, New York. 438pp.

Harker, D.E., C.E. Woodward, and D.H. Wooden, 2005. The dust grains from 9P/Tempel 1 before and after the encounter with Deep Impact. *Science*, **310**, 278–280.

Harmon, J.K., D.B. Campbell, A.A. Hine, I.I Shapiro, and B.G. Marsden, 1989. Radar observations of Comet IRAS–Araki–Alcock. *Astrophys. J.*, **338**, 1071–1093.

Harmon, J.K., + 5 co-authors, 1994. Radar mapping of Mercury's polar anomalies. *Nature*, **369**, 213–215.

Harrington, J., + 6 co-authors, 2004. Lessons from Shoemaker–Levy 9 about Jupiter and planetary impacts. In *Jupiter: Planet, Satellites and Magnetosphere*. Eds. F. Bagenal, T.E. Dowling, and W. McKinnon. Cambridge University Press, Cambridge, pp. 158–184.

Harrington, J., + 5 co-authors, 2007. The hottest planet. *Nature*, **447**, 691–693.

Hartmann, W.K., 1989. *Astronomy: The Cosmic Journey*. Wadsworth Publishing Company, Belmont, CA. 698pp.

Hartmann, W.K., 2005. *Moons and Planets*, 5th Edition. Brooks/Cole, Thomson Learning, Belmont, CA. 428pp.

Hedman, M.M., J.A. Burns, M.W. Evans, M.S. Tiscareno, and C.C. Porco, 2011. Saturn's curiously corrugated C ring. *Science*, **332**, 708–711.

Helfenstein, P.J., *et al.*, 1994. Galileo photometry of asteroid 951 Gaspra. *Icarus*, **107**, 37–60.

Helfenstein, P., J. Veverka, and J. Hiller, 1997. The lunar opposition effect: A test of alternative models. *Icarus*, **128**, 2–14.

Hendrix, A.R., R.M. Nelson, and D.L. Domingue, 2007. The Solar System at ultraviolet wavelengths. In *Encyclopedia of the Solar System*, 2nd Edition. Eds. L. McFadden, P.R. Weissman, and T.V. Johnson. Academic Press, San Diego, pp. 659–680.

Herzberg, G., 1944. *Atomic Spectra and Atomic Structure*. Dover Publications, New York. 257pp.

Hill, T.W., A.J. Dessler, and C.K. Goertz, 1983. Magnetospheric models. In *Physics of the Jovian Magnetosphere*. Ed. A.J. Dessler. Cambridge University Press, Cambridge, pp. 353–394.

Holman, M.J., 1997. A possible long-lived belt of objects between Uranus and Neptune. *Nature*, **387**, 785–788.

Holton, J.R., 1972. *An Introduction to Dynamic Meteorology*. Academic Press, New York. 319pp.

Hood, L., and J. Jones, 1987. Geophysical constraints on lunar bulk composition and structure: A reassessment. Proc. 17th Lunar Planet. Sci. Conf., Part 2. *J. Geophys. Res.*, **92**, E396–E410.

Hood, L., and M.T. Zuber, 2000. Recent refinements in geophysical constraints on lunar origin and evolution. In *Origin of the Earth and Moon*. Eds. R. Canup and K. Righter. University of Arizona Press, Tucson, pp. 397–409.

Horányi, M., J.A. Burns, M.M. Hedman, G.H. Jones, and S. Kempf, 2009. Diffuse Rings. In *Saturn from Cassini-Huygens*. Eds. M.K. Dougherlty, L.W. Esposito, and S.M. Krimigis, Springer-Verlag, Berlin, pp. 511–536.

Howard, A.D., 1967. Drainage analysis in geological interpretation: A summation. *Am. Ass. Petrol. Geol. Bull.*, **51**, 2246–2259.

Howard, A.W., + 9 co-authors, 2010. The occurrence and mass distribution of close-in super-Earths, Neptunes, and Jupiters. *Science*, **330**, 653–655.

Hubbard, W.B., 1984. *Planetary Interiors*. Van Nostrand Reinhold Company Inc., New York. 334pp.

Hubbard, W.B., M. Podolak, and D.J. Stevenson, 1995. The interior of Neptune. In *Neptune and Triton*. Ed. D.P. Cruikshank. University of Arizona Press, Tucson, pp. 109–138.

Huebner, W.F., Ed., 1990. *Physics and Chemistry of Comets*. Springer-Verlag, Berlin. 376pp.

Hueso, R., +16 co-authors, 2010. First Earth-based detection of a superbolide on Jupiter. *Astrophys. J. Lett.*, **721**, L129–L133.

Hughes, W.J., 1995. The magnetopause, magnetotail, and magnetic reconnection. In *Introduction to Space Physics*. Eds. M.G. Kivelson and C.T. Russell. Cambridge University Press, Cambridge, pp. 227–287.

Hundhausen, A.J., 1995. The solar wind. In *Introduction to Space Physics*. Eds. M.G. Kivelson, and C.T. Russell. Cambridge University Press, Cambridge, pp. 91–128.

Hunten, D.M., 2007. Venus: Atmosphere. In *Encyclopedia of the Solar System*, 2nd Edition. Eds. L. McFadden, P.R. Weissman, and T.V. Johnson. Academic Press, San Diego, pp. 139–148.

Hunten, D.M., and J. Veverka, 1976. Stellar and spacecraft occultations by Jupiter: A critical review of derived temperature profiles. In *Jupiter*. Ed. T. Gehrels. University of Arizona Press, Tucson, pp. 247–283.

Hunten, D.M., + 5 co-authors, 1984. Titan. In *Saturn*. Eds. T. Gehrels and M.S. Matthews. University of Arizona Press, Tucson, pp. 671–759.

Hunten, D.M., T.H. Morgan, and D.E. Shemansky, 1988. The Mercury atmosphere. In *Mercury*. Eds. F. Vilas, C.R. Chapman, and M.S. Matthews. University of Arizona Press, Tucson, pp. 562–612.

Hunten, D.M., T.M. Donahue, J.C.G. Walker, and J.F. Kasting, 1989. Escape of atmospheres and loss of water. In *Origin and Evolution of Planetary and Satellite Atmospheres*. Eds. S.K. Atreya, J.B. Pollack, and M.S. Matthews. University of Arizona Press, Tucson. pp. 386–422.

Huygens, C., 1659. *Systema Saturnia*.

Ingersoll, A.P., 1999. Atmospheres of the giant planets. In *The New Solar System*. 4th Edition. Eds. J.K. Beatty, C.C. Petersen and A. Chaikin. Cambridge University Press and Sky Publishing Corporation, pp. 201–220.

Ip, W.-H., and W.I. Axford, 1990. The plasma. In *Physics and Chemistry of Comets*. Ed. W.F. Huebner. Springer-Verlag, Berlin, pp. 177–233.

Israel, M.H., 2012. Cosmic rays: 1912–2012. *EOS*, **93**, 373–374.

Jackson, J.D., 1999. *Classical Electrodynamics*, 3rd Edition. John Wiley and Sons, New York. 641pp.

Jacobs, J.A., 1987. *The Earth's Core*, 2nd Edition. Academic Press, New York. 416pp.

Jacobson, M.Z., 1999. *Fundamentals of Atmospheric Modeling*. Cambridge University Press, New York. 656pp.

Jacobson, R., 2010. Orbits and masses of the Martian satellites and the libration of Phobos. *Astron. J.*, **139**, 668–679.

Jacobson, R.A., + 6 co-authors, 2008. Revised orbits of Saturn's small inner satellites. *Astron. J.*, **135**, 261–263.

Jakosky, B. 1998. *The Search for Life on Other Planets*. Cambridge University Press, New York. 326pp.

Jeanloz, R., 1989. Physical chemistry at ultrahigh pressures and temperatures. *Annu. Rev. Phys. Chem.*, **40**, 237–259.

Jewitt, D., L. Chizmadia, R. Grimm, and D. Prialnik, 2007. Water in the small bodies of the Solar System. In *Protostars and Planets V*. Eds. B. Reipurth, D. Jewitt, and K. Keil. University of Arizona Press, Tucson, pp. 863–878.

Jewitt, D., + 4 co-authors, 2013. The extraordinary multi-tailed main-belt comet P/2013 P5. *Astrophys. J.*, **778**, L21–L24.

Johnson, J.A., + 22 co-authors, 2011. HAT-P-30b: A transiting hot Jupiter on a highly oblique orbit. *Astrophys. J.*, **735**, 24.

Kaasalainen, M., S. Mottola, and M. Fulchignoni, 2002. Asteroid models from disk-integrated data. In *Asteroids III*. Eds. W.F. Bottke Jr., A. Cellino, P. Paolicchi, and R.P. Binzel. University of Arizona Press, Tucson, pp. 139–150.

Karkoschka, E., 1994. Spectrophotometry of the jovian planets and Titan at 300 to 1000 nm wavelength: The methane spectrum. *Icarus*, **111**, 174–192.

Kary, D.M., and L. Dones, 1996. Capture statistics of short-period comets: Implications for Comet D/Shoemaker–Levy 9. *Icarus*, **121**, 207–224.

Kerridge, J.F., 1993. What can meteorites tell us about nebular conditions and processes during planetesimal accretion? *Icarus*, **106**, 135–150.

Kivelson, M.G., and F. Bagenal, 1999. Planetary magnetospheres. In *Encyclopedia of the Solar System*. Eds. P.R. Weissman, L. McFadden, and T.V. Johnson. Academic Press, Inc., New York. pp. 477–498.

Kivelson, M.G. and F. Bagenal, 2007. Planetary magnetospheres. In *Encyclopedia of the Solar System*, 2nd Edition. Eds. L. McFadden, P.R. Weissman, and T.V. Johnson. Academic Press, San Diego, pp. 519–540.

Kivelson, M.G., and C.T. Russell, Eds., 1995. *Introduction to Space Physics*. Cambridge University Press, Cambridge. 568pp.

Kivelson, M.G., and G. Schubert, 1986. Atmospheres of the terrestrial planets. In *The Solar System: Observations and Interpretations*. Rubey Vol. IV. Ed. M.G. Kivelson. Prentice Hall, Englewood Cliffs, NJ, pp. 116–134.

Klein, M.J., and M.D. Hofstadter, 2006. Long-term variations in the microwave brightness temperature of the Uranus atmosphere. *Icarus*, **184**, 170–180.

Klein, M.J., S. Gulkis, and S.J. Bolton, 1995. Changes in Jupiter's 13-cm synchrotron radio emission following the impacts of Comet Shoemaker–Levy-9. *GRL*, **22**, 1797–1800.

Klein, M.J., + 6 co-authors, 2001. Cassini-Jupiter microwave observing campaign: DSN and GAVRT observations of jovian synchrotron radio emission. *Planetary Radio Emissions*, **V**, 221–228.

Kliore, A.J., + 6 co-authors, 2009. Midlatitude and high-latitude electron density profiles in the ionosphere of Saturn obtained by Cassini radio occultation observations. *J. Geophys. Res.*, **114**, CiteID A04315.

Kloosterman, J.L., B. Butler, and I. de Pater, 2007. VLA observations of Jupiter's synchrotron radiation at 15 GHz. *Icarus*, **193**, 644–648.

Knutson, H.A., + 8 co-authors, 2007. A map of the day–night contrast of the extrasolar planet HD 189733b. *Nature*, **447**, 183–186.

Kokubo, E., and S. Ida, 1999. Formation of protoplanets from planetesimals in the solar nebula. *Icarus*, **143**, 15–27.

Kokubo, E., R.M. Canup, and S. Ida, 2000. Lunar accretion from an impact-generated disk. In *Origin of the Earth and Moon*. Eds. R.M. Canup and K. Righter. University of Arizona Press, Tucson, pp. 145–163.

Kolokolova, L., M.S. Hanner, A.-C. Levasseur-Regourd, and B.A.S. Gustafson, 2004. Physical properties of cometary dust from light scattering and thermal emission. In *Comets II*. Eds. M.C. Festou, H.U. Keller, and H.A. Weaver. University of Arizona Press, Tucson, pp. 577–604.

Konackie, M. and A. Wolszczan, 2003. Masses and orbital inclinations of planets in the PSR B1257+12 system. *Astrophys. J.*, **597**, 1076–1091.

Konopliv, A.S., A.B. Binder, L.L. Hood, A.B. Kucinskas, W.L. Sjogren, and J.G. Williams, 1998. Improved gravity field of the Moon from Lunar Prospector. *Science*, **281**, 1476–1480.

Konopliv, A.S., W.B. Banerdt, and W.L. Sjogren, 1999. Venus gravity: 180th degree and order model. *Icarus*, **139**, 3–18.

Krasnopolsky, V.A., B.R. Sandel, F. Herbert, and R.J. Vervack Jr., 1993. Temperature, N_2, and N density profiles of Triton's atmosphere: Observations and model. *J. Geophys. Res.*, **98**(E2), 3065–3078.

Kraus, J.D., 1986. *Radio Astronomy*, 2nd Edition. Cygnus Books, Powell, OH. 719pp.

Kring, D., 2003. Environmental consequences of impact cratering events as a function of ambient conditions on Earth. *Astrobiology*, **3**, 133–152.

Krishna Swamy, K.S., 1986. *Physics of Comets*. World Scientific Publishing Co. Pte. Ltd., Singapore. 273pp.

Krot, A.N., E.R.D. Scott and B. Reipurth, Eds., 2005. *Chondrites and the Protoplanetary Disk*. ASP Conference Series, **341**, Astronomical Society of the Pacific, San Francisco. 1029pp.

Kurth, W.S., 1997. Whistler. In *Encyclopedia of Planetary Sciences*. Eds. J.H. Shirley and R.W. Fairbridge. Chapman and Hall, London, pp. 936–937.

Kurth, W.S., and D.A. Gurnett, 1991. Plasma waves in planetary magnetospheres. *J. Geophys. Res.*, **96**, 18 977–18 991.

Kurth, W.S., + 9 co-authors, 2005. High spectral and temporal resolution observations of Saturn kilometric radiation. *Geophys. Res. Lett.*, **32**, L20S07.

Lacerda, P., and D.C. Jewitt, 2007. Densities of Solar System objects from their rotational light curves. *Astron. J.*, **133**, 1393–1408.

Lambeck, K., 1988. *Geophysical Geodesy: The Slow Deformations of the Earth*. Oxford Science Publications, Oxford. 718pp.

Lamy, P.L., I. Toth, Y.R. Fernández, and H.A. Weaver, 2004. The sizes, shapes, albedos and colors of cometary nuclei. In *Comets II*. Eds. M.C. Festou, H.U. Keller, and H.A. Weaver. University of Arizona Press, Tucson, pp. 223–264.

Laskar, J., T. Quinn, and S. Tremaine, 1992. Confirmation of resonant structure in the Solar System. *Icarus*, **95**, 148–152.

Lauretta, D.S., and H.Y. McSween, Eds., 2006. *Meteorites and the Early Solar System II*. University of Arizona Press, Tucson. 942pp.

Lecacheux, A., 2001. Radio observations during the Cassini flyby of Jupiter. In *Planetary Radio Emissions V*. Eds. H.O. Rucker, M.L. Kaiser, and Y. Leblanc. Austrian Academy of Sciences Press, Vienna, pp. 1–13.

Lederer, S.M., + 14 co-authors, 2005. Physical characteristics of Hayabusa target asteroid 25143 Itokawa. *Icarus*, **173**, 153–165.

Lellouch, E., M.J.S. Belton, I. de Pater, S. Gulkis, and T. Encrenaz, 1990. Io's atmosphere from microwave detection of SO_2. *Nature*, **346**, 639–641.

Lellouch, E., D.F. Strobel, and M. Belton, 1995. Detection of SO in the atmosphere of Io. *BAAS*, **27**, 1155.

Lellouch, E., G. Paubert, D.F. Strobel, and M. Belton, 2000. Millimeter-wave observations of Io's atmosphere: The IRAM 1999 campaign. *BAAS*, **32**, #35.11.

Lellouch, E., G. Paubert, J.L. Moses, N.M. Schneider, and D.F. Strobel, 2003. Volcanically emitted sodium chloride as a source for Io's neutral clouds and plasma torus. *Nature*, **421**, 45–47.

Levison, H.F., and L. Dones, 2007. Comet populations and cometary dynamics. In *Encyclopedia of the Solar System*, 2nd Edition. Eds. L. McFadden, P.R. Weissman, and T.V. Johnson. Academic Press, San Diego, pp. 575–588.

Levison, H.F., J.J. Lissauer, and M.J. Duncan, 1998. Modeling the diversity of outer planetary systems. *Astron. J.*, **116**, 1998–2014.

Levy, E.H., 1986. The generation of magnetic fields in planets. In *The Solar System: Observations and Interpretations*. Rubey Vol. IV. Ed. M.G. Kivelson. Prentice Hall, Englewood Cliffs, NJ, pp. 289–310.

Lewis, J.S., 1995. *Physics and Chemistry of the Solar System*, Revised Edition. Academic Press, San Diego. 556pp.

Lewis, J.S., 2004. *Physics and Chemistry of the Solar System*, 2nd Edition. Elsevier, Academic Press, San Diego. 684pp.

Li, X.D., and B. Romanowicz, 1996. Global mantle shear velocity model developed using nonlinear asymptotic coupling theory. *J. Geophys. Res.*, **101**, pp. 22 245–22 272.

Lide, D.R., Ed., 2005. *CRC Handbook of Chemistry and Physics*, 86th Edition. CRC Press, Boca Raton, FL. 2544pp.

Lillis, R.J., + 5 co-authors, 2008. An improved crustal magnetic field map of Mars from electron reflectometry: Highland volcano magmatic history and the end of the martian dynamo. *Icarus*, **194**, 575–596.

Lin, D.N.C., 1986. The nebular origin of the Solar System. In *The Solar System: Observations and Interpretations*, Rubey Vol. IV. Ed. M.G. Kivelson. Prentice Hall, Englewood Cliffs, NJ, pp. 28–87.

Lindal, G.F., 1992. The atmosphere of Neptune: An analysis of radio occultation data acquired with Voyager 2. *Astron. J.*, **103**, 967–982.

Lindal, G.F., + 5 co-authors, 1987. The atmosphere of Uranus: Results of radio occultation measurements with Voyager 2. *J. Geophys. Res.*, **92**, 14 987–15 002.

Lipschutz, M.E., and L. Schultz, 2007. Meteorites. In *Encyclopedia of the Solar System*, 2nd Edition. Eds. L. McFadden, P.R. Weissman, and T.V. Johnson. Academic Press, San Diego, pp. 251–282.

Lis, D.C., + 10 co-authors, 1997. New molecular species in Comet C/1995 O1 (Hale-Bopp) observed with the Caltech Submillimeter Observatory. *Earth Moon Planets*, **78**, 13–20.

Lissauer, J.J., 1993. Planet formation. *Annu. Rev. Astron. Astrophys.*, **31**, 129–174.

Lissauer, J.J., 1995. Urey Prize lecture: On the diversity of plausible planetary systems. *Icarus*, **114**, 217–236.

Lissauer, J.J., 1999. How common are habitable planets? *Nature*, **402**, C11–C14.

Lissauer, J.J., and I. de Pater, 2013. *Fundamental Planetary Science: Physics, Chemistry and Habitability.* Cambridge University Press, Cambridge. 616pp.

Lissauer, J.J., J.B. Pollack, G.W. Wetherill, and D.J. Stevenson, 1995. Formation of the Neptune system. In *Neptune and Triton*. Ed. D.P. Cruikshank. University of Arizona Press, Tucson, pp. 37–108.

Lissauer, J.J., O. Hubickyj, G. D'Angelo, and P. Bodenheimer, 2009. Models of Jupiter's growth incorporating thermal and hydrodynamic constraints. *Icarus*, **199**, 338–350.

Lissauer, J.J., +16 co-authors, 2013. All six planets known to orbit Kepler-11 have low densities. *Astrophys. J.*, **770**, 131.

Lisse, C. M., K.E. Kraemer, J.A. Nuth, A. Li, and D. Joswiak, 2007. Comparison of the composition of the Tempel 1 ejecta to the dust in Comet C/Hale Bopp 1995 O1 and YSO HD100546. *Icarus*, **187**, 69–86.

Lithgow-Bertelloni, C., and M.A. Richards, 1998. The dynamics of cenozoic and mesozoic plate motions. *Rev. Geophys.*, **36**, 27–78.

Lodders, K., 2003. Solar System abundances and condensation temperatures of the elements. *Astrophys. J.*, **591**, 1220–1247.

Lodders, K., 2010. Solar System abundances of the elements. In *Principles and Perspectives in*

Cosmochemistry, Astrophysics and Space Science Proceedings. Springer-Verlag, Berlin, pp. 379–417.

Lopes, R.M.C., and T.K.P. Gregg, Eds., 2004. *Volcanic Worlds.* Springer-Praxis, New York. 236pp.

Lopes, R.M.C., and J.R. Spencer, Eds., 2007. *Io after Galileo: A New View of Jupiter's Volcanic Moon.* Springer, Praxis Publishing, Chichester, UK. 342pp.

Lovett, L., J. Horvath, and J. Cuzzi, 2006. *Saturn: A New View.* H.N. Abrams, New York. 192pp.

Luhmann, J.G., 1995. Plasma interactions with unmagnetized bodies. In *Introduction to Space Physics.* Eds. M.G. Kivelson and C.T. Russell. Cambridge University Press, Cambridge, pp. 203–226.

Luhmann, J.G, and S.C. Solomon, 2007. The Sun–Earth connection. In *Encyclopedia of the Solar System*, 2nd Edition. Eds. L. McFadden, P.R. Weissman, and T.V. Johnson. Academic Press, San Diego, pp. 213–226.

Luhmann, J.G., C.T. Russell, L.H. Brace, and O.L. Vaisberg, 1992. The intrinsic magnetic field and solar-wind interaction of Mars. In *Mars.* Eds. H.H. Kieffer, B.M. Jakosky, C.W. Snyder, and M.S. Matthews. University of Arizona Press, Tucson, pp. 1090–1134.

Lunine, J.I., 2005. *Astrobiology: A Multi-Disciplinary Approach.* Pearson Education, San Francisco. 586pp.

Lunine, J.I., and W.C. Tittemore, 1993. Origins of outer-planet satellites. In *Protostars and Planets III.* Eds. E.H. Levy and J.I. Lunine. University of Arizona Press, Tucson, pp. 1149–1176.

Lyons, L.R., and D.J. Williams, 1984. *Quantitative Aspects of Magnetospheric Physics.* Reidel Publishing Company, Dordrecht. 231pp.

Mahaffy, P.R., 1998. Mass spectrometers developed for planetary missions. In *Laboratory Astrophysics and Space Research.* Eds. P. Ehrenfreund, C. Krafft, H. Kochan, and V. Pironello. Kluwer Academic Publishing, Dordrecht, pp. 355–376.

Malin, M.C., and K.S. Edgett, 2000. Evidence for recent groundwater seepage and surface runoff on Mars. *Science*, **288**, 2330–2335.

Marchis, F., + 8 co-authors, 2002. High-resolution Keck adaptive optics imaging of violent volcanic activity on Io. *Icarus*, **160**, 124–131.

Marchis, F., P. Descamps, D. Hestroffer, J. Berthier, and I. de Pater, 2005a, Mass and density of Asteroid 121

Hermione from an analysis of its companion orbit. *Icarus*, **178**, 450–464.

Marchis, F., P. Descamps, D. Hestroffer and J. Berthier, 2005b. Discovery of the triple asteroidal system 87 Sylvia. *Nature*, **436**, 822–824.

Marchis, F., + 7 co-authors, 2006. Search of binary Jupiter-Trojan asteroids with laser guide star AO systems: A moon around 624 Hektor. *BAAS*, **38**, #65.07.

Marchis, F., + 7 co-authors, 2008a. Main belt asteroidal systems with eccentric mutual orbits. *Icarus*, **195**, 295–316.

Marchis, F., + 7 co-authors, 2008b. Main belt asteroidal systems with circular mutual orbits. *Icarus*, **196**, 97–118.

Marcy, G.W., + 6 co-authors, 2005. Observed properties of exoplanets: Masses, orbits and metallicities. *Prog. Theor. Phys. Supp.*, **158**, 24–42.

Margot, J.-L., + 7 co-authors, 2002. Binary asteroids in the near-Earth object population. *Science*, **296**, 1445–1448.

Marley, M.S., and J.J. Fortney, 2007. Interiors of the giant planets. *Encyclopedia of the Solar System*, 2nd Edition. Eds. L. McFadden, P. Weissman, and T.V. Johnson. Academic Press, San Diego, pp. 403–418.

Marsden, B.G., and G.V. Williams, 2008. *Catalogue of Cometary Orbits*, 17th Edition. The International Astronomical Union, Minor Planet Center and Smithsonian Astrophysical Observatory, Cambridge, MA.

Martin, S., I. de Pater, J. Kloosterman, and H.B. Hammel, 2008. Multi-wavelength Observations of Neptune's Atmosphere. EPSC2008-A-00277.

Mayor, M., and D. Queloz, 1995. A Jupiter-mass companion to a solar-type star. *Nature*, **378**, 355–359.

McClintock, W.E., + 8 co-authors, 2009. MESSENGER observations of Mercury's exosphere: Detection of magnesium and distribution of constituents. *Science*, **324**, 610–613.

McEwen, A.S., L.P. Keszthelyi, R. Lopes, P.M. Schenk, and J.R. Spencer, 2004. Lithosphere and surface of Io. In *Jupiter. The Planet, Satellites and Magnetosphere*. Eds. F. Bagenal, T. E. Dowling, and W.B. McKinnon. Cambridge Planetary Science, Vol. 1. Cambridge University Press, Cambridge, pp. 313–334.

McFadden, L., P.R. Weissman, and T.V. Johnson, Eds., 2007. *Encyclopedia of the Solar System*, 2nd Edition. Academic Press, San Diego. 982pp.

McKinnon, W.B., and R.L. Kirk, 2007. Triton, In *Encyclopedia of the Solar System*, 2nd Edition. Eds. L. McFadden, P.R. Weissman, and T.V. Johnson. Academic Press, San Diego, pp. 483–502.

McPherron, R.L., 1995. Magnetospheric dynamics. In *Introduction to Space Physics*. Eds. M.G. Kivelson and C.T. Russell. Cambridge University Press, Cambridge, pp. 400–458.

McSween, H.Y., Jr., 1999. *Meteorites and their Parent Planets*, 2nd Edition. Cambridge University Press, Cambridge. 322pp.

Meadows, V., and S. Seager, 2011. Terrestrial planet atmospheres and biosignatures. In *Exoplanets*. Ed. S. Seager. University of Arizona Press, Tucson, pp. 441–470.

Megnin, C., and B. Romanowicz, 2000. The 3D shear velocity structure of the mantle from the inversion of body, surface, and higher mode waveforms. *Geophys. J. Int.*, **143**, 709–728.

Melosh, H.J., 1989. *Impact Cratering: A Geologic Process*. Oxford Monographs on Geology and Geophysics, No. 11. Oxford University Press, New York. 245pp.

Mewaldt, R.A., R.S. Selesnick, and J.R. Cummings, 1997. Anomalous cosmic rays: The principal source of high energy heavy ions in the radiation belts. In *Radiation Belts: Models and Standards*. Geophysical Monograph **97**. Eds. J.F. Lemaire, D. Heynderickx, and D.N. Baker. American Geophysical Union, pp. 35–42.

Miller, R., and W.K. Hartmann, 2005. *The Grand Tour: A Traveler's Guide to the Solar System*, 3rd Edition. Workman Publishing, New York. 208pp.

Mitchell, D.L., 1993. *Microwave Imaging of Mercury's Thermal Emission: Observations and Models*. Ph.D. Thesis, University of California, Berkeley.

Mitchell, D.L., and I. de Pater, 1994. Microwave imaging of Mercury's thermal emission: Observations and models. *Icarus*, **110**, 2–32.

Mohanty, S., R. Jayawardhana, N. Hulamo, and E. Mamajek, 2007. The planetary mass companion 2MASS 1207-3932B: Temperature, mass, and evidence for an edge-on disk. *Astrophys. J.*, **657**, 1064–1091.

Moore, J.M. + 11 co-authors, 2004. Callisto. In *Jupiter: The Planet, Satellites and Magnetosphere*. Cambridge Planetary Science, Vol. 1. Eds. F. Bagenal, T.E. Dowling, and W.B. McKinnon. Cambridge University Press, Cambridge, pp. 397–426.

Moore, W.B., G. Schubert, J.D. Anderson, and J.R. Spencer, 2007. The interior of Io. In *Io after Galileo: A New View of Jupiter's Volcanic Moon*. Springer-Praxis, Chichester, UK, pp. 89–108.

Morbidelli, A., 2002. *Modern Celestial Mechanics: Aspects of Solar System Dynamics*. Taylor and Francis/Cambridge Scientific Publishers, London. 368pp. (Out of print: see http://www.oca.eu/morby/.)

Morbidelli, A., and H.F. Levison, 2007. Kuiper belt: Dynamics. In *Encyclopedia of the Solar System*, 2nd Edition. Eds. L. McFadden, P.R. Weissman, and T.V. Johnson. Academic Press, San Diego, pp. 589–604.

Morel, P., 1997. CESAM: A code for stellar evolution calculations. *A&A Supp. Ser.*, **124**, September 1997, 597–614.

Morgan, J., + 18 co-authors + the Chicxulub Working Group, 1997. Size and morphology of the Chicxulub impact crater. *Nature*, **390**, 472–476.

Moroz, V.I., 1983. Stellar magnitude and albedo data of Venus. In *Venus*. Eds. D.M. Hunten, L. Colin, T.M. Donahue, and V.I. Moroz. University of Arizona Press, Tucson, pp. 27–68.

Morrison, D., and T. Owen, 1996. *The Planetary System*. Addison-Wesley Publishing Company, New York.

Morrison, D., and T. Owen, 2003. *The Planetary System*, 3rd Edition. Addison-Wesley Publishing Company, New York. 531pp.

Moses, J.I., + 5 co-authors, 2005. Photochemistry and diffusion in Jupiter's stratosphere: Constraints from ISO observations and comparisons with other giant planets. *J. Geophys. Res.*, **110**, E08001.

Mottola, S., and F. Lahulla, 2000. Mutual eclipse events in asteroidal binary system 1996 FG_3: Observations and a numerical model. *Icarus*, **146**, 556–567.

Mueller, M., + 7 co-authors, 2010. Eclipsing binary Trojan asteroid Patroclus: thermal inertia from Spitzer observations. *Icarus*, **205**, 505–515.

Muinonen, K., J. Piironen, Y.G. Shkuratov, A. Ovcharenko, and B.E. Clark, 2002. Asteroid photometric and polarimetric phase effects. In *Asteroids III*. Eds. W.F. Bottke Jr., A. Cellino, P. Paolicchi, and

R.P. Binzel. University of Arizona Press, Tucson, pp. 123–138.

Mumma, M.J., R.E. Novak, M.A. DiSanti, B.P. Bonev, and N. Dello Russo, 2004. Detection and mapping of methane and water on Mars. *BAAS*, **36**, 1127.

Murchie, S.L., + 10 co-authors, 2008. Geology of Caloris basin, Mercury: A view from MESSENGER. *Science*, **321**, 73–76.

Murray, C., and S. Dermott, 1999. *Solar System Dynamics*. Cambridge University Press, Cambridge. 592pp.

Nair, H., M. Allen, A.D. Anbar, and Y.L. Yung, 1994. A photochemical model of the martian atmosphere. *Icarus*, **111**, 124–150.

Ness, N.F., J.E.P. Connerney, R.P. Lepping, M. Schulz, and G.-H. Voigt, 1991. The magnetic field and magnetospheric configuration of Uranus. In *Uranus*. Eds. J.T. Bergstrahl, E.D. Miner, and M.S. Matthews. University of Arizona Press, Tucson, pp. 739–779.

Nesvorný, D., W.F. Bottke, H.F. Levison, and L. Dones, 2003. Recent origin of the Solar System dust bands. *Astrophys. J.*, **591**, 486–497.

Neubauer, F.M., 1991. The magnetic field structure of the cometary plasma environment. In *Comets in the Post-Halley Era*. Eds. R.L. Newburn, M. Neugebauer, and J. Rahe. A book resulting from an international meeting on 'Comets in the Post-Halley Era,' Bamberg, April 24–28, 1989, pp. 1107–1124.

Nicholson, P.D., 2009. Natural satellites of the planets. In *Observer's Handbook*, Ed. P. Kelly, Royal Astron. Soc. Canada, pp. 24–30.

Niemann, H.B., + 11 co-authors, 1998. The composition of the jovian atmosphere as determined by the Galileo probe mass spectrometer. *J. Geophys. Res.*, **103**, 22 831–22 845.

Niemann, H.B., *et al.*, 2005. The abundances of constituents of Titan's atmosphere from the GCMS instrument on the Huygens probe. *Nature*, **438**, 779–784.

Noll, K.S., W.M. Grundy, E.I. Chiang, J.-L. Margot, and S.D. Kern, 2008. Binaries in the Kuiper belt. In *The Kuiper Belt*, Space Science Series, University of Arizona Press, Tucson, pp. 345–363.

O'Brien, D.P.. A. Morbidelli, and H.F. Levison, 2006. Terrestrial planet formation with strong dynamical friction. *Icarus*, **184**, 39–58.

Ockert-Bell, M.E., + 6 co-authors, 1999. The structure of the jovian ring system as revealed by the Galileo imaging experiment. *Icarus*, **138**, 188–213.

Opal, C.B., and G.R. Carruthers, 1977. Lyman-alpha observations of Comet West/1975n. *Icarus*, **31**, 503–508.

Ostro, S.J., 1989. Radar observations of asteroids. In *Asteroids II*. Eds. R.P. Binzel, T. Gehrels, and M.S. Matthews. University of Arizona Press, Tucson, pp. 192–212.

Ostro, S.J., 2007. Planetary radar. *Encyclopedia of the Solar System*, 2nd Edition. Eds. L. McFadden, P.R. Weissman, and T.V. Johnson. Academic Press, San Diego, pp. 735–764.

Ostro, S.J., + 6 co-authors, 1991. Asteroid 1986 DA: Radar evidence for a metallic composition. *Science*, **252**, 1399–1404.

Ostro, S.J., + 12 co-authors, 1996. Radar observations of asteroid 1620 Geographos. *Icarus*, **121**, 44–66.

Ostro, S.J., + 8 co-authors, 2000. Radar observations of asteroid 216 Kleopatra. *Science*, **288**, 836–839.

Ostro, S.J., + 6 co-authors, 2002. Asteroid radar astronomy. In *Asteroids III*. Eds. W.F. Bottke Jr., A. Cellino, P. Paolicchi, and R.P. Binzel. University of Arizona Press, Tucson, pp. 151–182.

Ostro, S.J., + 15 co-authors, 2006. Radar imaging of binary near-Earth asteroid (66391) 1999 KW_4. *Science*, **314**, 1276–1280.

Palme, H., and W.N. Boynton, 1993. Meteoritic constraints on conditions in the solar nebula. In *Protostars and Planets III*. Eds. E.H. Levy and J.I. Lunine. University of Arizona Press, Tucson, pp. 979–1004.

Parker, E.N., 1963. *Interplanetary Dynamical Processes*. Interscience, New York. 272pp.

Pasachoff, J.M., and M.L. Kutner, 1978. *University Astronomy*. W.B. Saunders Company, Philadelphia. 851pp.

Peale, S.J., 1976. Orbital resonances in the Solar System. *Annu. Rev. Astron. Astrophys.*, **14**, 215–246.

Pedlovsky, J., 1987. *Geophysical Fluid Dynamics*, 2nd Edition. Springer-Verlag, New York. 710pp.

Perly, R.A., F.R. Schwab, and A.H. Bridle, 1989. *Synthesis Imaging in Radio Astronomy*, NRAO Workshop No. 21, Astronomical Society of the Pacific. 509pp.

Perryman, M.A.C., + 22 co-authors, 1995. Parallaxes and the Hertzsprung–Russell diagram for the preliminary HIPPARCOS solution H30. *Astron. Astrophys.*, **304**, 69–81.

Phillips, O.M., 1968. *The Heart of the Earth*. Freeman, Cooper and Co., San Francisco. 236pp.

Pieri, D.C., and A.M. Dziewonski, 1999. Earth as a planet: Surface and interior. In *Encyclopedia of the Solar System*. Eds. P.R. Weissman, L. McFadden, and T.V. Johnson. Academic Press, Inc., New York, pp. 209–245.

Pieri, D.C., and A.M. Dziewonski, 2007. Earth as a planet: Surface and interior. In *Encyclopedia of the Solar System*, 2nd Edition. Eds. L. McFadden, P.R. Weissman, and T.V. Johnson. Academic Press, San Diego, pp. 189–212.

Pilcher, F., S. Mottola and T. Denk, 2012. Photometric lightcurve and rotation period of Himalia (Jupiter VI). *Icarus*, **219**, 741–742.

Podosek, F.A., and P. Cassen, 1994. Theoretical, observational, and isotopic estimates of the lifetime of the solar nebula. *Meteoritics*, **29**, 6–25.

Pollack, J.B., + 5 co-authors, 1994. Composition and radiative properties of grains in molecular clouds and accretion disks. *Astrophys. J.*, **421**, 615–639.

Porco, C.C., P.D. Nicholson, J.N. Cuzzi, J.J. Lissauer, and L.W. Esposito, 1995. Neptune's ring system. In *Neptune*. Ed. D.P. Cruikshank. University of Arizona Press, Tucson, pp. 703–804.

Porco, C.C., P.C. Thomas, J.W Weiss, and D.C. Richardson, 2007. Saturn's small inner satellites: Clues to their origins. *Science*, **318**, 1602–1607.

Pravec, P., A.W. Harris, and B.D. Warner, 2007. NEA rotations and binaries. *Near-Earth Objects: Our Celestial Neighbors – Opportunity and Risk*, Proceedings IAU Symposium No. 236. Eds. A. Milani, G.B. Valsecchi, and D. Vokrouhlický, pp. 167–176.

Press, F., and R. Siever, 1986. *Earth*. W.H. Freeman and Company, New York. 626pp.

Prettyman, T.H., 2007. Remote chemical sensing using nuclear spectroscopy. In *Encyclopedia of the Solar System*, 2nd Edition. Eds. L. McFadden, P.R. Weissman, and T.V. Johnson. Academic Press, San Diego, pp. 765–786.

Prinn, R.G., 1993. Chemistry and evolution of gaseous circumstellar disks. In *Protostars and Planets III*. Eds.

E.H. Levy and J.I. Lunine. University of Arizona Press, Tucson, pp. 1014–1028.

Prinn, R.G., and B. Fegley, Jr., 1989. Solar nebula chemistry: Origin of planetary, satellite and cometary volatiles. In *Origin and Evolution of Planetary and Satellite Atmospheres*. Eds. S.K. Atreya, J.B. Pollack, and M.S. Matthews. University of Arizona Press, Tucson, pp. 78–136.

Pudritz, R., P. Higgs, and J. Stone, 2007. *Planetary Systems and the Origins of Life*. Cambridge University Press, Cambridge, 315pp.

Putnis, A., 1992. *Mineral Science*. Cambridge University Press, Cambridge. 457pp.

Quinn, T.R., S. Tremaine, and M. Duncan, 1991. A three million year integration of the Earth's orbit. *Astron. J.*, **101**, 2287–2305.

Rauch, B.F., +17 co-authors, 2009. Cosmic-ray origin in OB associations and preferential acceleration of refractory elements: Evidence from abundances of elements 26Fe through 34Se. *Astrophys. J.*, **697**, 2083–2088.

Rayner, J.T., M.C. Cushing, and W.D. Vacca, 2009. The IRTF Spectral Library: Cool stars. *Astrophys. J. Supp.*, **185**, 289–432. See http://irtfweb.ifa.hawaii.edu/~spex/IRTF_Spectral_Library/.

Reipurth, B., D. Jewitt, and K. Keil, Eds., 2007. *Protostars and Planets V*. University of Arizona Press, Tucson. 951pp.

Retherford, K.D., H.W. Moos, and D.F. Strobel, 2003. Io's auroral limb glow: Hubble Space Telescope FUV observations. *J. Geophys. Res.* **108**, doi:10.1029/2002JA009710.

Rivera, E.J., G. Laughlin, R. P. Butler, S.S. Vogt, N. Hagigihipour, and S. Meschiari, 2010. The Lick-Carnegie Exoplanet Survey: a Uranus-mass fourth planet for GJ 876 in an extrasolar Laplace configuration. *Astrophys. J.*, **719**, 890–899.

Roe, H.G., I. de Pater, B.A. Macintosh, and C.P. McKay, 2002. Titan's clouds from Gemini and Keck adaptive optics imaging. *Astrophys. J.*, **581**, 1399–1406.

Roe, H.G., T.K. Greathouse, M.J. Richter, and J.H. Lacy, 2003. Propane on Titan. *Astrophys. J.*, **597**, L65–L68.

Roederer, J.G., 1970. *Physics and Chemistry in Space 2: Dynamics of Geomagnetically Trapped Radiation*. Springer-Verlag, Berlin. 166pp.

Roederer, J.G., 1972. Geomagnetic field distortions and their effects on radiation belt particles. *Rev. Geophys. Space Phys.*, **10**, 599–630.

Rosen, P.A., 1989. Waves in Saturn's rings probed by radio occultation. Ph.D. Thesis, Dept. of Electrical Engineering, Stanford University.

Rosen, P.A., G.L. Tyler, E.A. Marouf, and J.J. Lissauer, 1991. Resonance structures in Saturn's rings probed by radio occultation. II: Results and interpretation. *Icarus*, **93**, 25–44.

Russell, C.T., 1995. A brief history of solar-terrestrial physics. In *Introduction to Space Physics*. Eds. M.G. Kivelson and C.T. Russell. Cambridge University Press, Cambridge, pp. 1–26.

Russell, C.T., and M.G. Kivelson, 2001. Evidence of sulfur dioxide, sulfur monoxide, and hydrogen sulfide in the Io exosphere, *J. Geophys. Res.*, **106**, 33 267–33 272.

Russell, C.T., D.N. Baker, and J.A. Slavin, 1988. The magnetosphere of Mercury. In *Mercury*. Eds. F. Vilas, C.R. Chapman, and M.S. Matthews. University of Arizona Press, Tucson, pp. 514–561.

Rybicki, G.B., and A.P. Lightman, 1979. *Radiative Processes in Astrophysics*. John Wiley and Sons, New York. 382pp.

Salby, M.L., 1996. *Fundamentals of Atmospheric Physics*. Academic Press, New York. 624pp.

Sault, R.J., C. Engel, and I. de Pater, 2004. Longitude-resolved imaging of Jupiter at $\lambda = 2$ cm. *Icarus*, **168**, 336–343.

Schaller, E.L., and M.E. Brown, 2007. Detection of methane on Kuiper Belt Object (50000) Quaoar. *Astrophys. J. Lett.*, **670**, L49–L51.

Schenk, P.M., C.R. Chapman, K. Zahnle, and J.M. Moore, 2004. Ages, interiors, and the cratering record of the Galilean satellites. In *Jupiter: The Planet, Satellites and Magnetosphere*. Cambridge Planetary Science, Vol. 1. Eds. F. Bagenal, T.E. Dowling, and W.B. McKinnon. Cambridge University Press, Cambridge, pp. 427–456.

Schleicher, D.G., 1983. The fluorescence of cometary OH and CN. Ph.D. Dissertation, University of Maryland.

Schloerb, F.P., 1985. Millimeter-wave spectroscopy of Solar System objects: Present and future. Proceedings of the ESO-IRAM-Onsala workshop on (sub) millimeter astronomy, Aspenas, Sweden, 17–20 June 1985. Eds. P.A. Shaver and K. Kjar. *ESO Conference and Workshop Proceedings*, **22**, ESO, Garching, Munich, pp. 603–616.

Schneider, N.M., W.H. Smyth, and M.S. McGrath, 1987. Io's atmosphere and neutral clouds. In *Time Variable Phenomena in the Jovian System*. Eds. M.J.S. Belton, R.A. West, and J. Rahe. NASA SP-494, pp. 75–79.

Schowengerdt, R.A., 2007. *Remote Sensing, Models, and Methods for Image Processing*, 3rd Edition. Elsevier Academic Press, Burlington, MA. 515pp.

Schubert, G., D.L. Turcotte, and P. Olsen, 2001. *Mantle Convection in the Earth and Planets*. Cambridge University Press, Cambridge. 456pp.

Schubert, G., J.D. Anderson, T. Spohn, and W.B. McKinnon, 2004. Interior composition, structure and dynamics of the Galilean satellites. In *Jupiter: Planet, Satellites and Magnetosphere*. Eds. F. Bagenal, T.E. Dowling, and W. McKinnon. Cambridge University Press, Cambridge, pp. 281–306.

Schulz, M., and L.J. Lanzerotti, 1974. *Physics and Chemistry in Space. 7: Particle Diffusion in the Radiation Belts*. Springer-Verlag, Berlin. 215pp.

Scott, E.R.D., and A.N. Krot, 2005. Thermal processing of silicate dust in the solar nebula: Clues from primitive chondrite matricies. *Astrophys. J.*, **623**, 571–578.

Sears, D.W.G., and R.T. Dodd, 1988. In *Meteorites and the Early Solar System*. Eds. J.F. Kerridge and M.S. Matthews. University of Arizona Press, Tucson, pp. 3–31.

Seiff, A., 1983. Thermal structure of the atmosphere of Venus. In *Venus*. Eds. D.M. Hunten, L. Colin, T.M. Donahue, and V.I. Moroz. University of Arizona Press, Tucson, pp. 215–279.

Seiff, A., + 9 co-authors, 1998. Thermal structure of Jupiter's atmosphere near the edge of a 5-μm hot spot in the north equatorial belt. *J. Geophys. Res.*, **103**, 22 857–22 890.

Seinfeld, J.H., and S.N. Pandis, 2006. *Atmospheric Chemistry and Physics: From Air Pollution to Climate Change*, 2nd Edition. John Wiley and Sons, New York. 1203pp.

Sekanina, Z., 1976. Progress in our understanding of cometary dust tails. In *The Study of Comets*. Eds. B. Donn, M. Mumma, W. Jackson, M. A'Hearn, and R. Harrington. NASA SP-393, pp. 893–942.

Sekanina, Z., and J.A. Farrell, 1978. Comet West 1976. VI: Discrete bursts of dust, split nucleus, flare-ups, and particle evaporation. *Astron. J.*, **83**, 1675–1680.

Shoemaker, E.M., 1960. Penetration mechanics of high velocity meteorites, illustrated by Meteor Crater, Arizona. *Rep. of the Int. Geol. Congress*, XXI Session, Norden, Copenhagen, Part XVIII, pp. 418–434.

Showalter, M.R., 1991. Visual detection of 1981S13, Saturn's eighteenth satellite, and its role in the Encke gap. *Nature*, **351**, 709–713.

Showalter, M.R., 1996. Saturn's D ring in the Voyager images. *Icarus*, **124**, 677–689.

Showalter, M.R., and J.J. Lissauer, 2006. The second ring–moon system of Uranus: Discovery and dynamics. *Science*, **311**, 973–977.

Showalter, M.R., I. de Pater, G. Verbanac, D.P. Hamilton, and J.A. Burns, 2008. Properties and dynamics of Jupiter's gossamer rings from Galileo, Voyager, Hubble and Keck images. *Icarus*, **195**, 361–377.

Shu, F.H., 1982. *The Physical Universe: An Introduction to Astronomy*. University Science Books, Berkeley, CA. 584pp.

Shu, F.H., 1991. *The Physics of Astrophysics. Vol. I: Radiation*. University Science Books, Mill Valley, CA. 429pp.

Shu, F.H., J.N. Cuzzi, and J.J. Lissauer, 1983. Bending waves in Saturn's rings. *Icarus*, **53**, 185–206.

Shu, F.H., L. Dones, J.J. Lissauer, C. Yuan, and J.N. Cuzzi, 1985. Nonlinear spiral density waves: Viscous damping. *Astrophys. J.*, **299**, 542–573.

Shu, F.H., H. Shang, and T. Lee, 1996. Toward an astrophysical theory of chondrites. *Science*, **271**, 1545–1552.

Smith, B.A., + 28 co-authors, 1982. A new look at the Saturn system: The Voyager 2 images. *Science*, **215**, 504–537.

Smith, D.E., M.Y. Zuber, G.A. Neumann, and F.G. Lemoine, 1997. Topography of the Moon from the Clementine Lida. *J. Geophys. Res.*, **102**, 1591.

Smith, M.D., 2004. Interannual variability in TES atmospheric observations of Mars during 1999–2003. *Icarus*, **167**, 148–165.

Smrekar, S.E., and E.R. Stofan, 2007. Venus: Surface and interior. *Encyclopedia of the Solar System*, 2nd Edition. Eds. L. McFadden, P.R. Weissman, and T.V. Johnson. Academic Press, San Diego, pp. 149–168.

Solomon, S.C., + 10 co-authors, 2008. Return to Mercury: A global perspective on MESSENGER's first Mercury flyby. *Science*, **321**, 59–62.

Sotin, C., O. Grasset, and A. Mocquet, 2007. Mass-radius curve for extrasolar Earth-like planets and ocean planets. *Icarus*, **191**, 337–351.

Spencer, J.R., K.L. Jessup, M.A. McGrath, G.E. Ballester, and R. Yelle, 2000. Discovery of gaseous S_2 in Io's Pele plume. *Science*, **288**, 1208–1210.

Spencer, J.R., J.A. Stansberry, W.M. Grundy, and K.S. Noll, 2006. A low density for binary Kuiper Belt Object (26308) 1998 SM165. *BAAS*, **38**, #34.01.

Sprague, A.L., R.W.H. Kozlowski, D.M. Hunten, W.K. Wells, and F.A. Grosse, 1992. The sodium and potassium atmosphere of the Moon and its interaction with the surface. *Icarus*, **96**, 27–42.

Sromovsky, L.A., P.M. Frye, T. Dowling, K.H. Baines, and S.S. Limaye, 2001. Neptune's atmospheric circulation and cloud morphology: Changes revealed by 1998 HST imaging. *Icarus*, **150**, 244–260.

Sromovsky, L.A., + 7 co-authors, 2009. Uranus at Equinox: Cloud morphology and dynamics. *Icarus*, **203**, 265–286.

Stahler, S.W., and F. Palla, 2005. *The Formation of Stars*. Wiley-VCH, Weinheim, Germany. 865pp.

Stansberry, J., + 6 co-authors, 2008. Physical properties of Kuiper belt and Centaur objects: Constraints from Spitzer Space Telescope. In *The Solar System beyond Neptune*. Eds. M.A. Barucci, H. Boehnhardt, D.P. Cruikshank, and A. Morbidelli. University of Arizona Press, Tucson, pp. 161–179.

Stein, S., and M. Wysession, 2003. *An Introduction to Seismology, Earthquakes and Earth's Structure*. Wiley-Blackwell, Oxford. 498pp.

Stern, S.A., 2007. Pluto. In *Encyclopedia of the Solar System*, 2nd Edition. Eds. L. McFadden, P.R. Weissman,

and T.V. Johnson. Academic Press, San Diego, pp. 541–556.

Stevenson, D.J., 1982. Interiors of the giant planets. *Annu. Rev. Earth Planet. Sci.*, **10**, 257–295.

Stevenson, D.J., and E.E. Salpeter, 1976. Interior models of Jupiter. In *Jupiter*. Eds. T. Gehrels and M.S. Matthews. University of Arizona Press, Tucson, pp. 85–112.

Stix, M., 1987. In *Solar and Stellar Physics*, Lecture Notes Phys., **292**. Eds. E.H. Schröter and M. Schüssler. Springer, Berlin, Heidelberg, p. 15.

Stone, E.C., and E.D. Miner, 1989. The Voyager 2 encounter with the neptunian system. *Science*, **246**, 1417–1421.

Strobel, D.F., and B.C. Wolven, 2001. The atmosphere of Io: Abundances and sources of sulfur dioxide and atomic hydrogen. *Astrophys. Space Sci.*, **277**, 271–287.

Strom, R.G., 2007. Mercury. In *Encyclopedia of the Solar System*, 2nd Edition. Eds. L. McFadden, P.R. Weissman, and T.V. Johnson. Academic Press, San Diego, pp. 117–138.

Stuart, J.S., and R.P. Binzel, 2004. Bias-corrected population, size distribution, and impact hazard for the near-Earth objects. *Icarus*, **170**, 295–311.

Sutton, G.P., and O. Biblarz, 2001. *Rocket Propulsion Elements*. Wiley Europe, Chichester, UK. 772pp.

Svedhem, H., D.V. Tiov, F.W. Taylor, and O. Witasse, 2007. Venus is a more Earth-like planet. *Nature*, **450**, 629–632.

Taylor, F.W., + 5 co-authors, 2004. The composition of the atmosphere of Jupiter. In *Jupiter: Planet, Satellites and Magnetosphere*. Eds. F. Bagenal, T.E. Dowling, and W. McKinnon. Cambridge University Press, Cambridge, pp. 59–78.

Taylor, S.R., 1975. *Lunar Science: A Post-Apollo View*. Pergamon Press, New York. 372pp.

Taylor, S.R., 1992. *Solar System Evolution: A New Perspective*. Cambridge University Press, Cambridge. 307pp.

Taylor, S.R., 2001. *Solar System Evolution*, 2nd Edition. Cambridge University Press, Cambridge. 484pp.

Taylor, S.R., 2007. The Moon. In *Encyclopedia of the Solar System*, 2nd Edition. Eds. L. McFadden, P.

Weissman, and T.V. Johnson. Academic Press, San Diego, pp. 227–250.

Thomas, G.E., and K. Stamnes, 1999. *Radiative Transfer in the Atmosphere and Ocean*. Atmospheric and Space Science Series. Cambridge University Press, Cambridge. 517pp.

Thomas, P. C., 2010. Sizes, shapes, and derived properties of the saturnian satellites after the Cassini nominal mission. *Icarus*, **208**, 395–401.

Thomas, P.C., + 7 co-authors, 1998. Small inner satellites of Jupiter. *Icarus*, **135**, 360–371.

Thompson, A.R., J.M. Moran, and G.W. Swenson Jr., 2001. *Interferometry and Synthesis in Radio Astronomy*, 2nd Edition. John Wiley and Sons, New York. 692pp.

Tilton, G.R. 1988. Principles of radiometric dating. In *Meteorites and the Early Solar System*. Eds. J.F. Kerridge and M.S. Matthews. University of Arizona Press, Tucson, pp. 249–258.

Tiscareno, M.S., P.C. Thomas and J.A. Burns, 2009. The rotation of Janus and Epimetheus. *Icarus*, **204**, 254–261.

Tiscareno, M.S., + 10 co-authors, 2013. Observations of ejecta clouds produced by impacts onto Saturn's rings. *Science*, **340**, 460–464.

Tokunaga, A.T., and R. Jedicke, 2007. New generation ground-based optical/infrared telescopes. In *Encyclopedia of the Solar System*, 2nd Edition. Eds. L. McFadden, P.R. Weissman, and T.V. Johnson. Academic Press, San Diego, pp. 719–734.

Townes, C.H., and A.L. Schawlow, 1955. *Microwave Spectroscopy*. McGraw-Hill, New York. 698pp.

Turcotte, D.L., and G. Schubert, 2002. *Geodynamics*, 2nd Edition. Cambridge University Press, New York. 456pp.

Tyler, G.L., 1987. Radio propagation experiments in the outer Solar System with Voyager. *Proc. IEEE*, **75**, 1404–1431.

Tyler, G.L., I.R. Linscott, M.K. Bird, D.P. Hinson, D.F. Strobel, M. Pätzold, M.E. Summers, and K. Sivaramakrishan, 2008. The New Horizon radio science experiment (REX). *Space Sci. Rev.*, **140**, 217–259.

Udry, S., D. Fischer and D. Queloz, 2007. A decade of radial-velocity discoveries in the exoplanet domain. In *Protostars and Planets V*. Eds. B. Reipurth, D. Jewitt, and K. Keil. University of Arizona Press, Tucson, pp. 685–699.

Van de Hulst, H.C., 1957. *Light Scattering by Small Particles*. Wiley, New York. (Also Dover edition, 1981, 470pp.)

Vasavada, A.R., and A.P. Showman, 2005. Jovian atmospheric dynamics: An update after Galileo and Cassini. *Rep. Prog. Physics*, **68**, 1935–1996.

Veillet, C., + 8 co-authors, 2000. The binary Kuiper-belt object 1998 WW31. *Nature*, **416**, 711–713.

Veverka, J., P. Helfenstein, B. Hapke, and J.D. Goguen, 1988. Photometry and polarimetry of Mercury. In *Mercury*. Eds. F. Vilas, C.R. Chapman, and M.S. Matthews. University of Arizona Press, Tucson, pp. 37–58.

Veverka, J., + 10 co-authors, 1996. Dactyl: Galileo observations of Ida's satellite. *Icarus*, **120**, 200–211.

Waite, J.H., Jr., + 20 co-authors, 2005. Ion Neutral Mass Spectrometer results from the first flyby of Titan. *Science*, **308**, 982–986.

Waite, J.H., Jr., + 13 co-authors, 2006. Cassini ion and neutral mass spectrometer: Enceladus plume composition and structure. *Science*, **311**, 1419–1422.

Wasson, J.T., 1985. *Meteorites: Their Record of Early Solar-System History*. W.H. Freeman, New York. 274pp.

Weinberg, S., 1988. *The First Three Minutes*. Basic Books, New York. 198pp.

Weisberg, M.K., T.J. McCoy, and A.N. Krot, 2006. Systematics and evaluation of meteorite classification. In *Meteorites and the Early Solar System II*. Eds. D.S. Lauretta and H.Y. McSween Jr. University of Arizona Press, Tucson, pp. 19–52.

Weissman, P.R., 1986. Are cometary nuclei primordial rubble piles? *Nature*, **320**, 242–244.

Wilcox, J.M., and N.F. Ness, 1965. Quasi-stationary corotating structure in the interplanetary medium. *J. Geophys. Res.*, **70**, 5793–5805.

Williams, D.M., + 9 co-authors, 1997. Measurement of submicron grains in the coma of Comet Hale-Bopp C/1995 O1 during 1997 February 15–20 UT1997. *Astrophy. J. Lett.*, **489**, L91–L94.

Williams, J., 1992. *The Weather Book*. Vintage Books, New York. 212pp.

Wisdom, J., 1983. Chaotic behavior and the origin of the 3/1 Kirkwood Gap. *Icarus*, **56**, 51–74.

Wolf, R.A., 1995. Magnetospheric configuration. In *Introduction to Space Physics*. Eds. M.G. Kivelson and C.T. Russell. Cambridge University Press, Cambridge, pp. 288–329.

Wong, M.H., + 7 co-authors, 1996. Observations of Jupiter's 20-cm synchrotron emission during the impacts of Comet P/Shoemaker–Levy 9. *Icarus*, **121**, 457–468.

Wong, M.H., P.R. Mahaffy, S.K. Atreya, H.B. Niemann, and T.C. Owen, 2004. Updated Galileo probe mass spectrometer measurements of carbon, oxygen, nitrogen and sulfur on Jupiter. *Icarus*, **171**, 153–170.

Wooden, D., S. Desch, D. Harker, H.-P. Gail, and L. Keller, 2007. Comet grains and implications for heating and radial mixing in the protoplanetary disk. In *Protostars and Planets V*. Eds. B. Reipurth, D. Jewitt, and K. Keil. University of Arizona Press, Tucson, pp. 815–830.

Wright, M.C.H., + 10 co-authors, 1998. Mosaiced images and spectra of $J = 1 \rightarrow 0$ HCN and HCO^+ emission from Comet Hale–Bopp (1995 O1). *Astron. J.*, **116**, 3018–3028.

Yelle, R.V., 1991. Non-LTE models of Titan's upper atmosphere. *Astrophys. J.*, **383**, 380–400.

Yelle, R.V., and S. Miller, 2004. Jupiter's thermosphere and ionosphere,. In *Jupiter: Planet, Satellites and Magnetosphere*, Eds. F. Bagenal, T. E. Dowling, and W. McKinnon. Cambridge University Press, Cambridge, pp. 185–218.

Yelle, R.V., D.F. Strobel, E. Lellouch, and D. Gautier, 1997. Engineering models for Titan's atmosphere. In *Huygens Science, Payload and Mission*, ESA SP-1177, pp. 243–256.

Yoder, C.F., 1995. Astrometric and geodetic properties of Earth and the Solar System. In *GlobalEarth*

Physics: A Handbook of Physical Constants. AGU Reference Shelf 1, American Geophysical Union, pp. 1–31.

Youssef, A., and P.S. Marcus, 2003. The dynamics of jovian white ovals from formation to merger. *Icarus*, **162**, 74–93.

Zahnle, K., 1996. Dynamics and chemistry of SL9 plumes. In *The Collision of Comet Shoemaker–Levy 9 and Jupiter*. Eds. K.S. Noll, H.A. Weaver, and P.D. Feldman. Space Telescope Science Institute Symposium Series 9, IAU Colloquium 156. Cambridge University Press, Cambridge, pp. 183–212.

Zahnle, K.J., and N.H. Sleep, 1997. Impacts and the early evolution of life. In *Comets and the Origin and Evolution of Life*. Eds. P.J. Thomas, C.F. Chyba, and C.P. McKay. Springer, New York, pp. 175–208.

Zarka, P., and W. S. Kurth, 2005. Radio wave emission from the outer planets before Cassini. *Space Sci. Rev.*, **116**, 371–397.

Zarka, P., + 6 co-authors, 1995. Radio emissions from Neptune. In *Neptune and Triton*. Eds. D.P. Cruikshank and M.S. Matthews. University of Arizona Press, Tucson, pp. 341–387.

Zarnecki, J.C., + 25 co-authors, 2005. A soft solid surface on Titan as revealed by the Huygens Surface Science Package. *Nature*, **438**, 792–795.

Zhang, Z.-W., + 24 co-authors, 2008. First Results from the Taiwanese-American Occultation Survey (TAOS). *Astrophys. J.*, **685**, L157–L160.

Zinner, E., 1998. Stellar nucleosynthesis and the isotopic composition of presolar grains from primitive meteorites. *Annu. Rev. Earth Planet. Sci.*, **26**, 147–188.

Zuber, M.T., + 14 co-authors, 2000. Internal structure and early thermal evolution of Mars from Mars Global Surveyor topography and gravity. *Science*, **287**, 1788–1793.

Zurbuchen, T.H., +14 co-authors, 2011. MESSENGER observations of the spatial distribution of planetary ions near Mercury. *Science*, **333**, 1859–1862.

Index